Leah Epstein   Paolo Ferragina (Eds.)

Algorithms – ESA 2012

W0114696

20th Annual European Symposium
Ljubljana, Slovenia, September 10-12, 2012
Proceedings

 Springer

Volume Editors

Leah Epstein
University of Haifa
Department of Mathematics
Mount Carmel
31905 Haifa, Israel
E-mail: lea@math.haifa.ac.il

Paolo Ferragina
Università di Pisa
Dipartimento di Informatica
Largo B. Pontecorvo 3
56100 Pisa, Italy
E-mail: ferragin@di.unipi.it

ISSN 0302-9743   e-ISSN 1611-3349
ISBN 978-3-642-33089-6   e-ISBN 978-3-642-33090-2
DOI 10.1007/978-3-642-33090-2
Springer Heidelberg Dordrecht London New York

Library of Congress Control Number: : 2012945300

CR Subject Classification (1998): F.2, I.3.5, C.2, E.1, G.2, D.2, F.1

LNCS Sublibrary: SL 1 – Theoretical Computer Science and General Issues

Typesetting: Camera-ready by author, data conversion by Scientific Publishing Services, Chennai, India

Printed on acid-free paper

Springer is part of Springer Science+Business Media (www.springer.com)

Preface

This volume contains the papers presented at the 20th Annual European Symposium on Algorithms (ESA 2012) held in Ljubljana, Slovenia, during September 10–12, 2012. ESA 2012 was organized as part of ALGO 2012, which also included the Workshop on Algorithms for Bioinformatics (WABI), the International Symposium on Parameterized and Exact Computation (IPEC), the Workshop on Approximation and Online Algorithms (WAOA), the International Symposium on Algorithms for Sensor Systems, Wireless Ad Hoc Networks and Autonomous Mobile Entities (ALGOSENSORS), the Workshop on Algorithmic Approaches for Transportation Modeling, Optimization, and Systems (ATMOS), and the Workshop on Massive Data Algorithms (MASSIVE). The previous symposia were held in Saarbrücken (2011), Liverpool (2010), Copenhagen (2009), Karlsruhe (2008), Eilat (2007), Zürich (2006), Palma de Mallorca (2005), Bergen (2004), Budapest (2003), Rome (2002), Aarhus (2001), Saarbrücken (2000), Prague (1999), Venice (1998), Graz (1997), Barcelona (1996), Corfu (1995), Utrecht (1994), and Bad Honnef (1993).

The ESA symposia are devoted to fostering and disseminating the results of high-quality research on the design and evaluation of algorithms and data structures. The forum seeks original algorithmic contributions for problems with relevant theoretical and/or practical applications and aims at bringing together researchers in the computer science and operations research communities. Papers were solicited in all areas of algorithmic research, both theoretical and experimental, and were evaluated by two Program Committees (PC). The PC of Track A (Design and Analysis) selected contributions with a strong emphasis on the theoretical analysis of algorithms. The PC of Track B (Engineering and Applications) evaluated papers reporting on the results of experimental evaluations and on algorithm engineering contributions for interesting applications.

The conference received 285 submissions from 40 countries. Each submission was reviewed by at least three PC members and was carefully evaluated on quality, originality, and relevance to the conference. Overall, the PCs wrote more than 900 reviews with the help of more than 500 trusted external referees, who also participated in an extensive electronic discussion that led the committees of the two tracks to select 69 papers (56 out of 231 in Track A and 13 out of 54 in Track B), yielding an acceptance rate of about 24%. In addition to the accepted contributions, the symposium featured two distinguished plenary lectures by Yossi Matias (Google Israel and Tel Aviv University) and Jiří Sgall (Charles University, Prague).

The European Association for Theoretical Computer Science (EATCS) sponsored a best paper award and a best student paper award. The former award was shared by two papers: one by Clément Maria and Jean-Daniel Boissonnat for their contribution on "The Simplex Tree: An Efficient Data Structure for General

Simplicial Complexes," and the other by Sebastian Wild and Markus E. Nebel for their contribution titled "Average Case Analysis of Java 7's Dual Pivot Quicksort." The best student paper prize was awarded to Martin Groß, Jan-Philipp W. Kappmeier, Daniel R. Schmidt, and Melanie Schmidt for their contribution titled "Approximating Earliest Arrival Flows in Arbitrary Networks." Our warmest congratulations to them for these achievements.

We would like to thank all the authors who responded to the call for papers, the invited speakers, the members of the PCs, as well as the external referees and the Organizing Committee members. We also gratefully acknowledge the developers and maintainers of the EasyChair conference management system, which provided invaluable support throughout the selection process and the preparation of these proceedings. We hope that the readers will enjoy the papers published in this volume, sparking their intellectual curiosity and providing inspiration for their work.

July 2012   Leah Epstein
                                                                Paolo Ferragina

Organization

Program Committees

Track A (Design and Analysis)

Matthias Englert	University of Warwick, UK
Leah Epstein (Chair)	University of Haifa, Israel
Gregory Gutin	University of London, UK
Pinar Heggernes	University of Bergen, Norway
Martin Hoefer	RWTH Aachen University, Germany
Jochen Könemann	University of Waterloo, Canada
Petr Kolman	Charles University, Czech Republic
Kim Skak Larsen	University of Southern Denmark, Denmark
Asaf Levin	The Technion, Israel
Alejandro López-Ortiz	University of Waterloo, Canada
Krzysztof Onak	Carnegie Mellon University, USA
Dror Rawitz	Tel Aviv University, Israel
Günter Rote	Freie Universität Berlin, Germany
Andreas Schulz	MIT, USA
Ola Svensson	EPFL Lausanne, Switzerland
Marc Uetz	University of Twente, The Netherlands
Carola Wenk	University of Texas at San Antonio, USA
Peter Widmayer	ETH Zurich, Switzerland
Christian Wulff-Nilsen	University of Southern Denmark, Denmark
Raphael Yuster	University of Haifa, Israel

Track B (Engineering and Applications)

Susanne Albers	Humboldt University, Berlin, Germany
Alexandr Andoni	Microsoft Research, USA
Ioannis Z. Emiris	National Kapodistrian University of Athens, Greece
Paolo Ferragina (Chair)	University of Pisa, Italy
Irene Finocchi	Sapienza University of Rome, Italy
Johannes Fischer	Karlsruhe Institute of Technology, Germany
Michael T. Goodrich	University of California, Irvine, USA
Herman Haverkort	Eindhoven University of Technology, The Netherlands
Vahab Mirrokni	Google Research, USA
Gonzalo Navarro	University of Chile, Chile
Rina Panigrahy	Microsoft Research Silicon Valley, USA
Rajeev Raman	University of Leicester, UK

Jens Stoye   Bielefeld University, Germany
Oren Weimann   University of Haifa, Israel
Ke Yi   Hong Kong University of Science and
                                        Technology, China

Organizing Committee

Andrej Brodnik (Chair)   University of Ljubljana, Slovenia
Uroš Čibej   University of Ljubljana, Slovenia
Gašper Fele-Žorž   University of Ljubljana, Slovenia
Matevž Jekovec   University of Ljubljana, Slovenia
Jurij Mihelič   University of Ljubljana, Slovenia
Borut Robič (Co-chair)   University of Ljubljana, Slovenia
Andrej Tolič   University of Ljubljana, Slovenia

Additional Reviewers

Abed, Fidaa   Ada, Anil
Adamaszek, Anna   Adamaszek, Michał
Afshani, Peyman   Ahmed, Mahmuda
Aichholzer, Oswin   Al-Bawani, Kamal
Alaei, Saeed   Alon, Noga
Alt, Helmut   Altiparmak, Nihat
Ambainis, Andris   Amit, Mika
Angelopoulos, Spyros   Anshelevich, Elliot
Antoniadis, Antonios   Arge, Lars
Aronov, Boris   Arroyuelo, Diego
Ashlagi, Itai   Asinowski, Andrei
Atariah, Dror   Aurenhammer, Franz
Avin, Chen   Babaioff, Moshe
Bagchi, Amitabha   Bampis, Evripidis
Bansal, Nikhil   Bar-Noy, Amotz
Barbay, Jeremy   Barman, Siddharth
Bast, Hannah   Baswana, Surender
Bateni, Hossein   Beimel, Amos
Bein, Wolfgang   Belmonte, Rémy
Belovs, Aleksandrs   Bendich, Paul
Berry, Anne   Bhaskar, Umang
Bhawalkar, Kshipra   Bille, Philip
Bingmann, Timo   Birks, Martin
Björklund, Andreas   Bock, Adrian
Bodlaender, Hans L.   Boeckenhauer, Hans-Joachim
Bohmova, Katerina   Bonichon, Nicolas
Bonifaci, Vincenzo   Bose, Prosenjit
Boucher, Christina   Boutsidis, Christos

Boyar, Joan
Brandstaedt, Andreas
Brodal, Gerth Stølting
Bruhn, Henning
Călinescu, Gruia
Caragiannis, Ioannis
Carmi, Paz
Chalopin, Jérémie
Chazal, Frederic
Chen, Ning
Chowdhury, Rezaul
Clarkson, Ken
Cleve, Richard
Cook Iv, Atlas F.
Crowston, Robert
Czumaj, Artur
Dadush, Daniel
Danna, Emilie
Delling, Daniel
Dieckmann, Claudia
Dinitz, Yefim
Doerr, Benjamin
Dorrigiv, Reza
Driemel, Anne
Dvořák, Zdeněk
Dürr, Christoph
Ederer, Thorsten
Eisenbrand, Friedrich
Elmasry, Amr
Eppstein, David
Erten, Cesim
Even, Guy
Fagerberg, Rolf
Feige, Uriel
Feldman, Danny
Feldman, Moran
Fellows, Michael
Fiala, Jiří
Fischer, Johannes
Fotakis, Dimitris
Frangioni, Antonio
Frati, Fabrizio
Friedman, Eran
Fu, Hu
Fulek, Radoslav

Brandes, Ulrik
Brass, Peter
Broersma, Hajo
Buchbinder, Niv
Candogan, Ozan
Carlsson, John
Caskurlu, Bugra
Chan, Timothy M.
Chechik, Shiri
Cheng, Siu-Wing
Christodoulou, George
Claude, Francisco
Cohen, Ilan
Cooper, Colin
Cygan, Marek
Czyżowicz, Jurek
Dams, Johannes
Dell, Holger
Devillers, Olivier
Dinitz, Michael
Disser, Yann
Doerr, Daniel
Doty, David
Duan, Ran
Dósa, György
Edelkamp, Stefan
Ediger, David
Elberfeld, Michael
Emek, Yuval
Erlebach, Thomas
Escoffier, Bruno
Ezra, Esther
Favrholdt, Lene M.
Fekete, Sándor
Feldman, Michal
Feldmann, Andreas Emil
Fernau, Henning
Fineman, Jeremy
Fleischer, Lisa
Fountoulakis, Nikolaos
Fraser, Robert
Freund, Rob
Friggstad, Zac
Fukasawa, Ricardo
Furnon, Vincent

Fusco, Emanuele Guido
Gärtner, Bernd
Gairing, Martin
Garg, Naveen
Gavinsky, Dmitry
Ge, Rong
Georgiou, Konstantinos
Goaoc, Xavier
Goldberg, Ian
Grabowski, Szymon
Grigoriev, Alexander
Gudmundsson, Joachim
Guillemot, Sylvain
Gupta, Ankur
Haeupler, Bernhard
Hajiaghayi, Mohammadtaghi
Hanusse, Nicolas
Harks, Tobias
Hasan, Masud
Hazan, Elad
Henry, Kevin
Hirai, Hiroshi
Hoffmann, Frank
Horn, Silke
Hruz, Tomas
Hähnle, Nicolai
van Iersel, Leo
Immorlica, Nicole
Italiano, Giuseppe F.
Jakobi, Tobias
Jensen, Anders
Jaume, Rafel
Jelínek, Vít
Johnson, David
Jowhari, Hossein
Kaibel, Volker
Kannan, Rajgopal
Kaplan, Haim
Karavelas, Menelaos I.
Kellerer, Hans
Kesselheim, Thomas
Kimmel, Shelby
Kleiman, Elena
Kobourov, Stephen
Kontogiannis, Spyros
Korula, Nitish

Gao, Xiao-Shan
Gagie, Travis
Gamarnik, David
Gavenčiak, Tomáš
Gawrychowski, Paweł
Georgiadis, Loukas
Giannopoulos, Panos
Gog, Simon
Golovach, Petr
Grandoni, Fabrizio
Grossi, Roberto
Guha, Sudipto
Guo, Jiong
Gupta, Varun
Haghpanah, Nima
Halabi, Nissim
Har-Peled, Sariel
Hartline, Jason
Hassin, Refael
He, Meng
Hermelin, Danny
Hoeksma, Ruben
Hofri, Micha
Hornus, Samuel
Huang, Chien-Chung
Iacono, John
Ilie, Lucian
Imreh, Csanad
Ito, Takehiro
Jakoby, Andreas
Jansen, Klaus
Jeffery, Stacey
Jerrum, Mark
Jones, Mark
Jurdziński, Marcin
Kaminski, Marcin
Kantor, Ida
Karakostas, George
Karrenbauer, Andreas
Kern, Walter
Kim, Eun Jung
Kintali, Shiva
Klein, Philip
Kogan, Kirill
Kortsarz, Guy
Kothari, Robin

Koutecky, Martin
Koutis, Yiannis
Kowalski, Dariusz
Kratochvil, Jan
Kratsch, Stefan
Kreutzer, Stephan
Krivelevich, Michael
Kucherov, Gregory
Kumar, Ravi
Köhler, Ekkehard
Lang, Kevin
Larsen, Kasper Green
Lecroq, Thierry
van Leeuwen, Erik Jan
Leonardi, Stefano
Levy, Avivit
Li, Shi
Lidicky, Bernard
Loebl, Martin
Lokshtanov, Daniel
Lubiw, Anna
Lyu, Yu-Han
Luxen, Dennis
Mahabadi, Sepideh
Maheshwari, Anil
Malec, David
Manthey, Bodo
Mastrolilli, Monaldo
McClain, James
Meeks, Kitty
Meister, Daniel
Meyer, Ulrich
Mihalak, Matus
Miltzow, Tillmann
Mirrokni, Vahab
Mitra, Pradipta
Mnich, Matthias
Moitra, Ankur
Montanari, Sandro
Moseley, Benjamin
Muciaccia, Gabriele
Muller, Haiko
Munro, Ian
Moreno, Eduardo
Nagarajan, Viswanath
Nederlof, Jesper

Koucký, Michal
Kowalik, Łukasz
Král', Daniel
Kratsch, Dieter
Krauthgamer, Robert
Kriegel, Klaus
Krčál, Marek
Kumar, Piyush
Kynčl, Jan
Lampis, Michael
Langerman, Stefan
Le Gall, François
Lee, Troy
Lenzen, Christoph
Levin, Alex
Li, Fei
Liaghat, Vahid
Lingas, Andrzej
Löffler, Maarten
Lotker, Zvi
Lucier, Brendan
Lübbecke, Marco
Mach, Lukáš
Maher, Michael
Mahini, Hamid
Manlove, David
Mares, Martin
Matoušek, Jiří
Medina, Moti
Megow, Nicole
Mertzios, George
Michail, Dimitrios
Milanic, Martin
Mironov, Ilya
Misra, Neeldhara
Mittal, Shashi
Moemke, Tobias
Moldenhauer, Carsten
Morin, Pat
Mozes, Shay
Müller, Rudolf
Mulzer, Wolfgang
Mølhave, Thomas
Mądry, Aleksander
Naor, Seffi
Newman, Alantha

Nicholson, Patrick K.
Nikolova, Evdokia
Nitto, Igor
Nussbaum, Yahav
Okamoto, Yoshio
Olver, Neil
Orlin, James
Pagh, Rasmus
Panigrahi, Debmalya
Paschos, Vangelis
Patt-Shamir, Boaz
Pelc, Andrzej
Peng, Richard
Pettie, Seth
Philip, Geevarghese
Pietracaprina, Andrea
Pilaud, Vincent
Pilipczuk, Michal
Pinchasi, Rom
Pountourakis, Emmanouil
Proeger, Tobias
Pruhs, Kirk
Pérez-Lantero, Pablo
Rafiey, Arash
Raman, Venkatesh
Reich, Alexander
Reishus, Dustin
Rosamond, Frances
Rubin, Natan
Rutter, Ignaz
Röglin, Heiko
Salinger, Alejandro
Samet, Hanan
Sankaranarayanan, Jagan
Sarkar, Rik
Sauerwald, Thomas
Sayed, Umar
Schäfer, Guido
Schirra, Stefan
Schlotter, Ildikó
Schulz, Christian
Schwartz, Roy
Seco, Diego
Segal, Michael
Seki, Shinnosuke
Sheehy, Don

Niemeier, Martin
Nisgav, Aviv
Nonner, Tim
Nutov, Zeev
Oliveira, Fabiano
Omri, Eran
Otachi, Yota
Pangrac, Ondřej
Papadopoulou, Evanthia
Paterson, Mike
Paulusma, Daniel
Pendavingh, Rudi
Pergel, Martin
Pfetsch, Marc
Phillips, Jeff
Pignolet, Yvonne-Anne
Pilipczuk, Marcin
Pilz, Alexander
Polacek, Lukas
Price, Eric
Proietti, Guido
Prädel, Lars
Radzik, Tomasz
Raman, Rajeev
Ramanujan, M.S.
Reichardt, Ben
Ribó Mor, Ares
Rothvoss, Thomas
Ruozzi, Nicholas
Räcke, Harald
Sach, Benjamin
Šámal, Robert
Sanita, Laura
Sankowski, Piotr
Satti, Srinivasa Rao
Saurabh, Saket
Scalosub, Gabriel
Scharf, Ludmila
Schlipf, Lena
Schoengens, Marcel
Schwartz, Jarett
Schwiegelshohn, Uwe
Sedgewick, Bob
Segev, Danny
Sgall, Jiří
Sherette, Jessica

Sikdar, Somnath
Simon, Hans
Sivan, Balasubramanian
Snir, Sagi
Sommer, Christian
Spalek, Robert
Speckmann, Bettina
Sramek, Rastislav
van Stee, Rob
Stoddard, Greg
Subramani, K.
Suchy, Ondrej
Sviridenko, Maxim
Tamir, Tami
Tanigawa, Shin-Ichi
Tas, Baris
Terlecky, Peter
Thilikos, Dimitrios
Torng, Eric
Träff, Jesper Larsson
Tsur, Gilad
Tsirogiannis, Constantinos
Uhan, Nelson
Varadarajan, Kasturi
Ventre, Carmine
Vermeulen, Dries
Villanger, Yngve
Vredeveld, Tjark
Vyskočil, Tomáš
Wagner, Lisa
Wakrat, Ido
Wang, Ling
Watrous, John
Weibel, Christophe
Wenner, Cenny
Westermann, Matthias
Wiese, Andreas
Wittler, Roland
Wolf, Stefan
Xin, Qin
Yeo, Anders
Yiu, S.M.
Zaman, Tauhid
Zeh, Norbert

Silvestri, Francesco
Sitchinava, Nodari
Skopalik, Alexander
Snoeyink, Jack
Soto, Jose A.
Speck, Jochen
Spieksma, Frits
Stacho, Juraj
Still, Georg
Strehler, Martin
Subramanian, C.R.
Sundermann, Linda
Swamy, Chaitanya
Tancer, Martin
Tarjan, Robert
Telha, Claudio
Thévenin, Annelyse
Tomozei, Dan-Cristian
Toth, Csaba
Tůma, Vojtěch
Tsigaridas, Elias
Tzoumas, George M.
Ummels, Michael
Vaya, Shailesh
Venturini, Rossano
Vidali, Angelina
Volec, Jan
Vygen, Jens
Wagner, Fabian
Wahlström, Magnus
Waleń, Tomasz
Wang, Yusu
Watson, Bruce
Werneck, Renato
Werner, Daniel
Wieder, Udi
Williams, Ryan
Woeginger, Gerhard J.
Xiao, David
Yan, Li
Yi, Ke
Young, Maxwell
Zaroliagis, Christos
van Zuylen, Anke

Table of Contents

On Big Data Algorithmics 1
Yossi Matias

Open Problems in Throughput Scheduling 2
Jiří Sgall

Preemptive Coordination Mechanisms for Unrelated Machines 12
Fidaa Abed and Chien-Chung Huang

Hierarchical Hub Labelings for Shortest Paths 24
*Ittai Abraham, Daniel Delling, Andrew V. Goldberg, and
Renato F. Werneck*

Bottleneck Non-crossing Matching in the Plane 36
*A. Karim Abu-Affash, Paz Carmi, Matthew J. Katz, and
Yohai Trabelsi*

Lower Bounds for Sorted Geometric Queries in the I/O Model 48
Peyman Afshani and Norbert Zeh

Constructing Street Networks from GPS Trajectories 60
Mahmuda Ahmed and Carola Wenk

I/O-efficient Hierarchical Diameter Approximation 72
Deepak Ajwani, Ulrich Meyer, and David Veith

On the Value of Job Migration in Online Makespan Minimization 84
Susanne Albers and Matthias Hellwig

Simplifying Massive Contour Maps 96
*Lars Arge, Lasse Deleuran, Thomas Mølhave, Morten Revsbæk, and
Jakob Truelsen*

Explicit and Efficient Hash Families Suffice for Cuckoo Hashing
with a Stash... 108
Martin Aumüller, Martin Dietzfelbinger, and Philipp Woelfel

On Online Labeling with Polynomially Many Labels.................. 121
*Martin Babka, Jan Bulánek, Vladimír Čunát, Michal Koucký, and
Michael Saks*

A 5-Approximation for Capacitated Facility Location 133
Manisha Bansal, Naveen Garg, and Neelima Gupta

Weighted Geometric Set Multi-cover via Quasi-uniform Sampling 145
    Nikhil Bansal and Kirk Pruhs

A Bicriteria Approximation for the Reordering Buffer Problem......... 157
    Siddharth Barman, Shuchi Chawla, and Seeun Umboh

Time-Dependent Route Planning with Generalized Objective
Functions 169
    Gernot Veit Batz and Peter Sanders

New Lower and Upper Bounds for Representing Sequences 181
    Djamal Belazzougui and Gonzalo Navarro

Span Programs and Quantum Algorithms for st-Connectivity and Claw
Detection 193
    Aleksandrs Belovs and Ben W. Reichardt

The Stretch Factor of L_1- and L_∞-Delaunay Triangulations 205
    *Nicolas Bonichon, Cyril Gavoille, Nicolas Hanusse, and
    Ljubomir Perković*

Two Dimensional Range Minimum Queries and Fibonacci Lattices 217
    *Gerth Stølting Brodal, Pooya Davoodi, Moshe Lewenstein,
    Rajeev Raman, and Satti Srinivasa Rao*

Locally Correct Fréchet Matchings 229
    *Kevin Buchin, Maike Buchin, Wouter Meulemans, and
    Bettina Speckmann*

The Clique Problem in Ray Intersection Graphs 241
    Sergio Cabello, Jean Cardinal, and Stefan Langerman

Revenue Guarantees in Sponsored Search Auctions 253
    *Ioannis Caragiannis, Christos Kaklamanis,
    Panagiotis Kanellopoulos, and Maria Kyropoulou*

Optimizing Social Welfare for Network Bargaining Games in the Face
of Unstability, Greed and Spite 265
    T.-H. Hubert Chan, Fei Chen, and Li Ning

Optimal Lower Bound for Differentially Private Multi-party
Aggregation 277
    T-H. Hubert Chan, Elaine Shi, and Dawn Song

A Model for Minimizing Active Processor Time 289
    Jessica Chang, Harold N. Gabow, and Samir Khuller

Polynomial-Time Algorithms for Energy Games with Special Weight
Structures 301
    *Krishnendu Chatterjee, Monika Henzinger,
    Sebastian Krinninger, and Danupon Nanongkai*

Data Structures on Event Graphs 313
   Bernard Chazelle and Wolfgang Mulzer

Improved Distance Oracles and Spanners for Vertex-Labeled Graphs.... 325
   Shiri Chechik

The Quantum Query Complexity of Read-Many Formulas 337
   Andrew M. Childs, Shelby Kimmel, and Robin Kothari

A Path-Decomposition Theorem with Applications to Pricing and
Covering on Trees 349
   *Marek Cygan, Fabrizio Grandoni, Stefano Leonardi,
   Marcin Pilipczuk, and Piotr Sankowski*

Steiner Forest Orientation Problems 361
   Marek Cygan, Guy Kortsarz, and Zeev Nutov

A Dual-Fitting $\frac{3}{2}$-Approximation Algorithmfor Some Minimum-Cost
Graph Problems 373
   James M. Davis and David P. Williamson

Kinetic Compressed Quadtrees in the Black-Box Model with
Applications to Collision Detection for Low-Density Scenes 383
   Mark de Berg, Marcel Roeloffzen, and Bettina Speckmann

Finding Social Optima in Congestion Games with Positive
Externalities 395
   Bart de Keijzer and Guido Schäfer

Better Bounds for Graph Bisection 407
   Daniel Delling and Renato F. Werneck

On the Complexity of Metric Dimension 419
   Josep Díaz, Olli Pottonen, Maria Serna, and Erik Jan van Leeuwen

Embedding Paths into Trees: VM Placement to Minimize Congestion ... 431
   Debojyoti Dutta, Michael Kapralov, Ian Post, and Rajendra Shinde

Faster Geometric Algorithms via Dynamic Determinant
Computation 443
   Vissarion Fisikopoulos and Luis Peñaranda

Lines through Segments in 3D Space 455
   Efi Fogel, Michael Hemmer, Asaf Porat, and Dan Halperin

A Polynomial Kernel for PROPER INTERVAL VERTEX DELETION 467
   Fedor V. Fomin, Saket Saurabh, and Yngve Villanger

Knowledge, Level of Symmetry, and Time of Leader Election   479
    Emanuele G. Fusco and Andrzej Pelc

An Experimental Study of Dynamic Dominators   491
    Loukas Georgiadis, Giuseppe F. Italiano, Luigi Laura, and
    Federico Santaroni

Optimizing over the Growing Spectrahedron.   503
    Joachim Giesen, Martin Jaggi, and Sören Laue

Induced Disjoint Paths in Claw-Free Graphs.   515
    Petr A. Golovach, Daniël Paulusma, and Erik Jan van Leeuwen

On Min-Power Steiner Tree   527
    Fabrizio Grandoni

Maximum Multicommodity Flows over Time without Intermediate
Storage   539
    Martin Groß and Martin Skutella

Approximating Earliest Arrival Flows in Arbitrary Networks   551
    Martin Groß, Jan-Philipp W. Kappmeier, Daniel R. Schmidt, and
    Melanie Schmidt

Resource Buying Games   563
    Tobias Harks and Britta Peis

Succinct Data Structures for Path Queries   575
    Meng He, J. Ian Munro, and Gelin Zhou

Approximation of Minimum Cost Homomorphisms   587
    Pavol Hell, Monaldo Mastrolilli, Mayssam Mohammadi Nevisi, and
    Arash Rafiey

Property Testing in Sparse Directed Graphs: Strong Connectivity and
Subgraph-Freeness   599
    Frank Hellweg and Christian Sohler

Improved Implementation of Point Location in General
Two-Dimensional Subdivisions.   611
    Michael Hemmer, Michal Kleinbort, and Dan Halperin

Parameterized Complexity of Induced H-Matching on Claw-Free
Graphs   624
    Danny Hermelin, Matthias Mnich, and Erik Jan van Leeuwen

Solving Simple Stochastic Games with Few Coin Toss Positions   636
    Rasmus Ibsen-Jensen and Peter Bro Miltersen

Efficient Communication Protocols for Deciding Edit Distance   648
    Hossein Jowhari

Approximation Algorithms for Wireless Link Scheduling with Flexible
Data Rates   659
    Thomas Kesselheim

Extending Partial Representations of Function Graphs and Permutation
Graphs   671
    Pavel Klavík, Jan Kratochvíl, Tomasz Krawczyk, and
    Bartosz Walczak

A Fast and Simple Subexponential Fixed Parameter Algorithm for
One-Sided Crossing Minimization   683
    Yasuaki Kobayashi and Hisao Tamaki

Minimum Average Distance Triangulations   695
    László Kozma

Colouring AT-Free Graphs   707
    Dieter Kratsch and Haiko Müller

Routing Regardless of Network Stability   719
    Bundit Laekhanukit, Adrian Vetta, and Gordon Wilfong

The Simplex Tree: An Efficient Data Structure for General Simplicial
Complexes   731
    Jean-Daniel Boissonnat and Clément Maria

Succinct Posets   743
    J. Ian Munro and Patrick K. Nicholson

Polynomial-Time Approximation Schemes for Shortest Path
with Alternatives   755
    Tim Nonner

On Computing Straight Skeletons by Means of Kinetic
Triangulations   766
    Peter Palfrader, Martin Held, and Stefan Huber

A Self-adjusting Data Structure for Multidimensional Point Sets   778
    Eunhui Park and David M. Mount

TSP Tours in Cubic Graphs: Beyond 4/3   790
    José R. Correa, Omar Larré, and José A. Soto

FPT Algorithms for Domination in Biclique-Free Graphs   802
    Jan Arne Telle and Yngve Villanger

Maximum Flow Networks for Stability Analysis of LEGO®
Structures 813
    Martin Waßmann and Karsten Weicker

Average Case Analysis of Java 7's Dual Pivot Quicksort 825
    Sebastian Wild and Markus E. Nebel

Author Index 837

On Big Data Algorithmics

Yossi Matias

Israel R&D Center, Google

Abstract. The extensive use of Big Data has now become common in plethora of technologies and industries. From massive data bases to business intelligence and datamining applications; from search engines to recommendation systems; advancing the state of the art of voice recognition, translation and more. The design, analysis and engineering of Big Data algorithms has multiple flavors, including massive parallelism, streaming algorithms, sketches and synopses, cloud technologies, and more. We will discuss some of these aspects, and reflect on their evolution and on the interplay between the theory and practice of Big Data algorithmics.

An extended summary will be available at http://goo.gl/FbYh9

L. Epstein and P. Ferragina (Eds.): ESA 2012, LNCS 7501, p. 1, 2012.
© Springer-Verlag Berlin Heidelberg 2012

Open Problems in Throughput Scheduling

Jiří Sgall*

Computer Science Institute of Charles University, Faculty of Mathematics and
Physics, Malostranské nám. 25, CZ-11800 Praha 1, Czech Republic
sgall@iuuk.mff.cuni.cz

Abstract. In this talk we survey the area of scheduling with the objective to maximize the throughput, i.e., the (weighted) number of completed jobs. We focus on several open problems.

1 Introduction

Scheduling is a very wide research area with a multitude of variants that may appear confusing to non-specialists. The first rough classification is based on the objective that the algorithms try to optimize. Two most studied classes of problems try to minimize the costs generally depending on the completion times of jobs. One class minimizes a max-type objective, typically the length of the schedule (makespan). The other class minimizes a sum-type objective, typically the sum (average) of completion times or flow times (the time a job is in the system).

In this talk we focus on another area of scheduling. Our objective is to maximize the the benefit. The benefit it throughput, i.e., the number of scheduled jobs, or the weighted number of scheduled jobs, in case when jobs have profits individually given by their weights. This area received considerable attention in the last decade esp. in the online case, with motivation such as packet forwarding in network switches and power management.

In turns out that some of very basic and easily formulated problems in this area are still open. We center our talk around a few of such problems. Needless to say, any such selection is necessarily biased. Our choice is motivated mostly by simplicity and certain elegance of the problems.

In each of the considered variants the jobs are given by several parameters. We use the following notation: for a job j, p_j is its processing time also called length, r_j its release time, d_j its deadline, and w_j is its weight. There may be a single machine or m machines. A schedule assigns to each job a machine and a time slot of length p_j starting not before r_j and ending no later than d_j; some jobs may remain unscheduled. The time slots on each machine have to be non-overlapping. The objective is to maximize the sum of the weights w_j of all the scheduled jobs.

---

* Partially supported by the Center of Excellence – Inst. for Theor. Comp. Sci., Prague (project P202/12/G061 of GA ČR) and grant IAA100190902 of GA AV ČR.

L. Epstein and P. Ferragina (Eds.): ESA 2012, LNCS 7501, pp. 2–11, 2012.

We follow the standard convention that the job parameters are non-negative integers. Then we may assume that all jobs are started at integral time, as rounding the start times of all jobs down preserves feasibility.

We consider both offline and online problems. In the online setting, we let the time progress and at each time t we learn about jobs released at time t, i.e., those with $r_j = t$; at this time also all the job parameters of j become known. At time t, we may start some pending job (with $r_j \leq t$, $t + p_j \leq d_j$, and not running or completed) on one of the available machines (those not running any job). Sometimes we may also use preemption, i.e., stop the currently running job; in variants that we consider this job is then lost and cannot be even restarted.

We consider some variants of this general problem. Typically the set of possible job lengths is restricted. In the most restricted case, *unit-length jobs*, all job lengths are equal to 1. This corresponds to a special case of maximum matching on bipartite graph (where the vertices correspond to jobs and time slots), thus it can be solved to optimality offline. A less restricted case is that of *equal-length jobs* where all the job lengths are equal to some p (since release times and deadlines are integral, this is different from unit-length jobs). In several variants we have unit weights of jobs, which is equivalent to the objective of maximizing the number of scheduled jobs instead of the total weight. In some of the variants we allow parallel jobs. Then each job comes with an additional parameter $size_j$ and requires to be given the same time slot on $size_j$ machines.

2   Offline Scheduling

2.1   Instances with Length of Jobs Bounded by a Constant

If we allow arbitrary processing times, the problem is (strongly) NP-hard; it is even hard to check if all jobs can be scheduled in the unweighted version on a single machine by an easy reduction from 3-PARTITION. To avoid this reason for NP-hardness, a natural option is to reduce the number of job lengths. Surprisingly, very little is known about this case.

Open problem 1. *Consider scheduling on a single machine, with the restriction that $p_j \leq C$ for some constant C. Is there a polynomial-time algorithm to compute a schedule maximizing the number of completed jobs? Is it NP-hard to compute the maximal weight of scheduled jobs for some C?*

We do not even know a polynomial-time algorithm for the unweighted variant for $C = 2$, i.e., every job needs one or two time slots. The only positive result known is that for $C = 2$, it is possible to check in polynomial time if it is possible to schedule all jobs. (Since we are scheduling all the jobs, it makes no difference if the weights are unit or arbitrary.) This can be strengthened to the case when $p_j \in \{1, p\}$ for any fixed p. This result follows closely the methods from [12], but it appears to be new. We prove it in Section 4. Note that the restriction that one of the allowed job lengths equals 1 is essential in our proof. Even in the case $p_j \in \{2, 3\}$ it is not known whether checking feasibility of scheduling of all jobs is polynomial!

In case of equal length jobs, the problem of maximizing the number of scheduled jobs is still far from trivial. On a single machine it is polynomial [3,7], but on parallel machines, it is known to be polynomial only if the number of machines m is a constant [1]. The following is still open:

Open problem 2. *Consider scheduling of equal-length jobs on m parallel machines with m a part of the input. Is there a polynomial-time algorithm to compute a schedule maximizing the number of completed jobs? Is it NP-hard to compute the maximal weight of scheduled jobs for some?*

2.2   Unit-Length Parallel Jobs

If we allow parallel jobs but on the other had restrict the instances to the case of unit-length jobs, the situation is very similar to the case of restricted job lengths on a single machine. Indeed, instead of using m parallel machines, one can use a single machine and scale all the release times and deadline by a factor of m; the length of parallel jobs is also scaled to $p_j = m$. This transformation does not give an exact formal reduction between these two models, but intuitively they are very close to each other.

Open problem 3. *Consider scheduling of unit-length parallel jobs on m parallel machines. Is there a polynomial-time algorithm to compute a schedule maximizing the number of completed jobs for some fixed $m \geq 2$? Is it NP-hard to compute the maximal weight of scheduled jobs if m is a part of the input and the job sizes are bounded by $size_j \leq C$ for some constant C?*

We do not even know a polynomial-time algorithm for the unweighted variant for $m = 2$, i.e., there are only two machines and naturally every job needs one or two of them.

The best positive result known is that for the so called tall/small jobs, i.e., the case when $size_j \in \{1, m\}$, we can check the feasibility of scheduling of all jobs [2,12]. (The same proof actually works even if $size_j \in \{1, p\}$ for some p that divides m.) However, again, if $size_j \in \{2, 3\}$ it is not even known whether checking feasibility of scheduling of all jobs is polynomial.

Another interesting case is the restriction of job sizes to powers of two, which is motivated by scheduling parallel jobs on hypercubes. Even though the number of job sizes is unbounded, this restriction also avoids the hardness proof by the reduction from partition-type problems. Here it is known that the number of completed jobs can be maximized in polynomial time if all of them are released at the same time [23], or, more generally, if the intervals $[r_j, d_j]$ are nested [24]. It would be very nice to generalize this to general release times; this would be much stronger than a positive solution of the previous open problem for $m = 2$.

3   Online Scheduling

The most natural online algorithm is often some greedy strategy. For our problems, it means that at each time we choose to schedule the pending job with the maximal weight. For equal-length jobs, such an algorithm can be shown to be 2-competitive by a fairly generic argument: We "charge" each job in the optimal schedule to some job in the greedy schedule; either to the job that the greedy schedule runs at the same time (and machine), or to the same job if greedy schedule schedules it earlier than or at the same time as the optimum. Each job receives at most one charge of each type, furthermore all charges go to a job with weight at least the weight of the charged job. The challenge is then to improve the competitive ratio below 2.

3.1   Unit-Length Jobs: Buffer Management and Dynamic Queues

The online variant of unit-length jobs is probably the most extensively studied variant of throughput scheduling. It is motivated by buffer management in network elements, as the unit-length jobs represent packets of some fixed length and different weights represent their priorities.

The generic charging argument implies a 2-competitive algorithm, but to improve this upper bound, the algorithm has to deal with the choice of scheduling either a heavy job (possibly with a large deadline) or an urgent job (possibly with a small weight). The first such algorithm for this problem was randomized; its competitive ratio is $e/(e-1) \approx 1.582$ which is still the best upper bound [6,19]. An improved deterministic algorithm need a significantly more delicate argument; the first such algorithm has competitive ratio $64/33 \approx 1.939$ [8] and the currently best upper bound is $2\sqrt{2} - 1 \approx 1.828$ [14].

In both cases this is quite far from the lower bound, which is 1.25 for randomized algorithms and $\phi \approx 1.618$ for deterministic ones. In both cases, the lower bound instances are 2-bounded, which means that each job has $d_j - r_j \leq 2$, so that it has to be scheduled either immediately upon its arrival or in the next time step. This seems to be a very restricted case, so one would expect that larger lower bounds should be possible for more general instances.

On the other hand, the 2-bounded case is generalized by the case of agreeable deadlines (also called similarly ordered) which includes all instances which can be ordered so that both the sequence of release times and the sequence of deadlines are non-decreasing with j. In this substantially more general case there exist better algorithms, namely 4/3 [18,20] randomized and ϕ-competitive deterministic [21,20], which even matches the lower bound.

In light of this, the question is much less clear and we state it as our next problem. Of course, any improvement of any of the bounds would be nice, but the question whether the 2-bounded case is as hard as the general case is particularly interesting.

Open problem 4. *Does there exist an algorithm for scheduling of arbitrary unit-length jobs on a single machine that matches the performance of algorithms for the special case when $d_j - r_j \leq 2$? That is 1.618-competitive deterministic algorithm and/or 1.25-competitive algorithm?*

The randomized algorithms use only the relative order of the deadlines, not their actual values. This is not true for the deterministic algorithms, and for a long time the use of exact values of deadlines seemed to be essential. However, it turns out that even in the model of dynamic queues, which does allow the algorithm to use only the relative order of deadlines, a 1.897-competitive algorithm exists [5].

3.2   Equal Jobs

In this section we focus on the case when all the jobs have both equal length and unit weights. The generic charging argument again implies that any greedy strategy is 2-competitive [4], and in fact on a single machine no deterministic algorithm can be better [16].

One possibility to improve the algorithm is to use randomization. We know that there exists a 5/3-competitive randomized algorithm [9] for a single machine. This algorithm actually uses only a single bit of randomness: It generates online two schedules (connected by a lock mechanism) and randomly chooses to follow one of them from the beginning. This is still far from the best lower bound, which is only 4/3 for randomized algorithms on a single machine [16].

Open problem 5. *Suppose that all jobs have equal processing time $p_j = p$ and unit weights. What it the best competitive ratio of a randomized algorithm on a single machine with the objective to maximize the (expected) number of completed jobs?*

If we have more machines, the problem appears to be easier. At the first sight this may be surprising, but it is natural: If the online algorithm receives a single job with a large deadline, it has to decide to schedule it at some point. This blocks the only available machine, and the adversary may exploit this by releasing a job with a tight deadline in the next step. In contrast, already with two machines, the algorithm can try to keep one of them as a reserve for such tight jobs. This vaguely corresponds to generating two schedules and randomly choosing one of them, although we are not aware of any formal relationship of the two problems.

For two machines, it is known that the optimal competitive ratio of a deterministic algorithm is 3/2 [11,17]. For more machines, much less is known. The lower bound for deterministic algorithm approaches 6/5 from above for $m \to \infty$ [11]. The best algorithm has a competitive ratio that approaches $e/(e-1) \approx 1.582$ from above for $m \to \infty$ [10]. This algorithm actually works in a much more restricted model in which upon the release of each job it is immediately decided if and when it is scheduled. We note that in this restricted model with immediate decision, no better algorithm for $m \to \infty$ is possible even for unit-length jobs [13].

Open problem 6. *Suppose that all jobs have equal processing time $p_j = p$ and unit weights. For $m \to \infty$, find either a better than $e/(e-1)$-competitive deterministic algorithm or a lower bound larger than 6/5.*

3.3   Fixed Start Times

A special case of the general case is the problem of *interval scheduling*, where all the jobs are tight, i.e., $p_j = d_j - r_j$. This means that upon its release, a job must be started or it is lost. A classical study [22] shows that for variety of cases a 4-competitive preemptive deterministic algorithm exists and this is tight; this includes the case of unit-length jobs (with arbitrary weights) and jobs with weight equal to their length.

This 4-competitive algorithm was recently extended to a variant of parallel machines with speeds (uniformly related machines) [15]. Here each machine has a speed s_i and a job j needs time slot of length p_j/s_i to be run on this machine. Instead of formulating the problem as interval scheduling (which has no clear meaning, as tight jobs cannot be defined), we use the other equivalent formulation: Each job has to be started on one of the machines immediately upon its release, or else it is lost. We call this variant *scheduling with fixed start times*.

We conclude by a restricted variant of scheduling with fixed start times where we cannot even determine the competitive ratio of the greedy algorithm. In this case we have equal jobs once again (i.e., equal lengths and unit weights). Obviously, if we have a single machine or parallel machines with equal speeds, any greedy strategy is optimal. However, this is no longer true with speeds. Actually, with speeds there is no better than 3/2-competitive algorithm for $m \to \infty$ [15]. Any greedy algorithm is 2-competitive by the generic charging argument. However, no better upper bound is known for $m \geq 3$ for any algorithm. The most natural greedy algorithm always starts a job on the fastest available machine. This out to be 4/3-competitive and optimal for $m = 2$, but for $m \to \infty$ we only know that the competitive ratio is at least $25/16 = 1.5625$.

Open problem 7. *Suppose that all jobs have equal processing times and unit weights. For $m \to \infty$ machines with speeds, find a better than 2-competitive deterministic algorithm for scheduling with fixed start times.*

4   Checking Feasibility with Two Job Lengths

Now we return to the offline problem and give a polynomial-time algorithm for deciding if all the jobs can be scheduled if the lengths are restricted to $\{1, p\}$ for some p.

Theorem 1. *Suppose that our input consists of p and an instance with $p_j \in \{1, p\}$ for all jobs. Then there exists an algorithm which in polynomial time decides if there exists a feasible schedule that schedules all jobs.*

Proof. Let us call the jobs with $p_j = 1$ short and the jobs with $p_j = p$ long. For two times s, t let $A_{s,t} = \{j \mid p_j = 1 \wedge s \leq r_j \wedge d_j \leq t\}$, $a_{s,t} = |A_{s,t}|$, $B_{s,t} = \{j \mid p_j = 2 \wedge s \leq r_j \wedge d_j \leq t\}$, and $b_{s,t} = |B_{s,t}|$. I.e., $a_{s,t}$ and $b_{s,t}$ denote the number of short jobs and long jobs, respectively, that need to be scheduled (started and completed) between s and t.

We may assume that $r_j + p_j \leq d_j$ for all jobs; otherwise the instance is infeasible. We may also assume that

$$a_{s,t} \leq t - s \qquad (1)$$

for all t and s, as otherwise the instance is again infeasible.

We now formulate a linear program (2)–(5) that should describe the feasible schedules. It has no objective function, we are only interested in its feasibility. The intended meaning of the variable x_t is the number of long jobs started strictly before time t. For convenience, we set $x_t = 0$ for any $t \leq 0$. Let D be the maximal deadline.

$$\forall t \in \{1, \ldots, D\}: \qquad x_t - x_{t-1} \geq 0 \qquad (2)$$

$$\forall t \in \{p, \ldots, D\}: \qquad x_t - x_{t-p} \leq 1 \qquad (3)$$

$$\forall s, t \in \{0, \ldots, D\}, \ s + p \leq t: \qquad x_{t+1-p} - x_s \leq \left\lfloor \frac{t - s - a_{s,t}}{p} \right\rfloor \qquad (4)$$

$$\forall s, t \in \{0, \ldots, D\}, \ s + p \leq t: \qquad x_{t+1-p} - x_s \geq b_{s,t} \qquad (5)$$

Inequalities (2) make sure that x_t form a non-decreasing sequence, (3) that at most one long job is started in any p adjacent steps and thus the long jobs do not overlap. Inequalities (4) and (5) make sure that sufficiently many slots are available for short and long jobs in each interval.

The crucial observation is that the matrix of our linear program is totally unimodular. Indeed, every row contains at most single -1 and at most single $+1$ and such matrices are known to be totally unimodular. This implies that if the linear program is feasible, there exists an integral solution.

We now prove that the linear program is feasible if and only if there exist a feasible schedule.

Suppose we have a feasible schedule. As noted in the introduction, we may assume that all jobs are started at integral times. Then for (integral) s such that the schedule starts a long job at time s we put $x_{s+1} = x_s + 1$ and for all other s we put $x_{s+1} = x_s$. It is easy to check that this is a feasible solution of our linear program. Inequalities (2) and (3) are trivial. For (4) note that $x_{t+1-p} - x_s$ long jobs are started and completed in the slots $s, \ldots, t - 1$, thus they occupy $p(x_{t+1-p} - x_s)$ of these $t - s$ slots and at least $a_{s,t}$ of these slots is taken by the short jobs. Thus $p(x_{t+1-p} - x_s) \leq t - s - a_{s,t}$ and (4) follows from the integrality of x. Similarly, (5) follows from the fact that at least $b_{s,t}$ long jobs are started and completed in the slots $s, \ldots, t - 1$ in the feasible schedule.

Suppose now that the linear program is feasible and let us fix its integral solution x. The solution x determines at which time slots a long or short job can start. More precisely, let $S_1 = \{r \in \{0, \ldots, D - 1\} \mid x_{r+1-p} = x_{r+2-p} = \ldots = x_r = x_{r+1}\}$ and $S_2 = \{r \in \{0, \ldots, D - 1\} \mid x_{r+1} > x_r\}$. Note that S_2 are exactly the slots where a long job should start and S_1 are exactly the slots where no long job started at time $r \in S_2$ can be running. Now we construct the schedule greedily increasing r from $r = 0$. If $r \in S_1$ then we start a pending short job with the smallest deadline. If $r \in S_2$, then we start a pending long job with the smallest deadline. In both cases we break ties arbitrarily. The definitions of

S_1 and S_2 together with (2) and (3) guarantee that no two jobs overlap in the schedule.

It remains to show that the schedule completes all the jobs before their deadlines. For a contradiction, assume that j is a job with $d_j = t$ that is not completed before t and t is minimal. We distinguish two cases.

Case 1: j is a long job. Let $s - 1 \in S_2$ be the largest time before t such that at time $s - 1$ either no job is started or a job with the deadline larger than t is started. Let $s = 0$ if no such time exists. This implies that $r_j \geq s$ as otherwise we would have started j at time $s - 1$. Similarly, at any time in $S_2 \cap \{s, \ldots, t - p\}$, a long job with a deadline at most t is started (as otherwise we would have started j), and any of these jobs is released at s or later (as otherwise we would have started it at $s - 1$). Thus all these jobs and j belong to $B_{s,t}$. We have $b_{s,t} \geq 1 + |S_2 \cap \{s, \ldots, t - p\}| = 1 + x_{t+1-p} - x_s$, contradicting (5).

Case 2: j is a short job. Let t' be the first time in S_1 such that $t' \geq t$. Note that j' is not completed at time t'. Let $s - 1 \in S_1$ be largest time before t such that at time $s - 1$ either no job is scheduled or a job with the deadline larger than t is scheduled. Let $s = 0$ if no such time exists. This implies that $r_j \geq s$ as otherwise we would have started j at time $s - 1$. Similarly, at any time in $S_1 \cap \{s, \ldots, t - 1\} = S_1 \cap \{s, \ldots, t' - 1\}$, a short job with a deadline at most t is scheduled (as otherwise we would have scheduled j), and any of these job is released at s or later (as otherwise we would have started it at $s - 1$). Thus all these jobs and j belong to $A_{s,t} \subseteq A_{s,t'}$. Furthermore, since $s, t' \in S_1$, the number of time slots in $\{s, \ldots, t' - 1\} \setminus S_1$ is equal to $p(x_{t'+1-p} - x_s)$. We have $a_{s,t'} \geq 1 + |S_1 \cap \{s, \ldots, t' - 1\}| = 1 + (t' - s) - p(x_{t'+1-p} - x_s)$. It follows that

$$x_{t'+1-p} - x_s \geq \frac{1 + t' - s - a_{s,t'}}{p} > \frac{t' - s - a_{s,t'}}{p}.$$

This contradicts (4) if $s + p \leq t'$ or (1) if $s + p > t'$ (note that then $s \geq t' + 1 - p$ and $x_{t'+1-p} - x_s \leq 0$).

The last problem that we need to solve is that the linear program as we have formulated it may have a superpolynomial size if p or the release times or the deadlines are superpolynomial. This is only a technical issue, as there are clearly only polynomially many "important" times. More precisely, we can modify any feasible schedule so that each job starts either at its release time or at the completion time of some other job (the schedule with the lexicographically minimal sequence of start times will satisfy this). Then all the jobs are started at times $r_j + \alpha \cdot p + \beta$ for some job j and $\alpha, \beta \in \{0, \ldots, n\}$ where n is the number of jobs. There are $O(n^3)$ of such times. For all the other t, we set $x_{t+1} = x_t$. By substituting these identities, we get a linear program with only polynomially many variables and also polynomially many distinct constraints, and it can be even written down efficiently. Thus feasibility can be checked in polynomial time.

In fact, we have not been very efficient in our proof. As shown in [12], instead of solving a linear program, we can use a shortest path computation, as the matrix of the linear program is an adjacency matrix of a directed graph. $\square$

References

1. Baptiste, P., Brucker, P., Knust, S., Timkovsky, V.: Ten notes on equal-execution-time scheduling. 4OR 2, 111–127 (2004)
2. Baptiste, P., Schieber, B.: A note on scheduling tall/small multiprocessor tasks with unit processing time to minimize maximum tardiness. J. Sched. 6, 395–404 (2003)
3. Baptiste, P.: Polynomial time algorithms for minimizing the weighted number of late jobs on a single machine with equal processing times. J. Sched. 2, 245–252 (1999)
4. Baruah, S.K., Haritsa, J., Sharma, N.: On-line scheduling to maximize task completions. J. Comb. Math. Comb. Comput. 39, 65–78 (2001)
5. Bieńkowski, M., Chrobak, M., Dürr, C., Hurand, M., Jeż, A., Jeż, Ł., Stachowiak, G.: Collecting weighted items from a dynamic queue. To appear in ACM Trans. Algorithms
6. Chin, F.Y.L., Chrobak, M., Fung, S.P.Y., Jawor, W., Sgall, J., Tichý, T.: Online competitive algorithms for maximizing weighted throughput of unit jobs. Journal of Discrete Algorithms 4, 255–276 (2006)
7. Chrobak, M., Dürr, C., Jawor, W., Kowalik, Ł., Kurowski, M.: A note on scheduling equal-length jobs to maximize throughput. J. Sched. 9, 71–73 (2006)
8. Chrobak, M., Jawor, W., Sgall, J., Tichý, T.: Improved online algorithms for buffer management in QoS switches. ACM Trans. Algorithms 3(4), 19 (2007)
9. Chrobak, M., Jawor, W., Sgall, J., Tichý, T.: Online scheduling of equal-length jobs: Randomization and restarts help. SIAM J. Comput. 36, 1709–1728 (2007)
10. Ding, J., Ebenlendr, T., Sgall, J., Zhang, G.: Online Scheduling of Equal-Length Jobs on Parallel Machines. In: Arge, L., Hoffmann, M., Welzl, E. (eds.) ESA 2007. LNCS, vol. 4698, pp. 427–438. Springer, Heidelberg (2007)
11. Ding, J., Zhang, G.: Online Scheduling with Hard Deadlines on Parallel Machines. In: Cheng, S.-W., Poon, C.K. (eds.) AAIM 2006. LNCS, vol. 4041, pp. 32–42. Springer, Heidelberg (2006)
12. Dürr, C., Hurand, M.: Finding Total Unimodularity in Optimization Problems Solved by Linear Programs. In: Azar, Y., Erlebach, T. (eds.) ESA 2006. LNCS, vol. 4168, pp. 315–326. Springer, Heidelberg (2006)
13. Ebenlendr, T., Sgall, J.: A Lower Bound for Scheduling of Unit Jobs with Immediate Decision on Parallel Machines. In: Bampis, E., Skutella, M. (eds.) WAOA 2008. LNCS, vol. 5426, pp. 43–52. Springer, Heidelberg (2009)
14. Englert, M., Westerman, M.: Considering suppressed packets improves buffer management in QoS switches. In: Proc. 18th Symp. on Discrete Algorithms (SODA), pp. 209–218. ACM/SIAM (2007)
15. Epstein, L., Jeż, Ł., Sgall, J., van Stee, R.: Online Scheduling of Jobs with Fixed Start Times on Related Machines. In: Gupta, A., Jansen, K., Rolim, J., Servedio, R. (eds.) APPROX 2012 and RANDOM 2012. LNCS, vol. 7408, pp. 134–145. Springer, Heidelberg (2012)
16. Goldman, S.A., Parwatikar, J., Suri, S.: Online scheduling with hard deadlines. J. Algorithms 34, 370–389 (2000)
17. Goldwasser, M.H., Pedigo, M.: Online, Non-preemptive Scheduling of Equal-Length Jobs on Two Identical Machines. In: Arge, L., Freivalds, R. (eds.) SWAT 2006. LNCS, vol. 4059, pp. 113–123. Springer, Heidelberg (2006)
18. Jeż, Ł.: Randomized algorithm for agreeable deadlines packet scheduling. In: Proc. 27th Symp. on Theoretical Aspects of Computer Science (STACS), pp. 489–500 (2010)

19. Jeż, Ł.: One to Rule Them All: A General Randomized Algorithm for Buffer Management with Bounded Delay. In: Demetrescu, C., Halldórsson, M.M. (eds.) ESA 2011. LNCS, vol. 6942, pp. 239–250. Springer, Heidelberg (2011)
20. Jeż, Ł., Li, F., Sethuraman, J., Stein, C.: Online scheduling of packets with agreeable deadlines. To appear in ACM Trans. Algorithms
21. Li, F., Sethuraman, J., Stein, C.: An optimal online algorithm for packet scheduling with agreeable deadlines. In: Proc. 16th Symp. on Discrete Algorithms (SODA), pp. 801–802. ACM/SIAM (2005)
22. Woeginger, G.J.: On-line scheduling of jobs with fixed start and end times. Theoret. Comput. Sci. 130, 5–16 (1994)
23. Ye, D., Zhang, G.: Maximizing the throughput of parallel jobs on hypercubes. Inform. Process. Lett. 102, 259–263 (2007)
24. Zajíček, O.: A note on scheduling parallel unit jobs on hypercubes. Int. J. on Found. Comput. Sci. 20, 341–349 (2009)

Preemptive Coordination Mechanisms
for Unrelated Machines*

Fidaa Abed[1] and Chien-Chung Huang[2]

[1] Max-Planck-Institut für Informatik
fabed@mpi-inf.mpg.de
[2] Humboldt-Universität zu Berlin
villars@informatik.hu-berlin.de

Abstract. We investigate coordination mechanisms that schedule n jobs on m unrelated machines. The objective is to minimize the latest completion of all jobs, i.e., the makespan. It is known that if the mechanism is non-preemptive, the price of anarchy is $\Omega(\log m)$. Both Azar, Jain, and Mirrokni (SODA 2008) and Caragiannis (SODA 2009) raised the question whether it is possible to design a coordination mechanism that has constant price of anarchy using preemption. We give a negative answer.

All deterministic coordination mechanisms, if they are symmetric and satisfy the property of independence of irrelevant alternatives, even with preemption, have the price of anarchy $\Omega(\frac{\log m}{\log \log m})$. Moreover, all randomized coordination mechanisms, if they are symmetric and unbiased, even with preemption, have similarly the price of anarchy $\Omega(\frac{\log m}{\log \log m})$.

Our lower bound complements the result of Caragiannis, whose BCOORD mechanism guarantees $O(\frac{\log m}{\log \log m})$ price of anarchy. Our lower bound construction is surprisingly simple. En route we prove a Ramsey-type graph theorem, which can be of independent interest.

On the positive side, we observe that our lower bound construction critically uses the fact that the inefficiency of a job on a machine can be unbounded. If, on the other hand, the inefficiency is not unbounded, we demonstrate that it is possible to break the $\Omega(\frac{\log m}{\log \log m})$ barrier on the price of anarchy by using known coordination mechanisms.

1 Introduction

The input is a set $\mathcal{I}$ of jobs and a set $\mathcal{M}$ of machines. Each job $i \in \mathcal{I}$ has a processing time p_{ij} on a machine $j \in \mathcal{M}$. Each job is controlled by a selfish player, who aims to minimize the completion time of his job while disregarding the welfare of other players. Each machine, based on the information of the incoming jobs and a certain scheduling policy, decides the finishing time of each incoming job. The scheduling policy of the machines is referred to as the *coordination mechanism* [3] in algorithmic game theory literature. The objective is to

---

* Part of this work is based on the first author's Master thesis in Max-Planck-Institut für Informatik.

L. Epstein and P. Ferragina (Eds.): ESA 2012, LNCS 7501, pp. 12–23, 2012.
© Springer-Verlag Berlin Heidelberg 2012

minimize the latest finishing time of any job. Such an objective is conventionally called the *makespan*.

The above scenario is very similar to the *unrelated machine scheduling problem* $(R||C_{max})$ that has been extensively studied in the literature, e.g.,[9]. However, unlike the traditional setting where a central authority decides which job is to be assigned to which machine, here we assume that each job is controlled by a selfish player. Our setting captures certain real world situations, such as the Internet, where there is no central authority and users are self-interested.

We assume that in the coordination mechanism, a machine is allowed to use only the information of the incoming jobs, when deciding how to schedule them, while the information of the other jobs and how the other machines schedule them are irrelevant. Using the terminology of [1], such coordination mechanisms are *local policies*.[1]

This assumption is a natural one: the machines may not be able to communicate among themselves efficiently to make a scheduling decision, especially in a very fluid environment such as the Internet.

Coordination mechanisms can be *non-preemptive* or *preemptive*. In the former, the jobs on a machine are processed sequentially, and each job, once started, has to be processed in an uninterrupted manner; in the latter, a machine can interrupt an ongoing job and resume it later, or it can intentionally introduce delays, during which the machine just lies idling.

Our focus will be on the *pure Nash equilibrium* (PNE), where no player can unilaterally change his strategy, i.e., the machine, to reduce the completion time of his job. It can be expected that given the selfishness of the players, the makespan in a PNE can be sub-optimal. The worst ratio of the makespan in a PNE against that in an optimal schedule is called the *price of anarchy* (PoA) [8].

Table 1. Summary of various coordination mechanisms

Coordination mechanisms	PoA	PNE	Anonymous	Characteristics
ShortestFirst [7]	$\Theta(m)$	Yes	No	Strongly local, non-preemptive
LongestFirst [7]	Unbounded	No	No	Strongly local, non-preemptive
MAKESPAN [7]	Unbounded	Yes	Yes	Strongly local, preemptive
RANDOM [7]	$\Theta(m)$	No	Yes	Strongly local, non-preemptive
EQUI [4]	$\Theta(m)$	Yes	Yes	Strongly local, preemptive
AJM-1 [1]	$\Theta(\log m)$	No	No	Local, non-preemptive
AJM-2 [1]	$O(\log^2 m)$	Yes	No	Local, preemptive
ACOORD [2]	$O(\log m)$	Yes	No	Local, preemptive
BCOORD [2]	$\Theta(\frac{\log m}{\log \log m})$	?	Yes	Local, preemptive
CCOORD [2]	$O(\log^2 m)$	Yes	Yes	Local, preemptive

---

[1] A more stringent assumption proposed by Azar et al. [1] is that of the *strongly local policies*. In this case, a machine makes the scheduling decision only by the processing times of the incoming jobs on it, while the processing times of these jobs on other machines are irrelevant. Azar et al. [1] have shown that strongly local policies have much higher lower bound in terms of the price of anarchy. In this work, we consider only local policies.

It is desirable to have coordination mechanisms that have small PoA. Table 1 gives a summary of the various mechanisms that have been proposed so far in the literature. The "PNE" column shows whether the existance of a pure Nash equilibrium is guaranteed or not. For non-preemptive coordination mechanisms, Azar, Jain, and Mirrokni [1] designed a mechanism that achieves $O(\log m)$ PoA. This turns out to be optimal, since later Fleischer and Svitkina [6] showed that all non-preemptive coordination mechanisms have $\Omega(\log m)$ PoA.

Since non-preemptive mechanisms have the $\Omega(\log m)$ PoA barrier, an obvious question to ask is whether preemption can beat this lower bound. Caragiannis [2] showed that using preemption, his BCOORD mechanism achieves $O(\frac{\log m}{\log \log m})$ PoA. Both Azar et al. and Caragiannis raised the question whether it is possible to achieve constant PoA by preemption. We answer in the negative. (See the next section for the formal definitions of "symmetric", "IIA", and "unbiased.")

Theorem 1. *All deterministic coordination mechanisms, if they are symmetric and satisfy independence of irrelevant alternatives (IIA) property, even with preemption, have the price of anarchy $\Omega(\frac{\log m}{\log \log m})$. Moreover, all randomized coordination mechanisms, if they are symmetric and unbiased, even with preemption, have similarly the price of anarchy $\Omega(\frac{\log m}{\log \log m})$. These lower bounds hold even for the special case of* restricted assignment $(B||C_{max})$, *where each job can go to at most 2 machines on which it has the processing time of 1.*

Therefore, the BCOORD mechanism of Caragiannis [2] is essentially the best possible. We prove this theorem in Section 2.

In our proof, we use the fact that a job can be assigned to only a subset of all machines, i.e., the restricted assignment model $(B||C_{max})$. Let the *inefficiency* of a job i on a machine j be defined as $\frac{p_{ij}}{\min_{j' \in \mathcal{M}} p_{ij'}}$. The restricted assignment instances imply that the inefficiency of jobs on some machines is unbounded. This raises the issue whether it is possible to circumvent the $\Omega(\frac{\log m}{\log \log m})$ lower bound by assuming that inefficiency is bounded. We give a positive answer in Section 3. We show that the inefficiency-based mechanism [1] achieves $O(I)$ price of anarchy, where I is the largest possible inefficiency. Due to the space constraint, some proofs and a more detailed discussion on the related work are deferred to the full version.

1.1   Our Assumptions and Technique

In proving our lower bounds, it is critical to first state our assumptions and definitions precisely. Recall that each job $i \in \mathcal{I}$ is associated with a *load characteristic* $\mathbf{p}_i = \langle p_{i1}, p_{i2}, \cdots, p_{i|\mathcal{M}|} \rangle$. If a job i cannot be processed at a machine j, let $p_{ij} = \infty$. Each job may or may not have an ID. When each job has a unique ID, we say these jobs are *non-anonymous*. When jobs do not have (unique) IDs, we say they are *anonymous*.

Our lower bound construction shares similar ideas to those of Azar et al. [1] and Fleischer and Svitkina [6]. For each machine, we give a set of jobs that are *indistinguishable* so as to confuse it.

Definition 1. *Let $j \in \mathcal{M}$ be a machine. Then two jobs i, i' are* indistinguishable *to j if the following holds.*

1. $p_{ij} = p_{i'j} = 1$,
2. *there exists two different machines $j_i \neq j_{i'}$, $j \notin \{j_i, j_{i'}\}$ and $p_{ij_i} = p_{i'j_{i'}} = 1$,*
3. $p_{ij*} = \infty$ *for $j^* \in \mathcal{M}\backslash\{j, j_i\}$ and $p_{i'j*} = \infty$ for $j^* \in \mathcal{M}\backslash\{j, j_{i'}\}$.*

A set of jobs are indistinguishable to machine j if every two of them are indistinguishable to j.

Definition 2. *Let $\mathcal{C}$ be a deterministic coordination mechanism. $\mathcal{C}$ is said to be* symmetric *if the following holds.*

Let $j, j' \in \mathcal{M}$ be two different machines. Let $\mathcal{I}_1$ be a set of indistinguishable jobs to j and $\mathcal{I}_2$ a set of indistinguishable jobs to j'. Suppose that there exists a one-to-one correspondence $\gamma : \mathcal{I}_1 \rightarrow \mathcal{I}_2$ satisfying the following condition:

For every job $i \in \mathcal{I}_1$, there exists a job $\gamma(i) \in \mathcal{I}_2$ so that $p_{ij} = p_{\gamma(i)j'} = 1$. Furthermore, there exists a permutation $\sigma_i : \mathcal{M}\backslash\{j\} \rightarrow \mathcal{M}\backslash\{j'\}$ so that $p_{ij''} = p_{\gamma(i)\sigma_i(j'')}$ for all $j'' \in \mathcal{M}\backslash\{j\}$.

Then the set of the finishing times $t_{11} \leq t_{12} \leq \cdots \leq t_{1|\mathcal{I}_1|}$ for $\mathcal{I}_1$ on machine j and the set of the finishing times $t_{21} \leq t_{22} \leq \cdots \leq t_{2|\mathcal{I}_2|}$ for $\mathcal{I}_2$ on machine j' are the same. i.e., $t_{1l'} = t_{2l'}$ for $1 \leq l' \leq |\mathcal{I}_1|$.

Intuitively speaking, a coordination mechanism is symmetric, if two machines, when they are presented with two sets of jobs that *look essentially the same*, then the finishing times for these two sets of jobs are the same on both machines. All coordination mechanisms in Table 1 are symmetric.

As a clarification, the above assumption states nothing regarding the *order* of the jobs to be finished on the machines. It is only about the *set* of their finishing times.

Definition 3. *Let $\mathcal{C}$ be a deterministic coordination mechanism. $\mathcal{C}$ is said to satisfy the* independence of irrelevant alternative *(IIA) property if the following holds.*

Let $j \in \mathcal{M}$ be a machine and i, i' be two different jobs. Let $\{i, i'\} \subseteq \mathcal{I}' \subset \mathcal{I}$. If j is presented with the job set $\mathcal{I}'$ and it lets job i to be finished before i', then it also will let i to be finished before i' when it is presented with a job set $\mathcal{I}' \cup \{k\}$ for some job $k \in \mathcal{I}\backslash\mathcal{I}'$.

Informally speaking, the IIA property states that if job i is "preferred" over i' by machine j, then this "preference" should not change because of the availability of some other jobs $k \notin \{i, i'\}$. The IIA property appears as an axiom in voting theory, bargaining theory, and logic [10].

The next lemma states that if a mechanism satisfies the IIA property, then each machine must have some sort of "preference list" over a set of indistinguishable jobs.

Lemma 1. *Let $\mathcal{I}^*(j)$ be a set of indistinguishable jobs to machine j. A deterministic coordination mechanism satisfies the IIA property iff, each machine $j \in \mathcal{M}$ has a strict linear order $\mathbb{L}_j$ over jobs in $\mathcal{I}^*(j)$, so that when j is presented with a subset $\mathcal{I}' \in \mathcal{I}^*(j)$ of these indistinguishable jobs, a job $i \in \mathcal{I}'$ has smaller completion than i' only when i precedes i' in the order $\mathbb{L}_j$.*

Remark 1. We note that it is possible for a mechanism to satisfy the IIA property without "explicitly" having a strict linear order $\mathbb{L}_j$ over indistinguishable jobs: a machine can let all the indistinguishable jobs finish at the same time. This is indeed what several known deterministic mechanisms would have done, including MAKESPAN [7], BCOORD [2], and CCOORD [2]. In this case, the order $\mathbb{L}_j$ as stated in Lemma 1 can be just an arbitrary order over these indistinguishable jobs.

An IIA-satisfying deterministic mechanism could ask a machine to let all incoming indistinguishable jobs finish at the same time. But another possibility is that a machine j lets the incoming indistinguishable jobs finish at different times, when j does have an explicit linear order $\mathbb{L}_j$ over these indistinguishable jobs. An obvious candidate for $\mathbb{L}_j$ is an order over the job IDs when all jobs are non-anonymous. But even when jobs are anonymous, a machine still can use machine IDs of those machines on which these indistinguishable jobs can be processed to decide the order. To illustrate our point, assume that there are three machines, j_1, j_2, and j_3, and two jobs, i_1 and i_2. i_1 has the load characteristic $\langle 1, 1, \infty \rangle$ while i_2 has the load characteristic $\langle 1, \infty, 1 \rangle$. Even though these two jobs are indistinguishable to machine j_1, j_1 can deterministically choose to let i_1 finish first, if it prefers machine j_2 over j_3. By the above discussion, we make our assumption.

Definition 4. *Let $\mathcal{C}$ be a deterministic coordination mechanism satisfying the IIA property. Then the linear order $\mathbb{L}_j$ of each machine for a set of indistinguishable jobs as stated in Lemma 1 can take one of the following two forms.*

- ***Non-anonymous Case:*** *it is the preference list of machine j over the job IDs, or*
- ***Anonymous Case:*** *the preference list of machine j over the machine IDs. In particular, given two indistinguishable jobs i and i', i precedes i' in $\mathbb{L}_j$ if machine j prefers the machine j_i to $j_{i'}$, where j_i is the only other machine on which $p_{ij_i} = 1$ and $j_{i'}$ the only other machine on which $p_{i'j_{i'}} = 1$.*

Remark 2. Our assumptions stated in Definition 4 about the linear orders of the machines over the indistinguishable jobs are the same used as those by Azar et al. [1] and Fleischer and Svitkina [6] in their lower bound construction for non-preemptive coordination mechanisms. Suppose that machines do not have such preferences over the job IDs or the machine IDs, then a IIA-satisfying deterministic mechanism can only let all indistinguishable jobs finish at the same time (thus the linear order $\mathbb{L}_j$ of a machine j is an arbitrary order). We show that the same lower bound holds easily for this case in Section 2.3.

Sections 2.1 and 2.2 deal with the non-anonymous and anonymous cases respectively. The main technical challenge in our constructions is that the linear order $\mathbb{L}_j$ on the machines $j \in \mathcal{M}$ can differ from machine to machine. We need certain strategies to arrange the given set of jobs and machines so that in a PNE, the makespan is relatively high. Our lower bound construction for anonymous case is the most interesting part of this work. As a by-product, we derive a Ramsey-type graph theorem that has a similar flavor to the one obtained by Fleischer and Svitkina [6] when they proved their lower bound for non-preemptive mechanisms. In addition, our proof does not use Erdős-type probabilistic method; it is constructive and yields a polynomial time algorithm.

We now discuss our assumptions about randomized coordination mechanisms. It seems that there is not much work done concerning randomized mechanisms. The only one that we are aware of is the RANDOM mechanism of Immorlica et al. [7], which proceeds by ordering the incoming jobs of a machine uniformly at random and processing them non-preemptively. Cole et al. [5] used a randomized mechanism for the minimizing the weighted sum objective.

When randomized mechanisms are used, a PNE is an assignment in which no player can unilaterally change his machine to decrease the *expected* finishing time of his job.

Definition 5. *Let $\mathcal{C}$ be a randomized coordination mechanism.*

1. *$\mathcal{C}$ is* unbiased *if a machine $j \in \mathcal{M}$, when presented with a set of indistinguishable jobs, lets each of them have the same expected finishing time.*
2. *$\mathcal{C}$ is* symmetric *if two machines $j, j' \in \mathcal{M}$, when they are presented with the same number of indistinguishable jobs, let these two sets of indistinguishable jobs have the same set of expected finishing times.*

2   Lower Bounds

All of our three lower bound constructions are based on a specific tree structure, which we will call the *equilibrium tree*. In such a tree, the root has k children, each of its k children has $k-1$ children, and each of these $k(k-1)$ grandchildren has $k-2$ children and so on. Generally, a vertex whose distance to the root is l has $k-l$ children. See Figure 1 as an illustration for the case of $k = 3$. For convenience, we will use a bit unconventional terminology by referring to the root as the vertex in the k-th level, while its children are vertices in the $(k-1)$-st level and so on. Thus, a vertex in the l-th level has l children. We assume k to be some arbitrary large number.

In all our constructions, a vertex in the equilibrium tree corresponds to a machine, while an edge (j, j') in the tree corresponds to a job i. Such a job has processing time $p_{ij} = p_{ij'} = 1$, while $p_{ij''} = \infty$ for $j'' \notin \{j, j'\}$. Suppose that j is in level t while j' is in level $t-1$, we say j is the parent machine (vertex) and j' is the child machine (vertex) of job $i = (j, j')$.

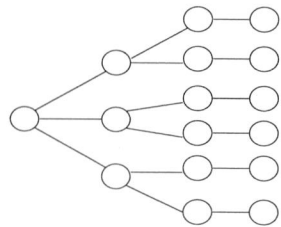

Fig. 1. The equilibrium tree with k=3

In our constructions, we will arrange the jobs and the machines corresponding to the equilibrium tree in such a way that in an optimal solution, all jobs will be assigned to their child machines, while in a PNE, all jobs are assigned to their parent machines. Clearly, in the optimal solution, the makespan is 1, while in the PNE, because there are k jobs on the root machines, the last job to be finished on it will have completion time at least k. Observe that the number of the vertices in the equilibrium tree is

$$\tilde{m} = 1 + \sum_{l=0}^{k} \prod_{s=0}^{l} (k - s) = 1 + k! \left(\sum_{s=0}^{k} \frac{1}{s!} \right) < 3 \left(\frac{k}{e} \right)^k \sqrt{2\pi k} < k^{2k}.$$

The function $f(x) = \frac{\ln x}{\ln \ln x}$ is strictly increasing when $x \geq e^e$. As we assume k to be large, both $\tilde{m}$ and k^{2k} are larger than e^e. So $f(\tilde{m}) < f(k^{2k})$, implying

$$\frac{\ln \tilde{m}}{\ln \ln \tilde{m}} < \frac{2k \ln k}{\ln 2 + \ln k + \ln \ln k} < \frac{2k \ln k}{\ln k} < 2k.$$

Thus, if the number of the machines initially given is m and $m = \theta(\tilde{m})$, by the above inequality, we can conclude that the PoA in the constructed instance is at least $k = \Omega(\frac{\log m}{\log \log m})$.

In the next two subsections, we consider two cases: deterministic mechanisms with non-anonymous jobs and with anonymus jobs. In the full version, we discuss deterministic mechanisms when neither jobs nor machines have IDs and randomized mechanisms.

2.1   Deterministic Mechanisms: Non-anonymous Case

In this section, we assume that jobs are non-anonymous and a machine, when faced with a set of indistinguishable jobs, uses its preferences over the job IDs to decide the order of these indistinguishable jobs.

Let $m = \tilde{m}$, i.e., all m machines given will be part of the equilibrium tree. We assign the machines arbitrarily to the equilibrium tree and we will create $m - 1$ jobs corresponding to their edges. Without loss of generality, let the job IDs to be from 1 to $m - 1$. Recall that each machine may have a different preference order over these IDs. In the following, let X denote the set of job IDs that have been used in the algorithm.

Let $X = \emptyset$.
For level l from $k - 1$ down to 0
  For each machine j in level l
    Let j' be the machine corresponding to j's parent vertex.
    Choose t to be the lowest ranking ID on j's preference list that are not included
    in X.
    Create a job i with ID t and let $p_{ij} = p_{ij'} = 1$ and $p_{ij''} = \infty$ for $j'' \in \mathcal{M} \backslash \{j, j'\}$.
    $X := X \cup \{t\}$.
  End
End

Fig. 2. An algorithm to construct the equilibrium tree in the non-anonymous case

We now apply the procedure in Figure 2 to construct the instance.

Observe that by the algorithm, a machine prefers all the jobs that can be assigned to its vertices corresponding to its children in the equilibrium tree over the job that can be assigned to its parent in the equilibrium tree. This property will be used in the proof.

Theorem 2. *In the constructed instance, the PoA is* $\Omega(\frac{\log m}{\log \log m})$.

Proof. Clearly in the optimal assignment, each job should be assigned to the child machine. We now argue that if each job is assigned to its parent machine, we have a PNE. If this is the case, then the PoA is at least $k = \Omega(\frac{\log m}{\log \log m})$ and we have the proof.

So suppose not. Then there exists some job i between machine j at level l and machine j' at level $l - 1$ and i has incentive to deviate from j to j'. Before the deviation, j has l incoming jobs that are indistinguishable; after the deviation, j' has similarly l incoming jobs that are indistinguishable. By Definition 2, the set of complete times for these l incoming jobs in both cases are identical $t_1 \leq t_2 \leq \cdots \leq t_l$. By our construction, job i would have the completion time t_l after its deviation since its ID ranks lower than the IDs of all other $l - 1$ jobs of machine j'. Before the deviation, job i has completion time $t_{l'}$ for some $1 \leq l' \leq l$. Since $t_{l'} \leq t_l$, we get a contradiction. □

2.2 Deterministic Mechanisms: Anonymous Case

In this section, we assume that jobs are anonymous and a machine, when faced with a set of indistinguishable jobs, uses its preferences over the machine IDs to decide the order of these indistinguishable jobs.

Assume that m machines are given, each with its own preference list over each other. (For convenience, the preference list over machine IDs can be interpreted as a preference order over other machines). We will choose a subset of machines ($\tilde{m}$ of them) and assign them to the equilibrium tree. Our goal is to make sure that each machine, if assigned to be a vertex in the equilibrium tree, ranks the machine corresponding to the parent vertex lower than all machines corresponding to its child vertices. We will discuss later how large m has to be (relative to $\tilde{m}$) so that such a construction is always possible. In the following, when the context is clear, we use the terms vertex and machine interchangeably.

Let n_s be the number of vertices in the s-th level in the equilibrium tree of totally k levels. Then

$$n_k = 1, \text{ and } n_{l-1} = l n_l, \quad \forall 1 \leq l \leq k.$$

We will define another sequence n'_s for $0 \leq s \leq k$. Roughly speaking, this sequence denotes the numbers of vertices we will need in each level s in our construction.

We now describe our algorithm. It proceeds in $k - 1$ iterations. In the beginning of each iteration l, we maintain n'_l equilibrium trees of l levels and n'_{l+1} equilibrium trees of $(l - 1)$ levels. Let the roots of the former set be A and the roots of the latter set be B. We discard all vertices of the latter set of equilibrium trees, except for their roots, i.e., B. Let the vertices in B be $v_1, v_2, \cdots v_{n'_{l+1}}$ and we process them in this order. For v_1, choose the $l + 1$ highest ranking vertices on v_1's preference list among all vertices in A. Make v_1 the parent of these roots. (So we have an equilibrium tree of $(l + 1)$ levels rooted at v_1.) Remove these roots from A and we process v_2, v_3, and so on, in the same manner. At the end, all vertices in B are roots of equilibrium trees of $l+1$ levels, while the remaining vertices in A are the roots of the equilibrium trees of l levels. Note that if we make sure that

$$n'_l - (l + 1)n'_{l+1} = n'_{l+2},$$

then in beginning of the next iteration, iteration $l+1$, we have the same condition as the the current iteration: n'_{l+1} equilibriums trees of $(l + 1)$ levels and n'_{l+2} equilibrium trees of l levels.

We now formally define the sequence $\{n'_s\}_{s=0}^k$.

$$n'_k = n_k.$$
$$n'_{k-1} = k n'_k.$$
$$n'_{k-s} = (k - s + 1)n'_{k-s+1} + n'_{k-s+2}, \quad \forall 2 \leq s \leq k.$$

We choose m to be $n'_0 + n'_1$. The full algorithm is presented in Figure 3.

Lemma 2. *The final outcome of the above algorithm is an equilibrium tree of k levels; moreover, in such a tree, every non-leaf/non-root vertex ranks all of its child vertices higher than its parent vertex.*

Proof. We prove by establishing the following claim.

Claim 1. *In the beginning of iteration l, $1 \leq l \leq k - 1$, there are n'_l equilibrium trees of l levels and n'_{l+1} equilibrium trees of $l - 1$ levels. Moreover, each root of the former ranks its child vertices higher than any of the roots of the latter.*

Proof. We prove by induction. The base case $l = 1$ holds trivially based on what the first for loop of the algorithm and the fact that $n'_0 - n'_1 = n'_2$. By induction hypothesis, in the beginning of the $(l-1)$-st iteration, there are n'_{l-1} equilibrium

Out of the m given vertices, choose n_1' arbitrary vertices and denote them as B and the rest as A.
For each vertex v in B
    Choose the highest ranking vertex $v' \in A$.
    Make v the parent of v'.
    $A = A \backslash \{v'\}$.
End   // The prepartion is done.
For level l from 1 up to $k - 1$
    Let the roots of the equilibrium trees of $(l - 1)$ levels be B; throw away all other vertices in these trees.
    Let the roots of the equilibrium trees of l levels be A.
    For each vertex v in B
        Choose $(l + 1)$ highest ranking vertices $v_1, v_2, \cdots, v_{l+1}$ among all vertices in A on v's preference list.
        Make v the parents of $v_1, v_2, \cdots, v_{l+1}$.
        $A = A \backslash \{v_i\}_{i=1}^{l+1}$.
    End
End

Fig. 3. An algorithm to construct the equilibrium tree in the anonymous case

trees of $l - 1$ levels, and n_l' equilibrium trees of $l - 2$ levels. At the end of $(l-1)$-st iteration, the latter set is thrown away except their roots. ln_l' of the former will be merged with these roots into n_l' equilibrium trees of l levels. So there are only $n_{l-1}' - ln_l' = n_{l+1}'$ equilibrium trees of $(l - 1)$ levels left. This completes the first part of the induction step. The second part of the induction step follows trivially from the way we choose to merge the equilibrium trees. $\square$

By the first part of the above claim, in the beginning of the last iteration, we have n_{k-1}' equilibrium trees of $k-1$ levels and n_k' equilibrium trees of $k-2$ levels. By the algorithm, at the end of the last iteration, we have $n_k' = 1$ equilibrium tree of k levels and $n_{k-1}' - kn_k' = 0$ equilibrium trees of $k - 1$ levels. So we are left with exactly an equilibrium tree of k levels. For the second part of the lemma, choose any vertex v at level l. Observe that such a vertex must be a root in the beginning of iteration l and its parent u must be one of the roots of those equilibrium trees of $l - 1$ levels. By the second part of the above claim, we conclude that v prefers its child vertices to u. The lemma follows. $\square$

We now bound m by establishing the following lemma.

Lemma 3. $n_l' < n_l + n_{l+1}'$, for each $0 \le l \le k - 1$.

Proof. We prove by induction. Let the base case be $k - 1$. Then $n_{k-1}' = kn_k' = kn_k = n_{k-1} < n_{k-1} + n_k'$. For the induction step,

$$n_l' = (l+1)n_{l+1}' + n_{l+2}' < (l+1)(n_{l+1} + n_{l+2}') + n_{l+2}' = n_l + (l+2)n_{l+2}' \le n_l + n_{l+1}',$$

where the first inequality follows from induction hypothesis. So the lemma follows. $\square$

Lemma 4. $\tilde{m} \le m \le 2\tilde{m}$.

Proof. The first inequality holds because by Lemma 3, after throwing away some vertices from the given m vertices, the algorithm ends up with an equilibrium tree of k levels, whose number of vertices is exactly $\tilde{m}$.

For the second inequality, by the definition n'_k and the previous lemma, we know that

$$n'_k \le n_k$$
$$n'_l \le n_l + n'_{l+1} \qquad \text{for all } 0 \le l \le k-1$$

Summing up the above inequalities, we have $n'_0 \le \sum_{l=0}^{k} n_l = \tilde{m}$. The lemma holds because

$$m = n'_0 + n'_1 < 2n'_0 \le 2\tilde{m}.$$

$\square$

Theorem 3. *In the constructed instance, the PoA is $\Omega(\frac{\log m}{\log \log m})$.*

Proof. Clearly in the optimal assignment, each job should be assigned to the child machine. We now argue that if each job is assigned to its parent machine, we have a PNE. If this is the case, then the PoA is at least $k = \Omega(\frac{\log \tilde{m}}{\log \log \tilde{m}}) = \Omega(\frac{\log m}{\log \log m})$, where the second equality follows from Lemma 4, and we would have the proof.

So suppose not. Then there exists some job i between machine j at level l and machine j' at level $l - 1$ and i has incentive to deviate from j to j'. Before the deviation, j has l incoming jobs that are indistinguishable; after the deviation, j' has similarly l incoming jobs that are indistinguishable. By Definition 2, the set of complete times for these l incoming jobs in both cases are identical $t_1 \le t_2 \le \cdots \le t_l$. By Lemma 3, machine j' prefers all child vertices over j, therefore, it also prefers all its other incoming jobs over i. This implies that job i would have the completion time t_l after its deviation. Before the deviation, job i has completion time $t_{l'}$ for some $1 \le l' \le l$. Since $t_{l'} \le t_l$, we arrive at a contradiction.

$\square$

The following corollary follows from Lemmas 3 and 4.

Corollary 1. *Let T be a tree of the following property: the root has k children, and the vertex whose distance to the root is l has $k - l$ children itself.*

Let $G = (V, E)$ be a graph, where each vertex in V has a strictly-ordered preference over other vertices. Suppose that $|V| \ge 2|T|$. Then we can always find a subset of vertices $V' \subset V$, $|V'| = |T|$, and assign these vertices to T so that a vertex $u \in V'$ prefers the vertices in V' corresponding to its children in T to the vertex in V' corresponding to its parent in T.

3   Upper Bound on Price of Anarchy When Inefficiency Is Bounded

In this section, we demonstrate that the $\Omega(\frac{\log m}{\log \log m})$ lower bound on PoA can be circumvented if the inefficiency of the jobs on the machines is bounded by I.

We analyze the upper bound of PoA of the inefficiency-based mechanism proposed by Azar et al. [1]. Let $p_i = \min_{j \in \mathcal{M}} p_{ij}$, the minimum processing time of a job on all machines. In this mechanism, each machine $j \in M$ non-preemptively processes the incoming jobs based on nondecreasing order of their inefficiency on it: that is, given two jobs i and i', if $\frac{p_{ij}}{p_i} < \frac{p_{i'j}}{p_{i'}}$, then job i should be processed before j (ties are broken by job IDs). This rather intuitive mechanism turns out to be optimal for non-preemptive mechanism. As shown by Azar et al. [1], its PoA is $O(\log m)$, matching the lower bound of non-preemptive mechanism.

Theorem 4. *The inefficiency-based mechanism has PoA at most $I + 2 \log I + 2$.*

References

1. Azar, Y., Jain, K., Mirrokni, V.S. (Almost) optimal coordination mechanisms for unrelated machine scheduling. In: SODA, pp. 323–332 (2008)
2. Caragiannis, I.: Efficient coordination mechanisms for unrelated machine scheduling. In: SODA, pp. 815–824 (2009)
3. Christodoulou, G., Koutsoupias, E., Nanavati, A.: Coordination Mechanisms. In: Díaz, J., Karhumäki, J., Lepistö, A., Sannella, D. (eds.) ICALP 2004. LNCS, vol. 3142, pp. 345–357. Springer, Heidelberg (2004)
4. Cohen, J., Dürr, C., Thang, N.K.: Non-clairvoyant scheduling games. Theory Comput. Syst. 49(1), 3–23 (2011)
5. Cole, R., Correa, J.R., Gkatzelis, V., Mirrokni, V.S., Olver, N.: Inner product spaces for minsum coordination mechanisms. In: STOC, pp. 539–548 (2011)
6. Fleischer, L., Svitkina, Z.: Preference-constrained oriented matching. In: Proceedings of the Seventh Workshop on Analytic Algorithmics and Combinatorics (ANALCO), pp. 66–73 (2010)
7. Immorlica, N., Li, L(Erran), Mirrokni, V.S., Schulz, A.S.: Coordination mechanisms for selfish scheduling. Theor. Comput. Sci. 410(17), 1589–1598 (2009)
8. Koutsoupias, E., Papadimitriou, C.H.: Worst-case equilibria. Computer Science Review 3(2), 65–69 (2009)
9. Lenstra, J.K., Shmoys, D.B., Tardos, É.: Approximation algorithms for scheduling unrelated parallel machines. Math. Program. 46, 259–271 (1990)
10. Wikipedia. Independence of irrelevant alternatives, http://en.wikipedia.org/wiki/

Hierarchical Hub Labelings for Shortest Paths

Ittai Abraham, Daniel Delling, Andrew V. Goldberg, and Renato F. Werneck

Microsoft Research Silicon Valley
{ittaia,dadellin,goldberg,renatow}@microsoft.com

Abstract. We study hierarchical hub labelings for computing shortest paths. Our new theoretical insights into the structure of hierarchical labels lead to faster preprocessing algorithms, making the labeling approach practical for a wider class of graphs. We also find smaller labels for road networks, improving the query speed.

1   Introduction

Computing point-to-point shortest paths in a graph is a fundamental problem with many applications. Dijkstra's algorithm [14] solves this problem in near-linear time [18], but for some graph classes sublinear-time queries are possible if preprocessing is allowed (e.g., [12,16]). In particular, Gavoille et al.'s *distance labeling* algorithm [16] precomputes a *label* for each vertex such that the distance between any two vertices s and t can be computed given only their labels. A special case is *hub labeling* (HL), where the label of each vertex u consists of a collection of vertices (the *hubs* of u) with their distances to or from u. Labels obey the *cover property*: for any two vertices s and t, there exists a vertex w on the shortest s–t path that belongs to both labels (of s and t). Cohen at al. [9] give a polynomial-time algorithm to approximate the smallest labeling within a factor of $O(\log n)$, where n is the number of vertices. The average label size can be quite large ($\Omega(\sqrt[3]{n})$ even for planar graphs [16]), but not always. On real-world DAG-like graphs, for instance, labels are quite small in practice [8].

Abraham et al. [4] conjecture that road networks have a small *highway dimension*, and show that if the highway dimension is polylogarithmic, so is the label size. This motivated the experimental study of labels for road networks. Unfortunately, preprocessing algorithms with theoretical guarantees on the label size run in $\Omega(n^4)$ time [9,4,1], which is impractical for all but small graphs. In previous work [2], we showed how to compute labels for road networks based on *contraction hierarchies* (CH) [17], an existing preprocessing-based shortest path algorithm. The resulting labels are unexpectedly small [2], with average size 85 on a graph of Western Europe [13] with 18 million vertices. The corresponding query algorithm is the fastest currently known for road networks.

In this paper we study *hierarchical* hub labelings, a natural special case where the relationship "vertex v is in the label of vertex w" defines a partial order on the vertices. We obtain theoretical insights into the structure of hierarchical labelings and their relationship to vertex orderings. In particular, we show that

L. Epstein and P. Ferragina (Eds.): ESA 2012, LNCS 7501, pp. 24–35, 2012.

for every total order there is a minimum hierarchical labeling. We use the theory to develop efficient algorithms for computing the minimum labeling from an ordering, and for computing orderings which yield small labelings. We also show that CH and hierarchical labelings are closely related and obtain new top-down CH preprocessing algorithms that lead to faster CH queries.

Our experimental study shows that our new label-generation algorithms are more efficient and compute labels for graphs that previous algorithms could not handle. For several graph classes (not only road networks), the labels are small enough to make HL the fastest distance oracle in practice. Furthermore, the new algorithms compute smaller labels; in particular, we reduce the average label size for Western Europe from 85 to 69, accelerating the fastest method by 8%.

This paper is organized as follows. After settling preliminaries in Section 2, Section 3 considers the relationship between vertex orderings and hierarchical labelings. Section 4 presents several techniques for computing vertex orderings, and Section 5 shows how to compute labels from orderings. Section 6 studies the relationship between labelings and CH, while Section 7 presents our experimental evaluation. Section 8 contains concluding remarks. Details omitted due to space constraints can be found in the full version of this paper [3].

2   Preliminaries

The input to the shortest path problem is a graph $G = (V, A)$, with $|V| = n$, $|A| = m$, and length $\ell(a) > 0$ for each arc a. The length of a path P in G is the sum of its arc lengths. The point-to-point problem (a *query*) is, given a source s and a target t, to find the distance $\text{dist}(s, t)$ between them, i.e., the length of the shortest path P_{st} between s and t in G. We assume that shortest paths are unique, which we can enforce by breaking ties consistently.

The standard solution to this problem is Dijkstra's algorithm [14]. It builds a shortest path tree by processing vertices in increasing order of distance from s. For every vertex v, it maintains the length $d(v)$ of the shortest s–v path found so far, as well as the predecessor $p(v)$ of v on the path. Initially $d(s) = 0$, $d(v) = \infty$ for all other vertices, and $p(v) = null$ for all v. At each step, a vertex v with minimum $d(v)$ value is extracted from a priority queue and *scanned*: for each arc $(v, w) \in A$, if $d(v) + \ell(v, w) < d(w)$, we set $d(w) = d(v) + \ell(v, w)$ and $p(v) = w$. The algorithm terminates when all vertices have been processed. Dijkstra's worst-case running time, denoted by $\text{Dij}(G)$, is $O(m + n \log n)$ [15] in the comparison model, and even better in weaker models [18].

For some applications, even linear time is too slow. For faster queries, *labeling* algorithms preprocess the graph and store a *label* with each vertex; the s–t distance can be computed from the labels of s and t. Our focus is on *hub labeling* (HL), a special case. For each vertex $v \in V$, HL builds a forward label $L_f(v)$ and a backward label $L_b(v)$. The forward label $L_f(v)$ consists of a sequence of pairs $(u, \text{dist}(v, u))$, where u is a vertex (a *hub* in this context). Similarly, $L_b(v)$ consists of pairs $(u, \text{dist}(u, v))$. Note that the hubs in the forward and backward labels of u may differ. Collectively, the labels obey the *cover property*: for any two vertices s and t, $L_f(s) \cap L_b(t)$ must contain at least one vertex on the shortest s–t path. For an s–t query, among all vertices $u \in L_f(s) \cap L_b(t)$ we pick the one minimizing $\mathrm{dist}(s,u) + \mathrm{dist}(u,t)$ and return this sum. If the entries in each label are sorted by hub ID, this can be done with a coordinated sweep over the two labels, as in mergesort.

We say that the forward (backward) *label size of* v, $|L_f(v)|$ ($|L_b(v)|$), is the number of hubs in $L_f(v)$ ($L_b(v)$). The time for an s–t query is $O(|L_f(s)|+|L_b(t)|)$. The *maximum label size* (denoted by M) is the size of the biggest label. The *labeling* $\mathcal{L}$ is the set of all forward and backward labels. We denote its *size* by $|\mathcal{L}| = \sum_v (|L_f(v)| + |L_b(v)|)$, while $L_a = |\mathcal{L}|/(2n)$ denotes the average label size.

Cohen et al. [9] show how to generate in $O(n^4)$ time labels whose average size is within a factor $O(\log n)$ of the optimum. Their algorithm maintains a set of U of uncovered shortest paths (initially all the paths) and labels $L_f(v)$ and $L_b(v)$ (initially empty) for every vertex v. Each iteration of the algorithm selects a vertex v and a set of labels to add v to so as to maximize the ratio between the number of paths covered and the increase in total label size.

Given two distinct vertices v, w, we say that $v \preceq w$ if $L_f(v) \cup L_b(v)$ contains w. A labeling is *hierarchical* if $\preceq$ is a partial order. We say that this order is *implied* by the labeling. Cohen et al.'s labels are not necessarily hierarchical.

Given a total order on vertices, the *rank function* $r : V \to [1 \dots n]$ ranks the vertices according to the order. We will call the corresponding order r.

3   Canonical Labelings

We say that a hierarchical labeling $\mathcal{L}$ *respects* a total order r if the implied partial order is consistent with r. Given a total order r, a *canonical labeling* is the labeling that contains only the following hubs. For every shortest path P_{st}, the highest ranked vertex $v \in P_{st}$ is in the forward label of s and in the backward label of t (with the corresponding distances). A canonical labeling is hierarchical by construction: the vertex v that we add to the labels of s and t has the highest rank on P_{st}, and therefore $r(v) \geq r(s)$ and $r(v) \geq r(t)$. This also implies that the canonical labeling respects r.

Lemma 1. *Let $\mathcal{L}$ be a hierarchical labeling. Then the set of vertices on any shortest path has a unique maximum element with respect to the partial order implied by $\mathcal{L}$.*

Proof. The proof is by induction on the number of vertices on the path. The result is trivial for paths with a single vertex. Consider a path $v_1 \dots v_k$ with $k > 1$. The subpath $v_2 \dots v_k$ has a maximum vertex v_i by the inductive assumption. Consider the subpath $v_1 \dots v_i$. The internal vertices v_j are not in $L_b(v_i)$ by the choice of v_i. Therefore either $v_i \in L_f(v_1)$ (and v_i is the maximum vertex on the path), or $v_1 \in L_b(v_i)$ (and v_1 is the maximum vertex).   □

Given a hierarchical labeling $\mathcal{L}$, the lemma implies that all total orders r that $\mathcal{L}$ respects have the same canonical labeling $\mathcal{L}'$.

Theorem 1. *Let $\mathcal{L}$ be a hierarchical labeling, r any total order such that $\mathcal{L}$ respects r, and $\mathcal{L}'$ the canonical labeling for r. Then $\mathcal{L}'$ is contained in $\mathcal{L}$.*

Proof. Consider the shortest path P from s to t, and let v be the maximum rank vertex on P. Let $\mathcal{L} = (L_f, L_b)$. We show that $v \in L_f(s)$; the case $v \in L_b(t)$ is similar. Consider the shortest path P' from s to v. Since shortest paths are unique, $P' \subseteq P$, and therefore v is the maximum rank vertex on P'. It follows that the only vertex of P' that is in $L_b(v)$ is v. Thus v must be in $L_f(s)$.   □

We now consider how to extract the canonical labeling $\mathcal{L}'$ from a hierarchical labeling $\mathcal{L}$. One approach is to first extract a total order r from $\mathcal{L}'$, then build the canonical label from the order. We can find r with a topological sort of the DAG representing the partial order induced by $\mathcal{L}$. We can then easily build canonical labels from r in $O(n\text{Dij}(G))$ time, as Section 5 will show.

We can often do better by simply *pruning* the labels in $\mathcal{L}$. We explain how to prune forward labels; backward labels can be dealt with similarly. Consider a vertex $w \in L_f(v)$. We must keep w in $L_f(v)$ if and only if it is the maximum-rank vertex on the shortest v–w path P_{vw}. Since we have not computed the ranks, we must test for maximality indirectly. We use the following observation: if the highest ranked vertex in P_{vw} is $u \neq w$, then u must be in both $L_f(v)$ and $L_b(w)$ (since it belongs to the canonical label). By running what is essentially a v–w HL query, we can determine if such a vertex u exists. If so, we delete w from $L_f(v)$. The algorithm takes $O(M)$ time to process each of the $|\mathcal{L}|$ vertex-hub pairs (v, w), for a total time of $O(|\mathcal{L}|M) = O(nM^2)$. (Recall that M is the maximum label size.) When labels are not too big ($M = O(\sqrt{m})$), this is faster than the $O(n\text{Dij}(G))$ approach mentioned above.

4   Vertex Orderings

Canonical labelings are the smallest hierarchical labelings defined by a vertex ordering. By the *quality* of an ordering we mean the size of its implied labeling. In this section, we discuss several ways of computing vertex orderings. We review a known approach, improve existing algorithms, and introduce new ones.

Contraction Hierarchies. CH [17] has been heavily studied in the context of point-to-point shortest paths in road networks. It is a preprocessing-based algorithm, but not a labeling algorithm. During preprocessing, CH orders all vertices and applies the *shortcut* operation to each vertex in that order. Intuitively, the ordering is from the least to the most important vertex. When applied to v, the shortcut operation temporarily removes v from the graph and adds as few arcs as needed to preserve the distances between the remaining vertices. More precisely, for any two neighbors u and w of v, it runs a *witness search* (Dijkstra) to compute $\text{dist}(u, w)$ in the current graph (without v). If $\ell(u, v) + \ell(v, w) < \text{dist}(u, w)$, it adds a *shortcut* arc (u, w) with $\ell(u, w) = \ell(u, v) + \ell(v, w)$. The output of CH preprocessing is a graph $G^+ = (V, A \cup A^+)$, where A^+ denotes the set of shortcuts, as well as the order in which vertices were shortcut. The CH query algorithm runs bidirectional Dijkstra on G^+, but considers only upward arcs.

The CH query performance and $|A^+|$ highly depend on the order in which vertices are shortcut. The best known orderings [17,19] use online heuristics to estimate how "important" each vertex is based on local graph properties (such as the net number of shortcuts added if the vertex were contracted).

The running time of CH preprocessing depends on the number of shortcuts. In the best case, when both G and G^+ have constant degree, it uses $O(n)$ space and runs in $O(nW)$ time, where W denotes the time for a witness search. In road networks, witness searches are local Dijkstra searches, which makes preprocessing quite fast. If G^+ is dense, however, shortcuts may need $O(n^2)$ witness searches, causing the standard implementation of CH preprocessing to run in $O(n^3W)$ time and $O(n^2)$ space.

Even for road networks, previous experiments [2] showed that the ordering computed by CH preprocessing can be improved. Next, we discuss ways to compute orderings with better worst-case time bounds that yield smaller labels.

Top-Down. We turn to an algorithm (which we call TD) that selects vertices top-down, from most to least important (as opposed to CH, which is bottom-up). Conceptually, each unpicked vertex v maintains the set U_v of all *uncovered paths*, i.e., shortest paths that contain v but do not contain any previously selected vertex. Each iteration selects the next most important vertex v^* according to some criterion that depends on U_{v^*}, then updates all sets $U_v = U_v \setminus U_{v^*}$. This is repeated until all paths are covered.

We consider two versions of this algorithm, with different selection criteria. The *covering* version (TDc) always picks the vertex v which maximizes $|U_v|$. The *weighted covering* version (TDwc) selects the v maximizing $|U_v|/(s_v + t_v)$, where s_v and t_v are the sizes of the sets consisting of the first and last vertices on the paths in U_v, respectively. TDwc is inspired by Cohen et al.'s algorithm [9].

An obvious implementation of TD is to compute every U_v from scratch in each round. This takes $O(n)$ space but $O(n^2\mathrm{Dij}(G))$ time, which is impractical even for mid-sized graphs. We therefore propose an *incremental* implementation of TDc that runs in $O(n\mathrm{Dij}(G))$ time. It can be extended to other TD algorithms as long as each iteration can pick the vertex v^* (given the updated U_v sets) in $O(\mathrm{Dij}(G))$ time. In particular, this holds for TDwc.

The incremental TDc implementation maintains a shortest path tree T_r for every vertex r, representing all uncovered shortest paths starting at r. Initially, these are full shortest path trees, computed in $O(n\mathrm{Dij}(G))$ time. Moreover, each vertex v also explicitly maintains a count $c(v) = |U_v|$ of the uncovered shortest paths containing v. For $r, v \in V$, the number of shortest paths starting at r and containing v is equal to the size of the subtree of T_r rooted at v. We can initialize all $c(\cdot)$ in $O(n^2)$ total time by adding the subtree sizes for all r.

Now consider an iteration in which a vertex v is selected. It hits several previously uncovered paths, and we must update the data structures accordingly. Consider a tree T_r. In the beginning of the iteration, it represents all uncovered paths that start at r. The ones that contain v are exactly those represented in the subtree of T_r rooted at v. We delete this subtree by setting to *null* the parent pointers of all of its vertices. Since each vertex can be removed from each tree once, the total cost of all subtree deletions over the entire algorithm is $O(nm)$.

While traversing the subtree, we also compute its size δ_r. Then we traverse the path in T_r from v to r and for every vertex w on the path, we subtract δ_r from $c(w)$. The total traversal cost (over all iterations of the algorithm) is $O(n^3)$ in the worst case (although much faster in practice). For provably better bounds, we need a more careful implementation of these *path updates*.

Lemma 2. *A* TD *iteration can update the* $c(\cdot)$ *values in* $O(\mathrm{Dij}(G))$ *time.*

Proof. Consider an iteration in which a vertex v is selected as the most important. For a vertex w, let Δ_w be the amount by which $c(w)$ must eventually be decreased. From the trivial algorithm above, Δ_w is the sum of δ_u over all u such that w is an ancestor of v in T_u. (Recall that δ_u is the size of the subtree of T_u rooted at v.) We claim we can compute all Δ_w's without explicitly traversing these paths. Consider the set S of all vertices whose $c(\cdot)$ values have to be updated; S is exactly the set of vertices in the union of the u–v paths on all trees T_u that contain v. Since all these paths end at v (and shortest paths are unique), their union is a shortest path tree I_v rooted at v. Also, it is easy to see that Δ_w is the sum of δ_u over all descendants u of w in I_v (including w itself). These values can thus be computed by a bottom-up traversal of I_v (from the leaves to the root v). The overall bottleneck is building I_v, which takes $O(\mathrm{Dij}(G))$ time.   □

The last aspect we need to consider is the time to select the next vertex in each iteration. For TDc, a simple $O(n)$ traversal of all $c(v)$ values suffices. For TDwc, we also need to know the values s_v and t_v for each candidate vertex v. The value of s_v is $|T_v|$, which we can maintain explicitly and update in $O(n)$ time per iteration. We keep t_v by maintaining the in-trees T_v', which are analogous to the out-trees T_v. Then $t_v = |T_v'|$. Maintaining the out-trees has no effect on the asymptotic complexity of the algorithm. We have the following result:

Theorem 2. *The* TDc *and* TDwc *algorithms can be implemented to run in* $\Theta(n\mathrm{Dij}(G))$ *time and* $\Theta(n^2)$ *space.*

The $\Theta(n^2)$ time and space requirements limit the size of the problems one can solve in practice. We are careful to minimize constant factors in space in our implementation. This is why we recompute subtree sizes instead of maintaining them explicitly. Moreover, we do not store distances within the trees (which we do not need); we only need the topology, as defined by parent pointers. We store the parent pointer of a vertex with in-degree $\deg(v)$ with only $\lceil \log_2(\deg(v)+1) \rceil$ bits, which is enough to decide which incoming arc—if any—is the parent. To find the children of a vertex v, we examine its neighbors w and check if v is the parent of w. Since we only look for the children of each vertex in each tree once (right before the vertex is deleted), the total time spent on traversals is $O(mn)$. Note that n, m, and the degree distribution fully determine the running time and space consumption of TD. This is not the case for CH, which can be much faster or slower than TD depending on the graph topology and cost function.

Range Optimization. Although TD yields better orderings then CH, its $\Theta(n^2)$ space and time requirements limit its applicability. We now discuss how to combine ideas from TD and CH to obtain better orderings for large graphs.

An obvious approach is to run CH preprocessing until the contracted graph is small enough, then use TD to order the remaining vertices. This idea has been used with the non-incremental implementation of TDc, and indeed improves the ordering [2], even though it can only optimize the most important vertices.

To improve the ordering among other vertices, we propose a *range optimization* algorithm. It takes an ordering r and parameters X and Y as input, and reorders all vertices v with $X < r(v) \leq Y$. It first shortcuts all vertices v with $r(v) \leq X$, creating a graph G' with $n - X$ vertices. It then runs Dijkstra's algorithm from each v in G' to compute all shortest paths U that are not covered by vertices w with $r(w) > Y$. The search from v is responsible for paths starting at v, and can stop as soon as all vertices in the priority queue have at least one vertex w with $r(w) > Y$ as an ancestor. Finally, we run TDwc with G' and U as input. Note that we only need to store a *partial* tree for each vertex v. If X and Y are chosen appropriately, the trees are small and can be stored explicitly.

This algorithm reoptimizes a range within a given vertex ordering. Intuitively, more important vertices in the range move up and less important ones move down. To allow a vertex to move between arbitrary positions, we use an *iterative range optimization algorithm*. It covers the interval $[1, n]$ by k overlapping intervals $[X_i, Y_i]$ with $X_1 = 0$, $Y_k = n$, $X_i < Y_i$, and $X_{i+1} \leq Y_i$. The algorithm starts with some vertex ordering, e.g., the one given by CH. It then proceeds in k steps, each reordering a different interval. In practice, it pays to process the intervals in decreasing order of importance; this is faster than processing them in increasing order, and the resulting labels are at least as small.

5   Computing the Labels

We now discuss how to build a canonical labeling from an ordering efficiently. As defined in Section 3, we must add the maximum-rank vertex on each shortest path P_{st} to the labels $L_f(s)$ and $L_b(t)$.

The most straightforward approach is to use Dijkstra to build a shortest path tree out of each vertex s. We then traverse the tree from s to the leaves, computing for each vertex v the maximum-rank vertex w on the s–v path. We add $(w, \text{dist}(s, w))$ to $L_f(s)$ and $(w, \text{dist}(s, v) - \text{dist}(s, w))$ to $L_b(v)$, if not already in the labels. Note that, if we use TD to order vertices, we can incorporate this approach and compute labels on the fly. However, the $O(n\text{Dij}(G))$ running time of this approach makes it infeasible for large networks.

A previous approach [2] is based on CH. Given G^+, we construct $L_f(s)$ for a given vertex s as follows ($L_b(s)$ is computed analogously). Run Dijkstra's algorithm from s in G^+ pruning arcs (u, v) with $r(u) > r(v)$ and add all scanned vertices to $L_f(s)$. To make the labeling canonical we apply label pruning (Section 3). Note that when pruning the label of v, we need labels of higher-ranked vertices, which we achieve by computing labels from high to low rank vertices.

We now introduce a *recursive* label-generation procedure, which is more efficient. It borrows from CH the shortcut operation, but not the query. Given a graph G_i where all vertices u with $r(u) < i$ are shortcut, and an ordering r, we pick the lowest-rank vertex v (with $r(v) = i$) and shortcut it, obtaining the graph G_{i+1}. Then we recursively compute the labels in G_{i+1}. (The basis of the recursion is G_n, with a single vertex s, when we just set $L_f(s) = L_b(s) = \{(s,0)\}$.) To extend the labeling to v, we *merge* the labels of its neighbors. We show how to construct $L_f(v)$; the construction of $L_b(v)$ is symmetric. We initialize $L_f(v)$ with $(v,0)$ and then, for every arc (v,w) in G_i and for every pair $(x, d_w(x)) \in L_f(w)$, we add to $L_f(v)$ a pair $(x, d_w(x) + \ell(v,w))$. If the same hub x appears in the labels of multiple neighbors w, we keep only the pair that minimizes $d_w(x) + \ell(v,w)$. Since labels are sorted by hub ID, we build the merged label by traversing all neighboring labels in tandem, as in mergesort.

Theorem 3. *The recursive algorithm computes a correct hierarchical labeling for an ordering r.*

We can make the labeling computed by this procedure canonical by pruning each label immediately after it is generated, as in the CH-based approach.

6   Building Contraction Hierarchies

A vertex ordering determines a canonical labeling. It also determines a contraction hierarchy (G^+): simply shortcut the vertices according to this order, which may take up to $O(n^3 W)$ time. We give an $O(n\text{Dij}(G))$ algorithm that does not use shortcutting. The key observation is that a shortcut (v,w) is added by CH preprocessing if and only if v and w are the two highest-ranked vertices on P_{vw}.

Given an ordering r, we first compute all shortcuts (u,v) with $r(u) < r(v)$. To do so, we run Dijkstra's algorithm from each vertex u of the graph. Whenever we scan a vertex v, we check whether v is the first vertex on the path from u to v with $r(v) > r(u)$. If so, we add (u,v) to G^+. We can stop the search as soon as all vertices v in the priority queue have an ancestor w with $r(w) > r(u)$. The shortcuts (v,u) with $r(v) > r(u)$ can be computed analogously. This algorithm builds G^+ in $O(n\text{Dij}(G))$ time.

Given a small canonical labeling, we can compute the shortcuts for G^+ even faster. We use $L_f(u)$ to compute all shortcuts (u,v) with $r(u) < r(v)$ as follows. For each pair $(v,d) \in L_f(u)$, we check whether there is at least one other pair $(v', d') \in L_f(u)$ such that $d = d' + \text{dist}(v', v)$. If there is none, we add (u,v) to G^+. Note that we need to run an HL query for $\text{dist}(v', v)$. We compute (v,u) with $r(v) < r(u)$ from $L_b(u)$ analogously. The overall running time of this approach is $O(nM^3)$, since for each label we must run $O(M^2)$ queries, each in $O(M)$ time. In practice, using the HL one-to-many query [11] may accelerate this approach.

7   Experiments

We implemented CH, HL, and TD in C++ and compiled them with Visual C++ 2010, using OpenMP for parallelization. We use a 4-ary heap as priority queue.

The experiments were conducted on a machine with two Intel Xeon X5680 CPUs and 96 GB of DDR3-133 RAM, running Windows 2008R2 Server. Each CPU has 6 cores (3.33 GHz, 6 x 64 kB L1, 6 x 256 kB L2, and 12 MB L3 cache). Preprocessing uses all 12 cores, but queries are sequential. Our implementation of CH follows [19] and uses $E_q(u) + O_q(u) + lev(u)$ as priority function. $E_q(u)$ is the *edge quotient* (number of shortcuts added divided by the number of elements removed if u were shortcut); $O_q(u)$ is the *original edges quotient* (number of original arcs in the shortcuts added divided by the number of original arcs in the shortcuts removed); and $lev(u)$ is the level of u. Initially, $lev(u) = 0$ for each vertex u. Whenever a vertex v is shortcut, the level of all its neighbors v in the graph with the shortcuts added so far is set to $\max\{lev(v), lev(u) + 1\}$. We implement HL queries as in [2], with no compression unless mentioned otherwise.

We first consider the orderings produced by TDWC on various inputs: road networks [13], meshes [21], communication and collaboration networks [5], and artificial inputs (Delaunay triangulations, random geometric graphs, small-world graphs [20,22]). For each graph, Table 1 shows its size ($|V|$), its density ($|A|/|V|$), the TDWC running time (TIME), the average resulting label size (L_a), and HL query time (QT). We then report the ratio of shortcuts to original arcs ($|A^+|/|A|$) in the graph G^+ produced from this order. Finally, we report query times and number of vertex scans for both CH and bidirectional Dijkstra queries (BD).

The results show that TDWC is very efficient for many graphs. Moreover, HL is practical for a wide variety graphs, with small labels and queries in microseconds or less. In fact, HL queries are always faster than BD, even when labels are large.

Table 1. TDWC labels on five groups of instances: meshes, road networks, artificial graphs, communication networks, and collaboration graphs. QT is the query time.

| | | | HL | | | CH | | | BD | |
| | | | TIME | SIZE | QT | $|A^+|$ | VERT | QT | VERT | QT |
INSTANCE	$|V|$	$|A|$/$|V|$	[s]	L_a	[μs]	/$|A|$	SCANS	[μs]	SCANS	[μs]
face	12530	5.8	20	48.8	0.3	2.2	143	55	3653	598
feline	20629	6.0	70	66.8	0.6	2.6	211	104	4529	810
horse	48485	6.0	362	108.4	0.8	2.8	355	250	13371	2154
bay-d	321270	2.5	14274	44.5	0.4	1.1	119	33	107813	13097
bay-t	321270	2.5	12739	29.3	0.3	1.0	78	17	93829	12352
G_pin_pout	99995	10.0	1833	8021.6	66.5	196.4	13166	1165430	354	160
smallworld	100000	10.0	2008	5975.0	49.7	102.8	9440	612682	523	203
rgg_18	262085	10.0	10694	225.6	2.2	1.0	723	668	85186	31612
del_18	262144	6.0	12209	97.1	0.7	2.0	304	183	49826	10818
klein_18	262144	6.0	10565	1718.7	12.8	13.6	4135	50844	4644	1590
as-22july06	22963	4.2	86	22.5	0.2	1.2	70	30	63	125
caidaRouter	190914	6.4	7399	343.8	3.1	2.7	2081	6527	377	298
astro-ph	14845	16.1	40	231.2	2.1	2.4	798	1529	151	112
cond-mat	36458	9.4	249	293.3	2.7	2.9	1145	2661	220	134
preferential	100000	10.0	1727	586.4	4.5	11.3	2265	10941	198	215
coAuthors	299067	6.5	19488	789.3	6.9	3.7	4047	26127	530	378

Due to better locality (it just merges two arrays), HL can still be slightly faster even when BD scans much fewer vertices than there are hubs in the labels, as in smallworld and G_n_pin_pout. The fact that labels are large for some graphs is not surprising, given known lower bounds [16].

CH queries are always slower than HL, and usually faster than BD. In several cases, however, CH queries are worse than BD in terms of both running times and number of scanned vertices. CH has a weaker termination condition, cannot balance the two sides of the search space, and works on a denser graph.

As predicted, TDwc preprocessing time depends mostly on the network size and not on its structure (unlike CH preprocessing). TDwc preprocessing uses $\Theta(n^2)$ space, limiting the size of the graphs we can handle in memory to a few hundred thousand vertices; at under six hours, running times are still acceptable.

Table 2 compares the TDwc orderings to those computed by TDc and CH preprocessing. In each case, we give ordering time, number of shortcuts in G^+, CH query search space, and label size. All values in the table are relative to those obtained from the TDwc ordering. Compared to TDwc, TDc produces larger labels (by 1% to 8%) and more shortcuts; since TDwc is not much slower, it is usually a better choice. Compared to TDwc, the CH ordering produces bigger labels (by 16% to 44%) and the CH query search space increases by 8% to 31%. Interestingly, this happens despite the fact that CH preprocessing adds fewer shortcuts (up to 30%). CH preprocessing is much faster than TDwc when G^+ is sparse, but much slower otherwise. In several cases, it did not finish in six hours.

We now consider the effects of range optimization on the road network of Western Europe [13] (18 million vertices and 42 million arcs) with travel times.

Table 2. Performance of orderings obtained from CH and TDc, relative to TDwc

INSTANCE	CH-PREPROCESSING				TDc			
	TIME	$|A^+|$	#SC	L_a	TIME	$|A^+|$	#SC	L_a
face	0.082	0.93	1.11	1.17	0.785	1.06	1.05	1.04
feline	0.115	0.95	1.11	1.16	0.651	1.07	1.05	1.05
horse	0.135	0.97	1.16	1.23	0.644	1.07	1.05	1.05
bay-d	0.000	0.88	1.31	1.36	0.589	1.04	1.06	1.06
bay-t	0.000	0.88	1.29	1.30	0.849	1.05	1.06	1.05
G_pin_pout	DNF	–	–	–	0.808	1.12	0.99	1.02
smallworld	DNF	–	–	–	0.752	1.09	1.00	1.02
rgg_18	0.045	0.96	1.16	1.21	0.948	1.07	1.05	1.08
del_18	0.003	0.94	1.08	1.16	0.521	1.06	1.04	1.05
klein_18	DNF	–	–	–	0.989	1.06	1.00	1.02
as-22july06	31.021	0.96	1.29	1.44	0.668	1.00	0.98	1.01
caidaRouter	3.623	0.70	1.17	1.23	0.798	1.01	0.97	1.02
astro-ph	56.873	0.79	1.25	1.20	0.689	1.04	1.00	1.03
cond-mat	24.771	0.81	1.20	1.19	0.658	1.05	1.00	1.03
preferential	DNF	–	–	–	0.826	1.15	0.98	1.06
coAuthors	DNF	–	–	–	0.761	1.09	0.99	1.05

Table 3 reports the label size after applying range optimization multiple (i) times, using the ranges $[n - 2^{17}, n]$, $[n - 2^{20}, n - 2^{15}]$, $[n - 2^{22}, n - 2^{17}]$, and $[0, n - 2^{20}]$. As initial ordering ($i = 0$), we use the one given by CH. We report the *total* time (in seconds) to obtain this ordering and the performance of CH queries. Each iteration takes about 70 minutes. The first one reduces label sizes by 25%, but later ones have a smaller impact. The effect on CH queries is similar. After five iterations the label size is stable. Experiments on smaller inputs indicate that the final ordering is close to the one obtained by TDwc.

Table 3. Range opt

i	TIME	L_a	#SC [μs]
0	108	95.52	288 96.4
1	4546	73.17	213 80.1
2	8925	70.32	206 77.8
3	12737	69.58	203 75.5
4	16730	69.17	201 74.2
5	20606	69.01	200 73.4
6	24512	69.01	200 73.4

For Western Europe, we also compare various versions of our algorithm with our previous HL implementations [2], Contraction Hierarchies, transit-node routing (TNR) [6], and its combination with arc-flags (TNR+AF) [7]. HL-0 uses pure CH preprocessing (label size 97.6), HL-15 uses TDwc on the topmost 32768 vertices (size 78.3), HL-17 optimizes the top 131072 (size 75.0), while HL-∞ uses five iterations of range optimization (size 69.0). In all cases, building G^+ takes 105 s, generating the labels takes 55 s, and the remaining time is spent on improving the ordering. As explained in [2], the "local" version uses 8/24 compression and an index, whereas the "global" version is index free with a partition oracle (cell size 20 000).

Table 4. Performance of various algorithms. HL and CH preprocessing use 12 cores, others are sequential.

method	prepro [h:m]	space [GB]	query [μs]
CH [17]	0:02	0.4	96.381
TNR [7]	0:58	3.7	1.775
TNR+AF [7]	2:00	5.7	0.992
HL local [2]	2:39	20.1	0.572
HL global [2]	2:45	21.3	0.276
HL-0 local	0:03	22.5	0.700
HL-15 local	0:05	18.8	0.556
HL-17 local	0:25	18.0	0.545
HL-∞ local	5:43	16.8	0.508
HL-∞ global	6:12	17.7	0.254

We observe that our new techniques (faster label generation, incremental TDwc) accelerate preprocessing by two orders of magnitude. With longer preprocessing, we improve the best previous query times (HL global [2]) by 8%. Preprocessing for HL-0 and HL-15 is fast enough for real-time traffic updates on large metropolitan areas, even when taking turn costs into account [10].

8   Conclusion

Our study of hierarchical labelings makes HL practical on a wider class of problems. The work raises several natural questions. It would be interesting to study the relationship between hierarchical labelings, which are more practical, and general hub labelings, which have theoretical guarantees on the label size. In preliminary experiments, we ran Cohen et al.'s algorithm [9] on significantly smaller graphs (given its $O(n^4)$ running time). It produces labels that are about as good as those created by TDwc for some graph classes (such as road networks), but asymptotically smaller for others (like small-world graphs). Another open question is how to efficiently compute (or approximate) optimal hierarchical labelings, or a good lower bound on their size. Finally, we would like to reduce the space consumption of HL further, beyond existing compression schemes [2].

Acknowledgments. We thank Ruslan Savchenko for implementing [9].

References

1. Abraham, I., Delling, D., Fiat, A., Goldberg, A.V., Werneck, R.F.: VC-Dimension and Shortest Path Algorithms. In: Aceto, L., Henzinger, M., Sgall, J. (eds.) ICALP 2011. LNCS, vol. 6755, pp. 690–699. Springer, Heidelberg (2011)
2. Abraham, I., Delling, D., Goldberg, A.V., Werneck, R.F.: A Hub-Based Labeling Algorithm for Shortest Paths in Road Networks. In: Pardalos, P.M., Rebennack, S. (eds.) SEA 2011. LNCS, vol. 6630, pp. 230–241. Springer, Heidelberg (2011)
3. Abraham, I., Delling, D., Goldberg, A.V., Werneck, R.F.: Hierarchical Hub Labelings for Shortest Paths. MSR-TR-2012-46. Microsoft Research (2012)
4. Abraham, I., Fiat, A., Goldberg, A.V., Werneck, R.F.: Highway Dimension, Shortest Paths, and Provably Efficient Algorithms. In: SODA, pp. 782–793 (2010)
5. Bader, D.A., Meyerhenke, H., Sanders, P., Wagner, D.: 10th DIMACS Implementation Challenge – Graph Partitioning and Graph Clustering (2011)
6. Bast, H., Funke, S., Matijevic, D., Sanders, P., Schultes, D.: In Transit to Constant Shortest-Path Queries in Road Networks. In: ALENEX, pp. 46–59 (2007)
7. Bauer, R., Delling, D., Sanders, P., Schieferdecker, D., Schultes, D., Wagner, D.: Combining Hierarchical and Goal-Directed Speed-Up Techniques for Dijkstra's Algorithm. ACM J. of Exp. Algo. 15(2.3), 1–31 (2010)
8. Cheng, J., Yu, J.X.: On-line Exact Shortest Distance Query Processing. In: EDBT, pp. 481–492. ACM Press (2009)
9. Cohen, E., Halperin, E., Kaplan, H., Zwick, U.: Reachability and Distance Queries via 2-hop Labels. SIAM J. Comput. 32 (2003)
10. Delling, D., Goldberg, A.V., Pajor, T., Werneck, R.F.: Customizable Route Planning. In: Pardalos, P.M., Rebennack, S. (eds.) SEA 2011. LNCS, vol. 6630, pp. 376–387. Springer, Heidelberg (2011)
11. Delling, D., Goldberg, A.V., Werneck, R.F.: Faster Batched Shortest Paths in Road Networks. In: ATMOS, OASIcs, vol. 20, pp. 52–63 (2011)
12. Delling, D., Sanders, P., Schultes, D., Wagner, D.: Engineering Route Planning Algorithms. In: Lerner, J., Wagner, D., Zweig, K.A. (eds.) Algorithmics of Large and Complex Networks. LNCS, vol. 5515, pp. 117–139. Springer, Heidelberg (2009)
13. Demetrescu, C., Goldberg, A.V., Johnson, D.S. (eds.): The Shortest Path Problem: Ninth DIMACS Implementation Challenge. DIMACS, vol. 74. AMS (2009)
14. Dijkstra, E.W.: A Note on Two Problems in Connexion with Graphs. Numerische Mathematik 1, 269–271 (1959)
15. Fredman, M.L., Tarjan, R.E.: Fibonacci Heaps and Their Uses in Improved Network Optimization Algorithms. J. Assoc. Comput. Mach. 34, 596–615 (1987)
16. Gavoille, C., Peleg, D., Pérennes, S., Raz, R.: Distance Labeling in Graphs. Journal of Algorithms 53, 85–112 (2004)
17. Geisberger, R., Sanders, P., Schultes, D., Vetter, C.: Exact Routing in Large Road Networks Using Contraction Hierarchies. Transportation Science (2012)
18. Goldberg, A.V.: A Practical Shortest Path Algorithm with Linear Expected Time. SIAM Journal on Computing 37, 1637–1655 (2008)
19. Kieritz, T., Luxen, D., Sanders, P., Vetter, C.: Distributed Time-Dependent Contraction Hierarchies. In: Festa, P. (ed.) SEA 2010. LNCS, vol. 6049, pp. 83–93. Springer, Heidelberg (2010)
20. Kleinberg, J.: Navigation in a Small World. Nature 406, 845 (2000)
21. Sander, P.V., Nehab, D., Chlamtac, E., Hoppe, H.: Efficient Traversal of Mesh Edges Using Adjacency Primitives. ACM Trans. on Graphics 27, 144:1–144:9 (2008)
22. Watts, D., Strogatz, S.: Collective Dynamics of "Small-World" Networks. Nature 393, 409–410 (1998)

Bottleneck Non-crossing Matching in the Plane[*]

A. Karim Abu-Affash, Paz Carmi, Matthew J. Katz, and Yohai Trabelsi

Department of Computer Science
Ben-Gurion University of the Negev, Beer-Sheva 84105, Israel
{abuaffas,carmip,matya,yohayt}@cs.bgu.ac.il

Abstract. Let P be a set of $2n$ points in the plane, and let M_C (resp., M_{NC}) denote a bottleneck matching (resp., a bottleneck non-crossing matching) of P. We study the problem of computing M_{NC}. We present an $O(n^{1.5} \log^{0.5} n)$-time algorithm that computes a non-crossing matching M of P, such that $bn(M) \leq 2\sqrt{10} \cdot bn(M_{NC})$, where $bn(M)$ is the length of a longest edge in M. An interesting implication of our construction is that $bn(M_{NC})/bn(M_C) \leq 2\sqrt{10}$. We also show that when the points of P are in convex position, one can compute M_{NC} in $O(n^3)$ time. (In the full version of this paper, we also prove that the problem is NP-hard and does not admit a PTAS.)

1 Introduction

Let P be a set of $2n$ points in the plane. A *perfect matching* M of P is a perfect matching in the complete Euclidean graph induced by P. Let $bn(M)$ denote the length of a longest edge of M. A *bottleneck matching* M_C of P is a perfect matching of P that minimizes $bn(\cdot)$. A *non-crossing matching* of P is a perfect matching whose edges are pairwise disjoint. In this paper, we study the problem of computing a bottleneck non-crossing matching of P; that is, a non-crossing matching M_{NC} of P that minimizes $bn(\cdot)$, where only non-crossing matchings of P are being considered.

The non-crossing requirement is natural, and indeed many researches have considered geometric problems concerning crossing-free configurations in the plane; see, e.g. [2–5, 14]. In particular, (bottleneck) non-crossing matching is especially important in the context of layout of VLSI circuits [12] and operations research. It is easy to see that there always exists a non-crossing matching of P (e.g., match each point in odd position with the first point to its right). Actually, any minimum weight matching of P is non-crossing. However, as shown

[*] Work by A.K. Abu-Affash was partially supported by a fellowship for doctoral students from the Planning & Budgeting Committee of the Israel Council for Higher Education, and by a scholarship for advanced studies from the Israel Ministry of Science and Technology. Work by A.K. Abu-Affash and Y. Trabelsi was partially supported by the Lynn and William Frankel Center for Computer Sciences. Work by P. Carmi was partially supported by grant 680/11 from the Israel Science Foundation. Work by M. Katz was partially supported by grant 1045/10 from the Israel Science Foundation. Work by M. Katz and P. Carmi was partially supported by grant 2010074 from the United States – Israel Binational Science Foundation.

L. Epstein and P. Ferragina (Eds.): ESA 2012, LNCS 7501, pp. 36–47, 2012.
© Springer-Verlag Berlin Heidelberg 2012

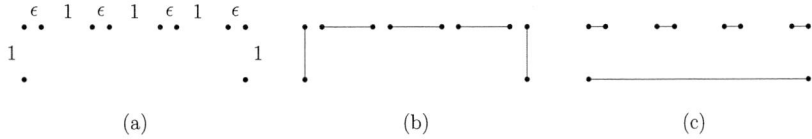

Fig. 1. (a) The set P. (b) A bottleneck non-crossing matching M_1 of P. (c) A minimum weight matching M_2 of P. Notice that $bn(M_2)/bn(M_1) \to n - 2$.

in Figure 1, which is borrowed from [6], the length of a longest edge of a minimum weight matching can be much larger than that of a bottleneck non-crossing matching.

In this paper we present a constat-factor approximation algorithm for the non-crossing bottleneck problem. Moreover, the length of the longest edge in the resulting non-crossing matching, is not much longer than the length of the longest edge in the bottleneck matching (i.e., when crossings are allowed). This implies that the ratio between the longest edge of a (crossing) bottleneck matching and a non-crossing bottleneck matching is bounded by a constant.

1.1   Related Work

Matching problems play an important role in graph theory, and thus have been studied extensively, see [13]. The various matching algorithms developed for general weighted graphs of course apply in our setting. However, it turns out that one can do better in the case of points in the plane. Vaidya [15] presented an $O(n^{5/2} \log^4 n)$-time algorithm for computing a minimum weight matching, based on Edmonds' $O(n^3)$ algorithm. Subsequently, Varadarajan [16] described an $O(n^{3/2} \log^5 n)$-time algorithm for this problem. For the bipartite version, Vaidya [15] presented an $O(n^{5/2} \log n)$-time algorithm and Agarwal et al. [1] presented an $O(n^{2+\varepsilon})$-time algorithm; both algorithms are based on the Hungarian method [13]. As for bottleneck matching, Chang et al. [7] obtained an $O(n^{3/2} \log^{1/2} n)$-time algorithm for computing a bottleneck matching, by proving that such a matching is contained in the 17RNG (17 relative neighborhood graph). Efrat and Katz [9] extended this result to higher dimensions. For the bipartite version, Efrat et al. [8] presented an $O(n^{3/2} \log n)$-time algorithm. Algorithms for other kinds of matchings, as well as approximation algorithms for the problems above, have also been developed.

Self-crossing configurations are often undesirable and might even imply an error condition; for example, a potential collision between moving objects, or inconsistency in the layout of a circuit. Many of the structures studied in computational geometry are non-crossing, for instance, minimum spanning tree, minimum weight matching, Voronoi diagram, etc. Jansen and Woeginger [11] proved that deciding whether there exists a non-crossing matching of a set of points with integer coordinates, such that all edges are of length *exactly d*, for a given integer $d \geq 2$, is NP-complete. Carlsson and Armbruster [6] proved that the bipartite version of the bottleneck non-crossing matching problem is NP-hard, and presented an $O(n^4 \log n)$-time algorithm (both for the bipartite and non-bipartite version), assuming the points are in convex position. Alon et al. [4] considered the problem of computing the longest (i.e., maximum weight) non-crossing matching of a set of points in the plane. They presented an approximation algorithm that computes a non-crossing matching of length at least $2/\pi$ of the length of the longest non-crossing matching. Aloupis et al. [5] considered the problem of finding a non-crossing matching between points and geometric objects in the plane. See also [2, 3, 14] for results related to non-crossing matching.

1.2   Our Results

In the full version of this paper, we prove that the problem of computing M_{NC} is NP-hard. The proof is based on a reduction from the planar 3-SAT problem, and is influenced by the proof of Carlsson and Armbruster mentioned above. It also implies that the problem does not admit a PTAS.

Our main result is described in Section 2. In this section, we present an $O(n \log n)$-time algorithm for converting *any* (crossing) matching $M_{\times}$ into a non-crossing matching $M_{=}$, such that $bn(M_{=}) \leq 2\sqrt{10} \cdot bn(M_{\times})$. The algorithm consists of two stages: converting $M_{\times}$ into an intermediate (crossing) matching $M_{\times}'$ with some desirable properties, and using $M_{\times}'$ as a "template" for the construction of $M_{=}$. The algorithm implies that (i) $bn(M_{\mathrm{NC}})/bn(M_{\mathrm{C}}) \leq 2\sqrt{10}$, and (ii) one can compute, in $O(n^{3/2} \log^{1/2} n)$-time, a non-crossing matching M, such that $bn(M) \leq 2\sqrt{10} \cdot bn(M_{\mathrm{NC}})$. We are not aware of any previous constant-factor approximation algorithm for the problem of computing M_{NC}, nor are we aware of previous results concerning the ratio $bn(M_{\mathrm{NC}})/bn(M_{\mathrm{C}})$. In Section 3, we present an $O(n^3)$-time algorithm, based on dynamic programming, for computing M_{NC} when the points of P are in convex position, thus improving the result of [6].

2   Approximation Algorithm

Let P be a set of $2n$ points in general position in the plane. The *bottleneck* of a perfect matching M of P, denoted $bn(M)$, is the length of a longest edge of M. Let $M_{\times}$ be a perfect matching of P. In this section we show how to convert $M_{\times}$ into a non-crossing perfect matching $M_{=}$ of P, such that $bn(M_{=}) \leq 2\sqrt{10} \cdot bn(M_{\times})$.

Set $\delta = bn(M_{\times})$. We begin by laying a grid of edge length $2\sqrt{2}\delta$. W.l.o.g. assume that each of the points in P lies in the interior of some grid cell. Consider an edge e of $M_{\times}$. Since e is of length at most δ, it is either contained in a single grid cell, or its endpoints lie in two adjacent cells (i.e., in two cells sharing a side or only a corner). In the former case, we say that e is *internal*, and in the latter case, we say that e is *external*. We distinguish between two types of external edges: a *straight* external edge (or s-edge for short) connects between a pair of points in two cells that share a side, while a *diagonal* external edge (or d-edge for short) connects between a pair of points in two cells that share only a corner. Finally, the *degree* of a grid cell C, denoted $deg(C)$, is the number of external edges with an endpoint in C.

Our algorithm consists of two stages. In the first stage we convert $M_\times$ into another perfect matching $M'_\times$, such that (i) each edge of $M'_\times$ is either contained in a single cell, or connects between a pair of points in two adjacent cells, (ii) for each grid cell C, $deg(C) \leq 4$ and these $deg(C)$ edges connect C to $deg(C)$ of its adjacent cells, and (iii) some additional properties hold (see below). In the second stage, we construct the matching $M_=$ according to $M'_\times$, such that there is a one-to-one correspondence between the external edges of $M'_\times$ and the external edges of $M_=$. That is, there exists an edge in $M'_\times$ connecting between two adjacent cells C_1 and C_2 if and only if there exists such an edge in $M_=$. However, the endpoints of an external edge of $M_=$ might be different than those of the corresponding edge of $M'_\times$.

The second stage itself consists of two parts. In the first part, we consider each non-empty grid cell separately. When considering such a cell C, we first determine the at most four points that will serve as endpoints of the external edges (as dictated by $M'_\times$). Next, we construct a non-crossing matching for the remaining points in C. In the second part of this stage, we add the external edges between the points that were chosen as endpoints for these edges in the first part.

We now describe each of the stages in detail.

2.1 Stage 1

In this stage we convert $M_\times$ into $M'_\times$. We do it by applying a sequence of reduction rules to the current matching, starting with $M_\times$. Each of the rules is applied multiple times, as long as there is an instance in the current matching to which it can be applied. When there are no more such instances, we move to the next rule in the sequence.

We associate a d-edge connecting between two cells with the corner shared by these cells.

Rule I is applied to a pair of d-edges associated with the same corner and connecting between the same pair of cells. The d-edges are replaced by a pair of internal edges.

Rule II is applied to a pair of d-edges associated with the same corner and connecting between different pairs of cells. The d-edges are replaced by a pair of s-edges.

Notice that when we are done with Rule I, each corner has at most two d-edges associated with it, and if it has two, then they connect between different pairs of cells. Moreover, when we are done with Rule II, each corner has at most one d-edge associated with it. Finally, since the length of a d-edge is at most δ, any edge created by Rule I or Rule II is contained in the disk $D_\delta(a)$ of radius δ centered at the appropriate corner a.

A d-edge associated with a corner a defines a *danger zone* in each of the two other cells sharing a, see Figure 2(a). A danger zone is an isosceles right triangle. The length of its legs is $\sqrt{2}\delta$ and it is semi-open, i.e., it does not include the hypotenuse (which is of length 2δ). Let $S_\delta(a)$ denote the square that two of its sides are the hypotenuses of the danger zones defined by the d-edge associated with a. Notice that if p is a point in cell C that is not in the danger zone in C, then it is impossible to draw a d-edge between C_1 and C_2, with endpoints in the interior of $S_\delta(a)$, that passes through p.

Rule III is applied to a d-edge e_1 and an edge e_2 with an endpoint in a danger zone defined by e_1, see Figure 2(b–d). We distinguish between two cases. If e_2 is an s-edge (Figure 2(b)), then, by the claim below, its other endpoint is in one of the cells C_1 or C_2. In this case, we replace e_1 and e_2 with an internal edge and an s-edge. If e_2 is an internal edge, then consider the other endpoint q of e_2. If q is not in a danger zone in C defined by another d-edge (Figure 2(c)), then replace e_1 and e_2 with two s-edges. If, however, q is in a danger zone in C defined by another d-edge e_3 (Figure 2(d)), then, by the claim below, e_3 is associated with one of the two corners of C adjacent to the corner a (to which e_1 is associated), and therefore, either C_1 or C_2 contains an endpoint of both e_1 and e_3. Replace e_1 and e_2 with two s-edges, such that one of them connects q to the endpoint of e_1 that is in the cell that also contains an endpoint of e_3.

Claim. Consider Figure 2. In an application of Rule III, (i) if e_2 is an s-edge, then its other endpoint q is in one of the cells C_1 or C_2, and (ii) if q is in a danger zone defined by another d-edge e_3 associated with corner b, then $\overline{ab}$ is a side of C and the danger zone containing q is the one contained in C.

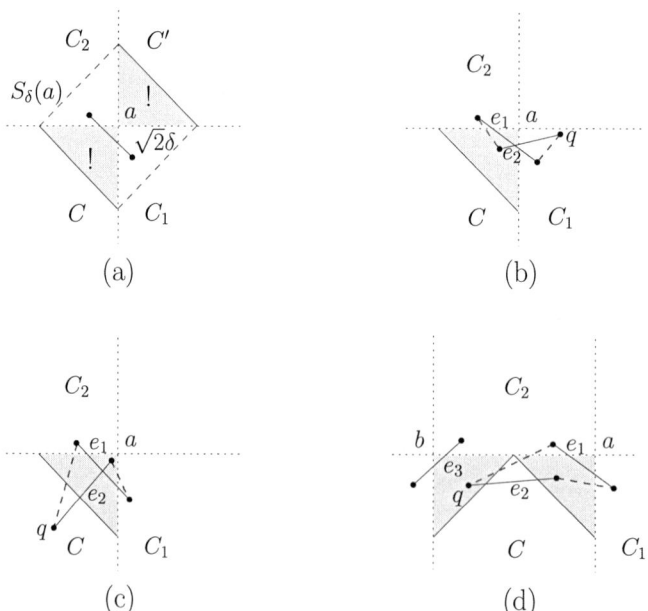

Fig. 2. (a) The danger zones defined by a d-edge. Rule III: (b) e_2 is an s-edge. (c) e_2 is an internal edge and q is not in another danger zone. (d) e_2 is an internal edge and q is in another danger zone.

Proof. Statement (i) is surely true just before the first application of Rule III, since the length of e_2 then is at most δ. (Notice that an s-edge created by Rule II cannot have an endpoint in a danger zone.) For the same reason, Statement (ii) is surely true just before the first application of Rule III. (Notice that if e_2 is an internal edge created by Rule I, then it is contained in one of the two danger zones defined by e_1.) It remains to verify that if e_2 was created by a previous application of Rule III, then both statements are still true. Indeed, if e_2 is an s-edge created by a previous application of Rule III, then it had to be an application of the type depicted in Figure 2(d), and the replacement instructions for this type ensure that Statement (i) is true. As for Statement (ii), if e_2 was created by an application of Rule III, then, since q was an endpoint of a d-edge that was removed by the application of Rule III, q is not in a danger zone.   □

Let $C(p)$ denote the cell containing point p.

Rule IV is applied to a d-edge $e_1 = (p_1, q_1)$ and to an s-edge $e_2 = (p_2, q_2)$, such that $C(p_1) = C(p_2)$ and $C(q_1)$ and $C(q_2)$ share a side. Edges e_1 and e_2 are replaced by an internal edge and an s-edge.

Rule V is applied to a pair of s-edges $e_1 = (p_1, q_1)$ and $e_2 = (p_2, q_2)$, such that $C(p_1) = C(p_2)$ and $C(q_1) = C(q_2)$. Edges e_1 and e_2 are replaced by a pair of internal edges.

Let $M'_\times$ be the matching that is obtained after applying Rules I-V. The following lemma summarizes some of the properties of $M'_\times$; its proof follows immediately from the discussion above.

Lemma 1. *$M'_\times$ has the following properties:*

1. *Each edge is either contained in a single cell, or connects between a pair of points in two adjacent cells.*
2. *A corner has at most one d-edge associated with it.*
3. *A d-edge is of length at most δ.*
4. *The two danger zones defined by a d-edge e are empty of points of P.*
5. *For each grid cell C, $deg(C) \leq 4$ and these $deg(C)$ edges connect C to $deg(C)$ of its adjacent cells.*
6. *If e is a d-edge in $M'_\times$ connecting between cells C_1 and C_2, and C is a cell sharing a side with both C_1 and C_2, then there is no s-edge in $M'_\times$ connecting between C and either C_1 or C_2.*

2.2   Stage 2

In this stage we construct $M_=$ according to $M'_\times$. This stage consists of two parts.

Part 1: Considering each cell separately. In this part we consider each non-empty grid cell separately. Let C be a non-empty grid cell and set $P_C = P \cap C$. We have to determine which of the points in C will serve as endpoints of external edges. The rest of the points will serve as endpoints of internal edges. We have to consider both types of external edges, d-edges and s-edges. We first consider the d-edges, then the s-edges, and, finally, after fixing the endpoints of the external edges, we form the internal edges.

d-edges. For each corner a of C that has a d-edge (with endpoint in C) associated with it in $M'_\times$, consider the line through a supporting C and parallel to its appropriate diagonal, and pick the point p_a in C that is closest to this line as the endpoint (in C) of the corresponding d-edge in $M_=$. By Lemma 1 (Property 3), $P_C \cap D_\delta(a) \neq \emptyset$, and therefore the distance between p_a and a is less than $\sqrt{2}\delta$. Moreover, since the length of a side of C is $2\sqrt{2}\delta$, each of the relevant corners is assigned a point of its own. Observe also that by picking the endpoints in this way, we ensure that in the final matching a d-edge with endpoint in C will not cross any of the s-edges with endpoint in C. This follows from Lemma 1 (Property 6). We thus may ignore the points in C that were chosen as endpoints of d-edges, and proceed to choose the endpoints of the s-edges.

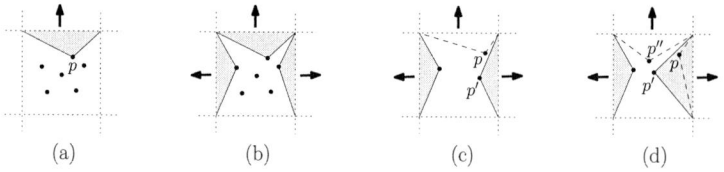

(a)   (b)   (c)   (d)

Fig. 3. (a) $\Delta(p,\text{"up"})$. Procedure 1: (b) $D = \{\text{"left"}, \text{"up"}, \text{"right"}\}$ and $|Q| = 3$. (c) $p \notin \Delta(p', \text{"right"})$ and p is assigned to "up". (d) $p \in \Delta(p', \text{"right"})$ and p is assigned to "right"; p'' will be assigned to "up".

s-edges. Let $\Delta(p, dir)$ denote the triangle whose corners are p and the two corners of C in direction dir, where $dir \in \{\text{"up"}, \text{"down"}, \text{"left"}, \text{"right"}\}$; see Figure 3(a). If there is only one s-edge in $M'_\times$ with endpoint in C, and its direction is dir, then we pick the extreme point in direction dir as the endpoint (in C) of the corresponding s-edge in $M_=$. Assume now that there are k, $2 \leq k \leq 4$, s-edges in $M'_\times$ with endpoints in C. We pick the k endpoints (in C) of the corresponding s-edges in $M_=$ according to the recursive procedure below.

Lemma 2. *The s-edges in $M_=$ with endpoints in C do not cross each other.*

Proof. By induction on k. If $k = 1$, then there is nothing to prove. Assume $k \geq 2$ and consider Procedure 1. If $|Q| = k$ (Line 5), then, for each $p_i \in Q$, the triangle $\Delta(p_i, dir_i)$ is empty, see Figure 3(b). Therefore, any two s-edges, one with endpoint p_i and direction dir_i and another with endpoint p_j and direction dir_j, $i \neq j$, do not cross each other. If $|Q| < k$ (Line 8), then, since $k \geq 2$, the directions dir_i and dir_j are not a pair of opposite directions. Now, if p, the extreme point in directions dir_i and dir_j, is not in $\Delta(p', dir_j)$ (Figure 3(c)), then $|Q' \cup \{p\}| = k$ and each of the corresponding k triangles is empty. If, however,

---

Procedure 1. Pick-s-Endpoints (a cell C, a set of directions $D = \{dir_1, \ldots, dir_k\}$)

1: $Q \leftarrow \emptyset$
2: **for** each direction $dir_i \in D$ **do**
3:     let p_i be the extreme point in direction dir_i
4:     $Q \leftarrow Q \cup \{p_i\}$
5: **if** $|Q| = k$ **then**
6:     assign p_i to direction dir_i, $i = 1, \ldots, k$
7:     Return(Q)
8: let $p \in Q$ be a point that is the extreme point in two directions $dir_i, dir_j \in D$
9: $Q' \leftarrow$ Pick-s-Endpoints($C \setminus \{p\}, D \setminus \{dir_i\}$)
10: let $p' \in Q'$ be the point assigned to direction dir_j
11: **if** $p \notin \Delta(p', dir_j)$ **then**
12:     assign p to direction dir_i
13: **else**
14:     $Q' \leftarrow$ Pick-s-Endpoints($C \setminus \{p\}, D \setminus \{dir_j\}$)
15:     assign p to direction dir_j
16: Return($Q' \cup \{p\}$)

$p \in \Delta(p', dir_j)$ (Figure 3(d)), then let p'' be the point assigned to direction dir_i by the call to Pick-s-Endpoints in Line 14. We claim that $p \notin \Delta(p'', dir_i)$. Indeed, if p were in $\Delta(p'', dir_i)$, then either $p'' \in \Delta(p', dir_j)$ or $p' \in \Delta(p'', dir_i)$, contradicting in both cases the induction hypothesis for $k - 1$.   □

Internal edges. We are now ready to form the internal edges. Let $P_C^E \subseteq P_C$ be the set of points that were chosen as endpoints of external edges, and set $P_C^I = P_C \setminus P_C^E$. We show below that P_C^I is contained in the interior of a convex region $R \subseteq C$, such that any external edge with endpoint in C does not intersect the interior of R. Hence, if we form the internal edges by visiting the points in P_C^I from left to right and matching each odd point with the next point in the sequence, then the resulting edges do not cross each other and do not cross any of the external edges.

It remains to define the convex region R. For each endpoint p_i of an s-edge, draw a line l_i through p_i that is parallel to the side of C crossed by the s-edge. Let h_i be the half-plane defined by l_i and not containing the s-edge, and set $R_i = h_i \cap C$. Similarly, for each endpoint p_a of a d-edge, draw a line l_a through p_a that is parallel to the appropriate diagonal of C. Let h_a be the half-plane defined by l_a and not containing the d-edge, and set $R_a = h_a \cap C$. Finally, set $R = (\cap\{R_a\}) \cap (\cap\{R_i\})$. It is clear that R is convex and that any external edge with endpoint in C does not intersect the interior of R. Moreover, by the way we chose the endpoints of the d-edges and s-edges, it is clear that P_C^I is contained in the interior of R.

Part 2: Putting everything together. In this part we form the external edges of $M_=$. For each external edge of $M_\times'$ connecting between cells C_1 and C_2, let p (resp., q) be the point that was chosen as the endpoint in C_1 (resp., in C_2) of the corresponding edge of $M_=$, and match p with q.

We have already shown that if e_1 and e_2 are two edges of $M_=$, for which there exists a grid cell containing an endpoint of both e_1 and e_2, then e_1 and e_2 do not cross each other. It remains to verify that a d-edge e of $M_=$ connecting between C_1 and C_2 cannot cause any trouble in the cell C through which it passes. Notice that $e \cap C$ is contained in the danger zone in C defined by e (or, more precisely, by the d-edge of $M'_\times$ corresponding to e). But, by Lemma 1 (Property 4), this danger zone is empty of points of P, thus e cannot cross an internal edge contained in C. Moreover, Lemma 1 (Property 2) guarantees that there is no d-edge in $M_=$ crossing e, and Lemma 1 (Property 6) guarantees that there is no s-edge crossing e. We conclude that e does not cause any trouble in C.

Finally, observe that $bn(M_=) \leq 2\sqrt{10}\delta = 2\sqrt{10} \cdot bn(M_\times)$. This is true since the length of a d-edge in $M_=$ is at most $2\sqrt{2}\delta$ (i.e., the diagonal of $S_\delta(a)$), the length of an internal edge in $M_=$ is at most 4δ (i.e., the diagonal of a single cell), and the length of an s-edge in $M_=$ is at most $2\sqrt{10}\delta$ (i.e., the diagonal of a pair of cells sharing a side).

The following theorem summarizes the main results of this section.

Theorem 1. *Let P be a set of $2n$ points in the plane. Let $M_\times$ be a perfect matching of P, and let M_C (resp., M_{NC}) be a bottleneck matching (resp., a bottleneck non-crossing matching) of P. Then,*

1. *One can compute in $O(n \log n)$ time a non-crossing perfect matching $M_=$, such that $bn(M_=) \leq 2\sqrt{10} \cdot bn(M_\times)$.*
2. $\frac{bn(M_{NC})}{bn(M_C)} \leq 2\sqrt{10}$.
3. *One can compute in $O(n^{1.5} \log^{0.5} n)$ time a non-crossing matching M of P, such that $bn(M) \leq 2\sqrt{10} \cdot bn(M_{NC})$.*

Proof. 1. It is easy to see that the time complexity of the algorithm of this section is only $O(n \log n)$. 2. By applying the algorithm of this section to M_C, we obtain a non-crossing matching M, such that $bn(M_{NC}) \leq bn(M) \leq 2\sqrt{10} \cdot bn(M_C)$. 3. Compute M_C in $O(n^{1.5} \log^{0.5} n)$ time, using the algorithm of Chang et al. [7]. Then, apply the algorithm of this section to M_C to obtain a non-crossing matching M, such that $bn(M) \leq 2\sqrt{10} \cdot bn(M_C) \leq 2\sqrt{10} \cdot bn(M_{NC})$.   □

Corollary 1. *For any $\varepsilon > 1$, one can compute in $\tilde{O}(n^{1+1/(1+\varepsilon)})$ time a non-crossing matching $M_=$ of P, such that $bn(M_=) \leq 4(1 + \varepsilon)\sqrt{10} \cdot bn(M_{NC})$.*

Proof. By a result of Goel et al. [10], for any $\varepsilon > 1$, one can compute in $\tilde{O}(n^{1+1/(1+\varepsilon)})$ time a perfect matching $M_\times$ of P, such that $bn(M_\times) \leq 2(1 + \varepsilon) \cdot bn(M_C)$.   □

Remarks. 1. There exists a set P of $2n$ points in the plane, for which $bn(M_{NC}) \geq \frac{\sqrt{85}}{8} \cdot bn(M_C)$; see Figure 4.

2. It is interesting to note that in the bipartite version, the ratio $bn(M_{NC})/bn(M_C)$ can be linear in n, even if the red and blue points are separated by a line; see Figures 5.

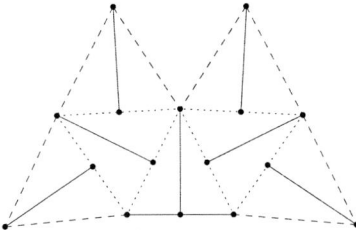

Fig. 4. The length of the solid edges is 1 and the length of the dashed edges is $\sqrt{85}/8$. The bottleneck (crossing) matching consists of the 8 solid edges, while any bottleneck non-crossing matching must use at least one of the dashed edges.

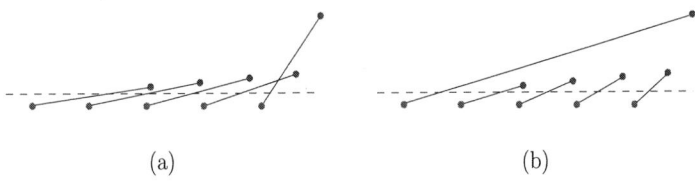

(a)   (b)

Fig. 5. (a) Bottleneck (crossing) and (b) bottleneck non-crossing matching of n blue and n red points separated by a line.

3   Points in Convex Position

In this section, we consider the special case where the points in P are in convex position, i.e., all the points in P are vertices of the convex hull of P. Let $(p_1, p_2, \ldots, p_{2n})$ be the sequence of points obtained by traversing the convex hull of P in clockwise order starting from an arbitrary point p_1; see Figure 6. Let M_{NC} be a bottleneck non-crossing matching of P. We first observe that, for each edge (p_i, p_j) of M_{NC}, $i+j$ is odd, since the number of points on each side of the line containing $\overline{p_i p_j}$ must be even. We thus associate with each edge (p_i, p_j) in the complete graph induced by P a weight $w_{i,j}$ as follows. $w_{i,j}$ is the length of the edge (p_i, p_j), if $i+j$ is odd, and $w_{i,j} = \infty$, if $i+j$ is even.

Notice that, for any $1 \leq i < j \leq 2n$, since the points in $\{p_1, p_2, \ldots, p_{2n}\}$ are in convex position, so are the points in $\{p_i, p_{i+1}, \ldots, p_j\}$ and the points

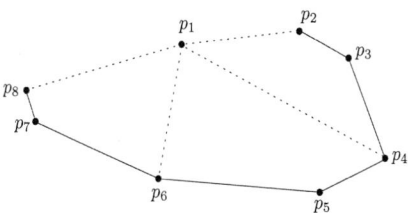

Fig. 6. The convex hull of P. p_1 can be matched only to p_2, p_4, p_6, or p_8.

in $\{p_j, p_{j+1}, \ldots, p_{2n}, p_1, \ldots, p_i\}$. For $1 \leq i, j \leq 2n$, $i + j$ odd, let $b_{i,j}$ denote the length of a longest edge of a bottleneck matching of $\{p_i, p_{i+1}, \ldots, p_j\}$. If p_k is the point matched to p_1 in M_{NC}, then $bn(M_{NC}) = b_{1,2n} = \max\{w_{1,k}, b_{2,k-1}, b_{k+1,2n}\}$. Therefore, in order to compute $b_{1,2n}$, we compute $\max\{w_{1,k}, b_{2,k-1}, b_{k+1,2n}\}$, for each even k between 2 and $2n$, and take the minimum over these values. In general, for any $1 \leq i < j \leq 2n$, $i + j$ odd, we have

$$b_{i,j} = \min_{k=i+1,i+3,\ldots,j} \begin{cases} w_{i,k} & \text{, if } k = i+1 = j \\ \max\{w_{i,k}, b_{k+1,j}\} & \text{, if } k = i+1 \\ \max\{w_{i,k}, b_{i+1,k-1}\} & \text{, if } k = j \\ \max\{w_{i,k}, b_{i+1,k-1}, b_{k+1,j}\} & \text{, otherwise.} \end{cases}$$

We compute $bn(M_{NC}) = b_{1,2n}$ via dynamic programming. The dynamic programming table M consists of $2n$ rows and $2n$ columns. The entry $M_{i,j}$ corresponds to a solution of the problem for the set $\{p_i, p_{i+1}, \ldots, p_j\}$. More precisely, $M_{i,j} = b_{i,j}$, if $i + j$ is odd, and $M_{i,j} = \infty$, if $i + j$ is even. We first fill the entries $M_{i,i+1} = w_{i,i+1}$, for $i = 1, \ldots, 2n$, corresponding to all chains of size 2. Next, we fill the entries $M_{i,i+3}$, for $i = 1, \ldots, 2n$, corresponding to all chains of size 4, etc. When considering an entry $M_{i,j}$, we fill it by consulting $O(n)$ entries that have already been filled, using the rule above. We thus obtain the following theorem.

Theorem 2. *Given a set P of $2n$ points in convex position, one can compute in $O(n^3)$ time and $O(n^2)$ space a bottleneck non-crossing matching of P.*

Corollary 2. *Let P be a set of $2n$ points on the boundary of a simple polygon Q. Let G be the graph over P, in which there is an edge between p_i and p_j if and only if $\overline{p_i p_j} \subseteq Q$. Then, one can compute a bottleneck non-crossing matching of P (if exists) in polynomial time.*

References

1. Agarwal, P.K., Efrat, A., Sharir, M.: Vertical decomposition of shallow levels in 3-dimensional arrangements and its applications. SIAM J. on Computing 29(3), 912–953 (1999)
2. Aichholzer, O., Bereg, S., Dumitrescu, A., García, A., Huemer, C., Hurtado, F., Kano, M., Márquez, A., Rappaport, D., Smorodinsky, S., Souvaine, D., Urrutia, J., Wood, D.R.: Compatible geometric matchings. Computational Geometry: Theory and Applications 42, 617–626 (2009)
3. Aichholzer, O., Cabello, S., Fabila-Monroy, R., Flores-Peñaloza, D., Hackl, T., Huemer, C., Hurtado, F., Wood, D.R.: Edge-removal and non-crossing configurations in geometric graphs. Discrete Mathematics and Theoretical Computer Science 12(1), 75–86 (2010)
4. Alon, N., Rajagopalan, S., Suri, S.: Long non-crossing configurations in the plane. In: Proceedings of the 9th ACM Symposium on Computational Geometry (SoCG 1993), pp. 257–263 (1993)

5. Aloupis, G., Cardinal, J., Collette, S., Demaine, E.D., Demaine, M.L., Dulieu, M., Fabila-Monroy, R., Hart, V., Hurtado, F., Langerman, S., Saumell, M., Seara, C., Taslakian, P.: Matching Points with Things. In: López-Ortiz, A. (ed.) LATIN 2010. LNCS, vol. 6034, pp. 456–467. Springer, Heidelberg (2010)
6. Carlsson, J.G., Armbruster, B.: A bottleneck matching problem with edge-crossing constraints,
   http://users.iems.northwestern.edu/~armbruster/2010matching.pdf
7. Chang, M.S., Tang, C.Y., Lee, R.C.T.: Solving the Euclidean bottleneck matching problem by k-relative neighborhood graphs. Algorithmica 8(1-6), 177–194 (1992)
8. Efrat, A., Itai, A., Katz, M.J.: Geometry helps in bottleneck matching and related problems. Algorithmica 31(1), 1–28 (2001)
9. Efrat, A., Katz, M.J.: Computing Euclidean bottleneck matchings in higher dimensions. Information Processing Letters 75(4), 169–174 (2000)
10. Goel, A., Indyk, P., Varadarajan, K.R.: Reductions among high dimensional proximity problems. In: Proceeding of the 12th ACM-SIAM Symposium on Discrete Algorithms (SODA 2001), pp. 769–778 (2001)
11. Jansen, K., Woeginger, G.J.: The complexity of detecting crossingfree configurations in the plane. BIT 33(4), 580–595 (1993)
12. Lengauer, T.: Combinatorial Algorithms for Integrated Circuit Layout. John Wiley & Sons, New York (1990)
13. Lovász, L., Plummer, M.D.: Matching Theory. Elsevier Science Ltd. (1986)
14. Rappaport, D.: Tight Bounds for Visibility Matching of f-Equal Width Objects. In: Akiyama, J., Kano, M. (eds.) JCDCG 2002. LNCS, vol. 2866, pp. 246–250. Springer, Heidelberg (2003)
15. Vaidya, P.M.: Geometry helps in matching. SIAM Journal on Computing 18(6), 1201–1225 (1989)
16. Varadarajan, K.R.: A divide-and-conquer algorithm for min-cost perfect matching in the plane. In: Proceedings of the 39th Symposium on Foundations of Computer Science (FOCS 1998), pp. 320–331 (1998)

Lower Bounds for Sorted Geometric Queries
in the I/O Model

Peyman Afshani[1,*] and Norbert Zeh[2,**]

[1] MADALGO[***], Department of Computer Science
Aarhus University, Denmark
peyman@cs.au.dk
[2] Faculty of Computer Science, Dalhousie University
Halifax, Canada
nzeh@cs.dal.ca

Abstract. We study sorted geometric query problems, a class of problems that, to the best of our knowledge and despite their applications, have not received much attention so far. Two of the most prominent problems in this class are angular sorting queries and sorted K-nearest neighbour queries. The former asks us to preprocess an input point set S in the plane so that, given a query point q, the clockwise ordering of the points in S around q can be computed efficiently. In the latter problem, the output is the list of K points in S closest to q, sorted by increasing distance from q. The goal in both problems is to construct a small data structure that can answer queries efficiently. We study sorted geometric query problems in the I/O model and prove that, when limited to linear space, the naïve approach of sorting the elements in S in the desired output order from scratch is the best possible. This is highly relevant in an I/O context because storing a massive data set in a superlinear-space data structure is often infeasible. We also prove that answering queries using $O(N/B)$ I/Os requires $\Omega(N \log_M N)$ space, where N is the input size, B is the block size, and M is the size of the main memory. This bound is unlikely to be optimal and in fact we can show that, for a particular class of "persistence-based" data structures, the space lower bound can be improved to $\Omega(N^2/M^{O(1)})$. Both these lower bounds are a first step towards understanding the complexity of sorted geometric query problems. All our lower bounds assume indivisibility of records and hold as long as $B = \Omega(\log_{M/B} N)$.

1 Introduction

We study the problem of storing a set of geometric objects in a data structure that supports the following type of queries efficiently: the query implicitly specifies a particular order of the input objects and the data structure has to report

---

* This research was done while the first author was a postdoctoral fellow at the Faculty of Computer Science of Dalhousie University and was supported by an NSERC postdoctoral fellowship.
** Research supported by NSERC and the Canada Research Chairs programme.
*** Center for Massive Data Algorithms, a Center of the Danish National Research Foundation.

L. Epstein and P. Ferragina (Eds.): ESA 2012, LNCS 7501, pp. 48–59, 2012.
© Springer-Verlag Berlin Heidelberg 2012

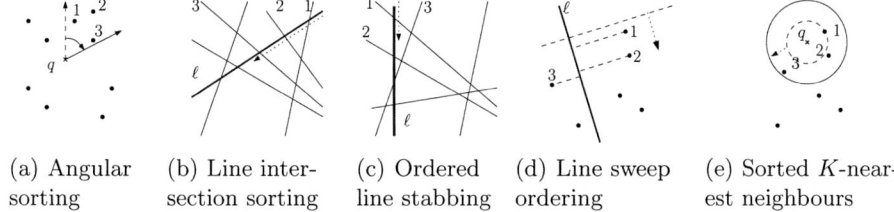

(a) Angular sorting   (b) Line intersection sorting   (c) Ordered line stabbing   (d) Line sweep ordering   (e) Sorted K-nearest neighbours

Fig. 1. Examples of sorted geometric query problems

the objects in this order. In most of the problems that we study, in contrast to typical data structure problems, the challenge is *not* to decide *which* objects to report: *all* objects in the data structure are to be reported. The challenge is to arrange the objects in the desired order more efficiently than sorting them from scratch. Specifically, we study the following problems.

Angular sorting: Given a set S of points in the plane and a query point q, report the points in S in the order they are passed by a ray shot upwards from q and then rotated in clockwise direction around q (Figure 1(a)).

Line intersection sorting: Given a set S of lines in the plane and a query line ℓ, sort the lines in S by the y-coordinates of their intersections with ℓ (Figure 1(b)). (We assume w.l.o.g. that ℓ is not horizontal and no line in S is parallel to ℓ.)

Ordered line stabbing: This is a special case of line intersection sorting where the query line ℓ is vertical (Figure 1(c)).

Line sweep ordering: Given a set S of points in the plane and a non-horizontal query line ℓ, sort the points in S by the y-coordinates of their projections onto ℓ. This corresponds to the order in which a sweep line perpendicular to ℓ passes over the points in S (Figure 1(d)).

Ordered K-nearest neighbours: Given a set S of points in the plane and a query point q, report the K points in S closest to q, sorted by their distances from q (Figure 1(e)).

These arise naturally in a number of geometric algorithms and applications. Angular sorting is used as a preprocessing step in a number of geometric algorithms, e.g., robust statistical measures of a point q in a point set, such as the Tukey depth [13], Oja depth [13] or simplicial depth [7, 10, 13], can be computed in linear time once the input points are sorted by their angles around q. Angular sorting is also used in the classical convex hull algorithm known as Graham's scan [8] and in the computation of visibility polygons (e.g., see [6,14]). Reporting the K nearest neighbours of a query point is a classical problem with many applications. In a number of these applications, it is natural to ask that these neighbours be reported in order of increasing distance from the query point. For example, when a user of a GPS device asks for the bookstores closest to their current location, it is natural to report the results by increasing distance.

Nevertheless, the existing literature on K-nearest neighbour queries considers exclusively the unordered version of this problem (e.g., see [2]). Line intersection sorting is the dual of angular sorting. Line sweep ordering is motivated by applications of the plane sweep technique in computational geometry.

The basic idea behind these problems is a natural one: many problems can be solved in linear time after an initial sorting step. Thus, presorting the input can help to speed up such algorithms. In one dimension, there is only one ordering of the input, and storing the input in sorted order uses linear space. In two dimensions, the ordering depends on the query and there are sufficiently many different orderings to make the naïve approach of storing each of them explicitly infeasible. On the other hand, the geometry of the input implies that not every ordering of the input objects corresponds to an ordering generated by a query. For instance, N input lines can be intersected by a query vertical line in $O(N^2)$ ways only. Furthermore, the total number of combinatorially different arrangements of lines is $N^{O(N)} = N!^{O(1)}$ (see e.g., [11, Chapter 6]), which is not much larger than the number of permutations of N objects in one dimensions. These raise the question whether the geometric constraints can be exploited to represent all possible output orderings compactly in a data structure while still being able to extract the ordering produced by any query using sub-sorting cost. This is the question we investigate in this paper, from a lower bound perspective.

We focus primarily on ordered line stabbing queries. Using fairly standard reductions, a lower bound for ordered line stabbing queries implies lower bounds for all the problems introduced previously (more details will follow in the full version of the paper).

We study these problems in the I/O model [3]. In this model, the computer has an internal memory capable of holding up to M data items (e.g., points) and an external memory of conceptually unlimited size. Computation has to happen on data in memory and is free. Data is transferred between the two memory levels using *I/O operations* (I/Os). Each such operation transfers a block of B consecutive data elements between internal and external memory. The cost of an algorithm is the number of such I/O operations it performs.

The I/O model is the most widely accepted model for designing algorithms for large data sets beyond the size of the computer's main memory. Statistical data sets can be very large, which motivates the study of methods to speed up the computation of statistical measures in the I/O model. Moreover, any result in the I/O model can also be applied to any two levels in a computer's cache hierarchy, making our results relevant to cache-efficient data structures and thus to data sets of more modest size. The most compelling reason to focus on the I/O model in this paper, however, is that it is the only well-established computing model with known lower bound properties that are sufficiently strong to obtain non-trivial space lower bounds for sorted query problems. Specifically, permuting and sorting have a superlinear cost in the I/O model.

Previous results. The problem of preprocessing an input set S of N points to answer angular sorting queries was considered as early as 1986, in the context of computing the visibility polygon [5]. Their data structure uses $O(N^2)$ space and answers queries in linear time. The techniques used in this structure are now standard: store the arrangement $\mathcal{A}$ formed by the dual of S and answer the query by finding the intersection of $\mathcal{A}$ with the dual line of the query point. Ghodsi and Nouri also studied angular sorting queries [12] but focused on obtaining a *sublinear* query time by returning a pointer to a simple data structure that stores the sorted order. Using $N^2 h$ space, for $1 \le h \le N^2$, they find such a pointer for any given query in $O((N \log h)/\sqrt{h})$ time. We are not aware of any other non-trivial results for angular sorting. For any fixed K, K-nearest neighbour queries can be answered using $O(\log_B N + K/B)$ I/Os by a linear-space data structure [2][1]; the output can then be sorted using $O((K/B) \log_{M/B}(K/B))$ I/Os.

To the best of our knowledge, the only relevant lower bound result is due to Afshani *et al.* [1], who considered the following one-dimensional problem: given an array A of numbers, preprocess A such that, given two query indices $i < j$, the numbers in the subarray $A[i \mathinner{.\,.} j]$ can be efficiently reported in sorted order. They proved that answering such queries using $O(\log^\alpha n + fK/B)$ I/Os requires $\Omega\left(N \frac{f^{-1} \log_M n}{\log(f^{-1} \log_M n)}\right)$ space, assuming $B = \Omega(\log N)$, where $n = N/B$, f is a parameter with $1 \le f \le \log_{M/B} n$, and α is any constant.

Our Results. We prove non-trivial lower bounds for sorted geometric query problems, which as a class of problems are novel. Our results provide the first insights into the complexity of answering such queries.

General lower bounds. Assuming $B = \Omega(\log_{M/B} N)$, we show that, to achieve a query bound of $O(N/B)$ I/Os, any data structure for angular sorting, line sweep ordering, ordered line stabbing or line intersection ordering has to use $\Omega(N \log_M N)$ space in the worst case. To achieve a query bound of $O(\log_B N + K/B)$ I/Os for ordered K-nearest neighbour queries in the plane, for a fixed $K = \Omega(\max(M, B \log_B N))$, $\Omega(N \log_M K)$ space is needed in the worst case. At the other end of the spectrum, we show that linear-space data structures for these problems can only achieve query bounds of $O((N/B) \log_M N)$ and $O(1 + (K/B) \log_M K)$ I/Os, respectively. The latter result is highly relevant in an I/O context because storing a massive data set in a superlinear-space data structure is often infeasible.

Lower bounds for "persistence-based" data structures. Using persistence to build data structures for two- or three-dimensional problems is one of the classical tools in algorithms [9]. For the ordered line stabbing problem, this easily gives a data structure that uses $O(N^2)$ space and can answer queries in optimal time: given an input S of N lines, sweep a vertical line q from $x = -\infty$ to $x = \infty$ and store the order in which the lines in S intersect q in a partially persistent B-tree. Each time q passes the intersection point of two lines in S, only the order of these two lines switches, which results in $O(N^2)$ updates overall and thus in an $O(N^2)$-space data structure [4]. Inspired by this, we consider "persistence-based" data structures that generalize the above solution (for details see Section 3.3). We prove that any persistence-based data structure with

---

[1] The space requirement increases to $O(N \log^*(N/B))$ if we want to answer K-nearest neighbour queries for any K, where $\log^*$ is the iterated log function.

a linear query bound for angular sorting, line sweep ordering, ordered line stabbing or line intersection ordering requires $\Omega(N^2/M^{O(1)})$ space in the worst case. For 2-d K-nearest neighbour queries, for a fixed $K = \Omega(\max(M, B \log_B N))$, the space bound is $\Omega(NK/M^{O(1)})$. This shows that, if small data structures for these problems exist, more sophisticated techniques are needed to obtain them.

Our lower bound proof is inspired by [1]: the idea is to construct sufficiently many queries that produce sufficiently different permutations of the input that the answers to these queries must be produced using nearly disjoint portions of the data structure if the query bound is to be linear. While the proof in [1] used purely combinatorial properties of integer sequences, we develop a fundamentally different, *geometric* construction of such a set of queries in this paper.

2   Lower Bound Model

The input to any sorted query problem is a set S of objects. Each query implicitly specifies a particular order of S. Our goal is to build a data structure $\mathcal{D}$ that is capable of generating the required ordering of the objects in S, for any query. We view $\mathcal{D}$ as a linear sequence of *cells* (disk locations). Each cell can hold one input object. We assume *indivisibility of records*: The only operations the query algorithm can perform is reading the contents of a block of B cells into memory, rearranging the objects in memory at no cost, and writing B objects currently in memory to a block of cells on disk. The query algorithm *cannot* create new objects. This implies that each object output by the query algorithm can be traced back to an *origin cell* in the data structure. For a query q, we use S_q to denote the output sequence of q. For any subsequence S' of S_q, we use $C(S')$ to denote the set of origin cells used to produce S'.

Our strategy is to show that there exist an input S, a query set $\{q_1, q_2, \ldots, q_t\}$, and subsequences $S_{q_1}^m, S_{q_2}^m, \ldots, S_{q_t}^m$ of the outputs of these queries such that each sequence $S_{q_i}^m$ has size $\Omega(N)$ and $C(S_{q_i}^m) \cap C(S_{q_j})$ is small, for all $i \neq j$ and any data structure $\mathcal{D}$. Given the right choice of parameters, this implies our lower bounds. Since we only consider the cost of permuting the objects, our lower bound holds even for an omnipotent data structure that is capable of identifying the cheapest set of origin cells for each query at no cost. The following lemma states an important property of this model. A very similar version was proved in [1] and we postpone the proof to the full version of the paper.

Lemma 1. *For a sequence σ of N elements and $1 \leq f \leq \log_M N$, the maximum number of permutations of σ that can be produced using $O(fN/B)$ I/Os is $M^{O(fN)}$, provided $B = \Omega(\log_M N)$.*

3   Lower Bounds for Ordered Query Problems

In this section, we prove our main result (Theorem 1). We prove this theorem for ordered line stabbing and as discussed, standard reductions give the same lower bound for other problems introduced in this paper.

Theorem 1. *For $B = \Omega(\log_{M/B} N)$, any data structure that answers angular sorting, line intersection sorting, line sweep ordering or ordered line stabbing queries using $O(N/B)$ I/Os requires $\Omega(N \log_M N)$ space in the worst case. Any data structure capable of answering ordered K-nearest neighbour queries using $O(\log_B N + K/B)$ I/Os, for any fixed $K = \Omega(\max(M, B \log_B N))$, requires $\Omega(N \log_M K)$ space in the worst case. The query bound of any linear-space data structure for these problems is $\Omega((N/B) \log_M N)$ I/Os or, in the case of K-nearest neighbour queries, $\Omega(\log_B N + (K/B) \log_M K)$ I/Os in the worst case.*

3.1   Input Instance and Geometric Properties

To prove Theorem 1, we construct a random input instance and a query set which, with non-zero probability, require a large part of the output of each query to be produced using a unique set of cells in the data structure (see Section 2). We present this construction in this section and prove a number of geometric properties that are used in Section 3.2 to prove Theorem 1.

Consider two squares $Q := [-N, N] \times [-N, N]$ and $Q^m := [-N/2, N/2] \times [-N/2, N/2]$. The set S of N lines to be stored in the data structure is generated using the following random process. For each line $\ell \in S$, we choose one point each uniformly at random from the left and right sides of Q; ℓ is the line through these two *anchor points*. We create t queries $q_1, q_2, \ldots, q_t$, for a parameter t to be chosen later, by partitioning the top side of Q^m into t subsegments of equal length; $q_1, q_2, \ldots, q_t$ are the vertical lines through the centers of these segments (Figure 2(a)). For each query q_i, let $q_i^m := q_i \cap Q^m$, and let $S_{q_i}^m \subseteq S_{q_i}$ be the sequence of lines in S that intersect q_i^m, sorted by the y-coordinates of their intersections with q_i^m. Our first lemma shows that $S_{q_i}^m$ is large, for each query q_i. From here on, we do not distinguish between a line segment and its length, and we use $I_{\ell,s}$ to denote the event that a random line $\ell \in S$ intersects a segment s.

Lemma 2. *Consider a query q_i and a subsegment $q_i' \subseteq q_i^m$. A random line in S intersects q_i' with probability $\Theta(q_i'/N)$.*

Proof. First we prove that $\Pr[I_{\ell,q_i'}] = \Omega(q_i'/N)$. Assume first that $q_i' \leq N/12$. Then there exists a subsegment s_L of the left side of Q whose length is at least $2N/3 - 4q_i' \geq N/3 = \Omega(N)$ and such that any line intersecting both s_L and q_i' intersects the right side of Q (see Figure 2(b)). Now let $s_L' \subseteq s_L$ be a subsegment of s_L, and let s_R' be the largest subsegment of the right side of Q such that every line with anchors on s_L' and s_R' intersects q_i' (Figure 2(b)). It is easy to verify that, for $s_L' \leq q_i'$, $q_i' \leq s_R' \leq 4q_i'$, that is, $s_R' = \Theta(q_i')$.

Now consider a random line ℓ in S. Its left anchor lies on s_L' with probability $s_L'/2N$. Its right anchor lies on s_R' with probability $s_R'/2N$. Since a line that intersects both s_L' and s_R' also intersects q_i', this shows that $\Pr[I_{\ell,s_L'} \cap I_{\ell,q_i'}] \geq s_L' s_R'/(4N^2) = \Omega(s_L' q_i'/N^2)$. Since we can cover s_L with $\lceil s_L/s_L' \rceil$ *disjoint* subsegments of length s_L', we have $\Pr[I_{\ell,q_i'}] \geq \Pr[I_{\ell,s_L} \cap I_{\ell,q_i'}] = \Omega(s_L q_i'/N^2) = \Omega(q_i'/N)$.

If $N/12 \leq q_i' \leq N$, we apply the same argument to a segment $q_i'' \subset q_i'$ of length $N/12 = \Omega(q_i')$. This proves that $\Pr[I_{\ell,q_i'}] \geq \Pr[I_{\ell,q_i''}] = \Omega(q_i''/N) = \Omega(q_i'/N)$.

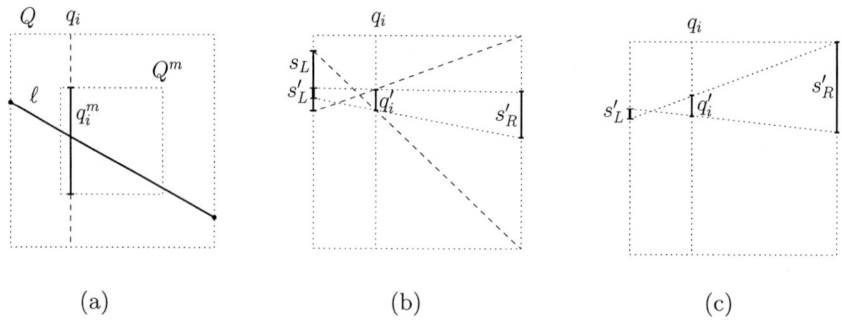

Fig. 2. (a) Q, Q^m, an input line ℓ and a query line q_i. (b,c) Proof of Lemma 2.

The upper bound proof is similar. For any subsegment $s'_L \leq q'_i$ of the left side of Q, any line $\ell \in S$ intersecting s'_L and q'_i has its right anchor on a subsegment s'_R of the right side of Q of length at most $4q'_i + 3s'_L = O(q'_i)$ (Figure 2(c)). Thus, $\Pr[I_{\ell,s'_L} \cap I_{\ell,q'_i}] = O(s'_L q'_i / N^2)$. Since we can cover the left side of Q using $\lceil 2N/s'_L \rceil$ subsegments of length s'_L, this proves that $\Pr[I_{\ell,q'_i}] = O(q'_i / N)$. □

Corollary 1. $|S^m_{q_i}| = \Omega(N)$ with probability at least $1 - 1/N^c$, for any constant c.

Proof. By Lemma 2, each line in S intersects q^m_i with constant probability, since q^m_i has length N. Thus, $\mathrm{E}[|S^m_{q_i}|] = \Omega(N)$. The high-probability bound follows by applying Chernoff bounds. □

To prove that any data structure needs to use largely disjoint subsets of cells to produce the sequences $S^m_{q_i}$ and S_{q_j}, we require that S_{q_i} and S_{q_j} are nearly independent random permutations of S. In other words, once we fix the output of query q_i, a large number of permutations remain that are possible outputs of query q_j, provided the distance between q_i and q_j is sufficiently large.

Intuitively, the argument is as follows (see Figure 3(a)): Conditioned on the event that $\ell \in S$ passes through a point $p \in q^m_i$, the intersection between ℓ and q_j is uniformly distributed inside an interval q'_j on q_j (because ℓ intersects the right side of Q). For such a conditional probability to make sense, however, the event it is conditional on must have a non-zero probability, whereas ℓ passes through p with probability zero. To overcome this problem, we consider a subsegment $q'_i \subseteq q^m_i$ instead of a point $p \in q^m_i$ and assume ℓ intersects q'_i (see Figure 3(b)). If q'_i is sufficiently small, this increases the length of q'_j only slightly and the distribution of the intersection between ℓ and q_j over q'_j remains almost uniform:

Lemma 3. *Let $f : q_j \to \mathbb{R}$ be the density function that describes the distribution of the intersection point between a random line $\ell \in S$ and q_j over q_j conditioned on the event that ℓ intersects a subsegment $q'_i \subseteq q^m_i$. Let d be the distance between q_i and q_j. If $q'_i = O(d)$, then f is zero except on a subsegment q'_j of q_j of length $\Theta(d)$, and $f(p) = O(1/d)$, for all $p \in q_j$.*

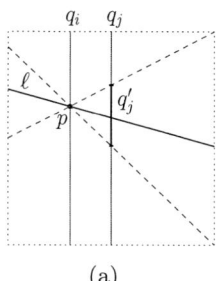 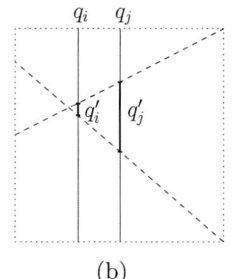 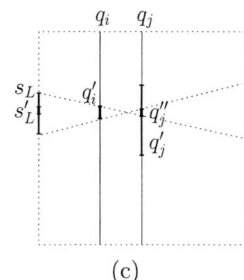

(a)   (b)   (c)

Fig. 3. Independence of $S_{q_i}^m$ and S_{q_j}

Proof. Since every line in S has slope between -1 and 1, the subsegment q_j' of possible intersection points between ℓ and q_j has size at most $q_i' + 2d = O(d)$. Clearly, f is zero outside of q_j'. To bound $f(p)$, for all $p \in q_j$, we consider a subsegment q_j'' of q_j such that $q_j'' \leq q_i'$ (see Figure 3(c)). It suffices to prove that, for any $q_j'' > 0$, $\Pr[I_{\ell,q_j''}|I_{\ell,q_i'}] = O(q_j''/d)$. By Lemma 2, $\Pr[I_{\ell,q_i'}] = \Theta(q_i'/N)$. Thus, to prove that $\Pr[I_{\ell,q_j''}|I_{\ell,q_i'}] = O(q_j''/d)$, it suffices to prove that $\Pr[I_{\ell,q_i'} \cap I_{\ell,q_j''}] = O(q_i'q_j''/(dN))$. The ratio between q_i's distance from Q's left side and its distance from q_j is $\Theta(N/d)$. Thus, any line in S intersecting q_i' and q_j'' has its left anchor on a subsegment s_L of Q's left side of length at most $(q_i' + q_j'')\Theta(N/d) = \Theta(q_i'N/d)$. Now consider a subsegment $s_L' \subseteq s_L$ with $s_L' \leq q_j''$. Line ℓ intersects s_L', q_i', and q_j'' only if it intersects s_L' and q_j''. The probability of the latter event is $O(s_L'q_j''/N^2)$ because the right anchor of ℓ must be chosen from a subsegment of Q's right side of length at most $3s_L' + 4q_j'' = O(q_j'')$ in this case. Now we cover s_L with $\lceil s_L/s_L' \rceil = O(q_i'N/(ds_L'))$ subsegments of length s_L'. The probability that a line $\ell \in S$ intersects both q_i' and q_j'' is upper bounded by the probability that it intersects s_L and q_i', which is $O(q_i'N/(ds_L') \cdot s_L'q_j''/N^2) = O(q_i'q_j''/(dN))$. $\square$

As we show next, Lemma 3 implies that fixing an order in which the lines in S intersect q_i does not significantly constrain the order in which they intersect q_j.

Lemma 4. *Let $I = [0, d]$ be an interval, and let $f_1, f_2, \ldots, f_X$ be density functions on I such that, for all $1 \leq i \leq X$ and every $p \in I$, $f_i(p) = O(1/d)$. Let $p_1, p_2, \ldots, p_X$ be random points, where each p_i is chosen according to the density function f_i. For a permutation π of $[X]$, the event E_π that $p_1, p_2, \ldots, p_X$ are sorted according to π happens with probability at most $2^{O(X)}/X!$.*

Proof. By relabeling the points, we can assume that $\pi = \langle 1, 2, \ldots, X \rangle$. E_π is the event that $p_X \in [0, d]$, $p_{X-1} \in [0, p_X]$, $p_{X-2} \in [0, p_{X-1}]$, and so on. For $1 \leq j \leq X$ and a value $0 \leq x \leq d$, let $F_j(x)$ be the probability that points $p_1, p_2, \ldots, p_{j-1}$ lie in the interval $[0, x]$ and are sorted according to π. Then

$$\Pr[E_\pi] = \int_0^d f_X(y)F_X(y)\,dy \quad \text{and} \quad F_j(x) = \int_0^x f_{j-1}(y)F_{j-1}(y)\,dy.$$

Using induction on j and the facts that $F_1(y) = 1$ and $f_j(y) \leq \alpha/d$, for all $y \in [0, d]$ and some constant $\alpha > 0$, we obtain the bound

$$F_j(y) \leq \frac{(\alpha y/d)^{j-1}}{(j-1)!}, \text{for all } 1 \leq j \leq X.$$

This implies that

$$\Pr[E_\pi] = \int_0^d f_X(y) F_X(y) \, dy \leq \int_0^d \frac{\alpha}{d} \cdot \frac{(\alpha y/d)^{X-1}}{(X-1)!} \, dy = \frac{\alpha^X}{X!} = \frac{2^{O(X)}}{X!}. \qquad \square$$

Finally, we use Lemmas 3 and 4 to prove a nearly uniform distribution over the possible orders in which (a subset of) the lines in S intersect a query line q_j, even once their intersections with a query line q_i are fixed.

Lemma 5. *Consider two queries q_i and q_j $(i < j)$, and let d be the horizontal distance between q_i and q_j. Let $q'_{i,1}, q'_{i,2}, \ldots, q'_{i,X}$ be pairwise disjoint subsegments of q_i^m of length at most d each, let I be the event $I := I_{\ell_1, q'_{i,1}} \cap I_{\ell_2, q'_{i,2}} \cap \cdots \cap I_{\ell_X, q'_{i,X}}$, where $\{\ell_1, \ell_2, \ldots, \ell_X\} \subseteq S$, and let π be a fixed permutation of the lines $\ell_1, \ell_2, \ldots, \ell_X$. The probability that the lines $\ell_1, \ell_2, \ldots, \ell_X$ intersect q_j in the order specified by π conditioned on event I is at most $\frac{2^{O(X)}}{(Xd/N)^{\Theta(X)}}$.*

Proof. Since $q'_{i,k} \leq d$, for all $1 \leq k \leq X$, and q_i^m has length N, we can partition q_i^m into $\Theta(N/d)$ disjoint subsegments $q''_{i,1}, q''_{i,2}, \ldots, q''_{i,Y}$ of length $\Theta(d)$ and such that either $q'_{i,k} \subseteq q''_{i,h}$ or $q'_{i,k} \cap q''_{i,h} = \emptyset$, for all $1 \leq k \leq X$ and $1 \leq h \leq Y$. Now consider two segments $q'_{i,k} \subseteq q''_{i,h}$ and the line $\ell_k \in S$. By Lemma 3, conditioned on ℓ_k intersecting $q'_{i,k}$, the potential intersection points between ℓ_k and q_j form a subsegment $q'_{j,k}$ of q_j of length $\Theta(d)$, and the distribution of ℓ_k's intersection with q_j over $q'_{j,k}$ is almost uniform. Since $q''_{i,h} = \Theta(d)$, the segments $q'_{j,k}$, for all $q'_{i,k} \subseteq q''_{i,h}$, are contained in a subsegment $q''_{j,h}$ of q_j of length $\Theta(d)$. Now let t_h be the number of segments $q'_{i,k}$ contained in $q''_{i,h}$. By Lemma 4, the lines in S intersecting these segments intersect q_j in the order specified by π with probability at most $2^{O(t_h)}/t_h!$. The lemma follows by observing that the probability that *all* the lines $\ell_1, \ell_2, \ldots, \ell_X$ intersect q_j in the order specified by π is at most $\prod_{h=1}^{Y} \frac{2^{O(t_h)}}{t_h!}$ which is maximized (over the reals) when $t_1 = \cdots = t_Y = X/Y = \Theta(Xd/N)$. $\qquad \square$

3.2 Proof of Theorem 1

For any two queries q_i and q_j, let R_{ij} be the event that there exists a data structure such that $C(S_{q_i}^m) \cap C(S_{q_j}) \geq X$, for a parameter X to be chosen later. To bound $\Pr[R_{ij}]$, we partition the segment $q_i \cap Q$ into a set S of N^4 subsegments of length $2N^{-3}$. Let S^N be the set of all sequences of N (not necessarily distinct) segments in S. For any $\sigma = \langle s_1, s_2, \ldots, s_N \rangle \in S^N$, let I_σ be the event that, for all $1 \leq k \leq N$, the kth line $\ell_k \in S$ intersects s_k. For every choice of lines in S, exactly one of the events I_σ, $\sigma \in S^N$, occurs. Thus, we have

$$\Pr[R_{ij}] = \sum_{\sigma \in S^N} \Pr[I_\sigma \cap R_{ij}] = \sum_{\sigma \in S^N} \Pr[I_\sigma] \Pr[R_{ij} | I_\sigma].$$

We bound this sum by partitioning $\mathcal{S}^N$ into two subsets $\mathcal{S}_1$ and $\mathcal{S}_2$. A sequence $\sigma = \langle s_1, s_2, \ldots, s_N \rangle \in \mathcal{S}^N$ belongs to $\mathcal{S}_1$ if $s_h \neq s_k$, for all $1 \leq h < k \leq N$, and, for a small enough constant $\beta > 0$, at least βN of its segments belong to Q^m. Otherwise $\sigma \in \mathcal{S}_2$. By Corollary 1, the sum of the probabilities $\Pr[I_\sigma]$, where σ has less than βN segments in Q^m, is $O(N^{-2})$. The probability that two lines in S intersect the same segment in $\mathcal{S}$ is at most $\binom{N}{2} N^{-4} = O(N^{-2})$. Thus, $\sum_{\sigma \in \mathcal{S}_2} \Pr[I_\sigma] \Pr[R_{ij} | I_\sigma] \leq \sum_{\sigma \in \mathcal{S}_2} \Pr[I_\sigma] = O(N^{-2})$ and

$$\Pr[R_{ij}] = O(N^{-2}) + \sum_{\sigma \in \mathcal{S}_1} \Pr[I_\sigma] \Pr[R_{ij} | I_\sigma].$$

Next we bound $\Pr[R_{ij} | I_\sigma]$, for $\sigma \in \mathcal{S}_1$. For a subset $S' \subseteq S$ of at least X lines, let $R_{S'}$ be the event that the cells in $C(S_{q_i}^m) \cap C(S_{q_j})$ store exactly the lines in S'. Then $\Pr[R_{ij} | I_\sigma] = \sum_{S' \in 2^S, |S'| \geq X} \Pr[R_{S'} | I_\sigma]$. Let S_i' and S_j' be the order of S' as reported by q_i and q_j, respectively, and let $C(S_i')$ and $C(S_j')$ be the set of origin cells used to generate S_i' and S_j', respectively. Event $R_{S'}$ occurs exactly when $C(S_i') = C(S_j')$. Thus, if the query cost is $O(fN/B)$ I/Os, it is possible to generate S_j' from S_i' using $O(fN/B)$ I/Os, by going via $C(S_i')$. Since $\sigma \in \mathcal{S}_1$, the segments in σ are pairwise distinct, that is, no two lines pass through a same segment in σ. Thus, conditioned on event I_σ, S_i' is fixed. Let $\mathcal{P}$ be the set of all possible permutations of S' that can be generated using $O(fN/B)$ I/Os starting from S_i'. Since S_i' is fixed, Lemma 1 shows that $|\mathcal{P}| = M^{O(fN)}$. Given event I_σ, $R_{S'}$ can happen only when $S_j' \in \mathcal{P}$. By Lemma 5, the probability that the lines in S' cross q_j according to any permutation in $\mathcal{P}$ is at most $\frac{2^{O(X)}}{(Xd/N)^{\Theta(X)}}$, where d is the distance between q_i and q_j. Thus, $\Pr[R_{S'} | I_\sigma] \leq \frac{2^{O(X)} M^{O(fN)}}{(Xd/N)^{\Theta(X)}}$. Since there are at most 2^N different subsets S', $\Pr[R_{ij} | I_\sigma] \leq 2^N \cdot \frac{2^{O(X)} M^{O(fN)}}{(Xd/N)^{\Theta(X)}} = \frac{2^{O(X)} M^{O(fN)}}{(Xd/N)^{\Theta(X)}}$. If we choose $X = \frac{\gamma f N}{\log_M N}$, for a large enough constant γ and $d \geq N^\varepsilon$, for a fixed constant $\varepsilon > 0$, it follows that

$$\Pr[R_{ij} | I_\sigma] = \frac{2^{O(X)} M^{O(fN)}}{(Xd/N)^{\Theta(X)}} \leq \frac{2^{O(fN \log M)}}{2^{\gamma \cdot \Omega(fN \log N / \log_M N)}} \leq \frac{2^{O(fN \log M)}}{2^{\gamma \Omega(fN \log M)}} = O(N^{-2})$$

and

$$\Pr[R_{ij}] = O(N^{-2}) + \sum_{\sigma \in \mathcal{S}_1} O\left(\frac{\Pr[I_\sigma]}{N^2}\right) = O(N^{-2}).$$

Thus, for $f = 1$ and $t = (\log_M N)/(2\gamma) = (N/2)/X$ queries at distance at least N^ε from each other, we have a non-zero probability that each query uses at least $N/2$ origin cells not used by any other query. This proves that any ordered line stabbing data structure with query bound $O(N/B)$ has to use $\Omega(tN) = \Omega(N \log_M N)$ space. For $f = o(\log_M N)$ and $t = \log_M N/(2\gamma f) = \omega(1)$ queries, we once again have a non-zero probability that each query uses at least $N/2$ origin cells not used by any other query. This proves that any data structure with query bound $o((N/B) \log_M N)$ must use $\Omega(tN) = \omega(N)$ space.

3.3 Lower Bound for Persistence-Based Data Structures

As discussed in the introduction, by using a partially persistent B-tree in a straightforward manner, it is possible to answer ordered line stabbing queries using $O(N/B)$ I/Os and $O(N^2)$ space. Here we prove that one cannot do much better, even using a more complex application of (full) persistence.

We consider the following model for persistence. Let $q_1, q_2, \ldots, q_t$ be a sequence of all the possible queries in an arbitrary order. A "persistence-based" structure is built in t stages. At the beginning of stage i the data structure stores a number of *versions*, $\mathcal{V}_1, \mathcal{V}_2, \ldots, \mathcal{V}_{i-1}$, each of which is a particular permutation of the input elements. The query q_j can be answered by retrieving $\mathcal{V}_j$ and transforming it into S_{q_j} using $O(N/B)$ I/Os. During stage i, we pick some version $\mathcal{V}_j$, $j < i$, and perform some number $k \geq 0$ of updates (insertions and deletions) on it to obtain a new version $\mathcal{V}_i$ that can be transformed into S_{q_i} using $O(N/B)$ I/Os. This new version is represented by storing only the k updates required to transform $\mathcal{V}_j$ into $\mathcal{V}_i$. The final space consumption is thus the total number of updates required to answer all queries, which is a lower bound on the size of any realistic persistent data structure that can answer queries $q_1, q_2, \ldots, q_t$.

Our lower bound exploits the property of the above model that every query q_i can share origin cells with at most one previous query q_j. (It may share origin cells also with other queries, but these are a subset of the cells it shares with q_j.)

Theorem 2. *Any persistence-based data structure with a query bound of $O(\frac{N}{B})$ I/Os for ordered line stabbing requires $\Omega(N^2/M^{O(1)})$ space.*

Proof. We follow the proof of Theorem 1 in Section 3.2 but choose different parameters. We pick $d = 4M^\gamma$, for a constant γ, $X = N/4$, and $f = 1$. We have,

$$\Pr[R_{ij}|I_\sigma] = \frac{2^{O(X)} M^{O(fN)}}{(Xd/N)^{\Theta(X)}} \leq \frac{2^{O(N \log M)}}{(M^\gamma)^{\Theta(X)}} = \frac{2^{O(N \log M)}}{2^{\Omega(\gamma N \log M)}} = O(N^{-2}).$$

Thus, with non-zero probability, every query q_i shares at most $N/4$ origin cells with any previous query q_j; so to obtain a version $\mathcal{V}_i$ that answers q_i using $O(N/B)$ I/Os, the data structure needs to perform at least $3N/4$ updates on $\mathcal{V}_j$, regardless of the choice of $\mathcal{V}_j$. Since $d = 4M^\gamma$, we can construct a set of $N/(4M^\gamma)$ queries with pairwise distance at least d. Each such query q_i contributes at least $\Omega(N)$ to the size of the data structure, the data structure has size $\Omega(N^2/M^\gamma)$. ☐

4 Conclusions

We provided the first space lower bounds for a number of sorted geometric query problems. We focused on the I/O model to exploit the non-trivial nature of permuting in this model as a basis for our results. Proving similar lower bounds in other models, such as pointer machine or RAM, is likely to be very difficult, as no suitable superlinear sorting lower bounds are known in these models.

We believe it is unlikely that any of the problems we studied admits an $O(N^{2-\varepsilon})$ space solution. Thus, a natural question is whether our lower bounds can be strengthened. There are a number of barriers that make this very difficult to achieve. For example, any two queries q_i and q_j can share $\Theta(N/\log_M N)$ origin cells, as sorting $\Theta(N/\log_M N)$ elements takes $O(N/B)$ I/Os. This makes our general $\Omega(N \log_M N)$ space lower bound difficult to improve without developing fundamentally different and substantially more difficult techniques. Similarly, it is not difficult to see that our $\Omega(N^2/M^{O(1)})$ space lower bound for persistence-based data structures is actually optimal.

Acknowledgements. We would like to thank Timothy Chan for bringing the type of problems we study in this paper to our attention.

References

1. Afshani, P., Brodal, G., Zeh, N.: Ordered and unordered top-K range reporting in large data sets. In: Proceedings of the 22nd ACM-SIAM Symposium on Discrete Algorithms, pp. 390–400 (2011)
2. Afshani, P., Chan, T.M.: Optimal halfspace range reporting in three dimensions. In: Proceedings of the 20th ACM-SIAM Symposium on Discrete Algorithms, pp. 180–186 (2009)
3. Aggarwal, A., Vitter, J.S.: The input/output complexity of sorting and related problems. Communications of the ACM 31(9), 1116–1127 (1988)
4. Arge, L., Danner, A., Teh, S.-M.: I/O-efficient point location using persistent B-trees. In: Proceedings of the 5th Workshop on Algorithm Engineering and Experiments, pp. 82–92 (2003)
5. Asano, T., Asano, T., Guibas, L., Hershberger, J., Imai, H.: Visibility of disjoint polygons. Algorithmica 1, 49–63 (1986)
6. Chazelle, B.: Filtering search: a new approach to query answering. SIAM Journal on Computin 15(3), 703–724 (1986)
7. Gil, J., Steiger, W., Wigderson, A.: Geometric medians. Discrete Mathematics 108(1-3), 37–51 (1992)
8. Graham, R.: An efficient algorith for determining the convex hull of a finite planar set. Information Processing Letters 1(4), 132–133 (1972)
9. Kaplan, H.: Persistent data structures. In: Handbook on Data Structures and Applications. CRC Press (2005)
10. Khuller, S., Mitchell, J.S.B.: On a triangle counting problem. Information Processing Letters 33(6), 319–321 (1990)
11. Matoušek, J.: Lectures on Discrete Geometry. Springer (2002)
12. Nouri, M., Ghodsi, M.: Space–Query-Time Tradeoff for Computing the Visibility Polygon. In: Deng, X., Hopcroft, J.E., Xue, J. (eds.) FAW 2009. LNCS, vol. 5598, pp. 120–131. Springer, Heidelberg (2009)
13. Rousseeuw, P.J., Ruts, I.: Bivariate location depth. Journal of Applied Statistics 45(4), 516–526 (1996)
14. Suri, S., O'Rourke, J.: Worst-case optimal algorithms for constructing visibility polygons with holes. In: Proceedings of the 2nd ACM Symposium on Computational Geometry, pp. 14–23. ACM (1986)

Constructing Street Networks
from GPS Trajectories*

Mahmuda Ahmed and Carola Wenk

Department of Computer Science,
University of Texas at San Antonio
{mahmed,carola}@cs.utsa.edu

Abstract. We consider the problem of constructing street networks from geo-referenced trajectory data: Given a set of trajectories in the plane, compute a street network that represents all trajectories in the set. We present a simple and practical incremental algorithm that is based on partial matching of the trajectories to the graph. We use minimum-link paths to reduce the complexity of the reconstructed graph. We provide quality guarantees and experimental results based on both real and synthetic data. For the partial matching we introduce a new variant of partial Fréchet distance.

1 Introduction

We study the task of developing geometric algorithms with quality and performance guarantees for constructing street networks from geo-referenced trajectory data. This is a new type of geometric reconstruction problem in which the task is to extract the underlying geometric structure described by a set of movement-constrained trajectories, or in other words reconstruct a geometric domain that has been sampled with continuous curves that are subject to noise. As a finite sequence of time-stamped position samples, each input trajectory represents a finite noisy sample of a continuous curve.

Due to the ubiquitous availability of geo-referenced trajectory data, the road network construction task has widespread applications ranging from a variety of location-based services on street maps to the analysis of tracking data for hiking trail map generation or for studying social behavior in animals. The underlying movement constraints can be in the form of an explicit road network that the trajectories move in. In other applications the movement might be constrained by non-geometric reasons such as behavioral patterns of animals, or the geometric domain may be implicitly given such as by wind or sea currents enabling efficient flight routes for birds or swim routes for sea turtles. In this scenario the "street" network represents a common path structure described by the set of input trajectories. From a theoretical point of view the street network construction task poses a new class of geometric shape-handling problems dealing with sets of continuous curves that are subject to noise.

---

* This work has been supported by the National Science Foundation grant NSF CAREER CCF-0643597.

L. Epstein and P. Ferragina (Eds.): ESA 2012, LNCS 7501, pp. 60–71, 2012.
© Springer-Verlag Berlin Heidelberg 2012

The problem of constructing digital maps from GPS traces has been considered in the Intelligent Transportation Systems and GIS communities [1,2,3,4], but the presented solutions are of a heuristic nature and do not provide quality guarantees. Recently the street network construction problem has received attention in the Computational Geometry community [5,6,7]. Chen et al. [5] reconstruct "good" portions of the edges (streets) and provide connectivities between these sections. They bound the complexity of the reconstructed graph and they guarantee a small directed Hausdorff distance between each original and corresponding reconstructed edge. Aanjaneya et al. [6] viewed street networks as metric graphs and they prove that their reconstructed structure is homeomorphic to the original street network. Their main focus is on computing an almost isometric space with lower complexity, therefore they focus on computing the combinatorial structure but they do not compute an explicit embedding of the edges or vertices. Ge et al. [7] employ a topological approach by modeling the reconstructed graph as a Reeb graph. They define an intrinsic function which respects the shape of the simplicial complex of a given unorganized data point set, and they provide partial theoretical guarantees that there is a one-to-one correspondence between cycles in the original graph and the reconstructed graph. All these approaches provide quality guarantees under certain assumptions on the original street network and the input trajectories, but they are based on sub-sampling the trajectory data and then working with an unorganized set of sample points and local neighborhood properties.

Our Contribution We present a simple and practical incremental algorithm that is based on partial matching of the trajectories to the graph. Our algorithm preserves the shape of the edges (streets) in the reconstructed graph by maintaining the continuous structure of the input curves. We provide quality guarantees based on the Fréchet distance, which is a distance measure suitable for comparing shapes of continuous curves. For the partial matching we introduce a new variant of partial Fréchet distance. We provide experimental results of our algorithm for real vehicle trajectory data as well as for generated data, and a statistical comparison between the original and reconstructed graph.

We prove that there is a one-to-one correspondence with bounded complexity between *well-separated* portions of the original and the reconstructed edges. However, giving quality guarantees for portions of the graph where edges come close together, in particular in regions around vertices, is a much more challenging task. We provide the first attempt at reconstructing vertex regions and providing quality guarantees, without assuming very clean and extremely densely sampled data. We reconstruct intersections as sets of vertices within bounded regions (*vertex regions*), where the size of each set is bounded by the degree and the region is bounded by the minimum incident angle of the streets at that intersection. We guarantee that if the vertices are sufficiently far apart so that the vertex regions do not overlap, then the vertices of each vertex region correspond to exactly one vertex in the original graph.

2   Problem Statement, Data Model and Assumptions

We model the *original* graph (street network) $G_o = (V_o, E_o)$ as an embedded undirected graph in $\mathbb{R}^2$. Each trajectory in the input curve set I is assumed to have sampled a connected sequence of edges in G_o(*street-path*). We model the error associated with each trajectory by a precision parameter ε. Given an input set I of polygonal curves in the plane and a precision parameter $\varepsilon > 0$, our goal is to compute an undirected *reconstructed* graph $G = (V, E)$ that represents all curves in the set. I.e., a *well-separated* portion of each edge in E_o corresponds to a sub-curve of an edge in E, and each vertex in V_o corresponds to a set of vertices in V.

We assume that V_o and V are sets of vertices with degree > 2 and each edge is represented as a polygonal curve. We assume that G_o is fully sampled i.e., for each street $\gamma \in E_o$ there is a sampled sub-curve in input curve set, I. We refer to each edge of G_o as a *street*.

An input trajectory is a finite sequence of time-stamped position samples, which represents a finite noisy sample of a continuous curve. Generally, the measurements of the position samples are only accurate within certain bounds (measurement error), and the movement transition in between position samples can be modeled with varying accuracies depending on the application (sampling error). We model trajectory data as piecewise linear curves, and we will in particular consider trajectories of vehicles driving on a street network. In this case, the input curves in fact sample an $\omega/2$-fattening of a street-path, where ω is the street width. The δ-fattening of a point set A is the Minkowski sum $A_\delta = A \oplus B(0, \delta)$, where $B(x, r)$ is the closed ball of radius r centered at x.

We work with a single precision parameter ε which captures the different kinds of noise as well as the street width. We use the Fréchet distance [8] to measure the similarity between the shape of an input curve and a street-path. For two planar curves $f, g : [0, 1] \to \mathbb{R}^2$, the Fréchet distance δ_f is defined as

$$\delta_F(f, g) = \inf_{\alpha, \beta : [0,1] \to [0,1]} \max_{t \in [0,1]} \| f(\alpha(t)) - g(\beta(t)) \| \tag{1}$$

where α, β range over continuous and non-decreasing reparametrizations, and $\|.\|$ denotes the Euclidean norm.

To define the well-separability of streets, we make use of the following definition from [5].

Definition 1 (α-Good). *A point p on G is α-good if $B(p, \alpha\varepsilon) \cap G$ is a 1-ball that intersects the boundary of $B(p, \alpha\varepsilon)$ in two points. A point p is α-bad if it is not α-good. A curve β is α-good if all points on β are α-good.*

2.1   Assumptions

The correctness of our algorithm depends on the following assumptions that we make about the original graph and input data, see Figure 1.

1. **Assumptions on G_o:** (a) Each street has a well-separated portion which is 3ε-good. We refer to this as a *good section*. (b) If for two streets γ_1, γ_2 there are points $p_1 \in \gamma_1$ and $p_2 \in \gamma_2$ with distance $\leq 3\varepsilon$, then γ_1 and γ_2 must share a vertex v, and the sub-curves $\gamma_1[p_1, v]$ and $\gamma_2[p_2, v]$ have Fréchet distance $\leq 3\varepsilon$ and they are fully contained in $B(v_0, 3\varepsilon/\sin\alpha)$. Here $\alpha = \angle p_1 v p_2$ (see Figure 1b).

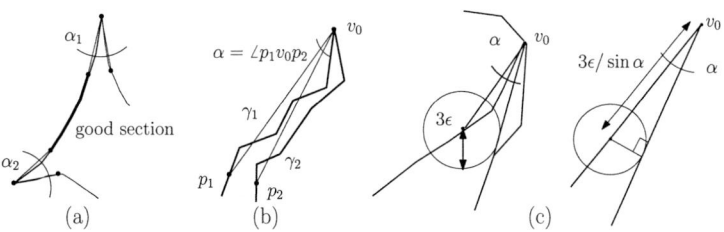

Fig. 1. (a) Assumption 1a (b) Assumption 1b (c) Assumptions yield minimum distance between two vertices

From both assumptions follows that the minimum distance between two intersections is $> 3\varepsilon/\sin\alpha_1 + 3\varepsilon/\sin\alpha_2$. Assumption 1a states the minimum requirement to justify the existence of a street based on it being well-separated from other streets. Assumption 1b requires streets that are close together to converge to a vertex, and it discards streets that are close but do not share a vertex, because the input curves do not clearly distinguish between such streets. Note that the bound $3\varepsilon/\sin\alpha$ can be large for small values of α. So, the assumptions allow streets to be close together for a long time but restrict them to go far off once they are close to a vertex.

2. **Assumptions on Input Data:** (a) Each input curve is within Fréchet distance $\varepsilon/2$ of a street-path in G_o. (b) All input curves sample an acyclic path in G_o.

Assumption 2a ensures that the street-path and the corresponding input curve have to be similar. Assumption 2b ensures that, during an incremental construction algorithm, the first curve in I that represents a new edge does not sample a cycle. Our algorithm in fact only needs the unmatched portion (defined in Subsection 3.1) to sample an acyclic path. And even if it samples a cyclic path, such a cycle can be split in order to maintain this assumption.

Note that if a curve is well-sampled then the sampled curve is naturally within a bounded Fréchet distance of the original curve, which fulfils Assumption 2a. In particular, for our test data of vehicle trajectories, the GPS device error is generally bounded, and data from OpenStreetMap.org is generally sampled every second, and it captures every feature of the shape of the original street very well.

3   Algorithm

3.1   Preliminaries

In our algorithm we employ the concept of the *free space* F_ε and the *free space surface* FS_ε of one curve and a graph to identify clusters of sub-curves which sample the same street-path. For two planar curves $f, g : [0,1] \to \mathbb{R}^2$, and $\varepsilon > 0$, *free space* is defined as $F_\varepsilon(f, g) := \{(s,t) \in [0,1]^2 |\ \|f(s) - g(t)\| \leq \varepsilon\}$. The *free space diagram* $FD_\varepsilon(f, g)$ is the partition of $[0,1]^2$ into regions belonging or not belonging to $F_\varepsilon(f, g)$. The *free space surface* $FS_\varepsilon(G, l)$ for a graph $G = (V, E)$ and a curve l is a collection of free space diagrams $FD_\varepsilon(e, l)$ for all $e \in E$ glued together according to the adjacency information of G (see Figure 3).

In [8] it has been shown that $\delta_F(f, g) \leq \varepsilon$ if and only if there exists a curve within F_ε from the lower left corner to the upper right corner, which is monotone in both coordinates, see Figure 2 for an illustration. $FS_\varepsilon(G, l)$ could be disconnected if the graph G is not connected.

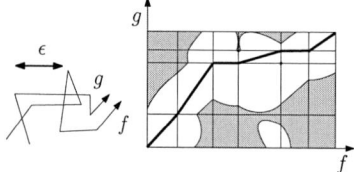

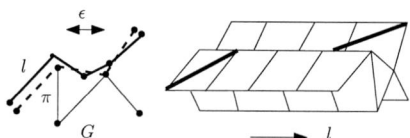

Fig. 2. FD_ε for two polygonal curves f, g. A monotone path is drawn in the free space.

Fig. 3. FS_ε for a graph G and a curve l. An example path π is shown in dashed line and an l-monotone path in FS_ε is highlighted in bold.

In our algorithm we combine the idea of *map matching* and *partial curve matching* to map a curve partially to a graph. Buchin et al. [9] solved a *partial matching* for two curves by finding a monotone shortest path on a weighted FD_ε from lower left to upper right end point, where free space (white region) has weight 0 and non-free space (black region) has weight 1 (see Figure 4a). Alt et al. [10] solved the *map-matching* problem of matching a curve l to a graph by finding an l-monotone path in the free space of FS_ε from any left to any right end point. Here, the path is not allowed to penetrate the black region. In our case we need to find an l-monotone shortest path on the weighted free-space surface from any left end point to any right end point. However, finding such a shortest path on a weighted non-manifold surface is hard. Moreover, as the path can begin and end at any left or right end point, in some cases such a path does not provide us with the mapping we are looking for (see Figure 4a). The bold path is the desired one but the shortest path based on L_2 is the dashed one.

We used a minimum-link chain stabbing algorithm [11] to lower the complexity, such that the complexity of the representative edge will depend only on the complexity of the original street rather than on the complexity of the input curves.

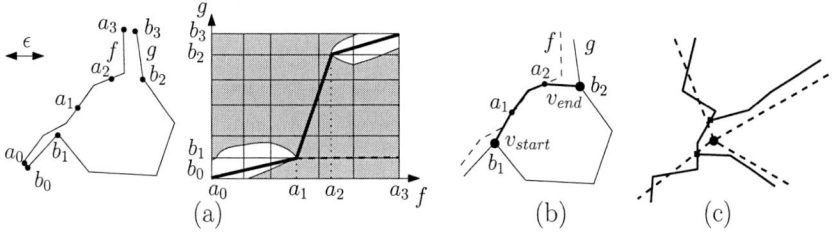

Fig. 4. (a) Partially similar f and g. $M_\varepsilon(f,g) = \{(f\,[a_0,a_1]\,,g\,[b_0,b_1]),(f\,[a_1,a_2]\,,$ $null),(null,g[b_1,b_2]),(f\,[a_2,a_3]\,,g\,[b_2,b_3])\}$. (b) Adding the unmatched portion of f as an edge. (c) G_o and G are in dashed and bold respectively.

3.2   Street Network Reconstruction Algorithm

Our algorithm can be divided into two phases, the first one involves computing a reconstructed graph (step 1, step 2) and in the second phase we compress the complexity of that graph (step 3). The simplicity of the algorithm relies on careful observation of how $FS_{1.5\varepsilon}(l,G)$ looks like when the original graph and the input curves satisfy the assumptions described in Subsection 2.1. In the first step it computes a partial curve-graph mapping.

Definition 2. *The Curve-Graph Partial Mapping* $M_\varepsilon(l,G)$ *consists of pairs* $(l\,[a_{j-1},a_j]\,,X)$ *where* $0 = a_0 \leq a_1 \leq \ldots \leq a_k = 1$ *such that every sub-curve* $l\,[a_{j-1},a_j]$ *in a partition* $l\,[a_0,a_1,\ldots,a_k]$ *of* l *is either mapped to a portion* $X = e\,[start,end]$ *of an edge* $e \in E$ *with* $\delta_F(l\,[a_{j-1},a_j]\,,e\,[start,end]) \leq \varepsilon$ *or* $X = null$ *if no such edge exists. A sub-curve is called a matched portion if it has a non-null mapping, and an unmatched portion otherwise. We assume that all unmatched portions as well as all matched portions along the same edge are maximal.*

Our iterative algorithm proceeds as follows: Let $G_i = (V_i,E_i)$ be the graph after the i-th iteration.

---

for each $l \in I$
    Step 1: Compute Curve-Graph Mapping $M_{1.5\varepsilon}(l,G_{i-1})$
    for each $(l\,[a_{j-1},a_j]\,,X) \in M_{1.5\varepsilon}(l,G_{i-1}))$
    if X is null
        // Add edge for the unmatched portion $l\,[a_{j-1},a_j]$
        Step 2: Create/split edges and create vertices
    else
        // Update $e\,[start,end]$ with matched portion $l\,[a_{j-1},a_j]$
        Step 3: Compute Minimum-Link Representative Edge

The total worst case runtime for adding the i-th curve is $O(N_{i-1}n_i + k_u(N_{i-1} + n_i)^2 \log(N_{i-1} + n_i) + k_c)$, where N_{i-1} is the complexity of G_{i-1}, n_i is the complexity of the ith curve $l \in I$, and k_u and k_c are the number of edges updated or created, respectively. The runtime for each individual step will be given with the description of the step below.

Step 1: Compute Curve-Graph Mapping $M_{1.5\varepsilon}(l_i, G_{i-1})$

In this step of the algorithm we compute a partial mapping $M_{1.5\varepsilon}(l_i, G_{i-1})$ which minimizes the total length of unmatched portions of l_i. First, we compute $FS_{1.5\varepsilon}(l_i, G_{i-1})$, and then we project the white regions onto the curve $(l\,[a_{j-1}, a_j],\ e\,[start, end])$ which yields *white intervals* (matched portions). Lastly, we fill the gap between non-overlapping white intervals with a *black interval* $(l\,[a_{j-1}, a_j],\ null)$ (unmatched portion).

In our setting, the objective is one-sided and we are interested in matching the curve maximally with the graph. Therefore the problem of computing the shortest monotone path reduces to computing the total length of the unmatched portions along the curve l. This allows the desired path to go through the black region along the graph direction without adding any additional cost. Therefore, we measure the length of the path **only within the unmatched portion and only along the curve direction**.

Our projection approach is justified by Assumption 1a on the street network that no two streets are less than or equal to 3ε close to each other if they do not share a vertex, which implies for each $l\,[a_{j-1}, a_j]$ if it samples a portion of a good section then only one $e\,[start, end]$ exists.

After this step we have the curve-graph mapping as a list of black and white intervals ordered by their start points on l_i. Such a list can be computed in $O(n_i N_{i-1})$ total time.

Step 2: Create/Split Edges and Create Vertices

In this step of the algorithm, for each $(l_i\,[a_{j-1}, a_j],\ null) \in M_{1.5\varepsilon}(l_i, G_{i-1})$ we create an edge in constant time. By construction, the previous and next intervals of a black-interval are either *null* (the interval is the first or last element in the list) or white. Assume them to be non-*null* (if one or both of them are *null* then we do not have to create a vertex on the *null*-end). Let $e_{prev} = p_0 p_1 p_2 \ldots p_{n_1}$ and $e_{next} = n_0 n_1 n_2 \ldots n_{n_2}$ be the edges of previous and next intervals $(l_i\,[a_{j-2}, a_{j-1}],\ e[start,\ end])$ and $(l_i\,[a_j, a_{j+1}],\ e\,[start,\ end])$, respectively. Then $v_{start} = e_{prev}\,[end]$ and $v_{end} = e_{next}\,[start]$ are two points on the line segments $p_{i-1}p_i$ and $n_{j-1}n_j$. We create new points on the segments if $start > 0$ or $end < 1$, otherwise we take the existing endpoints as v_{start}, v_{end} and insert them into the vertex list as new vertices (multiple vertices might be created for a single original one, see Figure 4c). Then we follow the steps below to split an existing edge and create a new one. a) split e_{prev} as $p_0 p_1 \ldots p_{i-1} v_{start}$ and $v_{start} p_i \ldots p_{n_1}$ b) split e_{next} as $n_0 n_1 \ldots n_{j-1} v_{end}$ and $v_{end} n_j \ldots n_{n_2}$ c) insert $v_{start} l_i\,[a_{j-1}, a_j]\, v_{end}$ as an edge in E_{i-1}. For example, in Figure 4a, consider g as an edge in G_{i-1} and f as l_i. Figure 4b shows the addition of the unmatched portion of f as an edge in the graph, here $e_{prev} = e_{next}$.

Step 3: Compute Minimum-Link Representative Edge

In first and second step we have computed the reconstructed graph for first i input curves. In this step, we compute a minimum-link representation for each $(l_i [a_{j-1}, a_j], X) \in M_{1.5\varepsilon}(l_i, G_{i-1})$, where $X \neq null$. The problem address the following: Given two polygonal curves f, g which both have bounded Fréchet distance to another polygonal curve γ such that $\delta_F(\gamma, f) \leq \varepsilon$ and $\delta_F(\gamma, g) \leq \varepsilon/2$, we have to find a minimum-link representative curve e of γ such that $\delta_F(\gamma, e) \leq \varepsilon$. First, we construct a combined vertex-sequence γ' of vertices of f and g by following a monotone mapping in $FD_{1.5\varepsilon}(f, g)$. According to Lemma 1, the polygonal curve associated with such sequence has Fréchet distance $\leq 2\varepsilon$ with γ. Then we apply the minimum-link algorithm [11] for the sequence of $B(v_{\gamma'}, 2\varepsilon)$ objects, with the additional restriction that the vertices must lie within $g^{\varepsilon/2}$, where $v_{\gamma'}$ are the vertices of γ'. The resulting path e obtained by our variant might not be the minimum-link path for the sequence of $B(v_{\gamma'}, 2\varepsilon)$, but the path has complexity less than or equal to the original curve γ, as $\gamma'^{2\varepsilon}$, f^{ε} and $g^{\varepsilon/2}$ all contain γ. Using the triangle inequality it can be proven that $\delta_F(e, \gamma) \leq \varepsilon$. In Step 3 of our algorithm, we update the edges of G_{i-1} for each white interval using the algorithm described above, where the polygonal curve f corresponds to $e[start, end]$ and g corresponds to $l_i [a_{j-1}, a_j]$. The time complexity to compute such a path is $O(n^2 \log n)$, where n is the number of vertices in γ'.

Lemma 1. *Let f and g be curves that sample a curve γ such that $\delta_F(f, \gamma) \leq \varepsilon$ and $\delta_F(g, \gamma) \leq \varepsilon/2$. Then any curve γ' comprised of vertices of f and g based on their order of a monotone mapping in $FD_{1.5\varepsilon}$ has $\delta_F(\gamma, \gamma') \leq 2\varepsilon$.*

4   Quality Analysis

In this section in Lemma 2 we prove that if an input curve $l \in I$ samples a good section of a street or a street-path, then that street-path is unique in G_o. It can be proven using a loop invariant that, if every edge in a path has exactly one good section, then after adding the i-th curve, the reconstruction graph G_i preserves all paths of G_o^i. By *preserving* a path we mean that all good sections have Fréchet distance less than or equal to ε to the original street and all vertices lie in the vertex region around the original vertices. Here, G_o^i is the sub-graph of G_o fully sampled by the first i curves in I.

Lemma 2. *For each $l \in I$ there exists a mapping $M_{\varepsilon_p}(l, G_o) = \{(l[0, a_1], \gamma_1[b_0, 1]), (l[a_1, a_2], \gamma_2[0, 1]), \ldots (l[a_{k-1}, 1], \gamma_k[0, b_k])\}$ for $\varepsilon/2 \leq \varepsilon_p < 2.5\varepsilon$. And for $k \geq 3$ if l samples a good section of γ_1 and γ_k then $\gamma_1 \gamma_2 \gamma_3 \ldots \gamma_k$ otherwise $\gamma_2 \gamma_3 \ldots \gamma_{k-1}$ is unique.*

4.1   Recovering Good Sections

In this subsection we prove that if a street $\gamma \in E_o$ is sampled by a set of input curves $I_\gamma = \{l_1, l_2, \ldots, l_k\}$, then our algorithm reconstructs each good section of γ as one edge $e \in E$.

Lemma 3. *For each edge $\gamma \in E_o$, if β is a good section of γ and there exists a non-empty set of input curves $I_\gamma = \{l_1, l_2, \ldots, l_k\} \subseteq I$ such that for every $l_i \in I_\gamma$, $\delta_F(l_i[start_i, end_i], \gamma) \leq \varepsilon/2$, then there exists only one $e \in E$ such that $\delta_F(e\,[start, end]\,, \beta) \leq \varepsilon$ and the complexity of $e\,[start, end]$ is less than or equal to the complexity of β.*

Proof. The proof follows from the construction of the curve-graph mapping $M_{1.5\varepsilon}(l_i, G_{i-1})$. When the first curve of the set is added to the graph, it is identified as an unmatched portion. And after that, all other sub-curves that sample γ will be identified as matched portions. Once we get the first matched portion (i.e., for the second curve in I_γ), we compute a center-line representative curve which ensures the minimum complexity and which has Fréchet distance $\leq \varepsilon$ from the original street. Thus, all the other sub-curves will also appear as matched portions in $M_{1.5\varepsilon}(l_i, G_{i-1})$, that means for all $l_i \in I_\gamma$ that sample β only one edge will be created in the graph. $\qquad \square$

4.2   Bounding Vertex Regions

In this section, we bound the vertex region, R_v for reconstructed vertices around the original vertex, v. We perform the analysis for a 3-way intersection which could easily be extended to an arbitrary n-way intersection.

Consider three streets γ_1, γ_2 and γ_3 incident to a vertex v_0. $v_i{}^{i-1}$ and $v_i{}^{(i+1)\%3}$, for $1 \leq i \leq 3$, are defined as the two points on γ_i which are farthest from v_0 along the curve (see Figure 5a). Here, $\angle v_i{}^{i+1} v_0 v_{i+1}{}^i = \alpha_i$ (see Figure 5b), and according to Assumption 2a, $v_0 v_i{}^{i+1}$ and $v_0 v_{i+1}{}^i$ are fully contained in $B(v_0, 3\varepsilon/\sin \alpha_i)$. For a 3-way intersection we can have input curves that sample three different street-paths and to reconstruct the vertex we need only two of them. Based on which two we use in which order, the reconstructed vertex can be in a different location in the vertex region. For analysis purpose, we define such a minimal input curve set, which is sufficient for reconstruction, considering all different choices. The idea is to compute the vertex region for each of these sets and then union them all to get R_{v_0}.

For a 3-way vertex we have six such sets. Let the first set I_1 contains l_i which samples street-path $\pi_1 = \gamma_1 \gamma_2$ ($\delta_F(l_i, \pi_1) \leq \varepsilon/2$) and l_j which samples $\pi_2 = \gamma_2 \gamma_3$ ($\delta_F(l_j, \pi_2) \leq \varepsilon/2$), where $i < j$. In the i-th iteration l_i will be inserted to G_{i-1} as an edge e, and in the j-th iteration l_j will be considered. To compute $M_{1.5\varepsilon}(l_j, G_{j-1})$, we compute intersection of $l_j{}^{1.5\varepsilon}$ and e and create a vertex with the intersection point which defines the mapping of a partition point of a matched and an unmatched portion of l_j. As e could be anywhere within $\pi_1{}^\varepsilon$ and l_j could be anywhere within $\pi_2{}^{\varepsilon/2}$. Considering all possibilities, we obtain the intersection region as $\pi_1{}^\varepsilon \cap \pi_2{}^{2\varepsilon}$. Again, as we are creating vertices only on the boundaries of the intersection region, no vertices would be created when the street-paths are $< 1.5\varepsilon$ close to each other, so the vertex region for I_1 is $R(v_0, I_1) = (\pi_1{}^\varepsilon \cap \pi_2{}^{2\varepsilon}) \setminus (\pi_1{}^\varepsilon \cap \pi_2{}^\varepsilon)$ (see Figure 5c). Consolidating all six sets, the vertex region is shown in Figure 5d.

The above analysis can be extended to an arbitrary n-way intersection, where there are $n(n-1)/2$ possible paths and $r = (n(n-1)/2)!/(n(n-1)/2 - (n-1))!$

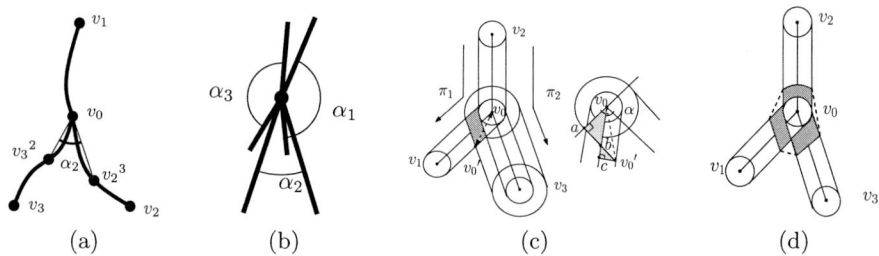

Fig. 5. (a) 3-way intersection. (b) 3-way intersection with illustration of α. (c) 3 way intersection region considering only I_1. (d) The shaded region is $R(v_0)$.

sets of input curves are involved. The region is then defined as $R(v_0) = R(v_0, I_1) \cup R(v_0, I_2) \cup R(v_0, I_3) \cup \cdots \cup R(v_0, I_r)$.

Lemma 4. *If $v_0' \in R(v_0)$ is the farthest point from a vertex $v_0 \in V_o$ with degree n, then $\varepsilon \leq d(v_0, v_0') \leq \varepsilon / \sin\alpha \sqrt{5 + 4\cos\alpha}$, where α is the minimum of all $\alpha_i s$.*

Proof. In Figure 5c, considering $\triangle v_0 ab$ we have that $v_0 b = v_0 a / \cos(\pi/2 - \alpha) = 2\varepsilon / \sin\alpha$. And considering $\triangle v_0' bc$ we have $bc = c v_0' / \tan\alpha = \varepsilon / \tan\alpha$. $d(v, v_0) = \sqrt{v_0 c^2 + v_0' c^2} = \sqrt{\varepsilon^2 + ((2\varepsilon + \varepsilon \cos\alpha) / \sin\alpha)^2} = \sqrt{\varepsilon^2 / \sin\alpha^2 (5 + 4\cos\alpha)} = \varepsilon / \sin\alpha \sqrt{5 + 4\cos\alpha}$.   □

5   Experimental Results

We implemented the first phase our algorithm using *java*. Instead of computing the minimum-link representative curve we used the unmatched portion of the first curve that defines an edge to represent the reconstructed edge. Experiments were conducted on a 2.93GHz Intel(R) Core(TM)2 Duo machine with 4.00GB of RAM. In this section we present our results obtained by running our algorithm on different datasets. It seems that, even using the noisy data our algorithm produces graphs with good accuracy.

We applied our algorithm on real tracking data obtained by sampling vehicle movements at a rate of 30 seconds. The dataset consists of $3,237$ vehicle trajectories consisting of a total of $57,109$ position samples. The data was collected from taxi cabs in the municipal area of Berlin, Germany, in 2007[1]. Figure 6 shows the reconstructed graph (dark) overlayed with the original graph. As we can see, almost all sampled good sections were reconstructed, but as our data was noisy the distortion is high. For this case our ε was 82.5 meters and the running time was 11 minutes. The reconstructed graph had $3,044$ vertices and $3,546$ edges. Total complexity of all edges was $12,434$. As the original graph was not fully sampled it was hard to compare the graphs.

---

[1] A lot of tracking data is available at openstreetmap.org also, but it is hard to extract the desired type, since they contain a lot of hiking trails as well as vehicle trajectories.

(a)   (b)

Fig. 6. Reconstructed graph of Berlin and a zoomed-in portion

Table 1. Experimental Results

GPS error	s.r. in sec	ε m	# of curves	# of points	G edges	G g.s.(m)	G vertices	G_o edges	G_o g.s.(m)	G_o vertices
	1	12	900	14515	805	105,062	56	560	76,248	37
	5	17	651	5934	731	66,755	46	560	67,959	32
± 5m	10	20	583	4318	681	56,379	32	560	65,270	28
	20	25	556	3323	733	62,522	28	560	58,810	26
	30	30	543	2793	696	65,784	22	560	53,200	20
	1	22	900	14515	912	76,991	47	560	63,128	28
	5	27	651	5934	501	46,031	31	560	56,251	24
± 10m	10	30	583	4318	499	31,816	31	560	53,200	20
	20	33	556	3323	605	37,709	20	560	49,643	17
	30	35	543	2793	635	50,112	16	560	48,183	14

We generated our second type of dataset by almost fully sampling a sub-graph of the Berlin graph which had 363 vertices and 560 edges/streets with a total length of 147,463 meters. We generated multiple datasets by varying sampling rates (s.r.) and device error, which effectively influence the choice of ε. We assumed the speed of the vehicle is 88.5km/h. Our results are summarized in Table 1. The performance of our algorithm is measured in terms of number of vertices, number of edges and length of good sections (g.s.) in G and G_o. For simplicity of implementation we approximated the vertex region assuming $\alpha = 2\pi/7$, and mapped every reconstructed vertex to an original one which is within 3.5ε distance. As we can see, the number of reconstructed vertices in G is consistently higher than the number of good vertices in G_o. We identified good sections only with respect to vertices, and for each line segment we classified it as either good or bad. As we did not compute minimum-link representative edges, when the sampling rate is very small the length of the good section in reconstructed graph became larger than the original one (because the reconstructed edges zig-zag within very small distance). Our code is publicly available at www.my.cs.utsa.edu/~mahmed/research.

6   Conclusions

We outlined an incremental algorithm for street network construction and update. Our algorithm is very practical to use in real life as it is very simple to implement and it involves only one parameter ε. Unlike existing algorithms, it does not require very densely sampled data, as long as the features are captured, the algorithm produces pretty accurate maps in reasonable time. Our algorithm assures reconstruction of good sections and bounds the vertex regions. Our future work includes investigating vertex regions to reduce redundant vertices.

Acknowledgment. We thank Dieter Pfoser and Maike Buchin for fruitful discussions.

References

1. Bruntrup, R., Edelkamp, S., Jabbar, S., Scholz, B.: Incremental map generation with GPS traces. In: Proc. IEEE Intelligent Transp. Systems, pp. 574–579 (2005)
2. Li, Z., Lee, J.-G., Li, X., Han, J.: Incremental Clustering for Trajectories. In: Kitagawa, H., Ishikawa, Y., Li, Q., Watanabe, C. (eds.) DASFAA 2010. LNCS, vol. 5982, pp. 32–46. Springer, Heidelberg (2010)
3. Guo, T., Iwamura, K., Koga, M.: Towards high accuracy road maps generation from massive GPS traces data. In: IEEE Int. Geoscience and Remote Sensing Symposium, pp. 667–670 (2007)
4. Cao, L., Krumm, J.: From GPS traces to a routable road map. In: Proc. of the 17th ACM SIGSPATIAL Int. Conf. on Advances in GIS. GIS 2009, pp. 3–12. ACM, New York (2009)
5. Chen, D., Guibas, L., Hershberger, J., Sun, J.: Road network reconstruction for organizing paths. In: Proc. ACM-SIAM Symp. on Discrete Algorithms (2010)
6. Aanjaneya, M., Chazal, F., Chen, D., Glisse, M., Guibas, L.J., Morozov, D.: Metric graph reconstruction from noisy data. In: Proc. ACM Symp. Computational Geometry, pp. 37–46 (2011)
7. Ge, X., Safa, I., Belkin, M., Wang, Y.: Data skeletonization via Reeb graphs. In: 25th Annual Conference on Neural Info. Processing Systems, pp. 837–845 (2011)
8. Alt, H., Godau, M.: Computing the Fréchet distance between two polygonal curves. Int. J. of Computational Geometry and Applications 5, 75–91 (1995)
9. Buchin, K., Buchin, M., Wang, Y.: Exact algorithm for partial curve matching via the Fréchet distance. In: Proc. ACM-SIAM Symp. on Discrete Algo. (SODA 2009), pp. 645–654 (2009)
10. Alt, H., Efrat, A., Rote, G., Wenk, C.: Matching planar maps. Journal of Algorithms, 262–283 (2003)
11. Guibas, L.J., Hershberger, J.E., Mitchell, J.S.B., Snoeyink, J.S.: Approximating polygons and subdivisions with minimum-link paths. Int. J. of Computational Geometry and Applications, 3–4 (1993)

I/O-efficient Hierarchical Diameter Approximation*

Deepak Ajwani[1,**], Ulrich Meyer[2], and David Veith[2]

[1] Centre for Unified Computing, University College Cork, Cork, Ireland
[2] Institut für Informatik, Goethe-Universität Frankfurt, Robert-Mayer-Str. 11–15,
D-60325 Frankfurt am Main, Germany

Abstract. Computing diameters of huge graphs is a key challenge in complex network analysis. As long as the graphs fit into main memory, diameters can be efficiently approximated (and frequently even exactly determined) using heuristics that apply a limited number of BFS traversals. If the input graphs have to be kept and processed on external storage, even a single BFS run may cause an unacceptable amount of time-consuming I/O-operations.

Meyer [17] proposed the first parameterized diameter approximation algorithm with fewer I/Os than that required for exact BFS traversal. In this paper we derive hierarchical extensions of this randomized approach and experimentally compare their trade-offs between actually achieved running times and approximation ratios. We show that the hierarchical approach is frequently capable of producing surprisingly good diameter approximations in shorter time than BFS. We also provide theoretical and practical insights into worst-case input classes.

1 Introduction

Massive graph data stemming from social networks or the world wide web have implicitly become part of our daily life and also means big business. Consequently, a whole branch of computer science (and industry) deals with network analysis [8]. For connected undirected unweighted graphs $G(V, E)$ where $n = |V|$ and $m = |E|$, the distance $d(u, v)$ between two nodes $u, v \in V$ is the number of edges in the shortest path connecting u and v. The eccentricity of a node v is defined as $ecc(v) = \max_u d(v, u)$. A fundamental step in the analysis of a massive graph is to compute its diameter $D := \max_{u,v} d(u, v) = \max_v ecc(v)$.

We focus on the case when G is sparse ($m = O(n)$) but nevertheless too big to fit into the main memory of a single computing device. In this situation one can either distribute the data over many computers and apply parallel algorithms (e. g., see [6]) and/or store the data on secondary memory like hard disks or flash memory. In this paper we will concentrate on improved diameter approximation algorithms for the second (external memory) approach.

---

* Partially supported by the DFG grant ME 3250/1-3, and by MADALGO – Center for Massive Data Algorithmics, a Center of the Danish National Research Foundation.
** Research of this author is supported by the EPS grant from IRCSET and IBM.

L. Epstein and P. Ferragina (Eds.): ESA 2012, LNCS 7501, pp. 72–83, 2012.
© Springer-Verlag Berlin Heidelberg 2012

External Memory Model. The huge difference in data access times between main memory and disks is captured in the external memory (EM) model (aka I/O model) by Aggarwal and Vitter [1]. It assumes a two level memory hierarchy: The internal memory is fast, but has a limited size of M elements (nodes/edges). The external memory (of potentially unlimited size) can only be accessed using I/Os that move B contiguous elements between internal and external memory. The computation can only use data currently kept in internal memory. The cost of an algorithm is the number of I/Os it performs – the less, the better. The number of I/Os required to scan n contiguously stored elements in the external memory is $scan(n) = O(n/B)$ and the number of I/Os required for sorting n elements is $sort(n) = O(\frac{n}{B} \log_{M/B} n/B)$. For realistic values of B, M, and n, $scan(n) < sort(n) \ll n$. Frequently, the goal of designing external memory algorithms for sparse graphs is to reduce the I/O complexity from $\Omega(n)$ to not much more than $O(sort(n))$.

2   Related Work

The traditional approaches for computing the exact diameter rely on computationally expensive primitives such as solving the all pair shortest path problem (APSP) or fast matrix multiplication algorithms. Thus, many heuristics (e.g., [11,15,7,10,12]) have been proposed to approximate the diameter. However, these heuristics still rely on BFS and SSSP traversals, which are very expensive when the graph does not fit in the main memory.

In the external memory, network analysis algorithms have been developed for specific graph classes (e.g., [14] for naturally sparse graphs). For general undirected graphs, algorithms for the exact computation of the diameter [5,9] require $\Theta(n \cdot sort(n))$ I/Os for sparse graphs ($m = O(n)$) and are thus impractical. We are interested in approaches that work for general undirected graphs and require even less I/Os than BFS, as even carefully tuned BFS implementations [4] require many hours on graphs with billions of edges. In this context, Meyer's parameterized approach [17] achieves efficient trade-offs between approximation quality and I/O-complexity in theory and can give good approximation bounds even when the desired I/O complexity is restricted to be less than that of external memory BFS algorithms [16].

Parameterized Approach. The parameterized approach decomposes the input undirected graph into many small-diameter clusters and then contracts each cluster to a single node forming a condensed graph. This condensed graph preserves the structure of the input graph and in particular, the diameter of the condensed graph approximates the diameter of the input graph. The condensed graph is typically much smaller than the input graph. If it fits internally, its diameter can be approximated using internal memory heuristics; otherwise a semi-external or fully external memory single source shortest path (SSSP) from an arbitrary vertex can be used to obtain a good approximation.

To compute small-diameter clusters, the parameterized algorithm first selects some vertices to be masters uniformly at random and then "grows" the clusters around the selected master vertices "in parallel". In a parallel cluster growing round, each cluster tries to capture all its unvisited neighbors, with ties being broken arbitrarily. Each such round is simulated in the external memory with a constant number of scanning and sorting steps.

For undirected, unweighted graphs with n nodes and $m = O(n)$ edges, an $O(\sqrt{k} \cdot \log n)$ approximation for the diameter can be obtained with high probability (whp) using $O(n \cdot \sqrt{\log k/(k \cdot B)} + k \cdot scan(n) + sort(n))$ I/Os by contracting the input graph to n/k vertices. Thus, if we strive for $O(n/B^{2/3})$ I/Os, we should contract the graph to $O(n/(B^{1/3} \cdot \log B))$ vertices and expect multiplicative errors of $O(B^{1/6} \cdot \sqrt{\log B} \cdot \log n)$ in the computed diameter whp.

In a recent workshop paper, Ajwani et al. [3] have shown that this parameterized approach produces much better approximation bounds on many different graph classes with varying diameters than its worst-case approximation guarantee would suggest. Also, if the number of vertices in the contracted graph is carefully chosen, the parameterized approach is significantly faster than the various internal memory heuristics for diameter approximation, even if the BFS routine in those heuristics is replaced by its external memory counterpart. This makes the diameter approximation of graphs with a few billion edges, a viable task in external memory on typical desktop machines.

3   Extending the Parameterized Approach

Hierarchical Extension. The scalability of the parameterized approach is still limited. For very large graph sizes, either one needs to pick a very large value of k to contract the graph enough to be able to run semi-external SSSP and thus settle for a poor approximation ratio or one has to depend on the more I/O-intensive primitive of fully-external SSSP. Also, we are not aware of any efficient implementation for computing SSSP in fully-external memory. A potential approach to scale the parameterized approach further is to extend it by contracting the graph recursively. The difficulty in such an extension lies in the fact that after the first iteration, the graph becomes weighted and it is not clear how to I/O-efficiently contract a weighted graph while minimizing the approximation ratio of the resultant diameter. We consider different ways to select master vertices (and their numbers) for weighted graphs and study the approximation quality of the resultant diameter on various graph classes.

Bad Graph Class. Our first attempt for selecting the master vertices in the weighted graphs is to completely ignore the edge weights occurring from graph shrinking and simply select masters uniformly at random and with the same probability in all recursive levels. Already for two levels of graph contraction the multiplicative approximation error for the diameter might grow to $\Omega(k \cdot \sqrt{k} \cdot \log n)$: a factor of $O(\sqrt{k} \cdot \log n)$ based on the analysis of [17] for unweighted graphs, exacerbated by the fact that after the first shrinking, edges may represent a weight of $\Omega(k)$. Surprisingly, it will turn out in Section 4 that this weight-oblivious master selection rule is often capable to yield reasonable approximation bounds. Nevertheless, in Section 4.2, we also present a sophisticated artificial graph class where the computed diameter is provably a factor $\Omega(k^{4/3-\epsilon})$ away from the actual diameter using two recursive levels and the simple master selection. Note that after the second level, the number of clusters is reduced to $O(n/k^2)$. On the other hand, if we would have directly chosen master vertices with probability $1/k^2$ using the non-recursive parameterized approach, we could have achieved an approximation factor of $O(k \cdot \log n)$. Thus, on this graph class, the diameter computed by the recursive extension of the parameterized approach is significantly worse than the one computed directly by the parameterized approach.

Adaptive Reduction in Graph Size. Our first approach described above reduces the number of vertices by the same (expected) multiplicative factor in each recursive level. Since in the later recursive levels, the size of the graph is smaller, it might be beneficial to perform more aggressive reduction by performing extra scanning rounds to minimize the number of recursive levels required to reduce the size of the graph enough for it to fit in the main memory. Ideally, the reduction factor should depend not only on the graph size and the available internal memory, but also on the graph type. This is because the condensed graphs from different classes have very different sizes. For instance, if we contract a linear chain with 2^{28} vertices and $2^{28} - 1$ edges to 2^{20} vertices, we get $2^{20} - 1$ edges (the size reducing by almost 256). On the other hand, we found that when we reduced a $\sqrt{n}$-level random graph (described in Section 4) of 2^{28} vertices and 2^{30} edges to 2^{20} vertices, there were still more than 2^{28} edges in the graph (the size reducing by a factor less than 4).

Our eventual goal is to design a software for approximating the diameter of large graphs without assuming any a priori information about the graph class. Thus our scheme needs to learn the graph class on-the-fly and use it to adapt the number of master vertices and consider weights, too.

Selecting the Master Vertices. To improve the approximation quality without increasing the I/O complexity and adapt the number of master vertices, we experimented with different probability distribution functions for selecting the master vertices. We found that selecting the ith vertex to be a master vertex in k^{th} round with a probability p_i as defined below, provides a good diameter approximation for a bad graph class of the two-level recursive approach. Let W_i be the set of weights of edges incident to a vertex i and let max_i be the maximum value in the set W_i. Let n_k (m_k) be the number of vertices (edges) in the graph before the kth round of contraction, $sum\_max = \sum_{i=1}^{n_k} max_i$, $max\_max = \max_{i=1}^{n_k} max_i$ and $min\_max = \min_{i=1}^{n_k} max_i$. Then the probability p_i is equal to $\alpha(\cdot) \cdot (((max_i/sum\_max) \cdot n_k) - min\_max)/(max\_max - min\_max)$. Here, α is an adaptability function that infers the rate of graph contraction based on the past contraction history (values of $m_0, \ldots, m_{k-1}, m_k; n_0, \ldots, n_{k-1}, n_k$) and then adapts the number of master vertices n_{k+1} in the kth recursive level based on the history and the expected graph size m_{k+1} from this level. We restrict the graph size in the final level to fit in the main memory.

Modeling the Graph Contraction. We considered various inference functions for modeling the graph contraction. Our first model assumed a direct proportionality between the number of vertices and edges, i.e., $n_{k+1} = n_k \cdot m_{k+1}/m_k$. Later, we integrated more complex linear and non-linear models such as $m_{k+1} = A \cdot n_{k+1} + B$ and $m_{k+1} = A \cdot n_{k+1} + B \cdot \sqrt{n_{k+1}}$ in our adaptability function, learning the values of parameters A and B based on the (limited) past contraction history of the graph.

Tie Breaking. In the parallel cluster growing approach, ties are broken arbitrarily among the clusters vying for the same vertex. To reduce the weighted diameter of the clusters in the second recursive level, ties can be broken in favor of the cluster that has the shortest distance from the vertex to its master.

Move Master Vertices. We use another trick to further reduce the weighted diameter of the clusters. Once the clustering is done using parallel cluster growing, we move the master vertices closer to the weighted center of their corresponding clusters and re-compute the distances between the clusters based on the new master vertices. Since the clusters fit in the internal memory, it requires only one scanning step to load all clusters in the main memory and compute the distances of all vertices from the master vertex of their cluster.

Once the contracted graph fits into the main memory, we can use any internal memory technique to approximate the diameter of the contracted graph. In this work, we use the double sweep lower bound [15] technique that first computes a single-source shortest path from an arbitrary source s and then returns the weighted eccentricity of a farthest node from s. Other techniques (e.g., [11]) can also be used for this purpose.

We study the relative merits of these ideas and show that by carefully tuning our implementation based on these ideas, we can approximate the diameter of graphs with many billions of edges in a few hours. Note that these graph sizes are significantly bigger than that of past experiments reported in the literature, even for external memory algorithms.

4   Experiments

Since external memory experiments on large graphs can take many hours and even days, a certain self-restraint in the number of such experiments is unavoidable. As such, we performed our initial experiments for analyzing the approximation quality of various extensions of the parameterized algorithm on a machine with 64 GB RAM, using an internal memory prototype. For the variant that gave the best approximation, we implemented it for optimized I/O performance in external memory. The external memory implementation relies on the STXXL library [13], exploiting the various features supported by STXXL such as pipelining and overlap of I/O and computation. The running time and I/O volume reported in the paper are based on external memory experiments.

Also, we restrict ourselves to extensions involving only two levels of recursion. We found that for the graph sizes that we considered, we could get acceptable running time and a good approximation ratio with two recursive levels.

4.1   Configuration

For experiments in internal memory we used a machine from the HPC cluster at Goethe University on graphs with 256 million vertices and about 1 billion edges. For our external memory experiments, we used an Intel dual core E6750 processor @ 2.66 GHz, 4 GB main memory (around 3.5 GB was available for the application) and four hard disks with 500 GB. Only the 250 GB from the outer tracks were used in a RAID-0.

4.2   Graph Classes

We chose four different graph classes: one real-world graph with logarithmic diameter, two synthetic graph classes with diameters $\Theta(\sqrt{n})$ and $\Theta(n)$ and a graph class that was designed to elicit poor performance from the simple extension of the parameterized approach. Recall that the simple extension (hereafter referred as Basic) chooses the master vertices at different recursive levels with the same probability.

The real-world graph sk-2005 has around 50 million vertices, about 1.8 billion edges and is based on a web-crawl. It was also used by Crescenzi et al. [11] and has a known diameter of 40. The synthetic x-level graphs are similar to the B-level random graphs in [2]. The graph consists of x levels, each having $\frac{n}{x}$ vertices (except the level 0 containing only one vertex). The edges are randomly distributed between consecutive levels, such that these x levels approximate the BFS levels if BFS were performed from the source vertex in level 0. The edges are evenly distributed between different levels. We selected $x = \sqrt{n}$ and $x = \Theta(n)$ to generate $\sqrt{n}$ and $\Theta(n)$-level graphs for our experiments. Figure 1a illustrates an example of such a graph.

While the basic recursive extension of the parameterized approach with uniform master probabilities already yields reasonable approximation ratios on many graph-classes including real-world data, it is possible to design artificial inputs that cause significant approximation errors. Before we sketch the construction of such a graph class (referred worse 2step), we consider worst-case inputs for the standard non-recursive approach with master probability $1/k$. Figure 1b displays a graph consisting of a main chain C_0 with x_1 nodes. Each of these x_1 nodes is connected to a side chain C_i of length x_2 that ends with a fan of size x_3. The diameter of this graph is $\Theta(x_1 + x_2)$.

For $x_1 \cdot x_2 \leq k^{1-\epsilon}$ and constant $0 < \epsilon \ll 1$, there is at least a constant probability that master vertices of the non-recursive approach only appear at the very ends of the side chains (i.e., in the fans, outside of the marked box). Furthermore, if the value of x_3 is chosen sufficiently large ($\Omega(k \cdot \log n)$), each fan receives at least one master with high probability. Thus, with at least constant probability, for each side chain a cluster is grown from its fan towards the main

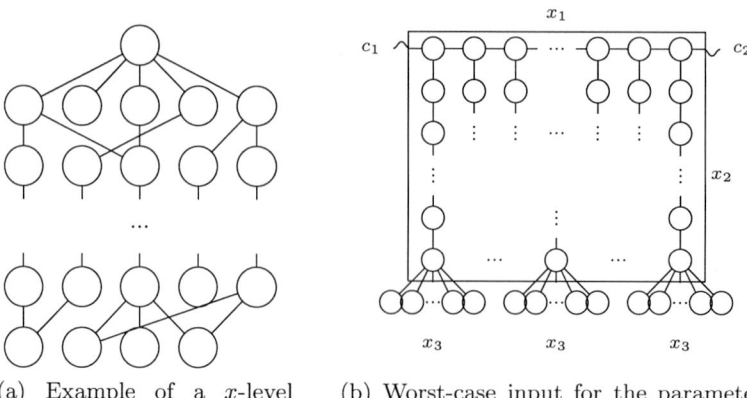

(a) Example of a x-level graph.

(b) Worst-case input for the parameterized approach.

Fig. 1. Various graph classes used in our experiments

chain C_0 and all these clusters reach C_0 simultaneously and then stop growing. Therefore, the shrunken weighted graph features a path with $x_1 - 1$ edges of weight $\Theta(x_2)$ each, resulting in a diameter of $\Theta(x_1 \cdot x_2)$. Choosing $x_1 = k^{1/2}$ and $x_2 = k^{1/2-\epsilon}$, the expected multiplicative approximation is therefore $\Omega((x_1 \cdot x_2)/(x_1 + x_2)) = \Omega(k^{1/2-\epsilon})$, asymptotically matching the upper bound proved in [17]. Larger graphs with similar behavior can be obtained by chaining several copies of the graph structure discussed above along the zigzag lines (using simple paths of length $\Theta(x_1 + x_2)$ each).

For the recursive graph shrinking approach with $i \geq 2$ levels of shrinking we would like to construct an input class that after the first $i - 1$ levels resembles the worst-case graph for the non-recursive version and has accumulated huge edge weights in the side chains but not on the main chain. For ease of exposition we only sketch the 2-level case with master probability $1/k$ for both shrinking phases: (1) edges of weight $O(1)$ in the shrunken graph G' can be obtained with high probability from two high-degree fans that are connected by an edge in the input graph G. (2) simple isolated paths of length $\Theta(z \cdot k)$ in G will result in paths of total weight $\Theta(z \cdot k)$ distributed over an *expected* number of $\Theta(z)$ edges in G'. Appropriate fans in G' are obtained with high probability from double fan structures in G at the cost of a quadratic node overhead concerning the fans.

While the base chain for G' can easily be generated using (1), there is a technical problem with the side chain generation via (2): since the number of vertices in those side chains in G' are only bounded in expectation, their actual numbers will vary and therefore during the second shrinking phase some clusters would reach the main chain earlier than others. As a consequence, the negative impact on the approximation error caused by those clusters reaching the main chain late would be lost and the overall theoretical analysis will be significantly hampered. We deal with this problem by reducing the number of side chains so that each side chain has a certain buffer area on the main chain and therefore with high probability side chains do not interfere with each other, even if they feature different number of vertices. By Chernoff bounds, for side chains with $E[x_2] = k^{2/3-\epsilon}$ in G', buffer areas of $\Theta(k^{1/3})$ between two consecutive side chains suffice with high probability. Filling in the details, it turns out that the expected diameter of the resulting graph G'' after the second shrinking phase exceeds the diameter of the input graph G by a factor of $\Omega(k^{4/3-\epsilon})$.

We randomize the layout of the synthetic graphs on the disk to ensure that the disk layout does not reveal any additional information that is exploitable. However, we use the ordering provided with sk-2005 graph for fair comparison with results reported in the literature.

4.3   Results

In this section, we first demonstrate that our bad graph class does elicit poor approximation results from the basic extension. Then, we show that a combination of techniques mentioned in Section 3 improves upon the approximation quality significantly, even for the bad graph class. We analyze the reasons for this improvement, and finally show that altogether our external memory code results in a good approximation on a wide range of graph classes in a few hours, even on graphs with billions of vertices and many billions of edges.

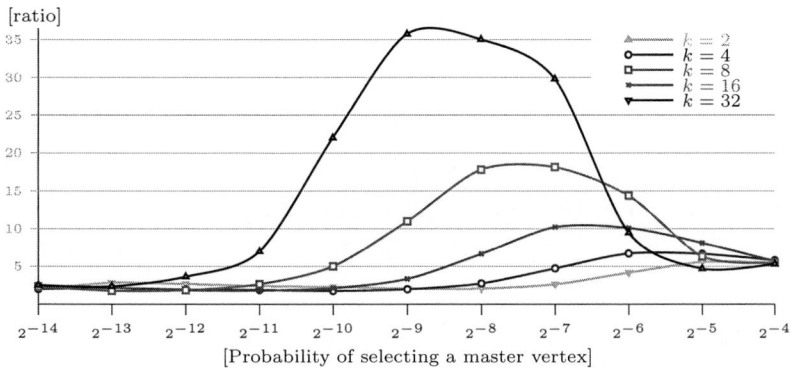

Fig. 2. Approximation ratio using the basic extension with varying probability of choosing the master vertex on five worse 2step graph instances generated with different parameters depending on k

Empirical Analysis of Worse Case Graphs. We first ran the basic extension of the parameterized approach on worse 2step graph instances that were generated using parameter settings suitable for different values of k. As can be seen in Figure 2, for larger values of k and within a certain range, smaller master probabilities cause higher approximation ratios and as k grows the highest achievable ratios start growing even faster. However, due to the constants hidden in the construction the maxima appear for somewhat lower master probabilities than k suggests. Nevertheless, already for rather small values of k, we experience significant ratios (larger than k) demonstrating that this graph class can indeed elicit poor approximation ratio from the basic extension.

Table 1. Diameters on different graph classes with various extensions

	Exact	Basic	Adaptive	Tie break	Move	All
sk-2005	40	185	196	173	182	168
ratio		4.625	4.900	4.325	4.550	4.200
$\sqrt{n}$-level	16385	16594	16416	16604	16597	16408
ratio		1.013	1.002	1.013	1.013	1.001
$\Theta(n)$-level	67108864	67212222	67131347	67212123	67174264	67131036
ratio		1.002	1.000	1.002	1.001	1.000
worse 2step	3867	138893	38643	137087	17321	13613
ratio		35.918	9.993	35.450	4.479	3.520

Approximation Quality. Next, we consider the various techniques described in Section 3 to determine if they can improve the approximation ratio on different graph classes. These techniques include (i) *Adaptive*, where we use the probability distribution function described in Section 3; (ii) *Tie break*, where the ties in the parallel cluster growing rounds are broken such that among all clusters competing to get a vertex v, the cluster that has the shortest distance between its master and v gets the vertex v; (iii) *Move*, where the masters are re-selected after the clustering to reduce the weighted diameter of the clusters and (iv) *All*, where all of the above techniques are combined. Table 1 presents the diameter computed by these techniques together with the exact diameter for different graph classes. This table is computed for graphs with 2^{28} vertices where in both recursive levels master vertices are chosen with a probability of 2^{-8}. Thus, the various extensions contract the number of vertices in the input graph by a factor of around $2^{16} = 65,536$, making it possible to handle graphs that are significantly bigger than the main memory size. Despite such a large reduction in the number of vertices, we get a fairly small approximation ratio with the All approach in our experiments, thus proving the efficacy of our techniques.

For graphs with small diameter such as sk-2005, various additive errors dominate the approximation ratio. However, this can be easily rectified as once it is determined that the input graph has a small diameter, one can run BFS based heuristics to compute better approximations. For such graphs, BFS can be efficiently computed in external memory. In contrast, the hardest case for the external memory BFS (in terms of running time) is the $\sqrt{n}$-level graph. For this hard graph class, however, even the basic extension yields an approximation ratio of 1.013 and the All approach improves it to 1.001. The case for $\Theta(n)$-level graph is similar – the Basic approach already gives a fairly good approximation ratio and the All approach improves it further. For the interesting case of the worse 2step graph where the basic variant gives a very poor approximation ratio of 35.92, the All approach manages to improve it to 3.52 – *a factor of more than 10.* These results imply that the additional techniques provide considerable robustness to the approximation quality by significantly improving the ratio for the very bad case. Most importantly, this improvement comes at little or no additional cost to the I/O complexity and the overall runtime of the external memory implementation.

Distance Distribution Within Clusters. Next, we analyze the reasons for the poor performance of basic extension on the worse 2step graph. After the first level of clustering, the resultant graph has a diameter of 10,067 with an approximation ratio of 2.6. It is in the (second level) clustering of this (weighted) contracted graph that the quality of the approximation deteriorates. Our techniques such as careful selection of master vertices closer to high weight edges, breaking the ties in favor of clusters with shorter distance to their masters, and moving the masters to weighted cluster centers reduce the weighted diameter of the second level clusters. The smaller diameter clusters, in turn, ensure that the resultant condensed graph better captures the diameter of the input graph.

The fact that vertices in the second level clusters are closer to their master vertices in the All scheme than in the basic extension is evident from Figure 3, where we plot the number of vertices at varying distance from their cluster masters. The number of vertices with distance at most 10 from the master vertex is 3638 for the basic approach and 5362 for the All approach, while the number of vertices with distance greater than 300 is 210 for the basic approach and 12 for the All approach.

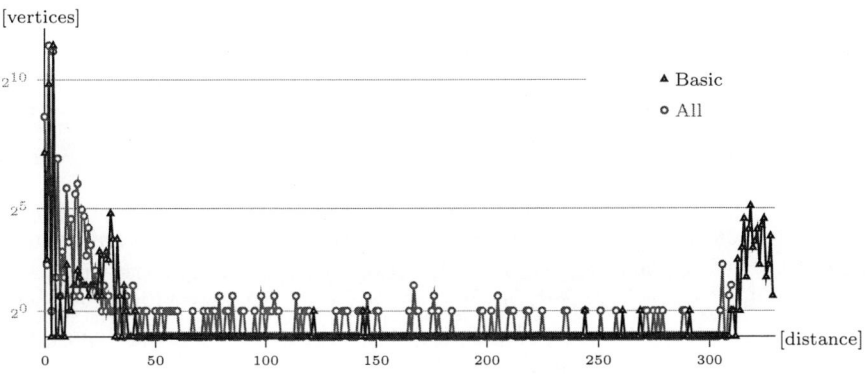

Fig. 3. Distance distribution on worst 2step graph after the second recursive level of the basic approach and all extensions

Running Time and I/O Complexity. Theoretically, our approach requires $O(n \cdot \sqrt{\log{(k_1 \cdot k_2)}/(k_1 \cdot k_2 \cdot B)} + (k_1 + k_2) \cdot scan(n) + sort(n))$ I/Os for contracting the graph to n/k_1 nodes in the first iteration and then to $n/(k_1 \cdot k_2)$ in the second iteration. This is better than the BFS complexity when $k_1 + k_2 < \sqrt{B}$.

Table 2. Results of our external memory implementation

	Size [GB]	Exact Diameter	Computed Diameter	Approx Ratio	Running Time [h]	I/O Time [h]	I/O Volume [TB]
sk-2005	27.0	40	133	3.325	0.7	0.1	0.3
$\sqrt{n}$-level	128.0	46342	46380	1.001	11.1	5.7	4.7
$\Theta(n)$-level	83.8	536870913	546560415	1.018	5.4	3.5	3.0
worse 2step	31.9	8111	25271	3.116	2.1	1.4	1.1

Empirically, Table 2 presents the result of using our external memory implementation to approximate the diameter on $\sqrt{n}$-level, $\Theta(n)$-level and worse 2step graphs with 2^{31} vertices and 2^{33} edges (except worse 2step which is a tree). On moderate to large diameter graphs such as these, the external memory BFS implementation based on parallel cluster growing requires *months* and even the BFS implementation based on chopping the Euler tour of a bidirectional spanning tree requires many hours [2] on graphs that are 8 times smaller than ours. Thus, the various diameter approximation techniques based on multiple calls to BFS routine are likely to take many *days*. On the other hand, our approach provides fairly good approximations of graph diameter in a few hours, that is *even less time than that required for one exact BFS call*. Alternatively, applying the single step approach [3] hardly yields better approximations in practice than the hierarchical method but results in significantly larger running times: on the $\Theta(n)$-level graph, e.g., 76 hours were needed to approximate the diameter with 2^{22} master vertices and still 21 hours with 2^{24} masters. Note that the running time of our external memory approach is still dominated by the time required for I/Os and the total volume of data moved between the two memory levels for approximating the diameter is still in the order of terabytes, thereby showing that the problem continues to remain I/O-bound, even with four parallel disks.

5   Conclusion

Using our new hierarchical extensions of the parameterized diameter approximation approach we managed to process significantly larger external-memory graphs than before while keeping approximation ratio and computation time reasonable. Open problems concern directed and dynamic versions, and the question whether the *worse 2step* graph class construction yields an asymptotically tight lower bound on the approximation error for the two level basic extension.

Our framework can as well be initialized directly from a weighted graph, but our analysis of approximation quality holds only when the input graph is unweighted. To make our hierarchical extension cache-oblivious, one might consider recursively contracting the number of vertices by a factor independent of B, till the graph reduces to a constant size.

Acknowledgements. The authors would like to thank Andreas Beckmann for his help with STXXL and the hardware resources and the anonymous reviewers for their feedback.

References

1. Aggarwal, A., Vitter, J.S.: The input/output complexity of sorting and related problems. Communications of the ACM 31(9), 1116–1127 (1988)
2. Ajwani, D.: Traversing large graphs in realistic setting. PhD thesis, Saarland University (2008)
3. Ajwani, D., Beckmann, A., Meyer, U., Veith, D.: I/O-efficient approximation of graph diameters by parallel cluster growing – A first experimental study. In: 10th Workshop on Parallel Systems and Algorithms, PASA (2012)
4. Ajwani, D., Meyer, U., Osipov, V.: Improved external memory BFS implementation. In: Proc. 9th ALENEX, pp. 3–12 (2007)
5. Arge, L., Meyer, U., Toma, L.: External Memory Algorithms for Diameter and All-Pairs Shortest-Paths on Sparse Graphs. In: Díaz, J., Karhumäki, J., Lepistö, A., Sannella, D. (eds.) ICALP 2004. LNCS, vol. 3142, pp. 146–157. Springer, Heidelberg (2004)
6. Bader, D.A., Madduri, K.: Snap, small-world network analysis and partitioning: An open-source parallel graph framework for the exploration of large-scale networks. In: Proc. 22nd IPDPS, pp. 1–12. IEEE (2008)
7. Boitmanis, K., Freivalds, K., Lediņš, P., Opmanis, R.: Fast and Simple Approximation of the Diameter and Radius of a Graph. In: Àlvarez, C., Serna, M. (eds.) WEA 2006. LNCS, vol. 4007, pp. 98–108. Springer, Heidelberg (2006)
8. Brandes, U., Erlebach, T. (eds.): Network Analysis. LNCS, vol. 3418. Springer, Heidelberg (2005)
9. Chowdury, R., Ramachandran, V.: External-memory exact and approximate all-pairs shortest-paths in undirected graphs. In: Proc. 16th SODA, pp. 735–744. ACM-SIAM (2005)
10. Corneil, D.G., Dragan, F.F., Habib, M., Paul, C.: Diameter determination on restricted graph families. Discrete Applied Mathematics 113(2-3), 143–166 (2001)
11. Crescenzi, P., Grossi, R., Imbrenda, C., Lanzi, L., Marino, A.: Finding the Diameter in Real-World Graphs. In: de Berg, M., Meyer, U. (eds.) ESA 2010. LNCS, vol. 6346, pp. 302–313. Springer, Heidelberg (2010)
12. Crescenzi, P., Grossi, R., Lanzi, L., Marino, A.: On Computing the Diameter of Real-World Directed (Weighted) Graphs. In: Klasing, R. (ed.) SEA 2012. LNCS, vol. 7276, pp. 99–110. Springer, Heidelberg (2012)
13. Dementiev, R., Sanders, P.: Asynchronous parallel disk sorting. In: Proc. 15th SPAA, pp. 138–148. ACM (2003)
14. Goodrich, M.T., Pszona, P.: External-Memory Network Analysis Algorithms for Naturally Sparse Graphs. In: Demetrescu, C., Halldórsson, M.M. (eds.) ESA 2011. LNCS, vol. 6942, pp. 664–676. Springer, Heidelberg (2011)
15. Magnien, C., Latapy, M., Habib, M.: Fast computation of empirically tight bounds for the diameter of massive graphs. Journal of Experimental Algorithmics 13, 1.10–1.9 (2009)
16. Mehlhorn, K., Meyer, U.: External-Memory Breadth-First Search with Sublinear I/O. In: Möhring, R.H., Raman, R. (eds.) ESA 2002. LNCS, vol. 2461, pp. 723–735. Springer, Heidelberg (2002)
17. Meyer, U.: On Trade-Offs in External-Memory Diameter-Approximation. In: Gudmundsson, J. (ed.) SWAT 2008. LNCS, vol. 5124, pp. 426–436. Springer, Heidelberg (2008)

On the Value of Job Migration
in Online Makespan Minimization

Susanne Albers and Matthias Hellwig

Department of Computer Science, Humboldt-Universität zu Berlin
{albers,mhellwig}@informatik.hu-berlin.de

Abstract. Makespan minimization on identical parallel machines is a classical scheduling problem. We consider the online scenario where a sequence of n jobs has to be scheduled non-preemptively on m machines so as to minimize the maximum completion time of any job. The best competitive ratio that can be achieved by deterministic online algorithms is in the range $[1.88, 1.9201]$. Currently no randomized online algorithm with a smaller competitiveness is known, for general m.

In this paper we explore the power of job migration, i.e. an online scheduler is allowed to perform a limited number of job reassignments. Migration is a common technique used in theory and practice to balance load in parallel processing environments. As our main result we settle the performance that can be achieved by deterministic online algorithms. We develop an algorithm that is α_m-competitive, for any $m \geq 2$, where α_m is the solution of a certain equation. For $m = 2$, $\alpha_2 = 4/3$ and $\lim_{m\to\infty} \alpha_m = W_{-1}(-1/e^2)/(1 + W_{-1}(-1/e^2)) \approx 1.4659$. Here W_{-1} is the lower branch of the Lambert W function. For $m \geq 11$, the algorithm uses at most $7m$ migration operations. For smaller m, $8m$ to $10m$ operations may be performed. We complement this result by a matching lower bound: No online algorithm that uses $o(n)$ job migrations can achieve a competitive ratio smaller than α_m. We finally trade performance for migrations. We give a family of algorithms that is c-competitive, for any $5/3 \leq c \leq 2$. For $c = 5/3$, the strategy uses at most $4m$ job migrations. For $c = 1.75$, at most $2.5m$ migrations are used.

1 Introduction

Makespan minimization on identical machines is a fundamental scheduling problem that has received considerable research interest over the last forty years. Let $\sigma = J_1, \ldots, J_n$ be a sequence of jobs that has to be scheduled non-preemptively on m identical parallel machines. Each job J_i is specified by a processing time p_i, $1 \leq i \leq n$. The goal is to minimize the makespan, i.e. the maximum completion time of any job in a schedule. In the offline setting all jobs are known in advance. In the online setting the jobs arrive one by one. Each job J_i has to be scheduled immediately on one of the machines without knowledge of any future jobs J_k, $k > i$. An online algorithm A is called c-competitive if, for any job sequence, the makespan of A's schedule is at most c times the optimum makespan for that sequence [24].

L. Epstein and P. Ferragina (Eds.): ESA 2012, LNCS 7501, pp. 84–95, 2012.

Early work on makespan minimization studied the offline setting. Already in 1966, Graham [14] presented the *List* scheduling algorithm that schedules each job on a least loaded machine. *List* can be used both as an offline and as an online strategy and achieves a performance ratio of $2 - 1/m$. Hochbaum and Shmoys devised a famous polynomial time approximation scheme [17]. More recent research, published mostly in the 1990s, investigated the online setting. The best competitive factor that can be attained by deterministic online algorithms is in the range $[1.88, 1.9201]$. Due to this relatively high factor, compared to *List*'s ratio of $2 - 1/m$, it is interesting to consider scenarios where an online scheduler has more flexibility to serve the job sequence.

In this paper we investigate the impact of job migration. At any time an online algorithm may perform *reassignments*, i.e. a job already scheduled on a machine may be removed and transfered to another machine. Process migration is a well-known and widely used technique to balance load in parallel and distributed systems. It leads to improved processor utilization and reduced processing delays. Migration policies have been analyzed extensively in theory and practice.

It is natural to investigate makespan minimization with job migration. In this paper we present a comprehensive study and develop tight upper and lower bounds on the competitive ratio that can be achieved by deterministic online algorithms. It shows that even with a very limited number of migration operations, significantly improved performance guarantees are obtained.

Previous Work: We review the most important results relevant to our work. As mentioned above, *List* is $(2 - 1/m)$-competitive. Deterministic online algorithms with a smaller competitive ratio were presented in [2,4,12,13,18]. The best algorithm currently known is 1.9201-competitive [12]. Lower bounds on the performance of deterministic strategies were given in [2,3,11,16,20]. The best bound currently known is 1.88, for general m, see [20]. Randomized online algorithms cannot achieve a competitive ratio smaller than $e/(e-1) \approx 1.58$ [7,22]. No randomized algorithm whose competitive ratio is provably below the deterministic lower bound is currently known, for general m. If job preemption is allowed, the best competitiveness of online strategies is equal to $e/(e-1) \approx 1.58$ [8].

Makespan minimization with job migration was first addressed by Aggarwal et al. [1]. They consider an offline setting. An algorithm is given a schedule, in which all jobs are already assigned, and a budget. The algorithm may perform job migrations up to the given budget. The authors design strategies that perform well with respect to the best possible solution that can be constructed with the budget. Online makespan minimization on $m = 2$ machines was considered in [19,23]. The best competitiveness is 4/3. Sanders et al. [21] study an online setting in which before the assignment of each job J_i, jobs up to a total processing volume of βp_i may be migrated, for some constant β. For $\beta = 4/3$, they present a 1.5-competitive algorithm. They also show a $(1 + \epsilon)$-competitive algorithm, for any $\epsilon > 0$, where β depends exponentially on $1/\epsilon$. The algorithms are robust in that the stated competitive ratios hold after each job assignment. However in this framework, over time, $\Omega(n)$ migrations may be performed and jobs of total processing volume $\beta \sum_{i=1}^{n} p_i$ may be moved.

Englert et al. [10] study online makespan minimization if an algorithm is given a buffer that may be used to partially reorder the job sequence. In each step an algorithm assigns one job from the buffer to the machines. Then the next job in σ is admitted to the buffer. Englert et al. show that, using a buffer of size $\Theta(m)$, the best competitive ratio is $W_{-1}(-1/e^2)/(1 + W_{-1}(-1/e^2))$, where W_{-1} is the Lambert W function. Dósa and Epstein [9] consider the corresponding problem assuming that job preemption is allowed and show that the competitiveness is $4/3$. The paper by Englert et al. [10] also addresses uniformly related machines. Finally, scheduling with job reassignments on two uniform machines has been considered, see [5,6] and references therein.

Our Contribution: We investigate online makespan minimization with limited migration. The number of job reassignments does not depend on the length of the job sequence. We determine the exact competitiveness achieved by deterministic algorithms, for general m.

In Section 2 we develop an optimal algorithm. For any $m \geq 2$, the strategy is α_m-competitive, where α_m is the solution of an equation representing load in an ideal machine profile for a subset of the jobs. For $m = 2$, the competitive ratio is $4/3$. The ratios are non-decreasing and converge to $W_{-1}(-1/e^2)/(1 + W_{-1}(-1/e^2)) \approx 1.4659$ as m tends to infinity. Again, W_{-1} is the lower branch of the Lambert W function. The algorithm uses at most $(\lceil (2 - \alpha_m)/(\alpha_m - 1)^2 \rceil + 4)m$ job migrations. For $m \geq 11$, this expression is at most $7m$. For smaller machine numbers it is $8m$ to $10m$. We note that the competitiveness of 1.4659 is considerably below the factor of roughly 1.9 obtained by deterministic algorithms in the standard online setting. It is also below the ratio of $e/(e - 1)$ attainable if randomization or job preemption are allowed.

In Section 3 we give a matching lower bound. We show that no deterministic algorithm that uses $o(n)$ job migrations can achieve a competitive ratio smaller than α_m, for any $m \geq 2$. Hence in order to beat the factor of α_m, $\Theta(n)$ reassignments are required. Finally, in Section 4 we trade migrations for performance. We develop a family of algorithms that is c-competitive, for any constant c with $5/3 \leq c \leq 2$. Setting $c = 5/3$ we obtain a strategy that uses at most $4m$ job migrations. For $c = 1.75$, the strategy uses no more than $2.5m$ migrations.

Our algorithms rely on a number of new ideas. All strategies classify incoming jobs into small and large depending on a careful estimate on the optimum makespan. The algorithms consist of a job arrival phase followed by a migration phase. The optimal algorithm, in the arrival phase, maintains a load profile on the machines with respect to jobs that are currently small. In the migration phase, the algorithm removes a certain number of jobs from each machine. These jobs are then rescheduled using strategies by Graham [14,15]. Our family of algorithms partitions the m machines into two sets A and B. In the arrival phase the algorithms prefer to place jobs on machines in A so that machines in B are available for later migration. In general, the main challenge in the analyses of the various algorithms is to bound the number of jobs that have to be migrated from each machine.

We finally relate our contributions to some existing results. First we point out that the goal in online makespan minimization is to construct a good schedule when jobs arrive one by one. Once the schedule is constructed, the processing of the jobs may start. It is not stipulated that machines start executing jobs while other jobs of σ still need to be scheduled. This framework is assumed in all the literature on online makespan minimization mentioned above. Consequently it is no drawback to perform job migrations when the entire job sequence has arrived. Nonetheless, as for the algorithms presented in this paper, the machines can start processing jobs except for the up to 10 largest jobs on each machine. A second remark is that the algorithms by Aggarwal et al. [1] cannot be used to achieve good results in the online setting. The reason is that those strategies are designed to perform well relative to the best possible makespan attainable from an initial schedule using a given migration budget. The strategies need not perform well compared to a globally optimal schedule.

Our results exhibit similarities to those by Englert et al. [10] where a reordering buffer is given. The optimal competitive ratio of α_m is the solution of an equation that also arises in [10]. This is due to the fact that our optimal algorithm and that in [10] maintain a certain load profile on the machines. Our strategy does so w.r.t. jobs that are currently small while the strategy in [10] considers all jobs assigned to machines. In our framework the profile is harder to maintain because of *shrinking jobs*, i.e. jobs that are large at some time t but small at later times $t' > t$. In the job migration phase our algorithm reschedules jobs removed from some machines. This operation corresponds to the "final phase" of the algorithm in [10]. However, our algorithm directly applies policies by Graham [14,15] while the algorithm in [10] computes a virtual schedule. In general, an interesting question is if makespan minimization with limited migration is equivalent to makespan minimization with a bounded reordering buffer. We cannot prove this in the affirmative. As for the specific algorithms presented in [10] and in this paper, the following relation holds. All our algorithms can be transformed into strategies with a reordering buffer. The competitive ratios are preserved and the number of job migrations is equal to the buffer size. On the other hand, despite careful investigation, we do not know how to translate the algorithms by Englert et al. [10] into strategies with job migration. Further details can be found in the full version of this paper.

2   An Optimal Algorithm

For the description of the algorithm and the attained competitive ratio we define a function $f_m(\alpha)$. Intuitively, $f_m(\alpha)$ represents accumulated normalized load in a "perfect" machine profile for a subset of the jobs. In such a profile the load ratios of the first $\lfloor m/\alpha \rfloor$ machines follow a Harmonic series of the form $(\alpha - 1)/(m - 1), \ldots, (\alpha - 1)/(m - \lfloor m/\alpha \rfloor)$ while the remaining ratios are α/m. Summing up these ratios we obtain $f_m(\alpha)$. Formally, let

$$f_m(\alpha) = (\alpha - 1)(H_{m-1} - H_{\lceil (1-1/\alpha)m \rceil - 1}) + \lceil (1 - 1/\alpha)m \rceil \alpha/m,$$

for any machine number $m \geq 2$ and real-valued $\alpha > 1$. Here $H_k = \sum_{i=1}^{k} 1/i$ denotes the k-th Harmonic number, for any integer $k \geq 1$. We set $H_0 = 0$. For any fixed $m \geq 2$, let α_m be the value satisfying $f_m(\alpha) = 1$. Lemma 1 below implies that α_m is well-defined. The algorithm we present is exactly α_m-competitive. By Lemma 2, the values α_m form a non-decreasing sequence. There holds $\alpha_2 = 4/3$ and $\lim_{m\to\infty} \alpha_m = W_{-1}(-1/e^2)/(1 + W_{-1}(-1/e^2)) \approx 1.4659$. This convergence was also stated by Englert et al. [10] but no thorough proof was presented. The following two technical lemmas are proven in the full version of the paper.

Lemma 1. *The function $f_m(\alpha)$ is continuous and strictly increasing in α, for any integer $m \geq 2$ and real number $\alpha > 1$. There holds $f_m(1 + 1/(3m)) < 1$ and $f_m(2) \geq 1$.*

Lemma 2. *The sequence $(\alpha_m)_{m\geq 2}$ is non-decreasing with $\alpha_2 = 4/3$ and $\lim_{m\to\infty} \alpha_m = W_{-1}(-1/e^2)/(1 + W_{-1}(-1/e^2))$.*

2.1   Description of the Algorithm

Let $m \geq 2$ and $M_1, \ldots, M_m$ be the available machines. Furthermore, let α_m be as defined above. The algorithm, called $ALG(\alpha_m)$, operates in two phases, a *job arrival phase* and a *job migration phase*. In the job arrival phase all jobs of $\sigma = J_1, \ldots, J_n$ are assigned one by one to the machines. In this phase no job migrations are performed. Once σ is scheduled, the job migration phase starts. First the algorithm removes some jobs from the machines. Then these jobs are reassigned to other machines.

Job Arrival Phase. In this phase $ALG(\alpha_m)$ classifies jobs into small and large and, moreover, maintains a load profile with respect to the small jobs on the machines. At any time the load of a machine is the sum of the processing times of the jobs currently assigned to it. Let *time t* be the time when J_t has to be scheduled, $1 \leq t \leq n$.

In order to classify jobs $ALG(\alpha_m)$ maintains a lower bound L_t on the optimum makespan. Let $p_t^+ = \sum_{i=1}^{t} p_i$ be the sum of the processing times of the first t jobs. Furthermore, for $i = 1, \ldots, 2m+1$, let p_t^i denote the processing time of the i-th largest job in $J_1, \ldots, J_t$, provided that $i \leq t$. More formally, if $i \leq t$, let p_t^i be the processing time of the i-th largest job; otherwise we set $p_t^i = 0$. Obviously, when t jobs have arrived, the optimum makespan cannot be smaller than the average load $\frac{1}{m}p_t^+$ on the m machines. Moreover, the optimum makespan cannot be smaller than $3p_t^{2m+1}$, which is three times the processing time of $(2m+1)$-st largest job seen so far. Define

$$L_t = \max\{\tfrac{1}{m}p_t^+, 3p_t^{2m+1}\}.$$

A job J_t is called *small* if $p_t \leq (\alpha_m - 1)L_t$; otherwise it is *large*. As the estimates L_t are non-decreasing over time, a large job J_t does not necessarily satisfy $p_t > (\alpha_m - 1)L_{t'}$ at times $t' > t$. Therefore we need a more refined notion of small

Algorithm ALG(α_m):

Job arrival phase. Each J_t, $1 \le t \le n$, is scheduled as follows.

- J_t is small: Assign J_t to an M_j with $\ell_s(j,t) \le \beta(j)L_t^*$.
- J_t is large: Assign J_t to a least loaded machine.

Job migration phase.

- Job removal: Set $R := \emptyset$. While there exists an M_j with $\ell(j) > \max\{\beta(j)L^*, (\alpha-1)L\}$, remove the largest job from M_j and add it to R.
- Job reassignment: $R' = \{J_i \in R \mid p_i > (\alpha_m - 1)L\}$. For $i = 1, \ldots, m$, set P_i contains J_r^i, if $i \le |R'|$, and J_r^{2m+1-i}, if $p_r^{2m+1-i} > p_r^i/2$ and $2m+1-i \le |R'|$. Number the sets in order of non-increasing total processing time. For $i = 1, \ldots, m$, assign P_i to a least loaded machine. Assign each $J_i \in R \setminus (P_1 \cup \ldots \cup P_m)$ to a least loaded machine.

Fig. 1. The algorithm $ALG(\alpha_m)$

and large. A job J_i, with $i \le t$, is *small at time t* if $p_i \le (\alpha_m - 1)L_t$; otherwise it is *large at time t*. We introduce a final piece of notation. In the sequence $p_t^1, \ldots, p_t^{2m}$ of the $2m$ largest processing times up to time t we focus on those that are large. More specifically, for $i = 1, \ldots, 2m$, let $\hat{p}_t^i = p_t^i$ if $p_t^i > (\alpha_m - 1)L_t$; otherwise let $\hat{p}_t^i = 0$. Define

$$L_t^* = \tfrac{1}{m}(p_t^+ - \textstyle\sum_{i=1}^{2m} \hat{p}_t^i).$$

Intuitively, L_t^* is the average machine load ignoring jobs that are large at time t. Since $\alpha_m \ge 4/3$, by Lemma 2, and $L_t \ge 3p_t^{2m+1}$, there can exist at most $2m$ jobs that are large at time t.

We describe the scheduling steps in the job arrival phase. Initially, the machines are numbered in an arbitrary way and this numbering $M_1, \ldots, M_m$ remains fixed throughout the execution of $ALG(\alpha_m)$. As mentioned above the algorithm maintains a load profile on the machines as far as small jobs are concerned. Define

$$\beta(j) = \begin{cases} (\alpha_m - 1)\frac{m}{m-j} & \text{if } j \le \lfloor m/\alpha_m \rfloor \\ \alpha_m & \text{otherwise.} \end{cases}$$

We observe that $f_m(\alpha_m) = \frac{1}{m}\sum_{j=1}^m \beta(j)$, because $m - \lfloor m/\alpha_m \rfloor = \lceil (1-1/\alpha_m)m \rceil$. For any machine M_j $1 \le j \le m$, let $\ell(j,t)$ denote its load at time t *before* J_t is assigned to a machine. Let $\ell_s(j,t)$ be the load caused by the jobs on M_j that are small at time t. $ALG(\alpha_m)$ ensures that at any time t there exists a machine M_j satisfying $\ell_s(j,t) \le \beta(j)L_t^*$.

For $t = 1, \ldots, n$, each J_t is scheduled as follows. If J_t is small, then it is scheduled on a machine with $\ell_s(j,t) \le \beta(j)L_t^*$. In Lemma 3 we show that such a machine always exists. If J_t is large, then it is assigned to a machine having the smallest load among all machines. At the end of the phase let $L = L_n$ and $L^* = L_n^*$.

Job Migration Phase. This phase consists of a *job removal step* followed by a *job reassignment step*. At any time during the phase, let $\ell(j)$ denote the current load of M_j, $1 \le j \le m$. In the removal step $ALG(\alpha_m)$ maintains a set R

of removed jobs. Initially $R = \emptyset$. During the removal step, while there exists a machine M_j whose load $\ell(j)$ exceeds $\max\{\beta(j)L^*, (\alpha_m-1)L\}$, $ALG(\alpha_m)$ removes the job with the largest processing time currently residing on M_j and adds the job to R.

If $R = \emptyset$ at the end of the removal step, then $ALG(\alpha_m)$ terminates. If $R \neq \emptyset$, then the reassignment step is executed. Let $R' \subseteq R$ be the subset of the jobs that are large at the end of σ, i.e. whose processing time is greater than $(\alpha_m - 1)L$. Again there can exist at most $2m$ such jobs. $ALG(\alpha_m)$ first sorts the jobs of R' in order of non-increasing processing time; ties are broken arbitrarily. Let J_r^i, $1 \leq i \leq |R'|$, be the i-th job in this sorted sequence and p_r^i be its processing time. For $i = 1, \ldots, m$, $ALG(\alpha_m)$ forms jobs pairs consisting of the i-th largest and the $(2m + 1 - i)$-th largest jobs provided that the processing time of the latter job is sufficiently high. A pairing strategy combining the i-th largest and the $(2m + 1 - i)$-th largest jobs was also used by Graham [15]. Formally, $ALG(\alpha_m)$ builds sets $P_1, \ldots, P_m$ that contain up to two jobs. Initially, all these sets are empty. In a first step J_r^i is assigned to P_i, for any i with $1 \leq i \leq \min\{m, |R'|\}$. In a second step J_r^{2m+1-i} is added to P_i provided that $p_r^{2m+1-i} > p_r^i/2$, i.e. the processing time of J_r^{2m+1-i} must be greater than half times that of J_r^i. This second step is executed for any i such that $1 \leq i \leq m$ and $2m+1-i \leq |R'|$. For any set P_i, $1 \leq i \leq m$, let π_i be the total summed processing time of the jobs in P_i. $ALG(\alpha_m)$ now renumbers the sets in order of non-increasing π_i values such that $\pi_1 \geq \ldots \geq \pi_m$. Then, for $i = 1, \ldots, m$, it takes the set P_i and assigns the jobs of P_i to a machine with the smallest current load. If P_i contains two jobs, then both are placed on the same machine. Finally, if $R \setminus (P_1 \cup \ldots \cup P_m) \neq \emptyset$, then $ALG(\alpha_m)$ takes care of the remaining jobs. These jobs may be scheduled in an arbitrary order. Each job of $R \setminus (P_1 \cup \ldots \cup P_m)$ is scheduled on a machine having the smallest current load. This concludes the description of $ALG(\alpha_m)$. A summary in pseudo-code is given in Figure 1.

Theorem 1. $ALG(\alpha_m)$ is α_m-competitive and uses at most $(\lceil (2 - \alpha_m)/(\alpha_m - 1)^2 \rceil + 4)m$ job migrations.

As we shall see in the analysis of $ALG(\alpha_m)$ in the job migration phase the algorithm has to remove at most $\mu_m = \lceil (2 - \alpha_m)/(\alpha_m - 1)^2 \rceil + 4$ jobs from each machine. Table 1 depicts the competitive ratios α_m (exactly and approximately) and the migration numbers μ_m, for small values of m. The ratios α_m are rational numbers, for any $m \geq 2$.

Table 1. The values of α_m and μ_m, for small m.

m	2	3	4	5	6	7	8	9	10	11
α_m	$\frac{4}{3}$	$\frac{15}{11}$	$\frac{11}{8}$	$\frac{125}{89}$	$\frac{137}{97}$	$\frac{273}{193}$	$\frac{586}{411}$	$\frac{1863}{1303}$	$\frac{5029}{3517}$	$\frac{58091}{40451}$
$\approx$		1.3636	1.375	1.4045	1.4124	1.4145	1.4258	1.4298	1.4299	1.4360
μ_m	10	9	9	8	8	8	8	8	8	7

2.2   Analysis of the Algorithm

We first show that the assignment operations in the job arrival phase are well defined. A corresponding statement was shown by Englert et al. [10]. The proof of Lemma 3, which is given in the full version of the paper, is more involved because we have to take care of large jobs in the current schedule.

Lemma 3. *At any time t there exists a machine M_j satisfying $\ell_s(j,t) \le \beta(j)L_t^*$.*

We next analyze the job migration phase.

Lemma 4. *In the job removal step $ALG(\alpha_m)$ removes at most $\lceil (2-\alpha_m)/(\alpha_m - 1)^2 \rceil + 4$ jobs from each of the machines.*

Proof. Consider any M_j, with $1 \le j \le m$. We show that it suffices to remove at most $\lceil (2-\alpha_m)/(\alpha_m - 1)^2 \rceil + 4$ jobs so that M_j's resulting load is upper bounded by $\max\{\beta(j)L^*, (\alpha_m - 1)L\}$. Since $ALG(\alpha_m)$ removes the largest jobs the lemma follows.

Let time $n+1$ be the time when the entire job sequence σ is scheduled and the job migration phase with the removal step starts. A job J_i, with $1 \le i \le n$, is *small at time $n+1$* if $p_i \le (\alpha_m - 1)L$; otherwise it is *large at time $n+1$*. Since $L = L_n$ any job that is small (large) at time $n+1$ is also small (large) at time n. Let $\ell(j, n+1)$ be the load of M_j at time $n+1$. Similarly, $\ell_s(j, n+1)$ is M_j's load consisting of the jobs that are small at time $n+1$. Throughout the proof let $k := \lceil (2-\alpha_m)/(\alpha_m - 1)^2 \rceil$.

First assume $\ell_s(j, n+1) \le \beta(j)L^*$. If at time $n+1$ machine M_j does not contain any jobs that are large at time $n+1$, then $\ell(j, n+1) = \ell_s(j, n+1) \le \beta(j)L^*$. In this case no job has to be removed and we are done. If M_j does contain jobs that are large at time $n+1$, then it suffices to remove these jobs. Let time l be the last time when a job J_l that is large at time $n+1$ was assigned to M_j. Since $L_l \le L$, J_l was also large at time l and hence it was assigned to a least loaded machine. This implies that prior to the assignment of J_l, M_j has a load of at most $p_l^+/m \le L_l \le L$. Hence it could contain at most $1/(\alpha_m - 1)$ jobs that are large at time $n+1$ because any such job has a processing time greater than $(\alpha_m - 1)L$. Hence at most $1/(\alpha_m - 1) + 1$ jobs have to be removed from M_j, and the latter expression is upper bounded by $k+4$.

Next assume $\ell_s(j, n+1) > \beta(j)L^*$. If $\ell_s(j, n) \le \beta(j)L^* = \beta(j)L_n^*$, then J_n was assigned to M_j. In this case it suffices to remove J_n and, as in the previous case, at most $1/(\alpha_m - 1) + 1$ jobs that are large at time $n+1$. Again $1/(\alpha_m - 1) + 2 \le k+4$.

In the remainder of this proof we consider the case that $\ell_s(j, n+1) > \beta(j)L^*$ and $\ell_s(j, n) > \beta(j)L_n^*$. Let t^* be the earliest time such that $\ell_s(j, t) > \beta(j)L_t^*$ holds for all times $t^* \le t \le n$. We have $t^* \ge 2$ because $\ell_s(j, 1) = 0 \le \beta(j)L_1^*$. Hence time $t^* - 1$ exists. We partition the jobs residing on M_j at time $n+1$ into three sets. Set T_1 is the set of jobs that were assigned to M_j at or before time $t^* - 1$ and are small at time $t^* - 1$. Set T_2 contains the jobs that were assigned to M_j at or before time $t^* - 1$ and are large at time $t^* - 1$. Finally T_3 is the set of jobs assigned to M_j at or after time t^*. We show a number of claims that we will use in the further proof.

Claim 4.1. Each job in $T_2 \cup T_3$ is large at the time it is assigned to M_j.

Claim 4.2. There holds $\sum_{J_i \in T_1 \setminus \{J_l\}} p_i \leq \beta(j) L_{t^*-1}^*$, where J_l is the job of T_1 that was assigned last to M_j.

Claim 4.3. There holds $|T_2| \leq 3$.

Claim 4.4. For any $J_l \in T_3$, M_j's load immediately before the assignment of J_l is at most L_l.

Claim 4.5. Let $J_l \in T_3$ be the last job assigned to M_j. If M_j contains at least k jobs, different from J_l, each having a processing time of at least $(\alpha_m - 1)^2 L$, then it suffices to remove these k jobs and J_l such that M_j's resulting load is upper bounded by $(\alpha_m - 1)L$.

Claim 4.6. If there exists a $J_l \in T_3$ with $p_l < (\alpha_m - 1)^2 L$, then M_j's load immediately before the assignment of J_l is at most $(\alpha_m - 1)L$.

Proof of Claim 4.1. The jobs of T_2 are large at time $t^* - 1$ and hence at the time they were assigned to M_j. By the definition of t^*, $\ell_s(j,t) > \beta(j) L_t^*$ for any $t^* \leq t \leq n$. Hence $ALG(\alpha_m)$ does not assign small jobs to M_j at or after time t^*.

Proof of Claim 4.2. All jobs of $T_1 \setminus \{J_l\}$ are small at time $t^* - 1$ and their total processing time is at most $\ell_s(j, t^* - 1)$. In fact, their total processing time is equal to $\ell_s(j, t^* - 1)$ if $l = t^* - 1$. By the definition of t^*, $\ell_s(j, t^* - 1) \leq \beta(j) L_{t^*-1}^*$.

Proof of Claim 4.3. We show that for any time t, $1 \leq t \leq n$, when J_t has been placed on a machine, M_j can contain at most three jobs that are large at time t. The claim then follows by considering $t^* - 1$. Suppose that when J_t has been scheduled, M_j contained more than three jobs that are large as time t. Among these jobs let J_l be the one that was assigned last to M_j. Immediately before the assignment of J_l machine M_j had a load greater than L_l because the total processing time of three large jobs is greater than $3(\alpha_m-1)L_t \geq 3(\alpha_m-1)L_l \geq L_l$ since $\alpha_m \geq 4/3$, see Lemma 2. This contradicts the fact that J_l is placed on a least loaded machine, which has a load of at most $p_l^+/m \leq L_l$.

Proof of Claim 4.4. By Claim 4.1 J_l is large at time l and hence is assigned to a least loaded machine, which has a load of at most $p_l^+/m \leq L_l$.

Proof of Claim 4.5. Claim 4.4 implies that immediately before the assignment of J_l machine M_j has a load of at most $L_l \leq L$. If M_j contains at least k jobs, different from J_l, with a processing time of at least $(\alpha_m - 1)^2 L$, then the removal of these k jobs and J_l from M_j leads to a machine load of at most $L - k(\alpha_m - 1)^2 L \leq L - \lceil (2 - \alpha_m)/(\alpha_m - 1)^2 \rceil (\alpha_m - 1)^2 L \leq (\alpha_m - 1)L$.

Proof of Claim 4.6. By Claim 4.1 J_l is large at time l and hence $p_l > (\alpha_m - 1)L_l$. Since $p_l < (\alpha_m - 1)^2 L$, it follows $L_l < (\alpha_m - 1)L$. By Claim 4.4, M_j's load prior to the assignment of J_l is at most L_l and hence at most $(\alpha_m - 1)L$.

We now finish the proof of the lemma and distinguish two cases depending on the cardinality of $T_2 \cup T_3$.

Case 1: If $|T_2 \cup T_3| < k + 4$, then by Claim 4.2 it suffices to remove the jobs of $T_2 \cup T_3$ and the last job of T_1 assigned to M_j.

Case 2: Suppose $|T_2 \cup T_3| \geq k+4$. By Claim 4.3, $|T_2| \leq 3$ and hence $|T_3| \geq k+1$. Among the jobs of T_3 consider the last $k + 1$ ones assigned to M_j. If each of them has a processing time of at least $(\alpha_m-1)^2 L$, then Claim 4.5 ensures that it suffices to remove these $k + 1$ jobs. If one of them, say J_l, has a processing time smaller than $(\alpha_m - 1)^2 L$, then by Claim 4.6 M_j's load prior to the assignment of J_l is at most $(\alpha_m - 1)L$. Again it suffices to remove these $k + 1$ jobs from M_j. $\square$

After the job removal step each machine M_j, $1 \leq j \leq m$, has a load of at most $\max\{\beta(j)L^*, (\alpha_m - 1)L\}$. By the definition of $\beta(j)$, L and L^*, this is at most $\alpha_m L \leq \alpha_m OPT$, where OPT denotes the value of the optimum makespan for the job sequence σ. By the following lemma, after the reassignment step, each machine still has a load of at most $\alpha_m OPT$. This establishes Theorem 1.

Lemma 5. *After the reassignment step each machine M_j, $1 \leq j \leq m$, has a load of at most $\alpha_m OPT$.*

Proof. (Sketch) A detailed proof is presented in the full paper and we only describe some of the main ideas. A first step is to show that each of the sets $P_1, \ldots, P_m$ has a total processing time of at most OPT. Moreover, each job of $R' \setminus (P_1 \cup \ldots \cup P_m)$ has a processing time of at most $OPT/3$. This implies that the processing time of each job in $R \setminus (P_1 \cup \ldots \cup P_m)$ is upper bounded by $(\alpha_m - 1)OPT$. Taking into account that before the reassignment step each machine has a load of at most $\max\{\beta(j)L^*, (\alpha_m - 1)L\}$, we can show that all further scheduling operations preserve a load of at most $\alpha_m OPT$ on each of the machines. $\square$

3   A Lower Bound

We present a lower bound showing that $ALG(\alpha_m)$ is optimal. The proof is given in the full version of the paper.

Theorem 2. *Let $m \geq 2$. No deterministic online algorithm can achieve a competitive ratio smaller than α_m if $o(n)$ job migrations are allowed.*

4   Algorithms Using Fewer Migrations

We present a family of algorithms $ALG(c)$ that uses a smaller number of job migrations. The family is defined for any constant c with $5/3 \leq c \leq 2$, where c is the targeted competitive ratio. An important feature of $ALG(c)$ is that it partitions the machines $M_1, \ldots, M_m$ into two sets $A = \{M_1, \ldots, M_{\lfloor m/2 \rfloor}\}$ and $B = \{M_{\lceil m/2 \rceil}, \ldots, M_m\}$ of roughly equal size. In a job arrival phase the jobs are preferably assigned to machines in A, provided that their load it not too high. In the job migration phase, jobs are mostly migrated from machines of A (preferably to machines in B) and this policy will allow us to achieve a smaller number of migrations. In the following let $5/3 \leq c \leq 2$.

Algorithm $ALG(c)$: Job arrival phase. At any time t $ALG(c)$ maintains a lower bound L_t on the optimum makespan, i.e. $L_t = \max\{\frac{1}{m}p_t^+, p_t^1, 2p_t^{m+1}\}$. As in Section 2, p_t^1 and p_t^{m+1} denote the processing times of the largest and $(m+1)$-st largest jobs in $J_1, \ldots, J_t$, respectively. A job J_t is *small* if $p_t \leq (2c - 3)L_t$; otherwise it is *large*. A job J_i, with $i \leq t$, is *small* at time t if $p_i \leq (2c - 3)L_t$.

For any machine M_j and any time t, $\ell(j,t)$ is M_j's load immediately before J_t is assigned and $\ell_s(j,t)$ is its load consisting of the jobs that are small at time t.

Any job J_t, $1 \leq t \leq n$, is processed as follows. If J_t is small, then $ALG(c)$ checks if there is a machine in A whose load value $\ell_s(j,t)$ is at most $(c-1)L_t$. If this is the case, then among the machines in A with this property, J_t is assigned to one having the smallest $\ell_s(j,t)$ value. If there is no such machine in A, then J_t is assigned to a least loaded machine in B. If J_t is large, then $ALG(c)$ checks if there is machine in A whose load value $\ell(j,t)$ is at most $(3-c)L_t$. If this is the case, then J_t is scheduled on a least loaded machine in A. Otherwise J_t is assigned to a least loaded machine in B. At the end of the phase let $L = L_n$.

Job Migration Phase. At any time during the phase let $\ell(j)$ denote the current load of M_j, $1 \leq j \leq m$. In the job removal step, for any machine $M_j \in B$, $ALG(c)$ removes the largest job from that machine. Furthermore, while there exists a machine $M_j \in A$ whose current load exceeds $(c-1)L$, $ALG(c)$ removes the largest job from the machine. Let R be the set of all removed jobs. In the job reassignment step $ALG(c)$ first sorts the jobs in order of non-increasing processing times. For any i, $1 \leq i \leq |R|$, let J_r^i be the i-th largest job in this sequence, and let p_r^i be the corresponding processing time. For $i = 1, \ldots, |R|$, J_r^i is scheduled as follows. If there exists a machine $M_j \in B$ such that $\ell(j) + p_r^i \leq cL$, i.e. J_r^i can be placed on M_j without exceeding a makespan of cL, then J_r^i is assigned to this machine. Otherwise the job is scheduled on a least loaded machine in A. A pseudo-code description of $ALG(c)$ is given in the full paper.

Theorem 3. *$ALG(c)$ is c-competitive, for any constant c with $5/3 \leq c \leq 2$.*

The proof of Theorem 3 is also presented in the full version of the paper. In order to obtain good upper bounds on the number of job migrations, we focus on specific values of c. First, set $c = 5/3$. In $ALG(5/3)$ a job J_t is small if $p_t \leq 1/3 \cdot L_t$. In the arrival phase a small job is assigned to a machine in A if there exists a machine in this set whose load consisting of jobs that are currently small is at most $2/3 \cdot L_t$. A large job is assigned to a machine in A if there exists a machine in this set whose load is at most $4/3L_t$.

Theorem 4. *$ALG(5/3)$ is $\frac{5}{3}$-competitive and uses at most $4m$ job migrations.*

In fact for any c, $5/3 \leq c \leq 2$, $ALG(c)$ uses at most $4m$ migrations. Finally, let $c = 1.75$.

Theorem 5. *$ALG(1.75)$ is 1.75-competitive and uses at most $2.5m$ job migrations.*

Again, for any c with $1.75 \leq c \leq 2$, $ALG(c)$ uses at most $2.5m$ job migrations. The proofs of Theorems 4 and 5 are contained in the full version of the paper.

References

1. Aggarwal, G., Motwani, R., Zhu, A.: The load rebalancing problem. Journal of Algorithms 60(1), 42–59 (2006)
2. Albers, S.: Better bounds for online scheduling. SIAM J. Comput. 29, 459–473 (1999)

3. Bartal, Y., Karloff, H., Rabani, Y.: A better lower bound for on-line scheduling. Infomation Processing Letters 50, 113–116 (1994)
4. Bartal, Y., Fiat, A., Karloff, H., Vohra, R.: New algorithms for an ancient scheduling problem. Journal of Computer and System Sciences 51, 359–366 (1995)
5. Cao, Q., Liu, Z.: Online scheduling with reassignment on two uniform machines. Theoretical Computer Science 411(31-33), 2890–2898 (2010)
6. Chen, X., Lan, Y., Benko, A., Dósa, G., Han, X.: Optimal algorithms for online scheduling with bounded rearrangement at the end. Theoretical Computer Science 412(45), 6269–6278 (2011)
7. Chen, B., van Vliet, A., Woeginger, G.J.: A lower bound for randomized on-line scheduling algorithms. Information Processing Letters 51, 219–222 (1994)
8. Chen, B., van Vliet, A., Woeginger, G.J.: A optimal algorithm for preemptive online scheduling. Operations Research Letters 18, 127–131 (1995)
9. Dósa, G., Epstein, L.: Preemptive Online Scheduling with Reordering. In: Fiat, A., Sanders, P. (eds.) ESA 2009. LNCS, vol. 5757, pp. 456–467. Springer, Heidelberg (2009)
10. Englert, M., Özmen, D., Westermann, M.: The power of reordering for online minimum makespan scheduling. In: Proc. 49th Annual IEEE Symposium on Foundations of Computer Science, pp. 603–612 (2008)
11. Faigle, U., Kern, W., Turan, G.: On the performance of on-line algorithms for partition problems. Acta Cybernetica 9, 107–119 (1989)
12. Fleischer, R., Wahl, M.: Online scheduling revisited. Journal of Scheduling 3, 343–353 (2000)
13. Galambos, G., Woeginger, G.: An on-line scheduling heuristic with better worst case ratio than Graham's list scheduling. SIAM J. on Computing 22, 349–355 (1993)
14. Graham, R.L.: Bounds for certain multi-processing anomalies. Bell System Technical Journal 45, 1563–1581 (1966)
15. Graham, R.L.: Bounds on multiprocessing timing anomalies. SIAM Journal of Applied Mathematics 17(2), 416–429 (1969)
16. Gormley, T., Reingold, N., Torng, E., Westbrook, J.: Generating adversaries for request-answer games. In: Proc. 11th ACM-SIAM Symposium on Discrete Algorithms, pp. 564–565 (2000)
17. Hochbaum, D.S., Shmoys, D.B.: Using dual approximation algorithms for scheduling problems: Theoretical and practical results. Journal of the ACM 34, 144–162 (1987)
18. Karger, D.R., Phillips, S.J., Torng, E.: A better algorithm for an ancient scheduling problem. Journal of Algorithms 20, 400–430 (1996)
19. Min, X., Liu, J., Wang, Y.: Optimal semi-online algorithms for scheduling problems with reassignment on two identical machines. Information Processing Letters 111(9), 423–428 (2011)
20. Rudin III, J.F.: Improved bounds for the on-line scheduling problem. Ph.D. Thesis. The University of Texas at Dallas (May 2001)
21. Sanders, P., Sivadasan, N., Skutella, M.: Online scheduling with bounded migration. Mathematics of Operations Research 34(2), 481–498 (2009)
22. Sgall, J.: A lower bound for randomized on-line multiprocessor scheduling. Information Processing Letters 63, 51–55 (1997)
23. Tan, Z., Yu, S.: Online scheduling with reassignment. Operations Research Letters 36(2), 250–254 (2008)
24. Sleator, D.D., Tarjan, R.E.: Amortized efficiency of list update and paging rules. Communications of the ACM 28, 202–208 (1985)

Simplifying Massive Contour Maps

Lars Arge[1], Lasse Deleuran[1], Thomas Mølhave[2,*],
Morten Revsbæk[1], and Jakob Truelsen[3]

[1] MADALGO**, Department of Computer Science, Aarhus University
[2] Department of Computer Science, Duke University
[3] SCALGO, Scalable Algorithmics, Denmark

Abstract. We present a simple, efficient and practical algorithm for constructing and subsequently simplifying contour maps from massive high-resolution DEMs, under some practically realistic assumptions on the DEM and contours.

1 Introduction

Motivated by a wide range of applications, there is extensive work in many research communities on modeling, analyzing, and visualizing terrain data. A (3D) digital elevation model (DEM) of a terrain is often represented as a planar triangulation $\mathbf{M}$ with heights associated with the vertices (also known as a triangulated irregular network or simply a TIN). The l-level set of $\mathbf{M}$ is the set of (2D) segments obtained by intersecting $\mathbf{M}$ with a horizontal plane at height l. A *contour* is a connected component of a level set, and a *contour map* $\mathcal{M}$ the union of multiple level sets; refer to Figure 1. Contour maps are widely used to visualize a terrain primarily because they provide an easy way to understand the topography of the terrain from a simple two-dimensional representation.

Early contour maps were created manually, severely limiting the size and resolution of the created maps. However, with the recent advances in mapping technologies, such as laser based LIDAR technology, billions of (x, y, z) points on a terrain, at sub-meter resolution with very high accuracy ($\sim$10-20 cm), can be acquired in a short period of time and with a relatively low cost. The massive size of the data (DEM) and the contour maps created from them creates problems, since tools for processing and visualising terrain data are often not designed to handle data that is larger than main memory. Another problem is that contours generated from high-resolution LIDAR data are very detailed, resulting in a large amount of excessively jagged and spurious contours; refer to Figure 1. This in turn hinders their primary applications, since it becomes difficult to interpret the maps and gain understanding of the topography of the terrain. Therefore we are interested in simplifying contour maps.

* This work is supported by NSF under grants CCF-06 -35000, CCF-09-40671, and CCF-1012254, by ARO grants W911NF-07-1-0376 and W911NF-08-1-0452, and by U.S. Army ERDC-TEC contract W9132V-11-C-0003.
** Center for Massive Data Algorithmics, a Center of the Danish National Research Foundation

L. Epstein and P. Ferragina (Eds.): ESA 2012, LNCS 7501, pp. 96–107, 2012.

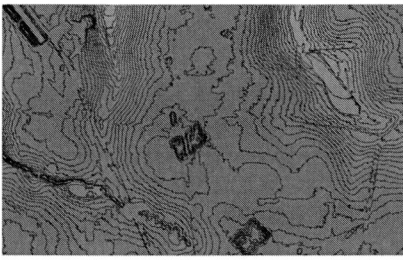 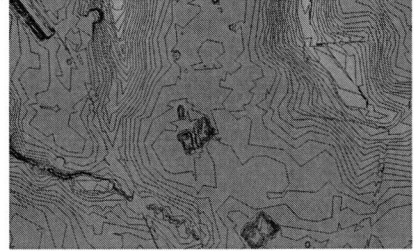

Fig. 1. A snapshot of contours generated from the Danish island of Als. The contour granularity is $\Delta = 0.5$m, the xy-constraints was $\varepsilon_{xy} = 4$m and the vertical constraint was $\varepsilon_z = 0.2$m. Left: Original map $\mathcal{M}$, right: simplified map.

Previous Work: The inefficiency of most tools when it comes to processing massive terrain data stems from the fact that the data is too large to fit in main memory and must reside on slow disks. Thus the transfer of data between disk and main memory is often a bottleneck (see e.g. [10]). To alleviate this bottleneck one needs algorithms designed in the I/O-model of computation [5]. In this model, the machine consists of a main memory of size M and an infinite-size disk. A block of B consecutive elements can be transferred between main memory and disk in one *I/O operation* (or simply *I/O*). Computation can only take place on elements in main memory, and the complexity of an algorithm is measured in terms of the number of I/Os it performs. Over the last two decades, I/O-efficient algorithms and data structures have been developed for several fundamental problems. See recent surveys [6,20] for a comprehensive review of I/O-efficient algorithms. Here we mention that scanning and sorting N elements takes $O(\mathrm{Scan}(N)) = O(N/B)$ and $O(\mathrm{Sort}(N)) = O(N/B \log_{M/B}(N/B))$ I/Os, respectively. The problem of computing contours and contour maps has previously been studied in the I/O-model [3,2]. Most relevant for this paper, Agarwal *et al.* [2] presents an optimal I/O-efficient algorithm that computes a contour map in $O(\mathrm{Sort}(N) + \mathrm{Scan}(|\mathcal{M}|))$ I/Os, where $|\mathcal{M}|$ is the number of segments in the contour map and N is the number of triangles in the DEM. It also computes the nesting of the contours and the segments around each contour are produced in sorted order. However, this algorithm is not practical.

Although the problem that contour maps generated from high-resolution LI-DAR data contain excessively jagged contours can be alleviated by contour map simplification (while also alleviating some of the scalability problems encountered when processing contour maps), the main issue is of course to guarantee the accuracy of the simplified contour map. There are two fundamental approaches to simplifying a contour map: The DEM **M** can be simplified before computing the contour map $\mathcal{M}$, or $\mathcal{M}$ can be simplified directly. There has been a lot of work on simplifying DEMs; refer e.g. to [9,15] and the references therein. However, most often the simple approaches do not provide a guarantee on the simplification accuracy, while the more advanced approaches are not I/O-efficient and therefore do not scale to large data sets. Developing I/O-efficient DEM simplification algorithms with simplification guarantees has shown to be a considerable challenge, although an $O(\mathrm{Sort}(N))$ I/O (topological persistence based) algorithm for removing "insignificant" features from a DEM (resulting in small contours) has recently been developed [4]. Simplifying the contour map $\mathcal{M}$ directly is very similar to simplifying a set of polygons (or polygonal lines) in the plane. Polygonal line simplification is a well studied problem; refer to [17] for a comprehensive survey. However, there are at least three important differences between contour maps and polygonal line simplification. Most noticeably, simplifying a contour line in the plane using a polygonal line simplification algorithms will, even if it guarantees simplification accuracy in the plane, not provide a z-accuracy guarantee. Furthermore, simplifying the contours individually may lead to intersections between the simplified contours. Finally, when simplifying contour maps its very important to preserve the relationships between the contours (the *homotopic* relationship), that is, maintain the nesting of the contours in the map. Note that intersections are automatically avoided and homotopy preserved when simplifying the DEM before computing the contour map.

One polygonal line simplification algorithm that is often favored for its simplicity and high subjective and objective quality on real life data is the Douglas-Peucker line simplification algorithm [13]. The algorithm simplifies a contour by removing points from the contour while ensuring that the distance between the original and the simplified contour is within a distance parameter ε_{xy} (but it does not guarantee that it removes the optimal number of endpoints under this constraint). Modifications of this algorithm have been developed, that remove self-intersections in the output [19], as well as ensure homotopy relative to a set of obstacles (polygons) [12,8]. However, these modified algorithms are complicated and/or not I/O-efficient (and do also not consider z-accuracy).

Our Results: In this paper we present a simple, efficient and practical algorithm for constructing and subsequently simplifying contour maps from massive high-resolution DEMs, under some practically realistic assumptions on the DEM and contours. The algorithm guarantees that the contours in the simplified contour map are homotopic to the unsimplified contours, non-intersecting, and within a distance of ε_{xy} of the unsimplified contours in the xy plane. Furthermore, it guarantees that for any point p on a contour in the l-level-set of the simplified contour map, the difference between l and the elevation of p in $\mathbf{M}$ (the z-error) is less than ε_z. We also present experimental results that show a significant improvement in the quality of the simplified contours along with a major (about 90%) reduction in size.

Overall, our algorithm has three main components. Given the levels $\ell_1, \ldots, \ell_d$, the *first component*, described in Section 3, computes the segments in the contour map $\mathcal{M}$. The component also computes level-sets for each of the levels $\ell_i \pm \varepsilon_z$, $1 \leqslant i \leqslant d$. The contours generated from these extra levels will be used to ensure that the z-error is bounded by ε_z. We call these contours *constraint* contours and mark the contours in $\mathcal{M}$ that are not constraint contours. The component also orders the segments around each contour and computes how the contours are nested. It uses $O(\mathrm{Sort}(|\mathcal{M}|))$ I/Os under the practically realistic assumptions that each contour, as well as the contour segments intersected by any horizontal line, fit in memory. This is asymptotically slightly worse than the theoretically optimal but complicated algorithm by Agarwal *et al.* [2]. The *second component*, described in Section 4, computes, for each of the marked contours P, the set $\mathcal{P}$ of contours that need to be considered when simplifying P. Intuitively, $\mathcal{P}$ contains all (marked as well as constraint) contours that can be reached from P without crossing any other contour of $\mathcal{M}$. Although each contour can be involved in many sets $\mathcal{P}$, the entire algorithm uses only $O(\text{Sort}(|\mathcal{M}|))$ I/Os. The *third component*, described in Section 5, simplifies each marked contour P within $\mathcal{P}$. This is done using a modified version of the Douglas-Peucker line simplification algorithm [13]. As with the Douglas-Peucker algorithm it guarantees that the simplified contour P' is within distance ε_{xy} of the original contour P, but it also guarantees that P' is homotopic to P (with respect to $\mathcal{P}$) and that P' does not have self intersections. The existence of the constraint contours in $\mathcal{P}$ together with the homotopy guarantee that the z-error is within ε_z. Under the practically realistic assumptions that each contour P along with $\mathcal{P}$ fits in internal memory, the algorithm does not use any extra I/Os.

Finally, the details on the implementation of our algorithm are given in Section 6 along with experimental results on a terrain data set of Denmark with over 12 billion points.

2   Preliminaries

Terrains: We consider the terrain $\mathbf{M}$ to be represented as a triangular irregular network, which is a planar triangulation whose vertices v are associated with heights $h(v)$. The height of a point interior to a triangle is determined by linearly interpolating from the triangle vertex heights; we use $h(p)$ to refer to the height of any point p in $\mathbb{R}^2$.

Paths and Polygons: Let $p_1, \ldots, p_n$, be a sequence of $n > 1$ points in $\mathbb{R}^2$. The *path* Q defined by these points is the set of line segments defined by pairs of consecutive points in the sequence. The points $p_1, \ldots, p_n$ are called the *vertices* of Q. A *simple path* is a path where only consecutive segments intersect, and only at the endpoints. Given integers $1 \leqslant i < j \leqslant n$, the *sub-path* Q_{ij} is the path defined by the vertices $p_i, p_{i+1}, \ldots, p_j$. We abuse notation slightly by using Q to denote both the sequence of vertices, and the path itself. We define the size of Q as its number of segments, i.e. $|Q| = n - 1$. A path Q' is a *simplification* of a path Q if $Q' \subseteq Q$ and the vertices of Q' appear in the same order as in Q.

A *polygon* (*simple polygon*) is a path (simple path) P where $p_1 = p_n$. A simple polygon P partitions $\mathbb{R}^2 \setminus P$ into two open sets — a bounded one called *inside* of P and denoted by P^i, and an unbounded one called *outside* of P and denoted by P^o. We define a *family of polygons* to be a set of non-intersecting and vertex-disjoint simple polygons. Consider two simple polygons P_1 and P_2 in a family of polygons A. P_1 and P_2 are called *neighbors* if no polygon $P \in A$ separates them, i.e., there is no P where one of P_1 and P_2 is contained in P^i and the other in P^o. If P_1 is a neighbor of P_2 and $P_1 \subset P_2^i$, then P_1 is called a *child* of P_2, and P_2 is called the *parent* of P_1; we will refer to the parent of P as $\hat{P}$. The *topology* of a family of polygons $\mathcal{M}$ describes how the polygons are nested i.e. the parent/child relationship between polygons. Given a polygon P in A, the

Fig. 2. (a) Polygonal domain $\mathcal{P}$ (solid lines) of P (dashed lines). (b) Polygon P' is homotopic to P in $\mathcal{P}$, Q is not.

polygonal domain of P, denoted $\mathcal{P}$, consists of the neighbors $P_1 \ldots P_k$ of P in A; refer to Figure 2(a). We define the size of $\mathcal{P}$ to be $|\mathcal{P}| = |P| + \sum_i |P_i|$.

Intuitively, two paths Q and Q' are *homotopic* with regards to a polygonal domain $\mathcal{P}$ if one can be continuously transformed into the other without intersecting any of the polygons of $\mathcal{P}$; refer to Figure 2. Path Q' is *strongly homotopic* to Q if Q' is a simplification of Q and if every segment $q_i'q_{i+1}'$ in Q' is homotopic to the corresponding sub-path $Q_{k,l}$ where $q_i' = q_k$ and $q_j' = q_l$. It follows that Q and Q' are also homotopic, but the reverse implication does not necessarily hold.

Given two indices $1 \leqslant i, j \leqslant n$ we define the distance $d(p, i, j)$ between any point $p \in \mathbb{R}^2$ and line segment $p_i p_j$ as the distance from p perpendicular to the line defined by $p_i p_j$. We define the error $\varepsilon(i, j)$ of replacing $P_{i,j}$ with the line segment $p_i p_j$ to be the maximum distance between the vertices of P_{ij} and $p_i p_j$, i.e. $\varepsilon(i, j) = \max\{d(p_i, i, j), d(p_{i+1}, i, j), \ldots, d(p_j, i, j)\}$. Let P' be a simplification of P. Given a *simplification threshold* ε, we say that P' is a *valid simplification* of P if it is a simple polygon homotopic to P in $\mathcal{P}$ and $\varepsilon(i, j) < \varepsilon$ for any segment $p_i p_j$ of P'.

Contours and Contour Maps: For a given terrain $\mathbf{M}$ and a *level* $\ell \in \mathbb{R}$, the ℓ-level set of $\mathbf{M}$, denoted by $\mathbf{M}_\ell$, is defined as $h^{-1}(\ell) = \{x \in \mathbb{R}^2 \mid h(x) = \ell\}$. A *contour* of a terrain $\mathbf{M}$ is a connected component of a level set of $\mathbf{M}$. Given a list of levels $\ell_1 < \ldots < \ell_d \in \mathbb{R}$, the *contour map* $\mathcal{M}$ of $\mathbf{M}$ is defined as the union of the level-sets $\mathbf{M}_{\ell_1}, \ldots, \mathbf{M}_{\ell_d}$. For simplicity, we assume that no vertex of $\mathbf{M}$ has height $\ell_1, \ldots, \ell_d$ and we assume that $\mathbf{M}$ is given such that $\mathbf{M}_{\ell_1}$ consists of only a single boundary contour $\mathcal{U}$. This implies that the collection of contours in the contour map form a family of polygons and that each polygon in the family, except $\mathcal{U}$, has a parent. It allows us to represent the topology of $\mathcal{M}$ as a tree $\mathcal{T} = (V, E)$ where the vertices V is the family of polygons and where E contains an edge from each polygon $P \neq \mathcal{U}$ to its parent polygon $\hat{P}$. The root of $\mathcal{T}$ is $\mathcal{U}$. We will refer to $\mathcal{T}$ as the *topology tree* of $\mathcal{M}$.

3   Building the Contour Map

In this section we describe our practical and I/O-efficient algorithm for constructing the contour map $\mathcal{M}$ of the terrain $\mathbf{M}$ and the topology tree $\mathcal{T}$ of $\mathcal{M}$, given a list of regular levels $\ell_1 < \ldots < \ell_d \in \mathbb{R}$. We will represent $\mathcal{M}$ as a sequence of line segments such that the clockwise ordered segments in each polygon P of

$\mathcal{M}$ appear consecutively in the sequence, and $\mathcal{T}$ by a sequence of edges (P_2,P_1) indicating that P_2 is the parent of P_1; all segments in $\mathcal{M}$ of polygon P will be augmented with (a label for) P and the BFS number of P in $\mathcal{T}$. We will use that the segments in any polygon P in $\mathcal{M}$, as well as the segments in $\mathcal{M}$ intersecting any horizontal line, fit in memory.

Computing Contour Map $\mathcal{M}$: To construct the line segments in $\mathcal{M}$, we first scan over all the triangles of $\mathbf{M}$. For each triangle f we consider each level ℓ_i within the elevation range of the three vertices of f and construct a line segment corresponding to the intersection of z_{ℓ_i} and f. To augment each segment with a polygon label, we then view the edges as defining a planar graph such that each polygon is a connected component in this graph. We find these connected components practically I/O-efficiently using an algorithm by Arge et al. [7], and then we use the connected component labels assigned to the segments by this algorithm as the polygon label. Next we sort the segments by their label. Then, since the segments of any one polygon fit in memory, we can in a simple scan load the segments of each polygon P into memory in turn and sort them in clock-wise order around the boundary of P.

Computing the Topology Tree $\mathcal{T}$ of $\mathcal{M}$: We use a plane-sweep algorithm to construct the edges of $\mathcal{T}$ from the sorted line segments in $\mathcal{M}$. During the algorithm we will also compute the BFS number of each polygon P in $\mathcal{T}$. After the algorithm it is easy to augment every segment in $\mathcal{M}$ with the BFS number of the polygon P it belongs to in a simple sorting and scanning step.

For a given $\mu \in \mathbb{R}$, let y_μ be the horizontal line through μ. Starting at $\mu = y_\infty$, our algorithm sweeps the line y_μ though the (pre-sorted) edges of $\mathcal{M}$ in the negative y-direction. We maintain a search tree $\mathcal{S}$ on the set of segments of $\mathcal{M}$ that intersect y_μ. For each edge in $\mathcal{S}$ we maintain the invariant that we have already computed its parent and that each edge also knows the identity of its own polygon. The set of edges in $\mathcal{S}$ changes as the sweep-line encounters endpoints of segments in $\mathcal{M}$, and each endpoint v from some polygon P has two adjacent edges s_1 and s_2. If the other two endpoints of s_1 and s_2 are both above y_μ we simply delete s_1 and s_2 from $\mathcal{S}$. If both endpoints are below y_μ and there is not already a segment from P in $\mathcal{S}$, then this is the first time P is encountered in the sweep. We can then use $\mathcal{S}$ to find the closest polygon segment s_3 from some other polygon P_1 to the left of v. Depending on the orientation of s_3 we know that the parent of P, $\hat{P}$ is either P_1, or the parent of P_1 which is stored with P_1 in $\mathcal{S}$, and we can output the corresponding edge $(P,\hat{P})$ of $\mathcal{T}$. It is easy to augment this algorithm so that it also computes the BFS number of each P in $\mathcal{T}$ by also storing, with each edge of $\mathcal{S}$, the BFS rank r of its parent, the rank of a child polygon is then $r+1$. Details will appear in the full version of this paper.

Analysis: The algorithm for computing the contour map $\mathcal{M}$ uses $O(\text{Sort}(|\mathcal{M}|))$ I/Os: First it scans the input triangles to produce the segments of $\mathcal{M}$ and invokes the practically efficient connected component algorithm of Arge et al. [7] that uses $O(\text{Sort}(|\mathcal{M}|))$ I/Os under the assumption that the segments in $\mathcal{M}$ intersecting any horizontal line fit in memory. Then it sorts the labeled segments using another $O(\text{Sort}(|\mathcal{M}|))$ I/Os. Finally, it scans the segments to sort each polygon in internal memory, utilizing that the segments in any polygon P in $\mathcal{M}$ fits in memory.

Also the algorithm for computing the topology tree $\mathcal{T}$ uses $O(\text{Sort}(|\mathcal{M}|))$: After scanning the segments of $\mathcal{M}$ to produce L, it performs one sort and one scan of L, utilizing the assumption that $\mathcal{S}$ fits in memory. In the full paper we show that augmenting each segment in $\mathcal{M}$ with the BFS number of the polygon P that it belongs to, can also be performed in $O(\text{Sort}(|\mathcal{M}|))$ in a simple way.

4   Simplifying Families of Polygons

In this section we describe our practical and I/O-efficient algorithm for simplifying a set of marked polygons in a family of polygons given an algorithm for simplifying a single polygon P within its polygonal domain $\mathcal{P}$. In essence the problem consist of computing $\mathcal{P}$ for each marked polygon P. We assume the family of polygons is given by a contour map $\mathcal{M}$ represented by a sequence of line segments such that the clockwise ordered segments in each polygon P of $\mathcal{M}$ appear consecutively in the sequence, and a topology tree $\mathcal{T}$ given as a sequence of edges $(P, \hat{P})$ indicating that $\hat{P}$ is the parent of P; all segments in $\mathcal{M}$ of polygon P are augmented with (a label for) P and the BFS number of P in $\mathcal{T}$.

To compute $\mathcal{P}$ for every marked P we need to retrieve the neighbors of P in $\mathcal{M}$. These are exactly the parent, siblings and children of P in $\mathcal{T}$. Once $\mathcal{P}$ and the simplification P' of P has been computed we need to update $\mathcal{M}$ with P'. We describe an I/O-efficient simplification algorithm that allows for retrieving the polygonal domains and updating polygons without spending a constant number of I/Os for each marked polygon. The algorithm simplifies the polygons across different BFS levels of $\mathcal{T}$ in order of increasing level, starting from the root. Within a given level the polygons are simplified in the same order as their parents were simplified. Polygons with the same parent can be simplified in arbitrary (label) order. Below we first describe how to reorder the polygons in $\mathcal{M}$ such that they appear in the order they will be simplified. Then we describe how to simplify $\mathcal{M}$.

Reordering: To reorder the polygons we first compute the *simplification rank* of every polygon P i.e. the rank of P in the simplification order described above. The simplification rank for the root of $\mathcal{T}$ is 0. To compute ranks for the remaining polygons of $\mathcal{T}$, we sort the edges $(P, \hat{P})$ of $\mathcal{T}$ in order of increasing BFS level of P. By scanning through the sorted list of polygons, we then assign simplification ranks to vertices one layer at a time. When processing a given layer we have already determined the ranks of the previous layer and can therefore order the vertices according to the ranks of their parents. After computing the simplification ranks we can easily reorder the polygons in a few sort and scan steps. Details will appear in the full paper.

Simplifying: Consider the sibling polygons $P_1, P_2 \ldots P_k$ in $\mathcal{M}$ all sharing the same parent P in $\mathcal{T}$. The polygonal domains of these sibling polygons all share the polygons $P, P_1, P_2 \ldots P_k$. We will refer to these shared polygons as the *open* *polygonal domain* of P and denote them $\mathcal{P}_{open}(P)$. It is easily seen that $\mathcal{P}$ for P_i where $i = 1 \ldots k$ is equal to $\mathcal{P}_{open}(P)$ together with the children of P_i.

We now traverse the polygons of $\mathcal{M}$ in the order specified by their simplification ranks, and refer to P as an *unfinished polygon* if we have visited P but not yet visited all the children of P. During the traversal we will maintain a queue Q containing an open polygonal domain for every unfinished polygon. The algorithm handles each marked polygon P as follows; if P is the root of $\mathcal{T}$ then $\mathcal{P}$ simply corresponds to the children of P which are at the front of $\mathcal{M}$. Given $\mathcal{P}$ we invoke the polygon simplification algorithm to get P'. Finally, we put $\mathcal{P}_{open}(P')$ at the back of Q. If P is not the root of $\mathcal{T}$, it will be contained in the open polygonal domain $\mathcal{P}_{open}(\hat{P})$. Since $\hat{P}$ is the unfinished polygon with lowest simplification rank, $\mathcal{P}_{open}(\hat{P})$ will be the front element of Q. If P is the first among its siblings to be visited, we retrieve $\mathcal{P}_{open}(\hat{P})$ from Q, otherwise it has already been retrieved and is available in memory. To get $\mathcal{P}$, we then retrieve the children of P from $\mathcal{M}$ and combine them with $\mathcal{P}_{open}(\hat{P})$ (if P is a leaf then $\mathcal{P} = \mathcal{P}_{open}(\hat{P})$). Finally, we invoke the polygon simplification algorithm on P and $\mathcal{P}$ to get P' and put the open polygonal domain of P' at the back of Q. It is easy to see that this algorithm simplifies $\mathcal{M}$, details will appear in the full version of this paper.

Analysis: Both reordering and simplifying is done with a constant number of sorts and scans of $\mathcal{M}$ and therefore require $O(\text{Sort}(|\mathcal{M}|))$ I/Os.

5   Internal Simplification Algorithm

In this section we describe our simplification algorithm, which given a single polygon P along with its polygonal domain $\mathcal{P}$ outputs a valid simplification P' of P.

Simplifying P: We first show how to compute a simplification Q^* of a path or polygon Q such that Q^* is homotopic to Q in $\mathcal{P}$, based on the Douglas-Peucker algorithm [13]. The recursive algorithm is quite simple. Given Q and a sub-path Q_{ij} for $i < j + 1$ to simplify, we find the vertex p_k in Q_{ij} maximizing the error $\varepsilon(i, j)$. Then we insert vertex p_k in the output path Q^* and recurse on Q_{ik} and Q_{kj}. When the error is sufficiently small, i.e. $\varepsilon(i, j) < \varepsilon_{xy}$, we check if the segment of Q^* is homotopic to P_{ij}. If this is the case the recursion stops. By construction every segment $p_i p_j$ of Q^* is homotopic to Q_{ij}. This implies that Q^* is strongly homotopic to Q, which again implies that Q^* and Q are homotopic.

When the input to the algorithm above is a polygon P the output is also a polygon P^*. However, even though P^* is homotopic to P, it is not necessarily a valid simplification since P^* may not be simple. Thus after using the algorithm above we may need to turn P^* into a simple polygon P'. This is done by finding all segments s of P^* that intersect and add more vertices from P to those segments using the same algorithm as above. Once this has been done we check for new intersections and keep doing this until no more intersection are found. Details will appear in the full paper.

Checking Segment-sub-path Homotopy: To check if a segment $p_i p_j$ is homotopic to P_{ij} in the above algorithm we need to be able to navigate the space around $\mathcal{P}$. Since $\mathcal{P}$ is given as a set of ordered simple polygons, we can efficiently and easily compute its trapezoidal decomposition[11] $\mathcal{D}$ using a simple sweep line algorithm on the segments of $\mathcal{P}$. To check homotopy we use the ideas of Cabello *et al.* [8] but arrive at a simpler algorithm by taking advantage of $\mathcal{D}$. We define the *trapezoidal sequence* $t(Q)$ of a path Q to be the sequence of trapezoids traversed by Q, sorted in the order of traversal. Using an argument similar to the ones used in [8,18] it is easy to show that if two paths have the same trapezoidal sequence then they are homotopic. Furthermore, in the full paper we show that if $t(Q)$ contains the subsequence $tt't$ for trapezoids $t, t' \in \mathcal{D}$ then this subsequence can be replaced by t without affecting Q's homotopic relationship to any other path; we call this a *contraction* of $t(Q)$. By repeatedly performing contractions on the sequence $t(Q)$ until no more contractions are possible we get a new sequence $t_c(Q)$, called the *canonical trapezoidal sequence* of Q. Q and Q' are homotopic if and only if $t_c(Q) = t_c(Q')$ [8,18].

Our algorithm for checking if two paths/polygons are homotopic simply computes and compares their canonical trapezoidal sequences. Note however that, for a path Q, the sizes of $t(Q')$ and $t_c(Q')$ are not necessarily linear in the size of the decomposition $\mathcal{D}$. In our case, we are interested in checking an instance of the strong homotopy condition, i.e. we want to check if a segment $s = p_i p_j$ is homotopic to $Q_{i,j}$. Since s is a line segment, we know that $t_c(s) = t(s)$, and we can thus simply traverse the trapezoids along in Q_i, tracing out $t_c(Q_{ij})$, and check that s intersects the same trapezoids as we go along, we do not need to precompute and store) $t(Q_{ij})$.

Analysis. We assume that $\mathcal{P}$ and Q fit in memory. Since the size of $\mathcal{D}$ is linear in the size of $\mathcal{P}$, it also fits in memory and the entire algorithm thus uses no I/Os. The I/Os needed to bring $\mathcal{P}$ and Q into memory were accounted for in the previous section.

6 Experiments

In this section we describe the experiments performed to verify that our algorithm for computing and simplifying a contour map $\mathcal{M}$ performs well in practice.

Implementation: We implemented our algorithm using the TPIE[1]: environment for efficient implementation of I/O-efficient algorithms, while taking care to handle all degeneracies (e.g. contours with height equal to vertices of $\mathbf{M}$, contour points with the same $x-$ or $y-$coordinate, and the existence of a single boundary contour). The implementation takes an input TIN $\mathbf{M}$ along with parameters ε_{xy} and ε_z, and Δ, and produces a simplified contour map $\mathcal{M}$ with equi-spaced contours at distance Δ.

We implemented one major internal improvement compared to algorithm described in Section 5, which results in a speed-up of an order of magnitude: As described in Section 5 we simplify a polygon P by constructing a trapezoidal decomposition of its entire polygonal domain $\mathcal{P}$. In practice, some polygons are

Table 1. Results for Funen with different ε_{xy} thresholds ($\varepsilon_z = 0.2m$ and with $\Delta = 0.5m$)

ε_{xy} in m.	0.2	0.5	1	2	3	5	10	15	20	25	50
Output points (%)	40.4	23.7	15.2	10.2	8.8	7.9	7.6	7.6	7.6	7.6	7.6
ε_z points		0.8	5.0	13.9	33.5	46.0	59.3	71.7	76.3	78.8	80.4	84.1
ε_{xy} points	99.2	95.0	86.1	66.5	54.0	40.7	28.3	23.7	21.2	19.6	15.9

very large and have many relatively small child polygons. In this case, even though a child polygon is small, its polygonal domain (and therefore also its trapezoidal decompositions) will include the large parent together with its siblings. However, it is easy to realize that for a polygon P it is only the subset of $\mathcal{P}$ within the bounding box of P that can constrain its simplification, and line segments outside the bounding box can be ignored when constructing the trapezoidal decomposition. We incorporate this observation into our implementation by building an internal memory R-tree [16] for each polygon P in the open polygonal domain $\mathcal{P}_{open}(\hat{P})$. These R-trees are constructed when loading large open polygonal domain into memory. To retrieve the bounding box of a given polygon P in $\mathcal{P}_{open}(\hat{P})$, we query the R-trees of its siblings and its parent, and retrieve the children of P as previously.

Data and Setup: All our experiments were performed on a machine with an 8-core Intel Xenon CPU running at 3.2GHz and 12GB of RAM. For our experiments we used a terrain model for the entire country of Denmark constructed from detailed LIDAR measurements (the data was generously provided to us by COWI A/S). The model is a $2m$ grid model giving the terrain height for every $2m \cdot 2m$ in the entire country, which amounts to roughly 12.4 billion grid cells. From this grid model we built a TIN by triangulating the grid cell center points. Before triangulating and performing our experiments, we used the concept of topological persistence [14] to compute the depth of depressions in the model. This can be done I/O-efficiently using an algorithm by Agarwal et. al [4]. For depressions that are not deeper than Δ it is coincidental whether the depression results in a contour or not. In case a contour is created it appears noisy and spurious in the contour map. For our experiments, we therefore raised depressions with a depth less than Δ. We removed small peaks similarly by simply inverting terrain heights. Our results show that this significantly reduces the size of the non-simplified contour map. Details will appear in the full paper.

Experimental Results: In all our experiments we generate contour maps with $\Delta = 0.5m$, and since the LIDAR measurements on which the terrain model of Denmark is based have a height accuracy of roughly $0.2m$, we used $\varepsilon_z = 0.2m$ in the experiments. In order to determine a good value of ε_{xy} we first performed experiments on a subset of the Denmark dataset consisting of $844,554,140$ grid cells and covering the island of Funen. Below we first describe the results of these experiments and then we describe the result of the experiments on the entire Denmark dataset. When discussing our results, we will divide the number of contour segment points in the simplified contour map (output points) into ε_z

Table 2. Results for Funen and Denmark with $\Delta = 0.5m$, $\varepsilon_z = 0.2m$ and $\varepsilon_{xy} = 5m$

Dataset	Funen	Denmark
Input points	365,641,479	4,046,176,743
Contours	636,973	16,581,989
Constraint factor	3	3
Running time (hours)	1.5	39
Output points (% of input points)	7.9	8.2
ε_z points (% of output points)	59.3	57.8
ε_{xy} points (% of output points)	40.7	42.2
Total number of intersections	38,992	585,813
Contours with intersections (% of input contours)	2.4	1.1

points and ε_{xy} *points*. These are the points that were not removed due to the constraint contours and the constraints of our polygon simplification algorithm (e.g. ε_{xy}), respectively.

Funen dataset: The non-simplified contour map generated from the triangulation of Funen consists of $636,973$ contours with $365,641,479$ points (not counting constraint contours). The results of our test runs are given in Table 1. The number of output points is given as a percentage of the number of points in the non-simplified contour map (not counting constraint contours). From the table it can be seen that the number of output points drops significantly as ε_{xy} is increased from $0.2m$ up to $5m$. However, for values larger than $5m$ the effect on output size of increasing ε_{xy} diminishes. This is most likely linked with high percentage of ε_z points in the output e.g. for $\varepsilon_{xy} = 10m$ we have that 71.7% of the output points are ε_z points (and increasing ε_{xy} will not have an effect on these).

Denmark dataset: When simplifying the contour map of the entire Denmark dataset we chose $\varepsilon_{xy} = 5m$, since our test runs on Funen had shown that increasing ε_{xy} further would not lead to a significant reduction in output points. Table 2 gives the results of simplifying the contour map of Denmark. The non-simplified contour map consists of $4,046,176,743$ points on $16,581,989$ contours. Adding constraint contours increases the contour map size with a factor 3 (the *constraint factor*) both in terms of points and contours. In total it took 39 hours to generate and simplify the contour map and the resulting simplified contour map contained 8.2% of the points in the non-simplified contour map (not counting constraint contours). Since 57.8% of the output points were ε_z points, it is unlikely that increasing ε_{xy} would reduce the size of the simplified contour map significantly. This corresponds to our observations on the Funen dataset. Table 2 also contains statistics on the number of self-intersections removed after the simplification (as discussed in Section 5); both the actual number of intersections and the percentage of the contours with self-intersections are given. As it can be seen these numbers are relatively small and their removal does not contribute significantly to the running time. The largest contour consisted of $8,924,584$

vertices while the largest family consisted of 19, 145, 568 vertices, which easily fit in memory.

References

1. TPIE - Templated Portable I/O-Environment, http://madalgo.au.dk/tpie
2. Agarwal, P., Arge, L., Mølhave, T., Sadri, B.: I/O-efficient algorithms for computing contours on a terrain. In: Proc. Symposium on Computational Geometry, pp. 129–138 (2008)
3. Agarwal, P.K., Arge, L., Murali, T.M., Varadarajan, K., Vitter, J.S.: I/O-efficient algorithms for contour line extraction and planar graph blocking. In: Proc. ACM-SIAM Symposium on Discrete Algorithms, pp. 117–126 (1998)
4. Agarwal, P.K., Arge, L., Yi, K.: I/O-efficient batched union-find and its applications to terrain analysis. ACM Trans. Algorithms 7(1), 11:1–11:21 (2010)
5. Aggarwal, A., Vitter, S., Jeffrey: The input/output complexity of sorting and related problems. Commun. ACM 31(9), 1116–1127 (1988)
6. Arge, L.: External memory data structures. In: Abello, J., Pardalos, P.M., Resende, M.G.C. (eds.) Handbook of Massive Data Sets, pp. 313–358 (2002)
7. Arge, L., Larsen, K., Mølhave, T., van Walderveen, F.: Cleaning massive sonar point clouds. In: Proc ACM SIGSPATIAL International Symposium on Advances in Geographic Information Systems, pp. 152–161 (2010)
8. Cabello, S., Liu, Y., Mantler, A., Snoeyink, J.: Testing homotopy for paths in the plane. In: Proc. Symposium on Computational Geometry, pp. 160–169 (2002)
9. Carr, H., Snoeyink, J., van de Panne, M.: Flexible isosurfaces: Simplifying and displaying scalar topology using the contour tree. In: Computational Geometry, pp. 42–58 (2010) (Special Issue on the 14th Annual Fall Workshop)
10. Danner, A., Mølhave, T., Yi, K., Agarwal, P., Arge, L., Mitasova, H.: TerraStream: From elevation data to watershed hierarchies. In: Proc. ACM International Symposium on Advances in Geographic Information Systems, pp. 28:1–28:8 (2007)
11. de Berg, M., van Kreveld, M., Overmars, M., Schwarzkopf, O.: Computational Geometry – Algorithms and Applications (1997)
12. de Berg, M., van Kreveld, M., Schirra, S.: A new approach to subdivision simplification. In: Proc. 12th Internat. Sympos. Comput.-Assist. Cartog., pp. 79–88 (1995)
13. Douglas, D., Peucker, T.: Algorithms for the reduction of the number of points required to represent a digitized line or its caricature (1973)
14. Edelsbrunner, H., Letscher, D., Zomorodian, A.: Topological persistence and simplification. In: Proc. IEEE Symposium on Foundations of Computer Science, pp. 454–463 (2000)
15. Garland, M., Heckbert, P.: Surface simplification using quadric error metrics. In: Proc. Computer Graphics and Interactive Techniques, pp. 209–216 (1997)
16. Guttman, A.: R-trees: A dynamic index structure for spatial searching. In: Proc. SIGMOD International Conference on Management of Data, pp. 47–57 (1984)
17. Heckbert, P.S., Garland, M.: Survey of polygonal surface simplification algorithms. Technical report, CS Dept., Carnegie Mellon U (to appear)
18. Hershberger, J., Snoeyink, J.: Computing minimum length paths of a given homotopy class. Comput. Geom. Theory Appl, 63–97 (1994)
19. Saalfeld, A.: Topologically consistent line simplification with the douglas peucker algorithm. In: Geographic Information Science (1999)
20. Vitter, J.: External memory algorithms and data structures: Dealing with MASSIVE data. ACM Computing Surveys, 209–271 (2001)

Explicit and Efficient Hash Families Suffice for Cuckoo Hashing with a Stash

Martin Aumüller[1], Martin Dietzfelbinger[1,*], and Philipp Woelfel[2,**]

[1] Fakultät für Informatik und Automatisierung, Technische Universität Ilmenau,
98694 Ilmenau, Germany
{martin.aumueller,martin.dietzfelbinger}@tu-ilmenau.de
[2] Department of Computer Science, University of Calgary,
Calgary, Alberta T2N 1N4, Canada
woelfel@cpsc.ucalgary.ca

Abstract. It is shown that for cuckoo hashing with a stash as proposed by Kirsch, Mitzenmacher, and Wieder (2008) families of very simple hash functions can be used, maintaining the favorable performance guarantees: with stash size s the probability of a rehash is $O(1/n^{s+1})$, and the evaluation time is $O(s)$. Instead of the full randomness needed for the analysis of Kirsch *et al.* and of Kutzelnigg (2010) (resp. $\Theta(\log n)$-wise independence for standard cuckoo hashing) the new approach even works with 2-wise independent hash families as building blocks. Both construction and analysis build upon the work of Dietzfelbinger and Woelfel (2003). The analysis, which can also be applied to the fully random case, utilizes a graph counting argument and is much simpler than previous proofs. As a byproduct, an algorithm for simulating uniform hashing is obtained. While it requires about twice as much space as the most space efficient solutions, it is attractive because of its simple and direct structure.

1 Introduction

Cuckoo hashing as proposed by Pagh and Rodler [17] is a popular implementation of a dictionary with guaranteed constant lookup time. To store a set S of n keys from a universe U (i.e., a finite set), cuckoo hashing utilizes two hash functions, $h_1, h_2 : U \to [m]$, where $m = (1+\varepsilon)n$, $\varepsilon > 0$. Each key $x \in S$ is stored in one of two hash tables of size m; either in the first table at location $h_1(x)$ or in the second one at location $h_2(x)$. The pair h_1, h_2 might not be suitable to accommodate S in these two tables. In this case, a *rehash* operation is necessary, which chooses a new pair h_1, h_2 and inserts all keys anew.

In their ESA 2008 paper [11], Kirsch, Mitzenmacher, and Wieder deplored the order of magnitude of the probability of a rehash, which is as large as $\Theta(1/n)$. They proposed adding a *stash*, an additional segment of storage that can hold up to s keys for some (constant) parameter s, and showed that this change

---
* Research supported in part by DFG grant DI 412/10-2.
** Research supported by a Discovery Grant from the National Sciences and Research Council of Canada (NSERC).

L. Epstein and P. Ferragina (Eds.): ESA 2012, LNCS 7501, pp. 108–120, 2012.

reduces the rehash probability to $\Theta(1/n^{s+1})$. However, the analysis of Kirsch *et al.* requires the hash functions to be fully random. In the journal version [12] Kirsch *et al.* posed "proving the above bounds for explicit hash families that can be represented, sampled, and evaluated efficiently" as an open problem.

Our Contribution. In this paper we generalize a hash family construction proposed by Dietzfelbinger and Woelfel [9] and show that the resulting hash functions have random properties strong enough to preserve the qualities of cuckoo hashing with a stash. The proof involves a new and simpler analysis of this hashing scheme, which also works in the fully random case. The hash functions we propose have a very simple structure: they combine functions from $O(1)$-wise independent families[1] with a few tables of size $n^\delta, 0 < \delta < 1$ constant, with random entries from $[m] = \{0, \ldots, m - 1\}$. An attractive version of the construction for stash capacity s has the following performance characteristics: the description of a hash function pair (h_1, h_2) consists of a table with $\sqrt{n}$ entries from $[m]^2$ and $2s+6$ functions from 2-wise independent classes. To evaluate $h_1(x)$ and $h_2(x)$ for $x \in U$, we must evaluate these $2s + 6$ functions, read $2s + 4$ table entries, and carry out $4s + 8$ additions modulo m. Our main result implies for these hash functions and for any set $S \subseteq U$ of n keys that with probability $1 - O(1/n^{s+1})$ S can be accommodated according to the cuckoo hashing rules.

In addition, we present a simple data structure for simulating a uniform hash function on S with range R, using our hash class and essentially a table with $2(1 + \varepsilon)n$ random elements from R.

Cuckoo Hashing with a Stash and Weak Hash Functions. In [12,14] it was noticed that for the analysis of cuckoo hashing with a stash of size s the properties of the so-called *cuckoo graph* $G(S, h_1, h_2)$ are central. Assume a set S and hash functions h_1 and h_2 with range $[m]$ are given. The associated cuckoo graph $G(S, h_1, h_2)$ is the bipartite multigraph whose two node sets are copies of $[m]$ and whose edge set contains the n pairs $(h_1(x), h_2(x))$, for $x \in S$. It is known that a single parameter of $G = G(S, h_1, h_2)$ determines whether a stash of size s is sufficient to store S using (h_1, h_2), namely the *excess* ex(G), which is defined as the minimum number of edges one has to remove from G so that all connected components of the remaining graph are acyclic or unicyclic.

Lemma 1 ([12]). *The keys from S can be stored in the two tables and a stash of size s using (h_1, h_2) if and only if* ex$(G(S, h_1, h_2)) \leq s$.

A proof of this lemma, together with proofs of other known facts, can be found in the online version of this paper [1].

Kirsch *et al.* [12] showed that with probability $1 - O(1/n^{s+1})$ a random bipartite graph with $2m = 2(1 + \varepsilon)n$ nodes and n edges has excess at most s. Their proof uses sophisticated tools such as Poissonization and Markov chain coupling. This result generalizes the analysis of standard cuckoo hashing [17] with no stash, in which the rehash probability is $\Theta(1/n)$. Kutzelnigg [14] refined

---

[1] κ-wise independent families of hash functions are defined in Section 2.

the analysis of [12] in order to determine the constant factor in the asymptotic bound of the rehash probability. His proof uses generating functions and differential recurrence equations. Both approaches inherently require that the hash functions h_1 and h_2 used in the algorithm are fully random.

Recently, Pătraşcu and Thorup [18] showed that simple tabulation hash functions are sufficient for running cuckoo hashing, with a rehash probability of $\Theta(1/n^{1/3})$, which is tight. Unfortunately, for these hash functions the rehash probability cannot be improved by using a stash.

Our main contribution is a new analysis that shows that explicit and efficient hash families are sufficient to obtain the $O(1/n^{s+1})$ bound on the rehash probability. We build upon the work of Dietzfelbinger and Woelfel [9]. For standard cuckoo hashing, they proposed hash functions of the form $h_i(x) = \left(f_i(x) + z^{(i)}[g(x)]\right) \bmod m$, for $x \in U$, for $i \in \{1, 2\}$, where f_i and g are from $2k$-wise independent classes with range $[m]$ and $[\ell]$, resp., and $z^{(1)}, z^{(2)} \in [m]^\ell$ are random vectors. They showed that with such hash functions the rehash probability is $O(1/n + n/\ell^k)$. Their proof has parts (i) and (ii). Part (i) already appeared in [4] and [17]: The rehash probability is bounded by the sum, taken over all minimal excess-1 graphs H of different sizes and all subsets T of S, of the probability that $G(T, h_1, h_2)$ is isomorphic to H. In Sect. 5 of this paper we demonstrate that for h_1 and h_2 fully random a similar counting approach also works for minimal excess-$(s + 1)$ graphs, whose presence in $G(S, h_1, h_2)$ determines whether a rehash is needed when a stash of size s is used. As in [17], this analysis also works for $O((s + 1) \log n)$-wise independent families.

Part (ii) of the analysis in [9] is a little more subtle. It shows that for each key set S of size n there is a part B_S^{conn} of the probability space given by (h_1, h_2) such that $\Pr(B_S^{\text{conn}}) = O(n/\ell^k)$ and in $\overline{B_S^{\text{conn}}}$ the hash functions act fully randomly on $T \subseteq S$ as long as $G(T, h_1, h_2)$ is connected. In Sect. 4 we show how this argument can be adapted to the situation with a stash, using subgraphs without leaves in place of the connected subgraphs. Woelfel [23] already demonstrated by applying functions in [9] to balanced allocation that the approach has more general potential to it.

A comment on the "full randomness assumption" and work relating to it seems in order. It is often quoted as an empirical observation that weaker hash functions like κ-wise independent families will behave almost like random functions. Mitzenmacher and Vadhan [15] showed that if the key set S has a certain kind of entropy then 2-wise independent hash functions will behave similar to fully random ones. However, as demonstrated in [7], there are situations where cuckoo hashing fails for a standard 2-wise independent family and even a random set S (which is "too dense" in U). The rather general "split-and-share" approach of [6] makes it possible to justify the full randomness assumption for many situations involving hash functions, including cuckoo hashing (with a stash) and further variants. However, for practical application this method is less attractive, since space consumption and failure probability are negatively affected by splitting the key set into "chunks" and treating these separately.

Simulating Uniform Hashing. Consider a universe U of keys and a finite set R. By the term *"simulating uniform hashing for U and R"* we mean an algorithm that does the following. On input $n \in \mathbb{N}$, a randomized procedure sets up a data structure DS_n that represents a hash function $h \colon U \to R$, which can then be evaluated efficiently for keys in U. For each set $S \subseteq U$ of cardinality n there is an event B_S with the property that conditioned on $\overline{B_S}$ the values $h(x)$, $x \in S$, are fully random. The quality of the algorithm is determined by the space needed for DS_n, the evaluation time for h, and the probability of the event B_S. It should be possible to evaluate h in constant time. The amount of entropy required for such an algorithm implies that at least $n \log |R|$ bits are needed to represent DS_n.

Pagh and Pagh [16] proposed a construction with $O(n)$ random words from R, based on Siegel's functions [19], which have constant, but huge evaluation time. They also gave a general method to reduce the space to $(1 + \varepsilon)n$, at the cost of an evaluation time of $O(1/\varepsilon^2)$. In [9] a linear-space construction with tables of size $O(n)$ was given that contain (descriptions of) $O(1)$-wise independent hash functions. The construction with the currently asymptotically best performance parameters, $(1 + \varepsilon)n$ words from R and evaluation time $O(\log(1/\varepsilon))$, as given in [6], is based on results of Calkin [2] and the "split-and-share" approach, involving the same disadvantages as mentioned above.

Our construction, to be described in Sect. 6, essentially results from the construction in [16] by replacing Siegel's functions with functions from our new class. The data structure consists of a hash function pair (h_1, h_2) from our hash class, a $O(1)$-wise independent hash function with range R, $O(s)$ small tables with entries from R, and two tables of size $m = (1 + \varepsilon)n$ each, filled with random elements from R. The evaluation time of h is $O(s)$, and for $S \subseteq U, |S| = n$, the event B_S occurs with probability $O(1/n^{s+1})$. The construction requires roughly twice as much space as the most space-efficient solutions [6,16]. However, it seems to be a good compromise combining simplicity with moderate space consumption.

2   Basics

Let U (the "universe") be a finite set. A mapping from U to $[r]$ is a *hash function with range* $[r]$. For an integer $\kappa \geq 2$, a set $\mathcal{H}$ of hash functions with range $[r]$ is called a κ-*wise independent* hash family if for arbitrary distinct keys $x_1, \ldots, x_\kappa \in U$ and for arbitrary $j_1, \ldots, j_\kappa \in [r]$ we have $\mathrm{Pr}_{h \in \mathcal{H}}\big(h(x_1) = j_1 \wedge \ldots \wedge h(x_\kappa) = j_\kappa\big) = 1/r^\kappa$. The classical κ-wise independent hash family construction is based on polynomials of degree $\kappa - 1$ over a finite field [22]. More efficient hash function evaluation can be achieved with tabulation-based constructions [9,20,21,13]. Throughout this paper, $\mathcal{H}_r^\kappa$ denotes an arbitrary κ-wise independent hash family with domain U and range $[r]$.

We combine κ-wise independent classes with lookups in tables of size ℓ in order to obtain pairs of hash functions from U to $[m]$:

Definition 1. *Let $c \geq 1$ and $\kappa \geq 2$. For integers m, $\ell \geq 1$, and given $f_1, f_2 \in \mathcal{H}_m^\kappa$, $g_1, \ldots, g_c \in \mathcal{H}_\ell^\kappa$, and vectors $z_j^{(i)} \in [m]^\ell$, $1 \leq j \leq c$, for $i \in \{1, 2\}$, let*

$$(h_1, h_2) = (h_1, h_2)\langle f_1, f_2, g_1, \ldots, g_c, z_1^{(1)}, \ldots, z_c^{(1)}, z_1^{(2)}, \ldots, z_c^{(2)}\rangle, \textit{ where}$$

$$h_i(x) = \left(f_i(x) + \sum_{1 \le j \le c} z_j^{(i)}[g_j(x)]\right) \bmod m, \textit{ for } x \in U, i \in \{1, 2\}.$$

Let $\mathcal{Z}_{\ell,m}^{\kappa,c}$ be the family of all these pairs (h_1, h_2) of hash functions.

While this is not reflected in the notation, we consider (h_1, h_2) as a structure from which the components $g_1, \ldots, g_c$ and $f_i, z_1^{(i)}, \ldots, z_c^{(i)}$, $i \in \{1, 2\}$, can be read off again. It is family $\mathcal{Z} = \mathcal{Z}_{\ell,m}^{2k,c}$, for some $k \ge 1$, made into a probability space by the uniform distribution, that we will study in the following. We usually assume that c and k are fixed and that m and ℓ are known.

2.1 Basic Facts

We start with some basic observations concerning the effects of compression properties in the "g-part" of (h_1, h_2), extending similar statements in [9].

Definition 2. *For $T \subseteq U$, define the random variable d_T, the "deficiency" of (h_1, h_2) with respect to T, by $d_T((h_1, h_2)) = |T| - \max\{k, |g_1(T)|, \ldots, |g_c(T)|\}$. (Note: d_T depends only on the g_j-components of (h_1, h_2).) Further, define*

- (i) bad_T *as the event that $d_T > k$;*
- (ii) good_T *as $\overline{\mathrm{bad}_T}$, i.e., the event that $d_T \le k$;*
- (iii) crit_T *as the event that $d_T = k$.*

Hash function pairs (h_1, h_2) in these events are called "T-bad", "T-good", and "T-critical", resp.

Lemma 2. *Assume $k \ge 1$ and $c \ge 1$. For $T \subseteq U$, the following holds:*

(a) $\Pr(\mathrm{bad}_T \cup \mathrm{crit}_T) \le \left(|T|^{2k}/\ell^k\right)^c.$

(b) *Conditioned on good_T (or on crit_T), the pairs $(h_1(x), h_2(x))$, $x \in T$, are distributed uniformly and independently in $[r]^2$.*

Proof. (a) Assume $|T| \ge 2k$ (otherwise the events bad_T and crit_T cannot occur). Since $g_1, \ldots, g_c$ are independent, it suffices to show that for a function g chosen randomly from $\mathcal{H}_\ell^{2k}$ we have $\Pr(|T| - |g(T)| \ge k) \le |T|^{2k}/\ell^k$.

We first argue that if $|T| - |g(T)| \ge k$ then there is a subset T' of T with $|T'| = 2k$ and $|g(T')| \le k$. Initialize T' as T. Repeat the following as long as $|T'| > 2k$: (i) if there exists a key $x \in T'$ such that $g(x) \ne g(y)$ for all $y \in T'\setminus\{x\}$, remove x from T'; (ii) otherwise, remove any key. Clearly, this process terminates with $|T'| = 2k$. It also maintains the invariant $|T'| - |g(T')| \ge k$: In case (i) $|T'| - |g(T')|$ remains unchanged. In case (ii) before the key is removed from T' we have $|g(T')| \le |T'|/2$ and thus $|T'| - |g(T')| \ge |T'|/2 > k$.

Now fix a subset T' of T of size $2k$ that satisfies $|g(T')| \le k$. The preimages $g^{-1}(u)$, $u \in g(T')$, partition T' into k' classes, $k' \le k$, such that g is constant on each class. Since g is chosen from a $2k$-wise independent class, the probability that g is constant on all classes of a given partition of T' into classes $C_1, \ldots, C_{k'}$, with $k' \le k$, is exactly $\ell^{-(2k-k')} \le \ell^{-k}$.

Finally, we bound $\Pr(|g(T)| \leq |T| - k)$. There are $\binom{|T|}{2k}$ subsets T' of T of size $2k$. Every partition of such a set T' into $k' \leq k$ classes can be represented by a permutation of T' with k' cycles, where each cycle contains the elements from one class. Hence, there are at most $(2k)!$ such partitions. This yields:

$$\Pr(|T| - |g(T)| \geq k) \leq \binom{|T|}{2k} \cdot (2k)! \cdot \frac{1}{\ell^k} \leq \frac{|T|^{2k}}{\ell^k}. \tag{1}$$

(b) If $|T| \leq 2k$, then h_1 and h_2 are fully random on T simply because f_1 and f_2 are $2k$-wise independent. So suppose $|T| > 2k$. Fix an arbitrary g-part of (h_1, h_2) so that $good_T$ occurs, i.e., $\max\{k, |g_1(T)|, \ldots, |g_c(T)|\} \geq |T| - k$. Let $j_0 \in \{1, \ldots, c\}$ be such that $|g_{j_0}(T)| \geq |T| - k$. Arbitrarily fix all values in the tables $z_j^{(i)}$ with $j \neq j_0$ and $i \in \{1, 2\}$. Let T^* be the set of keys in T colliding with other keys in T under g_{j_0}. Then $|T^*| \leq 2k$. Choose the values $z_{j_0}^{(i)}[g_{j_0}(x)]$ for all $x \in T^*$ and $i \in \{1, 2\}$ at random. Furthermore, choose f_1 and f_2 at random from the $2k$-wise independent family $\mathcal{H}_r^{2k}$. This determines $h_1(x)$ and $h_2(x)$, $x \in T^*$, as fully random values. Furthermore, the function g_{j_0} maps the keys $x \in T - T^*$ to distinct entries of the vectors $z_{j_0}^{(i)}$ that were not fixed before. Thus, the hash function values $h_1(x), h_2(x)$, $x \in T - T^*$, are distributed fully randomly as well and are independent of those with $x \in T^*$.   □

3   Graph Properties and Basic Setup

For $m \in \mathbb{N}$ let $\mathcal{G}_m$ denote the set of all bipartite (multi-)graphs with vertex set $[m]$ on each side of the bipartition. A set $\mathcal{A} \subseteq \mathcal{G}_m$ is called a *graph property*. For example, $\mathcal{A}$ could be the set of graphs in $\mathcal{G}_m$ that have excess larger than s. For a graph property $\mathcal{A} \subseteq \mathcal{G}_m$ and $T \subseteq U$, let $\mathcal{A}_T$ denote the event that $G(T, h_1, h_2)$ has property $\mathcal{A}$ (i.e., that $G(T, h_1, h_2) \in \mathcal{A}$). In the following, our main objective is to bound the probability $\Pr(\exists T \subseteq S \colon \mathcal{A}_T)$ for graph properties $\mathcal{A}$ which are important for our analysis.

For the next lemma we need the following definitions. For $S \subseteq U$ and a graph property $\mathcal{A}$ let $B_S^{\mathcal{A}} \subseteq \mathcal{Z}$ be the event $\exists T \subseteq S \colon \mathcal{A}_T \cap bad_T$ (see Def. 2). Considering fully random hash functions (h_1^*, h_2^*) for a moment, let $p_T^{\mathcal{A}} = \Pr(G(T, h_1^*, h_2^*) \in \mathcal{A})$.

Lemma 3. *For an arbitrary graph property $\mathcal{A}$ we have*

$$\Pr(\exists T \subseteq S \colon \mathcal{A}_T) \leq \Pr(B_S^{\mathcal{A}}) + \sum_{T \subseteq S} p_T^{\mathcal{A}}. \tag{2}$$

Proof. $\Pr(\exists T \subseteq S \colon \mathcal{A}_T) \leq \Pr(B_S^{\mathcal{A}}) + \Pr((\exists T \subseteq S \colon \mathcal{A}_T) \cap \overline{B_S^{\mathcal{A}}})$, and

$$\sum_{T \subseteq S} \Pr(\mathcal{A}_T \cap \overline{B_S^{\mathcal{A}}}) \underset{(i)}{\leq} \sum_{T \subseteq S} \Pr(\mathcal{A}_T \cap good_T) \leq \sum_{T \subseteq S} \Pr(\mathcal{A}_T \mid good_T) \underset{(ii)}{=} \sum_{T \subseteq S} p_T^{\mathcal{A}},$$

where (i) holds by the definition of $B_S^{\mathcal{A}}$, and (ii) holds by Lemma 2(b).   □

This lemma encapsulates our overall strategy for bounding $\Pr(\exists T \subseteq S \colon \mathcal{A}_T)$. The second summand in (2) can be bounded assuming full randomness. The task of bounding the first summand is tackled separately, in Section 4.

4   A Bound for Leafless Graphs

The following observation, which is immediate from the definitions, will be helpful in applying Lemma 3.

Lemma 4. *Let $m \in \mathbb{N}$, and let $\mathcal{A} \subseteq \mathcal{A}' \subseteq \mathcal{G}_m$. Then $\Pr(B_S^{\mathcal{A}}) \le \Pr(B_S^{\mathcal{A}'})$.* □

We define a graph property to be used in the role of $\mathcal{A}'$ in applications of Lemma 4. A node with degree 1 in a graph is called a *leaf*; an edge incident with a leaf is called a *leaf edge*. An edge is called a *cycle edge* if removing it does not disconnect any two nodes. A graph is called *leafless* if it has no leaves. Let $\mathsf{LL} \subseteq \mathcal{G}_m$ be the set of all leafless graphs. The 2-*core* of a graph is its (unique) maximum leafless subgraph. The purpose of the present section is to prove a bound on $\Pr(B_S^{\mathsf{LL}})$.

Lemma 5. *Let $\varepsilon > 0$, let $S \subseteq U$ with $|S| = n$, and let $m = (1+\varepsilon)n$. Assume (h_1, h_2) is chosen at random from $\mathcal{Z} = \mathcal{Z}_{\ell,m}^{2k,c}$. Then $\Pr(B_S^{\mathsf{LL}}) = O\big(n/\ell^{ck}\big)$.*

We recall a standard notion from graph theory (already used in [9]; cf. [1, App. A]): The *cyclomatic number* $\gamma(G)$ of a graph G is the smallest number of edges one has to remove from G to obtain a graph with no cycles. Also, let $\zeta(G)$ denote the number of connected components of G (ignoring isolated points).

Lemma 6. *Let $N(t,\ell,\gamma,\zeta)$ be the number of non-isomorphic (multi-)graphs with ζ connected components and cyclomatic number γ that have t edges, ℓ of which are leaf edges. Then $N(t,\ell,\gamma,\zeta) = t^{O(\ell+\gamma+\zeta)}$.*

Proof. In [9, Lemma 2] it is shown that $N(t,\ell,\gamma,1) = t^{O(\ell+\gamma)}$. Now note that each graph G with cyclomatic number γ, ζ connected components, $t-\ell$ non-leaf edges, and ℓ leaf edges can be obtained from some connected graph G' with cyclomatic number γ, $t - \ell + \zeta - 1$ non-leaf edges, and ℓ leaf edges by removing $\zeta - 1$ non-leaf, non-cycle edges. There are no more than $(t - \ell + \zeta - 1)^{\zeta-1}$ ways for choosing the edges to be removed. This implies, using [9, Lemma 2]:

$$N(t,\ell,\gamma,\zeta) \le N(t+\zeta-1,\ell,\gamma,1) \cdot (t - \ell + \zeta - 1)^{\zeta-1}$$
$$\le (t+\zeta)^{O(\ell+\gamma)} \cdot (t+\zeta)^{\zeta} = (t+\zeta)^{O(\ell+\gamma+\zeta)} = t^{O(\ell+\gamma+\zeta)}. \qquad \square$$

We shall need more auxiliary graph properties: A graph from $\mathcal{G}_m$ belongs to LCY if at most one connected component contains leaves (the *leaf component*); for $K \ge 1$ it belongs to $\mathsf{LCY}^{(K)}$ if it has the following four properties:

1. at most one connected component of G contains leaves (i. e., $\mathsf{LCY}^{(K)} \subseteq \mathsf{LCY}$);
2. the number $\zeta(G)$ of connected components is bounded by K;
3. if present, the leaf component of G contains at most K leaf and cycle edges;
4. the cyclomatic number $\gamma(G)$ is bounded by K.

Lemma 7. *If $T \subseteq U$ and (h_1, h_2) is from $\mathcal{Z}$ such that $G(T, h_1, h_2) \in \mathsf{LL}$ and (h_1, h_2) is T-bad, then there exists a subset T' of T such that $G(T', h_1, h_2) \in \mathsf{LCY}^{(4ck)}$ and (h_1, h_2) is T'-critical.*

Proof. Fix T and (h_1, h_2) as in the assumption. Initialize T' as T. We will remove ("peel") edges from $G(T', h_1, h_2)$ in four stages. Of course, by "removing edge $(h_1(x), h_2(x))$ from $G(T', h_1, h_2)$" we mean removing x from T'.

Stage 1: Initially, we have $d_{T'}((h_1, h_2)) > k$. Repeat the following step: If $G(T', h_1, h_2)$ contains a leaf, remove a leaf edge from it, otherwise remove a cycle edge. Clearly, such steps maintain the property that $G(T', h_1, h_2)$ belongs to LCY. Since $d_{T'}((h_1, h_2))$ can decrease by at most 1 when an edge is removed, we finally reach a situation where $d_{T'}((h_1, h_2)) = k$, i.e., (h_1, h_2) is T'-critical. Then $G(T', h_1, h_2)$ satisfies Property 1 from the definition of $\mathsf{LCY}^{(4ck)}$.

To prepare for the next stages, we define a set $T^* \subseteq T'$ with $2k \leq |T^*| \leq 2ck$, capturing keys that have to be "protected" during the following stages to maintain criticality of T'. If $|T'| = 2k$, we simply let $T^* = T'$. Then (h_1, h_2) is T^*-critical. If $|T'| > 2k$, a little more work is needed. By the definition of $d_T((h_1, h_2))$ we have $|T'| - \max\{|g_1(T')|, \ldots, |g_c(T')|\} = k$. For each $j \in \{1, \ldots, c\}$ Lemma 2(a) gives us a set $T_j^* \subseteq T'$ such that $|T_j^*| = 2k$ and $|g_j(T_j^*)| \leq k$. Let $T^* := T_1^* \cup \ldots \cup T_c^*$. Clearly, $2k \leq |T^*| \leq 2ck$. Since $T_j^* \subseteq T^*$, we have $|T^*| - |g_j(T^*)| \geq |T_j^*| - |g_j(T_j^*)| \geq k$, for $1 \leq j \leq c$, and hence $d_{T^*}((h_1, h_2)) \geq k$. On the other hand we know that there exists some j with $|T'| - |g_j(T')| = k$. Since $T^* \subseteq T'$, we have $|T^*| - |g_j(T^*)| \leq k$ for this j. Altogether we get $d_{T^*}((h_1, h_2)) = k$, which means that (h_1, h_2) is T^*-critical also in this case.

Now we "mark" all edges of $G = G(T', h_1, h_2)$ whose keys belong to T^*.

Stage 2: Remove all components of G without marked edges. Afterwards there are at most $2ck$ components left, and G satisfies Property 2.

Stage 3: If G has a leaf component C, repeatedly remove unmarked leaf and cycle edges from C, while C has such edges. The remaining leaf and cycle edges in C are marked, and thus there number is at most $2ck$; Property 3 is satisfied.

Stage 4: If there is a leaf component C with z marked edges (where $z \leq 2ck$), then $\gamma(C) \leq z-1$. Now consider a leafless component C' with cyclomatic number z. We need the following graph theoretic claim, which is proved in [1].

Claim. Every leafless connected graph with i marked edges has a leafless connected subgraph with cyclomatic number $\leq i+1$ that contains all marked edges.

This claim gives us a leafless subgraph C'' of C' with $\gamma(C'') \leq z + 1$ that contains all marked edges of C'. We remove from G all vertices and edges of C' that are not in C''. Doing this for all leafless components yields the final key set T' and the final graph $G = G(T', h_1, h_2)$. Summing contributions to the cyclomatic number of G over all (at most $2ck$) connected components, we see that $\gamma(G) \leq 4ck$; Property 4 is satisfied. $\square$

What have we achieved? By Lemma 7, we can bound $\Pr(B_S^{\mathsf{LL}})$ by just adding, over all $T' \subseteq S$, the probabilities $\Pr(\mathsf{LCY}_{T'}^{(4ck)} \cap \mathrm{crit}_{T'})$, that means, the terms $\Pr(\mathsf{LCY}_{T'}^{(4ck)} \mid \mathrm{crit}_{T'}) \cdot \Pr(\mathrm{crit}_{T'})$. Lemma 2(a) takes care of the second factor. By Lemma 2(b), we may assume that (h_1, h_2) acts fully random on T' for the first factor. The next lemma estimates this factor, using the notation from Section 3.

Lemma 8. *Let $T \subseteq U, |T| = t$, and $c, k \geq 1$. Then $p_T^{\mathsf{LCY}^{(4ck)}} \leq t! \cdot t^{O(1)} / m^{t-1}$.*

Proof. By Lemma 6, there are at most $t^{O(ck)} = t^{O(1)}$ ways to choose a bipartite graph G in $\mathsf{LCY}^{(4ck)}$ with t edges. Graph G cannot have more than $t + 1$ nodes, since cyclic components have at most as many nodes as edges, and in the single leaf component, if present, the number of nodes is at most one bigger than the number of edges. In each component of G, there are two ways to assign the vertices to the two sides of the bipartition. After such an assignment is fixed, there are at most m^{t+1} ways to label the vertices with elements of $[m]$, and there are $t!$ ways to label the edges of G with the keys in T. Assume now such labels have been chosen for G. Draw t edges $(h_1^*(x), h_2^*(x))$ from $[m]^2$ uniformly at random. The probability that they exactly fit the labeling of nodes and edges of G is $1/m^{2t}$. Thus, $p_T^{\mathsf{LCY}^{(4ck)}} \leq m^{t+1} \cdot t! \cdot t^{O(1)} / m^{2t} = t! \cdot t^{O(1)} / m^{t-1}$. □

We can now finally prove Lemma 5, the main lemma of this section.

Proof (of Lemma 5). By Lemma 7, and using the union bound, we get

$$\Pr(B_S^{\mathsf{LL}}) = \Pr(\exists T \subseteq S : \mathsf{LL}_T \cap \mathrm{bad}_T) \leq \Pr(\exists T' \subseteq S : \mathsf{LCY}_{T'}^{(4ck)} \cap \mathrm{crit}_{T'})$$
$$\leq \sum_{T' \subseteq S} \Pr(\mathsf{LCY}_{T'}^{(4ck)} \mid \mathrm{crit}_{T'}) \cdot \Pr(\mathrm{crit}_{T'}) =: \rho_S.$$

By Lemma 2(b), given the event that (h_1, h_2) is T'-critical, (h_1, h_2) acts fully random on T'. Using Lemma 8 and Lemma 2(a), this yields:

$$\Pr(\mathsf{LCY}_{T'}^{(4ck)} \mid \mathrm{crit}_{T'}) \cdot \Pr(\mathrm{crit}_{T'}) \leq (|T'|! \cdot |T'|^{O(1)} / m^{|T'|-1}) \cdot (|T'|^2 / \ell)^{ck}.$$

Summing up, collecting sets T' of equal size together, and using that ck is constant, we obtain

$$\rho_S \leq \sum_{2k \leq t \leq n} \binom{n}{t} \cdot \frac{t! \cdot t^{O(1)}}{m^{t-1}} \cdot \left(\frac{t^2}{\ell}\right)^{ck} \leq \frac{n}{\ell^{ck}} \cdot \sum_{2k \leq t \leq n} \frac{t^{O(1)}}{(1+\varepsilon)^{t-1}} = O\left(\frac{n}{\ell^{ck}}\right). \quad □$$

5   Cuckoo Hashing with a Stash

In this section we prove the desired bound on the rehash probability of cuckoo hashing with a stash when functions from $\mathcal{Z}$ are used. We focus on the question whether the pair (h_1, h_2) allows storing key set S in the two tables with a stash of size s. In view of Lemma 1, we identify minimal graphs with excess $s + 1$.

Definition 3. *An excess-$(s + 1)$ core graph is a leafless graph G with excess exactly $s + 1$ in which all connected components have at least two cycles. By $\mathsf{CS}^{(s+1)}$ we denote the set of all excess-$(s + 1)$ core graphs in $\mathcal{G}_m$.*

Lemma 9. *Let $G = G(S, h_1, h_2)$ be a cuckoo graph with $\mathrm{ex}(G) \geq s + 1$. Then G contains an excess-$(s + 1)$ core graph as a subgraph.*

Proof. We repeatedly remove edges from G. First we remove cycle edges until the excess is exactly $s + 1$. Then we remove components that are trees or unicyclic. Finally we remove leaf edges one by one until the remaining graph is leafless.   □

We are now ready to state our main theorem.

Theorem 1. *Let $\varepsilon > 0$ and $0 < \delta < 1$, let $s \geq 0$ and $k \geq 1$ be given. Assume $c \geq (s+2)/(\delta k)$. For $n \geq 1$ consider $m \geq (1+\varepsilon)n$ and $\ell = n^{\delta}$. Let $S \subseteq U$ with $|S| = n$. Then for (h_1, h_2) chosen at random from $\mathcal{Z} = \mathcal{Z}_{\ell,m}^{2k,c}$ the following holds:*

$$\Pr(\mathrm{ex}(G(S, h_1, h_2)) \geq s + 1) = O(1/n^{s+1}).$$

Proof. By Lemma 3 and Lemma 9, the probability that the excess of $G(S, h_1, h_2)$ is at least $s + 1$ is

$$\Pr(\exists T \subseteq S \colon \mathsf{CS}_T^{(s+1)}) \leq \Pr(B_S^{\mathsf{CS}^{(s+1)}}) + \sum_{T \subseteq S} p_T^{\mathsf{CS}^{(s+1)}}. \tag{3}$$

Since $\mathsf{CS}^{(s+1)} \subseteq \mathsf{LL}$, we can combine Lemmas 4 and 5 to obtain $\Pr(B_S^{\mathsf{CS}^{(s+1)}}) \leq \Pr(B_S^{\mathsf{LL}}) = O(n/\ell^{ck})$. It remains to bound the second summand in (3).[2]

Lemma 10. $\sum_{T \subseteq S} p_T^{\mathsf{CS}^{(s+1)}} = O(1/n^{s+1})$.

Proof. We start by counting (unlabeled) excess-$(s+1)$ core graphs with t edges. A connected component C of such a graph G with cyclomatic number $\gamma(C)$ (which is at least 2) contributes $\gamma(C) - 1$ to the excess of G. This means that if G has $\zeta = \zeta(G)$ components, then $s + 1 = \gamma(G) - \zeta$ and $\zeta \leq s + 1$, and hence $\gamma = \gamma(G) \leq 2(s+1)$. Using Lemma 6, there are at most $N(t, 0, \gamma, \zeta) = t^{O(\gamma+\zeta)} = t^{O(s)}$ such graphs G. If from each component C of such a graph G we remove $\gamma(C) - 1$ cycle edges, we get unicyclic components, which have as many nodes as edges. This implies that G has $t - (s + 1)$ nodes.

Now fix a bipartite (unlabeled) excess-$(s+1)$ core graph G with t edges and ζ components, and let $T \subseteq U$ with $|T| = t$ be given. There are $2^{\zeta} \leq 2^{s+1}$ ways of assigning the $t - s - 1$ nodes to the two sides of the bipartition, and then at most m^{t-s-1} ways of assigning labels from $[m]$ to the nodes. Thus, the number of bipartite graphs with property $\mathsf{CS}^{(s+1)}$, where each node is labeled with one side of the bipartition and an element of $[m]$, and where the t edges are labeled with distinct elements of T is smaller than $t! \cdot 2^{s+1} \cdot m^{t-s-1} \cdot t^{O(s)}$.

Now if G with such a labeling is fixed, and we choose t edges from $[m]^2$ uniformly at random, the probability that all edges $(h_1(x), h_2(x))$, $x \in T$, match the labeling is $1/m^{2t}$. For constant s, this yields the following bound:

$$\sum_{T \subseteq S} p_T^{\mathsf{CS}^{(s+1)}} \leq \sum_{s+3 \leq t \leq n} \binom{n}{t} \frac{2^{s+1} \cdot n^{t-s-1} \cdot t! \cdot t^{O(s)}}{m^{2t}} \leq \frac{2^{s+1}}{n^{s+1}} \cdot \sum_{s+3 \leq t \leq n} \frac{n^t \cdot t^{O(1)}}{m^t}$$

$$= O\left(\frac{1}{n^{s+1}}\right) \cdot \sum_{s+3 \leq t \leq n} \frac{t^{O(1)}}{(1+\varepsilon)^t} = O\left(\frac{1}{n^{s+1}}\right). \qquad \square$$

---

[2] We remark that the following calculations also give an alternative, simpler proof of [11, Theorem 2.1] for the fully random case, even if the effort needed to prove Lemma 6 and [9, Lemma 2] is taken into account.

Since $\ell = n^\delta$ and $c \geq (s+2)/(k\delta)$, we get

$$\Pr(\text{ex}(G) \geq s+1) = O(n/\ell^{ck}) + O(1/n^{s+1}) = O(1/n^{s+1}).$$ □

6   Simulating Uniform Hashing

In the following, let R be the range of the hash function to construct, and assume that $(R, \oplus)$ is a commutative group. (We could use $R = [t]$ with addition mod t.)

Theorem 2. *Let $n \geq 1, 0 < \delta < 1, \varepsilon > 0$, and $s \geq 0$ be given. There exists a data structure DS_n that allows us to compute a function $h\colon U \to R$ such that:*

(i) *For each $S \subseteq U$ of size n there is an event B_S of probability $O(1/n^{s+1})$ such that conditioned on $\overline{B_S}$ the function h is distributed uniformly on S.*
(ii) *For arbitrary $x \in U$, $h(x)$ can be evaluated in time $O(s/\delta)$.*
(iii) *DS_n comprises $2(1+\varepsilon)n + O(sn^\delta)$ words from R and $O(s)$ words from U.*

Proof. Let $k \geq 1$ and choose $c \geq (s+2)/(k\delta)$. Given U and n, set up DS_n as follows. Let $m = (1+\varepsilon)n$ and $\ell = n^\delta$, and choose and store a hash function pair (h_1, h_2) from $\mathcal{Z} = \mathcal{Z}_{\ell,m}^{2k,c}$ (see Definition 1), with component functions $g_1, \ldots, g_c$ from $\mathcal{H}_\ell^{2k}$. In addition, choose 2 random vectors $t_1, t_2 \in R^m$, c random vectors $y_1, \ldots, y_c \in R^\ell$, and choose f at random from a $2k$-wise independent family of hash functions from U to R.

Using DS_n, the mapping $h\colon U \to R$ is defined as follows:

$$h(x) = t_1[h_1(x)] \oplus t_2[h_2(x)] \oplus f(x) \oplus y_1[g_1(x)] \oplus \ldots \oplus y_c[g_c(x)].$$

DS_n satisfies (ii) and (iii) of Theorem 2. We show that it satisfies (i) as well.

First, consider only the hash functions (h_1, h_2) from $\mathcal{Z}$. By Lemma 5 we have $\Pr(B_S^{\mathsf{LL}}) = O(n/\ell^{ck}) = O(1/n^{s+1})$. Now fix $(h_1, h_2) \notin B_S^{\mathsf{LL}}$, which includes fixing the components $g_1, \ldots, g_c$. Let $T \subseteq S$ be such that $G(T, h_1, h_2)$ is the 2-core of $G(S, h_1, h_2)$, the maximal subgraph with minimum degree 2. Graph $G(T, h_1, h_2)$ is leafless, and since $(h_1, h_2) \notin B_S^{\mathsf{LL}}$, we have that (h_1, h_2) is T-good. Now we note that the part $f(x) \oplus \bigoplus_{1 \leq j \leq c} y_j[g_j(x)]$ of $h(x)$ acts exactly as one of our hash functions h_1 and h_2 (see Definition 1(a)), so that arguing as in the proof of Lemma 2 we see that $h(x)$ is fully random on T.

Now assume that f and the entries in the tables $y_1, \ldots, y_c$ are fixed. It is not hard to show that the random entries in t_1 and t_2 alone make sure that $h(x)$, $x \in S - T$, is fully random. (Such proofs were given in [9] and [16].) □

Concluding Remarks

We presented a family of efficient hash functions and showed that it exhibits sufficiently strong random properties to run cuckoo hashing with a stash, preserving the favorable performance guarantees of this hashing scheme. We also described a simple construction for simulating uniform hashing. We remark that the performance of our construction can be improved by using 2-universal hash families[3] (see, e. g., [3,5]) for the g_j-components. The proof of Lemma 2 can be adapted easily to these weaker families. It remains open whether generalized cuckoo hashing [10,8] can be run with efficient hash families.

References

1. Aumüller, M., Dietzfelbinger, M., Woelfel, P.: Explicit and efficient hash families suffice for cuckoo hashing with a stash. CoRR abs/1204.4431 (2012)
2. Calkin, N.J.: Dependent sets of constant weight binary vectors. Combinatorics, Probability and Computing 6(3), 263–271 (1997)
3. Carter, L., Wegman, M.N.: Universal classes of hash functions. J. Comput. Syst. Sci. 18(2), 143–154 (1979)
4. Devroye, L., Morin, P.: Cuckoo hashing: Further analysis. Inf. Process. Lett. 86(4), 215–219 (2003)
5. Dietzfelbinger, M., Hagerup, T., Katajainen, J., Penttonen, M.: A reliable randomized algorithm for the closest-pair problem. J. Algorithms 25(1), 19–51 (1997)
6. Dietzfelbinger, M., Rink, M.: Applications of a Splitting Trick. In: Albers, S., Marchetti-Spaccamela, A., Matias, Y., Nikoletseas, S., Thomas, W. (eds.) ICALP 2009. LNCS, vol. 5555, pp. 354–365. Springer, Heidelberg (2009)
7. Dietzfelbinger, M., Schellbach, U.: On risks of using cuckoo hashing with simple universal hash classes. In: 20th Ann. ACM-SIAM Symp. on Discrete Algorithms (SODA), pp. 795–804 (2009)
8. Dietzfelbinger, M., Weidling, C.: Balanced allocation and dictionaries with tightly packed constant size bins. Theor. Comput. Sci. 380(1-2), 47–68 (2007)
9. Dietzfelbinger, M., Woelfel, P.: Almost random graphs with simple hash functions. In: Proc. 35th ACM Symp. on Theory of Computing (STOC), New York, NY, USA, pp. 629–638 (2003)
10. Fotakis, D., Pagh, R., Sanders, P., Spirakis, P.G.: Space efficient hash tables with worst case constant access time. Theory Comput. Syst. 38(2), 229–248 (2005)
11. Kirsch, A., Mitzenmacher, M., Wieder, U.: More Robust Hashing: Cuckoo Hashing with a Stash. In: Halperin, D., Mehlhorn, K. (eds.) ESA 2008. LNCS, vol. 5193, pp. 611–622. Springer, Heidelberg (2008)
12. Kirsch, A., Mitzenmacher, M., Wieder, U.: More Robust Hashing: Cuckoo Hashing with a Stash. SIAM J. Comput. 39(4), 1543–1561 (2009)
13. Klassen, T.Q., Woelfel, P.: Independence of Tabulation-Based Hash Classes. In: Fernández-Baca, D. (ed.) LATIN 2012. LNCS, vol. 7256, pp. 506–517. Springer, Heidelberg (2012)
14. Kutzelnigg, R.: A further analysis of cuckoo hashing with a stash and random graphs of excess r. Discr. Math. and Theoret. Comput. Sci. 12(3), 81–102 (2010)
15. Mitzenmacher, M., Vadhan, S.P.: Why simple hash functions work: exploiting the entropy in a data stream. In: Proc. 19th Ann. ACM-SIAM Symp. on Discrete Algorithms (SODA), pp. 746–755 (2008)
16. Pagh, A., Pagh, R.: Uniform hashing in constant time and optimal space. SIAM J. Comput. 38(1), 85–96 (2008)
17. Pagh, R., Rodler, F.F.: Cuckoo hashing. J. Algorithms 51(2), 122–144 (2004)

---

[3] A family $\mathcal{H}$ of hash functions with range R is 2-universal if for each pair $x, y \in U$, $x \neq y$, and h chosen at random from $\mathcal{H}$ we have $\Pr(h(x) = h(y)) \leq 2/|R|$.

18. Pătraşcu, M., Thorup, M.: The power of simple tabulation hashing. In: Proc. 43rd ACM Symp. on Theory of Computing (STOC), pp. 1–10 (2011)
19. Siegel, A.: On universal classes of extremely random constant-time hash functions. SIAM J. Comput. 33(3), 505–543 (2004)
20. Thorup, M., Zhang, Y.: Tabulation based 4-universal hashing with applications to second moment estimation. In: Proc. 15th Ann. ACM-SIAM Symp. on Discrete Algorithms (SODA), pp. 615–624 (2004)
21. Thorup, M., Zhang, Y.: Tabulation based 5-universal hashing and linear probing. In: Proc. 12th ALENEX, pp. 62–76. SIAM (2010)
22. Wegman, M.N., Carter, L.: New classes and applications of hash functions. In: Proc. 20th Ann. Symp. on Foundations of Computer Science (FOCS), pp. 175–182. IEEE Computer Society (1979)
23. Woelfel, P.: Asymmetric balanced allocation with simple hash functions. In: Proc. 17th ACM-SIAM Symp. on Discrete Algorithms (SODA), pp. 424–433 (2006)

On Online Labeling with Polynomially Many Labels

Martin Babka[1], Jan Bulánek[1,2], Vladimír Čunát[1,*],
Michal Koucký[2,3,**], and Michael Saks[4,***]

[1] Faculty of Mathematics and Physics, Charles University, Prague
[2] Institute of Mathematics, Academy of Sciences, Prague
[3] Department of Computer Science, Aarhus University
[4] Department of Mathematics, Rutgers University
{babkys,vcunat}@gmail, {bulda,koucky}@math.cas.cz, saks@math.rutgers.edu

Abstract. In the online labeling problem with parameters n and m we are presented with a sequence of n *keys* from a totally ordered universe U and must assign each arriving key a label from the label set $\{1, 2, \ldots, m\}$ so that the order of labels (strictly) respects the ordering on U. As new keys arrive it may be necessary to change the labels of some items; such changes may be done at any time at unit cost for each change. The goal is to minimize the total cost. An alternative formulation of this problem is the *file maintenance problem*, in which the items, instead of being labeled, are maintained in sorted order in an array of length m, and we pay unit cost for moving an item.

For the case $m = cn$ for constant $c > 1$, there are known algorithms that use at most $O(n \log(n)^2)$ relabelings in total [9], and it was shown recently that this is asymptotically optimal [1]. For the case of $m = \theta(n^C)$ for $C > 1$, algorithms are known that use $O(n \log n)$ relabelings. A matching lower bound was claimed in [7]. That proof involved two distinct steps: a lower bound for a problem they call *prefix bucketing* and a reduction from prefix bucketing to online labeling. The reduction seems to be incorrect, leaving a (seemingly significant) gap in the proof. In this paper we close the gap by presenting a correct reduction to prefix bucketing. Furthermore we give a simplified and improved analysis of the prefix bucketing lower bound. This improvement allows us to extend the lower bounds for online labeling to the case where the number m of labels is superpolynomial in n. In particular, for superpolynomial m we get an asymptotically optimal lower bound $\Omega((n \log n)/(\log \log m - \log \log n))$.

Keywords: online labeling, file maintenance problem, lower bounds.

* The first three authors gratefully acknowledge a support by the Charles University Grant Agency (grant No. 265 111 and 344 711) and SVV project no. 265 314.
** Currently a visiting associate professor at Aarhus University, partially supported by the Sino-Danish Center CTIC (funded under the grant 61061130540). Supported in part by GA ČR P202/10/0854, grant IAA100190902 of GA AV ČR, Center of Excellence CE-ITI (P202/12/G061 of GA ČR) and RVO: 67985840.
*** The work of this author was done while on sabbatical at Princeton University and was also supported in part by NSF under grant CCF-0832787.

L. Epstein and P. Ferragina (Eds.): ESA 2012, LNCS 7501, pp. 121–132, 2012.
© Springer-Verlag Berlin Heidelberg 2012

1   Introduction

In the online labeling problem with parameters n, m, r, we are presented with a sequence of n *keys* from a totally ordered universe U of size r and must assign each arriving key a label from the label set $\{1, 2, \ldots, m\}$ so that the order of labels (strictly) respects the ordering on U. As new keys arrive it may be necessary to change the labels of some items; such changes may be done at any time at unit cost for each change. The goal is to minimize the total cost. An alternative formulation of this problem is the *file maintenance problem*, in which the items, instead of being labeled, are maintained in sorted order in an array of length m, and we pay unit cost for moving an item.

The problem, which was introduced by Itai, Konheim and Rodeh [9], is natural and intuitively appealing, and has had applications to the design of data structures (see for example the discussion in [7], and the more recent work on cache-oblivious data structures [3,5,4]). A connection between this problem and distributed resource allocation was recently shown by Emek and Korman [8].

The parameter m, the *label space* must be at least the number of items n or else no valid labeling is possible. There are two natural range of parameters that have received the most attention. In the case of *linearly many labels* we have $m = cn$ for some $c > 1$, and in the case of *polynomially many labels* we have $m = \theta(n^C)$ for some constant $C > 1$. The problem is trivial if the universe U is a set of size at most m, since then we can simply fix an order preserving bijection from U to $\{1, \ldots, m\}$ in advance. In this paper we will usually restrict attention to the case that U is a totally ordered set of size at least exponential in n (as is typical in the literature).

Itai et al. [9] gave an algorithm for the case of linearly many labels having worst case total cost $O(n \log(n)^2)$. Improvements and simplifications were given by Willard [10] and Bender et al. [2]. In the special case that $m = n$, algorithms with cost $O(\log(n)^3)$ per item were given [11,6]. It is also well known that the algorithm of Itai et al. can be adapted to give total cost $O(n \log(n))$ in the case of polynomially many labels. All of these algorithms make no restriction on the size of universe U of keys.

For lower bounds, a subset of the present authors recently proved [1] a tight $\Omega(n \log(n)^2)$ lower bound for the case of linearly many labels and tight bound $\Omega(n \log(n)^3)$ for the case $m = n$. These bounds hold even when the size of the universe U is only a constant multiple of m. The bound remains non-trivial (superlinear in n) for $m = O(n \log(n)^{2-\varepsilon})$ but becomes trivial for $m \in \Omega(n \log(n)^2)$.

For the case of polynomially many labels, Dietz at al. [7] (also in [11]) claim a matching lower bound for the $O(n \log(n))$ upper bound. Their result consists of two parts; a lower bound for a problem they call *prefix bucketing* and a reduction from prefix bucketing to online labeling. However, the argument giving the reduction seems to be incorrect, and we recently raised our concerns with one of the authors (Seiferas), who agrees that there is a gap in the proof.

This paper makes the following contributions:

- We provide a correct reduction from prefix bucketing to online labeling, which closes the gap in the lower bound proof for online labeling for the case of polynomially many labels.
- We provide a simpler and more precise lower bound for prefix bucketing which allows us to extend the lower bounds for online labeling to the case where the label size is as large as 2^{n^ε}. Specifically we prove a lower bound of $\Omega((n \log n)/(\log \log m - \log \log n))$ that is valid for m between $n^{1+\varepsilon}$ and 2^{n^ε}. Note that for polynomially many labels this reduces to $\Omega(n \log(n))$.

We remark that, unlike the bounds of [1] for the case of linearly many labels, our lower bound proof requires that the universe U is at least exponential in n. It is an interesting question whether one could design a better online labeling algorithm for U of size say $m \log n$. We summarize known results in Table 1. All the results are for deterministic algorithms. There are no known results for randomized algorithms except for what is implied by the deterministic case.

Table 1. Summary of known bounds for the online labeling problem

Array size (m)	Lower bound		Upper bound	
$m = n$	$\Omega(n \log(n)^3)$	[1]	$O(n \log(n)^3)$	[11]
$m = \Theta(n), m > n$	$\Omega(n \log(n)^2)$	[1]	$O(n \log(n)^2)$	[9]
$m = n^{1+o(1)}$	$\Omega\left(\frac{n \log(n)^2}{\log m - \log n}\right)$	[12]	$O\left(\frac{n \log(n)^2}{\log m - \log n}\right)$	[9]
$m = n^{1+\Theta(1)}$	$\Omega(n \log(n))$	[this paper]	$O(n \log(n))$	[9]
$m = n^{\Omega(\log(n))}$	$\Omega\left(\frac{n \log n}{\log \log m}\right)$	[this paper]	$O\left(\frac{n \log n}{\log \log m}\right)$	[1]

Our proof follows the high level structure of the proof from [7]. In the remainder of the introduction we sketch the two parts, and relate our proof to the one in [7].

1.1   Reducing Online Labeling to Prefix Bucketing

Dietz et al. [7] sketched a reduction from online labeling to prefix bucketing. In their reduction they describe an adversary for the labeling problem. They show that given any algorithm for online labeling, the behavior of the algorithm against the adversary can be used to construct a strategy for prefix bucketing. If one can show that the cost of the derived bucketing strategy is no more than a constant times the cost paid by the algorithm for relabelings then a lower bound on bucketing will give a similar lower bound on the cost of any relabeling algorithm. Unfortunately, their proof sketch does not show this. In particular, a single relabeling step may correspond to a bucketing step whose cost is $\Omega(\log(n))$, and this undermines the reduction. This may happen when inserting $\Theta(\log n)$ items into an empty segment of size n^ϵ without triggering any relabelings. We construct a different adversary for which one gets the needed correspondence between relabeling cost and bucketing steps.

Our goal in constructing an adversary is to force any online algorithm to perform many relabelings during insertion on n keys. The adversary is described in detail in Section 2 here we provide a high level description.

The adversary starts by inserting the minimal and maximal element of U, i.e. 1 and r, and five other keys uniformly spread in U. From then on the adversary will always pick a suitable pair of existing consecutive keys and the next inserted key will be the average of the pair. Provided that $r > 2^n$ there will always be an unused key between the pair.

It is illuminating to think of the problem in terms of the file maintenance problem mentioned earlier. In this reformulation we associate to the label space $[m]$ an array indexed by $\{1, \ldots, m\}$ and think of a key labeled by j as being stored in location j. Intuitively, the adversary wants to choose two consecutive keys that appear in a crowded region of this array. By doing this repeatedly, the adversary hopes to force the algorithm to move many keys within the array (which corresponds to relabeling them). The problem is to make precise the notion of "crowdedness". Crowding within the array occurs at different scales (so a small crowded region may lie inside a large uncrowded region) and we need to find a pair of consecutive keys with the property that all regions containing the pair are somewhat crowded.

With the array picture in mind, we call an interval of labels a *segment*, and say that a label is *occupied* if there is a key assigned to it. The *density* of a segment is the fraction of occupied labels.

In [7], the authors show that there is always a *dense point*, which is a point with the property that every segment containing it has density at least half the overall density of the label space. They use this as the basis for their adversary, but this adversary does not seem to be adequate.

We design a different adversary (which is related to the adversary constructed in [1] to handle the case of linearly many labels). The adversary maintains a sequence (*hierarchy*) of nested segments. Each successive segment in the hierarchy has size at most half the previous segment, and its density is within a constant factor of the density of the previous segment. The hierarchy ends with a segment having between 2 and 7 keys. The key chosen to be inserted next is the average (rounded down) of the two "middle" keys of the smallest segment.

For the first 8 insertions, the hierarchy consists only of the single segment $[m]$. After each subsequent insertion, the algorithm $\mathcal{A}$ specifies the label of the next key and (possibly) relabels some keys. The adversary then updates its hierarchy. For the hierarchy prior to the insertion, define the critical segment to be the smallest segment of the hierarchy that contains the label assigned to the inserted key and the old and new labels of all keys that were relabeled. The new hierarchy agrees with the previous hierarchy up to and including the critical segment. Beginning from the critical segment the hierarchy is extended as follows. Having chosen segment S for the hierarchy, define its *left buffer* to be the smallest subsegment of S that starts at the minimum label of S and includes at least $1/8$ of the occupied labels of S, and its *right buffer* to be the smallest subsegment that ends at the maximum label of S and includes at least $1/8$ of the occupied keys of S. Let S' be the segment obtained from S by deleting the left and right buffers. The successor segment of S in the hierarchy is the shortest subsegment of S' that contains exactly half (rounded down) of the occupied labels of S'. The hierarchy ends when we reach a segment with at most seven occupied labels; such a segment necessarily has at least two occupied labels.

It remains to prove that the algorithm will make lot of relabels on the sequence of keys produced by the adversary. For that proof we need a game introduced by Dietz et al. [7] that they call prefix bucketing.

A *prefix bucketing of n items into k buckets* (numbered 1 to k) is a one player game consisting of n steps. At the beginning of the game all the buckets are empty. In each step a new item arrives and the player selects an index $p \in [k]$. The new item as well as all items in buckets $p+1, \ldots, k$ are moved into bucket p at a cost equal to the total number of items in bucket p after the move. The goal is to minimize the total cost of n steps of the game. Notice that suffix bucketing would be a more appropriate name for our variant of the game, however we keep the original name as the games are equivalent.

We will show that if $\mathcal{A}$ is any algorithm for online labeling and we run $\mathcal{A}$ against our adversary then the behavior of $\mathcal{A}$ corresponds to a prefix bucketing of n items into $k = \lceil \log m \rceil$ buckets. The total cost of the prefix bucketing will be within a constant factor of the total number of relabelings performed by the online labeling algorithm. Hence, a lower bound on the cost of a prefix bucketing of n items into k buckets will imply a lower bound on the cost of the algorithm against our adversary.

Given the execution of $\mathcal{A}$ against our adversary we create the following prefix bucketing. We maintain a *level* for each key inserted by the adversary; one invariant these levels will satisfy is that for each segment in the hierarchy, the level of the keys inside the segment are at least the depth of the segment in the hierarchy. The level of a newly inserted key is initially k. After inserting the key, the algorithm does its relabeling, which determines the critical segment and the critical level. All keys in the current critical segment whose level exceeds the critical level have their levels reset to the current critical level.

The assignment of levels to keys corresponds to a bucketing strategy, where the level of a key is the bucket it belongs to. Hence, if p is the critical level, all the keys from buckets $p, p+1, \ldots, k$ will be merged into the bucket p.

We need to show that the cost of the merge operation corresponds to the number of relabelings done by the online algorithm at a given time step. For this we make the assumption (which can be shown to hold without loss of generality) that the algorithm is *lazy*, which means that at each time step the set of keys that are relabeled is a contiguous block of keys that includes the newly inserted keys. The cost of the bucketing merge step is at most the number of keys in the critical segment. One can argue that for each successor segment of the critical segment, either all labels in its left buffer or all labels in its right buffer were reassigned, and the total number of such keys is a constant fraction of the keys in the critical segment.

1.2   An Improved Analysis of Bucketing

It then remains to give a lower bound for the cost of prefix bucketing. This was previously given by Dietz et al. [7] for $k \in \Theta(\log n)$. We give a different and simpler proof that gives asymptotically optimal bound for k between $\log n$ and $O(n^\epsilon)$. We define a family of trees called k-admissible trees and show that the cost of bucketing for n and k, is between $dn/2$ and dn where d is the minimum depth of a k-admissible tree on n vertices. We further show that the minimum depth of a k-admissible tree on n vertices is equal $g_k(n)$ which is defined to be the smallest d such that $\binom{k+d-1}{k} \geq n$. This gives a characterization of the optimal cost of prefix bucketing (within a factor of 2). When we apply this characterization we need to use estimates of $g_k(n)$ in terms of more familiar functions (Lemma 11), and there is some loss in these estimates.

2   The Online Labeling Problem

In this paper, interval notation is used for sets of consecutive integers, e.g., $[a, b]$ is the set $\{k \in \mathbb{Z} : a \leq k \leq b\}$. Let m and $r \geq 1$ be integers. We assume without loss of generality $U = [r]$. An online labeling algorithm $\mathcal{A}$ with range m is an algorithm that on input sequence $y^1, y^2, \ldots, y^t$ of distinct elements from U gives an *allocation* $f : \{y^1, y^2, \ldots, y^t\} \to [m]$ that respects the natural ordering of $y^1, \ldots, y^t$ that is for any $x, y \in \{y^1, y^2, \ldots, y^t\}$, $f(x) < f(y)$ if and only if $x < y$. We refer to $y^1, y^2, \ldots, y^t$ as *keys*. The trace of $\mathcal{A}$ on a sequence $y^1, y^2, \ldots, y^n \in U$ is the sequence $f^0, f^1, f^2, \ldots, f^n$ of functions such that f^0 is the empty mapping and for $i = 1, \ldots, n$, f^i is the output of $\mathcal{A}$ on $y^1, y^2, \ldots, y^i$. For the trace $f^0, f^1, f^2, \ldots, f^n$ and $t = 1, \ldots, n$, we say that $\mathcal{A}$ *relocates* $y \in \{y^1, y^2, \ldots, y^t\}$ at time t if $f^{t-1}(y) \neq f^t(y)$. So each y^t is relocated at time t. For the trace $f^0, f^1, f^2, \ldots, f^n$ and $t = 1, \ldots, n$, Rel^t denotes the set of relocated keys at step t. The cost of $\mathcal{A}$ incurred on $y^1, y^2, \ldots, y^n \in U$ is $\chi_{\mathcal{A}}(y_1, \ldots, y_n) = \sum_{i=0}^{n} |Rel^i|$ where Rel is measured with respect to the trace of $\mathcal{A}$ on $y^1, y^2, \ldots, y^n$. The maximum cost $\chi_{\mathcal{A}}(y^1, \ldots, y^n)$ over all sequences $y^1, \ldots, y^n$ is denoted $\chi_{\mathcal{A}}(n)$. We write $\chi_m(n)$ for the smallest cost $\chi_{\mathcal{A}}(n)$ that can be achieved by any algorithm $\mathcal{A}$ with range m.

2.1   The Main Theorem

In this section, we state our lower bound results for $\chi_m(n)$.

Theorem 1. *There are positive constants C_0, C_1 and C_2 so that the following holds. Let m, n be integers satisfying $C_0 \leq n \leq m \leq 2^{n^{C_1}}$. Let the size of U be more than 2^{n+4}. Then $\chi_m(n) \geq C_2 \cdot \frac{n \log n}{3 + \log \log m - \log \log n}$.*

To prove the theorem for given algorithm $\mathcal{A}$ we will adversarially construct a sequence $y^1, y^2, \ldots, y^t$ of keys that will cause the algorithm to incur the desired cost. In the next section we will design the adversary.

2.2 Adversary Construction

Any interval $[a, b] \subseteq [m]$ is called a *segment*. Fix $n, m, r > 1$ such that $m \geq n$ and $r > 2^{n+4}$. Fix some online labeling algorithm $\mathcal{A}$ with range m. To pick the sequence of keys $y^1, \ldots, y^n$, the adversary will maintain a sequence of nested segments $S^t_{\mathbf{depth}(t)} \subseteq \cdots \subseteq S^t_2 \subseteq S^t_1 = [m]$, updating them after each time step t. The adversary will choose the next element y^t to fall between the keys in the smallest interval $S^t_{\mathbf{depth}(t)}$. In what follows, f^t is the allocation of $y^1, \ldots, y^t$ by the algorithm $\mathcal{A}$.

The *population of a segment* S *at time* t is $\mathbf{pop}^t(S) = (f^t)^{-1}(S)$ and the *weight of* S *at time* t is $\mathbf{weight}^t(S) = |\mathbf{pop}^t(S)|$. For $t = 0$, we extend the definition by $\mathbf{pop}^0(S) = \emptyset$ and $\mathbf{weight}^0(S) = 0$. The *density of* S *at time* t is $\rho^t(S) = \mathbf{weight}^t(S)/|S|$. For a positive integer b, let $\mathbf{densify}^t(S, b)$ be the smallest subsegment T of S of weight exactly $\lfloor(\mathbf{weight}^t(S) - 2b)/2\rfloor$ such that $\mathbf{pop}^t(T)$ does not contain any of the b largest and smallest elements of $\mathbf{pop}^t(S)$. Hence, $\mathbf{densify}^t(S, b)$ is the densest subsegment of S that contains the appropriate number of items but which is surrounded by a large population of S on either side. If $\mathbf{pop}^t(S) = \{x_1 < x_2 < \cdots < x_\ell\}$ then $\mathbf{midpoint}^t(S) = \lceil(x_{\lceil(\ell-1)/2\rceil} + x_{\lceil(\ell+1)/2\rceil})/2\rceil$.

Let $y^1, y^2, \ldots, y^t$ be the first t keys inserted and let Rel^t be the keys that are relabeled by $\mathcal{A}$ in response to the insertion of y^t. The *busy segment* $B^t \subseteq [m]$ at time t is the smallest segment that contains $f^t(Rel^t) \cup f^{t-1}(Rel^t \setminus \{y^t\})$. We say that the algorithm $\mathcal{A}$ is *lazy* if all the keys that are mapped to B^t are relocated at step t, i.e., $(f^t)^{-1}(B^t) = Rel^t$. By Proposition 4 in [1], when bounding the cost of $\mathcal{A}$ from below we may assume that $\mathcal{A}$ is lazy.

Adversary$(\mathcal{A}, n, m, r)$

Set $p^0 = 0$.

For $t = 1, 2, \ldots, n$ do

- If $t < 8$, let $S^t_1 = [m]$, $b^t_1 = 1$, $\mathbf{depth}(t) = 1$, and $p^t = 1$. Set $y^t = 1 + \lceil(t - 1) \cdot (r - 1)/6\rceil$, and run $\mathcal{A}$ on $y^1, y^2, \ldots, y^t$ to get f^t. Continue to next t.
- *Preservation Rule:* For $i = 1, \ldots, p^{t-1}$, let $S^t_i = S^{t-1}_i$ and $b^t_i = b^{t-1}_i$. (For $t \geq 8$, copy the corresponding segments from the previous time step.)
- *Rebuilding Rule:*
  Set $i = p^{t-1} + 1$.
  While $\mathbf{weight}^{t-1}(S^t_{i-1}) \geq 8$
    - Set $S^t_i = \mathbf{densify}^{t-1}(S^t_{i-1}, b^t_{i-1})$.
    - Set $b^t_i = \lceil\mathbf{weight}^{t-1}(S^t_i)/8\rceil$.
    - Increase i by one.
- Set $\mathbf{depth}(t) = i - 1$.
- Set $y^t = \mathbf{midpoint}^{t-1}(S^t_{\mathbf{depth}(t)})$.
- *The critical level:* Run $\mathcal{A}$ on $y^1, y^2, \ldots, y^t$ to get f^t. Calculate Rel^t and B^t. Set the critical level p^t to be the largest integer $j \in [\mathbf{depth}(t)]$ such that $B^t \subseteq S^t_j$.

Output: $y^1, y^2, \ldots, y^n$.

We make the following claim about the adversary which implies Theorem 1. We did not attempt to optimize the constants.

Lemma 1. *Let* m, n, r *be integers such that and* $2^{32} \leq n \leq m \leq 2^{\sqrt[4]{n}/8}$ *and* $2^{n+4} < r$. *Let* $\mathcal{A}$ *be a lazy online labeling algorithm with the range* m. *Let* $y^1, y^2, \ldots, y^n$ *be the output of* **Adversary**$(\mathcal{A}, n, m, r)$. *Then the cost*

$$\chi_{\mathcal{A}}(y^1, y^2, \ldots, y^n) \geq \frac{1}{512} \cdot \frac{n \log n}{3 + \log\lceil \log m \rceil - \log \log n} - \frac{n}{6}.$$

Notice, if $r > 2^{n+4}$ then for any $t \in [n-1]$ the smallest pair-wise difference between integers $y^1, y^2, \ldots, y^t$ is at least 2^{n+1-t} so y^{t+1} chosen by the adversary is different from all the previous y's. All our analysis will assume this.

To prove the lemma we will design a so called *prefix bucketing game* from the interaction between the adversary and the algorithm, we will relate the cost of the prefix bucketing to the cost $\chi_{\mathcal{A}}(y^1, y^2, \ldots, y^n)$, and we will lower bound the cost of the prefix bucketing.

In preparation for this, we prove several useful properties of the adversary.

Lemma 2. *For any* $t \in [n]$, $\mathbf{depth}(t) \leq \log m$.

Proof. The lemma is immediate for $t < 8$. For $t \geq 8$, it suffices to show that the hierarchy $S_1^t, S_2^t, \ldots$ satisfies that for each $i \in [1, \mathbf{depth}(t) - 1]$, $|S_{i+1}^t| \leq |S_i^t|/2$. Recall that S_i^t is obtained from S_{i-1}^t by removing the left and right buffer to get a subsegment S' and then taking S_{i+1}^t to be the shortest subsegment of S' that contains exactly half of the keys (rounded down) labeled in S'. Letting L' be the smallest $\lfloor |S'|/2 \rfloor$ labels of S' and R' be the largest $\lfloor |S'|/2 \rfloor$ labels of S', one of L' and R' contains at least half of the occupied labels (rounded down) in S', which implies $|S_{i+1}^t| \leq |S'|/2 < |S_i^t|/2$. $\qquad\square$

Lemma 3. *For any* $t \in [n]$ *and* $i \in [\mathbf{depth}(t) - 1]$, $64 \cdot b_{i+1}^t \geq \mathbf{weight}^{t-1}(S_i^t) - \mathbf{weight}^{t-1}(S_{i+1}^t)$.

The proof is omitted due to space constraints.

Corollary 1. *For any* $t \in [n]$ *and* $i \in [\mathbf{depth}(t) - 1]$, $8 + 64 \cdot \sum_{j=i+1}^{\mathbf{depth}(t)} b_j^t \geq \mathbf{weight}^{t-1}(S_i^t)$.

Lemma 4. *If* $\mathcal{A}$ *is lazy then for any* $t \in [n]$, $|Rel^t| \geq \sum_{i=p^t+1}^{\mathbf{depth}(t)} b_i^t$.

Proof. If $p^t = \mathbf{depth}(t)$ then the lemma is trivial. If $p^t = \mathbf{depth}(t) - 1$ then the lemma is trivial as well since $|Rel^t| \geq 1$ and $b_{\mathbf{depth}(t)}^t = 1$ always. So let us assume that $p^t < \mathbf{depth}(t) - 1$. By the definition of p^t we know that at least one of the following must happen: $f^{t-1}(\min(Rel^t)) \in S_{p^t}^t \setminus S_{p^t+1}^t$, $f^t(\min(Rel^t)) \in S_{p^t}^t \setminus S_{p^t+1}^t$, $f^{t-1}(\max(Rel^t)) \in S_{p^t}^t \setminus S_{p^t+1}^t$ or $f^t(\max(Rel^t)) \in S_{p^t}^t \setminus S_{p^t+1}^t$. Assume that $f^{t-1}(\min(Rel^t)) \in S_{p^t}^t \setminus S_{p^t+1}^t$ or $f^t(\min(Rel^t)) \in S_{p^t}^t \setminus S_{p^t+1}^t$, the other case is symmetric. Let $Y = \{y \in \{y^1, y^2, \ldots, y^{t-1}\}; \min(Rel^t) \le y < y^t\}$. Since $\mathcal{A}$ is lazy, $Y \subseteq Rel^t$. $S_{p^t+1}^t \setminus S_{\mathbf{depth}(t)}^t$ is the union of two subsegments, the left one S_L and the right one S_R. The population of S_L at time $t-1$ must be contained in Y. For any $i \in [\mathbf{depth}(t) - 1]$, the population of the left subsegment of $S_i^t \setminus S_{i+1}^t$ at time $t - 1$ is at least b_i^t, by the definition of S_{i+1}^t. Hence, $\sum_{i=p^t+1}^{\mathbf{depth}(t)-1} b_i^t \le |\mathbf{pop}^{t-1}(S_L)| \le |Y| < |Rel^t|$. Since $b_{\mathbf{depth}(t)}^t = 1$, the lemma follows. $\qquad\square$

Corollary 2. *Let $\mathcal{A}$ be a lazy algorithm. Then $64 \cdot \chi_{\mathcal{A}}(y^1, y^2, \ldots, y^n) + 8n \ge \sum_{t=1}^n \mathbf{weight}^{t-1}(S_{p^t}^t)$.*

3 Prefix Bucketing

A prefix bucketing of n items into k buckets is a sequence $a^0, a^1, \ldots, a^n \in \mathbb{N}^k$ of bucket configurations satisfying: $a^0 = (0, 0, \ldots, 0)$ and for $t = 1, 2, \ldots, n$, there exists $p^t \in [k]$ such that

1. $a_i^t = a_i^{t-1}$, for all $i = 1, 2, \ldots, p^t - 1$,
2. $a_{p^t}^t = 1 + \sum_{i \ge p^t} a_i^{t-1}$, and
3. $a_i^t = 0$, for all $i = p^t + 1, \ldots, k$.

The *cost of the bucketing* $a^0, a^1, \ldots, a^n$ is $c(a^0, a^1, \ldots, a^n) = \sum_{t=1}^n a_{p^t}^t$. In Section 3.1 we prove the following lemma.

Lemma 5. *Let $n \ge 2^{32}$ and k be integers where $\log n \le k \le \sqrt[4]{n}/8$. The cost of any prefix bucketing of n items into k buckets is greater than $\frac{n \log n}{8(\log 8k - \log \log n)} - n$.*

We want to relate the cost of online labeling to some prefix bucketing. We will build a specific prefix bucketing as follows. Set $k = \lceil \log m \rceil$. For a lazy on-line labeling algorithm $\mathcal{A}$ and $t = 1, \ldots, n$, let $f^t, S_i^t, B^t, p^t, y^t$ and f^0, p^0 be as defined by the **Adversary**$(\mathcal{A}, n, m, r)$ and the algorithm $\mathcal{A}$. Denote $Y = \{y^1, y^2, \ldots, y^n\}$. For $t = 0, 1, \ldots, n$ and $i = 1, \ldots, k$, define a sequence of sets $A_i^t \subseteq Y$ as follows: for all $i = 1, \ldots, k$, $A_i^0 = \emptyset$, and for $t > 0$:

 - $A_i^t = A_i^{t-1}$, for all $i = 1, \ldots, p^t - 1$,
 - $A_{p^t}^t = \{y^t\} \cup \bigcup_{i \ge p^t} A_i^{t-1}$, and
 - $A_i^t = \emptyset$, for all $i = p^t + 1, \ldots, k$.

The following lemma relates the cost of online labeling to a prefix bucketing.

Lemma 6. *Let the prefix bucketing $a^0, a^1, \ldots, a^n$ be defined by $a_i^t = |A_i^t|$, for all $t = 0, \ldots, n$ and $i = 1, \ldots, k$. The cost of the bucketing $a^0, a^1, \ldots, a^n$ is at most $64 \cdot \chi_{\mathcal{A}}(y^1, y^2, \ldots, y^n) + 9n$.*

Lemmas 5 and 6 together imply Lemma 1. The following lemma will be used to prove Lemma 6.

Lemma 7. *For any $t \in [n]$ and $i \in [\mathbf{depth}(t)]$, if $i \ne p^t$ then $f^{t-1}(A_i^t) \subseteq S_i^t$ otherwise $f^{t-1}(A_i^t \setminus \{y^t\}) \subseteq S_i^t$.*

The proof of this lemma is omitted due to space constraints.

Proof of Lemma 6. Using the previous lemma we see that for $t \in [n]$, $|A_{p^t}^t| \le \mathbf{weight}^{t-1}(S_{p^t}^t) + 1$. The lemma follows by Corollary 2. $\qquad\square$

3.1   Lower Bound for Bucketing

In this section we will prove Lemma 5. To do so we will associate with each prefix bucketing a k-tuple of ordered rooted trees. We prove a lower bound on the sum of depths of the nodes of the trees, and this will imply a lower bound for the cost of the bucketing.

An *ordered rooted tree* is a rooted tree where the children of each node are ordered from left to right. Since these are the only trees we consider, we refer to them simply as trees. The *leftmost principal subtree of a tree* T is the subtree rooted in the leftmost child of the root of T, the *i-th principal subtree of* T is the tree rooted in the i-th child of the root from the left. If the root has less than i children, we consider the i-th principal subtree to be empty. The number of nodes of T is called its *size* and is denoted $|T|$. The *depth* of a node is one more than its distance to the root, i.e., the root has depth 1. The cost of T, denoted $c(T)$, is the sum of depths of its nodes. The cost and size of an empty tree is defined to be zero.

We will be interested in trees that satisfy the following condition.

Definition 1 (k-admissible). *Let k be a non-negative integer. A tree T is k-admissible if it contains at most one vertex or*

- *its leftmost principal subtree is k-admissible and*
- *the tree created by removing the leftmost principal subtree from T is $(k-1)$-admissible.*

Notice that when a tree is k-admissible it is also k'-admissible, for any $k' > k$. The following easy lemma gives an alternative characterization of k-admissible trees:

Lemma 8. *A rooted ordered tree T is k-admissible if and only if for each $i \in [k]$, the i-th principal subtree of T is $(k-i+1)$-admissible.*

Proof. We show both directions at once by induction on k. For a single-vertex tree the statement holds. Let T be a tree with at least 2 vertices and let L be its leftmost principal subtree and T' be the subtree of T obtained by removing L. By definition T is k-admissible if and only if L is k-admissible and T' is $k-1$ admissible, and by induction on k, T' is $k-1$ admissible if and only if for each $2 \leq i \leq k$ the i-th principal subtree of T, which is the $(i-1)$-st principal subtree of T' is $(k-i+1)$-admissible.   □

We will assign a k-tuple of trees $T(\bar{a})_1, T(\bar{a})_2, \ldots, T(\bar{a})_k$ to each prefix bucketing $\bar{a} = a^0, a^1, \ldots, a^t$. The assignment is defined inductively as follows. The bucketing $\bar{a} = a^0$ gets assigned the k-tuple of empty trees. For bucketing $\bar{a} = a^0, a^1, \ldots, a^t$ we assign the trees as follows. Let p^t be as in the definition of prefix bucketing, so $0 \neq a_{p^t}^t \neq a_{p^t}^{t-1}$ and for all $i > p^t$, $a_i^t = 0$. Let $\bar{a}' = a^0, a^1, \ldots, a^{t-1}$. Then we let $T(\bar{a})_i = T(\bar{a}')_i$, for $1 \leq i < p^t$, and $T(\bar{a})_i$ be the empty tree, for $p^t < i \leq k$. The tree $T(\bar{a})_{p^t}$ consists of a root node whose children are the non-empty trees among $T(\bar{a}')_{p^t}, T(\bar{a}')_{p^t+1}, \ldots, T(\bar{a}')_k$ ordered left to right by the increasing index.

We make several simple observations about the trees assigned to a bucketing.

Proposition 1. *For any positive integer* k, *if* $\bar{a} = a^0, a^1, \ldots, a^t$ *is a prefix bucketing into* k *buckets then for each* $i \in [k]$, $|T(\bar{a})_i| = a_i^t$.

The proposition follows by a simple induction on t.

Proposition 2. *For any positive integer* k, *if* $\bar{a} = a^0, a^1, \ldots, a^t$ *is a prefix bucketing into* k *buckets then for each* $i \in [k]$, $T(\bar{a})_i$ *is* $(k+1-i)$-*admissible.*

Again this proposition follows by induction on t and the definition of k-admissibility. The next lemma relates the cost of bucketing to the cost of its associated trees.

Lemma 9. *For any positive integer* k, *if* $\bar{a} = a^0, a^1, \ldots, a^t$ *is a prefix bucketing into* k *buckets then* $\sum_{i=1}^{k} c(T(\bar{a})_i) = c(\bar{a})$.

Proof. By induction on t. For $t = 0$ the claim is trivial. Assume that the claim is true for $t - 1$ and we will prove it for t. Let $\bar{a}' = a^0, a^1, \ldots, a^{t-1}$ and p^t be as in the definition of prefix bucketing.

$$c(\bar{a}) = c(\bar{a}') + 1 + \sum_{i=p^t}^{k} a_i^{t-1} = \sum_{i=1}^{k} c(T(\bar{a}')_i) + 1 + \sum_{i=p^t}^{k} |T(\bar{a}')_i|$$

$$= \sum_{i=1}^{p^t-1} c(T(\bar{a})_i) + 1 + \sum_{i=p^t}^{k} (c(T(\bar{a}')_i + |T(\bar{a}')_i|)$$

where the first equality follows by the induction hypothesis and by Proposition 1, and the last equality follows by the definition of $T(\bar{a})_i$, for $i = 1, \ldots, p^t - 1$. For $i \geq p^t$, the depth of each node in $T(\bar{a}')_i$ increases by one when it becomes the child of $T(\bar{a})_{p^t}$ hence

$$c(T(\bar{a})_{p^t}) = 1 + \sum_{i=p^t}^{k} (c(T(\bar{a}')_i + |T(\bar{a}')_i|).$$

For $i > p^t$, $c(T(\bar{a})_i) = 0$ so the lemma follows.   □
Now, we lower bound the cost of any ordered rooted tree.

Lemma 10. *Let* $k, d \geq 1$ *and* T *be a* k-*admissible tree of depth* d. *Then* $c(T) \geq d \cdot |T|/2$ *and* $|T| \leq \binom{k+d-1}{k}$.

Proof. We prove the lemma by induction on $k + d$. Assume first that $k = 1$ and $d \geq 1$. The only 1-admissible tree T of depth d is a path of d vertices. Hence, $|T| = d = \binom{1+d-1}{1}$ and $c(T) = \sum_{i=1}^{d} i = d \cdot (d+1)/2$.

Now assume that $k > 1$ and that T is k-admissible. Denote the leftmost principal subtree of T by L and the tree created by removing L from T by R. By the induction hypothesis and definition of k-admissibility it follows that $|T| = |L| + |R| \leq \binom{k+d-2}{k} + \binom{k+d-2}{k-1} = \binom{k+d-1}{k}$. Furthermore, $c(T) = c(L) + |L| + c(R) \geq ((d-1) \cdot |L|/2) + |L| + (d \cdot (|T| - |L|)/2) \geq d \cdot |T|/2$.   □

Lemma 11. *Let n, k and d be integers such that $2^{32} \leq n$, $\log n \leq k \leq \sqrt[4]{n}/8$, and $d \leq \frac{\log n}{4(\log 8k - \log \log n)}$. Then $\binom{k+d}{k} < n$.*

The proof of Lemma 11 is omitted due to lack of space.

Proof of Lemma 5. Consider a prefix bucketing $\bar{a} = a^0, a^1, \ldots, a^t$ of n items into k buckets, where $\log n \leq k \leq \sqrt[4]{n}/8$. Let $a'^t = (n, 0, 0, \ldots, 0)$ be a k-tuple of integers and let $\bar{a}' = a^0, a^1, \ldots, a^{t-1}, a'^t$. Clearly, $\bar{a}'$ is also a prefix bucketing and $c(\bar{a}') \leq c(\bar{a}) + n - 1$. Hence, it suffices to show that $c(\bar{a}') \geq \frac{n \log n}{8(\log 8k - \log \log n)}$. Let T be $T(\bar{a}')_1$. By Proposition 1, $|T| = n$, and by Proposition 2, T is k-admissible. Furthermore, by Lemma 9, $c(\bar{a}) = \sum_{i=1}^{k} c(T(\bar{a})_i) = c(T)$. So we only need to lower bound $c(T)$. By Lemma 10 a k-admissible tree has size at most $\binom{k+d-1}{k}$, where d is its depth. For $d \leq \frac{\log n}{4(\log 8k - \log \log n)}$, $\binom{k+d-1}{k} \leq \binom{k+d}{k} < n$ by Lemma 11 so T must be of depth $d > \frac{\log n}{4(\log 8k - \log \log n)}$. By Lemma 10, $c(T) \geq \frac{n \log n}{8(\log 8k - \log \log n)}$. The lemma follows. $\square$

References

1. Bulánek, J., Koucký, M., Saks, M.: Tight lower bounds for the online labeling problem. In: Karloff, H.J., Pitassi, T. (eds.) Proc. of 66th Symp. of Theory of Computation (STOC 2012), pp. 1185–1198. ACM (2012)
2. Bender, M.A., Cole, R., Demaine, E.D., Farach-Colton, M., Zito, J.: Two Simplified Algorithms for Maintaining Order in a List. In: Möhring, R.H., Raman, R. (eds.) ESA 2002. LNCS, vol. 2461, pp. 152–164. Springer, Heidelberg (2002)
3. Bender, M.A., Demaine, E.D., Farach-Colton, M.: Cache-oblivious b-trees. Journal on Computing 35(2), 341–358 (2005)
4. Bender, M.A., Duan, Z., Iacono, J., Wu, J.: A locality-preserving cache-oblivious dynamic dictionary. Journal of Algorithms 53(2), 115–136 (2004)
5. Brodal, G.S., Fagerberg, R., Jacob, R.: Cache oblivious search trees via binary trees of small height. In: Eppstein, D. (ed.) Proc. of 13th ACM-SIAM Symp. on Discrete Algorithms. SODA, pp. 39–48. ACM/SIAM (2002)
6. Bird, R.S., Sadnicki, S.: Minimal on-line labelling. Information Processing Letters 101(1), 41–45 (2007)
7. Dietz, P.F., Seiferas, J.I., Zhang, J.: A tight lower bound for online monotonic list labeling. SIAM J. Discrete Mathematics 18(3), 626–637 (2004)
8. Emek, Y., Korman, A.: New bounds for the controller problem. Distributed Computing 24(3-4), 177–186 (2011)
9. Itai, A., Konheim, A.G., Rodeh, M.: A Sparse Table Implementation of Priority Queues. In: Even, S., Kariv, O. (eds.) ICALP 1981. LNCS, vol. 115, pp. 417–431. Springer, Heidelberg (1981)
10. Willard, D.E.: A density control algorithm for doing insertions and deletions in a sequentially ordered file in good worst-case time. Information and Computation 97(2), 150–204 (1992)
11. Zhang, J.: Density Control and On-Line Labeling Problems. PhD thesis, University of Rochester (1993)
12. Babka, M., Bulánek, J., Čunát, V., Koucký, M., Saks, M.: On Online Labeling with Superlinearly Many Labels (2012) (manuscript)

A 5-Approximation for Capacitated Facility Location

Manisha Bansal[1], Naveen Garg[2,*], and Neelima Gupta[1]

[1] University of Delhi
[2] Indian Institute of Technology Delhi

Abstract. In this paper, we propose and analyze a local search algorithm for the capacitated facility location problem. Our algorithm is a modification of the algorithm proposed by Zhang et al.[7] and improves the approximation ratio from 5.83 to 5. We achieve this by modifying the *close*, *open* and *multi* operations. The idea of taking linear combinations of inequalities used in Aggarwal et al.[1] is crucial in achieving this result. The example proposed by Zhang et al. also shows that our analysis is tight.

1 Introduction

In a facility location problem we are given a set of clients C and facility locations F. Each facility $i \in F$ has a facility cost $f(i)$ (the *facility cost*). The cost of servicing a client j by a facility i is given by $c(i, j)$ (the *service cost*) and these costs form a metric. The objective is to select a subset $S \subseteq F$, so that the total cost for opening the facilities and for serving all the clients is minimized.

When the number of clients that a facility i can serve is bounded by $u(i)$, we have the *capacitated facility location problem*. If $u(i)$ is same for all the facilities then the problem is called *uniform capacitated facility location*. We are interested in the setting when the capacities are non-uniform and do not require that all the demand from a client be served by a single facility.

For the problem of uniform capacities Korupolu et al.[3] showed that a simple local search heuristic proposed by Kuehn and Hamburger [4] is a constant factor approximation. Chudak and Williamson [2] improved the analysis of Korupolu et al. to obtain a (6,5)-approximation which implies that the solution returned by their algorithm had cost at most 6 times the facility cost plus 5 times the service cost of the optimum solution. Aggarwal et al.[1] improved this further to a (3,3)-approximation by introducing the idea of taking suitable linear combinations of inequalities which capture local optimality.

We apply some of the ideas of Aggarwal et al. to the setting of non-uniform capacities. To do this we need to modify the *close*, *multi* and *open* operations of Zhang et al.[7] so that apart from the demand of facilities being closed, some more demand, served by other facilities in the current solution, is assigned to the

---

* Supported by the Indo-German Max-Planck Center for Computer Science (IMPECS).

L. Epstein and P. Ferragina (Eds.): ESA 2012, LNCS 7501, pp. 133–144, 2012.

facilities being opened to utilize them more efficiently. The first local search algorithm for non-uniform capacities problem was due to Pal, Tardos and Wexler [6] who gave a (9,5)- approximation for this problem. Mahdian and Pal [5] reduced this to an (8,7)-approximation and simultaneously extended it to the more general *universal facility location* problem. The approximation guarantee for the capacitated facility location problem was further reduced by Zhang *et al.* [7] to (6,5).

In this paper, we present a (5,5)-approximation algorithm for this problem. The remainder of this paper is organized as follows. In Section 2 we describe the algorithm and analysis of Zhang *et al.*. In Section 3 we bound the facility cost of our solution. We also present a tight example in Section 4.

2   Preliminaries and Previous Work

For a given subset $S \subseteq F$, the optimal assignment of clients to the set of facilities in S can be computed by solving a mincost flow problem. Therefore to solve the capacitated facility location problem we only need to determine a good subset $S \subseteq F$ of facilities. We abuse notation and use S to denote both the solution and the set of open facilities in the solution. We denote the cost of solution S by $c(S) = c_f(S) + c_s(S)$, where $c_f(S)$ is the facility cost and $c_s(S)$ is the service cost of solution S.

Pal *et al.*[6] suggested a local search algorithm to find a good approximate solution for the capacitated facility location problem. Starting with a feasible solution S the following operations are performed to improve the solution if possible.

- **add(s):** $S \leftarrow S \cup \{s\}, s \notin S$. In this operation a facility s which is not in the current solution S is added if its addition improves the cost of the solution.
- **open(s,T):** $S \leftarrow S \cup \{s\} - T, s \notin S, T \subseteq S$. In this operation a facility $s \notin S$ is added and a subset of facilities $T \subseteq S$ is closed. Since the possibilities for the set T are exponentially large, to make the procedure polynomial, we do not compute the exact cost of the new solution but only an estimated cost which overestimates the exact cost. The operation is then performed if the estimated cost is less than the cost of the solution S. In computing the estimated cost we assume that for any $t \in T$, all clients served by t are assigned to s and that each such reassignment costs $c(s,t)$. Working with the estimated cost allows us to find in polynomial time, for a given s, the set T for which the estimated cost of $S - T \cup \{s\}$ is minimum [6].
- **close(s,T):** $S \leftarrow S \cup T - \{s\}, s \in S, T \subseteq F \setminus S$. In this operation a facility $s \in S$ is closed and a subset of facilities T, disjoint from S, is opened. Once again, we work with an estimated cost for this operation, in computing which we assume that a client which was assigned to s in the solution S will now be assigned to some $t \in T$ and that this reassignment costs $c(s,t)$. As before, working with the estimated cost allows us to find in polynomial time, for a given s, the set T for which the estimated cost of $S \cup T - \{s\}$ is minimum [6].

Zhang *et al.* added the following operation to the above set of operations:

- **multi(r,R,t,T):** $S \leftarrow S \cup R - \{r\} \cup \{t\} - T, r \in S, T \subseteq S, R \subseteq F \setminus S, t \notin S$.
  This operation is essentially a combination of a $\texttt{close}(r, R)$ and $\texttt{open}(t, T)$ with the added provision that clients served by r maybe assigned to facility t. For a choice of r, t the operation can be implemented by guessing the number of clients serviced by r that will be assigned to t and then determining the sets R, T which minimize the total expected cost.

S is locally optimum if none of the four operations improve the cost of the solution and at this point the algorithm stops. Polynomial running time can be ensured at the expense of an additive ϵ in the approximation factor by doing a local search operation only if the cost reduces by more than an $1 - \epsilon/n$ factor, for $\epsilon > 0$.

Let $S \subseteq F$ be a locally optimum solution and $O \subseteq F$ be the optimal solution. The add operation allows us to bound the service cost of the solution S.

Lemma 1 ([5,7]). $c_s(S) \leq c_s(O) + c_f(O) = c(O)$.

The local search procedure is analyzed by formulating a set of linear inequalities which arise from the consideration that the solution S is locally optimal. To formulate a suitable set of inequalities Pal *et al.* build an exchange graph G whose vertices are the set of facilities in S and the facilities in the optimum solution O; for a facility in $S \cap O$ we have two vertices in G. G has an edge $(s, o), s \in S, o \in O$ if there are clients served by s in the solution S and by o in the optimum solution and the value of this edge, $y(s, o)$, is the number of such clients. The cost of the edge (s, o) is the distance between the facilities s and o which is denoted by $c(s, o)$. Note that

1. $\sum_{s \in S, o \in O} c(s, o) y(s, o) \leq c_s(S) + c_s(O)$.
2. G is a bipartite graph with S and O defining the sets of the partition.
3. $\forall s \in S, \sum_{o \in O} y(s, o) \leq u(s)$ and $\forall o \in O, \sum_{s \in S} y(s, o) \leq u(o)$.

The graph G is now modified to make it acyclic. Consider a cycle in G and let C be the edges on the cycle. Partition the edges of C into sets C_1, C_2 such that the edges of C_1 (and C_2) are alternate edges on the cycle. Let ϵ be the minimum value of an edge in C. Consider two operations: one in which we increase the value of edges in C_1 and decrease the value of edges in C_2 by an amount ϵ and the other in which we do the inverse *i.e.*, decrease the value of the edges in C_1 and increase the value of the edges in C_2. Note that in one of these operations the total cost $\sum_{s \in S, o \in O} c(s, o) y(s, o)$ would not increase and the value of one of the edges would reduce to zero thereby removing it from the graph. This process is continued till the graph becomes acyclic. Note that the modified values of the edges continue to satisfy the three properties listed above although it is no more the case that value of an edge (s, o) is the number of clients which are served by $s \in S$ and $o \in O$.

Each tree in the graph G is rooted at an arbitrary facility in O. Figure 1 shows a subtree of height 2 with root $t \in O$. Recall that our aim is to formulate

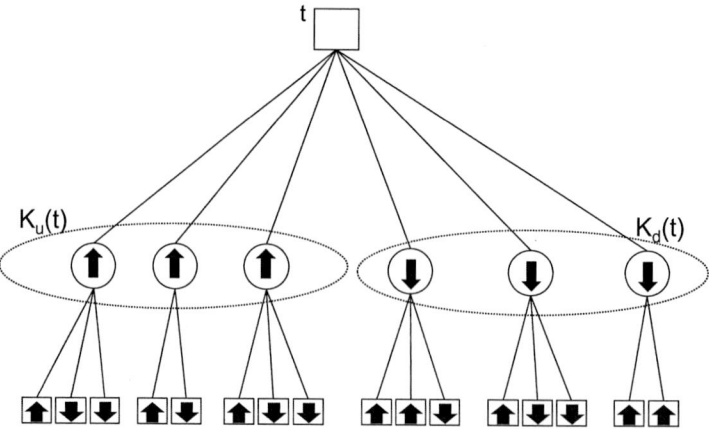

Fig. 1. The subtree of height 2 showing up-facilities and down-facilities. The square facilities are in the optimum solution while the circular facilities are in the locally optimum solution. The arrow in the facility identifies it as an up/down facility.

a set of inequalities that will let us bound the total facility cost of the solution S, $c_f(S)$. Each inequality is obtained by considering a potential local step and using the fact that S is a locally optimal solution. The inequalities are written such that

1. each facility in S is closed once.
2. each facility in O is opened at most thrice.
3. the total cost of reassigning clients is bounded by $2\sum_{s\in S, o\in O} c(s,o)y(s,o)$.

and when added yield

$$-c_f(S) + 3c_f(O) + 2(c_s(S) + c_s(O)) \geq 0 \qquad (1)$$

We now discuss the choice of inequalities as given by Zhang *et al.* in greater detail. For a facility i let $p(i)$ be the parent and $K(i)$ be the children of i. A facility i is an *up-facility* if $y(i, p(i)) \geq \sum_{j\in K(i)} y(i, j)$ and a *down-facility* otherwise. Let $K_u(i)$ (respectively $K_d(i)$) denote the children of i which are up-facilities (respectively down-facilities). For a facility $o \in O$ we further insist that

1. if o is an up-facility it is opened at most once in operations involving facilities which are descendants of o in the tree and is opened at most twice in other operations.
2. if o is a down-facility it is opened at most twice in operations involving facilities which are descendants of o in the tree and is opened at most once in other operations.

Consider a facility $s \in S$ which is a child of $t \in O$. Let s be a down-facility and let $o \in O$ be a child of s. When s is closed we can assign $2y(s,o)$ clients served

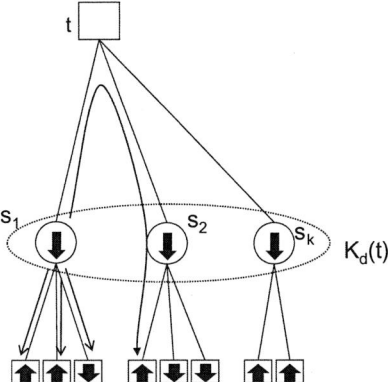

Fig. 2. The ordering of facilities in $K_d(t)$ and the reassignment of clients when one of these facilities is closed

by s to facility o if o is a down-facility. Else we can assign $y(s, o)$ clients served by s to o. Thus we need to assign

$$\sum_{j \in O} y(s, j) - \sum_{j \in K_u(s)} y(s, j) - 2 \sum_{j \in K_d(s)} y(s, j) = y(s, t) - \sum_{j \in K_d(s)} y(s, j)$$

to facilities other than the children of s; we refer to the above quantity as $\mathtt{rem}(s)$. Zhang et al. achieve this by ordering the facilities in $K_d(t)$ by increasing order of $\mathtt{rem}(s)$; let $s_1, s_2, \ldots, s_k$ be the order (Figure 2). They assign the remaining $\mathtt{rem}(s_i)$ clients of s_i to facilities in $K_u(s_{i+1})$ with $y(s_{i+1}, j)$ clients assigned to facility $j \in K_u(s_{i+1})$. This takes care of all the remaining demand because

$$\mathtt{rem}(s_i) \leq \mathtt{rem}(s_{i+1}) = y(s_{i+1}, t) - \sum_{j \in K_d(s_{i+1})} y(s_{i+1}, j) \leq \sum_{j \in K_u(s_{i+1})} y(s_{i+1}, j)$$

where the last inequality follows from the fact that s_{i+1} is a down-facility and hence

$$y(s_{i+1}, t) \leq \sum_{j \in K_d(s_{i+1})} y(s_{i+1}, j) + \sum_{j \in K_u(s_{i+1})} y(s_{i+1}, j).$$

However, the last facility in the order, say s_k needs to be handled differently. In fact, s_k is closed together with the facilities in $K_u(t)$.

To bound the cost of reassigning clients served by facilities in $K_d(t) \setminus \{s_k\}$ note that

1. since edge costs form a metric, $c(s_i, j), j \in K_u(s_{i+1})$ is at most $c(s_i, t) + c(t, s_{i+1}) + c(s_{i+1}, j)$.
2. the contribution of the edge (s_i, t) to the reassignment cost is at most $(\mathtt{rem}(s_i) + \mathtt{rem}(s_{i-1}))c(s_i, t)$. Since both $\mathtt{rem}(s_i)$ and $\mathtt{rem}(s_{i-1})$ are less than $y(s_i, t)$ the total contribution is at most $2y(s_i, t)c(s_i, t)$.

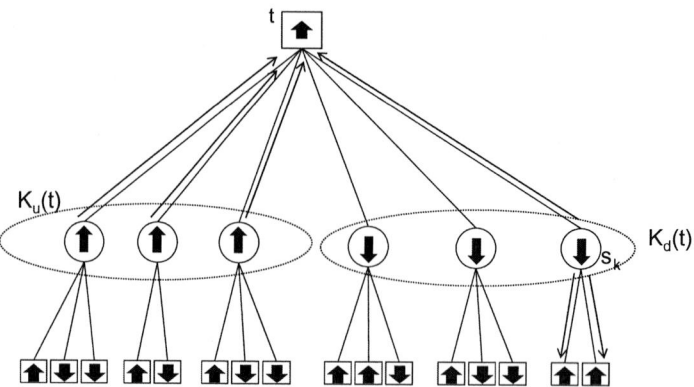

Fig. 3. The multi operation considered when t is an up-facility

3. the contribution of the edge $(s_i, j), j \in K_d(s_i)$ is at most $2y(s_i, j)c(s_i, j)$ since $2y(s_i, j)$ clients are assigned to j when s_i is closed.
4. the contribution of the edge $(s_i, j), j \in K_u(s_i)$ is at most $2y(s_i, j)c(s_i, j)$ since at most $y(s_i, j)$ clients are assigned to j when s_i is closed and when s_{i-1} is closed.

Now consider the facilities in $K_u(t) \cup \{s_k\}$. If t is an up-facility then Zhang *et al.* perform a multi operation $\mathtt{multi}(s_k, K(s_k), t, K_u(t))$ which can be viewed as a combination of $\mathtt{open}(t, K_u(t))$ and $\mathtt{close}(s_k, K(s_k))$ (Figure 3). Thus clients served by facilities in $K_u(t)$ are assigned to t. Clients served by s_k are assigned to facilities in $K(s_k) \cup \{t\}$ with $y(s_k, i)$ clients assigned to facility $i \in K(s_k) \cup \{t\}$. Note that

1. The total number of clients assigned to t is at most $2\sum_{s \in K_u(t)} y(s, t) + y(s_k, t)$ which is less than the capacity of t since t is an up-facility.
2. The contribution of the edges $(t, s), s \in K_u(t)$ to the reassignment cost is at most $2y(t, s)c(t, s)$.
3. The up-facility t is opened once when considering facilities of S which are descendants of t.

Next consider the case when t is a down-facility. Let h be the facility for which $y(s, t)$ is the maximum for $s \in K_u(t)$. Zhang *et al.* partition the facilities in $K_u(t) \setminus \{h\}$ into two sets A, B such that

$$\sum_{s \in A} 2y(t, s) + y(t, h) \leq u(t)$$

$$\sum_{s \in B} 2y(t, s) + y(t, s_k) \leq u(t)$$

This can be done as follows

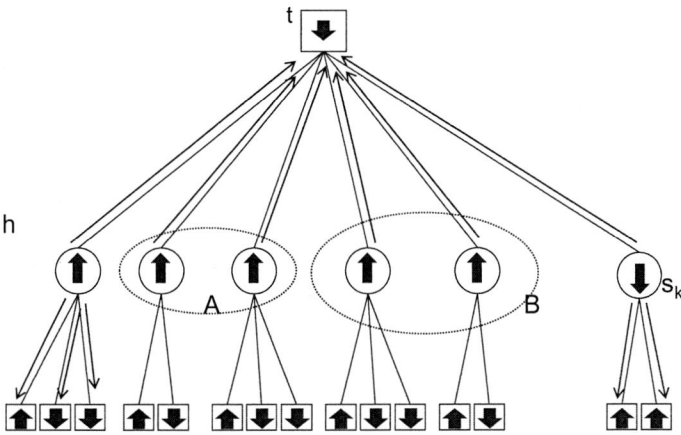

Fig. 4. The partition of $K_u(t)$ and the multi operations considered when t is a down-facility

1. Consider the facilities of $K_u(t) \setminus \{h\}$ in an arbitrary order and continue assigning them to the set A until $\sum_{s \in A} 2y(t, s) \geq u(t)$. The last facility considered and the remaining facilities are assigned to B. If $\sum_{s \in B} 2y(t, s) > u(t)$ then $\sum_{s \in A \cup B \cup \{h\}} 2y(t, s) > 2u(t)$ which is a contradiction.
2. Add h to A and s_k to B or h to B and s_k to A. One of these will ensure the property we need since if in both cases one of the constraints is violated then we again get $\sum_{s \in A \cup B \cup \{h\}} 2y(t, s) > 2u(t)$ which is a contradiction.

Zhang *et al.* now consider two multi operations (Figure 4). The first is a combination of close($s_k, K(s_k)$) and open(t, B) in which $y(s_k, s)$ clients are assigned to facilities $s \in K(s_k) \cup \{t\}$. The second operation is a combination of close($h, K(h)$) and open(t, A) in which $y(s, h)$ clients are assigned to facilities $s \in K(h) \cup \{t\}$. In both operations all clients served by facilities in A, B are assigned to t and the properties of the sets A, B ensure that the capacity of t is not violated. Once again

1. the down facility t is opened at most twice when considering facilities of S which are descendants of t.
2. the contribution of an edge $(t, s), s \in K_u(t)$ to the reassignment cost is at most $2y(t, s)c(t, s)$.

3   Improving the Operations

The key contribution of this paper is to modify the close and open operations (and consequently the multi operation) to exploit the following observation.

Claim. In the analysis of Zhang *et al.* a facility $o \in O$ is assigned a total of at most $2 \sum_s y(s, o) \leq 2u(o)$ clients over all operations considered.

Proof. We first consider the case when o is an up-facility. Then o would be part of a multi operation $\mathtt{multi}(s_k, K(s_k), o, K_u(o))$ and assigned $2\sum_{s\in K_u(o)} y(s,o) + y(s_k,o)$ clients where $s_k \in K_d(o)$. Note that this is at most $2\sum_{s\in K(o)} y(s,o)$.

We next consider the number of clients assigned to o when considering facilities of S which are not descendants of o. If the parent of o, $p(o)$, is an up-facility then o could be assigned at most $y(o,p(o))$ clients in a multi operation involving $p(o)$. If $p(o)$ is a down-facility then o would be assigned at most $2y(o,p(o))$ clients and this can be argued as follows. Consider the ordering of the down-facilities which are siblings of $p(o)$.

1. if $p(o)$ is the first facility in the ordering (referred to as s_1) then o is only part of $\mathtt{close}(s_1, K(s_1) \cup K_u(s_2))$ and is assigned $y(o,s_1)$ clients.
2. if $p(o)$ is the ith facility in the ordering (referred to as s_i) and is not the first nor the last facility then o is part of $\mathtt{close}(s_{i-1}, K(s_{i-1}) \cup K_u(s_i))$ and $\mathtt{close}(s_i, K(s_i) \cup K_u(s_{i+1}))$ and is assigned $y(o,s_i)$ clients in each of these operations.
3. if $p(o)$ is the last facility in the ordering (referred to as s_k) then o is part of $\mathtt{close}(s_{k-1}, K(s_{k-1}) \cup K_u(s_k))$ and a multi operation involving s_k. In both these operations o is assigned $y(o,s_k)$ clients.

Hence the total number of clients assigned to o when considering facilities of S which are not descendants of o is at most $2y(o,p(o))$ and so the total number of clients assigned to o when o is an up-facility is at most $2\sum_s y(s,o)$.

We next consider the case when o is a down-facility. Then o would be part of two multi operations $\mathtt{multi}(h, K(h), o, A)$ and $\mathtt{multi}(s_k, K(s_k), o, B)$ and the number of clients assigned to o in these operations is $2\sum_{s\in A} y(s,o) + y(h,o)$ and $2\sum_{s\in B} y(s,o) + y(s_k,o)$ respectively. Since $A \cup B \cup \{h\} = K_u(o)$ and $s_k \in K_d(o)$, the total number of clients assigned to o in these two multi operations is at most $2\sum_{s\in K(o)} y(s,o)$.

We next consider the number of clients assigned to o when considering facilities of S which are not descendants of o. If the parent of o, $p(o)$, is an up-facility then o would be assigned at most $y(o,p(o))$ clients in a multi operation involving $p(o)$. If $p(o)$ is a down-facility then o would be assigned at most $2y(o,p(o))$ clients and this can be argued as follows. As before, consider the ordering of the down-facilities which are siblings of $p(o)$.

1. if $p(o)$ is the ith facility in the ordering (referred to as s_i) and is not the last facility then o is part of $\mathtt{close}(s_i, K(s_i) \cup K_u(s_{i+1}))$ is assigned $2y(o,s_i)$ clients.
2. if $p(o)$ is the last facility in the ordering (referred to as s_k) then o is part of a multi operation involving s_k in which o is assigned $y(o,s_k)$ clients.

Hence the total number of clients assigned to o when considering facilities of S which are not descendants of o is at most $2y(o,p(o))$ and so the total number of clients assigned to o when o is a down-facility is at most $2\sum_s y(s,o)$.   □

Since each facility $o \in O$ is opened thrice in the analysis of Zhang *et al.* the above claim implies that when a facility is opened we do not use it to its full capacity.

3.1   The Modified Open

Recall that in the operation $\text{open}(s,T)$ we open a facility $s \notin S$ and close a subset of facilities $T \subseteq S$. Let S_j (respectively O_j) be the service cost of the client j in the solution S (respectively O). If the capacity of s is more than the total capacity of the facilities in T we use the remaining capacity of s to service clients j for which S_j is larger than $c(s,j)$.

Given a facility $s \notin S$ to determine the set T for which the estimated cost of the new solution is the minimum we proceed as follows.

1. Let k be a guess of the difference in the capacity of s and the total capacity of the facilities in T.
2. We reduce the capacity of s by k and using the procedure of Pal *et al.* find the subset $T \subseteq S$ which minimizes the estimated cost.
3. Order clients in decreasing order of $S_j - c(s,j)$ discarding those for which this quantity is negative. The first k clients in this ordering are assigned to s and the savings arising from this step is reduced from the estimated cost computed above.
4. The process is repeated for all values of k in $[0..u(s)]$ and the solution for which the cost is minimum gives the optimum set T.

Let $z(s,t)$ be the number of clients served by $t \in T$ which are assigned to s in the original operation $\text{open}(s,T)$. This operation then yields the inequality

$$f(s) - \sum_{t \in T} f(t) + \sum_{t \in T} z(s,t)c(s,t) \geq 0$$

For a facility s in the optimum solution let $N_O(s)$ be the set of clients served by s. If $z(s,t)$ is the number of clients served by $t \in T$ which are assigned to s in the modified operation $\text{mopen}(s,T)$ then we can formulate the inequality

$$f(s) - \sum_{t \in T} f(t) + \sum_{t \in T} z(s,t)c(s,t) + \left(1 - \sum_{t \in T} z(s,t)/u(s)\right) \sum_{j \in N_O(s)} (O_j - S_j) \geq 0 \quad (2)$$

The last term in the above inequality arises from the argument that instead of utilizing the remaining capacity, $u(s) - \sum_{t \in T} z(s,t)$, in the best possible way (which we did in determining the optimum set T) we could have assigned each client in $N_O(s)$ to the facility s to an extent $(1 - \sum_{t \in T} z(s,t)/u(s))$ and in doing this we would not have reduce the estimated cost of the new solution.

3.2   The Modified Close

In $\text{close}(s,T)$ we close a facility $s \in S$ and open a set of facilities $T \subseteq F \setminus S$. As in the modified open operation, if the total capacity of the facilities in T exceeds the number of clients served by s in the solution S, then we would like to use the remaining capacity in T to reduced the estimated cost of the operation.

However, to use the excess capacity in T we need to identify a *pivot* facility $t \in T$ which will be assigned the clients j for which S_j is larger than $c(j,t)$. Thus, given a facility $s \in S$ to determine the set T for which the estimated cost of the new solution is the minimum we proceed as follows.

1. Let $t^* \notin S$ be a guess of the pivot facility and let k be a guess of the difference in the capacity of s and the total capacity of the facilities in $T \setminus \{t^*\}$.
2. We reduce the capacity of s by k and using the procedure of Pal et $al.$ find the subset $T \subseteq F \setminus (S \cup \{t^*\})$ which minimizes the estimated cost.
3. The remaining k clients of s are assigned to t^* and the estimated cost is increased by $kc(s, t^*)$.
4. Order clients in decreasing order of $S_j - c(t^*, j)$ discarding those for which this quantity is negative. The first $u(t^*) - k$ clients in this ordering are assigned to t^* and the savings arising from this step is reduced from the estimated cost computed above.
5. The process is repeated for all choices of $t^* \notin S$ and values of k in $[0..u(t^*)]$ and the solution for which the cost is minimum gives the optimum set T.

Let $z(s, t)$ be the number of clients served by s which are assigned to t in the original operation $\texttt{close}(s, T)$. This operation then yields the inequality

$$-f(s) + \sum_{t \in T} f(t) + \sum_{t \in T} z(s, t)c(s, t) \geq 0$$

Claim. For the modified operation $\texttt{mclose}(s, T)$ we can instead formulate the inequality

$$-f(s) + \sum_{t \in T} f(t) + \sum_{t \in T} z(s, t)c(s, t) + \sum_{t \in T} \sum_{j \in N_O(t)} (1 - z(s, t)/u(t))(O_j - S_j) \geq 0$$

$$(3)$$

Proof. It is no loss of generality to assume that $u(s) = \sum_{t \in T} z(s, t)$. Consider the facilities of T in decreasing order of $z(s, t)/u(t)$ and keep including them into a set T' until the total capacity of the facilities in T' exceeds $u(s)$. Let t^* be the last facility to be included into T'. Then an $\texttt{mclose}(s, T')$ operation in which t^* is the pivot and is assigned $k = u(s) - \sum_{t \in T' - t^*} u(t)$ clients which are served by s would yield the inequality

$$-f(s) + \sum_{t \in T'} f(t) + \sum_{t \in T' - t^*} u(t)c(s, t) + kc(s, t^*) +$$

$$(1 - k/u(t^*)) \sum_{j \in N_O(t^*)} (O_j - S_j) \geq 0 \qquad (4)$$

The last term in the above inequality arises from the argument that instead of utilizing the remaining capacity, $u(t^*) - k$, in the best possible way we could have assigned each client in $N_O(t^*)$ to the facility t^* to an extent $(1 - k/u(t^*))$.

For all facilities $t \in T' \setminus \{t^*\}$ we reduce $z(s, t)$ by $\epsilon \cdot u(t)$ and reduce $z(s, t^*)$ by $\epsilon \cdot k$. At the same time we include the inequality 4 to an extent ϵ in a linear combination that would eventually yield the inequality 3.

This process can be viewed as sending $\epsilon \cdot u(s)$ units of flow from s to facilities in T' with facility $t \in T' \setminus \{t^*\}$ receiving $\epsilon \cdot u(t)$ flow and facility t^* receiving $\epsilon \cdot k$ flow. The edges (s, t) have capacity $z(s, t)$ which is reduced by the amount of flow sent. Initially the total capacity of all edges $\sum_{t \in T} z(s,t)$ equals the amount of flow that needs to be sent, $u(s)$, and this property is maintained with each step. By picking the facilities with the largest values of $z(s,t)/u(t)$ we are ensuring that the maximum of these quantities never exceeds the fraction of the flow that remains to be sent. This implies that when the procedure terminates all $z(s,t)$ are zero and $u(s)$ units of flow have been sent.

If a facility t was opened to an extent λ_t in this linear combination then it would contribute $\lambda_t f(t) + z(s,t)c(s,t) + (\lambda_t - z(s,t)/u(t)) \sum_{j \in N_O(t)} (O_j - S_j)$ to the left hand side of inequality 3. We add a $1 - \lambda_t$ multiple of the inequality

$$f(t) + \sum_{j \in N_O(t)} (O_j - S_j) \geq 0 \qquad (5)$$

which corresponds to the operation add(t), to the linear combination to match the contribution of t in inequality 3. □

3.3 Putting Things Together

The modification to the open and close operations also imply a modification to the multi operation which we refer to as *mmulti*. Inequalities 2 and 3 have an additional term due to these modifications which when taken over all operations involving facility $o \in O$ equals $(\alpha - \beta/u(o)) \sum_{j \in N_O(o)} (O_j - S_j)$ where α is the number of times o is opened and β is the total number of clients assigned to o in the operations defined by Zhang *et al.*. Recall that β is at most $2u(o)$ and α is at most 3. If a facility $o \in O$ is opened only twice in these operations then we add the inequality corresponding to add(o) to our linear combination. As a consequence, instead of the inequality 1 we obtain

$$-c_f(S) + 3c_f(O) + 2(c_s(S) + c_s(O)) + c_s(O) - c_s(S) \geq 0$$

which together with the bound on the service cost of S, $c_s(S) \leq c_f(O) + c_s(O)$, implies that

$$c_f(S) + c_s(S) \leq 5(c_f(O) + c_s(O)) = 5c(O).$$

Theorem 1. *The local search procedure with operations* add, mclose, mopen *and* mmulti *yields a locally optimum solution that is a (5,5)-approximation to the optimum solution.*

4 The Tight Example

Zhang *et al.*[7] provide an example (see Figure 5) to show that their analysis is tight. The rectangular boxes (respectively circles) in the Figure are the facilities in the optimal solution O (respectively S). The optimum solution has $2n$ facilities each with facility cost 0 and capacity $n - 1$ and one facility with facility cost

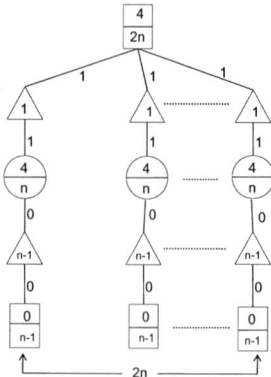

Fig. 5. The Tight example

4 and capacity $2n$. The solution S has $2n$ facilities each having a facility cost 4 and a capacity n. The clients are represented by triangles and there are $2n$ clients each having a demand of $n-1$ and $2n$ clients with 1 demand each. The numbers on the edges represent edge costs.

Zhang *et al.* argue that the solution S is locally optimal with respect to the add, open, close and multi operations. The cost of the optimal solution is $4+2n$ and that of S is $10n$ which gives a ratio of 5. By following the argument of Zhang *et al.* it is easy to confirm that this example is also tight with respect to the modified open, close and multi operations and the add operation. This establishes that our analysis is tight. Further, scaling costs, as done in Zhang*et al.* would destroy the local optimality of this solution.

References

1. Aggarwal, A., Anand, L., Bansal, M., Garg, N., Gupta, N., Gupta, S., Jain, S.: A 3-Approximation for Facility Location with Uniform Capacities. In: Eisenbrand, F., Shepherd, F.B. (eds.) IPCO 2010. LNCS, vol. 6080, pp. 149–162. Springer, Heidelberg (2010)
2. Chudak, F., Williamson, D.P.: Improved approximation algorithms for capacitated facility location problems. Math. Program. 102(2), 207–222 (2005)
3. Korupolu, M.R., Greg Plaxton, C., Rajaraman, R.: Analysis of a local search heuristic for facility location problems. J. Algorithms 37(1), 146–188 (2000)
4. Kuehn, A.A., Hamburger, M.J.: A heuristic program for locating warehouses. Management Science 9(4), 643–666 (1963)
5. Mahdian, M., Pál, M.: Universal Facility Location. In: Di Battista, G., Zwick, U. (eds.) ESA 2003. LNCS, vol. 2832, pp. 409–421. Springer, Heidelberg (2003)
6. Pál, M., Tardos, É., Wexler, T.: Facility location with nonuniform hard capacities. In: FOCS 2001: Proceedings of the 42nd IEEE Symposium on Foundations of Computer Science, p. 329. IEEE Computer Society, Washington, DC (2001)
7. Zhang, J., Chen, B., Ye, Y.: A multiexchange local search algorithm for the capacitated facility location problem. Math. Oper. Res. 30(2), 389–403 (2005)

Weighted Geometric Set Multi-cover
via Quasi-uniform Sampling

Nikhil Bansal[1] and Kirk Pruhs[2,*]

[1] Department of Mathematics and Computer Science,
Eindhoven University of Technology
n.bansal@tue.nl
[2] Computer Science Department, University of Pittsburgh,
Pittsburgh, PA 15260, USA
kirk@cs.pitt.edu

Abstract. We give a randomized polynomial time algorithm with approximation ratio $O(\log \phi(n))$ for weighted set multi-cover instances with a shallow cell complexity of at most $f(n,k) = n\phi(n)k^{O(1)}$. Up to constant factors, this matches a recent result of Könemann et al. for the set cover case, i.e. when all the covering requirements are 1. One consequence of this is an $O(1)$-approximation for geometric weighted set multi-cover problems when the geometric objects have linear union complexity; for example when the objects are disks, unit cubes or halfspaces in $\mathbb{R}^3$. Another consequence is to show that the real difficulty of many natural capacitated set covering problems lies with solving the associated priority cover problem only, and not with the associated multi-cover problem.

1 Introduction

In the *weighted set multi-cover problem* we are given a set $\mathcal{P}$ of n points and a collection $\mathcal{S}$ of m subsets of $\mathcal{P}$. Each element $p \in \mathcal{P}$ has a positive integer demand d_p, and each set $s \in \mathcal{S}$ has a positive weight w_s. A subset X of $\mathcal{S}$ is a feasible multi-cover if each $p \in \mathcal{P}$ lies in (or equivalently is covered by) at least d_p distinct sets in X, and the goal is to find a minimum weight feasible multi-cover. The case when all the demands are unit (i.e. $d_p = 1$ for all p) is known as the weighted set cover problem and has been studied extensively.

It is well known that the natural greedy algorithm is a $\ln n$-approximation algorithm for the weighted set multi-cover problem, and that no polynomial time algorithm can do better (up to lower order terms) even for unit demands and weights [1]. Thus, we focus here on special classes of instances where the underlying structure of the set system allows for an improved approximation. Such systems commonly arise in combinatorial optimization (e.g. several network design problems can be cast as covering a collection of cuts using edges). Another class of such problems arise in geometry, where the sets are geometric objects such as disks, rectangles or fat triangles and the elements are points in $\mathbb{R}^d$.

* Supported in part by an IBM Faculty Award, and NSF grants CCF-0830558 and CCF-1115575.

L. Epstein and P. Ferragina (Eds.): ESA 2012, LNCS 7501, pp. 145–156, 2012.

This paper is motivated by the following meta-question: If a particular set system (or a specific class of set systems) admits a good approximation algorithm for the set-cover problem, then does it also admit a good approximation for the multi-cover case?

In addition to its direct relevance (multi-cover problems arise naturally in many practical settings), there is another important theoretical motivation to study this question. Recently, [2] showed that the complexity of a capacitated set cover problem (defined formally later) can be reduced to that of the so-called priority set cover problem and the multi-cover problem on similar set systems. Thus in many cases, better bounds for multi-cover problems will directly improve upon known bounds for capacitated problems. We will discuss these consequences in more detail later.

While the set cover problem has been extensively studied for many classes of set systems, the corresponding multi-cover case has received less attention. At first glance, it seems that there should be a simple general method for extending a set cover result to the corresponding multi-cover case (after all, their natural IP/LP formulations differ only in the right hand side, i.e. $Ax \geq \mathbf{1}$ v.s. $Ax \geq d$). However, this seems unlikely in general. For example, consider the classic survivable network design problem (SNDP),[1] which can be viewed as a multi-cover problem where each cut S has covering requirement $d(S) = \max_{u \in S, v \notin S} r(u,v)$. Note that this problem seems much harder than the corresponding set cover case ($r(u,v) \in \{0,1\}$) which is the Steiner Network problem[2]. Similarly, while various techniques have been developed for geometric set cover problems (we will discuss these later), extending them to multi-cover poses several additional challenges (see for eg. [4] for a discussion). The only generic connection between set cover and multi-cover that we are aware of is the following (most likely folklore) result [5]: if a set system has an LP-based α-approximation for set cover, then it has an $O(\min(\log d_{\max}, \log \log n)\alpha)$-approximation for multi-cover.

In this paper, we study the weighted set multi-cover problem on geometric set systems, and extend the known set cover guarantees for such systems. Before we can describe our result, we must define some basic definitions and discuss the related previous work.

1.1 Previous Work

The goal in geometric set cover problems is to improve the $\ln n$ set-cover bound by exploiting the underlying geometric structure. This is an active area of research and various different techniques have been developed. However, until recently most of these techniques applied only to the unweighted case. A key idea is the connection between set covers and ϵ-nets due to [6], which implies that proving better bounds on sizes of ϵ-nets for various geometric systems (an active research

---

[1] Given a graph $G = (V, E)$ and demands $r(u,v)$ for pairs $u, v \in V$, find a minimum cost subset F such that $G' = (V, F)$ has $r(u,v)$ edge disjoint paths for each pair u, v.

[2] Indeed, a 2 approximation for Steiner Network was known long before Jain's breakthrough result on SNDP [3]. In fact, obtaining an $O(1)$ approximation for SNDP directly from the Steiner Network result would be a significant breakthrough.

area in discrete geometry) directly gives improved guarantees for the unweighted set-cover problem. In another direction, Clarkson and Varadarajan [7] related the guarantee for unweighted set-cover to the union complexity of the geometric objects. Here the complexity of a geometric shape refers to the size needed to describe it (which is roughly the number of vertices, edges, faces etc.). More precisely, [7] showed that if the complexity of the vertical decomposition of the complement of the union of any k sets is $O(kh(k))$, then there is an $O(h(n))$ approximation for unweighted set cover. This was subsequently improved to $O(\log(h(n))$ [8,9], and these results were further extended to the (unweighted) multi-cover case in certain "well-behaved" cases [4]. However, none of these techniques work with weights. Roughly speaking, the problem is the following: all these techniques begin by random sampling (say, according to the optimum LP solution), followed by an alteration phase where the points that are left uncovered during the sampling phase are now covered. They techniques are able to show that not many extra sets are needed in the alteration phase. However, they are unable the avoid the possibility that some sets may have a much higher chance of being picked than others, which is problematic if weights are present.

The first breakthrough on weighted geometric cover problems was made by Varadarajan [10]. He showed that for geometric set systems with union complexity[3] $O(kh(k))$, there is an efficient randomized algorithm with approximation ratio $O(\exp(O(\log^* n)) \log h(n))$. Further he showed that if the function $h(n)$ is mildly increasing, it suffices to have $h(n) = \omega(\log^{(j)} n)$ for some constant j, then there is an improved $O(\log h(n))$ approximation. The key idea behind this result was a new sampling approach (called quasi-uniform sampling), which gives a uniform bound on the probability of a set being sampled during the alteration phase. Recently, [11] refined this approach further and removed the mildly increasing requirement on $h(n)$. In particular they give an $O(\log h(n))$ guarantee for geometric set systems with union complexity $O(nh(n))$. This implies that if $h(n) = O(1)$, which for example is the case for disks, pseudo-disks, axis-aligned octants, unit cubes, or half-spaces in three dimensional space, their algorithm produces an $O(1)$ approximation [11].

Instead of using union complexity, [11] present their results using the notion of shallow cell complexity, which is also what we will do. To be more precise, let $f(n,k)$ be a function that is non-decreasing in both k and n. A set system has shallow cell complexity $f(n,k)$ if for all $1 \leq n \leq m$, if for any collection X of n sets from $\mathcal{S}$, the number of distinct regions covered by k or fewer sets in X is at most $f(n,k)$.[4] [11] gave an $O(\log \phi(n))$ approximation algorithm for weighted set cover instances with shallow cell complexity $n\phi(n)k^{O(1)}$. We remark that Varadarajan [10] also works with shallow cell complexity, without using this terminology directly. In particular, he uses that the fact that geometric sets in R^d with union complexity $n\phi(n)$ have shallow cell complexity $O(n\phi(n)k^{d-1})$.

---

[3] Unlike in [7,4], the union complexity here means the complexity of the shape of the union of sets, and not of the vertical decomposition of the complement of the union.

[4] Even though n refers to the number of points in the rest of the paper, we use n here intentionally as this parameter will later be related to the number of points.

1.2   Result and Consequences

We show the following result.

Theorem 1. *Given any weighted set multi-cover instance where the underlying set system has shallow cell complexity $f(n,k) = n\phi(n)k^c$, for some constant c, there is a randomized polynomial time algorithm that computes an $c' \log(\phi(n))$ approximation to the minimum weight multi-cover. Here c' is a universal constant that only depends on c.*

This matches the guarantee (up to constant factors) in [11] for the set cover case, and thus this extends all the geometric covering results in that paper from set cover to the multi-cover case.

An important consequence of our result is for capacitated covering problems. Here, each set s has a capacity c_s in addition to a weight w_s, and each point p has a demand d_p; a solution is feasible if the aggregate capacity of the selected sets covering a point is at least the point's demand. Perhaps surprising at first sight, capacities make the problem substantially more difficult. For eg., even if there is just one point to cover (a trivial problem without capacities), we obtain the NP-Hard (and non-trivial) Knapsack Cover problem.

Recently, Chakrabarty et al. [2] gave a general method to deal with capacities, and show that any weighted capacitated covering problem can be reduced to weighted multi-cover problem(s) and the so-called weighted priority cover problem on closely related systems[5]. In particular, they show that an $O(\alpha+\beta)$ approximation for capacitated cover problem follows from an α (resp. β)-approximation for the underlying multi-cover (resp. priority cover) problem

Our result implies that the real bottleneck for approximating capacitated covering problems in geometric settings is solving the associated priority cover problem. This already improves several results where previously the guarantee for the multi-cover problem was a bottleneck. One application is to Theorem 1.3 in [11], which says that any capacitated covering problem on an underlying network matrix has $O(\log\log n)$ approximation. Applying the improved bound in theorem 1 (together with the result in [11], that network matrices have $O(nk)$ shallow cell complexity), improves the approximation ratio to $O(1)$ for such problems. Another application is to a general scheduling problem introduced in [5]. They show that this problem can be cast as a capacitated covering problem where the sets are axis-parallel rectangles with the bottom side touching the x-axis. Applying Theorem 1 to such sets gives an $O(1)$ approximation for the associated multi-cover problem, improving upon the previous $O(\log\log n)$ bound. This improves the approximation for the scheduling problem in [5] from $O(\log\log nP)$ to $O(\log\log P)$, where P is the ratio of the maximum to minimum job size.

1.3   Overview of the Multicover Algorithm

As mentioned previously, most previous approaches for geometric covering problems work via random sampling. If α is the desired approximation, then first the

---

[5] We refer the reader to [2] for the definition of a priority cover problem.

sets are sampled with probability $O(\alpha x_s)$, where x_s is some optimum LP solution. After this step, a fixing procedure is applied (perhaps recursively) to cover the uncovered points. The key insight in [10] and [11] is to develop a fixing procedure where the probability of picking each set s remains bounded by αx_s. As our algorithm also uses this framework, we elaborate on this approach a bit more.

Instead of a single step, the sets are sampled in multiple steps (rounds). The process can be viewed as follows: We start with the solution x_s which wlog can be assumed to be an integer multiple of $1/M$ for some large integer M. At each round, x_s is rounded to an integer multiple of $2/M, 4/M, \ldots$, and so on, until it becomes 0 or 1, and in each round it is ensured that new rounded solution still covers each point fractionally to extent at least 1. This is done by a combination of careful random sampling and a forcing rule: the random sampling is designed to ensure that only a few points are left uncovered, and the forcing rule then ensures that each point is actually covered, while pick each set quasi-uniformly (with low probability).

To extend these ideas to multi-cover the main challenge is to design the above process such that no dependence on the covering requirement $d_{\max}$ is incurred in the approximation ratio. The main difficulty here is that we require that a point p must be covered by d_p *distinct* sets i.e. we cannot pick the same set twice[6]. For example, consider a scenario where a point p is covered by exactly d_p distinct sets in the LP solution. If the random sampling phase drops even one of these sets, the solution would be infeasible in later rounds. To avoid this, one would need to make either the sampling phase or the forcing rule more aggressive, leading to a dependence on $d_{\max}$ in the overall approximation guarantee.

To get around this problem, our basic idea is quite simple. Given an initial LP solution x_s, we first pick all sets s with $x_s \geq 1/Q$ for some constant $Q > 4$ and update the covering requirements accordingly. As the LP solution for the residual problem now satisfies $x_s < 1/Q$ for each s, there are at least $Q d_p$ distinct sets cover each point p in the support of the LP solution. We exploit this slack between $Q d_p$ and d_p. In particular, we start with a slack of Q initially, and design a sampling and forcing rule that ensures that even though the slack decreases in each round, this decrease is small enough so that the slack stays above 1 until the end of the algorithm. By allowing the slack to decrease slightly in each round, we can allow our forcing rule to be less aggressive, and avoid a dependence on d in the overall guarantee. Thus, the main technical contribution of this paper can be seen as further refining the quasi-uniform sampling procedure of [11] to incorporate these issues.

2   The Algorithm Description

Our algorithm proceeds in several rounds. We first introduce some notation and the initial setup. Then, we explain how a generic round works and state the

---

[6] Indeed, if a set can be picked multiple times, the set-cover analysis in [11] can be adapted very easily. See e.g. [4] for a more detailed discussion of this variant.

invariants that we will maintain at the end of each round. The analysis will be done in section 3.

Basic Setup: We assume that the given set system has shallow cell complexity $f(n,k) = n\phi(n)k^c$ for some constant c and some nondecreasing function $\phi(n) \geq 2$.

The exact integer programming formulation of the weighted set multicover problem is the following:

$$\min \sum_{s \in \mathcal{S}} w_s x_s \quad \text{s.t.} \quad \sum_{s:p \in s} x_s \geq d_p, \quad \forall p \in \mathcal{P} \quad \text{and} \quad x_s \in \{0,1\}, \quad \forall s \in \mathcal{S},$$

where x_s indicates whether the set s is selected or not.

Our algorithm begins by solving the natural linear programming (LP) relaxation where we relax the requirement $x_s \in \{0,1\}$ to $x_s \in [0,1]$. Let x_s denote some fixed basic optimum solution to this LP. As there are n constraints, at most n variables x_s lie strictly between 0 and 1. If $x_s > 1/Q$, where $Q > 4$ is a constant whose value we will specify later, the algorithm rounds $x_s = 1$ (meaning that set s is selected), at the expense of increasing the solution cost by at most Q times. The algorithm then considers the residual instance that ignores the selected sets, and where the demand d_p for point p is reduced by the number of selected sets that cover it. Clearly, the residual solution x_s is still feasible for the residual instance. Henceforth, we use d_p to denote this residual covering requirement.

Let $M = 2n$. We create a new instance by making $n_s = \lfloor M x_s \rfloor$ copies of each set s. To distinguish between the original sets and the copies in the new instance, we will use the term *replicas* to refer to the copies of a set s. Since the LP solution covers any point p to extent at least d_p, we have $\sum_{s:p \in s} M x_s \geq M d_p$ and hence p is covered by at least

$$\sum_{s:p \in s} n_s \geq \sum_{s:p \in s, x_s > 0} (M x_s - 1) \geq \left(\sum_{s:p \in s} M x_s\right) - n \geq M\left(d_p - \frac{1}{2}\right) \geq M\frac{d_p}{2}$$

replicas in the new instance.

As $x_s \leq 1/Q < 1/4$ for each set s, there are at most M/Q replicas of each set s in the new instance. We now double the number of replicas and redefine $n_s = 2n_s$ and $Q = Q/2$. So in this new instance, each point is covered by $M d_p$ replicas of sets containing p, and each set has at most M/Q replicas. Note that each point p is covered by replicas corresponding to at least $Q d_p$ distinct sets.

Rounds: The goal for the rest of this section is to describe a rounding procedure that covers the points with required demands and with the guarantee that for each replica, the probability that it is used is $O(\log(\phi(n))/M$.

The algorithm proceeds in rounds $1, \ldots, r$, where the value of $r = \log M - \log \log \phi(n) - O(1)$. At the beginning, when $r = 1$, we have the instance specified above. Recall the discussion in section 1.3. We view this instance as the scaled up version (by M times) of the LP solution where each variable x_s is an integer multiple of $1/M$. In each round, we will reduce the scaling factor M to $M/2$

(this process can be viewed as rounding x_s to an integer multiple of $2/M$) while maintaining various invariants. During these rounds we might also commit on picking certain sets, in which case we update the residual requirement of points accordingly. Eventually, when M reaches $O(\log \phi(n))$ the algorithm terminates, and we obtain our solution as each x_s is either 0 or at least $\Omega(1/\log(\phi(n)))$.

We now introduce some notation to track these rounds: Let $k(i) = M/2^{i-1}$. Let $n_s(i)$ be the number of replicas of set s at the start of round i. Let $d_p(i)$ be the residual covering requirement for point p at the start of round i. Let

$$\epsilon(i) = \Theta\left(\sqrt{\frac{\log k(i) + \log \phi(N)}{k(i)}}\right) \qquad \text{and let} \qquad b(i) = Q/\prod_{j=1}^{i-1}(1 + 4\epsilon(j)).$$

The Invariant: At the start of each round i, we will maintain the following invariant:

$$\sum_{s:p\in s} \min\left(n_s(i), \frac{k(i)}{b(i)}\right) \geq k(i)d_p(i) \qquad \forall p \tag{1}$$

Remark: Actually, our algorithm will also ensure that $n_s(i) \leq k(i)/b(i)$ for each set s, but for ease of algorithm description and analysis, we use the min in (1).

Note that at the beginning of round 1, we have that $k(1) = M$, $b(1) = Q > 2$ and $n_s(1) = \lfloor Mx_s \rfloor \leq M/Q$, and hence the setup discussion implies that (1) holds for $i = 1$.

The invariant (1) is useful for the following reason:

Fact 2. *If* (1) *holds, then each point p is covered by replicas of at least $b(i)d_p(i)$ distinct sets at the start of round i. In particular, if $b(i) \geq 1$, then there are replicas from at least $d_p(i)$ distinct sets.*

Proof. Invariant (1) implies that for each point p, $\sum_{s:p\in s, n_s(i)>0} \frac{k(i)}{b(i)} \geq k(i)d_p(i)$, and hence at least $b(i)d_p(i)$ terms contribute to the left side. $\qquad\square$

The Round i: The algorithm performs the following steps in each round i.

1. *Sampling:* For each set s, pick each of the $n_s(i)$ replicas for set s independently with probability $1/2 + \epsilon(i)$. Let $n_s^*(i)$ denote the number of replicas selected for set s.
2. *Fixing:* After the sampling step, there might be several points p for which invariant (1) for the next round is violated, i.e. $\sum_{s:p\in s} \min\left(n_s^*, \frac{k(i+1)}{b(i+1)}\right) \geq k(i+1)d_p(i)$. This is *fixed* via a forcing rule (described later), where we forcibly pick some sets to ensure that (1) holds for every point in the next round.
   If a set is forcibly picked, it is chosen forever. We remove all its replicas and update the residual covering requirement of a point accordingly.
3. *Clean up:* For each s, if $n_s^*(i) > k(i+1)/b(i+1)$, we arbitrarily discard the extra replicas of s until $n_s^*(i) \leq k(i+1)/b(i+1)$.

Note that the residual covering requirement of a point is modified only in the second step, when a set is picked by the forcing rule.

Termination: The algorithm terminates when $\epsilon(i)$ first exceeds $1/2$, which by the expression for $\epsilon(i)$ happens when $k(i)$ becomes $O(\log \phi(n))$. At this point the algorithm selects all the sets that have one or more replica left. In section 3 we will show that the invariants ensure that solution obtained is a valid multi-cover.

The Fixing Procedure: We now describe the fixing procedure for round i. We first need some notation.

Pseudo-depth: Define the *pseudo-depth* $q_p(i)$ of point p to be $\lfloor \sum_{s:p\in s} n_s(i)/k(i) \rfloor$, i.e. the number of replicas covering p at the start of round i, divided by $k(i)$.

Fact 3. *If invariant* (1) *holds at the start of round* i, *then for all* p, $q_p(i) \geq d_p(i)$.

Proof. Invariant (1) implies that for each p, $\sum_{s:p\in s} n_s(i) \geq k(i)d_p(i)$. Since $d_p(i)$ is an integer, this implies that $q_p(i) \geq d_p(i)$. □

For technical reasons, it will be easier to work with the pseudo-depth of a point rather than in terms of its residual covering requirement.

Constructing the list L_q: For $q = 1, \ldots,$ let P_q be the collection of all points with pseudo-depth q. Let C_q denote the collection of all replicas that cover some point in P_q. We construct an ordering L_q of the replicas in C_q as follows.

1. Initially L_q is empty.
2. Consider the regions (equivalence classes of points) formed by C_q. Select a replica $r \in C_q$ that covers at most $k(i)(q + 1)f(N, k(i)(q + 1))/N$ regions, where $N = |C_q|$. By the shallow cell complexity, this number is at most $(2qk(i))^{c+1}\phi(N)$. Since each set has most $2n$ replicas (at the start of round 1), and the support of the LP solution is at most n, we have that $N \leq 2n^2$. Such a replica must exist for the following reason. As the depth of any point in regions formed by C_q is at most $k(i)(q + 1)$ (by definition of the pseudo-depth q), the shallow cell complexity assumption implies that there are at most $f(N, k(i)(q + 1))$ regions. As each point lies in at most $k(i)(q + 1)$ replicas, the claim now follows by an averaging argument.
3. The replica r is appended to the end of the list L_q, and removed from C_q. Without loss of generality we can assume that all the replicas of a set appear consecutively in L_q (as removing a replica r does not change the regions formed by C_q unless r was the last remaining replica of its set).

As the replicas of a set appear consecutively in L_q, we will also view L_q as an ordering of the sets. Let $L_{p,q}$ denote the sublist of L_q consisting of those replicas that cover point p. Given a point p and a set $s \in L_{p,q}$ we define the rank $\rho_{p,q}(s)$ of s as the number of sets in $L_{p,q}$ that lie no later than s in the ordering. (i.e. there are exactly $\rho_{p,q}(s) - 1$ distinct sets before s in $L_{p,q}$).

Forcing Rule: For each value of pseudo-depth $q = 1, 2, \ldots$, we do the following.

Scan the sets in the list L_q starting from the end to the beginning of the list. Let s be the current set under consideration with rank $\rho_{p,q}(s)$. If there is some point $p \in s$ for which

$$\sum_{t \in L_{p,q} : \rho_{p,q}(t) \geq \rho_{p,q}(s)} \min\left(n_t^*, \frac{k(i)(1 + 4\epsilon(i))}{2b(i)}\right) < \frac{k(i)}{2}(q - (\rho_{p,q}(s) - 1)), \quad (2)$$

then we say that the forcing rule applies for p at set s, and mark all the sets in $L_{p,q}$ with rank $\rho_{p,q}(s)$ or less.

After all the lists L_q have been scanned, pick all the marked sets $L_{p,q}$. Update the residual requirements of points depending to the picked sets, discard the replicas of these sets and continue to the next round.

This completes the description of the algorithm.

Remark: We note here that a point p will not be forced due a set s which close to the end of the list $L_{p,q}$. This is because $\rho_{p,q}(s)$ is much larger than q for such sets s. This will be used crucially later and is stated as Fact 5.

The following claim illustrates why the forcing step is useful.

Claim. Let p be some point at pseudo-depth q. If condition (2) never applies for p at any set s, then the invariant (1) will hold for p at the start of round $i+1$. If condition (2) applies for p at some set s, then the forcing rule ensures that the invariant (1) will hold for p at the start of round $i + 1$. Moreover, applying the forcing rule at p does not violate the invariant (1) for any point $p' \neq p$.

Proof. First we note that $k(i+1) = k(i)/2$ and $b(i+1) = b(i)/(1+4\epsilon(i))$. If (2) never applies for p at any set s, then in particular it does not apply for s with $\rho_{p,q}(s) = 1$, and hence $\sum_{s : p \in s} \min\left(n_s^*, \frac{k(i+1)}{b(i+1)}\right) \geq k(i+1)d_p(i)$ and hence (1) is satisfied at the start of round $i + 1$ (note that $d_p(i+1) \leq d_p(i)$).

Now, suppose the forcing rule applies at set $s \in L_{p,q}$. Then, as the sets in $L_{p,q}$ of rank $\rho_{p,q}(s)$ or less are surely picked, the residual requirement for p reduces from $d_p(i)$ to at most $d_p(i) - \rho_{p,q}(s)$. Thus the invariant (1) for p will hold at the start of the next round $i + 1$ provided

$$\sum_{t : p \in t, \rho_{p,q}(t) > \rho_{p,q}(s)} \min\left(n_s^*, \frac{k(i+1)}{b(i+1)}\right) \geq k(i+1)(d_p(i) - \rho_{p,q}(s)). \quad (3)$$

Now, as the forcing rule did not apply for p at set s' with $\rho_{p,q}(s') = \rho_{p,q}(s) + 1$,

$$\sum_{t \in L_{p,q} : \rho_{p,q}(t) \geq \rho_{p,q}(s')} \min\left(n_t^*, \frac{k(i+1)}{b(i+1)}\right) \geq k(i+1)(q - (\rho_{p,q}(s') - 1)) \quad (4)$$

which is exactly the same as (3).

We now show that the invariant (1) at the start of round $i + 1$ is not violated for points $p' \neq p$. Suppose p forces a set u that is covering p'. This reduces the residual covering requirement $d_{p'}(i)$ of p' by 1 (and hence reduces the right hand side of (1) by $k(i+1)$), while reducing the contribution $\min(n_u^*, k(i+1)/b(i+1))$ on the left hand side of (1) by at most $k(i+1)/b(i+1)$. $\qquad\square$

3   The Algorithm Analysis

Feasibility: We claim that the algorithm produces a feasible solution provided $b(i) \geq 1$. This follows by fact 2 and noting that the forcing rule ensures that invariant (1) always holds. The fact that $b(i) \geq 1$, follows a simple calculation that we sketch below (a similar calculation can be found in Claim 2 in [11]). In fact, we can ensure that it is larger than any fixed constant by choosing Q large enough. We will choose Q so that $b(i) > 2$, which we will need later in our analysis.

Claim. Until the final round of the algorithm, it holds that $b(i) \geq 2$.

Proof. Recall that is is the case that $b(i) = Q / \prod_{j=1}^{i-1}(1 + 4\epsilon(j))$ and that $\epsilon(i) = \Theta\left(\sqrt{\frac{\log k(i) + \log \phi(N)}{k(i)}}\right)$. As $1 + x \leq \exp(x)$, we obtain $b(i) \geq Q \exp(-4 \sum_i \epsilon(i))$, where the summation is over all the rounds i until $\epsilon(i) \leq 1/2$. As $k(i)$ decreases geometrically, $\epsilon(i)$ increases geometrically and hence $\sum_{i:\epsilon(i) \leq 1/2} \epsilon(i)) = O(1)$. Choosing the constant Q large enough implies the result. □

Quasi-Uniformity: We now prove that the algorithm produces a quasi-uniform sample, that is, every set is sampled with probability at most $O(\log \phi(N))$ times[7] its LP contribution x_s. The key result, which is proved in lemma 1, is that the probability that a replica is forced in round i is most $\frac{1}{k(i)^2}$. Given lemma 1, the rest of the analysis follows essentially as in [11]. In particular, claim 4 from [11] shows that if the forcing probability is at most $1/k(i)^2$, then the probability that a particular replica r survives after the last phase is $O\left(\frac{\log \phi(N)}{M}\right)$. Since each set s has Mx_s replicas at the beginning of the algorithm, this implies that the probability that set s is picked is $O(x_s \log \phi(n))$.

We now focus on proving lemma 1. We first make two useful claims about the ordering relations for sets in $L_{p,q}$. Let $\sigma_{p,q}(s)$ denote the number of replicas of the first $\rho_{p,q}(s)$ sets in $L_{p,q}$.

Fact 4. *In round i, for all sets $s \in L_{p,q}$, it holds that $\rho_{p,q}(s) \geq b(i)\sigma_{p,q}(s)/k(i)$.*

Proof. This follows because each set s in $L_{p,q}$ has at most $\frac{k(i)}{b(i)}$ replicas. □

Fact 5. *If a set $s \in L_{p,q}$ is forced in round i, then $\sigma_{p,q}(s) \leq k(i)q/2$, provided $b(i) \geq 2$.*

Proof. Clearly, for a set s to be forced, the righthand size of inequality (2) must be positive, and thus $q > \rho_{p,q}(s)$. Thus by fact 4, $q \geq b(i)\sigma_{p,q}(s)/k(i)$, or equivalently $\sigma_{p,q}(s) < qk(i)/b(i) \leq qk(i)/2$. □

Lemma 1. *The probability that a set u is forced in round i is at most $\frac{1}{k(i)^2}$.*

---

[7] As $N = O(n^2)$ and $\phi(n) \leq n^d$ for geometric problem in dimension d, $\log(\phi(N)) = O(\log(\phi(n)))$, and hence we will use them interchangeably.

Proof. Consider some fixed round i and some set u. Let us denote $k(i)$ by k for notational convenience. The set u might potentially be forced when the forcing rule (2) holds for a particular point $p \in u$, and for a particular set s that appears later in $L_{p,q}$. By the construction of the list L_q, the set u was placed when there were at most $k(q+1)f(N,k(q+1))/N \leq \phi(N)(k(q+1))^{c+1}$ such possible points $p \in u$. For each such point p, there are at most $k(q+1)$ possibilities for the set s (by the definition of pseudo-depth). Thus it suffices to argue that the probability that a particular point p, with pseudo-depth q, goes tight at a particular set s is at most $(kq\phi(N))^{-g} = 1/\text{poly}(k,q,\phi(N))$, for some constant g that can be made arbitrarily large (in particular $g \geq c+3$ suffices). So let us fix a set s and a point p at pseudo-depth q. For notational simplicity we will drop p and q from the subscripts in the notation for $L_{p,q}(s), \rho_{p,q}(s)$ and $\sigma_{p,q}(s)$. Furthermore, we will drop i from $b(i)$.

From the definition of forcing in line (2), we need to upper bound

$$\Pr\left[\sum_{t \in L:\rho(t)>\rho(s)} \min\left(n_t^*, \frac{k(1+4\epsilon)}{2b}\right) < \frac{k}{2}(q - \rho(s)) \right] \tag{5}$$

Now by fact 4 and as $b > 1$ we have that $\frac{k}{2}(q - \rho(s)) \leq \frac{k}{2}\left(q - \frac{b\sigma(s)}{k}\right) = \frac{kq}{2} - \frac{b\sigma(s)}{2} \leq \frac{kq}{2} - \frac{\sigma(s)}{2}$. Thus to upper bound (5), it suffices to upper bound

$$\Pr\left[\sum_{t \in L:\rho(t)>\rho(s)} \min\left(n_t^*, \frac{k(1+4\epsilon)}{2b}\right) < \frac{kq}{2} - \frac{\sigma(s)}{2} \right] \tag{6}$$

Now, in order for the above event to occur, at least one of the following two events must occur:

$$\sum_{t \in L:\rho(t)>\rho(s)} n_t^* \leq \frac{kq}{2}\left(1 + \frac{\epsilon}{2}\right) - \frac{\sigma(s)}{2} \tag{7}$$

or

$$\sum_{t \in L:\rho(t)>\rho(s)} \left(1_{n_t^* > \frac{k(1+4\epsilon)}{2b}}\right)\left(n_t^* - \frac{k(1+4\epsilon)}{2b}\right) \geq \frac{\epsilon kq}{4} \tag{8}$$

Let us first bound (7). Note that $\sum_{t \in L:\rho(t)>\rho(s)} n_t^*$ is the sum of 0-1 random variables with mean equal to $\left(\frac{1}{2} + \epsilon\right)\sum_{t \in L:\rho(t)>\rho(s)} n_t \geq \left(\frac{1}{2} + \epsilon\right)(kq - \sigma(s)) = (1 + 2\epsilon)\left(\frac{kq}{2} - \frac{\sigma(s)}{2}\right)$. Thus by standard Chernoff bounds (Thm A.1.13 in [12]),

$$\Pr\left[\sum_{t \in L:\rho(t)>\rho(s)} n_t^* \leq (1+\epsilon)\left(\frac{kq}{2} - \frac{\sigma(s)}{2}\right) \right] \leq \exp(-(\epsilon/2)^2(kq - \sigma(s))/2) \tag{9}$$

Moreover, as $\sigma(s) < kq/2$ by fact 4, this probability is at most $\exp(-\epsilon^2 kq/8)$, which by our choice of ϵ, is at most $1/\text{poly}(k,q,\phi(N))$.

Finally, the probability of the event in (7) is upper bounded by the probability in (9) as $(1+\epsilon)\left(\frac{kq}{2} - \frac{\sigma(s)}{2}\right) = \left(1 + \frac{\epsilon}{2}\right)\frac{kq}{2} + \frac{kq\epsilon}{4} - \frac{\epsilon\sigma(s)}{2} - \frac{\sigma(s)}{2} \geq \frac{kq}{2}\left(1 + \frac{\epsilon}{2}\right) - \frac{\sigma(s)}{2}$ where the inequality above follows as $\sigma(s) \leq \frac{kq}{2}$, by Claim 4.

Let us now turn to upper bounding the probability of event (8). We make several simplifications. First, as $n_t^* \leq n_t$, clearly n_t^* cannot exceed $k(1 + 4\epsilon)/2b$ if $n_t \leq k(1 + 4\epsilon)/2b$. So it suffices to only consider sets t with $n_t \geq \frac{k(1+4\epsilon)}{2b}$. As $\sum_{t \in L} n_t \leq k(q + 1)$ (by definition of pseudo-depth) there can be at most $\frac{2b(q+1)}{1+4\epsilon} \leq 4bq$ such sets. Since $n_t^* \leq n_t \leq \frac{k}{b}$, to bound the probability of event (8), it suffices to upper bound $\Pr\left[\sum_{t:\rho(t)>\rho(s)}\left(1_{n_t^* > \frac{k(1+4\epsilon)}{2b}}\right)\frac{k}{b} \geq \frac{kq\epsilon}{4}\right]$ which can be written as

$$\Pr\left[\sum_{t:\rho(t)>\rho(s)}\left(1_{n_t^* > \frac{k(1+4\epsilon)}{2b}}\right) \geq \frac{bq\epsilon}{4}\right] \tag{10}$$

As $E[n_t^*] = n_t\left(\frac{1}{2} + \epsilon\right) \leq \frac{k}{b}\left(\frac{1}{2} + \epsilon\right) = \left(\frac{k(1+4\epsilon)}{2b}\right)\left(\frac{1+2\epsilon}{1+4\epsilon}\right)$, by the standard Chernoff bound (theorem A.1.12 in [12]), we have that $\Pr[n_t^* > k(1 + 4\epsilon)/2b] \leq \exp(-\epsilon^2(k/3b))$. By our choice of ϵ, this probability is at most $1/\text{poly}(k, \phi(n))$.

So, we can upper bound (10) as the probability that a sum of at most $2bq$ Bernoulli random variables, each having a probability of $1/\text{poly}(k, \phi(n))$ of being 1, exceeds $\frac{bq\epsilon}{4}$. By standard Chernoff bounds the probability of this event is $\exp(-\epsilon^2 \cdot q \cdot \text{poly}(k, \phi(N)))$, which is at most $\frac{1}{\text{poly}(k,q,\phi(N))}$.   □

References

1. Feige, U.: A threshold of ln n for approximating set cover. J. ACM 45(4), 634–652 (1998)
2. Chakrabarty, D., Grant, E., Könemann, J.: On Column-Restricted and Priority Covering Integer Programs. In: Eisenbrand, F., Shepherd, F.B. (eds.) IPCO 2010. LNCS, vol. 6080, pp. 355–368. Springer, Heidelberg (2010)
3. Jain, K.: A factor 2 approximation algorithm for the generalized steiner network problem. Combinatorica 21(1), 39–60 (2001)
4. Chekuri, C., Clarkson, K.L., Har-Peled, S.: On the set multi-cover problem in geometric settings. In: Symposium on Computational Geometry, pp. 341–350 (2009)
5. Bansal, N., Pruhs, K.: The geometry of scheduling. In: FOCS, pp. 407–414 (2010)
6. Brönnimann, H., Goodrich, M.T.: Almost optimal set covers in finite vc-dimension. Discrete & Computational Geometry 14(4), 463–479 (1995)
7. Clarkson, K.L., Varadarajan, K.R.: Improved approximation algorithms for geometric set cover. Discrete & Computational Geometry 37(1), 43–58 (2007)
8. Varadarajan, K.R.: Epsilon nets and union complexity. In: Symposium on Computational Geometry, pp. 11–16 (2009)
9. Aronov, B., Ezra, E., Sharir, M.: Small-size epsilon-nets for axis-parallel rectangles and boxes. In: STOC, pp. 639–648 (2009)
10. Varadarajan, K.: Weighted geometric set cover via quasi-uniform sampling. In: STOC, pp. 641–648 (2010)
11. Chan, T.M., Grant, E., Könemann, J., Sharpe, M.: Weighted capacitated, priority, and geometric set cover via improved quasi-uniform sampling. In: SODA, pp. 1576–1585 (2012)
12. Alon, N., Spencer, J.: The probabilistic method. John Wiley (2008)

A Bicriteria Approximation for the Reordering Buffer Problem[*]

Siddharth Barman, Shuchi Chawla, and Seeun Umboh

University of Wisconsin–Madison
{sid,shuchi,seeun}@cs.wisc.edu

Abstract. In the reordering buffer problem (RBP), a server is asked to process a sequence of requests lying in a metric space. To process a request the server must move to the corresponding point in the metric. The requests can be processed slightly out of order; in particular, the server has a buffer of capacity k which can store up to k requests as it reads in the sequence. The goal is to reorder the requests in such a manner that the buffer constraint is satisfied and the total travel cost of the server is minimized. The RBP arises in many applications that require scheduling with a limited buffer capacity, such as scheduling a disk arm in storage systems, switching colors in paint shops of a car manufacturing plant, and rendering 3D images in computer graphics.

We study the offline version of RBP and develop bicriteria approximations. When the underlying metric is a tree, we obtain a solution of cost no more than $9 \, \text{OPT}$ using a buffer of capacity $4k + 1$ where OPT is the cost of an optimal solution with buffer capacity k. Via randomized tree embeddings, this implies an $O(\log n)$ approximation to cost and $O(1)$ approximation to buffer size for general metrics. In contrast, when the buffer constraint is strictly enforced, constant-factor approximations are known only for the uniform metric (Avigdor-Elgrabli et al., 2012); the best known approximation ratio for arbitrary metrics is $O(\log^2 k \log n)$ (Englert et al., 2007).

1 Introduction

We consider the reordering buffer problem (RBP) where a server with buffer capacity k has to process a sequence of requests lying in a metric space. The server is initially stationed at a given vertex and at any point of time it can store at most k requests. In particular, if there are k requests in the buffer then the server must process one of them (that is, visit the corresponding vertex in the metric space) before reading in the next request from the input sequence. The objective is to process the requests in an order that minimizes the total distance travelled by the server.

RBP provides a unified model for studying scheduling with limited buffer capacity. Such scheduling problems arise in numerous areas including storage systems, computer graphics, job shops, and information retrieval (see [17,20,15,7]).

[*] This work was supported in part by NSF awards CCF-0643763 and CNS-0905134.

L. Epstein and P. Ferragina (Eds.): ESA 2012, LNCS 7501, pp. 157–168, 2012.
© Springer-Verlag Berlin Heidelberg 2012

For example, in a secondary storage system the overall performance critically depends on the response time of the underlying disk devices. Hence disk devices need to schedule their disk arm in a way that minimizes the *mean seek time*. Specifically, these devices receive read/write requests which are located on different cylinders and they must move the disk arm to the proper cylinder in order to serve a request. The device can buffer a limited number of requests and must deploy a scheduling policy to minimize the overall service time. Note that we can model this disk arm scheduling problem as a RBP instance by representing the disk arm as a server and the array of cylinders as a metric space over read/write requests.

The RBP can be seen to be NP-Hard via a reduction from the traveling salesperson problem. We study approximation algorithms. RBP has been considered in both online and offline contexts. In the online setting the entire input sequence is not known beforehand and the requests arrive one after the other. This setting was considered by Englert et al. [9], who developed an $O(\log^2 k \log n)$-competitive algorithm. To the best of our knowledge this is the best known approximation guarantee for RBP over arbitrary metrics both in the online and offline case.

RBP remains NP-Hard even when restricted to the uniform metric (see [8] and [2]). In fact the uniform metric is an interesting special case as it models scheduling of paint jobs in a car manufacturing plant. In particular, switching paint color is a costly operation; hence, paint shops temporarily store cars and process them out of order to minimize color switches. Over the uniform metric, RBP is somewhat related to paging. However, unlike for the latter, simple greedy strategies like First in First Out and Least Recently Used yield poor competitive ratios (see [17]). Even the offline version of the uniform metric case does not seem to admit simple approximation algorithms. The best known approximation for this setting, due to Avigdor-Elgrabli et al. [4], relies on intricate rounding of a linear programming relaxation in order to get a constant-factor approximation.

The hardness of the RBP appears to stem primarily from the strict buffer constraint; it is therefore natural to relax this constraint and consider bicriteria approximations. We say that an algorithm achieves an (α, β) bicriteria approximation if, given any RBP instance, it generates a solution of cost no more than α OPT using a buffer of capacity βk. Here OPT is the cost of an optimal solution with buffer capacity k. There are few bicriteria results known for the RBP. For the offline version of the uniform metric case, a bicriteria approximation of $\left(O(\frac{1}{\epsilon}), 2 + \epsilon\right)$ for every $\epsilon > 0$ was given by Chan et al. [8]. For the online version of this restricted case, Englert et al. [10] developed a $(4, 4)$-competitive algorithm. They further showed how to convert this bicriteria approximation into a true approximation with a logarithmic ratio. We show in the full version of the paper [6] that such a conversion from a bicriteria approximation to a true approximation is not possible at small loss in more general metrics, e.g. the evenly-spaced line metric. In more general metrics, relaxing the buffer constraint therefore gives us significant extra power in approximation.

We study bicriteria approximations for the offline version of RBP. When the underlying metric is a weighted tree we obtain a $\left(9, 4 + \frac{1}{k}\right)$ bicriteria approximation algorithm. Using tree embeddings of [11] this implies a $\left(O(\log n),\right.$ $\left. 4 + \frac{1}{k}\right)$ bicriteria approximation for arbitrary metrics over n points.

Other Related Work: Besides the work of Englert et al. [9], existing results address RBP over very specific metrics. RBP was first considered by Räcke et al. [17]. They focused on the uniform metric with online arrival of requests and developed an $O(\log^2 k)$-competitive algorithm. This was subsequently improved on by a number of results [10,3,1], leading to an $O(\sqrt{\log k})$-competitive algorithm [1]. With the disk arm scheduling problem in mind, Khandekar et al. [14] considered the online version of RBP over the evenly-spaced line metric (line graph with unit edge lengths) and gave an online algorithm with a competitive ratio of $O(\log^2 n)$. This was improved on by Gamzu et al. [12] to an $O(\log n)$-competitive algorithm.

On the lower bound side, the only known results, to the best of our knowledge, are for the online setting: $\Omega(\sqrt{\log k / \log \log k})$ for the uniform metric [1], and 2.154 for the line metric [12].

Bicriteria approximations have been studied previously in the context of resource augmentation (see [16] and references therein). In this paradigm, the algorithm is augmented with extra resources (usually faster processors) and the benchmark is an optimal solution without augmentation. This approach has been applied to, for example, paging [19], scheduling [13,5], and routing problems [18].

Techniques: We can assume without loss of generality that the server is lazy and services each request when it absolutely must—to create space in the buffer for a newly received request. Then after reading in the first k requests, the server must serve exactly one request for each new one received. Intuitively, adding extra space in the buffer lets us defer serving decisions. In particular, while the optimal server must serve a request at every step, we serve requests in batches at regular intervals. Partitioning requests into batches appears to be more tractable than determining the exact order in which requests appear in an optimal solution. This enables us to go beyond previous approaches (see [3,4]) that try to extract the order in which requests appear in an optimal solution. We enforce the buffer capacity constraint by placing lower bounds on the cardinalities of the batches. In particular, by ensuring that each batch is large enough, we make sure that the server "carries forward" few requests.

A crucial observation that underlies our algorithm is that when the underlying metric is a tree, we can find vertices $\{v_i\}_i$ that any solution with buffer capacity k must visit in order. This allows us to anchor the ith batch at v_i and equate the serving cost of a batch to the cost of the subtree spanning the batch and rooted at v_i. Overall, when the underlying metric is a tree, the problem of finding low-cost batches with cardinality constraints reduces to finding low-cost subtrees which are rooted at v_is, cover all the requests, and satisfy the same cardinality constraints. We formulate a linear programming relaxation, LP1, for this covering problem.

Rounding LP1 directly is difficult because of the cardinality constraints. To handle this we round LP1 partially to formulate another relaxation that is free of the cardinality constraints and is amenable to rounding. Specifically, using a fractional optimal solution to LP1, we determine for each request j an *interval* of indices, $\Gamma(j)$, such that any solution that assigns every request to a batch within its corresponding interval approximately satisfies the buffer constraint. This allows us to remove the cardinality constraints and instead formulate an interval-assignment relaxation LP2. In order to get the desired bicriteria approximation we show two things: first, the optimal cost achieved by LP2 is within a constant factor of the optimal cost for the given RBP instance; second, an integral feasible solution of LP2 can be transformed into a RBP solution using a bounded amount of extra buffer space. Finally we develop a rounding algorithm for LP2 which achieves an approximation ratio of 2.

2   Notation

An instance of RBP is specified by a metric space over a vertex set V, a sequence of n vertices (requests), an integer k, and a starting vertex v_0. The metric space is represented by a graph $G = (V, E)$ with distance function $d : E \to \mathbf{R}^+$ on edges. We index requests by j. We assume without loss of generality that requests are distinct vertices. Starting at v_0, the server reads requests from the input sequence into its buffer and clears requests from its buffer by visiting them in the graph (we say these requests are served). The goal is to serve all requests, having at most k buffered requests at any point in time, with minimum traveling distance. We denote the optimal solution as OPT. For the most part of this paper, we focus on the special case where G is a tree.

We break up the timeline into windows as follows. Without loss of generality, n is a multiple of $2k + 1$, i.e. $n = (2k + 1)m$. For $i \in [m]$, we define window W_i to be the set of requests from $(2k + 1)(i - 1) + 1$ to $(2k + 1)i$. Let $w(j)$ be the index of the window in which j belongs. The i-th *time window* is defined to be the duration in which the server reads W_i.

3   Reduction to Request Cover Problem

In this section we show how to use extra buffer space to convert the RBP into a new and simpler problem that we call Request Cover. The key tool for the reduction is the following lemma which states that we can find for every window a vertex in the graph G that must be visited by any feasible solution within the same window. We call these vertices *terminals*. This allows us to break up the server's path into segments that start and end at terminals.

Lemma 1. *For each i, there exists a vertex v_i such that all feasible solutions with buffer capacity k must visit v_i in the i-th time window.*

Proof. Fix a feasible solution and i. We orient the tree as follows. For each edge $e = (u, v)$, if after removing e from the tree, the component containing u contains at most k requests of W_i, then we direct the edge from u to v. Since $|W_i| = 2k+1$, there is exactly one directed copy of each edge.

An oriented tree is acyclic so there exists a vertex v_i with incoming edges only. We claim that the server must visit v_i during the i-th time window. During the i-th time window, the server reads all $2k+1$ requests of W_i. Since each component of the induced subgraph $G[V \setminus \{v_i\}]$ contains at most k requests of W_i and the server has a buffer of size k, it cannot remain in a single component for the entire time window. Therefore, the server must visit at least two components, passing by v_i, at some point during the i-th time window.   □

For the remainder of the argument, we will fix the terminals $v_1, \ldots, v_m$. Note that since G is a tree, there is a unique path visiting the terminals in sequence, and every solution must contain this path. For each i, let P_i denote the path from v_{i-1} to v_i. We can now formally define request covers.

Definition 1 (Request cover). *Let $\mathcal{B}$ be a partition of the requests into batches $B_1, \ldots, B_m$, and $\mathcal{E}$ be an ordered collection of m edge subsets $E_1, \ldots, E_m \subseteq E$. The pair $(\mathcal{B}, \mathcal{E})$ is a* request cover *if*

1. *For every request j, the index of the batch containing j is at least $w(j)$, i.e. the window in which j is released.*
2. *For all $i \in [m]$, $E_i \cup P_i$ is a connected subgraph spanning B_i.*
3. *There exists a constant β such that for all $i \in [m]$, we have $\sum_{l \leq i} |B_l| \geq (2k+1)i - \beta k$; we say that the request cover is β-feasible. We call the request cover* feasible *if $\beta = 1$.*

The length *of a request cover is $d(\mathcal{E}) = \sum_i d(E_i)$.*

Definition 2 (Request Cover Problem (RCP)). *In the RCP we are given a metric space $G = (V, E)$ with lengths $d(e)$ on edges, a sequence of n requests, buffer capacity constraint k, and a sequence of $m = n/(2k+1)$ terminals $v_1, \ldots, v_m$. Our goal is to find a feasible request cover of minimum length.*

We will now relate the request cover problem to the RBP. Let $(\mathcal{B}^*, \mathcal{E}^*)$ denote the optimal solution to the RCP. We show on the one hand (Lemma 2) that this solution has cost at most that of OPT, the optimal solution to RBP. On the other hand, we show (Lemma 3) that any β-feasible solution to RCP can be converted into a solution to the RBP that is feasible for a buffer of size $(2 + \beta)k + 1$ with a constant factor loss in length.

Lemma 2. $d(\text{OPT}) \geq d(\mathcal{E}^*)$.

Proof. For each i, let E_i be the edges traversed by the optimal server during the i-th time window and let $\mathcal{E}$ be the collection of edge subsets. We have $d(\text{OPT}) \geq \sum_i d(E_i) = d(\mathcal{E})$, so it suffices to show that $\mathcal{E} = (E_1, \ldots, E_m)$ is a feasible request cover. By Lemma 1, both E_i and P_i are connected subgraphs containing v_i for each i. Hence $\mathcal{E}$ is connected. Since E_l contains the requests served in the l-th time window for each l, and for each i the server has read $(2k+1)i$ requests and served all except at most k of them by the end of the i-th time window, we get that $\sum_{l \leq i} |B_l| \geq (2k + 1)i - k$. This proves that $\mathcal{E}$ is a feasible request cover.   □

Next, consider a request cover $(\mathcal{B}, \mathcal{E})$. We may assume without loss of generality that for all i, $E_i \cap P_i = \emptyset$. This observation implies that E_i can be partitioned into components $E_i(p)$ for each vertex $p \in P_i$, where $E_i(p)$ is the component of E_i containing p.

We will now define a server for the RBP, BATCH-SERVER$(\mathcal{B}, \mathcal{E})$, based on the solution $(\mathcal{B}, \mathcal{E})$. Recall that the server has to start at v_0. In the i-th iteration, it first buffers all requests in window W_i. Then it moves from v_{i-1} to v_i and serves requests of B_i as it passes by them.

Algorithm 1. BATCH-SERVER$(\mathcal{B}, \mathcal{E})$

1: Start at v_0
2: **for** $i = 1$ **to** m **do**
3:      (Buffering phase) Read W_i into buffer
4:      (Serving phase) Move from v_{i-1} to v_i along P_i, and for each vertex $p \in P_i$, perform an Eulerian tour of $E_i(p)$. Serve requests of B_i along the way.
5: **end for**

Lemma 3. *Given a β-feasible request cover $(\mathcal{B}, \mathcal{E})$, BATCH-SERVER$(\mathcal{B}, \mathcal{E})$ is a feasible solution to the RBP instance with a buffer of size $(2+\beta)k+1$, and has length at most $d(\mathrm{OPT}) + 2d(\mathcal{E})$.*

Proof. We analyze the length first. In iteration i, the server uses each edge of P_i exactly once. Since E_i is a disjoint union of $E_i(p)$ for $p \in P_i$, the server uses each edge of E_i twice during the Eulerian tours of E_i's components. The total length is therefore $\sum_i d(P_i) + \sum_i 2d(E_i) \leq d(\mathrm{OPT}) + 2d(\mathcal{E})$.

Next, we show that the server has at most $(2+\beta)k+1$ requests in its buffer at any point in time. We claim that all of B_i is served by the end of the i-th iteration. Consider a request j that belongs to a batch B_i. Since i is at least as large as $w(j)$, the request has already been received by the ith phase. The server visits j's location during the ith iteration and therefore services the request at that time if not earlier. This proves the claim.

The claim implies that the server begins the $(i+1)$-th iteration having read $(2k+1)i$ requests and served $\sum_{l \leq i} |B_l| \geq (2k+1)i - \beta k$ requests, that is, with at most βk requests in its buffer. It adds $2k+1$ requests to be the buffer during this iteration. So it uses at most $(2+\beta)k+1$ buffer space at all times. $\square$

4   Approximating the Request Cover Problem

We will now show how to approximate the request cover problem. Our approach is to start with an LP relaxation of the problem, and use the optimal fractional solution to the LP to further define a simpler covering problem which we then approximate in Section 4.2.

4.1 The Request Cover LP and the Interval Cover Problem

The integer linear program formulation of RCP is as follows. To obtain an LP relaxation we relax the last two constraints to $x(i, j), y(e, i) \in [0, 1]$.

$$
\begin{aligned}
&\text{minimize} \quad \sum_i \sum_e y(e, i) d_e \\
&\text{subject to} \quad \sum_{w(j) \leq i} x(j, i) = 1 \qquad\qquad\qquad\qquad \forall j \\
&\qquad\qquad \sum_{j: w(j) \leq i} \sum_{w(j) \leq i' \leq i} x(j, i') \geq (2k+1)i - k \qquad \forall i \\
&\qquad\qquad y(e, i) \geq x(j, i) \qquad\qquad\qquad\qquad \forall i, j, e \in R_{ji} \\
&\qquad\qquad x(j, i), y(e, i) \in \{0, 1\} \qquad\qquad\qquad \forall i, j, e
\end{aligned}
\qquad \text{(LP1)}
$$

Here the variable $x(j, i)$ indicates whether request j is assigned to batch B_i and the variable $y(e, i)$ indicates whether edge e is in E_i. Recall that the edge set E_i along with path P_i should span B_i. Let R_{ji} denote the (unique) path in G from j to P_i. The third inequality above captures the constraint that if j is assigned to B_i and $e \in R_{ji}$, then e must belong to E_i.

Let (x^*, y^*) be the fractional optimal solution to the linear relaxation of (LP1). Instead of rounding (x^*, y^*) directly to get a feasible request cover, we will show that it is sufficient to find request covers that "mimic" the fractional assignment x^* but do not necessarily satisfy the cardinality constraints on the batches (i.e. the second set of inequalities in the LP). To this end we define an *interval request cover* below.

Definition 3 (Interval request cover). *For each request j, we define the service deadline $h(j) = \min\{i \geq w(j) : \sum_{l \leq i} x^*(j, l) \geq 1/2\}$ and the service interval $\Gamma(j) = [w(j), h(j)]$. A request cover $(\mathcal{B}, \mathcal{E})$ is an interval request cover if it assigns every request to a batch within its service intervals.*

In other words, while x^* "half-assigns" each request no later than its service deadline, an interval request cover mimics x^* by integrally assigning each request no later than its service deadline. The following is a linear programming formulation for the problem of finding minimum length interval request covers.

$$
\begin{aligned}
&\text{minimize} \quad \sum_i \sum_e y(e, i) d_e \\
&\text{subject to} \quad \sum_{i \in \Gamma(j)} x(j, i) \geq 1 \qquad\qquad\qquad\qquad \forall j \\
&\qquad\qquad y(e, i) \geq x(j, i) \qquad\qquad \forall j, i \in \Gamma(j), e \in R_{ji} \\
&\qquad\qquad x(j, i), y(e, i) \in [0, 1] \qquad\qquad\qquad \forall i, j, e
\end{aligned}
\qquad \text{(LP2)}
$$

Let $(\tilde{x}, \tilde{y})$ be the fractional optimal solution of (LP2). We now show that interval request covers are 2-feasible request covers and that $d(\tilde{y}) \leq 2d(y^*)$. Since $d(y^*) \leq d(\mathcal{E}^*)$, it would then suffice to round (LP2).

Lemma 4. *Interval request covers are 2-feasible.*

Proof. Fix i. Let $H_i := \{j : h(j) \leq i\}$ denote the set of all requests whose service intervals end at or before the ith time window. We first claim that $|H_i| \geq (2k+1)i - 2k$. In particular, the second constraint of (LP1) and the definition of H_i gives us

$$(2k+1)i - k \leq \sum_{j:w(j)\leq i} \sum_{i'\leq i} x^*(j,i')$$

$$= \sum_{j\in H_i:w(j)\leq i} \sum_{i'\leq i} x^*(j,i') + \sum_{j\notin H_i:w(j)\leq i} \sum_{i'\leq i} x^*(j,i')$$

$$\leq \sum_{j\in H_i:w(j)\leq i} 1 + \sum_{j\notin H_i:w(j)\leq i} \frac{1}{2} = |H_i| + \frac{1}{2}\left((2k+1)i - |H_i|\right).$$

The claim now follows from rearranging the above inequality. Note that in an interval request cover, each request in H_i is assigned to some batch B_l with $l \leq i$. Therefore, $\sum_{l \leq i} |B_l| \geq |H_i| \geq (2k+1)i - 2k$.   □

We observe that multiplying all the coordinates of x^* and y^* by 2 gives us a feasible solution to (LP2). Thus we have the following lemma.

Lemma 5. *We have $d(\tilde{y}) \leq 2d(y^*)$.*

Note that the lemma says nothing about the integral optimal of (LP2) so a solution that merely approximates the optimal integral interval request cover may not give a good approximation to the RBP, and we need to bound the integrality gap of the LP. In the following subsection, we show that we can find an interval request cover of length at most $2d(\tilde{y})$.

4.2   Approximating the Interval Assignment LP

Before we describe the general approximation, we consider a special case.
Example: two edges. Suppose the tree is a line graph consisting of three vertices u_1, u_2 and v with unit-length edges $e_1 = (u_1, v)$ and $e_2 = (u_2, u_1)$. Requests reside at u_1 and u_2, and all terminals at v. For each j residing at u_1 and for all i, we have $R_{ji} = \{e_1\}$. For each j residing at u_2 and for all i, we have $R_{ji} = \{e_1, e_2\}$. Thus feasible solutions to (LP2) satisfy the constraints

$$\sum_{i\in\Gamma(j)} y(e_1, i) \geq 1 \quad \forall j, \qquad \sum_{i\in\Gamma(j)} y(e_2, i) \geq 1 \quad \forall j \in u_2.$$

The constraints suggest that the vector $y(e_1, \cdot)$ is a fractional hitting set for the collection of intervals $\mathcal{I}(e_1) := \{\Gamma(j)\}$, and $y(e_2, \cdot)$ for $\mathcal{I}(e_2) := \{\Gamma(j) : j \in u_2\}$. While the general hitting set problem is hard, the special case with interval sets is solvable in polynomial time and the relaxation has no integrality gap[1]. Thus,

---

[1] One way to see this is that the columns of the constraint matrix have consecutive ones, and thus the constraint matrix is totally unimodular.

a naive approach is to first compute minimum hitting sets $M(e_1)$ and $M(e_2)$ for $\mathcal{I}(e_1)$ and $\mathcal{I}(e_2)$, respectively. Then we add e_1 to E_i for $i \in M(e_1)$, and e_2 to E_i for $i \in M(e_2)$. However, the resulting edge sets may not be connected. Instead, we make use of the following observations: (1) we should include e_2 in E_i only if $e_1 \in E_i$, and, (2) minimal hitting sets are at most twice minimum fractional hitting sets (for our setting, we formally establish this in Lemma 9). These facts suggest that we should first compute a minimal hitting set $M(e_1)$ for $\mathcal{I}(e_1)$ and then compute a minimal hitting set $M(e_2)$ for $\mathcal{I}(e_2)$ with the constraint that $M(e_2) \subseteq M(e_1)$. This is a valid solution to (LP2) since $\mathcal{I}(e_2) \subseteq \mathcal{I}(e_1)$. We proceed as usual to compute $\mathcal{E}$. The resulting $\mathcal{E}$ is connected by observation (1), and $d(\mathcal{E}) \leq 2d(\tilde{y})$ by observation (2).

General case. Motivated by the two-edge example, at a high level, our approach for the general case is as follows:

1. We construct interval hitting set instances over each edge.
2. We solve these instances starting from the edges nearest to the paths P_i first.
3. We iteratively "extend" solutions for the instances nearer the paths to get minimal hitting sets for the instances further from the paths.

We then use Lemma 9 to argue a 2-approximation on an edge-by-edge basis.

Note that whether an edge is closer to some P_i along a path R_{ji} for some j depends on which direction we are considering the edge in. We therefore modify (LP2) to include directionality of edges, replacing each edge e with bidirected arcs and directing the paths R_{ji} from j to P_i.

$$
\begin{aligned}
&\text{minimize} && \sum_i \sum_a y(a,i)d_a \\
&\text{subject to} && \sum_{i \in \Gamma(j)} x(j,i) \geq 1 && \forall j \\
& && y(a,i) \geq x(j,i) && \forall j, i \in \Gamma(j), a \in R_{ji} \\
& && x(j,i), y(a,i) \in [0,1] && \forall i,j,a
\end{aligned}
\qquad \text{(LP2')}
$$

For every edge e and window i, there is a single orientation of edge e that belongs to R_{ji} for some j. So there is a 1-1 correspondence between the variables $y(e,i)$ in (LP2) and the variables $y(a,i)$ in (LP2'), and the two LPs are equivalent. Henceforth we focus on (LP2').

Before presenting our final approximation we need some more notation.

Definition 4. *For each request j, we define R_j to be the directed path from j to $\bigcup_{i \in \Gamma(j)} P_i$. For each arc a, we define $C(a) = \{j : a \in R_j\}$ and the set of intervals $\mathcal{I}(a) = \{\Gamma(j) : j \in C(a)\}$. We say that a is a cut arc if $C(a) \neq \emptyset$.*

We say that an arc a precedes arc a', written $a \prec a'$, if there exists a directed path in the tree containing both the arcs and a appears after a' in the path.

The proof of the following lemma uses an argument similar to the examples.

Lemma 6. *Feasible solutions (x, y) of (LP2') satisfy the following set of constraints for all arcs a: $\sum_{i \in \Gamma(j)} y(a, i) \geq 1 \quad \forall j \in C(a)$.*

Proof. Let (x, y) be a feasible solution of (LP2'). Fix an arc a and $j \in C(a)$. For each $i \in \Gamma(j)$, we have $a \in R_{ji}$ since R_j is a path from j to a connected subgraph containing P_i. By feasibility, we have $y(a, i) \geq x(j, i)$. Summing over $\Gamma(j)$, we get $\sum_{i \in \Gamma(j)} y(a, i) \geq \sum_{i \in \Gamma(j)} x(j, i) \geq 1$ where the last inequality follows from feasibility. $\qquad \square$

We are now ready to describe the algorithm. At a high level, Algorithm 2 does the following: initially, it finds a cut arc a with no cut arc preceding it and computes a minimal hitting set $M(a)$ for $\mathcal{I}(a)$; iteratively, it finds a cut arc a whose preceding cut arcs have been processed previously, and minimally "extends" the hitting sets $M(a')$ computed previously for the preceding arcs a' to form a minimal hitting set $M(a)$.

The next lemma shows that Algorithm 2 actually manages to process all cut arcs a and that $F(a)$ is a hitting set for $\mathcal{I}(a)$.

Lemma 7. *For each iteration, the following holds.*

1. *If $U \neq \emptyset$, the inner 'while' loop finds an arc.*
2. *$F(a)$ is a hitting set for the intervals $\mathcal{I}(a)$.*

Proof. The first statement follows from the fact that an arc does not precede its reverse arc and so the inner 'while' loop does not repeat arcs and hence it stops with some arc. We prove the second statement of the lemma by induction on the algorithm's iterations. In the first iteration, the set U consists of cut arcs so $a' \not\prec a$ for all cut arcs a'. Therefore, for all $\Gamma(j) \in \mathcal{I}(a)$, a is the arc on R_j closest to $\bigcup_{i \in \Gamma(j)} P_i$ and $v \in \bigcup_{i \in \Gamma(j)} P_i$. This proves the base case. Now we prove the inductive case. Fix an interval $\Gamma(j) \in \mathcal{I}(a)$. If a is the arc on R_j closest to $\bigcup_{i \in \Gamma(j)} P_i$ and $v \in \bigcup_{i \in \Gamma(j)} P_i$, then $F(a) \cap \Gamma(j) \neq \emptyset$. If not, then there exists a neighboring arc $(v, w) \in R_j$ closer to $\bigcup_{i \in \Gamma(j)} P_i$. We have that $\Gamma(j) \in \mathcal{I}((v, w))$ and $(v, w) \prec a$. Since the algorithm has processed all cut arcs preceding a, by the inductive hypothesis we have $F((v, w)) \cap \Gamma(j) \neq \emptyset$. This implies that $M((v, w))$ is a hitting set for $\mathcal{I}((v, w))$ and so $F(a) \cap \Gamma(j) \neq \emptyset$. Hence, $F(a)$ is a hitting set for $\mathcal{I}(a)$. $\qquad \square$

Let E_i be the set of edges whose corresponding arcs are in A_i and $\mathcal{E} = (E_1, \ldots, E_m)$, i.e. the undirected version of $\mathcal{A}$.

Lemma 8. *$(\mathcal{B}, \mathcal{E})$ is an interval request cover.*

Proof. The connectivity of $E_i \cup P_i$ follows from the fact that the algorithm starts with $A_i = \emptyset$, and in each iteration an arc $a = (u, v)$ is added to A_i only if $v \in P_i$ or v is incident to some edge previously added to A_i.

Now it remains to show that there exists $i \in \Gamma(j)$ such that j is incident to A_i or P_i. If $R_j = \emptyset$, then $j \in \bigcup_{i \in \Gamma(j)} P_i$. On the other hand if $R_j \neq \emptyset$, then let $a \in R_j$ be the arc incident to j. Since the algorithm processes all cut arcs, we have $a \in \bigcup_{i \in \Gamma(j)} A_i$ and thus j is incident to $\bigcup_{i \in \Gamma(j)} A_i$. $\qquad \square$

Algorithm 2. Greedy extension

1: $U \leftarrow \{a : C(a) \neq \emptyset\}$
2: $A_i \leftarrow \emptyset$ for all i
3: $M(a) \leftarrow \emptyset$ for all arcs a
4: **while** $U \neq \emptyset$ **do**
5:     Let a be any arc in U
6:     **while** there exists $a' \prec a$ in U **do**
7:         $a \leftarrow a'$
8:     **end while**
9:     Let $a = (u, v)$
10:    $F(a) \leftarrow \{i : v \in P_i\} \cup \bigcup_{w:(v,w) \prec a} M((v, w))$
11:    Set $M(a) \subseteq F(a)$ to be a minimal hitting set for the intervals $\mathcal{I}(a)$
12:    $A_i \leftarrow A_i \cup \{a\}$ for all $i \in M(a)$
13:    $U \leftarrow U \setminus \{a\}$
14: **end while**
15: $f(j) \leftarrow \min\{i \in \Gamma(j) : j \text{ incident to } A_i \text{ or } P_i\}$ for all j
16: $B_i \leftarrow \{j : f(j) = i\}$ for all i
17: **return** $\mathcal{A} = (A_1, \ldots, A_m)$, $\mathcal{B} = (B_1, \ldots, B_m)$

The following lemma lower bounds $D(a)$, the number of disjoint intervals in $\mathcal{I}(a)$.

Lemma 9. $D(a) \geq |M(a)|/2$ for all arcs a.

Proof. Say $i_1 < \ldots < i_{|M(a)|}$ are the elements in $M(a)$. For each $1 \leq l \leq |M(a)|$, there exists an interval $\Gamma(j_l) \in \mathcal{I}(a)$ such that $M(a) \cap \Gamma(j_l) = \{i_l\}$, because otherwise $M(a) \setminus \{i_l\}$ would still be a hitting set, contradicting the minimality of $M(a)$. We observe that the intervals $\Gamma(j_l)$ and $\Gamma(j_{l+2})$ are disjoint since $\Gamma(j_l)$ contains i_l and $\Gamma(j_{l+2})$ contains i_{l+2} but neither contains i_{l+1}. Therefore, the set of $\lceil |M(a)|/2 \rceil$ intervals $\{\Gamma(j_l) : 1 \leq l \leq |M(a)| \text{ and } l \text{ odd}\}$ are disjoint. $\square$

Finally, we analyze the cost of the algorithm.

Lemma 10. $d(\mathcal{E}) \leq 2d(\tilde{y})$.

Proof. Fix an arc a. From Lemmas 6 and 9, we get $\sum_i \tilde{y}(a, i) \geq D(a) \geq |M(a)|/2$. Since $d(\mathcal{E}) = d(\mathcal{A})$, we have $d(\mathcal{E}) = \sum_a |M(a)| \cdot d_a \leq \sum_a (2 \sum_i \tilde{y}(a, i)) \times d_a = 2d(\tilde{y})$. $\square$

Together with Lemmas 4 and 5, we have that $(\mathcal{B}, \mathcal{E})$ is a 2-strict request cover of length at most $4d(y^*) \leq 4d(\mathcal{E}^*)$.

Lemmas 2 and 3 imply that BATCH-SERVER$(\mathcal{B}, \mathcal{E})$ travels at most 9 OPT and uses a buffer of capacity $4k + 1$. This gives us the following theorem.

Theorem 1. *There exists an offline $(9, 4 + \frac{1}{k})$-bicriteria approximation for RBP when the underlying metric is a weighted tree.*

Using tree embeddings of [11], we get

Theorem 2. *There exists an offline $(O(\log n), 4 + \frac{1}{k})$-bicriteria approximation for RBP over general metrics.*

References

1. Adamaszek, A., Czumaj, A., Englert, M., Räcke, H.: Almost tight bounds for reordering buffer management. In: STOC (2011)
2. Asahiro, Y., Kawahara, K., Miyano, E.: Np-hardness of the sorting buffer problem on the uniform metric. Discrete Applied Mathematics (2012)
3. Avigdor-Elgrabli, N., Rabani, Y.: An improved competitive algorithm for reordering buffer management. In: SODA (2010)
4. Avigdor-Elgrabli, N., Rabani, Y.: A constant factor approximation algorithm for reordering buffer management. CoRR, abs/1202.4504 (2012)
5. Bansal, N., Pruhs, K.: Server scheduling in the l p norm: a rising tide lifts all boat. In: STOC 2003 (2003)
6. Barman, S., Chawla, S., Umboh, S.: A bicriteria approximation for the reordering buffer problem. CoRR, abs/1204.5823 (2012)
7. Blandford, D., Blelloch, G.: Index compression through document reordering. In: Proceedings of the Data Compression Conference, DCC 2002 (2002)
8. Chan, H.-L., Megow, N., Sitters, R., van Stee, R.: A note on sorting buffers offline. Theor. Comput. Sci. (2012)
9. Englert, M., Räcke, H., Westermann, M.: Reordering buffers for general metric spaces. Theory of Computing (2010)
10. Englert, M., Westermann, M.: Reordering Buffer Management for Non-uniform Cost Models. In: Caires, L., Italiano, G.F., Monteiro, L., Palamidessi, C., Yung, M. (eds.) ICALP 2005. LNCS, vol. 3580, pp. 627–638. Springer, Heidelberg (2005)
11. Fakcharoenphol, J., Rao, S., Talwar, K.: A tight bound on approximating arbitrary metrics by tree metrics. J. Comput. Syst. Sci. 69(3) (2004)
12. Gamzu, I., Segev, D.: Improved online algorithms for the sorting buffer problem on line metrics. ACM Transactions on Algorithms (2009)
13. Kalyanasundaram, B., Pruhs, K.: Speed is as powerful as clairvoyance. Journal of the ACM 47(4) (2000)
14. Khandekar, R., Pandit, V.: Online and offline algorithms for the sorting buffers problem on the line metric. Journal of Discrete Algorithms 8(1) (2010)
15. Krokowski, J., Räcke, H., Sohler, C., Westermann, M.: Reducing state changes with a pipeline buffer. In: VMV 2004 (2004)
16. Pruhs, K., Sgall, J., Torng, E.: Handbook of scheduling: Algorithms, models, and performance analysis (2004)
17. Räcke, H., Sohler, C., Westermann, M.: Online Scheduling for Sorting Buffers. In: Möhring, R.H., Raman, R. (eds.) ESA 2002. LNCS, vol. 2461, p. 820. Springer, Heidelberg (2002)
18. Roughgarden, T., Tardos, É.: How bad is selfish routing? Journal of the ACM 49(2) (2002)
19. Sleator, D.D., Tarjan, R.E.: Amortized efficiency of list update and paging rules. Communications of the ACM 28(2) (1985)
20. Spieckermann, S., Gutenschwager, K., Vosz, S.: A sequential ordering problem in automotive paint shops. International Journal of Production Research 42(9) (2004)

Time-Dependent Route Planning
with Generalized Objective Functions*

Gernot Veit Batz and Peter Sanders

Karlsruhe Institute of Technology (KIT), 76128 Karlsruhe, Germany
{batz,sanders}@kit.edu

Abstract. We consider the problem of finding routes in road networks
that optimize a combination of travel time and additional time-invariant
costs. These could be an approximation of energy consumption, distance,
tolls, or other penalties. The resulting problem is NP-hard, but if the additional cost is proportional to driving distance we can solve it optimally
on the German road network within 2.3 s using a multi-label A* search. A
generalization of time-dependent contraction hierarchies to the problem
yields approximations with negligible errors using running times below
5 ms which makes the model feasible for high-throughput web services.
By introducing tolls we get considerably harder instances, but still we
have running times below 41 ms and very small errors.

1 Introduction

In the last years time-dependent route planning in road networks has gained
considerable interest [1]. This has resulted in several algorithmic solutions that
are efficient in both time [2,3] and space [4,5]. Even *profile queries*, that compute
a result not only for a single but *all* departure times, can be answered fast [4].
However, all these techniques only deal with minimum *travel times*. Very little
work has been done on more general time-dependent objective functions so far.
This is disappointing from a practical perspective since real world route planners
based on static network models take into account many other aspects one would
not like to give up. For example, some approximation of energy consumption (just
distance traveled in the most simple case) is important to avoid large detours just
to save a few minutes of travel time – even environmentally insensitive users will
not appreciate such solutions due to their increased cost. Similarly, many users
want to avoid toll roads if this does not cost too much time. We might also want
additional penalties, e.g., for crossing residential areas or for using inconvenient
roads (narrow, steep, winding, bumpy,. . .). In Sect. 3 we show that this seemingly
trivial generalization already makes the problem NP-hard. Nevertheless, on road
networks computing such routes is feasible using a multi-label search algorithm
based on A* – at least for some instances (Sect. 4). However, this algorithm
is much too slow for applications like web-services that need running time in
the low milliseconds in order to guarantee low delays and in order to handle

---

* Partially supported by DFG project SA 933/5-1,2.

L. Epstein and P. Ferragina (Eds.): ESA 2012, LNCS 7501, pp. 169–180, 2012.

many user requests on a small number of servers. We therefore generalize time-dependent contraction hierarchies (TCHs) [2,4] to take additional constant costs into account. Although our adaptation gives up guaranteed optimality in favor of low query time, it turns out that the computed routes are nearly optimal in practice (Sect. 5). Just like original TCHs [6], the preprocessing of the adapted *heuristic TCHs* can be parallelized pretty well for shared memory. We support these claims by an experimental evaluation (Sect. 6).

Related Work. Time-dependent route planning started with classical results on minimum travel times [7,8] and generalized minimum objective functions [9]. Dean provided a introductory tutorial on minimum travel time paths [10]. The aforementioned TCHs have also been parallelized in distributed memory [11].

2   Preliminaries

We model road networks as directed graphs $G = (V, E)$, where nodes represent junctions and edges represent road segments. Every edge $u \to v \in E$ has two kinds of weights assigned, a *travel-time function (TTF)* $f : \mathbb{R} \to \mathbb{R}_{\geq 0}$ and an *additional constant cost* $c \in \mathbb{R}_{\geq 0}$. We often write $u \to_{f|c} v$.

The TTF f specifies the time $f(\tau)$ we need to travel from u to v via $u \to_{f|c} v$ when departing at time $\tau \in \mathbb{R}$. The constant cost c specifies some additional expenses that incur on traveling along $u \to_{f|c} v$ independently from the departure time τ. So, the *total (time-dependent) cost* of traveling along the edge $u \to_{f|c} v$ is $C(\tau) = f(\tau) + c$. In road networks we usually do not arrive earlier when we start later. So, all TTFs f fulfill the *FIFO-property*, that is $\tau' + f(\tau') \geq \tau + f(\tau)$ for $\tau' > \tau$. Moreover, we model TTFs as periodic piecewise linear functions, usually with a period of 24 hours. Given a path $\langle u \to_{f|c} v \to_{g|d} w \rangle$ in G the TTF of the entire path is denoted by $g * f := g \circ (f + \text{id}) + f$, that is $(g * f)(\tau) = g(f(\tau) + \tau) + f(\tau)$. Note that $f * (g * h) = (f * g) * h$ holds for TTFs f, g, h.

The time needed to travel along a path $P = \langle u_1 \to_{f_1|c_1} \cdots \to_{f_{k-1}|c_{k-1}} u_k \rangle$ instead of a single edge also depends on the departure time and is described by the TTF $f_P := f_{k-1} * \cdots * f_1$. The additional costs simply sum up along P and amount to $c_P := c_1 + \cdots + c_{k-1}$. Hence, the cost of traveling along P amounts to $C_P(\tau) := f_P(\tau) + c_P$ for departure time τ. As a lower and upper bound of this cost we further define $\underline{C}_P := \min f_1 + \cdots + \min f_{k-1} + c_P$ and $\overline{C}_P := \max f_1 + \cdots + \max f_{k-1} + c_P$ respectively.

Because of the FIFO-property there exists a minimum total cost for traveling from a node s to a node t for every departure time τ_0 in G, namely

$$\text{Cost}(s, t, \tau_0) := \min\{C_Q(\tau_0) \mid Q \text{ is path from } s \text{ to } t \text{ in } G\} \cup \{\infty\} .$$

The function $\text{Cost}(s, t, \cdot)$ is called the *cost profile* from s to t. If path P fulfills $C_P(\tau_0) = \text{Cost}(u_1, u_k, \tau_0)$, then it is called a *minimum total cost path* for departure time τ_0. Note that waiting is never beneficial in the described setting.

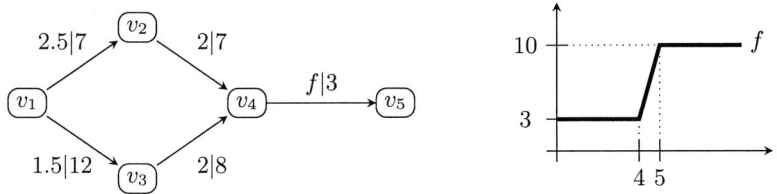

Fig. 1. A simple time-dependent graph. Most edges have constant TTFs. Only $v_4 \to v_5$ has the non-constant TTF f which is depicted on the right.

3 Complexity of Minimum Total Costs

On the first glance, computing minimum total cost paths looks related to the approach in static route planning with flexible objective functions [12] where a linear combination of objective functions is considered. However, this problem does not include any time-dependency but a time-*in*dependent parameter. Here, as we will show, we are stuck with a harder problem. Fig. 1 shows a simple example graph that demonstrates how quickly generalized time-dependent objective functions get unpleasant. There, $\langle v_1 \to v_3 \to v_4 \to v_5 \rangle$ is the only minimum total cost path from v_1 to v_5 for departure time 0.5. But its prefix path to v_4 has no minimal cost for departure time 0.5 and is suboptimal hence. This lack of *prefix optimality* implies that Dijkstra-like algorithms will not yield the correct result, as they throw suboptimal intermediate results away.

Our following proof shows that the problem is NP-hard. It uses a simple reduction to a partition problem[1] and is very much inspired by Ahuja et al. [13] who show the NP-hardness of a related problem with discrete time.

Theorem 1. *Computing a minimum total cost path for given start node, destination node, and departure time is NP-hard, even if there is only a single time-dependent edge with a constant number of bend points.*

Proof. To show our claim, we reduce the *number partitioning problem* to finding of a minimum total cost path: Given the numbers $b, a_1, \ldots, a_k \in \mathbb{N}_{>0}$ we ask whether $x_1, \ldots, x_k \in \{0, 1\}$ exist with $b = x_1 a_1 + \cdots + x_k a_k$. This question is an NP-complete problem [14]. Given an instance of number partitioning we construct the time-dependent graph depicted in Fig. 2. There are exactly 2^k paths from v_1 to v_{k+1}, all having the same total cost $2k + a_1 + \cdots + a_k$ but not necessary the same travel time. In particular, there is a path of travel time $b + k$ if and only if the underlying instance of number partitioning is answered yes. Hence, the answer is yes if and only if $\mathrm{Cost}(v_1, v_{k+2}, 0) = 1 + 2k + a_1 + \cdots + a_k$ holds. Note that the period of the TTF f has to be chosen sufficiently large. $\square$

---

[1] Indeed, using a slightly more complicated construction we can also directly reduce the static bicriteria shortest path problem to our minimum total cost time-dependent problem. This requires only a single time-dependent edge.

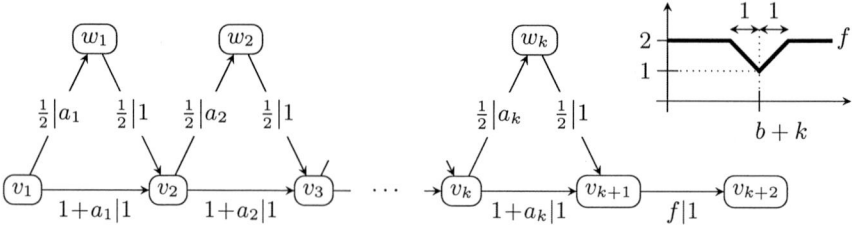

Fig. 2. A time-dependent graph encoding an instance of the number partitioning problem. All edges have constant TTFs except for $v_{k+1} \to v_{k+2}$ which has the TTF f depicted on the right.

Solving the problem for *all* departure times is also an interesting question in route planning – in order to find a convenient departure time for example. In other words, we want to compute cost profiles. The following statement suggests that cost profiles are harder to obtain than travel time profiles[2].

Theorem 2. *In a time-dependent network with additional constant costs and n nodes a cost profile can have $2^{\Omega(n)}$ bend points, even if there is only a single time-dependent edge with a constant number of bend points.*

Proof. Consider the graph in Fig. 2 with $a_i := 2^i$ for $1 \le i \le k$. Then

$$\text{Cost}(v_1, v_{k+2}, \tau) = \min_{I \subseteq \{1,\dots,k\}} \left\{ f\left(\tau + k + \sum_{i \in I} a_i\right) + 2k + 2^{k+1} - 1 \right\}$$

holds. So, $\text{Cost}(v_1, v_{k+2}, \cdot)$ has 2^k local minima and at least 2^k bend points.   □

Compared to cost profiles, travel time profiles are relatively simple. According to Foschini et al. [15] they can never have more than $Kn^{O(\log n)}$ bend points for a network with n nodes and K bend points in total. Note that our proof of Theorem 2 uses nearly the same construction as Hansen did to show that bicriterion settings can raise exponentially many Pareto optimal paths [16]. This strengthens the observation that the two problems are connected.

4   Exact Minimum Total Costs

Minimum total cost paths cannot be computed by Dijkstra-like algorithms as prefix optimality is violated. However, *suffix* optimality is always provided. So, can we not just search backward? But this would require us to know the arrival time which is part of what we want to compute. At least, on computing cost *profiles* this is not a problem. We utilize this in the following Dijkstra-like, label correcting, backward running algorithm that starts searching at the destination node. Theorem 2 implies that its execution is very expensive of course.

---

[2] If the additional constant costs are zero for all edges, then total cost and travel time is the same. In this special case a cost profile is called a *travel time profile*.

Backward Cost Profile Search. Given a start node s and a destination node t we want to compute the cost profile $\mathrm{Cost}(s, t, \cdot)$. The label of a node w is a pair $f_w | c_w$. There, f_w is a piecewise linear, *piecewise continuous* function that maps the departure time at w to the time one needs to travel from w to t using the cheapest path discovered for this departure time so far. Correspondingly, c_w is a *piecewise constant* function that maps the departure time to the sum of additional constant costs along the respective path. The function $f_w + c_w$ is a continuous and piecewise linear tentative cost profile for traveling from w to t. All nodes have $\infty | \infty$ as initial label, except for t which has $0|0$. With defining

$$\min(g|d,\, h|e) : \tau \mapsto \begin{cases} g(\tau) \,|\, d(\tau) & \text{if } (g+d)(\tau) \le (h+e)(\tau) \\ h(\tau) \,|\, e(\tau) & \text{otherwise} \end{cases}$$

we relax an edge $u \to_{f|c} v$ in backward direction as follows: If $f_v|c_v$ is the label of v we update the label $f_u|c_u$ of u by $\min(f_u|c_u,\; f_v * f \,|\, c_v \circ (f + \mathrm{id}) + c)$.   □

In forward direction Dijkstra-like algorithms are not applicable, but the following variant of *multi-label search* runs in forward direction and works. It is similar to the algorithm by Hansen [16] that finds all Pareto optimal paths in bicriteria settings. Our algorithm generalizes the time-dependent version [7] of Dijkstra's algorithm in a way that a node can have multiple labels at a time.

Plain Multi-Label Search. Given a start node s, a destination node t, and a departure time τ_0 we want to compute $\mathrm{Cost}(s, t, \tau_0)$. Nodes w are labeled with pairs $\tau_w | \gamma_w$ where τ_w is a time we arrive at w and γ_w is the sum of the additional constant costs of the corresponding path from s to w. The initial label of s is $\tau_0 | 0$. In every step we *expand* a label $\tau_u | \gamma_u$ of a node u such that the value $\tau_u + \gamma_u$ is currently minimal amongst all *non-expanded* labels of all nodes. Expanding $\tau_u | \gamma_u$ means, that we *relax* all outgoing edges $u \to_{f|c} v$ of u: If the new label $\tau_u + f(\tau_u) | \gamma_u + c$ is not *weakly dominated* by any existing label of u, it is added to the *label set* L_u of u.[3] Every old label in L_u *strictly dominated* by $\tau_u + f(\tau_u) | \gamma_u + c$ is removed from L_u, all other old labels in L_u are kept. The algorithm stops as soon as a label $\tau_t | \gamma_t$ of t is expanded for the first time. Then, with $\tau_t + \gamma_t - \tau_0 = \mathrm{Cost}(s, t, \tau_0)$, we have found the desired result.   □

In Sect. 3 we state that we cannot necessary throw intermediate results (i.e., labels) away. But in case of dominated labels we can. However, the running time of plain multi label search is not feasible as our experiments show (Sect. 6). This can be improved by switching over to A* search. The resulting algorithm is inspired by the multiobjective A* algorithm NAMOA [17].

Multi-Label A* Search. With the *heuristic* function $\underline{h}_t(w) := \min\{\underline{C}_Q \,|\, Q$ is path from w to $t\}$ which fulfills $0 \le \underline{h}_t(w) \le \mathrm{Cost}(w, t, \tau)$ for all $w \in V,\ \tau \in \mathbb{R}$ we modify the plain multi-label search described above as follows: We do not choose the label $\tau_u | \gamma_u$ of an node u with minimal $\tau_u + \gamma_u$ amongst all non-expanded

---

[3] $\tau | \gamma$ *weakly dominates* $\tau' | \gamma'$ if and only if $\tau \le \tau'$ and $\gamma \le \gamma'$ holds. If we also have $\tau \ne \tau'$ or $\gamma \ne \gamma'$, then we speak of *strict dominance*.

labels of all nodes, but the label with minimal $\tau_u + \gamma_u + \underline{h}_t(u)$. The final result remains unchanged, but the *order* in which the labels are expanded can change a lot. If a label of t is expanded earlier, we have to expand less labels before the computation stops. This can save much running time.   □

Our experiments show that $\underline{h}_t$ makes the computation more feasible on road networks (Sect. 6). To compute $\underline{h}_t$ we could run Dijkstra's algorithm in a backward manner starting from t as an initial step of the computation. But then we would process the whole graph as we would not know when to stop. Instead we perform an initial backward *interval search* [4] to compute the intervals $[\underline{h}_t(w), \overline{h}_t(w)]$ for all reached nodes w (with $\overline{h}_t(w) := \max\{\overline{C}_Q \,|\, Q$ is path from w to $t\}$). This enables us to stop the interval search as soon as the minimum key of the priority queue exceeds $\overline{h}_t(s)$. Also, multi-label A* search can use $\overline{h}_t$ to maintain an upper bound of the desired result $\mathrm{Cost}(s, t, \tau_0)$ and then desist from relaxing edges $u \rightarrow_{f|c} v$ where $(\tau_u + \gamma_u) + (f(\tau_u) + c) + \underline{h}_t(v)$ exceeds this upper bound – or where v has not even been reached by the interval search.

5   Heuristic Total Costs with TCHs

Essentially, a contraction hierarchy [18,19] orders the nodes by some notion of *importance* with more important nodes higher up in the hierarchy. The hierarchy is constructed bottom up during *preprocessing*, by successively *contracting* the least important remaining node. Contracting a node v means that v is removed from the graph while preserving all optimal routes. To do so, we have to insert a *shortcut* edge $u \rightarrow w$ for every deleted path $\langle u \rightarrow v \rightarrow w \rangle$ that is part of an optimal route. The result of this construction is the contraction hierarchy (CH). It is stored as a single graph in a condensed way: Every node is materialized exactly once and the original edges and shortcuts are put together and *merged* if necessary. The CH has the useful property that optimal paths from s to t can be constructed from an *upward path* $\langle s \rightarrow \cdots \rightarrow x \rangle$ and a *downward path* $\langle x \rightarrow \cdots \rightarrow t \rangle$.[4] This enables us to find an optimal route basically by performing two upward searches, a forward and a backward search each starting from s and t. As a well-constructed CH should be flat and sparse, such a *bidirectional* search should only take little running time.

Preprocessing. When contracting a node v we want to find out whether a shortcut $u \rightarrow w$ has to be inserted for a path $\langle u \rightarrow v \rightarrow w \rangle$. In other words, we have to discover whether $C_P(\tau) = \mathrm{Cost}(s, t, \tau)$ holds for some path $P := \langle s \rightarrow \cdots \rightarrow u \rightarrow v \rightarrow w \rightarrow \cdots \rightarrow t \rangle$ and some $\tau \in \mathbb{R}$. However, the lack of prefix-optimality imposed by the additional constant costs makes this question difficult to answer as non-optimal routes can now be part of optimal ones. To decide correctly, we could examine all *non-dominated* (i.e., Pareto optimal) paths from u to w. But as we had to do this for *all* departure times, we expect that this is too expensive

---

[4] Upward and downward paths use only edges leading from less important to more important nodes and from more important to less important nodes respectively.

with respect to time and space. Also, this might produce so many shortcuts that also the query times would be disappointing.

Therefore we switched over to a heuristic version which may loose the one or another optimal path. We insert a shortcut $u \rightarrow_{g*f \,|\, d\circ(f+\text{id})+c} w$ for a deleted path $\langle u \rightarrow_{f|c} v \rightarrow_{g|d} w \rangle$ when $(g * f + d \circ (f + \text{id}) + c)(\tau) = C_{\langle u \rightarrow_{f|c} v \rightarrow_{g|d} w \rangle}(\tau) = \text{Cost}(u, w, \tau)$ holds for some $\tau \in \mathbb{R}$ (note that c and d can now be piecewise constant functions and that f and g can now have points of discontinuity). To check this condition we compute $\text{Cost}(u, w, \cdot)$ using cost profile search (Sect. 4). If a shortcut $u \rightarrow_{g*f \,|\, d\circ(f+\text{id})+c} w$ has to be inserted for a path $\langle u \rightarrow_{f|c} v \rightarrow_{g|d} w \rangle$ it can happen that an edge $u \rightarrow_{h|e} w$ is already present. In this case we *merge* the two edges, that is we replace $u \rightarrow_{h|e} w$ by $u \rightarrow_{\min(g*f \,|\, d\circ(f+\text{id})+c,\ h|e)} w$. It is this kind of merging which introduces noncontinuous functions to the preprocessing and the resulting *heuristic* TCH.

Just like in case of the original travel time TCHs [2,4,6] preprocessing of heuristic TCHs is a computationally expensive task (or maybe even more expensive with respect to Theorem 2). So, to make the preprocessing feasible, we adopted many of the techniques applied to travel time TCHs.

Querying. Given a start node s, a destination node t, and a departure time τ_0, querying works similar as in case of travel time TCHs: In a first phase, we perform the aforementioned bidirectional search where forward and backward search only go upward. Here, the forward search is a plain multi-label search starting from s with initial label $\tau_0|0$. The backward search is a interval search as performed before running the multi-label A* search (Sect. 4). Every node x where the two searches meet is a *candidate* node. Both searches apply *stall on demand* known from the original travel time TCHs. But, in case of the forward search we stall a node based on strict dominance. In case of the backward search we stall a node based on upper and lower bounds.

In a second phase we perform a *downward search* which is again a plain multi-label search But this time it runs downward in the hierarchy starting from the candidate nodes processing only edges touched by the backward search. The labels taken from the forward search act as initial labels.

6 Experiments

Inputs and Setup. As input we use a road network of Germany provided by PTV AG for scientific use. It has 4.7 million nodes, 10.8 million edges, and TTFs reflecting the midweek (Tuesday till Thursday) traffic collected from historical data, i.e., a high traffic scenario with about 7.2 % non-constant TTFs. For all edges also the *driving distance* is available which we use as a basis to form the additional constant costs. Our idea is that the additional constant cost of an edge estimates the energy consumption raised by traveling along that edge. With typical gasoline prices we assume that driving 1 km costs 0.1 €. To prize the travel time we use rates of 5 €, 10 €, and 20 € per hour. So, using 0.1 s as unit of time and m as unit of distance we obtain time-dependent total costs of

$$time + \lambda \cdot distance$$

where λ has the values 0.72, 0.36, and 0.18 respectively. Accordingly, the additional constant edge costs are simply the driving distance scaled by the respective value of λ. Hence, we have three instances of Germany, one for every value of λ.

We also consider the effect of tools. To do so, we fix an extra price of 0.1 € per km on motorways. This means that edges belonging to motorways have double additional constant edge costs. This yields three further instances of Germany, one for every value of λ.

The experimental evaluation was done on different 64-bit machines with Ubunbtu Linux. All running times have been measured on one machine with four Core i7 Quad-Cores (2.67 Ghz) with 48 GiB of RAM with Ubuntu 10.4. There, all programs were compiled using GCC 4.4.3 with optimization level 3. Running times were always measured using one single thread except of the preprocessing where we used 8 threads.

The performance of the algorithms is evaluated in terms of running time, memory usage, error, and how often *deleteMin* is invoked. The latter comes from the fact, that we implemented all algorithms using priority queues. To measure the average running time of time-dependent cost queries we use 1 000 randomly selected start and destination pairs, together with a departure time randomly selected from $[0h, 24h)$ each. To measure the errors and the average number of invocations of *deleteMin* we do the same but with 10 000 test cases instead of 1 000. The memory usage is given in terms of the average *total* space usage of a node (not the overhead) in byte per node. For TCH-based techniques, all figures refer to the scenario that not only the total time-dependent costs but also the routes have to be determined. This increases the running time and the memory consumption a bit.

Results. For the different values of λ, Table 1 summarizes the behavior of the different algorithmic techniques. These are multi-label A* search (Sect. 4), heuristic TCHs (Sect. 5), and the original travel time TCHs [4]. In case of the travel time TCHs we did not use their original query algorithm, but the algorithm of the heuristic TCHs as described in Sect. 5. It turns out that with tolls we get much harder instances than without tolls, as the observed space usages and running times are the larger there. Without tolls we get the hardest instance at $\lambda = 0.72$, which corresponds to an hourly rate of 5 €. For smaller values of λ, which involve that time and total costs are more correlated, things seem to be easier. With tolls, in contrast, we get the hardest instance at $\lambda = 0.36$. This is because tolls are negatively correlated to travel time.

Without tolls, multi-label A* has running times similar to Dijkstra's algorithm, as its running time is mainly governed by the running time of the Dijkstra-like backward interval search in this case. This follows from the relatively low number of invocations of *deleteMin* (the invocations raised by the backward

Table 1. Behavior of time-dependent total cost queries for different values of λ. Node ordering is performed in parallel with 8 threads. delMins= number of invocations of deleteMin (without interval search in case of A*), MAX and AVG denote the maximum and average relative error, rate= percentage of routes with a relative error $\geq 0.05\%$.

| | | space | order | query | | error | | |
| | | usage | time | delMin | time | MAX | AVG | rate |
method	λ	[B/n]	[h:m]	#	[ms]	[%]	[%]	[%]
Germany midweek, no toll								
multi-label A*		130		253 933	2 328.72	–	–	–
heuristic TCH	0.72	1 481	0:28	2 142	4.92	0.09	0.00	0.00
travel time TCH		1 065	0:21	1 192	2.67	12.40	0.68	1.19
multi-label A*		130		184 710	2 208.76	–	–	–
heuristic TCH	0.36	1 316	0:26	1 774	4.22	0.03	0.00	0.00
travel time TCH		1 065	0:21	1 183	2.33	7.69	0.27	0.08
multi-label A*		130		150 970	2 234.04	–	–	–
heuristic TCH	0.18	1 212	0:25	1 464	3.51	0.01	0.00	0.00
travel time TCH		1 065	0:21	1 165	2.33	3.85	0.08	0.00
Germany midweek, with toll								
heuristic TCH	0.72	1 863	1:05	4 676	14.96			
travel time TCH		1 065	0:21	2 631	4.54			
heuristic TCH	0.36	2 004	1:16	10 725	40.96			
travel time TCH		1 065	0:21	2 634	4.46			
heuristic TCH	0.18	1 659	0:46	7 347	27.90			
travel time TCH		1 065	0:22	2 482	4.39			

interval search are not included in Table 1). Note that this low number of invocations implies that a more efficient realization of the heuristic function h_t would turn multi-label A* search into a pretty fast algorithm. Hub-labeling [20] may be an appropriate candidate. However, with tolls multi-label A* is no longer efficient. Accordingly, Table 1 omits some errors and error rates as we do not have the exact results for tolls.

With 5 ms and 41 ms heuristic TCHs have much faster query time than multi-label A*. Though being heuristic in theory, the method is practically exact for the kind of costs examined here, as there are nearly no routes with a relative error significantly away from 0 %. Admittedly, there are outliers, but they are not serious. However, the memory consumption is quite large. But it is likely that similar techniques as used for ATCHs [4] can reduce the memory consumption very much. Please note that our preprocessing is partly prototypical and takes considerably longer as in case of our original TCH implementation [4]. It may be possible to do the preprocessing faster hence.

Simply using the original TCHs with the adapted query algorithm of heuristic TCHs is not a good alternative to heuristic TCHs. Though the average error is relatively small, some more serious outliers spoil the result. This is supported by Fig. 3 and 4 which show the distribution of the relative error over the Dijkstra

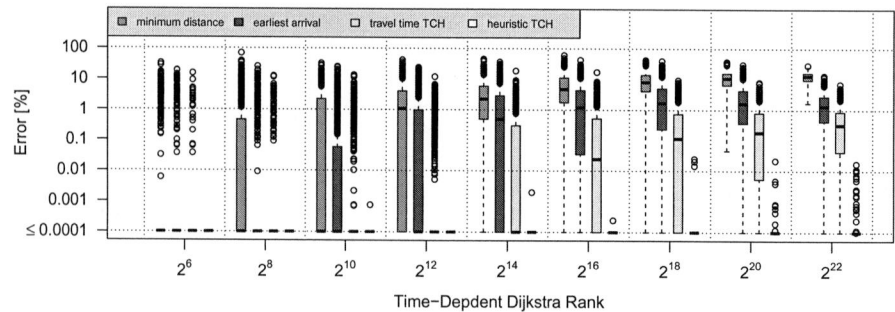

Fig. 3. Relative error of time-dependent total cost plotted over Dijkstra rank for different kinds of routes with $\lambda = 0.72$ and no tolls: minimum distance routes (green), earliest arrival routes (red), routes created using travel time TCHs (blue), and using heuristic TCHs (yellow). Number of queries per rank is 1000 for every algorithm.

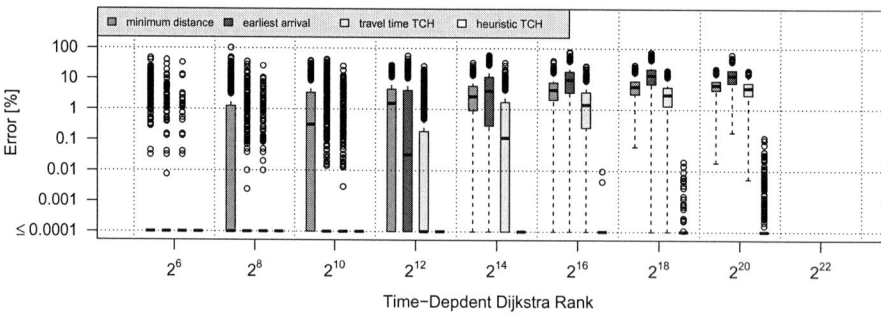

Fig. 4. Like Fig. 3 but with $\lambda = 0.36$ and with tolls. No error known for rank 2^{22}.

rank[5]. In practice, even few serious outliers may annoy some users which may already lead to a bad publicity. Of course, this can affect the success of a product or service.

Fig. 3 and 4 also report the quality of minimum distance and minimum travel time routes with respect to time-dependent total costs. Obviously, the quality gets worse with increasing Dijkstra rank. However, serious outliers occur for all ranks. Fig. 5 shows how the relative error behaves for the three different values of λ – with and without tolls. We are able to report errors for tolls for all Dijkstra ranks up to 2^{20}. For higher ranks the multi-label A* gets too slow. It turns out that with tolls the errors are larger but still small.

7   Conclusions and Future Work

We have shown that time-dependent route planning with additional constant costs is NP-hard in theory, but more than feasible on real-life road networks when

---

[5] For $i = 5..22$ we select a number of random queries such that the time-dependent version of Dijkstra's algorithm settles 2^i nodes. We call 2^i the *Dijkstra rank*.

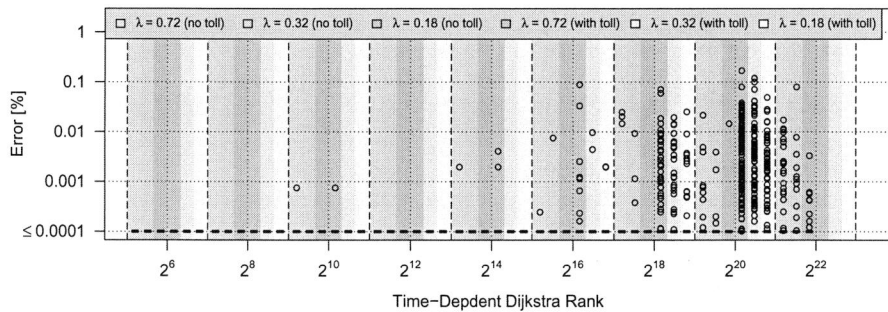

Fig. 5. Boxplot of relative error over Dijkstra rank with heuristic TCHs for $\lambda = 0.72, 0.36, 0.18$. As less than 25 % of the routes have greater error than 0.0001 %, all boxes and whiskers degenerate to small bars at the bottom. So, for readability we underlay all figures with colors as indicated in the legend. The number of queries per instance and rank is 1 000. With toll we do not know the error for rank 2^{22}.

we optimize travel time with a penalty proportional to driving distance: Our exact multi-label A* search finds optimal routes on the German road network within 2.3 s. But this may not be the end of the story. Using hub-labeling [20] as a heuristic oracle should yield average running times considerably smaller than 1 s. Our heuristic TCHs compute routes within 5 ms with mostly negligible errors. Motorway tolls, however, which are negatively correlated to travel time, make the problem considerably harder such that the multi-label A* search is no more feasible. But still our heuristic TCHs show running times below 41 ms and small errors. Our current implementation of heuristic TCHs is very space consuming. However, the careful use of approximation – which greatly reduced the space usage of original TCHs [4] – should also work well here. Faster computation of heuristic time-dependent cost profiles with very small error should also be possible with heuristic TCHs. Note that the very efficient *corridor contraction* [4] turns to be heuristic in this setting too. An A*-like backward cost profile search in the corridor with a preceding interval search in the corridor to provide a heuristic function may bring some speedup too.

Our ideas should also be applicable to additional costs that are themselves time-dependent as long as the overall cost function has the FIFO-property. Even waiting, which can be beneficial in some situations, may be translatable in such a setting. Whether all this runs fast enough in practice, has to be found out experimentally of course.

Acknowledgements. We thank Robert Geisberger for fruitful discussions.

References

1. Delling, D., Wagner, D.: Time-Dependent Route Planning. In: Ahuja, R.K., Möhring, R.H., Zaroliagis, C.D. (eds.) Robust and Online Large-Scale Optimization. LNCS, vol. 5868, pp. 207–230. Springer, Heidelberg (2009)

2. Batz, G.V., Delling, D., Sanders, P., Vetter, C.: Time-Dependent Contraction Hierarchies. In: Proceedings of the 11th Workshop on Algorithm Engineering and Experiments (ALENEX 2009), pp. 97–105. SIAM (April 2009)
3. Delling, D.: Time-Dependent SHARC-Routing. Algorithmica 60(1), 60–94 (2011)
4. Batz, G.V., Geisberger, R., Neubauer, S., Sanders, P.: Time-Dependent Contraction Hierarchies and Approximation. In: [21], pp. 166–177
5. Brunel, E., Delling, D., Gemsa, A., Wagner, D.: Space-Efficient SHARC-Routing. In: [21], pp. 47–58
6. Vetter, C.: Parallel Time-Dependent Contraction Hierarchies. Student Research Project (2009), http://algo2.iti.kit.edu/download/vetter_sa.pdf
7. Dreyfus, S.E.: An Appraisal of Some Shortest-Path Algorithms. Operations Research 17(3), 395–412 (1969)
8. Orda, A., Rom, R.: Shortest-Path and Minimum Delay Algorithms in Networks with Time-Dependent Edge-Length. Journal of the ACM 37(3), 607–625 (1990)
9. Orda, A., Rom, R.: Minimum Weight Paths in Time-Dependent Networks. Networks 21, 295–319 (1991)
10. Dean, B.C.: Shortest Paths in FIFO Time-Dependent Networks: Theory and Algorithms. Technical report, Massachusetts Institute of Technology (1999)
11. Kieritz, T., Luxen, D., Sanders, P., Vetter, C.: Distributed Time-Dependent Contraction Hierarchies. In: [21], pp. 83–93
12. Geisberger, R., Kobitzsch, M., Sanders, P.: Route Planning with Flexible Objective Functions. In: Proceedings of the 12th Workshop on Algorithm Engineering and Experiments (ALENEX 2010), pp. 124–137. SIAM (2010)
13. Ahuja, R.K., Orlin, J.B., Pallottino, S., Scutellà, M.G.: Dynamic Shortest Paths Minimizing Travel Times and Costs. Networks 41(4), 197–205 (2003)
14. Garey, M.R., Johnson, D.S.: Computers and Intractability. A Guide to the Theory of $\mathcal{NP}$-Completeness. W.H. Freeman and Company (1979)
15. Foschini, L., Hershberger, J., Suri, S.: On the Complexity of Time-Dependent Shortest Paths. In: Proceedings of the 22nd Annual ACM–SIAM Symposium on Discrete Algorithms (SODA 2011), pp. 327–341. SIAM (2011)
16. Hansen, P.: Bricriteria Path Problems. In: Fandel, G., Gal, T. (eds.) Multiple Criteria Decision Making–Theory and Application, pp. 109–127. Springer (1979)
17. Mandow, L., Pérez-de-la-Cruz, J.L.: Multiobjective A* Search with Consistent Heuristics. Journal of the ACM 57(5), 27:1–27:24 (2010)
18. Geisberger, R., Sanders, P., Schultes, D., Delling, D.: Contraction Hierarchies: Faster and Simpler Hierarchical Routing in Road Networks. In: McGeoch, C.C. (ed.) WEA 2008. LNCS, vol. 5038, pp. 319–333. Springer, Heidelberg (2008)
19. Geisberger, R., Sanders, P., Schultes, D., Vetter, C.: Exact Routing in Large Road Networks Using Contraction Hierarchies. Transportation Science (2012) (accepted for publication)
20. Abraham, I., Delling, D., Goldberg, A.V., Werneck, R.F.: A Hub-Based Labeling Algorithm for Shortest Paths in Road Networks. In: Pardalos, P.M., Rebennack, S. (eds.) SEA 2011. LNCS, vol. 6630, pp. 230–241. Springer, Heidelberg (2011)
21. Festa, P. (ed.): SEA 2010. LNCS, vol. 6049. Springer, Heidelberg (2010)

New Lower and Upper Bounds
for Representing Sequences*

Djamal Belazzougui[1] and Gonzalo Navarro[2]

[1] LIAFA, Univ. Paris Diderot - Paris 7, France
dbelaz@liafa.jussieu.fr
[2] Department of Computer Science, University of Chile
gnavarro@dcc.uchile.cl

Abstract. Sequence representations supporting queries *access*, *select* and *rank* are at the core of many data structures. There is a considerable gap between different upper bounds, and the few lower bounds, known for such representations, and how they interact with the space used. In this article we prove a strong lower bound for *rank*, which holds for rather permissive assumptions on the space used, and give matching upper bounds that require only a compressed representation of the sequence. Within this compressed space, operations *access* and *select* can be solved within almost-constant time.

1 Introduction

A large number of data structures build on sequence representations. In particular, supporting the following three queries on a sequence $S[1, n]$ over alphabet $[1, \sigma]$ has proved extremely useful:

- *access*(S, i) gives $S[i]$;
- *select*$_a(S, j)$ gives the position of the jth occurrence of $a \in [1, \sigma]$ in S; and
- *rank*$_a(S, i)$ gives the number of occurrences of $a \in [1, \sigma]$ in $S[1, i]$.

For example, Ferragina and Manzini's FM-index [9], a compressed indexed representation for text collections that supports pattern searches, is most successfully implemented over a sequence representation supporting *access* and *rank* [10]. Grossi et al. [18] had used earlier similar techniques for text indexing, and invented *wavelet trees*, a compressed sequence representation that solves the three queries in time $O(\lg \sigma)$. The time was reduced to $O(\frac{\lg \sigma}{\lg \lg n})$ with multiary wavelet trees [10,17].[1] Golynski et al. [16] used these operations for representing labeled trees and permutations, and proposed another representation that solved the operations in time $O(\lg \lg \sigma)$, and some even in constant time. This representation was made compressed by Barbay et al. [1]. Further applications of the three operations to multi-labeled trees and binary relations were uncovered by Barbay et al. [2]. Ferragina et al. [8] and Gupta et al. [20] devised new applications to

---

* Partially funded by Fondecyt Grant 1-110066, Chile. First author also partially supported by the French ANR-2010-COSI-004 MAPPI Project.
[1] For simplicity, throughout this paper we will assume that $\lg x$ means $\max(1, \lg x)$. Similarly, $O(x)$ will mean $O(\max(1, x))$ and $o(x)$ will mean $o(\max(1, x))$.

L. Epstein and P. Ferragina (Eds.): ESA 2012, LNCS 7501, pp. 181–192, 2012.
© Springer-Verlag Berlin Heidelberg 2012

XML indexing. Barbay et al. [3,1] gave applications to representing permutations and inverted indexes. Claude and Navarro [7] presented applications to graph representation. Mäkinen and Välimäki [29] and Gagie et al. [13] applied them to document retrieval on general texts.

The most basic case is that of bitmaps, when $\sigma = 2$. In this case obvious applications are set representations supporting membership and predecessor search. We assume throughout this article the RAM model with word size $w = \Omega(\lg n)$. Jacobson [21] achieved constant-time $rank$ using $o(n)$ extra bits on top of a plain representation of S, and Munro [23] and Clark [6] achieved also constant-time $select$. Golynski [14] showed a lower bound of $\Omega(n \lg \lg n / \lg n)$ extra bits for supporting both operations in constant time if S is to be represented in plain form (i.e., as an array of n bits), and gave matching upper bounds. When S can be represented arbitrarily, Patrascu [25] achieved $\lg \binom{n}{m} + O(n/\lg^c n)$ bits of space, where m is the number of 1s in S and c is any constant, and showed this is optimal [28].

For general sequences, a useful measure of compressibility is the *zero-order entropy* of S, $H_0(S) = \sum_{a \in [1,\sigma]} \frac{n_a}{n} \lg \frac{n}{n_a}$, where n_a is the number of occurrences of a in S. This can be extended to the *k-th order entropy*, $H_k(S) = \frac{1}{n} \sum_{A \in [1,\sigma]^k} |T_A| H_0(T_A)$, where T_A is the string of symbols following k-tuple A in S. It holds $0 \leq H_k(S) \leq H_{k-1}(S) \leq H_0(S) \leq \lg \sigma$ for any k, but the entropy measure is only meaningful for $k < \lg_\sigma n$. See Manzini [22] and Gagie [12] for a deeper discussion.

When representing sequences supporting these operations, we may aim at using $O(n \lg \sigma)$ bits of space, but frequently one aims for less space. We may aim at *succinct* representation of S, taking $n \lg \sigma + o(n \lg \sigma)$ bits, at a *zero-order* compressed representation, taking at most $nH_0(S) + o(n \lg \sigma)$ bits (we might also wish to compress the redundancy, $o(n \lg \sigma)$, to achieve for example $nH_0(S) + o(nH_0(S)))$, or at a *high-order* compressed representation, $nH_k(S) + o(n \lg \sigma)$.

Upper and lower bounds for sequence representations supporting the three operations are far less understood over larger alphabets. When $\sigma = O(\text{polylog } n)$, the three operations can be carried out in constant time over a data structure using $nH_0(S) + o(n)$ bits [10]. For larger alphabets, this solution requires the same space and answers the queries in time $O(\frac{\lg \sigma}{\lg \lg n})$ [10,17]. Another class of solutions [16,19,1], especially attractive for "large alphabets", achieves time $O(\lg \lg \sigma)$ for $rank$. For $access$ and $select$ they offer complementary complexities, where one of the operations is constant-time and the other requires $O(\lg \lg \sigma)$ time. They achieve zero-order compressed space, $nH_0(S) + o(nH_0(S)) + o(n)$ bits [1], and even high-order compressed space, $nH_k(S) + o(n \lg \sigma)$ for any $k = o(\lg_\sigma n)$ [19].

There are several curious aspects in the map of the current solutions for general sequences. On one hand, the times for $access$ and $select$ seem to be complementary, whereas that for $rank$ is always the same. On the other hand, there is no smooth transition between the complexity of one solution, $O(\frac{\lg \sigma}{\lg \lg n})$, and that of the other, $O(\lg \lg \sigma)$.

The complementary nature of $access$ and $select$ is not a surprise. Golynski [15] gave lower bounds that relate the time performance that can be achieved for these operations with the redundancy of an encoding of S on top of its information content. The lower bound acts on the product of both times, that is, if t and t' are the time complexities, and ρ is the bit-redundancy per symbol, then $\rho \cdot t \cdot t' = \Omega((\lg \sigma)^2/w)$ holds for a wide range of values of σ. The upper bounds for large alphabets [16,19] match this lower bound.

Although operation *rank* seems to be harder than the others (at least no constant-time solution exists except for polylog-sized alphabets), no general lower bounds on this operation have been proved. Only a recent result for the case in which S must be encoded in plain form states that if one solves *rank* within $a = O(\frac{\lg \sigma}{\lg \lg \sigma})$ access to the sequence, then the redundancy per symbol is $\rho = \Omega((\lg \sigma)/a)$ [19]. Since in the RAM model one can access up to $w/\lg \sigma$ symbols in one access, this implies a lower bound of $\rho \cdot t = \Omega((\lg \sigma)^2/w)$, similar to the one by Golynski [15] for the product of *access* and *select* times.

In this article we make several contributions that help close the gap between lower and upper bounds on sequence representation.

1. We prove the first general lower bound on *rank*, which shows that this operation is, in a sense, noticeably harder than the others: No structure using $O(n \cdot w^{O(1)})$ bits can answer *rank* queries in time $o(\lg \frac{\lg \sigma}{\lg w})$. Note the space includes the rather permissive $O(n \cdot \text{polylog } n)$. For this range of times our general bound is much stronger than the existing restricted one [19], which only forbids achieving it within $n \lg \sigma + O(n \lg^2 \sigma/(w \lg \frac{\lg \sigma}{\lg w}))$ bits. Our lower bound uses a reduction from predecessor queries.
2. We give a matching upper bound for *rank*, using $O(n \lg \sigma)$ bits of space and answering queries in time $O(\lg \frac{\lg \sigma}{\lg w})$. This is lower than any time complexity achieved so far for this operation within $O(n \cdot w^{O(1)})$ bits, and it elegantly unifies both known upper bounds under a single and lower time complexity. This is achieved via a reduction to a predecessor query structure that is tuned to use slightly less space than usual.
3. We derive succinct and compressed representations of sequences that achieve time $O(\frac{\lg \sigma}{\lg w})$ for *access*, *select* and *rank*, improving upon previous results [10]. This yields constant-time operations for $\sigma = w^{O(1)}$. Succinctness is achieved by replacing universal tables used in other solutions with bit manipulations on the RAM model. Compression is achieved by combining the succinct representation with existing compression boosters.
4. We derive succinct and compressed representations of sequences over larger alphabets, which achieve time $O(\lg \frac{\lg \sigma}{\lg w})$ for *rank*, which is optimal, and almost-constant time for *access* and *select*. The result improves upon almost all succinct and compressed representations proposed so far [16,2,1,19]. This is achieved by plugging our $O(n \lg \sigma)$-bit solutions into existing succinct and compressed data structures.

Our results assume a RAM model where bit shifts, bitwise logical operations, and arithmetic operations (including multiplication) are permitted. Otherwise we can simulate them with universal tables within $o(n)$ extra bits of space, but all $\lg w$ in our upper bounds become $\lg \lg n$.

2   Lower Bound for *rank*

Our technique is to reduce from a predecessor problem and apply the density-aware lower bounds of Patrascu and Thorup [26]. Assume that we have n keys from a universe of size $u = n\sigma$, then the keys are of length $\ell = \lg u = \lg n + \lg \sigma$. According to branch 2 of Patrascu and Thorup's result, the time for predecessor queries in this setting is lower bounded by $\Omega\left(\lg\left(\frac{\ell - \lg n}{a}\right)\right)$, where $a = \lg(s/n) +$ $\lg w$ and s is the space in words of our representation (the lower bound is in the cell probe model for word length w, so the space is always expressed in number of cells). The lower bounds holds even for a more restricted version of the predecessor problem in which one of two colors is associated with each element and the query only needs to return the color of the predecessor. We assume $\sigma = O(n)$; the other case will be considered at the end of the section.

The reduction is as follows. We divide the universe $n \cdot \sigma$ into σ intervals, each of size n. This division can be viewed as a binary matrix of n columns by σ rows, where we set a 1 at row r and column c iff element $(r-1) \cdot n + c$ belongs to the set. We will use four data structures.

1. A plain bitvector $L[1, n]$ which stores the color associated with each element. The array is indexed by the original ranks of the elements.
2. A partial sums structure R stores the number of elements in each row. It is a bitmap concatenating the σ unary representations, $1^{n_r}0$, of the number of 1s in each row $r \in [1, \sigma]$. Thus R is of length $n + \sigma$ and can give in constant time the number of 1s up to (and including) any row r, $count(r) = rank_1(R, select_0(R, r)) = select_0(R, r) - r$, in constant time and $O(n+\sigma) = O(n)$ bits of space [23,6].
3. A column mapping data structure C that maps the original columns into a set of columns where (*i*) empty columns are eliminated, and (*ii*) new columns are created when two or more 1s fall in the same column. C is a bitmap concatenating the n unary representations, $1^{n_c}0$, of the numbers n_c of 1s in each column $c \in [1, n]$. So C is of length $2n$. Note that the new matrix of mapped columns has also n columns (one per element in the set) and exactly one 1 per column. The original column c is then mapped to $col(c) = rank_1(C, select_0(C, c)) = select_0(C, c) - c$, using constant time and $O(n)$ bits. Note that $col(c)$ is the last of the columns to which the original column c might have been expanded.
4. A string $S[1, n]$ over alphabet $[1, \sigma]$, so that $S[c] = r$ iff the only 1 at column c (after column remapping) is at row r. Over this string we build a data structure able to answer queries $rank_r(S, c)$.

Queries are done in the following way. Given an element $x \in [1, u]$, we first deompose it into a pair (r, c) where $x = (r-1) \cdot n + c$. In a first step, we compute $count(r-1)$ in constant time. This gives us the count of elements up to point $(r-1) \cdot n$. Next we must compute the count of elements in the range $[(r-1) \cdot n + 1, (r-1) \cdot n + c]$. For doing that we first remap the column to $c' = col(c)$ in constant time, and finally compute $rank_r(S, c')$, which gives the number of 1s in row r up to column c'. Note that if column c was expanded to several ones, we are counting the 1s up to the last of the expanded columns, so that all the original 1s at column c are counted at their respective rows. Then the rank of the predecessor of x is $p = count(r - 1) + rank_r(S, col(c))$. Finally, the color associated with x is given by $L[p]$.

Theorem 1. *Given a data structure that supports rank queries on strings of length n over alphabet $[1, \sigma]$ in time $t(n, \sigma)$ and using $s(n, \sigma)$ bits of space, we can solve the colored predecessor problem for n integers from universe $[1, n\sigma]$ in time $t(n, \sigma) + O(1)$ using a data structure that occupies $s(n, \sigma) + O(n)$ bits.*

By the reduction above we get that any lower bound for predecessor search for n keys over a universe of size $n\sigma$ must also apply to *rank* queries on sequences of length n over alphabet of size σ. In our case, if we aim at using $O(n \cdot w^{O(1)})$ bits of space, this lower bound (branch 2 [26]) is $\Omega \left(\lg \frac{\ell - \lg n}{\lg(s/n) + \lg w} \right) = \Omega \left(\lg \frac{\lg \sigma}{\lg w} \right)$.

For $\sigma = \Theta(n)$ and $w = \Theta(\lg n)$, the bound is simply $\Omega(\lg \lg \sigma)$. In case $\sigma = \omega(n)$, $\Omega(\lg \frac{\lg \sigma}{\lg w})$ must still be a lower bound, as otherwise we could break it in the case $\sigma = O(n)$ by just declaring σ artificially larger.

Theorem 2. *Any data structure that uses space $O(n \cdot w^{O(1)})$ bits to represent a sequence of length n over alphabet $[1, \sigma]$, must use time $\Omega(\lg \frac{\lg \sigma}{\lg w})$ to answer rank queries.*

For simplicity, assume $w = \Theta(\lg n)$. This lower bound is trivial for small $\lg \sigma = O(\lg \lg n)$ (i.e., $\sigma = O(\text{polylog } n)$), where constant-time solutions for *rank* exist that require only $nH_0(S) + o(n)$ bits [10]. On the other hand, if σ is sufficiently large, $\lg \sigma = \Omega((\lg \lg n)^{1+\epsilon})$ for any constant $\epsilon > 0$, the lower bound becomes simply $\Omega(\lg \lg \sigma)$, where it is matched by known compact and compressed solutions [16,1,19] requiring as little as $nH_0(S) + o(nH_0(S)) + o(n)$ or $nH_k(S) + o(n \lg \sigma)$ bits.

The range where this lower bound has not yet been matched is $\omega(\lg \lg n) = \lg \sigma = o((\lg \lg n)^{1+\epsilon})$, for any constant $\epsilon > 0$. The next section presents a new matching upper bound.

3   Optimal Upper Bound for *rank*

We now show a matching upper bound with optimal time and space $O(n \lg \sigma)$ bits. In the next sections we make the space succinct and even compressed.

We reduce the problem to predecessor search and then use an existing solution for that problem. The idea is simply to represent the string $S[1, n]$ over alphabet $[1, \sigma]$ as a matrix of σ rows and n columns, and regard the matrix as the set of n points $\{(S[c] - 1) \cdot n + c, \ c \in [1, n]\}$ over the universe $[1, n\sigma]$. Then we store an array of n cells containing $\langle r, rank_r(S, c) \rangle$, where $r = S[c]$, for the point corresponding to column c in the set.

To query $rank_r(S, c)$ we compute the predecessor of $(r - 1) \cdot n + c$. If it is a pair $\langle r, v \rangle$, for some v, then the answer is v. Else the answer is zero.

This solution requires $n \lg \sigma + n \lg n$ bits for the pairs, on top of the space of the predecessor query. If $\sigma \leq n$ we can reduce this extra space to $2n \lg \sigma$ by storing the pairs $\langle r, rank_r(S, c) \rangle$ in a different way. We virtually cut the string into chunks of length σ, and store the pair as $\langle r, rank_r(S, c) - rank_r(S, c - (c \bmod \sigma)) \rangle$. The rest of the $rank_r$ information is obtained in constant time and $O(n)$ bits using Golynski et al.'s [16] reduction to chunks: They store a bitmap $A[1, 2n]$ where the matrix is traversed row-wise and we append to A a 1 for each 1 found in the matrix and a 0 each time we move to the next chunk (so we append n/σ 0s per row). Then the remaining information for $rank_r(S, c)$ is $rank_r(S, c - (c \bmod \sigma)) = select_0(A, p_1) - select_0(A, p_0) - (c \operatorname{div} \sigma)$, where $p_0 = (r-1) \cdot n/\sigma$ and $p_1 = p_0 + (c \operatorname{div} \sigma)$ (we have simplified the formulas by assuming σ divides n).

Theorem 3. *Given a solution for predecessor search on a set of n keys chosen from a universe of size u, that occupies space $s(n, u)$ and answers in time $t(n, u)$, there exists a solution for rank queries on a sequence of length n over an alphabet $[1, \sigma]$ that runs in time $t(n, n\sigma) + O(1)$ and occupies $s(n, n\sigma) + O(n \lg \sigma)$ bits.*

In the extended version of their article, Patrascu and Thorup [27] give an upper bound matching the lower bound of branch 2 and using $O(n \lg u)$ bits for n elements over a universe $[1, u]$, and give hints to reduce the space to $O(n \lg(u/n))$. For completeness, we do this explicitly in an extended version of the present paper [5, App. A]. By using this predecessor data structure, the following result on *rank* is immediate.

Theorem 4. *A string $S[1, n]$ over alphabet $[1, \sigma]$ can be represented in $O(n \lg \sigma)$ bits, so that operation rank is solved in time $O(\lg \frac{\lg \sigma}{\lg w})$.*

Note that, within this space, operations *access* and *select* can also be solved in constant time.

4   Optimal-Time *rank* in Succinct and Compressed Space

We start with a sequence representation using $n \lg \sigma + o(n \lg \sigma)$ bits (i.e., succinct) that answers *access* and *select* queries in almost-constant time, and *rank* in time $O(\lg \frac{\lg \sigma}{\lg w})$. This is done in two phases: a constant-time solution for $\sigma = w^{O(1)}$, and then a solution for general alphabets. Then we address compression.

4.1   Succinct Representation for Small Alphabets

Using multiary wavelet trees [10] we can obtain succinct space and $O(\frac{\lg \sigma}{\lg \lg n})$ time for *access*, *select* and *rank*. This is constant for $\lg \sigma = O(\lg \lg n)$. We start by extending this result to the case $\lg \sigma = O(\lg w)$, as a base case for handling larger alphabets thereafter. More precisely, we prove the following result.

Theorem 5. *A string $S[1, n]$ over alphabet $[1, \sigma]$ can be represented using $n \lg \sigma + o(n \lg \sigma)$ bits so that operations access, select and rank can be solved in time $O(\frac{\lg \sigma}{\lg w})$. If $\sigma = w^{O(1)}$, the space is $n\lceil \lg \sigma \rceil + o(n)$ bits and the times are $O(1)$.*

A multiary wavelet tree for $S[1,n]$ divides, at the root node v, the alphabet $[1,\sigma]$ into r contiguous regions of the same size. A sequence $R_v[1,n]$ recording the region each symbol belongs to is stored at the root node (note R_v is a sequence over alphabet $[1,r]$). This node has r children, each handling the subsequence of S formed by the symbols belonging to a given region. The children are decomposed recursively, thus the wavelet tree has height $O(\lg_r \sigma)$. Queries *access*, *select* and *rank* on sequence $S[1,n]$ are carried out via $O(\lg_r \sigma)$ similar queries on the sequences R_v stored at wavelet tree nodes [18]. By choosing r such that $\lg r = \Theta(\lg \lg n)$, it turns out that the operations on the sequences R_v can be carried out in constant time, and thus the cost of the operations on the original sequence S is $O(\frac{\lg \sigma}{\lg \lg n})$ [10].

In order to achieve time $O(\frac{\lg \sigma}{\lg w})$, we need to handle in constant time the operations over alphabets of size $r = w^\beta$, for some $0 < \beta < 1$, so that $\lg r = \Theta(\lg w)$. This time we cannot resort to universal tables of size $o(n)$, but rather must use bit manipulation on the RAM model.

The sequence $R_v[1,n]$ is stored as the concatenation of n fields of length $\lg r$, into consecutive machine words. Thus achieving constant-time *access* is trivial: To access $R_v[i]$ we simply extract the corresponding bits, from the $(1+(i-1) \cdot \lg r)$-th to the $(i \cdot \lg r)$-th, from one or two consecutive machine words, using bit shifts and masking.

Operations *rank* and *select* are more complex. We will proceed by cutting the sequence R_v into blocks of length $b = w^\alpha$ symbols, for some $\beta < \alpha < 1$. First we show how, given a block number i and a symbol a, we extract from $R[1,b] = R_v[(i-1) \cdot b + 1, i \cdot b]$ a bitmap $B[1,b]$ such that $B[j] = 1$ iff $R[j] = a$. Then we use this result to achieve constant-time *rank* queries. Next, we show how to solve predecessor queries in constant time, for several fields of length $\lg w$ bits fitting in a machine word. Finally, we use this result to obtain constant-time *select* queries.

Projecting a Block. Given sequence $R[1,b] = R_v[1 + (i-1) \cdot b, i \cdot b]$, which is of length $b \cdot \ell = w^\alpha \lg r < w^\alpha \lg w = o(w)$ bits, where $\ell = \lg r$, and given $a \in [1,r]$, we extract $B[1, b \cdot \ell]$ such that $B[j \cdot \ell] = 1$ iff $R[j] = a$.

To do so, we first compute $X = a \cdot (0^{\ell-1}1)^b$. This creates b copies of a within ℓ-bit long fields. Second, we compute $Y = R$ XOR X, which will have zeroed fields at the positions j where $R[j] = a$. To identify those fields, we compute $Z = (10^{\ell-1})^b - Y$, which will have a 1 at the highest bit of the zeroed fields in Y. Now $W = Z$ AND $(10^{\ell-1})^b$ isolates those leading bits.

Constant-time rank Queries. We now describe how we can do rank queries in constant time for $R_v[1,n]$. Our solution follows that of Jacobson [21]. We choose a superblock size $s = w^2$ and a block size $b = (\sqrt{w}-1)/\lg r$. For each $a \in [1,r]$, we store the accumulated values per superblock, $rank_a(R_v, i \cdot s)$ for all $1 \le i \le n/s$. We also store the within-superblock accumulated values per block, $rank_a(R_v, i \cdot b) - rank_a(R_v, \lfloor (i \cdot b)/s \rfloor \cdot s)$, for $1 \le i \le n/b$. Both arrays of counters require, over all symbols, $r((n/s) \cdot w + (n/b) \cdot \lg s) = O(nw^\beta (\lg w)^2/\sqrt{w})$ bits. Added over the $O(\frac{\lg \sigma}{\lg w})$ wavelet tree levels, the space required is $O(n \lg \sigma \lg w / w^{1/2-\beta})$ bits. This is $o(n \lg \sigma)$ for any $\beta < 1/2$, and furthermore it is $o(n)$ if $\sigma = w^{O(1)}$.

To solve a query $rank_a(R_v, i)$, we need to add up three values: (i) the superblock accumulator at position $\lfloor i/s \rfloor$, (ii) the block accumulator at position $\lfloor i/b \rfloor$, (iii), the bits set at $B[1, (i \bmod b) \cdot \ell]$, where B corresponds to the values equal to a in $R_v[\lfloor i/b \rfloor \cdot b + 1, \lfloor i/b \rfloor \cdot b + b]$. We have shown above how to extract $B[1, b \cdot \ell]$, so we count the number of bits set in $C = B$ AND $1^{(i \bmod b) \cdot \ell}$.

This counting is known as a popcount operation. Given a bit block of length $b\ell = \sqrt{w} - 1$, with bits set at positions multiple of ℓ, we popcount it using the following steps:

1. We first duplicate the block b times into b fields. That is, we compute $X = C \cdot (0^{b\ell-1}1)^b$.
2. We now isolate a different bit in each different field. This is done with $Y = X$ AND $(0^{b\ell}10^{\ell-1})^b$. This will isolate the ith aligned bit in field i.
3. We now sum up all those isolated bits using the multiplication $Z = Y \cdot (0^{b\ell+\ell-1}1)^b$. The end result of the popcount operation lies at the bits $Z[b^2\ell + 1, b^2\ell + \lg b]$.
4. We finally extract the result as $c = (Z \gg b^2\ell)$ AND $(1^{\lg b})$.

Constant-time select Queries. The solution to *select* queries is similar but more technical. For lack of space we describe it in the extended version [5, Sec. 4.1.3].

4.2 Succinct Representation for Larger Alphabets

We assume now $\lg \sigma = \omega(\lg w)$; otherwise the previous section achieves succinctness and constant time for all operations.

We build on Golynski et al.'s solution [16]. They first cut S into chunks of length σ. With bitvector $A[1, 2n]$ described in Section 3 they reduce all the queries, in constant time, to within a chunk. For each chunk they store a bitmap $X[1, 2\sigma]$ where the number of occurrences of each symbol $a \in [1, \sigma]$ in the chunk, n_a, is concatenated in unary, $X = 1^{n_1}01^{n_2}0 \ldots 1^{n_\sigma}0$. Now they introduce two complementary solutions.

Constant-time Select. The first one stores, for each consecutive symbol $a \in [1, \sigma]$, the chunk positions where it appears, in increasing order. Let π be the resulting permutation, which is stored with the representation of Munro et al. [24]. This requires $\sigma \lg \sigma (1 + 1/f(n, \sigma))$ bits and computes any $\pi(i)$ in constant time and any $\pi^{-1}(j)$ in time $O(f(n, \sigma))$, for any $f(n, \sigma) \geq 1$. With this representation they solve, within the chunk, $select_a(i) = \pi(select_0(X, a - 1) - (a - 1) + i)$ in constant time and $access(i) = 1 + rank_0(select_1(X, \pi^{-1}(i)))$ in time $O(f(n, \sigma))$.

For $rank_a(i)$, they basically carry out a predecessor search within the interval of π that corresponds to a: $[select_0(X, a - 1) - (a - 1) + 1, select_0(X, a) - a]$. They have a sampled predecessor structure with one value out of $\lg \sigma$, which takes just $O(\sigma)$ bits. With this structure they reduce the interval to size $\lg \sigma$, and a binary search completes the process, within overall time $O(\lg \lg \sigma)$.

To achieve optimal time, we sample one value out of $\frac{\lg \sigma}{\lg w}$ within chunks. We build the predecessor data structures of Patrascu and Thorup [27], mentioned in Section 3, over the sampled values. Added over all the chunks, these structures take $O((n/\frac{\lg \sigma}{\lg w}) \lg \sigma) = O(n \lg w) = o(n \lg \sigma)$ bits (as we assumed $\lg \sigma = \omega(\lg w)$). The predecessor structures take time $O(\lg \frac{\lg \sigma}{\lg w})$ (see Theorem 10 in the extended version [5, App. A]). The search is then completed with a binary search between two consecutive sampled values, which also takes time $O(\lg \frac{\lg \sigma}{\lg w})$.

Constant-time Access. This time we use the structure of Munro et al. on π^{-1}, so we compute any $\pi^{-1}(j)$ in constant time and any $\pi(i)$ in time $O(f(n,\sigma))$. Thus we get *access* in constant time and *select* in time $O(f(n,\sigma))$.

Now the binary search of *rank* needs to compute values of π, which is not anymore constant time. This is why Golynski et al. [16] obtained time slightly over $\lg \lg \sigma$ time for *rank* in this case. We instead set the sampling step to $(\frac{\lg \sigma}{\lg w})^{\frac{1}{f(n,\sigma)}}$. The predecessor structures on the sampled values still answer in time $O(\lg \frac{\lg \sigma}{\lg w})$, but they take $O((n/(\frac{\lg \sigma}{\lg w})^{\frac{1}{f(n,\sigma)}}) \lg \sigma)$ bits of space. This is $o(n \lg \sigma)$ provided $f(n,\sigma) = o(\lg \frac{\lg \sigma}{\lg w})$. On the other hand, the time for the binary search is $O(\frac{f(n,\sigma)}{f(n,\sigma)} \lg \frac{\lg \sigma}{\lg w})$, as desired.

The following theorem, which improves upon Golynski et al.'s [16] (not only as a consequence of a higher low-order space term), summarizes our result.

Theorem 6. *A string $S[1,n]$ over alphabet $[1,\sigma]$, $\sigma \leq n$, can be represented using $n \lg \sigma + o(n \lg \sigma)$ bits, so that, given any function $\omega(1) = f(n,\sigma) = o(\lg \frac{\lg \sigma}{\lg w})$, (i) operations access and select can be solved in time $O(1)$ and $O(f(n,\sigma))$, or vice versa, and (ii) rank can be solved in time $O(\lg \frac{\lg \sigma}{\lg w})$.*

For larger alphabets we must add a dictionary mapping $[1,\sigma]$ to the (at most) n symbols actually occurring in S, in the standard way.

4.3   Zero-Order Compression

Barbay et al. [1] showed how, given a representation $\mathcal{R}$ of a sequence in $n \lg \sigma + o(n \lg \sigma)$ bits, its times for *access*, *select* and *rank* can be maintained while reducing its space to $nH_0(S) + o(nH_0(S)) + o(n)$ bits. This can be done even if $\mathcal{R}$ works only for $\sigma \geq (\lg n)^c$ for some constant c. The technique separates the symbols according to their frequencies into $O(\lg n)$ classes. The sequence of classes is represented using a multiary wavelet tree [10], and the subsequences of the symbols of each class are represented with an instance of $\mathcal{R}$.

We can use this technique to compress the space of our succinct representations. By using Theorem 5 as our structure $\mathcal{R}$, we obtain the following result, which improves upon Ferragina et al. [10].

Theorem 7. *A string $S[1,n]$ over alphabet $[1,\sigma]$ can be represented using $nH_0(S) + o(nH_0(S)) + o(n)$ bits so that operations access, select and rank can* be solved in time $O(\frac{\lg \sigma}{\lg w})$. If $\sigma = w^{O(1)}$, the space is $nH_0(S) + o(n)$ and the operation times are $O(1)$.

To handle larger alphabets, we use Theorem 6 as our structure $\mathcal{R}$. The only technical problem is that the subsequences range over a smaller alphabet $[1, \sigma']$, and Theorem 6 holds only for $\lg \sigma' = \omega(\lg w)$. In subsequences with smaller alphabets we can use Theorem 5, which give *access*, *select* and *rank* times $O(\frac{\lg \sigma'}{\lg w})$. More precisely, we use that structure for $\frac{\lg \sigma'}{\lg w} \le f(n, \sigma)$, else use Theorem 6. This gives the following result, which improves upon Barbay et al.'s [1].

Theorem 8. *A string $S[1, n]$ over alphabet $[1, \sigma]$, $\sigma \le n$, can be represented using $nH_0(S) + o(nH_0(S)) + o(n)$ bits, so that, given any function $\omega(1) = f(n, \sigma) = o(\lg \frac{\lg \sigma}{\lg w})$, (i) operations access and select can be solved in time $O(f(n, \sigma))$, and (ii) rank can be solved in time $O(\lg \frac{\lg \sigma}{\lg w})$.*

4.4   High-Order Compression

Ferragina and Venturini [11] showed how a string $S[1, n]$ over alphabet $[1, \sigma]$ can be stored within $nH_k(S) + o(n \lg \sigma)$ bits, for any $k = o(\lg_\sigma n)$, so that it offers constant-time *access* to any $O(\lg_\sigma n)$ consecutive symbols.

We provide *select* and *rank* functionality on top of this representation by adding extra data structures that take $o(n \lg \sigma)$ bits, whenever $\lg \sigma = \omega(\lg w)$. The technique is similar to those used by Barbay et al. [2] and Grossi et al. [19]. We divide the text logically into chunks, as with Golynski et al. [16], and for each chunk we store a monotone minimum perfect hash function (mmphf) f_a for each $a \in [1, \sigma]$. Each f_a stores the positions where symbol a occurs in the chunk, so that given the position i of an occurrence of a, $f_a(i)$ gives $rank_a(i)$ within the chunk. All the mmphfs can be stored within $O(\sigma \lg \lg \sigma) = o(\sigma \lg \sigma)$ bits and can be queried in constant time [4]. With array X we can know, given a, how many symbols smaller than a are there in the chunk.

Now we have sufficient ingredients to compute π^{-1} in constant time: Let a be the ith symbol in the chunk (obtained in constant time using Ferragina and Venturini's structure), then $\pi^{-1}(i) = f_a(i) + select_0(X, a - 1) - (a - 1)$. Now we can compute *select* and *rank* just as done in the "constant-time *access*" branch of Section 4.2. The resulting theorem improves upon Barbay et al.'s results [2] (they did not use mmphfs).

Theorem 9. *A string $S[1, n]$ over alphabet $[1, \sigma]$, for $\sigma \le n$ and $\lg \sigma = \omega(\lg w)$, can be represented using $nH_k(S) + o(n \lg \sigma)$ bits for any $k = o(\lg_\sigma n)$ so that, given any function $\omega(1) = f(n, \sigma) = o(\lg \frac{\lg \sigma}{\lg w})$, (i) operation access can be solved in constant time, (ii) operation select can be solved in time $O(f(n, \sigma))$, and (ii) operation rank can be solved in time $O(\lg \frac{\lg \sigma}{\lg w})$.*

To compare with the corresponding result by Grossi et al. [19] (who do use mmphfs) we can fix the redundancy to $O(\frac{n \lg \sigma}{\lg \lg \sigma})$, where they obtain $O(\lg \lg \sigma)$ time for *select* and *rank*, whereas we obtain the same time for *select* and our improved time for *rank*, as long as $\lg \sigma = \Omega(\lg w \lg \lg w \lg \lg \lg w)$.

5   Conclusions

This paper considerably reduces the gap between upper and lower bounds for sequence representations providing *access*, *select* and *rank* queries. Most notably, we give matching lower and upper bounds $\Theta(\lg \frac{\lg \sigma}{\lg w})$ for operation *rank*, which was the least developed one in terms of lower bounds. The issue of the space related to this complexity is basically solved as well: we have shown it can be achieved even within compressed space, and it cannot be surpassed within space $O(n \cdot w^{O(1)})$. On the other hand, operations *access* and *select* can be solved, within the same compressed space, in almost constant time (i.e., as close to $O(1)$ as desired but not both reaching it, unless we double the space).

There are still some intriguing issues that remain unclear:

1. Golynski's lower bounds [15] leave open the door to achieving constant time for *access* and *select* simultaneously, with $O(n(\lg \sigma)^2 / \lg n)$ bits of redundancy. However, this has not yet been achieved for the interesting case $\omega(\lg w) = \lg \sigma = o(\lg n)$. We conjecture that this is not possible and a stronger lower bound holds.

2. While we can achieve constant-time *select* and almost-constant time for *access* (or vice versa) within zero-order entropy space, we can achieve only the second combination within high-order entropy space. If simultaneous constant-time *access* and *select* is not possible, then no solution for the first combination can build over a compressed representation of S giving constant-time access, as it has been the norm [2,1,19].

3. We have achieved high-order compression with almost-constant *access* and *select* times, and optimal *rank* time, but on alphabets of size superpolynomial in w. By using one Golynski's binary *rank/select* index [14] per symbol over Ferragina and Venturini's representation [11], we get high-order compression and constant time for all the operations for any $\sigma = o(\lg n)$. This leaves open the interesting band of alphabet sizes $\Omega(\lg n) = \sigma = w^{O(1)}$.

References

1. Barbay, J., Gagie, T., Navarro, G., Nekrich, Y.: Alphabet Partitioning for Compressed Rank/Select and Applications. In: Cheong, O., Chwa, K.-Y., Park, K. (eds.) ISAAC 2010, Part II. LNCS, vol. 6507, pp. 315–326. Springer, Heidelberg (2010)
2. Barbay, J., He, M., Munro, J.I., Rao, S.S.: Succinct indexes for strings, binary relations and multi-labeled trees. In: Proc. 18th SODA, pp. 680–689 (2007)
3. Barbay, J., Navarro, G.: Compressed representations of permutations, and applications. In: Proc. 26th STACS, pp. 111–122 (2009)
4. Belazzougui, D., Boldi, P., Pagh, R., Vigna, S.: Monotone minimal perfect hashing: searching a sorted table with $o(1)$ accesses. In: Proc. 20th SODA, pp. 785–794 (2009)
5. Belazzougui, D., Navarro, G.: New lower and upper bounds for representing sequences. CoRR, arXiv:1111.26211v1 (2011) http://arxiv.org/abs/1111.2621v1
6. Clark, D.: Compact Pat Trees. PhD thesis, University of Waterloo, Canada (1996)
7. Claude, F., Navarro, G.: Extended Compact Web Graph Representations. In: Elomaa, T., Mannila, H., Orponen, P. (eds.) Ukkonen Festschrift 2010. LNCS, vol. 6060, pp. 77–91. Springer, Heidelberg (2010)

8. Ferragina, P., Luccio, F., Manzini, G., Muthukrishnan, S.: Compressing and indexing labeled trees, with applications. Journal of the ACM 57(1) (2009)
9. Ferragina, P., Manzini, G.: Indexing compressed texts. Journal of the ACM 52(4), 552–581 (2005)
10. Ferragina, P., Manzini, G., Mäkinen, V., Navarro, G.: Compressed representations of sequences and full-text indexes. ACM Transactions on Algorithms 3(2), article 20 (2007)
11. Ferragina, P., Venturini, R.: A simple storage scheme for strings achieving entropy bounds. Theoretical Computer Science 372(1), 115–121 (2007)
12. Gagie, T.: Large alphabets and incompressibility. Information Processing Letters 99(6), 246–251 (2006)
13. Gagie, T., Navarro, G., Puglisi, S.J.: Colored Range Queries and Document Retrieval. In: Chavez, E., Lonardi, S. (eds.) SPIRE 2010. LNCS, vol. 6393, pp. 67–81. Springer, Heidelberg (2010)
14. Golynski, A.: Optimal lower bounds for rank and select indexes. Theoretical Computer Science 387(3), 348–359 (2007)
15. Golynski, A.: Cell probe lower bounds for succinct data structures. In: Proc. 20th SODA, pp. 625–634 (2009)
16. Golynski, A., Munro, I., Rao, S.: Rank/select operations on large alphabets: a tool for text indexing. In: Proc. 17th SODA, pp. 368–373 (2006)
17. Golynski, A., Raman, R., Rao, S.S.: On the Redundancy of Succinct Data Structures. In: Gudmundsson, J. (ed.) SWAT 2008. LNCS, vol. 5124, pp. 148–159. Springer, Heidelberg (2008)
18. Grossi, R., Gupta, A., Vitter, J.: High-order entropy-compressed text indexes. In: Proc. 14th SODA, pp. 841–850 (2003)
19. Grossi, R., Orlandi, A., Raman, R.: Optimal Trade-Offs for Succinct String Indexes. In: Abramsky, S., Gavoille, C., Kirchner, C., Meyer auf der Heide, F., Spirakis, P.G. (eds.) ICALP 2010. LNCS, vol. 6198, pp. 678–689. Springer, Heidelberg (2010)
20. Gupta, A., Hon, W.-K., Shah, R., Vitter, J.: Dynamic rank/select dictionaries with applications to XML indexing. Technical Report CSD TR #06-014, Purdue University (July 2006)
21. Jacobson, G.: Space-efficient static trees and graphs. In: Proc. 30th FOCS, pp. 549–554 (1989)
22. Manzini, G.: An analysis of the Burrows-Wheeler transform. Journal of the ACM 48(3), 407–430 (2001)
23. Munro, I.: Tables. In: Chandru, V., Vinay, V. (eds.) FSTTCS 1996. LNCS, vol. 1180, pp. 37–42. Springer, Heidelberg (1996)
24. Munro, J.I., Raman, R., Raman, V., Rao, S.S.: Succinct Representations of Permutations. In: Baeten, J.C.M., Lenstra, J.K., Parrow, J., Woeginger, G.J. (eds.) ICALP 2003. LNCS, vol. 2719, pp. 345–356. Springer, Heidelberg (2003)
25. Patrascu, M.: Succincter. In: Proc. 49th FOCS, pp. 305–313 (2008)
26. Patrascu, M., Thorup, M.: Time-space trade-offs for predecessor search. In: Proc. 38th STOC, pp. 232–240 (2006)
27. Patrascu, M., Thorup, M.: Time-space trade-offs for predecessor search. CoRR, cs/0603043v1 (2008), http://arxiv.org/pdf/cs/0603043v1
28. Patrascu, M., Viola, E.: Cell-probe lower bounds for succinct partial sums. In: Proc. 21st SODA, pp. 117–122 (2010)
29. Välimäki, N., Mäkinen, V.: Space-Efficient Algorithms for Document Retrieval. In: Ma, B., Zhang, K. (eds.) CPM 2007. LNCS, vol. 4580, pp. 205–215. Springer, Heidelberg (2007)

Span Programs and Quantum Algorithms for *st*-Connectivity and Claw Detection

Aleksandrs Belovs[1] and Ben W. Reichardt[2]

[1] University of Latvia
[2] University of Southern California

Abstract. We use span programs to develop quantum algorithms for several graph problems. We give an algorithm that uses $O(n\sqrt{d})$ queries to the adjacency matrix of an n-vertex graph to decide if vertices s and t are connected, under the promise that they either are connected by a path of length at most d, or are disconnected. We also give $O(n)$-query algorithms that decide if a graph contains as a subgraph a path, a star with two subdivided legs, or a subdivided claw. These algorithms can be implemented time efficiently and in logarithmic space. One of the main techniques is to modify the natural *st*-connectivity span program to drop along the way "breadcrumbs," which must be retrieved before the path from s is allowed to enter t.

1 Introduction

Span programs are a linear-algebraic model of computation, dating back to 1993 [KW93]. Recently, span programs have been shown to be equivalent to quantum query algorithms [Rei09, Rei11b]. The span program model therefore provides a new way of developing algorithms for quantum computers. The advantage of this approach is that span programs are a very simple model to understand. The model has no notions of entanglement, unitary operations, measurements, or any other of the potentially confounding features of quantum computation. After all, span programs were originally developed to study *classical* models such as branching programs, symmetric logspace and secret-sharing schemes. Although the extreme simplicity of span programs can be startling—the model even lacks any concept of dynamics—they have proven to be a powerful tool for developing new quantum algorithms. To date, they have been used to devise algorithms for formula evaluation [RŠ08, Rei11a, Rei11c, ZKH11, Kim11], matrix rank [Bel11] and graph collision [GI12]. A class of span programs known as learning graphs has been used to devise algorithms for subgraph detection and related problems [Bel12b, Zhu11, LMS11], and the k-distinctness problem [BL11, Bel12a].

In this paper, we present a new, span program-based quantum algorithm for the *st*-connectivity problem. The algorithm runs faster in many cases than the previous best algorithm [DHHM04], and uses exponentially less space—only $O(\log n)$ qubits, where n is the number of vertices. Space complexity is especially important for quantum algorithms because qubits and quantum RAM devices are expensive to implement, and yet many of the most-important quantum

L. Epstein and P. Ferragina (Eds.): ESA 2012, LNCS 7501, pp. 193–204, 2012.

algorithms are very space-inefficient [Amb07, MSS05]. The underlying span program is quite natural, and is somewhat analogous to the famous random walk algorithm of Aleliunas et al. [AKL+79]. Its complexity relates to the effective electrical resistance in the graph, a quantity that also arises in the analysis of random walks.

The st-connectivity span program has several applications. It can be seen as the key subroutine for quantum query algorithms based on learning graphs. Using the span program, we give a new quantum algorithm for detecting fixed paths in a graph, that is roughly $\sqrt{n}$ times faster than the previous best algorithms. We also extend the span program with a new "breadcrumb" technique: along the path from s, drop crumbs that must be retrieved before being allowed to enter t. This technique requires span programs, and does not seem to be analogous to anything that can be achieved using random or quantum walks on the graph. We use it to give optimal quantum algorithms for detecting subdivided claw subgraphs, and optimal quantum algorithms for detecting subdivided stars and triangles under a certain promise.

A disadvantage of the span program approach is that span programs are equivalent to quantum algorithms under query complexity only, not time complexity. Aside from the algorithms for formula evaluation, it is not known how to implement efficiently the previous span program-based quantum query algorithms. Nonetheless, by combining an algorithm for evaluating span programs with a Szegedy-style quantum walk [Sze04], we show how to implement our algorithms time efficiently, with only a poly-logarithmic overhead. A main open problem is whether this technique can be applied to learning graphs more broadly, thereby giving time-efficient quantum algorithms for a wide class of problems.

st-connectivity problem: In the undirected st-connectivity problem, we are given query access to the adjacency matrix of a simple, undirected, n-vertex graph G with two selected vertices s and t. The task is to determine whether there is a path from s to t. Classically, it can be solved in quadratic time by a variety of algorithms, e.g., depth- or breadth-first search. Its randomized query complexity is $\Theta(n^2)$. With more time, it can be solved in logarithmic space [Rei05, AKL+79].

Dürr *et al.* have given a quantum algorithm for st-connectivity that makes $O(n^{3/2})$ queries to the adjacency matrix [DHHM04]. The algorithm's time complexity is also $O(n^{3/2})$ up to logarithmic factors. It uses $O(\log n)$ qubits and requires coherently addressable access to $O(n \log n)$ classical bits, or quantum RAM.

Our algorithm has the same time complexity in the worst case, but only logarithmic space complexity. Moreover, the time complexity reduces to $\tilde{O}(n\sqrt{d})$, if it is known that the shortest path between s and t, if one exists, has length at most d. Note, though, that our algorithm only detects the presence of an s-t path. When $d = \omega(\log n)$, there is not even space to output the path.

The algorithm is based on a span program originally introduced in [KW93]. Consider an n-dimensional real vector space, with a coordinate for each vertex. For any edge (u, v) of G, you are allowed to use the vector $|u\rangle - |v\rangle$, i.e., the vector with coefficient $+1$ in the u coordinate, -1 in the v coordinate and 0 in all other coordinates. The goal is to reach the "target" vector, $|t\rangle - |s\rangle$. If s is connected to t, say via a path $u_0 = s, u_1, u_2, \ldots, u_m = t$, then the sum of the vectors $|u_{i+1}\rangle - |u_i\rangle$ equals the target vector. If s is disconnected from t, then no linear combination of the edge vectors will reach the target.

Subgraph containment problems: The promised upper bound on the length of an s-t path arises in naturally in applications such as learning graphs. As another example, we give an optimal, $O(n)$-query quantum algorithm for detecting the presence of a length-k path in a graph G given by its adjacency matrix. To our knowledge, this is the first linear-query quantum algorithm capable of detecting subgraphs with arbitrarily large, constant size. The previous best quantum query algorithms for deciding if a graph contains a length-k path use $\tilde{O}(n)$ queries for $k \leq 4$, $\tilde{O}(n^{3/2-1/(\lceil k/2 \rceil-1)})$ queries for $k \geq 9$, and certain intermediate polynomials for $5 \leq k \leq 8$ [CK11].

The reduction to *st*-connectivity is as follows. Assign to each vertex of G a color chosen from $\{0, 1, \ldots, k\}$, independently and uniformly at random. Discard the edges of G except those between vertices with consecutive colors. Add two new vertices s and t, and join s to all vertices of color 0, and t to all vertices of color k. If there is a path from s to t in the resulting graph H, then G contains a length-k path. Conversely, if G contains a length-k path, then with probability at least $2(k+1)^{-k-1} = \Omega(1)$ the vertices of the path are colored consecutively, and hence s and t are connected in H. The algorithm's query complexity is $O(n\sqrt{d}) = O(n)$, using $d = k + 2$.

Path detection is a special case of a *forbidden subgraph property*, i.e., a graph property characterized by a finite set of forbidden subgraphs. Another class of graph properties are *minor-closed* properties, i.e., properties that if satisfied by a graph G are also satisfied by all minors of G. Examples include whether the graph is acyclic, and whether it can be embedded in some surface. Robertson and Seymour have famously shown that any minor-closed property can be described by a finite set of forbidden minors [RS04]. Childs and Kothari have shown that the quantum query complexity of a minor-closed graph property is $\Theta(n^{3/2})$ unless the property is a forbidden subgraph property, in which case it is $o(n^{3/2})$ [CK11].

We make further progress on characterizing the quantum query complexity of minor-closed forbidden subgraph properties. We show that a minor-closed property can be solved by a quantum algorithm that uses $O(n)$ queries and $\tilde{O}(n)$ time if it is characterized by a *single* forbidden subgraph. The graph is then necessarily a collection of disjoint paths and subdivided claws. This query complexity is optimal. The algorithm for these cases is a generalization of the *st*-connectivity algorithm. It still checks connectivity in a certain graph built from G, but also pairs some of the edges together so that if one edge in the pair is used then so must be the other. Roughly, it is as though the algorithm drops breadcrumbs along the way that must be retrieved before the path is allowed to enter t.

For an example of the breadcrumb technique, consider the problem of deciding whether G contains a triangle. Triangle-finding has been studied extensively. It can be used as a subroutine in boolean matrix multiplication [WW10]. The best known quantum algorithm, based on span programs and learning graphs, uses $O(n^{1.296})$ queries [Bel12b], whereas the best known lower bound is $\Omega(n)$ queries.

As another approach to this problem, randomly color the vertices of G by $\{1, 2, 3\}$. Discard edges between vertices of the same color, make two copies of each vertex of color 1, the first connected to color-2 vertices and the second connected to color-3 vertices. Connect s to all the first vertices of color 1 and connect t to all the second vertices of color 1, and ask if s is connected to t. Unfortunately, this will give false positives. If G is a path of length four, with vertices colored $1, 2, 3, 1$, then s will be connected to t even though there is no triangle. To fix it, we can drop a breadcrumb at the first color-1 vertex and require that it be retrieved after the color-3 vertex; then the algorithm will no longer accept this path. The algorithm still does not work, though, because it will accept a cycle of length five colored $1, 2, 3, 2, 3$. Since the st-connectivity algorithm works in undirected graphs, it cannot see that the path backtracked; graph minors can fool the algorithm. What we can say is that in this particular case, the technique gives an $O(n)$-query and $\tilde{O}(n)$-time quantum algorithm that detects whether G contains a triangle or is acyclic, i.e., does not contain a triangle as a minor.

The full version of this article is available at [BR12].

2   Span Programs

Definition 1. *A span program $\mathcal{P} = (n, d, |\tau\rangle, V_{free}, \{V_{i,b}\})$ consists of a "target" vector $|\tau\rangle \in \mathbf{R}^d$ and finite sets V_{free} and $V_{1,0}, V_{1,1}, \ldots, V_{n,0}, V_{n,1}$ of "input" vectors from $\mathbf{R}^d$. To $\mathcal{P}$ corresponds a boolean function $f_{\mathcal{P}} : \{0, 1\}^n \to \{0, 1\}$, defined by $f_{\mathcal{P}}(x) = 1$ if and only if $|\tau\rangle$ lies in the span of the vectors in $V_{free} \cup \bigcup_{i=1}^n V_{i,x_i}$.*

For an input $x \in \{0, 1\}^n$, define the *available input vectors* as the vectors in $V_{free} \cup \bigcup_{i \in [n]} V_{i,x_i}$. Let V, $V(x)$ and V_{free} be matrices having as columns the input vectors, the available input vectors and the free input vectors in V_{free}, respectively. Then $f_{\mathcal{P}}(x) = 1$ if and only if $|\tau\rangle \in \mathcal{C}(V(x))$.

A useful notion of span program complexity is the *witness size*.

- If $\mathcal{P}$ evaluates to 1 on input x, a *witness* for this input is a pair of vectors $|w\rangle$ and $|w_{free}\rangle$ such that $V_{free}|w_{free}\rangle + V(x)|w\rangle = |\tau\rangle$. Its *witness size* is $\||w\rangle\|^2$.
- If $f_{\mathcal{P}}(x) = 0$, then a witness for x is any vector $|w'\rangle \in \mathbf{R}^d$ such that $\langle w'|\tau\rangle = 1$ and $|w'\rangle \perp \mathcal{C}(V(x))$. Since $|\tau\rangle \notin \text{span}(V(x))$, such a vector exists. The witness size of $|w'\rangle$ is defined as $\|V^\dagger|w'\rangle\|^2$. This equals the sum of the squares of the inner products of $|w'\rangle$ with all unavailable input vectors.

The witness size of span program $\mathcal{P}$ on input x, $\text{wsize}(\mathcal{P}, x)$, is defined as the minimal size among all witnesses for x. For $\mathcal{D} \subseteq \{0, 1\}^n$, let $\text{wsize}_b(\mathcal{P}, \mathcal{D}) = \max_{x \in \mathcal{D}: f_{\mathcal{P}}(x) = b} \text{wsize}(\mathcal{P}, x)$. Then the witness size of $\mathcal{P}$ on domain $\mathcal{D}$ is defined as

$$\text{wsize}(\mathcal{P}, \mathcal{D}) = \sqrt{\text{wsize}_0(\mathcal{P}, \mathcal{D})\,\text{wsize}_1(\mathcal{P}, \mathcal{D})}. \tag{1}$$

The bounded-error quantum query complexity of any function $f : \mathcal{D} \to \{0, 1\}$, $\mathcal{D} \subseteq \{0, 1\}^n$, is within a constant factor of the least witness size of a span program computing f [Rei09, Rei11b]. Thus, searching for good quantum query algorithms is equivalent to searching for span programs with small witness size.

3   Span Program for st-Connectivity

Theorem 2. *Consider the st-connectivity problem on a graph G given by its adjacency matrix. Assume there is a promise that if s and t are connected by a path, then they are connected by a path of length at most d. Then the problem can be decided in $O(n\sqrt{d})$ quantum queries. Moreover, the algorithm can be implemented in $\tilde{O}(n\sqrt{d})$ quantum time and $O(\log n)$ quantum space.*

Proof. The span program is as specified in the introduction.

When s is connected to t in G, let $t = u_0, u_1, \ldots, u_m = s$ be a path between them of length $m \leq d$. All vectors $|u_i\rangle - |u_{i+1}\rangle$ are available, and their sum is $|t\rangle - |s\rangle$. Thus the span program evaluates to 1. The witness size is at most $m \leq d$.

Next assume that t and s are in different connected components of G. Define $|w'\rangle$ by $\langle w', u\rangle = 1$ if u is in the connected component of t, and 0 otherwise. Then $\langle w', t - s\rangle = 1$ and $|w'\rangle$ is orthogonal to all available input vectors. Thus the span program evaluates to 0 with $|w'\rangle$ a witness. Since there are $O(n^2)$ input vectors, having inner products with $|w'\rangle$ at most one, the witness size is $O(n^2)$.

Thus the witness size is $O(n\sqrt{d})$. See Sec. 5 for the time and space claims.   □

Observe that when s and t are connected, the span program's witnesses correspond exactly to balanced unit flows from s to t in G. The witness size of a flow is the sum over all edges of the square of the flow across that edge. If there are multiple simple paths from s to t, then it is therefore beneficial to spread the flow across the paths in order to minimize the witness size. The optimal witness size is the same as the resistance distance between s and t, i.e., the effective resistance, or equivalently twice the energy dissipation of a unit electrical flow, when each edge in the graph is replaced by a unit resistor [DS84]. Spreading the flow to minimize its energy is the main technique used in the analysis of quantum query algorithms based on learning graphs [Bel12b, Zhu11, LMS11, BL11, Bel12a], for which this span program for st-connectivity can be seen to be the key subroutine.

4   Subgraph/Not-a-Minor Promise Problem

A natural strategy for deciding a minor-closed forbidden subgraph property is to take the list of forbidden subgraphs and test the input graph G for each subgraph one by one. Let T be a forbidden subgraph from the list. To simplify the problem of detecting T, we can add the promise that G either contains T as a subgraph or does not contain T *as a minor*. Call this promise problem the subgraph/not-a-minor promise problem for T.

In this section, we develop an approach to the subgraph/not-a-minor problem using span programs. We first show that the approach achieves the optimal $O(n)$ query complexity in the case that T is a subdivided star. As a special case, this implies an optimal quantum query algorithm for deciding minor-closed forbidden subgraph properties that are determined by a single forbidden subgraph. We extend the approach to give an optimal $O(n)$-query algorithm for the case that T is a triangle. However, the approach fails for the case $T = K_5$.

Theorem 3. *Let T be a subdivision of a star. Then there exists a quantum algorithm that, given query access to the adjacency matrix of a simple graph G with n vertices, makes $O(n)$ queries, and, with probability at least $2/3$, accepts if G contains T as a subgraph and rejects if G does not contain T as a minor. Furthermore, the algorithm can be implemented efficiently, in $\tilde{O}(n)$ quantum time and $O(\log n)$ quantum space.*

It can be checked that if T is a path or a subdivision of a claw, i.e., a star with three legs, then a graph G contains T as a minor if and only if it contains it as a subgraph. Disjoint collections of paths and subdivided claws are the only graphs T with this property. In these cases, the algorithm therefore solves the subgraph-detection problem. The previous best algorithm for detecting an $\{\ell_1, \ell_2, \ell_3\}$-subdivided claw subgraph uses $\tilde{O}(n^{3/2 - 2/(\ell_1 + \ell_2 + \ell_3 - 1)})$-query algorithm if all ℓ_js are even, with a similar expression if any of them is odd [CK11].

Proof. The proof uses the color-coding technique from [AYZ95]. Let T be a star with d legs, of lengths $\ell_1, \ldots, \ell_d > 0$. Denote the root vertex by r and the vertex at depth i along the jth leg by $v_{j,i}$. The vertex set of T is $V_T = \{r, v_{1,1}, \ldots, v_{1,\ell_1}, \ldots, v_{d,1}, \ldots, v_{d,\ell_d}\}$. Color every vertex u of G with an element $c(u) \in V_T$ chosen independently and uniformly at random. For $v \in V_T$, let $c^{-1}(v)$ be its preimage set of vertices of G. We design a span program that

- Accepts if there is a correctly colored T-subgraph in G, i.e., an injection ι from V_T to the vertices of G such that $c \circ \iota$ is the identity, and (t, t') being an edge of T implies that $(\iota(t), \iota(t'))$ is an edge of G;
- Rejects if G does not contain T as a minor, no matter the coloring c.

If G contains a T-subgraph, then the probability it is colored correctly is at least $|V_T|^{-|V_T|} = \Omega(1)$. Evaluating the span program for a constant number of independent colorings therefore suffices to detect T with probability at least $2/3$.

Span program. The span program we define works on the vector space with orthonormal basis

$$\{|s\rangle, |t\rangle\} \cup \left\{|u, b\rangle : (u, b) \in \left(c^{-1}(r) \times \{0, \ldots, d\}\right) \cup \bigcup_{v \in V_T \smallsetminus \{r\}} c^{-1}(v) \times \{0, 1\}\right\}. \quad (2)$$

The target vector is $|t\rangle - |s\rangle$. For $u \in c^{-1}(r)$, there are free input vectors $|u, 0\rangle - |s\rangle$ and $|t\rangle - |u, d\rangle$. For $j \in [d]$ and $u \in c^{-1}(v_{j,\ell_j})$, there are free input vectors $|u, 1\rangle - |u, 0\rangle$. For $j \in [d]$, there are the following input vectors:

- For $i \in [\ell_j - 1]$, $u \in c^{-1}(v_{j,i})$ and $u' \in c^{-1}(v_{j,i+1})$, the input vectors $|u', 0\rangle - |u, 0\rangle$ and $|u, 1\rangle - |u', 1\rangle$ are available when there is an edge (u, u') in G.
- For $u \in c^{-1}(r)$ and $u' \in c^{-1}(v_{j,1})$, the input vector $(|u', 0\rangle - |u, j - 1\rangle) + (|u, j\rangle - |u', 1\rangle)$ is available when there is an edge (u, u') in G.

For arguing about this span program, it is convenient to define a graph H whose vertices are the basis vectors in Eq. (2). Edges of H correspond to the available span program input vectors; for an input vector with two terms, $|\alpha\rangle - |\beta\rangle$, add an edge $(|\alpha\rangle, |\beta\rangle)$, and for the four-term input vectors $(|u', 0\rangle - |u, j - 1\rangle) + (|u, j\rangle - |u', 1\rangle)$ add two "paired" edges, $(|u, j - 1\rangle, |u', 0\rangle)$ and $(|u, j\rangle, |u', 1\rangle)$.

Positive case. Assume that there is a correctly colored T-subgraph in G, given by a map ι from V_T to the vertices of G. Then the target $|t\rangle - |s\rangle$ is achieved as the sum of the input vectors spanned by $|s\rangle$, $|t\rangle$ and the basis vectors of the form $|u, \cdot\rangle$ with $u \in \iota(V_T)$. All these vectors are available. This sum corresponds to a path from s to t in H, passing through $|\iota(r), 0\rangle, |\iota(v_{1,1}), 0\rangle, \ldots, |\iota(v_{1,\ell_1}), 0\rangle, |\iota(v_{1,\ell_1}), 1\rangle, |\iota(v_{1,\ell_1-1}), 1\rangle, \ldots, |\iota(v_{1,1}), 1\rangle, |\iota(r), 1\rangle, |\iota(v_{2,1}), 0\rangle, \ldots, |\iota(r), d\rangle$. Pulled back to T, the path goes from r out and back along each leg, in order. The positive witness size is $O(1)$, since there are $O(1)$ input vectors along the path.

This argument shows much of the intuition for the span program. T is detected as a path from $|s\rangle$ to $|t\rangle$, starting at a vertex in G with color r and traversing each leg of T in both directions. It is not enough just to traverse T in this manner, though, because the path might each time use different vertices of color r. The purpose of the four-term input vectors $(|u', 0\rangle - |u, j - 1\rangle) + (|u, j\rangle - |u', 1\rangle)$ is to enforce that if the path goes out along an edge $(|u, j - 1\rangle, |u', 0\rangle)$, then it must return using the paired edge $(|u', 1\rangle, |u, j\rangle)$.

Negative case. Assume that G does not contain T as a minor. It may still be that $|s\rangle$ is connected to $|t\rangle$ in H. We construct an ancillary graph H' from H by removing some vertices and adding some extra edges, so that $|s\rangle$ is disconnected from $|t\rangle$ in H'. Figure 1 shows an example.

H' is defined starting from H. Let $V_j = \{v_{j,1}, \ldots, v_{j,\ell_j}\}$, $H_j = \{|u, b\rangle : c(u) \in V_j, b \in \{0, 1\}\}$ and $R_j = \{|u, j\rangle : c(u) = r\}$. For $j \in [d]$ and $u \in c^{-1}(r)$,

- Add an edge $(|u, j - 1\rangle, |u, j\rangle)$ to H' if $|u, j - 1\rangle$ is connected to R_j in H via a path for which all internal vertices, i.e., vertices besides the two endpoints, are in H_j; and
- Remove all vertices in H_j that are connected both to R_{j-1} and R_j in H via paths with all internal vertices in H_j.

Note that in the second case, for each $u' \in c^{-1}(V_j)$, either both $|u', 0\rangle$ and $|u', 1\rangle$ are removed, or neither is. Indeed, if there is a path from $|u', 0\rangle$ to R_j, then it necessarily must pass along an edge $(|u'', 0\rangle, |u'', 1\rangle)$ with $c(u'') = v_{j,\ell_j}$. Then backtracking along the path before this edge, except with the second coordinate switched $0 \leftrightarrow 1$, gives a path from $|u', 0\rangle$ to $|u', 1\rangle$. Similarly $|u', 0\rangle$ is connected to $|u', 1\rangle$ if there is a path from $|u', 1\rangle$ to R_{j-1}.

Define the negative witness $|w'\rangle$ by $\langle v|w'\rangle = 1$ if $|s\rangle$ is connected to $|v\rangle$ in H', and $\langle v|w'\rangle = 0$ otherwise. Then $|w'\rangle$ is orthogonal to all available input vectors. In particular, it is orthogonal to any available four-term input vector $(|u', 0\rangle - |u, j - 1\rangle) + (|u, j\rangle - |u', 1\rangle)$, corresponding to two paired edges in H, because either the same edges are present in H', or $|u', 0\rangle$ and $|u', 1\rangle$ are removed and a new edge $(|u, j - 1\rangle, |u, j\rangle)$ is added.

To verify that $|w'\rangle$ is a witness for the span program evaluating to 0, with $(\langle t| - \langle s|)|w'\rangle = -1$, it remains to prove that $|s\rangle$ is disconnected from $|t\rangle$ in H'. Assume that $|s\rangle$ is connected to $|t\rangle$ in H', via a simple path p. Based on the path p, we will construct a minor of T, giving a contradiction.

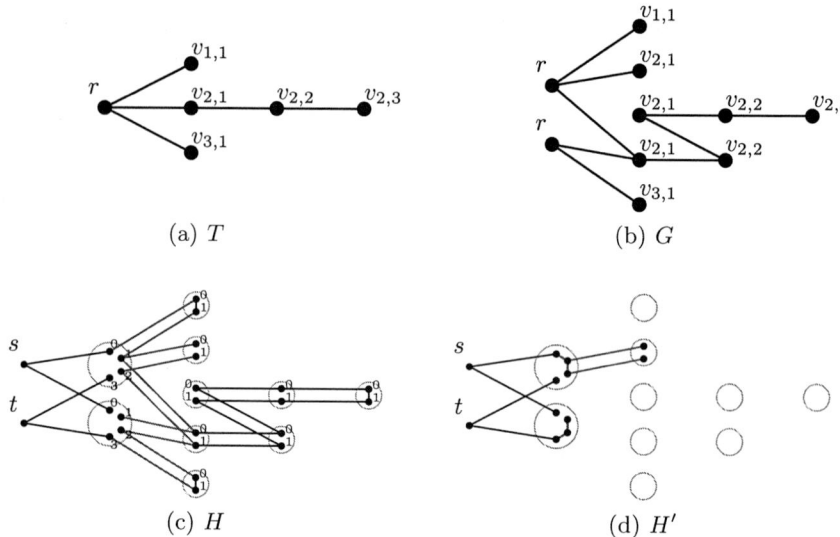

Fig. 1. An example to illustrate the constructions of the graphs H and H' in the negative case of the proof of Theorem 3. (a) A subdivided star T with $(\ell_1, \ell_2, \ell_3) = (1, 3, 1)$. (b) A graph G with vertices labeled by their colors, i.e., vertices of T. Although G contains T as a subgraph, the coloring is incorrect and the span program will reject. (c) The graph H, edges and paired edges of which correspond to available span program input vectors. Vertices of G have been split into two or four parts, and vertices s and t have been added. Paired edges are colored blue. Note that s is connected to t in H. (d) The graph H'. New edges are colored red. Note that s is disconnected from t.

The path p begins at $|s\rangle$ and next must move to some vertex $|u_0, 0\rangle$, where $c(u_0) = r$. The path ends by going from a vertex $|u_d, d\rangle$, where $c(u_d) = r$, to $|t\rangle$. By the structure of the graph H', p must also pass in order through some vertices $|u_1, 1\rangle, |u_2, 2\rangle, \ldots, |u_{d-1}, d-1\rangle$, where $c(u_j) = r$.

Consider the segment of the path from $|u_{j-1}, j-1\rangle$ to $|u_j, j\rangle$. Due to the construction, this segment must cross a new edge added to H', $(|u'_j, j-1\rangle, |u'_j, j\rangle)$ for some u'_j with $c(u'_j) = r$. Thus the path p has the form $|s\rangle, \ldots, |u'_1, 0\rangle, |u'_1, 1\rangle, \ldots, |u'_2, 0\rangle, |u'_2, 1\rangle, \ldots \ldots, |u'_d, 0\rangle, |u'_d, 1\rangle, \ldots, |t\rangle$. Based on this path, we can construct a minor for T. The branch set of the root r consists of all the vertices in G that correspond to vertices along p (by discarding the second coordinate). Furthermore, for each edge $(|u'_j, j-1\rangle, |u'_j, j\rangle)$, there is a path in H from $|u'_j, j-1\rangle$ to R_j, in which every internal vertex is in H_j. The first ℓ_j vertices along the path give a minor for the jth leg of T. It is vertex-disjoint from the minors for the other legs because the colors are different. It is also vertex-disjoint from the branch set of r because no vertices along the path are present in H'. Therefore, we obtain a minor for T, a contradiction.

Since each coefficient of $|w'\rangle$ is 0 or 1, the overlap of $|w'\rangle$ with any of the $O(n^2)$ input vectors is at most 2 in magnitude. Thus the witness size is $O(n^2)$.

By Eq. (1), the span program's overall witness size is $O(n)$.   □

Theorem 4. *Let T be a collection of subdivided stars. Then there is a quantum algorithm that, given query access to the adjacency matrix of a simple graph G with n vertices, makes $O(n)$ queries, and, with probability at least $2/3$, accepts if G contains T as a subgraph and rejects if G does not contain T as a minor.*

Proof. It is not enough to apply Theorem 3 once for each component of T, because some components might be subgraphs of other components. Instead, proceed as in the proof of Theorem 3, but for each fixed coloring of G by the vertices of T run the span program once for every component on the graph G restricted to vertices colored by that component. This ensures that in the negative case, if the span programs for all components accept, then there are vertex-disjoint minors for every component, which together form a minor for T.

$\square$

In this paper's full version, we show the above technique extends to triangles:

Theorem 5. *There is a bounded error $O(n)$-query quantum algorithm that, given query access to the adjacency matrix of a graph G with n vertices, accepts if G contains a triangle, and rejects if G is a forest.*

5   Time-Efficient Implementation

It is known that any problem can be evaluated query-efficiently by a Grover-like algorithm that alternates input-dependent and input-independent reflections [Rei11b]. We use this result, and additionally apply a Szegedy-type quantum walk to implement the input-independent reflection. We will show:

Theorem 6. *The algorithm from Theorem 2 can be implemented in $\tilde{O}(n\sqrt{d})$ time. Both algorithms from Theorems 3 and 5 can be implemented in $\tilde{O}(n)$ time. The algorithms from Theorems 2 and 3 use $O(\log n)$ quantum space.*

For example, the algorithm for st-connectivity has a very simple form. It works in the Hilbert space $\mathbf{C}^{\binom{n}{2}}$, with one dimension per possible edge in the graph. It alternates two reflections. The first reflection applies a coherent query to the adjacency matrix input in order to add a phase of -1 to all edges not present in G. The second reflection is a reflection about all balanced flows from s to t in the complete graph K_n. The second reflection is equivalent to reflecting about the constraints that for every vertex $v \notin \{s, t\}$, the net flow into v must be zero. This is difficult to implement directly because constraints for different vertices do not commute with each other. Our time-efficient procedure essentially works by inserting a new vertex in the middle of every edge. Balanced flows in the new graph correspond exactly to balanced flows in the old graph, but the new graph is bipartite. We reflect separately about the constraints in each half of the bipartition, and combine these reflections with a phase-estimation subroutine.

The main framework for the algorithms is the same as for Algorithm 1 from [Rei11b], which we extend from "canonical" span programs to general span programs. The analysis uses the Effective Spectral Gap Lemma from [LMR+11].

Let $\mathcal{P}$ be a span program with the target vector $|\tau\rangle$ and $m-1$ input vectors $\{|v_j\rangle\}$ in $\mathbf{R}^d$. Let W_1 and W_0 be the positive and the negative witness sizes, respectively, and let $W = \sqrt{W_0 W_1}$ be the witness size of $\mathcal{P}$. Also let $|\tilde{\tau}\rangle = |\tau\rangle/\alpha$, where $\alpha = C_1\sqrt{W_1}$ for some constant C_1 to be specified later.

Let V be the matrix containing the input vectors of $\mathcal{P}$, and also $|\tilde{\tau}\rangle$, as columns. Our quantum algorithm works in the vector space $\mathcal{H} = \mathbf{R}^m$, with standard basis elements $|j\rangle$ for $j = \{0, \ldots, m-1\}$. Basis vectors $|j\rangle$ for $j > 0$ correspond to the input vectors, and $|0\rangle$ corresponds to $|\tilde{\tau}\rangle$. Let Λ be the orthogonal projection onto the nullspace of V. For any input x of $\mathcal{P}$, let $\Pi_x = \sum |j\rangle\langle j|$ where the summation is over $j = 0$ and those indices $j > 0$ corresponding to the available input vectors.

Let $U = R_\Lambda R_\Pi$, where $R_\Lambda = 2\Lambda - I$ and $R_\Pi = 2\Pi_x - I$ are the reflections about the images of Λ and Π_x. Starting in $|0\rangle$, the algorithm runs phase estimation on U with precision $\Theta = \frac{1}{C_2 W}$, and accepts if and only if the measured phase is zero. In the full version of this paper, we show:

Theorem 7. *The algorithm is correct and uses $O(W)$ controlled calls to U.*

The reflection R_Π can be implemented efficiently in most cases, but implementing R_Λ is more difficult. We describe a general way of implementing R_Λ, which is efficient for relatively uniform span programs like those in this paper.

Essentially, we consider the $d \times m$ matrix V as the biadjacency matrix for a bipartite graph on $d + m$ vertices, and run a Szegedy-type quantum walk as in [ACR+10, RŠ08]. Such a quantum walk requires "factoring" V into two sets of unit vectors, vectors $|a_i\rangle \in \mathbf{R}^m$ for each row $i \in [k]$, and vectors $|b_j\rangle \in \mathbf{R}^d$ for each column $j = 0, \ldots, m-1$, satisfying $\langle a_i|j\rangle\langle i|b_j\rangle = V'_{ij}$, where V' differs from V only by a rescaling of the rows. Given such a factorization, let $A = \sum_{i \in [d]}(|i\rangle \otimes |a_i\rangle)\langle i|$ and $B = \sum_{j=0}^{m-1}(|b_j\rangle \otimes |j\rangle)\langle j|$, so $A^\dagger B = V'$. Let R_A and R_B be the reflections about the column spaces of A and B, respectively. Embed $\mathcal{H}$ into $\tilde{\mathcal{H}} = \mathbf{R}^k \otimes \mathbf{R}^m$ using the isometry B. R_Λ can be implemented on $B(\mathcal{H})$ as the reflection about the -1 eigenspace of $R_B R_A$. Indeed, this eigenspace equals $(\mathcal{C}(A)^\perp \cap \mathcal{C}(B)) \oplus (\mathcal{C}(A) \cap \mathcal{C}(B)^\perp)$, or $B(\ker V)$ plus a part that is orthogonal to $\mathcal{C}(B)$ and therefore irrelevant.

The reflection about the -1 eigenspace of $R_B R_A$ is implemented using phase estimation. The efficiency depends on two factors:

1. The implementation costs of R_A and R_B. They can be easier to implement than R_Λ directly because they decompose into local reflections.
2. The spectral gap around the -1 eigenvalue of $R_B R_A$. This gap is determined by the spectral gap of $A^\dagger B = V'$ around singular value zero.

So far the arguments have been general. Let us now specialize to the span programs in Theorems 2, 3 and 5. These span programs are sufficiently uniform that neither of the above two factors is a problem. Both reflections can be implemented efficiently, in polylogarithmic time, using quantum parallelism. Similarly, we can show that V' has an $\Omega(1)$ spectral gap around singular value zero. Therefore, approximating to within an inverse polynomial the reflection about the -1 eigenspace of $R_B R_A$ takes only polylogarithmic time.

The algorithms from Theorems 3 and 5 both look similar. In each case, the span program is based on a graph H, whose vertices form an orthonormal basis for the span program vector space. The vertices of H can be divided into a sequence of layers that are monochromatic according to the coloring c, such that edges only go between consecutive layers. Precisely, place the vertices s and t each on their own separate layer at the beginning and end, respectively, and set the layer of a vertex v to be the distance from s to $c(v)$ in the graph H for the case that $G = T$. For example, in the span program for detecting a subdivided star with branches of lengths $\{\ell_1, \ldots, \ell_d\}$, there are $2 + 2\sum_{j \in [d]}(\ell_j + 1)$ layers, because the s-t path is meant to traverse each branch of the star out and back.

In order to facilitate finding factorizations A and B such that R_A and R_B are easily implementable, we modify the span programs, by splitting vertices and adding dummy vertices and edges, so that between any two adjacent layers of H there is a complete bipartite graph. It is then easy to factor $V' = \frac{1}{2\sqrt{n}}V$. In the full version, we analyze the spectral gap around zero of V', using simple facts about the spectra of block matrices.

We thank Andrew Childs, Robin Kothari and Rajat Mittal for useful discussions, and Robert Špalek for comments on an earlier draft. Research conducted at the Institute for Quantum Computing, University of Waterloo. A.B. supported by the European Social Fund within the project "Support for Doctoral Studies at University of Latvia" and by FET-Open project QCS. B.R. acknowledges support from NSERC, ARO-DTO and Mitacs.

References

[ACR+10] Ambainis, A., Childs, A.M., Reichardt, B.W., Špalek, R., Zhang, S.: Any AND-OR formula of size N can be evaluated in time $N^{1/2+o(1)}$ on a quantum computer. SIAM J. Comput. 39(6), 2513–2530 (2010); Earlier version in FOCS 2007

[AKL+79] Aleliunas, R., Karp, R.M., Lipton, R.J., Lovasz, L., Rackoff, C.: Random walks, universal traversal sequences, and the complexity of maze problems. In: Proc. 20th IEEE FOCS, pp. 218–223 (1979)

[Amb07] Ambainis, A.: Quantum walk algorithm for element distinctness. SIAM J. Computing 37(1), 210–239 (2007), arXiv:quant-ph/0311001

[AYZ95] Alon, N., Yuster, R., Zwick, U.: Color-coding. J. ACM 42, 844–856 (1995); Earlier version in STOC 1994

[Bel11] Belovs, A.: Span-program-based quantum algorithm for the rank problem (2011), arXiv:1103.0842

[Bel12a] Belovs, A.: Learning-graph-based quantum algorithm for k-distinctness. To Appear in FOCS 2012 (2012), arXiv:1205.1534

[Bel12b] Belovs, A.: Span programs for functions with constant-sized 1-certificates. In: Proc. 44th ACM STOC, pp. 77–84 (2012), arXiv:1105.4024

[BL11] Belovs, A., Lee, T.: Quantum algorithm for k-distinctness with prior knowledge on the input (2011), arXiv:1108.3022

[BR12] Belovs, A., Reichardt, B.W.: Span programs and quantum algorithms for st-connectivity and claw detection (2012), arXiv:1203.2603

[CK11] Childs, A.M., Kothari, R.: Quantum query complexity of minor-closed graph properties. In: Proc. 28th STACS, pp. 661–672 (2011), arXiv:1011.1443

[DHHM04] Dürr, C., Heiligman, M., Høyer, P., Mhalla, M.: Quantum Query Complexity of Some Graph Problems. In: Díaz, J., Karhumäki, J., Lepistö, A., Sannella, D. (eds.) ICALP 2004. LNCS, vol. 3142, pp. 481–493. Springer, Heidelberg (2004), arXiv:quant-ph/0401091

[DS84] Doyle, P.G., Laurie Snell, J.: Random Walks and Electric Networks. Carus Mathematical Monographs, vol. 22. Mathematical Association of America (1984), arXiv:math/0001057 [math.PR]

[GI12] Gavinsky, D., Ito, T.: A quantum query algorithm for the graph collision problem (2012), arXiv:1204.1527

[Kim11] Kimmel, S.: Quantum adversary (upper) bound (2011), arXiv:1101.0797

[KW93] Karchmer, M., Wigderson, A.: On span programs. In: Proc. 8th IEEE Symp. Structure in Complexity Theory, pp. 102–111 (1993)

[LMR+11] Lee, T., Mittal, R., Reichardt, B.W., Špalek, R., Szegedy, M.: Quantum query complexity of state conversion. In: Proc. 52nd IEEE FOCS, pp. 344–353 (2011), arXiv:1011.3020

[LMS11] Lee, T., Magniez, F., Santha, M.: A learning graph based quantum query algorithm for finding constant-size subgraphs (2011), arXiv:1109.5135

[MSS05] Magniez, F., Santha, M., Szegedy, M.: Quantum algorithms for the triangle problem. In: Proc. 16th ACM-SIAM Symp. on Discrete Algorithms, SODA (2005), arXiv:quant-ph/0310134

[Rei05] Reingold, O.: Undirected ST-connectivity in log-space. In: Proc. 37th ACM STOC, pp. 376–385 (2005)

[Rei09] Reichardt, B.W.: Span programs and quantum query complexity: The general adversary bound is nearly tight for every boolean function. Extended Abstract in Proc. 50th IEEE FOCS, pp. 544–551 (2009), arXiv:0904.2759

[Rei11a] Reichardt, B.W.: Faster quantum algorithm for evaluating game trees. In: Proc. 22nd ACM-SIAM Symp. on Discrete Algorithms (SODA), pp. 546–559 (2011), arXiv:0907.1623

[Rei11b] Reichardt, B.W.: Reflections for quantum query algorithms. In: Proc. 22nd ACM-SIAM Symp. on Discrete Algorithms (SODA), pp. 560–569 (2011), arXiv:1005.1601

[Rei11c] Reichardt, B.W.: Span-program-based quantum algorithm for evaluating unbalanced formulas. In: 6th Conf. on Theory of Quantum Computation, Communication and Cryptography, TQC (2011), arXiv:0907.1622

[RS04] Robertson, N., Seymour, P.D.: Graph minors XX. Wagner's conjecture. J. Combin. Theory Ser. B 92, 325–357 (2004)

[RŠ08] Reichardt, B.W., Špalek, R.: Span-program-based quantum algorithm for evaluating formulas. In: Proc. 40th ACM STOC, pp. 103–112 (2008), arXiv:0710.2630

[Sze04] Szegedy, M.: Quantum speed-up of Markov chain based algorithms. In: Proc. 45th IEEE FOCS, pp. 32–41 (2004)

[WW10] Williams, V.V., Williams, R.: Subcubic equivalences between path, matrix, and triangle problems. In: Proc. 51st IEEE FOCS, pp. 645–654 (2010)

[Zhu11] Zhu, Y.: Quantum query complexity of subgraph containment with constant-sized certificates (2011), arXiv:1109.4165

[ZKH11] Zhan, B., Kimmel, S., Hassidim, A.: Super-polynomial quantum speed-ups for boolean evaluation trees with hidden structure (2011), arXiv:1101.0796

The Stretch Factor of L_1- and L_∞-Delaunay Triangulations

Nicolas Bonichon[1], Cyril Gavoille[2,*], Nicolas Hanusse[3],
and Ljubomir Perković[4,**]

[1] LaBRI - INRIA Bordeaux Sud-Ouest, Bordeaux, France
[2] LaBRI - University of Bordeaux, Bordeaux, France
[3] LaBRI - CNRS, Bordeaux, France
{bonichon,gavoille,hanusse}@labri.fr
[4] DePaul University, Chicago, USA
lperkovic@cs.depaul.edu

Abstract. In this paper we determine the stretch factor of L_1-Delaunay and L_∞-Delaunay triangulations, and we show that it is equal to $\sqrt{4+2\sqrt{2}} \approx 2.61$. Between any two points x, y of such triangulations, we construct a path whose length is no more than $\sqrt{4+2\sqrt{2}}$ times the Euclidean distance between x and y, and this bound is the best possible. This definitively improves the 25-year old bound of $\sqrt{10}$ by Chew (SoCG '86). This is the first time the stretch factor of the L_p-Delaunay triangulations, for any real $p \geq 1$, is determined exactly.

Keywords: :Delaunay triangulations, L_1-metric, L_∞-metric, stretch factor.

1 Introduction

Given a finite set of points P on the plane, the Voronoï diagram of P is the decomposition of the plane into polygonal regions, one for each point of P, such that the points in the region associated with a point are closer to it than to any other point of P. The Delaunay triangulation for P is a spanning subgraph of the complete Euclidean graph on P that is the dual of the Voronoï diagram of P. In some applications (including on-line routing [BM04]), the Delaunay triangulation is used as a spanner, defined as a spanning subgraph in which the distance between any pair of points is no more than a constant multiplicative ratio of the Euclidean distance between the points. The constant ratio is typically referred to as the stretch factor of the spanner. While Delaunay triangulations have been studied extensively, obtaining a tight bound on its stretch factor has been elusive even after decades of attempts.

In the mid-1980s, it was not known whether Delaunay triangulations were spanners at all. In order to gain an understanding of the spanning properties

* Member of the "Institut Universitaire de France". Supported by the ANR-11-BS02-014 "DISPLEXITY" project and the équipe-projet INRIA "CEPAGE".
** Supported by a Fulbright Aquitaine Regional grant and a DePaul University grant.

L. Epstein and P. Ferragina (Eds.): ESA 2012, LNCS 7501, pp. 205–216, 2012.
© Springer-Verlag Berlin Heidelberg 2012

Table 1. Key stretch factor upper bounds (optimal values are bold)

Paper	Graph	Stretch factor
[DFS87]	L_2-Delaunay	$\pi(1 + \sqrt{5})/2 \approx 5.08$
[KG92]	L_2-Delaunay	$4\pi/(3\sqrt{3}) \approx 2.41$
[Xia11]	L_2-Delaunay	1.998
[Che89]	TD-Delaunay	**2**
[Che86]	L_1-,L_∞-Delaunay	$\sqrt{10} \approx 3.16$
[this paper]	L_1-,L_∞-Delaunay	$\boldsymbol{\sqrt{4 + 2\sqrt{2}} \approx 2.61}$

of Delaunay triangulations, Chew considered related, "easier" structures. In his seminal 1986 paper [Che86], he proved that an L_1-Delaunay triangulation — the dual of the Voronoï diagram of P based on the L_1-metric rather than the L_2-metric — has a stretch factor bounded by $\sqrt{10}$. Chew then continued on and showed that a TD-Delaunay triangulation — the dual of a Voronoï diagram defined using a *Triangular Distance*, a distance function not based on a circle (L_2-metric) or a square (L_1-metric) but an equilateral triangle — has a stretch factor of 2 [Che89]. Finally, Dobkin et al. [DFS87] succeeded in showing that the (classical, L_2-metric) Delaunay triangulation of P is a spanner as well. The bound on the stretch factor they obtained was subsequently improved by Keil and Gutwin [KG92] as shown in Table 1. The bound by Keil and Gutwin stood unchallenged for many years until Xia improved the bound to below 2 [Xia11].

None of the techniques developed so far lead to a tight bound on the stretch factor of a Delaunay triangulation. There has been some progress recently on the lower bound side. The trivial lower bound of $\pi/2 \approx 1.5707$ has recently been improved to 1.5846 [BDL+11] and then to 1.5932 [XZ11].

While much effort has been made on studying the stretch factor of (classical) Delaunay triangulations, since Chew's original work little has been done on L_p-Delaunay triangulations for $p \neq 2$. It is known that L_p-Delaunay triangulations are spanners: Bose et al. [BCCS08] have shown that Delaunay triangulations that are based on any convex distance function are spanners whose stretch factor depends only on the shape of the associated convex body. However, due to the general approach, the bounds on the stretch factor that they obtain are loose: the bound for L_2-Delaunay triangulations, for example, is greater than 24.

The general picture is that, in spite of much effort, with the exception of the triangular distance the exact value of the stretch factor of Delaunay triangulations based on any convex function is unknown. In particular, the stretch factor of L_p-Delaunay triangulations is unknown for each $p \geq 1$.

Our contribution. We show that the exact stretch factor of L_1-Delaunay triangulations and L_∞-Delaunay triangulations is $\sqrt{4 + 2\sqrt{2}} \approx 2.61$, ultimately improving the 3.16 bound of Chew [Che86].

2   Preliminaries

Given a set P of points in the two-dimensional Euclidean space, the Euclidean graph $\mathcal{E}$ is the complete weighted graph embedded in the plane whose nodes are identified with the points. We assume a Cartesian coordinate system is associated with the Euclidean space and thus every point can be specified with x and y coordinates. For every pair of nodes u and w, the edge (u, w) represents the segment $[uw]$ and the weight of edge (u, w) is the Euclidean distance between u and w: $d_2(u, w) = \sqrt{d_x(u, w)^2 + d_y(u, w)^2}$ where $d_x(u, w)$ (resp. $d_y(u, w)$) is the difference between the x (resp. y) coordinates of u and w.

We say that a subgraph H of a graph G is a *t-spanner* of G if for any pair of vertices u, v of G, the distance between u and v in H is at most t times the distance between u and v in G; the constant t is referred to as the *stretch factor* of H (with respect to G). H is a t-spanner (or spanner for some t constant) of P if it is a t-spanner of $\mathcal{E}$.

In our paper, we deal with the construction of spanners based on Delaunay triangulations. As we saw in the introduction, the L_1-Delaunay triangulation is the dual of the Voronoï diagram based on the L_1-metric $d_1(u, w) = d_x(u, w) + d_y(u, w)$. A property of the L_1-Delaunay triangulations, actually shared by all L_p-Delaunay triangulations, is that all their triangles can be defined in terms of empty circumscribed convex bodies (squares for L_1 or L_∞ and circles for L_2). More precisely, let a *square* in the plane be a square whose sides are parallel to the x and y axis and let a *tipped square* be a square tipped at 45°. For every pair of points $u, v \in P$, (u, v) is an edge in the L_1-*Delaunay triangulation* of P iff there is a tipped square that has u and v on its boundary and contains no point of P in its interior (cf. [Che89]).

If a *square* with sides parallel to the x and y axes, rather than a tipped square, is used in this definition then a different triangulation is defined; it corresponds to the dual of the Voronoï diagram based on the L_∞-metric $d_\infty(u, w) = \max\{d_x(u, w), d_y(u, w)\}$. We refer to this triangulation as the L_∞-Delaunay triangulation. This triangulation is nothing more than the L_1-Delaunay triangulation of the set of points P after rotating all the points by 45° around the origin. Therefore Chew's bound of $\sqrt{10}$ on the stretch factor of the L_1-Delaunay triangulation ([Che86]) applies to L_∞-Delaunay triangulations as well. In the remainder of this paper, we will be referring to L_∞-Delaunay (rather than L_1) triangulations because we will be (mostly) using the L_∞-metric and squares, rather than tipped squares.

One issue with Delaunay triangulations is that there might not be a unique triangulation of a given set of points P. To insure uniqueness and keep our arguments simple, we make the usual assumption that the points in P are in *general position*, which for us means that no four points lie on the boundary of a square and no two points share the same abscissa or the same ordinate.

We end this section by giving a lower bound on the stretch factor of L_∞-Delaunay triangulations.

Proposition 1. *For every $\varepsilon > 0$, there is a set of points P in the plane such that the L_∞-Delaunay triangulation on P has stretch factor at least $\sqrt{4 + 2\sqrt{2}} - \varepsilon$.*

This lower bound applies, of course, to L_1-Delaunay triangulations as well. The proof of this proposition, omitted for lack of space, relies on the example shown in Fig. 1.

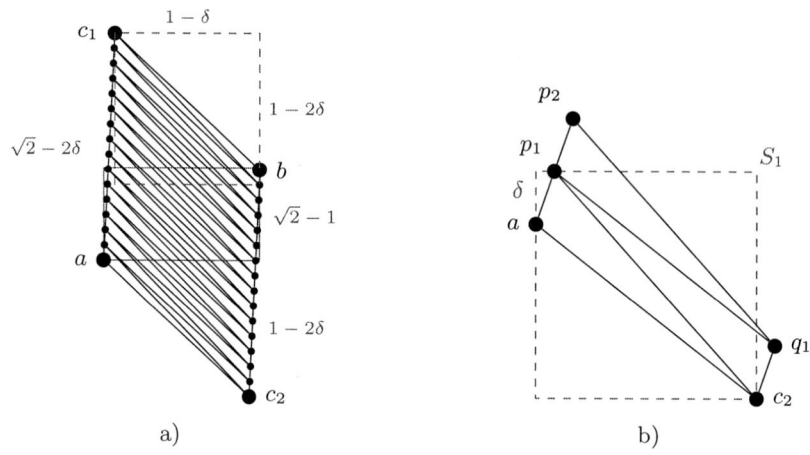

Fig. 1. a) An L_∞-Delaunay triangulation with stretch factor arbitrarily close to $\sqrt{4 + 2\sqrt{2}}$. The coordinates of points a, b, c_1, and c_2 are $(0,0)$, $(1, \sqrt{2}-1)$, $(\delta, \sqrt{2}-2\delta)$, and $(1 - \delta, 2\delta - 1)$, respectively. b) A closer look at the first few faces of this triangulation.

3   Main Result

In this section we obtain a tight upper bound on the stretch factor of an L_∞-Delaunay triangulation. It follows from this key theorem:

Theorem 1. *Let T be the L_∞-Delaunay triangulation on a set of points P in the plane and let a and b be any two points of P. If $x = d_\infty(a,b) = \max\{d_x(a,b), d_y(a,b)\}$ and $y = \min\{d_x(a,b), d_y(a,b)\}$ then*

$$d_T(a,b) \leq (1 + \sqrt{2})x + y$$

where $d_T(a,b)$ denotes the distance between a and b in triangulation T.

Corollary 1. *The stretch factor of the L_1- and the L_∞-Delaunay triangulation on a set of points P is at most*

$$\sqrt{4 + 2\sqrt{2}} \approx 2.6131259\ldots$$

Proof. By Theorem 1, an upper-bound of the stretch factor of an L_∞-Delaunay triangulation is the maximum of the function

$$\frac{(1+\sqrt{2})x + y}{\sqrt{x^2 + y^2}}$$

over values x and y such that $0 < y \le x$. The maximum is reached when x and y satisfy $y/x = \sqrt{2} - 1$, and the maximum is equal to $\sqrt{4 + 2\sqrt{2}}$. As L_1- and L_∞-Delaunay triangulations have the same stretch factor, this result also holds for L_1-Delaunay triangulations.   □

To prove Theorem 1, we will construct a bounded length path in T between two arbitrary points a and b of P. To simplify the notation and the discussion, we assume that point a has coordinates $(0,0)$ and point b has coordinates (x,y) with $0 < y \le x$. The line containing segment $[ab]$ divides the Euclidean plane into two half-planes; a point in the same half-plane as point $(0,1)$ is said to be *above* segment $[ab]$, otherwise it is *below*. Let $T_1, T_2, T_3, \ldots, T_k$ be the sequence of triangles of triangulation T that line segment $[ab]$ intersects when moving from a to b. Let h_1 and l_1 be the nodes of T_1 other than a, with h_1 lying above segment $[ab]$ and l_1 lying below. Every triangle T_i, for $1 < i < k$, intersects line segment $[ab]$ twice; let h_i and l_i be the endpoints of the edge of T_i that intersects segment $[ab]$ last, when moving on segment $[ab]$ from a to b, with h_i being above and l_i being below segment $[ab]$. Note that either $h_i = h_{i-1}$ and $T_i = \triangle(h_i, l_i, l_{i-1})$ or $l_i = l_{i-1}$ and $T_i = \triangle(h_{i-1}, h_i, l_i)$, for $1 < i < k$. We also set $h_0 = l_0 = a$, $h_k = b$, and $l_k = l_{k-1}$. For $1 \le i \le k$, we define S_i to be the square whose sides pass through the three vertices of T_i (see Fig. 2); since T is an L_∞-Delaunay triangulation, the interior of S_i is devoid of points of P. We will refer to the sides of the square using the notation: N (north), E (east), S (south), and W (west). We will also use this notation to describe the position of an edge connecting two points lying on two sides of a square: for example, a WN edge connects a point on the west and a point on the N side. We will say that an edge is *gentle* if the line segment corresponding to it in the graph embedding has a slope within $[-1, 1]$; otherwise we will say that it is *steep*.

We will prove Theorem 1 by induction on the distance, using the L_∞-metric, between a and b. Let $R(a,b)$ be the rectangle with sides parallel to the x and y axes and with vertices at points a and b. If there is a point of P inside $R(a,b)$, we will easily apply induction. The case when $R(a,b)$ does not contain points of P — and in particular the points h_i and l_i for $0 < i < k$ — is more difficult and we need to develop tools to handle it. The following Lemma describes the structure of the triangles $T_1, \ldots, T_k$ when $R(a,b)$ is empty. We need some additional terminology first though: we say that a point u is *above* (resp. *below*) $R(a,b)$ if $0 < x_u < x$ and $y_u > y$ (resp. $y_u < 0$).

Lemma 1. *If $(a,b) \notin T$ and no point of P lies inside rectangle $R(a,b)$, then point a lies on the W side of square S_1, point b lies on the E side of square S_k, points $h_1, \ldots, h_k$ all lie above $R(a,b)$, and points $l_1, \ldots, l_k$ all lie below $R(a,b)$. Furthermore, for any i such that $1 < i < k$:*

a) Either $T_i = \triangle(h_{i-1}, h_i, l_{i-1} = l_i)$, points h_{i-1}, h_i, and $l_{i-1} = l_i$ lie on the sides of S_i in clockwise order, and (h_{i-1}, h_i) is a WN, WE, or NE edge in S_i

b) Or $T_i = \triangle(h_{i-1} = h_i, l_{i-1}, l_i)$, points $h_{i-1} = h_i$, l_i, and l_{i-1} lie on the sides of S_i in clockwise order, and (l_{i-1}, l_i) is a WS, WE, or SE edge in S_i.

These properties are illustrated in Fig. 2.

Proof. Since points of P are in general position, points a, h_1, and l_1 must lie on 3 different sides of S_1. Because segment $[ab]$ intersects the interior of S_1 and since a is the origin and b is in the first quadrant, a can only lie on the W or S side of S_1. If a lies on the S side then $l_1 \neq b$ would have to lie inside $R(a, b)$, which is a contradiction. Therefore a lies on the W side of S_1 and, similarly, b lies on the E side of S_k.

Since points h_i ($0 < i < k$) are above segment $[ab]$ and points l_i ($0 < i < k$) are below segment $[ab]$, and because all squares S_i ($0 < i < k$) intersect $[ab]$, points $h_1, \ldots, h_k$ all lie above $R(a, b)$, and points $l_1, \ldots, l_k$ all lie below $R(a, b)$.

The three vertices of T_i can be either $h_i = h_{i-1}$, l_{i-1}, and l_i or h_{i-1}, h_i, and $l_{i-1} = l_i$. Because points of T are in general position, the three vertices of T_i must appear on three different sides of S_i. Finally, because h_{i-1} and h_i are above $R(a, b)$, they cannot lie on the S side of S_i, and because l_{i-1} and l_i are below $R(a, b)$, they cannot lie on the N side of S_i.

If $T_i = \triangle(h_{i-1}, h_i, l_{i-1} = l_i)$, points h_{i-1}, h_i, l_i must lie on the sides of S_i in clockwise order. The only placements of points h_{i-1} and h_i on the sides of S_i that satisfy all these constraints are as described in a). If $T_i = \triangle(h_{i-1} = h_i, l_{i-1}, l_i)$, points h_i, l_i, l_{i-1} must lie on the sides of S_i in clockwise order. Part b) lists the placements of points l_{i-1} and l_i that satisfy the constraints.   □

In the following definition, we define the points on which induction can be applied in the proof of Theorem 1.

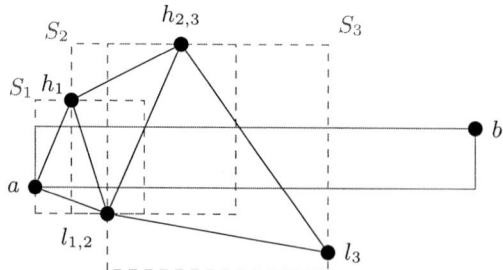

Fig. 2. Triangles T_1 (with points a, h_1, l_1), T_2 (with points h_1, h_2, and l_2), and T_3 (with points l_2, h_3, and l_3) and associated squares S_1, S_2, and S_3. When traveling from a to b along segment $[a, b]$, the edge that is hit when leaving T_i is (h_i, l_i).

Definition 1. *Let $R(a,b)$ be empty. Square S_j is* inductive *if edge (l_j, h_j) is gentle. The point $c = h_j$ or $c = l_j$ with the larger abscissa is the* inductive point *of inductive square S_j.*

The following lemma will be the key ingredient of our inductive proof of Theorem 1.

Lemma 2. *Assume that $R(a,b)$ is empty. If no square $S_1, \ldots, S_k$ is inductive then*

$$d_T(a,b) \leq (1 + \sqrt{2})x + y .$$

Otherwise let S_j be the first inductive square in the sequence $S_1, S_2, \ldots, S_k$. If h_j is the inductive point of S_j then

$$d_T(a, h_j) + (y_{h_j} - y) \leq (1 + \sqrt{2})x_{h_j} .$$

If l_j is the inductive point of S_j then

$$d_T(a, l_j) - y_{l_j} \leq (1 + \sqrt{2})x_{l_j} .$$

Given an inductive point c, we can use use Lemma 2 to bound $d_T(a,c)$ and then apply induction to bound $d_T(c,b)$, *but only if* the position of point c relative to the position of point b is *good*, i.e., if $x - x_c \geq |y - y_c|$. If that is not the case, we will use the following Lemma:

Lemma 3. *Let $R(a,b)$ be empty and let the coordinates of point $c = h_i$ or $c = l_i$ satisfy $0 < x - x_c < |y - y_c|$.*

  a) If $c = h_i$, and thus $0 < x - x_{h_i} < y_{h_i} - y$, then there exists j, with $i < j \leq k$ such that all edges in path $h_i, h_{i+1}, h_{i+2}, \ldots, h_j$ are NE edges in their respective squares and $x - x_{h_j} \geq y_{h_j} - y \geq 0$.
  b) If $c = l_i$, and thus $0 < x - x_{l_i} < y - y_{l_i}$, then there exists j, with $i < j \leq k$ such that all edges in path $l_i, l_{i+1}, l_{i+2}, \ldots, l_j$ are SE edges and $x - x_{l_j} \geq y - y_{l_j} \geq 0$.

Proof. We only prove the case $c = h_j$ as the case $c = l_i$ follows using a symmetric argument.

We construct the path $h_i, h_{i+1}, h_{i+2}, \ldots, h_j$ iteratively. If $h_i = h_{i+1}$, we just continue building the path from h_{i+1}. Otherwise, (h_i, h_{i+1}) is an edge of T_{i+1} which, by Lemma 1, must be a WN, WE, or NE edge in square S_{i+1}. Since the S side of square S_{i+1} is below $R(a,b)$ and because $x - x_{h_i} < y_{h_i} - y$, point h_i cannot be on the W side of S_{i+1} (otherwise b would be inside square S_{i+1}). Thus (h_i, h_{i+1}) is a NE edge. If $x - x_{h_{i+1}} \geq y_{h_{i+1}} - y$ we stop, otherwise we continue the path construction from h_{i+1}. $\qquad\square$

We can now prove the main theorem.

Proof of Theorem 1. The proof is by induction on the distance, using the L_∞-metric, between points of P (since P is finite there is only a finite number of distances to consider).

Let a and b be the two points of P that are the closest points, using the L_∞-metric. We assume w.l.o.g. that a has coordinates $(0,0)$ and b has coordinates (x, y) with $0 < y \le x$. Since a and b are the closest points using the L_∞-metric, the largest square having a as a southwest vertex and containing no points of P in its interior, which we call S_a must have b on its E side. Therefore (a, b) is an edge in T and $d_T(a, b) = d_2(a, b) \le x + y \le (1 + \sqrt{2})x + y$.

For the induction step, we again assume, w.l.o.g., that a has coordinates $(0,0)$ and b has coordinates (x, y) with $0 < y \le x$.

Case 1: $R(a, b)$ is not empty

We first consider the case when there is at least one point of P lying inside rectangle $R(a, b)$. If there is a point c inside $R(a, b)$ such that $y_c \le x_c$ and $y - y_c \le x - x_c$ (i.e., c lies in the region B shown in Fig. 3 then we can apply induction to get $d_T(a, c) \le (1+\sqrt{2})x_c + y_c$ and $d_T(c, b) \le (1+\sqrt{2})(x-x_c)+y-y_c$ and use these to obtain the desired bound for $d_T(a, b)$.

We now assume that there is no point inside region B. If there is still a point in $R(a, b)$ then there must be one that is on the border of S_a, the square we defined in the basis step, or S_b, defined as the largest square having b as a northeast vertex and containing no points of P in its interior. W.l.o.g., we assume the former and thus there is an edge $(a, c) \in T$ such that either $y_c > x_c$ (i.e., c is inside region A shown in Fig. 3 or $y - y_c > x - x_c$ (i.e., c is inside region C). Either way, $d_T(a, c) = d_2(a, c) \le x_c + y_c$. If c is in region A, since $x - x_c \ge y - y_c$, by induction we also have that $d_T(c, b) \le (1 + \sqrt{2})(x - x_c) + (y - y_c)$. Then

$$d_T(a, b) \le d_T(a, c) + d_T(c, b)$$
$$\le x_c + y_c + (1 + \sqrt{2})(x - x_c) + (y - y_c) \le (1 + \sqrt{2})x + y$$

In the second case, since $x - x_c < y - y_c$, by induction we have that $d_T(c, b) \le (1 + \sqrt{2})(y - y_c) + (x - x_c)$. Then, because $y < x$,

$$d_T(a, b) \le d_T(a, c) + d_T(c, b)$$
$$\le x_c + y_c + (1 + \sqrt{2})(y - y_c) + (x - x_c) \le (1 + \sqrt{2})x + y$$

Case 2: The interior of $R(a, b)$ is empty

If no square $S_1, S_2, \dots, S_k$ is inductive, $d_T(a, b) \le (1+\sqrt{2})x+y$ by Lemma 2. Otherwise, let S_i be the first inductive square in the sequence and suppose that h_i is the inductive point of S_i. By Lemma 3, there is a j, $i \le j \le k$, such that $h_i, h_{i+1}, h_{i+2}, \dots, h_j$ is a path in T of length at most $(x_{h_j} - x_{h_i}) + (y_{h_i} - y_{h_j})$

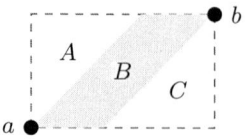

Fig. 3. Partition of $R(a, b)$ into three regions in Case 1 of the proof of Theorem 1

and such that $x - x_{h_j} \geq y_{h_j} - y \geq 0$. Since h_j is closer to b, using the L_∞-metric, than a is, we can apply induction to bound $d_T(h_j, b)$. Putting all this together with Lemma 2, we get:

$$d_T(a, b) \leq d_T(a, h_i) + d_T(h_i, h_j) + d_T(h_j, b)$$
$$\leq (1 + \sqrt{2})x_{h_i} - (y_{h_i} - y) + (x_{h_j} - x_{h_i}) + (y_{h_i} - y_{h_j})$$
$$+ (1 + \sqrt{2})(x - x_{h_j}) + (y_{h_j} - y) \leq (1 + \sqrt{2})x .$$

If l_i is the inductive point of S_i, by Lemma 3 there is a j, $i \leq j \leq k$, such that $l_i, l_{i+1}, l_{i+2}, \ldots, l_j$ is a path in T of length at most $(x_{h_j} - x_{h_i}) + (y_{h_j} - y_{h_i})$ and such that $x - x_{h_j} \geq y - y_{h_j} \geq 0$. Because the position of j with respect to b is good and since l_j is closer to b, using the L_∞-metric, than a is, we can apply induction to bound $d_T(l_j, b)$. Putting all this together with Lemma 2, we get:

$$d_T(a, b) \leq d_T(a, l_i) + d_T(l_i, l_j) + d_T(l_j, b)$$
$$\leq (1 + \sqrt{2})x_{l_i} + y_{l_i} + (x_{l_j} - x_{l_i}) + (y_{l_j} - y_{l_i}) + (1 + \sqrt{2})(x - x_{l_j})$$
$$+ (y - y_{l_j}) \leq (1 + \sqrt{2})x + y .$$

$\square$

What remains to be done is to prove Lemma 2. To do this, we need to develop some further terminology and tools. Let x_i, for $1 \leq i \leq k$, be the horizontal distance between point a and the E side of S_i, respectively. We also set $x_0 = 0$.

Definition 2. *A square S_i has* potential *if*

$$d_T(a, h_i) + d_T(a, l_i) + d_{S_i}(h_i, l_i) \leq 4x_i$$

where $d_{S_i}(h_i, l_i)$ is the Euclidean distance when moving from h_i to l_i along the sides of S_i, clockwise.

Lemma 4. *If $R(a, b)$ is empty then S_1 has potential. Furthermore, for any $1 \leq i < k$, if S_i has potential but is not inductive then S_{i+1} has potential.*

Proof. If $R(a, b)$ is empty then, by Lemma 1, a lies on the W side of S_1 and x_1 is the side length of square S_1. Also, h_1 lies on the N or E side of S_1, and l_1 lies on the S or E side of S_1. Then $d_T(a, h_1) + d_T(a, l_1) + d_{S_1}(h_1, l_1)$ is bounded by the perimeter of S_1 which is $4x_1$.

Now assume that S_i, for $1 \leq i < k$, has potential but is not inductive. Squares S_i and S_{i+1} both contain points l_i and h_i. Because S_i is not inductive, edge (l_i, h_i) must be steep and thus $d_x(l_i, h_i) < d_y(l_i, h_i)$. To simplify the arguments, we assume that l_i is to the W of h_i, i.e., $x_{l_i} < x_{h_i}$. The case $x_{l_i} > x_{h_i}$ can be shown using equivalent arguments.

By Lemma 1, $T_i = \triangle(h_{i-1}, h_i, l_{i-1} = l_i)$ or $T_i = \triangle(h_{i-1} = h_i, l_{i-1}, l_i)$ and there has to be a side of S_i between the sides on which l_i and h_i lie, when moving clockwise from l_i to h_i. Using the constraints on the position of h_i and l_i within S_i from Lemma 1 and using the assumptions that (l_i, h_i) is steep and

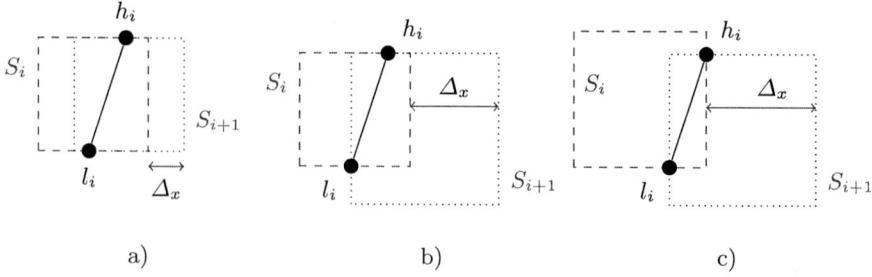

Fig. 4. The first, second and fourth case in the proof of Lemma 4. In each case, the difference $d_{S_{i+1}}(h_i, l_i) - d_{S_i}(h_i, l_i)$ is shown to be at most $4\Delta_x$, where $\Delta_x = x_{i+1} - x_i$.

that $x_{l_i} < x_{h_i}$, we deduce that l_i must be on the S side and h_i must be on the N or E side of S_i.

If h_i is on the N side of S_i then, because $x_{l_i} < x_{h_i}$, h_i must also be on the N side of S_{i+1} and either l_i is on the S side of S_{i+1} and

$$d_{S_{i+1}}(h_i, l_i) - d_{S_i}(h_i, l_i) = 2(x_{i+1} - x_i) \tag{1}$$

as shown in Fig. 4a) or l_i is on the W side of S_{i+1}, in which case

$$d_{S_{i+1}}(h_i, l_i) - d_{S_i}(h_i, l_i) \leq 4(x_{i+1} - x_i) \tag{2}$$

as shown in Fig. 4b).

If h_i is on the E side of S_i then, because $x_{i+1} > x_i$ (since (l_i, h_i) is steep), h_i must be on the N side of S_{i+1} and either l_i is on the S side of S_{i+1} and inequality (1) holds or l_i is on the W side of S_{i+1} and inequality (2) holds, as shown in Fig. 4c).

Since S_i has potential, in all cases we obtain:

$$d_T(a, h_i) + d_T(a, l_i) + d_{S_{i+1}}(h_i, l_i) \leq 4x_{i+1} . \tag{3}$$

Assume $T_{i+1} = \triangle(h_i, h_{i+1}, l_i = l_{i+1})$; in other words, (h_i, h_{i+1}) is an edge of T with h_{i+1} lying somewhere on the boundary of S_{i+1} between h_i and l_i, when moving clockwise from h_i to l_i. Then $d_T(a, h_{i+1}) \leq d_T(a, h_i) + d_2(h_i, h_{i+1})$. By the triangular inequality, $d_2(h_i, h_{i+1}) \leq d_{S_{i+1}}(h_i, h_{i+1})$ and we have that:

$$d_T(a, h_{i+1}) + d_T(a, l_{i+1}) + d_{S_{i+1}}(h_{i+1}, l_{i+1}) \leq d_T(a, h_i) + d_T(a, l_i) + d_{S_{i+1}}(h_i, l_i)$$
$$\leq 4x_{i+1} .$$

Thus S_{i+1} has potential. The argument for the case when $T_{i+1} = \triangle(h_i = h_{i+1}, l_i, l_{i+1})$ is symmetric. $\square$

Definition 3. *A vertex c (h_i or l_i) of T_i is promising in S_i if it lies on the E side of S_i.*

Lemma 5. *If square S_i has potential and $c = h_i$ or $c = l_i$ is a promising point in S_i then*

$$d_T(a,c) \leq 2x_c .$$

Proof. W.l.o.g., assume $c = h_i$. Since h_i is promising, $x_c = x_{h_i} = x_i$. Because S_i has potential, either $d_T(a,h_i) \leq 2x_{h_i}$ or $d_T(a,l_i) + d_{S_i}(l_i,h_i) \leq 2x_{h_i}$. In the second case, we can use edge (l_i,h_i) and the triangular inequality to obtain $d_T(a,h_i) \leq d_T(a,l_i) + |l_i h_i| \leq 2x_{h_i}$. □

We now define the *maximal high path* and the *minimal low path*.

Definition 4.

- If h_j is promising in S_j, the maximal high path ending at h_j is simply h_j; otherwise, it is the path $h_i, h_{i+1}, \ldots, h_j$ such that $h_{i+1}, \ldots, h_j$ are not promising and either $i = 0$ or h_i is promising in S_i.
- If l_j is promising in S_j, the maximal low path ending at l_j is simply l_j; otherwise, it is the path $l_i, l_{i+1}, \ldots, l_j$ such that $l_{i+1}, \ldots, l_j$ are not promising and either $i = 0$ or l_i is promising in S_i.

Note that by Lemma 1, all edges on the path $h_i, h_{i+1}, \ldots, h_j$ are WN edges and thus the path length is bounded by $(x_{h_j} - x_{h_i}) + (y_{h_j} - y_{h_i})$. Similarly, all edges in path $l_i, l_{i+1}, \ldots, l_j$ are WS edges and the length of the path is at most $(x_{l_j} - x_{l_i}) + (y_{l_i} - y_{l_j})$.

We now have the tools to prove Lemma 2.

Proof of Lemma 2. If $R(a,b)$ is empty then, by Lemma 1, b is promising. Thus, by Lemma 4 and Lemma 5, if no square $S_1, \ldots, S_k$ is inductive then $d_T(a,b) \leq 2x < (1 + \sqrt{2})x + y$.

Assume now that there is at least one inductive square in the sequence of squares $S_1, \ldots, S_k$. Let S_j be the first inductive square and assume, for now, that h_j is the inductive point in S_j. By Lemma 4, every square S_i, for $i < j$, is a potential square.

Since (l_j, h_j) is gentle, it follows that $d_2(l_j, h_j) \leq \sqrt{2}(x_{h_j} - x_{l_j})$. Let $l_i, l_{i+1}, \ldots, l_{j-1} = l_j$ be the maximal low path ending at l_j. Note that $d_T(l_i, l_j) \leq (x_{l_j} - x_{l_i}) + (y_{l_i} - y_{l_j})$. Either $l_i = l_0 = a$ or l_i is a promising point in potential square S_i; either way, by Lemma 5, we have that $d_T(a,l_i) \leq 2x_{l_i}$. Putting all this together, we get

$$d_T(a,h_j) + (y_{h_j} - y) \leq d_T(a,l_i) + d_T(l_i,l_j) + d_2(l_j,h_j) + y_{h_j}$$
$$\leq 2x_{l_i} + (x_{l_j} - x_{l_i}) + (y_{l_i} - y_{l_j}) + \sqrt{2}(x_{h_j} - x_{l_j}) + y_{h_j}$$
$$\leq \sqrt{2}x_{h_j} + x_{l_j} + y_{h_j} - y_{l_j}$$
$$\leq (1 + \sqrt{2})x_{h_j}$$

where the last inequality follows $x_{l_j} + y_{h_j} - y_{l_j} \leq x_{h_j}$, i.e., from the assumption that edge (l_j, h_j) is gentle.

The case when $c = l_j$ is the inductive point in square S_j is shown similarly. □

4   Conclusion and Perspectives

The L_1-Delaunay triangulation is the first type of Delaunay triangulation to be shown to be a spanner [Che86]. Progress on the spanning properties of the TD-Delaunay and the classical L_2-Delaunay triangulation soon followed. In this paper, we determine the precise stretch factor of an L_1- and L_∞-Delaunay triangulation and close the problem for good.

We believe that our proof techniques can be extended and that they will lead, yet again, to new insights on the stretch factor of other types of Delaunay triangulations. For example, let P_k denote the convex distance function defined by a regular k-gon. We observe that the stretch factor of P_k-Delaunay triangulations is known for $k = 3, 4$ since P_3 is the triangular distance function of [Che89], and P_4 is nothing else than the L_∞-metric. Determining the stretch factor of P_k-Delaunay triangulations for larger k would undoubtedly be an important step towards understanding the stretch factor of classical Delaunay triangulations.

References

[BCCS08]   Bose, P., Carmi, P., Collette, S., Smid, M.: On the Stretch Factor of Convex Delaunay Graphs. In: Hong, S.-H., Nagamochi, H., Fukunaga, T. (eds.) ISAAC 2008. LNCS, vol. 5369, pp. 656–667. Springer, Heidelberg (2008)

[BDL+11]   Bose, P., Devroye, L., Löffler, M., Snoeyink, J., Verma, V.: Almost all Delaunay triangulations have stretch factor greater than $\pi/2$. Comp. Geometry 44, 121–127 (2011)

[BFvRV12] Bose, P., Fagerberg, R., van Renssen, A., Verdonschot, S.: Competitive routing in the half-θ_6-graph. In: 23rd ACM Symp. on Discrete Algorithms (SODA), pp. 1319–1328 (2012)

[BM04]   Bose, P., Morin, P.: Competitive online routing in geometric graphs. Theoretical Computer Science 324(2-3), 273–288 (2004)

[Che86]   Paul Chew, L.: There is a planar graph almost as good as the complete graph. In: 2nd Annual ACM Symposium on Computational Geometry (SoCG), pp. 169–177 (August 1986)

[Che89]   Paul Chew, L.: There are planar graphs almost as good as the complete graph. Journal of Computer and System Sciences 39(2), 205–219 (1989)

[DFS87]   Dobkin, D.P., Friedman, S.J., Supowit, K.J.: Delaunay graphs are almost as good as complete graphs. In: 28th Annual IEEE Symposium on Foundations of Computer Science (FOCS), pp. 20–26. IEEE Computer Society Press (October 1987)

[KG92]   Mark Keil, J., Gutwin, C.A.: Classes of graphs which approximate the complete Euclidean graph. Discrete & Computational Geometry 7(1), 13–28 (1992)

[Xia11]   Xia, G.: Improved upper bound on the stretch factor of delaunay triangulations. In: 27th Annual ACM Symposium on Computational Geometry (SoCG), pp. 264–273 (June 2011)

[XZ11]   Xia, G., Zhang, L.: Toward the tight bound of the stretch factor of Delaunay triangulations. In: 23rd Canadian Conference on Computational Geometry (CCCG) (August 2011)

Two Dimensional Range Minimum Queries
and Fibonacci Lattices

Gerth Stølting Brodal[1], Pooya Davoodi[2,*], Moshe Lewenstein[3],
Rajeev Raman[4], and Satti Srinivasa Rao[5,**]

[1] MADALGO[***], Aarhus University, Denmark
gerth@cs.au.dk
[2] Polytechnic Institute of New York University, United States
pooyadavoodi@gmail.com
[3] Bar-Ilan University, Israel
moshe@cs.biu.ac.il
[4] University of Leicester, UK
r.raman@leicester.ac.uk
[5] Seoul National University, S. Korea
ssrao@cse.snu.ac.kr

Abstract. Given a matrix of size N, two dimensional range minimum
queries (2D-RMQs) ask for the position of the minimum element in a
rectangular range within the matrix. We study trade-offs between the
query time and the additional space used by indexing data structures
that support 2D-RMQs. Using a novel technique—the discrepancy properties of Fibonacci lattices—we give an indexing data structure for 2D-
RMQs that uses $O(N/c)$ bits additional space with $O(c \log c (\log \log c)^2)$
query time, for any parameter c, $4 \leq c \leq N$. Also, when the entries of
the input matrix are from $\{0, 1\}$, we show that the query time can be
improved to $O(c \log c)$ with the same space usage.

1 Introduction

The problem we consider is to preprocess a matrix (two dimensional array) of
values into a data structure that supports *range minimum queries (2D-RMQs)*,
asking for the position of the minimum in a rectangular range within the matrix.
More formally, an input is an $m \times n$ matrix A of $N = m \cdot n$ distinct totally ordered
values, and a range minimum query asks for the position of the minimum value
in a range $[i_1 \cdots i_2] \times [j_1 \cdots j_2]$, where $0 \leq i_1 \leq i_2 \leq m - 1$ and $0 \leq j_1 \leq j_2 \leq$
$n - 1$ (the case when $m = 1$ is referred to hereafter as 1D-RMQ). Both 1D-

* Research supported by NSF grant CCF-1018370 and BSF grant 2010437. Research
partly done while the author was a PhD student at MADALGO.
** Research supported by Basic Science Research Program through the National
Research Foundation of Korea (NRF) funded by MEST (Grant number 2012-
0008241).
*** Center for Massive Data Algorithmics, a Center of the Danish National Research
Foundation.

L. Epstein and P. Ferragina (Eds.): ESA 2012, LNCS 7501, pp. 217–228, 2012.

and 2D-RMQ problems are fundamental problems with a long history dating back nearly 25 years [13,5] that find applications in computer graphics, image processing, computational biology, databases, etc.

We consider this problem in two models [4]: the *encoding* and *indexing* models. In the encoding model, we preprocess A to create a data structure *enc* and queries have to be answered just using *enc*, *without* access to A. In the indexing model, we create an index *idx* and are able to refer to A when answering queries. The main measures we are interested in are (i) the time to answer queries (ii) the space (in bits) taken by *enc* and *idx* respectively and (iii) the preprocessing time (in general there may be a trade-off between (i) and (ii)). Note that an encoding is only interesting if it uses $o(N \log N)$ bits—otherwise we could store the sorted permutation of A in *enc* and trivially avoid access to A when answering queries. We defer to [4] for a further discussion of these models.

The 1D-RMQ problem seems to be well-understood. In the encoding model, it is known that *enc* must be of size at least $2N - O(\log N)$ bits [17,12]. Furthermore, 1D-RMQs can be answered in $O(1)$ time[1] using an encoding of $2N + o(N)$ bits after $O(N)$-time preprocessing [12,7]. In the indexing model, Brodal et al. [4] showed that an index of size $O(N/c)$ bits suffices to answer queries in $O(c)$ time, and that this trade-off is optimal.

Following Chazelle and Rozenberg's work [5], the study of 2D-RMQs was restarted by Amir et al. [1], who explicitly asked in what ways the 2D- and 1D-RMQ problems differ; this question turned out to have a complex answer:

1. In the indexing model, Atallah and Yuan [2] showed that, just as for 1D-RMQs, $O(N)$ preprocessing time suffices to get an index that answers 2D-RMQs in $O(1)$ time, improving upon the preprocessing time of [1]. Their index size of $O(N)$ words, or $O(N \log N)$ bits, was further reduced to $O(N)$ bits by Brodal et al. [4], while maintaining query and preprocessing times.
2. Demaine et al. [8] showed that, unlike 1D-RMQ, non-trivial (size $o(N \log N)$ bits) encodings for 2D-RMQs do not exist in general. Non-trivial encodings may only exist when $m \ll n$ [4], or in expectation for random A [14].
3. In the indexing model, Brodal et al. [4] showed that with an index of size $O(N/c)$ bits, $\Omega(c)$ query time is needed to answer queries, and gave an index of size $O(N/c)$ bits that answered queries in $O(c \log^2 c)$ time.

Thus, the question of whether 1D-RMQs and 2D-RMQs differ in the time-space trade-off for the indexing model remains unresolved, and we make progress towards this question in this paper.

Our Contributions. We show the following results:

– We give an index of size $O(N/c)$ bits that for any $4 \leq c \leq N$, takes $O(c \log c (\log \log c)^2)$ time to answer 2D-RMQs, improving upon Brodal et al.'s result by nearly a $\log c$ factor.

---

[1] This result uses, as do all the results in this paper, the word RAM model of computation with word size logarithmic in the size of inputs.

- For the case where A is a 0-1 matrix, and the problem is simplified to finding the location of *any* 0 in the query rectangle (or else report that there are no 0s in the query rectangle) we give an index of size $O(N/c)$ bits that supports 2D-RMQs in $O(c \log c)$ time, improving upon the Brodal et al.'s result by a $\log c$ factor. Note that in the 1D-RMQ case, the 0-1 case and the general case have the same time-space trade-off [4].

We recursively decompose the input matrix, which in turn partitions a 2D-RMQ into several disjoint sub-queries, some of which are *3-sided* or *2-sided* queries, which have one or two sides of the query rectangle ending at the edge of a recursive sub-matrix. To deal with 3-sided queries in the general case, we recursively partition the 3-sided query into a number of disjoint sub-queries including 2-sided queries (also called dominance queries), whereas for 0-1 matrices we make a simple non-recursive data structure.

The techniques used to answer the 2-sided queries in the general case are novel. We use a *Fibonacci lattice*, a 2D point set with applications in graphics and image processing [6,11] and parallel computing [10], defined as follows:

Definition 1. [15] *Let f_k be the k-th number in the Fibonacci sequence, where $f_1 = f_2 = 1$. In an $n \times n$ grid, the Fibonacci lattice of size $n = f_k$ is the two dimensional point set $\{(i, i \cdot f_{k-1} \bmod n) \mid i = 0, \ldots, n-1\}$.*

A Fibonacci lattice has several "low-discrepancy" properties including:

Lemma 1. [11] *In an $n \times n$ grid, there exists a constant α such that any axis-aligned rectangle whose area is larger than $\alpha \cdot n$, contains at least one point from the Fibonacci lattice of size n.*

We solve 2-sided queries via *weighted* Fibonacci lattices, in which every point is assigned a value from A. Such queries are reduced, via new properties of Fibonacci lattices (which may be useful in other contexts as well), to 2-sided queries on the weighted Fibonacci lattice. The weighted Fibonacci lattice is space-efficiently encoded using a succinct index of Farzan et al. [9].

2   2-Sided 2D-RMQs and Fibonacci Lattices

We present data structures that support 2-sided 2D-RMQs within an $n \times n$ matrix. Without loss of generality, assume that the query range is of the form $[0 \cdots i] \times [0 \cdots j]$, i.e. that 2-sided queries have a fixed corner on the top-left corner of the matrix. First we present a data structure of size $O(n \log n)$ bits additional space that supports 2-sided queries in $O(n \log n)$ time. Although this result is weaker than that of [4]), it gives the main ideas. We then reduce the query time to $O(n)$ with the same space usage.

A First Solution. Given n, we create a modified Fibonacci lattice as follows. Choose the Fibonacci lattice of smallest size $n' \geq n$, clearly $n' = O(n)$. Observe that no points in the Fibonacci lattice have the same x or y coordinate (this

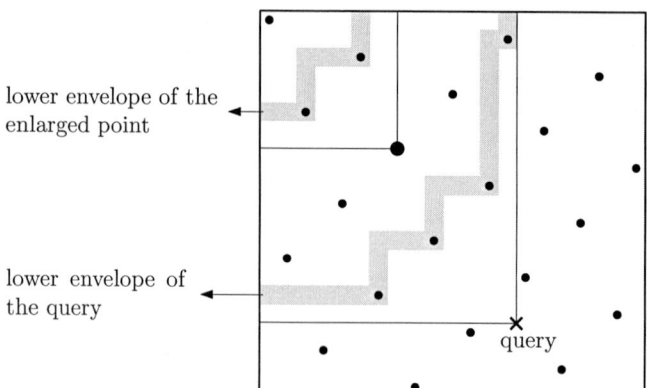

Fig. 1. The 2-sided query is divided into two sub-queries by its lower envelope. The enlarged point above the lower envelope is the FP with minimum priority within the query range, and its priority comes from an element below its lower envelope.

follows from the fact that $\gcd(f_k, f_{k-1}) = 1$). Eliminate all points that do not lie in $[0 \cdots n - 1] \times [0 \cdots n - 1]$ and call the resulting set F; note that F satisfies Lemma 1, albeit with a different constant in the $O(\cdot)$ notation. The points of F, which specify entries of the matrix, are called the *FPs*. For any point p in the matrix, its *2-sided region* is the region above and to the left of p. As shown in Figure 1, a 2-sided query is divided into two disjoint sub-queries by a polygonal chain (called the *lower envelope*) derived from all the FPs within the query range that are not in the 2-sided region of any other FP within the query range. We find the answer to each of the two sub-queries and return the smaller.

Upper Sub-query. Each FP $p \in F$ is assigned a "priority" which equals the value of the minimum element within its 2-sided region (i.e. p's priority is the value of the element returned by a 2-sided query at p). As the upper sub-query region is entirely covered by the 2-sided regions of all FPs within the query range, the minimum element in the upper sub-query is the smallest priority of all the FPs within the upper sub-query. The data structure for the upper sub-query comprises three parts: the first maps each FP p to the row and column of the matrix element containing the minimum value in the 2-sided query at p, using say a balanced search tree.

The next part is an *orthogonal range reporting* data structure that returns a list of all FPs that lie in a rectangle $[i_1 \cdots i_2] \times [j_1 \cdots j_2]$.

Lemma 2. *Given the Fibonacci lattice of size n, there exists a data structure of size $O(n)$ bits that supports 4-sided orthogonal range reporting queries in $O(k)$ time, where k is the number of reported points.*

Proof. We report all the FPs within by sweeping a horizontal line from the top of the rectangle towards the bottom in k steps, reporting one FP at each step in $O(1)$ time. The line initially jumps down from above the rectangle to the row in which the top-most FP within the rectangle is located. In subsequent steps, we move the sweep line from the row i containing the just-reported FP down to a row $i + x$ containing the next FP within the query rectangle. Notice that the function $L(i) = i \cdot f_{k-1} \bmod n$ is invertible, and thus row and column indexes in the lattice can be converted to each other in constant time (details omitted).

Let the columns j_ℓ and j_r be the left and right sides of the query. The point in row $i + x$ is within the rectangle if $j_\ell \leq L(i + x) = (L(i) + L(x)) \bmod n \leq j_r$; thus $L(x) \geq j_\ell - L(i) + n$ or $L(x) \leq j_r - L(i)$. Let $Z_1(i) = j_\ell - L(i) + n$ and $Z_2(i) = j_r - L(i)$. Since there is no point between the rows i and $i + x$ within the rectangle, therefore x must be the minimum positive number such that $L(x) \geq Z_1(i)$ or $L(x) \leq Z_2(i)$. We preprocess $L(i)$ for $i \in [1 \cdots n - 1]$ into a data structure to find x in $O(1)$ time.

Construct an array $L_{inv}[0 \cdots n - 1]$, where $L_{inv}[L(x)] = x$. We find the minimum in each of the ranges $[Z_1(i) \cdots n - 1]$ and $[0 \cdots Z_2(i)]$, and we return the smaller one as the right value for x. To do this, we encode L_{inv} with a 1D-RMQ structure of size $O(n)$ bits that answers queries in $O(1)$ time [16,12,7]. □

The final part solves the following problem: given a point q, find the point $p \in F$ in the 2-sided region specified by q with the smallest priority. For this, we use a *succinct index* by Farzan et al. [9, Corollary 1], which takes F and the priorities as input, preprocesses it and outputs a data structure D of size $O(n)$ bits. D encodes priority information about points in F, but *not* the coordinates of the points in F. Given a query point p, D returns the desired output q, but while executing the query, it calls the instance of Lemma 2 a number of times to retrieve coordinates of relevant points. The index guarantees that each call will report $O(\log n)$ points. The result is stated as follows:

Lemma 3. [9, Corollary 1] *Given n points in 2D rank space, where each point is associated with a priority, there exists a succinct index of size $O(n)$ bits that supports queries asking for the point with minimum priority within a 2-sided range in $O(\log \log n \cdot (\log n + T))$ time. Here T is the time taken to perform orthogonal range reporting on the n input points, given that the index guarantees that no query it asks will ever return more than $O(\log n)$ points. The space bound does not include the space for the data structure for orthogonal range reporting.*

We now put things together. To solve the upper sub-query, we use Lemmas 2 and 3[2] to find the FP with lowest priority in $O(\log n \log \log n)$ time (by Lemma 2 we can use $T = O(\log n)$ in Lemma 3). Since each FP is mapped to the location of the minimum-valued element that lies in its 2-sided region, we can return the minimum-valued element in the upper query region in $O(\log n)$ further time. The space usage is $O(n)$ bits for Lemmas 2 and 3, and $O(n \log n)$ bits for maintaining the location of minimum-valued elements (the mapping).

Corollary 1. *The upper sub-query can be solved in $O(\log n \log \log n)$ time using $O(n \log n)$ bits.*

---

[2] The points in F are technically not in 2D rank space, as some rows may have no points from F; this can be circumvented as in [3] (details omitted).

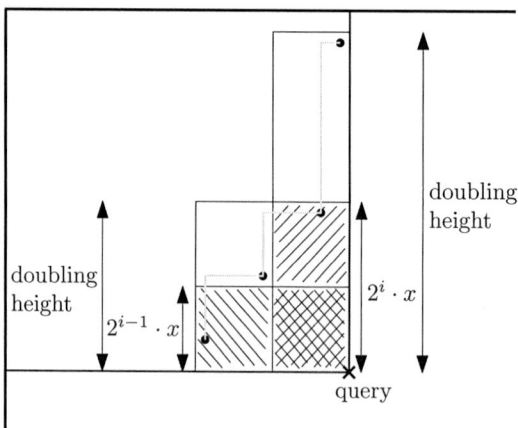

Fig. 2. The cross sign is the bottom-right corner of a 2-sided query

Lower Sub-Query. The lower sub-query is answered by scanning the whole region in time proportional to its size. The following lemma states that its size is $O(n \log n)$, so the lower sub-query also takes $O(n \log n)$ time (the region can be determined in $O(n)$ time by a sweep line similar to Lemma 2).

Lemma 4. *The area of the lower sub-query region is $O(n \log n)$.*

Proof. Consider $O(\log n)$ axis-aligned rectangles each with area $\alpha \cdot n$, with different aspect ratios, where the bottom-right corner of each rectangle is on the query point, where α is the constant of Lemma 1. By Lemma 1, each of the rectangles has at least one FP (see Figure 2).
Let $x \times (\alpha n/x)$ be the size of the first rectangle such that it contains the top-most FP on the lower envelope. Let $(2^i x) \times ((\alpha n)/(2^i x))$ be the size of the other rectangles, for $i = 1 \cdots \lceil \log(\alpha n/x) \rceil$. Doubling the height of each rectangle ensures that all the rectangles together cover the area below the lower envelope, and this does not increase the size of the rectangles asymptotically, thus the total area is $O(\alpha \cdot n \cdot \log n) = O(n \log n)$.   □

From Corollary 1 and Lemma 4 we obtain:

Lemma 5. *There exists an index of size $O(n \log n)$ bits that can answer 2-sided 2D-RMQs within an $n \times n$ matrix in $O(n \log n)$ query time.*

Improving Query Time to $O(n)$. The bottleneck in the query time of Lemma 5 is the brute force algorithm, which reads the entire region below the lower envelope. To reduce the size of this region to $O(n)$, we increase the number of FPs in the grid from n to at most $n \log n$.

Lower Sub-Query. We create a *dense* Fibonacci lattice of at most $n \log n$ FPs in the $n \times n$ grid A as follows. In the preprocessing of A, we expand A to an $n \log n \times n \log n$ grid A' by expanding each cell to a $\log n \times \log n$ *block*. We make a (modified) Fibonacci lattice of size $n \log n$ in A'. In the next step, we shrink the Fibonacci lattice as follows, such that it fits in A. Put a FP in a cell of A iff the block of A' corresponding to the cell has at least one FP. Notice that the number of FPs in A is at most $n \log n$. The following lemma states that the area below the lower envelope, after growing the number of FPs to $n \log n$, is $O(n)$.

Lemma 6. *In the dense FP set, the area of the lower sub-query is $O(n)$.*

Proof. Recall that A and A' are respectively the $n \times n$ and $n \log n \times n \log n$ grids. We can deduce from Lemma 4 that the area below the lower envelope in the Fibonacci lattice of size $n \log n$ in A' is $O(n \log^2 n)$. We prove that the area below the lower envelope in A is $O(n)$ by contradiction. Suppose that the area is $\omega(n)$. Since each element in A corresponds to a $\log n \times \log n$ block in A', the same region in A' has area $\omega(n \log^2 n)$, and it can be easily shown that this region is also below the lower envelope in A'. This is a contradiction.   □

Upper Sub-Query. We proceed as in the previous subsection, but with a space budget of $O(n \log n)$ bits. We begin by modifying Lemma 2 to work with the dense Fibonacci lattices (proof omitted):

Lemma 7. *Given the dense Fibonacci lattice of size at most $n \log n$ within an $n \times n$ grid A, there exists a data structure of size $O(n \log n)$ bits that supports orthogonal range reporting queries in $O(k \log^2 n)$ time, where k is the number of reported points.*

From Lemmas 7 and 3, we can find the FP with lowest priority in $O(n \log n)$ bits and $O(\log^3 n \log \log n)$ time[3]. Unfortunately, the mapping from FPs to minimum values now requires $O(n(\log n)^2)$ bits, which is over budget. For this, recall that the priority of each FP p was previously the minimum value in the 2-sided region of p. Instead of mapping p to the position of this minimum, we change the procedure of assigning priorities as follows. Suppose the minimum value in p's 2-sided region is x. If there is another FP q contained in p's 2-sided region, such that x is also the minimum in q's 2-sided region, then we assign p a priority of $+\infty$; otherwise we assign p a priority equal to x (ties between priorities of FPs are broken arbitrarily). This ensures that entry containing the value equal to the priority of p always comes from below the lower envelope of p: once we obtain a FP with minimum priority within the sub-query, we read the entire region of the lower envelope of the FP to find the position that its priority comes from (see Figure 1). This can be done in $O(n)$ time, by Lemma 6. We have thus shown:

Lemma 8. *There exists a data structure of size $O(n \log n)$ bits additional space that can answer 2-sided 2D-RMQs within an $n \times n$ matrix in $O(n)$ query time, excluding the space to perform range reporting on the dense FP set.*

---

[3] Again, when using Lemma 3, the fact that the dense FP points are not in 2D rank space is circumvented as in [3].

3   Improved Trade-off for 2D-RMQs

We present a data structure of size $O(N/c \cdot \log c \log \log c)$ bits additional space that supports 2D-RMQs in $O(c \log \log c)$ time in a matrix of size N, for any $c \leq N$. Substituting c with $O(c \log c \log \log c)$ gives a data structure of size $O(N/c)$ bits additional space with $O(c \log c (\log \log c)^2)$ query time.

We first reduce the problem to answering queries that are within small $c \times c$ matrices. Answering these queries is in fact the bottleneck of our solution. The reduction algorithm spends $O(N/c)$ bits additional space and $O(c)$ query time (later in Lemma 10). We show how to answer each query within a $c \times c$ matrix in $O(c \log \log c)$ time using $O(c \log c \log \log c)$ bits additional space. The reduction algorithm makes this sub-structure for $O(N/c^2)$ disjoint sub-matrices of size $c \times c$. We will make use of the following result in our solution.

Lemma 9. [4] *Given a matrix of size N, there exists a data structure supporting 2D-RMQs in the matrix in $O(1)$ time using $O(N)$ bits additional space.*

3.1   Reduction to Queries within Matrices of size $c \times c$

We show how to reduce general 2D-RMQs in a matrix of size N to 2D-RMQs in small matrices of size $c \times c$. We partition the matrix into blocks of size $1 \times c$, we build a data structure D_1 of Lemma 9 for the matrix M of size $m \times n/c$ containing the minimum element of each block, and then we delete M. The size of D_1 is $O(N/c)$ bits, and whenever its query algorithm wants to read an element from M, we read the corresponding block and find its minimum in $O(c)$ time. Similarly, we make another partitioning of the original matrix into blocks of size $c \times 1$, and build another data structure D_2 of Lemma 9 on the minimum of the blocks. Now, we explain how to make the reduction using D_1 and D_2.

The two partitionings together divide the matrix into $O(N/c^2)$ square blocks each of size $c \times c$. If a query is contained within one of these square blocks, the reduction is done. If a query is large, it spans over several square blocks; the first partitioning divides the query into three vertical sub-queries: the middle part that consists of full $1 \times c$ blocks, and the left and right sub-queries that contain partial $1 \times c$ blocks. We find the minimum element of the middle part in $O(c)$ query time using D_1. Each of the other two parts is similarly divided into three parts by the second partitioning: the middle part which consists of full $c \times 1$ blocks, and the other two parts that contain partial $c \times 1$ blocks. The middle part can be answered using D_2 in $O(c)$ query time, and the other two parts are contained in square blocks of size $c \times c$. We sum up in the following lemma:

Lemma 10. *If, for a matrix of size $c \times c$, there exists a data structure of size S bits additional space that answers 2D-RMQs in time T, then, for a rectangular matrix of size N, we can build a data structure of size $O(N/c + N/c^2 \cdot S)$ bits additional space that supports 2D-RMQs in time $O(c + T)$.*

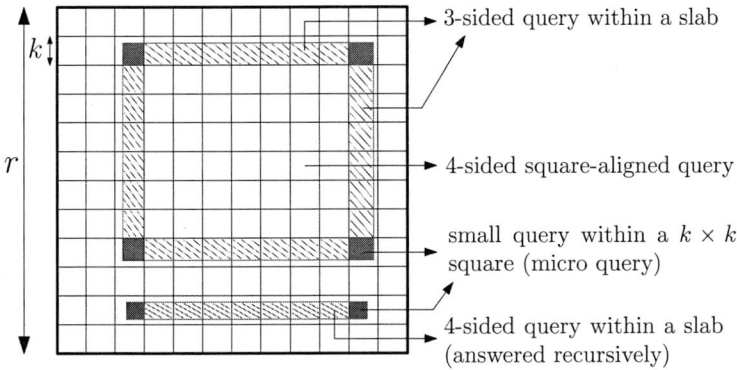

Fig. 3. The recursive decomposition. Two different queries which are decomposed in two different ways.

3.2   2D-RMQs within Matrices of Size $c \times c$

We present a recursive data structure of size $O(c \log c \log \log c)$ bits additional space that supports 2D-RMQs within a $c \times c$ matrix in $O(c \log \log c)$ time. Let r denote the size of the recursive problem with an input matrix of size $r \times r$ (at the top level $r = c$). We assume that c is a power of 2. We divide the matrix into r/k mutually disjoint horizontal slabs of size $k \times r$, and r/k mutually disjoint vertical slabs of size $r \times k$. A horizontal and a vertical slab intersect in a $k \times k$ square. We choose k the power of 2 such that $k/2 < \lceil \sqrt{r} \rceil \leq k$; observe that $r/k \leq k$, $k = \Theta(\sqrt{r})$, and $k^2 = \Theta(r)$.

We recurse on each horizontal or vertical slab. A horizontal slab of size $k \times r$ is *compressed* to a $k \times k$ matrix by dividing each row into k groups of k consecutive elements, and representing each group by its minimum element. A vertical slab is also compressed to a $k \times k$ matrix similarly. The compressed matrix is not represented explicitly. Instead, if a solution to a recursive $k \times k$ problem wants to read some entry x from its input (which is a compressed $k \times k$ matrix), we scan the entire sub-array that x represents to find the location of x in the $r \times r$ matrix; the size of the sub-array will be called the *weight* of x. Note that all values in a recursive sub-problem will have the same weight denoted by w. The recursion terminates when $r = O(1)$.

A given query rectangle is decomposed into a number of disjoint 2-sided, 3-sided, and 4-sided queries (see Figure 3). In addition to the (trivial) base case that $r = O(1)$, there are three kinds of *terminal* queries which do not generate further recursive problems: small queries contained within the $k \times k$ squares that are the intersection of slabs, also called *micro queries*; 4-sided *square-aligned* queries whose horizontal and vertical boundaries are aligned with slab boundaries; and 2-sided queries. To answer micro queries, we simply scan the entire query range in $O(k^2 w) = O(rw)$ time without using any additional space. For each of the recursive problems, we store a data structure of Lemma 8 that answers 2-sided queries in $O(rw)$ time using $O(r \log r)$ bits additional space.

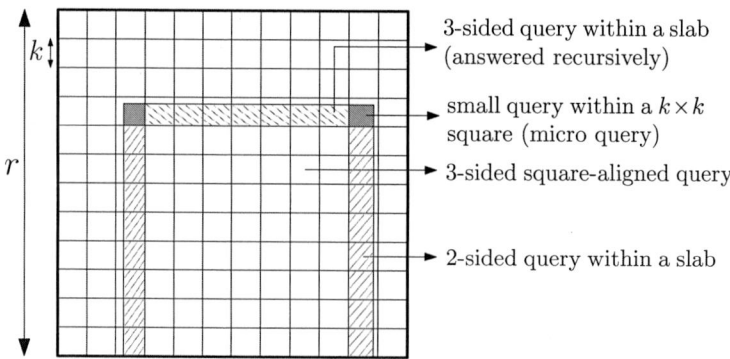

Fig. 4. The recursive decomposition of 3-sided queries

4-Sided Square-Aligned Queries. To answer 4-sided square-aligned queries in an $r \times r$ recursive problem, we divide the $r \times r$ matrix into k^2 disjoint blocks of size $k \times k$ in the preprocessing. We make a $k \times k$ *rank-matrix* out of the minimum element of the blocks, and then replace each element in the rank-matrix by its rank among all the elements in the rank-matrix. The rank-matrix is stored using $O(k^2 \log k) = O(r \log r)$ bits. A square-aligned query is first converted into a 4-sided query within the rank-matrix, which can be answered by brute force in $O(k^2) = O(r)$ time. This determines the block containing the minimum within the query range. Then we find the position of the minimum within the block in $O(k^2 w) = O(rw)$ time by scanning the block.

Query Algorithm. The recursion terminates when $r = O(1)$, or after $\log \log c - O(1)$ levels (recall that $k = \Theta(\sqrt{r})$); no data is stored with these terminal problems at the base of the recursion, and they are solved by brute-force.

A query, depending on its size and position, is decomposed in one of the following two ways (Figure 3): (1) decomposed into at most four micro queries, at most one 4-sided square-aligned query, and at most four 3-sided queries each within a slab; (2) decomposed into at most two micro queries, and at most one 4-sided query within a slab.

As previously described, micro queries and 4-sided square-aligned queries are terminals. A 4-sided query within a slab is answered recursively on the compressed matrix of the slab. A 3-sided query within a slab is converted to a 3-sided query within the compressed matrix of the slab, and that is answered using our decomposition as follows (see Figure 4). A 3-sided query within an $r \times r$ matrix is decomposed into at most two micro queries, at most one 3-sided square-aligned query, at most two 2-sided queries each within a slab, and at most one 3-sided query within a slab. As previously described, micro queries, and 4-sided (including 3-sided) square-aligned queries are terminals. Also a 2-sided query within a slab is converted into a 2-sided query within the compressed matrix of the slab, and thus becomes a terminal query. A 3-sided query within a slab is answered recursively on the compressed matrix of the slab.

Lemma 11. *Given a $c \times c$ matrix, there exists a data structure supporting 2D-RMQs in $O(c \log \log c)$ time using $O(c \log c \log \log c)$ bits additional space.*

Proof. The query time of the algorithm for an $r \times r$ problem is given by $T(r, w) = O(rw) + T(k, kw)$, The recurrence is terminated with $O(w)$ cost when $r = O(1)$. Since the product rw is maintained across the recursive calls, and the depth of the recursion is $O(\log \log c)$, it follows that $T(c, 1) = O(c \log \log c)$.

The space usage of the data structure for an $r \times r$ problem is given by $S(r) = O(r \log r) + (2r/k) \cdot S(k)$, which solves to $S(r) = O(r \log r \log \log r)$.   □

Theorem 1. *There exists a data structure of size $O(N/c)$ bits additional space that supports 2D-RMQs in a matrix of size N in $O(c \log c (\log \log c)^2)$ query time, for a parameter $c \leq N$.*

Proof. Lemmas 10 and 11 together imply a data structure of size $O(N/c' \cdot \log c' \log \log c')$ bits additional space with $O(c' \log \log c')$ query time, for a parameter $c' \leq N$. Substituting c' with $O(c \log c \log \log c)$ proves the claim.   □

4   2D-RMQs in 0-1 Matrices

We present a data structure of size $O(N/c)$ bits additional space that supports 2D-RMQs in $O(c \log c)$ time within a 0-1 matrix of size N. Analogous to Section 3, the problem is reduced to 2D-RMQs in $c \times c$ matrices using Lemma 10. We show that a 2D-RMQ in a $c \times c$ matrix can be answered in $O(c)$ query time using $O(c \log c)$ bits additional space. This implies a data structure of size $O(N/c \cdot \log c)$ bits with $O(c)$ query time, which then leads to our claim after substituting c with $c \log c$.

Lemma 12. *There exists a data structure of size $O(c \log c)$ bits additional space that can answer 2D-RMQs within a $c \times c$ matrix in $O(c)$ time.*

Proof. The data structure is recursive and similar to the one in Section 3.2, with the following differences. For 4-sided square-aligned queries, we make the data structure of Lemma 9 for the rank-matrix. We build a *non-recursive* data structure of size $O(c)$ bits that answers 3-sided queries in $O(c)$ time as follows. Assume w.l.o.g. that 3-sided queries are open to the left. We make an array A of size c such that $A[i]$ stores the column-number of the left most 0 within the i-th row of the matrix. We encode A with a 1D-RMQ structure of size $O(c)$ bits. To answer the 3-sided query, we first find the row j that contains the left most 0 within the query, using A in $O(1)$ time, and then we scan the row j in $O(c)$ time to find the position of the left-most zero within the query; if such a 0 does not exist, we return an arbitrary position.

To finish the proof, we show that the total area consisting all the micro sub-queries in all the recursive levels altogether is $O(c)$, which also holds for general matrices too. This is proved regarding the fact that the weight of entries along the levels is increasing exponentially, and the size of micro sub-queries in the last levels dominate the total size (details omitted).   □

Theorem 2. *For a matrix of size N with elements from $\{0,1\}$, there exists a data structure of size $O(N/c)$ bits additional space that supports 2D-RMQs in $O(c \log c)$ time, for a parameter $c \leq N$.*

References

1. Amir, A., Fischer, J., Lewenstein, M.: Two-Dimensional Range Minimum Queries. In: Ma, B., Zhang, K. (eds.) CPM 2007. LNCS, vol. 4580, pp. 286–294. Springer, Heidelberg (2007)
2. Atallah, M.J., Yuan, H.: Data structures for range minimum queries in multi-dimensional arrays. In: Proc. 20th Annual ACM-SIAM Symposium on Discrete Algorithms, pp. 150–160. SIAM (2010)
3. Bose, P., He, M., Maheshwari, A., Morin, P.: Succinct Orthogonal Range Search Structures on a Grid with Applications to Text Indexing. In: Dehne, F., Gavrilova, M., Sack, J.-R., Tóth, C.D. (eds.) WADS 2009. LNCS, vol. 5664, pp. 98–109. Springer, Heidelberg (2009)
4. Brodal, G.S., Davoodi, P., Rao, S.S.: On space efficient two dimensional range minimum data structures. Algorithmica 63(4), 815–830 (2012)
5. Chazelle, B., Rosenberg, B.: Computing partial sums in multidimensional arrays. In: Proc. 5th Annual Symposium on Computational Geometry, pp. 131–139. ACM (1989)
6. Chor, B., Leiserson, C.E., Rivest, R.L., Shearer, J.B.: An application of number theory to the organization of raster-graphics memory. Journal of ACM 33(1), 86–104 (1986)
7. Davoodi, P., Raman, R., Rao, S.S.: Succinct representations of binary trees for range minimum queries. In: Proc. 18th Annual International Conference on Computing and Combinatorics (to appear, 2012)
8. Demaine, E.D., Landau, G.M., Weimann, O.: On Cartesian Trees and Range Minimum Queries. In: Albers, S., Marchetti-Spaccamela, A., Matias, Y., Nikoletseas, S., Thomas, W. (eds.) ICALP 2009. LNCS, vol. 5555, pp. 341–353. Springer, Heidelberg (2009)
9. Farzan, A., Munro, J.I., Raman, R.: Succinct Indices for Range Queries with Applications to Orthogonal Range Maxima. In: Czumaj, A., Mehlhorn, K., Pitts, A., Wattenhofer, R. (eds.) ICALP 2012. LNCS, vol. 7391, pp. 327–338. Springer, Heidelberg (2012)
10. Fiat, A., Shamir, A.: Polymorphic arrays: A novel VLSI layout for systolic computers. Journal of Computer and System Sciences 33(1), 47–65 (1986)
11. Fiat, A., Shamir, A.: How to find a battleship. Networks 19, 361–371 (1989)
12. Fischer, J., Heun, V.: Space-efficient preprocessing schemes for range minimum queries on static arrays. SIAM J. Comput. 40(2), 465–492 (2011)
13. Gabow, H.N., Bentley, J.L., Tarjan, R.E.: Scaling and related techniques for geometry problems. In: Proc. 16th Annual ACM Symposium on Theory of Computing, pp. 135–143. ACM (1984)
14. Golin, M.J., Iacono, J., Krizanc, D., Raman, R., Rao, S.S.: Encoding 2D Range Maximum Queries. In: Asano, T., Nakano, S.-i., Okamoto, Y., Watanabe, O. (eds.) ISAAC 2011. LNCS, vol. 7074, pp. 180–189. Springer, Heidelberg (2011)
15. Matousek, J.: Geometric Discrepancy. Algorithms and Combinatorics. Springer (1999)
16. Sadakane, K.: Succinct data structures for flexible text retrieval systems. Journal of Discrete Algorithms 5(1), 12–22 (2007)
17. Vuillemin, J.: A unifying look at data structures. Communications of the ACM 23(4), 229–239 (1980)

Locally Correct Fréchet Matchings[*]

Kevin Buchin, Maike Buchin, Wouter Meulemans, and Bettina Speckmann

Dep. of Mathematics and Computer Science,
TU Eindhoven, The Netherlands
{k.a.buchin,m.e.buchin,w.meulemans,speckman}@tue.nl

Abstract. The Fréchet distance is a metric to compare two curves, which is based on monotonous matchings between these curves. We call a matching that results in the Fréchet distance a Fréchet matching. There are often many different Fréchet matchings and not all of these capture the similarity between the curves well. We propose to restrict the set of Fréchet matchings to "natural" matchings and to this end introduce *locally correct* Fréchet matchings. We prove that at least one such matching exists for two polygonal curves and give an $O(N^3 \log N)$ algorithm to compute it, where N is the total number of edges in both curves. We also present an $O(N^2)$ algorithm to compute a locally correct discrete Fréchet matching.

1 Introduction

Many problems ask for the comparison of two curves. Consequently, several distance measures have been proposed for the similarity of two curves P and Q, for example, the Hausdorff and the Fréchet distance. Such a distance measure simply returns a number indicating the (dis)similarity. However, the Hausdorff and the Fréchet distance are both based on matchings of the points on the curves. The distance returned is the maximum distance between any two matched points. The Fréchet distance uses *monotonous matchings* (and limits of these): if point p on P and q on Q are matched, then any point on P after p must be matched to q or a point on Q after q. The *Fréchet distance* is the maximal distance between two matched points minimized over all monotonous matchings of the curves. Restricting to monotonous matchings of only the vertices results in the *discrete Fréchet distance*. We call a matching resulting in the (discrete) Fréchet distance a *(discrete) Fréchet matching*. See Section 2 for more details.

There are often many different Fréchet matchings for two curves. However, as the Fréchet distance is determined only by the maximal distance, not all of these matchings capture the similarity between the curves well (see Fig. 1). There are applications that directly use a matching, for example, to map a GPS track to

---

[*] M. Buchin is supported by the Netherlands Organisation for Scientific Research (NWO) under project no. 612.001.106. W. Meulemans and B. Speckmann are supported by the Netherlands Organisation for Scientific Research (NWO) under project no. 639.022.707. A short abstract of the results presented in Section 3 and Section 4 appeared at the informal workshop EuroCG 2012. For omitted proofs we refer to [4].

L. Epstein and P. Ferragina (Eds.): ESA 2012, LNCS 7501, pp. 229–240, 2012.

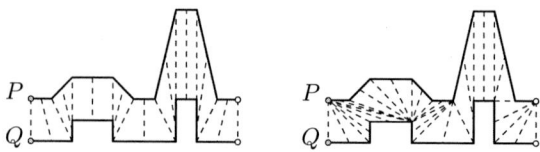

Fig. 1. Two Fréchet matchings for curves P and Q

a street network [9] or to morph between the curves [5]. In such situations a
"good" matching is important. Furthermore, we believe that many applications
of the (discrete) Fréchet distance, such as protein alignment [10] and detecting
patterns in movement data [3], would profit from good Fréchet matchings.

Results. We restrict the set of Fréchet matchings to "natural" matchings by introducing *locally correct* Fréchet matchings: matchings that for any two matched
subcurves are again a Fréchet matching on these subcurves. In Section 3 we prove
that there exists such a locally correct Fréchet matching for any two polygonal
curves. Based on this proof we describe in Section 4 an $O(N^3 \log N)$ algorithm to
compute such a matching, where N is the total number of edges in both curves.
We consider the discrete Fréchet distance in Section 5 and give an $O(N^2)$ algorithm to compute locally correct matchings under this metric.

Related Work. The first algorithm to compute the Fréchet distance was given
by Alt and Godau [1]. They also consider a non-monotone Fréchet distance
and their algorithm for this variant results in a locally correct non-monotone
matching (see Remark 3.5 in [7]). Eiter and Mannila gave the first algorithm
to compute the discrete Fréchet distance [6]. Since then, the Fréchet distance
has received significant attention. Here we focus on approaches that restrict the
allowed matchings. Efrat *et al.* [5] introduced Fréchet-like metrics, the geodesic
width and link width, to restrict to matchings suitable for curve morphing. Their
method is suitable only for non-intersecting polylines. Moreover, geodesic width
and link width do not resolve the problem illustrated in Fig. 1: both matchings
also have minimal geodesic width and minimal link width. Maheshwari *et al.* [8]
studied a restriction by "speed limits", which may exclude all Fréchet matchings
and may cause undesirable effects near "outliers" (see Fig. 2). Buchin *et al.* [2]
describe a framework for restricting Fréchet matchings, which they illustrate by
restricting slope and path length. The former corresponds to speed limits. We
briefly discuss the latter at the end of Section 4.

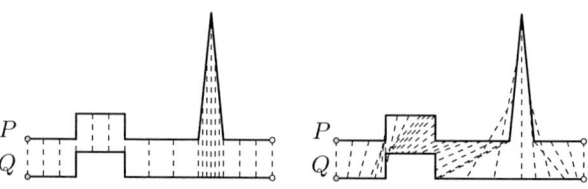

Fig. 2. Two Fréchet matchings. Right: the result of speed limits is not locally correct.

2   Preliminaries

Curves. Let P be a polygonal curve with m edges, defined by vertices $p_0, \ldots, p_m$. We treat a curve as a continuous map $P : [0, m] \to \mathbb{R}^d$. In this map, $P(i)$ equals p_i for integer i. Furthermore, $P(i + \lambda)$ is a parameterization of the $(i+1)$st edge, that is, $P(i + \lambda) = (1 - \lambda) \cdot p_i + \lambda \cdot p_{i+1}$, for integer i and $0 < \lambda < 1$. As a reparametrization $\sigma : [0, 1] \to [0, m]$ of a curve P, we allow any continuous, non-decreasing function such that $\sigma(0) = 0$ and $\sigma(1) = m$. We denote by $P_\sigma(t)$ the actual location according to reparametrization σ: $P_\sigma(t) = P(\sigma(t))$. By $P_\sigma[a, b]$ we denote the subcurve of P in between $P_\sigma(a)$ and $P_\sigma(b)$. In the following we are always given two polygonal curves P and Q, where Q is defined by its vertices $q_0, \ldots, q_n$ and is reparametrized by $\theta : [0, 1] \to [0, n]$. The reparametrized curve is denoted by Q_θ.

Fréchet Matchings. We are given two polygonal curves P and Q with m and n edges. A (monotonous) *matching* μ between P and Q is a pair of reparametrizations (σ, θ), such that $P_\sigma(t)$ matches to $Q_\theta(t)$. The Euclidean distance between two matched points is denoted by $d_\mu(t) = |P_\sigma(t) - Q_\theta(t)|$. The maximum distance over a range is denoted by $d_\mu[a, b] = \max_{a \leq t \leq b} d_\mu(t)$. The *Fréchet distance* between two curves is defined as $\delta_F(P, Q) = \inf_\mu d_\mu[0, 1]$. A *Fréchet matching* is a matching μ that realizes the Fréchet distance: $d_\mu[0, 1] = \delta_F(P, Q)$ holds.

Free Space Diagrams. Alt and Godau [1] describe an algorithm to compute the Fréchet distance based on the decision variant (that is, solving $\delta_F(P, Q) \leq \varepsilon$ for some given ε). Their algorithm uses a *free space diagram*, a two-dimensional diagram on the range $[0, m] \times [0, n]$. Every point (x, y) in this diagram is either "free" (white) or not (indicating whether $|P(x) - Q(y)| \leq \varepsilon$). The diagram has m columns and n rows; every cell (c, r) ($1 \leq c \leq m$ and $1 \leq r \leq n$) corresponds to the edges $p_{c-1}p_c$ and $q_{r-1}q_r$. We scale every row and column in the diagram to correspond to the (relative) length of the actual edge of the curve instead of using unit squares for cells. To compute the Fréchet distance, one finds the smallest ε such that there exists an x- and y-monotone path from point $(0, 0)$ to (m, n) in free space. For this, only certain *critical values* for the distance have to be checked. Imagine continuously increasing the distance ε starting at $\varepsilon = 0$. At so-called *critical events*, which are illustrated in Fig. 3, passages open in the free space. The critical values are the distances corresponding to these events.

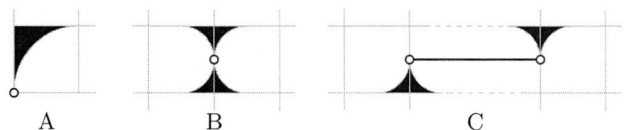

Fig. 3. Three event types. (A) Endpoints come within range of each other. (B) Passage opens on cell boundary. (C) Passage opens in row (or column).

3   Locally Correct Fréchet Matchings

We introduce *locally correct* Fréchet matchings, for which the matching between any two matched subcurves is a Fréchet matching.

Definition 1 (Local correctness). *Given two polygonal curves P and Q, a matching $\mu = (\sigma, \theta)$ is locally correct if for all a, b with $0 \leq a \leq b \leq 1$*

$$d_\mu[a, b] = \delta_F(P_\sigma[a, b], Q_\theta[a, b]).$$

Note that not every Fréchet matching is locally correct. See for example Fig. 2. The question arises whether a locally correct matching always exists and if so, how to compute it. We resolve the first question in the following theorem.

Theorem 1. *For any two polygonal curves P and Q, there exists a locally correct Fréchet matching.*

Existence. We prove Theorem 1 by induction on the number of edges in the curves. The proofs for the lemmata of this section have been omitted but can be found in [4]. First, we present the lemmata for the two base cases: one of the two curves is a point, and both curves are line segments. In the following, n and m again denote the number of edges of P and Q, respectively.

Lemma 1. *For two polygonal curves P and Q with $m = 0$, a locally correct matching is (σ, θ), where $\sigma(t) = 0$ and $\theta(t) = t \cdot n$.*

Lemma 2. *For two polygonal curves P and Q with $m = n = 1$, a locally correct matching is (σ, θ), where $\sigma(t) = \theta(t) = t$.*

For induction, we split the two curves based on events (see Fig. 4). Since each split must reduce the problem size, we ignore any events on the left or bottom boundary of cell $(1, 1)$ or on the right or top boundary of cell (m, n). This

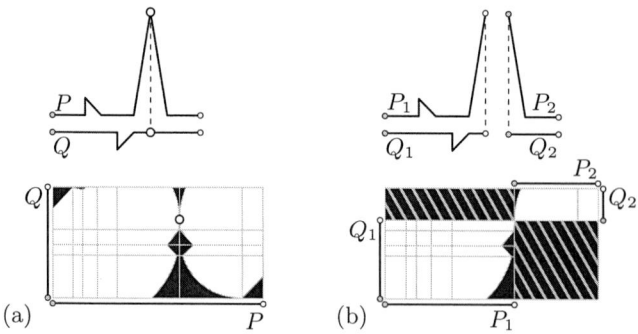

Fig. 4. (a) Curves with the free space diagram for $\varepsilon = \delta_F(P, Q)$ and the realizing event. (b) The event splits each curve into two subcurves. The hatched areas indicate parts that disappear after the split.

excludes both events of type A. A free space diagram is *connected* at value ε, if a monotonous path exists from the boundary of cell $(1,1)$ to the boundary of cell (m,n). A *realizing event* is a critical event at the minimal value ε such that the corresponding free space diagram is connected.

Let $\mathcal{E}$ denote the set of concurrent realizing events for two curves. A *realizing set* E_r is a subset of $\mathcal{E}$ such that the free space admits a monotonous path from cell $(1,1)$ to cell (m,n) without using an event in $\mathcal{E} \backslash E_r$. Note that a realizing set cannot be empty. When $\mathcal{E}$ contains more than one realizing event, some may be "insignificant": they are never required to actually make a path in the free space diagram. A realizing set is *minimal* if it does not contain a strict subset that is a realizing set. Such a minimal realizing set contains only "significant" events.

Lemma 3. *For two polygonal curves P and Q with $m > 1$ and $n \geq 1$, there exists a minimal realizing set.*

The following lemma directly implies that a locally correct Fréchet matching always exists. Informally, it states that curves have a locally correct matching that is "closer" (except in cell $(1,1)$ or (m,n)) than the distance of their realizing set. Further, this matching is linear inside every cell. In the remainder, we use realizing set to indicate a minimal realizing set, unless indicated otherwise. Now, the following lemma can be proven by induction: a realizing set is used to split the curves into pieces; combining the locally correct matchings of the pieces results in a single, locally correct matching.

Lemma 4. *If the free space diagram of two polygonal curves P and Q is connected at value ε, then there exists a locally correct Fréchet matching $\mu = (\sigma, \theta)$ such that $d_\mu(t) \leq \varepsilon$ for all t with $\sigma(t) \geq 1$ or $\theta(t) \geq 1$, and $\sigma(t) \leq m-1$ or $\theta(t) \leq n-1$. Furthermore, μ is linear in every cell.*

4 Algorithm for Locally Correct Fréchet Matchings

The existence proof directly results in a recursive algorithm, which is given by Algorithm 1. Fig. 1 (left), Fig. 2 (left), Fig. 5, Fig. 6, and Fig. 7 (left) illustrate matchings computed with our algorithm. This section is devoted to proving the following theorem.

Theorem 2. *Algorithm 1 computes a locally correct Fréchet matching of two polygonal curves P and Q with m and n edges in $O((m+n)mn \log(mn))$ time.*

Using the notation of Alt and Godau [1], $L_{i,j}^F$ denotes the interval of free space on the left boundary of cell (i,j); $L_{i,j}^R$ denotes the subset of $L_{i,j}^F$ that is reachable from point $(0,0)$ of the free space diagram with a monotonous path in the free space. Analogously, $B_{i,j}^F$ and $B_{i,j}^R$ are defined for the bottom boundary.

With a slight modification to the decision algorithm, we can compute the minimal value of ε such that a path is available from cell $(1,1)$ to cell (m,n). This requires only two changes: $B_{1,2}^R$ should be initialized with $B_{1,2}^F$ and $L_{2,1}^R$ with $L_{2,1}^F$; the answer should be "yes" if and only if $B_{m,n}^R$ or $L_{m,n}^R$ is non-empty.

Algorithm 1. ComputeLCFM(P, Q)

Require: P and Q are curves with m and n edges
Ensure: A locally correct Fréchet matching for P and Q
1: **if** $m = 0$ or $n = 0$ **then**
2:     **return** (σ, θ) where $\sigma(t) = t \cdot m$, $\theta(t) = t \cdot n$
3: **else if** $m = n = 1$ **then**
4:     **return** (σ, θ) where $\sigma(t) = \theta(t) = t$
5: **else**
6:     Find event e_r of a minimal realizing set
7:     Split P into P_1 and P_2 according to e_r
8:     Split Q into Q_1 and Q_2 according to e_r
9:     $\mu_1 \rightarrow$ ComputeLCFM(P_1, Q_1)
10:    $\mu_2 \rightarrow$ ComputeLCFM(P_2, Q_2)
11:    **return** concatenation of μ_1, e_r, and μ_2

Realizing Set. By computing the Fréchet distance using the modified Alt and Godau algorithm, we obtain an ordered, potentially non-minimal realizing set $\mathcal{E} = \{e_1, \ldots, e_l\}$. The algorithm must find an event that is contained in a realizing set. Let E_k denote the first k events of $\mathcal{E}$. For now we assume that the events in $\mathcal{E}$ end at different cell boundaries. We use a binary search on $\mathcal{E}$ to find the r such that E_r contains a realizing set, but E_{r-1} does not. This implies that event e_r is contained in a realizing set and can be used to split the curves. Note that r is unique due to monotonicity. For correctness, the order of events in $\mathcal{E}$ must be consistent in different iterations, for example, by using a lexicographic order. Set E_r contains only realizing sets that use e_r. Hence, E_{r-1} contains a realizing set to connect cell $(1,1)$ to e_r and e_r to cell (m,n). Thus any event found in subsequent iterations is part of E_{r-1} and of a realizing set with e_r.

To determine whether some E_k contains a realizing set, we check whether cells $(1,1)$ and (m,n) are connected without "using" the events of $\mathcal{E} \backslash E_k$. To do this efficiently, we further modify the Alt and Godau decision algorithm. We require only a method to prevent events in $\mathcal{E} \backslash E_k$ from being used. After $L_{i,j}^R$ is computed, we check whether the event e (if any) that ends at the left boundary of cell (i,j) is part of $\mathcal{E} \backslash E_k$ and necessary to obtain $L_{i,j}^R$. If this is the case, we replace $L_{i,j}^R$ with an empty interval. Event e is necessary if and only if $L_{i,j}^R$ is a singleton. To obtain an algorithm that is numerically more stable, we introduce

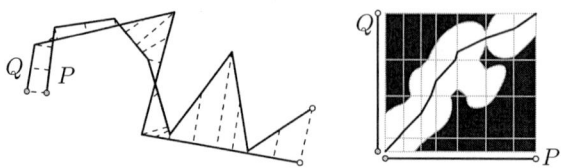

Fig. 5. Locally correct matching produced by Algorithm 1. Free space diagram drawn at $\varepsilon = \delta_F(P, Q)$.

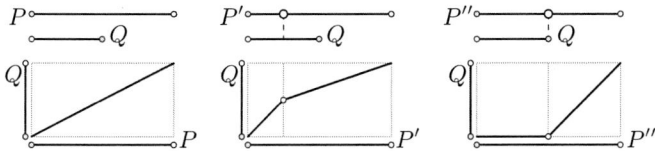

Fig. 6. Different sampling may result in different matchings

entry points. The *entry point* of the left boundary of cell (i, j) is the maximal $i' < i$ such that $B^R_{i',j}$ is non-empty. These values are easily computed during the decision algorithm. Assume the passage corresponding to event e starts on the left boundary of cell (i_s, j). Event e is necessary to obtain $L^R_{i,j}$ if and only if $i' < i_s$. Therefore, we use the entry point instead of checking whether $L^R_{i,j}$ is a singleton. This process is analogous for horizontal boundaries of cells.

Earlier we assumed that each event in $\mathcal{E}$ ends at a different cell boundary. If events end at the same boundary, then these occur in the same row (or column) and it suffices to consider only the event that starts at the rightmost column (or highest row). This justifies the assumption and ensures that $\mathcal{E}$ contains $O(mn)$ events. Thus computing e_r (Algorithm 1, line 6) takes $O(mn \log(mn))$ time, which is equal to the time needed to compute the Fréchet distance. Each recursion step splits the problem into two smaller problems, and the recursion ends when $mn \leq 1$. This results in an additional factor $m + n$. Thus the overall running time is $O((m + n)mn \log(mn))$.

Sampling and Further Restrictions. Two curves may still have many locally correct Fréchet matchings: the algorithm computes just one of these. However, introducing extra vertices may alter the result, even if these vertices do not modify the shape (see Fig. 6). This implies that the algorithm depends not only on the shape of the curves, but also on the sampling. Increasing the sampling further and further seems to result in a matching that decreases the matched distance as much as possible within a cell. However, since cells are rectangles, there is a slight preference for taking longer diagonal paths. Based on this idea, we are currently investigating "locally optimal" Fréchet matchings. The idea is to restrict to the locally correct Fréchet matching that decreases the matched distance as quickly as possible.

We also considered restricting to the "shortest" locally correct Fréchet matching, where "short" refers to the length of the path in the free space diagram. However, Fig. 7 shows that such a restriction does not necessarily improve the quality of the matching.

5   Locally Correct Discrete Fréchet Matchings

Here we study the discrete variant of Fréchet matchings. For the discrete Fréchet distance, only the vertices of curves are matched. The discrete Fréchet distance can be computed in $O(m \cdot n)$ time via dynamic programming [6]. Here, we show how to also compute a locally correct discrete Fréchet matching in $O(m \cdot n)$ time.

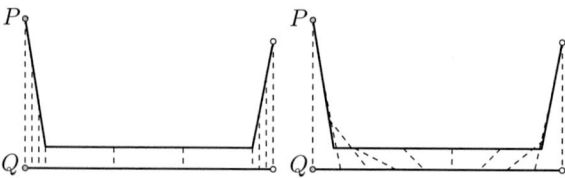

Fig. 7. Two locally correct Fréchet matchings for P and Q. Right: shortest matching.

Grids. Since we are interested only in matching vertices of the curves, we can convert the problem to a grid problem. Suppose we have two curves P and Q with m and n edges respectively. These convert into a grid G of non-negative values with $m + 1$ columns and $n + 1$ rows. Every column corresponds to a vertex of P, every row to a vertex of Q. Any node of the grid $G[i, j]$ corresponds to the pair of vertices (p_i, q_j). Its value is the distance between the vertices: $G[i, j] = |p_i - q_j|$. Analogous to free space diagrams, we assume that $G[0, 0]$ is the bottomleft node and $G[m, n]$ the topright node.

Matchings. A monotonous path π is a sequence of grid nodes $\pi(1), \ldots, \pi(k)$ such that every node $\pi(i)$ $(1 < i \le k)$ is the above, right, or above/right diagonal neighbor of $\pi(i-1)$. In the remainder of this section a path refers to a monotonous path unless indicated otherwise. A monotonous discrete matching of the curves corresponds to a path π such that $\pi(1) = G[0, 0]$ and $\pi(k) = G[m, n]$. We call a path π locally correct if for all $1 \le t_1 \le t_2 \le k$, $\max_{t_1 \le t \le t_2} \pi(t) = \min_{\pi'} \max_{1 \le t \le k'} \pi'(t)$, where π' ranges over all paths starting at $\pi'(1) = \pi(t_1)$ and ending at $\pi'(k') = \pi(t_2)$.

Algorithm. The algorithm needs to compute a locally correct path between $G[0, 0]$ and $G[m, n]$ in a grid G of non-negative values. To this end, the algorithm incrementally constructs a tree T on the grid such that each path in T is locally correct. The algorithm is summarized by Algorithm 2. We define a *growth node* as a node of T that has a neighbor in the grid that is not yet

---

Algorithm 2. ComputeDiscreteLCFM(P, Q)

Require: P and Q are curves with m and n edges
Ensure: A locally correct discrete Fréchet matching for P and Q
1: Construct grid G for P and Q
2: Let T be a tree consisting only of the root $G[0, 0]$
3: **for** $i \leftarrow 1$ **to** m **do**
4:    Add $G[i, 0]$ to T
5: **for** $j \leftarrow 1$ **to** n **do**
6:    Add $G[0, j]$ to T
7: **for** $i \leftarrow 1$ **to** m **do**
8:    **for** $j \leftarrow 1$ **to** n **do**
9:       AddToTree(T, G, i, j)
10: **return** path in T between $G[0, 0]$ and $G[m, n]$

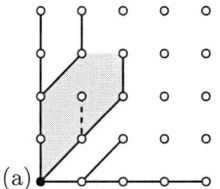 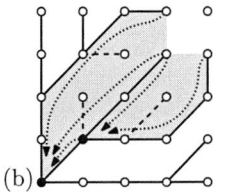 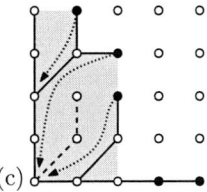

Fig. 8. (a) Face of tree (gray area) with its unique sink (solid dot). A dashed line represents a dead path. (b) Two adjacent faces with some shortcuts indicated. (c) Tree with 3 faces. Solid dots indicate growth nodes with a growth node as parent. These nodes are incident to at most one face. All shortcuts of these nodes are indicated.

part of T: a new branch may sprout from such a node. The growth nodes form a sequence of horizontally or vertically neighboring nodes. A *living node* is a node of T that is not a growth node but is an ancestor of a growth node. A *dead node* is a node of T that is neither a living nor a growth node, that is, it has no descendant that is a growth node. Every pair of nodes in this tree has a *nearest common ancestor* (NCA). When we have to decide what parent to use for a new node in the tree, we look at the maximum value on the path in the tree between the parents and their NCA (excluding the value of the latter). A *face* of the tree is the area enclosed by the segment between two horizontally or vertically neighboring growth nodes (without one being the parent of another) and the paths to their NCA. The unique *sink* of a face is the node of the grid that is in the lowest column and row of all nodes on the face. Fig. 8 (a-b) shows some examples of faces and their sinks.

Shortcuts. To avoid repeatedly walking along the tree to compute maxima, we maintain up to two *shortcuts* from every node in the tree. The segment between the node and its parent is incident to up to two faces of the tree. The node maintains shortcuts to the sink of these faces, associating the maximum value encountered on the path between the node and the sink (excluding the value of the sink). Fig. 8 (b) illustrates some shortcuts. With these shortcuts, it is possible to determine the maximum up to the NCA of two (potentially diagonally) neighboring growth nodes in constant time.

Note that a node g of the tree that has a growth node as parent is incident to at most one face (see Fig. 8 (c)). We need the "other" shortcut only when the parent of g has a living parent. Therefore, the value of this shortcut can be obtained in constant time by using the shortcut of the parent. When the parent of g is no longer a growth node, then g obtains its own shortcut.

Extending the Tree. Algorithm 3 summarizes the steps required to extend the tree T with a new node. Node $G[i, j]$ has three *candidate parents*, $G[i - 1, j]$, $G[i - 1, j - 1]$, and $G[i, j - 1]$. Each pair of these candidates has an NCA. For the actual parent of $G[i, j]$, we select the candidate c such that for any other candidate c', the maximum value from c to their NCA is at most the maximum value from c' to their NCA—both excluding the NCA itself. We must be consistent when breaking ties between candidate parents. To this end, we use

---

Algorithm 3. AddToTree(T, G, i, j)

Require: G is a grid of non-negative values; any path in tree T is locally correct

Ensure: node $G[i, j]$ is added to T and any path in T is locally correct

1: $parent(G[i, j]) \leftarrow$ candidate parent with lowest maximum value to NCA
2: **if** $G[i-1, j-1]$ is dead **then**
3:    Remove the dead path ending at $G[i-1, j-1]$ and extend shortcuts
4: Make shortcuts for $G[i-1, j]$, $G[i, j-1]$, and $G[i, j]$ where necessary

---

the preference order of $G[i-1, j] \succ G[i-1, j-1] \succ G[i, j-1]$. Since paths in the tree cannot cross, this order is consistent between two paths at different stages of the algorithm. Note that a preference order that prefers $G[i-1, j-1]$ over both other candidates or vice versa results in an incorrect algorithm.

When a dead path is removed from the tree, adjacent faces merge and a sink may change. Hence, shortcuts have to be extended to point toward the new sink. Fig. 9 illustrates the incoming shortcuts at a sink and the effect of removing a dead path on the incoming shortcuts. Note that the algorithm does not need to remove dead paths that end in the highest row or rightmost column.

Finally, $G[i-1, j]$, $G[i, j-1]$, and $G[i, j]$ receive shortcuts where necessary. $G[i-1, j]$ or $G[i, j-1]$ needs a shortcut only if its parent is $G[i-1, j-1]$. $G[i, j]$ needs two shortcuts if $G[i-1, j-1]$ is its parent, only one shortcut otherwise.

Correctness. To prove correctness of Algorithm 2, we require a stronger version of local correctness. A path π is *strongly locally correct* if for all paths π' with the same endpoints $\max_{1 < t \leq k} \pi(t) \leq \max_{1 < t' \leq k'} \pi'(t')$ holds. Note that the first node is excluded from the maximum. Since $\max_{1 < t \leq k} \pi(t) \leq \max_{1 < t' \leq k'} \pi'(t')$ and $\pi(1) = \pi'(1)$ imply $\max_{1 \leq t \leq k} \pi(t) \leq \max_{1 \leq t' \leq k'} \pi'(t')$, a strongly locally correct path is also locally correct. Lemma 5 implies the correctness of Algorithm 2. We only sketch the proof here. For the full proof see [4].

Lemma 5. *Algorithm 2 maintains the following invariant: any path in T is strongly locally correct.*

Proof (sketch). Initially any path in T spans either only the first row or only the first column. Hence, the path is unique between its endpoints and this path is strongly locally correct.

The algorithm extends tree T to T' by including a node g that has three candidate parents which are growth nodes. From the invariant, we derive that

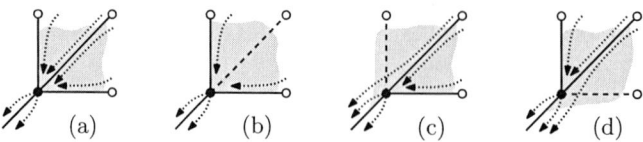

Fig. 9. (a) Each sink has up to four sets of shortcuts. (b-d) Removing a dead path (dashed) extends at most one set of shortcuts.

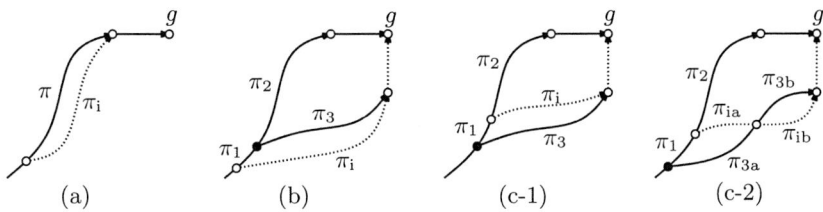

Fig. 10. The four cases for the proof of Lemma 5

if a path in T' is not strongly locally correct, then an invalidating path exists that ends at g and starts at an ancestor of g. The main observation is that this invalidating path uses one of the candidate parents of g as its before-last node. We distinguish three cases on how this path is situated compared to T'. The last case, however, needs two subcases in order to deal with candidate parents that have the same maximum value to their NCA. Fig. 10 illustrates these cases. For the first three cases, we derive that there is a path in T that is not strongly locally correct, contradicting the induction hypothesis. These paths are π, (π_1, π_3), and π_3 respectively. For case (c-2), either π_{3a} is not strongly locally correct or (π_1, π_{ia}) conflicts with the preference order.   □

Execution Time. When a dead path π_d is removed, we may need to extend a list of incoming shortcuts at $\pi_d(1)$, the node that remains in T. Let k denote the number of nodes in π_d. The lemma below relates the number of extended shortcuts to the size of π_d. The omitted proof can be found in [4]. The main observation is that the path requiring extensions starts at $\pi_d(1)$ and ends at either $G[i-1, j]$ or $G[i, j-1]$, since $G[i, j]$ has not yet received any shortcuts.

Lemma 6. *A dead path π_d with k nodes can result in at most $2 \cdot k - 1$ extensions.*

Hence, we can charge every extension to one of the $k - 1$ dead nodes (all but $\pi_d(1)$). A node gets at most 3 charges, since it is a (non-first) node of a dead path at most once. Because an extension can be done in constant time, the execution time of the algorithm is $O(mn)$. Note that shortcuts that originate from a living node with outdegree 1 could be removed instead of extended. We summarize the findings of this section in the following theorem.

Theorem 3. *Algorithm 2 computes a locally correct discrete Fréchet matching of two polygonal curves P and Q with m and n edges in $O(mn)$ time.*

6   Conclusion

We set out to find "good" matchings between two curves. To this end we introduced the local correctness criterion for Fréchet matchings. We have proven that there always exists at least one locally correct Fréchet matching between any two polygonal curves. This proof resulted in an $O(N^3 \log N)$ algorithm, where

N is the total number of edges in the two curves. Furthermore, we considered computing a locally correct matching using the discrete Fréchet distance. By maintaining a tree with shortcuts to encode locally correct partial matchings, we have shown how to compute such a matching in $O(N^2)$ time.

Future Work. Computing a locally correct discrete Fréchet matching takes $O(N^2)$ time, just like the dynamic program to compute only the discrete Fréchet distance. However, computing a locally correct continuous Fréchet matching takes $O(N^3 \log N)$ time, a linear factor more than computing the Fréchet distance. An interesting question is whether this gap in computation can be reduced as well.

Furthermore, it would be interesting to investigate the benefit of local correctness for other matching-based similarity measures, such as the geodesic width [5].

References

1. Alt, H., Godau, M.: Computing the Fréchet distance between two polygonal curves. Int. J. of Comp. Geometry and Appl. 5(1), 75–91 (1995)
2. Buchin, K., Buchin, M., Gudmundsson, J.: Constrained free space diagrams: a tool for trajectory analysis. Int. J. of GIS 24(7), 1101–1125 (2010)
3. Buchin, K., Buchin, M., Gudmundsson, J., Löffler, M., Luo, J.: Detecting Commuting Patterns by Clustering Subtrajectories. Int. J. of Comp. Geometry and Appl. 21(3), 253–282 (2011)
4. Buchin, K., Buchin, M., Meulemans, W., Speckmann, B.: Locally correct Fréchet matchings. CoRR, abs/1206.6257 (June 2012)
5. Efrat, A., Guibas, L., Har-Peled, S., Mitchell, J., Murali, T.: New Similarity Measures between Polylines with Applications to Morphing and Polygon Sweeping. Discrete & Comp. Geometry 28(4), 535–569 (2002)
6. Eiter, T., Mannila, H.: Computing Discrete Fréchet Distance. Technical Report CD-TR 94/65, Christian Doppler Laboratory (1994)
7. Har-Peled, S., Raichel, B.: Fréchet distance revisited and extended. CoRR, abs/1202.5610 (February 2012)
8. Maheshwari, A., Sack, J., Shahbaz, K., Zarrabi-Zadeh, H.: Fréchet distance with speed limits. Comp. Geometry: Theory and Appl. 44(2), 110–120 (2011)
9. Wenk, C., Salas, R., Pfoser, D.: Addressing the need for map-matching speed: Localizing global curve-matching algorithms. In: Proc. 18th Int. Conf. on Sci. and Stat. Database Management, pp. 379–388 (2006)
10. Wylie, T., Zhu, B.: A Polynomial Time Solution for Protein Chain Pair Simplification under the Discrete Fréchet Distance. In: Bleris, L., Măndoiu, I., Schwartz, R., Wang, J. (eds.) ISBRA 2012. LNCS, vol. 7292, pp. 287–298. Springer, Heidelberg (2012)

The Clique Problem in Ray Intersection Graphs[*]

Sergio Cabello[1], Jean Cardinal[2], and Stefan Langerman[2],[**]

[1] Department of Mathematics, IMFM, and Department of Mathematics, FMF,
University of Ljubljana, Slovenia
sergio.cabello@fmf.uni-lj.si
[2] Computer Science Department, Université Libre de Bruxelles (ULB),
Brussels, Belgium
{jcardin,slanger}@ulb.ac.be

Abstract. Ray intersection graphs are intersection graphs of rays, or
halflines, in the plane. We show that any planar graph has an even subdivision whose complement is a ray intersection graph. The construction
can be done in polynomial time and implies that finding a maximum
clique in a segment intersection graph is NP-hard. This solves a 21-year
old open problem posed by Kratochvíl and Nešetřil.

1 Introduction

The intersection graph of a collection of sets has one vertex for each set, and an
edge between two vertices whenever the corresponding sets intersect. Of particular interest are families of intersection graphs corresponding to geometric sets
in the plane. In this contribution, we will focus on *segment intersection graphs*,
intersection graphs of line segments in the plane.

In a seminal paper, Kratochvíl and Nešetřil [11] proposed to study the complexity of two classical combinatorial optimization problems, the maximum independent set and the maximum clique, in geometric intersection graphs. While
those problems are known to be hard to approximate in general graphs (see for
instance [5,13]), their restriction to geometric intersection graphs may be more
tractable. They proved that the maximum independent set problem remains
NP-hard for segment intersection graphs, even if those segments have only two
distinct directions. It was also shown that in that case, the maximum clique
problem can be solved in polynomial time. The complexity of the maximum
clique problem in general segment intersection graphs was left as an open problem, and remained so until now. In their survey paper "On six problems posed
by Jarik Nešetřil" [3], Bang-Jensen *et al.* describe this problem as being "among
the most tantalizing unsolved problem in the area".

Some progress has been made in the meanwhile. In 1992, Middendorf and
Pfeiffer [12] showed, with a simple proof, that the maximum clique problem was

[*] Research was supported in part by the Slovenian Research Agency, program P1-0297
and projects J1-4106, L7-4119, BI-BE/11-12-F-002, and within the EUROCORES
Programme EUROGIGA (projects GReGAS and ComPoSe) of the European Science
Foundation. Large part of the work was done while Sergio was visiting ULB.
[**] Maître de Recherches, Fonds de la Recherche Scientifique (F.R.S-FNRS).

L. Epstein and P. Ferragina (Eds.): ESA 2012, LNCS 7501, pp. 241–252, 2012.

NP-hard for intersection graphs of 1-intersecting curve segments that are either line segments or curves made of two orthogonal line segments. They also give a polynomial time dynamic programming algorithm for the special case of line segments with endpoints of the form $(x, 0), (y, i)$, with $i \in \{1, \dots k\}$ for some fixed k. Another step was made by Ambühl and Wagner [2] in 2005, who showed that the maximum clique problem was NP-hard for intersection graphs of ellipses of fixed, arbitrary, aspect ratio. Unfortunately, this ratio must be bounded, which excludes the case of segments.

Our results. We prove that the maximum clique problem in segment intersection graphs is NP-hard. In fact, we prove the stronger result that the problem is NP-hard even in *ray intersection graphs*, defined as intersection graphs of rays, or halflines, in the plane. This complexity result is a consequence of the following structural lemma: every planar graph has an even subdivision whose complement is a ray intersection graph. Furthermore, the corresponding set of rays has a natural polynomial size representation. Hence solving the maximum clique problem in this graph allows to recover the maximum independent set in the original planar graph, a task well known to be NP-hard [8]. The construction is detailed in Section 2.

Related work. We prove that the complement of some subdivision of any planar graph can be represented as a segment intersection graph. Whether the complement of every planar graph is a segment intersection graph remains an open question. In 1998, Kratochvíl and Kuběna [9] showed that the complement of any planar graph is the intersection graph of a set of convex polygons. More recently, Francis, Kratochvíl, and Vyskočil [7] proved that the complement of any partial 2-tree is a segment intersection graph. Partial 2-trees are planar, and in particular every outerplanar graph is a partial 2-tree. The representability of planar graphs by segment intersection graphs, formerly known as Scheinerman's conjecture, was proved recently by Chalopin and Gonçalves [4].

The maximum independent set problem in intersection graphs has been studied by Agarwal and Mustafa [1]. In particular, they proved that it could be approximated within a factor $n^{1/2+o(1)}$ in polynomial time for segment intersection graphs. This has been recently improved by Fox and Pach [6], who described, for any $\epsilon > 0$, an n^ϵ-approximation algorithm. In fact, their technique also applies to the maximum clique problem, and therefore n^ϵ is the best known approximation factor for this problem too.

In 1994, Kratochvíl and Matoušek [10] proved that the recognition problem for segment intersection graphs was in PSPACE, and was also NP-hard.

Notation. For any natural number m we use $[m] = \{1, \dots m\}$. In a graph G, a *rotation system* is a list $\pi = (\pi_v)_{v \in V(G)}$, where each π_v fixes the clockwise order of the edges of $E(G)$ incident to v. When G is an embedded planar graph, the embedding uniquely defines a rotation system, which is often called a combinatorial embedding. For the rest of the paper we use *ray* to refer to an *open ray*, that is, a ray does not contain its origin. Therefore, whenever two rays intersect they do so in the relative interior of both. Since our construction does not use degeneracies, we are not imposing any restriction by considering only open rays. A subdivision of a graph G is said to be *even* if each edge of G is subdivided an even number of times.

2   Construction

Let us start providing an overview of the approach. We first construct a set of curves that will form the *reference frame*. This construction is quite generic and depends only on a parameter $k \in \mathbb{N}$. We then show that the complement of any tree has a special type of representation, called *snooker representation*, which is constructed iteratively over the levels of the tree. The number of levels of the tree is closely related to k, the parameter used for the reference frame. We then argue that if G is a planar graph that consists of a tree T and a few, special paths of length two and three, then the complement of G can be represented as an intersection graph of rays by extending a snooker representation of T. Finally, we argue that any planar graph has an even subdivision that can be decomposed into a tree and a set of paths of length two and three with the required properties.

We first describe the construction using real coordinates. The construction does not rely on degeneracies, and thus we can slightly perturb the coordinates used in the description. This perturbation is enough to argue that a representation can be computed in polynomial time. Then, using a relation between the independence number of a graph G and an even subdivision of G, we obtain that computing a maximum clique in a ray intersection graph is NP-hard, even if a representation is given as part of the input.

2.1   Reference Frame

Let $k \geq 3$ be an *odd* number to be chosen later. We set $\theta = \frac{k-1}{k}\pi$, and define for $i = 0, \dots, k-1$ the points

$$p_i = (\cos(i \cdot \theta), \sin(i \cdot \theta)).$$

The points p_i lie on a unit circle centered at the origin; see Figure 1(a). For each $i \in [k-2]$ we construct a rectangle R_i as follows. Let q_i be the point $p_i + (p_i - p_{i+1})$, symmetric of p_{i+1} with respect to p_i, let m_i be the midpoint between p_i and q_i, and let t_i be, among the two points along the line through q_{i+1} and m_i with the property $|m_i t_i| = |m_i p_i|$, the one that is furthest from q_{i+1}. We define R_i to be the rectangle with vertices p_i, t_i, and q_i. The fourth vertex of R_i is implicit and is denoted by r_i. Any two rectangles R_i and R_j are congruent with respect to a rotation around the origin. We have constructed the rectangles R_i in such way that, for any $i \in [k-2]$, the line supporting the diagonal $p_i q_i$ of R_i contains p_{i+1} and the line supporting the diagonal $r_i t_i$ of R_i contains q_{i+1}.

For each $i \in [k-2]$, let α_i be the circular arc tangent to both diagonals of R_i with endpoints q_i and r_i (see Figure 1(b)). Note that the curves α_i and the

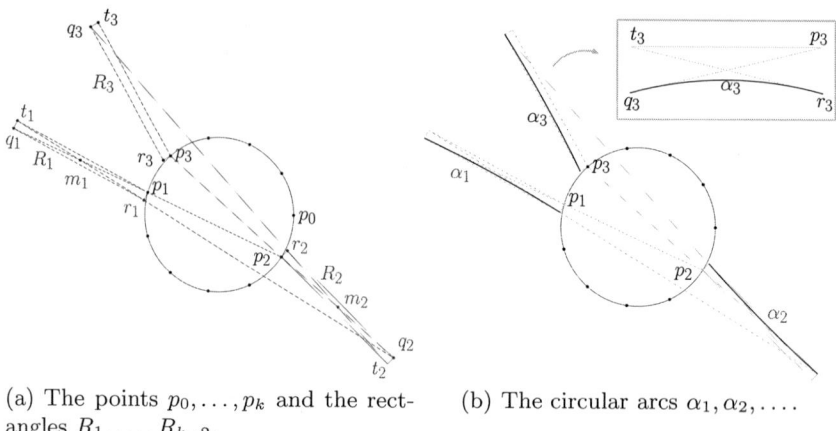

(a) The points $p_0, \ldots, p_k$ and the rectangles $R_1, \ldots, R_{k-2}$.

(b) The circular arcs $\alpha_1, \alpha_2, \ldots$.

Fig. 1. The reference frame

rectangles R_i have been chosen so that any line that intersects α_i twice or is tangent to the curve α_i must intersect the curve α_{i+1}. For any $i \in [k-2]$, let Γ_i be the set of rays that intersect α_i twice or are tangent to α_i and have their origins on α_{i+1}. We also define Γ_0 as the set of rays with origin on α_1 and passing through p_0. The rays of Γ_i that are tangent to α_i will play a special role. In fact, we will only use rays of Γ_i that are "near-tangent" to α_i.

Lemma 1. *When $|j - i| > 1$, any ray from Γ_i intersects any ray from Γ_j.*

Lemma 2. *Any ray tangent to α_{i+1} at the point $x \in \alpha_{i+1}$ intersects any ray from Γ_i, except those having their origin at x.*

Note that the whole construction depends only on the parameter k. We will refer to it as *reference frame*.

2.2   Complements of Trees

Let T be a graph with a rotation system π_T, let r be a vertex in T and let $rs \in E(T)$ be an arbitrary edge incident to r. The triple (π_T, r, rs) induces a natural linear order $\tau = \tau(\pi_T, r, rs)$ on the vertices of T. This order τ corresponds to the order followed by a breadth-first traversal of T from r with the following additional restrictions:

(i) s is the second vertex;
(ii) the children of any vertex v are visited according to the clockwise order π_v;
(iii) if $v \neq r$ has parent v', the first child of u of v is such that vu is the successor of vv' in the clockwise order π_v.

We say that vertices v and v' at the same level are *consecutive* when they are consecutive in τ. See Figure 2. The linear order will be fixed through our discussion, so we will generally drop it from the notation. Henceforth, whenever we

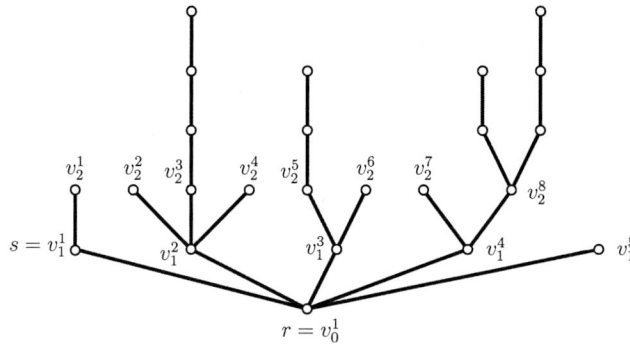

Fig. 2. In this example, assuming that π is as drawn, the linear order τ is $r, v_1^1, v_1^2, \ldots, v_1^5, v_2^1, \ldots$

talk about a tree T and a linear order τ on $V(T)$, we assume that τ is the natural linear order induced by a triple (π_T, r, rs). In fact, the triple (π_T, r, rs) is implicit in τ. For any vertex v we use v^+ for its successor and v^- for its predecessor.

A *snooker representation* of the complement of an embedded tree T with linear order τ is a representation of T with rays that satisfies the following properties:

(a) Each vertex v at level i in T is represented by a ray γ_v from Γ_i. Thus, the origin of γ_v, denoted by a_v, is on α_{i+1}. Note that this implies that k is larger than the depth of T.

(b) If a vertex u has parent v, then γ_u passes through the origin a_v of γ_v. (Here it is relevant that we consider all rays to be open, as otherwise γ_u and γ_v would intersect.) In particular, all rays corresponding to the children of v pass through the point a_v.

(c) The origins of rays corresponding to consecutive vertices u and v of level i are consecutive along α_{i+1}. That is, no other ray in the representation has its origin on α_{i+1} and between the origins γ_u and γ_v.

Lemma 3. *The complement of any embedded tree with a linear order τ has a snooker representation.*

Proof. Consider a reference frame with k larger than the depth of T. The construction we provide is iterative over the levels of T. Note that, since we provide a snooker representation, it is enough to tell for each vertex $v \neq r$ the origin a_v of the ray γ_v. Property (b) of the snooker representation provides another point on the ray γ_v, and thus γ_v is uniquely defined. The ray γ_r for the root r is the ray of Γ_0 with origin a_r in the center of α_1.

Consider any level $i > 1$ and assume that we have a representation of the vertices at level $i - 1$. Consider a vertex v at level $i - 1$ and let $u_1, \ldots, u_d$ denote its d children. If the successor v^+ of v is also at level $i - 1$, we take $a_v^+ = a_{v^+}$, and else we take a_v^+ to be an endpoint of α_i such that no other origin is between the endpoint and a_v. See Figure 3. Similarly, if the predecessor v^- of v is at level $i - 1$, we take $a_v^- = a_{v^-}$, and else we take a_v^- to be an endpoint of α_i such that no other origin is between the endpoint and a_v. (If v is the only one vertex at level i, we also make sure that $a_v^- \neq a_v^+$.) Let ℓ_v^+ be the line through a_v and a_v^+. Similarly, let ℓ_v^- be the line through a_v and a_v^-. We then choose the points $a_{u_1}, \ldots, a_{u_d}$ on the portion of α_{i+1} contained between ℓ_v^+ and ℓ_v^- such that the $d+2$ points $\ell_v^- \cap \alpha_{i+1}, a_{u_1}, \ldots, a_{u_d}, \ell_v^+ \cap \alpha_{i+1}$ are regularly spaced. Since the ray γ_{u_j} has origin a_{u_j} and passes through a_v, this finishes the description of the procedure. Because a_{u_j} lies between ℓ_v^+ and ℓ_v^-, the ray γ_{u_j} either intersects α_i twice or is tangent to α_i, and thus $\gamma_{u_j} \in \Gamma_i$.

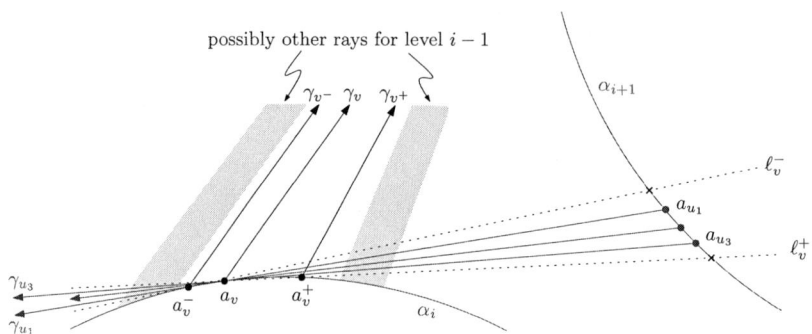

Fig. 3. An example showing the construction of a snooker representation when v has three children and v^+ and v^- are at the same level as v

Recall that any ray from Γ_i intersects any ray from Γ_j when $|j - i| > 1$. Therefore, vertices from levels i and j, where $|i - j| > 1$, intersect. For vertices u and v at levels $i - 1$ and i, respectively, the convexity of the curve α_i and the choices for a_v imply that γ_v intersects γ_u if and only if u is not the parent of v in T. For vertices u and v at the same level i, the rays γ_u and γ_v intersect: if they have the same parent w, then they intersect on a_w, and if they have different parents, the order of their origins a_u and a_v on α_{i+1} and the order of their intersections with α_i are reversed.   □

2.3   A Tree with a Few Short Paths

Let T be an embedded tree with a linear order τ. An *admissible extension* of T is a graph P with the following properties

- P is the union of vertex-disjoint paths (i.e., two paths don't share internal vertices but they are allowed to share endpoints);
- each maximal path in P has 3 or 4 vertices;
- the endpoints of each maximal path in P are leaves of T that are consecutive and at the same level;
- the internal vertices of any path in P are not vertices of $V(T)$.

Note that $T+P$ is a planar graph because we only add paths between consecutive leaves.

Lemma 4. *Let T be an embedded tree and let P be an admissible extension of T. The complement of $T + P$ is a ray intersection graph.*

Proof. We construct a snooker representation of T using Lemma 3 where k, the number of levels, is the depth of T plus 2 or 3, whichever is odd. We will use a local argument to represent each maximal path of P, and each maximal path of P is treated independently. It will be obvious from the construction that rays corresponding to vertices in different paths intersect. We distinguish the case where the maximal path has one internal vertex or two.

Consider first the case of a maximal path in P with one internal vertex. Thus, the path is uwv where u and v are consecutive leaves in T and $w \notin V(T)$ is not yet represented by a ray. The origins a_u and a_v of the rays γ_u and γ_v, respectively, are distinct and consecutive along α_{i+1} because we have a snooker representation. We thus have the situation depicted in Figure 4. We can then just take the γ_w to be the line through a_u and a_v. (This line can also be a ray with an origin sufficiently far away.) This line intersects the ray of any other vertex, different that γ_u and γ_v.

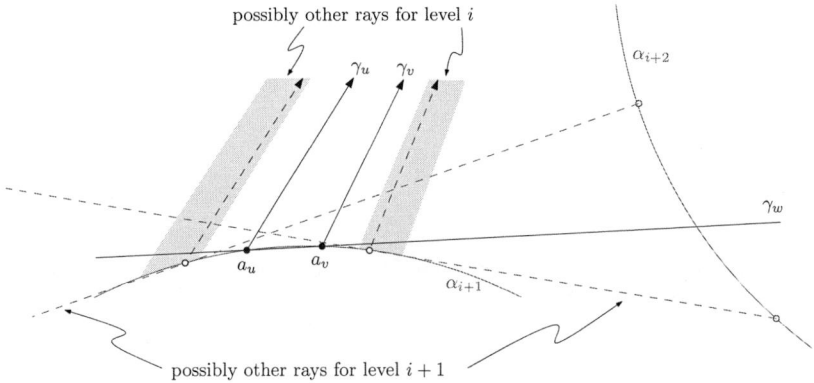

Fig. 4. Case 1 in the proof of Lemma 4: adding a path with one internal vertex

Consider now the case of a maximal path in P with two internal vertices. Thus, the path is $uww'v$ where u and v are consecutive leaves in T and $w, w' \notin V(T)$. In this case, the construction depends on the relative position of the origins a_u and a_v, and we distinguish two scenarios: (i) shifting the origin a_u of ray γ_u towards a_v while maintaining the slope introduces an intersection between γ_u and the ray for the parent of u or (ii) shifting the origin a_v of ray γ_v towards a_u while maintaining the slope introduces an intersection between γ_v and the ray for the parent of v. Note that exactly one of the two options must occur.

Let us consider only scenario (i), since the other one is symmetric; see Figure 5. We choose a point b_w in α_{i+1} between a_u and a_v very near a_u and represent w

with a ray γ_w parallel to γ_u with origin b_w. Thus γ_w does not intersect γ_u but intersects any other ray because we are in scenario (i). Finally, we represent w' with the line $\gamma_{w'}$ through points b_w and a_v. With this, $\gamma_{w'}$ intersects the interior of any other ray but γ_w and γ_v, as desired. Note that $\gamma_{w'}$ also intersects the rays for vertices in the same level because those rays are near-tangent to α_{i+2}, which is intersected by γ_w.

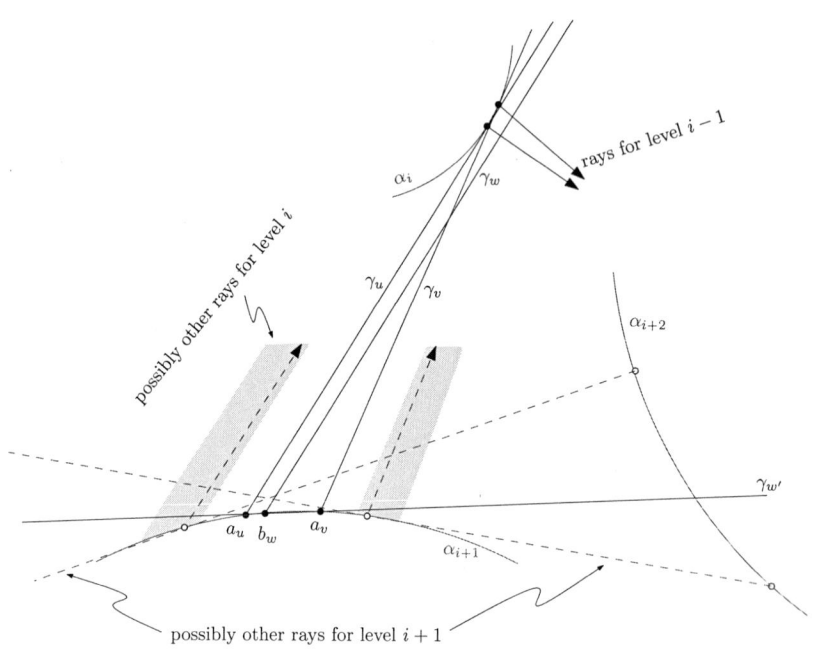

Fig. 5. Case 2(i) in the proof of Lemma 4: adding a path with two internal vertices

□

Lemma 5. *Any embedded planar graph G has an even subdivision $T + P$, where T is an embedded tree and P is an admissible extension of T. Furthermore, such T and P can be computed in polynomial time.*

Proof. Let r be an arbitrary vertex in the outer face f of G. Let B be the set of edges in a BFS tree of G from r. With a slight abuse of notation, we also use B for the BFS tree itself. Let $C = E(G) \setminus B$. In the graph G^* dual to G, the edges $C^* = \{c^* \mid c \in C\}$ are a spanning tree of G^*, which with a slight abuse of notation we also denote by C^*. We root C^* at the node f^*, corresponding to the outer face. This is illustrated on Figure 6.

We define for each edge $e \in C$ the number k_e of subdivisions it will undertake using a bottom-up approach. Any edge $e \in C$ that is incident to a leaf of C^* gets assigned $k_e = 4$. For any other edge $e \in C$, we define k_e as 2 plus the

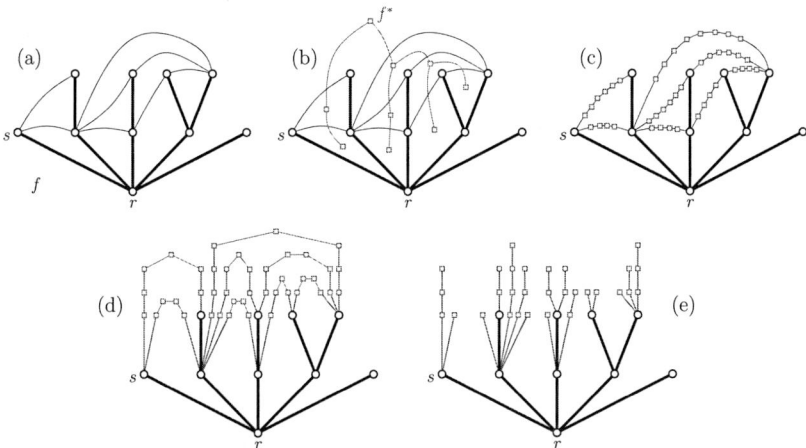

Fig. 6. Figure for the proof of Lemma 5. (a) A planar graph with a BFS from r in bold. (b) The spanning tree C^* of the dual graph. (c) The subdivision of the edges from C. (d) The resulting subdivided graph with P marked in red and thinner lines. (e) The resulting graph T after the removal of P, drawn such that the heights of the vertices correspond to their levels.

maximum $k_{e'}$ over all descendants $(e')^* \in C^*$ of e^* in C^*. Let H be the resulting subdivision. For any edge $e \in C$, let Q_e be the path in H that corresponds to the subdivision of e. We use in H the combinatorial embedding induced by G.

We can now compute the tree T and the paths P. Since B is a BFS tree, every edge $e \in C$ connects two vertices that are either at the same level in B, or at two successive levels. For every edge $e \in C$ that connects two vertices at the same level in B, let P_e be the length-three subpath in the middle of Q_e. For every edge $uv \in C$ that connects vertex u at level i to vertex v at level $i+1$ in B, let P_e be the length-two subpath obtained by removing the first edge of the length-three subpath in the middle of $Q_e - u$. We then take $P = \bigcup_{e \in C} P_e$ and take T to be the graph $H - P$, after removing isolated vertices. In T we use the rotation system inherited from H and use the edge rs to define the linear order, where rs is an edge in f.

It is clear from the construction that $T + P = H$ is an even subdivision of G. We have to check that P is indeed an admissible extension of T. The maximal paths of P are vertex disjoint and connect leaves of P because the paths Q_e, $e \in C$, are edge-disjoint and each P_e is strictly contained in the interior of Q_e. Since in $H - P$ we removed isolated vertices, it is clear that no internal vertex of a path of P is in T. The graph T is indeed a tree because, for every edge $e \in C$, we have removed some edges from its subdivision Q_e. Since B is a BFS tree and P_{uv} is centered within Q_{uv}, when u and v are at the same level, or within $Q_{uv} - u$, when u is one level below v, the maximal paths of P connect vertices that are equidistant from r in H.

It remains to show that the endpoints of any maximal path in P are consecutive vertices in T. This is so because of the inductive definition of k_e. The base case is when $e \in C$ is incident to a leaf of C^*, and is subdivided $k_e = 4$ times. The edge e either connects two vertices at the same level in B, or at two successive levels. In both cases it can be checked that P_e connects two consecutive vertices in T. The inductive case is as follows. We let k_e be equal to 2 plus the maximum $k_{e'}$ over all descendants $(e')^* \in C^*$ of e^* in C^*. By induction, all the corresponding $P_{e'}$ connect two consecutive vertices of T, say at level i in T. By definition, P_e will connect two consecutive vertices of T at level $i + 1$.

The construction of T and P only involves computing the BFS tree B, the spanning tree C^*, and the values k_e, which can clearly be done in polynomial time.   □

Combining lemmas 4 and 5 directly yields the following.

Theorem 1. *Any planar graph has an even subdivision whose complement is a ray intersection graph. Furthermore, this subdivision can be computed in polynomial time.*

2.4   Polynomial-Time Construction

In so far, the construction has used real coordinates. We next argue how to make a construction using integers each using a polynomial number of bits. Due to space constraints, some proof details are left out.

Lemma 6. *In the construction of Lemma 4, if $n = |V(T + P)|$,*

- *any two points are at distance at least $n^{-O(n)}$ and at most $O(n)$;*
- *the distance between any line through two origins and any other point is at least $n^{-O(n)}$, unless the three origins are collinear.*

Proof (sketch). Recall that the circle containing points $p_0, p_1, \ldots$ has radius 1. The rectangles R_i are all congruent and have two diagonals of length $|p_0 p_1| = \Theta(1)$. The small side of rectangle R_i has size $\Theta(1/n)$ and both diagonals of R_i form an angle of $\Theta(1/n)$. It follows that the center of the circles supporting α_i have coordinates $\Theta(n)$. For points from different curves α_i there is at least a separation of $\Theta(1/n)$.

We first bound the distance between the origins for the rays representing vertices of T. Let us refer by *features* on the curve α_i the origins of rays lying on α_i and the extremes of α_i. Let δ_i be the minimum separation between any two features on the curve α_i. In the curve α_1 there are three features: the two extremes of α_1 and the origin a_r of γ_r, which is in the middle of α_1. Since α_1 have length $\Omega(1)$, it follows that $\delta_1 = \Omega(1)$.

We will bound the ratio δ_{i+1}/δ_i for $i \geq 1$. Consider the construction of Lemma 3 to determine the features on α_{i+1}. By induction, any two consecutive features along α_i are separated at least by δ_i. Since α_i is supported by a circle of radius $\Theta(n)$, the lines ℓ_v^+ and ℓ_v^- form an angle of at least $\Omega(\delta_i/n)$.

This implies that the points $\ell_v^- \cap \alpha_{i+1}, a_{u_1}, \ldots, a_{u_d}, \ell_v^+ \cap \alpha_{i+1}$ have a separation of $\Omega(\delta_i/(nd))$. It follows that $\delta_{i+1} = \Omega(\delta_i/(nd)) = \Omega(\delta_i/n^2)$, and thus $\delta_{i+1}/\delta_i = \Omega(1/n^2)$. Since T has depth at most n, all features are at distance at least $n^{-O(n)}$.

We can now argue that the origins of the rays used in the construction of Lemma 4 also have a separation of $n^{-O(n)}$. In Case 1, we just add a line through previous features. In Case 2, it is enough to place b_w at distance $|a_u a_v|/n$ from a_u, and thus the features keep a separation of at least $n^{-O(n)}$.

The proof of the second item is omitted. $\qquad\square$

We can now give the following algorithmic version of Lemma 4.

Lemma 7. *Let T be an embedded tree and let P be an admissible extension of T. We can find in polynomial time a family of rays described with integer coordinates whose intersection graph is isomorphic to the complement of $T + P$.*

Proof. Recall that, after constructing the reference frame, each point in the construction is created either as (i) an intersection between a line through two existing points and a curve α_i, or (ii) by equally spacing $O(n)$ points between two existing points on a curve α_i. Consider that each point is moved by a distance $< \varepsilon$ after it is constructed. This causes any point further constructed from those to move as well. In case (ii), the new points would be moved by no more than ε as well. In case (i) however, the error could be amplified. Let a and b be two previously constructed points, and suppose they are moved to a' and b', within a radius of ε. By the previous lemma, the distance between a and b is at least $n^{-O(n)}$ and so the angle between the line ab and $a'b'$ is at most $\varepsilon n^{O(n)}$. Because the radius of the supporting circle of α_i is $\Theta(n)$, the distance between of $ab \cap \alpha_i$ and $a'b' \cap \alpha_i$ is at most $O(n)\varepsilon n^{O(n)} = \varepsilon n^{O(n)}$.

Therefore, an error in one construction step expands by a factor $n^{O(n)}$. Now observe that each point of type (i) is constructed on the next level, and a point of type (ii) is always constructed from points of type (i). Therefore, as there are $O(n)$ levels, an error is propagated at most $O(n)$ times, and therefore, the total error propagated from moving each constructed point by a distance $< \varepsilon$ is at most $\varepsilon n^{O(n^2)}$.

By the previous lemma, there is a constant c such that any origin in the construction of Lemma 4 is separated at least by a distance $A = 1/n^{cn}$ from any other origin and any ray not incident to it. Therefore, by choosing $\varepsilon = n^{-c'n^3}$ for c' large enough, the total propagation will never exceed A, and therefore perturbing the basic construction points of the reference frame and each further constructed point by a distance $< \varepsilon$ will not change the intersection graph of the modified rays.

Therefore, to construct the required set of rays in polynomial time, we multiply every coordinate by the smallest power of 2 larger than $1/\varepsilon$ and snap every constructed point to the nearest integer while following the construction of Lemma 4. Each coordinate can then be represented by $O(n^3 \log n)$ bits. $\qquad\square$

Let $\alpha(G)$ be the size of the largest independent set in a graph G. The following simple lemma can be deduced from the observation that subdividing an edge twice increases the independence number by exactly one.

Lemma 8. *If G' is an even subdivision of G where each edge $e \in E(G)$ is subdivided $2k_e$ times, then $\alpha(G') = \alpha(G) + \sum_e k_e$*

By combining Lemmas 5, 7, and 8, we obtain:

Theorem 2. *Finding a maximum clique in a ray intersection graph is NP-hard, even when the input is given by a geometric representation as a set of rays.*

References

1. Agarwal, P.K., Mustafa, N.H.: Independent set of intersection graphs of convex objects in 2D. Computational Geometry 34(2), 83–95 (2006)
2. Ambühl, C., Wagner, U.: The clique problem in intersection graphs of ellipses and triangles. Theory Comput. Syst. 38(3), 279–292 (2005)
3. Bang-Jensen, J., Reed, B., Schacht, M., Sámal, R., Toft, B., Wagner, U.: On six problems posed by Jarik Nešetřil. In: Klazar, M., Kratochvíl, J., Loebl, M., Matoušek, J., Thomas, R., Valtr, P. (eds.) Topics in Discrete Mathematics. Algorithms and Combinatorics, vol. 26, pp. 613–627. Springer (2006)
4. Chalopin, J., Gonçalves, D.: Every planar graph is the intersection graph of segments in the plane. In: STOC, pp. 631–638 (2009)
5. Engebretsen, L., Holmerin, J.: Clique Is Hard to Approximate within $n^{1-o(1)}$. In: Welzl, E., Montanari, U., Rolim, J.D.P. (eds.) ICALP 2000. LNCS, vol. 1853, pp. 2–12. Springer, Heidelberg (2000)
6. Fox, J., Pach, J.: Computing the independence number of intersection graphs. In: SODA, pp. 1161–1165 (2011)
7. Francis, M.C., Kratochvíl, J., Vyskocil, T.: Segment representation of a subclass of co-planar graphs. Discrete Mathematics 312(10), 1815–1818 (2012)
8. Garey, M.R., Johnson, D.S.: The rectilinear steiner tree problem is NP-complete. SIAM J. Appl. Math. 32, 826–834 (1977)
9. Kratochvíl, J., Kuběna, A.: On intersection representations of co-planar graphs. Discrete Math. 178(1-3), 251–255 (1998)
10. Kratochvíl, J., Matoušek, J.: Intersection graphs of segments. J. Comb. Theory, Ser. B 62(2), 289–315 (1994)
11. Kratochvíl, J., Nešetřil, J.: Independent set and clique problems in intersection-defined classes of graphs. Commentationes Mathematicae Universitatis Carolinae 31(1), 85–93 (1990)
12. Middendorf, M., Pfeiffer, F.: The max clique problem in classes of string-graphs. Discrete Math. 108(1-3), 365–372 (1992)
13. Håstad, J.: Clique is hard to approximate within $n^{1-\epsilon}$. In: Proceedings of the 37th Annual Symposium on Foundations of Computer Science (FOCS 1996), p. 627 (1996)

Revenue Guarantees in Sponsored Search Auctions*

Ioannis Caragiannis, Christos Kaklamanis,
Panagiotis Kanellopoulos, and Maria Kyropoulou

Computer Technology Institute and Press "Diophantus" &
Department of Computer Engineering and Informatics,
University of Patras, 26504 Rio, Greece

Abstract. Sponsored search auctions are the main source of revenue for search engines. In such an auction, a set of utility-maximizing advertisers compete for a set of ad slots. The assignment of advertisers to slots depends on bids they submit; these bids may be different than the true valuations of the advertisers for the slots. Variants of the celebrated VCG auction mechanism guarantee that advertisers act truthfully and, under mild assumptions, lead to revenue or social welfare maximization. Still, the sponsored search industry mostly uses generalized second price (GSP) auctions; these auctions are known to be non-truthful and suboptimal in terms of social welfare and revenue. In an attempt to explain this tradition, we study a Bayesian setting where the valuations of advertisers are drawn independently from a regular probability distribution. In this setting, it is well known by the work of Myerson (1981) that the optimal revenue is obtained by the VCG mechanism with a particular reserve price that depends on the probability distribution. We show that by appropriately setting the reserve price, the revenue over any Bayes-Nash equilibrium of the game induced by the GSP auction is at most a small constant fraction of the optimal revenue, improving recent results of Lucier, Paes Leme, and Tardos (2012). Our analysis is based on the Bayes-Nash equilibrium conditions and on the properties of regular probability distributions.

1 Introduction

The sale of advertising space is the main source of income for information providers on the Internet. For example, a query to a search engine creates advertising space that is sold to potential advertisers through auctions that are known as *sponsored search auctions* (or ad auctions). In their influential papers, Edelman et al. [6] and Varian [18] have proposed a (now standard) model for this process. According to this model, a set of utility-maximizing advertisers compete for a set of ad slots with non-increasing click-through rates. The auctioneer collects bids from the advertisers and assigns them to slots

---

* This work is co-financed by the European Social Fund and Greek national funds through the research funding program Thales on "Algorithmic Game Theory".

L. Epstein and P. Ferragina (Eds.): ESA 2012, LNCS 7501, pp. 253–264, 2012.

(usually, in non-increasing order of their bids). In addition, it assigns a payment per click to each advertiser. Depending on the way the payments are computed, different auctions can be defined. Typical examples are the Vickrey-Clark-Groves (VCG), the generalized second price (GSP), and the generalized first price (GFP) auction. Naturally, the advertisers are engaged as players in a strategic game defined by the auction; the bid submitted by each player is such that it maximizes her utility (i.e., the total difference of her valuation minus her payment over all clicks) given the bids of the other players. This behavior leads to equilibria, i.e., states of the induced game from which no player has an incentive to unilaterally deviate.

Traditionally, truthfulness has been recognized as an important desideratum in the Economics literature on auctions [11]. In truthful auctions, truth-telling is an equilibrium according to specific equilibrium notions (e.g., dominant strategy, Nash, or Bayes-Nash equilibrium). Such a mechanism guarantees that the social welfare (i.e., the total value of the players) is maximized. VCG is a typical example of a truthful auction [5,8,19]. In contrast, GSP auctions are not truthful [6,18]; still, they are the main auction mechanisms used in the sponsored search industry adopted by leaders such as Google and Yahoo!

In an attempt to explain this prevalence, several papers have provided bounds on the social welfare of GSP auctions [2,12,13,17] over different classes of equilibria (pure Nash, coarse-correlated, Bayes-Nash). The main message from these studies is that the social welfare is always a constant fraction of the optimal one. However, one would expect that revenue (as opposed to social welfare) maximization is the major concern from the point of view of the sponsored search industry. In this paper, following the recent paper by Lucier et al. [14], we aim to provide a theoretical justification for the wide adoption of GSP by focusing on the revenue generated by these auctions.

In order to model the inherent uncertainty in advertisers' beliefs, we consider a Bayesian setting where the advertisers have random valuations drawn independently from a common probability distribution. This is the classical setting that has been studied extensively since the seminal work of Myerson [15] for single-item auctions (which is a special case of ad auctions). The results of [15] carry over to our model as follows. Under mild assumptions, the revenue generated by a player in a Bayes-Nash equilibrium depends only on the distribution of the click-through rate of the ad slot the player is assigned to for her different valuations. Hence, two Bayes-Nash equilibria that correspond to the same allocation yield the same revenue even if they are induced by different auction mechanisms; this statement is known as *revenue equivalence*. The allocation that optimizes the expected revenue is one in which low-bidding advertisers are excluded and the remaining ones are assigned to ad slots in non-increasing order of their valuations. Such an allocation is a Bayes-Nash equilibrium of the variation of the VCG mechanism where an appropriate reserve price (the Myerson reserve) is set in order to exclude the low-bidding advertisers.

GSP auctions may lead to different Bayes-Nash equilibria [7] in which a player with a higher valuation is assigned with positive probability to a slot with lower click-through rate than another player with lower valuation. This implies that the revenue is suboptimal. Our purpose is to quantify the loss of revenue over all Bayes-Nash equilibria of GSP auctions by proving worst-case *revenue guarantees*. A revenue guarantee of ρ for an auction mechanism implies that, at any Bayes-Nash equilibrium, the revenue generated is at most ρ times smaller than the optimal one. Note that, it is not even clear whether Myerson reserve is the choice that minimizes the revenue guarantee in GSP auctions. This issue is the subject of existing experimental work (see [16]).

Recently, Lucier et al. [14] proved theoretical revenue guarantees for GSP auctions. Among other results for full information settings, they consider two different Bayesian models. When the advertisers' valuations are drawn independently from a common probability distribution with monotone hazard rate (MHR), GSP auctions with Myerson reserve have a revenue guarantee of 6. This bound is obtained by comparing the utility of players at the Bayes-Nash equilibrium with the utility they would have by deviating to a single alternative bid (and by exploiting the special properties of MHR distributions). The class of MHR distributions is wide enough and includes many common distributions (such as uniform, normal, and exponential). In the more general case where the valuations are regular, the same bound is obtained using a different reserve price. This reserve is computed using a prophet inequality [10]. Prophet inequalities have been proved useful in several Bayesian auction settings in the past [4,9].

In this work, we consider the same Bayesian settings with [14] and improve their results. We show that when the players have i.i.d. valuations drawn from a regular distribution, there is a reserve price so that the revenue guarantee is at most 4.72. For MHR valuations, we present a bound of 3.46. In both cases, the reserve price is either Myerson's or another one that maximizes the revenue obtained by the player allocated to the first slot. The latter is computed by developing new prophet-like inequalities that exploit the particular characteristics of the valuations. Furthermore, we show that the revenue guarantee of GSP auctions with Myerson reserve is at most 3.90 for MHR valuations. In order to analyze GSP auctions with Myerson reserve, we extend the techniques recently developed in [2,13] (see also [3]). The Bayes-Nash equilibrium condition implies that the utility of each player does not improve when she deviates to any other bid. This yields a series of inequalities which we take into account with different weights. These weights are given by families of functions that are defined in such a way that a relation between the revenue at a Bayes-Nash equilibrium and the optimal revenue is revealed; we refer to them as *deviation weight function families*.

The rest of the paper is structured as follows. We begin with preliminary definitions in Section 2. Our prophet-type bounds are presented in Section 3. The role of deviation weight function families in the analysis is explored in Section 4. Then, Section 5 is devoted to the proofs of our main statements for GSP auctions. We conclude with open problems in Section 6. Due to lack of space several proofs have been omitted.

2   Preliminaries

We consider a Bayesian setting with n players and n slots[1] where slot $j \in [n]$ has a click-through rate α_j that corresponds to the frequency of clicking an ad in slot j. We add an artificial $(n+1)$-th slot with click-through rate 0 and index the slots so that $\alpha_1 \geq \alpha_2 \geq \cdots \geq \alpha_n \geq \alpha_{n+1} = 0$. Each player's valuation (per click) is non-negative and is drawn from a publicly known probability distribution.

The auction mechanisms we consider use a reserve price t and assign slots to players according to the bids they submit. Player i submits a bid $b_i(v_i)$ that depends on her valuation v_i; the bidding function b_i is the strategy of player i. Given a realization of valuations, let $\mathbf{b} = (b_1, \ldots, b_n)$ denote a bid vector and define the random permutation π so that $\pi(j)$ is the player with the j-th highest bid (breaking ties arbitrarily). The mechanism assigns slot j to player $\pi(j)$ whenever $b_{\pi(j)} \geq t$; if $b_{\pi(j)} < t$, the player is not allocated any slot. In such an allocation, let $\sigma(i)$ denote the slot that is allocated to player i. This is well-defined when player i is assigned a slot; if this is not the case, we follow the convention that $\sigma(i) = n + 1$. Given $\mathbf{b}$, the mechanism also defines a payment $p_i \geq t$ for each player i that is allocated a slot. Then, the *utility* of player i is $u_i(\mathbf{b}) = \alpha_{\sigma(i)}(v_i - p_i)$. A set of players' strategies is a Bayes-Nash equilibrium if no player has an incentive to deviate from her strategy in order to increase her expected utility. This means that for every player i and every possible valuation x, $\mathbb{E}[u_i(\mathbf{b})|v_i = x] \geq \mathbb{E}[u_i(b_i', \mathbf{b}_{-i})|v_i = x]$ for every alternative bid b_i'. Note that the expectation is taken over the randomness of the valuations of the other players and the notation $(b_i', \mathbf{b}_{-i})$ is used for the bid vector where player i has deviated to b_i' and the remaining players bid as in $\mathbf{b}$. The *social welfare* at a Bayes-Nash equilibrium $\mathbf{b}$ is $\mathcal{W}_t(\mathbf{b}) = \mathbb{E}[\sum_i \alpha_{\sigma(i)} v_i]$, while the revenue generated by the mechanism is $\mathcal{R}_t(\mathbf{b}) = \mathbb{E}[\sum_i \alpha_{\sigma(i)} p_i]$.

We focus on the case where the valuations of players are drawn independently from a common probability distribution $\mathcal{D}$ with probability density function f and cumulative distribution function F. Given a distribution $\mathcal{D}$ over players' valuations, the *virtual valuation* function is $\phi(x) = x - \frac{1-F(x)}{f(x)}$. We consider *regular* probability distributions where $\phi(x)$ is non-decreasing. The work of Myerson [15] implies that the expected revenue from player i at a Bayes-Nash equilibrium $\mathbf{b}$ of any auction mechanism is $\mathbb{E}[\alpha_{\sigma(i)} \phi(v_i)]$, i.e., it depends only on the allocation of player i and her virtual valuation. Hence, the total expected revenue is maximized when the players with non-negative virtual valuations are assigned to slots in non-increasing order of their virtual valuations and players with negative virtual valuations are not assigned any slot. A mechanism that imposes this allocation as a Bayes-Nash equilibrium (and, hence, is revenue-maximizing) is the celebrated VCG mechanism with reserve price t such that $\phi(t) = 0$. We refer to this as Myerson reserve and denote it by r in the following. We use the notation μ to denote such an allocation. Note that, in μ, players with zero virtual

---

[1] Our model can simulate cases where the number of slots is smaller than the number of players by adding fictitious slots with zero click-through rate.

valuation can be either allocated slots or not; such players do not contribute to the optimal revenue.

A particular subclass of regular probability distributions are those with *monotone hazard rate* (MHR). A regular distribution $\mathcal{D}$ is MHR if its hazard rate function $h(x) = f(x)/(1 - F(x))$ is non-decreasing. These distributions have some nice properties (see [1]). For example, $F(r) \leq 1 - 1/e$ and $\phi(x) \geq x - r$ for every $x \geq r$.

In this paper, we focus on the GSP mechanism. For each player i that is allocated a slot (i.e., with bid at least t), GSP computes her payment as the maximum between the reserve price t and the next highest bid $b_{\pi(i+1)}$ (assuming that $b_{\pi(n+1)} = 0$). As it has been observed in [7], GSP may not admit the allocation μ as a Bayes-Nash equilibrium. This immediately implies that the revenue over Bayes-Nash equilibria would be suboptimal. In order to capture the revenue loss due to the selfish behavior of the players, we use the notion of *revenue guarantee*.

Definition 1. *The revenue guarantee of an auction game with reserve price t is* $\max_{\mathbf{b}} \frac{\mathcal{R}_{OPT}}{\mathcal{R}_t(\mathbf{b})}$, *where* $\mathbf{b}$ *runs over all Bayes-Nash equilibria of the game.*

In our proofs, we use the notation σ to refer to the random allocation that corresponds to a Bayes-Nash equilibrium. Note that, a player with valuation strictly higher than the reserve has always an incentive to bid at least the reserve and be allocated a slot. When her valuation equals the reserve, she is indifferent between bidding the reserve or not participating in the auction. For auctions with Myerson reserve, when comparing a Bayes-Nash equilibrium to the revenue-maximizing allocation μ, we assume that a player with valuation equal to the reserve has the same behavior in both σ and μ (this implies that $\mathbb{E}[\sum_i \alpha_{\sigma(i)}] = \mathbb{E}[\sum_i \alpha_{\mu(i)}]$). This assumption is without loss of generality since such a player contributes zero to the optimal revenue anyway. In our proofs, we also use the random variable $o(j)$ to denote the player with the j-th highest valuation (breaking ties arbitrarily). Hence, $\mu(i) = o^{-1}(i)$ if the virtual valuation of player i is positive and $\mu(i) = n + 1$ if it is negative. When the virtual valuation of player i is zero, it can be either $\mu(i) = o^{-1}(i)$ or $\mu(i) = n + 1$.

When considering GSP auctions, we make the assumption that players are *conservative*: whenever the valuation of player i is v_i, she only selects a bid $b_i(v_i) \in [0, v_i]$ at Bayes-Nash equilibria. This is a rather natural assumption since any bid $b_i(v_i) > v_i$ is weakly dominated by bidding $b_i(v_i) = v_i$ [17].

In the following, we use the notation x^+ to denote $\max\{x, 0\}$ while the expression $x\mathbb{1}\{E\}$ equals x when the event E is true and 0 otherwise.

3   Achieving Minimum Revenue Guarantees

Our purpose in this section is to show that by appropriately setting the reserve price, we can guarantee a high revenue from the advertiser that occupies the first slot at any Bayes-Nash equilibrium. Even though this approach will not give us a "standalone" result, it will be very useful later when we will combine it with the analysis of GSP auctions with Myerson reserve. These bounds are similar in spirit to prophet inequalities in optimal stopping theory [10].

We begin with a simple lemma.

Lemma 1. *Consider n random valuations v_1, ..., v_n that are drawn i.i.d. from a regular distribution $\mathcal{D}$. Then, for every $t \geq r$, it holds that*

$$\mathbb{E}[\max_i \phi(v_i)^+] \leq \phi(t) + \frac{n(1 - F(t))^2}{f(t)}.$$

We can use Lemma 1 in order to bound the revenue in the case of regular valuations.

Lemma 2. *Let $\mathbf{b}$ be a Bayes-Nash equilibrium for a GSP auction game with n players with random valuations v_1, ..., v_n drawn i.i.d. from a regular distribution $\mathcal{D}$. Then, there exists $r' \geq r$ such that $\mathcal{R}_{r'}(\mathbf{b}) \geq (1 - 1/e)\alpha_1 \mathbb{E}[\max_i \phi(v_i)^+]$.*

For MHR valuations, we show an improved bound.

Lemma 3. *Let $\mathbf{b}$ be a Bayes-Nash equilibrium for a GSP auction game with n players with random valuations v_1, ..., v_n drawn i.i.d. from an MHR distribution $\mathcal{D}$. Then, there exists $r' \geq r$ such that $\mathcal{R}_{r'}(\mathbf{b}) \geq (1 - e^{-2})\alpha_1 \mathbb{E}[\max_i \phi(v_i)^+] - (1 - e^{-2})\alpha_1 r(1 - F^n(r))$.*

Proof. We will assume that $\mathbb{E}[\max_i \phi(v_i)^+] \geq r(1 - F^n(r))$ since the lemma holds trivially otherwise. Let t^* be such that $F(t^*) = 1 - \eta/n$ where $\eta = 2 - (1 - 1/e)^n$. We will distinguish between two cases depending on whether $t^* \geq r$ or not.

We first consider the case $t^* \geq r$. We will use the definition of the virtual valuation, the fact that the hazard rate function satisfies $h(t^*) \geq h(r) = 1/r$, the definition of t^*, Lemma 1 (with $t = t^*$), and the fact that $F(r) \leq 1 - 1/e$ which implies that $1 - F^n(r) \geq \eta - 1$. We have

$$t^*(1 - F^n(t^*))$$
$$= \phi(t^*)(1 - F^n(t^*)) + \frac{1}{h(t^*)}(1 - F^n(t^*))$$
$$= \phi(t^*)(1 - F^n(t^*)) + \frac{\eta}{h(t^*)}(1 - F^n(t^*)) - \frac{\eta - 1}{h(t^*)}(1 - F^n(t^*))$$
$$\geq \phi(t^*)(1 - F^n(t^*)) + \frac{n(1 - F(t^*))^2}{f(t^*)} \cdot \frac{\eta(1 - F^n(t^*))}{n(1 - F(t^*))} - (\eta - 1)r(1 - F^n(t^*))$$
$$= (1 - F^n(t^*))\left(\phi(t^*) + \frac{n(1 - F(t^*))^2}{f(t^*)} - (\eta - 1)r\right)$$
$$\geq \left(1 - \left(1 - \frac{2 - (1 - 1/e)^n}{n}\right)^n\right)\left(\mathbb{E}[\max_i \phi(v_i)^+] - r(1 - F^n(r))\right).$$

Note that the left side of the above equality multiplied with α_1 is a lower bound on the revenue of GSP with reserve t^*. Also, $\left(1 - \frac{2-(1-1/e)^n}{n}\right)^n$ is non-decreasing in n and approaches e^{-2} from below as n tends to infinity. Furthermore, the right-hand side of the above inequality in non-negative. Hence,

$$\mathcal{R}_{t^*}(\mathbf{b}) \geq (1 - e^{-2})\alpha_1 \mathbb{E}[\max_i \phi(v_i)^+] - (1 - e^{-2})\alpha_1 r(1 - F^n(r))$$

as desired.

We now consider the case $t^* < r$. We have $1 - \eta/n = F(t^*) \leq F(r) \leq 1 - 1/e$ which implies that $n \leq 5$. Tedious calculations yield

$$\frac{1 - F^n(r)}{n(1 - F(r))} = \frac{1 + F(r) + \ldots + F^{n-1}(r)}{n} \geq \frac{1 - e^{-2}}{2 - e^{-2}}$$

for $n \in \{2, 3, 4, 5\}$ since $F(r) \geq 1 - \eta/n$. Hence,

$$\begin{aligned}
\mathcal{R}_r(\mathbf{b}) &\geq \alpha_1 r(1 - F^n(r)) \\
&\geq (1 - e^{-2})\alpha_1 n r(1 - F(r)) - (1 - e^{-2})\alpha_1 r(1 - F^n(r)) \\
&\geq (1 - e^{-2})\alpha_1 \mathbb{E}[\max_i \phi(v_i)^+] - (1 - e^{-2})\alpha_1 r(1 - F^n(r)),
\end{aligned}$$

where the last inequality follows by applying Lemma 1 with $t = r$.   □

4   Deviation Weight Function Families

The main idea we use for the analysis of Bayes-Nash equilibria of auction games with reserve price t is that the utility of player i with valuation $v_i = x \geq t$ does not increase when this player deviates to any other bid in $[t, x]$. This provides us with infinitely many inequalities on the utility of player i that are expressed in terms of her valuation, the bids of the other players, and the reserve price. Our technique combines these infinite lower bounds by considering their weighted average. The specific weights with which we consider the different inequalities are given by families of functions with particular properties that we call deviation weight function families.

Definition 2. *Let* $\beta, \gamma, \delta \geq 0$ *and consider the family of functions* $\mathcal{G} = \{g_\xi : \xi \in [0, 1)\}$ *where* g_ξ *is a non-negative function defined in* $[\xi, 1]$. $\mathcal{G}$ *is a* (β, γ, δ)-*DWFF (deviation weight function family) if the following two properties hold for every* $\xi \in [0, 1)$:

$$i) \quad \int_\xi^1 g_\xi(y)\, dy = 1,$$

$$ii) \quad \int_z^1 (1 - y)g_\xi(y)\, dy \geq \beta - \gamma z + \delta\xi, \quad \forall z \in [\xi, 1].$$

The next lemma is used in order to prove most of our bounds together with the deviation weight function family presented in Lemma 5.

Lemma 4. *Consider a Bayes-Nash equilibrium* **b** *for a GSP auction game with* n *players and reserve price* t. *Then, the following two inequalities hold for every player* i.

$$\mathbb{E}[u_i(\mathbf{b})] \geq \sum_{j=c}^{n} \mathbb{E}[\alpha_j(\beta v_i - \gamma b_{\pi(j)} + \delta t)\mathbb{1}\{\mu(i) = j\}], \tag{1}$$

$$\mathbb{E}[\alpha_{\sigma(i)}\phi(v_i)] \geq \sum_{j=c}^{n} \mathbb{E}[\alpha_j(\beta\phi(v_i) - \gamma b_{\pi(j)})\mathbb{1}\{\mu(i) = j\}], \tag{2}$$

where c *is any integer in* $[n]$, β, γ, *and* δ *are such that a* (β, γ, δ)-*DWFF exists, and* μ *is any revenue-maximizing allocation.*

Lemma 5. *Consider the family of functions* $\mathcal{G}_1$ *consisting of the functions* g_ξ : $[\xi, 1] \to \mathbb{R}_+$ *defined as follows for every* $\xi \in [0, 1)$:

$$g_\xi(y) = \begin{cases} \frac{\kappa}{1-y}, & y \in [\xi, \xi + (1-\xi)\lambda), \\ 0, & \text{otherwise,} \end{cases}$$

where $\lambda \in (0, 1)$ *and* $\kappa = -\frac{1}{\ln(1-\lambda)}$. *Then,* $\mathcal{G}_1$ *is a* $(\kappa\lambda, \kappa, \kappa(1-\lambda))$-*DWFF.*

We remark that the bound for GSP auctions with Myerson reserve (and players with MHR valuations) follows by a slightly more involved deviation weight function family. Due to lack of space, we omit it from this extended abstract; it will appear in the final version of the paper.

5   Revenue Guarantees in GSP Auctions

We will now exploit the techniques developed in the previous sections in order to prove our bounds for GSP auctions. Throughout this section, we denote by O_j the event that slot j is occupied in the revenue-maximizing allocation considered. The next lemma provides a lower bound on the revenue of GSP auctions.

Lemma 6. *Consider a Bayes-Nash equilibrium* **b** *for a GSP auction game with Myerson reserve price* r *and* n *players. It holds that*

$$\sum_{j \geq 2} \mathbb{E}[\alpha_j b_{\pi(j)} \mathbb{1}\{O_j\}] \leq \mathcal{R}_r(\mathbf{b}) - \alpha_1 r \cdot \Pr[O_1].$$

Proof. Consider a Bayes-Nash equilibrium **b** for a GSP auction game with Myerson reserve price r. Define $\Pr[O_{n+1}] = 0$. Consider some player whose valuation exceeds r and is thus allocated some slot. Note that the player's payment per click is determined by the bid of the player allocated just below her, if there is one, otherwise, the player's (per click) payment is set to r. It holds that

$$\mathcal{R}_r(\mathbf{b}) = \sum_{j} \alpha_j r(\Pr[O_j] - \Pr[O_{j+1}]) + \sum_{j} \mathbb{E}[\alpha_j b_{\pi(j+1)} \mathbb{1}\{O_{j+1}\}]$$

$$= \sum_{j\geq2} \mathbb{E}[\alpha_j b_{\pi(j)} \mathbb{1}\{O_j\}] + \sum_j \alpha_j r(\Pr[O_j] - \Pr[O_{j+1}])$$

$$+ \sum_j \mathbb{E}[(\alpha_j - \alpha_{j+1}) b_{\pi(j+1)} \mathbb{1}\{O_{j+1}\}]$$

$$\geq \sum_{j\geq2} \mathbb{E}[\alpha_j b_{\pi(j)} \mathbb{1}\{O_j\}] + \sum_j \alpha_j r(\Pr[O_j] - \Pr[O_{j+1}])$$

$$+ \sum_j (\alpha_j - \alpha_{j+1}) r \cdot \Pr[O_{j+1}]$$

$$= \sum_{j\geq2} \mathbb{E}[\alpha_j b_{\pi(j)} \mathbb{1}\{O_j\}] + \sum_j \alpha_j r \Pr[O_j] - \sum_j \alpha_{j+1} r \cdot \Pr[O_{j+1}]$$

$$= \sum_{j\geq2} \mathbb{E}[\alpha_j b_{\pi(j)} \mathbb{1}\{O_j\}] + \alpha_1 r \cdot \Pr[O_1].$$

The proof follows by rearranging the terms in the last inequality.   □

The next statement follows by Lemmas 2 and 4 using the DWFF defined in Lemma 5.

Theorem 1. *Consider a regular distribution $\mathcal{D}$. There exists some r^*, such that the revenue guarantee over Bayes-Nash equilibria of GSP auction games with reserve price r^* is 4.72, when valuations are drawn i.i.d. from $\mathcal{D}$.*

Proof. By Lemma 2, we have that there exists $r' \geq r$ such that the expected revenue over any Bayes-Nash equilibrium $\mathbf{b}'$ of the GSP auction game with reserve price r' satisfies

$$\mathcal{R}_{r'}(\mathbf{b}') \geq (1 - 1/e)\mathbb{E}[\alpha_1 \phi(v_{o(1)})^+]. \tag{3}$$

Now, let $\mathbf{b}''$ be any Bayes-Nash equilibrium of the GSP auction game with Myerson reserve and let β, γ, and δ be parameters so that a (β, γ, δ)-DWFF exists. Using inequality (2) from Lemma 4 with $c = 2$ and Lemma 6 we obtain

$$\mathcal{R}_r(\mathbf{b}'') = \sum_i \mathbb{E}[\alpha_{\sigma(i)} \phi(v_i)]$$

$$\geq \sum_i \sum_{j\geq2} \mathbb{E}[\alpha_j (\beta\phi(v_i) - \gamma b_{\pi(j)}) \mathbb{1}\{\mu(i) = j\}]$$

$$= \beta \sum_{j\geq2} \mathbb{E}[\alpha_j \phi(v_{o(j)})^+] - \gamma \sum_{j\geq2} \mathbb{E}[\alpha_j b_{\pi(j)} \mathbb{1}\{O_j\}]$$

$$\geq \beta \sum_{j\geq2} \mathbb{E}[\alpha_j \phi(v_{o(j)})^+] - \gamma \mathcal{R}_r(\mathbf{b}'').$$

In other words,

$$(1 + \gamma)\mathcal{R}_r(\mathbf{b}'') \geq \beta \sum_{j\geq2} \mathbb{E}[\alpha_j \phi(v_{o(j)})^+].$$

Using this last inequality together with inequality (3), we obtain

$$\left(1 + \gamma + \frac{e\beta}{e-1}\right) \max\{\mathcal{R}_r(\mathbf{b}''), \mathcal{R}_{r'}(\mathbf{b}')\} \geq (1+\gamma)\mathcal{R}_r(\mathbf{b}'') + \frac{e\beta}{e-1}\mathcal{R}_{r'}(\mathbf{b}')$$

$$\geq \beta \sum_j \mathbb{E}[\alpha_j \phi(v_{o(j)})^+]$$

$$= \beta \mathcal{R}_{OPT}.$$

We conclude that there exists some reserve price r^* (either r or r') such that for any Bayes-Nash equilibrium $\mathbf{b}$ it holds that

$$\frac{\mathcal{R}_{OPT}}{\mathcal{R}_{r^*}(\mathbf{b})} \leq \frac{1+\gamma}{\beta} + \frac{e}{e-1}.$$

By Lemma 5, the family $\mathcal{G}_1$ is a $(\beta, \gamma, 0)$-DWFF with $\beta = \kappa\lambda$ and $\gamma = \kappa$, where $\lambda \in (0,1)$ and $\kappa = -\frac{1}{\ln(1-\lambda)}$. By substituting β and γ with these values and using $\lambda \approx 0.682$, the right-hand side of our last inequality is upper-bounded by 4.72.   □

The next statement applies to MHR valuations. It follows by Lemmas 3 and 4 using the DWFF defined in Lemma 5.

Theorem 2. *Consider an MHR distribution $\mathcal{D}$. There exists some r^*, such that the revenue guarantee over Bayes-Nash equilibria of GSP auction games with reserve price r^* is 3.46, when valuations are drawn i.i.d. from $\mathcal{D}$.*

Proof. Let $\mathbf{b}'$ be any Bayes-Nash equilibrium of the GSP auction game with Myerson reserve and let β, γ, and δ be parameters so that a (β, γ, δ)-DWFF exists. Since $\mathcal{D}$ is an MHR probability distribution, we have

$$\mathbb{E}[\alpha_{\sigma(i)}r] \geq \mathbb{E}[\alpha_{\sigma(i)}(v_i - \phi(v_i))] = \mathbb{E}[u_i(\mathbf{b}')]$$

for every player i. By summing over all players and using inequality (1) from Lemma 4 with $c = 2$, we obtain

$$\sum_i \mathbb{E}[\alpha_{\sigma(i)}r] \geq \sum_i \mathbb{E}[u_i(\mathbf{b}')]$$

$$\geq \sum_i \sum_{j=2}^n \mathbb{E}[\alpha_j(\beta v_i - \gamma b_{\pi(j)} + \delta r)\mathbb{1}\{\mu(i) = j\}]$$

$$\geq \sum_{j\geq 2} \mathbb{E}[\alpha_j(\beta\phi(v_{o(j)})^+ - \gamma b_{\pi(j)} + \delta r)\mathbb{1}\{O_j\}]$$

$$\geq \beta \sum_{j\geq 2} \mathbb{E}[\alpha_j\phi(v_{o(j)})^+] - \gamma \sum_{j\geq 2} \mathbb{E}[\alpha_j b_{\pi(j)}\mathbb{1}\{O_j\}] + \delta \sum_{j\geq 2} \mathbb{E}[\alpha_j r\mathbb{1}\{O_j\}]$$

$$\geq \beta \sum_{j\geq 2} \mathbb{E}[\alpha_j\phi(v_{o(j)})^+] - \gamma\mathcal{R}_r(\mathbf{b}') + (\gamma - \delta)\mathbb{E}[\alpha_1 r\mathbb{1}\{O_1\}] + \delta \sum_j \mathbb{E}[\alpha_j r\mathbb{1}\{O_j\}]$$

$$= \beta \sum_{j\geq 2} \mathbb{E}[\alpha_j\phi(v_{o(j)})^+] - \gamma\mathcal{R}_r(\mathbf{b}') + (\gamma - \delta)\mathbb{E}[\alpha_1 r\mathbb{1}\{O_1\}] + \delta \sum_i \mathbb{E}[\alpha_{\mu(i)}r].$$

The last inequality follows by Lemma 6. Since $\sum_i \mathbb{E}[\alpha_{\mu(i)} r] = \sum_i \mathbb{E}[\alpha_{\sigma(i)} r]$, we obtain that

$$\gamma \mathcal{R}_r(\mathbf{b}') \geq \beta \sum_{j \geq 2} \mathbb{E}[\alpha_j \phi(v_{o(j)})^+] + (\gamma - \delta)\alpha_1 r \cdot \Pr[O_1] + (\delta - 1) \sum_i \mathbb{E}[\alpha_{\sigma(i)} r].$$

$$(4)$$

By Lemma 3, we have that there exists $r' \geq r$ such that the expected revenue over any Bayes-Nash equilibrium $\mathbf{b}''$ of the GSP auction game with reserve price r' satisfies

$$\mathcal{R}_{r'}(\mathbf{b}'') \geq (1 - e^{-2})\mathbb{E}[\alpha_1 \phi(v_{o(1)})^+] - (1 - e^{-2})\mathbb{E}[\alpha_1 r \mathbb{1}\{O_1\}].$$

Using this last inequality together with inequality (4), we obtain

$$\left(\gamma + \frac{e^2 \beta}{e^2 - 1} \right) \max\{\mathcal{R}_r(\mathbf{b}'), \mathcal{R}_{r'}(\mathbf{b}'')\}$$

$$\geq \gamma \mathcal{R}_r(\mathbf{b}') + \frac{e^2 \beta}{e^2 - 1} \mathcal{R}_{r'}(\mathbf{b}'')$$

$$\geq \beta \sum_j \mathbb{E}[\alpha_j \phi(v_{o(j)})^+] + (\gamma - \delta - \beta)\mathbb{E}[\alpha_1 r \mathbb{1}\{O_1\}] + (\delta - 1) \sum_i \mathbb{E}[\alpha_{\sigma(i)} r]$$

$$\geq \beta \mathcal{R}_{OPT} + (\gamma - \delta - \beta)\mathbb{E}[\alpha_1 r \mathbb{1}\{O_1\}] + (\delta - 1) \sum_i \mathbb{E}[\alpha_{\sigma(i)} r].$$

By Lemma 5, the family $\mathcal{G}_1$ is a (β, γ, δ)-DWFF with $\beta = \gamma - \delta = \kappa\lambda$, $\gamma = \kappa$, and $\delta = \kappa(1 - \lambda)$, where $\lambda \in (0, 1)$ and $\kappa = -\frac{1}{\ln(1-\lambda)}$. By setting $\lambda \approx 0.432$ so that $\delta = \kappa(1 - \lambda) = 1$, the above inequality implies that there exists some reserve price r^* (either r or r') such that for any Bayes-Nash equilibrium $\mathbf{b}$ of the corresponding GSP auction game, it holds that

$$\frac{\mathcal{R}_{OPT}}{\mathcal{R}_{r^*}(\mathbf{b})} \leq \frac{1}{\lambda} + \frac{e^2}{e^2 - 1} \approx 3.46,$$

as desired.   □

For GSP auctions with Myerson reserve, our revenue bound follows using a slightly more involved deviation weight function family.

Theorem 3. *Consider an MHR distribution $\mathcal{D}$. The revenue guarantee over Bayes-Nash equilibria of GSP auction games with Myerson reserve price r is 3.90, when valuations are drawn i.i.d. from $\mathcal{D}$.*

6   Conclusions

Even though we have significantly improved the results of [14], we conjecture that our revenue guarantees could be further improved. The work of Gomes and Sweeney [7] implies that the revenue guarantee of GSP auctions with Myerson reserve is in general higher than 1; however, no explicit lower bound is known. Due to the difficulty in computing Bayes-Nash equilibria analytically, coming up with a concrete lower bound construction is interesting and would reveal the gap of our revenue guarantees.

References

1. Barlow, R., Marshall, R.: Bounds for distributions with monotone hazard rate. Annals of Mathematicals Statistics 35(3), 1234–1257 (1964)
2. Caragiannis, I., Kaklamanis, C., Kanellopoulos, P., Kyropoulou, M.: On the efficiency of equilibria in generalized second price auctions. In: Proceedings of the 12th ACM Conference on Electronic Commerce (EC), pp. 81–90 (2011)
3. Caragiannis, I., Kaklamanis, C., Kanellopoulos, P., Kyropoulou, M., Lucier, B., Paes Leme, R., Tardos, É.: On the efficiency of equilibria in generalized second price auctions. arXiv:1201.6429 (2012)
4. Chawla, S., Hartline, J., Malec, D., Sivan, B.: Multi-parameter mechanism design and sequential posted pricing. In: Proceedings of the 41th ACM Symposium on Theory of Computing (STOC), pp. 311–320 (2010)
5. Clarke, E.H.: Multipart pricing of public goods. Public Choice 11, 17–33 (1971)
6. Edelman, B., Ostrovsky, M., Schwarz, M.: Internet advertizing and the generalized second-price auction: selling billions of dollars worth of keywords. The American Economic Review 97(1), 242–259 (2007)
7. Gomes, R., Sweeney, K.: Bayes-Nash equilibria of the generalized second price auction. Working paper, 2011. Preliminary version in Proceedings of the 10th ACM Conference on Electronic Commerce (EC), pp. 107–108 (2009)
8. Groves, T.: Incentives in teams. Econometrica 41(4), 617–631 (1973)
9. Hajiaghayi, M., Kleinberg, R., Sandholm, T.: Automated mechanism design and prophet inequalities. In: Proceedings of the 22nd AAAI Conference on Artificial Intelligence (AAAI), pp. 58–65 (2007)
10. Krengel, U., Sucheston, L.: Semiamarts and finite values. Bulletin of the American Mathematical Society 83(4), 745–747 (1977)
11. Krishna, V.: Auction Theory. Academic Press (2002)
12. Lahaie, S.: An analysis of alternative slot auction designs for sponsored search. In: Proceedings of the 7th ACM Conference on Electronic Commerce (EC), pp. 218–227 (2006)
13. Lucier, B., Paes Leme, R.: GSP auctions with correlated types. In: Proceedings of the 12th ACM Conference on Electronic Commerce (EC), pp. 71–80 (2011)
14. Lucier, B., Paes Leme, R., Tardos, É.: On revenue in generalized second price auctions. In: Proceedings of the 21st World Wide Web Conference (WWW), pp. 361–370 (2012)
15. Myerson, R.: Optimal auction design. Mathematics of Operations Research 6(1), 58–73 (1981)
16. Ostrovsky, M., Schwarz, M.: Reserve prices in Internet advertising auctions: a field experiment. In: Proceedings of the 12th ACM Conference on Electronic Commerce (EC), pp. 59–60 (2011)
17. Paes Leme, R., Tardos, É.: Pure and Bayes-Nash price of anarchy for generalized second price auction. In: Proceedings of the 51st Annual IEEE Symposium on Foundations of Computer Science (FOCS), pp. 735–744 (2010)
18. Varian, H.: Position auctions. International Journal of Industrial Organization 25, 1163–1178 (2007)
19. Vickrey, W.: Counterspeculation, auctions, and competitive sealed tenders. The Journal of Finance 16(1), 8–37 (1961)

Optimizing Social Welfare for Network Bargaining Games in the Face of Unstability, Greed and Spite

T.-H. Hubert Chan, Fei Chen, and Li Ning

Department of Computer Science, The University of Hong Kong
{hubert,fchen,lning}@cs.hku.hk

Abstract. Stable and balanced outcomes of network bargaining games have been investigated recently, but the existence of such outcomes requires that the linear program relaxation of a certain maximum matching problem has integral optimal solution.

We propose an alternative model for network bargaining games in which each edge acts as a player, who proposes how to split the weight of the edge among the two incident nodes. Based on the proposals made by all edges, a selection process will return a set of accepted proposals, subject to node capacities. An edge receives a commission if its proposal is accepted. The social welfare can be measured by the weight of the matching returned.

The node users, as opposed to being rational players as in previous works, exhibit two characteristics of human nature: greed and spite. We define these notions formally and show that the distributed protocol by Kanoria et. al can be modified to be run by the edge players such that the configuration of proposals will converge to a pure Nash Equilibrium, without the LP integrality gap assumption. Moreover, after the nodes have made their greedy and spiteful choices, the remaining ambiguous choices can be resolved in a way such that there exists a Nash Equilibrium that will not hurt the social welfare too much.

1 Introduction

Bargaining games have been studied with a long history, early in economics [19] and sociology, and recently in computer science, there has been a lot of attention on bargaining games in social exchange networks [16,1,4,15], in which users are modeled as nodes in an undirected graph $G = (V, E)$, whose edges are weighted. An edge $\{i, j\} \in E$ with weight $w_{ij} > 0$ means that users i and j can potentially form a contract with each other and split a profit of w_{ij}. A capacity vector $b \in \mathbb{Z}_+^V$ limits the maximum number b_i of contracts node i can form with its neighbors, and the set M of executed contracts form a b-matching in G.

In previous works, the nodes bargain with one another to form an *outcome* which consists of the set M of executed contracts and how the profit in each contract is distributed among the two participating nodes. The outside option of a node is the maximum profit the node can get from another node with whom there

L. Epstein and P. Ferragina (Eds.): ESA 2012, LNCS 7501, pp. 265–276, 2012.

is no current contract. An outcome is *stable* if for every contract a node makes, the profit the node gets from that contract is at least its outside option. Hence, under a stable outcome, no node has motivation to break its current contract to form another one. Extending the notion of Nash bargaining solution [19], Cook and Yamagishi [11] introduced the notion of balanced outcome. An outcome is balanced if, in addition to stability, for every contract made, after each participating node gets its outside option, the surplus is divided equally between the two nodes involved. For more notions of solutions, the reader can refer to [7].

Although stability is considered to be an essential property, as remarked in [4,15], a stable outcome exists *iff* the linear program (LP) relaxation (given in Section 4) for the b-matching problem on the given graph G has integrality gap 1. Hence, even for very simple graphs like a triangle with unit node capacities and unit edge weights, there does not exist a stable outcome. Previous works simply assumed that the LP has integrality gap 1 [15,8] or considered restriction to bipartite graphs [16,4], for which the LP always has integrality gap 1.

We think the integrality gap condition is a limitation to the applicability of such framework in practice. We would like to consider an alternative model for network bargaining games, and investigate different notions of equilibrium, whose existence does not require the integrality gap condition.

Our Contribution and Results. In this work, we let the edges take over the role of the "rational" players from the nodes. Each edge $e = \{i, j\} \in E$ corresponds to an agent, who proposes a way to divide up the potential profit w_{ij} among the two nodes. Formally, each edge $\{i, j\}$ has the action set $A_{ij} := \{(x, y) : x \geq 0, y \geq 0, x+y \leq w_e\}$, where a proposal (x, y) means that node i gets amount x and j gets amount y.[1] Based on the configuration $m \in A_E := \times_{e \in E} A_e$ of proposals made by all the agents, a selection process (which can be randomized) will choose a b-matching M, which is the set of contracts formed. An agent e will receive a commission if his proposal is selected; his payoff $u_e(m)$ is the probability that edge e is in the matching M returned.[2] Observe that once the payoff function u is defined, the notion of (pure or mixed) Nash Equilibrium is also well-defined. We measure the social welfare $\mathcal{S}(m)$ by the (expected) weight $w(M)$ of the matching M returned, which reflects the volume of transaction.

We have yet to describe the selection process, which will determine the payoff function to each agent, and hence will affect the corresponding Nash Equilibrium. We mention earlier that the rational players in our framework will be the edges, as opposed to the nodes in previous works; in fact, in the selection process we assume the node users will exhibit two characteristics of human nature: greed and spite.

Greedy Users. For a node i with capacity b_i, user i will definitely want an offer that is strictly better than his $(b_i + 1)$-st best offer. If this happens for both users forming an edge, then the edge will definitely be selected. We also say the resulting payoff function is greedy.

---

[1] In case $x + y < w_{ij}$ the remaining amount is lost and not gained by anyone.

[2] The actual gain of an agent could be scaled according to the weight w_e, but this will not affect the Nash Equilibrium.

Spiteful Users. Spite is an emotion that describes the situation that once a person has seen a better offer, he would not settle for anything less, even if the original better offer is no longer available. If user i with capacity b_i sees that an offer is strictly worse than his b_i-th best offer, then the corresponding edge will definitely be rejected. We also say the resulting payoff function is spiteful.

One can argue that greed is a rational behavior (hence the regime of greedy algorithms), but spite is clearly not always rational. In fact, we shall see in Section 2 that there exist a spiteful payoff function and a configuration of agent proposals that is a pure Nash Equilibrium, in which all proposals are rejected by the users out of spite, even though no single agent can change the situation by unilaterally offering a different proposal. The important question is that: can the agents follow some protocol that can avoid such bad Nash Equilibrium? In other words, can they collaboratively find a Nash Equilibrium that achieves good social welfare?

We answer the above question in the affirmative. We modify the distributed protocol of Kanoria et. al [15] to be run by edge players and allow general node capacities b. As before, the protocol is iterative and the configuration of proposals returned will converge to a fixed point m of some non-expansive function $\mathcal{T}$. In Section 3, we show that provided the payoff function u is greedy and spiteful, then any fixed point m of $\mathcal{T}$ is in the corresponding set $\mathcal{N}_u$ of pure Nash Equilibria.

In Section 4, we analyze the social welfare through the linear program (LP) relaxation of the maximum b-matching problem. As in [15], we investigate the close relationship between a fixed point of $\mathcal{T}$ and (LP). However, we go beyond previous analysis and do not need the integrality gap assumption, i.e., (LP) might not have an integral optimum. We show that when greedy users choose an edge, then all (LP) optimal solutions must set the value of that edge to 1; on the other hand, when users reject an edge out of spite, then all (LP) optimal solutions will set the value of that edge to 0. We do need some technical assumptions in order for our results to hold: either (1) (LP) has a unique optimum, or (2) the given graph G has no even cycle such that the sum of the weights of the odd edges equals that of the even edges; neither assumption implies the other, but both can be achieved by perturbing slightly the edge weights of the given graph. Unlike the case for simple 1-matching, we show (in the full version) that assumption (2) is necessary for general b-matching, which indicates that there is some fundamental difference between the two cases.

The greedy behavior states that some edges must be selected and the spiteful behavior requires that some edges must be rejected. However, there is still some freedom to deal with the remaining *ambiguous* edges.[3] Observe that a fixed point will remain a Nash Equilibrium (for the edge players) no matter how the ambiguous edges are handled, so it might make sense at this point to maximize the total number of extra contracts made from the ambiguous edges. However, optimizing the cardinality of a matching can be arbitrarily bad in terms of weight, but a maximum weight matching is a 2-approximation in terms of cardinality. Therefore, in Section 5, we consider a greedy and spiteful payoff function u that

---

[3] As a side note, we remark that our results implies that under the unique integral (LP) optimum assumption, there will be no ambiguous edges left.

corresponds to selecting a maximum weight matching (approximate or exact) among the ambiguous edges (subject to remaining node capacities b'); in reality, we can imagine this corresponds to a centralized clearing process or a collective effort performed by the users. We show that if a $(1+c)$-approximation algorithm for maximum weight matching is used for the ambiguous edges, then the social welfare is at least $\frac{2}{3(1+c)}$ fraction of the social optimum, i.e., the price of stability is $1.5(1+c)$. Finally, observe that the iterative protocol we mention will converge to a fixed point, but might never get there exactly; hence, we relax the notions of greed and spite in Section 5.1 and show that the same guarantee on the price of stability can be achieved eventually (and quickly).

We remark that if the topology of the given graph and the edge weights naturally indicate that certain edges should be selected while some should be rejected (both from the perspectives of social welfare and selfish behavior), then our framework of greed and spite can detect these edges. However, we do not claim that our framework is a silver bullet to all issues; in particular, for the triangle example given above, all edges will be ambiguous and our framework simply implies that one node will be left unmatched, but does not specify how this node is chosen. We leave as future research direction to develop notions of fairness in such situation.

Related Work. Kleinberg and Tardos [16] recently started the study of network bargaining games in the computer science community; they showed that a stable outcome exists *iff* a balanced outcome exists, and both can be computed in polynomial time, if they exist. Chakraborty et. al [9,10] explored equilibrium concepts and experimental results for bipartite graphs. Celis et. al [8] gave a tight polynomial bound on the rate of convergence for unweighted bipartite graphs with a unique balanced outcome. Kanoria [14] considered *unequal division* (UD) solutions for bargaining games, in which stability is still guaranteed while the surplus is split with ratio $r : 1 - r$, where $r \in (0, 1)$. They provided an FPTAS for the UD solutions assuming the existence of such solutions.

Azar et. al [1] considered a local dynamics that converges to a balanced outcome provided that it exists. Assuming that the LP relaxation for matching has a unique integral optimum, Kanoria et. al [15] designed a local dynamics that converges in polynomial time. Our distributed protocol is based on [15], but is generalized to general node capacities, run by edges and does not require the integrality condition on (LP).

Bateni et. al [4] also considered general node capacities; moreover, they showed that the network bargaining problem can be recast as an instance of the well-studied cooperative game [12]. In particular, a stable outcome is equivalent to a point in the core of a cooperative game, while a balanced outcome is equivalent to a point in the core and the prekernel. Azar et. al [2] also studied bargaining games from the perspective of cooperative games, and proved some monotonicity property for several widely considered solutions.

In our selection process, we assume that the maximum weight b'-matching problem is solved on the ambiguous edges. This problem is well-studied and can be solved exactly in polynomial time [22][Section 33.4]; moreover, the problem can be solved by a distributed algorithm [5], and $(1 + c)$-approximation for any $c > 0$ can be achieved in poly-logarithmic time [18,20,17].

2   Notation and Preliminaries

Consider an undirected graph $G = (V, E)$, with vertex set V and edge set E. Each node $i \in V$ corresponds to a *user* i (vertex player), and each edge $e \in E$ corresponds to an *agent* e (edge player). Agents arrange contracts to be formed between users where each agent $e = \{i, j\}$ gains a commission when users i and j form a contract. Each edge $e = \{i, j\} \in E$ has weight $w_e = w_{ij} > 0$, which is the maximum profit that can be shared between users i and j if a contract is made between them. Given a node i, denoted by $N(i) := \{j \in V : \{i, j\} \in E\}$ the set of its neighbors in G, there exists a capacity vector $b \in \mathbb{Z}_+^V$ such that each node i can make at most b_i contracts with its neighbors in $N(i)$, where at most one contract can be made between a pair of users; hence, the set M of edges on which contracts are made is a b-matching in G.

Agent Proposal. For each $e = \{i, j\} \in E$, agent e makes a proposal of the form $(m_{j \to i}, m_{i \to j})$ from an action set A_e to users i and j, where $A_e := \{(x, y) : x \geq 0, y \geq 0, x + y \leq w_{ij}\}$, such that if users i and j accepts the proposal and form a contract with each other, user i will receive $m_{j \to i}$ and user j will receive $m_{i \to j}$ from this contract.

Selection Procedure and Payoff Function u. Given a configuration $m \in A_E := \times_{e \in E} A_e$ of all agent's proposals, some selection procedure is run on m to return a b-matching M, where an edge $e = \{i, j\} \in M$ means that a contract is made between i and j. The procedure can be (1) deterministic or randomized, (2) centralized or (more preferably) distributed.

If i and j are *matched* in M, i.e., $e = \{i, j\} \in M$, agent e will receive a commission, which can either be fixed or a certain percentage of w_e; since an agent either gains the commission or not, we can assume that its payoff is 1 when a contract is made and 0 otherwise. Hence, the selection procedure defines a payoff function $u = \{u_e : A_E \to [0, 1] | e \in E\}$, such that for each $e \in E$, $u_e(m)$ is the probability that the edge e is in the b-matching M returned when the procedure is run on $m \in A_E$. We shall consider different selection procedures, which will lead to different payoff functions u. However, the selection procedure should satisfy several natural properties, which we relate to the human nature of the users as follows.

We use $\max^{(b)}$ to denote the b-th maximum value among a finite set of numbers (by convention it is 0 if there are less than b numbers). Given $m \in A_E$, we define $\hat{m}_i = \max_{j \in N(i)}^{(b_i)} m_{j \to i}$ and $\overline{m}_i = \max_{j \in N(i)}^{(b_i+1)} m_{j \to i}$.

Greedy Users. If both users i and j see that they cannot get anything better from someone else, then they will definitely make a contract with each other. Formally, we say that the payoff function u is *greedy* (or the users are greedy), if for each $e = \{i, j\} \in E$ and $m \in A_E$, if $m_{j \to i} > \overline{m}_i$ and $m_{i \to j} > \overline{m}_j$, then $u_e(m) = 1$.

Spiteful Users. It is human nature that once a person has seen the best, they will not settle for anything less. We try to capture this behavior formally. We say that the payoff function u is *spiteful* (or the users are spiteful) if for each $e = \{i, j\} \in E$ and $m \in A_E$, if $m_{j \to i} < \widehat{m}_i$, then $u_e(m) = 0$, i.e., if user i cannot get the b_i-th best offer from j, then no contract will be formed between i and j.

Game Theory and Social Welfare. We have described a game between the agents, in which agent e has the action set A_e, and has payoff function u (determined by the selection procedure). In this paper, we consider pure strategies and pure Nash Equilibria. A configuration $m \in A_E$ of actions is a Nash equilibrium if no single player can increase its payoff by unilaterally changing its action.

Given a payoff function u, we denote by $\mathcal{N}_u \subset A_E$ the set of Nash Equilibria. Given a configuration $m \in A_E$ of proposals and a payoff function u, we measure social welfare by $\mathcal{S}_u(m) := \sum_{e \in E} w_e \cdot u_e(m)$, which is the expected weight of the b-matching returned. When there is no ambiguity, the subscript u is dropped. The optimal social welfare $\mathcal{S}^* := \max_{m \in A_E} \mathcal{S}(m)$ is the maximum weight b-matching; to achieve the social optimum, given a maximum weight b-matching M, every agent $e \in M$ proposes $(\frac{w_e}{2}, \frac{w_e}{2})$, while other agents proposes $(0, 0)$. The weight of the b-matching can be an indicator of the volume of transactions or how active the market is. The Price of Anarchy (PoA) is defined as $\frac{\mathcal{S}^*}{\min_{m \in \mathcal{N}} \mathcal{S}(m)}$ and the Price of Stability (PoS) is defined as $\frac{\mathcal{S}^*}{\max_{m \in \mathcal{N}} \mathcal{S}(m)}$.

Proposition 1 (Infinite Price of Anarchy). *There exists an instance of the game such that when the users are spiteful, there exists a Nash Equilibrium $m \in A_E$ under which no contracts are made.*

We defer the proof of Propistion 1 to the full version.

3   A Distributed Protocol for Agents

We describe a distributed protocol for the agents to update their actions in each iteration. The protocol is based on the one by Kanoria et. al [15], which is run by nodes and designed for (1-)matchings. The protocol can easily be generalized to be run by edges and for general b-matchings. In each iteration, two agents only need to communicate if their corresponding edges share a node. Given a real number $r \in \mathbb{R}$, we denote $(r)_+ := \max\{r, 0\}$. Moreover, as described in [15,3] a damping factor $\kappa \in (0, 1)$ is used in the update; we can think of $\kappa = \frac{1}{2}$.

Although later on we will also consider the LP relaxation of b-matching, unlike previous works [21,6,15], we do not require the assumption that the LP relaxation has a unique integral optimum.

In Algorithm 1, auxiliary variables $\alpha^{(t)} \in \mathbb{R}_+^{2|E|}$ are maintained. Intuitively, the parameter $\alpha_{i \backslash j}$ is meant to represent the b_i-th best offer user i can receive if user j is removed. Suppose $W := \max_{e \in E} w_e$ and we define a function $\mathcal{T} : [0, W]^{2|E|} \to [0, W]^{2|E|}$ as follows.

Given $\alpha \in [0, W]^{2|E|}$, for each $\{i, j\} \in E$, define the following quantities.

$$S_{ij}(\alpha) = w_{ij} - \alpha_{i \backslash j} - \alpha_{j \backslash i} \tag{1}$$

Input: $G = (V, E, w)$
Initialization: For each $e = \{i, j\} \in E$, agent e picks arbitrary
$(m_{j \to i}^{(0)}, m_{i \to j}^{(0)}) \in A_e$.
for agent $e = \{i, j\} \in E$ **do**
$\quad \alpha_{i \backslash j}^{(1)} := \max_{k \in N(i) \backslash j}^{(b_i)} m_{k \to i}^{(0)} \; ; \; \alpha_{j \backslash i}^{(1)} := \max_{k \in N(j) \backslash i}^{(b_j)} m_{k \to j}^{(0)}$
end
for $t = 1, 2, 3, \ldots$ **do**
$\quad$ **for** agent $e = \{i, j\} \in E$ **do**
$\quad\quad S_{ij} := w_{ij} - \alpha_{i \backslash j}^{(t)} - \alpha_{j \backslash i}^{(t)}$
$\quad\quad m_{j \to i}^{(t)} := (w_{ij} - \alpha_{j \backslash i}^{(t)})_+ - \frac{1}{2}(S_{ij})_+ \; ; \; m_{i \to j}^{(t)} := (w_{ij} - \alpha_{i \backslash j}^{(t)})_+ - \frac{1}{2}(S_{ij})_+$
$\quad$ **end**
$\quad$ **for** agent $e = \{i, j\} \in E$ **do**
$\quad\quad \alpha_{i \backslash j}^{(t+1)} = (1 - \kappa) \cdot \alpha_{i \backslash j}^{(t)} + \kappa \cdot \max_{k \in N(i) \backslash j}^{(b_i)} m_{k \to i}^{(t)} \; ;$
$\quad\quad \alpha_{j \backslash i}^{(t+1)} = (1 - \kappa) \cdot \alpha_{j \backslash i}^{(t)} + \kappa \cdot \max_{k \in N(j) \backslash i}^{(b_j)} m_{k \to j}^{(t)}$
$\quad$ **end**
end

Algorithm 1. A Distributed Protocol for Agents. For each time t, agent $e = \{i, j\}$ computes its action $(m_{j \to i}^{(t)}, m_{i \to j}^{(t)}) \in A_e$; the first value is sent to other edges incident on i and the second to edges incident on j.

$$m_{j \to i}(\alpha) = (w_{ij} - \alpha_{j \backslash i})_+ - \frac{1}{2}(S_{ij}(\alpha))_+ \tag{2}$$

Then, we define $\mathcal{T}(\alpha) \in [0, W]^{2|E|}$ by $(\mathcal{T}(\alpha))_{i \backslash j} := \max_{k \in N(i) \backslash j}^{(b_i)} m_{k \to i}(\alpha)$. It follows that Algorithm 1 defines the sequence $\{\alpha^{(t)}\}_{t \geq 1}$ by $\alpha^{(t+1)} := (1 - \kappa)\alpha^{(t)} + \kappa \mathcal{T}(\alpha^{(t)})$.

Given a vector space D, a function $T : D \to D$ is *non-expansive* under norm $||\cdot||$ if for all $x, y \in D$, $||T(x) - T(y)|| \leq ||x - y||$; a point $\alpha \in D$ is a fixed point of T if $T(\alpha) = \alpha$. As in [15], it can be proved that the function $\mathcal{T}$ is non-expansive, and using a result by Ishikawa [13] (we defer the details to the full version), the next theorem follows.

Theorem 1 (Convergence to a Fixed Point). *The distributed protocol shown in Figure 1 maintains the sequence $\{\alpha^{(t)}\}$ which converges to a fixed point of the function $\mathcal{T}$ under the ℓ_∞ norm.*

Properties of a Fixed Point. Given a fixed point α of the function $\mathcal{T}$, the quantities $S \in \mathbb{R}^{|E|}$ and $m \in A_E$ are defined according to Equations (1) and (2). We also say that (m, α, S), (m, α) or m is a fixed point (of $\mathcal{T}$). Similar to [15], we give several important properties of a fixed point, whose details are given in the full version. Theorems 1 and 2 imply that as long as the payoff function is greedy and spiteful, the game defined between the agents (edge players) always has a pure Nash Equilibrium. We defer the proof of Theorem 2 to the full version.

Theorem 2 (Fixed Point is NE). *Suppose the payoff function u is greedy and spiteful. Then, any fixed point $m \in A_E$ of $\mathcal{T}$ is a Nash Equilibrium in $\mathcal{N}_u$.*

4   Analyzing Social Welfare via LP Relaxation

Theorem 2 states that a fixed point (m, α) of the function $\mathcal{T}$ is a Nash Equilibrium in $\mathcal{N}_u$, as long as the underlying payoff function is greedy and spiteful. Our goal is to show that there exists some greedy and spiteful u such that the fixed point m also achieves good social welfare $\mathcal{S}_u(m) = \sum_{e \in E} w_e \cdot u_e(m)$.

As observed by Kanoria et. al [15], the network bargain game is closely related to the linear program (LP) relaxation of the b-matching problem, which has the form $\mathcal{S}_{LP} := \max_{x \in \mathcal{L}} w(x)$, where $w(x) := \sum_{\{i,j\} \in E} x_{ij} w_{ij}$ and $\mathcal{L} := \{x \in [0,1]^E : \forall i \in V, \sum_{j:\{i,j\} \in E} x_{ij} \leq b_i\}$ is the set of feasible fractional solutions. Given $x \in \mathcal{L}$, we say a node i is *saturated* under x if $\sum_{j:\{i,j\} \in E} x_{ij} = b_i$, and otherwise *unsaturated*.

They showed that when the LP relaxation has a unique integral maximum, a fixed point (m, α, S) corresponds naturally to the unique maximum (1-)matching. However, their analysis cannot cover the case when the optimal solution is fractional or when the maximum matching is not unique.

In this section, we fully exploit the relationship between a fixed point and the LP relaxation, from which we show that good social welfare can be achieved. Note that we do not require the unique integral optimum assumption. On the other hand, we assume that either (1) the LP has a unique optimum or (2) the following technical assumption.

No Cycle with Equal Alternating Weight. We say that a cycle has *equal alternating weight* if it is even, and the sum of the odd edges equals that of the even edges. We assume that the given weighted graph G has no such cycle. The weights of any given graph can be perturbed slightly such that this condition holds. Observe that the optimum of (LP) might not be unique even with this assumption. For the 1-matching case, the conclusions listed in Theorem 3 have been derived by Kanoria et. al [15] without the "no cycle with equal alternating weight" assumption. However, for general b-matching, this assumption in Theorem 3 is necessary for cases (b) and (c). The reason is shown in the full version.

Theorem 3 (Fixed Point and LP). *Suppose (LP) has a unique optimum or the graph G has no cycle with equal alternating weight, and (m, α, S) is a fixed point of $\mathcal{T}$. Then, for any edge $\{i, j\} \in E$, the following holds.*

(a) *Suppose (LP) has a unique integral optimum corresponding to the maximum b-matching M^*. Then, $S_{ij} \geq 0$ implies that $\{i, j\} \in M^*$.*
(b) *Suppose $S_{ij} > 0$. Then, any optimal solution x to (LP) must satisfy $x_{ij} = 1$.*
(c) *Suppose $S_{ij} < 0$. Then, any optimal solution x to (LP) must satisfy $x_{ij} = 0$.*

Although the three statements in Theorem 3 look quite different, they can be implied by the three similar-looking corresponding statements in the following lemma.

Lemma 1 (Fixed Point and LP). *Suppose (m, α, S) is a fixed point of $\mathcal{T}$, and x is a feasible solution to (LP). Then, for each $\{i, j\} \in E$, the following properties hold.*

(a) If $S_{ij} \geq 0$ and $x_{ij} = 0$, then there is $\widehat{x} \in \mathcal{L}$ such that $\widehat{x} \neq x$ and $w(\widehat{x}) \geq w(x)$.
(b) If $S_{ij} > 0$ and $x_{ij} < 1$, then there is $\widehat{x} \in \mathcal{L}$ such that $\widehat{x} \neq x$ and $w(\widehat{x}) \geq w(x)$.
(c) If $S_{ij} < 0$ and $x_{ij} > 0$, then there is $\widehat{x} \in \mathcal{L}$ such that $\widehat{x} \neq x$ and $w(\widehat{x}) \geq w(x)$.

Moreover, strict inequality holds for (b) and (c), if in addition the graph G has no cycle with equal alternating weight.

4.1   Finding Alternative Feasible Solution via Alternating Traversal

Lemma 1 shows the existence of alternative feasible solutions under various conditions. We use the unifying framework of the *alternating traversal* to show its existence.

Alternating Traversal. Given a fixed point (m, α, S) of $\mathcal{T}$ and a feasible solution $x \in \mathcal{L}$, we define a structure called *alternating traversal* as follows.
(1) An alternating traversal $\mathcal{Q}$ (with respect to (m, α, S) and x) is a path or circuit (not necessarily simple and might contain repeated edges), which alternates between two disjoint edge sets $\mathcal{Q}^+$ and $\mathcal{Q}^-$ (hence $\mathcal{Q}$ can be viewed as a multiset which is the disjoint union of $\mathcal{Q}^+$ and $\mathcal{Q}^-$) such that $\mathcal{Q}^+ \subset S^+$ and $\mathcal{Q}^- \subset S^-$, where $S^+ := \{e \in E : S_e \geq 0\}$ and $S^- := \{e \in E : S_e \leq 0\}$.
   The alternating traversal is called *feasible* if in addition $\mathcal{Q}^+ \subset E^+$ and $\mathcal{Q}^- \subset E^-$, where $E^+ := \{e \in S^+ : x_e < 1\}$ and $E^- := \{e \in S^- : x_e > 0\}$.
   An edge e is called **critical** if e is in exactly one of E^+ and E^-, and is called **strict** if $S_e \neq 0$. Given an edge $e \in E$, we denote by $r_{\mathcal{Q}}(e)$ the number of times e appears in $\mathcal{Q}$, and by $\mathrm{sgn}_{\mathcal{Q}}(e)$ to be $+1$ if $e \in \mathcal{Q}^+$, -1 if $e \in \mathcal{Q}^-$ and 0 otherwise. Given a multiset $\mathcal{U}$ of edges, we denote by $w(\mathcal{U}) := \sum_{e \in \mathcal{U}} r_{\mathcal{U}}(e) w_e$ the sum of the weights of the edges in $\mathcal{U}$ in accordance with each edge's multiplicity.
(2) The following additional properties must be satisfied if the traversal $\mathcal{Q}$ is a path. If one end of the path has edge $\{i, j\} \in \mathcal{Q}^+$ and end node i, then i is unsaturated under x, i.e., $\sum_{e:i \in e} x_e < b_i$; if the end has edge $\{i, j\} \in \mathcal{Q}^-$ and end node i, then $\alpha_{i \backslash j} = 0$. Observe that there is a special case where the path starts and ends at the same node i; we still consider this as the path case as long as the end node conditions are satisfied for both end edges (which could be the same).

Lemma 2 (Alternative Feasible Solution.). *Suppose $\mathcal{Q}$ is a feasible alternating traversal with respect to some feasible $x \in \mathcal{L}$. Then, there exists feasible $\widehat{x} \neq x$ such that $w(\widehat{x}) - w(x)$ has the same sign ($\{-1, 0, +1\}$) as $w(\mathcal{Q}^+) - w(\mathcal{Q}^-)$.*

Proof. Suppose $\mathcal{Q}$ is a feasible alternating traversal. Then, for some $\lambda > 0$, we can define an alternative feasible solution $\widehat{x} \neq x$ by $\widehat{x}_e := x_e + \lambda \cdot \mathrm{sgn}_{\mathcal{Q}}(e) \cdot r_{\mathcal{Q}}(e)$. Moreover, $w(\widehat{x}) - w(x) = \lambda(w(\mathcal{Q}^+) - w(\mathcal{Q}^-))$. $\square$

Lemma 3 (Alternating Traversal Weight). *Suppose $\mathcal{Q}$ is an alternating traversal. Then, the following holds.*

(a) We have $w(\mathcal{Q}^+) \geq w(\mathcal{Q}^-)$, where strict inequality holds if $\mathcal{Q}$ contains a strict edge.

(b) If Q is a simple cycle with no strict edges, then $w(Q^+) = w(Q^-)$, i.e, Q is a cycle with equal alternating weight; in particular, with the "no cycle with alternating weight" assumption, any alternating traversal that is an even cycle must contain a strict edge.

Lemma 4 (Growing Feasible Alternating Traversal). *Suppose a fixed point (m, α, S) and a feasible $x \in \mathcal{L}$ are given as above.*

1. *Suppose $\{i, j\} \in E^+$ and node j is saturated (we stop if j is unsaturated). Then, there exists some node $k \in N(j) \setminus i$ such that $\{j, k\} \in E^-$.*
2. *Suppose $\{j, k\} \in E^-$ and $\alpha_{k \setminus j} > 0$ (we stop if $\alpha_{k \setminus j} = 0$). Then, there exists some node $l \in N(k) \setminus j$ such that $\{k, l\} \in E^+$.*

The proofs of Lemmas 3 and 4 are deferred to the full version.

Lemma 5 (Unifying Structural Lemma). *Suppose edge $e \in E$ is critical (with respect to some fixed point (m, α) and feasible $x \in \mathcal{L}$). Then, there exists a feasible alternating traversal Q; if in addition e is strict and there is no cycle with equal alternating weight, then Q contains a strict edge.*

Proof. To find a feasible alternating traversal Q, we apply a growing procedure (described in the full version) that starts from the critical edge $e = \{i, j\}$. Moreover, if Q is a simple even cycle, then by Lemma 3(b), Q contains a strict edge under the "no cycle with equal alternating weight" assumption; otherwise, Q contains the edge e, in which case e being strict implies that Q contains a strict edge.   □

Proof of **Lemma 1**: It suffices to check the given edge $\{i, j\}$ is critical in each of the three cases. Then, Lemma 4 promises the existence of a feasible alternating traversal, which contains a strict edge where appropriate. Then, Lemmas 3 and 2 guarantee the existence of feasible $\widehat{x} \neq x$ such that $w(\widehat{x}) \geq w(x)$, where strict inequality holds where appropriate.   □

5   Achieving Social Welfare with Greedy and Spiteful Users

We saw in Proposition 1 that a Nash Equilibrium m can result in zero social welfare if users are spiteful. In this section, we investigate under what conditions can a fixed point (m, α, S) of $\mathcal{T}$ achieve good social welfare, even if the underlying payoff function u is greedy and spiteful. Given $m \in A_E$, recall that for each node i, $\widehat{m}_i$ is the b_i-th best offer to i and $\overline{m}_i$ is the (b_i+1)-st best offer to i. Observe that each edge $e = \{i, j\} \in E$ falls into exactly one of the following three categories.

1. **Greedy Edges:** $m_{j \to i} > \overline{m}_i$ and $m_{i \to j} > \overline{m}_j$. Edge e will be selected and $u_e(m) = 1$.
2. **Spiteful Edges:** $m_{j \to i} < \widehat{m}_i$ or $m_{i \to j} < \widehat{m}_j$. Edge e will be rejected and $u_e(m) = 0$.
3. **Ambiguous Edges:** these are the remaining edges that are neither greedy nor spiteful.

Given a fixed point (m, α, S), by propositions shown in the full version, the category of an edge $e \in E$ can be determined by the sign of S_e: greedy (+1), ambiguous (0). Observe that after the greedy edges are selected and the spiteful edges are rejected, even if ambiguous edges are chosen arbitrarily (deterministic or randomized) to form a b-matching, the resulting payoff function is still greedy and spiteful. Since no agent (edge player) has motivation to unilaterally change his action for fixed point m, and any contract made for an ambiguous edge will be within the best b_i offers for a node i (i.e., if $\{i, j\} \in E$ is ambiguous, then $m_{j \to i} = \widehat{m}_i$ and $m_{i \to j} = \widehat{m}_j$), we can optimize the following, subject to remaining node capacity constraints b' (after greedy edges are selected).

- Find a maximum cardinality b'-matching among the ambiguous edges, hence optimizing the number of contracts made.
- Find a maximum weight b'-matching among the ambiguous edges, hence optimizing the social welfare.

Choosing Maximum Weight Matching among Ambiguous Edges. Observe that a maximum cardinality matching can be arbitrarily bad in terms of weight, but a maximum weight matching must be maximal and so is a 2-approximation for maximum cardinality. Hence, we argue that it makes sense to find a maximum weight b'-matching among the ambiguous edges. This step can be performed centrally or as a collective decision by the users. We give the main result in the following theorem and leave the details in the full version.

Theorem 4 (Price of Stability). *Suppose the given graph has no cycle with equal alternating weight or (LP) has a unique optimum. Then, there exists a greedy and spiteful payoff function u such that any fixed point m of $\mathcal{T}$ is a Nash Equilibrium in $\mathcal{N}_u$; moreover, the social welfare $\mathcal{S}_u(m) \geq \frac{2}{3}\mathcal{S}_{LP} \geq \frac{2}{3}\max_{m' \in A_E} \mathcal{S}_u(m')$, showing that the Price of Stability is at most 1.5*

5.1   Rate of Convergence: ϵ-Greedy and ϵ-Spiteful Users

Although the iterative protocol described in Figure 1 will converge to some fixed point (m, α, S), it is possible that a fixed point will never be exactly reached. However, results in Section 4 and 5 can be extended if we relax the notions of greedy and spiteful users, and relax also the technical assumption on no cycle with equal alternating weight. The details are given in the full version and we state the convergence result.

Theorem 5 (Rate of Convergence). *Suppose $\epsilon > 0$, and the given graph has maximum edge weight W and has no cycle with ϵ-equal alternating weight. Then, there exists an ϵ-greedy and ϵ-spiteful payoff function u such that the following holds. For any sequence $\{(m^{(t)}, \alpha^{(t)})\}$ produced by the iterative protocol in Figure 1, and for all $t \geq \Theta(\frac{W^2|V|^4}{\epsilon^2})$, the social welfare $\mathcal{S}_u(m^{(t)}) \geq \frac{2}{3}\mathcal{S}_{LP}$.*

References

1. Azar, Y., Birnbaum, B., Celis, L.E., Devanur, N.R., Peres, Y.: Convergence of local dynamics to balanced outcomes in exchange networks. In: FOCS 2009, pp. 293–302 (2009)

2. Azar, Y., Devanur, N.R., Jain, K., Rabani, Y.: Monotonicity in bargaining networks. In: SODA 2010, pp. 817–826 (2010)
3. Baillon, J., Bruck, R.E.: The rate of asymptotic regularity is $O(1/\sqrt{n})$. In: Theory and Applications of Nonlinear Operators of Accretive and Monotone Type. Lecture Notes in Pure and Appl. Math., pp. 51–81 (1996)
4. Bateni, M., Hajiaghayi, M., Immorlica, N., Mahini, H.: The Cooperative Game Theory Foundations of Network Bargaining Games. In: Abramsky, S., Gavoille, C., Kirchner, C., Meyer auf der Heide, F., Spirakis, P.G. (eds.) ICALP 2010. LNCS, vol. 6198, pp. 67–78. Springer, Heidelberg (2010)
5. Bayati, M., Borgs, C., Chayes, J., Zecchina, R.: Belief-propagation for weighted b-matchings on arbitrary graphs and its relation to linear programs with integer solutions. SIAM J. Discrete Math. 25, 989 (2011)
6. Bayati, M., Shah, D., Sharma, M.: Max-product for maximum weight matching: Convergence, correctness, and lp duality. IEEE Transactions on Information Theory 54(3), 1241–1251 (2008)
7. Binmore, K.: Game Theory and the Social Contract, vol. 2. MIT Press (1998)
8. Celis, L.E., Devanur, N.R., Peres, Y.: Local Dynamics in Bargaining Networks via Random-turn Games. In: Saberi, A. (ed.) WINE 2010. LNCS, vol. 6484, pp. 133–144. Springer, Heidelberg (2010)
9. Chakraborty, T., Kearns, M.: Bargaining Solutions in a Social Network. In: Papadimitriou, C., Zhang, S. (eds.) WINE 2008. LNCS, vol. 5385, pp. 548–555. Springer, Heidelberg (2008)
10. Chakraborty, T., Kearns, M., Khanna, S.: Network bargaining: algorithms and structural results. In: EC 2009, pp. 159–168 (2009)
11. Cook, K.S., Yamagishi, T.: Power in exchange networks: A power-dependence formulation. Social Networks 14, 245–265 (1992)
12. Driessen, T.S.H.: Cooperative Games: Solutions and Applications. Kluwer Academic Publishers (1988)
13. Ishikawa, S.: Fixed points and iteration of a nonexpansive mapping in a banach space. Proceedings of the American Mathematical Society 59(1), 65–71 (1976)
14. Kanoria, Y.: An FPTAS for Bargaining Networks with Unequal Bargaining Powers. In: Saberi, A. (ed.) WINE 2010. LNCS, vol. 6484, pp. 282–293. Springer, Heidelberg (2010)
15. Kanoria, Y., Bayati, M., Borgs, C., Chayes, J.T., Montanari, A.: Fast convergence of natural bargaining dynamics in exchange networks. In: SODA 2011, pp. 1518–1537 (2011)
16. Kleinberg, J., Tardos, É.: Balanced outcomes in social exchange networks. In: STOC 2008, pp. 295–304 (2008)
17. Koufogiannakis, C., Young, N.E.: Distributed Fractional Packing and Maximum Weighted b-Matching via Tail-Recursive Duality. In: Keidar, I. (ed.) DISC 2009. LNCS, vol. 5805, pp. 221–238. Springer, Heidelberg (2009)
18. Lotker, Z., Patt-Shamir, B., Pettie, S.: Improved distributed approximate matching. In: SPAA 2008, pp. 129–136. ACM (2008)
19. Nash, J.: The bargaining problem. Econometrica 18, 155–162 (1950)
20. Nieberg, T.: Local, distributed weighted matching on general and wireless topologies. In: Proceedings of the Fifth International Workshop on Foundations of Mobile Computing, DIALM-POMC 2008, pp. 87–92. ACM, New York (2008)
21. Sanghavi, S., Malioutov, D.M., Willsky, A.S.: Linear programming analysis of loopy belief propagation for weighted matching. In: NIPS 2007 (2007)
22. Schrijver: Combinatorial Optimizatin, vol. A-C. Springer (2004)

Optimal Lower Bound for Differentially Private Multi-party Aggregation

T-H. Hubert Chan[1], Elaine Shi[2], and Dawn Song[3]

[1] The University of Hong Kong
hubert@cs.hku.hk
[2] University of Maryland, College Park
elaine@cs.umd.edu
[3] UC Berkeley
dawnsong@cs.berkeley.edu

Abstract. We consider distributed private data analysis, where n parties each holding some sensitive data wish to compute some aggregate statistics over all parties' data. We prove a tight lower bound for the private distributed summation problem. Our lower bound is strictly stronger than the prior lower-bound result by Beimel, Nissim, and Omri published in CRYPTO 2008. In particular, we show that any n-party protocol computing the sum with sparse communication graph must incur an additive error of $\Omega(\sqrt{n})$ with constant probability, in order to defend against potential coalitions of compromised users. Furthermore, we show that in the client-server communication model, where all users communicate solely with an untrusted server, the additive error must be $\Omega(\sqrt{n})$, regardless of the number of messages or rounds. Both of our lower-bounds, for the general setting and the client-to-server communication model, are strictly stronger than those of Beimel, Nissim and Omri, since we remove the assumption on the number of rounds (and also the number of messages in the client-to-server communication model). Our lower bounds generalize to the (ϵ, δ) differential privacy notion, for reasonably small values of δ.

1 Introduction

Dwork *et al.* [DMNS06] proposed (information theoretical) differential privacy, which has become a de-facto standard privacy notion in private data analysis. In this paper, we investigate the setting of *distributed private data analysis* [BNO08], in which n parties each holds some private input, and they wish to jointly compute some statistic over all parties' inputs in a way that respects each party's privacy.

In a seminal work by Beimel, Nissim, and Omri [BNO08], they demonstrate a lower bound result for distributed private data analysis. Specifically, they consider the *distributed summation* problem, namely, computing the sum of all parties' inputs. They prove that any differentially-private multi-party protocol with *a small number of rounds* and *small number of messages* must have large error.

This paper proves a strictly stronger lower bound than the result by Beimel, Nissim, and Omri [BNO08]. We show that for the distributed summation

L. Epstein and P. Ferragina (Eds.): ESA 2012, LNCS 7501, pp. 277–288, 2012.
© Springer-Verlag Berlin Heidelberg 2012

problem, any differentially private multi-party protocol *with a sparse communi-cation graph* must have large error, where two nodes are allowed to communicate only if they are adjacent in the communication graph. In comparison with the previous lower bound by Beimel *et al.* [BNO08], our lower bound relaxes the constraint on the small number of messages or rounds. In this sense, our lower bound is strictly stronger than that of Beimel *et al.* [BNO08].

We also consider a special setting in which only client-server communication is allowed (i.e., the communication graph is the star graph with the server at the center). Beimel *et al.* [BNO08] referred to this communication model as *local model*. In the client-server communication setting, we prove a lower bound showing that any differentially-private protocol computing the sum must have large error. This lower bound has no restriction on the number of messages or the number of rounds, and is also strictly stronger than [BNO08], who showed that in the client-server setting, any differentially-private protocol with *a small number of rounds* must have large error.

Furthermore, our lower-bound results hold for (ϵ, δ)-differential privacy where δ is reasonably small. Since ϵ-differential privacy is a special case of this with $\delta = 0$, our lower bounds are also more general than those of Beimel *et al.* who considered ϵ differential privacy.

The lower bounds proven in this paper hold for information theoretic differential privacy. By contrast, previous works have demonstrated the possibility of constructing multi-party protocols with $O(1)$ error and small message complexity in the computational differential privacy setting [DKM+06, RN10, SCR+11]. Therefore, our lower-bound results also imply a gap between computational and information theoretic differential privacy in the multi-party setting.

1.1   Informal Summary of Main Results

Lower Bound for the General Setting (Corollary 2). Informally, we show that any n-party protocol computing the sum, which consumes at most $\frac{1}{4}n(t+1)$ messages must incur $\Omega(\sqrt{n})$ additive error (with constant probability), in order to preserve differentially privacy against coalitions of up to t compromised users.

Lower Bound for Client-Server Model (Corollary 1). Informally, we show that in the client-server model, an aggregator would make an additive error $\Omega(\sqrt{n})$ on the sum from any n-user protocol that preserves differential privacy. This lower-bound holds regardless of the number of messages or number of rounds.

Tightness of the Lower Bounds. Both of the above lower bounds are tight in the following sense. First, for the client-server model, there exists a naive protocol, in which each user perturbs their inputs using Laplace or geometric noise with standard deviation $O(\frac{1}{\epsilon})$, and reveals their perturbed inputs to the aggregator. Such a naive protocol has additive error $O(\sqrt{n})$; so in some sense, the naive protocol is the best one can do in the client-server model.

To see why the lower bound is tight for the general multi-party setting, we combine   standard   techniques   of   secure   function   evaluation   [CK93]   and distributed randomness [SCR+11] and state in Section 5 that there exists a protocol which requires only $O(nt)$ messages, but achieves $o(\sqrt{n})$ error.

Techniques. To prove the above-mentioned lower-bounds, we combine techniques from communication complexity and measure anti-concentration techniques used in the metric embedding literature. Our communication complexity techniques are inspired by the techniques adopted by McGregor *et al.* [MMP+10] who proved a gap between information-theoretic and computational differential privacy in the 2-party setting. The key observation is that independent inputs remain independent even after conditioning on the transcript of the protocol. This eliminates the dependence on the number of rounds of communication in the lower bound.

As argued by in [BNO08], if a party communicates with only a small number of other parties, then there must still be sufficient randomness in that party's input. Then, using anti-concentration techniques, we show that the sum of these independent random variables is either much smaller or much larger than the mean, both with constant probability, thereby giving a lower bound on the additive error. The anti-concentration techniques are inspired by the analysis of the square of the sum of independent sub-Gaussian random variables [IN07], which generalizes several Johnson-Lindenstrauss embedding constructions [DG03, Ach03]. Moreover, we generalize the techniques to prove the lower bound for (ϵ, δ)-differentially private protocols (as opposed to just ϵ-differential privacy). The challenge is that for $\delta > 0$, it is possible for some transcript to break a party's privacy and there might not be enough randomness left in its input. However, we show that for small enough δ, the probability that such a transcript is encountered is small, and hence the argument is still valid.

2   Related Work

Differential privacy [DMNS06, Dwo06, Dwo10] was traditionally studied in a setting where a trusted curator, with access to the entire database in the clear, wishes to release statistics in a way that preserves each individual's privacy. The trusted curator is responsible for introducing appropriate perturbations prior to releasing any statistic. This setting is particularly useful when a company or a government agency, in the possession of a dataset, would like to share it with the public.

In many real-world applications, however, the data is distributed among users, and users may not wish to entrust their sensitive data to a centralized party such as a cloud service provider. In these cases, we can employ distributed private data analysis – a problem proposed and studied in several recent works [MMP+10, BNO08, DKM+06, RN10, SCR+11] – where participating parties are mutually distrustful, but wish to learn some statistics over their joint datasets. In particular, the client-server communication model [BNO08, RN10, SCR+11] where all users communicate solely with an untrusted server, is especially desirable in real-world settings.

This work subsumes the distributed private data analysis setting previously studied by Beimel, Nissim, and Omri [BNO08], and improves their lower-bounds for information-theoretic differentially private multi-party protocols.

While this work focuses on lower bounds for information theoretic differential privacy, computational differential privacy is an alternative notion first formalized by Mironov *et al.* [MPRV09], aiming to protect individual user's sensitive data against polynomially-bounded adversaries. Previous works have shown the possibility of constructing protocols with $O(1)$ error and small message complexity in the computational differential privacy setting [DKM+06, RN10, SCR+11]. This demonstrates a gap between information theoretic and computational differential privacy in the multi-party setting. In particular, the constructions by Rastogi *et al.* [RN10] and Shi *et al.* [SCR+11] require only client-server communication, and no peer-to-peer interactions.

3   Problem Definition and Assumptions

Consider a group of n parties (or nodes), indexed by the set $[n] := \{1, 2, \ldots n\}$. Each party $i \in [n]$ has private data $x_i \in \mathcal{U}$, where $\mathcal{U} := \{0, 1, 2, \ldots, \Delta\}$ for some positive integer Δ. We use the notation $\mathbf{x} := (x_1, x_2, \ldots, x_n) \in \mathcal{U}^n$ to denote the vector of all parties' data, also referred to as an *input configuration*. The n parties participate in a protocol such that at the end at least one party learns or obtains an estimate of the sum, denoted $\mathsf{sum}(\mathbf{x}) := \sum_{i \in [n]} x_i$. For a subset $S \subseteq [n]$, we denote $\mathsf{sum}(x_S) := \sum_{i \in S} x_i$.

Given a protocol Π and an input $\mathbf{x} \in \mathcal{U}^n$, we use $\Pi(\mathbf{x})$ to denote the execution of the protocol on the input. A *coalition* is a subset T of nodes that share their information with one another in the hope of learning the other parties' input. The view $\Pi(\mathbf{x})_{|T}$ of the coalition T consists of the messages, any input and private randomness viewable by the nodes in T. In contrast, we denote by $\pi(\mathbf{x})$ the transcript of the messages and use $\pi(\mathbf{x})_{|T}$ to mean the messages sent or received by nodes in T.

Trust and Attack Model. As in Beimel *et al.* [BNO08], we assume that all parties are semi-honest. A subset T of parties can form a *coalition* and share their input data, private randomness and view of the transcript with one another in order to learn the input data of other parties. Since we adopt the semi-honest model, all parties, whether within or outside the coalition, honestly use their true inputs and follow the protocol. The *data pollution* attack, where parties inflate or deflate their input values, is out of the scope of this paper. Defense against the data pollution attack can be considered as orthogonal and complementary to our work, and has been addressed by several works in the literature [PSP03].

Communication Model. Randomized *oblivious protocols* are considered in [CK93, BNO08], where the communication pattern (i.e., which node sends message to which node in which round) is independent of the input and the randomness. We relax this notion by assuming that for a protocol Π, there is a *communication graph* G_Π (independent of input and randomness) on the nodes such that only adjacent nodes can communicate with each other. For a node i, we denote by $N_\Pi(i)$ its set of neighbors in G_Π. The subscript Π is dropped when there is no risk of ambiguity. Observe that the number of messages sent in each round is only limited by the number of edges in the communication graph, and to simply our proofs, we only assume that there is some finite upper bound on the number of rounds for all possible inputs and randomness used by the protocol.

3.1 Preliminaries

Intuitively, differential privacy against a coalition guarantees that if an individual outside the coalition changes its data, the view of the coalition in the protocol will not be affected too much. In other words, if two input configurations $\mathbf{x}$ and $\mathbf{y}$ differ only in 1 position outside the coalition, then the distribution of $\Pi(\mathbf{x})_{|T}$ is very close to that of $\Pi(\mathbf{y})_{|T}$. This intuition is formally stated in the following definition.

Definition 1 (Differential Privacy Against Coalition). *Let $\epsilon > 0$ and $0 \leq \delta < 1$. A (randomized) protocol Π preserves (ϵ, δ)-differential privacy against coalition T if for all vectors $\mathbf{x}$ and $\mathbf{y}$ in $\mathcal{U}^n$ that differ by only 1 position corresponding to a party outside T, for all subsets S of possible views by T, $\Pr[\Pi(\mathbf{x})_{|T} \in S] \leq \exp(\epsilon) \cdot \Pr[\Pi(\mathbf{y})_{|T} \in S] + \delta$.*

A protocol Π preserves ϵ-differential privacy against a coalition if it preserves $(\epsilon, 0)$-differential privacy against the same coalition.

Two noise distributions are commonly used to perturb the data and ensure differential privacy, the Laplace distribution [DMNS06], and the Geometric distribution [GRS09]. The advantage of using the geometric distribution over the Laplace distribution is that we can keep working in the domain of integers.

Definition 2 (Geometric Distribution). *Let $\alpha > 1$. We denote by $\mathsf{Geom}(\alpha)$ the symmetric geometric distribution that takes integer values such that the probability mass function at k is $\frac{\alpha-1}{\alpha+1} \cdot \alpha^{-|k|}$.*

Proposition 1. *Let $\epsilon > 0$. Suppose u and v are two integers such that $|u - v| \leq \Delta$. Let r be a random variable having distribution $\mathsf{Geom}(\exp(\frac{\epsilon}{\Delta}))$. Then, for any integer k, $Pr[u + r = k] \leq \exp(\epsilon) \cdot Pr[v + r = k]$.*

The above property of Geom distribution is useful for designing differentially private mechanisms that output integer values. In our setting, changing one party's data can only affect the sum by at most Δ. Hence, it suffices to consider $\mathsf{Geom}(\alpha)$ with $\alpha = e^{\frac{\epsilon}{\Delta}}$. Observe that $\mathsf{Geom}(\alpha)$ has variance $\frac{2\alpha}{(\alpha-1)^2}$. Since $\frac{\sqrt{\alpha}}{\alpha-1} \leq \frac{1}{\ln \alpha} = \frac{\Delta}{\epsilon}$, the magnitude of the error added is $O(\frac{\Delta}{\epsilon})$.

Naive Scheme. As a warm-up exercise, we describe a Naive Scheme, where each party generates an independent $\mathsf{Geom}(e^{\frac{\epsilon}{\Delta}})$ noise, adds the noise to its data, and sends the perturbed data to one special party called an aggregator, who then computes the sum of all the noisy data. As each party adds one copy of independent noise to its data, n copies of noises would accumulate in the sum. It can be shown that the accumulated noise is $O(\frac{\Delta\sqrt{n}}{\epsilon})$ with high probability. In comparison with our lower-bound, this shows that under certain mild assumptions, if one wishes to guarantee small message complexity, the Naive Scheme is more or less the best one can do in the information theoretic differential privacy setting.

4   Lower Bound for Information-Theoretic Differential Privacy

This section proves lower-bounds for differentially private distributed summation protocols. We consider two settings, the general settings, where all nodes are allowed to interact with each other; and the client-server communication model, where all users communicate only with an untrusted server, but not among themselves.

We will prove the following main result, and then show how to extend the main theorem to the afore-mentioned two communication models.

Theorem 1 (Lower Bound for Size-t Coalitions). *Let $0 < \epsilon \leq \ln 99$ and $0 \leq \delta \leq \frac{1}{4n}$. There exists some $\eta > 0$ (depending on ϵ) such that the following holds. Suppose n parties, where party i ($i \in [n]$) has a secret bit $x_i \in \{0, 1\}$, participate in a protocol Π to estimate $\sum_{i \in [n]} x_i$. Suppose further that the protocol is (ϵ, δ)-differentially private against any coalition of size t, and there exists a subset of m parties, each of whom has at most t neighbors in the protocol's communication graph. Then, there exists some configuration of the parties' bits x_i's such that with probability at least η (over the randomness of the protocol), the additive error is at least $\Omega(\frac{\sqrt{\gamma}}{1+\gamma} \cdot \sqrt{m})$, where $\gamma = 2e^\epsilon$.*

Note that the assumption that $0 \leq \delta \leq \frac{1}{4n}$ is not a limitation. Typically, when we adopt (ϵ, δ) differential privacy, we wish to have $\delta = o(\frac{1}{n})$, to ensure that no individual user's sensitive data is leaked with significant probability.

The following corollaries are special cases of Theorem 1, corresponding to the client-server communication model, and the general model respectively. In both settings, our results improve upon the lower bounds by Beimel *et al.* [BNO08]. We will first show how to derive these corollaries from Theorem 1. We then present a formal proof for Theorem 1.

Corollary 1 (Lower Bound for Client-Server Communication Model). *Let $0 < \epsilon \leq \ln 99$ and $0 \leq \delta \leq \frac{1}{4n}$. Suppose n parties, each having a secret bit, participate in a protocol Π with a designated party known as the aggregator, with no peer-to-peer communication among the n parties. Suppose further that the protocol is (ϵ, δ)-differentially private against any single party (which forms a coalition on its own). Then, with constant probability (depending on ϵ), the aggregator estimates the sum of the parties' bits with additive error at least at least $\Omega(\frac{\sqrt{\gamma}}{1+\gamma} \cdot \sqrt{n})$, where $\gamma = 2e^\epsilon$.*

Proof. The communication graph is a star with the aggregator at the center. The protocol is also differentially private against any coalition of size 1, and there are n parties, each of which has only 1 neighbor (the aggregator). Therefore, the result follows from Theorem 1.   □

Corollary 2 (Lower Bound for General Setting). *Let $0 < \epsilon \leq \ln 99$ and $0 \leq \delta \leq \frac{1}{4n}$. Suppose n parties participate in a protocol that is (ϵ, δ)-differentially private against any coalition of size t. If there are at most $\frac{1}{4}n(t+1)$ edges in the* *communication graph of the protocol, then with constant probability (depending on ϵ), the protocol estimates the sum of the parties' bits with additive error at least $\Omega(\frac{\sqrt{\gamma}}{1+\gamma} \cdot \sqrt{n})$, where $\gamma = 2e^\epsilon$.*

Proof. Since there are at most $\frac{1}{4}n(t+1)$ edges in the communication graph, there are at least $\frac{n}{2}$ nodes with at most t neighbors (otherwise the sum of degrees over all nodes is larger than $\frac{1}{2}n(t+1)$). Hence, the result follows from Theorem 1.   □

Proof Overview for Theorem 1. We fix some $\epsilon > 0$ and $0 \le \delta \le \frac{1}{4n}$, and consider some protocol Π that preserves (ϵ, δ)-differential privacy against any coalition of size t.

Suppose that the bits X_i's from all parties are all uniform in $\{0,1\}$ and independent. Suppose M is the subset of m parties, each of whom has at most t neighbors in the communication graph. For each $i \in M$, we consider a set $\mathcal{P}^{(i)}$ of *bad* transcripts for i, which intuitively is the set of transcripts π under which the view of party i's neighbors can compromise party i's privacy.

We consider the set $\mathcal{P} := \cup_{i \in M} \mathcal{P}^{(i)}$ of bad transcripts (which we define formally later), and show that the probability that a bad transcript is produced is at most $\frac{3}{4}$. Conditioning on a transcript $\pi \notin \mathcal{P}$, for $i \in M$, each X_i still has enough randomness, as transcript π does not break the privacy of party i. Therefore, the conditional sum $\sum_{i \in M} X_i$ still has enough variance like the sum of $m = |M|$ independent uniform $\{0,1\}$-random variables. Using anti-concentration techniques, we can show that the sum deviates above or below the mean by $\Omega(\sqrt{m})$, each with constant probability. Since the transcript determines the estimation of the final answer, we conclude that the error is $\Omega(\sqrt{m})$ with constant probability.

Notation. Suppose that each party i's bit X_i is uniform in $\{0,1\}$ and independent. We use $\mathbf{X} := (X_i : i \in [n])$ to denote the collection of the random variables. We use a probabilistic argument to show that the protocol must, for some configuration of parties' bits, make an additive error of at least $\Omega(\sqrt{m})$ on the sum with constant probability.

For convenience, given a transcript π (or a view of the transcript by certain parties) we use $\Pr[\pi]$ to mean $\Pr[\pi(\mathbf{X}) = \pi]$ and $\Pr[\cdot|\pi]$ to mean $\Pr[\cdot|\pi(\mathbf{X}) = \pi]$; given a collection $\mathcal{P}$ of transcripts (or collection of views), we use $\Pr[\mathcal{P}]$ to mean $\Pr[\pi(\mathbf{X}) \in \mathcal{P}]$.

We can assume that the estimate made by the protocol is a deterministic function on the whole transcript of messages, because without loss of generality we can assume that the last message sent in the protocol is the estimate of the sum.

We will define some event $\mathcal{E}$ where the protocol makes a large additive error.

Bad Transcripts. Denote $\gamma := 2e^\epsilon$. For $i \in M$, define $\mathcal{P}_0^{(i)} := \{\pi : \Pr[\pi_{|N(i)}|X_i = 0] > \gamma \cdot \Pr[\pi_{|N(i)}|X_i = 1]\}$ and $\mathcal{P}_1^{(i)} := \{\pi : \Pr[\pi_{|N(i)}|X_i = 1] > \gamma \cdot \Pr[\pi_{|N(i)}|X_i = 0]\}$. We denote by $\mathcal{P}^{(i)} := \mathcal{P}_0^{(i)} \cup \mathcal{P}_1^{(i)}$ the set of *bad* transcripts with respect to party i. Let $\mathcal{P} := \cup_{i \in M} \mathcal{P}^{(i)}$.

Proposition 2 (Projection of Events). *Suppose U is a subset of the views of the transcript by the neighbors of i, and define the subset of transcripts by $\mathcal{P}_U := \{\pi : \pi_{|N(i)} \in U\}$. Then, it follows that $\Pr_{\mathbf{X},\Pi}[\pi(\mathbf{X}) \in \mathcal{P}_U] = \Pr_{\mathbf{X},\Pi}[\pi(\mathbf{X})_{|N(i)} \in U]$.*

Lemma 1 (Most Transcripts Behave Well). *Let $\epsilon > 0$ and $0 \leq \delta \leq \frac{1}{4n}$. Suppose the protocol is (ϵ, δ)-differentially private against any coalition of size t, and $\mathcal{P}$ is the union of the bad transcripts with respect to parties with at most t neighbors in the communication graph. Then, $\mathrm{Pr}_{\mathbf{X}, \Pi}[\mathcal{P}] \leq \frac{3}{4}$.*

Proof. From definition of $\mathcal{P}_0^{(i)}$ and using Proposition 2, we have $\mathrm{Pr}[\mathcal{P}_0^{(i)} | X_i = 0] > \gamma \cdot \mathrm{Pr}[\mathcal{P}_0^{(i)} | X_i = 1]$. Since the protocol is (ϵ, δ)-differentially private against any coalition of size t, we have for each $i \in M$, $\mathrm{Pr}[\mathcal{P}_0^{(i)} | X_i = 0] \leq e^\epsilon \mathrm{Pr}[\mathcal{P}_0^{(i)} | X_i = 1] + \delta$. Hence, we have $(\gamma - e^\epsilon) \mathrm{Pr}[\mathcal{P}_0^{(i)} | X_i = 1] \leq \delta$, which implies that $\mathrm{Pr}[\mathcal{P}_0^{(i)} | X_i = 1] \leq e^{-\epsilon} \delta$, since $\gamma = 2e^\epsilon$.

Hence, we also have $\mathrm{Pr}[\mathcal{P}_0^{(i)} | X_i = 0] \leq e^\epsilon \mathrm{Pr}[\mathcal{P}_0^{(i)} | X_i = 1] + \delta \leq 2\delta$. Therefore, we have $\mathrm{Pr}[\mathcal{P}_0^{(i)}] = \frac{1}{2}(\mathrm{Pr}[\mathcal{P}_0^{(i)} | X_i = 0] + \mathrm{Pr}[\mathcal{P}_0^{(i)} | X_i = 1]) \leq \frac{3\delta}{2}$.

Similarly, we have $\mathrm{Pr}[\mathcal{P}_1^{(i)}] \leq \frac{3\delta}{2}$. Hence, by the union bound over $i \in M$, we have $\mathrm{Pr}[\mathcal{P}] \leq 3n\delta \leq \frac{3}{4}$, since we assume $0 \leq \delta \leq \frac{1}{4n}$. □

We perform the analysis by first conditioning on some transcript $\pi \notin \mathcal{P}$. The goal is to show that $\mathrm{Pr}_{\mathbf{X}}[\mathcal{E} | \pi] \geq \eta$, for some $\eta > 0$. Then, since $\mathrm{Pr}[\mathcal{P}] \leq \frac{3}{4}$, we can conclude $\mathrm{Pr}_{\mathbf{X}}[\mathcal{E}] \geq \frac{\eta}{4}$, and hence for some configuration $\mathbf{x}$, we have $\mathrm{Pr}[\mathcal{E} | \mathbf{x}] \geq \frac{\eta}{4}$, as required.

Conditioning on Transcript π. The first step (Lemma 2) is analogous to the techniques of [MMP+10, Lemma 1]. We show that conditioning on the transcript $\pi \notin \mathcal{P}$, the random variables X_i's are still independent and still have enough randomness remaining.

Definition 3 (γ-random). *Let $\gamma \geq 1$. A random variable X in $\{0, 1\}$ is γ-random if $\frac{1}{\gamma} \leq \frac{\mathrm{Pr}[X = 1]}{\mathrm{Pr}[X = 0]} \leq \gamma$.*

Lemma 2 (Conditional Independence and Randomness). *Suppose each party's bit X_i is uniform and independent, and consider a protocol to estimate the sum that is (ϵ, δ)-differentially private against any coalition of size t, where $0 \leq \delta \leq \frac{1}{4n}$. Then, conditioning on the transcript $\pi \notin \mathcal{P}$, the random variables X_i's are independent; moreover, for each party $i \in M$ that has at most t neighbors in the communication graph, the conditional random variable X_i is γ-random, where $\gamma = 2e^\epsilon$.*

Proof. The proof is similar to that of [MMP+10, Lemma 1]. Since our lower bound does not depend on the number of rounds, we can without loss of generality sequentialize the protocol and assume only one node sends a message in each round. The conditional independence of the X_i's can be proved by induction on the number of rounds of messages. To see this, consider the first message m_1 sent by the party who has input X_1, and suppose X' is the joint input of all other parties. Observe that (X_1, m_1) is independent of X'. Hence, we have $\mathrm{Pr}[X_1 = a, X' = b | m_1 = c] = \frac{\mathrm{Pr}[X_1 = a, X' = b, m_1 = c]}{\mathrm{Pr}[m_1 = c]} = \frac{\mathrm{Pr}[X_1 = a, m_1 = c] \mathrm{Pr}[X' = b]}{\mathrm{Pr}[m_1 = c]} = \mathrm{Pr}[X_1 = a | m_1 = c] \cdot \mathrm{Pr}[X' = b | m_1 = c]$, which means conditioning on m_1, the random variables X_1 and X' are independent. After conditioning on m_1, one can view the remaining protocol as one that has one less round of messages. Therefore, by induction, one can argue that conditioning on the whole transcript, the inputs of the parties are independent.

For each party i having at most t neighbors, the γ-randomness of each conditional X_i can be proved by using the uniformity of X_i and that $\pi \notin \mathcal{P}^{(i)}$ is not bad for i.

We first observe that the random variable X_i has the same conditional distribution whether we condition on π or $\pi_{|N(i)}$, because as long as we condition on the messages involving node i, everything else is independent of X_i.

We next observe that if party $i \in M$ has at most t neighbors in the communication graph and $\pi \notin \mathcal{P}^{(i)}$, then by definition we have $\frac{\Pr[\pi_{|N(i)} | X_i=1]}{\Pr[\pi_{|N(i)} | X_i=0]} \in [\gamma^{-1}, \gamma]$.

Hence, $\frac{\Pr[X_i=1|\pi]}{\Pr[X_i=0|\pi]} = \frac{\Pr[X_i=1| \pi_{|N(i)}]}{\Pr[X_i=0| \pi_{|N(i)}]} = \frac{\Pr[\pi_{|N(i)} | X_i=1] \cdot \Pr[X_i=1]}{\Pr[\pi_{|N(i)} | X_i=0] \cdot \Pr[X_i=0]} = \frac{\Pr[\pi_{|N(i)} | X_i=1]}{\Pr[\pi_{|N(i)} | X_i=0]}$
$\in [\gamma^{-1}, \gamma]$. $\qquad\square$

We use the superscripted notation X' to denote the version of the random variable X conditioning on some transcript π. Hence, Lemma 2 states that the random variables X_i''s are independent, and each X_i' is γ-random for $i \in M$. It follows that the sum $\sum_{i \in M} X_i'$ has variance at least $\frac{m\gamma}{(1+\gamma)^2}$.

The idea is that conditioning on the transcript π, the sum of the parties' bits (in M) has high variance, and so the protocol is going to make a large error with constant probability. We describe the precise properties we need in the following technical lemma, whose proof appears in Section 4.1, from which Theorem 1 follows.

Lemma 3 (Large Variance Dichotomy). *Let $\gamma \geq 1$. There exists $\eta > 0$ (depending on γ) such that the following holds. Suppose Z_i's are m independent random variables in $\{0,1\}$ and are all γ-random, where $i \in [n]$. Define $Z := \sum_{i \in [m]} Z_i$ and $\sigma^2 := \frac{m\gamma}{2(1+\gamma)^2}$. Then, there exists an interval $[a,b]$ of length $\frac{\sigma}{2}$ such that the probabilities $\Pr[Z \geq b]$ and $\Pr[Z \leq a]$ are both at least η.*

Proof of Theorem 1: Using Lemma 3, we set $\gamma := \exp(\epsilon)$ and $Z_i := X_i'$, for each $i \in M$. Suppose $\eta > 0$ (depending on γ and hence on ϵ), $\sigma^2 := \frac{m\gamma}{2(1+\gamma)^2}$ and the interval $[a,b]$ are as guaranteed from the lemma. Suppose s is the sum of the bits of parties outside M. Let $c := \frac{a+b}{2} + s$.

Suppose the protocol makes an estimate that is at most c. Then, conditioning on π, the system still has enough randomness among parties in M, and with probability at least η, the real sum is at least $b + s$, which means the additive error is at least $\frac{\sigma}{4}$. The case when the protocol makes an estimate greater than c is symmetric. Therefore, conditioning on $\pi \notin \mathcal{P}$, the protocol makes an additive error of at least $\frac{\sigma}{4}$ with probability at least η in any case. Note that this is true even if the protocol is randomized.

Let $\mathcal{E}$ be the event that the protocol makes an additive error of at least $\frac{\sigma}{4}$. We have just proved that for $\pi \notin \mathcal{P}$, $\Pr_{\mathbf{X}, \Pi}[\mathcal{E}|\pi] \geq \eta$, where the probability is over the $\mathbf{X} = (X_i : i \in [n])$ and the randomness of the protocol Π.

Observe that $\Pr_{\mathbf{X}, \Pi}[\mathcal{E}|\pi] \geq \eta$ for all transcripts $\pi \notin \mathcal{P}$, and from Lemma 1, $\Pr[\mathcal{P}] \leq \frac{3}{4}$. Hence, we conclude that $\Pr_{\mathbf{X}, \Pi}[\mathcal{E}] \geq \frac{\eta}{4}$. It follows that there must exist some configuration $\mathbf{x}$ of the parties' bits such that $\Pr_\Pi[\mathcal{E}|\mathbf{x}] \geq \frac{\eta}{4}$. This completes the proof of Theorem 1. $\square$

4.1   Large Variance Dichotomy

We prove Lemma 3. For $i \in M$, let $p_i := \Pr[Z_i = 1]$. From the γ-randomness of Z_i, it follows that $\frac{1}{1+\gamma} \leq p_i \leq \frac{\gamma}{1+\gamma}$. Without loss of generality, we assume that there are at least $\frac{m}{2}$ indices for which $p_i \geq \frac{1}{2}$; otherwise, we consider $1 - Z_i$. Let $J \subseteq M$ be a subset of size $\frac{m}{2}$ such that for each $i \in J$, $p_i \geq \frac{1}{2}$.

Define for $i \in J$, $Y_i := Z_i - p_i$. Let $Y := \sum_{i \in J} Y_i$, and $Z' := \sum_{i \in J} Z_i$. It follows that $E[Y_i] = 0$ and $E[Y_i^2] = p_i(1 - p_i) \geq \frac{\gamma}{(1+\gamma)^2}$. Denote $\sigma^2 := \frac{m\gamma}{2(1+\gamma)^2}$, $\mu := E[Z'] = \sum_{i \in J} p_i$ and $\nu^2 := E[Y^2] = \sum_{i \in J} p_i(1 - p_i)$. We have $\nu^2 \geq \sigma^2$.
The required result can be achieved from the following lemma.

Lemma 4 (Large Deviation). *There exists $\eta_0 > 0$ (depending only on γ) such that $\Pr[|Y| \geq \frac{9\sigma}{10}] \geq \eta_0$.*

We show how Lemma 4 implies the conclusion of Lemma 3. Since $\Pr[|Y| \geq \frac{9\sigma}{10}] = \Pr[Z' \geq E[Z'] + \frac{9\sigma}{10}] + \Pr[Z' \leq E[Z'] - \frac{9\sigma}{10}]$, at least one of the latter two terms is at least $\frac{\eta_0}{2}$. We consider the case $\Pr[Z' \geq E[Z'] + \frac{9\sigma}{10}] \geq \frac{\eta_0}{2}$; the other case is symmetric.

By Hoeffding's Inequality, for all $u > 0$, $\Pr[Z' \geq E[Z'] + u] \leq \exp(-\frac{2u^2}{n})$. Setting $u := \frac{2\sigma}{5}$, we have $\Pr[Z' < E[Z'] + \frac{2\sigma}{5}] \geq 1 - \exp(-\frac{8\gamma}{25(1+\gamma)^2}) =: \eta_1$.

We set $\eta := \frac{1}{2}\min\{\frac{\eta_0}{2}, \eta_1\}$. Let $\widehat{Z} := \sum_{i \in M \setminus J} Z_i$. Observe that $\widehat{Z}$ and Z are independent. Hence we can take the required interval to be $[\mathsf{median}(\widehat{Z}) + E[Z] + \frac{2\sigma}{5}, \mathsf{median}(\widehat{Z}) + E[Z] + \frac{9\sigma}{10}]$, which has width $\frac{\sigma}{2}$.

Hence, it remains to prove Lemma 4.
Proof of Lemma 4: We use the method of sub-Gaussian moment generating function in the way described in [IN07, Remark 3.1].

First, for each $i \in M$, for any real h,

$$E[e^{hY_i}] = p_i \cdot e^{h(1-p_i)} + (1 - p_i) \cdot e^{h(0-p_i)}$$
$$= \exp(-p_i h) \cdot (1 + p_i(e^h - 1)) \leq \exp(p_i h^2),$$

where the last inequality follows from $1 + p(e^h - 1) \leq \exp(ph^2 + ph)$, for all real h and $\frac{1}{2} \leq p \leq 1$.

Let g be a standard Gaussian random variable, i.e., it has density function $x \mapsto \frac{1}{\sqrt{2\pi}}e^{\frac{1}{2}x^2}$. It is assumed that g is independent of all other randomness in the proof. Recall that $E[e^{hg}] = e^{\frac{1}{2}h^2}$ and for $h < \frac{1}{2}$, $E[e^{hg^2}] = \frac{1}{\sqrt{1-2h}}$.

For $0 \leq h \leq \frac{1}{8\mu}$, we have

$$E[e^{hY^2}] = E_Y[E_g[e^{\sqrt{2h}gY}]] = E_g[E_Y[e^{\sqrt{2h}g\sum_i Y_i}]]$$
$$= E_g[\prod_i E_{Y_i}[e^{\sqrt{2h}gY_i}]] \leq E_g[\prod_i e^{2hp_ig^2}]$$
$$= E_g[\exp(2\mu hg^2)] = \frac{1}{\sqrt{1 - 4\mu h}} \leq \sqrt{2}.$$

For $-\frac{1}{8\mu} \leq h \leq \frac{1}{8\mu}$, we have

$$E[e^{hY^2}] \leq 1 + hE[Y^2] + \sum_{m \geq 2} \frac{1}{m!}(8\mu|h|)^m (\frac{1}{8\mu})^m E[Y^{2m}]$$

$$\leq 1 + h\nu^2 + (8\mu h)^2 \sum_{m \geq 2} \frac{1}{m!}(\frac{1}{8\mu})^m E[Y^{2m}]$$

$$\leq 1 + h\nu^2 + (8\mu h)^2 E[\exp(\frac{Y^2}{8\mu})]$$

$$\leq 1 + h\nu^2 + 100\mu^2 h^2 \leq \exp(h\nu^2 + 100\mu^2 h^2)$$

Let $0 < \beta < 1$. For $-\frac{1}{8\mu} \leq h < 0$, we have

$$\Pr[Y^2 \leq (1-\beta)\nu^2] = \Pr[hY^2 \geq h(1-\beta)\nu^2]$$

$$\leq \exp(-h(1-\beta)\nu^2) \cdot E[\exp(hY^2)]$$

$$\leq \exp(h\beta\nu^2 + 100\mu^2 h^2).$$

Observe that $\frac{1}{1+\gamma} \leq \frac{\nu^2}{\mu} = \frac{\sum_i p_i(1-p_i)}{\sum_i p_i} \leq \frac{\gamma}{1+\gamma}$.

We can set $h := -\frac{\beta\nu^2}{200\mu^2} \geq -\frac{1}{8\mu}$, and we have $\Pr[Y^2 \leq (1-\beta)\nu^2] \leq \exp(-\frac{\beta^2\nu^4}{400\mu^2})$
$\leq \exp(-\frac{\beta^2}{400(1+\gamma)^2})$.

Setting $\beta := \frac{19}{100}$ and observing that $\nu^2 \geq \sigma^2$, we have $\Pr[|Y| \geq \frac{9}{10}\sigma] \geq 1 - \exp(-(\frac{19}{2000(1+\gamma)})^2)$. $\qquad\square$

5   Differentially Private Protocols against Coalitions

We show that the lower bound proved in Section 4 is essentially tight. As noted by Beimel *et al.* [BNO08], one can generally obtain differentially private multi-party protocols with small error, by combining general (information theoretic) Secure Function Evaluation (SFE) techniques with differential privacy. Although our upper-bound constructions use standard techniques from SFE and differential privacy, we include the main result here for completeness. The details are given in the full version.

Theorem 2 (Differentially Private Protocols Against Coalitions). *Given $\epsilon > 0$, $0 < \delta < 1$ and a positive integer t, there exists an oblivious protocol among n parties each having a secret input $x_i \in \mathcal{U} := \{0, 1, 2, \ldots, \Delta\}$, such that the protocol uses only $O(nt)$ messages to estimate the sum $\sum_{i \in [n]} x_i$; the differential privacy guarantees and error bounds of the protocols are given as follows.*

(a) *For ϵ-differential privacy against any coalition of size t, with probability at least $1 - \eta$, the additive error is at most $O(\frac{\Delta}{\epsilon} \cdot \exp(\frac{\epsilon}{2\Delta})\sqrt{t+1}\log\frac{1}{\eta})$.*

(b) *For (ϵ, δ)-differential privacy against any coalition of size t, with probability at least $1 - \eta$, the additive error is at most $O(\frac{\Delta}{\epsilon} \cdot \exp(\frac{\epsilon}{2\Delta})\sqrt{\frac{n}{n-t}\log\frac{1}{\delta}}\log\frac{1}{\eta})$.*

References

[Ach03]   Achlioptas, D.: Database-friendly random projections: Johnson-Lindenstrauss with binary coins. J. Comput. Syst. Sci. 66(4), 671–687 (2003)

[BNO08]   Beimel, A., Nissim, K., Omri, E.: Distributed Private Data Analysis: Simultaneously Solving How and What. In: Wagner, D. (ed.) CRYPTO 2008. LNCS, vol. 5157, pp. 451–468. Springer, Heidelberg (2008)

[CK93]   Chor, B., Kushilevitz, E.: A communication-privacy tradeoff for modular addition. Information Processing Letters 45, 205–210 (1993)

[DG03]   Dasgupta, S., Gupta, A.: An elementary proof of a theorem of Johnson and Lindenstrauss. Random Struct. Algorithms 22(1), 60–65 (2003)

[DKM+06]   Dwork, C., Kenthapadi, K., McSherry, F., Mironov, I., Naor, M.: Our Data, Ourselves: Privacy Via Distributed Noise Generation. In: Vaudenay, S. (ed.) EUROCRYPT 2006. LNCS, vol. 4004, pp. 486–503. Springer, Heidelberg (2006)

[DMNS06]   Dwork, C., McSherry, F., Nissim, K., Smith, A.: Calibrating Noise to Sensitivity in Private Data Analysis. In: Halevi, S., Rabin, T. (eds.) TCC 2006. LNCS, vol. 3876, pp. 265–284. Springer, Heidelberg (2006)

[Dwo06]   Dwork, C.: Differential Privacy. In: Bugliesi, M., Preneel, B., Sassone, V., Wegener, I. (eds.) ICALP 2006. LNCS, vol. 4052, pp. 1–12. Springer, Heidelberg (2006)

[Dwo10]   Dwork, C.: A firm foundation for private data analysis. In: Communications of the ACM (2010)

[GRS09]   Ghosh, A., Roughgarden, T., Sundararajan, M.: Universally utility-maximizing privacy mechanisms. In: STOC 2009 (2009)

[IN07]   Indyk, P., Naor, A.: Nearest-neighbor-preserving embeddings. ACM Transactions on Algorithms 3(3) (2007)

[MMP+10]   McGregor, A., Mironov, I., Pitassi, T., Reingold, O., Talwar, K., Vadhan, S.: The limits of two-party differential privacy. In: FOCS (2010)

[MPRV09]   Mironov, I., Pandey, O., Reingold, O., Vadhan, S.: Computational Differential Privacy. In: Halevi, S. (ed.) CRYPTO 2009. LNCS, vol. 5677, pp. 126–142. Springer, Heidelberg (2009)

[PSP03]   Przydatek, B., Song, D., Perrig, A.: Sia: secure information aggregation in sensor networks. In: ACM Sensys (2003)

[RN10]   Rastogi, V., Nath, S.: Differentially private aggregation of distributed time-series with transformation and encryption. In: SIGMOD 2010, pp. 735–746 (2010)

[SCR+11]   Shi, E., Chan, H., Rieffel, E., Chow, R., Song, D.: Privacy-preserving aggregation of time-series data. In: NDSS (2011)

A Model for Minimizing Active Processor Time

Jessica Chang[1,*], Harold N. Gabow[2], and Samir Khuller[3,**]

[1] University of Washington, Seattle WA 98195
jschang@cs.washington.edu
[2] University of Colorado, Boulder CO 80309
hal@cs.colorado.edu
[3] University of Maryland, College Park MD 20742
samir@cs.umd.edu

Abstract. We introduce the following elementary scheduling problem. We are given a collection of n jobs, where each job J_i has an integer length ℓ_i as well as a set T_i of time intervals in which it can be feasibly scheduled. Given a parameter B, the processor can schedule up to B jobs at a timeslot t so long as it is "active" at t. The goal is to schedule all the jobs in the fewest number of active timeslots. The machine consumes a fixed amount of energy per active timeslot, *regardless* of the number of jobs scheduled in that slot (as long as the number of jobs is non-zero). In other words, subject to ℓ_i units of each job i being scheduled in its feasible region and at each slot at most B jobs being scheduled, we are interested in minimizing the total time during which the machine is active. We present a linear time algorithm for the case where jobs are unit length and each T_i is a single interval. For general T_i, we show that the problem is NP-complete even for $B = 3$. However when $B = 2$, we show that it can be solved. In addition, we consider a version of the problem where jobs have arbitrary lengths and can be preempted at any point in time. For general B, the problem can be solved by linear programming. For $B = 2$, the problem amounts to finding a triangle-free 2-matching on a special graph. We extend the algorithm of Babenko et. al. [3] to handle our variant, and also to handle non-unit length jobs. This yields an $O(\sqrt{L}m)$ time algorithm to solve the preemptive scheduling problem for $B = 2$, where $L = \sum_i \ell_i$. We also show that for $B = 2$ and unit length jobs, the optimal non-preemptive schedule has at most 4/3 times the active time of the optimal preemptive schedule; this bound extends to several versions of the problem when jobs have arbitrary length.

1 Introduction

Power management strategies have been widely studied in the scheduling literature [1,20,4]. Many of the models are motivated by the energy consumption of the processor. Consider, alternatively, the energy consumed by the operation of large storage systems. Data is stored in memory which may be turned on and off [2],

---

* Supported by NSF CCF-1016509, and GRF.
** Supported by NSF CCF-0728839, CCF-0937865 and a Google Award.

L. Epstein and P. Ferragina (Eds.): ESA 2012, LNCS 7501, pp. 289–300, 2012.
© Springer-Verlag Berlin Heidelberg 2012

and each task or job needs to access a subset of data items to run. At each time step, the scheduler can perform at most B jobs. The only requirement is that the memory banks containing the required data from these jobs be turned on. The problem studied in this paper is the special case where all the data is in one memory bank. (For even special cases involving multiple memory banks, the problem often becomes NP-complete.)

We propose a simple model for measuring energy usage on a parallel machine. Instead of focusing on solution quality from the scheduler's perspective, we focus on the energy savings of "efficient" schedules.

In many applications, a job has many intervals of availability because, e.g., it interfaces with an external event like a satellite reading or a recurring broadcast. The real-time and period scheduling literatures address this. More broadly, tasks may be constrained by user availability, introducing irregularity in the feasible intervals. Our model is general enough to capture jobs of this nature.

More formally, we are given a collection of n jobs, each job J_i having integer length ℓ_i and a set T_i of time intervals with integer boundaries in which it can be feasibly scheduled. In particular, $T_i = \{I_k^i = [r_k^i, d_k^i]\}_{k=1}^{m_i}$ is a non-empty set of disjoint intervals, where $|I_k^i| = d_k^i - r_k^i$ is the size of I_k^i. Note that if $m_i = 1$, then we can think of job J_i as having a single release time and a single deadline.

For ease of notation, we may refer to J_i as job i. Additionally, time is divided into unit length timeslots and for a given parallelism parameter B, the system (or machine) can schedule up to B jobs in a single timeslot. If the machine schedules any jobs at timeslot t, we say that it is "active at t". The goal is to schedule all jobs within their respective feasible regions, while minimizing the number of slots during which the machine is active. The machine consumes a fixed amount of energy per active slot. In other words, subject to each job J_i being scheduled in its feasible region T_i, and subject to at most B jobs being scheduled at any time, we would like to minimize the total active time spent scheduling the jobs. There may be instances when there is no feasible schedule for all the jobs. However, this case is easy to check, as we show in Sec. 2. Note that for a timeslot significantly large (e.g. on the order of an hour), any overhead cost for starting a memory bank is negligible compared to the energy spent being "on" for that unit of time.

To illustrate this model in other domains, consider the following operational problem. Suppose that a ship can carry up to B cargo containers from one port to another. Jobs have delivery constraints, i.e. release times and deadlines. An optimal schedule yields the minimum number of times the ship must be sent to deliver all the packages on time. The motivating assumption is that it costs roughly the same to send the ship, regardless of load and that there is an upper bound on the load.

We could also consider this as a basic form of "batch" processing similar to the work of Ikura and Gimple [19], which presented an algorithm minimizing completion time for batch processing on a single machine in the case of *agreeable*[1]

---

[1] When the ordering of jobs by release times is the same as the ordering of jobs by deadlines.

release times and deadlines. The natural extension to general release times and deadlines has an efficient algorithm as well [7]. Both of these works focus on finding a feasible schedule (which can be used to minimize maximum lateness). However in our problem, in addition we wish to minimize the number of batches.

For the cases of unit length jobs or those in which jobs can be preempted at integral time points, our scheduling problem can be modeled as a bipartite matching problem in which each node on the left needs to be matched with a node on the right. Each node on the right can be matched to up to B nodes on the left, and we are interested in minimizing the number of nodes on the right that have non-zero degree in the matching selected. This problem can easily be shown to be NP-hard. For unit length jobs and single intervals T_i, we can develop a fast algorithm to obtain an optimal solution to this scheduling problem. [2]

In particular, even for $B = 2$, there is plenty of structure due to the connection with matchings in graphs. We anticipate that this structure will be useful in the design of improved approximation algorithms for $B > 2$.

Main Results

For proof details, we refer the reader to the full version of this paper [5].

1. For the case where jobs have unit length and each T_i is one interval (i.e. $m_i = 1$), we develop an algorithm whose running time is linear. Our algorithm takes as input n jobs with release times and deadlines and outputs a non-preemptive schedule with the smallest number of active slots. The algorithm has the additional property that it schedules the maximum number of jobs. We also note that without loss of generality, time is slotted when job lengths, release times and deadlines are integral (Sec. 2). When the release times and deadlines are not integral, minimizing the number of batches can be solved optimally in polynomial time via dynamic programming. This objective is more restrictive than active time: a batch must start all its jobs at the same time and the system may work on at most one batch at any given point in time. Even so, scheduling unit length jobs with integer release times and deadlines to minimize active time is a special case of this. We extend this to the case with a budget on the number of active slots. The running time of the former algorithm was recently improved in [22].

2. We consider the generalization to arbitrary T_i, which is closely related to capacitated versions of vertex cover, k-center and facility location, all classic covering problems. In particular, for the special case where every job is feasible in exactly two timeslots, a 2-approximation is implied from the vertex cover result in [17]. The complexity of the problem depends on the value of B, since for any fixed $B \geq 3$, the problem is NP-hard. When $B = 2$ this problem can be solved optimally in $O(m\sqrt{n})$ time where m is the total number of timeslots which are feasible for some job (Sec. 3). We show that

---

[2] The problem can be solved in $O(n^2T^2(n+T))$ time using Dynamic Programming as was shown by Even et. al. [12], albeit the complexity of their solution is high. Their algorithm solves in the problem of stabbing a collection of horizontal intervals with the smallest number of vertical stabbers, each stabber having a bounded capacity.

this problem is essentially equivalent to the maximum matching problem computationally. In addition, we show that this algorithm can be extended to the case of non-unit length jobs where a job can be preempted at integer time points, i.e. scheduled in unit sized pieces.

3. We also consider the case when jobs have arbitrary lengths and can be preempted[3] at any time, i.e. not just at integer time points. For general B, the problem can be solved by linear programming. For $B = 2$, the problem amounts to finding a maximum triangle-free 2-matching on a special graph. Babenko et. al. [3] present an elegant algorithm that essentially shows a maximum cardinality triangle-free 2-matching can be found in the same time as a maximum cardinality matching. We extend it for our scheduling problem to show that when $B = 2$ and jobs can be scheduled in arbitrary given time slots, an optimal preemptive schedule can be found in $O(\sqrt{L}m)$ time, for L the total length of all jobs. The proof is sketched in Sec. 4.2.

4. We give a tight bound on the gain from arbitrary preemption: an optimal schedule allowing preemption only at integral times uses at most 4/3 the active time of the optimal preemptive schedule. This bound is the best possible. The proof draws on the Edmonds-Gallai decomposition of matching theory. A sketch is provided in Sec. 4.3.

Related Work

Problems of scheduling unit jobs have a rich history [29,18]. For scheduling unit length jobs with arbitrary release times and deadlines on B processors to minimize the sum of completion times, Simons and Warmuth [29] extended the work by Simons [28] giving an algorithm with running time $O(n^2 B)$ to find a feasible solution. For constant B, the running time is improved in [25].

In the closely related busy time problem [13,21], jobs of arbitrary length have release times and deadlines and are scheduled in batches, so that the job demand in a batch is never more than a given value. The goal is to minimize the total busy time. Unlike our problem, this model permits an unbounded number of machines, making every instance feasible. The "min gap" problem [4] is to find a schedule of unit length jobs minimizing the number of idle intervals. We refer the reader to [18] for results on unit job scheduling and to surveys [20,1] for a more comprehensive overview of scheduling results for power management problems.

2   Unit Jobs and Single Execution Windows

If time is not slotted and release times and deadlines are integral, then w.l.o.g., the optimal solution schedules jobs in "slots" implied by integer time boundaries.

We only provide a high level description of the algorithm. We assume at first that the instance is feasible, since this is easy to check by an EDF computation. Denote the distinct deadlines by $d_{i_1} < d_{i_2} < \ldots < d_{i_k}$, and let S_p be the set of jobs with deadline d_{i_p}. Then $S_1 \cup S_2 \ldots \cup S_k$ is the entire set of jobs where the deadlines range from time 1 to $T = d_{i_k}$.

---

[3] For the non-preemptive case minimizing active time is strongly NP-complete, even for $B = 2$.

We process the jobs in two phases. In Phase I we scan the jobs in order of *decreasing* deadline. We do not schedule any jobs, but modify their deadlines to create a new instance which has an equivalent optimal solution. The desired property of the new instance is that at most B jobs have the same deadline. Process the timeslots from right to left. At slot D, let S be the set of jobs that currently have deadline D. From S, select $\max(0, |S| - B)$ jobs with earliest release times and decrement their deadlines by one. If $|S| \leq B$ then we do not modify the deadlines of jobs in S. (Note that a job may have its deadline reduced multiple times since it may be processed repeatedly.)

Assume for simplicity's sake that after Phase I, S_p refers to the jobs of (modified) deadline d_{i_p}. In Phase II, jobs are actually scheduled. Initially, all jobs are labeled *unscheduled*. As the algorithm assigns a job to an active timeslot, its status changes to *scheduled*. Once a job is scheduled, it remains scheduled. Once a slot is declared active, it remains active for the entire duration of the algorithm.

We schedule the slots d_{i_p} from left to right. To schedule d_{i_p}, if there remain unscheduled jobs with that deadline, we schedule them. If there are fewer than B such jobs, we schedule additional jobs that are available. Job j is available if it is currently unscheduled and has $d_{i_p} \in [r_j, d_j)$. We schedule these available jobs EDF, until the slot is full or no more jobs are available.

Analysis of the Algorithm
It is easy to implement our algorithm in $O(n \log n)$ time using standard data structures. Suppose the initial instance I is transformed by Phase I to a modified instance I'. We prove the following properties about an optimal solution for I'.

Proposition 1. *An optimal solution for I' has the same number of active slots as an optimal solution for the original instance I.*

Proposition 2. *W.l.o.g., an optimal solution for I' uses a subset of slots that are deadlines.*

Theorem 1. *The algorithm computes an optimal solution for I' (i.e., with the fewest active slots).*

Proposition 3. *On infeasible instances, the algorithm maximizes the number of scheduled jobs and does so in the fewest number of active slots.*

Linear Time Implementation: We define Phase I' to be equivalent to Phase I: initially each deadline value has no jobs assigned to it. Process jobs j in order of decreasing release time, assigning j to a new deadline equal to the largest value $D \leq d_j$ that has $< B$ jobs assigned to it. We now argue their equivalence.

Let D (D^*, respectively) be the deadline assigned to job j in Phase I' (Phase I). Suppose for jobs i, $i < j$, i's Phase I and Phase I' deadlines are equal. Then $D^* \leq D$ by Phase I''s choice of D. Phase I first assigns D to the $< B$ jobs currently assigned D in Phase I', since they have release time $\geq r_j$. Among all jobs which can still be assigned D, j has the latest release time. So $D^* = D$. We implement Phase I' in $O(n+T)$ time using a disjoint set merging algorithm [16].

Let $D_1 < D_2 < \ldots < D_\ell$ be the distinct deadline values assigned to jobs in Phase I. Phase II compresses the active interval from $[0, T]$ to $[0, \ell]$, as follows: job j with Phase I deadline D_k gets deadline k; its release time r_j changes to the largest integer i with $D_i \leq r_j$ (let $D_0 = 0$). The algorithm returns the EDF schedule on this modified instance, with slot i changed back to D_i. Phase II assigns new release times in time $O(n + T)$ and constructs an EDF schedule via disjoint set merging, achieving time $O(n + T)$ [16].

3   Disjoint Collection of Windows

We discuss the generalization where each job's feasible timeslots are an arbitrary set of slots (as opposed to a single interval). For any fixed $B \geq 3$ the problem is NP-hard (see the full paper). There is an $O(\log n)$ approximation for this problem by an easy reduction to a submodular covering problem [30].

3.1   Connections to Other Problems

Capacitated Vertex Cover. Given graph $G = (V, E)$ and vertex capacities $k(v)$, the goal is to pick a subset $S \subseteq V$ and assign edges to vertices in S, so that each edge is assigned to an incident vertex and each vertex v is assigned up to $k(v)$ edges. Approximation algorithms are given in [17,6].

In the special case of the activation problem where each job has exactly two feasible timeslots, there is an equivalence with VCHC with uniform capacities $k(v) = B$: the slots are the vertices and the jobs are the edges. One implication of our result is that we can solve VCHC optimally when $k(v) = 2$.

Capacitated K-center Problem. Given a graph, pick K nodes (called centers) and assign each vertex to a center that is "close" to it [23]. No more than L nodes should be assigned to a center, minimize d such that each node is assigned to a center within distance d of it. Create an unweighted graph G^d induced by edges of weight $\leq d$. Create job J_i and timeslot t_i for each node i. Let J_i be feasible at t_j if and only if edge (i, j) is in G^d or $j = i$. Then assigning all vertices to K centers in G^d is equivalent to scheduling all jobs in K timeslots so that no slot is assigned more than $B = L$ jobs. Since we can solve the latter problem for $B = 2$, we can solve the K-center problem when $L = 2$.

Capacitated Facility Location. Given a set of facilities (with capacities) and clients, the goal is to open a subset of facilities and find a capacity-respecting assignment of clients to open facilities that minimizes the sum of facility opening costs and connection costs. When capacities are uniformly two, the problem can be solved exactly. Details are described at the end of Sec. 3.2.

3.2   Polynomial Solution for B=2

We consider scheduling jobs in the fewest number of active slots, when there are two processors and each job can be scheduled in a specified subset of timeslots. We use the following graph G. The vertex set is $J \cup T$, where J is the set of all jobs and T the set of all timeslots. The edge set is $\{(j,t) : \text{job } j \text{ can be scheduled}$ in timeslot $t\} \cup \{(t,t) : t \in T\}$. A degree-constrained subgraph (DCS) problem is defined on G by the degree constraints $d(j) \leq 1$ for every $j \in J$ and $d(t) \leq 2$ for every $t \in T$. By definition a loop (t,t) contributes two to the degree of t. Any DCS gives a schedule, and any schedule gives a DCS that contains a loop in every nonactive timeslot. Also any DCS D that covers ι jobs and contains λ loops has cardinality $|D| = \iota + \lambda$.

Lemma 1. *A maximum cardinality DCS (MC-DCS) D of G minimizes the number of active slots used by any schedule of $|V(D) \cap J|$ jobs.*

Let n be the number of jobs and m the number of given pairs (j,t) where job j can be scheduled at time t. Observe that G has $O(m)$ vertices and $O(m)$ edges, since we can assume every timeslot t is on some edge (j,t). In any graph, the majority of edges in a simple non-trivial path are not loops. So in the algorithms for MC-DCS and maximum matching, an augmenting path has length $O(n)$. An MC-DCS on G can be found in time $O(\sqrt{n}m)$. The approach is via non-bipartite matching. We describe two ways to handle the loops.

The first approach reduces the problem to maximum cardinality matching in a graph called the MG graph. To do this modify G, replacing each timeslot vertex t by two vertices t_1, t_2. Replace each edge (j,t) by two edges from j to the two replacement vertices for t. Finally replace each loop (t,t) by an edge joining the two corresponding replacement vertices.

A DCS D corresponds to a matching M in a natural way: if slot t is active in D then M matches corresponding edges of the form (j,t_i). If loop (t,t) is in D then M matches the replacement edge (t_1, t_2). Thus it is easy to see that an MC-DCS corresponds to a maximum cardinality matching of the same size.

A second approach is to use an algorithm for MC-DCS on general graphs. If it does not handle loops directly, modify G by replacing each loop (t,t) by a triangle (t, t_1, t_2, t), where each t_i is a new vertex with degree constraint $d(t_i) = 1$. A DCS in G corresponds to a DCS in the new graph that contains exactly $|T|$ more edges, one from each triangle. The cardinality algorithm of Gabow and Tarjan [15] gives the desired time bound. Let ι^* be the greatest number of jobs that can be scheduled. One can compute ι^* in time $O(\sqrt{n}m)$ by finding a MC-DCS on the bipartite graph formed by removing all loops from G [11].

Theorem 2. *A schedule for the greatest possible number of jobs (ι^*) that minimizes the number of active slots can be found in time $O(\sqrt{n}m)$.*

The idea is to show that the choice of ι^* jobs is irrelevant: the minimum number of active slots for a schedule of ι^* jobs can be achieved using *any* set of ι^* jobs that can be scheduled.

We note that matching is a natural tool for power minimization. Given a maximum cardinality matching problem, create a job for each vertex. If two vertices are adjacent, create a common slot in which they can be scheduled. A maximum cardinality matching corresponds to a schedule of minimum active time. Thus if $T(m)$ is the optimal busy time, a maximum cardinality matching on a graph of m edges can be found in $O(T(m))$ time. For bipartite graphs, the number of slots can be reduced from m to n. Thus a maximum cardinality matching on a bipartite graph of n vertices and m edges can be found in $O(T(n,m))$ time, where $T(n,m)$ is the optimal time to find a minimum active time schedule for $O(n)$ jobs, $O(n)$ timeslots, and m pairs (j,t).

Our algorithm can be extended to other versions of the power minimization problem. For example the following corollary models a situation where power is limited.

Corollary 1. *For any given integer α, a schedule for the greatest possible number of jobs using at most α active slots can be found in time $O(\sqrt{n}m)$.*

Now consider capacitated facility location with capacities being two. With clients as jobs and facilities as slots, denote by MG_w the weighted equivalent of the graph MG. Edge (j,t_i) has weight equal to the cost of connecting client j to facility t, for $i=1,2$. Edge (t_1,t_2) has weight $-C(t)$ where $C(t)$ is the cost of opening facility t. Then, solving the facility location problem amounts to determining a minimum weight matching in MG_w where each job is matched, which can be achieved by generalizing our algorithm to the weighted case in a natural way.

4   Active Time and Arbitrary Preemption

This section considers jobs j of arbitrary lengths ℓ_j. There are B processors that operate in a collection of unit length "timeslots" $s \in S$. Each job has a set of timeslots in which it may be scheduled. Preemptions are allowed at any time, i.e., job j must be scheduled in a total of ℓ_j units of time but it can be started and stopped arbitrarily many times, perhaps switching processors. The only constraint is that it cannot execute on more than one processor at a time. We seek a schedule minimizing the active time.

4.1   Linear Program Formulation

The problem can be written as a linear program. Form a bipartite graph where edge $js \in E$ if and only if job j can be scheduled in slot s. A variable x_{js} gives the amount of time job j is scheduled in slot s, and a variable i_s gives the amount of idletime in slot s. The problem is equivalent to the following LP:

$$
\begin{aligned}
\text{max.} \quad & \textstyle\sum_{s\in S} i_s \\
\text{s.t.} \quad & \textstyle\sum_{js\in E} x_{js} \geq \ell_j \quad j \in J && (1.a) \\
& \textstyle\sum_{js\in E} x_{js} + Bi_s \leq B \quad s \in S && (1.b) \\
& x_{js} + i_s \leq 1 \quad js \in E && (1.c) \\
& i_s, x_{js} \geq 0 \quad s \in S,\ js \in E
\end{aligned}
\qquad (1)
$$

Observe that the inequalities (1.b)–(1.c) are necessary and sufficient for scheduling x_{js} units of job j, $js \in E$, in $1-i_s$ units of time (on B processors). Necessity is clear. For sufficiency, order jobs and processors arbitrarily. Repeatedly schedule the next $1 - i_s$ units of jobs on the next processor, until all jobs are scheduled. Even though a job may be split across processors, it would be scheduled last on one processor and first on the next, so (1.c) guarantees it is not executed simultaneously on two processors. A variant of this LP gives a polynomial-time solution to the problem where each job has a single interval $[r_j, d_j]$ in which it can be scheduled. The idea is to redefine S as the collection of minimal intervals formed by the $2n$ points $\{r_j, d_j : j \in J\}$.

4.2   The Case $B = 2$ and 2-Matchings

It is convenient to use a variant of the above LP based on the matching graph MG of Sec. 3. Each edge of MG has a linear programming variable. Specifically each edge $js \in E(G)$ gives rise to two variables x_e, $e = js_i$, $i \in \{1, 2\}$, that give the amount of time job j is scheduled on processor i in timeslot s. Also each timeslot $s \in S$, has a variable x_e, $e = s_1 s_2$ that gives the amount of inactive time in slot s. The problem is equivalent to the following LP:

$$
\begin{aligned}
\text{max.} \quad & \sum\{x_e : e \in E(MG)\} \\
\text{s.t.} \quad & \sum\{x_e : e \text{ incident to } j\} && = \ell_j \; j \in J && \text{(2.a)} \\
& \sum\{x_e : e \text{ incident to } s_i\} && \leq 1 \; s \in S, \; i \in \{1, 2\} && \text{(2.b)} \quad \text{(2)} \\
& \sum\{x_e : e \in \{js_1, js_2, s_1 s_2\}\} && \leq 1 \; js \in E(G) && \text{(2.c)} \\
& x_e && \geq 0 \; e \in E(MG)
\end{aligned}
$$

To see that this formulation is correct note that any schedule supplies feasible x_e values. Conversely any feasible x_e's give a feasible solution to LP (1) (e.g., set $x_{js} = x_{js_1} + x_{js_2}$ and $i_s = x_{s_1 s_2}$) and so corresponds to a schedule. Finally the objective of (2) is the sum over job lengths plus the total inactive time, so like (1) it maximizes the total inactive time.

Now consider an arbitrary graph $G = (V, E)$ and the polyhedron defined by the following system of linear inequalities:

$$
\begin{aligned}
& \sum\{x_e : e \text{ incident to } v\} \leq 1 && v \in V && \text{(3.a)} \\
& \sum\{x_e : e \text{ an edge of } T\} \leq 1 && T \text{ a triangle of } G && \text{(3.b)} \quad \text{(3)} \\
& x_e \geq 0 && e \in E
\end{aligned}
$$

Call $\sum\{x_e : e \in E\}$ the *size* of a solution to (3). Say vertex v is *covered* if equality holds in (3.a). Define a *2-matching* M to be an assignment of weight 0,1 or $\frac{1}{2}$ to each edge of G so that each vertex is incident to edges of total weight ≤ 1.[4] M is *basic* if its edges of weight $\frac{1}{2}$ form a collection of (vertex-disjoint) odd cycles. The basic 2-matchings are precisely the vertices of the polyhedron determined by the constraints (3) with (3.b) omitted. A basic 2-matching is *triangle-free* if no triangle has positive weight on all three of its edges. Cornuéjols and Pulleyblank [8] showed the triangle-free 2-matchings are precisely the vertices of the polyhedron determined by constraints (3).

---

[4] This definition of a 2-matching scales the usual definition by a factor $\frac{1}{2}$, i.e., the weights are usually 0,1 or 2.

When job lengths are one, the inequalities of (2) are system (3) for graph MG, with the requirement that every job vertex is covered. It is clear that the result of Cornuéjols and Pulleyblank implies our scheduling problem is solved by any maximum-size triangle-free 2-matching on MG subject to the constraint that it covers each job vertex. Interestingly, there is always a solution where each job is scheduled either completely in one slot or split into two pieces of size $1/2$.

Cornuéjols and Pulleyblank give two augmenting path algorithms for triangle-free 2-matching: they find such a matching that is perfect (i.e., every vertex is fully covered) in time $O(nm)$ [9], and such a matching that has minimum cost in time $O(n^2m)$ [8]. (The latter bound clearly applies to our scheduling problem, since it can model the constraint that every job vertex is covered.) Babenko et. al. [3] showed that a maximum cardinality triangle-free 2-matching can be found in time $O(\sqrt{n}m)$. This is done by reducing the problem to ordinary matching, with the help of the Edmonds-Gallai decomposition.

The full paper gives two easy extensions of [3]: a simple application of the Mendelsohn-Dulmage Theorem [24] shows that a maximum cardinality triangle-free 2-matching can be found in the same asymptotic time as a maximum cardinality matching (e.g., the algebraic algorithm of Mucha and Sankowski [27] can be used). This result extends to maximum cardinality triangle-free 2-matchings that are constrained to cover a given set of vertices (this is the variant needed for our scheduling problem). The development in the full paper is self-contained and based on standard data structures for blossoms. For instance, it is based on a simple definition of an *augmenting cycle* (analog of an augmenting path), leading to an *augmenting blossom* (which models the *triangle cluster*[5] of Cornuéjols and Pulleyblank [8,9]); we show a maximum cardinality matching with the greatest number of augmenting blossoms gives a maximum cardinality cardinality triangle-free 2-matching. [6]

4.3   Applications to Preemptive Scheduling for $B = 2$

Our algorithm for finding a maximum triangle-free 2-matching and LP (2) solve the preemptive scheduling problem for unit jobs. We extend this to jobs of arbitrary length by reducing to unit lengths, using vertex substitutes from matching theory. A sketch follows; details are in the full paper.

A graph UG is defined wherein a job of length ℓ that can be preemptively scheduled in s given time slots is represented by a complete bipartite graph $K_{s-\ell,s}$. The $s - \ell$ vertices on the small side will be matched to $s - \ell$ vertices on the big side, leaving ℓ vertices to be matched to time slots where the job is scheduled. The triangles of UG are similar to those of MG. Lemma 4 (*ii*) in the full paper (or [9]) implies that a solution to the scheduling problem is given by a maximum cardinality triangle-free 2-matching on UG that covers all vertices in $K_{s-\ell,s}$ graphs. For efficiency we use the sparse vertex substitute technique of [14] which works on (various) graphs of $O(m)$ edges rather than UG itself.

---

[5] A triangle cluster is a connected subgraph whose edges can be partitioned into triangles, such that every vertex shared by two or more triangles is a cut vertex.

[6] Our algorithm was developed independently of the authoritative algorithm of [3].

Theorem 3. *Let J be a set of unit jobs that can be scheduled on $B = 2$ processors. A preemptive schedule for J minimizing the active time can be found in $O(\sqrt{n}m)$ time. The result extends to jobs of arbitrary integral length ℓ_j, where the time is $O(\sqrt{L}m)$ for L the sum of all job lengths.*

What can be gained by allowing arbitrary preemption? For example when $B = 2$, three unit jobs that may be scheduled in slots 1 or 2 require 2 units of active time under preemption at integral points, and only $3/2$ units under arbitrary preemption. We now show this example gives the greatest disparity between the two preemption models for $B = 2$.

Recall that we construct an optimal preemptive schedule $\mathcal{P}$ by finding a maximum size triangle-free 2-matching (on MG for unit jobs, and UG for arbitrary length jobs) and converting it to $\mathcal{P}$ in the obvious way via LP (2). The following lemma is proved by examining the structure of blossoms in our special graphs, e.g., the triangles all have the form js_1s_2 for j a job vertex and s_1s_2 the edge representing a time slot; also vertices s_1 and s_2 are isomorphic. The matching terminology in the following lemma is from [10; or see 24].

Lemma 2. *(i) A blossom in MG (UG) is a triangle cluster iff it contains exactly one vertex of J (J_U).(ii) Let H be a Hungarian subgraph in MG (UG). Any slot s either has both its vertices inner, or both its vertices outer (and in the same blossom) or neither vertex in V_H. (iii) Let the optimal preemptive schedule $\mathcal{P}$ be constructed from $T(M)$, the 2-matching corresponding to matching M, and let $\mathcal{B}$ be an augmenting blossom of M. The slots with both vertices in $\mathcal{B}$ have $\geq 3/2$ units of active time in $\mathcal{P}$.*

Theorem 4. *For $B = 2$ and a set of jobs J, the minimum active time in a schedule allowing preemption at integral points is $\leq 4/3$ times that of a schedule permitting arbitrary preemption.*

Remark: The theorem holds for unit jobs, and for jobs of arbitrary length ℓ_j when time is slotted, i.e., a schedule is allowed to execute a length ℓ_j job in ℓ_j distinct timeslots, but no more.

References

1. Albers, S.: Energy-efficient algorithms. CACM 53(5), 86–96 (2010)
2. Amur, H., Cipra, J., Gupta, V., Kozuch, M., Ganger, G., Schwan, K.: Robust and flexible power-proportional storage. In: Proceedings of 1st SoCC, pp. 217–218 (2010)
3. Babenko, M., Gusakov, A., Razenshteyn, I.: Triangle-Free 2-Matchings Revisited. In: Thai, M.T., Sahni, S. (eds.) COCOON 2010. LNCS, vol. 6196, pp. 120–129. Springer, Heidelberg (2010)
4. Baptiste, P.: Scheduling unit tasks to minimize the number of idle periods: a polynomial time algorithm for offline dynamic power management. In: SODA 2006 (2006)
5. Chang, J., Gabow, H.N., Khuller, S.: A Model for Minimizing Active Processor Time (2012), http://www.cs.umd.edu/~samir/grant/active.pdf

6. Chuzhoy, J., Naor, S.: Covering problems with hard capacities. SIAM J. on Computing 36(2), 498–515 (2006)
7. Condotta, A., Knust, S., Shakhlevich, N.: Parallel batch scheduling of equal-length jobs with release and due dates. J. of Scheduling 13(5), 463–677 (2010)
8. Cornuéjols, G., Pulleyblank, W.: A matching problem with side conditions. Discrete Math. 29(2), 135–159 (1980)
9. Cornuéjols, G., Pulleyblank, W.: Perfect triangle-free 2-matchings. Math. Programming Study 13, 1–7 (1980)
10. Edmonds, J.: Paths, trees and flowers. Canadian J. Math. 17, 447–467 (1965)
11. Even, S., Tarjan, R.E.: Network flow and testing graph connectivity. SIAM J. on Computing 4(4), 507–581 (1975)
12. Even, G., Levi, R., Rawitz, D., Schieber, B., Shahar, S., Sviridenko, M.: Algorithms for capacitated rectangle stabbing and lot sizing with joint set up costs. ACM Trans. on Algorithms 4(3), 34–51 (2008)
13. Flammini, M., Monaco, G., Moscardelli, L., Shachnai, H., Shalom, M., Tamir, T., Zaks, S.: Minimizing total busy time in parallel scheduling with application to optical networks. In: IPDPS Conference, pp. 1–12 (2009)
14. Gabow, H.N.: An efficient reduction technique for degree-constrained subgraph and bidirected network flow problems. In: Proc. 15th Annual STOC, pp. 448–456 (1983)
15. Gabow, H.N., Tarjan, R.E.: Faster scaling algorithms for general graph matching problems. J. ACM 38(4), 815–853 (1991)
16. Gabow, H.N., Tarjan, R.E.: A linear-time algorithm for a special case of disjoint set union. J. Computer and System Sciences 30(2), 209–221 (1985)
17. Gandhi, R., Halperin, E., Khuller, S., Kortsarz, G., Srinivasan, A.: An improved approximation algorithm for vertex cover with hard capacities. J. of Computer and System Sciences 72(1), 16–33 (2006)
18. Garey, M., Johnson, D., Simons, B., Tarjan, R.: Scheduling unit-time jobs with arbitrary release times and deadlines. SIAM J. on Computing 10(2), 256–269 (1981)
19. Ikura, Y., Gimple, M.: Efficient scheduling algorithms for a single batch processing machine. Operations Research Letters 5(2), 61–65 (1986)
20. Irani, S., Pruhs, K.: Algorithmic problems in power management. SIGACT News 36(2), 63–76 (2005)
21. Khandekar, R., Schieber, B., Shachnai, H., Tamir, T.: Minimizing busy time in multiple machine real-time scheduling. In: FST&TCS Conference, pp. 169–180 (2010)
22. Koehler, F., Khuller, S.: Quick and efficient: fast algorithms for completion time and batch minimization on multiple machines (manuscript)
23. Khuller, S., Sussman, Y.J.: The capacitated k-center problem. SIAM J. on Discrete Mathematics 13(30), 403–418 (2000)
24. Lawler, E.: Combinatorial Optimization. In: Holt, Rinehart and Winston (1976)
25. López-Ortiz, A., Quimper, C.-G.: A fast algorithm for multi-machine scheduling problems with jobs of equal processing times. In: Proc. of Symposium on Theoretical Aspects of Computer Science (2011)
26. Lovász, L., Plummer, M.D.: Matching Theory. North-Holland (1986)
27. Mucha, M., Sankowski, P.: Maximum Matchings via Gaussian Elimination. In: FOCS, pp. 248–255 (2004)
28. Simons, B.: Multiprocessor scheduling of unit-time jobs with arbitrary release times and deadlines. SIAM J. on Computing 12(2), 294–299 (1983)
29. Simons, B., Warmuth, M.: A fast algorithm for multiprocessor scheduling of unit length jobs. SIAM J. on Computing 18(4), 690–710 (1989)
30. Wolsey, L.A.: An analysis of the greedy algorithm for the submodular set covering problem. Combinatorica 2(4), 385–393 (1982)

Polynomial-Time Algorithms for Energy Games with Special Weight Structures*

Krishnendu Chatterjee[1],[**], Monika Henzinger[2],[***], Sebastian Krinninger[2],[***], and Danupon Nanongkai[2]

[1] IST Austria (Institute of Science and Technology, Austria)
[2] University of Vienna, Austria

Abstract. Energy games belong to a class of *turn-based two-player infinite-duration games* played on a weighted directed graph. It is one of the rare and intriguing combinatorial problems that lie in NP∩co-NP, but are not known to be in P. While the existence of polynomial-time algorithms has been a major open problem for decades, there is no algorithm that solves any non-trivial subclass in polynomial time.

In this paper, we give several results based on the weight structures of the graph. First, we identify a notion of *penalty* and present a polynomial-time algorithm when the penalty is large. Our algorithm is the first polynomial-time algorithm on a large class of weighted graphs. It includes several counter examples that show that many previous algorithms, such as value iteration and random facet algorithms, require at least subexponential time. Our main technique is developing the first non-trivial *approximation* algorithm and showing how to convert it to an exact algorithm. Moreover, we show that in a practical case in verification where weights are clustered around a constant number of values, the energy game problem can be solved in polynomial time. We also show that the problem is still as hard as in general when the clique-width is bounded or the graph is strongly ergodic, suggesting that restricting graph structures need not help.

1 Introduction

Consider a restaurant A having a budget of e competing with its rival B across the street who has an unlimited budget. Restaurant B can observe the food price at A, say p_0, and responds with a price p_1, causing A a loss of $w(p_0, p_1)$, which could potentially put A out of business. If A manages to survive, then it can respond to B with a price p_2, gaining itself a profit of $w(p_1, p_2)$. Then B will

---

* Full version available at http://eprints.cs.univie.ac.at/3455/
** Supported by the Austrian Science Fund (FWF): P23499-N23, the Austrian Science Fund (FWF): S11407-N23 (RiSE), an ERC Start Grant (279307: Graph Games), and a Microsoft Faculty Fellows Award.
*** Supported by the Austrian Science Fund (FWF): P23499-N23, the Vienna Science and Technology Fund (WWTF) grant ICT10-002, the University of Vienna (IK I049-N), and a Google Faculty Research Award.

L. Epstein and P. Ferragina (Eds.): ESA 2012, LNCS 7501, pp. 301–312, 2012.
© Springer-Verlag Berlin Heidelberg 2012

try to put A out of business again with a price p_3. How much initial budget e does A need in order to guarantee that its business will survive forever? This is an example of a perfect-information turn-based infinite-duration game called an *energy game*, defined as follows.

In an energy game, there are two players, Alice and Bob, playing a game on a finite directed graph $G = (V, E)$ with weight function $w : E \to \mathbb{Z}$. Each node in G belongs to either Alice or Bob. The game starts by placing an imaginary car on a specified starting node v_0 with an imaginary energy of $e_0 \in \mathbb{Z}^{\geq 0} \cup \{\infty\}$ in the car (where $\mathbb{Z}^{\geq 0} = \{0, 1, \ldots\}$). The game is played in *rounds*: at any round $i > 0$, if the car is at node v_{i-1} and has energy e_{i-1}, then the owner of v_{i-1} moves the car from v_{i-1} to a node v_i along an edge $(v_{i-1}, v_i) \in E$. The energy of the car is then updated to $e_i = e_{i-1} + w(v_{i-1}, v_i)$. The goal of Alice is to sustain the energy of the car while Bob will try to make Alice fail. That is, we say that Alice *wins* the game if the energy of the car is never below zero, i.e. $e_i \geq 0$ for all i; otherwise, Bob wins. The problem of *computing the minimal sufficient energy* is to compute the minimal initial energy e_0 such that Alice wins the game. (Note that such e_0 always exists since it could be ∞ in the worst case.) The important parameters are the number n of nodes in the graph, the number m of edges in the graph, and the weight parameter W defined as $W = \max_{(u,v) \in E} |w(u, v)|$.

The class of energy games is a member of an intriguing family of *infinite-duration turn-based games* which includes alternating games [22], and has applications in areas such as computer-aided verification and automata theory [3,7], as well as in online and streaming problems [24]. The energy game is polynomial-time equivalent to the *mean-payoff game* [11,16,24,5], includes the *parity game* [18] as a natural subproblem, and is a subproblem of the *simple stochastic game* [9,24]. While the energy game is relatively new and interesting in its own right, it has been implicitly studied since the late 80s, due to its close connection with the mean-payoff game. In particular, the seminal paper by Gurvich et al. [16] presents a simplex-like algorithm for the mean-payoff game which computes a "potential function" that is essentially the energy function. These games are among the rare combinatorial problems, along with *Graph Isomorphism*, that are unlikely to be NP-complete (since they are in UP ∩ co-UP ⊆ NP ∩ co-NP) but not known to be in P. It is a major open problem whether any of these games are in P or not. The current fastest algorithms run in pseudopolynomial ($O(nmW)$) and randomized subexponential ($O(2^{\sqrt{n \log n}} \log W)$) time [6,2]. We are not aware of any polynomial-time algorithms for non-trivial special cases of energy games or mean-payoff games.

These games also have a strong connection to *Linear Programming* and *(mixed) Nash equilibrium computation* (see, e.g., [23,10]). For example, along with the problem of Nash Equilibrium computation, they are in a low complexity class lying very close to P called CCLS [10] which is in PPAD ∩ PLS, implying that, unlike many problems in Game Theory, these problems are *unlikely* to be PPAD-complete. Moreover, they are related to the question whether there exists a pivoting rule for the simplex algorithm that requires a polynomial number of pivoting steps on any linear program, which is perhaps one of the most important problems in the field of linear programming. In fact, several randomized pivoting rules have been conjectured to solve linear programs in polynomial time until recent breakthrough results (see [14,12,13]) have rejected these conjectures. As noted in [14], infinite-duration turn-based games played an important role in this breakthrough as the lower bounds were first developed for these games and later extended to linear programs. Moreover, all these games are LP-type problems which generalize linear programming [17].

Our Contributions. In this paper we identify several classes of graphs (based on weight structures) for which energy games can be solved in polynomial time. Our first contribution is an algorithm whose running time is based on a parameter called *penalty*. For the sake of introduction, we define penalty as follows[1]. For any starting node s, let $e^*_{G,w}(s)$ be the minimal sufficient energy. We say that s *has a penalty of at least D* if Bob has a strategy[2] τ such that (1) if Alice plays her best strategy, she will need $e^*_{G,w}(s)$ initial energy to win the game, and (2) if Alice plays a strategy σ that makes her lose the game for any initial energy, then she still loses the game against τ even if she can add an additional energy of D to the car *in every turn*. Intuitively, the penalty of D means that either Alice does not need additional energy in any turn or otherwise she needs an additional energy of at least D in every turn. Let the *penalty of graph* (G, w), denoted by $P(G, w)$, be the supremum of all D such that every node s has penalty of at least D.

Theorem 1. *Given a graph (G, w) and an integer M we can compute the minimal initial energies of all nodes in $O(mn(\log \frac{M}{n})(\log \frac{M}{n\lceil P(G,w)\rceil}) + m\frac{M}{\lceil P(G,w)\rceil})$ time,[3] provided that for all v, $e^*_{G,w}(v) < \infty$ implies that $e^*_{G,w}(v) \leq M$.*

We note that in addition to (G, w), our algorithm takes M as an input. If M is unknown, we can simply use the universal upper bound $M = nW$ due to [6]. Allowing different values of M will be useful in our proofs. We emphasize that the algorithm can run without knowing $P(G, w)$. Our algorithm is as efficient as the fastest known pseudopolynomial algorithm in the general case (where $M = nW$ and $P(G, w) = 1/n$), and solves several classes of problems that are not known to be solvable in polynomial time. As an illustration, consider the class of graphs where each cycle has total weight either positive or less than $-W/2$. In this case, our algorithm runs in polynomial time. *No previously known algorithm* (including the value iteration algorithm [6] and simplex-style algorithms with several pivoting rules [14,12,13]) can do this because previous worst-case instances (e.g., [13,16]) fall in this class of graphs (see Sect. 6).

Our second contribution is an algorithm that approximates the minimal energy within some *additive error* where the size of the error depends on the penalty. This result is the main tool in proving Theorem 1 where we show how to use the approximation algorithm to compute the minimal energy *exactly*.

---

[1] We note that the precise definition is slightly more complicated (see Sect. 2).
[2] To be precise, the strategy must be a so-called *positional strategy* (see Sect. 2).
[3] For simplicity we assume that logarithms in running times are always at least 1.

Theorem 2. *Given a graph (G, w) with $P(G, w) \geq 1$, an integer M, and an integer c such that $n \leq c \leq nP(G, w)$, we can compute an energy function e such that $e(v) \leq e^*_{G,w}(v) \leq e(v) + c$ for all nodes v in $O(mnM/c)$ time, provided that for any node v, $e^*_{G,w}(v) < \infty$ implies that $e^*_{G,w}(v) \leq M$.*

The main technique in proving Theorem 2 is rounding weights appropriately. We note that a similar idea of approximation has been explored earlier in the case of mean-payoff games [4]. Roth et al. [22] show an *additive* FPTAS for rational weights in $[-1, 1]$. This implies an additive error of ϵW for any $\epsilon > 0$ in our setting. This does not help in general since the error depends on W. Boros et al [4] later achieved a *multiplicative* error of $(1 + \epsilon)$. This result holds, however, only when the edge weights are non-negative integers. In fact, it is shown that if one can approximate the mean-payoff within a small multiplicative error in the general case, then the exact mean-payoff can be found [15]. There is currently *no* polynomial-time approximation algorithm for general energy games.

Our third contribution is a variant of the *Value Iteration Algorithm* by Brim et al [6] which runs faster in many cases. The running time of the algorithm depends on a concept that we call *admissible list* (defined in Sect. 3) which uses the weight structure. One consequence of this result is used to prove Theorem 2. The other consequence is an algorithm for what we call the *fixed-window* case.

Theorem 3. *If there are d values $w_1, \ldots, w_d$ and a window size δ such that for every edge $(u, v) \in G$ we have $w(u, v) \in \{w_i - \delta, \ldots, w_i + \delta\}$ for some $1 \leq i \leq d$, then the minimal energies can be computed in $O(m\delta n^{d+1} + dn^{d+1} \log n)$ time.*

The fixed-window case, besides its theoretical attractiveness, is also interesting from a practical point of view. Energy and mean-payoff games have many applications in the area of verification, mainly in the synthesis of reactive systems with resource constraints [3] and performance aware program synthesis [7]. In most applications related to synthesis, the resource consumption is through only a few common operations, and each operation depending on the current state of the system consumes a related amount of resources. In other words, in these applications there are d groups of weights (one for each operation) where in each group the weights differ by at most δ (i.e, δ denotes the small variation in resource consumption for an operation depending on the current state), and d and δ are typically constant. Theorem 3 implies a polynomial time algorithm for this case.

We also show that the problem is still as hard as the general case even when the clique-width is bounded or the graph is strongly ergodic (see Sect. 6). This suggests that restricting the graph structures might not help in solving the problem, which is in sharp contrast to the fact that parity games can be solved in polynomial time in these cases.

2   Preliminaries

Energy Games. An energy game is played by two players, Alice and Bob. Its input instance consists of a finite weighted directed graph (G, w) where all nodes have out-degree at least one[4]. The set of nodes V is partitioned into V_A and V_B, which belong to Alice and Bob respectively, and every edge $(u, v) \in E$ has an integer weight $w(u, v) \in \{-W, \ldots, W\}$. Additionally, we are given a node s and an initial energy e_0. We have already described the energy game informally in Sect. 1. To define this game formally, we need the notion of *strategies*. While general strategies can depend on the history of the game, it has been shown (see, e.g., [8]) that we can assume that if a player wins a game, a *positional strategy* suffices to win. Therefore we only consider positional strategies. A positional strategy σ of Alice is a mapping from each node in V_A to one of its out-neighbors, i.e., for any $u \in V_A$, $\sigma(u) = v$ for some $(u, v) \in E$. This means that Alice sends the car to v every time it is at u. We define a positional strategy τ of Bob similarly. We simply write "strategy" instead of "positional strategy" in the rest of the paper.

A *pair of strategies* (σ, τ) consists of a strategy σ of Alice and τ of Bob. For any pair of strategies (σ, τ), we define $G(\sigma, \tau)$ to be the subgraph of G having only edges corresponding to the strategies σ and τ, i.e., $G(\sigma, \tau) = (V, E')$ where $E' = \{(u, \sigma(u)) \mid u \in V_A\} \cup \{(u, \tau(u)) \mid u \in V_B\}$. In $G(\sigma, \tau)$ every node has a unique successor.

Now, consider an energy game played by Alice and Bob starting at node s with initial energy e_0 using strategies σ and τ, respectively. We use $G(\sigma, \tau)$ to determine who wins the game as follows. For any i, let P_i be the (unique) directed path of length i in $G(\sigma, \tau)$ originating at s. Observe that P_i is exactly the path that the car will be moved along for i rounds and the energy of the car after i rounds is $e_i = e_0 + w(P_i)$ where $w(P_i)$ is the sum of the edge weights in P_i. Thus, we say that *Bob wins* the game if there exists i such that $e_0 + w(P_i) < 0$ and *Alice wins* otherwise. Equivalently, we can determine who wins as follows. Let C be the (unique) cycle reachable by s in $G(\sigma, \tau)$ and let $w(C)$ be the sum of the edge weights in C. If $w(C) < 0$, then Bob wins; otherwise, Bob wins if and only if there exists a *simple* path P_i of some length i such that $e_0 + w(P_i) < 0$.

This leads to the following definition of the *minimal sufficient energy* at node s *corresponding to strategies* σ and τ, denoted by $e^*_{G(\sigma,\tau),w}(s)$: If $w(C) < 0$, then $e^*_{G(\sigma,\tau),w}(s) = \infty$; otherwise, $e^*_{G(\sigma,\tau),w}(s) = \max\{0, -\min w(P_i)\}$ where the minimization is over all *simple* paths P_i in $G(\sigma, \tau)$ originating at s. We then define the *minimal sufficient energy* at node s to be

$$e^*_{G,w}(s) = \min_{\sigma} \max_{\tau} e^*_{G(\sigma,\tau),w}(s) \tag{1}$$

where the minimization and the maximization are over all positional strategies σ of Alice and τ of Bob, respectively. We note that it follows from Martin's determinacy theorem [20] that $\min_{\sigma} \max_{\tau} e^*_{G(\sigma,\tau),w}(s) = \max_{\tau} \min_{\sigma} e^*_{G(\sigma,\tau),w}(s)$, and thus it does not matter which player picks the strategy first. We say that a strategy σ^* of Alice is an *optimal strategy* if for any strategy τ of Bob, $e^*_{G(\sigma^*,\tau),w}(s) \leq e^*_{G,w}(s)$. Similarly, τ^* is an optimal strategy of Bob if for any strategy σ of Alice, $e^*_{G(\sigma,\tau^*),w}(s) \geq e^*_{G,w}(s)$.

---

[4] (G, w) is usually called a "game graph" in the literature. We will simply say "graph".

We call any $e : V \rightarrow \mathbb{Z}^{\geq 0} \cup \{\infty\}$ an *energy function*. We call $e_{G,w}^*$ in Eq. (1) a *minimal sufficient energy function* or simply a *minimal energy function*. If $e(s) \geq e_{G,w}^*(s)$ for all s, then we say that e is a *sufficient* energy function. The goal of the energy game problem is to compute $e_{G,w}^*$.

We say that a natural number M is an *upper bound on the finite minimal energy* if for every node v either $e_{G,w}^*(v) = \infty$ or $e_{G,w}^*(v) \leq M$. A universal upper bound is $M = nW$ [6].

Penalty. Let (G, w) be a weighted graph. For any node s and real $D \geq 0$, we say that s has a *penalty of at least D* if there exists an optimal strategy τ^* of Bob such that for any strategy σ of Alice, the following condition holds for the (unique) cycle C reachable by s in $G(\sigma, \tau^*)$: if $w(C) < 0$, then the average weight on C is at most $-D$, i.e. $\sum_{(u,v) \in C} w(u,v)/|C| \leq -D$. Intuitively, this means that either Alice wins the game using a finite initial energy, or she loses *significantly*, i.e., even if she constantly receive an extra energy of a little less than D per round, she still needs an infinite initial energy in order to win the game. We note that $\sum_{(u,v) \in C} w(u,v)/|C|$ is known in the literature as the *mean-payoff* of s when Alice and Bob play according to σ and τ^*, respectively. Thus, the condition above is equivalent to saying that either the mean-payoff of s (when (σ, τ^*) is played) is non-negative or otherwise it is at most $-D$. We define the penalty of s, denoted by $P_{G,w}(s)$, as the supremum of all D such that s has a penalty of at least D. We say that the graph (G, w) has a penalty of at least D if every node s has a penalty of at least D, and define $P(G, w) = \min_{s \in G} P_{G,w}(s)$. Note that for any graph (G, w), $P(G, w) \geq 1/n$.

3   Value Iteration Algorithm with Admissible List

In this section we present a variant of the *Value Iteration Algorithm* for computing the minimal energy (see, e.g., [6] for the fastest previous variant). In addition to the graph (G, w), our algorithm uses one more parameter A which is a sorted list containing all possible minimal energy values. That is, the algorithm is promised that $e_{G,w}^*(v) \in A$ for every node v. We call A an *admissible list*.

Lemma 4. *There is an algorithm that, given a (sorted) admissible list A, computes the minimal energies of all nodes in (G, w) in $O(m|A|)$ time.*

We note that in some cases space can be saved by giving an algorithm that generates A instead of giving A explicitly. Before we present the idea of the theorem, we note that the simplest choice of an admissible list is $A = \{0, 1, \ldots, nW, \infty\}$. In this case the algorithm works like the current fastest pseudopolynomial algorithm by Brim et al [6] and has a running time of $O(mnW)$. If we consider certain special cases, then we can give smaller admissible lists. Our first example are graphs where every edge weight is a multiple of an integer $B > 0$.

Corollary 5. *Let (G, W) be a graph for which there is an integer $B > 0$ such that the weight of every edge $(u, v) \in G$ is of the form $w(u, v) = iB$ for some integer i, and M is an upper bound on the finite minimal energy* *(i.e., for any node v, if $e^*_{G,w}(v) < \infty$, then $e^*_{G,w}(v) \leq M$). There is an admissible list of size $O(M/B)$ which can be computed in $O(M/B)$ time. Thus there is an algorithm that computes the minimal energies of (G, w) in $O(mM/B)$ time.*

The above corollary will be used later in this paper. Our second example are graphs in which we have a (small) set of values $\{w_1, \ldots, w_d\}$ of size d and a window size δ such that *every* weight lies in $\{w_i - \delta, \ldots, w_i + \delta\}$ for one of the values w_i. This is exactly the situation described in Theorem 3 which is a consequence of Lemma 4. As noted in Sect. 1, in some applications d is a constant and δ is polynomial in n. In this case Theorem 3 implies a polynomial time algorithm. We now sketch the proofs of all these results

Proof (Proof idea of Lemma 4). The value iteration algorithm relies on the following characterization of the minimal energy (see full version for details).

Lemma 6 ([6]). *An energy function e is the minimal energy function if for every node $u \in V_A$ there is an edge $(u, v) \in E$ such that $e(u) + w(u, v) \geq e(v)$, and for every node $u \in V_B$ and edge $(u, v) \in E$ we have $e(u) + w(u, v) \geq e(v)$. Moreover, for any e' that satisfies this condition, $e(v) \leq e'(v)$ for every node v.*

The basic idea of the modified value iteration algorithm is as follows. The algorithm starts with an energy function $e(v) = \min A$ for every node v and keeps increasing e slightly in an attempt to satisfy the condition in Lemma 6. That is, as long as the condition is not fulfilled for some node u, it increases $e(u)$ to the next value in A, which could also be ∞. This updating process is repeated until e satisfies the condition in Lemma 6 (which will eventually happen at least when all $e(u)$ becomes ∞). Based on the work of Brim et al [6], it is straightforward to show the correctness of this algorithm. To get a fast running time we use their speed-up trick that avoids unnecessary checks for updates [6].   □

Proof (Proof idea of Corollary 5 and Theorem 3). To see how to get the results for our two special cases, we give suitable formulations of admissible lists based on a list that is always admissible. Given an upper bound M on the finite minimal energy, we define $U_M = \{0, \ldots, M, \infty\}$. We denote the set of different weights of a graph (G, w) by $R_{G,w} = \{w(u, v) \mid (u, v) \in E\}$. The set of all combinations of edge weights is defined as

$$C_{G,w} = \left\{ -\sum_{i=1}^{k} x_i \mid x_i \in R_{G,w} \text{ for all } i, 0 \leq k \leq n \right\} \cup \{\infty\}.$$

Our key observation is the following lemma (which we prove in the full version).

Lemma 7. *For every graph (G, w) with an upper bound M on the finite minimal energy we have $e^*_{G,w}(v) \in C_{G,w} \cap U_M$ for every $v \in G$.*

Now, for graphs with upper bound M on the finite minimal energy and edge weights that are multiples of B we define a list $A = \{i \cdot B \mid 0 \leq i \leq \lceil M/B \rceil\} \cup \{\infty\}$ which is admissible since $C_{G,w} \cap U_M \subseteq A$. Similarly, for graphs with values

$w_1, \ldots, w_d$ and a window size δ we define an admissible list $A' = \{x - \sum_{j=1}^{k} w_{i_j} \mid 1 \le i_j \le d, 0 \le k \le n, -n\delta \le x \le n\delta\} \cup \{\infty\}$. To prove the claimed running times we note that A has $O(M/B)$ elements and can be computed in $O(M/B)$ time, and A' has $O(\delta n^{d+1})$ elements and can be computed in $O(\delta n^{d+1} + d n^{d+1} \log n)$ time (see full version for details). $\square$

4   Approximating Minimal Energies for Large Penalties

This section is devoted to proving Theorem 2. We show that we can approximate the minimal energy of nodes in high-penalty graphs (see Sect. 2 for the definition of penalty). The key idea is *rounding edge weights*, as follows. For an integer $B > 0$ we denote the weight function resulting from rounding up every edge weight to the nearest multiple of B by w_B. Formally, the function w_B is given by

$$w_B(u, v) = \left\lceil \frac{w(u, v)}{B} \right\rceil \cdot B$$

for every edge $(u, v) \in E$. Our algorithm is as follows. We set $B = \lfloor c/n \rfloor$ (where c is as in Theorem 2). Since weights in (G, w_B) are multiples of B, e^*_{G, w_B} can be found faster than $e^*_{G, w}$ due to Corollary 5: we can compute e^*_{G, w_B} in time $O(mM/B) = O(mnM/c)$ provided that M is an upper bound on the finite minimal energy. This is the running time stated in Theorem 2. We complete the proof of Theorem 2 by showing that e^*_{G, w_B} is a good approximation of $e^*_{G, w}$ (i.e., it is the desired function e).

Proposition 8. *For every node v with penalty $P_{G, w}(v) \ge B = \lfloor c/n \rfloor$ (where $c \ge n$) we have*

$$e^*_{G, w_B}(v) \le e^*_{G, w}(v) \le e^*_{G, w_B}(v) + nB \le e^*_{G, w_B}(v) + c.$$

The first inequality is quite intuitive: We are doing Alice a favor by *increasing* edge weights from w to w_B. Thus, Alice should not require more energy in (G, w_B) than she needs in (G, w). As we show in the full version, this actually holds for *any* increase in edge weights: For any w' such that $w'(u, v) \ge w(u, v)$ for all $(u, v) \in G$, we have $e^*_{G, w'}(v) \le e^*_{G, w}(v)$. Thus we get the first inequality by setting $w' = w_B$.

We now show the second inequality in Proposition 8. Unlike the first inequality, we do not state this result for general increases of the edge weights as the bound depends on our rounding procedure. At this point we also need the precondition that the graph we consider has penalty at least B. We first show that the inequality holds when both players play some "nice" pair of strategies.

Lemma 9. *Let (σ, τ) be a pair of strategies. For any node v, if $e^*_{G(\sigma, \tau), w}(v) = \infty$ implies $e^*_{G(\sigma, \tau), w_B}(v) = \infty$, then $e^*_{G(\sigma, \tau), w}(v) \le e^*_{G(\sigma, \tau), w_B}(v) + nB$.*

The above lemma needs strategies (σ, τ) to be nice in the sense that if Alice needs infinite energy at node v in the original graph $(G(\sigma, \tau), w)$ then she also needs infinite energy in the rounded-weight graph $(G(\sigma, \tau), w_B)$. Our second crucial fact shows that if v has penalty at least B then there is a pair of strategies that has this nice property required by Lemma 9. *This is where we exploit the fact that the penalty is large.*

Lemma 10. *Let v be a node with penalty $P_{G,w}(v) \geq B$. Then, there is an optimal strategy τ^* of Bob such that for every strategy σ of Alice we have that $e^*_{G(\sigma,\tau^*),w}(v) = \infty$ implies $e^*_{G(\sigma,\tau^*),w_B}(v) = \infty$.*

To prove Lemma 9 we only have to consider a special graph where all nodes have out-degree one. Lemma 10 is more tricky as we need to come up with the right τ^*. We use τ^* that comes from the definition of the penalty (cf. Sect. 2). We give detailed proofs of Lemmas 9 and 10 in the full version.

The other tricky part of the proof is translating our result from graphs with fixed strategies to general graphs in order to prove the second inequality in Proposition 8. We do this as follows. Let σ^* be an optimal strategy of Alice for (G, w) and let (σ_B^*, τ_B^*) be a pair of optimal strategies for (G, w_B). Since v has penalty $P_{G,w}(v) \geq B$, Lemma 10 tells us that the preconditions of Lemma 9 are fulfilled. We use Lemma 9 and get that there is an optimal strategy τ^* of Bob such that $e^*_{G(\sigma_B^*,\tau^*),w}(v) \leq e^*_{G(\sigma_B^*,\tau^*),w_B}(v) + nB$. We now arrive at the chain of inequalities

$$
e^*_{G,w}(v) \overset{(a)}{=} e^*_{G(\sigma^*,\tau^*),w}(v) \overset{(b)}{\leq} e^*_{G(\sigma_B^*,\tau^*),w}(v) \leq e^*_{G(\sigma_B^*,\tau^*),w_B}(v) + nB
$$

$$
\overset{(c)}{\leq} e^*_{G(\sigma_B^*,\tau_B^*),w_B}(v) + nB \overset{(d)}{=} e^*_{G,w_B}(v) + nB
$$

that can be explained as follows. Since (σ^*, τ^*) and (σ_B^*, τ_B^*) are pairs of *optimal* strategies we have (a) and (d). Due to the optimality we also have $e^*_{G(\sigma^*,\tau^*),w}(v) \leq e^*_{G(\sigma,\tau^*),w}(v)$ for *any* strategy σ of Alice, and in particular σ_B^*, which implies (b). A symmetric argument gives (c).

5   Exact Solution by Approximation

We now use our result of the previous sections to prove Theorem 1. As the first step, we provide an algorithm that computes the minimal energy given a lower bound on the penalty of the graph. For this algorithm, we show how we can use the approximation algorithm in Sect. 4 to find an *exact* solution.

Lemma 11. *There is an algorithm that takes a graph (G, w), a lower bound D on the penalty $P(G, w)$, and an upper bound M on the finite minimal energy of (G, w) as its input and computes the minimal energies of (G, w) in $O(mn \log D + m \cdot \frac{M}{\lceil D \rceil})$ time. Specifically, for $D \geq \frac{M}{2n}$ it runs in $O(mn \log (M/n))$ time.*

Proof (Sketch). To illustrate the main idea, we focus on the case $D = M/(2n)$ where we want to show an $O(mn \log(M/n))$ running time. A proof for the general case is given in the full version. Let $\mathcal{A}$ be the approximation algorithm given in Theorem 2. Recall that $\mathcal{A}$ takes c as its input and returns $e(v)$ such that $e(v) \leq e^*_{G,w}(v) \leq e(v) + c$ provided that $n \leq c \leq nP(G, w)$. Our exact algorithm, will run $\mathcal{A}$ with parameter $c = \lceil M/2 \rceil$ which satisfies $c \leq nD \leq nP(G, w)$. By Theorem 2, this takes $O(mnM/c) = O(mn)$ time. Using the energy function e returned by $\mathcal{A}$, our algorithm produces a new graph (G, w') defined by $w'(u, v) = w(u, v) + e(u) - e(v)$ for all $(u, v) \in E$. It can be proved that this graph has the following crucial properties:

1. The penalty does not change, i.e., $P_{G,w}(v) = P_{G,w'}(v)$ for every node v.
2. We have $e^*_{G,w}(v) = e^*_{G,w'}(v) + e(v)$ for every node v.
3. The largest finite minimal energy of nodes in (G, w') is at most c (this follows from property 2 and the inequality $e^*_{G,w}(v) \leq e(v) + c$ of Theorem 2).

The algorithm then recurses on input (G, w'), D and $M' = c = M/2$. Properties 1 and 3 guarantee that the algorithm will return $e^*_{G,w'}(v)$ for every node v. It then outputs $e^*_{G,w'}(v) + e(v)$ which is guaranteed to be a correct solution (i.e., $e^*_{G,w}(v) = e^*_{G,w'}(v) + e(v)$) by the second property. The running time of this algorithm is $T(n, m, M) \leq T(n, m, M/2) + O(mn)$. We stop the recursion if $M \leq 2n$ is reached because in this case the value iteration algorithm runs in $O(mn)$ time. Thus we get $T(n, m, M) = O(mn \log(M/n))$ as desired.   □

We now prove Theorem 1 by extending the previous result to an algorithm that does not require the knowledge of a lower bound of the penalty. We repeatedly guess a lower bound for the penalty $P_{G,w}$ from $M/(2n), M/(4n), \ldots$. We can check whether the values returned by our algorithm are indeed the minimal energies in linear time (see Lemma 6). If our guess was successful we stop. Otherwise we guess a new lower bound which is half of the previous one. Eventually, our guess will be correct and we will stop before the guessed value is smaller than $P(G, w)/2$ or one (in the latter case we simply run the value iteration algorithm). Therefore we get a running time of

$$O\left(mn\left(\log \tfrac{M}{2n} + \log \tfrac{M}{4n} + \ldots + \log(\lceil P(G,w)\rceil)\right) + m\left(2n + 4n + \ldots + \tfrac{M}{\lceil P(G,w)\rceil}\right)\right)$$

which solves to $O(mn(\log \tfrac{M}{n})(\log \tfrac{M}{n\lceil P(G,w)\rceil}) + \tfrac{mM}{\lceil P(G,w)\rceil})$. In the worst case, i.e., when $P(G, w) = 1/n$ and $M = nW$, our algorithm runs in time $O(mnW)$ which matches the current fastest pseudopolynomial algorithm [6]. The result also implies that graphs with a penalty of at least $W/\text{poly}(n)$ form an interesting class of polynomial-time solvable energy games.

6   Discussions

Hardness for Bounded Clique-width and Strongly Ergodic Cases. We note the fact that energy games on complete bipartite graphs are polynomial-time equivalent to the general case. This implies that energy games on graphs of bounded clique-width [21] and strongly ergodic[5] graphs [19] are as hard as the general case. It also indicates that, in sharp contrast to parity games (a natural subclass of energy and mean-payoff games), structural properties of the input graphs might not yield efficiently solvable subclasses.

Let (G, w) be any input graph. For our reduction we first make the graph bipartite and then add two types of edges. (1) For any pair of nodes u and v such

---

[5] There are many notions of ergodicity [19,4]. Strong ergodicity is the strongest one as it implies other ergodicity conditions.

that $u \in V_A$, $v \in V_B$ and $(u, v) \notin E$, we add an edge (u, v) of weight $-2nW - 1$. (2) For any pair of nodes u and v such that $u \in V_B$, $v \in V_A$ and $(u, v) \notin E$, we add an edge (u, v) of weight $2n^2W + n + 1$. Let (G', w') be the resulting complete bipartite graph which is strongly ergodic and has clique width two.[6]

The polynomial-time reduction follows straightforwardly from the following claim: For any node v, if $e^*_{G,w}(v) < \infty$ then $e^*_{G',w'}(v) = e^*_{G,w}(v)$ (which is at most nW [6]); otherwise, $e^*_{G',w'}(v) > nW$. We now sketch the idea of this claim (see full version for more detail). Let (G'', w'') be the graph resulting from adding edges of the first type only. It can be shown that any strategy that uses such an edge of weight $-2nW - 1$ will require an energy of at least $nW + 1$ since we can gain energy of at most nW before using this edge. If $e^*_{G,w}(u) < \infty$, Alice will use her optimal strategy of (G, w) also in (G'', w'') which gives a minimal energy of at most nW. If $e^*_{G,w}(u) = \infty$, Alice might use a new edge but in that case $e^*_{G'',w''}(u) > nW$. This implies that the claim holds if we add edges of the first type only. Using the same idea we can show that the claim also holds when we add the second type of edges. The argument is slightly more complicated as we have to consider two cases depending on whether $e^*_{G,w}(v) = \infty$ for all v.

Previous Hard Examples Have Large Penalties. Consider the class of graphs where each cycle has total weight either positive or less than $-W/2$. Clearly, graphs in this class have penalty at least $-W/(2n)$. We observe that, for this class of graphs, the following algorithms need at least subexponential time while our algorithm runs in polynomial time: the algorithm by Gurvich et al [16], the value iteration algorithm by Brim et al [6], the algorithm by Zwick and Paterson [24] and the random facet algorithm (the latter two algorithms are for the decision version of mean-payoff and parity games, respectively). An example of such graphs for the first two algorithms is from [1] (for the second algorithm, we exploit the fact that it is deterministic and there exists a bad ordering in which the nodes are processed). Examples of the third and fourth algorithms are from [24] and [13], respectively. We note that the examples from [1] and [13] have one small cycle. One can change the value of this cycle to $-W$ to make these examples belong to the desired class of graphs without changing the worst-case behaviors of the mentioned algorithms.

References

1. Beffara, E., Vorobyov, S.: Is Randomized Gurvich-Karzanov-Khachiyan's Algorithm for Parity Games Polynomial? Tech. Rep. 2001-025, Department of Information Technology, Uppsala University (October 2001)
2. Björklund, H., Vorobyov, S.G.: A combinatorial strongly subexponential strategy improvement algorithm for mean payoff games. Discrete Applied Mathematics 155(2), 210–229 (2007)
3. Bloem, R., Chatterjee, K., Henzinger, T.A., Jobstmann, B.: Better Quality in Synthesis through Quantitative Objectives. In: Bouajjani, A., Maler, O. (eds.) CAV 2009. LNCS, vol. 5643, pp. 140–156. Springer, Heidelberg (2009)

---

[6] Note that every graph has clique-width at least two.

4. Boros, E., Elbassioni, K., Fouz, M., Gurvich, V., Makino, K., Manthey, B.: Stochastic Mean Payoff Games: Smoothed Analysis and Approximation Schemes. In: Aceto, L., Henzinger, M., Sgall, J. (eds.) ICALP 2011. LNCS, vol. 6755, pp. 147–158. Springer, Heidelberg (2011)
5. Bouyer, P., Fahrenberg, U., Larsen, K.G., Markey, N., Srba, J.: Infinite Runs in Weighted Timed Automata with Energy Constraints. In: Cassez, F., Jard, C. (eds.) FORMATS 2008. LNCS, vol. 5215, pp. 33–47. Springer, Heidelberg (2008)
6. Brim, L., Chaloupka, J., Doyen, L., Gentilini, R., Raskin, J.F.: Faster algorithms for mean-payoff games. Formal Methods in System Design 38(2), 97–118 (2011)
7. Černý, P., Chatterjee, K., Henzinger, T.A., Radhakrishna, A., Singh, R.: Quantitative Synthesis for Concurrent Programs. In: Gopalakrishnan, G., Qadeer, S. (eds.) CAV 2011. LNCS, vol. 6806, pp. 243–259. Springer, Heidelberg (2011)
8. Chakrabarti, A., de Alfaro, L., Henzinger, T.A., Stoelinga, M.: Resource Interfaces. In: Alur, R., Lee, I. (eds.) EMSOFT 2003. LNCS, vol. 2855, pp. 117–133. Springer, Heidelberg (2003)
9. Condon, A.: The Complexity of Stochastic Games. Information and Computation 96(2), 203–224 (1992)
10. Daskalakis, C., Papadimitriou, C.H.: Continuous Local Search. In: SODA, pp. 790–804 (2011)
11. Ehrenfeucht, A., Mycielski, J.: Positional strategies for mean payoff games. International Journal of Game Theory 8(2), 109–113 (1979)
12. Friedmann, O.: A Subexponential Lower Bound for Zadeh's Pivoting Rule for Solving Linear Programs and Games. In: Günlük, O., Woeginger, G.J. (eds.) IPCO 2011. LNCS, vol. 6655, pp. 192–206. Springer, Heidelberg (2011)
13. Friedmann, O., Hansen, T.D., Zwick, U.: A subexponential lower bound for the Random Facet algorithm for Parity Games. In: SODA, pp. 202–216 (2011)
14. Friedmann, O., Hansen, T.D., Zwick, U.: Subexponential lower bounds for randomized pivoting rules for the simplex algorithm. In: STOC, pp. 283–292 (2011)
15. Gentilini, R.: A Note on the Approximation of Mean-Payoff Games. In: CILC (2011)
16. Gurvich, V.A., Karzanov, A.V., Khachiyan, L.G.: Cyclic games and an algorithm to find minimax cycle means in directed graphs. USSR Computational Mathematics and Mathematical Physics 28(5), 85–91 (1990)
17. Halman, N.: Simple Stochastic Games, Parity Games, Mean Payoff Games and Discounted Payoff Games Are All LP-Type Problems. Algorithmica 49(1), 37–50 (2007)
18. Jurdzinski, M.: Deciding the Winner in Parity Games is in UP∩co-UP. Inf. Process. Lett. 68(3), 119–124 (1998)
19. Lebedev, V.N.: Effectively Solvable Classes of Cyclical Games. Journal of Computer and Systems Sciences International 44(4), 525–530 (2005)
20. Martin, D.A.: Borel determinacy. Annals of Mathematics 102(2), 363–371 (1975)
21. Obdržálek, J.: Clique-Width and Parity Games. In: Duparc, J., Henzinger, T.A. (eds.) CSL 2007. LNCS, vol. 4646, pp. 54–68. Springer, Heidelberg (2007)
22. Roth, A., Balcan, M.F., Kalai, A., Mansour, Y.: On the Equilibria of Alternating Move Games. In: SODA, pp. 805–816 (2010)
23. Vorobyov, S.: Cyclic games and linear programming. Discrete Applied Mathematics 156(11), 2195–2231 (2008)
24. Zwick, U., Paterson, M.: The Complexity of Mean Payoff Games on Graphs. Theoretical Computer Science 158(1&2), 343–359 (1996), also in COCOON 1995

Data Structures on Event Graphs

Bernard Chazelle[1] and Wolfgang Mulzer[2]

[1] Department of Computer Science, Princeton University, USA
chazelle@cs.princeton.edu
[2] Institut für Informatik, Freie Universität Berlin, Germany
mulzer@inf.fu-berlin.de

Abstract. We investigate the behavior of data structures when the input and operations are generated by an *event graph*. This model is inspired by the model of Markov chains. We are given a fixed graph G, whose nodes are annotated with operations of the type *insert, delete,* and *query*. The algorithm responds to the requests as it encounters them during a (adversarial or random) walk in G. We study the limit behavior of such a walk and give an efficient algorithm for recognizing which structures can be generated. We also give a near-optimal algorithm for successor searching if the event graph is a cycle and the walk is adversarial. For a random walk, the algorithm becomes optimal.

1 Introduction

In contrast with the traditional adversarial assumption of worst-case analysis, many data sources are modeled by Markov chains (e.g., in queuing, speech, gesture, protein homology, web searching, etc.). These models are very appealing because they are widely applicable and simple to generate. Indeed, locality of reference, an essential pillar in the design of efficient computing systems, is often captured by a Markov chain modeling the access distribution. Hence, it does not come as a surprise that this connection has motivated and guided much of the research on self-organizing data structures and online algorithms in a Markov setting [1,7–11,15–18]. That body of work should be seen as part of a larger effort to understand algorithms that exploit the fact that input distributions often exhibit only a small amount of entropy. This effort is driven not only by the hope for improvements in practical applications (e.g., exploiting coherence in data streams), but it is also motivated by theoretical questions: for example, the key to resolving the problem of designing an optimal deterministic algorithm for minimum spanning trees lies in the discovery of an optimal heap for constant-entropy sources [2]. Markov chains have been studied intensively, and there exists a huge literature on them (e.g., [12]). Nonetheless, the focus has been on state functions (such as stationary distribution or commute/cover/mixing times) rather than on the behavior of complex objects evolving over them. This leads to a number of fundamental questions which, we hope, will inspire further research.

Let us describe our model in more detail. Our object of interest is a structure $\mathcal{T}(X)$ that evolves over time. The structure $\mathcal{T}(X)$ is defined over a finite subset

L. Epstein and P. Ferragina (Eds.): ESA 2012, LNCS 7501, pp. 313–324, 2012.
© Springer-Verlag Berlin Heidelberg 2012

X of a universe $\mathcal{U}$. In the simplest case, we have $\mathcal{U} = \mathbb{N}$ and $\mathcal{T}(X) = X$. This corresponds to the classical dictionary problem where we need to maintain a subset of a given universe. We can also imagine more complicated scenarios such as $\mathcal{U} = \mathbb{R}^d$ with $\mathcal{T}(X)$ being the Delaunay triangulation of X. An *event graph* $G = (V, E)$ specifies restrictions on the queries and updates that are applied to $\mathcal{T}(X)$. For simplicity, we assume that G is undirected and connected. Each node $v \in V$ is associated with an item $x_v \in \mathcal{U}$ and corresponds to one of three possible requests: (i) `insert`(x_v); (ii) `delete`(x_v); (iii) `query`(x_v). Requests are specified by following a walk in G, beginning at a designated start node of G and hopping from node to neighboring node. We consider both *adversarial* walks, in which the neighbors can be chosen arbitrarily, and *random* walks, in which the neighbor is chosen uniformly at random. The latter case corresponds to the classic Markov chain model. Let v^t be the node of G visited at time t and let $X^t \subseteq \mathcal{U}$ be the set of *active* elements, i.e., the set of items inserted prior to time t and not deleted after their last insertions. We also call X^t an *active set*. For any $t > 0$, $X^t = X^{t-1} \cup \{x_{v^t}\}$ if the operation at v^t is an insertion and $X^t = X^{t-1} \setminus \{x_{v^t}\}$ in the case of a deletion. The query at v depends on the structure under consideration (successor, point location, ray shooting, etc.). Another way to interpret the event graph is as a finite automaton that generates words over an alphabet with certain cancellation rules.

Markov chains are premised on forgetting the past. In our model, however, the structure $\mathcal{T}(X^t)$ can remember quite a bit. In fact, we can define a secondary graph over the much larger vertex set $V \times 2^{\mathcal{U}_{|V}}$, where $\mathcal{U}_{|V} = \{x_v \mid v \in V\}$ denotes those elements in the universe that occur as labels in G. We call this larger graph the *decorated event graph*, $\text{dec}(G)$, since the way to think of this secondary graph is to picture each node v of G being "decorated" with any realizable $X \subseteq \mathcal{U}_{|V}$. An edge (v, w) in the original graph gives rise to up to 2^n edges $(v, X)(w, Y)$ in the decorated graph, with Y derived from X in the obvious way. A trivial upper bound on the number of states is $n2^n$, which is essentially tight. If we could afford to store all of $\text{dec}(G)$, then any of the operations at the nodes of the event graph could be precomputed and the running time would be constant. However, the required space might be huge, so the main question is

Can the decorated graph be compressed with no loss of performance?

This seems a difficult question to answer in general. In fact, even counting the possible active sets in decorated graphs seems highly nontrivial, as it reduces to counting words in regular languages augmented with certain cancellation rules. Hence, in this paper we will focus on basic properties and special cases that highlight the interesting behavior of the decorated graph. Beyond the results themselves, the main contribution of this work is to draw the attention of algorithm designers to a more realistic input model that breaks away from worst-case models.

Our Results. The paper has two main parts. In the first one, we investigate some basic properties of decorated graphs. We show that the decorated graph $\text{dec}(G)$

has a unique strongly connected component that corresponds to the limiting phase of a walk on the event graph G, and we give characterizations for when a set $X \subseteq \mathcal{U}_{|V}$ appears as an active set in this limiting phase. We also show that whether X is an active set can be decided in linear time (in the size of G).

In the second part, we consider the problem of maintaining a dictionary that supports successor searches during a one-dimensional walk on a cycle. We show how to achieve linear space and constant expected time for a random walk. If the walk is adversarial, we can achieve a similar result with near-linear storage. The former result is in the same spirit as previous work by the authors on randomized incremental construction (RIC) for Markov sources [3]. RIC is a fundamental algorithmic paradigm in computational geometry that uses randomness for the construction of certain geometric objects, and we showed that there is no significant loss of efficiency if the randomness comes from a Markov chain with sufficiently high conductance.

2 Basic Properties of Decorated Graphs

We are given a labeled, connected, undirected graph $G = (V, E)$. In this section, we consider only labels of the form $\mathtt{i}x$ and $\mathtt{d}x$, where x is an element from a finite universe $\mathcal{U}$ and $\mathtt{i}$ and $\mathtt{d}$ stand for $\mathtt{insert}$ and $\mathtt{delete}$. We imagine an adversary that maintains a subset $X \subseteq \mathcal{U}$ while walking on G and performing the corresponding operations on the nodes. Since the focus of this section is the evolution of X over time, we ignore queries for now.

Recall that $\mathcal{U}_{|V}$ denotes the elements that appear on the nodes of G. For technical convenience, we require that for every $x \in \mathcal{U}_{|V}$ there is at least one node with label $\mathtt{i}x$ and at least one node with label $\mathtt{d}x$. The walk on G is formalized through the *decorated graph* $\mathrm{dec}(G)$. The graph $\mathrm{dec}(G)$ is a directed graph on vertex set $V' := V \times 2^{\mathcal{U}_{|V}}$. The pair $((u, X), (v, Y))$ is an edge in E' if and only if $\{u, v\}$ is an edge in G and $Y = X \cup \{x_v\}$ or $Y = X \backslash \{x_v\}$ depending on whether v is labeled $\mathtt{i}x_v$ or $\mathtt{d}x_v$.

By a *walk W* in a (directed or undirected) graph, we mean any finite sequence of nodes such that the graph contains an edge from each node in W to its successor in W. Let A be a walk in $\mathrm{dec}(G)$. By taking the first components of the nodes in A, we obtain a walk in G, the *projection* of A, denoted by $\mathrm{proj}(A)$. Similarly, let W be a walk in G with start node v, and let $X \subseteq 2^{\mathcal{U}_{|V}}$. Then the *lifting* of W with respect to X is the walk in $\mathrm{dec}(G)$ that begins at node (v, X) and follows the steps of W in $\mathrm{dec}(G)$. We denote this walk by $\mathrm{lift}(W, X)$.

Since $\mathrm{dec}(G)$ is a directed graph, it can be decomposed into strongly connected components that induce a directed acyclic graph D. We call a strongly connected component of $\mathrm{dec}(G)$ a *sink component* (also called essential class in Markov chain theory), if it corresponds to a sink (i.e., a node with out-degree 0) in D. First, we will show that to understand the behaviour of a walk on G in the limit, it suffices to focus on a single sink component of $\mathrm{dec}(G)$.

Lemma 2.1. *In $\mathrm{dec}(G)$ there exists a unique sink component $\mathcal{C}$ such that for every node $(v, \emptyset)$ in $\mathrm{dec}(G)$, $\mathcal{C}$ is the only sink component that $(v, \emptyset)$ can reach.*

Proof. Suppose there is a node v in G such that $(v, \emptyset)$ can reach two different sink components $\mathcal{C}$ and $\mathcal{C}'$ in $\mathrm{dec}(G)$. Since G is connected and since $\mathcal{C}$ and $\mathcal{C}'$ are sink components, both $\mathcal{C}$ and $\mathcal{C}'$ must contain at least one node with first component v. Call these nodes (v, X) (for $\mathcal{C}$) and (v, X') (for $\mathcal{C}'$). Furthermore, by assumption $\mathrm{dec}(G)$ contains a walk A from $(v, \emptyset)$ to (v, X) and a walk A' from $(v, \emptyset)$ to (v, X'). Let $W := \mathrm{proj}(A)$ and $W' := \mathrm{proj}(A')$. Both W and W' are closed walks in G that start and end in v, so their concatenations $WW'W$ and $W'W'W$ are valid walks in G, again with start and end vertex v. Consider the lifted walks $\mathrm{lift}(WW'W, \emptyset)$ and $\mathrm{lift}(W'W'W, \emptyset)$ in $\mathrm{dec}(G)$. We claim that these two walks have the same end node (v, X''). Indeed, for each $x \in \mathcal{U}_{|V}$, whether x appears in X'' or not depends solely on whether the label $\mathrm{i}x$ or the label $\mathrm{d}x$ appears last on the original walk in G. This is the same for both $WW'W$ and $W'W'W$. Hence, $\mathcal{C}$ and $\mathcal{C}'$ must both contain (v, X'), a contradiction to the assumption that they are distinct sink components. Thus, each node $(v, \emptyset)$ can reach exactly one sink component.

Now consider two distinct nodes $(v, \emptyset)$ and $(w, \emptyset)$ in $\mathrm{dec}(G)$ and assume that they reach the sink components $\mathcal{C}$ and $\mathcal{C}'$, respectively. Let W be a walk in G that goes from v to w and let $W' := \mathrm{proj}(A)$, where A is a walk in $\mathrm{dec}(G)$ that connects w to $\mathcal{C}'$. Since G is undirected, the reversed walk W^R is a valid walk in G from w to v. Now consider the walks $Z_1 := WW^RWW'$ and $Z_2 := W^RWW'$. The walk Z_1 begins in v, the walk Z_2 begins in w, and they both have the same end node. Furthermore, for each $x \in \mathcal{U}_{|V}$, the label $\mathrm{i}x$ appears last in Z_1 if and only if it appears last in Z_2. Hence, the lifted walks $\mathrm{lift}(Z_1, \emptyset)$ and $\mathrm{lift}(Z_2, \emptyset)$ have the same end node in $\mathrm{dec}(G)$, so $\mathcal{C} = \mathcal{C}'$. The lemma follows. □

Since the unique sink component $\mathcal{C}$ from Lemma 2.1 represents the limit behaviour of the set X during a walk in G, we will henceforth focus on this component $\mathcal{C}$. Let us begin with a few properties of $\mathcal{C}$. First, as we already observed in the proof of Lemma 2.1, it is easy to see that every node of G is represented in $\mathcal{C}$.

Lemma 2.2. *For each vertex v of G, there is at least one node in $\mathcal{C}$ whose first component is v.*

Proof. Let (w, X) be any node in $\mathcal{C}$. Since G is connected, there is a walk W in G from w to v, so $\mathrm{lift}(W, X)$ ends in a node in $\mathcal{C}$ whose first component is v. □

Next, we characterize the nodes in $\mathcal{C}$.

Lemma 2.3. *Let v be a node of G and $X \subseteq \mathcal{U}_{|V}$. We have $(v, X) \in \mathcal{C}$ if and only if there exists a closed walk W in G with the following properties:*

1. *the walk W starts and ends in v;*
2. *for each $x \in \mathcal{U}_{|V}$, there is at least one node in W with label $\mathrm{i}x$ or $\mathrm{d}x$;*
3. *We have $x \in X$ exactly if the last node in W referring to x is an insertion and $x \notin X$ exactly if the last node in W referring to x is a deletion.*

We call the walk W from Lemma 2.3 a *certifying walk* for the node (v, X) of $\mathcal{C}$.

Proof. First, suppose there is a walk with the given properties. By Lemma 2.2, there is at least one node in $\mathcal{C}$ whose first component is v, say (v, Y). The properties of W immediately imply that the walk lift(W, Y) ends in (v, X), which proves the first part of the lemma.

Now suppose that (v, X) is a node in $\mathcal{C}$. Since $\mathcal{C}$ is strongly connected, there exists a closed walk A in $\mathcal{C}$ that starts and ends at (v, X) and visits every node of $\mathcal{C}$ at least once. Let $W := \text{proj}(A)$. By Lemma 2.2 and our assumption on the labels of G, the walk W contains for every element $x \in \mathcal{U}_{|V}$ at least one node with label ix and one node with label dx. Therefore, the walk W meets all the desired properties.   □

This characterization of the nodes in $\mathcal{C}$ immediately implies that the decorated graph can have only one sink component.

Corollary 2.1 *The component $\mathcal{C}$ is the only sink component of* $\text{dec}(G)$.

Proof. Let (v, X) be a node in $\text{dec}(G)$. By Lemmas 2.2 and 2.3, there exists in $\mathcal{C}$ a node of the form (v, Y) and a corresponding certifying walk W. Clearly, the walk lift(W, X) ends in (v, Y). Thus, every node in $\text{dec}(G)$ can reach $\mathcal{C}$, so there can be no other sink component.   □

Next, we give a bound on the length of certifying walks, from which we can deduce a bound on the diameter of $\mathcal{C}$.

Theorem 2.2. *Let (v, X) be a node of $\mathcal{C}$ and let W be a corresponding certifying walk of minimum length. Then W has length at most $O(n^2)$, where n denotes the number of nodes in G. There are examples where any certifying walk needs $\Omega(n^2)$ nodes. It follows that $\mathcal{C}$ has diameter $O(n^2)$ and that this is tight.*

Proof. Consider the reversed walk W^R. We subdivide W^R into *phases*: a new phase starts when W^R encounters a node labeled ix or dx for an $x \in \mathcal{U}_{|V}$ that it has not seen before. Clearly, the number of phases is at most n. Now consider the i-th phase and let V_i be the set of nodes in G whose labels refer to the i distinct elements of $\mathcal{U}_{|V}$ that have been encountered in the first i phases. In phase i, W^R can use only vertices in V_i. Since W has minimum cardinality, the phase must consist of a shortest path in V_i from the first node of phase i to the first node of phase $i+1$. Hence, each phase consists of at most n vertices and W has total length $O(n^2)$.

For the lower bound, consider the path P in Fig. 1. It consists of $2n+1$ vertices. Let v be the middle node of P and $\mathcal{C}$ be the unique sink component of $\text{dec}(P)$. It is easy to check that (v, X) is a node of $\mathcal{C}$ for every $X \subseteq \{1, \ldots, n-1\}$ and that the length of a shortest certifying walk for the node $(v, \{2k \mid k = 0, \ldots, \lfloor n/2 \rfloor - 1\})$ is $\Theta(n^2)$.

We now show that any two nodes in $\mathcal{C}$ are connected by a path of length $O(n^2)$. Let (u, X) and (v, Y) be two such nodes and let Q be a shortest path from u to v in G and W be a certifying walk for (v, Y). Then lift(QW, X) is a walk of length $O(n^2)$ in $\mathcal{C}$ from (u, X) to (v, Y). Hence, the diameter of $\mathcal{C}$ is $O(n^2)$, and Fig. 1 again provides a lower bound: the length of a shortest path in $\mathcal{C}$ between $(v, \emptyset)$ and $(v, \{2k \mid k = 0, \ldots, \lfloor n/2 \rfloor - 1\})$ is $\Theta(n^2)$.   □

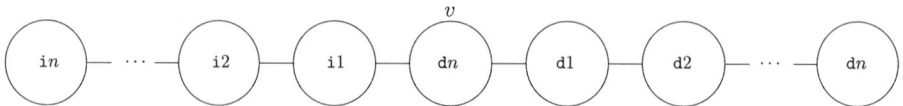

Fig. 1. The lower bound example

We now describe an algorithm that is given G and a set $X \subseteq \mathcal{U}_{|V}$ and then decides whether (v, X) is a node of the unique sink or not. For $W \subseteq V$, let $\mathcal{U}_{|W}$ denote the elements that appear in the labels of the nodes in W. For $U \subseteq \mathcal{U}$, let $V_{|U}$ denote the nodes of G whose labels contain an element of U.

Theorem 2.3. *Given an event graph G, a node v of G and a subset $X \subseteq \mathcal{U}_{|V}$, we can decide in $O(|V| + |E|)$ steps whether (v, X) is a node of the unique sink component $\mathcal{C}$ of $\mathrm{dec}(G)$.*

Proof. The idea of the algorithm is to construct a certifying walk for (v, X) through a modified breadth first search.

In the preprocessing phase, we color a vertex w of G *blue* if w is labeled $\mathtt{i}x$ and $x \in X$ or if w is labeled $\mathtt{d}x$ and $x \notin X$. Otherwise, we color w *red*. If v is colored red, then (v, X) cannot be in $\mathcal{C}$ and we are done. Otherwise, we perform a directed breadth first search that starts from v and tries to construct a reverse certifying walk. Our algorithm maintains several queues. The main queue is called the *blue fringe* B. Furthermore, for every $x \in \mathcal{U}_{|V}$, we have a queue R_x, the *red fringe* for x. At the beginning, we consider each neighbor w of v. If w is colored blue, we append it to the blue fringe B. If w is colored red, we append it to the appropriate red fringe R_{x_w}, where x_w is the element that appears in w's label.

The main loop of the algorithm takes place while B is not empty. We pull the next node w out of B, and we process w as follows: if we have not seen the element $x_w \in \mathcal{U}_{|V}$ for w before, we color all the nodes in $V_{\{x_w\}}$ blue, append all the nodes of R_{x_w} to B, and we delete R_{x_w}. Next, we process the neighbors of w as follows: if a neighbor w' of w is blue, we append it to B if w' is not in B yet. If w' is red and labeled with the element $x_{w'}$, we append w' to $R_{x_{w'}}$, if necessary.

The algorithm terminates after at most $|V|$ iterations. In each iteration, the cost is proportional to the degree of the current vertex w and (possibly) the size of one red fringe. The latter cost can be charged to later rounds, since the nodes of the red fringe are processed later on. Let V_{red} be the union of the remaining red fringes after the algorithm terminates.

If $V_{\mathrm{red}} = \emptyset$, we obtain a certifying walk for (v, X) by walking from one newly discovered vertex to the next inside the current blue component and reversing it. Now suppose $V_{\mathrm{red}} \neq \emptyset$. Let A be the set of all vertices that were traversed during the BFS. Then $G \setminus V_{\mathrm{red}}$ has at least two connected components (since there must be blue vertices outside of A). Furthermore, $\mathcal{U}_{|A} \cap \mathcal{U}_{|V_{\mathrm{red}}} = \emptyset$. We claim that a certifying walk for (v, X) cannot exist. Indeed, assume that W is such a certifying walk. Let $x_w \in \mathcal{U}_{|V_{\mathrm{red}}}$ be the element in the label of the last node w in W whose label refers to an element in $\mathcal{U}_{|V_{\mathrm{red}}}$. Suppose that the label of w is of the form $\mathtt{i}x_w$. Since W is a certifying walk, we have $x_w \in X$, so w was colored blue during the initialization phase. Furthermore, all the nodes on W that come after w are also blue at the end. This implies that $w \in A$, because by assumption a neighbor of w was in B, and hence w must have been added to B when this neighbor was processed. Hence, we get a contradiction to the fact that $\mathcal{U}_{|A} \cap \mathcal{U}_{|V_{\mathrm{red}}} = \emptyset$, so W cannot exist. Therefore, $(v, X) \notin \mathcal{C}$. □

The proof of Theorem 2.3 gives an alternative characterization of whether a node appears in the unique sink component or not.

Corollary 2.4 *The node (v, X) does not appear in $\mathcal{C}$ if and only if there exists a set $A \subseteq V(G)$ with the following properties:*

1. *$G \backslash A$ has at least two connected components.*
2. *$\mathcal{U}_{|A} \cap \mathcal{U}_{|B} = \emptyset$, where B denotes the vertex set of the connected component of $G \setminus A$ that contains v.*
3. *For all $x \in \mathcal{U}$, A contains either only labels of the form $\mathtt{i}x$ or only labels of the form $\mathtt{d}x$ (or neither). If A has a node with label $\mathtt{i}x$, then $x \notin X$. If A has a node with label $\mathtt{d}x$, then $x \in X$.*

A set A with the above properties can be found in polynomial time. □

Lemma 2.4. *Given $k \in \mathbb{N}$ and a node $(v, X) \in \mathcal{C}$, it is NP-complete to decide whether there exists a certifying walk for (v, X) of length at most k.*

Proof. The problem is clearly in NP. To show completeness, we reduce from Hamiltonian path in undirected graphs. Let G be an undirected graph with n vertices, and suppose the vertex set is $\{1, \ldots, n\}$. We let $\mathcal{U} = \mathbb{N}$ and take two copies G_1 and G_2 of G. We label the copy of node i in G_1 with $\mathtt{i}i$ and the copy of node i in G_2 with $\mathtt{d}i$. Then we add two nodes v_1 and v_2 and connect them to all nodes in G_1 and G_2, We label v_1 with $\mathtt{i}(n+1)$ and v_2 with $\mathtt{d}(n+1)$. The resulting graph G' has $2n + 2$ nodes and meets all our assumptions about an event graph. Furthermore, G has a Hamiltonian path if and only if the node $(v_1, \{1, \ldots, n+1\})$ has a certifying walk of length $n + 1$. This completes the reduction. □

3   Successor Searching on Cycle Graphs

We now consider the case that the event graph G is a simple cycle $v_1, \ldots, v_n, v_1$ and the item x_{v_i} at node v_i is a real number. Again, the structure $\mathcal{T}(X)$ is X itself, and we now have three types of nodes: insertion, deletion, and query. A query at time t asks for $\mathtt{succ}_{X^t}(x_{v^t}) = \min\{ x \in X^t \,|\, x \geq x_{v^t} \}$ (or ∞). Again, an example similar to the one of Fig. 1 shows that the decorated graph can be of exponential size: take $x_{v_{n+1-i}} = x_{v_i}$ and define the operation at v_i as: $\mathtt{i}(x_{v_i})$ if $i \leq n/2$ and $\mathtt{d}(x_{v_{n+1-i}})$ if $i > n/2$. It is easy to design a walk that produces *any* set[1] $X_{v_{n/2}}$ at the median node consisting of *any* subset of $\mathcal{U}_{|V}$, which implies a lower bound of $\Omega(2^{n/2})$ on the size of the decorated graph.

We consider two different walks on G. The *random* walk starts at v_1 and hops from a node to one of its neighbors with equal probability. The main result of this section is that for random walks maximal compression is possible.

Theorem 3.1. *Successor searching in a one-dimensional random walk can be done in constant expected time per step and linear storage.*

First, however, we consider an adversarial walk on G. Note that we always achieve a log-logarithmic running time per step by maintaining a van Emde Boas search structure dynamically [5,6], so the interesting question is how little storage we need if we are to perform each operation in constant time.

Theorem 3.2. *Successor searching along an n-node cycle in the adversarial model can be performed in constant time per operation, using $O(n^{1+\varepsilon})$ storage, for any fixed $\varepsilon > 0$.*

Before addressing the walk on G, we must consider the following range searching problem (see also [4]). Let $Y = \{y_1, \ldots, y_n\}$ be n distinct numbers. A query is a pair (i, j) and its "answer" is the smallest index $k > i$ such that $y_i < y_k < y_j$ (or $\emptyset$). Fig. 2(a) suggests 5 other variants of the query: the point (i, y_i) defines four quadrants and the horizontal line $Y = y_j$ two halfplanes (when the rectangle extends to the left, of course, we would be looking for the largest k, not the smallest). So, we specify a query by a triplet (i, j, σ), with σ to specify the type. We need the following result, which, as a reviewer pointed out to us, was also discovered earlier by Crochemore et al. [4]. We include our proof below for completeness.

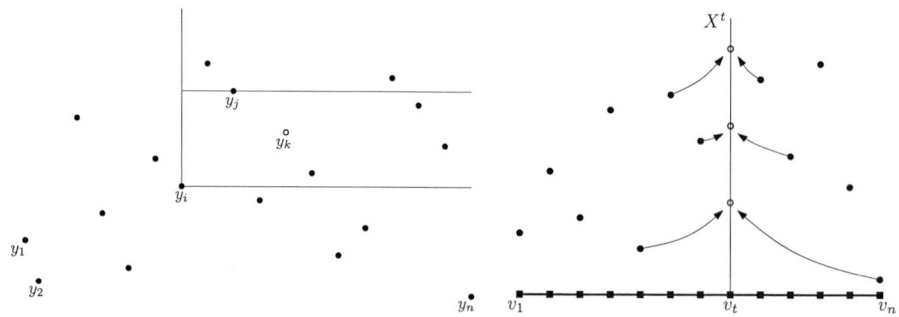

Fig. 2. (a) The query (i, j); (b) the successor data structure

---

[1] We abuse notation by treating $n/2$, $\sqrt{n}$, etc., as integers.

Lemma 3.1. *Any query can be answered in constant time with the help of a data structure of size $O(n^{1+\varepsilon})$, for any $\varepsilon > 0$.*

Using Lemma 3.1, we can prove Theorem 3.2.

Proof (of Theorem 3.2). At any time t, the algorithm has at its disposal: (i) a sorted doubly-linked list of the active set X^t (augmented with ∞); (ii) a (bidirectional) pointer to each $x \in X^t$ from the first v_k clockwise from v^t, if it exists, such that $\operatorname{succ}_{X^t}(x_{v_k}) = x$ (same thing counterclockwise)—see Fig. 2(b). Assume now that the data structure of Lemma 3.1 has been set up over $Y = \{x_{v_1}, \ldots, x_{v_n}\}$. As the walk enters node v^t at time t, $\operatorname{succ}_{X^t}(x_{v^t})$ is thus readily available and we can update X^t in $O(1)$ time. The only question left is how to maintain (ii). Suppose that the operation at node v^t is a successor request and that the walk reached v^t clockwise. If x is the successor, then we need to find the first v_k clockwise from v^t such that $\operatorname{succ}_{X^t}(x_{v_k}) = x$. This can be handled by two range search queries (i, j, σ): for i, use v^t; and, for j, use x and its predecessor in X^t. An insert can be handled by two such queries (one on each side of v_t), while a delete requires pointer updating but no range search queries. □

Proof (of Lemma 3.1). Given any (i, j), we add four more queries to our repertoire by considering the four quadrants cornered at (i, y_j). We define a single data structure to handle all 10 types simultaneously. We restrict our discussion to the type in Fig. 2(a) but kindly invite the reader to check that all other 9 types can be handled in much the same way. We prove by induction that with $c^s n^{1+1/s}$ storage, for a large enough constant c, any query can be answered in at most s table lookups. The case $s = 1$ being obvious (precompute all queries), we assume that $s > 1$. Sort and partition Y into consecutive groups $Y_1 < \cdots < Y_{n^{1/s}}$ of size $n^{1-1/s}$ each.

- **Ylinks:** for each $y_i \in Y$, link y_i to the highest-indexed y_j $(j < i)$ within each group $Y_1, \ldots, Y_{n^{1/s}}$ (left pointers in Fig. 3(a)).
- **Zlinks:** for each $y_i \in Y$, find the group Y_{ℓ_i} to which y_i belongs and, for each k, define Z_k as the subset of Y sandwiched (inclusively) between y_i and the smallest (resp. largest) element in Y_k if $k \leq \ell_i$ (resp. $k \geq \ell_i$). Note that this actually defined two sets for Z_{ℓ_i}, so that the total number of Z_k's is really $n^{1/s} + 1$. Link y_i to the lowest-indexed y_j $(j > i)$ in each Z_k (right pointers in Fig. 3(a)).
- Prepare a data structure of type $s - 1$ recursively for each Y_i.

Given a query (i, j) of type Fig. 3(a), we first check whether it fits entirely within Y_{ℓ_i} and, if so, solve it recursively. Otherwise, we break it down into two subqueries: one of them can be handled directly by using the relevant Zlink. The other one fits entirely within a single Y_k. By following the corresponding Ylink, we find $y_{i'}$ and solve the subquery recursively by converting it into another (i', j) of different type (Fig.2(b)). By induction, this requires s lookups and storage

$$dn^{1+1/s} + c^{s-1} n^{1/s + (1-1/s)(1+1/(s-1))} \leq c^s n^{1+1/s},$$

for some constant d and c large enough. □

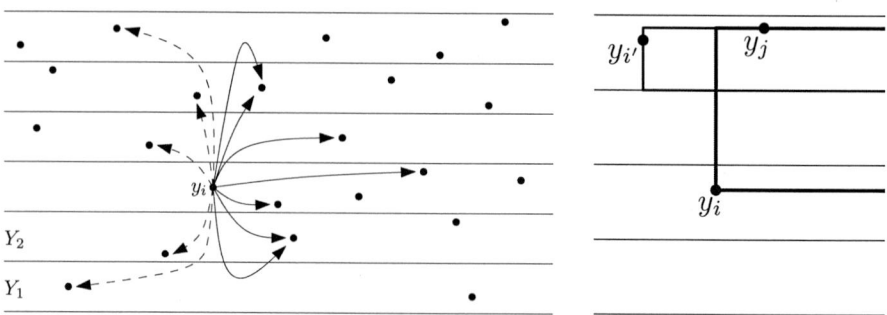

Fig. 3. (a) the recursive data structure; (b) decomposing a query

Using Theorem 3.2 together with the special properties of a random walk on G, we can quickly derive the algorithm for Theorem 3.1.

Proof (of Theorem 3.1). The idea is to divide up the cycle into $\sqrt{n}$ equal-size paths $P_1, \ldots, P_{\sqrt{n}}$ and prepare an adversarial data structure for each one of them right upon entry. The high cover time of a one-dimensional random walk is then invoked to amortize the costs. De-amortization techniques are then used to make the costs worst-case. The details follow. As soon as the walk enters a new P_k, the data structure of Lemma 3.1 is built from scratch for $\varepsilon = 1/3$, at a cost in time and storage of $O(n^{2/3})$. By merging $L_k = \{ x_{v_i} \mid v_i \in P_k \}$ with the doubly-linked list storing X^t, we can set up all the needed successor links and proceeds just as in Theorem 3.2. This takes $O(n)$ time per interpath transition and requires $O(n^{2/3})$ storage. There are few technical difficulties that we now address one by one.

- Upon entry into a new path P_k, we must set up successor links from P_k to X^t, which takes $O(n)$ time. Rather than forcing the walk to a halt, we use a "parallel track" idea to de-amortize these costs. (Fig. 4). Cover the cycle

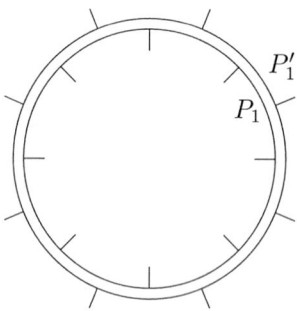

Fig. 4. The parallel tracks on the cycle

with paths P_i' shifted from P_i clockwise by $\frac{1}{2}\sqrt{n}$. and carry on the updates in parallel on both tracks. As we shall see below, we can ensure that updates do not take place simultaneously on both tracks. Therefore, one of them is always available to answer successor requests in constant time.

- Upon entry into a new path P_k (or P_k'), the relevant range search structure must be built from scratch. This work does not require knowledge of X^t and, in fact, the only reason it is not done in preprocessing is to save storage. Again, to avoid having to interrupt the walk, while in P_k we ensure that the needed structures for the two adjacent paths P_{k-1}, P_{k+1} are already available and those for P_{k-2}, P_{k+2} are under construction. (Same with P_k'.)
- The range search structure can only handle queries (i, j) for which *both* y_i and y_j are in the ground set. Unfortunately, j may not be, for it may correspond to an item of X^t inserted prior to entry into the current P_k. There is an easy fix: upon entering P_k, compute and store $\operatorname{succ}_{L_k}(x_{v_i})$ for $i = 1, \ldots, n$. Then, simply replace a query (i, j) by (i, j') where j' is the successor (or predecessor) in L_k.

The key idea now is that a one-dimensional random walk has a quadratic cover time [13]; therefore, the expected time between any change of paths on one track and the next change of paths on the other track is $\Theta(n)$. This means that if we dovetail the parallel updates by performing a large enough number of them per walking step, we can keep the expected time per operation constant. This proves Theorem 3.1.   □

4   Conclusion

We have presented a new approach to model and analyze restricted query sequences that is inspired by Markov chains. Our results only scratch the surface of a rich body of questions. For example, even for the simple problem of the adversarial walk on a path, we still do not know whether we can beat van Emde Boas trees with linear space. Even though there is some evidence that the known lower bounds for successor searching on a pointer machine give the adversary a lot of leeway [14], our lower bound technology does not seem to be advanced enough for this setting. Beyond paths and cycles, of course, there are several other simple graph classes to be explored, e.g., trees or planar graphs.

Furthermore, there are more fundamental questions on decorated graphs to be studied. For example, how hard is it to count the number of distinct active sets (or the number of nodes) that occur in the unique sink component of $\operatorname{dec}(G)$? What can we say about the behaviour of the active set in the limit as the walk proceeds randomly? And what happens if we go beyond the dictionary problem and consider the evolution of more complex structures during a walk on the event graph?

Acknowledgments. We would like to thank the anonymous referees for their thorough reading of the paper and their many helpful suggestions, as well as for pointing out [4] to us.

References

[1] Chassaing, P.: Optimality of move-to-front for self-organizing data structures with locality of references. Ann. Appl. Probab. 3(4), 1219–1240 (1993)

[2] Chazelle, B.: The discrepancy method: randomness and complexity. Cambridge University Press, New York (2000)

[3] Chazelle, B., Mulzer, W.: Markov incremental constructions. Discrete Comput. Geom. 42(3), 399–420 (2009)

[4] Crochemore, M., Iliopoulos, C.S., Kubica, M., Rahman, M.S., Walen, T.: Improved algorithms for the range next value problem and applications. In: Proc. 25th Sympos. Theoret. Aspects Comput. Sci (STACS), pp. 205–216 (2008)

[5] van Emde Boas, P.: Preserving order in a forest in less than logarithmic time and linear space. Inform. Process. Lett. 6(3), 80–82 (1977)

[6] van Emde Boas, P., Kaas, R., Zijlstra, E.: Design and implementation of an efficient priority queue. Math. Systems Theory 10(2), 99–127 (1976)

[7] Hotz, G.: Search trees and search graphs for Markov sources. Elektronische Informationsverarbeitung und Kybernetik 29(5), 283–292 (1993)

[8] Kapoor, S., Reingold, E.M.: Stochastic rearrangement rules for self-organizing data structures. Algorithmica 6(2), 278–291 (1991)

[9] Karlin, A.R., Phillips, S.J., Raghavan, P.: Markov paging. SIAM J. Comput. 30(3), 906–922 (2000)

[10] Konneker, L.K., Varol, Y.L.: A note on heuristics for dynamic organization of data structures. Inform. Process. Lett. 12(5), 213–216 (1981)

[11] Lam, K., Leung, M.Y., Siu, M.K.: Self-organizing files with dependent accesses. J. Appl. Probab. 21(2), 343–359 (1984)

[12] Levin, D.A., Peres, Y., Wilmer, E.L.: Markov chains and mixing times. American Mathematical Society. Providence, RI (2009)

[13] Motwani, R., Raghavan, P.: Randomized algorithms. Cambridge University Press, Cambridge (1995)

[14] Mulzer, W.: A note on predecessor searching in the pointer machine model. Inform. Process. Lett. 109(13), 726–729 (2009)

[15] Phatarfod, R.M., Pryde, A.J., Dyte, D.: On the move-to-front scheme with Markov dependent requests. J. Appl. Probab. 34(3), 790–794 (1997)

[16] Schulz, F., Schömer, E.: Self-organizing Data Structures with Dependent Accesses. In: Meyer auf der Heide, F., Monien, B. (eds.) ICALP 1996. LNCS, vol. 1099, pp. 526–537. Springer, Heidelberg (1996)

[17] Shedler, G.S., Tung, C.: Locality in page reference strings. SIAM J. Comput. 1(3), 218–241 (1972)

[18] Vitter, J.S., Krishnan, P.: Optimal prefetching via data compression. J. ACM 43(5), 771–793 (1996)

Improved Distance Oracles and Spanners for Vertex-Labeled Graphs

Shiri Chechik

Department of Computer Science, The Weizmann Institute, Rehovot, Israel
shiri.chechik@weizmann.ac.il

Abstract. Consider an undirected weighted graph $G = (V, E)$ with $|V| = n$ and $|E| = m$, where each vertex $v \in V$ is assigned a *label* from a set of labels $L = \{\lambda_1, ..., \lambda_\ell\}$. We show how to construct a compact distance oracle that can answer queries of the form: "what is the distance from v to the closest λ-labeled vertex" for a given vertex $v \in V$ and label $\lambda \in L$.

This problem was introduced by Hermelin, Levy, Weimann and Yuster [ICALP 2011] where they present several results for this problem. In the first result, they show how to construct a vertex-label distance oracle of expected size $O(kn^{1+1/k})$ with stretch $(4k - 5)$ and query time $O(k)$. In a second result, they show how to reduce the size of the data structure to $O(kn\ell^{1/k})$ at the expense of a huge stretch, the stretch of this construction grows exponentially in k, $(2^k - 1)$. In the third result they present a dynamic vertex-label distance oracle that is capable of handling label changes in a sub-linear time. The stretch of this construction is also exponential in k, $(2 \cdot 3^{k-1} + 1)$.

We manage to significantly improve the stretch of their constructions, reducing the dependence on k from exponential to polynomial $(4k - 5)$, without requiring any tradeoff regarding any of the other variables.

In addition, we introduce the notion of vertex-label spanners: subgraphs that preserve distances between every vertex $v \in V$ and label $\lambda \in L$. We present an efficient construction for vertex-label spanners with stretch-size tradeoff close to optimal.

1 Introduction

An approximate distance oracle for a given graph $G = (V, E)$ is a processed data structure that, given two vertices s and t, can quickly return an approximation of $\mathbf{dist}(s, t, G)$, the distance between s and t in G. [To ease notation, we let $\mathbf{dist}(s, t) = \mathbf{dist}(s, t, G)$. In other words, when we refer to the distance between s and t in some subgraph H of G, we will always state the subgraph explicitly and write $\mathbf{dist}(s, t, H)$. Otherwise, if we write $\mathbf{dist}(s, t)$ we mean $\mathbf{dist}(s, t, G)$.]

The approximate distance oracle is said to be of stretch k, or a k-approximate distance oracle, if for every two vertices, s and t, the reported distance $\tilde{\mathbf{dist}}(s, t)$ between s and t satisfies $\mathbf{dist}(s, t) \leq \tilde{\mathbf{dist}}(s, t) \leq k \cdot \mathbf{dist}(s, t)$.

Usually, the key concerns in designing approximate distance oracles are to minimize the size of the data structure, to minimize the stretch, and to minimize the query time.

L. Epstein and P. Ferragina (Eds.): ESA 2012, LNCS 7501, pp. 325–336, 2012.
© Springer-Verlag Berlin Heidelberg 2012

Distance oracles have been extensively studied. They were first introduced by Thorup and Zwick in a seminal paper [17]. Thorup and Zwick showed how to construct for a given integer $k \geq 1$, a $(2k - 1)$-approximate distance oracle of size $O(kn^{1+1/k})$ that can answer distance queries in $O(k)$ time. Thorup and Zwick [17] showed that their space requirements are essentially optimal assuming the girth conjecture of Erdős relating the number of edges of a graph and its girth [6]. Thorup and Zwick also showed how to derandomize this construction, but at the cost of increasing the preprocessing time and slightly increasing the size. Roditty, Thorup, and Zwick [14] later improved this result, presenting a faster deterministic construction and reducing the size of the data structure to $O(kn^{1+1/k})$ as in the randomized construction. Further improvements on the construction time were later introduced in [4,5,3]. For further results and lower bounds see also [9,15,10,1].

In this paper, we consider a natural variant of the approximate distance oracle problem for vertex-labeled graphs. We are given an undirected weighted graph, $G = (V, E)$, where each vertex, v, is assigned a *label*, $\lambda(v)$, where $\lambda(v)$ belongs to a set $L = \{\lambda_1, ..., \lambda_\ell\}$ of $\ell \leq n$ distinct labels. The goal is to construct a compact data structure that, given a vertex v and a label $\lambda \in L$, can quickly return an approximation to the distance $\mathbf{dist}(v, \lambda)$, where $\mathbf{dist}(v, \lambda)$ is the minimal distance between v and a λ-labeled vertex in G. This interesting variant of distance oracles was introduced by Hermelin, Levy, Weimann and Yuster [8]. The labels of the vertices often represent some functionality (or resources). In some settings, the natural question is not what is the distance between two given vertices, but rather what is the distance between a given vertex and some desired resource. For example, the vertices may represent cities and the labels may represent some public resources such as hospitals, courts, universities, and so on.

Hermelin et al. [8] mention that there is a simple solution for this problem: store a table of size $n \cdot \ell$, where the entry (v, λ) represents the distance $\mathbf{dist}(v, \lambda)$. This data structure is of size $O(n\ell)$, the query time is $O(1)$, and the stretch is 1 (exact distances). As shown in [8] this table can be constructed in $O(m\ell)$ time. However, as is also mentioned in [8], this data structure suffers from two main drawbacks. First, in many applications, $O(n\ell)$ might be still too large, and in such applications it might be preferable to store a more compact data structure at the price of approximate distances. Second, in some settings it might be desirable to allow label changes and it is not clear how to efficiently handle label changes using the above mentioned data structure.

Hermelin et al. [8] present several results for distance oracles for vertex-labeled graphs problem. In their first result, Hermelin et al. show how to construct a vertex-label distance oracle of expected size $O(kn^{1+1/k})$ with stretch $(4k - 5)$ and query time $O(k)$. This result is unsatisfactory when ℓ is very small, especially when $\ell = o(n^{1/k})$. In this case, the trivial $O(n\ell)$ solution gives a smaller data structure with exact distances. To overcome this issue, they propose a second data structure of size $O(kn\ell^{1/k})$. This, however, comes at the price of a huge stretch factor of $2^k - 1$. In a third result, they present a dynamic vertex-label distance oracle that is capable of handling label changes in sub-linear time. More specifically, they show how to construct a vertex-label distance oracle of expected size $O(kn^{1+1/k})$ and with stretch $(2 \cdot 3^{k-1} + 1)$ that can support label changes in $O(kn^{1/k} \log \log n)$ time and queries in $O(k)$ time.

Note that in the latter two results, the stretch depends exponentially on k. In this paper, we address an important question they left open, namely, is it possible to improve this dependence on k from exponential to polynomial. More specifically, we prove the following theorems.

Theorem 1. *A vertex-label distance oracle of expected size $O(kn\ell^{1/k})$ with stretch $(4k - 5)$ and query time $O(k)$ can be constructed in $O(m \cdot \min\{n^{k/(2k-1)}, \ell\})$ time.*

Theorem 2. *A vertex-label distance oracle of expected size $\tilde{O}(n^{1+1/k})$ with stretch $(4k - 5)$ and query time $O(k)$ can be constructed in $O(kmn^{1/k})$ time and can support label changes in $\tilde{O}(n^{1/k})$ time.*

A closely related notion of distance oracles is that of *spanners*. A subgraph H is said to be a k-spanner (or a spanner with stretch k) of the graph G if $\mathbf{dist}(u, v, H) \leq k \cdot \mathbf{dist}(u, v, G)$ for every $u, v \in V(G)$. Here and throughout, $V(G')$ denotes the set of vertices of graph G', and similarly, $E(G')$ denotes the set of edges of graph G'. A well-known theorem on spanners is that one can efficiently construct a $(2k - 1)$-spanner with $O(n^{1+1/k})$ edges [2]. This size-stretch tradeoff is conjectured to be optimal . The notion of spanners was introduced in the late 80's [11,12], and has been extensively studied. Spanners are used as a fundamental ingredient in many distributed applications (e.g., synchronizers [12], compact routing [13,16], broadcasting [7], etc.).

This paper also introduces a natural extension of spanners, spanners for vertex-labeled graphs, and presents efficient constructions for such spanners. Consider an undirected weighted graph $G = (V, E)$, where each vertex $v \in V$ is assigned a label from a set of labels $L = \{\lambda_1, ..., \lambda_\ell\}$. We say that a subgraph H is a *vertex-labeled k-spanner (VL k-spanner)* of G if $\mathbf{dist}(u, \lambda, H) \leq k \cdot \mathbf{dist}(u, \lambda, G)$ for every vertex $u \in V$ and label $\lambda \in L$. It is not hard to verify that every k-spanner is also a VL k-spanner. However, one may hope to find sparser VL-spanners when the number of labels is small. A naive approach would be to create for each label λ an auxiliary graph G_λ by adding a new vertex s_λ and then connect s_λ to all λ-labeled vertices with edges of weight 0. It is not hard to verify that by invoking a shortest path algorithm in every G_λ from s_λ and taking the union of all these shortest-paths trees (removing the vertices s_λ and their incident edges), the resulting subgraph is a VL 1-spanner (preserving the exact distances) with $O(n\ell)$ edges. However, the $O(n\ell)$ spanner's size may still be too large in many settings, and one may wish to reduce the size of the spanner at the price of approximated distances. Ideally, one would wish to find a VL $(2k - 1)$-spanner with $O(n\ell^{1/k})$ number of edges (beating these bounds yields improved trade-off for the standard spanners). We managed to come close this goal, presenting an efficient construction for VL spanners with stretch close to $(4k + 1)$ and with $\tilde{O}(n\ell^{1/k})$ number of edges. More specifically, we prove the following theorem.

Theorem 3. *For every weighted graph G with minimal edge weight 1 and fixed parameter $\epsilon > 0$, one can construct in polynomial time a vertex-label $(4k+1)(1+\epsilon)$-spanner with $O(\log n \cdot \log D \cdot n\ell^{1/k})$ edges, where D is the weighted diameter of the graph.*

We note that our constructions for vertex-labeled distance oracles and the constructions presented in [8] do not seem to transform well to also give spanners. Therefore, our vertex-labeled spanner constructions use different techniques and require some new ideas.

The rest of the paper is organized as follows. In Section 2 we prove Theorem 1. In Section 3 we prove Theorem 2. In Section 4 we prove Theorem 3, for simplicity, we first present a construction for unweighted graphs in Subsection 4.1 and then show how to generalize it to weighted graphs in Subsection 4.2.

Due to space limitations, all formal codes and most of the proofs are deferred to the full version.

2   Compact Vertex-Label Distance Oracles

In this section we prove Theorem 1. In Subsection 2.1 we present the construction of our data structure, in Subsection 2.2 we present our query answering algorithm, and in Subsection 2.3 we analyze the construction time.

2.1   The Data Structure

The first step of the construction of the data structure is similar to the algorithm presented in [8]. For a given positive integer k, construct the sets $V = A_0 \supseteq A_1 \supseteq \cdots \supseteq A_{k-1}$ as follows: The i-th level A_i is constructed by sampling the vertices of A_{i-1} independently at random with probability $\ell^{-1/k}$ for $1 \le i \le k - 1$.

Next, for every vertex v, define the bunch of v exactly as the Thorup-Zwick definition, but with a small change: that is, we have one level less, namely

$$B(v) = \bigcup_{i=0}^{k-2} \{u \in A_i \setminus A_{i+1} \mid \mathbf{dist}(v, u) < \mathbf{dist}(v, A_{i+1})\}.$$

The pivot $p_i(v)$ is also exactly as Thorup-Zwick's definition, namely $p_i(v)$ is the closest vertex to v in A_i (break ties arbitrarily).

The data structure is as follows. For every vertex v store $B(v)$ and for every vertex $x \in B(v)$ store $\mathbf{dist}(x, v)$. In addition, for every vertex v and every index i for $1 \le i \le k-1$ store $p_i(v)$ and $\mathbf{dist}(v, p_i(v))$. For every vertex $v \in A_{k-1}$, store its distance for every label $\lambda \in L$, namely $\mathbf{dist}(v, \lambda)$ in a hash table. Finally, for every label $\lambda \in L$, store $B(\lambda) = \bigcup_{v \in V_\lambda} B(v)$ in a hash table and for every vertex $x \in B(\lambda)$ store $\mathbf{dist}(x, \lambda)$, where $V_\lambda = \{v \in V \mid \lambda(v) = \lambda\}$. This completes the description of our data structure. Below, we bound the size of the data structure.

Lemma 1. $\mathbb{E}[|B(v)|] = (k-1)\ell^{1/k}$ *for every* $v \in V$.

Lemma 2. *The expected size of our data structure is* $O(kn\ell^{1/k})$.

Proof: By Lemma 1, the total expected size of the bunches of all vertices is $(k - 1)n\ell^{1/k}$. In addition, for every vertex in A_{k-1} we also store its distance to every label, $\lambda \in L$. That is, for every vertex in A_{k-1}, we store additional data of size ℓ. The expected size of A_{k-1} is $n\ell^{-(k-1)/k}$. To see this, note that the probability that a vertex v belongs to A_i is $\ell^{-i/k}$. Therefore, the total additional expected size stored for all vertices in A_{k-1} is $n\ell^{1/k}$. Finally, storing $B(\lambda)$ for every $\lambda \in L$ does not change the asymptotic size since $\sum_{\lambda \in L} |B(\lambda)| \leq \sum_{\lambda \in L} \sum_{v \in V_\lambda} |B(v)| = \sum_{v \in V} |B(v)|$. The lemma follows.   ∎

2.2   Vertex-Label Queries

We now describe our query answering algorithm, with the input vertex-label query, $(v \in V, \lambda \in L)$.

The query answering algorithm is done as follows. For every index i from 0 to $k - 2$, check if $p_i(v) \in B(\lambda)$, and if so return $\mathbf{dist}(v, p_i(v)) + \mathbf{dist}(p_i(v), \lambda)$. Otherwise, if no such index exists, return $\mathbf{dist}(v, p_{k-1}(v)) + \mathbf{dist}(p_{k-1}(v), \lambda)$. This completes the query answering algorithm.

We now turn to the stretch analysis. If there exists an index i such that $0 \leq i \leq k - 2$ and $p_i(v) \in B(\lambda)$, set i to be the first such index. If no such index exists, set $i = k - 1$. Let u be the λ-labeled vertex closest to v, namely $\mathbf{dist}(v, u) = \mathbf{dist}(v, \lambda)$. Note that $p_j(v) \notin B(u)$ for every $j < i$. This is due to the facts that $p_j(v) \notin B(\lambda)$ and that $B(u) \subseteq B(\lambda)$. Using the same analysis as in [16] (Lemma A.1), one can show that $\mathbf{dist}(v, p_i(v)) \leq 2i \cdot \mathbf{dist}(v, u)$ and $\mathbf{dist}(p_i(v), \lambda) \leq \mathbf{dist}(p_i(v), u) \leq (2i + 1)\mathbf{dist}(v, u)$. We get that the returned distance $\tilde{\mathbf{dist}}(v, \lambda)$ satisfies $\tilde{\mathbf{dist}}(v, \lambda) = \mathbf{dist}(v, p_i(v)) + \mathbf{dist}(p_i(v), \lambda) \leq (4k - 3)\mathbf{dist}(v, u) = (4k - 3)\mathbf{dist}(v, \lambda)$.

Note that if $i \leq k - 2$, then the distance $\mathbf{dist}(p_i(v), \lambda)$ is stored in $B(\lambda)$, or, if $i = k - 1$ then $p_i(v) \in A_{k-1}$ and recall that $\mathbf{dist}(u, \lambda)$ is stored for every $u \in A_{k-1}$ and therefore also $\mathbf{dist}(p_i(v), \lambda)$ is stored and can be retrieved in $O(1)$ time.

Finally, using the same method as in [16] (Lemma A.2) the stretch can be reduced to $4k - 5$ as required.

We note that Hermelin et al. [8] have to check all indices. Namely, their query algorithm is to return the minimal distance $\mathbf{dist}(v, w) + \mathbf{dist}(w, w_\lambda)$ for all $w = p_i(v)$ such that $w \in B(\lambda)$, where we define w_λ to be the λ-labeled vertex closest to w that satisfies $w \in B(w_\lambda)$. Let u be the λ-labeled vertex that satisfies $\mathbf{dist}(v, u) = \mathbf{dist}(v, \lambda)$. Hermelin et al. [8] note that the first $w = p_i(v) \in B(\lambda)$, does not necessarily satisfy $\mathbf{dist}(w, w_\lambda) \leq \mathbf{dist}(w, u)$ since there is a possibility that $w \notin B(u)$. Therefore, they have to iterate over all indices $1 \leq i \leq k - 1$ and take the one that gives the minimal distance. We bypass this issue by simply explicitly storing the distance $\mathbf{dist}(w, \lambda)$ —rather than $\mathbf{dist}(w, w_\lambda)$— for every

$w \in B(\lambda)$. This does not increase the asymptotic size and it simplifies the query algorithm and its analysis.

2.3   Construction Time

The preprocessing time of our construction is composed of the time it takes to construct the different components of our data structure. Recall that our data structure is composed of four components. The first component is the pivots: for every vertex v we store $p_i(v)$ for $1 \leq i \leq k - 1$. The second component is the bunches of the vertices: for every vertex v we store $B(v)$ and the distances $\mathbf{dist}(v, x)$ for every $x \in B(v)$. The third component is the bunches of the labels: for every label λ we store $B(\lambda)$ and the distances $\mathbf{dist}(x, \lambda)$ for every $x \in B(\lambda)$. The fourth part is the distances of the vertices in A_{k-1} to all labels: store $\mathbf{dist}(v, \lambda)$ for every $v \in A_{k-1}$ and $\lambda \in L$.

Using the same analysis as in [17], one can show that the time complexity for constructing the first component is $O(k \cdot m)$ and the time complexity for the second component is $O(km\ell^{1/k})$.

Constructing $B(\lambda)$ for every $\lambda \in L$ (the first part of the third component) can be done easily in $O(kn\ell^{1/k})$ time [just go over all vertices v, and add $B(v)$ to $B(\lambda(v))$].

We are left with computing $\mathbf{dist}(x, \lambda)$ for every $x \in B(\lambda)$ and then for every $x \in A_{k-1}$ and $\lambda \in L$. This can be done by invoking Dijkstra's Algorithm ℓ times (for every label $\lambda \in L$ add a source vertex s and connect all λ-labeled vertices to s with an edge of weight 0 and then invoke Dijkstra's Algorithm from s) and thus the running time for this part is $O(m\ell)$.

We get that the total running time for the preprocessing phase is $O(m\ell)$.

We note here that if $\ell > n^{k/(2k-1)}$, then it is possible to reduce the preprocessing running time to $O(mn^{k/(2k-1)})$. This can be done by storing $\mathbf{dist}(v, v_\lambda)$ as suggested in [8] instead of storing $\mathbf{dist}(v, \lambda)$ for every $v \in B(\lambda)$. This change forces checking all indices in the query algorithm as explained above. The analysis of the preprocessing time in this case is similar to the one presented in [8].

3   Dynamic Labels

In this section, we consider the problem of constructing a dynamic vertex-label distance oracle and prove Theorem 2. Namely, we show how to construct a vertex-label distance oracle that supports label changes of the form $update(v, \lambda)$ for $v \in V$ and $\lambda \in L$. This update changes the label of v to be λ and leaves all other vertices unchanged. Our data structure in this section is a slight adaptation of Thorup-Zwick's construction and is also similar to the one presented in [8] for static vertex-label distance oracles.

In Subsection 3.1 we present the data structure, Subsection 3.2 presents the query answering algorithm and in Subsection 3.3 we analyze the construction time.

3.1   The Data Structure

For a given positive integer k, construct the sets $V = A_0 \supseteq A_1 \supseteq \cdots \supseteq A_{k-1} \supseteq A_k = \emptyset$ as follows. The i-th level A_i is constructed by sampling the vertices of A_{i-1} independently at random with probability p to be specified shortly for $1 \le i \le k-1$.

The bunch of v is defined as in Thorup-Zwick as follows:

$$B(v) = \bigcup_{i=0}^{k-1} \{u \in A_i \setminus A_{i+1} \mid \mathbf{dist}(v,u) < \mathbf{dist}(v, A_{i+1})\}.$$

The pivot $p_i(v)$ is also defined exactly as Thorup-Zwick's definition, namely $p_i(v)$ is the closest vertex to v in A_i (break ties arbitrarily).

In order to allow fast updates, the size of every bunch $B(v)$ for $v \in V$ must be small. In order to ensure this property, we set the sampling probability to be $p = (n/\ln n)^{-1/k}$. It was proven in [17] that by setting $p = (n/\ln n)^{-1/k}$, the size of every bunch $B(v)$ is $O(n^{1/k} \log^{1-1/k} n)$ with high probability.

In addition, for every $\lambda \in L$, store $B(\lambda) = \bigcup_{v \in V_\lambda} B(v)$ in a hash table. Recall that in the static setting we store $\mathbf{dist}(v, \lambda)$ when $v \in B(\lambda)$. In the dynamic setting, we do not store this data as it is too costly to update it for two reasons. First, notice that a single label change, say from λ_1 to λ_2, might require updating $\mathbf{dist}(v, \lambda_1)$, $\mathbf{dist}(u, \lambda_2)$ for many vertices $v \in B(\lambda_1)$ and $u \in B(\lambda_2)$. As both $B(\lambda_1)$ and $B(\lambda_2)$ might be very large, this may take a long time. Second, even a single update of $\mathbf{dist}(v, \lambda)$ might be too costly as it might require invoking a shortest path algorithm during the update phase.

To avoid the need of updating $\mathbf{dist}(v, \lambda)$ for a vertex $v \in B(\lambda)$, we do the following two things. First, rather than maintaining the value $\mathbf{dist}(v, \lambda)$, we instead maintain the value $\mathbf{dist}(v, v_\lambda)$ where v_λ is defined to be the closest λ-labeled vertex such that $v \in B(v_\lambda)$. Second, we use the method of [8] and iterate on all indices $1 \le i \le k-1$ and return the minimal distance $\mathbf{dist}(v, w) + \mathbf{dist}(w, w_\lambda)$ for $w = p_i(v)$ in the answering query algorithm.

In order to maintain the value $\mathbf{dist}(v, w_\lambda)$ for a vertex $v \in B(\lambda)$, we store the set of λ-labeled vertices x such that v belongs to $B(x)$ in a heap, $\mathbf{Heap}(v, \lambda)$, namely the set of vertices in the $\mathbf{Heap}(v, \lambda)$ is $V(\mathbf{Heap}(v, \lambda)) = \{x \in V \mid v \in B(x) \text{ and } \lambda(x) = \lambda\}$ where the key, $\mathbf{key}(x)$, of a vertex, $x \in V(\mathbf{Heap}(v, \lambda))$, is the distance, $\mathbf{dist}(v, x)$. The heap, $\mathbf{Heap}(v, \lambda)$, supports the standard operations of [$\mathbf{insert}(x)$ - insert a vertex x to the heap], [$\mathbf{remove}(x)$ - remove a vertex x from the heap] and [$\mathbf{minimum}()$ - return the vertex x in the heap with minimal $\mathbf{key}(x)$]. For this purpose, we use any standard construction of heaps (e.g. [18]) that allow $\mathbf{insert}$ and $\mathbf{remove}$ operations in $O(\log \log n)$ time and $\mathbf{minimum}()$ operations in constant time.

We now summarize the different components of our data structure to make it clear what parts of the data structure need to be updated as a result of a label change.

(1) For every vertex v, store $B(v)$ and for every vertex $x \in B(v)$, store $\mathbf{dist}(v, x)$. This data is stored in a hash table, which allows checking if a vertex $x \in B(v)$ and, if so, finding $\mathbf{dist}(v, x)$ in $O(1)$ time.

(2) For every vertex v and index $1 \le i \le k - 1$, store $p_i(v)$.

(3) For every $\lambda \in L$, store $B(\lambda)$ in a hash table where the entry in the hash table of a vertex $v \in B(\lambda)$ points to the heap, $\mathbf{Heap}(v, \lambda)$.

It is not hard to see that only component (3) in our data structure needs to be modified as a result of a label change. Moreover, if the label of some vertex $v \in V$ is changed from $\lambda_1 \in L$ to $\lambda_2 \in L$, then only $B(\lambda_1)$ and $B(\lambda_2)$ need to be updated. The update is relatively simple. For every vertex $x \in B(v)$, do the following: Remove v from $\mathbf{Heap}(x, \lambda_1)$. If $\mathbf{Heap}(x, \lambda_1)$ becomes empty, then also remove x from the hash table of $B(\lambda_1)$. In addition, if $x \in B(\lambda_2)$, then add v to $\mathbf{Heap}(x, \lambda_2)$, otherwise add x to the hash table $B(\lambda_2)$ and create a new heap, $\mathbf{Heap}(x, \lambda_2)$, containing only v. Each such operation takes $O(\log \log n)$ time for every $x \in B(v)$. Recall that the size of $B(v)$ is $O(n^{1/k} \log^{1-1/k} n)$; thus we get that the update requires $O(n^{1/k} \log^{1-1/k} n \log \log n)$ time.

It is not hard to verify that the size of the data structure is

$$O(\sum_{v \in V} |B(v)|) = O(n^{1+1/k} \log^{1-1/k} n).$$

3.2   Vertex-Label Queries

The query answering algorithm is similar to the one presented in Section 2. Let $(v \in V, \lambda \in L)$ be the input vertex-label query.

The query answering algorithm is done by checking all indices $1 \le i \le k - 1$ and returning the minimal $\mathbf{dist}(v, p_i(v)) + \mathbf{dist}(p_i(v), w_\lambda)$ such that $p_i(v) \in B(\lambda)$. w_λ is the vertex returned by $\mathbf{Heap}(p_i(v), \lambda).\mathbf{minimum}()$, namely, w_λ is the λ-labeled vertex such that $p_i(v) \in B(\lambda)$ with minimal $\mathbf{dist}(p_i(v), w_\lambda)$.

Note that here we must check all indices and cannot stop upon reaching the first index j, such that $p_j(v) \in B(\lambda)$. Let u be the λ-labeled vertex closest to v, namely $\mathbf{dist}(v, u) = \mathbf{dist}(v, \lambda)$. As mentioned by Hermelin et al. [8] (and discussed above), the first $w = p_i(v) \in B(\lambda)$, does not necessarily satisfy $\mathbf{dist}(w, w_\lambda) \le \mathbf{dist}(w, u)$ since it may be that $w \notin B(u)$. Therefore, we also have to iterate over all indices $1 \le i \le k - 1$ and take the one that gives the minimal distance.

It is not hard to verify that the query algorithm takes $O(k)$ time.

Finally, as mentioned in Section 2, using the same analysis as in [16] (Lemma A.1), the stretch is $(4k - 3)$, and using the same method as in [16] (Lemma A.2), the stretch can be reduced to $4k - 5$ as required.

3.3   Construction Time

The first two components of the data structure are exactly the same construction as Thorup-Zwick's and thus can be constructed in $O(kmn^{1/k})$ time.

The third component can be constructed in $O(|\cup_{v\in V} B(v)| \cdot \log\log n) = O(kn^{1+1/k}\log\log n)$ time (the $\log\log n$ comes from the insertion to the heaps). We conclude that the total preprocessing time is $O(kmn^{1/k})$.

4   Sparse Vertex-Label Spanners

In this section, we shall address the question of finding low stretch sparse vertex-label spanners. More specifically, we show how to find a subgraph H with expected number of edges $\tilde{O}(n\ell^{1/k})$ such that for every vertex v and label λ, $\mathbf{dist}(v,\lambda,H) \leq (4k+1)(1+\epsilon)\mathbf{dist}(v,\lambda,G)$ for any fixed $\epsilon > 0$. Note that it is unclear how to transform the construction of Section 2 into a vertex-label spanner. To see this, recall that for every vertex v in A_{k-1} and for every label λ we store the distance $\mathbf{dist}(v,\lambda)$. However, in order to allow a low-stretch spanner, we need to add a shortest path P from $v \in A_{k-1}$ to its closest λ-labeled vertex. This path could be very long and, of course, may contain many vertices not in A_{k-1}. Thus, adding all these paths may result in a subgraph with too many edges. Therefore, transforming the construction of Section 2 into a vertex-label spanner seems challenging. We hence suggest a different construction for spanners.

For simplicity, we first present (Subsection 4.1) a construction for unweighted graphs and then (Subsection 4.2) we show how to generalize this construction to weighted graphs.

4.1   Unweighted Graphs

For a vertex v, radius r, and subgraph H, let $B(v,r,H) = \{x \in V(H) \mid \mathbf{dist}(v,x,H) \leq r\}$.

We start by describing an algorithm named **VL_Cover**, that when given a distance d, returns a spanner H with the following property. For every vertex $v \in V$ and label $\lambda \in L$, such that $\mathbf{dist}(v,\lambda,G) \leq d$, $\mathbf{dist}(v,\lambda,H) \leq (4k+1)d$.

Loosely speaking, the algorithm proceeds as follows: It consists of two stages. The first stage handles "sparse" areas, namely, balls around some vertex v such that $|B(v,kd,G)| < d \cdot \ell^{1-1/k}$. In this case, we show that we can tolerate adding a BFS tree $T(v)$ spanning $B(v,i\cdot d,G)$ for some $1 \leq i \leq k$, charging the vertices in $B(v,(i-1)d,G)$. It is not hard to verify that every path P of length at most d that contains a vertex in $B(v,(i-1)d,G)$ satisfies $V(P) \subseteq B(v,i\cdot d,G)$. Using the tree $T(v)$, we have a "short" alternative path to P. The second stage handles "dense" areas, namely, balls around some vertex v such that $|B(v,kd,G)| \geq d \cdot \ell^{1-1/k}$. The algorithm picks a set $C \subseteq V$ such that the distance between every two vertices in C is at least $2kd$, and such that every vertex in a "dense" area has a "close" vertex in C. In this case, we can tolerate adding $O(d \cdot \ell)$ edges for every vertex $c \in C$, charging the vertices in $B(c,kd,G)$. The algorithm connects every vertex $u \in V$ to some "close" vertex $c \in C$ by a shortest path. In addition, for every label λ such that the distance from c to λ is $O(d)$, we add a "short" path from c to λ. In this case for every vertex u and label λ such that $\mathbf{dist}(u,\lambda) \leq d$, we have a "short" alternative path by concatenating the path from u to its "close" vertex $c \in C$ and the path from c to λ.

We now describe the algorithm more formally.

Initially, set $G' = G$, $H_d = (V, \emptyset)$, and $C = \emptyset$. The algorithm consists of two stages. The first stage of the algorithm is done as follows. As long as there exists a vertex $v \in V(G')$ such that $|B(v, kd, G')| < d \cdot \ell^{1-1/k}$, pick v to be such a vertex. Let i be the minimal index such that $|B(v, id, G')| < d \cdot \ell^{(i-1)/k}$. Construct a shortest-path tree $T(v)$ rooted at v and spanning $B(v, id, G')$ and add the edges of $T(v)$ to H_d. If $i > 1$, then remove the vertices $B(v, (i-1)d, G')$ from G'; if $i = 1$ remove the vertices $B(v, d, G')$.

The second stage of the algorithm is done as follows. As long as there is a vertex $v \in V$ such that $B(v, 2k \cdot d, G') \cap C = \emptyset$, pick v to be such a vertex, and add it to C. For every vertex $c \in C$, do the following. First, let $B(c)$ be all vertices u in G' such that c is closer to u than any other $c' \in C$. (We assume that there is a unique such node c. This is without loss of generality, as one can just order the vertices and say that a vertex c_1 is closer to z than the vertex c_2 if either $\mathbf{dist}(c_1, z) < \mathbf{dist}(c_2, z)$ or $\mathbf{dist}(c_1, z) = \mathbf{dist}(c_2, z)$ and c_1 appears before c_2 in the ordering of the vertices.) Second, construct a BFS tree rooted at c and spanning the vertices in $B(c)$ and add the edges of the BFS tree to H_d. Third, for every label $\lambda \in L$, if there exists a vertex $v \in B(c)$ such that $\mathbf{dist}(v, \lambda, G) \leq d$, then pick v to be such a vertex and add a shortest path $P(v, \lambda)$ from v to its closest λ-labeled vertex, and add the edges of the path $P(v, \lambda)$ to H_d.

This completes the construction of the spanner.

We now turn to analyze the stretch and the number of edges in the resulting spanner H_d.

Lemma 3. *The number of edges in H_d is $O(n\ell^{1/k})$.*

Lemma 4. *For every vertex $u \in V$ and label $\lambda \in L$ such that $\mathbf{dist}(u, \lambda) \leq d$, $\mathbf{dist}(u, \lambda, H_d) \leq (4k+1)d$.*

The main algorithm for constructing our spanner operates in $\log n$ iterations. For a given fixed parameter ϵ, for every index $1 \leq i \leq \log n$, invoke Algorithm **VL_Cover** with parameter $d(i) = (1+\epsilon)^i$. Let $H_{d(i)}$ be the subgraph returned by the Algorithm **VL_Cover**. Let H be the union of all subgraphs $H_{d(i)}$ for $1 \leq i \leq \log n$. This completes our spanner construction.

It is not hard to verify that by Lemmas 3 and 4, we have the following.

Theorem 4. *For every unweighted graph G and fixed parameter ϵ, one can construct in polynomial time a vertex-label $(4k+1)(1+\epsilon)$-spanner with $O(\log n \cdot n\ell^{1/k})$ edges.*

4.2   Weighted Graphs

In this section we generalize our spanner construction to weighted graphs.

Note that in the unweighted case we exploit the fact that a path P of length d contains d edges. In the weighted case, this is no longer the case. A path of length d could potentially contain a much smaller or larger number of edges. We thus need to be more careful with the paths we add to the spanner. Roughly speaking, for every potential distance d and index j, we consider vertices that have at least 2^j vertices at distance d from them. In this case, we can tolerate adding paths with $O(2^j)$ number of edges.

For a path P, let $|P|$ be the number of edges in P and let $\mathbf{dist}(P)$ be the length of P. Let $\tilde{\mathbf{dist}}(v, u, x', H)$ be the minimal length of a path from u to v in H among all paths with at most x' edges. Let $\tilde{P}(u, v, x', H)$ be the shortest path in H between u and v among all paths with at most x' edges. We say that a vertex v is (x', d)-relevant in H if $x' \leq |B(v, d, H)|$. We say that a path P is (x', d)-relevant if $x' \leq |P| \leq 2x'$ and $\mathbf{dist}(P) \leq d$.

As in the unweighted case, we first describe an algorithm named **WVL_Cover** that given a distance d, an integer x, and a graph G, returns a subgraph $H_{d,x}$ that satisfy the following. For every vertex v and label $\lambda \in L$ such that there exists an (x, d)-relevant path P from v to a λ-labeled vertex, $\mathbf{dist}(v, \lambda, H_{d,x}) \leq (4k+1)d$.

The algorithm proceeds as follows. Initially, set $G' \leftarrow G$, $H_{d,x} \leftarrow (V, \emptyset)$, and $C \leftarrow \emptyset$. There are two stages. The first stage of the algorithm is as follows. As long as there exists an (x, d)-relevant vertex v in G' such that $|B(v, kd, G')| < x \cdot \ell^{1-1/k}$, pick v to be such a vertex. Let i be the minimal index such that $|B(v, id, G')| < x \cdot \ell^{(i-1)/k}$. Construct a shortest-path tree $T(v)$ rooted at v and spanning $B(v, id, G')$, and then add the edges of $T(v)$ to $H_{d,x}$. Finally, remove the vertices $B(v, (i-1)d, G')$ from G'.

The second stage of the algorithm is done as follows. As long as there exists an (x, d)-relevant vertex v in G' such that $B(v, 2k \cdot d, G') \cap C = \emptyset$, add v to C. For every vertex $c \in C$ do the following. First, let $B(c) = \{u \in V(G') \mid \mathbf{dist}(u, c) = \mathbf{dist}(u, C)\}$. Second, construct a shortest-path tree $T(c)$ rooted at c and spanning $B(c)$, and then add the edges of $T(c)$ to $H_{d,x}$. Finally, for every label $\lambda \in L$ such that $\exists y \in B(c)$ and $\tilde{\mathbf{dist}}(y, \lambda, 2x) \leq d$, add $E(\tilde{P}(y, \lambda, 2x, G))$ to $H_{d,x}$. This completes the construction of our spanner.

Lemma 5. *For every vertex $u \in V$ and label $\lambda \in L$ such that there exists an (x, d)-relevant path P from u to a λ-labeled vertex, $\mathbf{dist}(u, \lambda, H_{d,x}) \leq (4k+1)d$.*

Lemma 6. *The number of edges in $H_{d,x}$ is $O(n\ell^{1/k})$.*

The main algorithm for constructing our spanner operates in $\log n \cdot \log D$ iterations, where D is the diameter of the graph. For a given fixed parameter ϵ, for every index $1 \leq i \leq \log D$ and $1 \leq j \leq \log n$, invoke Algorithm **WVL_Cover** with parameters $d(i) = (1 + \epsilon)^i$ and $x(j) = 2^j$. Let $H_{d(i),x(j)}$ be the subgraph returned by the Algorithm **WVL_Cover**. Finally, let H be the union of all subgraphs $H_{d(i),x(j)}$ for $1 \leq i \leq \log D$ and $1 \leq j \leq \log n$. This completes our spanner construction.

The following lemma shows that the stretch of the spanner is $(4k+1)(1+\epsilon)$.

Lemma 7. *For every vertex $u \in V$ and label $\lambda \in L$, $\mathbf{dist}(u, \lambda, H) \leq (4k+1)(1+\epsilon)\mathbf{dist}(u, \lambda, G)$.*

Proof: Consider a vertex $u \in V$ and a label $\lambda \in L$. Let $P = P(u, \lambda)$ be the shortest path from u to λ in G. Let i and j be the indices such that $(1 + \epsilon)^{i-1} \leq$ $\mathbf{dist}(P) \leq (1 + \epsilon)^i$ and $2^j \leq |P| \leq 2^{j+1}$. By Lemma 5, $\mathbf{dist}(u, \lambda, H_{d,x}) \leq$ $(4k + 1)(1 + \epsilon)^i$, for $d = (1 + \epsilon)^i$ and $x = 2^j$. The lemma follows.   ∎

We thus conclude Theorem 3.

References

1. Abraham, I., Gavoille, C.: On Approximate Distance Labels and Routing Schemes with Affine Stretch. In: Peleg, D. (ed.) Distributed Computing. LNCS, vol. 6950, pp. 404–415. Springer, Heidelberg (2011)
2. Althöfer, I., Das, G., Dobkin, D., Joseph, D., Soares, J.: On sparse spanners of weighted graphs. Discrete & Computational Geometry 9, 81–100 (1993)
3. Baswana, S., Gaur, A., Sen, S., Upadhyay, J.: Distance Oracles for Unweighted Graphs: Breaking the Quadratic Barrier with Constant Additive Error. In: Aceto, L., Damgård, I., Goldberg, L.A., Halldórsson, M.M., Ingólfsdóttir, A., Walukiewicz, I. (eds.) ICALP 2008, Part I. LNCS, vol. 5125, pp. 609–621. Springer, Heidelberg (2008)
4. Baswana, S., Kavitha, T.: Faster algorithms for approximate distance oracles and all-pairs small stretch paths. In: FOCS, pp. 591–602 (2006)
5. Baswana, S., Sen, S.: Approximate distance oracles for unweighted graphs in expected $O(n^2)$ time. ACM Transactions on Algorithms, 557–577 (2006)
6. Erdős, P.: Extremal problems in graph theory. In: Theory of Graphs and Its Applications, pp. 29–36 (1964)
7. Farley, A.M., Proskurowski, A., Zappala, D., Windisch, K.: Spanners and message distribution in networks. Discrete Applied Mathematics 137(2), 159–171 (2004)
8. Hermelin, D., Levy, A., Weimann, O., Yuster, R.: Distance Oracles for Vertex-Labeled Graphs. In: Aceto, L., Henzinger, M., Sgall, J. (eds.) ICALP 2011, Part II. LNCS, vol. 6756, pp. 490–501. Springer, Heidelberg (2011)
9. Mendel, M., Naor, A.: Ramsey partitions and proximity data structures. Journal of the European Mathematical Society, 253–275 (2007)
10. Pătraşcu, M., Roditty, L.: Distance oracles beyond the thorup-zwick bound. In: FOCS, pp. 815–823 (2010)
11. Peleg, D., Scháffer, A.A.: Graph spanners. J. Graph Theory, 99–116 (1989)
12. Peleg, D., Ullman, J.D.: An optimal synchronizer for the hypercube. SIAM J. Computing 18(4), 740–747 (1989)
13. Peleg, D., Upfal, E.: A trade-off between space and efficiency for routing tables. J. ACM 36(3), 510–530 (1989)
14. Roditty, L., Thorup, M., Zwick, U.: Deterministic Constructions of Approximate Distance Oracles and Spanners. In: Caires, L., Italiano, G.F., Monteiro, L., Palamidessi, C., Yung, M. (eds.) ICALP 2005. LNCS, vol. 3580, pp. 261–272. Springer, Heidelberg (2005)
15. Sommer, C., Verbin, E., Yu, W.: Distance oracles for sparse graphs. In: FOCS, pp. 703–712 (2009)
16. Thorup, M., Zwick, U.: Compact routing schemes. In: SPAA, pp. 1–10 (2001)
17. Thorup, M., Zwick, U.: Approximate distance oracles. Journal of the ACM, 1–24 (2005)
18. van Emde Boas, P., Kaas, R., Ziljstra, E.: Design and implementation of an efficient priority queue. Mathematical Systems Theory 10, 99–127 (1977)

The Quantum Query Complexity
of Read-Many Formulas

Andrew M. Childs[1,2], Shelby Kimmel[3], and Robin Kothari[2,4]

[1] Department of Combinatorics & Optimization, University of Waterloo
[2] Institute for Quantum Computing, University of Waterloo
[3] Center for Theoretical Physics, Massachusetts Institute of Technology
[4] David R. Cheriton School of Computer Science, University of Waterloo

Abstract. The quantum query complexity of evaluating any read-once formula with n black-box input bits is $\Theta(\sqrt{n})$. However, the corresponding problem for read-many formulas (i.e., formulas in which the inputs can be repeated) is not well understood. Although the optimal read-once formula evaluation algorithm can be applied to any formula, it can be suboptimal if the inputs can be repeated many times. We give an algorithm for evaluating any formula with n inputs, size S, and G gates using $O(\min\{n, \sqrt{S}, n^{1/2}G^{1/4}\})$ quantum queries. Furthermore, we show that this algorithm is optimal, since for any n, S, G there exists a formula with n inputs, size at most S, and at most G gates that requires $\Omega(\min\{n, \sqrt{S}, n^{1/2}G^{1/4}\})$ queries. We also show that the algorithm remains nearly optimal for circuits of any particular depth $k \geq 3$, and we give a linear-size circuit of depth 2 that requires $\tilde{\Omega}(n^{5/9})$ queries. Applications of these results include a $\tilde{\Omega}(n^{19/18})$ lower bound for Boolean matrix product verification, a nearly tight characterization of the quantum query complexity of evaluating constant-depth circuits with bounded fanout, new formula gate count lower bounds for several functions including PAR- ITY, and a construction of an AC^0 circuit of linear size that can only be evaluated by a formula with $\Omega(n^{2-\epsilon})$ gates.

1 Introduction

A major problem in query complexity is the task of evaluating a Boolean formula on a black-box input. In this paper, we restrict our attention to the standard gate set {AND, OR, NOT}. A formula is a rooted tree of NOT gates and unbounded-fanin AND and OR gates, where each leaf represents an input bit and each internal vertex represents a logic gate acting on its children. The depth of a formula is the length of a longest path from the root to a leaf.

The quantum query complexity of evaluating read-once formulas is now well understood. (In this paper, the "query complexity" of f always refers to the bounded-error quantum query complexity, denoted $Q(f)$; for a formal definition, see for example [9].) A formula is read-once if each input bit appears at most once in it. Grover's algorithm shows that a single OR gate on n inputs can be evaluated in $O(\sqrt{n})$ queries [15], which is optimal [6]. It readily follows that balanced constant-depth read-once formulas can be evaluated in $\tilde{O}(\sqrt{n})$ queries [9] (where

L. Epstein and P. Ferragina (Eds.): ESA 2012, LNCS 7501, pp. 337–348, 2012.
© Springer-Verlag Berlin Heidelberg 2012

we use a tilde to denote asymptotic bounds that neglect logarithmic factors), and in fact $O(\sqrt{n})$ queries are sufficient [16]. A breakthrough result of Farhi, Goldstone, and Gutmann showed how to evaluate a balanced binary AND-OR formula with n inputs in time $O(\sqrt{n})$ [13] (in the Hamiltonian oracle model, which easily implies an upper bound of $n^{1/2+o(1)}$ queries [11]). Subsequently, it was shown that any read-once formula whatsoever can be evaluated in $n^{1/2+o(1)}$ queries [1], and indeed $O(\sqrt{n})$ queries suffice [26]. This result is optimal, since any read-once formula requires $\Omega(\sqrt{n})$ queries to evaluate [3] (generalizing the lower bound of [6]).

Now that the quantum query complexity of evaluating read-once formulas is tightly characterized, it is natural to consider the query complexity of evaluating more general formulas, which we call "read-many" formulas to differentiate them from the read-once case. Note that we cannot expect to speed up the evaluation of arbitrary read-many formulas on n inputs, since such formulas are capable of representing arbitrary functions, and some functions, such as parity, require as many quantum queries as classical queries up to a constant factor [4,14]. Thus, to make the task of read-many formula evaluation nontrivial, we must take into account other properties of a formula besides the number of inputs.

Two natural size measures for formulas are formula size and gate count. The size of a formula, which we denote by S, is defined as the total number of inputs counted with multiplicity (i.e., if an input bit is used k times it is counted as k inputs). The gate count, which we denote by G, is the total number of AND and OR gates in the formula. (By convention, NOT gates are not counted.) Note that $G < S$ since the number of internal vertices of a tree is always less than the number of leaves.

A formula of size S can be viewed as a read-once formula on S inputs by neglecting the fact that some inputs are identical, thereby giving a query complexity upper bound from the known formula evaluation algorithm. Thus, an equivalent way of stating the results on read-once formulas that also applies to read-many formulas is the following:

Theorem 1 (Formula evaluation algorithm [26, Cor. 1.1]). *The bounded-error quantum query complexity of evaluating a formula of size S is $O(\sqrt{S})$.*

However, this upper bound does not exploit the fact that some inputs may be repeated, so it can be suboptimal for formulas with many repeated inputs. We study such formulas and tightly characterize their query complexity in terms of the number of inputs n (counted *without* multiplicity), the formula size S, and the gate count G. By preprocessing a given formula using a combination of classical techniques and quantum search before applying the algorithm for read-once formula evaluation, we show that any read-many formula can be evaluated in $O(\min\{n, \sqrt{S}, n^{1/2}G^{1/4}\})$ queries (Theorem 2). Furthermore, we show that for any n, S, G, there exists a read-many formula with n inputs, size at most S, and at most G gates such that any quantum algorithm needs $\Omega(\min\{n, \sqrt{S}, n^{1/2}G^{1/4}\})$ queries to evaluate it (Theorem 3). We construct these formulas by carefully composing formulas with known quantum lower bounds and then applying recent results on the behavior of quantum query complexity under composition [26,25].

To refine our results on read-many formulas, it is natural to consider the query complexity of formula evaluation as a function of the depth k of the given formula in addition to its number of input bits n, formula size S, and gate count G. Beame and Machmouchi showed that constant-depth formulas (in particular, formulas of depth $k = 3$ with size $\tilde{O}(n^2)$) can require $\tilde{\Omega}(n)$ queries [5]. However, to the best of our knowledge, nothing nontrivial was previously known about the case of depth-2 formulas (i.e., CNF or DNF expressions) with polynomially many gates, or formulas of depth 3 and higher with gate count $o(n^2)$ (in particular, say, with a linear number of gates).

Building upon the results of [5], we show that the algorithm of Theorem 2 is nearly optimal even for formulas of any fixed depth $k \geq 3$. While we do not have a tight characterization for depth-2 formulas, we improve upon the trivial lower bound of $\Omega(\sqrt{n})$ for depth-2 formulas of linear gate count (a case arising in an application), giving an example of such a formula that requires $\tilde{\Omega}(n^{5/9}) = \Omega(n^{0.555})$ queries (Corollary 1). It remains an open question to close the gap between this lower bound and the upper bound of $O(n^{0.75})$ provided by Theorem 2, and in general, to better understand the quantum query complexity of depth-2 formulas.

Aside from being a natural extension of read-once formula evaluation, read-many formula evaluation has potential applications to open problems in quantum query complexity. For example, we apply our results to better understand the quantum query complexity of Boolean matrix product verification. This is the task of verifying whether $AB = C$, where A, B, C are $n \times n$ Boolean matrices provided by a black box, and the matrix product is computed over the Boolean semiring, in which OR plays the role of addition and AND plays the role of multiplication. Buhrman and Špalek gave an upper bound of $O(n^{1.5})$ queries for this problem, and their techniques imply a lower bound of $\Omega(n)$ [10]. We improve the lower bound to $\tilde{\Omega}(n^{19/18}) = \Omega(n^{1.055})$ (Theorem 5), showing in particular that linear query complexity is not achievable.

Our results can be viewed as a first step toward understanding the quantum query complexity of evaluating general circuits. A circuit is a directed acyclic graph in which source vertices represent black-box input bits, sink vertices represent outputs, and internal vertices represent AND, OR, and NOT gates. The size of a circuit is the total number of AND and OR gates, and the depth of a circuit is the length of a longest (directed) path from an input bit to an output bit.

The main difference between a formula and a circuit is that circuits allow fanout for gates, whereas formulas do not. In a circuit, the value of an input bit can be fed into multiple gates, and the output of a gate can be fed into multiple subsequent gates. A read-once formula is a circuit in which neither gates nor inputs have fanout. A general formula allows fanout for inputs (equivalently, inputs may be repeated), but not for gates. Note that by convention, the size of a circuit is the number of gates, whereas the size of a formula is the total number of inputs counted with multiplicity—even though formulas are a special case of circuits—so one must take care to avoid confusion. For example, the circuit that has 1 OR gate with n inputs has circuit size 1, formula size n, and formula gate count 1. To help clarify this distinction, we consistently use the symbol S for formula size and the symbol G for formula gate count and circuit size.

As another preliminary result on circuit evaluation, we provide a nearly tight characterization of the quantum query complexity of evaluating constant-depth circuits with bounded fanout. We show that any constant-depth bounded-fanout circuit of size G can be evaluated in $O(\min\{n, n^{1/2}G^{1/4}\})$ queries, and that there exist such circuits requiring $\tilde{\Omega}(\min\{n, n^{1/2}G^{1/4}\})$ queries to evaluate.

Finally, our results on the quantum query complexity of read-many formula evaluation have two purely classical applications. First, we give lower bounds on the number of gates (rather than simply the formula size) required for a formula to compute various functions. For example, while a classic result of Khrapchenko shows that parity requires a formula of size $\Omega(n^2)$ [19], we show that in fact any formula computing parity must have $\Omega(n^2)$ gates, which is a stronger statement. We also give similar results for several other functions. Second, for any $\epsilon > 0$, we give an example of an explicit constant-depth circuit of linear size that requires $\Omega(n^{2-\epsilon})$ gates to compute with a formula of any depth (Theorem 7).

2   Algorithm

In this section, we describe an algorithm for evaluating any formula on n inputs with G gates using $O(n^{1/2}G^{1/4})$ queries to the inputs, which gives an optimal upper bound when combined with Theorem 1.

We wish to evaluate a formula f with n inputs and G gates, using $O(n^{1/2}G^{1/4})$ queries. If the bottommost gates of f have a large fanin, say $\Omega(n)$, then the formula size can be as large as $\Omega(nG)$. If we directly apply the formula evaluation algorithm (Theorem 1) to this formula, we will get an upper bound of $O(\sqrt{nG})$, which is larger than claimed. Therefore, suppose the formula size of f is large.

Large formula size implies that there must be some inputs that feed into a large number of gates. Among the inputs that feed into many OR gates, if any input is 1, then this immediately fixes the output of a large number of OR gates, reducing the formula size. A similar argument applies to inputs that are 0 and feed into AND gates. Thus the first step in our algorithm is to find these inputs and eliminate them, reducing the formula size considerably. Then we use Theorem 1 on the resulting formula to achieve the claimed bound.

More precisely, our algorithm converts a formula f on n inputs and G gates into another formula f' of size $n\sqrt{G}$ on the same input. The new formula f' has the same output as f (on the given input), and this stage makes $O(n^{1/2}G^{1/4})$ queries. We call this the formula pruning algorithm.

The formula pruning algorithm uses a subroutine to find a marked entry in a Boolean string of length n, with good expected performance when there are many marked entries, even if the number of marked items is not known [8]. (We say an input is marked when it has value 1.)

Lemma 1. *Given an oracle for a string $x \in \{0,1\}^n$, there exists a quantum algorithm that outputs with certainty the index of a marked item in x if it exists, making $O(\sqrt{n/t})$ queries in expectation when there are t marked items. If there are no marked items the algorithm runs indefinitely.*

Lemma 2. *Given a formula f with n inputs and G gates, and an oracle for the input x, there exists a quantum algorithm that makes $O(n^{1/2}G^{1/4})$ queries and returns a formula f' on the same input x, such that $f'(x) = f(x)$ and f' has formula size $O(n\sqrt{G})$.*

Proof (Lemma 2). Consider the set of inputs that feed into OR gates. From these inputs and OR gates, we construct a bipartite graph with the inputs on one side and OR gates on the other. We put an edge between an input and all of the OR gates it feeds into. We call an input *high-degree* if it has degree greater than $\sqrt{G}$. Now repeat the following process. First, use Lemma 1 to find any high-degree input that is a 1. All OR gates connected to this input have output 1, so we delete these gates and their input wires and replace the gates with the constant 1. This input is removed from the set of inputs for the next iteration. If all the OR gates have been deleted or if there are no more high-degree inputs, the algorithm halts.

Say the process repeats $k-1$ times and in the kth round there are no remaining high-degree 1-valued inputs, although there are high-degree 0-valued inputs. Then the process will remain stuck in the search subroutine (Lemma 1) in the kth round. Let the number of marked high-degree inputs at the jth iteration of the process be m_j. Note that $m_j > m_{j+1}$ since at least 1 high-degree input is eliminated in each round. However it is possible that more than 1 high-degree input is eliminated in one round since the number of OR gates reduces in each round, which affects the degrees of the other inputs. Moreover, in the last round there must be at least 1 marked item, so $m_{k-1} \geq 1$. Combining this with $m_j > m_{j+1}$, we get $m_{k-r} \geq r$.

During each iteration, at least $\sqrt{G}$ OR gates are deleted. Since there are fewer than G OR gates in total, this process can repeat at most $\sqrt{G}$ times before we learn the values of all the OR gates, which gives us $k \leq \sqrt{G} + 1$.

In the jth iteration, finding a marked input requires $O(\sqrt{n/m_j})$ queries in expectation by Lemma 1. Thus the total number queries made in expectation until the kth round, but not counting the kth round itself, is

$$\sum_{j=1}^{k-1} O\left(\sqrt{\frac{n}{m_j}}\right) \leq \sum_{j=1}^{k-1} O\left(\sqrt{\frac{n}{j}}\right) \leq O(\sqrt{nk}) \leq O(n^{1/2}G^{1/4}). \tag{1}$$

In expectation, $O(n^{1/2}G^{1/4})$ queries suffice to reach the kth round, i.e., to reach a stage with no high-degree marked inputs. To get an algorithm with worst-case query complexity $O(n^{1/2}G^{1/4})$, halt this algorithm after it has made some constant times its expected number of queries. This gives a bounded-error algorithm with the same worst-case query complexity. After halting the algorithm, remaining high-degree inputs are known to have value 0, so we learn the values of all high-degree inputs, leaving only low-degree inputs unevaluated.

Using the same number of queries, we repeat the process with AND gates while searching for high-degree inputs that are 0. At the end of both these steps, each input has at most $\sqrt{G}$ outgoing wires to OR gates and at most $\sqrt{G}$ outgoing wires to AND gates. This yields a formula f' of size $O(n\sqrt{G})$ on the same inputs.

We can now upper bound the query complexity of Boolean formulas:

Theorem 2. *The bounded-error quantum query complexity of evaluating a formula with n inputs, size S, and G gates is $O(\min\{n, \sqrt{S}, n^{1/2}G^{1/4}\})$.*

Proof. We present three algorithms, with query complexities $O(n)$, $O(\sqrt{S})$, and $O(n^{1/2}G^{1/4})$, which together imply the desired result. Reading the entire input gives an $O(n)$ upper bound, and Theorem 1 gives an upper bound of $O(\sqrt{S})$. Finally, we can use Lemma 2 to convert the given formula to one of size $O(n\sqrt{G})$, at a cost of $O(n^{1/2}G^{1/4})$ queries. Theorem 1 shows that this formula can be evaluated using $O(n^{1/2}G^{1/4})$ queries.

We observed that our algorithm does better than the naive strategy of directly applying Theorem 1 to the given formula with G gates on n inputs, since its formula size could be as large as $O(nG)$, yielding a sub-optimal algorithm. Nevertheless, one might imagine that for every formula with G gates on n inputs, there exists another formula f' that represents the same function and has formula size $n\sqrt{G}$. This would imply Theorem 2 directly using Theorem 1.

However, this is not the case: there exists a formula with G gates on n inputs such that any formula representing the same function has formula size $\Omega(nG/\log n)$. This shows that in the worst case, the formula size of such a function might be close to $O(nG)$.

Proposition 1 ([24]). *There exists a function f that can be represented by a formula with G gates on n inputs, and any formula representing it must have formula size $S = \Omega(nG/\log n)$.*

3   Lower Bounds

Recent work on the quantum adversary method has shown that quantum query complexity behaves well with respect to composition of functions [26,25]. Here we apply this property to characterize the query complexity of composed circuits. By composing circuits appropriately, we construct a circuit whose depth is one less than the sum of the depths of the constituent circuits, but whose query complexity is still the product of those for the constituents. This construction can be used to give tradeoffs between circuit size and quantum query complexity: in particular, it shows that good lower bounds for functions with large circuits can be used to construct weaker lower bounds for functions with smaller circuits.

Lemma 3. *Let f be a circuit with n_f inputs, having depth k_f and size G_f; and let g be a circuit with n_g inputs, having depth k_g and size G_g. Then there exists a circuit h with $n_h = 4n_f n_g$ inputs, having depth $k_h = k_f + k_g - 1$ and size $G_h \leq 2G_f + 4n_f G_g$, such that $Q(h) = \Omega(Q(f)Q(g))$. Furthermore, if f is a formula and $k_g = 1$, then h is a formula of size $S_h = S_f S_g$.*

We now use this lemma to analyze formulas obtained by composing PARITY with AND gates. We show that for these formulas, the algorithm of Theorem 2 is optimal.

It is well known that the parity of n bits, denoted PARITY$_n$, can be computed by a formula of size $O(n^2)$. An explicit way to construct this formula is by recursion. The parity of two bits x and y can be expressed by a formula of size 4: $x \oplus y = (x \wedge \bar{y}) \vee (\bar{x} \wedge y)$. When n is a power of 2, given formulas of size $n^2/4$ for the parity of each half of the input, we get a formula of size n^2 for the parity of the full input. When n is not a power of 2 we can use the next largest power of 2 to obtain a formula size upper bound. Thus PARITY has a formula of size $O(n^2)$. Consequently, the number of gates in this formula is $O(n^2)$.

This observation, combined with Lemma 3 and known quantum lower bounds for PARITY, gives us the following theorem.

Theorem 3. *For any n, S, G, there is a read-many formula with n inputs, size at most S, and at most G gates with bounded-error quantum query complexity $\Omega(\min\{n, \sqrt{S}, n^{1/2}G^{1/4}\})$.*

Proof. If $\min\{n, \sqrt{S}, n^{1/2}G^{1/4}\} = n$ (i.e., $S \geq n^2$ and $G \geq n^2$), then consider the PARITY function, which has $Q(\text{PARITY}_n) = \Omega(n)$ [4,14]. Since the formula size and gate count of parity are $O(n^2)$, this function has formula size $O(S)$ and gate count $O(G)$. By adjusting the function to compute the parity of a constant fraction of the inputs, we can ensure that the formula size is at most S and the gate count is at most G with the same asymptotic query complexity.

In the remaining two cases, we obtain the desired formula by composing a formula for PARITY with AND gates. We apply Lemma 3 with $f = \text{PARITY}_m$ and $g = \text{AND}_{n/m}$ for some choice of m. The resulting formula has $\Theta(n)$ inputs, size $O(m^2(n/m)) = O(nm)$, and gate count $O(m^2)$. Its quantum query complexity is $\Omega(m\sqrt{n/m}) = \Omega(\sqrt{nm})$.

If $\min\{n, \sqrt{S}, n^{1/2}G^{1/4}\} = \sqrt{S}$ (i.e., $S \leq n^2$ and $S \leq n\sqrt{G}$), let $m = S/n$. Then the formula size is $O(S)$ and the gate count is $O(S^2/n^2) \leq O(G)$. By appropriate choice of constants, we can ensure that the formula size is at most S and the gate count is at most G. In this case, the query complexity is $\Omega(\sqrt{S})$.

Finally, if $\min\{n, \sqrt{S}, n^{1/2}G^{1/4}\} = n^{1/2}G^{1/4}$ (i.e., $G \leq n^2$ and $S \geq n\sqrt{G}$), let $m = \sqrt{G}$. Then the gate count is $O(G)$ and the formula size is $O(n\sqrt{G}) \leq O(S)$. Again, by appropriate choice of constants, we can ensure that the formula size is at most S and the gate count is at most G. In this final case, the query complexity is $\Omega(n^{1/2}G^{1/4})$.

4   Constant-Depth Formulas

Theorem 2 and Theorem 3 together completely characterize the quantum query complexity of formulas with n inputs, formula size S, and gate count G. However, while there exists such a formula for which the algorithm is optimal, particular formulas can sometimes be evaluated more efficiently. Thus it would be useful to have a finer characterization that takes further properties into account, such as the depth of the formula. In this section we consider formulas of a given depth, number of inputs, size, and gate count.

Since the algorithm described in Section 2 works for formulas of any depth, we know that any depth-k formula with G gates can be evaluated with $O(\min\{n, \sqrt{S}, n^{1/2}G^{1/4}\})$ queries. However, the lower bound in Section 3 uses a formula with non-constant depth.

However, instead of composing the PARITY function with AND gates, we can use the ONTO function (defined in [5]), which has a depth-3 formula and has nearly maximal query complexity. This observation leads to the following:

Theorem 4. *For any N, S, G, there is a depth-3 formula with N inputs, size at most S, and at most G gates with quantum query complexity $\tilde{\Omega}(\min\{N, \sqrt{S}, N^{1/2}G^{1/4}\})$.*

This gives us a matching lower bound, up to log factors, for any $k \geq 3$. Thus we have a nearly tight characterization of the query complexity of evaluating depth-k formulas with n inputs, size S, and G gates for all $k \geq 3$. Furthermore, since any depth-1 formula is either the AND or OR of a subset of the inputs, the query complexity of depth-1 formulas is easy to characterize.

It remains to consider the depth-2 case. Since all depth-2 circuits are formulas, we will refer to depth-2 formulas as depth-2 circuits henceforth and use circuit size (i.e., the number of gates) as the size measure.

There are also independent reasons for considering the query complexity of depth-2 circuits. Improved lower bounds on depth-2 circuits of size n imply improved lower bounds for the Boolean matrix product verification problem. We explain this connection and exhibit new lower bounds for the Boolean matrix product verification problem in Section 5.1.

Furthermore, some interesting problems can be expressed with depth-2 circuits, and improved upper bounds for general depth-2 circuits would give improved algorithms for such problems. For example, the graph collision problem [22] for a graph with G edges can be written as a depth-2 circuit of size G. Below, we exhibit a lower bound for depth-2 circuits that does not match the upper bound of Theorem 2. If there exists an algorithm that achieves the query complexity of the lower bound, this would improve the best known algorithm for graph collision, and consequently the triangle problem [22].

To use Lemma 3 to prove lower bounds for depth-2 circuits, we need a function that can be expressed as a depth-2 circuit and has high query complexity. The best function we know of with these properties is the element distinctness function, which has a depth-2 circuit of size $O((N/\log N)^3)$ that requires $\Omega((N/\log N)^{2/3})$ quantum queries to evaluate. Composing this function with AND gates, as in the proof of Theorem 3, we can prove the following lower bound, which also arises in an application to Boolean matrix product verification:

Corollary 1. *There exists a depth-2 circuit on n inputs of size n that requires $\tilde{\Omega}(n^{5/9}) = \Omega(n^{0.555})$ quantum queries to evaluate.*

The results of this section also allow us to characterize (up to log factors) the quantum query complexity of bounded-fanout AC^0 circuits, which we call AC^0_b

circuits. AC_b^0 circuits are like AC^0 circuits where the gates are only allowed to have $O(1)$ fanout, as opposed to the arbitrary fanout that is allowed in AC^0. Note that as complexity classes, AC^0 and AC_b^0 are equal, but the conversion from an AC^0 circuit to an AC_b^0 circuit will in general increase the circuit size.

The reason that formula size bounds apply to AC_b^0 circuits is that an AC_b^0 circuit can be converted to a formula whose gate count is at most a constant factor larger than the circuit size of the original circuit. (However, this constant depends exponentially on the depth.) Thus we have the following corollary:

Corollary 2. *Any language L in AC_b^0 has quantum query complexity $O(\min\{n, n^{1/2}G^{1/4}\})$ where $G(n)$ is the size of the smallest circuit family that computes L. Furthermore, for every $G(n)$, there exists a language in AC_b^0 that has circuits of size $G(n)$ and that requires $\tilde{\Omega}(\min\{n, n^{1/2}G^{1/4}\})$ quantum queries to evaluate.*

5   Applications

5.1   Boolean Matrix Product Verification

A decision problem closely related to the old and well-studied matrix multiplication problem is the matrix product verification problem. This is the task of verifying whether the product of two matrices equals a third matrix. The Boolean matrix product verification (BMPV) problem is the same problem where the input comprises Boolean matrices and the matrix product is performed over the Boolean semiring, i.e., the "sum" of two bits is their logical OR and the "product" of two bits is their logical AND. Recently, there has been considerable interest in the quantum query complexity of Boolean matrix multiplication [27,21,17].

More formally, the input to BMPV consists of three $n \times n$ matrices A, B, and C. We have to determine whether $C_{ij} = \bigvee_k A_{ik} \wedge B_{kj}$ for all $i, j \in [n]$. Note that the input size is $3n^2$, so $O(n^2)$ is a trivial upper bound on the query complexity of this problem. Buhrman and Špalek [10] show that BMPV can be solved in $O(n^{3/2})$ queries. This bound can be obtained by noting that checking the correctness of a single entry of C requires $O(\sqrt{n})$ queries, so one can use Grover's algorithm to search over the n^2 entries for an incorrect entry. Another way to obtain the same bound is to show that there exists a formula of size $O(n^3)$ that expresses the statement $AB = C$ and then apply Theorem 1.

While Buhrman and Špalek do not explicitly state a lower bound for this problem, their techniques yield a lower bound of $\Omega(n)$ queries. This leaves a gap between the best known upper and lower bounds. The following theorem summarizes known facts about Boolean matrix product verification.

Theorem 5. *If A, B, and C are $n \times n$ Boolean matrices available via oracles for their entries, checking whether the Boolean matrix product of A and B equals C, i.e., checking whether $C_{ij} = \bigvee_k A_{ik} \wedge B_{kj}$ for all $i, j \in [n]$, requires at least $\Omega(n)$ quantum queries and at most $O(n^{3/2})$ quantum queries.*

We can use the results of the previous section to improve the lower bound to $\Omega(n^{1.055})$, via an intermediate problem we call the Boolean vector product verification problem.

Theorem 6. *The bounded-error quantum query complexity of the Boolean matrix product verification problem is $\tilde{\Omega}(n^{19/18}) = \Omega(n^{1.055})$.*

5.2   Applications to Classical Circuit Complexity

In this section we present some classical applications of our results. In particular, we prove lower bounds on the number of gates needed in any formula representing certain functions. The main tool we use is the following corollary of Theorem 2.

Corollary 3. *For a function f with n inputs and quantum query complexity $Q(f)$, any (unbounded-fanin) formula representing f requires $\Omega(Q(f)^4/n^2)$ gates.*

Almost immediately, this implies that functions such as PARITY and MAJORITY, which have quantum query complexity of $\Omega(n)$, require $\Omega(n^2)$ gates to be represented as a formula. Similarly, this implies that functions with query complexity $\Omega(n^{3/4})$, such as GRAPH CONNECTIVITY [12], GRAPH PLANARITY [2], and HAMILTONIAN CYCLE [7], require formulas with $\Omega(n)$ gates.

To compare this with previous results, it is known that PARITY requires formulas of size $\Omega(n^2)$ [19]. This result is implied by our result since the number of gates is less than the size of a formula.

We can also use these techniques to address the question "Given a constant-depth circuit of size G, how efficiently can this circuit be expressed as a formula?" The best result of this type that we are aware of is the following: There exists a constant-depth circuit of linear size such that any formula expressing the same function has size at least $n^{2-o(1)}$. The result appears at the end of Section 6.2 in [18], where the function called $V^{\mathrm{OR}}(x,y)$ is shown to have these properties. Indeed, the function has a depth-3 formula with $O(n)$ gates. The idea of using such functions, also called universal functions, is attributed to Nechiporuk [23].

We can construct an explicit constant-depth circuit of linear size that requires $\Omega(n^{2-\epsilon})$ gates to be expressed as a formula. To the best of our knowledge, this result is new. Our result is incomparable to the previous result since we lower bound the number of gates, which also lower bounds the formula size, but we use a constant-depth circuit as opposed to a depth-3 formula.

Improving our lower bound to $\Omega(n^2)$ seems difficult, since we do not even know an explicit constant-depth circuit of linear size that requires formulas of *size* (as opposed to number of gates) $\Omega(n^2)$, which is a weaker statement. In fact, we do not know any explicit function in AC^0 with a formula size lower bound of $\Omega(n^2)$ (for more information, see [20]).

Theorem 7. *For every $\epsilon > 0$, there exists a constant-depth unbounded-fanin circuit (i.e., an AC^0 circuit) of size $O(n)$ such that any (unbounded-fanin) formula computing the same function must have $\Omega(n^{2-\epsilon})$ gates.*

6   Conclusions and Open Problems

We have given a tight characterization of the query complexity of read-many formulas in terms of their number of inputs n, formula size S, and gate count G. In particular, we showed that the query complexity of evaluating this class of formulas is $\Theta(\min\{n, \sqrt{S}, n^{1/2}G^{1/4}\})$. Our results suggest several new avenues of research, looking both toward refined characterizations of the query complexity of evaluating formulas and toward a better understanding of the quantum query complexity of evaluating circuits.

In Section 4 we showed that our query complexity bounds are nearly tight for all formulas of a given depth except for the case of depth 2. We made partial progress on this remaining case by giving a depth-2 circuit of size n that requires $\Omega(n^{0.555})$ queries, whereas our algorithm gives an upper bound of $O(n^{0.75})$.

While we have made progress in understanding the quantum query complexity of evaluating read-many formulas, we would also like to understand the query complexity of evaluating general circuits. It would be interesting to find upper and lower bounds on the query complexity of evaluating circuits as a function of various parameters such as their number of inputs, gate count, fanout, and depth. In particular, the graph collision problem can also be expressed using a circuit of depth 3 and linear size (in addition to the naive depth-2 circuit of quadratic size mentioned in Section 4), so it would be interesting to focus on the special case of evaluating such circuits.

Acknowledgments. We thank Noam Nisan for the proof of Proposition 1 (via the website cstheory.stackexchange.com). R. K. thanks Stasys Jukna for helpful discussions about circuit complexity.

This work was supported in part by MITACS, NSERC, the Ontario Ministry of Research and Innovation, QuantumWorks, and the US ARO/DTO. This work was done while S. K. was visiting the Institute for Quantum Computing at the University of Waterloo. S. K. also received support from NSF Grant No. DGE-0801525, *IGERT: Interdisciplinary Quantum Information Science and Engineering*, and from the U.S. Department of Energy under cooperative research agreement Contract Number DE-FG02-05ER41360.

References

1. Ambainis, A., Childs, A.M., Reichardt, B.W., Špalek, R., Zhang, S.: Any AND-OR formula of size N can be evaluated in time $N^{1/2+o(1)}$ on a quantum computer. SIAM J. Comput. 39(6), 2513–2530 (2010)
2. Ambainis, A., Iwama, K., Nakanishi, M., Nishimura, H., Raymond, R., Tani, S., Yamashita, S.: Quantum Query Complexity of Boolean Functions with Small On-Sets. In: Hong, S.-H., Nagamochi, H., Fukunaga, T. (eds.) ISAAC 2008. LNCS, vol. 5369, pp. 907–918. Springer, Heidelberg (2008)
3. Barnum, H., Saks, M.: A lower bound on the quantum query complexity of read-once functions. J. Comput. System Sci. 69(2), 244–258 (2004)

4. Beals, R., Buhrman, H., Cleve, R., Mosca, M., de Wolf, R.: Quantum lower bounds by polynomials. J. ACM 48(4), 778–797 (2001)
5. Beame, P., Machmouchi, W.: The quantum query complexity of AC^0. arXiv:1008.2422
6. Bennett, C.H., Bernstein, E., Brassard, G., Vazirani, U.: Strengths and weaknesses of quantum computing. SIAM J. Comput. 26, 1510–1523 (1997)
7. Berzina, A., Dubrovsky, A., Freivalds, R., Lace, L., Scegulnaja, O.: Quantum Query Complexity for Some Graph Problems. In: Van Emde Boas, P., Pokorný, J., Bieliková, M., Štuller, J. (eds.) SOFSEM 2004. LNCS, vol. 2932, pp. 140–150. Springer, Heidelberg (2004)
8. Boyer, M., Brassard, G., Høyer, P., Tapp, A.: Tight bounds on quantum searching. Fortschr. Phys. 46(4-5), 493–505 (1998)
9. Buhrman, H., Cleve, R., Wigderson, A.: Quantum vs. classical communication and computation. In: 30th STOC, pp. 63–68 (1998)
10. Buhrman, H., Špalek, R.: Quantum verification of matrix products. In: 17th SODA, pp. 880–889 (2006)
11. Childs, A.M., Cleve, R., Jordan, S.P., Yonge-Mallo, D.: Discrete-query quantum algorithm for NAND trees. Theory Comput. 5, 119–123 (2009)
12. Dürr, C., Heiligman, M., Høyer, P., Mhalla, M.: Quantum query complexity of some graph problems. SIAM J. Comput. 35(6), 1310–1328 (2006)
13. Farhi, E., Goldstone, J., Gutmann, S.: A quantum algorithm for the Hamiltonian NAND tree. Theory Comput. 4(1), 169–190 (2008)
14. Farhi, E., Goldstone, J., Gutmann, S., Sipser, M.: Limit on the speed of quantum computation in determining parity. Phys. Rev. Lett. 81(24), 5442–5444 (1998)
15. Grover, L.K.: Quantum mechanics helps in searching for a needle in a haystack. Phys. Rev. Lett. 79(2), 325–328 (1997)
16. Høyer, P., Mosca, M., de Wolf, R.: Quantum Search on Bounded-Error Inputs. In: Baeten, J.C.M., Lenstra, J.K., Parrow, J., Woeginger, G.J. (eds.) ICALP 2003. LNCS, vol. 2719, pp. 291–299. Springer, Heidelberg (2003)
17. Jeffery, S., Kothari, R., Magniez, F.: Improving quantum query complexity of Boolean matrix multiplication using graph collision, arXiv:1112.5855
18. Jukna, S.: Boolean Function Complexity. Springer (2012)
19. Khrapchenko, V.M.: Complexity of the realization of a linear function in the class of Π-circuits. Math. Notes 9(1), 21–23 (1971)
20. Kothari, R.: Formula size lower bounds for AC^0 functions. Theoretical Computer Science Stack Exchange, http://cstheory.stackexchange.com/q/7156
21. Le Gall, F.: Improved output-sensitive quantum algorithms for Boolean matrix multiplication. In: 23rd SODA, pp. 1464–1476 (2012)
22. Magniez, F., Santha, M., Szegedy, M.: Quantum algorithms for the triangle problem. SIAM J. Comput. 37, 413–424 (2007)
23. Nechiporuk, E.I.: A Boolean function. Soviet Mathematics Doklady 7, 999–1000 (1966)
24. Nisan, N.: Shortest formula for an n-term monotone CNF. Theoretical Computer Science Stack Exchange, http://cstheory.stackexchange.com/a/7087
25. Reichardt, B.: Span programs and quantum query complexity: The general adversary bound is nearly tight for every Boolean function. In: 50th FOCS, pp. 544–551 (2009)
26. Reichardt, B.: Reflections for quantum query algorithms. In: 22nd SODA, pp. 560–569 (2011)
27. Vassilevska Williams, V., Williams, R.: Subcubic equivalences between path, matrix, and triangle problems. In: 51st FOCS, pp. 645–654 (2010)

A Path-Decomposition Theorem with Applications to Pricing and Covering on Trees*

Marek Cygan[1], Fabrizio Grandoni[1], Stefano Leonardi[2],
Marcin Pilipczuk[3], and Piotr Sankowski[3]

[1] IDSIA, University of Lugano, Switzerland
{marek,fabrizio}@idsia.ch
[2] Department of Computer and System Science,
Sapienza University of Rome, Italy
leon@dis.uniroma1.it
[3] Institute of Informatics, University of Warsaw, Poland
{malcin,sank}@mimuw.edu.pl

Abstract. In this paper we focus on problems characterized by an input n-node tree and a collection of subpaths. Motivated by the fact that some of these problems admit a very good approximation (or even a poly-time exact algorithm) when the input tree is a path, we develop a decomposition theorem of trees into paths. Our decomposition allows us to partition the input problem into a collection of $O(\log \log n)$ subproblems, where in each subproblem either the input tree is a path or there exists a *hitting* set F of edges such that each path has a non-empty, small intersection with F. When both kinds of subproblems admit constant approximations, our method implies an $O(\log \log n)$ approximation for the original problem.

We illustrate the above technique by considering two natural problems of the mentioned kind, namely UNIFORM TREE TOLLBOOTH and UNIQUE TREE COVERAGE. In UNIFORM TREE TOLLBOOTH each subpath has a budget, where budgets are within a constant factor from each other, and we have to choose non-negative edge prices so that we maximize the total price of subpaths whose budget is not exceeded. In UNIQUE TREE COVERAGE each subpath has a weight, and the goal is to select a subset X of edges so that we maximize the total weight of subpaths containing exactly one edge of X. We obtain $O(\log \log n)$ approximation algorithms for both problems. The previous best approximations are $O(\log n / \log \log n)$ by Gamzu and Segev [ICALP'10] and $O(\log n)$ by Demaine et al. [SICOMP'08] for the first and second problem, respectively, however both previous results were obtained for much more general problems with arbitrary budgets (weights).

* The first two authors are partially supported by the ERC Starting Grant NEWNET 279352, whereas the third and fifth author are partially supported by the ERC Starting Grant PAAl 259515.

L. Epstein and P. Ferragina (Eds.): ESA 2012, LNCS 7501, pp. 349–360, 2012.

1   Introduction

Several natural graph problems are characterized by an input n-node tree $T = (V, E)$, and a (multi-)set $\mathcal{P} = \{P_1, \ldots, P_m\}$ of subpaths of T (the *requests*). Typically one has to make decisions on the edges or nodes of T, and based on these decisions one obtains some profit from each request. Some of the mentioned problems admit a very good approximation (or are even poly-time solvable) when T itself is a path, but get substantially harder in the tree case. In this paper we present a tool to reduce the complexity gap between the path and tree case to $O(\log \log n)$ for problems which have some extra natural properties. We illustrate the application of this method by providing an $O(\log \log n)$ approximation for one pricing problem and for one covering problem, improving on previous best logarithmic approximations.

1.1   Our Results

The main technical contribution of this paper is a simple decomposition theorem of trees into paths.

Theorem 1. *Given an n-node rooted tree T, and a collection of paths $\mathcal{P} = \{P_1, \ldots, P_m\}$, there is a polynomial time algorithm which partitions $\mathcal{P}$ into $\ell = O(\log \log n)$ subsets $\mathcal{P}_1, \ldots, \mathcal{P}_\ell$ such that one of the following two properties holds for each $\mathcal{P}_i$:*

(a) The algorithm provides a partition $\mathcal{L}$ of T into edge disjoint paths, such that each request of $\mathcal{P}_i$ is fully contained in some path of $\mathcal{L}$.

(b) The algorithm provides a collection of edges F_i, such that each $P \in \mathcal{P}_i$ contains at least one and at most 6 edges from F_i.

Intuitively, each subset $\mathcal{P}_i$ naturally induces a subproblem, and at least one of the subproblems carries a fraction $1/O(\log \log n)$ of the original profit. For subproblems of type (a), each L_i induces an independent instance of the problem on a path graph: here the problem becomes typically easier to solve. For subproblems of type (b), a good approximation can be obtained if the objective function and constraints allow us to retrieve a big fraction of the optimal profit from each path P by making a decision among a few options on a small, arbitrary subset of edges of P. In particular, it is sufficient to make decisions (even randomly and obliviously!) on F_i.

We illustrate the above approach by applying it to two natural problems on trees and subpaths. The first one is a pricing problem. In the well-studied TREE TOLLBOOTH problem (TT), we are given a tree $T = (V, E)$ with n vertices, and a set of paths $\mathcal{P} = \{P_1, \ldots, P_m\}$, where P_i has budget $b_i > 0$. The goal is to find a pricing function $p : E \to \mathbb{R}_+$ maximizing $\sum_{1 \le i \le m, p(P_i) \le b_i} p(P_i)$, where $p(P_i) := \sum_{e \in P_i} p(e)$. Here we consider the UNIFORM TREE TOLLBOOTH problem (UTT), which is the special case where the ratio of the largest to smallest budget is bounded by a constant. Interestingly enough, the best-known approximation for UTT is the same as for TT, i.e. $O(\log n / \log \log n)$ [7]. In this paper we obtain the following improved result.

Theorem 2. *There is an $O(\log \log n)$-approximation for the* UNIFORM TREE TOLLBOOTH *problem.*

The second application is to a covering problem. In the UNIQUE COVERAGE ON TREES problem (UCT) we are given a tree T and subpaths $\mathcal{P} = \{P_1, \ldots, P_m\}$ as in TT, plus a profit function $p : \mathcal{P} \to \mathbb{R}_+$. The goal is to compute a subset of edges $X \subseteq E$ so that we maximize the total profit of paths $P \in \mathcal{P}$ which share exactly one edge with X.

Theorem 3. *There is an $O(\log \log n)$-approximation for the* UNIQUE COVERAGE ON TREES *problem.*

1.2   Related Work

The tollbooth problem belongs to a wider family of *pricing* problems, which attracted a lot of attention in the last few years. In the *single-minded* version of these problems, we are given a collection of m clients and n item types. Each client wishes to buy a bundle of items provided that the total price of the bundle does not exceed her budget. Our goal is to choose item prices so that the total profit is maximized. In the *unlimited supply* model, there is an unbounded amount of copies of each item. For this problem an $O(\log n + \log m)$ approximation is given in [9]. This bound was refined in [3] to $O(\log L + \log B)$, where L denotes the maximum number of items in a bundle and B the maximum number of bundles containing a given item. A $O(L)$ approximation is given in [1]. On the negative side, Demaine et al. [4] show that this problem is hard to approximate within $\log^d n$, for some $d > 0$, assuming that $NP \not\subseteq BPTIME(2^{n^\varepsilon})$ for some $\varepsilon > 0$.

An interesting special case is when each bundle contains $k = O(1)$ items. The case $k = 2$ is also know as the VERTEX PRICING problem. VERTEX PRICING is APX-hard even on bipartite graphs [5], and it is hard to approximate better than a factor 2 under the Unique Game Conjecture [12]. On the positive side, there exists a 4-approximation for VERTEX PRICING, which generalizes to an $O(k)$-approximation for bundles of size k [1].

A $O(\log n)$ approximation for the TREE TOLLBOOTH problem was developed in [5]. This was recently improved to $O(\log n / \log \log n)$ by Gamzu and Segev [7]. TREE TOLLBOOTH is APX-hard [9]. One might consider a generalization of the problems on arbitrary graphs (rather than just trees). In that case the problem is APX-hard even when the graph has bounded degree, the paths have constant length and each edge belongs to a constant number of paths [3].

The HIGHWAY problem is the special case of TREE TOLLBOOTH where the input graph is a path. It was shown to be weakly NP-hard by Briest and Krysta [3], and strongly NP-hard by Elbassioni, Raman, Ray, and Sitters [5]. Balcan and Blum [1] give an $O(\log n)$ approximation for the problem. The result in [7] implies as a special case an $O(\log n / \log \log n)$ for the problem. Elbassioni, Sitters, and Zhang [6] developed a QPTAS, exploiting the profiling technique introduced by Bansal et al. [2]. Finally, a PTAS was given by Grandoni and Rothvoß [8].

FPTASs are known for some special case: for example when the graph has constant length [11], the budgets are upper bounded by a constant [9], the paths have constant length [9] or they induce a laminar family [1,3]. We will (crucially) use the PTAS in [8] as a black box to derive an $O(\log \log n)$ approximation for UNIQUE TREE TOLLBOOTH.

One can also consider variants of the above problems. For example, in the *coupon* version, prices can be negative and the profit of a bundle is zero if its total price is above the budget or below zero. The coupon version is typically harder than the classical one. For example, COUPON HIGHWAY is APX-hard [5]. Recently Wu and Popat [13] showed that COUPON HIGHWAY and COUPON VERTEX PRICING are not approximable within any constant factor given the Unique Game Conjecture.

Versions of the above problems with a limited supply are much less studied. Here, part of the problem is to assign the available copies to the clients who can afford to pay them. In the latter case, one can consider the *envy-free* version of the problem, where prices must satisfy the condition that all the clients who can afford the price of their bundle actually get it [9].

UNIQUE COVERAGE ON TREES is a special case of the UNIQUE COVERAGE problem: given a universe U a collection $\mathcal{F} \subseteq 2^U$ of subsets of U, the goal is to find a subset of the universe $X \subseteq U$, such that the number of sets in $\mathcal{F}$ containing exactly one element of X is maximized. Demaine et al. [4] show that this problem is hard to approximate within $\log^d n$, for some $d > 0$, assuming that $NP \not\subseteq BPTIME(2^{n^\varepsilon})$ for some $\varepsilon > 0$. However, if there is a solution covering all the sets in $\mathcal{F}$, then an e-approximation is known [10].

2   A Decomposition Theorem

In this section we prove Theorem 1. In order to introduce ideas in a smooth way, in Section 2.1 we start by proving a simplified version of the theorem in the case that the input graph has small (namely, logarithmic) diameter. In Section 2.2 we generalize the approach to arbitrary diameters.

For a positive integer a, by $\text{POWER}_2(a)$ we denote the smallest integer i such that 2^i does not divide a; in other words, $\text{POWER}_2(a)$ is the index of the lowest non-zero bit in the binary representation of a.

2.1   Small Diameter

Theorem 4. *Given an n-node tree $T = (V, E)$, and a collection of paths $\mathcal{P} = \{P_1, \ldots, P_m\}$, there is a polynomial time algorithm which partitions $\mathcal{P}$ into $\ell = O(\log D)$ subsets $\mathcal{P}_1, \ldots, \mathcal{P}_\ell$, where D is the diameter of T, and provides a collection of edges F_i for each $\mathcal{P}_i$, such that each $P \in \mathcal{P}_i$ contains at least one and at most two edges of F_i.*

Proof. Let $\ell = \lceil \log(D + 1) \rceil$. Fix any vertex $r \in V$ as a root of the tree T. Let $H_j \subseteq E$ $(1 \leq j \leq D)$ be the set of edges which are at distance exactly j from r (the edges incident to r are in H_1). For each $1 \leq i \leq \ell$ we take

$$F_i = \bigcup_{j:1\leq j\leq D,\texttt{POWER}_2(j)=i} H_j.$$

Note that as $1 \leq j \leq D$ we have $1 \leq \texttt{POWER}_2(j) \leq \lceil \log(D+1) \rceil = \ell$, and $\{F_1, \ldots, F_\ell\}$ indeed is a partition of the set of edges (see Figure 1 for an illustration). Now we partition the set of requests $\mathcal{P}$ into subsets $\mathcal{P}_1, \ldots, \mathcal{P}_\ell$. We put a request $P \in \mathcal{P}$ to the set $\mathcal{P}_i$ if P intersects F_i but does not intersect F_j for any $j > i$.

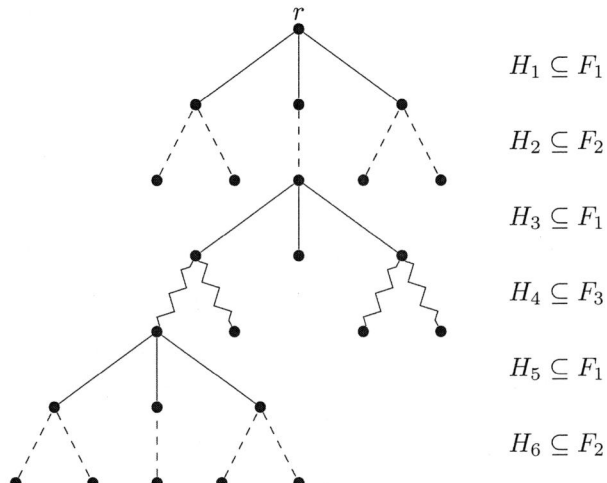

$H_1 \subseteq F_1$

$H_2 \subseteq F_2$

$H_3 \subseteq F_1$

$H_4 \subseteq F_3$

$H_5 \subseteq F_1$

$H_6 \subseteq F_2$

Fig. 1. Regular edges belong to F_1, dashed edges belong to F_2, whereas zigzagged edges belong to F_3

As each request $P \in \mathcal{P}$ is a nonempty path, each request belongs to some set $\mathcal{P}_i$. By the definition, each request $P \in \mathcal{P}_i$ intersects the set F_i. It remains to prove that each $P \in \mathcal{P}_i$ contains at most two edges from the set F_i. For a request $P \in \mathcal{P}_i$ let j be an index, such that P contains an edge of H_j and $\texttt{POWER}_2(j) = i$ (note that such index j exists, since $P \in \mathcal{P}_i$). If P contains an edge from a set $H_{j'}$ where $1 \leq j' \leq D$, $\texttt{POWER}_2(j') = i$, $j' \neq j$, then P also contains an edge from some $H_{j''}$, where j'' is between j' and j and $\texttt{POWER}_2(j'') > \texttt{POWER}_2(j) = i$. But this is a contradiction with the assumption that P is in the set $\mathcal{P}_i$.

Finally, any path can contain at most two edges from any fixed H_j, since H_j is the set of edges equidistant from the root. Therefore each request $P \in \mathcal{P}_i$ contains at least one and at most two edges of F_i. $\square$

2.2   Large Diameter

Now we want to adapt ideas from the previous section to handle the case of large diameters, hence proving Theorem 1. To that aim, we exploit the following two technical lemmas. A subpath is *upward* if one endpoint is an ancestor of the other endpoint.

Lemma 5. *Given an n-node rooted tree T, there is a polynomial-time algorithm which partitions the edge set of T into upward paths, and groups such paths into $s = O(\log n)$ collections (called levels) $L_1, \ldots, L_s$ so that, for any path P of T the following two properties hold:*

1. **(separation)** *If P shares an edge with $P' \in L_i$ and $P'' \in L_j$, $i < j$, P must share an edge with some $P''' \in L_k$ for any $i < k < j$.*
2. **(intersection)** *P shares an edge with at most two paths in each level L_i, and if it shares an edge with two such paths P' and P'', then it must contain the topmost edges of the latter two paths.*

Proof. For each vertex v that is not a root nor a leaf, consider all edges of T that connect v with a child of v, and denote as e_v the edge that leads to a subtree with the largest number of vertices (breaking ties arbitrarily). Now, for each leaf w of T, consider a path P_w constructed as follows: start from w, traverse the tree towards the root r and stop when a vertex v is reached via an edge different than e_v. Let v_w be the topmost (i.e., the last, closest to the root r) vertex on the path P_w .

Since for each non-leaf and non-root vertex v, exactly one edge connecting v with its child is denoted e_v, each edge of T belongs to exactly one path P_w. For each vertex u of T, by depth(u) we denote the number of different paths P_w that share an edge with the unique path connecting u with the root r. If a path P_w ends at a vertex $v_w \neq r$ and $v'v_w$ is the last edge of P_w, then, by the definition of the edge e_v, the subtree of T rooted at v_w has at least twice as many vertices than the subtree rooted at v'. We infer that for each vertex u we have $0 \leq \text{depth}(u) \leq \lceil \log n \rceil$; depth($u$) = 0 iff $u = r$.

For each path P_w we denote depth(P_w) = depth(v_w) + 1; note that $1 \leq \text{depth}(P_w) \leq \lceil \log n \rceil + 1$. Let $s = \lceil \log n \rceil + 1$ and let $L_j = \{P_w : \text{depth}(P_w) = j\}$ for $1 \leq j \leq s$. Clearly the family $\{L_j : 1 \leq j \leq s\}$ can be constructed in polynomial (even linear) time and partitions the edges into levels, each level consisting of a set of upward paths (see Figure 2 for an illustration). Moreover, if paths P_{w_1} and P_{w_2} share a vertex, then $|\text{depth}(P_{w_1}) - \text{depth}(P_{w_2})| \leq 1$, which proves the separation property in the claim. It remains to prove the intersection property.

Consider any upward path P in the tree T. Clearly, the values depth(v) for vertices v on the path P are ordered monotonously, with the lowest value in the topmost (closest to the root r) vertex of P. Therefore, for each layer L_j, the path P may contain edges of at most one path of L_j. Moreover, as all paths P_w are upward paths, if P contains an edge of P_w, then either P contains the topmost edge of P_w, or the topmost vertex of P lies on P_w, but is different from v_w. Since any path in T can be split into two upward paths, intersection property and the whole theorem follow.   □

The following lemma provides a way to partition requests.

Lemma 6. *Given a given rooted tree T and a set of requests $\mathcal{P}$, in polynomial time we can construct a collection of $\ell = O(\log \log n)$ families of paths*

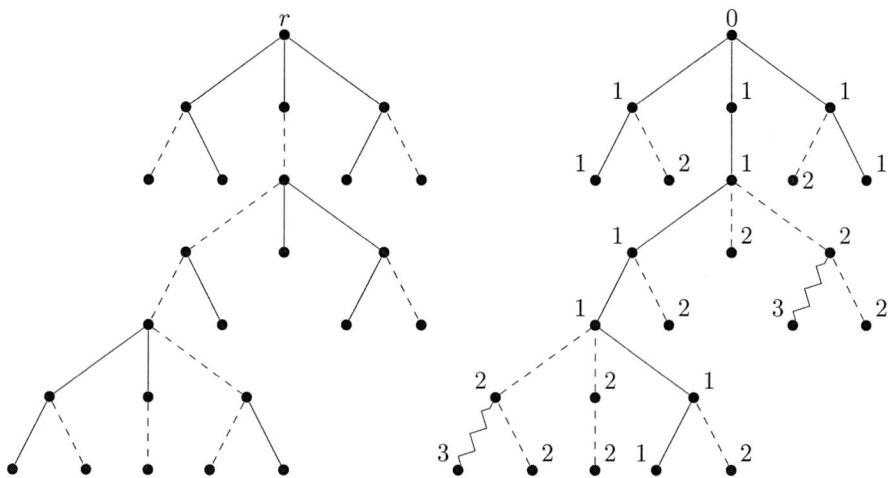

Fig. 2. In the left tree edges e_v are dashed. In the right tree nodes are labeled with their depth, regular edges belong to paths of level L_1, dashed edges belong to paths of level L_2, whereas zigzagged edges belong to paths of level L_3.

$G_1, G_2, \ldots, G_\ell$ in T (called groups) and a partition $K_1, K_2, \ldots, K_\ell$ of $\mathcal{P}$, such that:

1. Each request $P \in K_i$ shares an edge with at at least one and at most two paths from the group G_i.
2. Each request $P \in K_i$ contains at most 4 edges incident to paths of G_i.
3. If a request $P \in K_i$ shares an edge with two paths $Q_1, Q_2 \in G_i$, then it contains the topmost edges of the latter two paths.

Proof. First, invoke Lemma 5 on the tree T and obtain a partition into levels $L_1, \ldots, L_s$ with $s = O(\log n)$. Let $\ell = \lceil \log(s+1) \rceil = O(\log \log n)$ and, for $1 \le i \le \ell$, define G_i as follows:

$$G_i = \bigcup_{j:1 \le j \le s, \mathtt{POWER}_2(j)=i} L_j.$$

Now partition the set of requests $\mathcal{P}$ into ℓ subsets K_i, for $1 \le i \le \ell$, in such a way that a request $P \in \mathcal{P}$ is assigned to the subset K_i iff P contains an edge of some path of G_i and does not contain any edge of any path of G_j for $j > i$. The paths of the groups G_i cover all edges from the tree T, so each request $P \in \mathcal{P}$ is included in some layer.

Observe, that due to the separation property ensured by Lemma 5, if a path $P \in \mathcal{P}$ belongs to K_i and P shares an edge with some path from L_j, $\mathtt{POWER}_2(j) = i$, then it does not share an edge with any path of level $L_{j'}$, $j' \ne j$, $\mathtt{POWER}_2(j') = i$. Indeed, otherwise P shares an edge of some path from $L_{j''}$, $\mathtt{POWER}_2(j'') > i$ and j'' is between j' and j. Thus, due to the intersection property of Lemma 5 the path P shares an edge with at most two paths of G_i. Analogously, if we split P into two upward paths P_1 and P_2, then each of them contains at most two edges incident to some path of G_i, and therefore P contains at most 4 edges incident to paths of G_i.

Finally, if a request $P \in K_i$ shares an edge with two paths $Q_1, Q_2 \in G_i$, then those two paths belong to the same level L_j and consequently by the intersection property of Lemma 5 the request P contains the topmost edges of both Q_1 and Q_2.   □

We now have all the ingredients to prove Theorem 1.

Proof (of Theorem 1). First, invoke Lemma 6 to obtain groups $G_1, G_2, \ldots, G_\ell$ and a partition $K_1, K_2, \ldots, K_\ell$ of the set $\mathcal{P}$.

Next, consider each subset K_i independently. For a fixed subset of requests K_i we split the requests of K_i into two sets $\mathcal{P}_{2i-1}, \mathcal{P}_{2i}$:

- We put $P \in K_i$ into $\mathcal{P}_{2i-1}$ if it is entirely contained in some path Q of G_i;
- Otherwise we put P into $\mathcal{P}_{2i}$.

Note that for $\mathcal{P}_{2i-1}$ the levels give the desired partition $\mathcal{L} = \bigcup_{1 \leq i \leq \ell} G_i$ of the tree T such that each request in $\mathcal{P}_{2i-1}$ is entirely contained in some path of $\mathcal{L}$. Therefore, it remains to construct a set F_i such that each path of $\mathcal{P}_{2i}$ contains at least one and at most 6 edges of F_i.

As the set F_i we take all the topmost edges of paths in G_i as well as all edges in T incident to some path of G_i. Observe that, by Lemma 6, if a request $P \in \mathcal{P}_{2i}$ shares an edge with two paths Q_1 and Q_2 of G_i, then P contains the topmost edges of the latter two paths. Since no $P \in \mathcal{P}_{2i}$ is entirely contained in G_i, if P shares an edge with only one path of G_i, then it contains at least one incident edge to some path of G_i. Consequently, each path $P \in \mathcal{P}_{2i}$ contains at least one edge of F_i.

Finally, since each $P \in \mathcal{P}_{2i}$ contains at most 4 edges incident to paths of G_i and at most two topmost edges of paths of G_i, it contains at most 6 edges of F_i.   □

3   Applications of the Decomposition Theorem

In this section we present two applications of our decomposition Theorem 1.

3.1   Uniform Tollbooth on Trees

Since we assume that the ratio between the smallest and greatest budget in an instance of UNIFORM TOLLBOOTH ON TREES is bounded by a constant and our goal is to obtain an $O(\log \log n)$-approximation, we can assume that all the budgets are equal to 1 and therefore we are going to work with the following definition.

UNIFORM TOLLBOOTH ON TREES (UTT)

Input: A tree $T = (V, E)$ with n vertices, and a set of paths $\mathcal{P} = \{P_1, \ldots, P_m\}$.

Task: Find a pricing function $p : E \to \mathbb{R}_+$ maximizing

$$\sum_{1 \leq i \leq m, p(P_i) \leq 1} p(P_i) .$$

For an instance $\mathcal{I} = ((V, E), \mathcal{P})$ of UTT by $\mathrm{opt}(\mathcal{I})$ we denote the revenue obtained by an optimum solution. When the underlying tree does not change and we consider subsets $\mathcal{P}_i \subseteq \mathcal{P}$, then by $\mathrm{opt}(\mathcal{P}_i)$ we denote $\mathrm{opt}(((V, E), \mathcal{P}_i))$.

Theorem 7. *There is a polynomial time $O(\log \log n)$-approximation algorithm for* UNIFORM TOLLBOOTH ON TREES.

Proof. We use Theorem 1 and independently consider each of the subsets $\mathcal{P}_i$. We will obtain a constant factor approximation for each of the sets $\mathcal{P}_i$, which altogether gives an $O(\log \log n)$-approximation for UTT.

If a set $\mathcal{P}_i$ is of type (a), that is we are given a decomposition of the tree T into edge disjoint paths and each request of $\mathcal{P}_i$ is fully contained in one of the paths of the decomposition, then we can use a PTAS of Grandoni and Rothvoß [8] to obtain revenue at least $\mathrm{opt}(\mathcal{P}_i)/(1 + \epsilon)$.

If a set $\mathcal{P}_i$ is of type (b), then we are additionally given a set of edges F_i, such that each each path of $\mathcal{P}_i$ contains at least one and at most 6 edges of F_i. Consequently, the pricing function defined as:

$$p(e) = \begin{cases} 1/6 & \text{if } e \in F_i \\ 0 & \text{otherwise} \end{cases}$$

gives at least $|\mathcal{P}_i|/6 \geq \mathrm{opt}(\mathcal{P}_i)/6$ revenue.   □

3.2   Unique Coverage on Trees

UNIQUE COVERAGE ON TREES (UCT)

Input: A tree $T = (V, E)$ with n vertices, a set of paths $\mathcal{P} = \{P_1, \ldots, P_m\}$ and a profit function $p : \mathcal{P} \to \mathbb{R}_+$.

Task: Find a set of edges $X \subseteq E$, which maximizes the sum of profits of paths containing exactly one edge of X.

We start by presenting an exact polynomial-time algorithm for the UCT problem for the case when T is a path.

Lemma 8. *With an additional assumption that T is a path,* UCT *can be solved optimally in polynomial time.*

Proof. Let $\mathcal{I} = (T = (V, E), \mathcal{P}, p)$ be a UCT instance where T is a path and let $e_1, \ldots, e_{n-1}$ be the consecutive edges of T. Since each $P \in \mathcal{P}$ is a path, we can represent it as an interval $[a, b] \in P$, where $1 \leq a \leq b \leq n - 1$, e_a is the leftmost edge and e_b is the rightmost edge of P. We use a dynamic programming approach where we consider edges one by one and in a state we store the last two edges selected to the set X. Formally, for $1 \leq i < j < n$ we define $t[i, j]$ as a maximum profit of paths covered exactly once by a subset $X \subseteq E$, satisfying $X \cap \{e_i, \ldots, e_{n-1}\} = \{e_i, e_j\}$:

$$t[i, j] = \max_{X \subseteq E, X \cap \{e_i, \ldots, e_{n-1}\} = \{e_i, e_j\}} p(\{P \in \mathcal{P} : |P \cap X| = 1\}).$$

Observe, that with this definition of entries of the 2-dimensional table t, the optimum profit is the maximum value of $t[i, j]$ over $1 \leq i < j < n$. It remains to show how to compute all the values $t[i, j]$ in polynomial time. We use the following recursive formula, where we either decide that $X = \{e_i, e_j\}$, or we iterate over the first edge in X to the left of e_i:

$$t[i, j] = \max(p(\{P \in \mathcal{P} : |P \cap \{e_i, e_j\}| = 1\}),$$
$$\max_{1 \leq k < i} (t[k, i] + p(A) - p(B_k))).$$

where $A \subseteq \mathcal{P}$ is the set of paths $[a, b] \in \mathcal{P}$ with $i < a \leq j \leq b$ and $B_k \subseteq \mathcal{P}$ is the set of paths $[a, b] \in \mathcal{P}$ with $k < a \leq i < j \leq b$. Note that when adding e_j to the set corresponding to $t[k, i]$ the paths in A start being covered uniquely, and the paths in B are being covered for the second time.

By the standard method of extending the table t with backlinks one can in polynomial time retrieve the set X corresponding to each of the values $t[i, j]$. □

Theorem 9. *There is a polynomial time $O(\log \log n)$-approximation algorithm for the UCT problem.*

Proof. By using Theorem 1 for the tree T and the set of paths $\mathcal{P}$ it is enough to obtain a constant factor approximation for each set $\mathcal{P}_i$. Since for $\mathcal{P}_i$ with paths contained entirely in some path of the decomposition $\mathcal{L}$ we can use Lemma 8, it remains to handle type (b) of the set $\mathcal{P}_i$ given by Theorem 1.

Therefore we assume that each path of $\mathcal{P}_i$ contains at least one and at most 6 edges of the given set F_i. To the set X we independently take each edge of F_i with probability $1/6$. Note, that with constant probability each path in $\mathcal{P}_i$ contains exactly one edge of X and therefore the expected profit given by X is a constant fraction of $p(\mathcal{P})$, which is a trivial upper bound on $\mathrm{opt}(\mathcal{P}_i)$. By the standard method of conditional expectation we can derandomize this procedure, obtaining a deterministic constant factor approximation for the second type of the set $\mathcal{P}_i$, which proves the theorem. □

4   Conclusions

We have presented $O(\log \log n)$-approximation algorithms for UNIFORM TOLL-BOOTH ON TREES. A natural question is whether a constant factor approxima-tion is possible.

Moreover, obtaining a constant-, or even poly(log log n)-approximation for TOLLBOOTH ON TREES with general budgets remains open. As a corollary of our techniques one can prove, that with a loss of a factor of $O(\log \log n)$ in the approximation ratio, one can assume that each request starts and ends in a leaf of the tree. We believe, that it is worthwhile to investigate the TOLLBOOTH ON TREES problem with general budgets, but with the additional assumption that the tree is a full binary tree and all the requests have their endpoints in the leaves of the tree.

Acknowledgements. We thank Guy Kortsarz for mentioning the tree case of the UNIQUE COVERAGE problem as an interesting special case for which nothing better than the general $O(\log n)$ approximation was known.

Moreover we are thankful to anonymous referees for their helpful comments and remarks.

References

1. Balcan, M.-F., Blum, A.: Approximation algorithms and online mechanisms for item pricing. In: ACM Conference on Electronic Commerce, pp. 29–35 (2006)
2. Bansal, N., Chakrabarti, A., Epstein, A., Schieber, B.: A quasi-PTAS for unsplittable flow on line graphs. In: ACM Symposium on Theory of Computing (STOC), pp. 721–729 (2006)
3. Briest, P., Krysta, P.: Single-minded unlimited supply pricing on sparse instances. In: ACM-SIAM Symposium on Discrete Algorithms (SODA), pp. 1093–1102 (2006)
4. Demaine, E.D., Feige, U., Hajiaghayi, M.T., Salavatipour, M.R.: Combination can be hard: Approximability of the unique coverage problem. SIAM Journal on Computing 38(4), 1464–1483 (2008)
5. Elbassioni, K., Raman, R., Ray, S., Sitters, R.: On Profit-Maximizing Pricing for the Highway and Tollbooth Problems. In: Mavronicolas, M., Papadopoulou, V.G. (eds.) SAGT 2009. LNCS, vol. 5814, pp. 275–286. Springer, Heidelberg (2009)
6. Elbassioni, K.M., Sitters, R.A., Zhang, Y.: A Quasi-PTAS for Profit-Maximizing Pricing on Line Graphs. In: Arge, L., Hoffmann, M., Welzl, E. (eds.) ESA 2007. LNCS, vol. 4698, pp. 451–462. Springer, Heidelberg (2007)
7. Gamzu, I., Segev, D.: A Sublogarithmic Approximation for Highway and Tollbooth Pricing. In: Abramsky, S., Gavoille, C., Kirchner, C., Meyer auf der Heide, F., Spirakis, P.G. (eds.) ICALP 2010. Part I. LNCS, vol. 6198, pp. 582–593. Springer, Heidelberg (2010)
8. Grandoni, F., Rothvoß, T.: Pricing on paths: A ptas for the highway problem. In: ACM-SIAM Symposium on Discrete Algorithms (SODA), pp. 675–684 (2011)
9. Guruswami, V., Hartline, J.D., Karlin, A.R., Kempe, D., Kenyon, C., McSherry, F.: On profit-maximizing envy-free pricing. In: ACM-SIAM Symposium on Discrete Algorithms (SODA), pp. 1164–1173 (2005)
10. Guruswami, V., Trevisan, L.: The Complexity of Making Unique Choices: Approximating 1-in-k SAT. In: Chekuri, C., Jansen, K., Rolim, J.D.P., Trevisan, L. (eds.) APPROX 2005 and RANDOM 2005. LNCS, vol. 3624, pp. 99–110. Springer, Heidelberg (2005)

M. Cygan et al.

11. Hartline, J.D., Koltun, V.: Near-Optimal Pricing in Near-Linear Time. In: Dehne,
F., López-Ortiz, A., Sack, J.-R. (eds.) WADS 2005. LNCS, vol. 3608, pp. 422–431.
Springer, Heidelberg (2005)

12. Khandekar, R., Kimbrel, T., Makarychev, K., Sviridenko, M.: On Hardness of Pricing Items for Single-Minded Bidders. In: Dinur, I., Jansen, K., Naor, J., Rolim, J.
(eds.) APPROX and RANDOM 2009. LNCS, vol. 5687, pp. 202–216. Springer,
Heidelberg (2009)

13. Popat, P., Wu, Y.: On the hardness of pricing loss-leaders. In: ACM-SIAM Symposium on Discrete Algorithms (SODA), pp. 735–749 (2012)

Steiner Forest Orientation Problems

Marek Cygan[1,*], Guy Kortsarz[2,**], and Zeev Nutov[3]

[1] IDSIA, University of Lugano, Switzerland
marek@idsia.ch
[2] Rutgers University, Camden
guyk@camden.rutgers.edu
[3] The Open University of Israel
nutov@openu.ac.il

Abstract. We consider connectivity problems with orientation constraints. Given a directed graph D and a collection of ordered node pairs P let $P[D] = \{(u,v) \in P : D \text{ contains a } uv\text{-path}\}$. In the **Steiner Forest Orientation** problem we are given an undirected graph $G = (V, E)$ with edge-costs and a set $P \subseteq V \times V$ of ordered node pairs. The goal is to find a minimum-cost subgraph H of G and an orientation D of H such that $P[D] = P$. We give a 4-approximation algorithm for this problem.

In the **Maximum Pairs Orientation** problem we are given a graph G and a multi-collection of ordered node pairs P on V. The goal is to find an orientation D of G such that $|P[D]|$ is maximum. Generalizing the result of Arkin and Hassin [DAM'02] for $|P| = 2$, we will show that for a mixed graph G (that may have both directed and undirected edges), one can decide in $n^{O(|P|)}$ time whether G has an orientation D with $P[D] = P$ (for undirected graphs this problem admits a polynomial time algorithm for any P, but it is NP-complete on mixed graphs). For undirected graphs, we will show that one can decide whether G admits an orientation D with $|P[D]| \geq k$ in $O(n + m) + 2^{O(k \cdot \log \log k)}$ time; hence this decision problem is fixed-parameter tractable, which answers an open question from Dorn et al. [AMB'11]. We also show that **Maximum Pairs Orientation** admits ratio $O(\log |P| / \log \log |P|)$, which is better than the ratio $O(\log n / \log \log n)$ of Gamzu et al. [WABI'10] when $|P| < n$.

Finally, we show that the following node-connectivity problem can be solved in polynomial time: given a graph $G = (V, E)$ with edge-costs, $s, t \in V$, and an integer ℓ, find a min-cost subgraph H of G with an orientation D such that D contains ℓ internally-disjoint st-paths, and ℓ internally-disjoint ts-paths.

* Partially supported by National Science Centre grant no. N206 567140, Foundation for Polish Science, ERC Starting Grant NEWNET 279352, NSF CAREER award 1053605, DARPA/AFRL award FA8650-11-1-7162 and ONR YIP grant no.N000141110662.
** Partially supported by NSF support grant award number 0829959.

L. Epstein and P. Ferragina (Eds.): ESA 2012, LNCS 7501, pp. 361–372, 2012.
© Springer-Verlag Berlin Heidelberg 2012

1   Introduction

1.1   Problems Considered and Our Results

We consider connectivity problems with orientation constraints. Unless stated otherwise, graphs are assumed to be undirected (and may not be simple), but we also consider directed graphs, and even *mixed graphs*, which may have both directed and undirected edges. Given a mixed graph H, an *orientation of H* is a directed graph D obtained from H by assigning to each undirected edge one of the two possible directions. For a mixed graph H on node set V and a multi-collection of ordered node pairs (that is convenient to consider as a set of directed edges) P on V let $P[H]$ denote the subset of the pairs (or edges) in P for which H contains a uv-path. We say that H *satisfies* P if $P[H] = P$, and that H is *P-orientable* if H admits an orientation D that satisfies P. We note that for undirected graphs it is easy to check in polynomial time whether H is P-orientable, cf. [8] and Section 3 in this paper. Let $n = |V|$ denote the number of nodes in H and $m = |E(H)| + |P|$ the total number of edges and arcs in H and ordered pairs in P.

Our first problem is the classic Steiner Forest problem with orientation constraints.

Steiner Forest Orientation

Instance: A graph $G = (V, E)$ with edge-costs and a set $P \subseteq V \times V$ of ordered node pairs.

Objective: Find a minimum-cost subgraph H of G with an orientation D that satisfies P.

Theorem 1. Steiner Forest Orientation *admits a 4-approximation algorithm.*

Our next bunch of results deals with maximization problems of finding an orientation that satisfies the maximum number of pairs in P.

Maximum Pairs Orientation

Instance: A graph G and a multi-collection of ordered node pairs (i.e., a set of directed edges) P on V.

Objective: Find an orientation D of G such that the number $|P[D]|$ of pairs satisfied by D is maximum.

Let k Pairs Orientation be the decision problem of determining whether Maximum Pairs Orientation has a solution of value at least k. Let P-Orientation be the decision problem of determining whether G is P-orientable (this is the k Pairs Orientation with $k = |P|$). As was mentioned, for undirected graphs P-Orientation can be easily decided in polynomial time [8]. Arkin and Hassin [1] proved that on mixed graphs, P-Orientation is NP-complete, but it is polynomial-time solvable for $|P| = 2$. Using new techniques, we widely generalize the result of [1] as follows.

Theorem 2. *Given a mixed graph G and $P \subseteq V \times V$ one can decide in $n^{O(|P|)}$ time whether G is P-orientable; namely, P-Orientation with a mixed graph G can be decided in $n^{O(|P|)}$ time. In particular, the problem can be decided in polynomial time for any instance with constant $|P|$.*

In several papers, for example in [11], it is stated that any instance of Maximum Pairs Orientation admits a solution D such that $|P[D]| \geq |P|/(4 \log n)$. Furthermore Gamzu et al. [6] show that Maximum Pairs Orientation admits an $O(\log n/ \log \log n)$-approximation algorithm. In [3] it is shown that k Pairs Orientation is fixed-parameter tractable[1] when parameterized by the maximum number of pairs that can be connected via one node. They posed an open question if the problem is fixed-parameter tractable when parameterized by k (the number of pairs that should be connected), namely, whether k Pairs Orientation can be decided in $f(k)\text{poly}(n)$ time, for some computable function f. Our next result answers this open question, and for $|P| < n$ improves the approximation ratio $O(\log n/ \log \log n)$ for Maximum Pairs Orientation of [11,6].

Theorem 3. *Any instance of* Maximum Pairs Orientation *admits a solution D, that can be computed in polynomial time, such that $|P[D]| \geq |P|/(4 \log_2(3|P|))$. Furthermore*

(i) k Pairs Orientation *can be decided in $O(n + m) + 2^{O(k \cdot \log \log k)}$ time; thus it is fixed-parameter tractable when parameterized by k.*

(ii) Maximum Pairs Orientation *admits an $O(\log |P|/ \log \log |P|)$-approximation algorithm.*

Note that $|P|$ may be much smaller than n, say $|P| = 2^{\sqrt{\log n}}$. While this size of P does not allow exhaustive search in time polynomial in n, we do get an approximation ratio of $O(\sqrt{\log n}/ \log \log n)$, which is better than the ratio $O(\log n/ \log \log n)$ of Gamzu et al. [6].

One may also consider "high-connectivity" orientation problems, to satisfy prescribed connectivity demands. Several papers considered min-cost edge-connectivity orientation problems, cf. [10]. Almost nothing is known about min-cost node-connectivity orientation problems. We consider the following simple but still nontrivial variant.

ℓ Disjoint Paths Orientation
Instance: A graph $G = (V, E)$ with edge-costs, $s, t \in V$, and an integer ℓ.
Objective: Find a minimum-cost subgraph H of G with an orientation D such that D contains ℓ internally-disjoint st-paths, and ℓ internally-disjoint ts-paths.

Checking whether ℓ Disjoint Paths Orientation admits a feasible solution can be done in polynomial time using the characterization of feasible solutions of Egawa, Kaneko, and Matsumoto; we use this characterization to prove the following.

Theorem 4. ℓ Disjoint Paths Orientation *can be solved in polynomial time.*

---

[1] "Fixed-parameter tractable" means the following. In the parameterized complexity setting, an instance of a decision problem comes with an integer parameter k. A problem is said to be *fixed-parameter tractable* (w.r.t. k) if there exists an algorithm that decides any instance (I, k) in time $f(k)\text{poly}(|I|)$ for some (usually exponential) computable function f.

Theorems 1, 2, 3 are proved in sections 2, 3, 4, respectively. Theorem 4 is proved in the full version, due to space limitation.

1.2   Previous and Related Work

Let $\lambda_H(u,v)$ denote the (u,v)-edge-connectivity in a graph H, namely, the maximum number of pairwise edge-disjoint uv-paths in H. Similarly, let $\kappa_H(u,v)$ denote the (u,v)-node-connectivity in H, namely, the maximum number of pairwise internally node-disjoint uv-paths in H. Given an edge-connectivity demand function $r = \{r(u,v) : (u,v) \in V \times V\}$, we say that H *satisfies* r if $\lambda_H(u,v) \geq r(u,v)$ for all $(u,v) \in V \times V$; similarly, for node connectivity demands, we say that H *satisfies* r if $\kappa_H(u,v) \geq r(u,v)$ for all $(u,v) \in V \times V$.

Survivable Network Orientation

Instance: A graph $G = (V,E)$ with edge-costs and edge/node-connectivity
demand function $r = \{r(u,v) : (u,v) \in V \times V\}$.

Objective: Find a minimum-cost subgraph H of G with orientation D that
satisfies r.

So far we assumed that the orienting costs are symmetric; this means that orienting an undirected edge connecting u and v in each one of the two directions is the same, namely, that $c(u,v) = c(v,u)$. This assumption is reasonable in practical problems, but in a theoretic more general setting, we might have non-symmetric costs $c(u,v) \neq c(v,u)$. Note that the version with non-symmetric costs includes the min-cost version of the corresponding directed connectivity problem, and also the case when the input graph G is a mixed graph, by assigning large/infinite costs to non-feasible orientations. For example, Steiner Forest Orientation with non-symmetric costs includes the Directed Steiner Forest problem, which is Label-Cover hard to approximate [2]. This is another reason to consider the symmetric costs version.

Khanna, Naor, and Shepherd [10] considered several orientation problems with non-symmetric costs. They showed that when D is required to be k-edge-outconnected from a given roots s (namely, D contains k edge-disjoint paths from s to every other node), then the problem admits a polynomial time algorithm. In fact they considered a more general problem of finding an orientation that covers an intersecting supermodular or crossing supermodular set-function. See [10] for precise definitions. Further generalization of this result due to Frank, T. Király, and Z. Király was presented in [5]. For the case when D should be strongly connected, [10] obtained a 4-approximation algorithm; note that our Steiner Forest Orientation problem has much more general demands, that are *not* captured by intersecting supermodular or crossing supermodular set-functions, but we consider symmetric edge-costs (otherwise the problem includes the Directed Steiner Forest problem). For the case when D is required to be k-edge-connected, $k \geq 2$, [10] obtained a pseudo-approximation algorithm that computes a $(k-1)$-edge-connected subgraph of cost at most $2k$ times the cost of an optimal k-connected subgraph.

We refer the reader to [4] for a survey on characterization of graphs that admit orientations satisfying prescribed connectivity demands, and here mention only the following central theorem, that can be used to obtain a pseudo-approximation for edge-connectivity orientation problems.

Theorem 5 (Well-Balanced Orientation Theorem, Nash-Williams [12]). *Any undirected graph $H = (V, E_H)$ has an orientation D for which $\lambda_D(u, v) \geq \lfloor \frac{1}{2}\lambda_H(u, v) \rfloor$ for all $(u, v) \in V \times V$.*

We note that given H, an orientation as in Theorem 5 can be computed in polynomial time. It is easy to see that if H has an orientation D that satisfies r then H satisfies the demand function q defined by $q(u, v) = r(u, v) + r(v, u)$. Theorem 5 implies that edge-connectivity Survivable Network Orientation admits a polynomial time algorithm that computes a subgraph H of G and an orientation D of H such that $c(H) \leq 2\text{opt}$ and

$$\lambda_D(u, v) \geq \lfloor (r(u, v) + r(v, u))/2 \rfloor \geq \lfloor \max\{r(u, v), r(v, u)\}/2 \rfloor \ \forall (u, v) \in V \times V \ .$$

This is achieved by applying Jain's [9] algorithm to compute a 2-approximate solution H for the corresponding undirected edge-connectivity Survivable Network instance with demands $q(u, v) = r(u, v) + r(v, u)$, and then computing an orientation D of H as in Theorem 5. This implies that if the costs are symmetric, then by cost at most 2opt we can satisfy almost half of the demand of every pair, and if also the demands are symmetric then we can satisfy all the demands. The above algorithm also applies for non-symmetric edge-costs, invoking an additional cost factor of $\max_{uv \in E} c(v, u)/c(u, v)$. Summarizing, we have the following observation, which we failed to find in the literature.

Corollary 6. *Edge-connectivity Survivable Network Orientation (with non-symmetric costs) admits a polynomial time algorithm that computes a subgraph H of G and an orientation D of H such that $c(H) \leq 2\text{opt} \cdot \max_{uv \in E} c(v, u)/c(u, v)$ and $\lambda_D(u, v) \geq \lfloor \frac{1}{2}(r(u, v) + r(v, u)) \rfloor$ for all $(u, v) \in V \times V$. In particular, the problem admits a 2-approximation algorithm if both the costs and the demands are symmetric.*

2   Algorithm for Steiner Forest Orientation (Theorem 1)

In this section we prove Theorem 1. For a mixed graph or an edge set H on a node set V and $X, Y \subseteq V$ let $\delta_H(X, Y)$ denote the set of all (directed and undirected) edges in H from X to Y and let $d_H(X, Y) = |\delta_H(X, Y)|$ denote their number; for brevity, $\delta_H(X) = \delta_H(X, \bar{X})$ and $d_H(X) = d_H(X, \bar{X})$, where $\bar{X} = V \setminus X$.

Given an integral set-function f on subsets of V we say that H covers f if $d_H(X) \geq f(X)$ for all $X \subseteq V$. Define a set-function f_r by $f_r(\emptyset) = f_r(V) = 0$ and for every $\emptyset \neq X \subset V$

$$f_r(X) = \max\{r(u, v) : u \in X, v \in \bar{X}\} + \max\{r(v, u) : u \in X, v \in \bar{X}\} \ . \quad (1)$$

Note that the set-function f_r is symmetric, namely, that $f_r(X) = f_r(\bar{X})$ for all $X \subseteq V$.

Lemma 7. *If an undirected edge set H has an orientation D that satisfies an edge-connectivity demand function r then H covers f_r.*

Proof. Let $X \subseteq V$. By Menger's Theorem, any orientation D of H that satisfies r has at least $\max\{r(u,v) : u \in X, v \in \bar{X}\}$ edges from X to $\bar{X}$, and at least $\max\{r(v,u) : u \in X, v \in \bar{X}\}$ edges from $\bar{X}$ to X. The statement follows. □

Recall that in the Steiner Forest Orientation problem we have $r(u,v) = 1$ if $(u,v) \in P$ and $r(u,v) = 0$ otherwise. We will show that if $r_{max} = \max\limits_{u,v \in V} r(u,v) = 1$ then the inverse to Lemma 7 is also true, namely, if H covers f_r then H has an orientation that satisfies r; for this case, we also give a 4-approximation algorithm for the problem of computing a minimum-cost subgraph that covers f_r. We do not know if these results can be extended for $r_{max} \geq 2$.

Lemma 8. *For $r_{max} = 1$, if an undirected edge set H covers f_r then H has an orientation that satisfies r.*

Proof. Observe that if $(u,v) \in P$ (namely, if $r(u,v) = 1$) then u,v belong to the same connected component of H. Hence it sufficient to consider the case when H is connected. Let D be an orientation of H obtained as follows. Orient every 2-edge-connected component of H to be strongly connected (recall that a directed graph is strongly connected if there is a directed path from any of its nodes to any other); this is possible by Theorem 5. Now we orient the bridges of H. Consider a bridge e of H. The removal of e partitions V into two connected components $X, \bar{X}$. Note that $\delta_P(X, \bar{X}) = \emptyset$ or $\delta_P(\bar{X}, X) = \emptyset$, since $f_r(X) \leq d_H(X) = 1$. If $\delta_P(X, \bar{X}) \neq \emptyset$, we orient e from X to $\bar{X}$; if $\delta_P(\bar{X}, X) \neq \emptyset$, we orient e from $\bar{X}$ to X; and if $\delta_P(X, \bar{X}), \delta_P(\bar{X}, X) = \emptyset$, we orient e arbitrarily. It is easy to see that the obtained orientation D of H satisfies P. □

We say that an edge-set or a graph H *covers* a set-family $\mathcal{F}$ if $d_H(X) \geq 1$ for all $X \in \mathcal{F}$. A set-family $\mathcal{F}$ is said to be *uncrossable* if for any $X, Y \in \mathcal{F}$ the following holds: $X \cap Y, X \cup Y \in \mathcal{F}$ or $X \setminus Y, Y \setminus X \in \mathcal{F}$. The problem of finding a minimum-cost set of undirected edges that covers an uncrossable set-family $\mathcal{F}$ admits a primal-dual 2-approximation algorithm, provided the inclusion-minimal members of $\mathcal{F}$ can be computed in polynomial time [7]. It is known that the undirected Steiner Forest problem is a particular case of the problem of finding a min-cost cover of an uncrossable family, and thus admits a 2-approximation algorithm.

Lemma 9. *Let $H = (V, J \cup P)$ be a mixed graph, where edges in J are undirected and edges in P are directed, such that for every $uv \in P$ both u,v belong to the same connected component of the graph (V, J). Then the set-family $\mathcal{F} = \{S \subseteq V : d_J(S) = 1 \wedge d_P(S), d_P(\bar{S}) \geq 1\}$ is uncrossable, and its inclusion minimal members can be computed in polynomial time.*

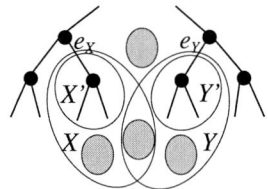

Fig. 1. Illustration to the proof of Lemma 9; components distinct from C_X, C_Y are shown by gray ellipses

Proof. Let C be the set of connected components of the graph (V, J). Let $C \in C$. Any bridge e of C partitions C into two parts $C'(e), C''(e)$ such that e is the unique edge in J connecting them. Note that the condition $d_J(S) = 1$ is equivalent to the following condition (C1), while if condition (C1) holds then the condition $d_P(S), d_P(\bar{S}) \geq 1$ is equivalent to the following condition (C2) (since no edge in P connects two distinct connected components of (V, J)).

(C1) There exists $C_S \in C$ and a bridge e_S of C_S, such that S is a union of one
 of the sets $X' = C'_S(e_S), X'' = C''_S(e_S)$ and sets in $C \setminus \{C_S\}$.
(C2) $d_P(X', X''), d_P(X'', X') \geq 1$.

Hence we have the following characterization of the sets in $\mathcal{F}$: $S \in \mathcal{F}$ *if, and only if, conditions (C1), (C2) hold for S*. This implies that every inclusion-minimal member of $\mathcal{F}$ is $C'(e)$ or $C''(e)$, for some bridge e of $C \in C$. In particular, the inclusion-minimal members of $\mathcal{F}$ can be computed in polynomial time.

Now let $X, Y \in \mathcal{F}$ (so conditions (C1), (C2) hold for each one of X, Y), let $C_X, C_Y \in C$ be the corresponding connected components and e_X, e_Y the corresponding bridges (possibly $C_X = C_Y$, in which case we also may have $e_X = e_Y$), and let X', X'' and Y', Y'' be the corresponding partitions of C_X and C_Y, respectively. Since e_X, e_Y are bridges, at least one of the sets $X' \cap Y', X' \cap Y'', X'' \cap Y', X'' \cap Y''$ must be empty, say $X' \cap Y' = \emptyset$. Note that the set-family $\mathcal{F}$ is symmetric, hence to prove that $X \cap Y, X \cup Y \in \mathcal{F}$ or $X \setminus Y, Y \setminus X \in \mathcal{F}$, it is sufficient to prove that $A \setminus B, B \setminus A \in \mathcal{F}$ for some pair A, B such that $A \in \{X, \bar{X}\}, B \in \{Y, \bar{Y}\}$. E.g., if $A = X$ and $B = \bar{Y}$, then $A \setminus B = X \cap Y$ and $B \setminus A = V \setminus (X \cup Y)$, hence $A \setminus B, B \setminus A \in \mathcal{F}$ together with the symmetry of $\mathcal{F}$ implies $X \cap Y, X \cup Y \in \mathcal{F}$. Similarly, if $A = \bar{X}$ and $B = \bar{Y}$, then $A \setminus B = Y \setminus X$ and $B \setminus A = X \setminus Y$, hence $A \setminus B, B \setminus A \in \mathcal{F}$ implies $Y \setminus X, X \setminus Y \in \mathcal{F}$. Thus w.l.o.g. we may assume that $X' \subseteq X$ and $Y' \subseteq Y$, see Figure 1, and we show that $X \setminus Y, Y \setminus X \in \mathcal{F}$. Recall that $X' \cap Y' = \emptyset$ and hence $X \cap Y$ is a (possibly empty) union of some sets in $C \setminus \{C_X, C_Y\}$. Thus $X \setminus Y$ is a union of X' and some sets in $C \setminus \{C_X, C_Y\}$. This implies that conditions (C1), (C2) hold for $X \setminus Y$, hence $X \setminus Y \in \mathcal{F}$; the proof that $Y \setminus X \in \mathcal{F}$ is similar. This concludes the proof of the lemma. □

Lemma 10. *Given a* Steiner Forest Orientation *instance, the problem of computing a minimum-cost subgraph H of G that covers f_r admits a 4-approximation algorithm.*

Proof. The algorithm has two phases. In the first phase we solve the corresponding undirected Steiner Forest instance with the same demand function r. The Steiner Forest problem admits a 2-approximation algorithm, hence $c(J) \leq 2\text{opt}$. Let J be a subgraph of G computed by such a 2-approximation algorithm. Note that $f_r(S) - d_J(S) \leq 1$ for all $S \subseteq V$. Hence to obtain a cover of f_r it is sufficient to cover the family $\mathcal{F} = \{S \subseteq V : f_r(S) - d_J(S) = 1\}$ of the deficient sets w.r.t. J. The key point is that the family $\mathcal{F}$ is uncrossable, and that the inclusion-minimal members of $\mathcal{F}$ can be computed in polynomial time. In the second phase we compute a 2-approximate cover of this $\mathcal{F}$ using the algorithm of [7]. Observe that the set $E(H) \setminus E(J)$, that is the set of edges of the optimum solution with edges of J removed, covers the family $\mathcal{F}$ and therefore the cost of the second phase is at most 2opt. Consequently, the problem of covering f_r is reduced to solving two problems of covering an uncrossable set-family.

To show that $\mathcal{F}$ is uncrossable we use Lemma 9. Note that for any $(u,v) \in P$ both u,v belong to the same connected component of (V,J), and that $f_r(S) - d_J(S) = 1$ if, and only if, $d_J(S) = 1$ and $d_P(S), d_P(\bar{S}) \geq 1$, hence $\mathcal{F} = \{S \subseteq V : d_J(S) = 1 \wedge d_P(S), d_P(\bar{S}) \geq 1\}$. Consequently, by Lemma 9, the family $\mathcal{F}$ is uncrossable and its inclusion-minimal members can be computed in polynomial time. This concludes the proof of the lemma.   □

The proof of Theorem 1 is complete.

3   Algorithm for P-Orientation on Mixed Graphs (Theorem 2)

In this section we prove Theorem 2. The following (essentially known) statement is straightforward.

Lemma 11. *Let G be a mixed graph, let P be a set of directed edges on V, and let C be a subgraph of G that admits a strongly connected orientation. Let G', P' be obtained from G, P by contracting C into a single node. Then G is P-orientable if, and only if, G' is P'-orientable. In particular, this is so if C is a cycle.*   □

Corollary 12. P-orientation *(with an undirected graph G) can be decided in polynomial time.*

Proof. By repeatedly contracting a cycle of G, we obtain an equivalent instance, by Lemma 11. Hence we may assume that G is a tree. Then for every $(u,v) \in P$ there is a unique uv-path in G, which imposes an orientation on all the edges of this path. Hence if suffices to check that no two pairs in P impose different orientations of the same edge of the tree.   □

Our algorithm for mixed graphs is based on a similar idea. We say that a mixed graph is an *ori-cycle* if it admits an orientation that is a directed simple cycle. We need the following statement.

Lemma 13. *Let $G = (V, E \cup A)$ be a mixed graph, where edges of E are undirected and A contains directed arcs, and let G' be obtained from G by contracting* *every connected component of the undirected graph (V, E) into a single node. If there is a directed cycle (possibly a self-loop) C' in G' then there is an ori-cycle C in G, and such C can be found in polynomial time.*

Proof. If C' is also a directed cycle in G, then we take $C = C'$. Otherwise, we replace every node v_X of C' that corresponds to a contracted connected component X of (V, E) by a path, as follows. Let a_1 be the arc entering v_X in C' and let a_2 be the arc leaving v_X in C'. Let v_1 be the head of a_1 and similarly let v_2 be a the tail of a_2. Since X is a connected component in (V, E), there is a v_1v_2-path in (V, E), and we replace X by this path. The result is the required ori-cycle C (possibly a self-loop) in G. It is easy to see that such C can be obtained from C' in polynomial time. □

By Lemmas 13 and 11 we may assume that the directed graph G' obtained from G by contracting every connected component of (V, E), is a directed acyclic multigraph (with no self-loops). This preprocessing step is similar to the one used by Silverbush et al. [13]. Let $p = |P|$. Let $f : V \rightarrow V(G')$ be the function which for each node v of G assigns a node $f(v)$ in G' that represents the connected component of (V, E) that contains v (in other words the function f shows a correspondence between nodes before and after contractions).

The first step of our algorithm is to guess the first and the last edge on the path for each of the p pairs in P, by trying all $n^{O(p)}$ possibilities. If for the i-th pair an undirected edge is selected as the first or the last one on the corresponding path, then we orient it accordingly and move it from E to A. Thus by functions last, first : $\{1, \ldots, p\} \rightarrow A$ we denote the guessed first and last arc for each of the p paths.

Now we present a branching algorithm with exponential time complexity which we later convert to $n^{O(p)}$ time by applying a method of memoization. Let π be a topological ordering of G'. By cur : $\{1, \ldots, p\} \rightarrow A$ we denote the most recently chosen arc from A for each of the p paths (initially cur(i) = first(i)). In what follows we consider subsequent nodes v_C of G' with respect to π and branch on possible orientations of the connected component C of G. We use this orientation to update the function cur for all the arguments i such that cur(i) is an arc entering a node mapped to v_C.

Let $v_C \in V(G')$ be the first node w.r.t. to π which was not yet considered by the branching algorithm. Let $I \subseteq \{1, \ldots, p\}$ be the set of indices i such that cur(i) = $(u, v) \in A$ for $f(v) = v_C$, and cur(i) $\neq$ last(i). If $I = \emptyset$ then we skip v_C and proceed to the next node in π. Otherwise for each $i \in I$ we branch on choosing an arc $(u, v) \in A$ such that $f(u) = v_C$, that is we select an arc that the i-path will use just after leaving the connected component of G corresponding to the node v_C (note that there are at most $|A|^{|I|} = n^{O(p)}$ branches). Before updating the arcs cur(i) for each $i \in I$ in a branch, we check whether the connected component C of (V, E) consisting of nodes $f^{-1}(v_C)$ is accordingly orientable by using Corollary 12 (see Figure 2). Finally after considering all the nodes in π we check whether for each $i \in \{1, \ldots, p\}$ we have cur(i) = last(i). If this is the case our algorithm returns YES and otherwise it returns NO in this branch.

The correctness of our algorithm follows from the invariant that each node v_C is considered at most once, since all the updated values cur(i) are changed to arcs that are to the right with respect to π.

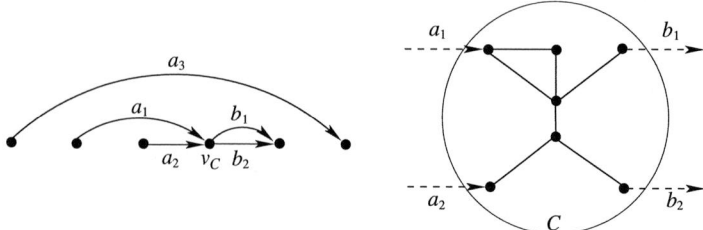

Fig. 2. Our algorithm considers what orientation the connected component C (of (V, E)) will have. Currently we have $\mathrm{cur}(1) = a_1$, $\mathrm{cur}(2) = a_2$ and $\mathrm{cur}(3) = a_3$, hence $I = \{1, 2\}$. If in a branch we set new values $\mathrm{cur}(1) = b_1$ and $\mathrm{cur}(2) = b_2$ then by Corollary 12 we can verify that it is possible to orient C, so that there is a path from the end-point of a_1 to the start-point of b_1 and from the end-point of a_2 to the start-point of b_2. However the branch with new values $\mathrm{cur}(1) = b_2$ and $\mathrm{cur}(2) = b_1$ will be terminated, since it is not possible to orient C accordingly.

Observe that when considering a node v_C it is not important what orientations previous nodes in π have, because all the relevant information is contained in the cur function. Therefore to improve the currently exponential time complexity we apply the standard technique of memoization, that is, store results of all the previously computed recursive calls. Consequently for any index of the currently considered node in π and any values of the function cur, there is at most one branch for which we compute the result, since for the subsequent recursive calls we use the previously computed results. This leads to $n^{O(p)}$ branches and $n^{O(p)}$ total time and space complexity.

4   Algorithms for Maximum Pairs Orientation (Theorem 3)

In this section we prove Theorem 3.

Lemma 14. *There exists a linear time algorithm that given an instance of* Maximum Pairs Orientation *or of* k Pairs Orientation, *transforms it into an equivalent instance such that the input graph is a tree with at most* $3p - 1$ *nodes.*

Proof. As is observed in [8], and also follows from Lemma 11, we can assume that the input graph G is a tree; such a tree can be constructed in linear time by contracting the 2-edge-connected components of G. For each edge e of G we compute an integer $P(e)$, that is the number of pairs (s, t) in P, such that e belongs to the shortest path between s and t in G. If for an edge e we have $P(e) \leq 1$, we contract this edge, since we can always orient it as desired by at most one st-path. Note that after this operation each leaf belongs to at least two pairs in P. If a node v has degree 2 in the tree and does not belong to any pair, we contract one of the edges incident to v; this is since in any inclusion minimal solution, one of the two edges enters v if, and only if, the other leaves v.

The linear time implementation of the presented reductions is as follows. First, using a linear time algorithm for computing 2-edge-connected components, and by scanning every edge in $E \cup P$, we can see which components it connects, thus obtaining an equivalent instance where G is a tree. The values $P(e)$ can be computed in linear time by standard methods, however we postpone the details to the full version. After computing the values $P(e)$ contract all the edges with $P(e) = 1$. Note that after the contractions are done we need to relabel pairs in P, since some nodes may have their labels changed, but this can be also done in linear time by storing a new label for each node in a table. Finally using a graph search algorithm we find maximal paths such that each internal node is of degree two and does not belong to any pair. For each such path we contract all but one edge.

We claim that after these reductions are implemented, the tree G' obtained has at most $3p - 1$ nodes. Let ℓ be the number of leaves and t the number of nodes of degree 2 in G'. As each node of degree less than 3 in G' is an s_i or t_i, $\ell + t \leq 2p$. Since each leaf belongs to at least two pairs in P, $\ell \leq p$. The number of nodes of degree at least 3 is at most $\ell - 1$ and so $|V(G')| \leq 2\ell + t - 1 \leq 3p - 1$. This concludes the proof of the lemma. □

After applying Lemma 14, the number of nodes n of the returned tree is at most $3p - 1$. Therefore, by [11], one can find in polynomial time a solution D, such that $|P[D]| \geq p/(4 \log_2 n) \geq p/(4 \log_2(3p))$. Therefore, if for a given k Pairs Orientation instance we have $k \leq p/(4 \log_2(3p))$, then clearly it is a YES instance. However if $k > p/(4 \log_2(3p))$, then $p = \Theta(k \log k)$. In order to solve the k Pairs Orientation instance we consider all possible $\binom{p}{k}$ subsets P' of exactly k pairs from P, and check if the graph is P'-orientable. Observe that

$$\binom{p}{k} \leq \frac{p^k}{k!} \leq \frac{p^k}{(k/e)^k} \leq \frac{p^k}{(p/(4e \log_2(3p)))^k} = (4e \log_2(3p))^k = 2^{O(k \log \log k)}$$

where the second inequality follows from Stirling's formula. Therefore the running time is $O(m + n) + 2^{O(k \log \log k)}$, which proves (i).

Combining Lemma 14 with the $O(\log n / \log \log n)$-approximation algorithm of Gamzu et al. [6] proves (ii). Thus the proof of Theorem 3 is complete.

5   Conclusions and Open Problems

In this paper we considered minimum-cost and maximum pairs orientation problems. Our main results are a 4-approximation algorithm for Steiner Forest Orientation, an $n^{O(|P|)}$ time algorithm for P-Orientation on mixed graphs, and an $O(n + m) + 2^{O(k \cdot \log \log k)}$ time algorithm for k Pairs Orientation (which implies that k Pairs Orientation is fixed-parameter tractable when parameterized by k, solving an open question from [3]). We now mention some open problems, most of them related to the work of Khanna, Naor, and Shepherd [10].

We have shown that the P-Orientation problem on mixed graphs parameterized by $|P|$ belongs to XP, however to the best of our knowledge it is not known whether this problem is fixed parameter tractable or $W[1]$-hard.

As was mentioned, [10] showed that the problem of computing a minimum-cost k-edge-outconnected orientation can be solved in polynomial time, even for non-symmetric edge-costs. To the best of our knowledge, for *node-connectivity*, and even for the simpler notion of *element-connectivity*, no non-trivial approximation ratio is known even for symmetric costs and $k = 2$. Moreover, even the decision version of determining whether an undirected graph admits a 2-outconnected orientation is not known to be in P nor NP-complete.

For the case when the orientation D is required to be k-edge-connected, $k \geq 2$, [10] obtained a pseudo-approximation algorithm that computes a $(k - 1)$-edge-connected subgraph of cost at most $2k$ times the cost of an optimal k-connected subgraph. It is an open question if the problem admits a non-trivial true approximation algorithm even for $k = 2$.

References

1. Arkin, E., Hassin, R.: A note on orientations of mixed graphs. Discrete Applied Mathematics 116(3), 271–278 (2002)
2. Dodis, Y., Khanna, S.: Design networks with bounded pairwise distance. In: STOC, pp. 750–759 (1999)
3. Dorn, B., Hüffner, F., Krüger, D., Niedermeier, R., Uhlmann, J.: Exploiting bounded signal flow for graph orientation based on cause-effect pairs. Algorithms for Molecular Biology 6(21) (2011)
4. Frank, A., Király, T.: Combined connectivity augmentation and orientation problems. Discrete Applied Mathematics 131(2), 401–419 (2003)
5. Frank, A., Király, T., Király, Z.: On the orientation of graphs and hypergraphs. Discrete Applied Mathematics 131, 385–400 (2003)
6. Gamzu, I., Segev, D., Sharan, R.: Improved Orientations of Physical Networks. In: Moulton, V., Singh, M. (eds.) WABI 2010. LNCS, vol. 6293, pp. 215–225. Springer, Heidelberg (2010)
7. Goemans, M., Goldberg, A., Plotkin, S., Shmoys, D., Tardos, E., Williamson, D.: Improved approximation algorithms for network design problems. In: SODA, pp. 223–232 (1994)
8. Hassin, R., Megiddo, N.: On orientations and shortest paths. Linear Algebra and Its Applications, 589–602 (1989)
9. Jain, K.: A factor 2 approximation algorithm for the generalized Steiner network problem. Combinatorica 21(1), 39–60 (2001)
10. Khanna, S., Naor, J., Shepherd, B.: Directed network design with orientation constraints. In: SODA, pp. 663–671 (2000)
11. Medvedovsky, A., Bafna, V., Zwick, U., Sharan, R.: An Algorithm for Orienting Graphs Based on Cause-Effect Pairs and Its Applications to Orienting Protein Networks. In: Crandall, K.A., Lagergren, J. (eds.) WABI 2008. LNCS (LNBI), vol. 5251, pp. 222–232. Springer, Heidelberg (2008)
12. Nash-Williams: On orientations, connectivity and odd vertex pairings in finite graphs. Canad. J. Math. 12, 555–567 (1960)
13. Silverbush, D., Elberfeld, M., Sharan, R.: Optimally Orienting Physical Networks. In: Bafna, V., Sahinalp, S.C. (eds.) RECOMB 2011. LNCS, vol. 6577, pp. 424–436. Springer, Heidelberg (2011)

A Dual-Fitting $\frac{3}{2}$-Approximation Algorithm for Some Minimum-Cost Graph Problems

James M. Davis* and David P. Williamson**

School of Operations Research and Information Engineering,
Cornell University, Ithaca, NY 14853, USA
jmd388@cornell.edu, dpw@cs.cornell.edu

Abstract. In an ESA 2011 paper, Couëtoux [2] gives a beautiful $\frac{3}{2}$-approximation algorithm for the problem of finding a minimum-cost set of edges such that each connected component has at least k vertices in it. The algorithm improved on previous 2-approximation algorithms for the problem. In this paper, we reanalyze Couëtoux's algorithm using dual-fitting and show how to generalize the algorithm to a broader class of graph problems previously considered in the literature.

1   Introduction

Consider the following graph problem: given an undirected graph $G = (V, E)$ with nonnegative edge weights $c(e) \geq 0$ for all $e \in E$, and a positive integer k, find a set F of edges such that each connected component of (V, F) has at least k vertices in it. If $k = 2$, the problem is the minimum-weight edge cover problem, and if $k = |V|$, the problem is the minimum spanning tree problem; both problems are known to be polynomial-time solvable. However, for any other constant value of k, the problem is known to be NP-hard (see Imielińska, Kalantari, and Khachiyan [6] for $k \geq 4$ and Bazgan, Couëtoux, and Tuza [1] for $k = 3$).

For this reason, researchers have considered approximation algorithms for the problem. An α-approximation algorithm is a polynomial-time algorithm that produces a solution of cost at most α times the cost of an optimal solution. Imielińska, Kalantari, and Khachiyan [6] give a very simple 2-approximation algorithm for this problem, which they call the *constrained forest problem*. The algorithm is a variant of Kruskal's algorithm [7] for the minimum spanning tree problem that considers edges in order of increasing cost, and adds an edge to the solution as long as it connects two different components (as in Kruskal's algorithm) and one of the two components has fewer than k vertices. Other 2-approximation algorithms based on edge deletion (instead of, or in addition to, edge insertion) are given by Laszlo and Mukherjee [8,9], who also perform some experimental comparisons.

---

* This material is based upon work supported by the National Science Foundation under Grant No. DGE-0707428.
** This work is supported in part by NSF grant CCF-1115256.

L. Epstein and P. Ferragina (Eds.): ESA 2012, LNCS 7501, pp. 373–382, 2012.
© Springer-Verlag Berlin Heidelberg 2012

In an ESA 2011 paper, Couëtoux [2] introduces a beautiful little twist to the Imielińska et al. algorithm and shows that the result is a $\frac{3}{2}$-approximation algorithm. In any partial solution constructed by the algorithm, let us call a connected component *small* if it has fewer than k vertices in it, and *big* otherwise. As the algorithm considers whether to add edge e to the solution (at cost $c(e)$), it also considers all edges e' of cost $c(e') \leq 2c(e)$, and checks whether adding such an edge e' would join two small components into a big component. If any such edge e' exists, the algorithm adds the edge to the solution (if there is more than one, it adds the cheapest such edge), otherwise it returns to considering the addition of edge e.

In another thread of work, Goemans and Williamson [3] show that the Imielińska et al. algorithm can be generalized to provide 2-approximation algorithms to a large class of graph problems. The graph problems are specified by a function $h : 2^V \rightarrow \{0,1\}$ (actually, [3] considers a more general case of functions $h : 2^V \rightarrow \mathbb{N}$, but we will consider only the 0-1 case here). Given a graph G and a subset of vertices S, let $\delta(S)$ be the set of all edges with exactly one endpoint in S. Then a set of edges F is a feasible solution to the problem given by h if $|F \cap \delta(S)| \geq h(S)$ for all nontrivial subsets of vertices S. Thus, for instance, the problem of finding components of size at least k is given by the function $h(S) = 1$ iff $|S| < k$. Goemans and Williamson [3] consider functions h that are *downwards monotone*; that is, if $h(S) = 1$ for some subset S, then $h(T) = 1$ for all $T \subseteq S$, $T \neq \emptyset$. They then show that the natural generalization of the Imielińska et al. algorithm is a 2-approximation algorithm for any downwards monotone function h; in particular, the algorithm considers edges in order of increasing cost, and adds an edge to the solution if it joins two different connected components C and C' and either $h(C) = 1$ or $h(C') = 1$ (or both). Laszlo and Mukherjee [10] have shown that their edge-deletion algorithms also provide 2-approximation algorithms for these problems. Goemans and Williamson show that a number of graph problems can be modeled with downwards monotone functions, including some location-design and location-routing problems; for example, they consider a problem in which every component not only must have at least k vertices and also must have an open depot from a subset $D \subseteq V$, where there is a cost $c(d)$ for opening the depot $d \in D$ to serve the component.

In this paper, we show that the natural generalization of Couëtoux's algorithm to downwards monotone functions gives $\frac{3}{2}$-approximation algorithms for this class of problems. In the process, we give a different and (we believe) somewhat simpler analysis than Couëtoux. We note that the algorithm can be viewed as a dual-growing algorithm similar to many primal-dual approximation algorithms for network design problems (see Goemans and Williamson [5] for a survey). However, in this case, the dual is not a feasible solution to the dual of a linear programming relaxation of the problem; in fact, it gives an overestimate on the cost of the tree generated by the algorithm. But we show that a dual-fitting style analysis works; namely, if we scale the dual solution down by a factor of 2/3, it gives a lower bound on the cost of any feasible solution. This leads directly to the performance guarantee of $\frac{3}{2}$.

Our result is interesting for another reason: we know of very few classes of network design problems such as this one that have a performance guarantee with constant strictly smaller than 2. For individual problems such as Steiner tree and prize-collecting Steiner tree there are approximation algorithms known with performance guarantees smaller than 2, but these results are isolated, and do not extend to well-defined classes of problems. It would be very interesting, for instance, to give an approximation algorithm for the class of proper functions (defined in Goemans and Williamson [4]) with a constant performance guarantee smaller than 2.

The paper is structured as follows. In Section 2, we give the algorithm in more detail. In Section 3, we turn to the analysis of the algorithm. We conclude with some open questions in Section 4. Some proofs are omitted and will appear in the final version of the paper.

2   The Algorithm

In this section we give the algorithm in slightly more detail; it is summarized in Algorithm 1. As stated in the introduction, we start with an infeasible solution $F = \emptyset$. In each iteration, we first look for the cheapest *good* edge e with respect to F; a good edge with respect to F has endpoints in two different components C_1 and C_2 of (V, F) such that $h(C_1) = h(C_2) = 1$ and $h(C_1 \cup C_2) = 0$; we then look for the cheapest edge e' with respect to F that has endpoints in two different components C_1 and C_2 such that $\max(h(C_1), h(C_2)) = 1$. If the cost of the good edge e is less than twice the cost of the edge e', we add e to F, otherwise we add e' to F. We continue until we have a feasible solution. Note that if the step of considering good edges is removed, then we have the previous 2-approximation algorithm of Goemans and Williamson.

3   Analysis of the Algorithm

3.1   Some Preliminaries

To give the analysis of the algorithm, we implicitly have the algorithm construct a dual solution y as it runs. While we call y a dual solution, this is a bit misleading; unlike typical primal-dual style analyses, we are not constructing a feasible dual solution to a linear programming relaxation of the problem. However, we will guarantee that it is feasible for a particular set of constraints, which we now describe. Suppose we take the original undirected graph $G = (V, E)$ and create a mixed graph $G_m = (V, E \cup A)$ by *bidirecting* every edge; that is, for each edge $e = (u, v) \in E$ of cost $c(e)$ we create two arcs $a = (u, v)$ and $a' = (v, u)$ of cost $c(a) = c(a') = c(e)$ and add them to the original set of undirected edges. Let $\delta^+(S)$ be the set of arcs of G_m whose tails are in S and whose heads are not in S; note that no undirected edge is in $\delta^+(S)$. Then the dual solution we construct will be feasible for the constraints

$$\sum_{S: a \in \delta^+(S)} y(S) \leq c(a) \qquad \text{for all } a \in A. \tag{1}$$

$F \leftarrow \emptyset$;
// Set dual solution $y(S) \leftarrow 0$ for all sets S, $t \leftarrow 0$
while F is not a feasible solution **do**
    Let e be the cheapest (good) edge joining two components C_1, C_2 of F with
        $h(C_1) = h(C_2) = 1$ and $h(C_1 \cup C_2) = 0$ (if such an edge exists);
    Let e' be the cheapest edge joining two components C_1, C_2 of F such that
        $\max(h(C_1), h(C_2)) = 1$;
    if e exists and $c(e) \leq 2c(e')$ **then**
        // Implicitly increase $y(C)$ by $h(C) \max(0, \frac{1}{2}c(e) - t)$ for all
        connected components C, t increases by $\max(0, \frac{1}{2}c(e) - t)$
        $F \leftarrow F \cup \{e\}$;
    else
        // Implicitly increase $y(C)$ by $h(C)(c(e') - t)$ for all connected
        components C, t increases by $c(e') - t$
        $F \leftarrow F \cup \{e'\}$;
  Return F;

Algorithm 1. The $\frac{3}{2}$-approximation algorithm.

Note that the constraints are only over the directed arcs of the mixed graph and not the undirected edges. We say that a constraint is *tight* for some arc a if $\sum_{S:a \in \delta^+(S)} y(S) = c(a)$. We will sometimes simply say that the arc a is tight.

As given in the comments of Algorithm 1, initially $y(S) \leftarrow 0$ for all $S \subseteq V$. We keep track of a quantity t, initially zero; we call t the *time*. If we add a good edge e joining two components, we increase $y(C)$ by $h(C)(\max(0, \frac{1}{2}c(e) - t))$ for all current connected components of F (before adding edge e) and update t to $\max(t, \frac{1}{2}c(e))$. If we add a nongood edge e' joining two components, we increase $y(C)$ by $h(C)(c(e') - t)$ for all connected components C of F (before adding edge e') and update t to $c(e')$.

For a component C at some iteration of the algorithm, we say C is *active* if $h(C) = 1$ and inactive otherwise. We observe that because the algorithm repeatedly merges components, and because the function h is downwards monotone, any vertex u in an inactive component C is never again part of an active component; this follows since at later iterations $u \in C'$ implies $C \subseteq C'$, and since $h(C) = 0$ it must be the case that $h(C') = 0$ also.

Lemma 1. *Consider any component C at the start of an iteration of the main loop of the algorithm. For any arc (u,v) with $u \in C$, $v \notin C$, and C active, $\sum_{S:(u,v) \in \delta^+(S)} y(S) = t$ at the start of the iteration.*

Lemma 2. *At the end of the algorithm, the dual solution y is feasible for the constraints (1).*

Corollary 1. *For any nongood edge $e' = (u,v)$ added by the algorithm, if $u \in C$ such that $h(C) = 1$, then the arc (u,v) is tight. For any good edge e added by the algorithm, $\sum_{S:(u,v) \in \delta^+(S)} y(S) + \sum_{S:(v,u) \in \delta^+(S)} y(S) \geq c(e)$.*

We now need to show that the cost of the algorithm's solution F is at most the sum of the dual variables. We will prove this on a component-by-component basis. To do this it is useful to think about directing most of the edges of the component. We now explain how we direct edges.

Consider component C of F; overloading notation, let C stand for both the set of vertices and the set of edges of the component. At the start of the algorithm, each vertex of C is in its own component, and these components are repeatedly merged until C is a component in the algorithm's solution F; at any iteration of the algorithm call the connected components whose vertices are subsets of C the *subcomponents* of C. We say that a subcomponent has *h-value* of 0 (1, respectively) if for the set of vertices S of the subcomponent $h(S) = 0$ ($h(S) = 1$ respectively). We claim that at any iteration there can be at most one subcomponent of h-value 0. Note that the algorithm never merges components both of h-value 0, and any merging of two components, one of which has h-value 1 and the other 0, must result in a component of h-value 0 by the properties of h. So if there were in any iteration two subcomponents of C both with h-value 0, the algorithm would not in any future iteration add an edge connecting the vertices in the two subcomponents, which contradicts the connectivity of C. It also follows from this reasoning that at some iteration (perhaps initially) the first subcomponent appears with h-value 0, and there is a single such subcomponent from then on. If there is a vertex $v \in C$ with $h(\{v\}) = 0$, then there is such a subcomponent initially; otherwise, such a subcomponent is formed by adding a good edge e^* to merge two subcomponents of h-value 1. In the first case, we consider directing all the edges in the component towards v and we call v the root of C. In the second case, we think about directing all the edges to the two endpoints of e^* and we call these vertices the roots of C; the edge e^* remains undirected. We say that directing the edges of the component in this way makes the component *properly birooted*; let the corresponding set of arcs (plus perhaps the one undirected edge) be denoted $\vec{C}$.

Lemma 3. *At the end of the algorithm* $\sum_{e \in F} c(e) \leq \sum_S y(S)$.

3.2   Proof of the Performance Guarantee

We use a dual fitting argument to show the performance guarantee. For a feasible solution F^*, let C^* be some connected component. We say that $C^* \in \delta(S)$ for a subset S of vertices if there is some edge e in C^* such that $e \in \delta(S)$. We will show the following lemma.

Lemma 4. *Let F^* be a feasible solution to the problem, and let C^* be any component of F^*. Let $y(S)$ be the dual variables returned by the algorithm. Then*

$$\sum_{S:C^* \in \delta(S)} y(S) \leq \frac{3}{2} \sum_{e \in C^*} c(e).$$

From the lemma, we can easily derive the performance guarantee.

Theorem 1. *Algorithm 1 is a $\frac{3}{2}$-approximation algorithm.*

Proof. Let F^* be an optimal solution to the problem, and let $\mathcal{C}^*$ be its connected components. Then

$$\frac{3}{2} \sum_{e \in F^*} c(e) = \frac{3}{2} \sum_{C^* \in \mathcal{C}^*} \sum_{e \in C^*} c(e) \geq \sum_{C^* \in \mathcal{C}^*} \sum_{S:C^* \in \delta(S)} y(S),$$

where the last inequality follows by Lemma 4. Let F be the solution returned by the algorithm. By Lemma 3, we have that

$$\sum_{e \in F} c(e) \leq \sum_{S} y(S).$$

Since we only increased variables $y(S)$ for subsets S with $h(S) = 1$, it is clear that if $y(S) > 0$, then there must exist some $C^* \in \mathcal{C}^*$ with $C^* \in \delta(S)$ in order for the solution to be feasible. Thus

$$\sum_{e \in F} c(e) \leq \sum_{S} y(S) \leq \sum_{C^* \in \mathcal{C}^*} \sum_{S:C^* \in \delta(S)} y(S) \leq \frac{3}{2} \sum_{e \in F^*} c(e).$$

$\square$

We now turn to proving Lemma 4. The essence of the analysis is that the feasibility of the dual solution shows that the sum of most of the duals is no greater than the cost of all but one edge e in the component (the edge e that gives the birooted property). We then need to account for the duals that intersect this edge e or that contain both of its endpoints. If the sum of these duals is sufficiently small, then Lemma 4 follows. If the sum of these duals is not small, then we can show that there must be another edge e' in the component that has a large cost, and we can charge these duals to the cost of e'.

As a warmup to our techniques for proving Lemma 4, we prove a simple special case of the lemma by orienting the arcs of the component and using the dual feasibility (1).

Lemma 5. *Given any connected component C^* of a feasible solution F^*, if there is a vertex $v \in C^*$ such that $h(\{v\}) = 0$, then $\sum_{S:C^* \in \delta(S)} y(S) \leq \sum_{e \in C^*} c(e)$.*

Proof. If there is a vertex $v \in C^*$ such that $h(\{v\}) = 0$, we consider a directed version of C^* which we call $\vec{C}^*$ in which all the edges are directed towards v. Because h is downwards monotone and $h(\{v\}) = 0$, any set S containing v has $h(S) = 0$ and therefore $y(S) = 0$ since we only increase y on sets S' with $h(S') = 1$. We say that $\vec{C}^* \in \delta^+(S)$ if there is some arc of $\vec{C}^*$ with a tail in S and head not in S. Then by the previous discussion $\sum_{S:C^* \in \delta(S)} y(S) = \sum_{S:\vec{C}^* \in \delta^+(S)} y(S)$.

Then by (1),

$$\sum_{e\in C^*} c(e) = \sum_{a\in \vec{C}^*} c(a) \geq \sum_{a\in \vec{C}^*} \sum_{S:a\in\delta^+(S)} y(S)$$

$$= \sum_{S:\vec{C}^*\in\delta^+(S)} \sum_{a\in \vec{C}^*:a\in\delta^+(S)} y(S)$$

$$\geq \sum_{S:\vec{C}^*\in\delta^+(S)} y(S) = \sum_{S:C^*\in\delta(S)} y(S).$$

$\square$

We would like to prove something similar in the general case; however, if we do not have a vertex v with $h(\{v\}) = 0$, then there might not be any orientation of the arcs such that if $C^* \in \delta(S)$ and $y(S) > 0$ then $\vec{C}^* \in \delta^+(S)$, as there is in the previous case. Instead, we will use a birooted component as we did previously, and argue that we can make the lemma hold for this case. To orient the arcs of C^*, we order the edges of C^* by increasing cost, and consider adding them one-by-one, repeatedly merging subcomponents of C^*. The first edge that merges two subcomponents of h-value 1 to a subcomponent of h-value 0 we call the undirected edge of C^*, and we designate it $e^* = (u^*, v^*)$. We now let $\vec{C}^*$ be the orientation of all the edges of C^* except e^* towards the two roots u^*, v^*; e^* remains undirected. Let

$$t(u^*) = \sum_{S:u\in S, v\notin S} y(S),$$

$$t(v^*) = \sum_{S:u\notin S, v\in S} y(S), \text{ and}$$

$$t(e^*) = \sum_{S:u,v\in S, C^*\in\delta(S)} y(S).$$

We begin with an observation and a few lemmas. Throughout what follows we will use the quantity t as an index for the algorithm. When we say "at time t", we mean the last iteration of the algorithm at which the time is t (since the algorithm may have several iterations where t is unchanged). For a set of edges X, let $\max(X) = \max_{e\in X} c(e)$.

Observation 2. *At time t, the algorithm has considered adding all edges of cost at most t to the solution and any active component must consist of edges of cost at most t. At time t, if component C is active, then any edge $e \in \delta(C)$ must have cost greater than t.*

Lemma 6. $\max\{t(u^*) + t(e^*), t(v^*) + t(e^*)\} \leq c(e^*)$.

Proof. Since $\vec{C}^*$ is birooted, there is some component $X \subseteq C^*$ that contains u^* and v^* with $\max(X) \leq c(e^*)$ and $h(X) = 0$. At time $c(e^*)$, u^* and v^* must be in inactive components (possibly the same component); if either u^* or v^* was not in an inactive component then, since $\max(X) \le c(e^*)$, the algorithm would have connected some superset of the vertices of X which, because $h(X) = 0$, is inactive. Thus at time $c(e^*)$ the algorithm is no longer increasing duals $y(S)$ for S containing either u^* or v^*, and thus the inequality of the lemma must be true.   □

We will make extensive use of the following lemma, which tells us that the dual growing procedure finds components of small maximum edge cost. Let C_u^t be the connected component constructed by the algorithm at time t that contains u (similarly, C_v^t).

Lemma 7. *Consider two vertices u and v and suppose the components C_u^t, C_v^t are active for all times $t < t'$ (they need not be disjoint). Then for any connected set of edges $C(u)$ in the original graph containing u (similarly $C(v)$) if $h(C(u) \cup C(v)) = 0$ while $h(C_u^t \cup C_v^t) = 1$ for all times $t < t'$, then $\max(C(u) \cup C(v)) \ge t'$.*

Proof. Consider the algorithm at any time $t < t'$. Since C_u^t and C_v^t are active, by Observation 2, any edge in $\delta(C_u^t)$ or $\delta(C_v^t)$ must have cost greater than t; also, any edge in the components must have cost at most t. If $\max(C(u) \cup C(v)) \le t$, it must be that the vertices of $C(u)$ are a subset of those of C_u^t, and similarly the vertices of $C(v)$ are a subset of those of C_v^t. This implies $C(u) \cup C(v) \subseteq C_u^t \cup C_v^t$, and since $h(\cdot)$ is downward monotone, $h(C(u) \cup C(v)) \ge h(C_u^t \cup C_v^t)$, which is a contradiction. So it must be the case that $\max(C(u) \cup C(v)) > t$. Since this is true for any time $t < t'$, it follows that $\max(C(u) \cup C(v)) \ge t'$.   □

We now split the proof into two cases, depending on whether $\min(t(u^*) + t(e^*), t(v^*) + t(e^*)) > \frac{1}{2}c(e^*)$ or not. We first suppose that it is true.

Lemma 8. *If $\min(t(u^*)+t(e^*), t(v^*)+t(e^*)) > \frac{1}{2}c(e^*)$, then C^* must have some edge e' other than e^* of cost $c(e') \ge \min(t(u^*) + t(e^*), t(v^*) + t(e^*))$.*

Proof. First assume that $t(e^*) > 0$; this implies that $t(u^*) = t(v^*)$ since u^* and v^* must be in active components until the point in time (at time $t(u^*) = t(v^*)$) when they are merged into a single component, which is then active until time $t(u^*)+t(e^*)$. Then for all times $t < \min(t(u^*)+t(e^*), t(v^*)+t(e^*))$, u^* and v^* are in active components $C_{u^*}^t$ and $C_{v^*}^t$, and for times t with $t(u^*) \le t < t(u^*)+t(e^*)$, in which u^* and v^* are in the same component, $h(C_{u^*}^t \cup C_{v^*}^t) = 1$. Let $C(u^*)$ be the component containing u^* in $C^* - e^*$ and $C(v^*)$ be the component containing v^* in $C^* - e^*$. Since $h(C(u^*) \cup C(v^*)) = h(C^*) = 0$, we can apply Lemma 7, and the lemma follows.

Now assume that $t(e^*) = 0$. In this case, u^* and v^* are never in an active component together for any positive length of time (since $y(S) = 0$ for any S containing both u^* and v^*). Assume $t(u^*) = \min(t(u^*) + t(e^*), t(v^*) + t(e^*)) = \min(t(u^*), t(v^*)) > \frac{1}{2}c(e^*)$. For all $t < t(u^*)$, $C_{u^*}^t$ and $C_{v^*}^t$ are active components. Furthermore, since the algorithm did not add edge e^* during time $\frac{1}{2}c(e^*) \le t < t(u^*)$, it must have been the case that e^* was not a good edge during that period of time: otherwise e^* would have been cheaper than whatever edge(s) the algorithm was adding in that period of time. Since both $C_{u^*}^t$ and $C_{v^*}^t$ are active components for $t < t(u^*)$, if e^* was not good, it must have been the case that $h(C_{u^*}^t \cup C_{v^*}^t) = 1$. Thus we can apply Lemma 7: again, let $C(u^*)$ be the component containing u^* in $C^* - e^*$ and $C(v^*)$ be the component containing v^* in $C^* - e^*$. We know that $h(C(u^*) \cup C(v^*)) = 0$ since $h(C^*) = 0$. Thus there must be an edge e' in $C(u^*) \cup C(v^*)$ of cost at least $t(u^*)$.   □

Finally, we can prove Lemma 4.

Proof of Lemma 4:   If there is $v \in C^*$ with $h(\{v\}) = 0$, then the statement follows from Lemma 5. If not, then we can get a directed, birooted version $\vec{C}^*$ of C^* with undirected edge $e^* = (u^*, v^*)$. Without loss of generality, suppose that $t(u^*) + t(e^*) = \min(t(u^*) + t(e^*), t(v^*) + t(e^*))$. If $t(u^*) + t(e^*) \leq \frac{1}{2}c(e^*)$, then

$$\sum_{S:C^* \in \delta(S)} y(S) \leq \sum_{S:\vec{C}^* \in \delta^+(S)} y(S) + \sum_{S:u^* \in S, v^* \notin S} y(S) + \sum_{S:u^* \notin S, v^* \in S} y(S)$$

$$+ \sum_{S:u^*, v^* \in S} y(S),$$

$$\leq \sum_{a \in \vec{C}^*} \sum_{S:a \in \delta^+(S)} y(S) + t(u^*) + t(v^*) + t(e^*),$$

$$\leq \sum_{a \in \vec{C}^*} c(a) + t(u^*) + t(e^*) + t(v^*) + t(e^*),$$

$$\leq \sum_{a \in \vec{C}^*} c(a) + \frac{1}{2}c(e^*) + c(e^*),$$

$$\leq \frac{3}{2} \sum_{e \in C^*} c(e^*).$$

where the penultimate inequality follows by our assumption and by Lemma 6. If $t(u^*) + t(e^*) > \frac{1}{2}c(e^*)$, then by Lemma 8, there is an edge $e' \in C^*$, $e' \neq e^*$, of cost at least $t(u^*) + t(e^*)$. Let a' be the directed version of e' in $\vec{C}^*$. Since $c(a') \geq t(u^*) + t(e^*)$ and by Lemma 6 $c(e^*) \geq t(v^*) + t(e^*) \geq t(u^*) + t(e^*)$, then $t(u^*) + t(e^*) \leq \frac{1}{2}(c(a') + c(e^*))$. Then following the inequalities above, we have

$$\sum_{S:C^* \in \delta(S)} y(S) \leq \sum_{a \in \vec{C}^*} c(a) + t(u^*) + t(e^*) + t(v^*) + t(e^*)$$

$$= \sum_{a \in \vec{C}^*, a \neq a'} c(a) + c(a') + \frac{1}{2}(c(a') + c(e^*)) + c(e^*)$$

$$= \sum_{a \in \vec{C}^*, a \neq a'} c(a) + \frac{3}{2}c(a') + \frac{3}{2}c(e^*)$$

$$\leq \frac{3}{2} \sum_{e \in C^*} c(e),$$

and we are done.   □

4   Conclusion

As we have already mentioned, it would be very interesting to extend this algorithm from the class of downwards monotone functions to the class of proper functions defined by Goemans and Williamson [4]. A function $f : 2^V \to \{0, 1\}$ is *proper* if $f(S) = f(V - S)$ for all $S \subseteq V$ and $f(A \cup B) \leq \max(f(A), f(B))$ for all disjoint $A, B \subset V$. This class includes problems such as Steiner tree and generalized Steiner tree; currently no ρ-approximation algorithm for constant $\rho < 2$ is known for the latter problem. However, our algorithm makes extensive use of the property that we grow duals around active components until they are inactive, then no further dual is grown around that component; whereas the algorithm of [4] may start growing duals around components that were previously inactive. The first step in extending the algorithm of this paper to proper functions might well be to consider the Steiner tree problem, since the dual-growing algorithm of [4] in this case does have the property that once a component becomes inactive, no further dual is grown around it.

Acknowledgements. We thank the anonymous reviewers for useful comments.

References

1. Bazgan, C., Couëtoux, B., Tuza, Z.: Complexity and approximation of the Constrained Forest problem. Theoretical Computer Science 412, 4081–4091 (2011)
2. Couëtoux, B.: A $\frac{3}{2}$ Approximation for a Constrained Forest Problem. In: Demetrescu, C., Halldórsson, M.M. (eds.) ESA 2011. LNCS, vol. 6942, pp. 652–663. Springer, Heidelberg (2011)
3. Goemans, M.X., Williamson, D.P.: Approximating minimum-cost graph problems with spanning tree edges. Operations Research Letters 16, 183–189 (1994)
4. Goemans, M.X., Williamson, D.P.: A general approximation technique for constrained forest problems. SIAM Journal on Computing 24, 296–317 (1995)
5. Goemans, M.X., Williamson, D.P.: The primal-dual method for approximation algorithms and its application to network design problems. In: Hochbaum, D.S. (ed.) Approximation Algorithms for NP-hard Problems, ch. 4. PWS Publishing, Boston (1996)
6. Imielińska, C., Kalantari, B., Khachiyan, L.: A greedy heuristic for a minimum-weight forest problem. Operations Research Letters 14, 65–71 (1993)
7. Kruskal, J.: On the shortest spanning subtree of a graph and the traveling salesman problem. Proceedings of the American Mathematical Society 7, 48–50 (1956)
8. Laszlo, M., Mukherjee, S.: Another greedy heuristic for the constrained forest problem. Operations Research Letters 33, 629–633 (2005)
9. Laszlo, M., Mukherjee, S.: A class of heuristics for the constrained forest problem. Discrete Applied Mathematics 154, 6–14 (2006)
10. Laszlo, M., Mukherjee, S.: An approximation algorithm for network design problems with downwards-monotone demand functions. Optimization Letters 2, 171–175 (2008)

Kinetic Compressed Quadtrees in the Black-Box Model with Applications to Collision Detection for Low-Density Scenes*

Mark de Berg, Marcel Roeloffzen, and Bettina Speckmann

Dept. of Computer Science, TU Eindhoven, The Netherlands
{mdberg,mroeloff,speckman}@win.tue.nl

Abstract. We present an efficient method for maintaining a compressed quadtree for a set of moving points in $\mathbb{R}^d$. Our method works in the *black-box KDS model*, where we receive the locations of the points at regular time steps and we know a bound $d_{\max}$ on the maximum displacement of any point within one time step. When the number of points within any ball of radius $d_{\max}$ is at most k at any time, then our update algorithm runs in $O(n \log k)$ time. We generalize this result to constant-complexity moving objects in $\mathbb{R}^d$. The compressed quadtree we maintain has size $O(n)$; under similar conditions as for the case of moving points it can be maintained in $O(n \log \lambda)$ time per time step, where λ is the density of the set of objects. The compressed quadtree can be used to perform broad-phase collision detection for moving objects; it will report in $O((\lambda + k)n)$ time a superset of all intersecting pairs of objects.

1 Introduction

(Kinetic) Collision Detection. Collision detection [17,19] is an important problem in computer graphics, robotics, and N-body simulations. One is given a set S of n objects, some or all of which are moving, and the task is to detect the collisions that occur. The most common way to perform collision detection is to use *time-slicing* and test for collisions at regular time steps; for graphics applications this is typically every frame. This approach can be wasteful, in particular if computations are performed from scratch every time: if the objects moved only a little, then much of the computation may be unnecessary. An alternative is to use the *kinetic-data-structure (KDS) framework* introduced by Basch *et al.* [4]. A KDS for collision detection maintains a collection of certificates (elementary geometric tests) such that there is no collision as long as the certificates remain true. The failure times of the certificates—these can be computed from the motion equations of the objects—are stored in an event queue. When the next event happens, it is checked whether there is a real collision and the set of certificates and the event queue are updated. (In addition, if there is a collision the motion equations of the objects involved are changed based on the collision response.)

---

* M. Roeloffzen and B. Speckmann were supported by the Netherlands' Organisation for Scientific Research (NWO) under project no. 600.065.120 and 639.022.707, respectively.

L. Epstein and P. Ferragina (Eds.): ESA 2012, LNCS 7501, pp. 383–394, 2012.

KDSs for collision detection have been proposed for 2D collision detection among polygonal objects [2,15], for 3D collision detection among spheres [14], and for 3D collision detection among fat convex objects [1].

The KDS framework is elegant and can lead to efficient algorithms, but it has its drawbacks. One is that it requires knowledge of the exact trajectories (motion equations) to compute when certificates fail. Such knowledge is not always available. Another disadvantage is that some KDSs are complicated and may not be efficient in practice—the collision-detection KDS for fat objects [1] is an example. We therefore study collision detection in the more practical *black-box model* [9,12], where we receive new locations for the objects of S at regular time steps $t = 1, 2, \ldots$ This brings us back to the time-slicing approach. The goal is now to use *temporal coherence* to speed up the computations—that is, not to perform the computations from scratch at every time step—and to *prove* under which conditions this leads to an efficient solution.

Following a previous paper [9] in which we studied Delaunay triangulations and convex hulls in the black-box model we make the following[1] assumption. Let $A(t)$ denote the location of $A \in S$ at time t and let $S(t) = \{A(t) \mid A \in S\}$.

Displacement Assumption:

- *There is a maximum displacement $d_{\max}$ such that for each object $A \in S$ and any time step t we have* $\operatorname{dist}(A(t), A(t+1)) \leqslant d_{\max}$.
- *Any ball of radius $d_{\max}$ intersects at most $k \ll n$ objects from $S(t)$, at any time step t.*

In the above, $\operatorname{dist}(A(t), A(t+1))$ denotes the distance between $A(t)$ and $A(t+1)$. When the objects are points this simply refers to the Euclidean distance between them. For non-point objects the distance is defined in Section 3.

The Displacement Assumption bounds how far an object can move in one time step, relative to the distances between the objects. In practise one would expect that the time intervals are such that $k = O(1)$. Note that the parameter k is not known to the algorithm, it is used only in the analysis.

Collision Detection and Spatial Decompositions. In practice collision detection is often performed in two phases: a *broad phase* that serves as a filter and reports a (hopefully small) set of potentially colliding pairs of objects—this set should include all pairs that actually collide—and a *narrow phase* that tests each of these pairs to determine if there is indeed a collision. Our paper is concerned with broad-phase collision detection; more information on the narrow phase can be found in a survey by Kockara *et al.* [16].

A natural way to perform broad-phase collision detection is to use a decomposition of the space into cells [10,18]. One then reports, for each cell in the decomposition, all pairs of objects intersecting the cell. This is also the approach that we take. For this to be efficient, one needs a decomposition with few cells such that each cell is intersected by few objects. In general this is impossible for

---

[1] The Displacement Assumption was phrased slightly differently in [9], but it is essentially the same.

convex cells: Chazelle [11] showed that there are sets of disjoint objects in $\mathbb{R}^d$ such that any convex decomposition in which each cell intersects $O(1)$ objects must consist of $\Omega(n^2)$ cells. However, if we take the *density* λ [6,8]—see Section 3—into account then a decomposition with a linear number of cells, each intersecting $O(\lambda)$ objects, exists and can be constructed in $O(n \log n)$ time [5]. (Practical applications typically have $\lambda \ll n$; in fact many papers assume that $\lambda = O(1)$.) Our main goal is thus to maintain a decomposition with these properties in $o(n \log n)$ time per time step under the Displacement Assumption.

Our Results. The known linear-size decomposition for low-density scenes [5] is a binary space partition (BSP), but that BSP seems difficult to maintain efficiently as the objects move. We therefore use a different decomposition, namely a compressed quadtree. In Section 2 we study the problem of maintaining a compressed quadtree for a set of moving points. We show that this can be done in $O(n \log k)$ time per time step for any (fixed) dimension d, where k is the parameter from the Displacement Assumption. (The dependency of our bounds on the dimension d is exponential, as is usually the case when quadtrees are used.) Note that $k \leqslant n$ and hence our result is never worse than recomputing the quadtree from scratch. Since hierarchical space decompositions such as (compressed) quadtrees have many uses, this result is of independent interest.

In Section 3 we turn our attention to compressed quadtrees for low-density scenes. It is known that a compressed quadtree on the bounding-box vertices of a set S of n objects in the plane has $O(n)$ cells that each intersect $O(\lambda)$ objects, where λ is the density of S [7]. We first prove that, similar to the planar case, a compressed quadtree on the bounding-box vertices of a set S of objects in $\mathbb{R}^d$ with density λ, has $O(n)$ cells each intersecting $O(\lambda)$ objects. We also show, using our result on compressed quadtrees for points, how to maintain a quadtree on a (well chosen) subset of $O(n/\lambda)$ bounding-box vertices such that each of the $O(n/\lambda)$ cells in the quadtree is intersected by $O(\lambda + k)$ objects. This quadtree can be maintained in $O(n \log \lambda)$ time per time step and can be used to report $O((\lambda + k)n)$ potentially overlapping pairs of objects. It is known that the density a set of disjoint fat objects is $O(1)$ [8]. Hence, our approach is particularly efficient for collision detection between fat objects.

2   Maintaining a Compressed Quadtree for Moving Points

A *quadtree subdivision* for a set P of n points in $\mathbb{R}^d$ is a recursive subdivision of an initial hypercube containing the points of P into 2^d equal-sized sub-hypercubes ("quadrants"). The subdivision process continues until a stopping criterion is met. We use a commonly employed criterion, namely that each final hypercube contains at most one point. A *quadtree* is a tree structure representing such a recursive subdivision. (In higher dimensions, especially in $\mathbb{R}^3$, the term *octree* is often used instead of quadtree. As our figures and examples will always be in the plane, we prefer to use the term quadtree.) A hypercube that can result from such a recursive partitioning of the initial hypercube is called a *canonical hypercube*.

In a quadtree subdivision there can be splits where only one of the resulting quadrants contains points. In the quadtree this corresponds to a node with only one non-empty child. There can even be many consecutive splits of this type, leading to a long path of nodes with only one non-empty child. As a result, a quadtree does not necessarily have linear size. A *compressed quadtree* replaces such paths by *compressed nodes*, which have two children: a child for the *hole* repre-

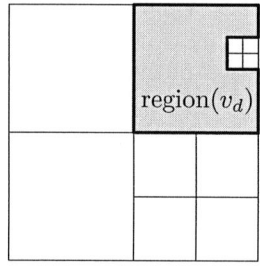

Fig. 1. A quadtree subdivion with a compressed node. The gray region denotes the region of the donut.

senting the smallest canonical hypercube containing all points, and a child for the *donut* representing the rest of the hypercube. Thus a donut is the set-theoretic difference of two hypercubes, one contained in the other. Note that the inner hypercube can share part of its boundary with the outer one; in the plane the donut then becomes a U-shape—see Fig. 1—or an L-shape. An internal node that is not compressed is called a *regular node*. In the following, we use region(v) to denote the region corresponding to a node v. For a node v whose region is a hypercube we use size(v) to denote the edge length of region(v); when v is a donut, size(v) refers to the edge length of its outer hypercube.

A compressed quadtree for a set of points has linear size and it can be constructed in $O(n \log n)$ time [3] in the appropriate model of computation [13, Chapter 2]. In this model we can compute the smallest canonical hypercube containing two points in $O(1)$ time. In this model and under the Displacement Assumption, our goal is to update the compressed quadtree in $O(n \log k)$ time.

Let $P(t)$ denote the set of points at time t, and let $\mathcal{T}(t)$ denote the compressed quadtree on $P(t)$. Our task is thus to compute $\mathcal{T}(t+1)$ from $\mathcal{T}(t)$. The root of $\mathcal{T}(t)$ represents a hypercube that contains all points of $P(t)$. We assume without loss of generality that both the size of the initial hypercube and $d_{\max}$ are powers of 2, so that there exists canonical hypercubes of size $d_{\max}$. As the points move, it can happen that the initial hypercube no longer contains all the points. When this happens, we enlarge the initial hypercube. The easy details are omitted here.

Since we assume that points move only a small distance—at most $d_{\max}$—we would like to maintain $\mathcal{T}$ by moving points to neighboring cells and then locally recomputing the quadtree. To this end we first prune the tree by removing all nodes whose region has size less than $d_{\max}$; moreover for compressed nodes u with size(u) $\geqslant d_{\max}$ but size(u_h) $< d_{\max}$ for the hole u_h of u, we replace u_h by a canonical hypercube of size $d_{\max}$ containing region(u_h). We use $\mathcal{T}_0$ to denote this reduced quadtree. Note that $\mathcal{T}_0$ is the compressed version of the quadtree for $P(t)$ that uses the following stopping criterion for the recursive subdivision process: a hypercube σ becomes a leaf when (i) σ contains at most one point from $P(t)$, or (ii) σ has edge length $d_{\max}$. Each leaf in $\mathcal{T}_0$ has a list of all points contained in its region. By the Displacement Assumption, this list contains $O(k)$ points. The reduced quadtree $\mathcal{T}_0$ can easily be computed from $\mathcal{T}(t)$ in $O(n)$ time.

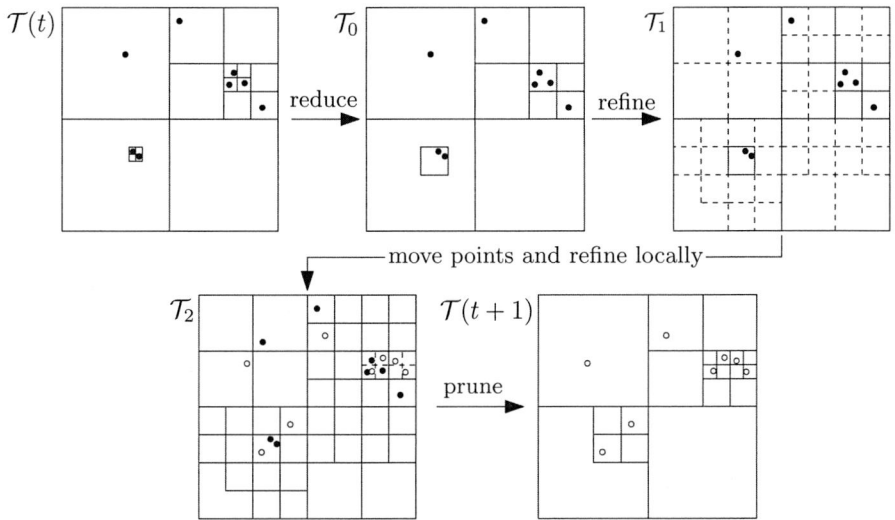

Fig. 2. The various intermediate quadtrees while updating $\mathcal{T}(t)$ to $\mathcal{T}(t+1)$. New edges in each step are illustrated by dashed edges, solid disks are points at time t and open disks are points at time $t+1$.

We do not lose any useful information in this step as within some region(v) with size(v) = $d_{\max}$ the points can move (nearly) arbitrarily and, hence, the subtree of u at time $t+1$ need not have any relation to the subtree of u at time t.

Now we would like to move the points from the leaf regions of $\mathcal{T}_0$ to neighboring leaf regions according to their new positions at time $t+1$, and then locally rebuild the compressed quadtree. The main problem in this approach is that a leaf region can border many other leaf regions and many points may move into it. Constructing the subtree replacing that leaf from scratch is then too expensive. We solve this by first building an intermediate tree $\mathcal{T}_1$, which is a suitable refinement of $\mathcal{T}_0$. We then move the points from $\mathcal{T}_0$ into $\mathcal{T}_1$—the definition of $\mathcal{T}_1$ guarantees that any node receives only $O(k)$ points—and we locally refine any nodes that contain more than one point, giving us another intermediary tree $\mathcal{T}_2$. Finally, we prune $\mathcal{T}_2$ to obtain $\mathcal{T}(t+1)$. An example of the different stages of updating a quadtree of eight points is shown in Fig. 2. The individual steps are described in more detail below.

Refining the Tree: Computing $\mathcal{T}_1$ from $\mathcal{T}_0$. The intermediary tree $\mathcal{T}_1$ has the property that each region in $\mathcal{T}_1$ has $O(1)$ neighbors in $\mathcal{T}_0$. To make this precise we define for a node v in $\mathcal{T}_1$ some related internal or leaf nodes in $\mathcal{T}_0$.

- original(v): The lowest node u in $\mathcal{T}_0$ such that region(v) $\subseteq$ region(u).
- $\mathcal{N}(v)$: The collection $\mathcal{N}(v)$ of *neighbors* of v consists of all nodes $w \in \mathcal{T}_0$ that are at least as large as v (this to avoid multiple neighbors on one "side" of region(v)) and are adjacent to region(v). So $\mathcal{N}(v)$ consists of all nodes $w \in \mathcal{T}_0$ such that:

(i) size(w) ⩾ size(v),

(ii) the regions region(w) and region(v) touch (that is, their boundaries intersect but their interiors are disjoint), and

(iii) w does not have a proper descendant w' with properties (i) and (ii) (so that w is a lowest node with properties (i) and (ii)).

Note that for every node in $\mathcal{T}_1$ we have $|\mathcal{N}(v)| \leqslant 3^d - 1$, hence $|\mathcal{N}(v)| = O(1)$ for any fixed dimension. We next modify $\mathcal{N}(v)$ and original(v) slightly, as follows. Any node $w_d \in \mathcal{N}(v) \cup \{\text{original}(v)\}$ that is a donut with a hole w_h that touches the boundary of region(v) and with size(w_h) < size(v) is replaced by its parent (which is a compressed node). The set $\mathcal{N}(v)$ and original(v) will be used in our algorithm to determine if region(v) needs to be split—this is the case when it has a neighbor that is not a leaf or when original(v) is not a leaf—and later we need $\mathcal{N}(v)$ to know from which regions points can move into region(v).

It is important to remember that original(v) and the nodes in $\mathcal{N}(v)$ are nodes in $\mathcal{T}_0$, while v is a node in $\mathcal{T}_1$. Thus the neighbors never get refined which is needed to ensure that $\mathcal{T}_1$ has linear size. We now express the conditions on $\mathcal{T}_1$:

(i) For every node u in $\mathcal{T}_0$ such that region(u) is a hypercube, there is a node v in $\mathcal{T}_1$ with region(v) = region(u). Thus, the subdivision induced by $\mathcal{T}_1$ is a refinement of the subdivision induced by $\mathcal{T}_0$.

(ii) For each leaf v in $\mathcal{T}_1$, every node $u \in \mathcal{N}(v)$ is a leaf of $\mathcal{T}_0$.

We construct $\mathcal{T}_1$ top-down. Whenever we create a new node v of $\mathcal{T}_1$, we also store pointers from v to original(v) and to the nodes in $\mathcal{N}(v)$. (This is not explicitly stated in Algorithm 1 below.) We obtain these pointers from the parent of v and the original and neighbors of that parent in $O(1)$ time. How we refine each node v in $\mathcal{T}_1$ depends on original(v) and the neighbors in $\mathcal{N}(v)$. We distinguish three cases: in Case 1 at least one of the neighbors or original(v) is a regular node, in Case 2 no neighbor or original(v) is a regular node and at least one is a compressed node, and in Case 3 all neighbors as well as original(v) are leaf nodes. How we refine v in Case 2 depends on the holes of the compressed nodes in $\mathcal{N}(v) \cup \{\text{original}(v)\}$. To determine how to refine v we use a set H_v which is defined as follows. Let R be a hypercube and f a facet of any dimension of R. Let mirror$_f(R)$ denote a hypercube R^* for which size(R^*) = size(R) and $R^* \cap R = f$. We say that mirror$_f(R)$ is a mirrored version of R along f. The set H_v^* then consists of the mirrored hypercubes of every hole w_h of a compressed node in $\mathcal{N}(v)$ as well as the holes themselves. The set H_v is then defined as

$$H_v = \{R \mid R \in H_v^* \text{ and } R \subset \text{region}(v)\}.$$

Note that H_v is never empty, because a compressed node in $\mathcal{N}(v)$ must have a hole smaller than region(v) along the boundary of region(v). The different cases are described in more detail in Algorithm 1 and illustrated in Fig. 3.

Lemma 1. *The tree $\mathcal{T}_1$ created by Algorithm 1 is a compressed quadtree with properties (i) and (ii) stated above.*

Algorithm 1. REFINE($\mathcal{T}_0$)

---

1 Create a root node v for $\mathcal{T}_1$ with original(v) = root($\mathcal{T}_0$) and $\mathcal{N}(v) = \emptyset$.
2 Initialize a queue Q and add v to Q.
3 **while** Q *is not empty* **do**
4     $v \leftarrow$ pop(Q)
5   ***Case 1:*** original(v) *or a neighbor in* $\mathcal{N}(v)$ *is a regular node.*
6   Make v into a regular node, create 2^d children for it and add these to Q.
7   ***Case 2:*** *Case 1 does not apply, and* original(v) *is a compressed node or a neighbor* $w \in \mathcal{N}(v)$ *is a compressed node.*
8        $\sigma \leftarrow$ the smallest canonical hypercube containing the regions of H_v.
9   ***Case 2a:*** $\sigma =$ region(v).
10   Make v into a regular node, create 2^d children for it and add these to Q.
11   ***Case 2b:*** $\sigma \subset$ region(v).
12   Make v into a compressed node by creating two children for it, one child v_h corresponding to the hole in region(v) and one child v_d corresponding to the donut region, with region(v_h) = σ and region(v_d) = region(v) \ σ. Add v_h to Q.
13   ***Case 3:*** *Cases 1 and 2 do not apply;* original(v) *and every neighbor* $w \in \mathcal{N}(v)$ *are leaves in* $\mathcal{T}_0$.
14   The node v remains a leaf.

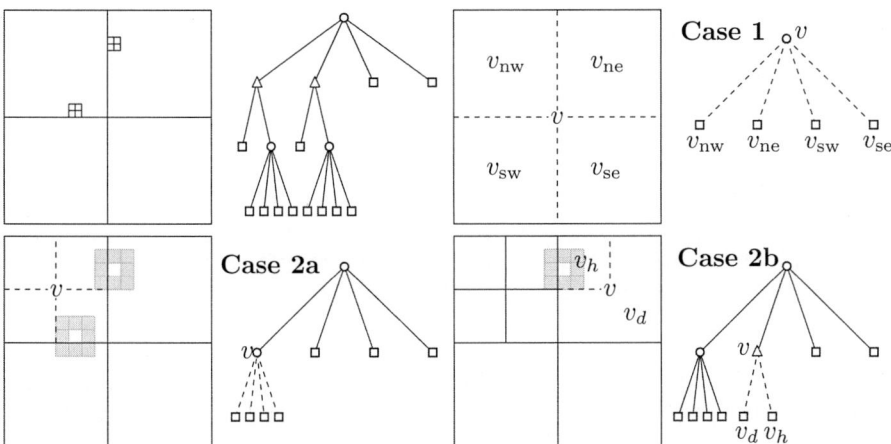

Fig. 3. Several steps from Algorithm 1. The tree on the top left is $\mathcal{T}_0$, the others illustrate consecutive steps in the construction of $\mathcal{T}_1$. The gray squares are the mirrored versions of the holes as used in Case 2.

Proof. First note that Algorithm 1 initially constructs a quadtree consisting of a single leaf. In the subsequent steps of the algorithm a leaf may be replaced by a compressed node or regular node, hence $\mathcal{T}_1$ is also a valid compressed quadtree.

For property (ii) it is sufficient to note that only in Case 3, where by definition the neighbors are leaves, the node v remains a leaf.

To prove property (i), suppose for a contradiction that there is a node u in $\mathcal{T}_0$ for which region(u) is a hypercube and there is no node v in $\mathcal{T}_1$ with region$(v) =$ region(u). Let v' be the lowest node in $\mathcal{T}_1$ such that region$(v') \supset$ region(u). Then original(v') is an ancestor of u. Hence, when v' is handled by Algorithm 1, either Case 1, 2a or 2b applies. Thus v' will be subdivided such that one of its children is a hypercube containing region(u), contradicting the fact that region(v') is the lowest node v' such that region$(v') \supset$ region(u).   □

Note that Algorithm 1 must terminate, since every node in $\mathcal{T}_1$ is treated at most once and $\mathcal{T}_1$ will never contain nodes whose region is smaller than the smallest node in $\mathcal{T}_0$. We can actually prove the following stronger statement.

Lemma 2. *The tree $\mathcal{T}_1$ contains $O(n)$ nodes and can be constructed in $O(n)$ time, where n is the number of points in P.*

Moving the Points: Computing $\mathcal{T}_2$ from $\mathcal{T}_1$. For a node v we define $P_t(v)$ as the set of points in P contained in region(v) at time t. For each leaf v in $\mathcal{T}_1$ we find $P_{t+1}(v)$ by inspecting all nodes $w \in \{\text{original}(v)\} \cup \mathcal{N}(v)$—these nodes must be leaves of $\mathcal{T}_0$. Recall that $\mathcal{N}(v)$ contains a neighbor for every j-facet of region(v) $(0 \leqslant j < d)$, and that these neighbors have size at least $d_{\max}$. This implies that the regions region(w) for $w \in \{\text{original}(v)\} \cup \mathcal{N}(v)$, contain all the points that can possibly move into region(v). For each such node w we then check for all points in $P_t(w)$ if they are inside region(v) at time $t + 1$; if so we add them to $P_{t+1}(v)$. Using a charging scheme we can show that this takes $O(n)$ time in total. Afterwards some cells may contain more than one point and we compute the quadtree for these cells from scratch. Since every cell contains $O(k)$ points this can be done in $O(n \log k)$ time in total.

As a last step we prune the tree. We replace regular nodes in which only one child contains points by compressed nodes and merge nested compressed nodes. This is straightforward to do in linear time by traversing the tree in a bottom-up manner. The following theorem summarizes the result from this section.

Theorem 1. *Let P be a set of n moving points in $\mathbb{R}^d$. Under the Displacement Assumption, we can maintain a compressed quadtree for P in $O(n \log k)$ time per time step, where k is the parameter in the Displacement Assumption.*

Note that the $\log k$ factor comes purely from computing subtrees within a cell of size $d_{\max}$. If we are satisfied with maintaining $\mathcal{T}_0$, in which cells of size $d_{\max}$ are not subdivided, then updating takes only $O(n)$ time per time step.

3   Maintaining a Compressed Quadtree for Moving Objects

Let S be a set of n moving objects in $\mathbb{R}^d$. In this section we show how to maintain a compressed quadtree for S, based on our results from the previous section. We assume that each object $A \in S$ has constant complexity, so that we can test in $O(1)$ time whether A intersects a given region (hypercube or donut) in a quadtree, and we can compute its axis-aligned bounding box $\mathrm{bb}(A)$ in $O(1)$ time. (For complex objects, we can either treat their constituent surface patches as the elementary objects, or we can work with a suitable simplification of the object.) Recall that the Displacement Assumption stipulates that $\mathrm{dist}(A(t), A(t+1)) \leqslant d_{\max}$ for any object $A \in S$ and any time t. For objects we interpret this as follows. If the objects only translate then the distance refers to the length of the translation vector. Otherwise, we require that the Hausdorff distance from $A(t+1)$ to $A(t)$ should not be more than $d_{\max}$—no point on $A(t+1)$ should be more than distance $d_{\max}$ from its nearest point in $A(t)$—and that no bounding-box vertex should move by more than $d_{\max}$. (Note that the bounding box is allowed to deform due to rotation or deformation of A.)

The idea of our approach is simple: We find a suitable subset $P^*(t)$—which is defined later—of the set $P(t)$ of $2^d n$ bounding-box vertices of the objects, such that every cell in the quadtree on $P^*(t)$ is intersected by a small number of objects. Our analysis will be done in terms of the so-called *density* of S, which is defined as follows. The density of a set S of objects is the smallest number λ such that any ball B intersects at most λ objects $A \in S$ with $\mathrm{diam}(A) \geqslant \mathrm{diam}(B)$ [6]. In the following, we assume that the density of $S(t)$ is at most λ, for any t. For $d = 2$ De Berg *et al.* [7] prove that each region in the subdivision induced by a compressed quadtree on the bounding-box vertices is intersected by $O(\lambda)$ objects. We extend this result to higher dimensions.

Lemma 3. *Let S be a set of n objects in $\mathbb{R}^d$, and let $\mathcal{T}$ be a compressed quadtree on the set of bounding-box vertices of S. Then $\mathrm{region}(v)$ intersects $O(\lambda)$ objects $A \in S$ for any leaf v of $\mathcal{T}$, where λ is the density of S.*

Proof (sketch). Assume v is a donut. (The case where v is a hypercube is easier.) It is known that any hypercube that does not contain any bounding-box vertices is intersected by $O(\lambda)$ objects [8]. Hence, it suffices to show that $\mathrm{region}(v)$ can always be covered by $O(1)$ hypercubes that lie inside the donut.

Let σ be the outer hypercube defining the donut $\mathrm{region}(v)$ and let σ_{in} be the inner hypercube (the hole). Our method to cover $\mathrm{region}(v)$ is as follows. A cell in a d-dimensional grid is surrounded by $3^d - 1$ other cells, that is, there are $3^d - 1$ other cells whose boundaries intersect the boundary of the given cell. We want to cover $\mathrm{region}(v) = \sigma \setminus \sigma_{\mathrm{in}}$ using at most $3^d - 1$ hypercubes that surround σ_{in} in a somewhat similar manner. Consider two canonical hypercubes $\sigma_A := [a_1 : a_1'] \times \cdots \times [a_d : a_d']$ and $\sigma_B := [b_1 : b_1'] \times \cdots \times [b_d : b_d']$ of the same size, and let $(\alpha_1, \ldots, \alpha_d) \in \{-, 0, +\}^d$. We say that σ_A is an $(\alpha_1, \ldots, \alpha_d)$-*neighbor* of σ_B if the following holds:

- if $\alpha_i = -$ then $a_i' = b_i$
- if $\alpha_i = 0$ then $[a_i : a_i'] = [b_i : b_i']$
- if $\alpha_i = +$ then $a_i = b_i'$

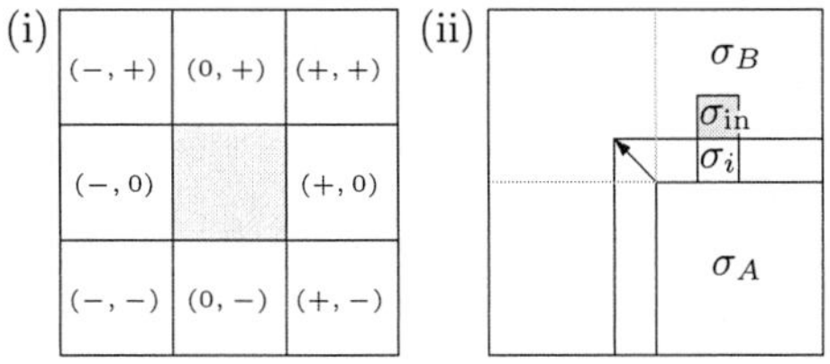

Fig. 4. (i) Labeling the neighbors of a cell. (ii) Growing σ_A until it contains σ_i.

In other words, α_i indicates the relative position of σ_A and σ_B in the i-th dimension. Fig. 4 illustrates this in the plane.

Now consider the donut $\sigma \setminus \sigma_{in}$ that we want to cover, and let j be such that $\operatorname{diam}(\sigma) = 2^j \cdot \operatorname{diam}(\sigma_{in})$. We will prove by induction on j that we can cover any donut region in a compressed quadtree by a collection C of hypercubes. The hypercubes in the constructed collection C will have a label from $\{-, 0, +\}^d$, where each label is used at most once—except $(0, \ldots, 0)$ which will not be used at all—and the following invariant is maintained:

> *Invariant:* If a hypercube $\sigma_i \in C$ has label $(\alpha_1, \ldots, \alpha_d)$, then there are canonical hypercubes σ_A and σ_B of the same size such that $\sigma_A \subset \sigma_i$ and $\sigma_{in} \subset \sigma_B$ and σ_A is an $(\alpha_1, \ldots, \alpha_d)$-neighbor of σ_B.

The invariant is maintained by stretching the hypercubes of octants that do not contain σ_{in} until they hit σ_{in}. These stretched hypercubes then replace the ones with the same labels in C. □

In the lemma above we consider a compressed quadtree on the set $P(t)$ of all bounding-box vertices. In our method we use a subset $P^*(t) \subset P(t)$ to define the compressed quadtree. Furthermore, to update the quadtree efficiently we cannot afford to refine the quadtree until each node contains only one point—even when the objects adhere to the Displacement Assumption, there can be $\Omega(n)$ bounding-box vertices in a hypercube of size d_{max}. Instead we use the same stopping criterion as we did for $\mathcal{T}_0$ for a set of points: a hypercube σ is a leaf in the compressed quadtree when (i) σ contains at most one point from $P^*(t)$, or (ii) σ has edge length d_{max}. Next we define $P^*(t)$.

We define the *Z-order* of $P(t)$ as the order in which the points are visited while doing an in-order treewalk on a non-compressed quadtree on $P(t)$ where for each interior node the children are visited in a specific order. (In two dimensions this order is: top left, top right, bottom left and bottom right.) For any set of n points such a Z-order can be computed in $O(n \log n)$ time [13] as long as a smallest canonical square of two points can be computed in $O(1)$ time. Ideally our subset $P^*(t)$ consists of every λ^{th} point in this Z-order, but for cells of size d_{max} (which can contain many points) we cannot afford to compute the exact Z-order. Instead we define an *approximate Z-order* as a Z-order where points within a cell of size d_{max} have an arbitrary ordering. The set $P^*(t)$ then consists of every λ^{th} point on such an approximate Z-order. This implies that $P^*(t)$ is not uniquely defined, but each choice of $P^*(t)$ leads to the same quadtree, since cells of size d_{max} are not refined further. In the rest of the section we use $\mathcal{T}^*(t)$ to denote the compressed quadtree on $P^*(t)$ with the above mentioned stopping criterion. Since $P^*(t)$ contains $O(n/\lambda)$ points, $\mathcal{T}^*(t)$ has $O(n/\lambda)$ nodes. Next we bound the number of objects that intersect a cell in $\mathcal{T}^*(t)$.

Lemma 4. *Every cell in the subdivision induced $\mathcal{T}^*(t)$ is intersected by $O(\lambda+k)$ objects from $S(t)$.*

Proof. Let σ be an arbitrary cell in the subdivision induced by $\mathcal{T}^*(t)$. If σ has edge length d_{max} then by the Displacement Assumption it intersects $O(k)$ objects. If σ is larger than d_{max} it can either be a hypercube or a donut. If σ is a hypercube, then by definition all points of $P(t)$ contained in σ must be consecutive in the approximate Z-order. It follows that σ contains at most 2λ points of $P(t)$. If σ is a donut, then the points of $P(t)$ contained in σ can be divided into two sets of consecutive points from the approximate Z-order. Since the donut does not contain any point of $P^*(t)$ it contains at most 2λ points of $P(t)$. In either case, by Lemma 3, σ is intersected by $O(\lambda)$ objects for which there is no bounding-box vertex in σ and, hence, by $O(\lambda)$ objects in total. □

We maintain $\mathcal{T}^*(t)$ in the same way as we did for points, but we also maintain for each cell a list of objects that intersect that cell and we have to update $P^*(t)$ at each time step. This leads to two additional steps in our algorithm.

We compute $P^*(t+1)$ just after we have moved the points of $P(t+1)$, but before we do any local refining. We obtain $P^*(t+1)$ by traversing $\mathcal{T}_1$ (in Z-order) and for each cell larger than $d_{\max}$ computing the local Z-order of the points of $P(t+1)$ that are contained in the cell. Since each such cell contains $O(\lambda)$ points this takes $O(n \log \lambda)$ time in total. We then locally refine each cell that contains more than one point of $P^*(t+1)$ until it adheres to the stopping criterion.

We also have to construct for each leaf v a list $S_{t+1}(v)$ of objects that intersect region(v) at time $t+1$. We compute $S_{t+1}(v)$ in the same way as we computed $P_{t+1}(v)$, by inspecting the objects that intersect neighbors and then testing which of these objects intersect region(v). To do this in $O(1)$ time per object we need the assumption that objects have constant complexity. Since $\mathcal{T}_2$ still has $O(n/\lambda)$ nodes we can do this in $O(n)$ time.

The resulting tree adheres to the stopping criterion with respect to $P^*(t+1)$ and each leaf has a list of objects that intersect it. We then prune the tree to obtain $\mathcal{T}^*(t+1)$. We can conclude with the following theorem.

Theorem 2. *Let S be a set of n moving objects in $\mathbb{R}^d$. Under the Displacement Assumption, we can maintain a compressed quadtree for S in $O(n \log \lambda)$ time per time step, such that each leaf region intersects $O(\lambda + k)$ objects, where λ is an upper bound on the density of S at any time and k is the parameter in the Displacement Assumption. The compressed quadtree has $O(n/\lambda)$ nodes, and the sum of the number of objects over all the leafs is $O(n)$.*

Notice that the parameter k does not show up in the update time, unlike for points. The reason for this is that we stop the construction when the size of a region drops below $d_{\max}$. The parameter k does however show up when we look at the number of potentially colliding objects. For each leaf v in $\mathcal{T}^*(t)$ we report all pairs of objects in $S_t(v)$. Observe that $|S_t(v)| = O(\lambda + m_v)$ where m_v is the number of bounding-box vertices in region(v). Summing $|S_t(v)|^2$ over all nodes v in $\mathcal{T}^*(t)$ we get $\sum_{v \in \mathcal{T}_0} |S_t(v)|^2 = O(\lambda + k) \sum_{v \in \mathcal{T}_0} O(\lambda + m_v) = O((\lambda + k)n)$.

Corollary 1. *Let S be a set of n objects with density λ. Under the Displacement Assumption, we can maintain a compressed quadtree on S in $O(n)$ time per time step that can be used for broad-phase collision detection, reporting $O((\lambda + k)n)$ pairs of potentially colliding objects.*

References

1. Abam, M.A., de Berg, M., Poon, S.-H., Speckmann, B.: Kinetic collision detection for convex fat objects. Algorithmica 53(4), 457–473 (2009)
2. Agarwal, P.K., Basch, J., Guibas, L.J., Hershberger, J., Zhang, L.: Deformable free-space tilings for kinetic collision detection. Int. J. Robotics Research 21(3), 179–197 (2002)
3. Aluru, S., Sevilgen, F.E.: Dynamic compressed hyperoctrees with application to the N-body problem. In: Proc. 19th Conf. Found. Softw. Tech. Theoret. Comput. Sci., pp. 21–33 (1999)
4. Basch, J., Guibas, L.J., Hershberger, J.: Data structures for mobile data. In: Proc. 8th ACM-SIAM Symp. Discr. Alg., pp. 747–756 (1997)
5. de Berg, M.: Linear size binary space partitions for uncluttered scenes. Algorithmica 28, 353–366 (2000)
6. de Berg, M., Cheong, O., van Kreveld, M., Overmars, M.: Computational Geometry: Algorithms and Applications, 3rd edn. Springer (2008)
7. de Berg, M., Haverkort, H., Thite, S., Toma, L.: Star-quadtrees and guard-quadtrees: I/O-efficient indexes for fat triangulations and low-density planar subdivisions. Comput. Geom. Theory Appl. 43, 493–513 (2010)
8. de Berg, M., Katz, M.J., van der Stappen, A.F., Vleugels, J.: Realistic input models for geometric algorithms. Algorithmica 34(1), 81–97 (2002)
9. de Berg, M., Roeloffzen, M., Speckmann, B.: Kinetic convex hulls and Delaunay triangulations in the black-box model. In: Proc. 27th ACM Symp. Comput. Geom., pp. 244–253 (2011)
10. Borro, D., Garcia-Alonso, A., Matey, L.: Approximation of optimal voxel size for collision detection in maintainability simulations within massive virtual environments. Comp. Graph. Forum 23(1), 13–23 (2004)
11. Chazelle, B.: Convex partitions of polyhedra: a lower bound and worst-case optimal algorithm. SIAM J. Comput. 13, 488–507 (1984)
12. Gao, J., Guibas, L.J., Nguyen, A.: Deformable spanners and applications. In: Proc. 20th ACM Symp. Comput. Geom., pp. 190–199 (2004)
13. Har-Peled, S.: Geometric Approximation Algorithms. American Mathematical Society (2011)
14. Kim, D.-J., Guibas, L.J., Shin, S.Y.: Fast collision detection among multiple moving spheres. IEEE Trans. Vis. Comp. Gr. 4, 230–242 (1998)
15. Kirkpatrick, D., Snoeyink, J., Speckmann, B.: Kinetic collision detection for simple polygons. Int. J. Comput. Geom. Appl. 12(1-2), 3–27 (2002)
16. Kockara, S., Halic, T., Iqbal, K., Bayrak, C., Rowe, R.: Collision detection: A survey. In: Proc. of Systems, Man and Cybernetics, pp. 4046–4051 (2007)
17. Lin, M., Gottschalk, S.: Collision detection between geometric models: A survey. In: Proc. of IMA Conf. Math. Surfaces, pp. 37–56 (1998)
18. Moore, M., Wilhelms, J.: Collision detection and response for computer animation. SIGGRAPH Comput. Graph. 22, 289–298 (1988)
19. Teschner, M., Kimmerle, S., Heidelberger, B., Zachmann, G., Raghupathi, L., Fuhrmann, A., Cani, M., Faure, F., Thalmann, M.N., Strasser, W., Volino, P.: Collision detection for deformable objects. Comp. Graph. Forum 24, 119–140 (2005)

Finding Social Optima in Congestion Games with Positive Externalities

Bart de Keijzer[1] and Guido Schäfer[1,2]

[1] Centrum Wiskunde & Informatica (CWI), Amsterdam, The Netherlands
{B.de.Keijzer,G.Schaefer}@cwi.nl
[2] VU University Amsterdam, The Netherlands

Abstract. We consider a variant of congestion games where every player i expresses for each resource e and player j a positive *externality*, i.e., a value for being on e together with player j. Rather than adopting a game-theoretic perspective, we take an optimization point of view and consider the problem of optimizing the social welfare.

We show that this problem is NP-hard even for very special cases, notably also for the case where the players' utility functions for each resource are affine (contrasting with the tractable case of linear functions [3]). We derive a 2-approximation algorithm by rounding an optimal solution of a natural LP formulation of the problem. Our rounding procedure is sophisticated because it needs to take care of the dependencies between the players resulting from the pairwise externalities. We also show that this is essentially best possible by showing that the integrality gap of the LP is close to 2.

Small adaptations of our rounding approach enable us to derive approximation algorithms for several generalizations of the problem. Most notably, we obtain an $(r + 1)$-approximation when every player may express for each resource externalities on player sets of size r. Further, we derive a 2-approximation when the strategy sets of the players are restricted and a $\frac{3}{2}$-approximation when these sets are of size 2.

1 Introduction

Congestion games constitute an important class of non-cooperative games that model situations in which n players compete for the usage of m resources. Every player i selects a subset of resources from a collection of resource subsets that are available to him. The utility $u_{i,e}(x)$ that player i receives for resource e depends on the total number x of players who have chosen this resource. Rosenthal's original definition of congestion games [13] assumes that the utility functions of the players are identical for each resource, i.e., $u_{i,e} = u_{j,e}$ for every two players i, j and every resource e. Milchtaich [11] introduced the more general congestion game model where every player i has a *player-specific* utility functions $u_{i,e}$ as described above.

Ever since their introduction in 1973, congestion games have been the subject of intensive research in game theory and, more recently, in algorithmic game

L. Epstein and P. Ferragina (Eds.): ESA 2012, LNCS 7501, pp. 395–406, 2012.

theory. Most of these studies adopt a distributed viewpoint and focus on issues like the existence and inefficiency of Nash equilibria, the computational complexity of finding such equilibria, etc. (see, e.g., [12] for an overview). Much less attention has been given to the study of congestion games from a centralized viewpoint.

Studying these congestion games from a centralized viewpoint is important in situations where a centralized authority has influence over the players in the game. Also, adopting a centralized perspective may help in acquiring insights about the decentralized setting: if it is hard to find an (approximate) optimum or near-optimum in the centralized case where all the players are completely coordinated, it certainly will be hard for the players to reach such a solution in the decentralized case, where besides lack of coordinated computation additional issues related to selfishness and stability arise. Lastly, we believe that studying this optimization problem is interesting for its own sake, as it can be seen as a generalization of various fundamental optimization problems (see paragraph on related work).

In this paper, we are interested in the problem of computing an assignment of players to resources such that the *social welfare*, i.e., the sum of the utilities of the players, is maximized. We are aware only of two related articles [3,4] (see paragraph on related work for more details) that study player-specific congestion games from an optimization perspective. Both articles assume that the players are *anonymous* [3] in the sense that the utility function $u_{i,e}$ of a player i only depends on the number of players using resource e, but not on their identities.

The assumption that all players are anonymous is overly simplistic in many situations. We therefore extend the player-specific congestion game model of Milchtaich [11] to incorporate *non-anonymous* players. More specifically, let $N = [n]$ be the set of players[1] and suppose that player i's value for sharing resource e with player j is v_{ije} (possibly $j = i$). We define the utility of player i for resource e and player set $S \subseteq N$ as $u_{i,e}(S) = \sum_{j \in S} v_{ije}$. We refer to these games as *generalized congestion games*. The externality v_{ije} that player j imposes on player i on resource e can be negative or positive. We speak of a generalized congestion game with *positive, negative* or *mixed externalities*, respectively, when the v_{ije}'s are positive, negative or arbitrary.

Our Contributions. We study the problem of computing a welfare maximizing assignment for generalized congestion games. As in [3,4], we concentrate on the case where each of the n players has to choose one of m available resources (symmetric players, singleton resource sets).

We first consider the problem with mixed externalities and show that it is strongly NP-hard and $n^{1-\varepsilon}$-inapproximable for every $\varepsilon > 0$ even for $m = 2$ resources. We also give a polynomial-time algorithm that solves the problem when the number of players is constant.

In light of this inapproximability result, we then focus on the problem of computing an optimal assignment for generalized congestion games with positive externalities (MAX-CG-POS-EXT). We derive a polynomial-time algorithm that

---

[1] We use the notation $[k]$ to denote the set $\{1, \ldots, k\}$ for a positive integer k.

solves MAX-CG-POS-EXT for $m = 2$ resources. We show that MAX-CG-POS-EXT is strongly NP-hard for $m \geq 3$ resources and therefore focus on approximation algorithms.

We derive a deterministic 2-approximation algorithm for MAX-CG-POS-EXT. Our algorithm computes an optimal solution to a natural LP-relaxation of the problem and then iteratively rounds this solution to an integer solution, thereby losing at most a factor 2 in the objective function value. We also show that the integrality gap of the underlying LP is close to 2 and therefore the approximation factor of our algorithm is essentially best possible.

Our rounding procedure is sophisticated because it needs to take care of the dependencies between the players resulting from the pairwise externalities. The key of our analysis is a probabilistic argument showing that these dependencies can always be resolved in each iteration. We believe that this approach might be applicable to similar problems and is therefore of independent interest.

Our approach is flexible enough to extend the algorithm to more general settings. One such generalization is to incorporate resource-restrictions for players (non-symmetric players). We show that our 2-approximation algorithm for MAX-CG-POS-EXT can be adapted to the case where the resources available to each player are restricted. We also obtain an improved $\frac{3}{2}$-approximation algorithm when every player is restricted to two resources. The proof of the $\frac{3}{2}$-approximation factor crucially exploits a characterization of the extreme point solutions of the LP relaxation.

A natural extension of our model are *r-generalized congestion games* where each player i specifies externalities v_{iTe} for all player subsets $T \subseteq N \setminus \{i\}$ of size at most r. The utility function of player i for resource e and player set $S \subseteq N$ is defined as $u_{i,e}(S) = \sum_{T \subseteq S \setminus \{i\}, |T| \leq r} v_{iTe}$. Using this terminology, generalized congestion games correspond to 1-generalized congestion games. We extend our rounding procedure to r-generalized congestion games with positive externalities and derive an $(r + 1)$-approximation algorithm.

Finally, we settle a question left open by Blumrosen and Dobzinski [3]. The authors showed that an optimal assignment for player-specific congestion games with non-negative linear utility functions $u_{i,e}(x) = a_{i,e}x$ can be computed efficiently. We show that this problem becomes NP-hard for affine utility functions $u_{i,e}(x) = a_{i,e}x + b_{i,e}$.

Related Work. There are various papers that study congestion games with negative or positive externalities. For example, negative externalities are studied in routing [14], scheduling and load balancing [2]. Positive externalities are studied in the context of cost sharing [7], facility location [1] and negotiations [5,6].

Meyers and Schulz [10] studied the complexity of finding a minimum cost solution in congestion games (according to Rosenthal's classical congestion game model [13]). They consider several variants of the problem and prove hardness results for some cases and polynomial time computability results for some other cases.

Chakrabarty, Mehta, Nagarajan and Vazirani [4] were the first to study player-specific congestion games from a centralized optimization perspective. The authors study the cost-minimization variant of the problem where each player has a non-negative and non-decreasing cost function associated with each resource.[2] They show that computing an assignment of minimum total cost is NP-hard. The authors also derive some positive results for certain special cases of the problem (see [4] for details).

Most related to our work is the paper by Blumrosen and Dobzinski [3]. They study the problem of welfare maximization in player-specific congestion games with non-negative utility functions. Among other results, they give NP-hardness and inapproximability results for positive and negative externalities. They also provide a randomized 18-approximation algorithm for arbitrary (non-negative) utility functions.

The problem of computing a welfare maximizing assignment for generalized congestion games can also be interpreted as the following graph coloring problem: We are given a complete undirected graph on n vertices and m colors $[m]$. Every edge (i, j) (including self-loops) has a weight $w_{ije} = v_{ije} + v_{jie}$ for each color $e \in [m]$. The goal is to assign a color to every node such that the total weight of all *monochromatic* edges, i.e., edges whose endpoints have the same color, is maximized. The weight of a monochromatic edge (i, j) is defined as w_{ije}, where e is the color of the endpoints. The minimization variant of this problem with identical weights $w_{ije} = w_{ij}$ for all $e \in [m]$ and every edge (i, j) is also known as the generalized graph coloring problem [9], graph k-partitioning [8], and k-min cluster [15].

2   Preliminaries

A *generalized congestion game* is given by $\Gamma = (N, E, \{\Sigma_i\}_{i \in N}, \{u_{i,e}\}_{i \in N, e \in E})$, where $N = [n]$ is the set of players, $E = [m]$ is the set of facilities (or resources), $\Sigma_i \subseteq 2^E$ is the strategy set of player i, and $u_{i,e} : 2^N \to \mathbb{R}$ is the utility function that player i associates with facility e. Unless stated otherwise, we assume throughout this paper that $\Sigma_i = E$ for every player i (symmetric players, singleton resource sets). The set $\Sigma = \times_{i=1}^{n} \Sigma_i$ is called the *joint strategy set*, and elements in it are *strategy profiles*. Each player chooses a strategy $s_i \in \Sigma_i$ from his strategy set, which gives rise to a strategy profile $s = (s_1, \ldots, s_n) \in \Sigma$. A player tries to maximize his utility $u_i(s) = u_{i,s_i}(\{j : s_j = s_i\})$.

The utility functions $u_{i,e}$ that we consider in this paper are of a specific form: Each player i associates with each facility e and player j a non-negative value v_{ije}. Then $u_{i,e}$ is given by $S \mapsto \sum_{j \in S} v_{ije}$. The idea behind this is that for $(i, j) \in N^2$ and $e \in E$, v_{ije} specifies player i's value of being on facility e together with player j. Here, v_{iie} is the value of player i for facility e, independently of the other players. We speak of a generalized congestion game with *positive, negative* or *mixed externalities*, respectively, when the v_{ije}'s are positive, negative or

---

[2] Equivalently, the utility functions are assumed to be non-positive and non-increasing.

arbitrary. Note that allowing players to remain unassigned would not result in a more expressive model, as this is equivalent to assigning a player to an artificial facility for which the player always has utility 0.

Associated to Γ is a *social welfare function* $\Pi : \Sigma \to \mathbb{R}$ defined by $s \mapsto \sum_{i=1}^{n} u_i(s)$. We are interested in the problem of finding an *optimal* strategy profile for a given generalized congestion games Γ, i.e., a strategy profile $s^* \in \Sigma$ that maximizes Π. We use MAX-CG-XXX-EXT as a short to refer to this optimization problem, where $\text{xxx} \in \{\text{POS}, \text{NEG}, \text{MIX}\}$ indicates the respective type of externalities.

Due to space limitations, many proofs are omitted from this extended abstract and will be given in the full version of the paper.

3   Mixed Externalities

We start off by studying the problem MAX-CG-MIX-EXT of optimizing the social welfare in generalized congestion games with mixed externalities. It turns out that this problem is highly inapproximable, even for 2 facilities.

Theorem 1. MAX-CG-MIX-EXT *is strongly* NP-*hard and is not* $n^{1-\epsilon}$-*approximable in polynomial time for every* $\epsilon > 0$, *unless* P $\neq$ NP, *even for* $m = 2$ *facilities.*

As it turns out, the problem can be solved efficiently if the number of players is fixed.

Proposition 1. MAX-CG-MIX-EXT *can be solved in polynomial time for a fixed number of players.*

4   Positive Externalities

Given the strong inapproximability result for mixed externalities (Theorem 1), we focus on the case of positive externalities in the remainder of this paper. Central to our study of this problem is the following integer program (IP) for the problem:

$$\max \sum_{e=1}^{m} \left(\sum_{i=2}^{n} \sum_{j=1}^{i-1} (v_{ije} + v_{jie}) x_{\{i,j\},e} + \sum_{i=1}^{n} v_{iie} x_{i,e} \right) \tag{1}$$

$$\text{s.t.} \sum_{e=1}^{m} x_{i,e} = 1 \qquad\qquad \forall i \in N \tag{2}$$

$$x_{\{i,j\},e} - x_{i,e} \leq 0 \qquad\qquad \forall i, j \in N, \forall e \in E \tag{3}$$

$$x_{\{i,j\},e} \in \{0,1\} \qquad\qquad \forall i, j \in N, \forall e \in E \tag{4}$$

Recall that $N = [n]$ and $E = [m]$. The variables are interpreted as follows: $x_{i,e}$ is the 0-1 variable that indicates whether player $i \in N$ is assigned to resource e; $x_{\{i,j\},e}$ is the 0-1 variable that indicates whether both players $i, j \in N$, $i \neq j$, are assigned to resource $e \in E$. There are thus $(n^2 + n)/2$ variables in total. Note that we exploit in the above formulation that all externalities are positive.

In the LP relaxation the constraints (4) are replaced by "$0 \leq x_{\{i,j\},e} \leq 1, \forall i,j \in N, \forall e \in E$". We can show that the LP relaxation of IP (1) is totally unimodular for $m = 2$. This implies the following positive result, which is in stark contrast with Theorem 1.

Theorem 2. MAX-CG-POS-EXT *can be solved in polynomial time for* $m = 2$ *facilities.*

Unfortunately, the problem becomes strongly NP-hard for $m \geq 3$.

Theorem 3. MAX-CG-POS-EXT *is strongly* NP-*hard for* $m \geq 3$ *facilities.*

4.1 A 2-Approximate LP-Rounding Algorithm

In this section, we derive a 2-approximation algorithm for MAX-CG-POS-EXT. We first need to introduce some more notation.

We extend Π to the domain of fractional solutions of the LP relaxation of IP (1) and use the term *social welfare* to refer to the objective function of IP (1).

For a congestion game $\Gamma = (N, E, \{\Sigma_i\}_{i \in N}, \{u_{i,e}\}_{i \in N, e \in E})$ with positive externalities $\{v_{ije}\}_{i,j \in N, e \in E}$, we define for $e \in E$ and $a \in \mathbb{Q}_{\geq 1}$ the (e, a)-*boosted game* as the game obtained by introducing a copy e' of facility e which all players value a times as much as the original e. Formally, the (e, a)-boosted game of Γ is the game $\Gamma' = (N, E \cup \{e'\}, \{\Sigma_i \cup \{e'\}\}_{i \in N}, \{u_{i,f}\}_{i \in N, f \in E \cup \{e'\}})$ where $u_{i,e'}(S) = \sum_{j \in S} a v_{ije}$ for all $i \in N$. In an (e, a)-boosted game, we refer to the introduced facility e' as *the boosted facility*.

We fix for each facility $e \in E$ a total order $\preceq_e$ on N which satisfies $i \prec_e j$ whenever $x_{ie} < x_{je}$. Using $\preceq_e$, we define $P(e, k)$ as the player set N' for which it holds that $|N'| = k$ and $i \in N'$ if $j \in N'$ and $i \succ_e j$. Informally, $P(e, k)$ consists of the k players with the highest fractional assignments on e.

Finally, for a fractional solution x for Γ, we define the (e, a, k)-*boosted assignment* as the fractional solution to the (e, a)-boosted game where players $P(e, k)$ are assigned integrally to the boosted facility and the remaining players are assigned according to x.

Our algorithm, BOOST(a), is a rounding algorithm that takes as its starting point the fractional optimum of the relaxation of IP (1), and iteratively picks some facility $e \in E$ and assigns a set of players to the boosted facility in the (e, a)-boosted game. The formal description of BOOST(a) is given in Algorithm 1.

Theorem 4. *Algorithm* BOOST(2) *is a deterministic polynomial time 2-approximation algorithm for* MAX-CG-POS-EXT.

Proof. The hard part of the proof is to show that in each iteration there exist e and k such that Step 2.1 is feasible. This is done in Lemma 1 given below.
The algorithm clearly outputs a feasible solution. It is straightforward to check that the algorithm runs in polynomial time: For Step 1, solving an LP to optimality can be done in polynomial time using the ellipsoid method or an interior

point method. For Step 2, there are only nm (e,k)-pairs to check. Lastly, in each iteration, at least one player will be assigned to a boosted facility, hence removed from $[m]$, so there are at most n iterations.

It is also easy to prove that Algorithm 1 outputs a solution for which the social welfare is within a factor $\frac{1}{2}$ from the optimal social welfare: In the solution x' at the beginning of Step 3, every player $i \in [n]$ is assigned to a copy e' of a facility $e \in [m]$ for which it holds that $v_{ije'} = 2v_{ije}$ for all j. So by assigning all players on such facilities e' to the original facility e decreases the social welfare at most factor 2. If we denote by SOL the social welfare of the strategy profile found by Algorithm 1, and if we denote by OPT the optimizal social welfare, then the following sequence of inequalities proves that the algorithm outputs a 2-approximate solution:[3]

$$SOL = \Pi_\Gamma(x'') \geq \frac{1}{2}\Pi_{\Gamma'}(x') \geq \frac{1}{2}\Pi_\Gamma(x) \geq \frac{1}{2}OPT.$$

$\square$

We come to the proof of the following lemma.

Lemma 1. *Suppose that x' is a solution to IP (1) for a congestion game Γ' with positive externalities. Denote by E the facility set of Γ' and assume that $E = [m] \cup E'$, where E' is a set of facilities such that $x_{\{i,j\},e'} \in \{0,1\}$ for all $i,j \in [n], e' \in E'$. Then there is a facility $e \in [m]$ and a number $k \in [n]$ such that the social welfare of the $(e,2,k)$-boosted assignment is at least the social welfare of x'. Moreover, e and k can be found in polynomial time.*

For the proof of Lemma 1, we need the following technical result:

Lemma 2. *Let $a_1, a_2, \ldots, a_n \in \mathbb{R}_{\geq 0}$, be a non-increasing sequence of non-negative numbers with a_1 positive, and let $b_1, \ldots, b_n \in \mathbb{R}$. Suppose that $\sum_{i=1}^{n} a_i b_i \geq 0$. Then, there is a $k \in [n]$ such that $\sum_{i=1}^{k} b_i \geq 0$.*

Proof. Let n' be the highest index such that $a_{n'} > 0$. There are two cases: either there is a $k < n'$ for which the claim holds, or there is not such a k. For the latter case, we show that the claim must hold for $k = n'$. It follows from the following derivation:

$$0 \leq \sum_{i=1}^{n'} a_i b_i = \sum_{j=1}^{n'-1} (a_j - a_{j+1}) \sum_{i=1}^{j} b_i + a_{n'} \sum_{i=1}^{n'} b_i \leq a_{n'} \sum_{i=1}^{n'} b_i.$$

$\square$

Proof (of Lemma 1). It suffices to only show existence of the appropriate e and k, as finding them in polynomial time can then simply be done by complete enumeration of all (e,k)-pairs (because there are only mn such pairs).

---

[3] The function Π is subscripted with the game of which it is the social welfare function.

Algorithm 1. BOOST(a): An LP rounding algorithm for MAX-CG-POS-EXT

---

1. Solve the relaxation of IP (1) for congestion game Γ, and let x be the optimal fractional solution. Let $\Gamma' := \Gamma$ and $x' = x$.
2. Repeat the following until x' is integral:

   2.1. Find a facility $e \in [m]$ and a number $k \in [n]$ such that the social welfare of the (e, a, k)-boosted assignment is at least the social welfare of x'.

   2.2. Let Γ' be the (e, a)-boosted game and let x' be the (e, a, k)-boosted assignment.

3. For every player $i \in [n]$ let e_i be the facility of $[m]$ such that i is assigned to a boosted copy of e_i in x'. Let x'' be the integral solution of IP (1) for Γ' obtained by assigning every i integrally to the original non-boosted facility e_i. Output the strategy profile for Γ that corresponds to x''.

For $e \in E$ and $k \in [n]$, let $\Delta(e, k)$ denote the amount by which social welfare increases when comparing the $(e, 2, k)$-boosted assignment of x' to the $(e, 2, k-1)$-boosted assignment of x'. Let $p(e, k)$ be the single player in $P(e, k) \backslash P(e, k-1)$. We can express $\Delta(e, k)$ as $\Delta^+(e, k) - \Delta^-(e, k)$, where $\Delta^+(e, k)$ is the increase in social welfare due to the additional utility on the boosted facility, and $\Delta^-(e, k)$ is the loss in utility due to setting the assignment for player $p(e, k)$ to 0 on all facilities in $[m]$. For notational convenience, we define $w_{\{i,j\},e} = v_{ije} + v_{jie}$ and $w_{i,e} = v_{iie}$. Then we can write $\Delta^+(e, k)$ and $\Delta^-(e, k)$ as follows:

$$\Delta^+(e, k) = 2w_{p(e,k),e} + \sum_{j:j \succ_e p(e,k)} 2w_{\{p(e,k),j\},e}$$

$$\Delta^-(e, k) = \sum_{f \in [m]} \left(x_{p(e,k),f} w_{p(e,k),f} + \sum_{j:j \prec_e p(e,k)} x_{\{p(e,k),j\},f} w_{\{p(e,k),j\},f} \right).$$

Clearly, if we move for some $e \in [m]$ and $k \in [n]$ the players $P(e, k)$ to the boosted facility e' in the $(e, 2)$-boosted game, then the change in utility is $\sum_{i=1}^{k} \Delta(e, i)$. We therefore need to show that there is a facility $e \in E$ and $k \in [n]$ such that $\sum_{i=1}^{k} \Delta(e, i) \geq 0$.

To show this, let X be a random variable that takes on the values $\{\Delta(e, k) : e \in [m], k \in [n]\}$, of which the distribution is given by

$$\mathbf{Pr}[X = \Delta(e, k)] = \frac{x_{p(e,k),e}}{\sum_{e \in [m], i \in [n]} x_{p(e,i),e}} \qquad \forall e \in [m], k \in [n].$$

Define $Y = \sum_{e \in [m], i \in [n]} x_{p(e,i),e}$.

We derive the following bound on the expectation of X:

$$\mathbf{E}[X] = \sum_{(e,k):e \in [m], k \in [n]} \mathbf{Pr}[X = \Delta(e, k)] \Delta(e, k)$$

$$= \frac{1}{Y} \sum_{e \in [m], k \in [n]} x_{p(e,k),e} \left(2w_{p(e,k),e} + \sum_{j:j \succ_e p(e,k)} 2w_{\{p(e,k),j\},e} \right.$$

$$
-\sum_{f\in[m]}\left(x_{p(e,k),f}w_{p(e,k),f}+\sum_{j:j\prec_e p(e,k)}x_{\{p(e,k),j\},f}w_{\{p(e,k),j\},f}\right)\Bigg)\Bigg)
$$

$$
=\frac{1}{Y}\Bigg(\sum_{e\in[m],k\in[n]}2x_{p(e,k),e}w_{p(e,k),e}
$$

$$
+\sum_{e\in[m],k\in[n]}\sum_{j:j\succ_e p(e,k)}2x_{p(e,k),e}w_{\{p(e,k),j\},e}
$$

$$
-\sum_{f\in[m],k\in[n]}\left(\sum_{e\in[m]}x_{p(e,k),e}\right)x_{p(e,k),f}w_{p(e,k),f}
$$

$$
-\sum_{e\in[m],k\in[n]}x_{p(e,k),e}\sum_{f\in[m]}\left(\sum_{j:j\prec_e p(e,k)}x_{\{p(e,k),j\},f}w_{\{p(e,k),j\},f}\right)\Bigg)
$$

$$
=\frac{1}{Y}\Bigg(\sum_{e\in[m],i\in[n]}2x_{i,e}w_{i,e}+\sum_{\{i,j\},i\neq j}\sum_{e\in[m]}2x_{\{i,j\},e}w_{\{i,j\},e}
$$

$$
-\sum_{e\in[m],i\in[n]}x_{i,e}w_{i,e}-\sum_{e\in[m],i\in[n]}x_{i,e}\sum_{f\in[m]}\sum_{j:j\prec_e i}x_{\{i,j\},f}w_{\{i,j\},f}\Bigg)
$$

$$
=\frac{1}{Y}\Bigg(\sum_{e\in[m],i\in[n]}x_{i,e}w_{i,e}+\sum_{\{i,j\},i\neq j}\sum_{e\in[m]}2x_{\{i,j\},e}w_{\{i,j\},e}
$$

$$
-\sum_{\{i,j\},i\neq j}\left(\sum_{f\in[m]}x_{\{i,j\},e}w_{\{i,j\},e}\right)\left(\sum_{e\in[m]}\max\{x_{i,e},x_{j,e}\}\right)\Bigg) \qquad (5)
$$

$$
\geq\frac{1}{Y}\sum_{e\in[m],i\in[n]}x_{i,e}w_{i,e}\geq 0.
$$

In the above derivation we make use of the following facts that hold for all $e\in[m], i,j\in[n]$: (i) $\sum_{f\in[m]}x_{i,f}=1$ (for the fourth equality), (ii) $x_{i,e}=\max\{x_{i,e},x_{j,e}\}$ if $i\succ_e j$ (for the fifth equality), (iii) $\sum_{f\in[m]}\max\{x_{i,f},x_{j,f}\}\leq 2$ (for the first inequality).

We can express $\mathbf{E}[X]$ as the sum of the m terms $\{T_e=\sum_{k\in[n]}\Delta(e,k)\mathbf{Pr}[X=\Delta(e,k)]:e\in[m]\}$. Because the expectation is non-negative, it holds that T_e is non-negative for at least one $e\in[m]$. We take this e and apply Lemma 2 to T_e (take $\Delta(e,i)$ for b_i and $\mathbf{Pr}[X=\Delta(e,i)]$ for a_i). We conclude that there is a $k\in[n]$ such that $\sum_{i=1}^{k}\Delta(e,k)\geq 0$. □

4.2   A Matching Integrality Gap Example

In this section, we show that the integrality gap of IP (1) is very close to 2. This implies that no algorithm based on the LP relaxation of IP (1) can achieve an approximation factor better than 2 and thus the approximation factor of BOOST(2) is essentially best possible.

We construct a family of generalized congestion games, parameterized by the number of players n, for which we determine optimal integral and fractional solutions to IP (1) and its LP relaxation, respectively. It can be verified computationally that the integrality gap approaches 2 as n increases. We are convinced that the integrality gap can be shown to approach 2 analytically, but at this point of time lack a formal proof of this.

Fix two parameters m and k with $k \leq m$ and let $H(m,k)$ be the k-uniform hypergraph on $[m]$. The instance $I(m,k)$ is then defined as follows: there are m facilities and $\binom{m}{k}$ players. We identify each player with a distinct edge of $H(m,k)$. The externality v_{ije} is set to 1 if the hyperedge of i and the hyperedge of j both contain facility e, and 0 otherwise. The value v_{iie} is set to 1 if the hyperedge of i contains facility e.

We define $OPT_{\mathrm{frac}}(m,k)$ as the feasible fractional solution for $I(m,k)$, where each player is assigned with value $1/k$ to each of its facilities. For each facility $e \in [m]$ there are $\frac{k}{m}\binom{m}{k} = \binom{m-1}{k-1}$ players that have e in their hyperedge and thus the social welfare of $OPT_{\mathrm{frac}}(m,k)$ is $\frac{m}{k}\binom{m-1}{k-1}^2$.

Using induction on m, one can straightforwardly prove that the social welfare of an optimal integral solution $OPT_{\mathrm{int}}(m,k)$ for this instance is $\sum_{i=k-1}^{m-1}\binom{i}{k-1}^2$. Subsequently, evaluating by computer the expression $OPT_{\mathrm{frac}}(m,k)/OPT_{\mathrm{int}}$ for particular choices of m and k indicates that the integrality gap approaches $\frac{2k-1}{k}$ as m gets larger. The largest integrality gap that we computed explicitly is 1.972013 for $m = 5000$ and $k = 71$.

5   Variations on the Problem

We study in this section two generalizations and one special case of the problem.

Restricted Strategy Sets.   Up to this point, we assumed that the strategy set Σ_i of each player i equals the facility set E (symmetric players, singleton strategy sets). We now show that our algorithm BOOST(2) generalizes to the case where every player i has a strategy set $\Sigma_i \subseteq E$ (non-symmetric players, singleton strategy sets). The modification is easy: in Step 1, the algorithm solves a modified version of IP (1) with the additional constraints $x_{i,e} = 0$ for all $i \in N, e \notin \Sigma_i$. The rounding procedure that follows after Step 1 of the algorithm then clearly puts every player i on a facility for which $x_{i,e} \neq 0$, so it produces a feasible integral solution. The rest of the proof of Theorem 4 remains valid.

Corollary 1. *There is a polynomial time 2-approximation algorithm for* MAX-CG-POS-EXT *with restricted strategy sets.*

For the special case that each player's strategy set is of size 2, we can improve the approximation factor.

Proposition 2. *There is a polynomial time $\frac{3}{2}$-approximation algorithm for* MAX-CG-POS-EXT *with restricted strategy sets of size 2.*

The proof of this proposition relies on a characterization result for the extreme point solutions of the LP relaxation of the modified IP, that we omit here. Moreover, one can show by a simple example that the integrality gap of the modified IP for strategy sets of size 2 is $\frac{3}{2}$.

Externalities on Bigger Sets of Players. Instead of restricting the players to express their externalities on pairs consisting of a single player and a facility, we can fix an arbitrary number $r \in \mathbb{N}$ and allow players to express their externalities on pairs consisting of a set of at most r players and a facility. We refer to such congestion games as *r-generalized congestion games*. When each player i is allowed to express externalities v_{iTe} for sets $T \subseteq N\backslash\{i\}, |T| \leq r$, we will show that a simple adaptation of the algorithm BOOST$(r + 1)$ returns an $(r + 1)$-approximate solution. The only change that needs to be made is that the relaxation of the following generalization of IP (1) is solved in Step 1 of BOOST$(r + 1)$: For each player i and facility e, this generalized LP has again the variable $x_{i,e}$ that indicates whether i is assigned to e. Moreover, for every facility e and set S of at most $r + 1$ players, there is a variable $x_{S,e}$ that indicates whether all players in S are assigned to e. For notational convencience, let $w_{S,e} = \sum_{i \in S} v_{i,S\backslash\{i\},e}$ for all sets $S \subseteq N, |S| \leq r$. The generalized LP reads as follows:

$$\max \sum_{e=1}^{m} \sum_{S:S\subseteq[n],|S|\leq r+1} w_{S,e}x_{S,e}$$

$$\text{s.t.} \sum_{e=1}^{m} x_{i,e} \leq 1 \qquad\qquad \forall i \in N$$

$$x_{S,e} - x_{i,e} \leq 0 \qquad\qquad \forall S \subseteq [n], S \leq r + 1, \forall i \in S, \forall e \in E,$$

$$x_{\{i,j\},e} \in \{0,1\} \qquad\qquad \forall i, j \in [n], \forall e \in E$$

We can show that the adapted version of BOOST$(r + 1)$ is a valid polynomial time $(r + 1)$-approximation algorithm.

Proposition 3. *There is a polynomial time $(r+1)$-approximation algorithm for computing a social welfare maximizing strategy profile for r-generalized congestion games with positive externalities.*

A Special Case: Affine Externalities. In this section, we study the special case of our problem where the utility functions $u_{i,e}(S)$ are *affine* non-negative functions of $|S|$, i.e., for every $i \in N$ and each $e \in E$, $u_{i,e}(S) = a_{ie}|S| + b_{ie}$ for non-negative rational numbers a_{ie}, b_{ie}. It is straightforward to see that this type utility functions falls within our positive externalities model. We refer to the respective optimization problem as MAX-CG-AFF-EXT.

The motivation for studying this is that Blumrosen and Dobzinski show in [3] that if $b_{ie} = 0$ for all $i \in N$, $e \in E$ then the optimal solution can be found in polynomial time. Allowing b_{ie} to be non-zero is thus one of the simplest generalizations that comes to mind.

In contrast to the polynomial time result in [3], we show that this problem is strongly NP-hard.

Theorem 5. MAX-CG-AFF-EXT *is strongly NP-hard.*

One could use the algorithm BOOST(2) to find a 2-approximate solution to this affine special case of the problem. However, it is easy to see that there is a much simpler and faster 2-approximation algorithm for this case: The algorithm simply chooses the maximum among the strategy profiles s and t defined as follows: for $i \in N$, $s_i = \arg_e \max\{b_{ie} : e \in E\}$ and $t_i = \arg_e \max\{\sum_{j \in N} na_{je} : e \in E\}$.

It is shown in Proposition 4.3 of [3] that if $b_{ie} = 0$ for all $i \in N$, $e \in E$, then the strategy profile t is optimal. It is obvious that if $a_{ie} = 0$ for all $i \in N$, $e \in E$, then the strategy profile s is optimal. Therefore, for the case that neither a_{ie} nor b_{ie} is necessarily 0, the maximum of these two joint strategies must be within a factor of 2 from the optimal strategy profile.

References

1. Anshelevich, E., Dasgupta, A., Kleinberg, J.M., Tardos, E., Wexler, T., Roughgarden, T.: The price of stability for network design with fair cost allocation. SIAM Journal on Computing 38(4), 1602–1623 (2008)
2. Awerbuch, B., Azar, Y., Grove, E.F., Kao, M.-Y., Krishnan, P., Vitter, J.S.: Load balancing in the L_p norm. In: IEEE Annual Symposium on Foundations of Computer Science (1995)
3. Blumrosen, L., Dobzinski, S.: Welfare maximization in congestion games. In: Proceedings of the 7th ACM Conference on Electronic Commerce, pp. 52–61 (2006)
4. Chakrabarty, D., Mehta, A., Nagarajan, V., Vazirani, V.: Fairness and optimality in congestion games. In: Proceedings of the 6th ACM Conference on Electronic Commerce, pp. 52–57. ACM, New York (2005)
5. Conitzer, V., Sandholm, T.: Expressive negotiation over donations to charities. In: Proceedings of the 5th ACM Conference on Electronic Commerce, EC 2004, pp. 51–60. ACM, New York (2004)
6. Conitzer, V., Sandholm, T.: Expressive negotiation in settings with externalities. In: Proceedings of the 20th National Conference on Artificial Intelligence - Volume 1, AAAI 2005, pp. 255–260. AAAI Press (2005)
7. Feigenbaum, J., Papadimitriou, C.H., Shenker, S.: Sharing the cost of multicast transmissions. Journal of Computer and System Sciences 63(1), 21–41 (2001)
8. Kann, V., Khanna, S., Lagergren, J., Panconesi, A.: On the hardness of approximating max k-cut and its dual. Chicago Journal of Theoretical Computer Science 2 (1997)
9. Kolen, A.W.J., Lenstra, J.K.: Combinatorics in operations research. In: Graham, R., Grötschel, M., Lovász, L. (eds.) Handbook of Combinatorics. Elsevier, Amsterdam (1995)
10. Meyers, C.A., Schulz, A.S.: The complexity of welfare maximization in congestion games. Networks, 252–260 (2012)
11. Milchtaich, I.: Congestion games with player-specific payoff functions. Games and Economic Behavior 13(1), 111–124 (1996)
12. Nisan, N., Roughgarden, T., Tardos, É., Vazirani, V.V. (eds.): Algorithmic Game Theory. Cambridge University Press (2007)
13. Rosenthal, R.W.: A class of games possessing pure-strategy nash equilibria. International Journal of Game Theory 2(1), 65–67 (1973)
14. Roughgarden, T.: Selfish Routing and the Price of Anarchy. MIT Press (2005)
15. Sahni, S., Gonzalez, T.: P-complete approximation problems. Journal of the ACM 23(3), 555–565 (1976)

Better Bounds for Graph Bisection

Daniel Delling and Renato F. Werneck

Microsoft Research Silicon Valley
{dadellin,renatow}@microsoft.com

Abstract. We introduce new lower bounds for the minimum graph bi-section problem. Within a branch-and-bound framework, they enable the solution of a wide variety of instances with tens of thousands of vertices to optimality. Our algorithm compares favorably with the best previous approaches, solving long-standing open instances in minutes.

1 Introduction

We study the minimum graph bisection problem: partition the vertices of a graph into two cells of equal vertex weight so as to minimize the number of edges between them. This classical NP-hard problem [10] has applications in areas as diverse as VLSI design, load-balancing, and compiler optimization. Fast and good heuristics exist [14,19], but provide no quality guarantee. Our focus is on *exact* algorithms, which normally rely on the branch-and-bound framework [16]. Traditional approaches apply sophisticated techniques to find lower bounds, such as multicommodity flows [20,21], or linear [3,5,9], semidefinite [1,3,13,17], and quadratic programming [11]. The bounds they obtain tend to be very good, but quite expensive to compute. As a result, they can handle relatively small graphs, typically with hundreds of vertices. (See Armbruster [1] for a survey.)

Delling et al. [8] have recently proposed a branch-and-bound algorithm using only combinatorial bounds. Their *packing bound* involves building collections of disjoint paths and arguing that any bisection must cut a significant fraction of those. This approach offers a different trade-off: branch-and-bound trees tend to be bigger (since the bounds are weaker), but each node can be processed much faster. This pays off for very sparse graphs with small bisections, and allowed them to solve (for the first time) instances with tens of thousands of vertices.

In this work, we propose a new combinatorial bound that follows the same principle, but is much stronger. Section 3 explains our new *edge-based packing bound* in detail: it finds a much larger collection of paths by allowing them to overlap in nontrivial ways. Section 4 shows how to actually build this collection efficiently, which requires sophisticated algorithms and significant engineering effort. We then show, in Section 5, how to fix some vertices to one of the cells without actually branching on them. After explaining additional details of the branch-and-bound algorithm (Section 6), we present extensive experimental results in Section 7. It shows that our new algorithm outperforms any previous technique on a wide range of inputs, and is almost never much worse. In fact, it can solve several benchmark instances that have been open for decades, often in minutes or even seconds.

L. Epstein and P. Ferragina (Eds.): ESA 2012, LNCS 7501, pp. 407–418, 2012.
© Springer-Verlag Berlin Heidelberg 2012

2 Preliminaries

We take as input a graph $G = (V, E)$, with $|V| = n$ and $|E| = m$. Each vertex $v \in V$ has an integral weight $w(v)$. By extension, for any set $S \subseteq V$, let $w(S) = \sum_{v \in S} w(v)$. Let $W = w(V)$ denote the total weight of all vertices. For a given input parameter $\epsilon \geq 0$, we are interested in computing a *minimum ϵ-balanced bisection* of G, i.e., a partition of V into exactly two sets (*cells*) such that (1) the weight of each cell is at most $W_+ = \lfloor (1 + \epsilon) \lceil W/2 \rceil \rfloor$ and (2) the number of edges between cells (*cut size*) is minimized. Conversely, $W_- = W - W_+$ is the *minimum allowed cell size*. Our algorithms assume edges are unweighted; it deals with small integral edge weights by creating parallel unweighted edges.

To find the optimum solution, we use the *branch-and-bound* technique [16]. It implicitly enumerates all solutions by dividing the original problem into slightly simpler subproblems, solving them recursively, and picking the best solution found. Each node of the branch-and-bound tree corresponds to a distinct subproblem. At all times, we keep a global *upper bound U* on the solution of the original problem, which is updated as better solutions are found. To process a node in the tree, we compute a *lower bound L* on any solution to the corresponding subproblem. If $L \geq U$, we *prune* the node: it cannot lead to a better solution. Otherwise, we *branch*, creating two or more simpler subproblems.

Concretely, for graph bisection each node of the branch-and-bound tree represents a *partial assignment* (A, B), where $A, B \subseteq V$ and $A \cap B = \emptyset$. The vertices in A or B are said to be *assigned*, and all others are *free* (or *unassigned*). This node implicitly represents all valid bisections (A^+, B^+) that are *extensions* of (A, B), i.e., such that $A \subseteq A^+$ and $B \subseteq B^+$. In particular, the *root* node has the form $(A, B) = (\{v_0\}, \emptyset)$ and represents all valid bisections. (Note that we fix an arbitrary vertex v_0 to one cell to break symmetry.) To process an arbitrary node (A, B), we must compute a lower bound $L(A, B)$ on the value of any extension (A^+, B^+) of (A, B). If $L(A, B) \geq U$, we prune. Otherwise, we choose a free vertex v and *branch* on it, generating subproblems $(A \cup \{v\}, B)$ and $(A, B \cup \{v\})$.

The number of nodes in the branch-and-bound tree depends crucially on the quality of the lower bound. As a starting point, we use the well-known [7] *flow bound*: the minimum s–t cut between A and B. It is a valid lower bound because any extension (A^+, B^+) must separate A from B. It also functions as a *primal heuristic*: if the minimum cut happens to be balanced, we can update U. Unfortunately, the flow bound can only work well when A and B have similar sizes, and even in such cases the corresponding cuts are often far from balanced, with one side containing almost all vertices. Because the flow bound does not use the fact that the final solution must be balanced, it is rather weak by itself.

3 Edge-Based Packing Bound

To take the balance constraint into account, we propose the *edge-based packing bound*, a novel lower bounding technique. Consider a partial assignment (A, B). Let f be the value of the maximum A–B flow, and G_f be the graph obtained by removing all flow edges from G. Without loss of generality, assume that A is the *main side*, i.e., that the set of vertices reachable from A in G_f has higher total weight than those reachable from B. We will compute our new bound on G_f, since this allows us to simply add it to f to obtain a unified lower bound.

To compute the bound, we need a *tree packing* $\mathcal{T}$. This is a collection of trees (acyclic connected subgraphs of G_f) such that: (1) the trees are edge-disjoint; (2) each tree contains exactly one edge incident to A; and (3) the trees are maximal (no edge can be added to $\mathcal{T}$ without violating the previous properties). Given a set $S \subseteq V$, let $\mathcal{T}(S)$ be the subset of $\mathcal{T}$ consisting of all trees that contain a vertex in S. By extension, let $\mathcal{T}(v) = \mathcal{T}(\{v\})$.

For now, assume a tree packing $\mathcal{T}$ is given (Section 4 shows how to build it). With $\mathcal{T}$, we can reason about any extension (A^+, B^+) of (A, B). By definition, a tree $T_i \in \mathcal{T}$ contains a path from each of its vertices to A; if a vertex in B^+ is in T_i, at least one edge from T_i must be cut in (A^+, B^+). Since each tree $T_i \in \mathcal{T}(B^+)$ contains a separate path from A to B^+ in G_f, the following holds:

Lemma 1. *If B^+ is an extension of B, then $f + |\mathcal{T}(B^+)|$ is a lower bound on the cost of the corresponding bisection $(V \setminus B^+, B^+)$.*

This applies to a fixed extension B^+ of B; we need a lower bound that applies to *all* (exponentially many) possible extensions. We must therefore reason about a *worst-case extension* B^*, i.e., one that minimizes the bound given by Lemma 1.

First, note that $w(B^*) \geq W_-$, since $(V \setminus B^*, B^*)$ must be a valid bisection.

Second, let $D_f \subseteq V$ be the set of all vertices that are *unreachable* from A in G_f (in particular, $B \subseteq D_f$). Without loss of generality, we can assume that B^* contains D_f. After all, regarding Lemma 1, any vertex $v \in D_f$ is *deadweight*: since there is no path from v to A, it does not contribute to the lower bound.

To reason about other vertices in B^*, we first establish a relationship between $\mathcal{T}$ and vertex weights by predefining a *vertex allocation*, i.e., a mapping from vertices to trees. We allocate each reachable free vertex v (i.e., $v \in V \setminus (D_f \cup A)$) to one of the trees in $\mathcal{T}(v)$. (Section 4 will discuss how.) The *weight* $w(T_i)$ of a tree $T_i \in \mathcal{T}$ is the sum of the weights of all vertices allocated to T_i.

Given a fixed allocation, we can assume without loss of generality that, if B^* contains a single vertex allocated to a tree T_i, it will contain *all* vertices allocated to T_i. To see why, note that, according to Lemma 1, the first vertex increases the lower bound by one unit, but the other vertices in the tree are free.

Moreover, B^* must contain a *feasible* set of trees $\mathcal{T}' \subseteq \mathcal{T}$, i.e., a set whose total weight $w(\mathcal{T}')$ (defined as $\sum_{T_i \in \mathcal{T}'} w(T_i)$) is at least as high as the *target weight* $W_f = W_- - w(D_f)$. Since B^* is the worst-case extension, it must pick a feasible set $\mathcal{T}'$ of minimum cardinality. Formally, given a partial assignment (A, B), a flow f, a tree packing $\mathcal{T}$, and an associated vertex allocation, we define the *packing bound* as $p(\mathcal{T}) = \min_{\mathcal{T}' \subseteq \mathcal{T}, w(\mathcal{T}') \geq W_f} |\mathcal{T}'|$.

Note that this bound can be computed by a *greedy algorithm*, which picks trees in decreasing order of weight until their accumulated weight is at least W_f.

We can strengthen this bound further by allowing *fractional allocations*. Instead of allocating v's weight to a single tree, we can distribute $w(v)$ arbitrarily among all trees in $\mathcal{T}(v)$. For v's allocation to be *valid*, each tree must receive a nonnegative fraction of v's weight, and these fractions must add up to one. The weight of a tree T is defined in the natural way, as the sum of all fractional weights allocated to T. Fractional allocations can improve the packing bound by making trees more balanced. They are particularly useful when the average number of vertices per tree is small, or when some vertices have high degree. The fact that the packing bound is valid is our main theoretical result.

Theorem 1. *Consider a partial assignment (A, B), a flow f, a tree packing $\mathcal{T}$, and a valid fractional allocation of weights. Then $f + p(\mathcal{T})$ is a lower bound on the cost of any valid extension of (A, B).*

Proof. Let (A^*, B^*) be a minimum-cost extension of (A, B). Let $\mathcal{T}^* = \mathcal{T}(B^*)$ be the set of trees in $\mathcal{T}$ that contain vertices in B^*. The cut size of (A^*, B^*) must be at least $f + |\mathcal{T}^*|$, since at least one edge in each tree must be cut. We must prove that $p(\mathcal{T}) \leq |\mathcal{T}^*|$. (Since we only consider G_f, the flow bound stays valid.) It suffices to show that $\mathcal{T}^*$ is feasible, i.e., that $w(\mathcal{T}^*) \geq W_f$ (since the packing bound minimizes over all feasible sets, it cannot be higher than any one of them). Let $R^* = B^* \setminus D_f$ be the set of vertices in B^* that are reachable from A in G_f; clearly, $w(R^*) = w(B^* \setminus D_f) \geq w(B^*) - w(D_f)$. Moreover, $w(\mathcal{T}^*) \geq w(R^*)$ must hold because (1) every vertex $v \in R^*$ must hit some tree in $\mathcal{T}^*$ (the trees are maximal); (2) although $w(v)$ may be arbitrarily split among several trees in $\mathcal{T}(v)$, all these must be in $\mathcal{T}^*$; and (3) vertices of $\mathcal{T}^*$ that are in A^* (and therefore not in R^*) can only contribute nonnegative weights to the trees. Finally, since B^* is a valid bisection, we must have $w(B^*) \geq W_-$. Putting everything together, we have $w(\mathcal{T}^*) \geq w(R^*) \geq w(B^*) - w(D_f) \geq W_- - w(D_f) = W_f$.   □

Comparison. Our packing bound is a generalization of the bound proposed by Delling et al. [8], which also creates a set of disjoint trees and uses a greedy packing algorithm to compute a lower bound. The crucial difference is that, while we only need the trees to be *edge-disjoint*, Delling et al. [8] also require them to be *vertex-disjoint*. We therefore refer to their method as VBB (for *vertex-based bound*). Dropping vertex-disjointness not only allows our method to balance the trees more effectively (since it can allocate vertex weights more flexibly), but also increases the number of available trees. This results in significantly better lower bounds, leading to much smaller branch-and-bound trees. As Section 7 will show, our method is particularly effective on instances with high-degree vertices, where it can be several orders of magnitude faster than VBB. The only drawback of our approach relative to VBB is that finding overlapping trees efficiently requires significantly more engineering effort, as the next section will show.

4   Bound Computation

Theorem 1 applies to any valid tree packing $\mathcal{T}$, but the quality of the bound it provides varies. Intuitively, we should pick $\mathcal{T}$ (and an associated weight allocation) so as to avoid heavy trees: this improves $p(\mathcal{T})$ by increasing the number of trees required to achieve the target weight W_f. Since $|\mathcal{T}|$ and $w(\mathcal{T})$ are both fixed (for any $\mathcal{T}$), in the ideal tree packing all trees would have the same weight. It is easy to show that finding the best such packing is NP-hard, so in practice we must resort to (fast) heuristics. Ideally, the trees and weight allocations should be computed simultaneously (to account for the interplay between them), but it is unclear how to do so efficiently. Instead, we use a two-stage approach: first compute a valid tree packing, then allocate vertex weights to these trees appropriately. We discuss each stage in turn.

Generating trees. The goal of the first stage is to generate maximal edge-disjoint trees rooted at A that are as balanced and intertwined as possible. We do so by growing these trees simultaneously, trying to balance their sizes.

More precisely, each tree starts with a single edge (the one adjacent to A) and is marked *active*. In each step, we pick an active tree with minimum size (number of edges) and try to expand it by one edge in DFS fashion. A tree that cannot be expanded is marked as inactive. We stop when there are no active trees left. We call this algorithm *SDFS* (for *simultaneous depth-first search*).

An efficient implementation of SDFS requires a careful choice of data structures. In particular, a standard DFS implementation associates information (such as parent pointers and status within the search) with vertices, which are the entities added and removed from the DFS stack. In our setting, however, the same vertex may be in several trees (and stacks) simultaneously. We get around this by associating information with *edges* instead. Since each edge belongs to at most one tree, it has at most one parent and is inserted into at most one stack. This takes $O(m)$ total space regardless of the number of trees.

Given this representation, we now describe the basic step of SDFS in more detail. First, pick an active tree T_i of minimum size (using buckets). Let (u, v) be the edge on top of S_i (the stack associated with T_i), and assume v is farther from T_i's root than u is. Scan vertex v, looking for an *expansion edge*. This is an edge (v, w) such that (1) (v, w) is free (not assigned to any tree yet) and (2) no edge incident to w belongs to T_i. The first condition ensures that the final trees are disjoint, while the second makes sure they have no cycles. If no such expansion edge exists, we pop (u, v) from S_i; if S_i becomes empty, T_i can no longer grow, so we mark it as inactive. If expansion edges do exist, we pick one such edge (v, w), push it onto S_i, and add it to T_i by setting $parent(v, w) \leftarrow (u, v)$. The algorithm repeats the basic step until there are no more active trees.

We must still define which expansion edge (v, w) to select when processing (u, v). We prefer an edge (v, w) such that w has several free incident edges (to help keep the tree growing) and is as far as possible from A (to minimize congestion around the roots, which is also why we do DFS). Note that we can precompute the distances from A to all vertices with a single BFS.

To bound the running time of SDFS, note that a vertex v can be scanned $O(\deg(v))$ times (each scan either eliminates a free edge or backtracks). When scanning v, we can process each outgoing edge (v, w) in $O(1)$ time using a hash table to determine whether w is already incident to v's tree. The worst-case time is therefore $\sum_{v \in V} (\deg(v))^2 = O(m\Delta)$, where Δ is the maximum degree.

Weight allocation. Once a tree packing $\mathcal{T}$ is built, we must allocate the weight of each vertex v to the trees $\mathcal{T}(v)$ it is incident to. Our final goal is to have the weights as evenly distributed among the trees as possible. We work in two stages: *initial allocation* and *local search*. We discuss each in turn.

The first stage allocates each vertex to a single tree. We maintain, for each tree T_i, its maximum *potential weight* $\Pi(i)$, defined as the sum of the weights of all vertices that are adjacent to T_i and have not yet been allocated to another tree. To keep the trees balanced, we allocate weight to trees with smaller $\Pi(\cdot)$ values first. More precisely, initially all vertices in $\mathcal{T}$ are *available* (not allocated), all trees T_i are *active*, and $\Pi(i)$ is the sum of the weights of all available vertices incident to T_i. In each step, the algorithm picks an active tree T_i such that $\Pi(i)$ is minimum. If there is an available vertex v incident to T_i, we allocate it to T_i; otherwise, we mark the tree as inactive. We stop when no active tree remains.

To implement this, we maintain the active trees in a priority queue (according to $\Pi(i)$), and each tree T_i keeps a list of all available vertices it is incident to; these lists have combined size $O(m)$. When v is allocated to a tree T_i, we decrease $\Pi(j)$ for all trees $T_j \neq T_i$ that are incident to v ($\Pi(i)$ does not change), remove v from the associated lists, and update the priority queue. The total time is $O(m \log m)$ with a binary heap or $O(m+W)$ with buckets (with integral weights).

Given an initial allocation, we then run a *local search* to rebalance the trees. Unlike the constructive algorithm, it allows fractional allocations. We process one vertex at a time (in arbitrary order) by reallocating v's weight among the trees in $\mathcal{T}(v)$ in a locally optimal way. More precisely, v is processed in two steps. First, we reset v's existing allocation by removing v from all trees it is currently allocated to, thus reducing their weights. We then distribute v's weight among the trees in $\mathcal{T}(v)$ (from lightest to heaviest), evening out their weights as much as possible. In other words, we add weight to the lightest tree until it is as heavy as the second lightest, then add weight to the first two trees (at the same rate) until each is as heavy as the third, and so on. We stop as soon as v's weight is fully allocated. The entire local search runs in $O(m \log m)$ time, since it must sort (by weight) the adjacency lists of each vertex in the graph once. In practice, we run the local search three times to further refine the weight distribution.

5   Forced Assignments

Consider a partial assignment (A, B). As observed by Delling et al. [8], if the current lower bound for (A, B) is close enough to the upper bound U, one can often infer that certain free vertices v must be assigned to A (or B) with no need to branch, reducing the size of the branch-and-bound tree. This section studies how these *forced assignments* can be generalized to work with our stronger edge-based bounds. As usual, assume A is the main side, let $\mathcal{T}$ be a tree packing with weight allocations, and let $f + p(A)$ be the current lower bound.

First, we consider *flow-based forced assignments*. Let v be a free vertex reachable from A in G_f, and consider what would happen if it were assigned to B. The flow bound would immediately increase by $|\mathcal{T}(v)|$ units, since each tree in $\mathcal{T}(v)$

contains a different path from v to A. We cannot, however, simply increase the overall lower bound to $f + p(\mathcal{T}) + |\mathcal{T}(v)|$, since the packing bound may already be "using" some trees in $\mathcal{T}(v)$. Instead, we must compute a new packing bound $p(\mathcal{T}')$, where $\mathcal{T}' = \mathcal{T} \setminus \mathcal{T}(v)$ but the weights originally assigned to the trees $\mathcal{T}(v)$ are treated as deadweight (unreachable). If the updated bound $f + p(\mathcal{T}') + |\mathcal{T}(v)|$ is U or higher, we have proven that no solution that extends $(A, B \cup \{v\})$ can improve the best known solution. Therefore, we can safely assign v to A.

Note that we can make a symmetric argument for vertices w that are reachable from B in G_f, as long as we also compute an edge packing $\mathcal{T}'_B$ on B's side. Assigning such a vertex w to A would increase the overall bound by $|\mathcal{T}'_B(w)|$ (because the extra flow is on B's side, it does not affect $p(\mathcal{T})$). If the new bound $f + p(\mathcal{T}) + |\mathcal{T}'_B(w)|$ is U or higher, we can safely assign w to B.

Another strategy we use is *subdivision-based forced assignments*, which subdivides heavy trees in $\mathcal{T}$. Let v be a free vertex reachable from A in G_f. If v were assigned to A, we could obtain a new tree packing $\mathcal{T}'$ by splitting each tree $T_i \in \mathcal{T}(v)$ into multiple trees, one for each edge of T_i that is incident to v. If $f + p(\mathcal{T}') \geq U$, we can safely assign v to B.

Some care is required to implement this test efficiently. In particular, to recompute the packing bound we need to compute the total weight allocated to each of the newly-created trees. To do so efficiently, we use some precomputation. For each edge e, let $T(e) \in \mathcal{T}$ be the tree to which e belongs. Define $s(e)$ as the weight of the subtree of $T(e)$ rooted at e: this is the sum, over all vertices descending from e in $T(e)$, of the (fractional) weights allocated to $T(e)$. (If e belongs to no tree, $s(e)$ is undefined.) The $s(e)$ values can be computed with a bottom-up traversal of all trees, which takes $O(m)$ total time.

These precomputed values are useful when the forced assignment routine processes a vertex v. Each edge $e = (v, u)$ is either a *parent* or a *child* edge, depending on whether u is on the path from v to $T(e)$'s root or not. If e is a child edge, it will generate a tree of size $s(e)$. If e is a parent edge, the new tree will have size $s(r(e)) - s(e)$, where $r(e)$ is the root edge of $T(e)$.

Note that both forced-assignment techniques (flow-based and subdivision-based) must compute a new packing bound $p(\mathcal{T}')$ for each vertex v they process. Although they need only $O(\deg(v))$ time to transform $\mathcal{T}$ into $\mathcal{T}'$, actually computing the packing bound from scratch can be costly. Our implementation uses an *incremental algorithm* instead. When computing the original $p(\mathcal{T})$ bound, we remember the entire state of its computation (including the sorted list of all original tree weights). To compute $p(\mathcal{T}')$, we can start from this initial state, discarding trees that are no longer valid and considering new ones appropriately.

6 The Full Algorithm

We test our improved bounds by incorporating them into Delling et al.'s branch-and-bound routine [8]. We process each node of the branch-and-bound tree as follows. We first compute the flow bound, then add to it our new edge-based packing bound (which fully replaces their vertex-based bound). If the result is not smaller than the best known upper bound U, we prune. Otherwise, we try both types of forced assignment, then branch. The remainder of this section discusses branching rules, upper bounds, and an optional decomposition technique.

VBB branches on the free vertex v that maximizes a certain *score* based on three parameters: the degree of v, the *distance* from v to $A \cup B$, and the average *weight* of the trees $\mathcal{T}(v)$ that v belongs to. Together, these criteria aim to maximize the overall (flow and packing) bound. Besides these, we propose a fourth criterion: whenever the current minimum cut A–B is almost balanced, we prefer to branch on vertices that already carry some flow in order to increase the number of reachable vertices in G_f.

Like VBB, we only update the best *upper bound* U when the minimum A–B cut happens to be balanced. Moreover, we do not use any heuristics to try to find a good initial bound U. Instead, we just call the branch-and-bound algorithm repeatedly, with increasing values of U, and stop when the bound it proves is better than the input. We use $U_1 = 1$ for the first call, and set $U_i = \lceil 1.05 U_{i-1} \rceil$ for call $i > 1$. (Delling et al. suggest using 1.5 instead of 1.05, but this is too aggressive for nontrivial instances; we therefore run VBB with 1.05 in our experiments.)

Finally, we consider *decomposition*. As in VBB, the quality of our lower bounds depend on the degrees of the vertices already assigned to A or B (which limit both the A–B flow and the number of trees). If a graph with small degrees has a large bisection, the branch-and-bound tree can get quite deep. Delling et al. [8] propose a decomposition-based preprocessing technique to get around this. Let U be an upper bound on the optimum bisection of the input graph $G = (V, E)$. Partition E into $U + 1$ sets $E_0, E_1, \ldots, E_U$, and for each i create a new graph G_i by taking G and *contracting* all edges in E_i. To solve G, we simply solve each G_i to optimality (with our standard branch-and-bound algorithm) and return the best solution found. At least one subproblem must preserve the optimum solution, since none of the solution edges will be contracted. Delling et al. propose partitioning the edges into clumps (paths with many neighbors) to ensure that, after contraction, each graph G_i will have at least a few high-degree vertices. We generate clumps similarly (see [8] for details), adjusting a few parameters to better suit our stronger lower bounds: we allow clumps to be twice as long, and randomize the distribution of clumps among subproblems.

7   Experiments

We implemented our algorithm in C++ and compiled it with full optimization on Visual Studio 2010. All experiments were run on a single core of an Intel Core 2 Duo E8500 with 4 GB of RAM running Windows 7 Enterprise at 3.16 GHz.

We test the effectiveness of our approach by comparing our branch-and-bound algorithm with other exact approaches proposed in the literature. We consider a set of benchmarks compiled by Armbruster [2] containing instances used in VLSI design (alue, alut, diw, dmxa, gap, taq), meshes (mesh), random graphs (G), random geometric graphs (U), and graphs deriving from sparse symmetric linear systems (KKT) and compiler design (cb). In each case, we use the same value of ϵ tested by Armbruster, which is either 0 or 0.05 (but note that his definition of ϵ differs slightly).

Table 1. Performance of our algorithm compared with the best available times obtained by Armbruster [1], Hager et al. [11], and Delling et al.'s VBB [8]. Columns indicate number of nodes (n), number of edges (m), allowed imbalance (ϵ), optimum bisection value (*opt*), number of branch-and-bound nodes (BB), and running times in seconds; "—" means "not tested" and *DNF* means "not finished in at least 5 hours".

NAME	n	m	ϵ	*opt*	BB	TIME	[Arm07]	[HPZ11]	VBB
G124.02	124	149	0.00	13	426	0.08	13.91	4.21	0.06
G124.04	124	318	0.00	63	204999	52.80	4387.67	953.63	768.20
G250.01	250	331	0.00	29	41754	10.58	1832.25	10106.14	16.34
KKT_capt09	2063	10936	0.05	6	33	0.16	1164.88	4658.59	0.10
KKT_skwz02	2117	14001	0.05	567	891	7.87	*DNF*	—	*DNF*
KKT_plnt01	2817	24999	0.05	46	12607	72.29	*DNF*	—	*DNF*
KKT_heat02	5150	19906	0.05	150	9089	120.13	*DNF*	—	614.04
U1000.05	1000	2394	0.00	1	456	0.33	53.62	—	0.27
U1000.10	1000	4696	0.00	39	2961	8.83	1660.63	—	17.38
U1000.20	1000	9339	0.00	222	38074	276.05	*DNF*	—	17469.18
U500.05	500	1282	0.00	2	138	0.12	19.81	—	0.30
U500.10	500	2355	0.00	26	967	1.32	495.91	—	2.42
U500.20	500	4549	0.00	178	66857	225.02	*DNF*	—	*DNF*
alut2292.6329	2292	13532	0.05	154	24018	251.05	391.76	—	3058.29
alue6112.16896	6112	36476	0.05	272	378320	13859.72	4774.15	—	*DNF*
cb.47.99	47	3906	0.00	765	270	1.24	5.28	0.29	*DNF*
cb.61.187	61	33281	0.00	2826	793	91.81	81.35	0.80	*DNF*
diw681.1494	681	3081	0.05	142	5362	12.38	*DNF*	—	*DNF*
diw681.3103	681	18705	0.05	1011	6673	162.85	*DNF*	—	*DNF*
diw681.6402	681	7717	0.05	330	2047	9.22	4579.12	—	*DNF*
dmxa1755.10867	1755	13502	0.05	150	2387	28.59	*DNF*	—	57.10
dmxa1755.3686	1755	7501	0.05	94	10390	63.42	1972.22	—	371.99
gap2669.24859	2669	29037	0.05	55	74	1.31	348.95	—	0.52
gap2669.6182	2669	12280	0.05	74	3225	31.26	651.03	—	105.87
mesh.138.232	138	232	0.00	8	124	0.05	10.22	6.91	0.04
mesh.274.469	274	469	0.00	7	79	0.05	8.52	24.62	0.38
taq170.424	170	4317	0.05	55	193	0.53	28.68	—	8.14
taq334.3763	334	8099	0.05	341	1379	4.88	*DNF*	—	*DNF*
taq1021.2253	1021	4510	0.05	118	3373	11.93	169.65	—	283.04

Table 1 reports the results obtained by our algorithm (using decomposition for U, alue, alut, and KKT_heat02). For comparison, we also show the running times obtained by recent state-of-the-art algorithms by Armbruster et al. [3,1] (using linear or semidefinite programming, depending on the instance), Hager et al. [11] (quadratic programming), and Delling et al.'s VBB approach. Our method and VBB were run on a 3.16 GHz Core 2 Duo, while Armbruster used a 3.2 GHz Pentium 4 540 and Hager et al. used a 2.66 GHz Xeon X5355. Since these machines are not identical, small differences in running times (a factor of two or so) should be disregarded.

The table includes all instances that can be solved by at least one method in less than 5 hours (the time limit set by Armbruster [1]), except those that can be solved in less than 10 seconds by both our method and Armbruster's. Note that we can solve every instance in the table to optimality. Although slightly slower than Armbruster's in a few cases (notably alue6112.16896), our method is usually much faster, often by orders of magnitude. We can solve in minutes (or even seconds) several instances no other method can handle in 5 hours.

Our approach is significantly faster than Hager et al.'s for mesh, KKT, and random graphs (G), but somewhat slower for the cb instances, which are small but have heavy edges. Since we convert them to parallel (unweighted) edges, we end up dealing with much denser graphs (as the m column indicates). On such dense instances, fractional allocations help the most: without them, we would need almost 1000 times as many branch-and-bound nodes to solve cb.47.99.

Compared to VBB, our method is not much better for graphs that are very sparse or have small bisections—it can be even slower, since it often takes 50% more time per branch-and-bound node due to its costlier packing computation. As in VBB, however, flow computations still dominate. For denser instances, however, our edge-based approach is vastly superior, easily handling several instances that are beyond the reach of VBB.

With longer runs, both Armbruster and Hager et al. [11] can solve random graphs G124.08 and G124.16 (not shown in the table) in a day or less. We would take about 3 days on G124.08, and a month or more for G124.16. Here the ratio between the solution value (449) and the average degree (22) is quite large, so we can only start pruning very deep in the tree. Decomposition would contract only about three edges per subproblem, which does not help. This shows that there are classes of instances in which our method is clearly outperformed.

For real-world instances, however, we can actually solve much larger instances than those shown on Table 1. In particular, we consider instances from the 10th DIMACS Implementation Challenge [4] representing social and communication networks (class clustering), road networks (streets), Delaunay triangulations (delaunay), random geometric graphs (rgg), and assorted graphs (walshaw) from the Walshaw benchmark [22] (mostly finite-element meshes). We also consider triangulations representing three-dimensional objects in computer graphics (mesh) [18] and grid graphs with holes (vlsi) representing VLSI circuits [15].

Table 2 compares our algorithm with VBB (results are not available for other exact methods). Both use decomposition for all classes but clustering. Since our focus is on lower bound quality, here we ran both algorithms directly with $U = opt + 1$ as an initial upper bound. For each instance, we report the number of branch-and-bound nodes and the running time of our algorithm, as well as its speedup (SPD) relative to VBB, i.e., the ratio between VBB's running time and ours. Note that VBB runs that would take more than a day were actually executed on a cluster using DryadOpt [6], which roughly doubles the total CPU time. The table includes most nontrivial instances tested by Delling et al. [8], and additional instances that could not be solved before. On instances marked *DNF*, VBB would be at least 200 times slower than our method.

Our new algorithm is almost always faster than VBB, which is only competitive for very sparse inputs, such as road networks. For denser graphs, our algorithm can be orders of magnitude faster, and can solve a much greater range

Table 2. Performance on various large instances with $\epsilon = 0$; BB is the number of branch-and-bound nodes, TIME is the total CPU time, and SPD is the speedup over Delling et al.'s VBB algorithm [8]; *DNF* means the speedup would be at least 200.

CLASS	NAME	n	m	*opt*	BB	TIME [S]	SPD
clustering	lesmis	77	820	61	21	0.02	12975.4
	as-22july06	22963	48436	3515	7677	417.27	*DNF*
delaunay	delaunay_n11	2048	6127	86	4540	18.87	9.3
	delaunay_n12	4096	12264	118	13972	140.64	19.3
	delaunay_n13	8192	24547	156	34549	759.67	49.5
	delaunay_n14	16384	49122	225	635308	30986.82	*DNF*
mesh	cow	2903	8706	79	2652	13.04	5.1
	fandisk	5051	14976	137	6812	81.81	63.4
	blob	8036	24102	205	623992	12475.18	*DNF*
	gargoyle	10002	30000	175	46623	1413.62	33.0
	feline	20629	61893	148	43944	1474.22	3.1
	dragon-043571	21890	65658	148	1223289	53352.66	109.7
	horse	48485	145449	355	121720	21527.24	*DNF*
rgg	rgg15	32768	160240	181	5863	1111.35	148.1
	rgg16	65536	342127	314	43966	24661.47	32.2
streets	luxembourg	114599	119666	17	844	101.21	0.9
vlsi	alue7065	34046	54841	80	9650	350.05	1.4
walshaw	data	2851	15093	189	29095	265.12	21689.9
	crack	10240	30380	184	19645	605.12	479.2
	fe_4elt2	11143	32818	130	1324	42.89	5.2
	4elt	15606	45878	139	4121	187.51	4.1
	fe_pwt	36519	144794	340	2310	394.50	39.7
	fe_body	45087	163734	262	147424	17495.68	*DNF*
	finan512	74752	261120	162	1108	339.70	2.2

of instances. Note that several instances could not be solved with VBB even after days of computation. In contrast, in a few hours (or minutes) we can solve to optimality a wide variety of graphs with tens of thousands of vertices.

8   Conclusion

We have introduced new lower bounds that provide excellent results in practice. They outperform previous methods on a wide variety of instances, and help find provably optimum bisections for several long-standing open instances (such as U500.20 [12]). While most previous approaches keep the branch-and-bound tree small by computing very good (but costly) bounds at the root, our bounds are only useful if some vertices have already been assigned. This causes us to branch more, but we usually make up for it with a faster lower bound computation.

References

1. Armbruster, M.: Branch-and-Cut for a Semidefinite Relaxation of Large-Scale Minimum Bisection Problems. PhD thesis, Technische Universität Chemnitz (2007)

2. Armbruster, M.: Graph Bisection and Equipartition (2007),
   http://www.tu-chemnitz.de/mathematik/discrete/armbruster/diss/
3. Armbruster, M., Fügenschuh, M., Helmberg, C., Martin, A.: A Comparative Study
   of Linear and Semidefinite Branch-and-Cut Methods for Solving the Minimum
   Graph Bisection Problem. In: Lodi, A., Panconesi, A., Rinaldi, G. (eds.) IPCO
   2008. LNCS, vol. 5035, pp. 112–124. Springer, Heidelberg (2008)
4. Bader, D.A., Meyerhenke, H., Sanders, P., Wagner, D.: 10th DIMACS Implementation Challenge: Graph Partitioning and Graph Clustering (2011),
   http://www.cc.gatech.edu/dimacs10/index.shtml
5. Brunetta, L., Conforti, M., Rinaldi, G.: A branch-and-cut algorithm for the equicut
   problem. Mathematical Programming 78, 243–263 (1997)
6. Budiu, M., Delling, D., Werneck, R.F.: DryadOpt: Branch-and-bound on distributed data-parallel execution engines. In: IPDPS, pp. 1278–1289 (2011)
7. Bui, T.N., Chaudhuri, S., Leighton, F., Sipser, M.: Graph bisection algorithms with
   good average case behavior. Combinatorica 7(2), 171–191 (1987)
8. Delling, D., Goldberg, A.V., Razenshteyn, I., Werneck, R.F.: Exact combinatorial
   branch-and-bound for graph bisection. In: ALENEX, pp. 30–44 (2012)
9. Ferreira, C.E., Martin, A., de Souza, C.C., Weismantel, R., Wolsey, L.A.: The
   node capacitated graph partitioning problem: A computational study. Mathematical Programming 81, 229–256 (1998)
10. Garey, M.R., Johnson, D.S.: Computers and Intractability. A Guide to the Theory
    of $\mathcal{NP}$-Completeness. W.H. Freeman and Company (1979)
11. Hager, W.W., Phan, D.T., Zhang, H.: An exact algorithm for graph partitioning.
    Mathematical Programming, 1–26 (2011)
12. Johnson, D.S., Aragon, C.R., McGeoch, L.A., Schevon, C.: Optimization by simu-
    lated annealing: an experimental evaluation; part I, graph partitioning. Operations
    Research 37(6), 865–892 (1989)
13. Karisch, S.E., Rendl, F., Clausen, J.: Solving graph bisection problems with
    semidefinite programming. INFORMS Journal on Computing 12, 177–191 (2000)
14. Karypis, G., Kumar, G.: A fast and high quality multilevel scheme for partitioning
    irregular graphs. SIAM J. Scientific Computing 20(1), 359–392 (1999)
15. Koch, T., Martin, A., Voß, S.: SteinLib: An updated library on Steiner tree prob-
    lems in graphs. Technical Report 00-37, Konrad-Zuse-Zentrum Berlin (2000)
16. Land, A.H., Doig, A.G.: An automatic method of solving discrete programming
    problems. Econometrica 28(3), 497–520 (1960)
17. Rendl, F., Rinaldi, G., Wiegele, A.: Solving max-cut to optimality by intersecting
    semidefinite and polyhedral relaxations. Math. Programming 121, 307–335 (2010)
18. Sander, P.V., Nehab, D., Chlamtac, E., Hoppe, H.: Efficient traversal of mesh edges
    using adjacency primitives. ACM Trans. on Graphics 27, 144:1–144:9 (2008)
19. Sanders, P., Schulz, C.: Distributed evolutionary graph partitioning. In: ALENEX,
    pp. 16–29. SIAM (2012)
20. Sellmann, M., Sensen, N., Timajev, L.: Multicommodity Flow Approximation Used
    for Exact Graph Partitioning. In: Di Battista, G., Zwick, U. (eds.) ESA 2003.
    LNCS, vol. 2832, pp. 752–764. Springer, Heidelberg (2003)
21. Sensen, N.: Lower Bounds and Exact Algorithms for the Graph Partitioning Prob-
    lem Using Multicommodity Flows. In: Meyer auf der Heide, F. (ed.) ESA 2001.
    LNCS, vol. 2161, pp. 391–403. Springer, Heidelberg (2001)
22. Soper, A.J., Walshaw, C., Cross, M.: The Graph Partitioning Archive (2004),
    http://staffweb.cms.gre.ac.uk/~c.walshaw/partition/

On the Complexity of Metric Dimension*

Josep Díaz[1], Olli Pottonen[1], Maria Serna[1], and Erik Jan van Leeuwen[2]

[1] Departament de Llenguatges i Sistemes Informatics, UPC, Barcelona, Spain
{diaz,mjserna}@lsi.upc.edu, olli.pottonen@iki.fi
[2] Dept. Computer, Control, Managm. Eng., Sapienza University of Rome, Italy
E.J.van.Leeuwen@dis.uniroma1.it

Abstract. The metric dimension of a graph G is the size of a smallest subset $L \subseteq V(G)$ such that for any $x, y \in V(G)$ there is a $z \in L$ such that the graph distance between x and z differs from the graph distance between y and z. Even though this notion has been part of the literature for almost 40 years, the computational complexity of determining the metric dimension of a graph is still very unclear. Essentially, we only know the problem to be NP-hard for general graphs, to be polynomial-time solvable on trees, and to have a $\log n$-approximation algorithm for general graphs. In this paper, we show tight complexity boundaries for the METRIC DIMENSION problem. We achieve this by giving two complementary results. First, we show that the METRIC DIMENSION problem on bounded-degree planar graphs is NP-complete. Then, we give a polynomial-time algorithm for determining the metric dimension of outerplanar graphs.

1 Introduction

Given a graph $G = (V, E)$, its *metric dimension* is the cardinality of a smallest set $L \subseteq V$ such that for every pair $x, y \in V$, there is a $z \in L$ such that the length of a shortest path from z to x is different from the length of a shortest path from z to y. In this case we say that x, y are *resolved* by z and L. Elements of the set L are called *landmarks*. A set $L \subseteq V$ that resolves all pairs of vertices is called a *resolving set*. The problem of finding the metric dimension of a given graph G is called METRIC DIMENSION, but is also known as *Harary's problem*, and the *locating number* or *rigidity* problem. The problem was defined independently by Harary and Melter [15] and Slater [21].

The reasons for studying this problem are three-fold. First, even though the problem is part of Garey and Johnson's famous book on computational intractability [14], very little is known about the computational complexity of this problem. Garey and Johnson proved thirty years ago that the decision version of METRIC DIMENSION is NP-complete on general graphs [18] (another proof appears in [19]). It was shown that there exists a $2 \log n$-approximation algorithm

* J. Díaz and M. Serna are partially supported by TIN-2007-66523 (FORMALISM) and SGR 2009-2015 (ALBCOM). O. Pottonen was supported by the Finnish Cultural Foundation. E.J. van Leeuwen is supported by ERC StG project PAAl no. 259515.

L. Epstein and P. Ferragina (Eds.): ESA 2012, LNCS 7501, pp. 419–430, 2012.
© Springer-Verlag Berlin Heidelberg 2012

on arbitrary graphs [19], which is best possible within a constant factor under reasonable complexity assumptions [2,16]. Hauptmann et al. [16] show hardness of approximation on sparse graphs and on complements of sparse graphs. On the positive side, fifteen years ago, Khuller et al. [19] gave a linear-time algorithm to compute the metric dimension of a tree, as well as characterizations for graphs with metric dimension 1 and 2. Epstein et al. [10] provide hardness results for split graphs, bipartite graphs, co-bipartite graphs, and line graphs of bipartite graphs, together with polynomial-time algorithms for a weighted variant of metric dimension for several simple graphs including paths, trees, and cycles. To the best of our knowledge, no further results are known about the complexity of this problem. It is thus interesting if the substantial, long-standing gap on the tractability of this problem (between trees and general graphs) can be closed.

Second, the problem has received a lot of attention from researchers in different disciplines and it is frequently studied from an analytical point of view (see e.g. [1,16] and references therein). For example, a recent survey by Bailey and Cameron [1] notes an interesting connection to group theory and graph isomorphism. It was also shown to be applicable to certain cop-and-robber games [11]. Therefore it makes sense to reopen the investigation on the computational complexity of METRIC DIMENSION and close the above-mentioned complexity gap.

The third reason for studying METRIC DIMENSION, particularly on (outer) planar graphs, is that known techniques in the area do not seem to apply to it. Crucially, it seems difficult to formulate the problem as an MSOL-formula in a natural way, implying that Courcelle's Theorem [5] does not apply. Hence there is no easy way to show that the problem is polynomial-time solvable on graphs of bounded treewidth. Also, the line of research pioneered by Baker [3], which culminated in the recent meta-theorems on planar graphs using the framework of bidimensionality [8,13], does not apply, as METRIC DIMENSION does not exhibit the required behavior. For example, the metric dimension of a grid is two [19], whereas bidimensionality requires it to be roughly linear in the size of the grid. Moreover, the problem is not closed under contraction. This behavior of METRIC DIMENSION contrasts that of many other problems, even of nonlocal problems such as FEEDBACK VERTEX SET. Hence by studying the METRIC DIMENSION problem, there is an opportunity to extend the toolkit that is available to us on planar graphs.

Our Results. In the present work, we significantly narrow the tractability gap of METRIC DIMENSION. From the hardness side, we show that METRIC DIMENSION on planar graphs, called PLANAR METRIC DIMENSION, is NP-hard, even for bounded-degree planar graphs. From the algorithmic side, we show that there is a polynomial-time algorithm to find the metric dimension of outerplanar graphs, significantly improving the known algorithm on trees.

The crux of both of these results is our ability to deal with the fact that the METRIC DIMENSION problem is extremely *nonlocal*. In particular, a landmark can resolve vertices that are very far away from it. The paper thus focusses on constraining the effects of a landmark to a small area. The NP-hardness proof does this by constructing a specific family of planar graphs for which METRIC

DIMENSION is essentially a local problem. The algorithm on outerplanar graphs uses a tree structure to traverse the graph, together with several data structures that track the influence of landmarks on other vertices. As we show later, this is sufficient to keep the nonlocality of the problem in check. We believe that our algorithmic techniques are of independent interest, and could lead to (new) algorithms for a broad class of nonlocal problems.

Overview of the Algorithm. Observe that the standard dynamic programming approach using a tree decomposition fails here as the amount of information one needs to maintain seems to depend exponentially on n, rather than on the width of the decomposition. To overcome this fact we take a different approach.

First, we characterize resolving sets in outerplanar graphs by giving two necessary and sufficient requirements for an arbitrary vertex set to be a resolving set. Then, taking as a base the duals of the biconnected components of the graph G, we define a tree T. Vertices of T correspond to faces and cut vertices of G, and edges to inner edges and bridges of G. Note that each vertex of T is a separator and splits the graph into two parts. The algorithm uses dynamic programming to process T, starting at the leaves and advancing towards the root.

At first sight, this decomposition has the same problem as we had with tree decompositions. Moreover, the size of a face might be arbitrarily big, leading to a decomposition of arbitrary 'width'. To overcome these obstacles, we introduce two data structures, called *boundary conditions* and *configurations*.

Boundary conditions track the effects of landmarks placed in the already processed part of the graph and the possible effects of sets of landmarks to be placed in the unexplored parts of the graphs. The main algorithmic novelty lies in the configurations, which control the process of combining the boundary conditions on edges towards children of the current vertex $v' \in V(T)$ into a boundary condition on the edge towards the parent of v'. The configurations depend on the vertices of G represented by v'. Even though the number of vertices of G represented by v' may be unbounded, we show that the total number of relevant configurations is only polynomial.

The use of configurations presents a stark contrast with the techniques used in bounded treewidth algorithms, where the combination process commonly is a simple static procedure. A similar contrast is apparent in our tree structure: whereas outerplanar graphs have constant treewidth [4], the tree structure used in our approach actually leads to a decomposition that can have arbitrary width. Our methods also provide a simpler algorithm to solve the problem on trees.

Preliminaries. For basic notions and results in graph theory, we refer the reader to any textbook on the topic, e.g. [9]. All graphs are finite, undirected, and unless otherwise stated, connected. We use (u, v) to denote an edge from u to v. By distance $d(u, v)$ we mean the graph distance. The vertex and edge sets of G are denoted by $V(G)$ and $E(G)$, respectively.

Recall that a graph is *outerplanar* if and only if it does not contain a subdivision of K_4 or $K_{2,3}$. A graph G has a *cut vertex* if the removal of that vertex disconnects the graph into two components. A graph is a *biconnected* if it has no cut vertices. If G is a biconnected outerplanar graph, G has a planar embedding in which the edges on the boundary of the outer face form a Hamiltonian cycle. We call those edges *outer edges* and the other edges are called *inner edges*. Given G and $v \in V(G)$, $\mathcal{N}(v)$ denotes the set of neighbors of v in G. Given a set S, we denote by $\mathcal{P}(S)$ the power set of S. Due to the space restrictions many technical details and proofs are omitted. A full version of the paper is available on arXiv.

2 NP-Hardness on Planar Graphs

We reduce from a variation of the 3-SAT problem. We first require some notation. Let Ψ be a boolean formula on a set V of variables and a set C of clauses. The *clause-variable graph* of Ψ is defined as $G_\Psi = (V \cup C, E)$, where $E = \{(v, c) \mid v \in V, c \in C, v \in c\}$. The notation $v \in c$ means that variable v (or its negation) occurs in clause C. Observe that G_Ψ is always bipartite.

It is well known that 3-SAT is NP-complete [14], and that it remains NP-complete even with additional restrictions, such as requiring G_Ψ to be planar [7, p. 877]. As a starting point for our work, we consider the following restrictions. Let 1-NEGATIVE PLANAR 3-SAT be the problem of deciding the satisfiability of a boolean formula Ψ with the following properties: (1) every variable occurs exactly once negatively and once or twice positively, (2) every clause contains two or three distinct variables, (3) every clause with three distinct variables contains at least one negative literal, and (4) G_Ψ is planar.

Lemma 1. 1-NEGATIVE PLANAR 3-SAT *is NP-complete.*

To prove that PLANAR METRIC DIMENSION is NP-hard, we give a reduction from 1-NEGATIVE PLANAR 3-SAT. The idea behind the graph constructed in this reduction is the following. Given an instance Ψ of 1-NEGATIVE PLANAR 3-SAT, we first find a planar embedding of its clause-variable graph G_Ψ. We then replace each variable vertex of G_Ψ by a *variable gadget*, and each clause vertex of G_Ψ by a *clause gadget*. By identifying vertices of variable gadgets and vertices of clause gadgets in an appropriate way, we obtain a planar graph H_Ψ that will be our instance of PLANAR METRIC DIMENSION.

Let n be the number of variables. Each variable gadget is constructed such that it must contain 4 landmarks: 3 at known, specific locations, but for the fourth we have three different choices. They correspond to the variable being true, false, or undefined. These $4n$ landmarks are a resolving set if and only if they resolve all pairs of vertices in clause gadgets, which happens only if they correspond to a satisfying truth assignment of the SAT-instance. Then we get the following theorem (technical details are deferred to the full version).

Theorem 2. PLANAR METRIC DIMENSION *is NP-complete, even on bounded-degree graphs.*

3 Algorithm for Outerplanar Graphs

In this section, we prove that METRIC DIMENSION can be solved in polynomial time on outerplanar graphs. So let G be an outerplanar graph, given together with an outerplanar embedding. Note that a metric base of a disconnected graph is the union of metric bases of its components[1]. So we can safely assume that G is a connected graph. We can also assume that it has at least three vertices.

As a technical trick we sometimes treat the midpoint of an inner edge $e = (v_1, v_2) \in E(G)$ as an actual vertex. The distances from this *midpoint vertex* v_e are such that $d(v_e, v_1) = d(v_e, v_2) = \frac{1}{2}$ and $d(v_e, x) = \min(d(v_e, v_1) + d(v_1, x), d(v_e, v_2) + d(v_2, x))$.

3.1   Characterization of Resolving Sets in Outerplanar Graphs

We first give a characterization of the effects of resolving sets in vertices and faces of outerplanar graphs. To this end, we introduce some notation. A *bifurcation point* associated with z, x, y is a vertex v farthest from z such that v is on shortest path from z to both x and y. More formally, v is a bifurcation point if it is on shortest paths $z \rightsquigarrow x$, $z \rightsquigarrow y$, and if any shortest paths $v \rightsquigarrow x$, $v \rightsquigarrow y$ intersect only in v. Notice that in an outerplanar graph the bifurcation point for each triple of vertices is unique.

We define the function $g : V(G) \times \mathcal{P}(V(G)) \to \mathcal{P}(V(G))$ as

$$g(v, L) = \{w \in \mathcal{N}(v) : d(z, w) = d(z, v) + 1 \text{ for all } z \in L\}.$$

In other words, a neighbor w of v is in $g(v, L)$ if for every $z \in L$, v is on some shortest path $z \rightsquigarrow w$ (but not necessarily on every shortest path.)

Any pair $x, y \in g(v, L)$ is left unresolved by landmarks in L. So any resolving set L satisfies the following:

Requirement 1. *Any vertex $v \in V(G)$ must have $|g(v, L)| \leq 1$.*

If G is a tree, then the requirement is sufficient. The following lemma gives an alternative and simpler correctness proof for the algorithm by Khuller et al. [19].

Lemma 3. *If G is a tree with at least three vertices, then a set $L \subseteq V(G)$ is a resolving set if and only if it satisfies Requirement 1.*

Proof. We have already seen that any resolving set L satisfies Requirement 1. Now assume that Requirement 1 is satisfied. We pick any two vertices $x, y \in V(G)$ and show that they are resolved.

Since G has at least 3 vertices, there is at least one vertex $v \in V(G)$ with degree at least 2. Since $|g(v, L)| \leq 1 < |\mathcal{N}(v)|$, L is not empty.

Choose any landmark $z \in L$. If z resolves x, y, we are done. Otherwise, let v be the bifurcation point associated with z, x, y, and let v_1, v_2 be the successors of v on the shortest paths $v \rightsquigarrow x, v \rightsquigarrow y$. Since $d(z, x) = d(z, y)$, we have $d(v, x) = d(v, y)$. By assumption, $g(v, L)$ can not contain both v_1 and v_2. Without loss of generality $v_1 \notin g(v, L)$. Then there is a landmark z_2 with $d(z_2, v_1) < d(z_2, v)$, and furthermore $d(z_2, x) < d(z_2, y)$.   □

---

[1] With one exception: isolated vertices. An edgeless graph of n vertices has metric dimension $n - 1$.

As stated earlier, the major difficulty of the metric dimension problem is that it is non-local. This is why Lemma 3 is useful. Although stopping short of giving an actual local characterization of resolving sets, it does make the effects of a resolving set local enough to devise a polynomial-time algorithm for trees.

Our algorithm relies on a generalization of Lemma 3 to outerplanar graphs. In this case Requirement 1 no longer implies that L is a resolving set. For example, if G is an even cycle and L contains two antipodal vertices, then Requirement 1 is satisfied, but L is not a resolving set. Therefore we need another requirement that provides an additional condition for the effects of a resolving set on the faces of outerplanar graphs. First, we need some auxiliary definitions and lemmas.

Definition 4. *Let $z \in V(G)$ be a landmark, and let C be either a single edge or a cycle—in particular, it may be a face. The* representative *of z on C is the element of $V(C)$ closest to z, if it is unique. Otherwise outerplanarity implies that there are two closest vertices, which are adjacent. In this case the representative is the midpoint of those two vertices.*

We can make the following observations. Let C, C' be cycles such that $V(C') \subseteq V(C)$ and C' is a face. If two vertices have the same representative on C, then they must have the same representative on C' as well.

Lemma 5. *Let G be a graph, and let $x, y, z, z_2 \in V(G)$. If neither z nor z_2 resolves the pair x, y and there exist intersecting shortest paths $z \rightsquigarrow x$, $z_2 \rightsquigarrow y$, then a bifurcation point of z, x, y is also a bifurcation point of z_2, x, y.*

We are now ready to present the other necessary requirement for a set of landmarks L to be a resolving set.

Requirement 2. *Let v' be a face of an outerplanar graph G in which L has exactly two representatives $\hat{z}_1, \hat{z}_2$ on v'. Let z_1^f and z_1^l be the landmarks with representative $\hat{z}_1$ which occur first and last on the walk along the outer face starting at $\hat{z}_2$, and define z_2^f, z_2^l analogously. Assume that neither z_1^f nor z_2^f resolves a pair x, y, and that shortest paths $z_1^f \rightsquigarrow x$, $z_2^f \rightsquigarrow y$ do not intersect. Let v be the bifurcation point of z_1^f, x, y, let u be the bifurcation point of x, z_1^f, z_2^f, let s be the bifurcation point of y, z_1^f, z_2^f and let t be the bifurcation point of z_2^f, x, y. By assumption, $v \neq t$. Therefore the shortest paths $s \rightsquigarrow t$, $t \rightsquigarrow u$, $u \rightsquigarrow v$, $v \rightsquigarrow s$ form a cycle C. If v' is inside the cycle C, then one of z_1^l, z_2^l must resolve the pair x, y.*

Note that the representatives of z_1^f and z_2^f on C are v and t, respectively.

The assumption that G is outerplanar is essential for Requirement 2 and Lemma 6 below. In particular, the definition of $z_1^f, z_2^l, z_2^f, z_2^l$, as well as the use of representatives in the proof of Lemma 6, relies on outerplanarity.

Lemma 6. *If G is an outerplanar graph, then any resolving set $L \subseteq V(G)$ satisfies Requirement 2.*

Now we are ready to generalize Lemma 3 to outerplanar graphs. This is a crucial result, since it characterizes resolving sets in a manner that allows dynamic programming.

Theorem 7. *If G is an outerplanar graph, then a set $L \subseteq V(G)$ is a resolving set if and only if it satisfies Requirements 1 and 2.*

Proof. We have already seen that any resolving set satisfies the requirements. So assume that $L \subseteq V(G)$ satisfies the Requirements, and choose any $x, y \in V(G)$. We show that there exists a $z \in L$ that resolves the pair x, y.

As in Lemma 3, L is non-empty. Choose $z \in L$ arbitrarily. If z resolves x and y, we are done; so assume that it does not. As in Lemma 3, let v be the bifurcation point of z, x, y, and let v_1, v_2 be successors of v on some shortest paths $v \rightsquigarrow x, v \rightsquigarrow y$ respectively. By Requirement 1, there is a $z_2 \in L$ such that, without loss of generality, $d(z_2, v_1) \leq d(z_2, v)$. If z_2 resolves x and y, we are done. So assume otherwise.

If there exist intersecting shortest paths $z \rightsquigarrow x$, $z_2 \rightsquigarrow y$, then by Lemma 5, v is on a shortest path $z_2 \rightsquigarrow x$. Then v_1 is also on such a path, and $d(z_2, v_1) = d(z_2, v) + 1$, a contradiction. Therefore no such pair of intersecting shortest paths exists. Define v, t, C as in Requirement 2, and let v' be a face inside C. If there exists a $z_3 \in L$ whose representative on v' is distinct from the representatives of z_1 and z_2, then its representative on C is neither v nor t. Hence z_3 resolves x, y. If such a landmark z_3 does not exist, then Requirement 2 implies the claim. □

3.2 Data Structures

We now introduce the complex data structures that we need in our algorithms. The *weak dual* of an outerplanar graph G is a graph which has the faces of G, except the unbounded outer face, as vertices. Two faces are adjacent if they share an edge. The weak dual of a biconnected outerplanar graph is a tree.

The *generalized dual tree* $T = (V', E')$ of an outerplanar graph G is defined as follows. T contains the weak dual of G, and also contains the subgraph of G induced by all cut vertices and vertices of degree one. For any cut vertex contained in a biconnected component, there is an edge from the vertex to an arbitrary incident face of the component. Observe that the resulting graph is a tree. According to this definition, a cut vertex is a vertex of both G and T. Let any vertex of T be the root, denoted by v'_r.

We now associate a subset of $V(G)$ with any vertex or edge of T. If $v' \in V(T)$ is a face, the set $s(v')$ consists of the vertices on the face. If v' is a cut vertex or a leaf, $s(v')$ consists of that vertex. If both endpoints of $e' \in E(T)$ are vertices of G, then $s(e')$ consists of those vertices. Otherwise, let either endpoint of e' is a face. Let $e' = (v'_1, v'_2)$, and define $s(e') = s(v'_1) \cap s(v'_2)$.

Removing an edge (v', w') divides T into two components, $T_{v'}$ and $T_{w'}$, where $T_{v'}$ is the one containing v'. Define $B(v', w')$ as the subgraph of G corresponding to $T_{v'}$. Formally, it is the subgraph of G induced by $\bigcup_{u' \in V(T_{v'})} s(u')$. Thus G is divided into two subgraphs $B(v', w')$, $B(w', v')$. If v', w' are adjacent faces, the subgraphs share two vertices. If v' is a face and w' a cut vertex (or the other way around), then the subgraphs share one vertex. Otherwise they do not intersect. Define $B^-(v', w')$ as the subgraph of G induced by $V(G) \setminus V(B(w', v'))$. Then we can divide G into two nonintersecting subgraphs, $B^-(v', w')$ and $B(w', v')$.

The following lemma is immediate from the definitions.

Lemma 8. *Given neighbors* $v', w' \in V(T)$, $B(v', w')$ *and* $B(w', v')$ *are connected subgraphs of* G, *and any path from* $B(v', w')$ *to* $B(w', v')$ *intersects* $s((v', w'))$.

Given a landmark z, let $S(z, W) = \{\{x, y\} \in W \times W : d(z, y) \neq d(z, x)\}$ be the set of pairs in $W \subseteq V(G)$ resolved by z.

Lemma 9. *Let* $e' = (v', w') \in E(T)$. *Assume* $z \in V(B(v', w'))$ *and denote* $W = V(B(w', v'))$. *Then* $S(z, W) = S(\hat{z}, W)$, *where* $\hat{z}$ *is the representative of* z *on* e'.

Let v' be a dual vertex and p' its parent. We use *boundary conditions* to describe the relation of landmarks in $B(v', p')$ to $B(p', v')$, and vice versa. Boundary conditions can be seen as an abstract data structure which depends on v', p', L and satisfies the following:

1. It consists of two components, one of which depends on landmarks in $B(p', v')$, and the other on landmarks in $B^-(v', p')$. The components are called *upper* and *lower* boundary conditions, respectively.
2. It determines which pairs in $B(p', v')$ are resolved by landmarks in $B^-(v', p')$, and vice versa.
3. If $B(v', p')$ contains landmarks with a representative v on v', then the boundary condition specifies which such landmarks occur first and last on a walk along the outer boundary of $B(v', p')$ starting at v.
4. For any $v \in s((p', v'))$, it specifies whether the set $g(v, L) \cap V(B^-(v', p'))$ is empty or not.
5. The number of possible boundary conditions is polynomial.

The first and the second properties are necessary to be able to run a dynamic programming algorithm along T. The third and fourth properties are needed when verifying Requirements 2 and 1, respectively. The last property is needed to ensure that the algorithm runs in polynomial time.

While a boundary condition describes how the landmarks are placed in relation to a dual edge e', a *configuration* describes their relation to a dual vertex v'. There are three quite different cases, depending on whether v' is a cut vertex of G, a face with two representatives, or a face with at least three representatives. A configuration L associated with v' is designed to satisfy following:

1. For any $w' \in \mathcal{N}(v')$, it determines which pairs of vertices in $B(w', v')$ are resolved by landmarks in $B(v', w')$.
2. It contains enough information to verify that v' satisfies Requirement 2.
3. The total number of configurations is polynomial.

Essentially, a configuration determines which vertices of $s(v')$ are landmarks, and which boundary conditions are allowed for edges (v', w').

Even if both boundary conditions for edges (v', w'_1), (v', w'_2) are allowed by a specific configuration, they may contradict each other. This happens when there is $v \in s((v', w'_1)) \cap s((v', w'_2))$, and the boundary conditions disagree about the value of $g(v, L)$. Hence, the algorithm will only process boundary conditions that agree with each other and with the configuration.

As a tool for avoiding such disagreements, we define a coarser variant of the function $g(\cdot, \cdot)$. The function $h : V(G) \times V(T) \times \mathcal{P}(V(T)) \to V(T) \cup \{\emptyset\}$ describes which part of the generalized dual contains $g(v, L)$. Let $v \in s(v')$. Then

$$h(v, v', L) = \begin{cases} v' & \text{if } g(v, L) \cap s(v') \neq \emptyset, \\ w' & \text{if } w' \text{ is neighbor of } v' \text{ and } g(v, L) \cap V(B^-(w', v')) \neq \emptyset, \\ \emptyset & \text{if } g(v, L) = \emptyset. \end{cases}$$

Notice that as long as Requirement 1 holds, the function is well defined.

Lemma 10. *The configuration associated with v' and L determines for any $v \in s(v')$ whether the equation $h(v, v', L) = v'$ holds.*

3.3 Algorithm

The algorithm works in a dynamic programming manner, by finding optimal resolving sets of subgraphs $B(v', p')$ with given boundary conditions. Formally, assume that we are given a vertex $v' \in V(T)$ and boundary condition $\mathbf{r}$ on the edge $e' = (v', p')$, where p' is the parent of v'. Let $m(v', \mathbf{t}) \subseteq V(B^-(v', p'))$ be a set $L \cap V(B^-(v', p'))$, where L is a minimal resolving set with boundary condition $\mathbf{t}$, if such a set exists. Otherwise $m(v', \mathbf{t}) = \text{NIL}$. For notational convenience, we let $|\text{NIL}| = \infty$ and $\text{NIL} \cup A = \text{NIL}$ for any A.

The values of $m(v', \mathbf{t})$ are computed in a recursive manner: the computation of $m(v', \mathbf{t})$ uses the values of $m(w', \mathbf{r})$, where w' is a child of v'. The basic idea is to iterate over all configurations on v'. For every configuration, we then find an optimal function h and an optimal set of landmarks.

First, we introduce several subroutines. Algorithm 1 (Intermediate-sol) returns, given a configuration (v', R), a boundary condition $\mathbf{t}$, and a function h, an optimal set of landmarks.

Given a configuration (v', R) and a boundary condition $\mathbf{t}$, the next subroutine finds an optimal function h. By Lemma 10, the configuration determines whether $h(v, v', L) = v'$ holds or not. Also $\mathbf{t}$ restricts some values of h. Otherwise, the value of $h(v, v', L)$ may be $\emptyset$ or w', where w' is a suitable neighbor of v', and the task is to determine which one of these is optimal. It turns out that the optimum can be found by a greedy algorithm (Algorithm 2).

Lemma 11. *Algorithm 2 runs in polynomial time and returns a smallest resolving set of $B(v', p')$ that agrees with the parameters.*

We just emphasize that the greedy algorithm relies on the following observation about cardinalities of optimal solutions L_1, L_2 with $h(v, v', L_1) = \emptyset$,

---

Algorithm 1. Intermediate-sol

Input: Configuration (v', R), boundary condition **t**, function h
  if the parameters are inconsistent or (v', R) violates Requirement 2 **then**
    return NIL
  end if
  Initialize W to the set of landmarks on $s(v')$ as described by (v', R)
  for all w' that are children of v' **do**
    $\mathbf{r} \leftarrow \arg\min_{\mathbf{r}} m[w', \mathbf{r}]$ such that $\mathbf{r}$ agrees with $C, h, \mathbf{t}$
    $W \leftarrow W \cup m[w', \mathbf{r}]$
  end for
  return $W \cap V(B^-(v', p'))$

$h(v, v', L_2) = w'$. Let $g(v, L_2) = \{w\}$ and note that $M_1 = L_2 \cup \{w\}$ is a solution with $h(v, v', M_1) = \emptyset$. Therefore $|L_1| \leq |M_1| = |L_2| + 1$.

Finally, we are ready to present the main algorithm (Algorithm 3) and its correctness proof.

Theorem 12. *Algorithm 3 correctly computes the values of $m[v', \mathbf{r}]$, and returns a metric base in polynomial time.*

Proof. The algorithm runs in polynomial time, because there is a polynomial number of dual vertices, boundary conditions, and configurations.

We prove by induction that $m[v', \mathbf{r}]$ is computed correctly. Assume that the values of $m[w', \mathbf{t}]$ have been computed correctly for any child w' of v'. Then, because Opt (Algorithm 2) works correctly, the value of $m[v', \mathbf{r}]$ will be computed correctly. Similarly, the return value will be computed correctly, since Opt works correctly and uses correct values of $m[w', \mathbf{r}]$. □

4   Conclusions and Open Problems

We have showed that METRIC DIMENSION is NP-hard for planar graphs, even when the graph has bounded degree (an open problem from 1976). We also gave a polynomial-time algorithm to solve the problem on outerplanar graphs. Our algorithm is based on an innovative use of dynamic programming which allows us to deal with the non-bidimensional, global problem of METRIC DIMENSION.

We pose some open problems about METRIC DIMENSION. First, it would be nice to extend our results to k-outerplanar graphs. The main obstacle to extending the result is that the separators to be associated with nodes of the computation tree should include faces and edges between consecutive levels. For such separators we lose the relevant property that shortest paths between nodes in different parts cross the separator only once.

Even if the problem turns out to be solvable on k-outerplanar graphs by a polynomial-time algorithm, it is not clear that such an algorithm could be used to derive a polynomial-time approximation scheme for PLANAR METRIC DIMENSION. The quest for such an approximation scheme or even for a constant approximation algorithm is an interesting challenge in its own right.

Algorithm 2. Opt

Input: Configuration (v', R), boundary condition $\mathbf{t}$

   $Q \leftarrow \emptyset$ {Q is the set of vertices for which $h[v]$ is already fixed}

   for all $v \in s(v')$ **do**

      if R or $\mathbf{r}$ determine $h(v, v', L)$ **then**

         $h[v] \leftarrow$ the appropriate value

         $Q \leftarrow Q \cup \{v\}$

      else

         $h[v] \leftarrow \emptyset$

      end if

   end for

   for all w' that are children of v', in clockwise order, starting on the successor of p'
   do

      $P \leftarrow s((v', w')) \setminus Q$

      Find k that minimizes |Intermediate-sol$(v', R, \mathbf{r}, k)$| such that

      $h[v]$ and $k[v]$ differ only for $v \in P$. If possible, choose a solution with $k[v_l] = \emptyset$ for

      the last (in clockwise order) element $v_l \in P$

      $h \leftarrow k$

      $Q \leftarrow Q \cup \{v : h[v] \neq \emptyset\}$

   end for

   return Intermediate-sol$(v', R, \mathbf{r}, h)$

Algorithm 3. Metric-D

Input: Graph G, its generalized dual tree T

   for all $v' \in V(T) \setminus \{v'_r\}$, boundary condition $\mathbf{r}$ **do** {recall that v'_r is the root}

      $m[v', \mathbf{r}] \leftarrow$ NIL

   end for

   for all $v' \in V(T) \setminus \{v'_r\}$, in postorder **do**

      $p' \leftarrow$ the parent of v'

      for all configuration (v', R) **do**

         for all boundary condition condition $\mathbf{r}$ on (v', p') **do**

            if |Opt$(v', R, \mathbf{r})$| $\leq$ |$m[v', \mathbf{r}]$| **then**

               $m[v', \mathbf{r}] \leftarrow$ Opt$(v', R, \mathbf{r})$

            end if

         end for

      end for

   end for

   $W \leftarrow V(G)$

   for all configuration (v'_r, R) **do**

      if |Opt(v'_r, R, NIL)| $\leq |W|$ **then**

         $W \leftarrow$ Opt(v'_r, R, NIL)

      end if

   end for

   return W

Another interesting line of research is the *parameterized complexity* of METRIC DIMENSION. Daniel Lokshtanov [20] posed this problem at a recent Dagstuhl seminar on parametrized complexity. Moreover, he conjectured that the problem could be $W[1]$-complete. We hope that the insights of this paper can help to obtain results in this direction.

Acknowledgment. The authors thank David Johnson for sending a copy of the NP-completeness proof.

References

1. Bailey, R.F., Cameron, P.J.: Base size, metric dimension and other invariants of groups and graphs. Bulletin London Math. Society 43(2), 209–242 (2011)
2. Beerliova, Z., Eberhard, T., Erlebach, T., Hall, A., Hoffmann, M., Mihalak, M., Ram, L.: Network Discovery and Verification. IEEE J. Selected Areas in Communication 24, 2168–2181 (2006)
3. Baker, B.: Approximation algorithms for NP-complete problems on planar graphs. JACM 41, 153–180 (1994)
4. Bodlaender, H.L.: Classes of Graphs with Bounded Treewidth. Bulletin of the EATCS 36, 116–125 (1988)
5. Courcelle, B.: Graph rewriting: An algebraic and logic approach. In: Handbook of Theoretical Computer Science, vol. B, pp. 194–242. Elsevier Science (1990)
6. Chartrand, G., Eroh, L., Johnson, M.A., Oellemann, O.R.: Resolvability in graphs and the metric dimension of a graph. Discrete Applied Math. 105, 99–113 (2000)
7. Dahlhaus, E., Johnson, D.S., Papadimitriou, C.H., Seymour, P.D., Yannakakis, M.: The Complexity of Multiterminal Cuts. SIAM J. Computing 23, 864–894 (1994)
8. Demaine, E.D., Hajiaghayi, M.T.: Bidimensionality: new connections between FPT algorithms and PTASs. In: SODA 2005, pp. 590–601 (2005)
9. Diestel, R.: Graph Theory. Springer (2000)
10. Epstein, L., Levin, A., Woeginger, G.J.: The (weighted) metric dimension of graphs: hard and easy cases. In: Proc. WG 2012 (to appear, 2012)
11. Erickson, L.H., Carraher, J., Choi, I., Delcourt, M., West, D.B.: Locating a robber on a graph via distance queries (preprint)
12. Eppstein, D.: Diameter and treewidth in minor-closed graph families. Algorithmica 27, 275–291 (2000)
13. Fomin, F.V., Lokshtanov, D., Raman, V., Saurabh, S.: Bidimensionality and EP-TAS. In: SODA 2011, pp. 748–759 (2011)
14. Garey, M.R., Johnson, D.S.: Computers and Intractability: A Guide to the Theory of NP-Completeness. Freeman (1979)
15. Harary, F., Melter, R.A.: The metric dimension of a graph. Ars Combinatoria 2, 191–195 (1976)
16. Hauptmann, M., Schmied, R., Viehmann, C.: On approximation complexity of metric dimension problem. In: J. Discrete Algorithms (2011) (in press)
17. Hopcroft, J., Tarjan, R.E.: Efficient planarity testing. JACM 21, 549–568 (1974)
18. Johnson, D.S.: Personal communication
19. Khuller, S., Raghavachari, B., Rosenfeld, A.: Landmarks in Graphs. Discrete Applied Math. 70, 217–229 (1996)
20. Lokshtanov, D.: Metric Dimension. In: Demaine, E.D., Hajiaghayi, M.T., Marx, D. (eds.) Open Problems from Dagstuhl Seminar 09511 (2009), http://erikdemaine.org/papers/DagstuhlFPT2009Open/paper.pdf
21. Slater, P.: Leaves of trees. Congressus Numerantium 14, 549–559 (1975)

Embedding Paths into Trees: VM Placement to Minimize Congestion

Debojyoti Dutta[1], Michael Kapralov[2,*], Ian Post[2,**],
and Rajendra Shinde[2,* * *]

[1] Cisco Systems
ddutta@gmail.com
[2] Stanford University, Stanford, CA, USA
{kapralov,itp,rbs}@stanford.edu

Abstract. Modern cloud infrastructure providers allow customers to rent computing capability in the form of a network of virtual machines (VMs) with bandwidth guarantees between pairs of VMs. Typical requests are in the form of a chain of VMs with an uplink bandwidth to the *gateway* node of the network (*rooted path requests*), and most data center architectures route network packets along a spanning tree of the physical network. VMs are instantiated inside servers which reside at the leaves of this network, leading to the following optimization problem: given a rooted tree network T and a set of rooted path requests, find an embedding of the requests that minimizes link congestion.

Our main result is an algorithm that, given a rooted tree network T with n leaves and set of weighted rooted path requests, embeds a $1 - \epsilon$ fraction of the requests with congestion at most $\text{poly}(\log n, \log \theta, \epsilon^{-1})$ · OPT (approximation is necessary since the problem is NP-hard). Here OPT is the congestion of the optimal embedding and θ is the ratio of the maximum to minimum weights of the path requests. We also obtain an $O(H \log n/\epsilon^2)$ approximation if node capacities can be augmented by a $(1 + \epsilon)$ factor (here H is the height of the tree). Our algorithm applies a randomized rounding scheme based on Group Steiner Tree rounding to a novel LP relaxation of the set of subtrees of T with a given number of leaves that may be of independent interest.

1 Introduction

Many web applications today use cloud computing services offered by data center providers like Amazon (EC2) to buy computing capabilities in the form of a group of virtual machines(VM) and require bandwidth guarantees between pairs of VMs. A request for computing capabilities can be thought of as a graph, where nodes are VMs and edge weights represent bandwidth requirements. To service such a request, the VMs are mapped to servers in the data center and network links are used to satisfy communication requirement between the VMs. Since

* Research supported by NSF grant 0904314.
** Research supported by NSF grants 0915040 and 0904314.
* * * Research supported by NSF grants 0915040.

network resources are currently more expensive than computing resources, we study the problem of finding the *minimum congestion embedding* of the requests with a goal of improving network utilization.

In most data center architectures, routing protocols employed are usually based on a spanning tree of the physical network. VMs are instantiated inside servers that reside at the leaves of this tree. Also, requests for common workloads like web services are often in the form of a chain of VMs with an uplink bandwidth requirement from the first VM to the *gateway* node (root of the spanning tree) of the network. We refer to such requests as *rooted path requests*. For instance, a web service request usually consists of a chain of the following VMs: web, app, database, and firewalls in between with bandwidth requirements between pairs of VMs and an uplink bandwidth requirement from the web server VM to the gateway node of the network.

Motivated by these practical simplifications, we study the following problem: given a tree network T of height H and a set of rooted path requests, find an embedding of the requests minimizing the link congestion in T. We start by showing that the problem is NP-hard, and then propose a polynomial time algorithm which satisfies $1 - \epsilon$ fraction of the requests incurring a congestion of $\mathrm{poly}(\log n, \log \theta, \epsilon^{-1}) \cdot \mathrm{OPT}$, where OPT is the congestion of the optimal embedding, θ is the ratio of maximum and minimum request bandwidth requirements and n is the number of leaves in T. We also obtain an $O(H \log n / \epsilon^2)$ approximation if node capacities can be augmented by a $1 + \epsilon$ factor.

1.1   Related Work

A number of heuristic methods have been proposed that perform VM allocation in data centers [4], [12], [11], [9]. However, to the best of our knowledge, the only work that provides solutions with provable guarantees is that of Bansal et al[3].

Comparison to [3]. In [3] the authors considered a closely related problem, in which an arbitrary set of request topologies need to be mapped to nodes of a network G while respecting edge and node capacities. Their problem is different from ours in the following two aspects. First, the request topologies can be arbitrary, as opposed to rooted path requests that we consider (in particular, in [3] there is no notion of a gateway node to which the request topologies must maintain connectivity). Second, their objective is to minimize edge *and node* congestion, while we can either handle hard capacity constraints with bicriteria approximation(Theorem 1) or minimize edge congestion with $1+\epsilon$ augmentation of node capacities (Theorem 2).

It is also interesting to contrast the algorithmic techniques. The authors of [3] show that embedding multiple requests in their setting reduces to a generalized problem of embedding a *single request* with a polylogarithmic loss in congestion. The resulting embedding problem for a single request turns out to be quite challenging. Nevertheless, the authors show how to solve this problem via an intricate algorithm that uses an LP relaxation inspired by the Sherali-Adams hierarchy, for the case when the request is a tree of depth d. Their solution runs in time $n^{O(d)}$, and thus is only polynomial when the depth is constant. On the other hand, our problem turns out to admit a polynomial time approximation algorithm with a polylogarithmic factor *independent* of the size of request graphs. The approximation to congestion that we achieve improves that of [3] for nodes and matches their approximation for edges when the height of the tree T is constant. Our bicriteria approximation is incomparable to any of the results of [3].

Other Related Work. Routing and admission control problems with a view to minimizing edge congestion have been studied in detail in the point-to-point [1] and multicast settings [2], [6], and oblivious routing schemes have been developed with polylogarithmic competitive ratios for arbitrary networks [10]. These problems deal with requests having *fixed source and destination pairs*, and routing involves choosing paths connecting the source to the destination while satisfying flow requirements. Minimizing link congestion has also been studied for quorum placement involving simultaneous mapping of "quorum" nodes and paths from them to the "client" nodes in a "product multicommodity flow" setting. On the other hand, in our problem the algorithm can choose which nodes handle requests, and the bandwidth requirements could be arbitrary. This makes our problem similar to the quadratic assignment problem(QAP) [7], but our objective is congestion as opposed to minimizing the assignment cost, and it is not clear whether known techniques for QAP can be applied in our setting.

1.2   Our Results and Techniques

Our main result is the following:

Theorem 1. *There exists a polynomial time algorithm that, given a capacitated tree $T = (V, E)$ and a set of weighted requests $\mathcal{P}$, embeds at least $(1-O(\epsilon))|\mathcal{P}|$ requests, incurring congestion at most $\mathrm{poly}(\log n, \log \theta, 1/\epsilon)$ of the optimum, where θ is the ratio of the largest to smallest weight in $\mathcal{P}$.*

Theorem 1 bounds congestion under *hard constraints on node capacities*, but only guarantees embedding a $1 - \epsilon$ fraction of requests. If node constraints can be relaxed, we obtain

Theorem 2. *For any $\epsilon \in (0,1)$ an $O(H \log n/\epsilon^2)$-approximation to congestion with $1 + \epsilon$ node capacity augmentation can be found in polynomial time.*

Note that augmenting a node of (integral) capacity a_u means increasing its capacity to $\lceil (1+\epsilon)a_u \rceil$, which could be a factor of 2 increase when a_u is small. The main technical contribution of our algorithm and analysis is the LP relaxation and rounding scheme used. Our LP formulation uses a novel relaxation of the set of subtrees of a given tree spanning at most k leaves that we term the k-branching polytope and which may be of independent interest. This polytope is a strict generalization of the flow polytope on a tree. The rounding scheme that we employ is a generalization of Group Steiner tree rounding of [5] to the k-branching polytope. In order to highlight the most interesting aspects of our contribution, we now give an overview of our algorithm and analysis.

   Given a set of rooted path requests $\mathcal{P}$ and a host tree T, we first reduce the problem of embedding path requests to that of embedding *subtree requests* with at most a factor 2 loss in congestion (Section 2.2). We then formulate an LP relaxation of the subtree embedding problem by defining an *allocation polytope* $\mathcal{A}(\mathcal{P}, z)$, which represents a fractional allocation of subtrees of T to requests $p \in \mathcal{P}$. The allocation polytope uses a novel relaxation of the set of subtrees of T spanning k leaves for $k = k_p, p \in \mathcal{P}$, which we refer to as the *k-branching polytope*. The polytope is a generalization of the flow polytope on a tree and includes crucial *bounded branching* constraints that allow us to control variance in our rounding scheme. Note that the allocation polytope $\mathcal{A}(\mathcal{P}, z)$ represents a *fractional allocation* of *fractional subtrees* to requests $p \in \mathcal{P}$. Given a point in $\mathcal{A}(\mathcal{P}, z)$, we show how to round it to obtain the desired integral allocation. It is convenient to describe the rounding process in terms of the *reachability graph*. The reachability graph, defined for a fractional solution, is a bipartite graph with requests on one side, leaves of T on the other side and an edge between a request $p \in \mathcal{P}$ and a leaf $l \in L^0$ if l is reachable from p in the current (possibly fractional) solution. Our rounding scheme can be viewed as the process of sparsifying the reachability graph G so as to preserve a large fractional assignment. Theorem 2 will essentially follow from our rounding scheme, while Theorem 1 requires additional preprocessing steps. We now state the main technical steps involved in the proof of Theorem 1:

1. **Decompose $\mathcal{P}$ into a geometric collection $\mathcal{P}'$.** For an arbitrary set of requests $\mathcal{P}$, we decompose $\mathcal{P}$ into a set of requests with weights and lengths in powers of two and $1/\epsilon$ respectively.
2. **Find a point in the allocation polytope $\mathcal{A}(\mathcal{P}', z)$.**
3. **Round the fractional solution.** The input to the rounding scheme is a point in $\mathcal{A}(\mathcal{P}', z)$. The output is a sampled reachability graph $\hat{G}$ and a collection of *integral* subtrees $\{T(p)\}_{p \in \mathcal{P}}$ of T such that *any integral allocation* in $\hat{G}$ can be realized using this collection with congestion at most $O(H \cdot \log n / \epsilon^2)$. Additionally, $\hat{G}$ supports a *fractional* allocation that satisfies at least $(1 - O(\epsilon))$ fraction of the demand of every $p \in \mathcal{P}$.
4. **Sample an integral allocation in $\hat{G}$.**
5. **Recombine allocation for $\mathcal{P}'$ into an allocation for $\mathcal{P}$.**

Since steps 1, 4 and 5 are mostly technical, and steps 2 and 3 are the main contribution of the paper, we first describe our LP relaxation (step 2), then state and analyze the rounding procedure (step 3) for a general set of requests $\mathcal{P}$ and prove Theorem 2. Details of steps 1, 4 and 5 and the complete proof of Theorem 1 are deferred to the full version of the paper.

The rest of the paper is organized as follows. In Section 2, we formalize the notation, define the problem, and show it is NP-complete. In Section 3, we formulate the LP relaxation of the problem. Section 4 discusses our rounding scheme.

2   Preliminaries

We are given a host tree $T = (V, E)$ and a set of rooted path requests $\mathcal{P}$. The tree T has n leaves, a root node r, and height $H \leq \log n$. Each edge $e \in E$

has a capacity c_e. For a node $v \in V$, we denote its parent in T by $\texttt{parent}(v)$ and the capacity of the edge $(\texttt{parent}(v), v)$ by c_v. By adding dummy nodes if necessary, we may assume that all leaves of T are at distance H from r. For $j = 0, \ldots, H$ we define L^j as the set of all nodes at distance j from the leaves, so L^0 is the set of leaves, and $L^H = \{r\}$. For a vertex v, we will also use the notation T_v for the subtree of T rooted at v, L^j_v for the nodes in T_v at level j, and $p^j(v)$ for the parent of v at level j when $v \in L^i$ and $i \leq j$. For a leaf $l \in L^0$, denote its capacity, i.e. the number of VMs it can contain, by a_l. Note that the case of general a_l can be reduced to the case of $a_l = 1, \forall l$ by replacing a leaf l of capacity a_l with a star with l as the root and infinite capacity edges to a_l children. Capacities a_l will be useful in the LP formulation, however.

Each of the path requests $p \in \mathcal{P}$ consists of a weight w_p, a length k_p, and a path V_p on $k_p + 1$ nodes labeled $0, 1, \ldots, k_p$. We want to map the nodes of V_p for all $p \in \mathcal{P}$ onto the host tree T, so that T can support a flow of w_p between adjacent nodes of V_p with as little congestion as possible. Node 0 must always be assigned to the root r in every path, and the remaining path nodes must be assigned to distinct leaves. More formally,

Definition 1. *For a tree T and a set of path requests $\mathcal{P}$, an* embedding *of $\mathcal{P}$ in T is a mapping $\pi_p : V_p \to L^0 \cup \{r\}$ for each $p \in \mathcal{P}$ such that (1) $\pi_p(0) = r$, for all $p \in \mathcal{P}$, (2) $\pi_p(i) \in L^0$ for all $i > 0$ and $p \in \mathcal{P}$ and (3) $\pi_p(i) \neq \pi_{p'}(j)$ if $i, j > 0$ and $(p, i) \neq (p', j)$.*

For two nodes u and v in T, let $E_{u,v}$ be the unique path from u to v in T. Each path p routes w_p flow from $\pi_p(i)$ to $\pi_p(i+1)$ for $0 \leq i \leq k_p - 1$, and the quality of an embedding π is defined by its congestion:

Definition 2. *Given a tree $T = (V, E)$, a set of path requests $\mathcal{P}$, and an embedding π of $\mathcal{P}$ in T the* congestion *of an edge $e = (\texttt{parent}(v), v) \in E$ under π, denoted by $\operatorname{cong}(\pi, e)$, is $\operatorname{cong}(\pi, e) = \frac{1}{c_v} \sum_{p \in \mathcal{P}} \left(\sum_{i \text{ s.t. } e \in E_{\pi_p(i), \pi_p(i+1)}} w_p \right)$, and the* congestion *of π is defined as $\operatorname{cong}(\pi) = \max_{e \in E} \operatorname{cong}(\pi, e)$.*

The problem that we are interested in is:

$$\begin{aligned} &\text{minimize } \operatorname{cong}(\pi) \\ &\text{subject to: } \pi \text{ is an embedding of } \mathcal{P} \text{ into } T \end{aligned} \tag{1}$$

2.1   NP-Hardness

We first show that problem (1) is NP-hard even with the restricted topologies of the host and request graphs. The proof is a simple reduction from 3-partition and is deferred to the full version of the paper.

Theorem 3. *Problem (1) is NP-hard to approximate to a factor better than $4/3$.*

2.2   Reduction to Embedding Subtrees

We first show that problem (1) can be converted into a similar problem where the requests in $\mathcal{P}$ are trees, which is equivalent to the original up to a factor of 2 in the congestion. Consider an embedding π_p of some path p of length k_p. The union of all the edges in the paths $E_{\pi_p(i),\pi_p(i+1)}$ $0 \leq i \leq k_p - 1$ forms a subtree T^{π_p} spanning r and k_p leaves, and each edge in T^{π_p} carries at least w_p flow for path p. In particular, if we remove p from $\mathcal{P}$, then the edges of T^{π_p} support a flow of w_p from r to any single leaf $\pi_p(i)$ with congestion of at most cong(π).

We will solve the following problem. We replace each path request p with a tree request p^t with weight w_p and k_p leaves. Let $\mathcal{P}^t$ be the set of tree requests. Our goal is to assign k_p leaves $\pi_{p^t}(1), \ldots, \pi_{p^t}(k_p)$ to each p^t, so as to minimize the congestion if each p^t requires w_p capacity on each edge along the paths from r to the $\pi_{p^t}(i)$. Note that the flow requirements are now different: previously we simultaneously routed w_p flow along all the paths $E_{\pi_p(i),\pi_p(i+1)}$ $0 \leq i \leq k_p - 1$, but now we only require reserving capacity so that each p^t can simultaneously route w_p flow from r to *one* of its leaves $\pi_{p^t}(i)$. Formally, the congestion of edge $e = (\texttt{parent}(v), v)$ is defined as cong(π, e) $= \frac{1}{c_v} \sum_{p \text{ s.t. } e \in \cup_i E_{r,\pi_p(i)}} w_p$.

We show the two problems are equivalent to within a factor of 2:

Lemma 1. *Let T be a host tree and let $\mathcal{P}$ be a set of requests. Then if there exists a solution of (1) with congestion z, then there exists an embedding of $\mathcal{P}^t$ with congestion at most z. Conversely, if there exists an embedding of $\mathcal{P}^t$ with congestion at most z, then there exists a solution of (1) with congestion at most $2z$.*

The proof is deferred to the full version. For the remainder of the paper, we will consider the embedding problem on $\mathcal{P}^t$, but to simplify notation we will drop the superscript and use the notation p and $\mathcal{P}$ to refer to path/tree requests.

3   Linear Programming Formulation

In this section we introduce the LP relaxation that our algorithm uses, and then give an outline of the algorithm. We begin by defining the k-branching polytope $\mathcal{T}_k$, which is a relaxation of the set of subtrees of T on k leaves.

3.1   The k-Branching Polytope

We now introduce the k-branching polytope, which is a relaxation of the set of subtrees of T with at most k leaves and serves as a basis of our LP relaxation. Intuitively, we represent a tree with at most k leaves as 1 unit of flow from the root r of T to the leaves that can be augmented at internal nodes of T by at most $k - 1$ extra units, thus allowing the flow to reach up to k leaves. We refer to such a flow as *augmented flow*. A key constraint is the *bounded branching constraint*, which states that the amount of extra flow injected into any subtree T_u rooted at $u \in T$ cannot be more than $k - 1$ times larger than the flow that enters u from $\texttt{parent}(u)$. This constraint is crucially used to control the variance in our rounding procedure in Section 4. A similar constraint is used in [8]. The k-branching polytope $\mathcal{T}_k$ is

$$
\mathcal{T}_k = \left\{ \begin{array}{l} (f, \Delta) \\ \in \mathbb{R}^{2|V|} \end{array} \middle| \begin{array}{ll} \sum_{v:\texttt{parent}(v)=u} f_v = f_u + \Delta_u \; \forall u \notin L^0 \text{ (flow conservation)} \\ \sum_{v \in T_u} \Delta_v \leq (k-1) f_u & \forall u \qquad \text{(bounded branching)} \\ f_u \leq f_{\texttt{parent}(u)} & \forall u \\ f_r = 1 \\ f, \Delta \geq 0 \end{array} \right\}
\tag{2}
$$

Any subtree T' of T with at most k leaves rooted at r satisfies (2):

Lemma 2. *Let T' be a subtree of $T = (V, E)$ with at most k leaves. Let $f_u = 1$ if T' contains the edge $(u, \texttt{parent}(u))$ and 0 otherwise. For each u, let Δ_u equal to the number of children of u in T' minus 1. Then $(f, \Delta) \in \mathcal{T}_k$.*

Proof. The flow conservation constraints are satisfied by definition of Δ. For each $u \in V$, we have that $\sum_{v \in T_u} \Delta_v$ is equal to the number of leaves in the subtree of u minus 1. Hence, since T' has at most k leaves, the bounded branching constraints are satisfied. □

We now describe the allocation polytope $\mathcal{A}(\mathcal{P}, z)$. For each path $p \in \mathcal{P}$ we introduce variables f_u^p, Δ_u^p that give a point in the k_p-branching polytope $\mathcal{T}_{k_p}$. We will slightly abuse notation by writing $(f, \Delta) = (f(p), \Delta(p))_{p \in |\mathcal{P}|} \in \mathbb{R}^{2|V| \cdot |\mathcal{P}|}$ to denote the vector (f, Δ) which is a concatenation of $(f(p), \Delta(p)), p \in \mathcal{P}$. We include separate $(f(p), \Delta(p))$ variables in the appropriate branching polytope for each $p \in \mathcal{P}$ and impose the additional constraints that (a) no leaf can accept more than (fractionally) one request, and (b) each fractional tree $(f(p), \Delta(p))$ should reach (fractionally) exactly k_p leaves. The allocation polytope is parameterized by a variable z that upper bounds congestion. In particular, a path p of weight w_p can only use edges $(u, \texttt{parent}(u))$ of T for which $c_u \cdot z \geq w_p$. Formally,

$$
\mathcal{A}(\mathcal{P}, z) = \left\{ \begin{array}{l} (f, \Delta) = \\ (f(p), \Delta(p))_{p \in \mathcal{P}} \end{array} \middle| \begin{array}{ll} \sum_{p \in \mathcal{P}} f_l(p) \leq a_l & \forall l \in L^0 \\ \sum_{l \in L^0} f_l(p) = k_p & \forall p \in \mathcal{P} \\ \sum_{p \in \mathcal{P}} w_p f_u(p) \leq c_u z \; \forall u \neq r \\ f_u(p) = 0 & \forall p, u \in V \text{ s.t. } c_u z < w_p \\ f(p) \in \mathcal{T}_{k_p} \end{array} \right\}
\tag{3}
$$

The following relation between the optimum congestion and the polytope $\mathcal{A}(\mathcal{P}, z)$ is easy to prove and is deferred to the full version of the paper.

Lemma 3. *If there exists an embedding of requests $\mathcal{P}$ into tree $T = (V, E)$ with congestion at most z, then $\mathcal{A}(\mathcal{P}, z) \neq \emptyset$.*

In what follows we will use the concept of a *reachability graph*, which is an unweighted bipartite graph with requests $p \in \mathcal{P}$ on one side, leaves L^0 on the other side, and edges (p, l) for pairs (p, l) such that l is reachable from the root r of T in the (fractional) subtree defined by $(f(p), \Delta(p))$. In fact, we will define two reachability graphs—a fractional reachability graph G and a sampled reachability graph $\hat{G}$. The former will reflect the fractional solution obtained from $\mathcal{A}(\mathcal{P}, z)$, while the latter will be the result of our rounding procedure and, in particular, will use integral as opposed to fractional subtrees. Formally,

Definition 3. *Let* $(f, \Delta) = (f(p), \Delta(p))_{p \in \mathcal{P}}$ *be a point in* $\mathcal{A}(\mathcal{P}, z)$. *The fractional reachability graph is the unweighted bipartite graph* $G = (\mathcal{P}, L^0, E)$ *defined by including an edge* (p, l) *in* E *if* $f_l(p) > 0$.

The core of our algorithm, which we will describe below, is a randomized rounding procedure that, given a point in $\mathcal{A}(\mathcal{P}, z)$, produces an fractional allocation that satisfies at least a $1 - \epsilon$ fraction of the requests at the expense of increasing the congestion by most a factor of $O(H \log n / \epsilon^2)$. We describe the rounding procedure in the next section. The proof of Theorem 2 follows directly by an application of the rounding procedure to the allocation polytope with augmented constraints, as we will show at the end of Section 4. A preprocessing step is needed to prove the bicriteria approximation guarantees in Theorem 1. The details of this step are deferred to the full version of the paper.

4   Rounding Scheme

In this section we describe our rounding scheme, which takes as input a point $(f, \Delta) \in \mathcal{A}(\mathcal{P}, z)$ and produces the following: (1) an integral subtree $\hat{T}(p)$ for each $p \in \mathcal{P}$, such that the overall congestion from embedding $\{\hat{T}(p)\}_{p \in \mathcal{P}}$ is at most $\text{poly}(\log n, \log \theta, 1/\epsilon) \cdot z$ (the subtrees $\hat{T}(p)$ need not have disjoint leaves for different p); (2) a *sampled reachability graph* $\hat{G}$, a bipartite graph with $\mathcal{P}$ on one side and leaves L^0 of T on the other side, such that any allocation in $\hat{G}$ can be realized using $\{\hat{T}(p)\}_{p \in \mathcal{P}}$; and (3) a *fractional allocation* in $\hat{G}$ that satisfies at least $(1 - O(\epsilon))$ of the demand of every request. Since any allocation of leaves of T to requests $p \in \mathcal{P}$ naturally corresponds to a flow in $\hat{G}$ where $p \in \mathcal{P}$ have capacity k_p and leaves $l \in L^0$ have capacity 1, we will also refer to such allocations as *flows* in $\hat{G}$. Also, recall that we assume the tree T to be balanced, and hence the height H of T satisfies $H \leq \log n$, where $n = |L^0|$.

The rounding scheme is based on the rounding scheme for Group Steiner Tree in [5]. In order to give intuition about the rounding and the analysis, it is useful to first recall Group Steiner Tree rounding. Let f denote a flow of unit value from the root of a tree T to the leaves of T, so that f_u denotes the flow on the edge $(\texttt{parent}(u), u)$. Note that such a flow emerges in the analysis of [5] as a flow to a specific group of leaves, but for the purposes of giving intuition we are only concerned with the fact that it is a flow from the root to the leaves. The flow is rounded in [5] by first keeping each edge $e = (u, v)$ of T with probability f_u / f_v independently, and then outputting the connected component that the root of T belongs to in the sampled graph. In [5] the authors show that $O(\log^2 n)$ independent repetitions of this procedure ensure that all groups of vertices are reached. With some simplifications our sampling procedure can be described in a similar way. Instead of a single flow from the root to the leaves we have *augmented flows* $f(r,p) \in \mathcal{T}_{k_p}$ for each $p \in \mathcal{P}$. We do Group Steiner Tree rounding $O(H \cdot \log n/\epsilon^2)$ times independently for each $p \in \mathcal{P}$, and then output the union of the sampled subtrees. While this is an intuitive description of the rounding procedure, it misses some important constraints of our problem. In particular, it is important for us that the rounding must output a fractional allocation that satisfies a $1 - \epsilon$ fraction of demand of each request simultaneously (we use this property in the preprocess step outlined in the full version of the paper). Thus, we will need to introduce more notation and describe it somewhat differently since (a) we are dealing with more general objects than flows and (b) we need to prove stronger guarantees on the output than in [5]. We now outline the main difficulties, and then define the rounding scheme formally.

The following difficulties arise in our setting. First, instead of a flow from the root to the leaves we need to deal with *augmented flows* $f(r,p)$ for each request $p \in \mathcal{P}$. Second, we need to prove stronger guarantees on the output of the scheme. In particular, we are starting from a *fractional allocation* of requests to *fractional subtrees* and need to show that the *integral subtrees* that emerge from the scheme still contain a nearly-perfect allocation of the requests. For example, the fact that flow can be augmented at intermediate nodes in the tree causes complications by increasing variance in the rounding scheme since the outcome of a coin flip at an intermediate node influences significantly more flow mass than the flow that comes into the vertex. However, the *bounded branching constraints* in the k-branching polytope let us control the variance in this step.

For each node $u \in V$ we now define *sampled solutions* $(f(u), \Delta(u))$ restricted to the subtree T_u. We start with $f(r) = f, \Delta(r) = \Delta$ and proceed inductively top-down from r in the tree T. For a sampled solution $f(u)$ we denote by $f(u,p)$ the flow corresponding to path $p \in \mathcal{P}$ that goes through u.

Base: $j = -1$. Let $f(r,p) := f(p), \Delta(r,p) = \Delta(p), \forall p \in \mathcal{P}$.

Inductive Step: $j \to j + 1$. For each u on level j do the following. Let $\{c_1, \ldots, c_{k_u}\}$ denote the children of u. For each $p \in \mathcal{P}$ let $Z_{u,p,i}, i = 1, \ldots, k_u$ denote independent Bernoulli random variables with on probability $q_{u,p,i} = \min\{Cf_{c_i}(u,p), 1\}$. Here $C = \Theta(H \log n/\epsilon^2)$ is an upsampling parameter and $H \leq \log n$ is the height of the tree T.

If $q_{u,p,i} = 0$, let $f(c_i, p) = \mathbf{0}, \Delta(c_i, p) = \mathbf{0}$. Otherwise define $f(c_i, p)$ by setting for all vertices $v \in T_{c_i}$

$$f_v(c_i, p) = \begin{cases} \frac{f_v(u,p)}{q_{u,p,i}}, & \text{if } Z_{u,p,i} = 1 \\ 0 & \text{o.w.} \end{cases} , \Delta_v(c_i, p) = \begin{cases} \frac{\Delta_v(u,p)}{q_{u,p,i}}, & \text{if } Z_{u,p,i} = 1 \\ 0 & \text{o.w.} \end{cases} \quad (4)$$

Note that $f_v(c_i, p), \Delta_v(c_i, p)$ are defined for $v \in T_{c_i}$. The output of the rounding scheme will be given by the fractional allocation $f_u(u,p), u \in L^0, p \in \mathcal{P}$. We will use the notation $f^*_{(u,p)} := f_u(u,p)$. Finally, for each $p \in \mathcal{P}$, let $\hat{T}(p)$ be the tree formed by edges $(u, \texttt{parent}(u)), u \in V$ such that $f_u(u,p) > 0$. The bound on the congestion from embedding $\{\hat{T}(p)\}_{p \in \mathcal{P}}$ essentially follows by Chernoff bounds and is given in the full version:

Lemma 4. *The congestion incurred after embedding $\{\hat{T}(p)\}_{p \in \mathcal{P}}$ is at most $C \cdot OPT$ with probability at least $1 - n^{-2}$.*

Definition 4. *Define the* sampled reachability graph *$\hat{G} = (\mathcal{P}, L^0, E)$ by including an edge $(p, l), p \in \mathcal{P}, l \in L^0$ iff $f^*_{(l,p)} > 0$.*

Note that any allocation in $\hat{G}$ can be realized using $\{\hat{T}(p)\}_{p \in \mathcal{P}}$. The rest of the section is devoted to proving that $f^*_{(l,p)}, p \in \mathcal{P}, l \in L^0$ yields a large flow from $\mathcal{P}$ to L^0 in $\hat{G}$. Proofs of the following lemmas follow directly from the definition of the k-branching polytope and the rounding scheme (see full version for details).

Lemma 5. *For all $p \in \mathcal{P}$ and all $u \in V$ if $f(u, p) \neq 0$, then $\left(\frac{f(u,p)}{f_u(u,p)}, \frac{\Delta(u,p)}{f_u(u,p)} \right) \in \mathcal{T}_{k_p}$, where $\mathcal{T}_{k_p}$ is the k-branching polytope on the subtree T_u of T.*

Lemma 6. *For all $p \in \mathcal{P}$ and all $u \in V$ one has $\sum_{l \in L^0_u} f_l(u, p) \leq k_p f_u(u, p)$.*

We now demonstrate a fractional flow from $\mathcal{P}$ to L in $\hat{G}$ that is at least $1 - \epsilon$ fraction of the optimum. We will need the following auxiliary lemmas.

Lemma 7. *Consider a path $p \in \mathcal{P}$ and a node $u \in L^j$ for some $j \in [0 : H]$ and denote the children of u in T by $\{c_1, \ldots, c_x\}$. Then if $f_{c_i}(u, p) \geq 1/C$, then $f_{c_i}(c_i, p) \equiv f_{c_i}(u, p)$. Otherwise $f_{c_i}(c_i, p)$ is a Bernoulli random variable bounded by $1/C$.*

Proof. Suppose that $f_{c_i}(u, p) \geq 1/C$. Then (4) implies that $q_{u,p,i} = 1$ and hence $f_{c_i}(c_i, p) \equiv f_{c_i}(u, p)$. Otherwise $q_{u,p,i} = C f_{c_i}(u, p)$, and hence, if nonzero, one has $f_{c_i}(c_i, p) = f_{c_i}(u, p)/q_{u,p,i} = 1/C$. □

Lemma 8. *Setting $C = \Theta(H \log n / \epsilon^2)$, we have that $\sum_{p \in \mathcal{P}} f^*_{(l,p)} \leq (1 + \epsilon) a_l$ for each $l \in L^0$ with probability at least $1 - n^{-2}$.*

Proof. We give an outline of the main ideas, leaving the details to the full version. We will prove a high probability bound for a fixed $l \in L^0$, and the statement of the lemma will follow by a union bound. First note that by (3) $\sum_{p \in \mathcal{P}} f_l(r, p) \leq a_l$, i.e. the fractional degree of l in the original solution is at most a_l. Consider the j-th level of T for some j. Let $Q_j = \sum_{p \in \mathcal{P}} f_l(\text{parent}^j(l), p)$, where $\text{parent}^j(l)$ is the parent of l at level j. We have $\mathbf{E}[Q_j] = Q_{j+1}$ for all $j = 0, \ldots, H$, and $Q_H \leq a_l$. We first give the intuition behind the proof. Informally, we show that for each j, by definition of our rounding procedure, conditional on the rounded solution at levels above j, Q_{j+1} is a sum of Bernoulli random variables bounded by $1/C$ that add up to at most a_l. This allows us to show that the difference $Q_j - Q_{j+1} = Q_j - \mathbf{E}[Q_j]$ is stochastically dominated by a simple random variable with expectation smaller than $\frac{a_l \epsilon}{10 \sqrt{H \cdot \log n}}$. Finally, we show that in order to have $Q_0 > (1 + \epsilon) Q_H$, the summation of these random variables over H levels of the tree would have to be significantly higher than its expectation, which is a low probability event by standard concentration inequalities. We now outline the proof.

First note that, $f_l(u,p)$ are conditionally independent given $\{f_l(w,p)\}_{w \in L^{j+1}}$ by the definition of our rounding scheme. Define the event $\mathcal{E}_j = \{Q_i < (1+\epsilon)Q_H, i = j, \ldots, H\}$. We will prove by induction on j that $\mathbf{Pr}[\mathcal{E}_j] \geq 1 - (H-j) \cdot n^{-4}$ for all j. The base is trivial since $\mathbf{Pr}[\mathcal{E}_H] = 1$. We now prove the inductive step. Let $u^* = \mathtt{parent}^j(l)$ denote the parent of l at level j. Let $\mathcal{P}^{0,j} = \{p \in \mathcal{P} : f_{u^*}(\mathtt{parent}(u^*),p) > 1/C\}$ and let $\mathcal{P}^{1,j} = \mathcal{P} \setminus \mathcal{P}^{0,j}$ and let $Q_{j,0} = \sum_{p \in \mathcal{P}^{0,j}} f_l(u^*,p)$ and $Q_{j,1} = \sum_{p \in \mathcal{P}^{1,j}} f_l(u^*,p)$, i.e. we split the paths into those that are chosen deterministically and those that are sampled. All variables in $Q_{j,1}$ are independent Bernoulli variables bounded by $1/C$, as follows from the definition of $\mathcal{P}^{1,j}$, the fact that $f_l(\mathtt{parent}(u^*),p) \leq f_{u^*}(\mathtt{parent}(u^*),p)$ by the definition of the k-branching polytope and Lemma 7.

It follows via a series of estimates given in the full version that for all $\gamma > 0$ one has $\mathbf{Pr}\left[Q_j > Q_{j+1} + \frac{\gamma a_l \epsilon}{10\sqrt{H \cdot \log n}} \middle| \mathcal{E}_{j+1} \wedge \bar{\mathcal{B}}\right] \leq e^{-32\gamma^2}$ (here $\mathcal{B}$ is a low probability bad event that we define properly in the full version). We next exhibit a simple random variable W_j such that $\mathbf{Pr}[W_j > \gamma] \geq e^{-32\gamma^2}$, and then use concentration inequalities to show that $\mathbf{Pr}\left[\mathcal{E}_j | \mathcal{E}_{j+1} \wedge \bar{\mathcal{B}}\right] \geq 1 - \mathbf{Pr}\left[\sum_{i=j}^H W_i > 10\sqrt{H \log n}\right] \geq 1 - n^{-5}$, concluding that $\mathbf{Pr}[\mathcal{E}_j | \mathcal{E}_{j+1} \wedge \bar{\mathcal{B}}] = 1 - \mathbf{Pr}[Q_j > 1 + \epsilon | \mathcal{E}_{j+1} \wedge \bar{\mathcal{B}}] \geq 1 - n^{-5}$.

Hence, $\mathbf{Pr}[\mathcal{E}_j] = (\mathbf{Pr}[\mathcal{E}_j | \mathcal{E}_{j+1} \wedge \bar{\mathcal{B}}]/\mathbf{Pr}[\bar{\mathcal{B}}]) \cdot \mathbf{Pr}[\mathcal{E}_{j+1}] \geq 1 - (H-j)n^{-4}$, which proves the inductive step. We refer the reader to the full version for details. □

We now prove a similar bound on the fractional degree of a request $p \in \mathcal{P}$.

Lemma 9. *Setting $C = \Theta(H \cdot \log n/\epsilon^2)$, we have that $(1-\epsilon)k_p \leq \sum_{l \in L^0} f^*_{(l,p)} \leq (1+\epsilon)k_p$ for each $p \in \mathcal{P}$ with probability at least $1 - n^{-2}$.*

Proof. The general outline of the proof is similar to Lemma 8, except that we use the bounded branching constraints of the k-matching polytope to bound the variance in rounding. First note that by (3) $\sum_{l \in L^0} f_l(r,p) = k_p$. Let $Q_j = \sum_{u \in L^j} X_u$, where $X_u = \sum_{l \in L^0_u} f_l(u,p)$. We have, as before, $\mathbf{E}[Q_j] = Q_{j+1}$ for all $j = 0, \ldots, H$, and $Q_H = k_p$.

Define the event $\mathcal{E}_j = \{Q_j < (1+\epsilon)Q_H \wedge Q_j > (1-\epsilon)Q_H\}$. Let $U^{(0,j)} = \{u \in L^j : f_u(\mathtt{parent}(u),p) > 1/C\}$ and let $U^{(1,j)} = L^j \setminus U^{(0,j)}$ and let $Q_{j,0} = \sum_{u \in U^{(0,j)}} X_u$, and $Q_{j,1} = \sum_{u \in U^{(1,j)}} X_u$. Unlike Lemma 8, it is not true that $f_l(u,p)$ are conditionally independent given $f_l(w,p), w \in L^{j+1}$, but $X_u = \sum_{l \in L^0_u} f_l(u,p), u \in L^j$ are. However, we have by Lemma 6 that for every $u \in V \sum_{l \in L^0_u} f_l(u,p) \leq k_p f_u(u,p)$. Thus, one has $X_u \leq k_p/C$ for all $u \in U^{(1,j)}$. Similarly to Lemma 8, we now get $\mathbf{Pr}\left[|Q_j - Q_{j+1}| > \frac{\gamma \epsilon k_p}{10\sqrt{H \cdot \log n}} \middle| \mathcal{E}_{j+1} \wedge \bar{\mathcal{B}}\right] \leq e^{-32\gamma^2}$ for a low probability bad event $\mathcal{B}$. We then prove by induction on j that $\mathbf{Pr}[\mathcal{E}_j] \geq 1 - (H-j) \cdot n^{-4}$ for all j.   □

Lemma 8 and Lemma 9 together imply (see the full version for proof)

Lemma 10. *There exists a fractional flow in the sampled reachability graph $\hat{G}$ that for each $p \in \mathcal{P}$ sends at least $(1-2\epsilon)k_p$ flow through p.*

We have described a rounding procedure that, given a point in $\mathcal{A}(\mathcal{P}, z)$, produces a sampled reachability graph $\hat{G}$ that supports a fractional flow of at least $(1 - 2\epsilon)k_p$ out of every $p \in \mathcal{P}$, together with a set of subtrees $\hat{T}(p)$ of T for each p, such that the embedding of $\hat{T}(p)$ incurs congestion at most a factor of $O(H \log n/\epsilon^2)$ of the optimum. Moreover, the set $\hat{T}(p)$ is such that any allocation of leaves L^0 to paths $p \in |\mathcal{P}|$ that is admissible in $\hat{G}$ can be realized using $\{\hat{T}(p)\}_{p\in\mathcal{P}}$. In the full version, we show how to convert the large *fractional* flow that exists in $\hat{G}$ to an *integral* allocation at the expense of an extra factor of $\text{poly}(\log n, \log \theta, 1/\epsilon)$ in the congestion, leading to a proof of Theorem 1. Before that, we can show:

Proof of Theorem 2: Let $c_u^\epsilon := c_u(1 + \epsilon)$, $a_l^\epsilon := \lceil a_l(1 + \epsilon) \rceil$ and impose $1 + \epsilon$ flow out of the root in the definition of the k-branching polytope $\mathcal{T}_k$. Use $c_u^\epsilon, a_l^\epsilon$ in the definition of the allocation polytope $\mathcal{A}(\mathcal{P}, z)$ (denote the modified polytope by $\mathcal{A}^\epsilon(\mathcal{P}, z)$). Denote the optimal congestion for the original problem by OPT. Note that $\mathcal{A}^\epsilon(\mathcal{P}, (1 + \epsilon)OPT) \neq \emptyset$. It now follows similarly to Lemma 10, that the sampled reachability graph contains a fractional flow that satisfies each request fully, and hence there also exists an *integral* flow that gives the desired allocation by the integrality of the flow polytope. $\qquad\square$

References

1. Aspnes, J., Azar, Y., Fiat, A., Plotkin, S., Waarts, O.: Online load balancing with applications to machine scheduling and virtual circuit routing. In: STOC (1993)
2. Awerbuch, B., Singh, T.: Online algorithms for selective multicast and maximal dense trees. In: STOC (1997)
3. Bansal, N., Lee, K.W., Nagarajan, V., Zafer, M.: Minimum congestion mapping in a cloud. In: PODC, pp. 267–276 (2011)
4. Chowdhury, N.M.M.K., Rahman, M.R., Boutaba, R.: Virtual network embedding with coordinated node and link mapping. In: INFOCOM (2009)
5. Garg, N., Konjevod, G., Ravi, R.: A polylogarithmic approximation algorithm for the group steiner tree problem. In: SODA (1998
6. Goel, A., Hezinger, M., Plotkin, S.: Online throughput-comptetitive algorithm for multicast routing and admission control. In: SODA (1998)
7. Hassin, R., Levin, A., Sviridenko, M.: Approximating the minimum quadratic assignment problems. ACM Transactions on Algorithms (2009)
8. Konjevod, G., Ravi, R., Srinivasan, A.: Approximation algorithms for the covering steiner problem. Random Structures & Algorithms 20(3), 465–482 (2002)
9. Meng, X., Pappas, V., Zhang, L.: Improving the scalability of data center networks with traffic-aware virtual machine placement. In: INFOCOM (2010)
10. Racke, H.: Minimizing congestion in general networks. In: FOCS (2002)
11. Rastogi, R., Silberschatz, A., Yener, B.: Secondnet: a data center network virtualization architecture with bandwidth guarantees. In: Co-NEXT Workshop (2010)
12. Yu, M., Yi, Y., Rexford, J., Chiang, M.: Rethinking virtual network embedding: Substrate support for path splitting and migration. In: SIGCOMM (2008)

Faster Geometric Algorithms
via Dynamic Determinant Computation

Vissarion Fisikopoulos and Luis Peñaranda

University of Athens, Dept. of Informatics & Telecommunications, Athens, Greece
{vfisikop,lpenaranda}@di.uoa.gr

Abstract. Determinant computation is the core procedure in many important geometric algorithms, such as convex hull computations and point locations. As the dimension of the computation space grows, a higher percentage of the computation time is consumed by these predicates. In this paper we study the sequences of determinants that appear in geometric algorithms. We use dynamic determinant algorithms to speed-up the computation of each predicate by using information from previously computed predicates.

We propose two dynamic determinant algorithms with quadratic complexity when employed in convex hull computations, and with linear complexity when used in point location problems. Moreover, we implement them and perform an experimental analysis. Our implementations outperform the state-of-the-art determinant and convex hull implementations in most of the tested scenarios, as well as giving a speed-up of 78 times in point location problems.

Keywords: computational geometry, determinant algorithms, orientation predicate, convex hull, point location, experimental analysis.

1 Introduction

Determinantal predicates are in the core of many important geometric algorithms. Convex hull and regular triangulation algorithms use *orientation* predicates, the Delaunay triangulation algorithms also involve the *in-sphere* predicate. Moreover, algorithms for exact volume computation of a convex polytope rely on determinantal volume formulas. In general dimension d, the orientation predicate of $d + 1$ points is the sign of the determinant of a matrix containing the homogeneous coordinates of the points as columns. In a similar way, the volume determinant formula and in-sphere predicate of $d + 1$ and $d + 2$ points respectively can be defined. In practice, as the dimension grows, a higher percentage of the computation time is consumed by these core procedures. For this reason, we focus on algorithms and implementations for the exact computation of the determinant. We give particular emphasis to division-free algorithms. Avoiding divisions is crucial when working on a ring that is not a field, *e.g.*, integers or polynomials. Determinants of matrices whose elements are in a ring arise in combinatorial problems [21], in algorithms for lattice polyhedra [4] and secondary polytopes [23] or in computational algebraic geometry problems [12].

L. Epstein and P. Ferragina (Eds.): ESA 2012, LNCS 7501, pp. 443–454, 2012.

Our main observation is that, in a sequence of computations of determinants that appear in geometric algorithms, the computation of one predicate can be accelerated by using information from the computation of previously computed predicates. In this paper, we study orientation predicates that appear in convex hull computations. The convex hull problem is probably the most fundamental problem in discrete computational geometry. In fact, the problems of regular and Delaunay triangulations reduce to it.

Our main contribution is twofold. First, we propose an algorithm with quadratic complexity for the determinants involved in a convex hull computation and linear complexity for those involved in point location problems. Moreover, we nominate a variant of this algorithm that can perform computations over the integers. Second, we implement our proposed algorithms along with division-free determinant algorithms from the literature. We perform an experimental analysis of the current state-of-the-art packages for exact determinant computations along with our implementations. Without taking the dynamic algorithms into account, the experiments serve as a case study of the best implementation of determinant algorithms, which is of independent interest. However, dynamic algorithms outperform the other determinant implementations in almost all the cases. Moreover, we implement our method on top of the convex hull package `triangulation` [6] and experimentally show that it attains a speed-up up to 3.5 times, results in a faster than state-of-the-art convex hull package and a competitive implementation for exact volume computation, as well as giving a speed-up of 78 times in point location problems.

Let us review previous work. There is a variety of algorithms and implementations for computing the determinant of a $d \times d$ matrix. By denoting $O(d^\omega)$ their complexity, the best current ω is 2.697263 [20]. However, good asymptotic complexity does not imply good behavior in practice for small and medium dimensions. For instance, LinBox [13] which implements algorithms with state-of-the-art asymptotic complexity, introduces a significant overhead in medium dimensions, and seems most suitable in very high dimensions (typically > 100). Eigen [18] and CGAL [10] implement decomposition methods of complexity $O(n^3)$ and seem to be suitable for low to medium dimensions. There exist algorithms that avoid divisions such as [24] with complexity $O(n^4)$ and [5] with complexity $O(nM(n))$ where $M(n)$ is the complexity of matrix multiplication. In addition, there exists a variety of algorithms for determinant sign computation [8,1]. The problem of computation of several determinants has also been studied. TOPCOM [23], the reference software for computing triangulations of a set of points, efficiently precomputes all orientation determinants that will be needed in the computation and stores their signs. In [15], a similar problem is studied in the context of computational algebraic geometry. The computation of orientation predicates is accelerated by maintaining a hash table of computed minors of the determinants. These minors appear many times in the computation. Although, applying that method to the convex hull computation does not lead to a more efficient algorithm.

Our main tools are the Sherman-Morrison formulas [27,3]. They relate the inverse of a matrix after a small-rank perturbation to the inverse of the original matrix. In [25] these formulas are used in a similar way to solve the dynamic transitive closure problem in graphs.

The paper is organized as follows. Sect. 2 introduces the dynamic determinant algorithms and the following section presents their application to the convex hull problem. Sect. 4 discusses the implementation, experiments, and comparison with other software. We conclude with future work.

2 Dynamic Determinant Computations

In the *dynamic determinant problem*, a $d \times d$ matrix A is given. Allowing some preprocessing, we should be able to handle updates of elements of A and return the current value of the determinant. We consider here only non-singular updates, *i.e.*, updates that do not make A singular. Let $(A)_i$ denote the i-th column of A, and e_i the vector with 1 in its i-th place and 0 everywhere else.

Consider the matrix A', resulting from replacing the i-th column of A by a vector u. The Sherman-Morrison formula [27,3] states that $(A + wv^T)^{-1} = A^{-1} - \frac{(A^{-1}w)(v^T A^{-1})}{1 + v^T A^{-1} w}$. An i-th column update of A is performed by substituting $v = e_i$ and $w = u - (A)_i$ in the above formula. Then, we can write A'^{-1} as follows.

$$A'^{-1} = (A + (u - (A)_i)e_i^T)^{-1} = A^{-1} - \frac{(A^{-1}(u - (A)_i))\,(e_i^T A^{-1})}{1 + e_i^T A^{-1}(u - (A)_i)} \qquad (1)$$

If A^{-1} is computed, we compute A'^{-1} using Eq. 1 in $3d^2 + 2d + O(1)$ arithmetic operations. Similarly, the matrix determinant lemma [19] gives Eq. 2 below to compute $det(A')$ in $2d + O(1)$ arithmetic operations, if $det(A)$ is computed.

$$det(A') = det(A + (u - (A)_i)e_i^T) = (1 + e_i^T A^{-1}(u - (A)_i)det(A) \qquad (2)$$

Eqs. 1 and 2 lead to the following result.

Proposition 1. [27] *The dynamic determinant problem can be solved using $O(d^\omega)$ arithmetic operations for preprocessing and $O(d^2)$ for non-singular one column updates.*

Indeed, this computation can also be performed over a ring. To this end, we use the adjoint of A, denoted by A^{adj}, rather than the inverse. It holds that $A^{adj} = det(A)A^{-1}$, thus we obtain the following two equations.

$$A'^{adj} = \frac{1}{det(A)}(A^{adj} det(A') - (A^{adj}(u - (A)_i))\,(e_i^T A^{adj})) \qquad (3)$$

$$det(A') = det(A) + e_i^T A^{adj}(u - (A)_i) \qquad (4)$$

The only division, in Eq. 3, is known to be exact, *i.e.*, its remainder is zero. The above computations can be performed in $5d^2 + d + O(1)$ arithmetic operations for Eq. 3 and in $2d + O(1)$ for Eq. 4. In the sequel, we will call *dyn_inv* the dynamic determinant algorithm which uses Eqs. 1 and 2, and *dyn_adj* the one which uses Eqs. 3 and 4.

3   Geometric Algorithms

We introduce in this section our methods for optimizing the computation of sequences of determinants that appear in geometric algorithms. First, we use dynamic determinant computations in incremental convex hull algorithms. Then, we show how this solution can be extended to point location in triangulations.

Let us start with some basic definitions from discrete and computational geometry. Let $\mathcal{A} \subset \mathbb{R}^d$ be a pointset. We define the *convex hull* of a pointset $\mathcal{A}$, denoted by $\mathrm{conv}(\mathcal{A})$, as the smallest convex set containing $\mathcal{A}$. A hyperplane *supports* $\mathrm{conv}(\mathcal{A})$ if $\mathrm{conv}(\mathcal{A})$ is entirely contained in one of the two closed half-spaces determined by the hyperplane and has at least one point on the hyperplane. A *face* of $\mathrm{conv}(\mathcal{A})$ is the intersection of $\mathrm{conv}(\mathcal{A})$ with a supporting hyperplane which does not contain $\mathrm{conv}(\mathcal{A})$. Faces of dimension $0, 1, d-1$ are called vertices, edges and facets respectively. We call a face f of $\mathrm{conv}(\mathcal{A})$ *visible* from $a \in \mathbb{R}^d$ if there is a supporting hyperplane of f such that $\mathrm{conv}(\mathcal{A})$ is contained in one of the two closed half-spaces determined by the hyperplane and a in the other. A k-simplex of $\mathcal{A}$ is an affinely independent subset S of $\mathcal{A}$, where $dim(\mathrm{conv}(S)) = k$. A *triangulation* of $\mathcal{A}$ is a collection of subsets of $\mathcal{A}$, the *cells* of the triangulation, such that the union of the cells' convex hulls equals $\mathrm{conv}(\mathcal{A})$, every pair of convex hulls of cells intersect at a common face and every cell is a simplex.

Denote $\boldsymbol{a}$ the vector $(a, 1)$ for $a \in \mathbb{R}^d$. For any sequence C of points $a_i \in \mathcal{A}$, $i = 1 \ldots d+1$, we denote A_C its orientation $(d+1) \times (d+1)$ matrix. For every a_i, the column i of A_C contains $\boldsymbol{a_i}$'s coordinates as entries. For simplicity, we assume general position of $\mathcal{A}$ and focus on the Beneath-and-Beyond (BB) algorithm [26]. However, our method can be extended to handle degenerate inputs as in [14, Sect. 8.4], as well as to be applied to any incremental convex hull algorithm by utilizing the dynamic determinant computations to answer the predicates appearing in point location (see Cor. 2). In what follows, we use the dynamic determinant algorithm *dyn_adj*, which can be replaced by *dyn_inv* yielding a variant of the presented convex hull algorithm.

The BB algorithm is initialized by computing a d-simplex of $\mathcal{A}$. At every subsequent step, a new point from $\mathcal{A}$ is inserted, while keeping a triangulated convex hull of the inserted points. Let t be the number of cells of this triangulation. Assume that, at some step, a new point $a \in \mathcal{A}$ is inserted and T is the triangulation of the convex hull of the points of $\mathcal{A}$ inserted up to now. To determine if a facet F is visible from a, an orientation predicate involving a and the points of F has to be computed. This can be done by using Eq. 4 if we know the adjoint matrix of points of the cell that contains F. But, if F is visible, this cell is unique and we can map it to the adjoint matrix corresponding to its points.

Our method (Alg. 1), as initialization, computes from scratch the adjoint matrix that corresponds to the initial d-simplex. At every incremental step, it computes the orientation predicates using the adjoint matrices computed in previous steps and Eq. 4. It also computes the adjoint matrices corresponding to the new cells using Eq. 3. By Prop. 1, this method leads to the following result.

Algorithm 1: Incremental Convex Hull $(\mathcal{A})$

Input   : pointset $\mathcal{A} \subset \mathbb{R}^d$
Output : convex hull of $\mathcal{A}$
sort $\mathcal{A}$ by increasing lexicographic order of coordinates, *i.e.*, $\mathcal{A} = \{a_1, \ldots, a_n\}$;
$T \leftarrow \{\, d\text{-face of } conv(a_1, \ldots, a_{d+1})\}$; // $conv(a_1, \ldots, a_{d+1})$ is a d-simplex
$Q \leftarrow \{\, \text{facets of } conv(a_1, \ldots, a_{d+1})\}$;
compute $A^{adj}_{\{a_1,\ldots,a_{d+1}\}}, \det(A_{\{a_1,\ldots,a_{d+1}\}})$;
foreach $a \in \{a_{d+2}, \ldots, a_n\}$ **do**
    $Q' \leftarrow Q$;
    foreach $F \in Q$ **do**
        $C \leftarrow$ the unique d-face s.t. $C \in T$ and $F \in C$;
        $u \leftarrow$ the unique vertex s.t. $u \in C$ and $u \notin F$;
        $C' \leftarrow F \cup \{a\}$;
        $i \leftarrow$ the index of u in A_C;
        // both $\det(A_C)$ and A^{adj}_C were computed in a previous step
        $\det(A_{C'}) \leftarrow \det(A_C) + (A^{adj}_C)^i(\boldsymbol{u} - \boldsymbol{a})$;
        if $\det(A_{C'})\det(A_C) < 0$ *and* $\det(A_{C'}) \neq 0$ **then**
            $A^{adj}_{C'} \leftarrow \frac{1}{\det(A_C)}(A^{adj}_C \det(A_{C'}) - A^{adj}_C(\boldsymbol{u} - \boldsymbol{a})(e_i^T A^{adj}_C))$;
            $T \leftarrow T \cup \{d\text{-face of } conv(C')\}$;
            $Q' \leftarrow Q' \ominus \{(d-1)\text{-faces of } C'\}$; // symmetric difference
    $Q \leftarrow Q'$;
return Q;

Theorem 1. *Given a d-dimensional pointset all, except the first, orientation predicates of incremental convex hull algorithms can be computed in $O(d^2)$ time and $O(d^2 t)$ space, where t is the number of cells of the constructed triangulation.*

Essentially, this result improves the computational complexity of the determinants involved in incremental convex hull algorithms from $O(d^\omega)$ to $O(d^2)$. To analyze the complexity of Alg. 1, we bound the number of facets of Q in every step of the outer loop of Alg. 1 with the number of $(d-1)$-faces of the constructed triangulation of $conv(\mathcal{A})$, which is bounded by $(d+1)t$. Thus, using Thm. 1, we have the following complexity bound for Alg. 1.

Corollary 1. *Given n d-dimensional points, the complexity of BB algorithm is $O(n \log n + d^3 nt)$, where $n \gg d$ and t is the number of cells of the constructed triangulation.*

Note that the complexity of BB, without using the method of dynamic determinants, is bounded by $O(n \log n + d^{\omega+1} nt)$. Recall that t is bounded by $O(n^{\lfloor d/2 \rfloor})$ [28, Sect.8.4], which shows that Alg. 1, and convex hull algorithms in general, do not have polynomial complexity. The schematic description of Alg. 1 and its coarse analysis is good enough for our purpose: to illustrate the application of dynamic determinant computation to incremental convex hulls and to quantify the improvement of our method. See Sect. 4 for a practical approach to incremental convex hull algorithms using dynamic determinant computations.

The above results can be extended to improve the complexity of geometric algorithms that are based on convex hulls computations, such as algorithms for regular or Delaunay triangulations and Voronoi diagrams. It is straightforward to apply the above method in orientation predicates appearing in point location algorithms. By using Alg. 1, we compute a triangulation and a map of adjoint matrices to its cells. Then, the point location predicates can be computed using Eq. 4, avoiding the computation of new adjoint matrices.

Corollary 2. *Given a triangulation of a d-dimensional pointset computed by Alg. 1, the orientation predicates involved in any point location algorithm can be computed in $O(d)$ time and $O(d^2 t)$ space, where t is the number of cells of the triangulation.*

4   Implementation and Experimental Analysis

We propose the *hashed dynamic determinants* scheme and implement it in C++. The design of our implementation is modular, that is, it can be used on top of either geometric software providing geometric predicates (such as orientation) or algebraic software providing dynamic determinant algorithm implementations. The code is publicly available from http://hdch.sourceforge.net.

The hashed dynamic determinants scheme consists of efficient implementations of algorithms *dyn_inv* and *dyn_adj* (Sect. 2) and a hash table, which stores intermediate results (matrices and determinants) based on the methods presented in Sect. 3. Every $(d - 1)$-face of a triangulation, *i.e.*, a common facet of two neighbor cells (computed by any incremental convex hull package which constructs a triangulation of the computed convex hull), is mapped to the indices of its vertices, which are used as keys. These are mapped to the adjoint (or inverse) matrix and the determinant of one of the two adjacent cells. Let us illustrate this approach with an example, on which we use the *dyn_adj* algorithm.

Example 1. Let $A = \{a_1 = (0,1), a_2 = (1,2), a_3 = (2,1), a_4 = (1,0), a_5 = (2,2)\}$ where every point a_i has an index i from 1 to 5. Assume we are in some step of an incremental convex hull or point location algorithm and let $T = \{\{1,2,4\}, \{2,3,4\}\}$ be the 2-dimensional triangulation of A computed so far. The cells of T are written using the indices of the points in A. The hash table will store as keys the set of indices of the 2-faces of T, *i.e.*, $\{\{1,2\},\{2,4\},\{1,4\}\}$ mapping to the adjoint and the determinant of the matrix constructed by the points a_1, a_2, a_4. Similarly, $\{\{2,3\},\{3,4\},\{2,4\}\}$ are mapped to the adjoint matrix and determinant of a_2, a_3, a_4. To insert a_5, we compute the determinant of a_2, a_3, a_5, by querying the hash table for $\{2,3\}$. Adjoint and determinant of the matrix of a_2, a_3, a_4 are returned, and we perform an update of the column corresponding to point a_4, replacing it by a_5 by using Eqs. 3 and 4.

To implement the hash table, we used the Boost libraries [7]. To reduce memory consumption and speed-up look-up time, we sort the lists of indices that form the hash keys. We also use the *GNU Multiple Precision arithmetic library* (GMP), the current standard for multiple-precision arithmetic, which provides integer and rational types mpz_t and mpq_t, respectively.

We perform an experimental analysis of the proposed methods. All experiments ran on an Intel Core i5-2400 3.1GHz, with 6MB L2 cache and 8GB RAM, running 64-bit Debian GNU/Linux. We divide our tests in four scenarios, according to the number type involved in computations: (a) rationals where the bit-size of both numerator and denominator is 10000, (b) rationals converted from doubles, that is, numbers of the form $m \times 2^p$, where m and p are integers of bit-size 53 and 11 respectively, (c) integers with bit-size 10000, and (d) integers with bit-size 32. However, it is rare to find in practice input coefficients of scenarios (a) and (c). Inputs are usually given as 32 or 64-bit numbers. These inputs correspond to the coefficients of scenario (b). Scenario (d) is also very important, since points with integer coefficients are encountered in many combinatorial applications (see Sect. 1).

We compare state-of-the-art software for exact computation of the determinant of a matrix. We consider LU decomposition in CGAL [10], optimized LU decomposition in Eigen [18], LinBox asymptotically optimal algorithms [13] (tested only on integers) and Maple 14 LinearAlgebra[Determinant] (the default determinant algorithm). We also implemented two division-free algorithms: Bird's [5] and Laplace expansion [22, Sect.4.2]. Finally, we consider our implementations of *dyn_inv* and *dyn_adj*.

We test the above implementations in the four coefficient scenarios described above. When coefficients are integer, we can use integer exact division algorithms, which are faster than quotient-remainder division algorithms. In this case, Bird, Laplace and *dyn_adj* enjoy the advantage of using the number type mpz_t while the rest are using mpq_t. The input matrices are constructed starting from a random $d \times d$ matrix, replacing a randomly selected column with a random d vector. We present experimental results of the most common in practice input scenarios (b), (d) (Tables 1, 2). The rest will appear in the full version of the paper. We stop testing an implementation when it is slow and far from being the fastest (denoted with '-' in the Tables).

On one hand, the experiments show the most efficient determinant algorithm implementation in the different scenarios described, without considering the dynamic algorithms. This is a result of independent interest, and shows the efficiency of division-free algorithms in some settings. The simplest determinant algorithm, Laplace expansion, proved to be the best in all scenarios, until dimension 4 to 6, depending on the scenario. It has exponential complexity, thus it is slow in dimensions higher than 6 but it behaves very well in low dimensions because of the small constant of its complexity and the fact that it performs no divisions. Bird is the fastest in scenario (c), starting from dimension 7, and in scenario (d), in dimensions 7 and 8. It has also a small complexity constant, and performing no divisions makes it competitive with decomposition methods (which have better complexity) when working with integers. CGAL and Eigen implement LU decomposition, but the latter is always around two times faster. Eigen is the fastest implementation in scenarios (a) and (b), starting from di-

Table 1. Determinant tests, inputs of scenario (b): rationals converted from `double`. Each timing (in milliseconds) corresponds to the average of computing 10000 (for $d < 7$) or 1000 (for $d \geq 7$) determinants. Light blue highlights the best non-dynamic algorithm while yellow highlights the dynamic algorithm if it is the fastest over all.

d	Bird	CGAL	Eigen	Laplace	Maple	dyn_inv	dyn_adj
3	.013	.021	.014	.008	.058	.046	.023
4	.046	.050	.033	.020	.105	.108	.042
5	.122	.110	.072	.056	.288	.213	.067
6	.268	.225	.137	.141	.597	.376	.102
7	.522	.412	.243	.993	.824	.613	.148
8	.930	.710	.390	–	1.176	.920	.210
9	1.520	1.140	.630	–	1.732	1.330	.310
10	2.380	1.740	.940	–	2.380	1.830	.430
11	–	2.510	1.370	–	3.172	2.480	.570
12	–	3.570	2.000	–	4.298	3.260	.760
13	–	4.960	2.690	–	5.673	4.190	1.020
14	–	6.870	3.660	–	7.424	5.290	1.360
15	–	9.060	4.790	–	9.312	6.740	1.830

mension 5 and 6 respectively, as well as in scenario (d) in dimensions between 9 and 12. It should be stressed that decomposition methods are the current standard to implement determinant computation. Maple is the fastest only in scenario (d), starting from dimension 13. In our tests, Linbox is never the best, due to the fact that it focuses on higher dimensions.

On the other hand, experiments show that dyn_adj outperforms all the other determinant algorithms in scenarios (b), (c), and (d). On each of these scenarios, there is a threshold dimension, starting from which dyn_adj is the most efficient, which happens because of its better asymptotic complexity. In scenarios (c) and (d), with integer coefficients, division-free performs much better, as expected, because integer arithmetic is faster than rational. In general, the sizes of the coefficients of the adjoint matrix are bounded. That is, the sizes of the operands of the arithmetic operations are bounded. This explains the better performance of dyn_adj over the dyn_inv, despite its worse arithmetic complexity.

For the experimental analysis of the behaviour of dynamic determinants used in convex hull algorithms (Alg. 1, Sect. 3), we experiment with four state-of-the-art exact convex hull packages. Two of them implement incremental convex hull algorithms: `triangulation` [6] implements [11] and `beneath-and-beyond` (bb) in `polymake` [17]. The package `cdd` [16] implements the double description method, and `lrs` implements the gift-wrapping algorithm using reverse search [2]. We propose and implement a variant of `triangulation`, which we will call `hdch`, implementing the hashed dynamic determinants scheme for dimensions higher than 6 (using Eigen for initial determinant and adjoint or inverse matrix computation) and using Laplace determinant algorithm for lower dimensions. The main difference between this implementation and Alg. 1 of Sect. 3 is that it does not sort the points and, before inserting a point, it performs a point location. Thus, we can take advantage of our scheme in two places: in the orientation predicates appearing in the point location procedure and in the ones that

Table 2. Determinant tests, inputs of scenario (d): integers of bit-size 32. Times in milliseconds, averaged over 10000 tests. Highlighting as in Table 1.

d	Bird	CGAL	Eigen	Laplace	Linbox	Maple	dyn_inv	dyn_adj
3	.002	.021	.013	.002	.872	.045	.030	.008
4	.012	.041	.028	.005	1.010	.094	.058	.015
5	.032	.080	.048	.016	1.103	.214	.119	.023
6	.072	.155	.092	.040	1.232	.602	.197	.033
7	.138	.253	.149	.277	1.435	.716	.322	.046
8	.244	.439	.247	–	1.626	.791	.486	.068
9	.408	.689	.376	–	1.862	.906	.700	.085
10	.646	1.031	.568	–	2.160	1.014	.982	.107
11	.956	1.485	.800	–	10.127	1.113	1.291	.133
12	1.379	2.091	1.139	–	13.101	1.280	1.731	.160
13	1.957	2.779	1.485	–	–	1.399	2.078	.184
14	2.603	3.722	1.968	–	–	1.536	2.676	.222
15	3.485	4.989	2.565	–	–	1.717	3.318	.269
16	4.682	6.517	3.391	–	–	1.850	4.136	.333

appear in construction of the convex hull. We design the input of our experiments parametrized on the number type of the coefficients and on the distribution of the points. The number type is either rational or integer. From now on, when we refer to rational and integer we mean scenario (b) and (d), respectively. We test three uniform point distributions: (i) in the d-cube $[-100, 100]^d$, (ii) in the origin-centered d-ball of radius 100, and (iii) on the surface of that ball.

We perform an experimental comparison of the four above packages and hdch, with input points from distributions (i)-(iii) with either rational or integer co-efficients. In the case of integer coefficients, we test hdch using mpq_t (hdch_q) or mpz_t (hdch_z). In this case hdch_z is the most efficient with input from distribution (ii) (Fig. 1(a); distribution (i) is similar to this) while in distribution (iii) both hdch_z and hdch_q perform better than all the other packages (see Fig. 1(b)). In the rational coefficients case, hdch_q is competitive to the fastest package (not shown for space reasons). Note that the rest of the packages cannot perform arithmetic computations using mpz_t because they are lacking division-free determinant algorithms. Moreover, we perform experiments to test the improvements of hashed dynamic determinants scheme on triangulation and their memory consumption. For input points from distribution (iii) with integer coefficients, when dimension ranges from 3 to 8, hdch_q is up to 1.7 times faster than triangulation and hdch_z up to 3.5 times faster (Table 3). It should be noted that hdch is always faster than triangulation. The sole modification of the determinant algorithm made it faster than all other implementations in the tested scenarios. The other implementations would also benefit from applying the same determinant technique. The main disadvantage of hdch is the amount of memory consumed, which allows us to compute up to dimension 8 (Table 3). This drawback can be seen as the price to pay for the obtained speed-up.

A large class of algorithms that compute the exact volume of a polytope is based on triangulation methods [9]. All the above packages compute the

Table 3. Comparison of hdch_z, hdch_q and triangulation using points from distribution (iii) with integer coefficients; swap means that the machine used swap memory

|A|	d	hdch_q		hdch_z		triangulation	
		time (sec)	memory (MB)	time (sec)	memory (MB)	time (sec)	memory (MB)
260	2	0.02	35.02	0.01	33.48	0.05	35.04
500	2	0.04	35.07	0.02	33.53	0.12	35.08
260	3	0.07	35.20	0.04	33.64	0.20	35.23
500	3	0.19	35.54	0.11	33.96	0.50	35.54
260	4	0.39	35.87	0.21	34.33	0.82	35.46
500	4	0.90	37.07	0.47	35.48	1.92	37.17
260	5	2.22	39.68	1.08	38.13	3.74	39.56
500	5	5.10	45.21	2.51	43.51	8.43	45.34
260	6	14.77	1531.76	8.42	1132.72	20.01	55.15
500	6	37.77	3834.19	21.49	2826.77	51.13	83.98
220	7	56.19	6007.08	32.25	4494.04	90.06	102.34
320	7	swap	swap	62.01	8175.21	164.83	185.87
120	8	86.59	8487.80	45.12	6318.14	151.81	132.70
140	8	swap	swap	72.81	8749.04	213.59	186.19

Fig. 1. Comparison of convex hull packages for 6-dimensional inputs with integer coefficients. Points are uniformly distributed (a) inside a 6-ball and (b) on its surface.

volume of the polytope, defined by the input points, as part of the convex hull computation. The volume computation takes place at the construction of the triangulation during a convex hull computation. The sum of the volumes of the cells of the triangulation equals the volume of the polytope. However, the volume of the cell is the absolute value of the orientation determinant of the points of the cell and these values are computed in the course of the convex hull computation. Thus, the computation of the volume consumes no extra time besides the convex hull computation time. Therefore, hdch yields a competitive implementation for the exact computation of the volume of a polytope given by its vertices (Fig. 1).

Finally, we test the efficiency of hashed dynamic determinants scheme on the point location problem. Given a pointset, triangulation constructs a data structure that can perform point locations of new points. In addition to that, hdch constructs a hash table for faster orientation computations. We perform tests with

Table 4. Point location time of 1K and 1000K (1K=1000) query points for hdch_z and triangulation (triang), using distribution (iii) for preprocessing and distribution (i) for queries and integer coefficients

	d	$|\mathcal{A}|$	preprocess time (sec)	data structures memory (MB)	# of cells in triangulation	query time (sec)	
						1K	1000K
hdch_z	8	120	45.20	6913	319438	0.41	392.55
triang	8	120	156.55	134	319438	14.42	14012.60
hdch_z	9	70	45.69	6826	265874	0.28	276.90
triang	9	70	176.62	143	265874	13.80	13520.43
hdch_z	10	50	43.45	6355	207190	0.27	217.45
triang	10	50	188.68	127	207190	14.40	14453.46
hdch_z	11	39	38.82	5964	148846	0.18	189.56
triang	11	39	181.35	122	148846	14.41	14828.67

triangulation and hdch using input points uniformly distributed on the surface of a ball (distribution (iii)) as a preprocessing to build the data structures. Then, we perform point locations using points uniformly distributed inside a cube (distribution (i)). Experiments show that our method yields a speed-up in query time of a factor of 35 and 78 in dimension 8 to 11, respectively, using points with integer coefficients (scenario (d)) (see Table 4).

5   Future Work

It would be interesting to adapt our scheme for gift-wrapping convex hull algorithms and implement it on top of packages such as [2]. In this direction, our scheme should also be adapted to other important geometric algorithms, such as Delaunay triangulations.

In order to overcome the large memory consumption of our method, we shall exploit hybrid techniques. That is, to use the dynamic determinant hashing scheme as long as there is enough memory and subsequently use the best available determinant algorithm (Sect. 4), or to clean periodically the hash table.

Another important experimental result would be to investigate the behavior of our scheme using filtered computations.

Acknowledgments. The authors are partially supported from project "Computational Geometric Learning", which acknowledges the financial support of the Future and Emerging Technologies (FET) programme within the Seventh Framework Programme for Research of the European Commission, under FET-Open grant number: 255827. We thank I.Z. Emiris for his advice and support, as well as M. Karavelas and E. Tsigaridas for discussions and bibliographic references.

References

1. Abbott, J., Bronstein, M., Mulders, T.: Fast deterministic computation of determinants of dense matrices. In: ISSAC, pp. 197–203 (1999)
2. Avis, D.: lrs: A revised implementation of the reverse search vertex enumeration algorithm. In: Polytopes - Combinatorics and Computation, Oberwolfach Seminars, vol. 29, pp. 177–198. Birkhäuser-Verlag (2000)

3. Bartlett, M.S.: An inverse matrix adjustment arising in discriminant analysis. The Annals of Mathematical Statistics 22(1), 107–111 (1951)
4. Barvinok, A., Pommersheim, J.E.: An algorithmic theory of lattice points in polyhedra. New Perspectives in Algebraic Combinatorics, 91–147 (1999)
5. Bird, R.: A simple division-free algorithm for computing determinants. Inf. Process. Lett. 111, 1072–1074 (2011)
6. Boissonnat, J.D., Devillers, O., Hornus, S.: Incremental construction of the Delaunay triangulation and the Delaunay graph in medium dimension. In: SoCG, pp. 208–216 (2009)
7. Boost: peer reviewed C++ libraries, http://www.boost.org
8. Brönnimann, H., Emiris, I., Pan, V., Pion, S.: Sign determination in Residue Number Systems. Theor. Comp. Science 210(1), 173–197 (1999)
9. Büeler, B., Enge, A., Fukuda, K.: Exact volume computation for polytopes: A practical study (1998)
10. CGAL: Computational geometry algorithms library, http://www.cgal.org
11. Clarkson, K., Mehlhorn, K., Seidel, R.: Four results on randomized incremental constructions. Comput. Geom.: Theory & Appl. 3, 185–212 (1993)
12. Cox, D.A., Little, J., O'Shea, D.: Using Algebraic Geometry. Graduate Texts in Mathematics. Springer, Heidelberg (2005)
13. Dumas, J.G., Gautier, T., Giesbrecht, M., Giorgi, P., Hovinen, B., Kaltofen, E., Saunders, B., Turner, W., Villard, G.: Linbox: A generic library for exact linear algebra. In: ICMS, pp. 40–50 (2002)
14. Edelsbrunner, H.: Algorithms in combinatorial geometry. Springer-Verlag New York, Inc., New York (1987)
15. Emiris, I., Fisikopoulos, V., Konaxis, C., Peñaranda, L.: An output-sensitive algorithm for computing projections of resultant polytopes. In: SoCG, pp. 179–188 (2012)
16. Fukuda, K.: cddlib, version 0.94f (2008), http://www.ifor.math.ethz.ch/~fukuda/cdd_home
17. Gawrilow, E., Joswig, M.: Polymake: a framework for analyzing convex polytopes, pp. 43–74 (1999)
18. Guennebaud, G., Jacob, B., et al.: Eigen v3 (2010), http://eigen.tuxfamily.org
19. Harville, D.A.: Matrix algebra from a statistician's perspective. Springer, New York (1997)
20. Kaltofen, E., Villard, G.: On the complexity of computing determinants. Computational Complexity 13, 91–130 (2005)
21. Krattenthaler, C.: Advanced determinant calculus: A complement. Linear Algebra Appl. 411, 68 (2005)
22. Poole, D.: Linear Algebra: A Modern Introduction. Cengage Learning (2006)
23. Rambau, J.: TOPCOM: Triangulations of point configurations and oriented matroids. In: Cohen, A., Gao, X.S., Takayama, N. (eds.) Math. Software: ICMS, pp. 330–340. World Scientific (2002)
24. Rote, G.: Division-free algorithms for the determinant and the Pfaffian: algebraic and combinatorial approaches. Comp. Disc. Math., 119–135 (2001)
25. Sankowski, P.: Dynamic transitive closure via dynamic matrix inverse. In: Proc. IEEE Symp. on Found. Comp. Sci., pp. 509–517 (2004)
26. Seidel, R.: A convex hull algorithm optimal for point sets in even dimensions. Tech. Rep. 81-14, Dept. Comp. Sci., Univ. British Columbia, Vancouver (1981)
27. Sherman, J., Morrison, W.J.: Adjustment of an inverse matrix corresponding to a change in one element of a given matrix. The Annals of Mathematical Statistics 21(1), 124–127 (1950)
28. Ziegler, G.: Lectures on Polytopes. Springer (1995)

Lines through Segments in 3D Space[*]

Efi Fogel, Michael Hemmer, Asaf Porat, and Dan Halperin

The Blavatnik School of Computer Science, Tel Aviv University
{efifogel,mhsaar,asafpor}@gmail.com, danha@post.tau.ac.il

Abstract. Given a set S of n line segments in three-dimensional space, finding all the lines that simultaneously intersect at least four line segments in S is a fundamental problem that arises in a variety of domains. We refer to this problem as *the lines-through-segments problem*, or LTS for short. We present an efficient output-sensitive algorithm and its implementation to solve the LTS problem. The implementation is exact and properly handles all degenerate cases. To the best of our knowledge, this is the first implementation for the LTS problem that is (i) output sensitive and (ii) handles all degenerate cases. The algorithm runs in $O((n^3 + I) \log n)$ time, where I is the output size, and requires $O(n \log n + J)$ working space, where J is the maximum number of output elements that intersect two fixed line segments; I and J are bounded by $O(n^4)$ and $O(n^2)$, respectively. We use CGAL arrangements and in particular its support for two-dimensional arrangements in the plane and on the sphere in our implementation. The efficiency of our implementation stems in part from careful crafting of the algebraic tools needed in the computation. We also report on the performance of our algorithm and its implementation compared to others. The source code of the LTS program as well as the input examples for the experiments can be obtained from http://acg.cs.tau.ac.il/projects/lts.

1 Introduction

Given a set S of line segments in $\mathbb{R}^3$, we study the *lines-through-segments* (LTS) problem, namely, the problem of computing all lines that simultaneously intersect at least four line segments in S. LTS is a fundamental problem that arises in a variety of domains. For instance, solving the LTS problem can be used as the first step towards solving the more general problem of finding all lines tangent to at least four geometric objects taken from a set of geometric objects. The latter is ubiquitous in many fields of computation such as computer graphics (visibility computations), computational geometry (line transversal), robotics and automation (assembly planning), and computer vision. Computing visibility information, for example, is crucial to many problems in computer graphics,

---

[*] This work has been supported in part by the 7th Framework Programme for Research of the European Commission, under FET-Open grant number 255827 (CGL—Computational Geometry Learning), by the Israel Science Foundation (grant no. 1102/11), by the German-Israeli Foundation (grant no. 969/07), and by the Hermann Minkowski–Minerva Center for Geometry at Tel Aviv University.

L. Epstein and P. Ferragina (Eds.): ESA 2012, LNCS 7501, pp. 455–466, 2012.

vision, and robotics, such as computing umbra and penumbra cast by a light source [7].

The number of lines that intersect four lines in $\mathbb{R}^3$ is 0, 1, 2, or infinite. Brönnimann et al. [6] showed that the number of lines that intersect four arbitrary line segments in $\mathbb{R}^3$ is 0, 1, 2, 3, 4, or infinite. They also showed that the lines lie in at most four maximal *connected components*.[1]

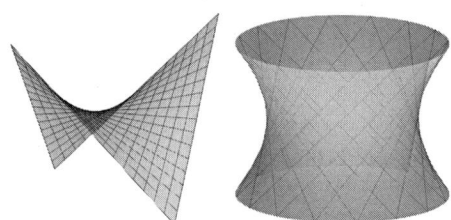

(a) A hyperbolic (b) A hyperboloid of paraboloid.   one sheet.

Fig. 1. Two configurations of lines segments in which an infinite number of lines intersect four line segments

A straightforward method to find all the lines that intersect four lines, given a set of n lines, examines each quadruplet of lines using, for example, the Plücker coordinate representation. This method has been used by Hohmeyer and Teller [17] and also described by Redburn [15]. It was later used by Everett et al. [10] as a building block for the problem of finding line transversals (the set of lines that intersect all given line segments). The running time of this method is $O(n^4)$.

The combinatorial complexity of the lines that intersect four line segments of a set of n line segments is $\Theta(n^4)$ (counting maximal connected components). The lower bound can be established by placing two grids of $n/2$ line segments each in two parallel planes and passing a line through every two intersection points, one from each grid. However, in many cases the number of output lines is considerably smaller. The size of the output tends to be even smaller, when the input consists of line segments (as opposed to lines), which is typically the case in practical problems, and it is expected to decrease with the decrease of the input line-segment lengths.

We present an efficient output-sensitive algorithm, and its complete and robust implementation, that solves the LTS problem in three-dimensional Euclidean space.[2] The implementation is complete in the sense that it handles all degenerate cases and guarantees exact results. Examples of degenerate cases are: A line segment may degenerate to a point, several segments may intersect, be coplanar, parallel, concurrent, lie on the same supporting line, or even overlap. To the best of our knowledge, this is the first implementation for the LTS problem that is (i) output sensitive and (ii) handles all degenerate cases. The algorithm utilizes the idea of McKenna and O'Rouke [13] to represent the set of lines that intersect three lines as a rectangular hyperbola with a vertical and a horizontal asymptotes in $\mathbb{R}^2$. However, as opposed to their algorithm, which takes $O(n^4\alpha(n))$ time, our algorithm is output sensitive and its asymptotic time and space complexities are $O((n^3 + I)\log n)$ and $O(n\log n + J)$, respectively,

---

[1] Two lines tangent to the same four line segments are in the same connected component iff one of the lines can be continuously moved into the other while remaining tangent to the same four line-segments.

[2] A short version of the extended abstract was presented in EuroCG this year.

where n is the input size, I is the output size, and J is the maximum number of output elements that intersect two fixed line segments; I and J are bounded by $O(n^4)$ and $O(n^2)$, respectively. The algorithm can be trivially altered to accept a constant $c \geq 4$ and compute all lines that simultaneously intersect exactly, or at least, c line segments from the input set. In addition, the algorithm can easily be changed to compute transversals to line segments in $\mathbb{R}^3$ [6].

A related problem to the problem at hand is the *the lines-tangent-to-polytopes problem*, or LTP for short. Formally, given a set $\mathcal{P}$ of n convex polytopes in three-dimensional space, the objective is to find all the lines that are simultaneously tangent to quadruples of polytopes in $\mathcal{P}$. This, in turn, can be generalized to the problem of determining the visibility between objects. In many cases a solution to the visibility or LTP problems can also serve as a solution to the LTS problem. Brönnimann et al. [5] provide a non-output sensitive solution to the visibility problem. It runs in time $O(n^2 k^2 \log n)$ for a scene of k polyhedra of total complexity n (although their algorithm is sensitive to the size of the 2D visibility skeletons, calculated during the process). Devillers et al. [8] introduce efficient algebraic methods to evaluate the geometric predicates required during the visibility computation process.

Our algorithms are implemented on top of the Computational Geometry Algorithm Library (CGAL) [18]. The implementation is mainly based on the *2D Arrangements* package of the library [20]. This package supports the robust and efficient construction and maintenance of arrangements induced by curves embedded on certain orientable two-dimensional parametric surfaces in three-dimensional space [2, 19], and robust operations on them.[3] The implementation uses in particular 2D arrangements of rectangular hyperbolas with vertical and horizontal asymptotes in the plane and 2D arrangements of geodesic arcs on the sphere [1].

The rest of this paper is organized as follows. In Section 2 we introduce the necessary terms and definitions and the theoretical foundation of the algorithm that solves the LTS problem. In Section 3 we present a limited version of the algorithm. In Section 4 we describe the specially tuned algebraic tools used in the implementation. We report on experimental results in Section 5 and suggest future directions in Section 6. Because of space limitation many details of the analysis (Section 2) and the algorithm (Section 3) are omitted.

2   Representation

This section discusses the encoding of all lines that intersect two fixed line segments, S_1 and S_2, and a third line segment, S_3. We represent a line $L \subset \mathbb{R}^3$ by a point $p \in \mathbb{R}^3$ and a direction $d \in \mathbb{R}^3 \setminus \{\mathcal{O}\}$ as $L(t) = p + t \cdot d$, where $\mathcal{O}$ denotes the origin and $t \in \mathbb{R}$. Clearly, this representation is not unique. A segment $S \subset L \subset \mathbb{R}^3$ is represented by restricting t to the interval $[a, b] \subset \mathbb{R}$. We refer to $S(a)$ and $S(b)$ as the source and target points, respectively, and set

---

[3] Arrangements on surfaces are supported as of CGAL version 3.4, albeit not documented yet.

$a = 0$ and $b = 1$. We denote the underlying line of a line segment S by $L(S)$. Two lines are *skew* if they are not coplanar. Three or more lines are *concurrent* if they all intersect at a common point.

2.1 S_1 and S_2 Are Skew

Given two lines L_1 and L_2 we define a map $\Psi_{L_1 L_2}$, which maps a point in $\mathbb{R}^3$ to a set in $\mathbb{R}^2$, as follows: $\Psi_{L_1 L_2}(q) = \{(t_1, t_2) \in \mathbb{R}^2 \,|\, L_1(t_1),\ L_2(t_2),\ \text{and } q \text{ are co-}$ llinear$\}$. The set $\Psi_{L_1 L_2}(q)$, which might be empty, corresponds to all lines that contain q and intersect L_1 and L_2. Now, consider the pair $(t_1, t_2) \in \mathbb{R}^2$. If $L_1(t_1) \neq L_2(t_2)$, then this pair uniquely defines the line that intersects L_1 and L_2 at $L_1(t_1)$ and $L_2(t_2)$, respectively. Thus, for skew lines L_1 and L_2 there is a canonical bijective map between $\mathbb{R}^2$ and all lines that intersect L_1 and L_2. It follows that for disjoint lines L_1 and L_2 and a third line L_3 the set $\Psi_{L_1 L_2}(L_3)$ is sufficient to represent all lines that intersect L_1, L_2, and L_3, where $\Psi_{L_1 L_2}(L_3) = \{\Psi_{L_1 L_2}(q) \,|\, q \in L_3\}$. Similarly, we define $\Psi_{s_1 s_2}(q) = \{(t_1, t_2) \,|\, S_1(t_1),\ S_2(t_2),\ \text{and } q \text{ are collinear}, (t_1, t_2) \in [0,1]^2\}$ for two line segments S_1 and S_2. The characterization of $\Psi_{S_1 S_2}(S_3) = \{\Psi_{S_1 S_2}(q) \,|\, q \in S_3\}$ serves as the theoretical foundation of the algorithm that solves the LTS problem presented in Section 3. As $\Psi_{S_1 S_2}(x) = \Psi_{L(S_1) L(S_2)}(x) \cap [0,1]^2$, it is sufficient to analyze $\Psi_{L_1 L_2}(S_3)$ for a line segment S_3. We omit the complete characterization due to limited space, and only mention the generic case where S_1, S_2, and S_3 are pairwise skew below. The characterization is summerized by the following corollary.

Corollary 1. *$\Psi_{S_1 S_2}(S_3) \subset \mathbb{R}^2$ is either a point, a one dimensional set consisting of line segments or arcs of a rectangular hyperbola with a vertical and a horizontal asymptotes, or a two-dimensional set bounded by linear segments or arcs of such hyperbolas.*

Consider the case in which the direction vectors of the underlying lines of the segments S_1, S_2, and S_3 are pairwise skew. From the complete analysis omitted here it follows that $\Psi_{L(S_1) L(S_2)}(L(S_3))$ is a rectangular hyperbola with a vertical and a horizontal asymptotes, and $\Psi_{S_1 S_2}(S_3)$ consists of at most three maximal con-

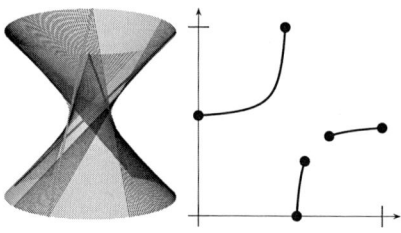

nected components, where each component represents a patch of a ruled surface. The figure depicted above shows three surface patches the lines of which intersect three skew line segments, S_1, S_2, and S_3, in $\mathbb{R}^3$. These surface patches are contained in a hyperboloid of one sheet; see Figure 1b. Also depicted is the point set $\Psi_{S_1 S_2}(S_3)$.

2.2 S_1 and S_2 Intersect

Assume L_1 and L_2 intersect, and let $q = L_1(\tilde{t_1}) = L_2(\tilde{t_2})$ be the intersection point. The point $(\tilde{t_1}, \tilde{t_2})$ represents all lines that contain q. We represent these lines by points on a semi open upper hemisphere centered at q. We define the additional map $\Xi_q : \mathbb{R}^3 \setminus \{q\} \to \mathbb{H}^2$ and $\Xi_q(p) \longmapsto d = s(p-q)/|p-q|$, with $s \in \{\pm 1\}$, such that $d \in \mathbb{H}^2 = \{p \,|\, p \in \mathbb{S}^2$ and p is lexicographically larger than $\mathcal{O}\}$.[4]

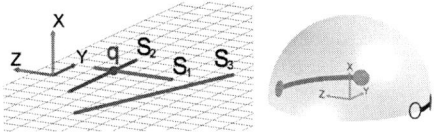

In the generic case a segment S maps to one or two geodesic arcs on $\mathbb{H}^2$. If S_3 is a point, or $L(S_3)$ contains q and S_3 does not, $\Xi_q(S)$ consists of a single point. If $q \in S_3$, we define $\Xi_q(S_3) = \mathbb{H}^2$. The left image of the figure above depicts three line segments, S_1, S_2, and S_3, such that S_1 and S_2 intersect at q (and S_3 does not). The right image depicts the mapping $\Xi_q(S_3)$, where $\Xi_q(S_3) = \{\Xi_q(p) \,|\, p \in S_3\}$. It consists of two geodesic arcs on $\mathbb{H}^2$.

2.3 S_1 and S_2 Are Collinear

The case where S_1 and S_2 are collinear completes the list of possible cases. If S_1 and S_2 do not overlap, the only line that can possibly intersect S_1 and S_2 is the line containing S_1 and S_2. Otherwise, the number of degrees of freedom of all the lines that intersect S_1 and S_2 is three. In any case S_1 and S_2 are handled separately. The handling does not involve a mapping to a two-dimensional surface.

We are now ready to describe our algorithm for solving the LTS problem in its full generality. Further details related to the variant cases handled are available at http://acg.cs.tau.ac.il/projects/lts.

3   The Algorithm

The input is a set $\mathcal{S} = \{S_1, \ldots, S_n\}$ of n line segments in $\mathbb{R}^3$. The output is a set of at most $O(n^4)$ elements. Assuming that an input line imposes a single intersection constraint, each element is either (i) a (one-dimensional) line that abides by at least four intersection constraints imposed by the line segments in $\mathcal{S}$, (ii) a (two-dimensional) ruled surface patch in $\mathbb{R}^3$, such that all lines that lie on the surface abide by at least four such intersection constraints, (iii) the set of all lines that contain a given point, which is, for example, the intersection of at least four concurrent lines, or (iv) the set of all lines that intersect a given line segment, which is the intersection of at least four input line segments. By default, however, we assume that a sub-segment that is the intersection of several overlapping line segments imposes a single intersection constraint, and a point that is either the intersection of several concurrent line segments, or simply a degenerate line segment, imposes two intersection constraints. The line segments that impose the constraints of an output element are referred to as the *generating line segments* of that element. It is possible to couple the output lines with their generating line segments.

---

[4] $\mathbb{H}^2$ is the union of an open half sphere, an open half circle, and a point.

To simplify the exposition of the algorithm, we assume that the line segments are full-dimensional, may intersect pairwise only at discrete and distinct points, and no three line segments are coplanar. Due to limited space we only partially describe the algorithm that handles this case. The detailed description of the complete algorithm that handles all cases can be found at `http://acg.cs.tau.ac.il/projects/lts`. The complete algorithm also respects several different settings selected by the user. They are also listed in the appendix.

We transform the original three-dimensional LTS problem into a collection of two-dimensional problems and use two-dimensional arrangements to solve them, exploiting the plane-sweep algorithmic framework, which is output sensitive. We go over unordered pairs of lines segments in S. For each pair, (S_i, S_j), we find all lines that intersect S_i, S_j, and two other line segments in S, that have not been found yet in previous iterations; see Algorithm 1 for pseudo code.

Algorithm 1. Compute lines that intersect line segments $S_1, \ldots, S_n$.

---
1 **for** $i = 1, \ldots, n - 3$,
2   **for** $j = n, \ldots, i + 3$,
3   Construct arrangement $\mathcal{A}_{S_i S_j}$ induced by $\{\Psi_{S_i S_j}(S_k) \mid k = i+1, \ldots, j-1\}$.
4   Extract lines that intersect S_i and S_j from $\mathcal{A}_{S_i S_j}$.
5   **if** S_i and S_j intersect,
6   Construct arrangement $\mathcal{A}^s_{S_i \cap S_j}$ induced by $\{\Xi_{S_i \cap S_j}(S_k) \mid k = i+1, \ldots, j-1\}$.
7   Extract lines that intersect S_i and S_j from $\mathcal{A}^s_{S_i \cap S_j}$.

In Line 3 of Algorithm 1 we construct the arrangement $\mathcal{A}_{S_i S_j}$ induced by the set of (possibly degenerate) hyperbolic arcs $\mathcal{C}_{ij} = \{\Psi_{S_i S_j}(S_k) \mid k = i+1, \ldots, j-1\}$. We process the line segments $S_{i+1}, \ldots, S_{j-1}$ one at a time to produce the inducing set $\mathcal{C}_{ij}$. Next, using a plane-sweep algorithm, we construct the arrangement $\mathcal{A}_{S_i S_j}$ induced by $\mathcal{C}_{ij}$. We store with each vertex and edge of the arrangement $\mathcal{A}_{S_i S_j}$ the sorted sequence of line segments that are mapped through $\Psi_{S_i S_j}$ to the points and curves that induce that cell. The segments are sorted by their indices. Observe that curves in $\mathcal{C}_{ij}$ may overlap; see Figure 2b.

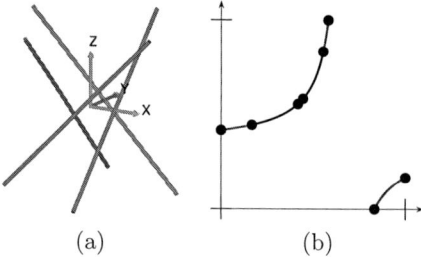

(a)   (b)

Fig. 2. (a) Four line segments, S_1, S_2, S_3, and S_4, supported by four lines of one ruling of a *hyperbolic paraboloid*, respectively; see also Figure 1a. (b) The arrangement $\mathcal{A}_{S_1 S_2}$. The edges drawn in purple are induced by two overlapping curves, one in $\Psi_{S_1 S_2}(S_3)$ and the other in $\Psi_{S_1 S_2}(S_4)$.

The generating line segments of every output element are immediately available from the sequences of line segments stored with vertices and edges. However, the role of these sequences extends beyond reporting. It turns out that some intersection points do not represent lines that intersect four line segments. An example of such a case occurs when either S_i or S_j intersects a third line segment, S_k. In such a case $\Psi_{S_i S_j}(S_k)$ consists of horizontal and vertical line segments. intersection point of the vertical and horizontal line segments does not represent a line that intersects four line segments and, thus, must be ignored. This case is detected by examining the sorted sequences of line segments.

Consider the case where S_i and S_j intersect at a point, say p. In this case we must output every line that contains p and abides by two additional intersection constraints. This information is not present in the planar arrangements constructed in Line 3. We can change the algorithm to construct an arrangement $\mathcal{A}_{S_i S_j}$ for every ordered pair (i, j) of indices of line segments in $\mathcal{S}$, where $\mathcal{A}_{S_i S_j}$ is induced by $\{\Psi_{S_i S_j}(S) \mid S \in \mathcal{S} \setminus \{S_i, S_j\}\}$, and filter out redundancies. While this modification does not increase the asymptotic complexity of the algorithm, our experiments show that constructing arrangements on the sphere instead exhibits much better performance in practice.

In Line 6 of Algorithm 1 we construct an arrangement on the sphere centered at the intersection point of S_i and S_j. The arrangement is induced by the set of (possibly degenerate) geodesic arcs $\mathcal{C}_{ij}^s = \{\Xi_{S_i \cap S_j}(S_k) \mid k = i + 1, \ldots, j - 1\}$. We process the line segments $S_{i+1}, \ldots, S_{j-1}$ one at a time to produce the inducing set $\mathcal{C}_{ij}^s$. When the underlying line of a line segment S_k contains the sphere center, $\Xi_{S_i \cap S_j}(S_k)$ consists of a single point. For each k, $i < k < j$, $\Xi_{S_i \cap S_j}(S_k)$ consists of either an isolated point or at most two geodesic arcs on the sphere; see Section 2.2. The pairwise intersections of the points and arcs in $\mathcal{C}_{ij}^s$ represent lines that intersect at least four input segments. Next, using a plane-sweep algorithm on the sphere, we construct the arrangement $\mathcal{A}_{S_i \cap S_j}^s$ induced by $\mathcal{C}_{ij}^s$. When $\Xi_{S_i \cap S_j}(S_k)$ consists of a single point it induces a single vertex in the arrangement. We store with each vertex and edge of the arrangement $\mathcal{A}_{S_i \cap S_j}^s$ the sorted sequence of line segments that are mapped through $\Xi_{S_i \cap S_j}$ to the points and geodesic arcs that induce that cell.

In Line 4 and Line 7 we extract the information and provide it to the user in a usable format. We refer to an arrangement cell that represents a valid output element as an eligible cell. We provide the user with the ability to iterate over eligible cells of different dimensions separately. This way, for example, a user can choose to obtain only the vertices that represent valid output lines. The eligibility of a given cell is immediately established from the sequence of line segments stored in that cell, and so are the generating line segments.

Constructing the arrangement $\mathcal{A}_{S_i S_j}$ and $\mathcal{A}_{S_i \cap S_j}^s$ are the dominant steps in the algorithm. For a given pair of indices (i, j), they are carried out in $O((n + k_{ij}) \log n)$ and $O((n + k_{ij}^s) \log n)$ time, respectively (using plane-sweep algorithms) where k_{ij} and k_{ij}^s are the numbers of intersections of the inducing hyperbolic and geodesic arcs, respectively. Thus, the entire process can be performed in $O((n^3 + I) \log n)$ running time, and $O(n + J)$ working space. Where n is the input size, I is the output size, and J is the maximum number of intersections in a single arrangement. I and J are bounded by $O(n^4)$ and $O(n^2)$, respectively.

4   Lazy Algebraic Tools

Our implementations are exact and complete. In order to achieve this, CGAL in general, and the CGAL *2D Arrangements* package [20] in particular, follows the *exact geometric-computation (EGC) paradigm* [22]. A naive attempt could realize this by carrying out each and every arithmetic operation using an expensive unlimited-precision number type. However, only the discrete decisions in an algorithm, namely the predicates, must be correct. This is a significant relaxation from the naive concept of numerical exactness. This way it is possible to use fast inexact arithmetic (e.g., double-precision floating-point arithmetic [9]), while analyzing the correctness. If the computation reaches a stage of uncertainty, the computation is redone using unlimited precision. In cases where such a state is never reached, expensive computation is avoided, while the result is still certified. In this context CGAL's Lazy kernel [14] is the state of the art, as it not only provides filtered predicates, but also delays the exact construction of coordinates and objects. While arithmetic is only carried out with (floating-point) interval arithmetic [4], each constructed object stores its construction history in a directed acyclic graph (DAG). Only in case the result of a predicate evaluated using interval arithmetic is uncertain, the DAG is evaluated using unlimited precision.

CGAL follows the *generic programming paradigm*; that is, algorithms are formulated and implemented such that they abstract away from the actual types, constructions, and predicates. Using the C++ programming language this is realized by means of class and function templates. CGAL's arrangement data structure is parameterized by a *traits* class that defines the set of curves types and the operations on curves of these types.

For the arrangement of geodesic arcs on the sphere we use the existing and efficient traits class that we have used in the past [1]. As this only requires a linear kernel, it uses CGAL's efficient Lazy Kernel [4]. However, in order to compute the planar arrangements of rectangular hyperbolic arcs with horizontal and vertical asymptotes, CGAL offered only a general traits class for rational functions, which was introduced in [16]. The class uses the general univariate algebraic kernel [3] of CGAL, which does not offer lazy constructions.

We modified the aforementioned traits class such that it is specialized for rectangular hyperbolic arcs with horizontal and vertical asymptotes. This way, it was possible to take advantage of the fact that in our case coordinates of intersection points are of algebraic degree at most 2. In particular, CGAL already offers a specific type[5] that represents such a number as $a+b\sqrt{c}$, where $a, b, c \in \mathbb{Q}$, This type can be used in conjunction with the lazy mechanism. Thus, the specialized traits constructs all point coordinates in a lazy fashion, which considerably speeds up the computation as shown in Section 5.

---

[5] CGAL::Sqrt_extension, part of the *Number Types* package of CGAL [12].

5   Experimental Results

We have conducted several experiments on three types of data sets. The first produces the worst-case combinatorial output and has many degeneracies. The second consists of transformed versions of the first and has many near-degeneracies. The third comprises random input. We report on the time consumption of our implementation, and compare it to those of other implementations. All experiments were performed on a Pentium PC clocked at 2.40 GHz.

5.1   Grid

The Grid data set comprises 40 line segments arranged in two grids of 20 lines segments each lying in two planes parallel to the yz-plane; see Section 1. Each grid consists of ten vertical and ten horizontal line segments. The output

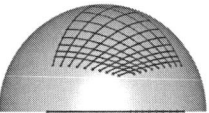

consists of several planar patches each lying in one of the two planes and exactly 10,000 lines, such that each contains one intersection point in one plane and one in the other plane.

The table to the right lists all output elements. 703 arrangements in the plane and 200 arrangements on the sphere were

Lines	Planar Curves	Spherical Arcs	Planar Regions	Time (secs)
10,000	36	1,224	17,060	20.74

constructed during the process. All single output lines are represented by vertices of arrangements on the sphere. Such an arrangement is depicted in the figure above. The origin of the sphere is the intersection point of two line segments S_1 and S_{40} lying in the same plane. The arrangement is induced by the point set $\{\Xi_{S_1 \cap S_{40}}(S_i) \mid i = 2, \ldots, 39\}$.

The figure to the right depicts an arrangement in the plane constructed during the process. The arrangement is induced by the point set $\{\Psi_{S_1 S_{39}}(S_i) \mid i = 2, \ldots, 38\}$. The line segments S_1 and S_{39} are parallel segments lying in the same plane. Each face of the arrangement represents a ruled surface patch, such that each line lying in the surface intersects at least 6 and up to 20 line segments. Different colors represent different number of generating line segments.

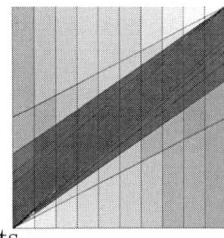

5.2   Transformed Grid

We have conducted three more experiments using a transformed version of the Grid data set. First, we slightly perturbed the input line segments, such that ev-

Input	Unlimited Precision			Double Precision	
	Time (secs)		Lines	Time	Lines
	LTS	Redburn		Redburn	
PG	23.72	140.17	12,139	0.70	12,009
TG 1	11.83	132.80	5,923	0.69	5,927
TG 2	6.90	128.80	1,350	0.70	1,253

ery two line segments became skew and the directions of every three line segments became linearly independent (referred to as **PG**). Secondly, we translated the (perturbed) horizontal line segments of one grid along the plane that contains this grid (referred to as **TG 1**), increasing the distance between the (perturbed) vertical and horizontal line segments of that grid. This drastically reduced the number of output lines. Thirdly, we translated the (perturbed) horizontal line segments of the other grid along the plane that contains this grid (referred to as **TG 2**), further reducing the number of output lines. The table above shows the number of output lines and the time it took to perform the computation using our implementation, referred to as **LTS**. The monotonic relation between the output size and time consumption of our implementation is prominent. The table also shows the time it took to perform the computation using two instances of a program developed by J. Redburn [15], which represents lines by their Plücker coordinates and exhaustively examines every quadruple of input line segments. One instance, relies on a number type with unlimited precision, while the other resorts to double-precision floating-point numbers. As expected, when limited precision numbers were used, the output was only an approximation. Notice that the influence of the output size on the time consumption of Redburn's implementation is negligible.[6]

5.3   Random Input

The Random data set consists of 50 line segments drawn uniformly at random. In particular, the endpoints are selected uniformly at random within a sphere. We experimented with three different radii,

Input	Time (secs)			Lines
	LTS	LLTS	Redburn	
Short	3.04	1.06	300.4	0
Medium	6.80	2.82	314.0	20,742
Long	12.36	5.15	327.0	64,151

namely, **Short**, **Medium**, and **Long** listed in increasing lengths. We verified that the line segments are in general position; that is, the directions of every three are linearly independent and they are pairwise skew. The table above shows the number of output lines and the time it took to perform the computation using (i) our implementation referred to as **LTS**, (ii) our implementation enhanced with the lazy mechanism referred to as **LLTS** (see Section 4), and (iii) the instance of Redburn's implementation that relies on unlimited precision. Once again, one can clearly observe how the time consumption of our implementation decreases with the decrease of the output size, which in turn decreases with the decrease in the line-segment lengths. Adversely, the time consumption of Redburn's implementation hardly changes.

6   Future Work

We are, in particular, motivated by assembly-partitioning problems, where a given collection of pairwise interior disjoint polyhedra in some relative position

---

[6] Our attempts to run Redburn's code on degenerate input failed, as it seems not to well handle such cases; thus, we were unable to experiment with the original Grid data set using this implementation.

in $\mathbb{R}^3$, referred to as an assembly, has to be partitioned into its basic polyhedra through the applications of a sequence of transforms applied to subsets of the assembly [11, 21]. The LTP problem is a building block in a solution that we forsee to certain assembly-partitioning problems. We strive to develop an output-sensitive algorithm that solves the LTP problem, and provide an exact implementation of it. We have already conceived the general framework for such an algorithm and implemented a raw version that handles the general position case. However, we still need to enhance the algorithm and its implementation to handle all cases and carefully analyze them.

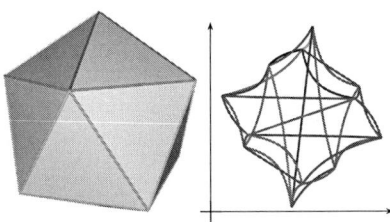

A glimpse at this future development can be seen in the figure to the right. It shows an icosahedron P and the arrangement induced by the point set $\Psi_{S_1 S_2}(E(P))$, where $E(P)$ is the set of the edges of P, and S_1 and S_2 are two skew segments (omitted in the figure). The color of each edge of the arrangement is the same as the color of its generating icosahedron edge. The boundary of the hole in the unbounded face contains points that represent lines that intersect S_1 and S_2 and are tangent to P.

Acknowledgement. We thank Michael Hoffmann for helpful discussions on assembly partitioning, which inspired us to conduct the research discussed in this article. We also thank Linqiao Zhang who provided us with Redburn's code that was used for the experiments. Zhang used it as part of an implementation of an algorithm that constructs the visibility skeleton [23].

References

1. Berberich, E., Fogel, E., Halperin, D., Kerber, M., Setter, O.: Arrangements on parametric surfaces II: Concretizations and applications. Math. in Comput. Sci. 4, 67–91 (2010)
2. Berberich, E., Fogel, E., Halperin, D., Mehlhorn, K., Wein, R.: Arrangements on parametric surfaces I: General framework and infrastructure. Math. in Comput. Sci. 4, 45–66 (2010)
3. Berberich, E., Hemmer, M., Kerber, M.: A generic algebraic kernel for non-linear geometric applications. In: Proc. 27th Annu. ACM Symp. Comput. Geom., pp. 179–186. ACM Press, New York (2011)
4. Brönnimann, H., Burnikel, C., Pion, S.: Interval arithmetic yields efficient dynamic filters for computational geometry. Disc. Appl. Math. 109, 25–47 (2001)
5. Brönnimann, H., Devillers, O., Dujmovic, V., Everett, H., Glisse, M., Goaoc, X., Lazard, S., Suk Na, H.: Lines and free line segments tangent to arbitrary three-dimensional convex polyhedra. SIAM J. on Computing 37, 522–551 (2006)
6. Brönnimann, H., Everett, H., Lazard, S., Sottile, F., Whitesides, S.: Transversals to line segments in three-dimensional space. Disc. Comput. Geom. 34, 381–390 (2005), doi:10.1007/s00454-005-1183-1

7. Demouth, J., Devillers, O., Everett, H., Glisse, M., Lazard, S., Seidel, R.: On the complexity of umbra and penumbra. Comput. Geom. Theory Appl. 42, 758–771 (2009)
8. Devillers, O., Glisse, M., Lazard, S.: Predicates for line transversals to lines and line segments in three-dimensional space. In: Proc. 24th Annu. ACM Symp. Comput. Geom., pp. 174–181. ACM Press (2008)
9. Devillers, O., Pion, S.: Efficient exact geometric predicates for Delaunay triangulations. In: Proc. 5th Workshop Alg. Eng. Experiments, pp. 37–44 (2003)
10. Everett, H., Lazard, S., Lenhart, W., Redburn, J., Zhang, L.: On the degree of standard geometric predicates for line transversals. Comput. Geom. Theory Appl. 42(5), 484–494 (2009)
11. Fogel, E., Halperin, D.: Polyhedral Assembly Partitioning with Infinite Translations or The Importance of Being Exact. In: Chirikjian, G.S., Choset, H., Morales, M., Murphey, T. (eds.) Algorithmic Foundation of Robotics VIII. STAR, vol. 57, pp. 417–432. Springer, Heidelberg (2009)
12. Hemmer, M., Hert, S., Kettner, L., Pion, S., Schirra, S.: Number types. CGAL User and Reference Manual. CGAL Editorial Board, 4.0 edn. (2012), http://www.cgal.org/Manual/4.0/doc_html/cgal_manual/ackages.html#Pkg:NumberTypes
13. McKenna, M., O'Rourke, J.: Arrangements of lines in 3-space: a data structure with applications. In: Proc. 4th Annu. ACM Symp. Comput. Geom., pp. 371–380. ACM Press, New York (1988)
14. Pion, S., Fabri, A.: A Generic Lazy Evaluation Scheme for Exact Geometric Computations. Sci. Comput. Programming 76(4), 307–323 (2011)
15. Redburn, J.: Robust computation of the non-obstructed line segments tangent to four amongst n triangles. PhD thesis, Williams College, Massachusetts (2003)
16. Salzman, O., Hemmer, M., Raveh, B., Halperin, D.: Motion planning via manifold samples. In: Proc. 19th Annu. Eur. Symp. Alg., pp. 493–505 (2011)
17. Teller, S., Hohmeyer, M.: Determining the lines through four lines. J. of Graphics, Gpu, and Game Tools 4(3), 11–22 (1999)
18. The CGAL Project. CGAL User and Reference Manual. CGAL Editorial Board, 4.0 edn. (2012), http://www.cgal.org/Manual/4.0/doc_html/cgal_manual/title.html
19. Wein, R., Fogel, E., Zukerman, B., Halperin, D.: Advanced programming techniques applied to CGAL's arrangement package. Comput. Geom. Theory Appl. 38(1-2), 37–63 (2007)
20. Wein, R., Fogel, E., Zukerman, B., Halperin, D.: 2D arrangements. CGAL User and Reference Manual. Cgal Editorial Board, 4.0 edn. (2012), http://www.cgal.org/Manual/4.0/doc_html/cgal_manual/packages.html#Pkg:Arrangement2
21. Wilson, R.H., Kavraki, L., Latombe, J.-C., Lozano-Pérez, T.: Two-handed assembly sequencing. Int. J. of Robotics Research 14, 335–350 (1995)
22. Yap, C.-K., Dubé, T.: The exact computation paradigm. In: Du, D.-Z., Hwang, F.K. (eds.) GI 1973, 2nd edn. LNCS, vol. 1, pp. 452–492. World Scientific, Singapore (1973)
23. Zhang, L., Everett, H., Lazard, S., Weibel, C., Whitesides, S.H.: On the Size of the 3D Visibility Skeleton: Experimental Results. In: Halperin, D., Mehlhorn, K. (eds.) ESA 2008. LNCS, vol. 5193, pp. 805–816. Springer, Heidelberg (2008)

A Polynomial Kernel for PROPER INTERVAL VERTEX DELETION

Fedor V. Fomin[1,*], Saket Saurabh[2], and Yngve Villanger[1]

[1] Department of Informatics, University of Bergen, N-5020 Bergen, Norway
{fomin,yngvev}@ii.uib.no
[2] The Institute of Mathematical Sciences, Chennai, India
saket@imsc.res.in

Abstract. It is known that the problem of deleting at most k vertices to obtain a proper interval graph (PROPER INTERVAL VERTEX DELETION) is fixed parameter tractable. However, whether the problem admits a polynomial kernel or not was open. Here, we answer this question in affirmative by obtaining a polynomial kernel for PROPER INTERVAL VERTEX DELETION. This resolves an open question of van Bevern, Komusiewicz, Moser, and Niedermeier.

1 Introduction

Study of graph editing problems cover a large part of parmeterized complexity. The problem of editing (adding/deleting vertices/edges) to ensure a graph to have some property is a well studied problem in theory and applications of graph algorithms. When we want the *edited graph* to be in a hereditary (that is, closed under induced subgraphs) graph class, the optimization version of the corresponding vertex deletion problem is known to be NP-hard by a classical result of Lewis and Yannakakis [13]. In this paper we study the problem of deleting vertices to get into proper interval graph in the realm of kernelization complexity.

A graph G is a proper (unit) interval graph if it is an intersection graph of unit-length intervals on a real line. Proper interval graphs form a well studied and well structured hereditary class of graphs. The parameterized study of the following problem of deleting vertices to get into proper interval graph was initiated by van Bevern et al. [1].

---

p-PROPER INTERVAL VERTEX DELETION (PIVD)   **Parameter:** k
Input: An undirected graph G and a positive integer k
Question: Decide whether G has a vertex set X of size at most k such that $G \setminus X$ is a proper interval graph

---

Wegner [20] (see also [2]) showed that proper interval graphs are exactly the class of graphs that are {*claw,net,tent,hole*}-*free*. *Claw, net*, and *tent* are graphs

---

* Supported by the European Research Council (ERC) advanced grant PREPRO-CESSING, reference 267959.

L. Epstein and P. Ferragina (Eds.): ESA 2012, LNCS 7501, pp. 467–478, 2012.

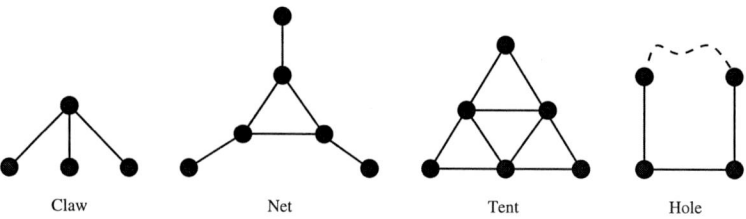

| Claw | Net | Tent | Hole |

Fig. 1. Excluded Subgraphs

containing at most 6 vertices depicted in Fig. 1, and *hole* is an induced cycle of length at least four. Combining results of Wegner, Cai, and Marx [3,15,20], it can be shown that PIVD is FPT. That is one can obtain an algorithm for PIVD running in time $\tau(k)n^{O(1)}$ where τ is a function depending only on k and n is the number of vertices in the input graph. Van Bevern et al. [1] presented a faster $O(k(14k+14)^{k+1}n^6)$ time algorithm for PIVD using the structure of a problem instance that is already $\{claw, net, tent, C_4, C_5, C_6\}$-free. The running time was recently improved by Villanger down to $O(6^k kn^6)$ [19]. However, the question, whether the problem has a polynomial kernel or not was not resolved. This question was explicitly asked by Van Bevern et al. [1]. This is precisely the problem we address in this paper.

Here, we study PIVD from kernelization perspective. A parameterized problem is said to admit a *polynomial kernel* if every instance (I, k) can be reduced in polynomial time to an equivalent instance with both size and parameter value bounded by a polynomial in k. In other words, it is possible in polynomial time to "compress" every instance of the problem to a new instance of size $k^{O(1)}$.

Our interest to PIVD is also motivated by the following more general problem. Let $\mathcal{G}$ be an arbitrary class of graphs. We denote by $\mathcal{G} + kv$ the class of graphs that can be obtained from a member of $\mathcal{G}$ by adding at most k vertices. For an example, PIVD is equivalent to deciding if G is in $\mathcal{G} + kv$, where $\mathcal{G}$ is the class of proper interval graphs. There is a generic criteria providing sufficient conditions on the properties of class $\mathcal{G}$ to admit a polynomial kernel for $\mathcal{G} + kv$ recognition problem. A graph class is called *hereditary* if every induced subgraph of every graph in the class also belongs to the class. Let Π be a hereditary graph class characterized by forbidden induced subgraphs of size at most d. Cai [3] showed that the $\Pi + kv$ problem, where given an input graph G and positive integers k, the question is to decide whether there exists a k sized vertex subset S such that $G[V \setminus S] \in \Pi$, is FPT when parameterized by k. The $\Pi + kv$ problem can be shown to be equivalent to p-d-HITTING SET and thus it admits a polynomial kernel [7], [12]. In the p-d-HITTING SET problem, we are given a family $\mathcal{F}$ of sets of size at most d over a universe $\mathcal{U}$ and a positive integer k and the objective is to find a subset $S \subseteq \mathcal{U}$ of size at most k intersecting, or *hitting*, every set of $\mathcal{F}$.

However, the result of Cai does not settle the parameterized complexity of $\Pi + kv$ when Π cannot be characterized by a finite number of forbidden induced subgraphs. Here even for graph classes with well-understood structure and very simple infinite set of forbidden subgraphs, the situation becomes challenging. In particular, for the "closest relatives" of proper interval graphs, chordal and interval graphs, the current situation is still obscure. For example, the FPT algorithm of Marx [15] for the problem of vertex deletion into a chordal graph, i.e. a graph without induced cycles of length at least four, requires heavy algorithmic machinery. The question if CHORDAL+kv admits a polynomial kernel is still open. Situation with INTERVAL+kv is even more frustrating, in this case we even do not know if the problem is in FPT or not.

In this paper we make a step towards understanding the kernelization behaviour for $\mathcal{G} + kv$ recognition problems, where $\mathcal{G}$ is well understood and the infinite set of forbidden subgraphs is simple. A generic strategy to obtain FPT algorithm for many $\mathcal{G} + kv$ recognition problems is to first take care of small forbidden subgraphs by branching on them. When these small subgraphs are not present, the structure of a graph is utilised to take care of infinite family of forbidden subgraphs. However, to apply a similar strategy for kernelization algorithm we need to obtain a polynomial kernel for a variant of p-d-HITTING SET that preserves all minimal solutions of size at most k along with a "witness" for the minimality, rather than the kernel for p-d-HITTING SET which was sufficient when we only had finitely many forbidden induced subgraphs. Preserving the witness for the minimality is crucial here as this "insulates" small constant size forbidden induced subgraphs from the large and infinite forbidden induced subgraph. In some way it mimics the generic strategy used for the FPT algorithm. Towards this we show that indeed one can obtain a kernel for a variant of p-d-HITTING SET that preserves all minimal solutions of size at most k along with a witness for the minimality (Section 3). Finally, using this in combination with reduction rules that shrinks "clique and clique paths" in proper interval graphs we resolve the kernelization complexity of PIVD. We show that PIVD admits a polynomial kernel and thus resolve the open question posed in [1]. We believe that our strategy to obtain polynomial kernel for PIVD will be useful in obtaining polynomial kernels for various other $\mathcal{G} + kv$ recognition problems.

2   Definitions and Notations

We consider simple, finite, and undirected graphs. For a graph G, $V(G)$ is the *vertex set* of G and $E(G)$ is the *edge set* of G. For every edge $uv \in E(G)$, vertices u and v are *adjacent* or *neighbours*. The *neighbourhood* of a vertex u in G is $N_G(u) = \{v \mid uv \in E\}$, and the *closed neighbourhood* of u is $N_G[u] = N_G(u) \cup \{u\}$. When the context will be clear we will omit the subscript. A set $X \subseteq V$ is called a *clique* of G if the vertices in X are pairwise adjacent. A *maximal* clique is a clique that is not a proper subset of any other clique. For $U \subseteq V$, the *subgraph of G induced by U* is denoted by $G[U]$ and it is the graph with vertex set U and edge set equal to the set of edges $uv \in E$ with $u, v \in U$. For every $U \subseteq V$, $G' = G[U]$ is an *induced subgraph* of G. By $G \setminus X$ for $X \subseteq V$, we denote the graph $G[V \setminus X]$.

Parameterized Problems and Kernels. A parameterized problem Π is a subset of $\Gamma^* \times \mathbb{N}$ for some finite alphabet Γ. An instance of a parameterized problem consists of (x, k), where k is called the parameter. The notion of kernelization is formally defined as follows. A *kernelization algorithm*, or in short, a *kernelization*, for a parameterized problem $\Pi \subseteq \Gamma^* \times \mathbb{N}$ is an algorithm that, given $(x, k) \in \Gamma^* \times \mathbb{N}$, outputs in time polynomial in $|x| + k$ a pair $(x', k') \in \Gamma^* \times \mathbb{N}$ such that (a) $(x, k) \in \Pi$ if and only if $(x', k') \in \Pi$ and (b) $|x'|, k' \le g(k)$, where g is some computable function depending only on k. The output of kernelization (x', k') is referred to as the *kernel* and the function g is referred to as the size of the kernel. If $g(k) \in k^{O(1)}$, then we say that Π admits a polynomial kernel. For general background on the theory, the reader is referred to the monographs [5,7,16].

Interval Graphs. A graph G is an *interval graph* if and only if we can associate with each vertex $v \in V(G)$ an open interval $I_v = (l_v, r_v)$ on the real line, such that for all $v, w \in V(G)$, $v \ne w$: $vw \in E(G)$ if and only if $I_v \cap I_w \ne \emptyset$. The set of intervals $\mathcal{I} = \{I_v\}_{v \in V}$ is called an (interval) *representation* of G. By the classical result of Gilmore and Hoffman [10], and Fulkerson and Gross [9], for every interval graph G there is a linear ordering of its maximal cliques such that for every vertex v, the maximal cliques containing v occur consequently. We refer to such an ordering of maximal cliques $C_1, C_2, \ldots, C_p$ of interval graph G as a *clique path* of G. Note that an interval graph can have several different clique paths. A clique path of an interval graph can be constructed in linear time [9].

A *proper interval graph* is an interval graph with an interval model where no interval is properly contained in any other interval. There are several equivalent definitions of proper interval graphs. Graph G is a *unit interval graph* if G is an interval graph with an interval model of unit-length intervals. By the result of Roberts [18], G is a unit interval graph if and only if it is a proper interval graph. A *claw* is a graph that is isomorphic to $K_{1,3}$, see Fig. 1. A graph is *claw-free* if it does not have a claw as an induced subgraph. Proper interval graphs are exactly the claw-free interval graphs [18].

A vertex ordering $\sigma = \langle u_1, \ldots, u_n \rangle$ of graph $G = (V, E)$ is called *interval ordering* if for every $1 \le i < j < k \le n$, $v_i v_k \in E$ implies $v_j v_k \in E$. A graph is an interval graph if and only if it admits an interval ordering [17]. A vertex ordering σ for G is called a *proper interval ordering* if for every for every $1 \le i < j < k \le n$, $v_i v_k \in E$ implies $v_i v_j, v_j v_k \in E$. A graph is a proper interval graph if and only if it admits a proper interval ordering [14]. Interval orderings and proper interval orderings can be computed in linear time, if they exist. We will need the following properties of proper interval graphs.

Proposition 1 ([20,2]). *A graph G is a proper interval graph if and only if it contains neither claw, net, tent, nor induced cycles (holes) of length at least 4 as induced subgraphs.*

A *circular-arc graph* is the intersection graph of a set of arcs on the circle. A circular-arc graph is a *proper circular-arc graph* if no arc is properly contained in any other arc.

Proposition 2 ([19]). *Every connected graph G that does not contain either tent, net or claw or induced cycles (holes) of length 4, 5 and 6 as an induced subgraph is a proper circular-arc graph. Moreover, for such a graph there is a polynomial time algorithm computing a set X of minimum size such that $G \setminus X$ is a proper interval graph.*

The following proposition of proper interval orderings of proper interval graphs follows almost directly from the definition.

Proposition 3. *Let $\sigma = \langle v_1, \ldots, v_n \rangle$ be a proper interval ordering of $G = (V, E)$.*

1. *For every maximal clique K of G, there exist integers $1 \leq i < j \leq n$ such that $K = \{v_i, v_{i+1}, \ldots, v_{j-1}, v_j\}$. That is, vertices of K occur consecutively.*
2. *For a vertex v_ℓ let i, j be the smallest and the largest numbers such that $v_i v_\ell, v_\ell v_j \in E$, then $N[v_\ell] = \{v_i, \ldots, v_j\}$ and the sets $\{v_i, \ldots, v_\ell\}$ and $\{v_\ell, \ldots, v_j\}$ are cliques;*
3. *Let $C_1, C_2, \ldots, C_p$ be a clique path of G. If $v_i \in C_j$ then $v_i \notin C_{j+\ell+1}$, where $\ell \geq |N[v_i]|$.*

3   Sunflower Lemma and Minimal Hitting Sets

In this section we obtain a kernel for a variant of p-d-HITTING SET that preserves all minimal solutions of size at most k along with a witness for the minimality. Towards this we introduce the notion of *sunflower*. A *sunflower* S with k petals and a *core* Y is a collection of sets $\{S_1, S_2, \ldots, S_k\}$ such that $S_i \cap S_j = Y$ for all $i \neq j$; the sets $S_i \setminus Y$ are petals and we require that none of them be empty. Note that a family of pairwise disjoint sets is a sunflower (with an empty core). We need the following algorithmic version of the classical result of Erdős and Rado [6].

Lemma 1 ([7]). [Sunflower Lemma] *Let $\mathcal{F}$ be a family of sets over a universe $\mathcal{U}$ each of cardinality at most d. If $|\mathcal{F}| > d!(k-1)^d$ then $\mathcal{F}$ contains a sunflower with k petals and such a sunflower can be found in $O(k + |\mathcal{F}|)$ time.*

A subset X of $\mathcal{U}$ intersecting every set in $\mathcal{F}$ is referred to as a *hitting set* for $\mathcal{F}$. Sunflower Lemma is a common tool used in parameterized complexity to obtain a polynomial kernel for p-d-HITTING SET [7]. The observation is that if $\mathcal{F}$ contains a sunflower $S = \{S_1, \ldots, S_{k+1}\}$ of cardinality $k+1$ then every hitting set of of size at most k of $\mathcal{F}$ must have a nonempty intersection with the core Y. However, for our purposes it is crucial that kernelization algorithm preserves *all* small minimal hitting sets. The following application of Sunflower Lemma is very similar to its use for kernelization for p-d-HITTING SET. However, it does not seem to exist in the literature in the form required for our kernelization and thus we give its proof here.

Lemma 2. *Let $\mathcal{F}$ be a family of sets of cardinality at most d over a universe $\mathcal{U}$ and k be a positive integer. Then there is an $\mathcal{O}(|\mathcal{F}|(k + |\mathcal{F}|))$ time algorithm that finds a non-empty set $\mathcal{F}' \subseteq \mathcal{F}$ such that*

1. For every $Z \subseteq \mathcal{U}$ of size at most k, Z is a minimal hitting set of $\mathcal{F}$ if and only if Z is a minimal hitting set of $\mathcal{F}'$; and
2. $|\mathcal{F}'| \leq d!(k+1)^d$.[1]

Proof. The algorithm iteratively constructs sets $\mathcal{F}_t$, where $0 \leq t \leq |\mathcal{F}|$. We start with $t = 0$ and $\mathcal{F}_0 = \mathcal{F}$. For $t \geq 1$, we use Lemma 1 to check if there is a sunflower of cardinality $k + 2$ in $\mathcal{F}_{t-1}$. If there is no such sunflower, we stop, and output $\mathcal{F}' = \mathcal{F}_{t-1}$. Otherwise, we use Lemma 1 to construct a sunflower $\{S_1, S_2, \ldots, S_{k+2}\}$ in $\mathcal{F}_{t-1}$. We put $\mathcal{F}_t = \mathcal{F}_{t-1} \setminus \{S_{k+2}\}$. At every step, we delete one subset of $\mathcal{F}$. Thus the algorithm calls the algorithm from Lemma 1 at most $|\mathcal{F}|$ times and hence its running time is $\mathcal{O}(|\mathcal{F}|(k+|\mathcal{F}|))$. Since $\mathcal{F}'$ has no sunflower of cardinality $k + 2$, by Lemma 1, $|\mathcal{F}'| \leq d!(k+1)^d$.

Now we prove that for each $t \geq 1$ and for every set $Z \subseteq \mathcal{U}$, it holds that Z is a minimal hitting set for $\mathcal{F}_{t-1}$ of size k if and only if Z is a minimal hitting set for $\mathcal{F}_t$. Since for $t = 1$, $\mathcal{F}_{t-1} = \mathcal{F}$, and for some $t \leq |\mathcal{F}|$, $\mathcal{F}_t = \mathcal{F}'$, by transitivity this is sufficient for proving the first statement of the lemma.

The set $\mathcal{F}_t$ is obtained from $\mathcal{F}_{t-1}$ by removing the set S_{k+2} of the sunflower $\{S_1, S_2, \ldots, S_{k+2}\}$ in $\mathcal{F}_{t-1}$. Let Y be the core of this sunflower. If $Y = \emptyset$, then $\mathcal{F}_{t-1}$ has no hitting set of size k. In this case, $\mathcal{F}_t$ contains pairwise disjoint sets $S_1, S_2, \ldots, S_{k+1}$ and hence $\mathcal{F}_t$ also has no hitting set of size k. Thus the interesting case is when $Y \neq \emptyset$.

Let Z be a minimal hitting set for $\mathcal{F}_{t-1}$ of size k. Since $\mathcal{F}_t \subseteq \mathcal{F}_{t-1}$, we have that set Z is a hitting set for $\mathcal{F}_t$. We claim that Z is a *minimal* hitting set for $\mathcal{F}_t$. Targeting towards a contradiction, let us assume that Z is not a minimal hitting set for $\mathcal{F}_t$. Then there is $u \in Z$, such that $Z' = Z \setminus \{u\}$ is a hitting set for $\mathcal{F}_t$. Sets $S_1, S_2, \ldots, S_{k+1}$ form a sunflower in $\mathcal{F}_t$, and thus every hitting set of size at most k, including Z', intersects its core Y. Thus Z' hits all sets of $\mathcal{F}_{t-1}$, as it hits all the sets of $\mathcal{F}_t$ and it also hits S_{k+2} because $Y \subset S_{k+2}$. Therefore, Z is not a minimal hitting set in $\mathcal{F}_{t-1}$, which is a contradiction. This shows that Z is a minimal hitting set for $\mathcal{F}_t$.

Let Z be a minimal hitting set for $\mathcal{F}_t$ of size k. Every hitting set of size k for $\mathcal{F}_t$ should contain at least one vertex of the core Y. Hence $Y \cap Z \neq \emptyset$. But then $Z \cap S_{k+2} \neq \emptyset$ and thus Z is a hitting set for $\mathcal{F}_{t-1}$. Because $\mathcal{F}_t \subseteq \mathcal{F}_{t-1}$, Z is a minimal hitting set for $\mathcal{F}_{t-1}$. □

Given a family $\mathcal{F}$ of sets over a universe $\mathcal{U}$ and a subset $T \subseteq \mathcal{U}$, we define $\mathcal{F}_T$ as the subset of $\mathcal{F}$, containing all sets $Q \in \mathcal{F}$ such that $Q \subseteq T$.

4   Proper Interval Vertex Deletion

In this section, we apply results from the previous section to obtain a polynomial kernel for PIVD. Let (G, k) be an instance to PIVD, where G is a graph on n vertices and k is a positive integer. The kernel algorithm is given in four steps.

---

[1] The set $\mathcal{F}'$ is not the same as the *full* kernels in [4] as we also preserve a witness for the minimality.

First we take care of small forbidden sets using Lemma 2, the second and the third steps reduce the size of maximal cliques and shrink the length of induced paths. Finally, we combine three previous steps into a kernelization algorithm.

4.1 Small Induced Forbidden Sets

In this section we show how we could use Lemma 2 to identify a vertex subset of $V(G)$, which allows us to forget about small induced subgraphs in G and to concentrate on long induced cycles in the kernelization algorithm for PIVD. We view vertex subsets of G inducing a forbidden subgraph as a set family. We prove the following lemma.

Lemma 3 ($\star^2$). *Let (G, k) be an instance to PIVD. Then there is a polynomial time algorithm that either finds a non-empty set $T \subseteq V(G)$ such that*

1. $G \setminus T$ *is a proper interval graph;*
2. *Every set $Y \subseteq V(G)$ of size at most k is a minimal hitting set for nets, tents, claws and induced cycles C_ℓ, $4 \leq \ell \leq 8$, in G if and only if it is a minimal hitting set for nets, tents, claws and induced cycles C_ℓ, $4 \leq \ell \leq 8$, contained in $G[T]$; and*
3. $|T| \leq 8 \cdot 8!(k+1)^8 + k$.

or concludes that (G, k) is a NO instance.

In the rest of the coming subsections we assume that

$$\boxed{G_T = G \setminus T \text{ is a proper interval graph and } |T| \leq \delta(k) = 8 \cdot 8!(k+1)^8 + k.}$$

4.2 Finding Irrelevant Vertices in G_T

In this subsection we show that if the maximum size of a clique in G_T is larger than $(k+1)(\delta(k)+2)$, then we can find some irrelevant vertex $v \in V(G_T)$ and delete it without altering the answer to the problem. More precisely, we prove the following result.

Lemma 4. *Let G and T be as described before. Furthermore, let the size of a maximum clique in G_T be greater than $\epsilon(k) = (k+1)(\delta(k)+2)$. Then in polynomial time we can find a vertex $v \in V(G_T)$ such that (G, k) is a YES instance to PIVD if and only if $(G \setminus v, k)$ is a YES instance.*

Proof. We start by giving a procedure to find an irrelevant vertex v. Let K be a maximum clique of G_T, it is well known that a maximum clique can be found in linear time in proper interval graphs [11]. Let $\sigma = \langle u_1, \dots, u_n \rangle$ be a linear vertex ordering of G_T. By Proposition 3, vertices of K form an interval in σ, we denote this interval by $\sigma(K)$. Suppose that $|K| > \epsilon(k)$. The following procedure marks vertices in the clique K and helps to identify an irrelevant vertex.

---

[2] Proofs of results marked with $\star$ are removed due to space restrictions. See [8] for a full version of the paper.

Set $Z = \emptyset$. For every vertex $v \in T$, pick $k + 1$ arbitrary neighbours of v in K, say S_v, and add them to Z. If v has at most k neighbors in K, then add all of them to Z. Furthermore, add V^F, the first $k+1$ vertices, and V^L, the last $k + 1$ vertices in $\sigma(K)$ to Z. Return Z.

Observe that the above procedure runs in polynomial time and adds at most $k + 1$ vertices for any vertex in T. In addition, the procedure also adds some other $2(k+1)$ vertices to Z. Thus the size of the set Z containing marked vertices is at most $(k + 1)(\delta(k) + 2) = \epsilon(k)$. By our assumption on the size of the clique we have that $K \setminus Z \neq \emptyset$. We show that any vertex in $K \setminus Z$ is irrelevant. Let $v \in (K \setminus Z)$. Now we show that (G, k) is a YES instance to PIVD if and only if $(G \setminus v, k)$ is a YES instance to PIVD. Towards this goal we first prove the following auxiliary claim.

Claim 1. *Let H be a proper interval graph, and $P = p_1, \ldots, p_\ell$ be an induced path in H. Let $u \notin \{p_1, \ldots, p_\ell\}$ be some vertex of H and let $N_P(u)$ be the set of its neighbours in P. Then, the vertices of $N_P(u)$ occur consecutively on the path P, and furthermore, $|N_P(u)| \leq 4$.*

Proof. The first statement follows from the fact that H has no induced cycle of length more than three and the second statement from the fact that H contains no claw. □

Let (G, k) be a YES instance and let $X \subseteq V(G)$ be a vertex set such that $|X| \leq k$ and $G \setminus X$ is a proper interval graph. Then clearly $(G \setminus v, k)$ is a YES instance of PIVD as $|X \setminus \{v\}| \leq k$ and $G \setminus (\{v\} \cup X)$ is a proper interval graph.

For the opposite direction, let $(G \setminus v, k)$ be a YES instance for PIVD and let X be a vertex set such that $|X| \leq k$ and $G \setminus (\{v\} \cup X)$ is a proper interval graph. Towards a contradiction, let us assume that $G \setminus X$ is not a proper interval graph. Thus $G \setminus X$ contains one of the forbidden induced subgraphs for proper interval graphs. We first show that this cannot contain forbidden induced subgraphs of size at most 8. Let Y be the subset of X such that it is a minimal hitting set for nets, tents, claws and induced cycles C_ℓ, $4 \leq \ell \leq 8$, contained in $G[T]$. By the definition of T and the fact that $v \notin T$ we know that Y is also a minimal hitting set for nets, tents, claws and induced cycles C_ℓ, $4 \leq \ell \leq 8$, contained in G. Thus, the only possible candidate for the forbidden subgraph in $G \setminus X$ is an induced cycle C_ℓ, where $\ell \geq 9$. Now since $G \setminus (X \cup \{v\})$ is a proper interval graph, the vertex v is part of the cycle $C_\ell = \{v = w_1, w_2, \ldots, w_\ell\}$. Furthermore, w_2 and w_ℓ are the neighbors of v on C_ℓ.

Next we show that using C_ℓ we can construct a forbidden induced cycle in $G \setminus (\{v\} \cup X)$, contradicting that $G \setminus (\{v\} \cup X)$ is a proper interval graph. Towards this we proceed as follows. For vertex sets V^F and V^L (the first and the last $k+1$ vertices of $\sigma(K)$), we pick up vertex $v^F \in V^F \setminus X$ and vertex $v^L \in V^L \setminus X$. Because $|X| \leq k$, such vertices always exist.

Claim 2. *Vertices $w_2, w_\ell \in T \cup N[v^F] \cup N[v^L]$.*

Proof. Suppose $w_a \in (T \cup K)$, $a \in \{2, \ell\}$, then because $K \subseteq N[v^F]$, we are done. Otherwise w_a is a vertex of the proper interval graph $G_T \setminus K$. Then w_a occurs either before or after the vertices of K in σ. If w_a occurs before then $w_a < v^F < v$ on σ. Now since w_a has an edge to v and σ is a proper interval ordering of G_T, we have that $w_a v^F$ is an edge and hence $w_a \in N[v^F]$. The case where w_a occurs after K is symmetric. In this case we could show that $w_a \in N[v^L]$.   □

Now with $w_a, a \in \{2, \ell\}$, we associate a partner vertex $p(w_a)$. If $w_a \in T$, then $Z \cap N(w_a)$ contains at least $k + 1$ vertices as $v \in K \cap N(w_a)$ is not in Z. Thus there exists $z^a \in (Z \cap N(w_a)) \setminus X$. In this case we define $p(w_a)$ to be z^a. If $w_a \notin T$ then by Claim 2 we know that either v^F or v^L is a neighbor to w_a. If v^F is neighbor to w_a then we define $p(w_a) = v^F$, else $p(w_a) = v^L$. Observe that $p(w_a) \in K \setminus \{v\}$ for $a \in \{2, \ell\}$.

Now consider the closed walk $W = \{p(w_2), w_2, \ldots, w_\ell, p(w_\ell)\}$ in $G \setminus (\{v\} \cup X)$. First of all W is a closed walk because $p(w_2)$ and $p(w_\ell)$ are adjacent. In fact, we would like to show that W is a simple cycle in $G \setminus (\{v\} \cup X)$ (not necessarily an induced cycle). Towards this we first show that $p(w_a) \notin \{w_2, \ldots, w_\ell\}$. Suppose $p(w_2) \in \{w_2, \ldots, w_\ell\}$, then it must be w_3 as the only neighbors of w_2 on C_ℓ are $v = w_1$ and w_3. However, v and $p(w_2)$ are part of the same clique K. This implies that v has w_2, w_3 and w_ℓ as its neighbors on C_ℓ, contradicting to the fact that C_ℓ is an induced cycle of length at least 9 in G. Similarly, we can also show that $p(w_\ell) \notin \{w_2, \ldots, w_\ell\}$. Now, the only reason W may not be a simple cycle is that $p(w_2) = p(w_\ell)$. However, in that case $W = \{p(w_2), w_2, \ldots, w_\ell\}$ is a simple cycle in $G \setminus (\{v\} \cup X)$.

Notice that $G[\{w_2, w_3, \ldots, w_\ell\}]$ is an induced path, where $\ell \geq 9$. Let i be the largest integer such that $w_i \in N(p(w_2))$ and let j be the smallest integer such that $w_j \in N(p(w_\ell))$. By Claim 1 and the conditions that $G[\{w_2, w_3, \ldots, w_\ell\}]$ is an induced path, $w_2 \in N(p(w_2))$, and $w_\ell \in N(p(w_\ell))$, we get that $i \leq 5, j \geq \ell - 3$. As $\ell \geq 9$ this implies that $i < j$, and hence $G[\{p(w_2), w_i, \ldots, w_j, p(w_\ell)\}]$ is an induced cycle of length at least four in $G \setminus (\{v\} \cup X)$, which is a contradiction. Therefore, $G \setminus X$ is a proper interval graph.   □

4.3   Shrinking G_T

Let (G, k) be a YES instance of PIVD, and let T be a vertex subset of G of size at most $\delta(k)$ such that $G_T = G \setminus T$ is a proper interval graph with the maximum clique size at most $\epsilon(k) = (k + 1)(\delta(k) + 2)$. The following lemma argues that if G_T has sufficiently long clique path, then a part of this path can be shrunk without changing the solution.

Lemma 5 ($\star$). *Let us assume that every claw in G contains at least two vertices from T and that there is a connected component of G_T with at least $\zeta(k) = \left(\delta(k)(8\epsilon(k) + 2) + 1\right)\left(2[\epsilon(k)]^2 + 32\epsilon(k) + 3\right)$ maximal cliques. Then there is a polynomial time algorithm transforming G into a graph G' such that*

- *(G, k) is a YES instance if and only if (G', k) is a YES instance;*
- *$|V(G')| < |V(G)|$.*

4.4 Putting All Together: Final Kernel Analysis

We need some auxiliary reduction rules to give the kernel for PIVD. Let $\mathcal{F}$ be the family consisting of all nets, tents, claws and induced cycles C_ℓ for $\ell \in \{4, 5, \ldots, 8\}$ of the input graph G.

Lemma 6 ($\star$). *Let (G, k) be an instance to* PIVD *and T be as defined before. Let X be a subset of T such that for every $x \in X$ we have a set $S_x \in \mathcal{F}$ such that $S_x \backslash \{x\} \subseteq (V(G) \backslash T)$. If $|X| > k$ then we conclude that G can not be transformed into proper interval graph by deleting at most k vertices. Else, (G, k) is a YES instance if and only if $(G[V \backslash X], k - |X|)$ is a YES instance.*

Now we are ready to state the main result of this paper.

Theorem 1. PIVD *admits a polynomial kernel.*

Before proceedings with the proof of the theorem, let us remind the definitions of all functions used so far.

Size of T:	$\leq \delta(k) = 8 \cdot 8!(k+1)^8 + k$
Maximum clique size in G_T:	$\leq \epsilon(k) = (k+1)(\delta(k)+2)$
# of maximal cliques in a connected component of G_T:	$\leq \zeta(k) = (\delta(k)(8\epsilon(k)+2)+1)(2[\epsilon(k)]^2 + 32\epsilon(k)+3)$

Proof. Let (G, k) be an instance to PIVD. We first show that if G is not connected then we can reduce it to the connected case. If there is a connected component $\mathcal{C}$ of G such that $\mathcal{C}$ is a proper interval graph then we delete this component. Clearly, (G, k) is a YES instance if and only if $(G \backslash \mathcal{C}, k)$ is a YES instance. We repeat this process until every connected component of G is not a proper interval graph. At this stage if the number of connected components is at least $k + 1$, then we conclude that G cannot be made into a proper interval graph by deleting at most k vertices. Thus, we assume that G has at most k connected components. Now we show how to obtain a kernel for the case when G is connected, and for the disconnected case we just run this algorithm on each connected component. This only increases the kernel size by a factor of k. From now onwards we assume that G is connected.

Now we apply Lemma 3 on G and in polynomial time either find a non-empty set $T \subseteq V(G)$ such that

1. $G \backslash T$ is a proper interval graph;
2. $Y \subseteq V(G)$ of size at most k is a minimal hitting set for nets, tents, claws and induced cycles C_ℓ for $\ell \in \{4, 5, \ldots, 8\}$ contained in G if and only if it is a minimal hitting set for nets, tents, claws and induced cycles C_ℓ for $\ell \in \{4, 5, \ldots, 8\}$ contained in $G[T]$; and
3. $|T| \leq \delta(k)$,

or conclude that G cannot be made into a proper interval graph by deleting at most k vertices. If Lemma 3 concludes that G cannot be transformed into a proper interval graph by deleting at most k vertices, then the kernelization algorithm returns the same.

If the size of a maximum clique in G_T is more than $\epsilon(k)$, then we apply Lemma 4 and obtain a vertex $v \in V(G_T)$ such that (G,k) is a YES instance if and only if $(G \setminus v, k)$ is a YES instance. We apply Lemma 4 repeatedly until the size of a maximum clique in G_T is at most $\epsilon(k)$. So, from now onwards we assume that the size of a maximum clique in G_T is at most $\epsilon(k)$.

Now we apply Lemma 6 on (G,k). If Lemma 6 concludes that G cannot be made into a proper interval graph by deleting at most k vertices, then (G,k) is a NO instance and the kernelization algorithm returns a trivial NO instance. Otherwise, we find a set $X \subseteq T$ such that (G,k) is a YES instance if and only if $(G \setminus X, k - |X|)$ is a YES instance. If $|X| \geq 1$ then $(G \setminus X, k - |X|)$ is a smaller instance and we start all over again with this as new instance to PIVD.

If we cannot apply Lemma 6 anymore, then every claw in G contains at least two vertices from T. Thus if the number of maximal cliques in a connected component of G_T is more than $\zeta(k)$, we can apply Lemma 5 on (G,k) and obtain an equivalent instance (G',k) such that $|V(G')| < |V(G)|$ and then we start all over again with instance (G',k).

Finally, we are in the case, where G_T is a proper interval graph and none of conditions of Lemmata 4, 5 and 6 can be applied. This implies that the number of maximal cliques in each connected component of G_T is at most $\zeta(k)$ and the size of each maximal clique is at most $\epsilon(k)$. Thus we have that every connected component of G_T has at most $\zeta(k)\epsilon(k)$ vertices. Since G is connected, we have that every connected component of G_T has some neighbour in T. However because Lemma 6 cannot be applied, we have that every vertex in T has neighbours in at most 2 connected components. The last assertion follows because of the following reason. If a vertex v in T has neighbours in at least 3 connected components of G_T then v together with a neighbour from each of the components of G_T forms a claw in G, with all the vertices except v in G_T, which would imply that Lemma 6 is applicable. This implies that the total number of connected components in G_T is at most $2\delta(k)$. Thus the total number of vertices in G is at most $2\delta(k)\zeta(k)\epsilon(k)$.

Recall that G may not be connected. However, we argued that G can have at most k connected components and we apply the kernelization procedure on each connected component. If the kernelization procedure returns that some particular component cannot be made into a proper interval graph by deleting at most k vertices, then we return the same for G. Else, the total number of vertices in the reduced instance is at most $2k \cdot \delta(k)\zeta(k)\epsilon(k)$, which is a polynomial.

Observe that the above procedure runs in polynomial time, as with every step of the algorithm, the number of vertices in the input graph reduces. This together with the fact that Lemmata 4, 5 and 6 run in polynomial time, we have that the whole kernelization algorithm runs in polynomial time. This concludes the proof. $\square$

References

1. van Bevern, R., Komusiewicz, C., Moser, H., Niedermeier, R.: Measuring Indifference: Unit Interval Vertex Deletion. In: Thilikos, D.M. (ed.) WG 2010. LNCS, vol. 6410, pp. 232–243. Springer, Heidelberg (2010) 1,2,3
2. Brandstädt, A., Le, V.B., Spinrad, J.P.: Graph classes: a survey. Society for Industrial and Applied Mathematics, Philadelphia (1999) 1,4
3. Cai, L.: Fixed-parameter tractability of graph modification problems for hereditary properties. Inform. Process. Lett. 58, 171–176 (1996) 2
4. Damaschke, P.: Parameterized enumeration, transversals, and imperfect phylogeny reconstruction. Theor. Comp. Sc. 351, 337–350 (2006) 6
5. Downey, R.G., Fellows, M.R.: Parameterized complexity. Springer, New York (1999) 4
6. Erdős, P., Rado, R.: Intersection theorems for systems of sets. J. London Math. Soc. 35, 85–90 (1960) 5
7. Flum, J., Grohe, M.: Parameterized Complexity Theory. In: Texts in Theoretical Computer Science. An EATCS Series, Springer, Berlin (2006) 2,4,5
8. Fomin, F.V., Saurabh, S., Villanger, Y.: A polynomial kernel for proper interval vertex deletion. CoRR, abs/1204.4880 (2012) 7
9. Fulkerson, D.R., Gross, O.A.: Incidence matrices and interval graphs. Pacific J. Math. 15, 835–855 (1965) 4
10. Gilmore, P.C., Hoffman, A.J.: A characterization of comparability graphs and of interval graphs. Canad. J. Math. 16, 539–548 (1964) 4
11. Golumbic, M.C.: Algorithmic Graph Theory and Perfect Graphs. Academic Press, New York (1980) 8
12. Kratsch, S.: Polynomial kernelizations for min f+1 and max np. Algorithmica 63, 532–550 (2012) 2
13. Lewis, J.M., Yannakakis, M.: The Node-Deletion Problem for Hereditary Properties is NP-Complete. J. Comput. Syst. Sci. 20, 219–230 (1980) 1
14. Looges, P.J., Olariu, S.: Optimal greedy algorithms for indifference graphs. Computers & Mathematics with Applications 25, 15–25 (1993) 4
15. Marx, D.: Chordal deletion is fixed-parameter tractable. Algorithmica 57, 747–768 (2010) 2,3
16. Niedermeier, R.: Invitation to fixed-parameter algorithms. Oxford Lecture Series in Mathematics and its Applications, vol. 31. Oxford University Press, Oxford (2006) 4
17. Olariu, S.: An optimal greedy heuristic to color interval graphs. Inf. Process. Lett. 37, 21–25 (1991) 4
18. Roberts, F.S.: Indifference graphs. In: Harary, F. (ed.) Proof Techniques in Graph Theory, pp. 139–146 (1969) 4
19. Villanger, Y.: Proper Interval Vertex Deletion. In: Raman, V., Saurabh, S. (eds.) IPEC 2010. LNCS, vol. 6478, pp. 228–238. Springer, Heidelberg (2010) 2,5
20. G. Wegner, Eigenschaften der Nerven homologisch-einfacher Familien im R^n. PhD thesis, Dissertation Göttingen (1967) 1,2,4

Knowledge, Level of Symmetry, and Time of Leader Election

Emanuele G. Fusco[1] and Andrzej Pelc[2,*]

[1] Computer Science Department, Sapienza, University of Rome, 00198 Rome, Italy
fusco@di.uniroma1.it
[2] Département d'informatique, Université du Québec en Outaouais,
Gatineau, Québec J8X 3X7, Canada
pelc@uqo.ca

Abstract. We study the time needed for deterministic leader election in the $\mathcal{LOCAL}$ model, where in every round a node can exchange any messages with its neighbors and perform any local computations. The topology of the network is unknown and nodes are unlabeled, but ports at each node have arbitrary fixed labelings which, together with the topology of the network, can create asymmetries to be exploited in leader election. We consider two versions of the leader election problem: strong LE in which exactly one leader has to be elected, if this is possible, while all nodes must terminate declaring that leader election is impossible otherwise, and weak LE, which differs from strong LE in that no requirement on the behavior of nodes is imposed, if leader election is impossible.

We show that the time of leader election depends on three parameters of the network: its diameter D, its size n, and its *level of symmetry* λ, which, when leader election is feasible, is the smallest depth at which some node has a unique view of the network. It also depends on the knowledge by the nodes, or lack of it, of parameters D and n. Optimal time of weak LE is shown to be $\Theta(D + \lambda)$ if either D or n is known to the nodes. (If none of these parameters is known, even weak LE is impossible.) For strong LE, knowing only D is insufficient to perform it. If only n is known then optimal time is $\Theta(n)$, and if both n and D are known, then optimal time is $\Theta(D + \lambda)$.

1 Introduction

The Model and the Problem. Leader election is one of the fundamental problems in distributed computing. Every node of a network has a boolean variable initialized to 0 and, after the election, exactly one node, called the *leader*, should change this value to 1. All other nodes should know which node is the leader. If nodes of the network have distinct labels, then leader election is always possible (e.g., the node with the largest label can become a leader). However, nodes may refrain from revealing their identities, e.g., for security or privacy

---

* Partially supported by NSERC discovery grant and by the Research Chair in Distributed Computing at the Université du Québec en Outaouais.

L. Epstein and P. Ferragina (Eds.): ESA 2012, LNCS 7501, pp. 479–490, 2012.

reasons. Hence it is desirable to have leader election algorithms that do not rely on node identities but exploit asymmetries of the network due to its topology and to port labelings. With unlabeled nodes, leader election is impossible in symmetric networks, e.g., in a ring in which ports at each node are 0,1, in the clockwise direction.

A network is modeled as an undirected connected graph. We assume that nodes are unlabeled, but ports at each node have arbitrary fixed labelings $0, \ldots, d-1$, where d is the degree of the node. Throughout the paper, we will use the term "graph" to mean a graph with the above properties. We do not assume any coherence between port labelings at various nodes. Nodes can read the port numbers. When sending a message through port p, a node adds this information to the message, and when receiving a message through port q, a node is aware that this message came through this port. The topology of the network is unknown to the nodes, but depending on the specification of the problem, nodes may know some numerical parameters of the network, such as the number n of nodes (size), and/or the diameter D. We consider two versions of the leader election problem (LE): *strong leader election* (SLE) and *weak leader election* (WLE). In SLE one leader has to be elected whenever this is possible, while all nodes must terminate declaring that leader election is impossible otherwise. WLE differs from SLE in that no requirement on the behavior of nodes is imposed, if leader election is impossible. In both cases, upon election of the leader, every non-leader is required to know a path (coded as a sequence of ports) from it to the leader.

In this paper we investigate the time of leader election in the extensively studied $\mathcal{LOCAL}$ model [15]. In this model, communication proceeds in synchronous rounds and all nodes start simultaneously. In each round each node can exchange arbitrary messages with all its neighbors and perform arbitrary local computations. The time of completing a task is the number of rounds it takes. In particular, the time of WLE is the number of rounds it takes to all nodes to elect a leader if this is possible, and the time of SLE is the number of rounds it takes to all nodes to elect a leader if this is possible, and to terminate by declaring that this is impossible, otherwise.

It should be observed that the synchronous process of the $\mathcal{LOCAL}$ model can be simulated in an asynchronous network, by defining for each node separately its asynchronous round i in which it performs local computations, then sends messages stamped i to all neighbors, and waits until getting messages stamped i from all neighbors. (To make this work, every node is required to send a message with all consecutive stamps, until termination, possibly empty messages for some stamps.) All our results concerning time can be translated for asynchronous networks by replacing "time of completing a task" by "the maximum number of asynchronous rounds to complete it, taken over all nodes".

If nodes have distinct labels, then time D in the $\mathcal{LOCAL}$ model is enough to solve any problem solvable on a given network, as after this time all nodes have an exact copy of the network. By contrast, in our scenario of unlabeled nodes, time D is often not enough, for example to elect a leader, even if this task is feasible. This is due to the fact that after time t each node learns only all paths of length t originating at it and coded as sequences of port numbers. This is far less information than having a picture of the radius t neighborhood. A node v may not know if two paths originating at it have the same other endpoint or not. It turns out that these ambiguities may force time much larger than D to accomplish leader election.

We show that the time of leader election depends on three parameters of the network: its diameter D, its size n, and its *level of symmetry* λ. The latter parameter is defined for any network (see Section 2 for the formal definition) and, if leader election is feasible, this is the smallest depth at which some node has a unique view of the network. The view at depth t from a node (formally defined in Section 2) is the tree of all paths of length t originating at this node, coded as sequences of port numbers on these paths. It is the maximal information a node can gain after t rounds of communication in the $\mathcal{LOCAL}$ model.

It turns out that the time of leader election also crucially depends on the *knowledge* of parameters n and/or D by the nodes. On the other hand, it does not depend on *knowing* λ, although it often depends on this parameter as well.

Our Results. Optimal time of WLE is shown to be $\Theta(D + \lambda)$, if either D or n is known to the nodes. More precisely, we give two algorithms, one working for the class of networks of given diameter D and the other for the class of networks of given size n, that elect a leader in time $O(D + \lambda)$ on networks with diameter D and level of symmetry λ, whenever election is possible. Moreover, we prove that this complexity cannot be improved. We show, for any values D and λ, a network of diameter D and level of symmetry λ on which leader election is possible but takes time at least $D + \lambda$, even when D, n and λ are known. If neither D nor n is known, then even WLE is impossible [20].

For SLE, we show that knowing only D is insufficient to perform it. Then we prove that, if only n is known, then optimal time is $\Theta(n)$. We give an algorithm working for the class of networks of given size n, which performs SLE in time $O(n)$ and we show, for arbitrarily large n, a network G_n of size n, diameter $O(\log n)$ and level of symmetry 0, such that any algorithm performing correctly SLE on all networks of size n must take time $\Omega(n)$ on G_n. Finally, if both n and D are known, then optimal time is $\Theta(D + \lambda)$. Here we give an algorithm, working for the class of networks of given size n and given diameter D which performs SLE in time $O(D + \lambda)$ on networks with level of symmetry λ. In this case the matching lower bound carries over from our result for WLE.

Table 1. Summary of the results

		KNOWLEDGE			
		NONE	DIAMETER	SIZE	DIAMETER & SIZE
TASK	WLE	IMPOSSIBLE [20]	$\Theta(D + \lambda)$ (fast)		
	SLE	IMPOSSIBLE		$\Theta(n)$ (slow)	$\Theta(D + \lambda)$ (fast)

Table 1 gives a summary of our results. The main difficulty of this study is to prove lower bounds on the time of leader election, showing that the complexity of the proposed algorithms cannot be improved for any of the considered scenarios.

The comparison of our results for various scenarios shows two exponential gaps. The first is between the time of strong and weak LE. When only the size n is known, SLE takes time $\Omega(n)$ on some graphs of logarithmic diameter and level of symmetry 0, while WLE is accomplished in time $O(\log n)$ for such graphs. The second exponential gap is for the time of SLE, depending on the knowledge provided. While knowledge of the diameter alone does not help to accomplish SLE, when this knowledge is added to the knowledge of the size, it may exponentially decrease the time of SLE. Indeed, SLE with the knowledge of n alone takes time $\Omega(n)$ on some graphs of logarithmic diameter and level of symmetry 0, while SLE with the knowledge of n and D is accomplished in time $O(\log n)$, for such graphs.

Due to the lack of space, proofs of several results are omitted.

Related Work. Leader election was first studied for rings, under the assumption that all labels are distinct. A synchronous algorithm, based on comparisons of labels, and using $O(n \log n)$ messages was given in [8]. It was proved in [5] that this complexity is optimal for comparison-based algorithms. On the other hand, the authors showed an algorithm using a linear number of messages but requiring very large running time. An asynchronous algorithm using $O(n \log n)$ messages was given, e.g., in [16]. Deterministic leader election in radio networks has been studied, e.g., in [9,10,13] and randomized leader election, e.g., in [18].

Many authors [1,3,11,12,17,20] studied various computing problems in anonymous networks. In particular, [2,20] characterize networks in which leader election can be achieved when nodes are anonymous. In [19] the authors study the problem of leader election in general networks, under the assumption that labels are not unique. They characterize networks in which this can be done and give an algorithm which performs election when it is feasible. They assume that the number of nodes of the network is known to all nodes. In [4] the authors study feasibility and message complexity of sorting and leader election in rings with nonunique labels. In [7] the leader election problem is approached in a model based on mobile agents. Memory needed for leader election in unlabeled networks has been studied in [6]. To the best of our knowledge, the problem of time of leader election in arbitrary unlabeled networks has never been studied before.

2 Preliminaries

We say that leader election is *possible* for a given graph, or that this graph is a *solvable graph*, if there exists an algorithm which performs LE for this graph.

We consider two versions of the leader election task for a class $\mathcal{C}$ of graphs :

- Weak leader election. Let G be any graph in class $\mathcal{C}$. If leader election is possible for the graph G, then it is accomplished.

– Strong leader election. Let G be any graph in class $\mathcal{C}$. If leader election is possible for the graph G, then it is accomplished. Otherwise, all nodes eventually declare that the graph is not solvable and stop.

Hence WLE differs from SLE in that, in the case of impossibility of leader election, no restriction on the behavior of nodes is imposed: they can, e.g., elect different leaders, or no leader at all, or the algorithm may never stop.

We will use the following notion from [20]. Let G be a graph and v a node of G. The *view* from v is the infinite rooted tree $\mathcal{V}(v)$ with labeled ports, defined recursively as follows. $\mathcal{V}(v)$ has the root x_0 corresponding to v. For every node v_i, $i = 1, \ldots, k$, adjacent to v, there is a neighbor x_i in $\mathcal{V}(v)$ such that the port number at v corresponding to edge $\{v, v_i\}$ is the same as the port number at x_0 corresponding to edge $\{x_0, x_i\}$, and the port number at v_i corresponding to edge $\{v, v_i\}$ is the same as the port number at x_i corresponding to edge $\{x_0, x_i\}$. Node x_i, for $i = 1, \ldots, k$, is now the root of the view from v_i. The following proposition directly follows from [20] and expresses the feasibility of leader election in terms of views.

Proposition 1. *Let G be a graph. The following conditions are equivalent:*
1. *Leader election is possible in G;*
2. *Views of all nodes are different;*
3. *There exists a node with a unique view.*

By $\mathcal{V}^t(v)$ we denote the view $\mathcal{V}(v)$ truncated to depth t. We call it the view of v at depth t. In particular, $\mathcal{V}^0(v)$ consists of the node v, together with its degree. The following proposition was proved in [14].

Proposition 2. *For a n-node graph, $\mathcal{V}(u) = \mathcal{V}(v)$, if and only if $\mathcal{V}^{n-1}(u) = \mathcal{V}^{n-1}(v)$.*

Define the following equivalence relations on the set of nodes of a graph. $u \sim v$ if and only if $\mathcal{V}(u) = \mathcal{V}(v)$, and $u \sim_t v$ if and only if $\mathcal{V}^t(u) = \mathcal{V}^t(v)$. Let Π be the partition of all nodes into equivalence classes of $\sim$, and Π_t the corresponding partition for $\sim_t$. It was proved in [20] that all equivalence classes in Π are of equal size σ. In view of Proposition 2 this is also the case for Π_t, where $t \geq n-1$. On the other hand, for smaller t, equivalence classes in Π_t may be of different sizes. Every equivalence class in Π_t is a union of some equivalence classes in $\Pi_{t'}$, for $t < t'$. The following result was proved in [14]. It says that if the sequence of partitions Π_t stops changing at some point, it will never change again.

Proposition 3. *If $\Pi_t = \Pi_{t+1}$, then $\Pi_t = \Pi$.*

For any graph G we define its *level of symmetry* λ as the smallest integer t, for which there exists a node v satisfying the condition $\{u : \mathcal{V}^t(u) = \mathcal{V}^t(v)\} = \{u : \mathcal{V}(u) = \mathcal{V}(v)\}$. By Proposition 1, for solvable graphs, the level of symmetry is the smallest t for which there is a node with a unique view at depth t. In general, the level of symmetry is the smallest integer t for which some equivalence class in Π_t has size σ.

Define Λ to be the smallest integer t for which $\Pi_t = \Pi_{t+1}$. We have

Proposition 4. $\Lambda \leq D + \lambda$.

We fix a canonical linear order on all finite rooted trees with unlabeled nodes and labeled ports, e.g., as the lexicographic order of DFS traversals of these trees, starting from the root and exploring children of a node in increasing order of ports. For any subset of this class, the term "smallest" refers to this order. Since views at a depth t are such trees, we will elect as leader a node whose view at some depth is the smallest in some class of views. The difficulty is to establish when views at some depth are already unique for a solvable graph, and to decide fast if the graph is solvable, in the case of SLE.

All our algorithms are written for a node u of the graph. In their formulations we will use the procedure $COM(i)$, with a non-negative integer parameter i. This procedure is executed in a single round and consists of sending to all neighbors the view $\mathcal{V}^i(u)$ and receiving messages from all neighbors.

3   Weak Leader Election

In this section we show that the optimal time of weak leader election is $\Theta(D+\lambda)$, if either the diameter D of the graph or its size n is known to the nodes. We first give two algorithms, one working for the class of graphs of given diameter D and the other for the class of graphs of given size n, that elect a leader in time $O(D + \lambda)$ on graphs with diameter D and level of symmetry λ, whenever election is possible.

Our first algorithm works for the class of graphs of given diameter D.

Algorithm WLE-known-diameter(D).

for $i := 0$ **to** $D - 1$ **do** $COM(i)$
construct Π_0; $j := 0$
repeat
    $COM(D + j)$; $j := j + 1$; construct Π_j
until $\Pi_j = \Pi_{j-1}$
$V :=$ the set of nodes v in $\mathcal{V}^{D+j}(u)$ having the smallest $\mathcal{V}^{j-1}(v)$
elect as leader the node in V having the lexicographically smallest
    path from u

Theorem 1. *Algorithm WLE-known-diameter(D) elects a leader in every solvable graph of diameter D, in time $O(D + \lambda)$, where λ is the level of symmetry of the graph.*

Proof. All nodes of the graph construct Π_0 after D rounds and then they construct consecutive partitions Π_j, for $j = 1, \ldots, \Lambda + 1$. At this time the exit condition of the "repeat" loop is satisfied. All nodes stop simultaneously and elect a leader. Since the graph is solvable, by the definition of Λ and in view of Propositions 1 and 3, all elements of the partition Π_Λ are singletons. Hence all nodes in V correspond to the same node in the graph and consequently all nodes elect as leader the same node. All nodes stop in round $D + \Lambda$, which is at most $2D + \lambda$ by Proposition 4. $\square$

Our second algorithm works for the class of graphs of given size n.

Algorithm WLE-known-size(n).

$i := 0;\ x := 1;$
while $x < n$ **do**
$\qquad COM(i);\ i := i + 1$
$\qquad$ **for** $j := 0$ **to** i **do**
$\qquad\qquad L_j :=$ the set of nodes in $\mathcal{V}^i(u)$ at distance at most j
$\qquad\qquad\qquad$ from u (including u)
$\qquad\qquad num_j :=$ the number of nodes in L_j with distinct views
$\qquad\qquad\qquad$ at depth $i - j$
$\qquad x := \max \{num_j : j \in \{0, 1, \ldots, i\}\}$
compute Λ and D
$V :=$ the set of nodes v in $\mathcal{V}^i(u)$ having the smallest $\mathcal{V}^\Lambda(v)$
elect as leader the node in V having the lexicographically smallest
$\qquad$ path from u
while $i \leq D + \Lambda$ **do**
$\qquad COM(i);\ i := i + 1$

Theorem 2. *Algorithm WLE-known-size(n) elects a leader in every solvable graph of size n, in time $O(D + \lambda)$, where λ is the level of symmetry of the graph.*

In order to show that Algorithm WLE-known-diameter(D) is optimal for the class of graphs of diameter D and Algorithm WLE-known-size(n) is optimal for the class of graphs of size n we prove the following theorem. It shows a stronger property: both above algorithms have optimal complexity even among WLE algorithms working only when all three parameters n, D and λ are known.

Theorem 3. *For any $D \geq 1$ and any $\lambda \geq 0$, with $(D, \lambda) \neq (1, 0)$, there exists an integer n and a solvable graph G of size n, diameter D and level of symmetry λ, such that every algorithm for WLE working for the class of graphs of size n, diameter D and level of symmetry λ takes time at least $D + \lambda$ on the graph G.* [1]

Before proving Theorem 3 we present a construction of a family Q_k of complete graphs (cliques) that will be used in the proof of our lower bound. The construction consists in assigning port numbers. In order to facilitate subsequent analysis we also assign labels to nodes of the constructed cliques.

$\qquad Q_1$ is the single node graph. Q_2 is a 4-node clique with port numbers defined as follows. For a node u, we say that edge (i, j) is *incident* to u, if the edge corresponding to port i at node u corresponds to port j at some node v. Nodes in Q_2 are uniquely identifiable by the set of their incident edges. Below we assign distinct labels to the four nodes of the clique depending on the sets of their incident edges:

---

[1] Notice that there is no solvable graph with $D = 1$ and $\lambda = 0$, because the latter condition, for solvable graphs, means that there is a node of a unique degree.

- set of edges $\{(0,0), (1,1), (2,2)\}$ – corresponding to label a;
- set of edges $\{(0,0), (1,1), (2,0)\}$ – corresponding to label b;
- set of edges $\{(0,2), (1,1), (2,0)\}$ – corresponding to label c;
- set of edges $\{(0,2), (1,1), (2,2)\}$ – corresponding to label d.

We additionally assign colors 0, 1, and 2 to edges of Q_2 as follows: edges $\{a,b\}$ and $\{c,d\}$ get color 0, edges $\{a,c\}$ and $\{b,d\}$ get color 1, and edges $\{a,d\}$ and $\{b,c\}$ get color 2.

Q_3 is constructed starting from two disjoint copies of clique Q_2 as follows. Denote by Q_2 one of the copies and by $\overline{Q}_2$ the other one. Nodes from the overlined copy will be distinguished from corresponding nodes of the other copy by adding overlines.

Ports 3, 4, 5, and 6 are used to connect each node in Q_2 to all nodes in $\overline{Q}_2$ to construct an 8-node clique. All edges connecting nodes in Q_2 with nodes in $\overline{Q}_2$ will have the same port number at both endpoints, hence we can collapse pairs of ports (i,i) to a single edge color $i \in \{3,4,5,6\}$.

More precisely, edges of color 4 connect nodes $\{a,\overline{c}\}$, $\{\overline{a},c\}$, $\{b,\overline{d}\}$, and $\{\overline{b},d\}$. Edges of color 5 connect nodes $\{a,\overline{a}\}$, $\{b,\overline{b}\}$, $\{c,\overline{c}\}$, and $\{d,\overline{d}\}$.

Notice that until now, this construction results in a graph for which each node has exactly one other node with the same view, e.g., a and $\overline{a}$ have the same view. Uniqueness of all views at depth 2 is guaranteed in clique Q_3 by the definitions of edges of colors 3 and 6. In particular, edges of color 3 connect nodes: $\{a,\overline{b}\}$, $\{b,\overline{c}\}$ $\{c,\overline{d}\}$, and $\{d,\overline{a}\}$. On the other hand, nodes $\{\overline{a},b\}$, $\{\overline{b},c\}$ $\{\overline{c},d\}$, and $\{\overline{d},a\}$ are connected by edges of color 6. This concludes the construction of Q_3.

A node x in Q_3 connected to a node y by an edge with color 3 receives as its label the concatenation of the labels of nodes x and y in their respective copies of Q_2 (removing all overlines).

The complete labeling of nodes in Q_3 is: ab, bc, ad, ba, cd, da, cb, and dc. We partition this set of nodes into two blocks of size four: $\{ab, bc, ad, ba\}$ and $\{cd, da, cb, dc\}$.

For $k \geq 3$, the clique Q_{k+1} is produced starting from disjoint copies Q_k and $\overline{Q}_k$ of clique Q_k as follows. By the inductive hypothesis, let $\pi_1, \ldots, \pi_{2^{k-2}}$ be blocks partitioning nodes in Q_k such that:

- each block π_i, for $i \in [1, 2^{k-2}]$, has size 4;
- labels of nodes in each block π_i are $\alpha_i\beta_i$, $\beta_i\gamma_i$, $\alpha_i\delta_i$, and $\beta_i\alpha_i$, where $\alpha_i, \beta_i, \gamma_i$, and δ_i are strings of length 2^{k-2} over the alphabet $\{a,b,c,d\}$.

The same partition applies to nodes in $\overline{Q}_k$. We will call the corresponding blocks $\overline{\pi}_1, \ldots, \overline{\pi}_{2^{k-2}}$.

The set of nodes of the clique Q_{k+1} is the union of the sets of nodes of Q_k and $\overline{Q}_k$. Nodes in Q_{k+1} are partitioned into blocks as follows. Consider blocks $\pi_i = \{\alpha_i\beta_i, \beta_i\gamma_i, \alpha_i\delta_i, \beta_i\alpha_i\}$ and $\overline{\pi}_i = \{\overline{\alpha_i\beta_i}, \overline{\beta_i\gamma_i}, \overline{\alpha_i\delta_i}, \overline{\beta_i\alpha_i}\}$. Using these blocks we construct two blocks in Q_{k+1}: $\{\alpha_i\beta_i, \beta_i\gamma_i, \alpha_i\beta_i, \beta_i\gamma_i\}$ and $\{\alpha_i\delta_i, \beta_i\alpha_i, \overline{\alpha_i\delta_i}, \overline{\beta_i\alpha_i}\}$.

Edges connecting nodes in Q_k to nodes in $\overline{Q}_k$ have the same port number at both endpoints. As before, this port number is called the color of the edge. Colors are assigned to edges as follows.

Let $f(k) = 2^k - 1$. Consider a node x in a block π_i. For any $h \in [1, 2^k - 3]$, if x is connected in Q_k to a node y by an edge of color h, then we connect x to node $\overline{y}$ by an edge of color $h + f(k)$. Moreover we connect each node x to $\overline{x}$ by an edge of color $2^{k+1} - 3$.

Uniqueness of each view at depth k in Q_{k+1} is guaranteed by the asymmetries caused by edges of colors $2^k - 1$ and $2^{k+1} - 2$. We connect nodes $\{\alpha_i\beta_i, \overline{\beta_i\gamma_i}\}$, $\{\beta_i\gamma_i, \overline{\alpha_i\delta_i}\}$, $\{\alpha_i\delta_i, \overline{\beta_i\alpha_i}\}$, and $\{\beta_i\alpha_i, \overline{\alpha_i\beta_i}\}$ by edges of color $2^k - 1$. We connect nodes $\{\overline{\alpha_i\beta_i}, \beta_i\gamma_i\}$, $\{\overline{\beta_i\gamma_i}, \alpha_i\delta_i\}$, $\{\overline{\alpha_i\delta_i}, \beta_i\alpha_i\}$, and $\{\overline{\beta_i\alpha_i}, \alpha_i\beta_i\}$ by edges of color $2^{k+1} - 2$. This completes the construction of Q_{k+1}.

We call *small* the colors in $[0, 2^k - 2]$, i.e., all the edge colors inherited from the copies of Q_k in the construction of clique Q_{k+1}. Moreover, we call *symmetric* all edge colors in $[2^k, 2^{k+1} - 4]$ and *mirror* the edge color $2^{k+1} - 3$. Finally, we call *skews* the edge colors $2^k - 1$ and $2^{k+1} - 2$. We also call small, symmetric, and skews the edges having respectively a small, symmetric, or skew color.

The label of a node x in Q_{k+1} connected to a node y by an edge of color $2^k - 1$ is the concatenation of the labels of nodes x and y in their respective copies of Q_k (removing all overlines).

A node of the clique Q_k is said to be of *type* a, b, c, or d if it is obtained from a node with label a, b, c, or d (respectively) in a copy of Q_2 in the construction of Q_k. Consider a path p defined as a sequence of consecutive colored edges. We call p a *distinguishing* path for nodes x and y in Q_k, if it yields two different sequences of node types a, b, c, and d traversed proceeding along p, depending on whether the origin of p is x or y.

Lemma 1. *The clique Q_k has level of symmetry $k - 1$.*

We will also use the following family of cliques $\widetilde{Q}_k$. $\widetilde{Q}_1$ is the clique on 2 nodes, with port number 0. $\widetilde{Q}_2$ is a clique on 4 nodes, where all nodes have the same set of incident edges $\{(0,0), (1,1), (2,2)\}$. For $k \geq 3$, $\widetilde{Q}_k$ is a clique obtained from two disjoint copies of Q_{k-1}. The construction of $\widetilde{Q}_k$ mimics the construction of Q_k for all small, symmetric, and mirror edges. The only differences in the construction concern skew edges. Edges of color $2^k - 1$ connect nodes $\{\alpha_i\beta_i, \overline{\beta_i\gamma_i}\}$, $\{\overline{\alpha_i\beta_i}, \beta_i\gamma_i\}$, $\{\alpha_i\delta_i, \overline{\beta_i\alpha_i}\}$, and $\{\overline{\alpha_i\delta_i}, \beta_i\alpha_i\}$, while nodes $\{\alpha_i\beta_i, \overline{\beta_i\alpha_i}\}$, $\{\overline{\alpha_i\beta_i}, \beta_i\alpha_i\}$, $\{\beta_i\gamma_i, \overline{\alpha_i\delta_i}\}$, and $\{\overline{\beta_i\gamma_i}, \alpha_i\delta_i\}$ are connected by edges of color $2^{k+1} - 2$, for $i \in [1, 2^{k-2}]$. Notice that, in graph $\widetilde{Q}_k$, nodes x and $\overline{x}$ have identical views. Nevertheless, in order to describe our construction, we artificially assign to nodes x and $\overline{x}$ the label they would respectively receive in the construction of graph Q_k.

We finally define a family of graphs that allow us to prove Theorem 3. For any pair of integers (D, λ), with $D \geq 2$ and $\lambda \geq 2$, the graph $R_{D,\lambda}$ is obtained using one copy of graph $Q_{\lambda+1}$ and $2D - 1$ copies of graph $\widetilde{Q}_{\lambda+1}$. The construction of graph $R_{D,\lambda}$ proceeds as follows. Arrange $2D - 1$ disjoint copies of $\widetilde{Q}_{\lambda+1}$ and one copy of $Q_{\lambda+1}$ in a cyclic order. Connect each node in a clique with all nodes in the subsequent clique. Let i be the color of the edge (x, y) inside clique $Q_{\lambda+1}$. Assign port numbers $(i + 2^{\lambda+1} - 1, i + 2^{\lambda+2} - 1)$ to the edge connecting node x' in some clique to node y'' in the subsequent clique, where x' has label x and y''

has label y. Assign port numbers $(2^{\lambda+2} - 2, 2^{\lambda+2} + 2^{\lambda+1} - 2)$ to edges (x', x''), where x', x'' have label x and x'' belongs to the clique subsequent to that of x'.

A distinguishing path in $R_{D,\lambda}$ is defined in the same way as in $Q_{\lambda+1}$, which is possible since in both graphs each node has type a, b, c, or d.

Lemma 2. *Let $\lambda \geq 2$ and $D \geq 2$. Let x and y be two nodes in $Q_{\lambda+1}$ and let $\ell \leq \lambda - 1$ be the maximum depth at which views of x and y in $Q_{\lambda+1}$ are identical. Then the views at depth ℓ of nodes x and y belonging to the copy of $Q_{\lambda+1}$ in $R_{D,\lambda}$ are identical.*

The following lemma proves Theorem 3 in the case when either D or λ are small.

Lemma 3. *For $D = 1$ and any $\lambda \geq 1$, and for any $D \geq 2$ and $0 \leq \lambda \leq 1$, there exists an integer n and a solvable graph G of size n, diameter D and level of symmetry λ, such that every algorithm for WLE working for the class of graphs of size n, diameter D and level of symmetry λ takes time at least $D + \lambda$ on the graph G.*

Proof of Theorem 3: Lemma 3 proves the theorem if either D or λ are less than 2. It remains to give the argument for $D, \lambda \geq 2$. Consider the clique $\widetilde{Q}_{\lambda+1}$ antipodal to the clique $Q_{\lambda+1}$ in graph $R_{D,\lambda}$. Consider nodes x and $\overline{x}$ from this clique. Any distinguishing path for nodes x and $\overline{x}$ in $R_{D,\lambda}$ must contain a node from $Q_{\lambda+1}$. Let q be a minimum length distinguishing path for nodes x and $\overline{x}$ and assume without loss of generality that y and $\overline{y}$ are the first nodes from $Q_{\lambda+1}$ found along path q, if starting from x and $\overline{x}$, respectively. By Lemmas 1 and 2, nodes y and $\overline{y}$ have the same views at depth $\lambda - 1$ and different views at depth λ in graph $R_{D,\lambda}$. Thus the minimum length distinguishing path in $R_{D,\lambda}$ for y and $\overline{y}$ has length $\lambda - 1$. Since nodes y and $\overline{y}$ are at distance D from x and $\overline{x}$, the views at depth $D + \lambda - 1$ of x and $\overline{x}$ are identical. This concludes the proof. □

4   Strong Leader Election

For strong leader election more knowledge is required to accomplish it, and even more knowledge is needed to perform it fast. We first prove that knowledge of the diameter D is not sufficient for this task. The idea is to show, for sufficiently large D, one solvable and one non-solvable graph of diameter D, such that both graphs have the same sets of views.

Theorem 4. *For any $D \geq 4$ there is no SLE algorithm working for all graphs of diameter D.*

By contrast, knowledge of the size n alone is enough to accomplish strong leader election, but (unlike for WLE), it may be slow. We will show that optimal time for SLE is $\Theta(n)$ in this case. We first show an algorithm working in time $O(n)$.

Algorithm SLE-known-size(n).

for $i := 0$ **to** $2n - 3$ **do** $COM(i)$
$L :=$ the set of nodes in $\mathcal{V}^{2n-2}(u)$ at distance at most $n - 1$ from u
   (including u)
$num :=$ the number of nodes in L with distinct views at depth $n - 1$
if $num < n$ **then** report "LE impossible"
else
      $V :=$ the set of nodes v in $\mathcal{V}^{2n-2}(u)$ having the smallest $\mathcal{V}^{n-1}(v)$
      elect as leader the node in V having the lexicographically smallest
         path from u

Theorem 5. *Algorithm SLE-known-size(n) performs SLE in the class of graphs of size n, in time $O(n)$.*

Our next result shows that Algorithm SLE-known-size(n) is optimal if only n is known. Compared to Theorem 2 it shows that the optimal time of SLE with known size can be exponentially slower than that of WLE with known size. Indeed, it shows that SLE may take time $\Omega(n)$ on some graphs of diameter logarithmic in their size and having level of symmetry 0, while WLE takes time $O(\log n)$, on any solvable graph of diameter $O(\log n)$ and level of symmetry 0.

The high-level idea of proving that Algorithm SLE-known-size(n) is optimal if only n is known is the following. For arbitrarily large n, we show one solvable and one non-solvable graph of size n, such that nodes in one graph have the same view at depth $\Omega(n)$ as some nodes of the other.

Theorem 6. *For arbitrarily large n there exist graphs H_n of size n, level of symmetry 0 and diameter $O(\log n)$, such that every SLE algorithm working for the class of graphs of size n takes time $\Omega(n)$ on graph H_n.*

We finally show that if both D and n are known, then the optimal time of SLE is $\Theta(D + \lambda)$, for graphs with level of symmetry λ. The upper bound is given by Algorithm SLE-known-size-and-diameter(n, D), obtained from Algorithm WLE-known-diameter(D) by adding a test on the size of the partition Π_j after exiting the "repeat" loop. If this partition has size less than n, then report "LE impossible", otherwise elect the leader as in Algorithm WLE-known-diameter(D).

The following result says that Algorithm SLE-known-size-and-diameter(n, D) is fast. Compared to Theorems 4 and 6, it shows that while knowledge of the diameter alone does not help to accomplish SLE, when this knowledge is added to the knowledge of the size, it may exponentially decrease the time of SLE.

Theorem 7. *Algorithm SLE-known-size-and-diameter(n, D) performs SLE in the class of graphs of size n and diameter D, in time $O(D + \lambda)$, for graphs with level of symmetry λ.*

Since the lower bound in Theorem 3 was formulated for known n, D and λ, it implies a matching lower bound for the optimal time of SLE with known n and D, showing that this time is indeed $\Theta(D+\lambda)$ for graphs with level of symmetry λ.

References

1. Attiya, H., Snir, M., Warmuth, M.: Computing on an Anonymous Ring. Journal of the ACM 35, 845–875 (1988)
2. Boldi, P., Shammah, S., Vigna, S., Codenotti, B., Gemmell, P., Simon, J.: Symmetry Breaking in Anonymous Networks: Characterizations. In: Proc. 4th Israel Symposium on Theory of Computing and Systems (ISTCS 1996), pp. 16–26 (1996)
3. Boldi, P., Vigna, S.: Computing anonymously with arbitrary knowledge. In: Proc. 18th ACM Symp. on Principles of Distributed Computing, pp. 181–188 (1999)
4. Flocchini, P., Kranakis, E., Krizanc, D., Luccio, F.L., Santoro, N.: Sorting and election in anonymous asynchronous rings. JPDC 64, 254–265 (2004)
5. Fredrickson, G.N., Lynch, N.A.: Electing a leader in a synchronous ring. Journal of the ACM 34, 98–115 (1987)
6. Fusco, E., Pelc, A.: How much memory is needed for leader election. Distributed Computing 24, 65–78 (2011)
7. Haddar, M.A., Kacem, A.H., Métivier, Y., Mosbah, M., Jmaiel, M.: Electing a leader in the local computation model using mobile agents. In: Proc. 6th ACS/IEEE Int. Conference on Computer Systems and Applications (AICCSA 2008), pp. 473–480 (2008)
8. Hirschberg, D.S., Sinclair, J.B.: Decentralized extrema-finding in circular configurations of processes. Communications of the ACM 23, 627–628 (1980)
9. Jurdzinski, T., Kutylowski, M., Zatopianski, J.: Efficient algorithms for leader election in radio networks. In: Proc., 21st ACM Symp. on Principles of Distr. Comp. (PODC 2002), pp. 51–57 (2002)
10. Kowalski, D., Pelc, A.: Leader Election in Ad Hoc Radio Networks: A Keen Ear Helps. In: Albers, S., Marchetti-Spaccamela, A., Matias, Y., Nikoletseas, S., Thomas, W. (eds.) ICALP 2009, Part II. LNCS, vol. 5556, pp. 521–533. Springer, Heidelberg (2009)
11. Kranakis, E.: Symmetry and Computability in Anonymous Networks: A Brief Survey. In: Proc. 3rd Int. Conf. on Structural Information and Communication Complexity, pp. 1–16 (1997)
12. Kranakis, E., Krizanc, D., van der Berg, J.: Computing Boolean Functions on Anonymous Networks. Information and Computation 114, 214–236 (1994)
13. Nakano, K., Olariu, S.: Uniform leader election protocols for radio networks. IEEE Transactions on Parallel and Distributed Systems 13, 516–526 (2002)
14. Norris, N.: Universal Covers of Graphs: Isomorphism to Depth $N - 1$ Implies Isomorphism to All Depths. Discrete Applied Mathematics 56(1), 61–74 (1995)
15. Peleg, D.: Distributed Computing, A Locality-Sensitive Approach. SIAM Monographs on Discrete Mathematics and Applications, Philadelphia (2000)
16. Peterson, G.L.: An $O(n \log n)$ unidirectional distributed algorithm for the circular extrema problem. ACM Transactions on Programming Languages and Systems 4, 758–762 (1982)
17. Sakamoto, N.: Comparison of Initial Conditions for Distributed Algorithms on Anonymous Networks. In: Proc. 18th ACM Symp. on Principles of Distributed Computing (PODC), pp. 173–179 (1999)
18. Willard, D.E.: Log-logarithmic selection resolution protocols in a multiple access channel. SIAM J. on Computing 15, 468–477 (1986)
19. Yamashita, M., Kameda, T.: Electing a Leader when Procesor Identity Numbers are not Distinct. In: Bermond, J.-C., Raynal, M. (eds.) WDAG 1989. LNCS, vol. 392, pp. 303–314. Springer, Heidelberg (1989)
20. Yamashita, M., Kameda, T.: Computing on anonymous networks: Part I - characterizing the solvable cases. IEEE Trans. Parallel and Distributed Systems 7, 69–89 (1996)

An Experimental Study of Dynamic Dominators

Loukas Georgiadis[1], Giuseppe F. Italiano[2],
Luigi Laura[3], and Federico Santaroni[2]

[1] Department of Computer Science, University of Ioannina, Greece
loukas@cs.uoi.gr
[2] Dipartimento di Informatica, Sistemi e Produzione
Università di Roma "Tor Vergata", via del Politecnico 1, 00133 Roma, Italy
{italiano,santaroni}@disp.uniroma2.it
[3] Dip. di Ingegneria Informatica, Automatica e Gestionale
"Sapienza" Università di Roma, via Ariosto 25, 00185, Roma, Italy
laura@dis.uniroma1.it

Abstract. Motivated by recent applications of dominator computations
we consider the problem of dynamically maintaining the dominators of
a flowgraph through a sequence of insertions and deletions of edges. We
present new algorithms and efficient implementations of existing algorithms and exhibit their efficiency in practice.

1 Introduction

A flowgraph $G = (V, E, r)$ is a directed graph with a distinguished root vertex
$r \in V$. A vertex v is *reachable* in G if there is a path from r to v; v is *unreachable*
if no such path exists. The *dominance relation* on G is defined for the set of
reachable vertices as follows. A vertex w *dominates* a vertex v if every path from
r to v includes w. We let $Dom(v)$ denote the set of all vertices that dominate v.
If v is reachable then $Dom(v) \supseteq \{r, v\}$; otherwise $Dom(v) = \emptyset$. For a reachable
vertex v, r and v are its *trivial dominators*. A vertex $w \in Dom(v) - v$ is a *proper
dominator* of v. The *immediate dominator* of a vertex $v \neq r$, denoted $idom(v)$,
is the unique vertex $w \neq v$ that dominates v and is dominated by all vertices
in $Dom(v) - v$. The dominance relation is transitive and can be represented
in compact form as a tree D, called the *dominator tree* of G, in which every
reachable vertex $v \neq r$ is made a child of $idom(v)$. The dominators of a vertex
are its ancestors in D.

The problem of finding dominators has been extensively studied, as it occurs
in several applications, including program optimization, code generation, circuit
testing and theoretical biology (see, e.g., [17]). Allen and Cocke showed that the
dominance relation can be computed iteratively from a set of data-flow equations [1]. A direct implementation of this method has an $O(mn^2)$ worst-case time
bound, for a flowgraph with n vertices and m edges. Cooper et al. [8] presented
a clever tree-based space-efficient implementation of the iterative algorithm. Although it does not improve the $O(mn^2)$ worst-case time bound, the tree-based
version is much more efficient in practice. Purdom and Moore [23] gave an algorithm, based on reachability, with complexity $O(mn)$. Improving on previous

L. Epstein and P. Ferragina (Eds.): ESA 2012, LNCS 7501, pp. 491–502, 2012.

work by Tarjan [28], Lengauer and Tarjan [19] proposed an $O(m \log n)$-time algorithm and a more complicated $O(m\alpha(m, n))$-time version, where $\alpha(m, n)$ is an extremely slow-growing functional inverse of the Ackermann function [29]. There are also more complicated truly linear-time algorithms [2,4,5,14].

An experimental study of static algorithms for computing dominators was presented in [17], where careful implementations of both versions of the Lengauer-Tarjan algorithm, the iterative algorithm of Cooper et al., and a new hybrid algorithm (SNCA) were given. In these experimental results the performance of all these algorithms was similar, but the simple version of the Lengauer-Tarjan algorithm and the hybrid algorithm were most consistently fast, and their advantage increased as the input graph got bigger or more complicated. The graphs used in [17] were taken from the application areas mentioned above and have moderate size (at most a few thousand vertices and edges) and simple enough structure that they can be efficiently processed by the iterative algorithm. Recent experimental results for computing dominators in large graphs are reported in [10,13] and the full version of [16]. There it is apparent that the simple iterative algorithms are not competitive with the more sophisticated algorithms based on Lengauer-Tarjan. The graphs used in these experiments were taken from applications of dominators in memory profiling [20,21], testing 2-vertex connectivity and computing sparse 2-vertex connected spanning subgraphs [12,13], and computing strong articulation points and strong bridges in directed graphs [18], which typically involve much larger and complicated graphs.

Here we consider the problem of dynamically maintaining the dominance relation of a flowgraph that undergoes both insertions and deletions of edges. Vertex insertions and deletions can be simulated using combinations of edge updates. We recall that a dynamic graph problem is said to be fully dynamic if it requires to process both insertions and deletions of edges, incremental if it requires to process edge insertions only and decremental if it requires to process edge deletions only. The fully dynamic dominators problem arises in various applications, such as data flow analysis and compilation [7]. Moreover, [12,18] imply that a fully dynamic dominators algorithm can be used for dynamically testing 2-vertex connectivity, and maintaining the strong articulation points of a digraph. The decremental dominators problem appears in the computation of 2-connected components in digraphs [18].

The problem of updating the dominance relation has been studied for few decades (see, e.g., [3,6,7,22,24,26]). However, a complexity bound for a single update better than $O(m)$ has been only achieved for special cases, mainly for incremental or decremental problems. Specifically, the algorithm of Cicerone et al. [7] achieves $O(n \max\{k, m\} + q)$ running time for processing a sequence of k edge insertions interspersed with q queries of the type "does x dominate y?", for a flowgraph with n vertices and initially m edges. The same bound is also achieved for a sequence of k deletions, but only for a reducible flowgraph[1]. This algorithm does not maintain the dominance relation in a tree but in an $n \times n$ matrix, so a query can be answered in constant time. Alstrup and Lauridsen describe in a

---

[1] A flowgraph with root r is reducible if every loop has a single entry vertex from r.

technical report [3] an algorithm that maintains the dominator tree through a sequence of k edge insertions interspersed with q queries in $O(m \min\{k,n\} + q)$ time. In this bound m is the number of edges after all insertions. However, the description of this algorithm is incomplete and contains some incorrect arguments, as it will be shown in the full paper.

Although theoretically efficient solutions to the fully dynamic dominators problem appear still beyond reach, there is a need for practical algorithms and fast implementations in several application areas. In this paper, we present new fully dynamic dominators algorithms and efficient implementations of known algorithms, such as the algorithm by Sreedhar, Gao and Lee [26]. We evaluate the implemented algorithms experimentally using real data taken from the application areas of dominators. To the best of our knowledge, the only previous experimental study of (fully) dynamic dominators algorithms appears in [22]; here we provide new algorithms, improved implementations, and an experimental evaluation using bigger graphs taken from a larger variety of applications. Other previous experimental results, reported in [25] and the references therein, are limited to comparing incremental algorithms against the static computation of dominators.

2   Basic Properties

The algorithms we consider can be stated in terms of two structural properties of dominator trees that we discuss next. Let T be a tree with the same vertices as G and rooted at r. Let $p_T(v)$ denote the parent of v in T. We call T *valid* if, for every edge $(v,w) \in E$, v is a descendant of $p_T(w)$. If T is valid, $p_T(v)$ dominates v for every vertex $v \neq r$ [16]. For a valid tree T we define the *support* $sp_T(v,w)$ of an edge (v,w) with respect to T as follows: if $v = p_T(w)$, $sp_T(v,w) = v$; otherwise, $sp_T(v,w)$ is the child of $p_T(w)$ that is an ancestor of v. We call T *co-valid* if, for every ordered pair of siblings v, w in T, v does not dominate w.

Theorem 1. *[16] Tree T is the dominator tree ($T = D$) if and only if it is valid and co-valid.*

Now consider the effect that a single edge update (insertion or deletion) has on the dominator tree D. Let (x, y) be the inserted or deleted edge. We let G' and D' denote the flowgraph and its dominator tree after the update. Similarly, for any function f on V we let f' be the function after the update. By definition, $D' \neq D$ only if x is reachable before the update. We say that a vertex v is *affected* by the update if $idom'(v) \neq idom(v)$. (Note that we can have $Dom'(v) \neq Dom(v)$ even if v is not affected.)

The difficulty in updating the dominance relation lies on two facts: (i) An affected vertex can be arbitrarily far from the updated edge, and (ii) a single update may affect many vertices. A pathological example is shown in Figure 1, where each single update affects every vertex except r. Moreover, we can construct sequences of $\Theta(n)$ edge insertions (deletions) such that each single insertion (deletion) affects $\Theta(n)$ vertices. (Consider, for instance, the graph of

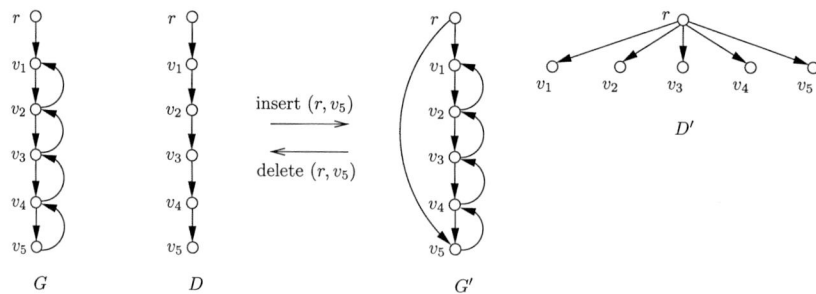

Fig. 1. Pathological updates: each update (insertion or deletion) affects $n - 1$ vertices

Figure 1 and the sequence of insertions $(v_3, v_5), (v_2, v_5), (v_1, v_5), (r, v_5)$.) This implies a lower bound of $\Omega(n^2)$ time for any algorithm that maintains D explicitly through a sequence of $\Omega(n)$ edge insertions or a sequence of $\Omega(n)$ edge deletions, and a lower bound of $\Omega(mn)$ time for any algorithm that maintains D through an intermixed sequence of $\Omega(m)$ edge insertions and deletions.

Using the structural properties of dominator trees stated above we can limit the number of vertices and edges processed during the search for affected vertices. Note that an edge insertion can violate the validity of D, while an edge deletion can violate the co-validity of D. Throughout the rest of this section and Section 3, (x, y) is the inserted or deleted edge and x is reachable.

2.1   Edge Insertion

We consider two cases, depending on whether y was reachable before the insertion. Suppose first that y was reachable. Let $nca_D(x, y)$ be the nearest (lowest) common ancestor of x and y in D. If either $nca_D(x, y) = idom(y)$ or $nca_D(x, y) = y$ then, by Theorem 1, the inserted edge has no effect on D. Otherwise, the validity of D implies that $nca_D(x, y)$ is a proper dominator of $idom(y)$. As shown in [24], for all affected vertices v, $idom'(v) = nca_D(x, y)$. Moreover, [3,26] observe that all affected vertices v satisfy $depth(nca_D(x, y)) < depth(idom(v)) < depth(v) \leq depth(y)$, where $depth(w)$ is the depth of vertex w in D. (These facts also follow from the validity of D.) Based on the above observations we can show the following lemma which is a refinement of a result in [3].[2]

Lemma 1. *Suppose x and y are reachable vertices in G. A vertex v is affected after the insertion of (x, y) if and only if $depth(nca_D(x, y)) < depth(idom(v))$ and there is a path P from y to v such that $depth(idom(v)) < depth(w)$ for all $w \in P$. If v is affected then $idom'(v) = nca_D(x, y)$.*

Now suppose y was unreachable before the insertion of (x, y). Then we have $x = idom'(y)$. Next, we need to process any other vertex that became reachable

---

[2] Due to limited space all proofs are deferred to the full paper.

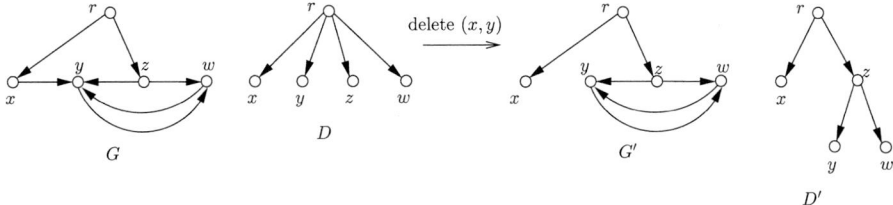

Fig. 2. After the deletion of (x, y), y still has two entering edges (z, y) and (w, y) such that $sp_D(z, y) \neq sp_D(w, y)$

after the insertion. To that end, we have two main options. The first is to process each edge leaving y as a new insertion, and continue this way until all edges adjacent to newly reachable vertices are processed. The second is to compute the set $R(y)$ of the vertices that are reachable from y and were not reachable from r before the insertion of (x, y). We can build the dominator tree $D(y)$ for the subgraph induced by $R(y)$, rooted at y, using any static algorithm. After doing that we connect $D(y)$ to form the complete dominator tree by adding the edge (x, y) in D. Finally we process every edge (x', y') where $x' \in R(y)$ and $y' \notin R(y)$ as a new insertion. Note that the effect on D of each such edge is equivalent to adding the edge (x, y') instead.

2.2   Edge Deletion

We consider two cases, depending on whether y becomes unreachable after the deletion of (x, y). Suppose first that y remains reachable. Deletion is harder than insertion because each of the affected vertices may have a different new immediate dominator. (Refer to Figure 1.) Also, unlike the insertion case, we do not have a simple test, as the one stated in Lemma 1, to decide whether the deletion affects any vertex. Consider, for example, the following (necessary but not sufficient) condition stated in [15,16] when $u = idom(y)$: if u is not a predecessor[3] of y then y has predecessors z and w, such that $sp_D(z, y) \neq sp_D(w, y)$. Unfortunately, this condition may still hold in D after the deletion even when $D' \neq D$ (see Figure 2). Despite this difficulty, in Section 3.1 we give a simple but conservative test (i.e., it allows false positives) to decide if there are any affected vertices. Also, as observed in [26], we can limit the search for affected vertices and their new immediate dominators as follows. Since the edge deletion may violate the co-validity of D but not validity, it follows that the new immediate dominator of an affected vertex v is a descendant of some sibling of v in D. This implies the following lemma that provides a necessary (but not sufficient) condition for a vertex v to be affected.

Lemma 2. *Suppose x is reachable and y does not becomes unreachable after the deletion of (x, y). A vertex v is affected only if $idom(v) = idom(y)$ and there is a path P from y to v such that $depth(idom(v)) < depth(w)$ for all $w \in P$.*

---

[3] A vertex u is a *predecessor* of a vertex v if $(u, v) \in E$.

Next we examine the case where y becomes unreachable after (x, y) is deleted. This happens if and only if y has no entering edge (z, y) in G' such that $sp_D(z, y) \neq y$. Furthermore, all descendants of y in D also become unreachable. To deal with this, we can collect all edges (u, v) such that u is a descendant of y in D ($y \in Dom(u)$) but v is not, and process (u, v) as a new deletion. (Equivalently we can substitute (u, v) with (y, v) and process (y, v) as a new deletion.) An alternative choice is to use a static algorithm to compute the immediate dominators of all possibly affected vertices, which can be identified by the following lemma.

Lemma 3. *Suppose x is reachable and y becomes unreachable after the deletion of (x, y). A vertex v is affected only if there is a path P from y to v such that $depth(idom(v)) < depth(w)$ for all $w \in P$.*

3   Algorithms

Here we present new algorithms for the dynamic dominators problem. We begin in Section 3.1 with a simple dynamic version of the SNCA algorithm (DSNCA), which also provides a necessary (but not sufficient) condition for an edge deletion to affect the dominator tree. Then, in Section 3.2, we present a depth-based search (DBS) algorithm which uses the results of Section 2. We improve the efficiency of deletions in DBS by employing some ideas used in DSNCA. Due to lack of space we defer some of implementation details to the full version of the paper. In the description below we let (x, y) be the inserted or deleted edge and assume that x is reachable.

3.1   Dynamic SNCA (DSNCA)

We develop a simple method to make the (static) SNCA algorithm [17] dynamic, in the sense that it can respond to an edge update faster (by some constant factor) than recomputing the dominator tree from scratch. Furthermore, by storing some side information we can test if the deletion of an edge satisfies a necessary condition for affecting D.

The SNCA algorithm is a hybrid of the simple version of Lengauer-Tarjan (SLT) and the iterative algorithm of Cooper et al. The Lengauer-Tarjan algorithm uses the concept of *semidominators*, as an initial approximation to the immediate dominators. It starts with a depth-first search on G from r and assigns preorder numbers to the vertices. Let T be the corresponding depth-first search tree, and let $pre(v)$ be the preorder number of v. A path $P = (u = v_0, v_1, \ldots, v_{k-1}, v_k = v)$ in G is a *semidominator path* if $pre(v_i) > pre(v)$ for $1 \leq i \leq k-1$. The semidominator of v, $sdom(v)$, is defined as the vertex u with minimum $pre(u)$ such that there is a semidominator path from u to v. Semidominators and immediate dominators are computed by executing path-minima computations, which find minimum $sdom$ values on paths of T, using an appropriate data-structure. Vertices are processed in reverse preorder, which ensures that all the necessary values are available when needed. With a simple implementation of the path-minima data structure, the algorithm SLT runs in $O(m \log n)$ time. With a more sophisticated strategy the algorithm runs in $O(m\alpha(m, n))$ time.

The SNCA algorithm computes dominators in two phases:

(a) Compute $sdom(v)$ for all $v \neq r$, as done by SLT.
(b) Build D incrementally as follows: Process the vertices in preorder. For each vertex w, ascend the path from $p_T(w)$ to r in D, where $p_T(w)$ is the parent of w in T (the depth-first search tree), until reaching the deepest vertex x such that $pre(x) \leq pre(sdom(w))$. Set x to be the parent of w in D.

With a naïve implementation, the second phase runs in $O(n^2)$ worst-case time. However, as reported in [10,17] and the full version of [16], it performs much better in practice. In fact, the experimental results reported in [10] and the full version of [16], show that the running times of SNCA and SLT are very close in practice even for large graphs. SNCA is simpler than SLT in that it requires fewer arrays, eliminates some indirect addressing, and there is one fewer pass over the vertices. This makes it easier to produce a dynamic version of SNCA, as described below. We note, however, that the same ideas can be applied to produce a dynamic version of SLT as well.

Edge insertion. Let T be the depth-first search tree used to compute semidominators. Let $pre(v)$ be the preorder number of v in T and $post(v)$ be the postorder number of v in T; if $v \notin T$ then $pre(v)(= post(v)) = 0$. The algorithm runs from scratch if

$$pre(y) = 0 \text{ or, } pre(x) < pre(y) \text{ and } post(x) < post(y). \tag{1}$$

If condition (1) does not hold then T remains a valid depth-first seach tree for G. If this is indeed the case then we can repeat the computation of semidominators for the vertices v such that $pre(v) \leq pre(y)$. To that end, for each such v, we initialize the value of $sdom(v)$ to $p_T(v)$ and perform the path-minima computations for the vertices v with $pre(v) \in \{2, \ldots, pre(y)\}$. Finally we perform the nearest common ancestor phase for all vertices $v \neq r$.

Edge deletion. The idea for handling deletions is similar to the insertion case. In order to test efficiently if the deletion may possibly affect the dominator tree we use the following idea. For any $v \in V - r$, we define $t(v)$ to be a predecessor of v that belongs to a semidominator path from $sdom(v)$ to v. Such vertices can be found easily during the computation of semi-dominators [15].

Lemma 4. *The deletion of (x, y) affects D only if $x = p_T(y)$ or $x = t(y)$.*

If $x = p_T(y)$ we run the whole SNCA algorithm from scratch. Otherwise, if $x = t(y)$, then we perform the path-evaluation phase for the vertices v such that $pre(v) \in \{2, \ldots, pre(y)\}$. Finally we perform the nearest common ancestor phase for all vertices $v \neq r$. We note that an insertion or deletion takes $\Omega(n)$ time, since our algorithm needs to reset some arrays of size $\Theta(n)$. Still, as the experimental results given in Section 4 show, the algorithm offers significant speedup compared to running the static algorithm from scratch.

3.2   Depth-Based Search (DBS)

This algorithm uses the results of Section 2 and ideas from [3,26] and DSNCA. Our goal is to search for affected vertices using the depth of the vertices in the dominator tree, and improve batch processing in the unreachable cases (when y is unreachable before the insertion or y becomes unreachable after a deletion).

Edge insertion. We use Lemma 1 to locate the affected vertices after inserting (x, y). As in the Sreedhar-Gao-Lee algorithm, we maintain a set A of affected vertices, sorted by their depth in D. Initially $A = \{y\}$, and while A is not empty, a vertex in A with maximum depth is extracted and processed. Let $\hat{v}$ be the most recently extracted affected vertex. During the search for affected vertices we visit a current vertex v. Initially $\hat{v} = v = y$. To process v, we look at the edges (v, w) leaving v. If $depth(w) > depth(\hat{v})$ then we recursively visit w. If $depth(nca(x, y)) + 1 < depth(w) \leq depth(\hat{v})$ then w is inserted into A. The last inequality makes it easy to extract in amortized $O(1)$ time a vertex in A with maximum depth to be processed next, using bucket sorting.

If y was unreachable before the insertion then we execute a depth-first search from y to find the set of vertices $R(y)$ that are reachable from y but were previously unreachable from r. During this search we also compute the edge sets $E(y) = \{(u, v) \in E \mid u \in R(y) \text{ and } v \in R(y)\}$ and $E^+(y) = \{(u, v) \in E \mid u \in R(y) \text{ and } v \notin R(y)\}$. We built the dominator tree $D(y)$ for the subgraph induced by $R(y)$ using SNCA. After computing $D(y)$ we set $idom(y) = x$ and process the edges in $E^+(y)$. For all $(u, v) \in E^+(y)$ both u and v are now reachable, so we use the insertion algorithm for the reachable case to update the dominator tree.

The running time for a single insertion is proportional to the number of visited vertices and their adjacent edges, which is $O(m)$. However, Lemma 1 allows us to amortize the running time in the insertions-only case: Any vertex w on the path P above is either affected or a descendant of an affected vertex, therefore $depth'(w) < depth(w)$. Thus, all vertices on the path decrease in depth, which implies the following result.

Theorem 2. *Algorithm DBS maintains the dominator tree of a flowgraph through a sequence of k edge insertions in $O(m \min\{k, n\} + kn)$ time, where n is the number of vertices and m is the number of edges after all insertions.*

We note that the $O(kn)$ term in the above bound corresponds to the time required to compute $nca_D(x, y)$ for all inserted edges (x, y). As observed in [3], this term can be reduced to $O(k)$ by applying a data structure of Gabow [11].

Edge deletion. We describe a method that applies SNCA. If y is still reachable after the deletion then we execute SNCA for the subgraph induced by $idom(y)$. Now suppose y becomes unreachable. Let $E^+(y) = \{(u, v) \in E \mid y \in Dom(u) \text{ and } y \notin Dom(v)\}$ and let $V^+(y) = \{v \in V \mid \text{there is an edge } (u, v) \in E^+(y)\}$. We compute $V^+(y)$ by executing a depth-first search from y, visiting only vertices w with $depth(w) \geq depth(y)$. At each visited vertex w we examine the edges (w, v) leaving w; v is included in $V^+(y)$ if $depth(v) \leq depth(y)$. Next, we find a vertex

$v \in V^+(y)$ of minimum depth such that $v \notin Dom(y)$. Finally we execute SNCA for the subgraph induced by $idom(v)$. Lemma 3 implies the correctness of this method. A benefit of this approach is that we can apply Lemma 4 to test if the deletion may affect D. Here we have the additional complication that we can maintain the $p_T(y)$ and $t(y)$ values required for the test only for the vertices y such that $idom(y)$ was last computed by SNCA. Therefore, when an insertion affects a reachable vertex y we set $p_T(y)$ and $t(y)$ to *null* and cannot apply the deletion test if some edge entering y is deleted.

4   Experimental Evaluation

4.1   Implementation and Experimental Setup

We evaluate the performance of four algorithms: the simple version of Lengauer-Tarjan (SLT), dynamic SNCA (DSNCA), an efficient implementation of Sreedhar-Gao-Lee (SGL) and the depth-based search algorithm (DBS). In this setting, SLT runs the simple version of the Lengauer-Tarjan algorithm after each edge (x, y) update, but only if x is currently reachable from r. We do not report running times for static SNCA as they are very close to those of SLT. We implemented all algorithms in C++. They take as input the graph and its root, and maintain an n-element array representing immediate dominators. Vertices are assumed to be integers from 1 to n. The code was compiled using g++ v. 3.4.4 with full optimization (flag -O4). All tests were conducted on an Intel Core i7-920 at 2.67GHz with 8MB cache, running Windows Vista Business Edition. We report CPU times measured with the getrusage function. Running times include allocation and deallocation of arrays and linked lists, as required by each algorithm, but do not include reading the graph from an input file. Our source code is available upon request.

4.2   Instances and Evaluation

Our test set consists of a sample of graphs used in [17], graphs taken from the Stanford Large Network Dataset Collection [27], and road networks [9]. We report running times for a representative subset of the above test set, which consist of the following: the control-flow graph uloop from the SPEC 2000 suite created by the IMPACT compiler, the foodweb baydry, the VLSI circuit s38584 from the ISCAS'89 suite, the peer-to-peer network p2p-Gnutella25, and the road network rome99. We constructed a sequence of update operations for each graph by simulating the update operations as follows. Let m be the total number of edges in the graph. We define parameters i and d which correspond, respectively, to the fraction of edges to be inserted and deleted. This means that $m_i = i * m$ edges are inserted and $m_d = d * m$ edges are deleted, and the flowgraph initially has $m' = m - m_i$ edges. The algorithms build (in static mode) the dominator tree for the first m' edges in the original graph and then they run in dynamic mode. For $i = d = 0$, SGL reduces to the iterative algorithm of Cooper et al. [8], whilst

Table 1. Average running times in seconds for 10 seeds. The best result in each row is bold.

GRAPH	INSTANCE		INSERTIONS	DELETIONS	SLT	DSNCA	SGL	DBS
	i	d						
uloop	10	0	315	0	0.036	0.012	0.008	**0.006**
$n = 580$	0	10	0	315	0.039	**0.012**	0.059	**0.012**
$m = 3157$	10	10	315	315	0.065	0.017	0.049	**0.014**
	50	0	1578	0	0.129	0.024	0.013	**0.012**
	0	50	0	1578	0.105	0.024	0.121	**0.023**
	50	50	1578	1578	0.067	**0.015**	0.041	0.033
	100	0	3157	0	0.178	0.033	0.023	**0.019**
	0	100	0	3157	0.120	**0.026**	0.136	0.050
baydry	10	0	198	0	0.010	**0.002**	0.004	0.003
$n = 1789$	0	10	0	198	0.014	**0.003**	0.195	0.004
$m = 1987$	10	10	198	198	0.020	**0.003**	0.012	0.005
	50	0	993	0	0.034	0.009	**0.007**	0.008
	0	50	0	993	0.051	**0.007**	0.078	0.012
	50	50	993	993	0.056	**0.006**	0.033	0.020
	100	0	1987	0	0.048	**0.004**	0.013	0.015
	0	100	0	1987	0.064	**0.010**	0.095	0.022
rome99	10	0	887	0	0.261	0.106	0.027	**0.017**
$n = 3353$	0	10	0	887	0.581	**0.252**	1.861	0.291
$m = 8870$	10	10	887	887	0.437	0.206	0.863	**0.166**
	50	0	4435	0	0.272	0.106	0.049	**0.031**
	0	50	0	4435	1.564	**0.711**	4.827	0.713
	50	50	4435	4435	0.052	**0.016**	0.074	0.065
	100	0	8870	0	0.288	0.103	0.068	**0.056**
	0	100	0	8870	1.274	**0.613**	4.050	0.639
s38584	10	0	3449	0	6.856	2.772	0.114	**0.096**
$n = 20719$	0	10	0	3449	7.541	**4.416**	15.363	4.803
$m = 34498$	10	10	3449	3449	6.287	3.586	5.131	**2.585**
	50	0	17249	0	9.667	3.950	0.228	**0.150**
	0	50	0	17249	10.223	**5.671**	17.543	5.835
	50	50	17249	17249	0.315	**0.107**	0.342	0.291
	100	0	34498	0	10.477	4.212	0.301	**0.285**
	0	100	0	34498	10.931	6.056	18.987	**6.016**
p2p-Gnutella25	10	0	5470	0	38.031	9.295	0.167	**0.123**
$n = 22687$	0	10	0	5470	38.617	**13.878**	38.788	16.364
$m = 54705$	10	10	5470	5470	72.029	21.787	37.396	**14.767**
	50	0	27352	0	129.668	37.206	0.415	**0.256**
	0	50	0	27352	133.484	**49.730**	131.715	51.764
	50	50	27352	27352	60.776	27.996	28.478	**19.448**
	100	0	54705	0	136.229	39.955	0.724	**0.468**
	0	100	0	54705	128.738	54.405	139.449	**44.064**

DSNCA and DBS reduce to SNCA. The remaining edges are inserted during the updates. The type of each update operation is chosen uniformly at random, so that there are m_i insertions interspersed with m_d deletions. During this simulation that produces the dynamic graph instance we keep track of the edges currently in the graph. If the next operation is a deletion then the edge to be deleted is chosen uniformly at random from the edges in the current graph.

The experimental results for various combinations of i and d are shown in Table 1. The reported running times for a given combination of i and d is the average of the total running time taken to process ten update sequences obtained from different seed initializations of the **srand** function. With the exception of **baydry** with $i = 50, d = 0$, in all instances DSNCA and DBS are the fastest. In most cases, DSNCA is by a factor of more than 2 faster than SLT. SGL and DBS are much more efficient when there are only insertions ($d = 0$), but their performance deteriorates when there are deletions ($d > 0$). For the $d > 0$ cases, DBS and DSNCA have similar performance for most instances, which is due to employing the deletion test of Lemma 4. On the other hand, SGL can be even worse than SLT when $d > 0$. For all graphs except **baydry** (which is extremely sparse) we observe a decrease in the running times for the $i = 50, d = 50$ case. In this case, many edge updates occur in unreachable parts of the graph. (This effect is more evident in the $i = 100, d = 100$ case, so we did not include it in Table 1.) Overall, DBS achieves the best performance. DSNCA is a good choice when there are deletions and is a lot easier to implement.

References

1. Allen, F.E., Cocke, J.: Graph theoretic constructs for program control flow analysis. Technical Report IBM RC 3923, IBM T.J. Watson Research (1972)
2. Alstrup, S., Harel, D., Lauridsen, P.W., Thorup, M.: Dominators in linear time. SIAM Journal on Computing 28(6), 2117–2132 (1999)
3. Alstrup, S., Lauridsen, P.W.: A simple dynamic algorithm for maintaining a dominator tree. Technical Report 96-3, Department of Computer Science, University of Copenhagen (1996)
4. Buchsbaum, A.L., Georgiadis, L., Kaplan, H., Rogers, A., Tarjan, R.E., Westbrook, J.R.: Linear-time algorithms for dominators and other path-evaluation problems. SIAM Journal on Computing 38(4), 1533–1573 (2008)
5. Buchsbaum, A.L., Kaplan, H., Rogers, A., Westbrook, J.R.: A new, simpler linear-time dominators algorithm. ACM Trans. on Programming Languages and Systems 27(3), 383–387 (2005)
6. Carroll, M.D., Ryder, B.G.: Incremental data flow analysis via dominator and attribute update. In: Proc. 15th ACM POPL, pp. 274–284 (1988)
7. Cicerone, S., Frigioni, D., Nanni, U., Pugliese, F.: A uniform approach to semi-dynamic problems on digraphs. Theor. Comput. Sci. 203, 69–90 (1998)
8. Cooper, K.D., Harvey, T.J., Kennedy, K.: A simple, fast dominance algorithm. Software Practice & Experience 4, 110 (2001)
9. Demetrescu, C., Goldberg, A., Johnson, D.: 9th DIMACS Implementation Challenge - Shortest Paths (2006)

10. Firmani, D., Italiano, G.F., Laura, L., Orlandi, A., Santaroni, F.: Computing Strong Articulation Points and Strong Bridges in Large Scale Graphs. In: Klasing, R. (ed.) SEA 2012. LNCS, vol. 7276, pp. 195–207. Springer, Heidelberg (2012)
11. Gabow, H.N.: Data structures for weighted matching and nearest common ancestors with linking. In: Proc. 1st ACM-SIAM SODA, pp. 434–443 (1990)
12. Georgiadis, L.: Testing 2-Vertex Connectivity and Computing Pairs of Vertex-Disjoint s-t Paths in Digraphs. In: Abramsky, S., Gavoille, C., Kirchner, C., Meyer auf der Heide, F., Spirakis, P.G. (eds.) ICALP 2010, Part I. LNCS, vol. 6198, pp. 738–749. Springer, Heidelberg (2010)
13. Georgiadis, L.: Approximating the Smallest 2-Vertex Connected Spanning Subgraph of a Directed Graph. In: Demetrescu, C., Halldórsson, M.M. (eds.) ESA 2011. LNCS, vol. 6942, pp. 13–24. Springer, Heidelberg (2011)
14. Georgiadis, L., Tarjan, R.E.: Finding dominators revisited. In: Proc. 15th ACM-SIAM SODA, pp. 862–871 (2004)
15. Georgiadis, L., Tarjan, R.E.: Dominator tree verification and vertex-disjoint paths. In: Proc. 16th ACM-SIAM SODA, pp. 433–442 (2005)
16. Georgiadis, L., Tarjan, R.E.: Dominators, Directed Bipolar Orders, and Independent Spanning Trees. In: Czumaj, A., Mehlhorn, K., Pitts, A., Wattenhofer, R. (eds.) ICALP 2012, Part I. LNCS, vol. 7391, pp. 375–386. Springer, Heidelberg (2012)
17. Georgiadis, L., Tarjan, R.E., Werneck, R.F.: Finding dominators in practice. Journal of Graph Algorithms and Applications (JGAA) 10(1), 69–94 (2006)
18. Italiano, G.F., Laura, L., Santaroni, F.: Finding strong bridges and strong articulation points in linear time. Theoretical Computer Science (to appear, 2012)
19. Lengauer, T., Tarjan, R.E.: A fast algorithm for finding dominators in a flowgraph. ACM Trans. on Programming Languages and Systems 1(1), 121–141 (1979)
20. Maxwell, E.K., Back, G., Ramakrishnan, N.: Diagnosing memory leaks using graph mining on heap dumps. In: Proc. 16th ACM SIGKDD Int. Conf. on Knowledge Discovery and Data Mining, KDD 2010, pp. 115–124 (2010)
21. Mitchell, N.: The Runtime Structure of Object Ownership. In: Hu, Q. (ed.) ECOOP 2006. LNCS, vol. 4067, pp. 74–98. Springer, Heidelberg (2006)
22. Patakakis, K., Georgiadis, L., Tatsis, V.A.: Dynamic dominators in practice. In: Proc. 16th Panhellenic Conference on Informatics, pp. 100–104 (2011)
23. Purdom Jr., P.W., Moore, E.F.: Algorithm 430: Immediate predominators in a directed graph. Communications of the ACM 15(8), 777–778 (1972)
24. Ramalingam, G., Reps, T.: An incremental algorithm for maintaining the dominator tree of a reducible flowgraph. In: Proc. 21st ACM POPL, pp. 287–296 (1994)
25. Sreedhar, V.C.: Efficient program analysis using DJ graphs. PhD thesis, School of Computer Science, McGill University (September 1995)
26. Sreedhar, V.C., Gao, G.R., Lee, Y.: Incremental computation of dominator trees. ACM Trans. Program. Lang. Syst. 19, 239–252 (1997)
27. Stanford network analysis platform (SNAP), http://snap.stanford.edu/
28. Tarjan, R.E.: Finding dominators in directed graphs. SIAM Journal on Computing 3(1), 62–89 (1974)
29. Tarjan, R.E.: Efficiency of a good but not linear set union algorithm. Journal of the ACM 22(2), 215–225 (1975)

Optimizing over the Growing Spectrahedron

Joachim Giesen[1], Martin Jaggi[2], and Sören Laue[1]

[1] Friedrich-Schiller-Universität Jena, Germany
[2] CMAP, École Polytechnique, Palaiseau, France

Abstract. We devise a framework for computing an approximate solution path for an important class of parameterized semidefinite problems that is guaranteed to be ε-close to the exact solution path. The problem of computing the entire regularization path for matrix factorization problems such as maximum-margin matrix factorization fits into this framework, as well as many other nuclear norm regularized convex optimization problems from machine learning. We show that the combinatorial complexity of the approximate path is independent of the size of the matrix. Furthermore, the whole solution path can be computed in *near linear* time in the size of the input matrix.

The framework employs an approximative semidefinite program solver for a fixed parameter value. Here we use an algorithm that has recently been introduced by Hazan. We present a refined analysis of Hazan's algorithm that results in improved running time bounds for a single solution as well as for the whole solution path as a function of the approximation guarantee.

1 Introduction

We provide an algorithm for tracking an approximate solution of a parameterized semidefinite program (SDP) along the parameter path. The algorithm is very simple and comes with approximation quality- and running time guaranties. It computes at some parameter value a slightly better approximate solution than required, and keeps this solution along the path as long as the required approximation quality can be guaranteed. Only when the approximation quality is no longer sufficient, a new solution of the SDP is computed. Hence, the complexity of the algorithm is determined by the time to compute a single approximate SDP solution and the number of solution updates along the path. We show that, if an approximation guarantee of $\varepsilon > 0$ is required, then the number of updates is in $O(1/\varepsilon)$, independent of the size of the problem.

Any SDP solver can be used within the framework to compute an approximate solution of the SDP at fixed parameter values. Here we use Hazan's algorithm [8], which is a Frank-Wolfe type algorithm (also known as conditional gradient descent) applied to SDPs. Hazan's algorithm scales well to large inputs, provides low-rank approximate solutions with guarantees, and only needs a simple approximate eigenvector computation in each of its iterations. A refined analysis of this algorithm, that we present here, shows that its running time can be improved to $\tilde{O}(N/\varepsilon^{1.5})$, where N is the number of non-zeros in the input problem. In the $\tilde{O}$ notation we are ignoring polylogarithmic factors in N.

L. Epstein and P. Ferragina (Eds.): ESA 2012, LNCS 7501, pp. 503–514, 2012.

Motivation. Our work is motivated by the problem of completing a matrix $Y \in \mathbb{R}^{m \times n}$ from which only a small fraction of the entries is known. This problem has been approached in a number of ways in recent years, see for example [2,16,10]. Here we will consider the maximum-margin matrix factorization approach that computes a completion X of Y as a solution of the following optimization problem:

$$\min_{X \in \mathbb{R}^{m \times n}} \sum_{(i,j) \in \Omega} (X_{ij} - Y_{ij})^2 + \lambda \cdot \|X\|_* \tag{1}$$

where Ω is the set of indices at which Y has been observed, and $\lambda > 0$ is the regularization parameter. The solution X whose nuclear norm (also known as trace norm) is $\|X\|_*$ does not need to coincide with Y on the index set Ω. It can be seen as a low rank approximation of Y, where the regularization parameter λ controls the rank of X.

We consider the following equivalent constraint formulation of Problem (1),

$$\begin{aligned} \min_X \ & \sum_{(i,j) \in \Omega} (X_{ij} - Y_{ij})^2 \\ \text{s.t.} \quad & \|X\|_* \leq t \\ & X \in \mathbb{R}^{m \times n}, \end{aligned} \tag{2}$$

such that there is a one-to-one mapping between the solutions at the parameters t and λ. In fact, Problem (1) is the Lagrangian of Problem (2). The latter is a convex optimization problem that can be solved for fixed values of t using standard convex optimization techniques. By the properties of the nuclear norm, it is known that Problem (2) can be equivalently re-formulated as

$$\begin{aligned} \min_X \ & f(X) \\ \text{s.t.} \quad & X \succeq 0 \\ & \operatorname{Tr}(X) \leq t \\ & X \in \mathbb{R}^{(m+n) \times (m+n)}, \end{aligned} \tag{3}$$

where f has to be chosen properly, and $\operatorname{Tr}(X)$ denotes the trace of X. We call the set defined by the constraints of Problem (3) a spectrahedron that is growing with t.

It remains to choose a good value for t. We use the simple algorithmic framework that we have described above to track an ε-approximate solution along the whole parameter range. It is interesting to note that in doing so, we follow the standard approach for choosing a good value for t, namely, computing an approximate solution for Problem (2) at a finite number of values for t, and then picking the best out of these finitely many solutions. However, in contrast to previous work, we automatically pick the values for t such that the whole parameter range is covered by an ε-approximate solution, for a specified accuracy ε. We show that the number t-values at which a solution needs to be computed such that the ε-approximation guarantee holds, is in $O(1/\varepsilon)$, independent of the size of the matrix Y. At the chosen t-values we use Hazan's algorithm [8] to compute an approximate solution for Problem (3).

Related Work. There have been algorithms proposed for computing the regularization path of matrix factorization problems. d'Aspremont et al. [4] consider the regularization path for sparse principal component analysis. However, here it is known at which parameters one should compute a solution, i.e., all integral values from 1 to n, where n is the number of variables.

Mazumder et al. [12] consider a very similar problem (regularized matrix factorization) and provide an algorithm for computing a solution at a fixed parameter t. They suggest to approximate the regularization path by computing solutions at various values for t. However, the parameter values at which the solutions are computed are chosen heuristically, e.g. on a uniform grid, and therefore no continuous approximation guarantee can be given.

Our solution path algorithm is motivated by the approach of [6] for parameterized convex optimization problems over the simplex, such as e.g., support vector machines. A direct but rather ad hoc extension of this approach to parameterized SDPs has appeared in [7].

Computing an approximate solution for a fixed parameter t simplifies to solving an SDP. The most widely-known implementations of SDP solvers are interior point methods, which provide high-accuracy solutions. However, with a running time that is a low-order polynomial in the number of variables, they do not scale to medium/large problems. Proximal methods have been proposed to overcome the disadvantages of interior point methods. These methods achieve better running times at the expense of less accurate solutions [13,1,14]. Alternating direction methods form yet another approach. On specific, well-structured SDP problems, they achieve very good speed in practice [17].

Notation. For arbitrary real matrices, the standard *inner product* is defined as $A \bullet B := Tr(A^T B)$, and the (squared) *Frobenius matrix norm* $\|A\|_{Fro}^2 := A \bullet A$ is the sum of all squared entries in the matrix. By $\mathbb{S}^{n \times n}$ we denote the set of *symmetric* $n \times n$ matrices. $A \in \mathbb{S}^{n \times n}$ is called *positive semidefinite* (PSD), written as $A \succeq 0$, iff $v^T A v \geq 0 \; \forall v \in \mathbb{R}^n$. Note that $v^T A v = A \bullet v v^T$. $\lambda_{\max}(A) \in \mathbb{R}$ denotes the largest eigenvalue of A. $\|A\|_*$ is the *nuclear norm* of the matrix A, also known as the *trace norm* (sum of the singular values).

2   Convex Optimization over the Spectrahedron

We consider convex optimization problems of the form

$$\min_{X \in \mathcal{S}_t} \; f(X) \tag{4}$$

where $f : \mathbb{S}^{n \times n} \to \mathbb{R}$ is symmetric, convex, and continuously differentiable such that $-\nabla f(X)$ is not negative definite for all $X \in \mathcal{S}_t$, and the domain $\mathcal{S}_t := \{X \in \mathbb{S}^{n \times n} \mid X \succeq 0, \; Tr(X) \leq t\}$ is the set of symmetric PSD matrices whose trace is at most t. This set generalizes the set of all symmetric PSD matrices of trace 1, i.e., the convex hull of all rank-1 matrices of unit trace, which is also known as the *spectrahedron*. The spectrahedron can be seen as a generalization of the unit simplex to the space of symmetric matrices.

By the convexity of f, we have the following linearization, for any $X, X_0 \in \mathcal{S}_t$:

$$f(X) \geq \nabla f(X_0) \bullet (X - X_0) + f(X_0).$$

This allows us to define a lower bound function on (4) for any fixed matrix $X_0 \in \mathcal{S}_t$ as follows,

$$\omega_t(X_0) := \min_{X \in \mathcal{S}_t} \nabla f(X_0) \bullet (X - X_0) + f(X_0) = f(X_0) - \max_{X \in \mathcal{S}_t} -\nabla f(X_0) \bullet (X - X_0).$$

The function $\omega_t(X_0)$ will be called here the Wolfe dual function[1]. Hence we can define the duality gap as

$$g_t(X_0) := f(X_0) - \omega_t(X_0) = \max_{X \in \mathcal{S}_t} -\nabla f(X_0) \bullet (X - X_0) \geq 0,$$

where non-negativity holds because of

$$f(X_0) \geq \min_{X \in \mathcal{S}_t} f(X) \geq \min_{X \in \mathcal{S}_t} \nabla f(X_0) \bullet (X - X_0) + f(X_0) = \omega_t(X_0).$$

By the definition of the objective function f, the gradient $\nabla f(X_0)$ is always a symmetric matrix (not necessarily PSD), and therefore has real eigenvalues, which is important in the following. Furthermore, this allows to equip any iterative SDP solver with guarantees for the duality gap by running until the gap is smaller than a prescribed bound. The latter can be easily checked using the following lemma.

Lemma 1. *The Wolfe dual of Problem (4) can be written as*

$$\omega_t(X_0) = f(X_0) - t \cdot \lambda_{\max} (-\nabla f(X_0)) - \nabla f(X_0) \bullet X_0,$$

and the duality gap can be written as

$$g_t(X_0) = t \cdot \lambda_{\max} (-\nabla f(X_0)) + \nabla f(X_0) \bullet X_0.$$

Proof. It is well-known that any matrix $X \in \mathbb{S}^{n \times n}$ can be written as $X = \sum_{i=1}^n \lambda_i u_u u_i^T$ with the orthonormal system of eigenvectors u_i, $\|u_i\| = 1$ of X and corresponding eigenvalues λ_i. Moreover $\text{Tr}(X) = \sum_{i=1}^n \lambda_i$. So we have

$$\max_{X \in \mathcal{S}_t} G \bullet X = \max \sum_{i=1}^n \alpha_i (G \bullet u_i u_i^T) = \max \sum_{i=1}^n \alpha_i u_i^T G u_i$$

$$= t \cdot \max_{v \in \mathbb{R}^n, \|v\|=1} v^T G v \quad [\text{since } G \text{ is not negative definite}]$$

$$= t \cdot \lambda_{\max} (G),$$

where the last equality is the variational characterization of the largest eigenvalue. The equality above can be interpreted as, any linear function must attain its maximum at a "vertex" (extreme point) of $\mathcal{S}_t$. Finally, both claims follow by plugging in $-\nabla f(X_0)$ for G. □

---

[1] Strictly speaking, this is not a proper "dual" function. However, we follow here the terminology of Clarkson [3].

Remark 1. If we alternatively define $\mathcal{S}_t := \{X \in \mathbb{S}^{n \times n} \mid X \succeq 0, \ \mathrm{Tr}(X) = t\}$, i.e., replacing "$\leq t$" by "$= t$", then Lemma 1 also holds even if $-\nabla f(X)$ can become negative definite.

From the definition of the duality gap we derive the definition of an approximate solution.

Definition 1. *A matrix $X \in \mathcal{S}_t$ is an ε-approximation to Problem (4) if $g_t(X)$ $\leq \varepsilon$.*

Obviously, the duality gap is always an upper bound for the *primal error* $h_t(X)$ of the approximate solution X, where $h_t(X) := f(X) - f(X^*)$ and X^* is an optimal solution of Problem (4) at parameter value t.

3   Hazan's Algorithm Revisited

We adapt the algorithm by Hazan [8] for approximating SDPs over the spectrahedron to our needs, i.e., approximating SDPs over $\mathcal{S}_t$. The proofs that we provide in the following fix some minor errors in the original paper. We also tighten the analysis and get improved bounds on the running time, $\tilde{O}(N/\varepsilon^{1.5})$ instead of $\tilde{O}(N/\varepsilon^2)$. In particular, we show that the eigenvector computation that is employed by the algorithm only needs to be computed up to an accuracy of ε instead of ε^2.

We start with the simple observation that every SDP in the form of Problem (4) be converted into an equivalent SDP where the inequality constraint $\mathrm{Tr}(X) \leq t$ is replaced by the equality constraint $\mathrm{Tr}(X) = t$, i.e., the optimization problem $\min_{X \in \mathcal{S}_t} f(X)$ is equivalent to the optimization problem

$$
\begin{aligned}
\min_{\hat{X}} \ & \hat{f}(\hat{X}) \\
\text{s.t.} \quad & \hat{X} \succeq 0 \\
& \mathrm{Tr}(\hat{X}) = t \\
& \hat{X} \in \mathbb{S}^{(n+1) \times (n+1)},
\end{aligned}
\tag{5}
$$

where $\hat{f}$ is the same function as f but defined on the larger set of symmetric matrices,

$$
\hat{f}(\hat{X}) = \hat{f}\left(\begin{pmatrix} X & X_2 \\ X_2^T & X_3 \end{pmatrix} \right) := f(X)
$$

for $X \in \mathbb{S}^{n \times n}$, $X_2 \in \mathbb{R}^{n \times 1}$, $X_3 \in \mathbb{R}^{1 \times 1}$. Every feasible solution X for Problem (4) can be converted into a feasible solution $\hat{X}$ for Problem (5) with the same objective function value and vice versa.

Hazan's algorithm for computing an ε-approximate solution for Problem (5) is summarized in pseudo-code in Algorithm 1.

The function ApproxEV used in Algorithm 1 returns an approximate eigenvector to the largest eigenvalue: given a square matrix M it returns a vector

---

Algorithm 1. ApproxSDP

Input: Convex function f, trace t, k
Output: Approximate solution of Problem 5.
Initialize $X_0 = t \cdot v_0 v_0^T$ for some arbitrary rank-1 PSD matrix $t \cdot v_0 v_0^T$ with trace t and $\|v_0\| = 1$.
for $i = 0$ **to** k **do**
$\quad$ Let $\alpha_i = \frac{2}{i+2}$.
$\quad$ Let $\varepsilon_i = \frac{\alpha_i \cdot C_f}{t}$.
$\quad$ Compute $v_i = \text{ApproxEV}(-\nabla f(X_i), \varepsilon_i)$.
$\quad$ Set $X_{i+1} = X_i + \alpha_i(t \cdot v_i v_i^T - X_i)$.
end for

$v = \text{ApproxEV}(M, \varepsilon)$ of unit length that satisfies $v^T M v \geq \lambda_{\max}(M) - \varepsilon$. The *curvature constant* C_f used in the algorithm is defined as

$$C_f := \sup_{\substack{X, Z \in \mathcal{S}_t, \\ \alpha \in [0,1], \\ Y = X + \alpha(Z-X)}} \frac{1}{\alpha^2} \left(f(Y) - f(X) - (Y - X) \bullet \nabla f(X) \right).$$

The curvature constant is a measure of how much the function $f(X)$ deviates from a linear approximation in X. The boundedness of this curvature is a widely used in convex optimization, and can be upper bounded by the Lipschitz constant of the gradient of f and the diameter of $\mathcal{S}_t$. Now we can prove the following theorem.

Theorem 1. *For each $k \geq 1$, the iterate X_k of Algorithm 1 satisfies $f(X_k) - f(X^*) \leq \varepsilon$, where $f(X^*)$ is the optimal value for the minimization Problem (5), and $\varepsilon = \frac{8C_f}{k+2}$.*

Proof. For each iteration of the algorithm, we have that

$$f(X_{i+1}) = f(X_i + \alpha_i(t \cdot v_i v_i^T - X_i))$$
$$\leq f(X_i) + \alpha_i(t \cdot v_i v_i^T - X_i) \bullet \nabla f(X_i) + \alpha_i^2 C_f, \tag{6}$$

where the last inequality follows from the definition of the curvature C_f. Furthermore,

$$(t \cdot v_i v_i^T - X_i) \bullet \nabla f(X_i) = (X_i - t \cdot v_i v_i^T) \bullet (-\nabla f(X_i))$$
$$= X_i \bullet (-\nabla f(X_i)) - t \cdot v_i^T(-\nabla f(X_i))v_i$$
$$\leq -X_i \bullet \nabla f(X_i) - t \cdot (\lambda_{\max}(-\nabla f(X_i)) - \varepsilon_i)$$
$$= -g_t(X_i) + t \cdot \varepsilon_i$$
$$= -g_t(X_i) + \alpha_i \cdot C_f.$$

The last two equalities follow from the remark after Lemma 1 and from setting $\varepsilon_i = \frac{\alpha_i \cdot C_f}{t}$ within Algorithm 1. Hence, Inequality (6) evaluates to

$$f(X_{i+1}) \leq f(X_i) - \alpha_i g_t(X_i) + \alpha_i^2 C_f + \alpha_i^2 C_f$$
$$= f(X_i) - \alpha_i g_t(X_i) + 2\alpha_i^2 C_f. \tag{7}$$

Subtracting $f(X^*)$ on both sides of Inequality (7), and denoting the current primal error by $h(X_i) = f(X_i) - f(X^*)$, we get

$$h(X_{i+1}) \leq h(X_i) - \alpha_i g_t(X_i) + 2\alpha_i^2 C_f \,, \tag{8}$$

which by using the fact that $g_t(X_i) \geq h(X_i)$ gives

$$h(X_{i+1}) \leq h(X_i) - \alpha_i h(X_i) + 2\alpha_i^2 C_f \,. \tag{9}$$

The claim of this theorem is that the primal error $h(X_i) = f(X_i) - f(X^*)$ is small after a sufficiently large number of iterations. Indeed, we will show by induction that $h(X_i) \leq \frac{8C_f}{i+2}$. In the first iteration ($i = 0$), we know from (9) that the claim holds, because of the choice of $\alpha_0 = 1$.

Assume now that $h(X_i) \leq \frac{8C_f}{i+2}$ holds. Using $\alpha_i = \frac{2}{i+2}$ in our inequality (9), we can now bound $h(X_{i+1})$ as follows,

$$
\begin{aligned}
h(X_{i+1}) &\leq h(X_i)(1 - \alpha_i) + 2\alpha_i^2 C_f \\
&\leq \frac{8C_f}{i+2}\left(1 - \frac{2}{i+2}\right) + \frac{8C_f}{(i+2)^2} \\
&\leq \frac{8C_f}{i+2} - \frac{8C_f}{(i+2)^2} \\
&\leq \frac{8C_f}{i+1+2} \,,
\end{aligned}
$$

which proves the theorem.   □

Theorem 1 states that after k iterations of Algorithm 1, the approximate solution is within $\varepsilon = \frac{8C_f}{k+2}$ of the optimum. However, for our framework of computing approximate solution paths we need a stronger theorem, namely, we also need the duality gap to be small. Though this cannot be guaranteed after k iterations, it can be guaranteed after at most $2k + 1$ iterations as the next theorem states (whose proof follows along the lines of [3]). Note that the theorem does not guarantee that the solution after $2k + 1$ iterations has a small enough duality gap, but only that the gap is small enough after one of the iterations between k and $2k + 1$. In practice one can run the algorithm for $2k + 1$ iterations and keep the solution with the smallest duality gap after any iteration.

Theorem 2. *After at most $2k + 1$ iterations, Algorithm 1 has computed an approximate solution X_i whose duality gap $g_t(X_i)$ is at most $\frac{8C_f}{k+2}$.*

Proof. Theorem 1 shows that k iterations suffice for Algorithm 1 to provide an approximate solution X_k with $h(X_k) \leq \frac{8C_f}{k+2}$. For the subsequent $k+1$ iterations we change Algorithm 1 slightly by fixing the step-length α_i to $\frac{2}{k+2}$ for all $i : k \leq i \leq 2k + 1$, i.e., we do not decrease the step-size anymore.

For a contradiction assume that $g_t(X_i) > \frac{8C_f}{k+2}$ for all $i : k \leq i \leq 2k + 1$. As we have seen in the proof of Theorem 1 (Equation (8)), the bound $h(X_{i+1}) \leq$

$h(X_i) - \alpha_i g_t(X_i) + 2\alpha_i^2 C_f$ in fact holds for any step size $\alpha_i \in [0,1]$. Using our assumption, we get

$$
\begin{aligned}
h(X_{i+1}) &\leq h(X_i) - \alpha_i g_t(X_i) + 2\alpha_i^2 C_f \\
&< h(X_i) - \frac{2}{k+2} \cdot \frac{8C_f}{k+2} + 2C_f \frac{2^2}{(k+2)^2} \\
&\leq h(X_i) - \frac{16C_f}{(k+2)^2} + \frac{8C_f}{(k+2)^2} \\
&= h(X_i) - \frac{8C_f}{(k+2)^2} .
\end{aligned}
$$

Since $h(X_k) \leq \frac{8C_f}{k+2}$ and $h(X_i) \geq 0$ for all $i \geq 0$, we must have that $g_t(X_i) \leq \frac{8C_f}{k+2}$ holds after at most $k+1$ such additional iterations, since otherwise the inequality from above for $h(X_{i+1})$ implies $h(X_{2k+1}) < 0$, which is a contradiction.   □

Note that matrix X_i is stored twice in Algorithm 1, once as a low rank factorization and once as a sparse matrix. The low rank factorization of X_i is obtained in the algorithm via $X_i = \sum_i \beta_i v_i v_i^T$, where v_i is the eigenvector chosen in iteration i, and β_i are the appropriate factors computed via all α_i. Collecting all $O(\frac{1}{\varepsilon})$ eigenvectors amounts to $O(\frac{n}{\varepsilon})$ time and space in total. The sparse representation of X_i stores only the N entries of X_i that are necessary for computing $f(X_i)$ and $\nabla f(X_i)$. Depending on the application, we often have $N \ll n^2$, as for instance in the case of the matrix completion problem (cf. Section 5). The sparse representation of X_i is computed in a straightforward way from the sparse representation of X_{i-1} and the current sparse representation of $v_i v_i^T$, where only the N necessary entries of $v_i v_i^T$ are computed. Hence, computing the sparse representation for all X_i amounts to $O(N)$ operations per iteration and $O(\frac{N}{\varepsilon})$ operations in total.

Furthermore, it is known that the function ApproxEV(M, ε) that computes an approximate eigenvector with guarantee ε can be implemented using the Lanczos method that runs in time $\tilde{O}\left(\frac{N\sqrt{L}}{\sqrt{\varepsilon}}\right)$ and returns a valid approximation with high probability, where N is the number of non-zero entries in the matrix $M \in \mathbb{R}^{n \times n}$ and L is a bound on the largest eigenvalue of M, see [11]. Hence, we can conclude with the following corollary.

Corollary 1. *Algorithm 1 computes an approximate solution X for Problems (5) whose duality gap $g_t(X)$ is at most ε, with high probability, and its running time is in $\tilde{O}\left(\frac{N}{\varepsilon^{1.5}}\right)$.*

4   Optimizing over the Growing Spectrahedron

We are interested in ε-approximations for Problem (4) for all parameter values of $t \in \mathbb{R}, t \geq 0$.

Definition 2. *The ε-approximation path complexity of Problem (4) is the minimum number of sub-intervals over all possible partitions of the parameter range*

$t \in \mathbb{R}, t \geq 0$, *such that for each individual sub-interval there is a single solution of Problem (4) which is an ε-approximation for that entire sub-interval.*

The following simple stability lemma is at the core of our discussion and characterizes all parameter values t' such that a given $\frac{\varepsilon}{\gamma}$-approximate solution X (for $\gamma > 1$) at t is at least an ε-approximate solution at t'.

Lemma 2. *Let $X \in \mathcal{S}_t$ be an $\frac{\varepsilon}{\gamma}$-approximate solution of Problem (4) for some fixed parameter value t, and for some $\gamma > 1$. Then for all $t' \geq t \in \mathbb{R}$ that satisfy*

$$(t' - t) \cdot \lambda_{\max}\left(-\nabla f(X)\right) \leq \varepsilon \left(1 - \tfrac{1}{\gamma}\right), \tag{10}$$

the solution X is still an ε-approximation to Problem (4) at the parameter value t'.

Proof. Any feasible solution X for the problem at parameter t is also a feasible solution at parameter t' since $\mathcal{S}_t \subseteq \mathcal{S}_{t'}$. We have to show that

$$g_{t'}(X) = t' \cdot \lambda_{\max}\left(-\nabla f(X)\right) + \nabla f(X) \bullet X \leq \varepsilon.$$

To do so, we add to Inequality (10) the inequality stating that X is an $\frac{\varepsilon}{\gamma}$-approximate solution at value t, i.e.,

$$t \cdot \lambda_{\max}\left(-\nabla f(X)\right) + \nabla f(X) \bullet X \leq \frac{\varepsilon}{\gamma},$$

and obtain

$$t' \cdot \lambda_{\max}\left(-\nabla f(X)\right) + \nabla f(X) \bullet X \leq \varepsilon,$$

which is the claimed bound on the duality gap at the parameter value t'.   □

We assume that the unconstrained minimization problem $\min_{X \in \mathbb{S}^{n \times n}} f(X)$ has a bounded solution X^*. This assumption holds in general for all practical problems and the applications that we will consider later.

Theorem 3. *The path complexity of Problem (4) over the parameter range $t \in \mathbb{R}, t \geq 0$ is in $O\left(\frac{1}{\varepsilon}\right)$.*

Proof. From Lemma 2, we obtain intervals of constant ε-approximate solutions, where each interval is of length as least $\frac{\varepsilon\left(1-\frac{1}{\gamma}\right)}{\lambda_{\max}(-\nabla f(X))}$. Let $t_{\max}$ be the trace of the optimal solution X^* to the unconstrained minimization problem $\min_{X \in \mathbb{S}^{n \times n}} f(X)$. Then X^* has a zero duality gap for Problem (4) for all parameters $t \geq t_{\max}$. On the other hand, for $t < t_{\max}$, we have $\mathcal{S}_t \subseteq \mathcal{S}_{t_{\max}}$ and can therefore bound the number of intervals by $\lceil \frac{\gamma_f}{\varepsilon} \rceil$, where

$$\gamma_f := t_{\max} \cdot \max_{X \in \mathcal{S}_{t_{\max}}} \lambda_{\max}\left(-\nabla f(X)\right) \frac{\gamma}{\gamma - 1}$$

is an absolute constant depending only on the function f, like its Lipschitz constant.   □

Lemma 2 immediately suggests a simple algorithm (Algorithm 2) to compute an ε-approximate solution path for Problem (4). The running time of this algorithm is by Theorem 3 in $O(T(\varepsilon)/\varepsilon)$, where $T(\varepsilon)$ is the time to compute a single ε-approximate solution for Problem (4) at a fixed parameter value.

Algorithm 2. SDP-Path

Input: convex function $f, t_{min}, t_{max}, \varepsilon, \gamma$
Output: ε-approximate solution path for Problem (4).
Set $t = t_{min}$.
repeat
    Compute approximate solution $X = \text{ApproxSDP}(f, t, \varepsilon/\gamma)$.
    Update $t = t + \frac{\varepsilon(1-1/\gamma)}{\lambda_{max}(-\nabla f(X))}$
until $t > t_{max}$
---

5 Applications

5.1 Nuclear Norm Regularized Problems and Matrix Factorization

A nuclear norm regularized problem

$$\min_{X \in \mathbb{R}^{m \times n}} f(X) + \lambda \|X\|_* \tag{11}$$

for *any* loss function f in its constrained formulation

$$\min_{X \in \mathbb{R}^{m \times n}, \|X\|_* \leq t/2} f(X)$$

is by a straightforward transformation, see [5,9,16], equivalent to the optimization problem

$$\begin{aligned} \min_X \hat{f}(\hat{X}) \\ \text{s.t.} \quad \hat{X} \in \mathcal{S}_t \end{aligned} \tag{12}$$

over the scaled spectrahedron $\mathcal{S}_t \subseteq \mathbb{S}^{(m+n) \times (m+n)}$, which is our original problem (4). Here $\hat{f}$ is defined using f as

$$\hat{f}(\hat{X}) = \hat{f}\left(\begin{pmatrix} X_1 & X_2 \\ X_2^T & X_3 \end{pmatrix}\right) := f(X_2)$$

for $\hat{X} \in \mathbb{S}^{(m+n) \times (m+n)}$, $X_2 \in \mathbb{R}^{m \times n}$. Observe that this is now a convex problem whenever the loss function f is convex.

The gradient of the transformed function $\hat{f}$ is

$$\nabla \hat{f}(\hat{X}) = \begin{pmatrix} 0 & \nabla f(X) \\ (\nabla f(X))^T & 0 \end{pmatrix}, \tag{13}$$

which is not negative definite because if $(v, w)^T$ is an eigenvector of $\nabla \hat{f}(\hat{X})$ for the eigenvalue λ, then $(-v, w)^T$ is an eigenvector for the eigenvalue $-\lambda$. We can now use this gradient in Algorithm 2 to compute the entire regularization path for Problem (12).

A *matrix factorization* can be obtained directly from the low-rank representation of X_i of Algorithm 1.

5.2   Matrix Completion

The matrix completion Problem (1) from the introduction is the probably the most commonly used instance of Problem (11) with the standard squared loss function

$$f(X) = \frac{1}{2} \sum_{(i,j) \in \Omega} (X_{ij} - Y_{ij})^2. \tag{14}$$

The gradient of this loss function is

$$(\nabla f(X))_{ij} = \begin{cases} X_{ij} - Y_{ij} & : (i,j) \in \Omega, \\ 0 & : \text{otherwise.} \end{cases}$$

Using the notation $(A)_\Omega$ for the matrix that coincides with A on the indices Ω and is zero otherwise, $\nabla f(X)$ can be written as $\nabla f(X) = (X - Y)_\Omega$. This implies that the square gradient matrix $\nabla \hat{f}(\hat{X})$ that we use in our algorithm (see Equation (13)) is also of this simple form. As this matrix is sparse–it has only $N = |\Omega|$ non-zero entries, storage and approximate eigenvector computations can be performed much more efficiently than for dense problems. Also note that the curvature constant $C_{\hat{f}}$ that appears in the bound for the running time of Hazan's algorithm equals t^2 for the squared loss function from Equation (14), see [9]. Hence, the full regularization path can be computed in time $\tilde{O}\left(\frac{N}{\varepsilon^{2.5}}\right)$.

Finally, note that our framework applies to matrix completion problems with any convex differentiable loss function, such as the smoothed hinge loss or the standard squared loss, and includes the classical maximum-margin matrix factorization variants [16].

6   Conclusion

We have presented a simple and efficient framework that allows to approximate solution paths for parameterized semidefinite programs with guarantees. Many well known regularized matrix factorization and completion problems from machine learning fit into this framework. Even weighted nuclear norm regularized convex optimization problems, see e.g., [15], fit into the framework though we have not shown this here for lack of space. We also have not shown experimental results, and just want to state here that they support the theory , i.e., the running time is near linear in the size of the input matrix. Finally, we have also improved the running time of Hazan's algorithm by a refined analysis.

Acknowledgments. The authors would like to thank the reviewers for their useful comments. The research of J. Giesen and S. Laue has been supported by the Deutsche Forschungsgemeinschaft (DFG) under grant GI-711/3-2. The research of M. Jaggi has been supported by a Google Research Award and by the Swiss National Science Foundation (SNF grant 20PA21-121957).

References

1. Arora, S., Hazan, E., Kale, S.: Fast Algorithms for Approximate Semidefinite Programming using the Multiplicative Weights Update Method. In: Proceedings of the Annual IEEE Symposium on Foundations of Computer Science (FOCS), pp. 339–348 (2005)
2. Candès, E.J., Tao, T.: The Power of Convex Relaxation: Near-Optimal Matrix Completion. IEEE Transactions on Information Theory 56(5), 2053–2080 (2010)
3. Clarkson, K.L.: Coresets, sparse greedy approximation, and the Frank-Wolfe algorithm. ACM Transactions on Algorithms 6(4) (2010)
4. d'Aspremont, A., Bach, F.R., El Ghaoui, L.: Full Regularization Path for Sparse Principal Component Analysis. In: Proceedings of the International Conference on Machine Learning (ICML), pp. 177–184 (2007)
5. Fazel, M., Hindi, H., Boyd, S.P.: A Rank Minimization Heuristic with Application to Minimum Order System Approximation. In: Proceedings of the American Control Conference, vol. 6, pp. 4734–4739 (2001)
6. Giesen, J., Jaggi, M., Laue, S.: Approximating Parameterized Convex Optimization Problems. In: de Berg, M., Meyer, U. (eds.) ESA 2010, Part I. LNCS, vol. 6346, pp. 524–535. Springer, Heidelberg (2010)
7. Giesen, J., Jaggi, M., Laue, S.: Regularization Paths with Guarantees for Convex Semidefinite Optimization. In: Proceedings International Conference on Artificial Intelligence and Statistics (AISTATS) (2012)
8. Hazan, E.: Sparse Approximate Solutions to Semidefinite Programs. In: Laber, E.S., Bornstein, C., Nogueira, L.T., Faria, L. (eds.) LATIN 2008. LNCS, vol. 4957, pp. 306–316. Springer, Heidelberg (2008)
9. Jaggi, M., Sulovský, M.: A Simple Algorithm for Nuclear Norm Regularized Problems. In: Proceedings of the International Conference on Machine Learning (ICML), pp. 471–478 (2010)
10. Koren, Y., Bell, R.M., Volinsky, C.: Matrix Factorization Techniques for Recommender Systems. IEEE Computer 42(8), 30–37 (2009)
11. Kuczyński, J., Woźniakowski, H.: Estimating the Largest Eigenvalue by the Power and Lanczos Algorithms with a Random Start. SIAM Journal on Matrix Analysis and Applications 13(4), 1094–1122 (1992)
12. Mazumder, R., Hastie, T., Tibshirani, R.: Spectral Regularization Algorithms for Learning Large Incomplete Matrices. Journal of Machine Learning Research 11, 2287–2322 (2010)
13. Nemirovski, A.: Prox-method with Rate of Convergence O(1/T) for Variational Inequalities with Lipschitz Continuous Monotone Operators and Smooth Convex-concave Saddle Point Problems. SIAM Journal on Optimization 15, 229–251 (2004)
14. Nesterov, Y.: Smoothing Technique and its Applications in Semidefinite Optimization. Math. Program. 110(2), 245–259 (2007)
15. Salakhutdinov, R., Srebro, N.: Collaborative Filtering in a Non-Uniform World: Learning with the Weighted Trace Norm. In: Proceedings of Advances in Neural Information Processing Systems (NIPS), vol. 23 (2010)
16. Srebro, N., Rennie, J.D.M., Jaakkola, T.: Maximum-Margin Matrix Factorization. In: Proceedings of Advances in Neural Information Processing Systems (NIPS), vol. 17 (2004)
17. Wen, Z., Goldfarb, D., Yin, W.: Alternating Direction Augmented Lagrangian Methods for Semidefinite Programming. Technical report (2009)

Induced Disjoint Paths in Claw-Free Graphs[*]

Petr A. Golovach[1], Daniël Paulusma[1], and Erik Jan van Leeuwen[2]

[1] School of Engineering and Computer Science, Durham University, England
{petr.golovach,daniel.paulusma}@durham.ac.uk
[2] Dept. Computer, Control, Managm. Eng., Sapienza University of Rome, Italy
e.j.van.leeuwen@dis.uniroma1.it

Abstract. Paths $P_1, \ldots, P_k$ in a graph $G = (V, E)$ are said to be mutually induced if for any $1 \leq i < j \leq k$, P_i and P_j have neither common vertices nor adjacent vertices (except perhaps their end-vertices). The INDUCED DISJOINT PATHS problem is to test whether a graph G with k pairs of specified vertices (s_i, t_i) contains k mutually induced paths P_i such that P_i connects s_i and t_i for $i = 1, \ldots, k$. This problem is known to be NP-complete already for $k = 2$, but for n-vertex claw-free graphs, Fiala et al. gave an $n^{O(k)}$-time algorithm. We improve the latter result by showing that the problem is fixed-parameter tractable for claw-free graphs when parameterized by k. Several related problems, such as the k-IN-A-PATH problem, are shown to be fixed-parameter tractable for claw-free graphs as well. We prove that an improvement of these results in certain directions is unlikely, for example by noting that the INDUCED DISJOINT PATHS problem cannot have a polynomial kernel for line graphs (a type of claw-free graphs), unless NP $\subseteq$ coNP/poly. Moreover, the problem becomes NP-complete, even when $k = 2$, for the more general class of $K_{1,4}$-free graphs. Finally, we show that the $n^{O(k)}$-time algorithm of Fiala et al. for testing whether a claw-free graph contains some k-vertex graph H as a topological induced minor is essentially optimal by proving that this problem is W[1]-hard even if G and H are line graphs.

1 Introduction

The problem of finding disjoint paths of a certain type in a graph has received considerable attention in recent years. The regular DISJOINT PATHS problem is to test whether a graph G with k pairs of specified vertices (s_i, t_i) contains a set of k mutually vertex-disjoint paths $P_1, \ldots, P_k$. This problem is included in Karp's list of NP-compete problems [20], provided that k is part of the input. If k is a *fixed* integer, i.e. not part of the input, the problem is called k-DISJOINT PATHS and can be solved in $O(n^3)$ time for n-vertex graphs, as shown by Robertson and Seymour [30] in one of their keystone papers on graph minor theory.

In this paper, we study a generalization of the DISJOINT PATHS problem by considering its *induced* version. We say that paths $P_1, \ldots, P_k$ in a graph

[*] This work is supported by EPSRC (EP/G043434/1), Royal Society (JP100692), and ERC StG project PAAl no. 259515. A full version of this paper can be found on arXiv: http://arxiv.org/abs/1202.4419

$G = (V, E)$ are *mutually induced* if for any $1 \leq i < j \leq k$, P_i and P_j have neither common vertices, i.e. $V(P_i) \cap V(P_j) = \emptyset$, nor adjacent vertices, i.e. $uv \notin E$ for any $u \in V(P_i), v \in V(P_j)$, except perhaps their end-vertices. We observe that the paths $P_1, \ldots, P_k$ are not required to be induced paths in G. However, this may be assumed without loss of generality, because we can replace non-induced paths by shortcuts. We can now define the following problem, where we call the vertex pairs specified in the input *terminal pairs* and their vertices *terminals*.

INDUCED DISJOINT PATHS
Instance: a graph G with k terminal pairs (s_i, t_i) for $i = 1, \ldots, k$.
Question: does G contain k mutually induced paths P_i such that P_i connects
terminals s_i and t_i for $i = 1, \ldots, k$?

When k is fixed, we call this the k-INDUCED DISJOINT PATHS problem. Observe that the INDUCED DISJOINT PATHS problem can indeed be seen as a generalization of the DISJOINT PATHS problem, since the latter can be reduced to the former by subdividing every edge of the graph. This generalization makes the problem significantly harder. In contrast to the original, non-induced version, the k-INDUCED DISJOINT PATHS problem is NP-complete even for $k = 2$ [9,1]. The hardness of the k-INDUCED DISJOINT PATHS problem motivates an investigation into graph classes for which it may still be tractable. For planar graphs, INDUCED DISJOINT PATHS stays NP-complete, as it generalizes DISJOINT PATHS for planar graphs, which is NP-complete as shown by Lynch [26]. However, Kobayashi and Kawarabayashi [23] presented an algorithm for k-INDUCED DISJOINT PATHS on planar graphs that runs in linear time for any fixed k. For AT-free graphs, IN-DUCED DISJOINT PATHS is polynomial-time solvable [14]. For claw-free graphs (graphs where no vertex has three pairwise nonadjacent neighbors), Fiala et al. [11] recently showed that the INDUCED DISJOINT PATHS problem is NP-complete. This holds even for line graphs, a subclass of the class of claw-free graphs. They also gave a polynomial-time algorithm for k-INDUCED DISJOINT PATHS for any fixed k. Their approach is based on a modification of the claw-free input graph to a special type of claw-free graph, namely to a quasi-line graph, in order to use the characterization of quasi-line graphs by Chudnovsky and Seymour [4]. This transformation may require $\Omega(n^{2k})$ time due to some brute-force guessing, in particular as claw-freeness must be preserved.

In this paper, we improve on the result of Fiala et al. [11] using the theory of parameterized complexity (cf. [29]). In this theory, we consider the problem input as a pair (I, k), where I is the main part and k the parameter. A problem is *fixed-parameter tractable* if an instance (I, k) can be solved in time $f(k) |I|^c$, where f denotes a computable function and c a constant independent of k. We consider INDUCED DISJOINT PATHS when the parameter is the number of terminals and show that it is fixed-parameter tractable for claw-free graphs.

A study on induced disjoint paths can also be justified from another direction, one that focuses on detecting induced subgraphs such as cycles, paths, and trees that contain some set of k specified vertices, which are also called terminals. The corresponding decision problems are called k-IN-A-CYCLE, k-IN-A-PATH, and k-IN-A-TREE, respectively. For general graphs, even 2-IN-A-CYCLE and 3-IN-A-PATH are NP-complete [9,1], whereas the k-IN-A-TREE problem is polynomial-time solvable for $k = 3$ [5], open for any fixed $k \geq 4$, and NP-complete when k is part of the input [7]. Several polynomial-time solvable cases are known for graph classes, see e.g. [8,22,25]. For claw-free graphs, the problems k-IN-A-TREE and k-IN-A-PATH are equivalent and polynomial-time solvable for any fixed integer k [11]. Consequently, the same holds for the k-IN-A-CYCLE problem. In this paper, we improve on these results by showing that all three problems are fixed-parameter tractable for claw-free graphs when parameterized by k.

As a final motivation for our work, we note that just as disjoint paths are important for (topological) graph minors, one may hope that induced disjoint paths are useful for finding induced (topological) minors in polynomial time on certain graph classes. Whereas the problems of detecting whether a graph contains some fixed graph H as a minor or topological minor can be solved in cubic time for any fixed graph H [30,15], complexity classifications of both problems with respect to some fixed graph H as induced minor or induced topological minor are still wide open. So far, only partial results that consist of both polynomial-time solvable and NP-complete cases are known [10,13,24].

Our Results. In Section 3, we prove that INDUCED DISJOINT PATHS is fixed-parameter tractable on claw-free graphs when parameterized by k. Our approach circumvents the time-consuming transformation to quasi-line graphs of Fiala et al. [11], and is based on an algorithmic application of the characterization for claw-free graphs by Chudnovsky and Seymour. Hermelin et al. [18] recently applied such an algorithmic structure theorem to DOMINATING SET on claw-free graphs. However, their algorithm reduces the strip-structure to have size polynomial in k and then follows an exhaustive enumeration strategy. For k-INDUCED DISJOINT PATHS, such an approach seems unsuitable, and our arguments thus differ substantially from those in [18]. We also show that the problems k-IN-A-PATH (or equivalently k-IN-A-TREE) and k-IN-A-CYCLE are fixed-parameter tractable when parameterized by k. This gives some answer to an open question of Bruhn and Saito [3]. They gave necessary and sufficient conditions for the existence of a path through three given vertices in a claw-free graph and asked whether such conditions also exist for k-IN-A-PATH with $k \geq 4$. But as this problem is NP-complete even for line graphs when k is part of the input [11], showing that it is fixed-parameter tractable may be the best answer we can hope for.

Fiala et al. [11] use their algorithm for the k-INDUCED DISJOINT PATHS problem to obtain an $n^{O(k)}$-time algorithm that solves the problem of testing whether a claw-free graph on n vertices contains a graph H on k vertices as a topological induced minor. However, we prove in Section 4 that this problem is W[1]-hard when parameterized by $|V(H)|$, even if G and H are line graphs. Finally, we show that our results for the INDUCED DISJOINT PATHS problem for claw-free graphs are best possible in the following ways. First, we show that the problem does not allow a polynomial kernel even for line graphs, unless NP $\subseteq$ coNP/poly. Second, we observe that a result from Derhy and Picouleau [7] immediately implies that 2-INDUCED DISJOINT-PATHS is NP-complete on $K_{1,4}$-free graphs (no vertex has 4 pairwise nonadjacent neighbors).

2   Preliminaries

We only consider finite undirected graphs that have no loops and no multiple edges.

Mutually Induced Paths. We consider a slight generalization of the standard definition. We say that paths $P_1, \ldots, P_k$ in a graph $G = (V, E)$ are *mutually induced* if (i) each P_i is an induced path in G; (ii) any distinct P_i, P_j may only share vertices that are ends of both paths; (iii) no inner vertex u of any P_i is adjacent to a vertex v of some P_j for $j \neq i$, except when v is an end-vertex of both P_i and P_j. This more general definition is exactly what we need later when detecting induced topological minors. In particular, this definition allows terminal pairs $(s_1, t_1), \ldots, (s_k, t_k)$ to have the following two properties: 1) for all $i < j$, a terminal of (s_i, t_i) may be adjacent to a terminal of (s_j, t_j); 2) for all $i < j$, it holds that $0 \leq |\{s_i, t_i\} \cap \{s_j, t_j\}| \leq 1$. Property 2 means that terminal pairs may overlap but not coincide, i.e. the set of terminal pairs is not a multiset.

For the remainder of the paper, we assume that the input of INDUCED DIS-JOINT PATHS consists of a graph G with a set of terminal pairs $(s_1, t_1), \ldots, (s_k, t_k)$ having Properties 1 and 2, and that the desired output is a set of paths $P_1, \ldots, P_k$ that are mutually induced, such that P_i has end-vertices s_i and t_i for $i = 1, \ldots, k$. We say that P_i is the $s_i t_i$-*path* and also call it a *solution path*. We say that the subgraph of G induced by the vertices of such paths forms a *solution for* G.

We also need the following terminology. Let $G = (V, E)$ be a graph with terminal pairs $(s_1, t_1), \ldots, (s_k, t_k)$. By Property 2, a vertex v can be a terminal in more than one terminal pair, e.g., $v = s_i$ and $v = s_j$ is possible for some $i \neq j$. For clarity reasons, we will view s_i and s_j as two different terminals *placed on* vertex v. We then say that a vertex $u \in V$ *represents* terminal s_i or t_i if $u = s_i$ or $u = t_i$, respectively. We call such a vertex a *terminal vertex*; the other vertices of G are called *non-terminal vertices*. We let T_u denote the set of terminals represented by u and observe that $|T_u| \geq 2$ is possible. We call two terminals that belong to the same terminal pair *partners*.

In our algorithm, we sometimes have to solve the INDUCED DISJOINT PATHS problem on a graph that contain no terminals as a subproblem. We consider such instances Yes-instances (that have an empty solution).

Graph Classes. The graph $K_{1,k}$ denotes the star with k rays. In particular, the graph $K_{1,3} = (\{a_1, a_2, a_3, b\}, \{a_1 b, a_2 b, a_3 b\})$ is called a *claw*. A graph is $K_{1,k}$-*free* if it has no induced subgraph isomorphic to $K_{1,k}$. If $k = 3$, then we usually call such a graph *claw-free*.

The *line graph* of a graph G with edges $e_1, \ldots, e_p$ is the graph $L(G)$ with vertices $u_1, \ldots, u_p$ such that there is an edge between any two vertices u_i and u_j if and only if e_i and e_j share one end vertex in H. Every line graph is claw-free. We call G the *preimage* of $L(G)$.

Structure of Claw-Free Graphs. Chudnovsky and Seymour have given a structural characterization for claw-free graphs, the proof of which can be found in a series of seven papers called *Claw-free graphs I* through *VII*. We refer to their survey [4] for a summary. Hermelin et al. [18] gave an algorithmic version of their result. This version plays an important role in the proof of our main result in Section 3. In order to state it we need the following additional terminology. We denote the maximum size of an independent set in a graph G by $\alpha(G)$.

Two adjacent vertices u and v in a graph G are called *(true) twins* if they share the same neighbors, i.e. $N[u] = N[v]$, where $N[u] = N(u) \cup \{u\}$. The equivalence classes of the twin relation are called *twin sets*. Two disjoint cliques A and B form a *proper W-join* in a graph G if $|A| \geq 2$, $|B| \geq 2$, every vertex $v \in V(G) \setminus (A \cup B)$ is either adjacent to all vertices of A or to no vertex of A, every vertex in A is adjacent to at least one vertex in B and non-adjacent to at least one vertex in B, and the above also holds with A and B reversed.

A graph is an *interval graph* if intervals of the real line can be associated with its vertices such that two vertices are adjacent if and only if their corresponding intervals intersect. An interval graph is *proper* if it has an interval representation in which no interval is properly contained in any other interval. Analogously, we can define the class of *circular-arc graphs* and *proper circular-arc graphs* by considering a set of intervals (arcs) on the circle instead of a real line.

A *hypergraph* is a pair $R = (V_R, E_R)$ where V_R is a set of elements called *vertices* and E_R is a collection of subsets of V_R called *hyperedges*. Two hyperedges e_1 and e_2 are *parallel* if they contain the same vertices of V_R. Graphs can be seen as hypergraphs in which all hyperedges have size two.

A *strip-structure* $(R, \{(J_e, Z_e) \mid e \in E(R)\})$ for a claw-free graph G is a hypergraph R, with possibly parallel and empty hyperedges, and a set of tuples (J_e, Z_e) for each $e \in E(R)$ called *strips* such that

- J_e is a claw-free graph and $Z_e \subseteq V(J_e)$,
- $\{V(J_e) \setminus Z_e \mid e \in E(R)\}$ is a partition of $V(G)$ and each $V(J_e) \setminus Z_e$ is nonempty,
- $J_e[V(J_e) \setminus Z_e]$ equals $G[V(J_e) \setminus Z_e]$,
- each $v \in e$ corresponds to a unique $z_v \in Z_e$ and vice versa.
- for each $v \in V(R)$, the set $C_v := \bigcup_{z_v \in Z_e : v \in e} N_{J_e}(z_v)$ induces a clique in G,
- each edge of G is either in $G[C_v]$ for some $v \in V(R)$ or in $J_e[V(J_e) \setminus Z_e]$ for some $e \in E(R)$.

When there is no confusion about e, we just talk about strips (J, Z). A strip (J, Z) is called a *stripe* if the vertices of Z are pairwise nonadjacent and any vertex in $V(J) \setminus Z$ is adjacent to at most one vertex of Z. A strip (J, Z) is called a *spot* if J is a three-vertex path and Z consists of both ends of this path.

We can now state the required lemma, which is easily derived from Lemma C.20 in [19] or Theorem 1 in [18].

Lemma 1 ([18,19]). *Let G be a connected claw-free graph, such that G does not admit twins or proper W-joins and $\alpha(G) > 4$. Then either*

1. *G is a proper circular-arc graph, or*
2. *G admits a strip-structure such that each strip (J, Z) either is*
   (a) *a spot, or*
   (b) *a stripe with $|Z| = 1$ and J is proper circular-arc or has $\alpha(J) \leq 3$, or*

(c) a stripe with $|Z| = 2$, and J is proper interval or has $\alpha(J) \leq 4$.

Moreover, it is possible to distinguish the cases and to find the strip-structure in polynomial time.

3   Mutually Induced Disjoint Paths

In this section we present the following result.

Theorem 1. *The* INDUCED DISJOINT PATHS *problem is fixed-parameter tractable on claw-free graphs when parameterized by* k.

Below, we outline the general approach of our algorithm, and then give the details of the subroutines that we use. First, we need some definitions. Let G be a graph with terminal pairs $(s_1, t_1), \ldots (s_k, t_k)$ that forms an instance of INDUCED DISJOINT PATHS. We call this instance *claw-free, twin-free,* or *proper W-join-free* if G is claw-free, twin-free, or proper W-join-free, respectively. In addition, we call the instance *independent* if the terminal vertices form an independent set, and no terminal vertex represents two terminals of the same terminal pair. Note that in this definition it is still possible for a terminal vertex to represent more than one terminal. However, no terminal vertex in an independent instance can represent more than two terminals if the instance has a solution and G is claw-free. Otherwise, any solution would induce a claw in the neighborhood of this terminal vertex.

In our algorithm, it may happen that the graph under consideration gets disconnected. In that case we make the following implicit check. We stop considering this graph if there is a terminal pair of which the terminals are in two different connected components. Otherwise we consider each connected component separately. Hence, we may assume that the graph is connected.

THE ALGORITHM AND PROOF OF THEOREM 1

Let a claw-free graph G on n vertices with terminal pairs $(s_1, t_1), \ldots, (s_k, t_k)$ for some $k \geq 1$ form an instance.

Step 1. Reduce to an independent instance.
We apply Lemma 2 (stated on page 7) and obtain in $O(k^2 n + n^2)$ time an independent and equivalent instance that consists of a claw-free graph with at most k terminal pairs. For simplicity, we denote this graph and these terminals pairs by G and $(s_1, t_1), \ldots, (s_k, t_k)$ as well.

Step 2. Solve the problem if α is small.
Because all terminal vertices are independent, we find that $k \leq \alpha$ holds. Hence, if $\alpha \leq 4$, we can solve the problem by applying the aforementioned $n^{O(k)}$ time algorithm of Fiala et al. [11]. From now on we assume that $\alpha > 4$.

Step 3. Remove twins and proper W-joins.
We apply Lemma 3 and obtain in $O(n^5)$ time an independent and equivalent instance that consists of a claw-free, twin-free, proper W-join-free graph with at

most k terminal pairs. For simplicity, we denote this graph and these terminals pairs by G and $(s_1, t_1), \ldots, (s_k, t_k)$ as well.

Step 4. Solve the problem for a proper circular-arc graph.
We can check in linear time if G is a proper circular-arc graph [6]. If so, then we apply Lemma 4 to solve the problem in linear time. From now on we assume that G is not a proper circular-arc graph.

Step 5. Reduce to a collection of line graphs.
By Lemma 1 we find in polynomial time a strip-structure of G, in which each strip (J, Z) is either a spot, or a stripe with $|Z| = 1$, and J is proper circular-arc or has $\alpha(J) \leq 3$, or a stripe with $|Z| = 2$, and J is proper interval or has $\alpha(J) \leq 4$. We apply Lemma 5, and in $6^k n^{O(1)}$ time either find that the instance has no solution, or obtain at most 6^k line graphs on at most n vertices and with at most k terminals each, such that G has a solution if and only if at least one of these line graphs has a solution.

Step 6. Solve the problem for each line graph.
For each of the 6^k line graphs G' created we can do this in $g(k) |V_G'|^6$ time due to Lemma 6. Here, $g(k)$ is a function that only depends on k. We conclude that our algorithm runs in $6^k g(k) n^{O(1)}$ time, as desired.

To finish the correctness proof and running time analysis of our algorithm, we state the missing lemmas (proofs omitted). Let $G = (V, E)$ be a claw-free graph with terminal pairs $(s_1, t_1), \ldots, (s_k, t_k)$ for $k \geq 1$. To obtain an independent instance we apply the following rules in order. We then show that these rules can be performed in polynomial time and yield an independent instance.

1. Remove every non-terminal vertex u from G that is adjacent to two adjacent terminal vertices v and w.
2. Remove every non-terminal vertex u from G that is adjacent to a terminal vertex v that only represents terminals whose partner is represented by a vertex of $N_G[v]$.
3. Remove the terminals of every terminal pair (s_i, t_i) with $s_i \in T_u$ for some $u \in V(G)$ and $t_i \in T_v$ for some $v \in N_G[u]$ from T.
4. For every pair of adjacent terminal vertices, remove the edge between them.

Lemma 2. *There is an $O(k^2 n + n^2)$-time algorithm that transforms an instance consisting of an n-vertex, claw-free graph G with k terminal pairs into an equivalent instance that is independent and claw-free.*

After the instance has become independent, we remove twins and proper W-joins. To remove twins, we remove $A \setminus \{v\}$ for each twin set A, where v is a vertex in A that is a terminal if possible. Removing proper W-joins is slightly more complicated.

Lemma 3. *There is an $O(n^5)$-time algorithm that transforms an independent instance consisting of an n-vertex, claw-free graph G with k terminal pairs into an equivalent instance that is independent, claw-free, twin-free, and proper W-join-free.*

Next, we consider (proper) circular-arc graphs.

Lemma 4. *The INDUCED DISJOINT PATHS problem can be solved in linear time for independent instances consisting of a proper circular-arc graph on n vertices with k terminal pairs.*

The proof of the above lemma can be modified to work even if G is a circular-arc graph (i.e. it is not necessarily proper circular-arc). This quickly leads to the following corollary, which contrasts the complexity of the DISJOINT PATHS problem for circular-arc graphs, as it is NP-complete even on interval graphs [28].

Corollary 1. *The INDUCED DISJOINT PATHS problem can be solved in polynomial time for circular-arc graphs on n vertices with k terminal pairs.*

We now reduce a strip-structure to a collection of line graphs.

Lemma 5. *Let G be a graph that together with a set S of k terminal pairs forms a claw-free, independent, and twin-free instance of INDUCED DISJOINT PATHS. Let $(R, \{J_e, Z_e\} \mid e \in E(R)\})$ be a strip-structure for G, in which each strip (J, Z) either is*

1. *a spot, or*
2. *a stripe with $|Z| = 1$, and J is proper circular-arc or has $\alpha(J) \leq 3$, or*
3. *a stripe with $|Z| = 2$, and J is proper interval or has $\alpha(J) \leq 4$.*

There is a $6^k n^{O(1)}$-time algorithm that either shows that (G, S) has no solution, or produces a set $\mathcal{G}$ of at most 6^k graphs, such that each $G' \in \mathcal{G}$ is a line graph with at most $|V(G)|$ vertices and at most k terminal pairs, and such that G has a solution if only if at least one graph in $\mathcal{G}$ has a solution.

Proof (Sketch). Let G be an n-vertex graph with a set of k terminal pairs that has the properties as described in the statement of the lemma. Our algorithm is a branching algorithm that applies a sequence of graph modifications to G until a line graph remains in each leaf of the branching tree. While branching, the algorithm keeps the terminal set and the strip structure up to date with the modifications being performed. This is possible dynamically, i.e. without needing to recompute a strip structure from scratch, and no new strips are created in the algorithm. Moreover, the modifications ensure that all intermediate instances are claw-free and independent, i.e. it is not necessary to reapply Lemma 2. Finally, we note that the modifications may remove some or all of the vertices of a strip. However, at any time during the algorithm, a strip (J, Z) still is of one of the types as in the lemma statement. It is also worth noting that such a deletion may create twins in an intermediate instance. However, the algorithm only relies on the original instance being twin-free.

The algorithm considers each strip at most once in any path of the branching tree. The branching strategy that the algorithm follows for a strip (J, Z) depends on a complex case analysis. The main distinction is between $|Z| = 1$ and $|Z| = 2$. We do not have to branch in the first case. However, in the second case we may have to do so. After processing a strip and possibly branching, we obtain for each branch a new, intermediate instance of the problem that consists of the induced subgraph G' of remaining vertices of G together with those terminal pairs of G that are represented by terminal vertices in G'. We call this *reducing to* G'. The algorithm then considers the next strip of the updated strip structure. Below we focus on one particular subcase that is indicative of our type of branching.

Case 2ciii. We assume that (J, Z) is a stripe with $|Z| = 2$. We write $H = G[J \setminus Z]$ and $F = G - H$. Assume that $Z = Z_{e_1}$ with $e_1 = \{v_1, v_2\}$. Let $e_2^h, \ldots, e_{p_h}^h$ be the other hyperedges of R that contain v_h for $h = 1, 2$. For $h = 1, 2$ and $i = 1, \ldots, p_h$, we let $z_v(e_i^h)$ denote the vertex in $Z_{e_i^h}$ corresponding to v_h. For $h = 1, 2$, let $X_h = N_{J_{e_1^h}}(z_v(e_1^h))$ and $Y_h = N_{J_{e_2^h}}(z_v(e_2^h)) \cup \cdots \cup N_{J_{e_p^h}}(z_v(e_{p_h}^h))$. Because (J, Z) is a stripe, $X_1 \cap X_2 = \emptyset$. Also by definition, we have that for $h = 1, 2$, the sets X_h and Y_h are both nonempty, $(X_1 \cup X_2) \cap (Y_1 \cup Y_2) = \emptyset$, and $X_h \cup Y_h$ is a clique in G. Moreover, $Y_1 \cup Y_2$ separates $V(H)$ from $V(F) \setminus (Y_1 \cup Y_2)$, should $V(F) \setminus (Y_1 \cup Y_2)$ be nonempty.

We assume that H contains at least one terminal vertex and that neither X_1 nor X_2 contains a terminal vertex. We branch in five directions. In the first four directions, we check whether G has a solution that contains no vertex from X_1, X_2, Y_1, Y_2, respectively. In these cases we may remove X_1, X_2, Y_1, or Y_2, respectively, from G and return to Case 1. In the remaining branch we check whether G has a solution in which a solution path uses a vertex from each of the sets X_1, X_2, Y_1, and Y_2. Note that these four vertices will be inner vertices of one or more solution paths. This is the branch that we analyze below.

We say that a terminal that is represented by a vertex in H but whose partner is represented by a vertex in F is *unpaired* in H. Assume that exactly two terminals are unpaired in H and denote them by s_i and s_j, respectively. Note that they may be represented by the same vertex.

Define the following two graphs. Let H'' be the graph obtained from H by adding a new vertex v_1' adjacent to all vertices in X_1 and a new vertex v_2' adjacent to all vertices in X_2. Note that H'' is isomorphic to J. This implies that INDUCED DISJOINT PATHS can be solved in polynomial time on H'' (as $\alpha(H'') \leq 4$ or H'' is an interval graph). We let F^* denote the graph obtained from F by adding a new vertex u_1', a new vertex adjacent to all vertices of Y_1 and to u_1', a new vertex u_2', and a new vertex adjacent to all vertices of Y_2 and to u_2'.

We now branch into two further directions. First, we check whether H'' has a solution after letting v_1', v_2' represent new terminals t_i', t_j', respectively, that are the new partners of s_i and s_j, respectively, in H''. If the answer is No, then we stop considering this branch. Otherwise, we reduce to F^* after letting u_1', u_2' represent new terminals s_i', s_j', respectively, that are the new partners of t_i and t_j in F^*. In the second branch, we do the same with the roles of i and j reversed.

This finishes Case 2ciii. Note that we branched in six directions in total.

After our branching algorithm we have either found in $k^6 n^{O(1)}$ time that G has no solution, or a set $\mathcal{G}$ of at most 6^k graphs. The latter follows because the search tree of our branching algorithm has depth k and we branch in at most six directions (as illustrated in Case 2ciii above). Because we only removed vertices from G, we find that every graph in $\mathcal{G}$ has at most n vertices. Since we only removed terminal pairs from S or replaced a terminal pair by another terminal pair, we find that every graph in $\mathcal{G}$ has at most k terminal pairs. Moreover, for each graph $G' \in \mathcal{G}$, the neighborhood of each of its vertices is the disjoint union of at most two cliques. Hence, G' is a line graph (see e.g. Ryjáček [32]).   □

Lemma 6. *The* INDUCED DISJOINT PATHS *problem can be solved in* $g(k)n^6$ *time for line graphs on n vertices and with k terminal pairs, where g is a function that only depends on k.*

Lemma 6 also appears in Fiala et al. [11], but we provide a better analysis here. This completes the proof of Theorem 1. We then obtain the following corollary.

Corollary 2. *The problems k-IN-A-CYCLE, k-IN-A-PATH, and k-IN-A-TREE are fixed-parameter tractable for claw-free graphs when parameterized by k.*

4   Conclusions

We showed that the INDUCED DISJOINT PATHS problem is fixed-parameter tractable in k for claw-free graphs. As a consequence, we also proved that the problems k-IN-A-CYCLE, k-IN-A-PATH, and k-IN-A-TREE are fixed-parameter tractable in k. We can also show that ANCHORED INDUCED TOPOLOGICAL MINOR is fixed-parameter tractable when parameterized by the number of vertices in the target graph. We omit the proof of this result and of the following result.

Theorem 2. *The* INDUCED TOPOLOGICAL MINOR *problem is* W[1]*-hard for pairs (G, H) where G and H are line graphs, and $|V(H)|$ is the parameter.*

Below, we show that our result for INDUCED DISJOINT PATHS is tight, from two other perspectives; proofs are omitted. First, it natural to ask whether our results generalize to $K_{1,\ell}$-free graphs for $\ell \geq 4$, but this is unlikely.

Proposition 1. *The problems* 2-INDUCED DISJOINT PATHS, 2-IN-A-CYCLE, *and* 3-IN-A-PATH *are* NP*-complete even for $K_{1,4}$-free graphs.*

It is well known that a parameterized problem is fixed-parameter tractable if and only if it is decidable and kernelizable (cf. [29]). Hence, the next step would be to try to construct a *polynomial* kernel. However, we show that for our problems this is not likely even for line graphs.

Proposition 2. *The* INDUCED DISJOINT PATHS *problem restricted to line graphs has no polynomial kernel when parameterized by k, unless* NP $\subseteq$ coNP/poly.

The question whether the same result as in Proposition 2 holds for k-IN-A-CYCLE and k-IN-A-PATH restricted to line graphs is open.

Instead of improving our result for the INDUCED DISJOINT PATHS problem, we could also work towards solving a more general problem. In the definition of induced disjoint paths, we explicitly disallowed duplicate terminal pairs, i.e. the set of terminal pairs is not a multiset. If we generalize to allow duplicate terminal pairs, then we can solve the k-INDUCED DISJOINT PATHS problem for claw-free graphs in polynomial time for fixed k as follows. In a nutshell, we replace two terminal pairs (s_i, t_i) and (s_j, t_j) with $s_i = s_j$ and $t_i = t_j$ by two new pairs (s'_i, t'_i) and (s'_j, t'_j), where s'_i, s'_j are two distinct nonadjacent neighbors of $s_i = s_j$ and t'_i, t'_j are two distinct nonadjacent neighbors of $t_i = t_j$. However, determining the parameterized complexity of the general case is still an open problem.

As a partial result towards answering this more general question, we consider the variation of INDUCED DISJOINT PATHS where all terminal pairs coincide. In general, this problem generalizes the 2-IN-A-CYCLE problem, which is NP-complete [1,9]. For claw-free graphs, we observe that no terminal vertex can represent more than two terminals in any Yes-instance. Hence the problem can be reduced to the 2-IN-A-CYCLE problem, which is polynomial-time solvable on claw-free graphs [11].

References

1. Bienstock, D.: On the complexity of testing for odd holes and induced odd paths. Discrete Mathematics 90, 85–92 (1991); See also Corrigendum. Discrete Mathematics 102, 109 (1992)
2. Bodlaender, H.L., Thomassé, S., Yeo, A.: Kernel bounds for disjoint cycles and disjoint paths. Theoretical Computer Science 412, 4570–4578 (2011)
3. Bruhn, H., Saito, A.: Clique or hole in claw-free graphs. Journal of Combinatorial Theory, Series B 102, 1–13 (2012)
4. Chudnovsky, M., Seymour, P.D.: The structure of claw-free graphs. In: Webb, B.S. (ed.) Surveys in Combinatorics. London Mathematical Society Lecture Notes Series, vol. 327, pp. 153–171. Cambridge University Press, Cambridge (2005)
5. Chudnovsky, M., Seymour, P.D.: The three-in-a-tree problem. Combinatorica 30, 387–417 (2010)
6. Deng, X., Hell, P., Huang, J.: Linear time representation algorithm for proper circular-arc graphs and proper interval graphs. SIAM Journal on Computing 25, 390–403 (1996)
7. Derhy, N., Picouleau, C.: Finding induced trees. Discrete Applied Mathematics 157, 3552–3557 (2009)
8. Derhy, N., Picouleau, C., Trotignon, N.: The four-in-a-tree problem in triangle-free graphs. Graphs and Combinatorics 25, 489–502 (2009)
9. Fellows, M.R.: The Robertson-Seymour theorems: A survey of applications. In: Richter, R.B. (ed.) Proceedings of the AMS-IMS-SIAM Joint Summer Research Conference. Contemporary Mathematics, vol. 89, pp. 1–18. American Mathematical Society, Providence (1989)
10. Fellows, M.R., Kratochvíl, J., Middendorf, M., Pfeiffer, F.: The Complexity of Induced Minors and Related Problems. Algorithmica 13, 266–282 (1995)

11. Fiala, J., Kamiński, M., Lidicky, B., Paulusma, D.: The k-in-a-path problem for claw-free graphs. Algorithmica 62, 499–519 (2012)
12. Fiala, J., Kamiński, M., Paulusma, D.: A note on contracting claw-free graphs (manuscript, submitted)
13. Fiala, J., Kamiński, M., Paulusma, D.: Detecting induced star-like minors in polynomial time (manuscript, submitted)
14. Golovach, P.A., Paulusma, D., van Leeuwen, E.J.: Induced Disjoint Paths in AT-Free Graphs. In: Fomin, F.V., Kaski, P. (eds.) SWAT 2012. LNCS, vol. 7357, pp. 153–164. Springer, Heidelberg (2012)
15. Grohe, M., Kawarabayashi, K., Marx, D., Wollan, P.: Finding topological subgraphs is fixed-parameter tractable. In: Proc. STOC 2011, pp. 479–488 (2011)
16. Habib, M., Paul, C., Viennot, L.: A Synthesis on Partition Refinement: A Useful Routine for Strings, Graphs, Boolean Matrices and Automata. In: Meinel, C., Morvan, M. (eds.) STACS 1998. LNCS, vol. 1373, pp. 25–38. Springer, Heidelberg (1998)
17. Harary, F.: Graph Theory. Addison-Wesley, Reading MA (1969)
18. Hermelin, D., Mnich, M., van Leeuwen, E.J., Woeginger, G.J.: Domination When the Stars Are Out. In: Aceto, L., Henzinger, M., Sgall, J. (eds.) ICALP 2011, Part I. LNCS, vol. 6755, pp. 462–473. Springer, Heidelberg (2011)
19. Hermelin, D., Mnich, M., van Leeuwen, E.J., Woeginger, G.J.: Domination when the stars are out. arXiv:1012.0012v1 [cs.DS]
20. Karp, R.M.: On the complexity of combinatorial problems. Networks 5, 45–68 (1975)
21. King, A., Reed, B.: Bounding χ in terms of ω and δ for quasi-line graphs. Journal of Graph Theory 228, 215–228 (2008)
22. Kobayashi, Y., Kawarabayashi, K.: Algorithms for finding an induced cycle in planar graphs and bounded genus graphs. In: Mathieu, C. (ed.) Proc. SODA 2009, pp. 1146–1155. ACM Press, New York (2009)
23. Kobayashi, Y., Kawarabayashi, K.: A linear time algorithm for the induced disjoint paths problem in planar graphs. Journal of Computer and System Sciences 78, 670–680 (2012)
24. Lévêque, B., Lin, D.Y., Maffray, F., Trotignon, N.: Detecting induced subgraphs. Discrete Applied Mathematics 157, 3540–3551 (2009)
25. Liu, W., Trotignon, N.: The k-in-a-tree problem for graphs of girth at least k. Discrete Applied Mathematics 158, 1644–1649 (2010)
26. Lynch, J.F.: The equivalence of theorem proving and the interconnection problem. SIGDA Newsletter 5, 31–36 (1975)
27. McConnell, R.M.: Linear-Time Recognition of Circular-Arc Graphs. Algorithmica 37, 93–147 (2003)
28. Natarajan, S., Sprague, A.P.: Disjoint paths in circular arc graphs. Nordic Journal of Computing 3, 256–270 (1996)
29. Niedermeier, R.: Invitation to Fixed-Parameter Algorithms. Oxford Lecture Series in Mathematics and its Applications. Oxford University Press (2006)
30. Robertson, N., Seymour, P.D.: Graph minors. XIII. The disjoint paths problem. Journal of Combinatorial Theory, Series B 63, 65–110 (1995)
31. Roussopoulos, N.D.: A max$\{m, n\}$ algorithm for determining the graph H from its line graph G. Information Processing Letters 2, 108–112 (1973)
32. Ryjáček, Z.: On a closure concept in claw-free graphs. Journal of Combinatorial Theory, Series B 70, 217–224 (1997)

On Min-Power Steiner Tree*

Fabrizio Grandoni

IDSIA, University of Lugano, Switzerland
fabrizio@idsia.ch

Abstract. In the classical (min-cost) Steiner tree problem, we are given an edge-weighted undirected graph and a set of terminal nodes. The goal is to compute a min-cost tree S which spans all terminals. In this paper we consider the min-power version of the problem (a.k.a. symmetric multicast), which is better suited for wireless applications. Here, the goal is to minimize the total power consumption of nodes, where the power of a node v is the maximum cost of any edge of S incident to v. Intuitively, nodes are antennas (part of which are terminals that we need to connect) and edge costs define the power to connect their endpoints via bidirectional links (so as to support protocols with ack messages). Observe that we do not require that edge costs reflect Euclidean distances between nodes: this way we can model obstacles, limited transmitting power, non-omnidirectional antennas etc. Differently from its min-cost counterpart, min-power Steiner tree is NP-hard even in the spanning tree case (a.k.a. symmetric connectivity), i.e. when all nodes are terminals. Since the power of any tree is within once and twice its cost, computing a $\rho_{st} \leq \ln(4) + \varepsilon$ [Byrka et al.'10] approximate min-cost Steiner tree provides a $2\rho_{st} < 2.78$ approximation for the problem. For min-power spanning tree the same approach provides a 2 approximation, which was improved to $5/3 + \varepsilon$ with a non-trivial approach in [Althaus et al.'06].

In this paper we present an improved approximation algorithm for min-power Steiner tree. Our result is based on two main ingredients. We present the first decomposition theorem for min-power Steiner tree, in the spirit of analogous structural results for min-cost Steiner tree and min-power spanning tree. Based on this theorem, we define a proper LP relaxation, that we exploit within the iterative randomized rounding framework in [Byrka et al.'10]. A careful analysis of the decrease of the power of nodes at each iteration provides a $3\ln 4 - \frac{9}{4} + \varepsilon < 1.91$ approximation factor. The same approach gives an improved $1.5 + \varepsilon$ approximation for min-power spanning tree as well. This matches the approximation factor in [Nutov and Yaroshevitch'09] for the special case of min-power spanning tree with edge weights in $\{0, 1\}$.

1 Introduction

Consider the following basic problem in wireless network design. We are given a set of antennas, and we have to assign the transmitting power of each antenna. Two antennas can exchange messages (directly) if they are within the transmission range

---

* This research is partially supported by the ERC Starting Grant NEWNET 279352.

L. Epstein and P. Ferragina (Eds.): ESA 2012, LNCS 7501, pp. 527–538, 2012.

of each other (this models protocols with ack messages). The goal is to find a minimum total power assignment so that a given subset of antennas can communicate with each other (using a multi-hop protocol).

We can formulate the above scenario as a *min-power Steiner tree* problem (a.k.a. *symmetric multicast*). Here we are given an undirected graph $G = (V, E)$, with edge costs $c : E \to \mathbb{Q}_{\geq 0}$, and a subset R of terminal nodes. The goal is to compute a Steiner tree S spanning R, of minimum *power* $p(S) := \sum_{v \in V(S)} p_S(v)$, with $p_S(v) := \max_{uv \in E(S)}\{c(uv)\}$. In words, the power of a node v with respect to tree S is the largest cost of any edge of S incident to v, and the power of S is the sum of the powers of its nodes[1]. The *min-power spanning tree* problem (a.k.a. *symmetric connectivity*) is the special case of min-power Steiner tree where $R = V$, i.e. all nodes are terminals. Let us remark that, differently from part of the literature on related topics, we do not require that edge costs reflect Euclidean distances between nodes. This way, we are able to model obstacles, limited transmitting power, antennas which are not omnidirectional, etc.

The following simple approximation-preserving reduction shows that min-power Steiner tree is at least as hard to approximate as its min-cost counterpart: given a min-cost Steiner tree instance, replace each edge e with a path of 3 edges, where the boundary edges have cost zero and the middle one has cost $c(e)/2$. Hence the best we can hope for in polynomial time is a c approximation for some constant $c > 1$. It is known [1,11] that, for any tree S of cost $c(S) := \sum_{e \in E(S)} c(e)$,

$$c(S) \leq p(S) \leq 2c(S). \tag{1}$$

As a consequence, a ρ_{st} approximation for min-cost Steiner tree implies a $2\rho_{st}$ approximation for min-power Steiner tree. In particular, the recent $\ln(4) + \varepsilon < 1.39$ approximation[2] in [3] for the first problem, implies a 2.78 approximation for the second one: no better approximation algorithm is known to the best of our knowledge.

Differently from its min-cost version, min-power spanning tree is NP-hard (even in quite restricted subcases) [1,11]. By the above argument, a min-cost spanning tree is a 2 approximation. However, in this case non-trivial algorithms are known. A $1 + \ln 2 + \varepsilon < 1.69$ approximation is given in [5]. This was improved to $\frac{5}{3} + \varepsilon$ in [1]. If edge costs are either 0 or 1, the approximation factor can be further improved to $\frac{3}{2} + \varepsilon$ [13]. Indeed, the same factor can be achieved if edge costs are either a or b, with $0 \leq a < b$: this models nodes with two power states, low and high. All these results exploit the notion of k-decomposition. The first result is obtained with a greedy algorithm, while the latter two use (as a black box) the FPTAS in [14] for the *min-cost connected spanning hypergraph* problem in 3-hypergraphs. We will also use k-decompositions, but our algorithms are rather different (in particular, they are LP-based).

Our Results. In this paper we present an improved approximation algorithm for min-power Steiner tree.

---

[1] When S is clear from the context, we will simply write $p(v)$.
[2] Throughout this paper ε denotes a small positive constant.

Theorem 1. *There is an expected* $3\ln 4 - \frac{9}{4} + \varepsilon < 1.909$ *approximation algorithm for min-power Steiner tree.*

Our result is based on two main ingredients. Informally, a k-decomposition of a Steiner tree S is a collection of (possibly overlapping) subtrees of S, each one containing at most k terminals, which together span S. The power/cost of a decomposition is the sum of the powers/costs of its components[3]. It is a well-known fact (see [2] and references therein) that, for any constant $\varepsilon > 0$, there is a constant $k = k(\varepsilon)$ such that there exists a k-decomposition of cost at most $(1+\varepsilon)$ times the cost of S. A similar result holds for min-power spanning tree [1]. The first ingredient in our approximation algorithm is a similar decomposition theorem for min-power Steiner tree, which might be of independent interest. This extends the qualitative results in [1,2] since min-power Steiner tree generalizes the other two problems. However, the dependence between ε and k is worse in our construction.

Theorem 2. (Decomposition) *For any* $h \geq 3$ *and any Steiner tree* S, *there exists a* h^h-*decomposition of* S *of power at most* $(1 + \frac{14}{h})p(S)$.

Based on this theorem, we are able to compute a $1 + \varepsilon$ approximate solution for a proper component-based LP-relaxation for the problem. We exploit this relaxation within the iterative randomized rounding algorithmic framework in [3]: we sample one component with probability proportional to its fractional value, set the corresponding edge costs to zero and iterate until there exists a Steiner tree of power zero. The solution is given by the sampled components plus a subset of edges of cost zero in the original graph. A careful analysis of the decrease of node powers at each iteration provides a $3\ln 4 - \frac{9}{4} + \varepsilon < 1.91$ approximation. We remark that, to the best of our knowledge, this is the only other known application of iterative randomized rounding to a natural problem.

The same basic approach also provides an improved approximation for min-power spanning tree.

Theorem 3. *There is an expected* $\frac{3}{2} + \varepsilon$ *approximation algorithm for min-power spanning tree.*

This improves on [14], and matches the approximation factor achieved in [13] (with a drastically different approach!) for the special case of 0-1 edge costs.

Preliminaries and Related Work. Min-power problems are well studied in the literature on wireless applications. Very often here one makes the assumption that nodes are points in $\mathbb{R}^2$ or $\mathbb{R}^3$, and that edge costs reflect the Euclidean distance d between pairs of nodes, possibly according to some power law (i.e., the cost of the edge is d^c for some constant c typically between 2 and 4). This assumption is often not realistic for several reasons. First of all, due to obstacles, connecting in a direct way geographically closer nodes might be more expensive (or impossible). Second, the power of a given antenna might be upper bounded

---

[3] Due to edge duplication, the cost of the decomposition can be larger than $c(S)$. Its power can be larger than $p(S)$ even for edge disjoint components.

(or even lower bounded) for technological reasons. Third, antennas might not be omnidirectional. All these scenarios are captured by the undirected graph model that we consider in this paper. A relevant special case of the undirected graph model is obtained by assuming that there are only two edge costs a and b, $0 \leq a < b$. This captures the practically relevant case that each node has only two power states, low and high. A typical goal is to satisfy a given connectivity requirement at minimum total power, as we assume in this paper.

Several results are known in the *asymmetric* case, where a unidirectional link is established from u to v iff v is within the transmission range of u (and possibly the vice versa does not hold). For example in the *asymmetric unicast* problem one wants to compute a min-power (or, equivalently, min-cost) directed path from node s to node t. In the *asymmetric connectivity* problem one wants to compute a min-power spanning arborescence rooted at a given root node r. This problem is NP-hard even in the 2-dimensional Euclidean case [7], and a minimum spanning tree provides a 12 approximation for the Euclidean case (while the general case is log-hard to approximate) [15]. The *asymmetric multicast* problem is the generalization of asymmetric connectivity where one wants a min-power arborescence rooted at r which contains a given set R of terminals. As observed in [1], the same approach as in [15] provides a $12\rho_{st}$ approximation for the Euclidean case, where ρ_{st} is the best-known approximation for Steiner tree in graphs. In the *complete range assignment* problem one wants to establish a strongly connected spanning subgraph. The authors of [11] present a 2-approximation which works for the undirected graph model, and show that the problem is NP-hard in the 3-dimensional Euclidean case. The NP-hardness proof was extended to 2 dimensions in [8]. Recently, the approximation factor was improved to $2 - \delta$ for a small constant $\delta > 0$ with a highly non-trivial approach [4]. The same paper presents a 1.61 approximation for edge costs in $\{a, b\}$, improving on the 9/5 factor in [6]. For more general edge activation models and higher connectivity requirements, see e.g. [9,10,12] and references therein.

In this paper we consider the *symmetric* case, where links must be bidirectional (i.e. u and v are not adjacent if one of the two is not able to reach the other). This is used to model protocols with ack messages, where nodes which receive a message have to send back a notification. The symmetric unicast problem can be solved by applying Dijkstra's algorithm to a proper auxiliary graph [1]. The symmetric connectivity and multicast problems are equivalent to min-power spanning tree and min-power Steiner tree, respectively.

Proofs which are omitted due to lack of space will be given in the full version of the paper. The min-power Steiner tree is denoted by S^*.

2   A Decomposition Theorem for Min-Power Steiner Tree

A *k-component* is a tree which contains at most k terminals. If internal nodes are non-terminals, the component is *full*. A *k-decomposition* of a Steiner tree S over terminals R is a collection of k-components on the edges of S which span S and such that the following auxiliary *component graph* is a tree: replace each component C with a star, where the leaves are the terminals of C and the central node is a distinct, dummy non-terminal v_C. Observe that, even if the component graph is a tree, the actual components might share edges. When the value of k is irrelevant, we simply use the terms component and decomposition. We will consider k-decompositions with $k = O(1)$. This is useful since a min-power component C on a constant number of terminals can be computed in polynomial time[4]. The assumption on the component graph is more technical, and it will be clearer later. Intuitively, when we compute a min-power component on a subset of terminals, we do not have full control on the internal structure of the component. For this reason, the connectivity requirements must be satisfied independently from that structure.

Assume w.l.o.g. that S consists of one full component. This can be enforced by appending to each terminal v a dummy node v' with a dummy edge of cost 0, and replacing v with v' in the set of terminals. Any decomposition into (full) k-components of the resulting tree can be turned into a k-decomposition of the same power for S by contracting dummy edges, and vice versa.

Next lemma shows that one can assume that the maximum degree of the components in a decomposition can be upper bounded by a constant while losing a small factor in the approximation.

Lemma 1. *For any $\Delta \geq 3$, there exists a decomposition of S of power at most $(1 + \frac{2}{\lceil \Delta/2 \rceil - 1})p(S)$ whose components have degree at most Δ.*

Proof. The rough idea is to split S at some node v not satisfying the degree constraint, so that the duplicated copies of v in each obtained component have degree (less than) Δ. Then we add a few paths between components, so that the component graph remains a tree. All the components but one will satisfy the degree constraint: we iterate the process on the latter component.

In more detail, choose any leaf node r as a *root*. The decomposition initially consists of S only. We maintain the invariant that all the components but possibly the component C_r containing r have degree at most Δ. Assume that C_r has degree larger than Δ (otherwise, we are done). Consider any *split node* v of degree $d(v) = d + 1 \geq \Delta + 1$ such that all its descendants have degree at most Δ. Let $u_1, \ldots, u_d$ be the children of v, in increasing order of $c(vu_i)$. Define $\Delta' := \lceil \Delta/2 \rceil \in [2, \Delta - 1]$. We partition the u_i's by iteratively removing the first Δ' children, until there are at most $\Delta - 2$ children left: let $V_1, \ldots, V_h$ be the resulting subsets of children. In particular, for $i < h$, $V_i = \{u_{(i-1)\Delta'+1}, \ldots, u_{i\Delta'}\}$. For $i = 1, \ldots, h - 1$, we let C_i be a new component induced by v, V_i, and the descendants of V_i. The new root component C_h is obtained by removing from C_r the nodes $\cup_{i<h} V(C_i) - \{v\}$ and the corresponding edges. In order to maintain

---

[4] One can guess (by exhaustive enumeration) the non-terminal nodes of degree at least 3 in C, and the structure of the tree where non-terminals of degree 2 are contracted. Each edge vu of the contracted tree corresponds to a path P whose internal nodes are non-terminals of degree 2: after guessing the boundary edges uu' and $v'v$ of P (which might affect the power of u and v, respectively), the rest of P is w.l.o.g. a min-power path between u' and v' (which can be computed in polynomial time [1]).

the connectivity of the component graph (which might be lost at this point), we expand C_{i+1}, $i \geq 1$, as follows: let P_j be any path from v to some leaf which starts with edge vu_j. We append to C_{i+1} the path $P_{m(i)}$ which minimizes $p(P_j) - c(vu_j)$ over $j \in V_i$. After this step, the component graph is a tree. The invariant is maintained: in fact, the degree of any node other than v can only decrease. In each C_i, $i < h$, v has degree either Δ' or $\Delta' + 1$, which is within 2 and Δ. Since $\Delta - 2 - \Delta' < |V_h| \leq \Delta - 2$, the cardinality $|V_h| + 2$ of v in C_h is also in $[2, \Delta]$. By the choice of v, all the components but C_r have maximum degree Δ. Observe that C_r loses at least $\Delta' - 1 \geq 1$ nodes, hence the process halts.

In order to bound the power of the final decomposition, we use the following charging argument, consisting of two charging rules. When we split C_r at a given node v, the power of v remains $p(v)$ in C_h and becomes $c(vu_{i\Delta'})$ in the other C_i's. We evenly charge the extra power $c(vu_{i\Delta'})$ to nodes V_{i+1}: observe that each $u_j \in V_{i+1}$ is charged by $\frac{c(vu_{i\Delta'})}{|V_{i+1}|} \leq \frac{c(vu_{i\Delta'})}{\Delta' - 1} \leq \frac{c(vu_j)}{\Delta' - 1} \leq \frac{p(u_j)}{\Delta' - 1}$.

Furthermore, we have an extra increase of the power by $p(P_{m(i)}) - c(vu_{m(i)})$ for every $i < h$: this is charged to the nodes of the paths $P_j - \{v\}$ with $u_j \in V_i - \{u_{m(i)}\}$, in such a way that no node w is charged by more than $\frac{1}{\Delta' - 1} p(w)$. This is possible since there are $\Delta' - 1$ such paths, and the nodes of each such path have total power at least $p(P_{m(i)}) - c(vu_{m(i)})$ by construction.

Each node w can be charged with the second charging rule at most once, since when this happens w is removed from C_r and not considered any longer. When w is charged with the first charging rule, it must be a child of some split node v. Since no node is a split node more than once, also in this case we charge w at most once. Altogether, each node v is charged by at most $\frac{2}{\Delta' - 1} p(v)$. □

Proof. (Theorem 2) Apply Lemma 1 with $\Delta = h$ to S, hence obtaining a decomposition of power at most $\frac{\lceil h/2 \rceil + 1}{\lceil h/2 \rceil - 1} p(S)$ whose components have degree at most h. We describe an h^h decomposition of each such component C with more than h^h terminals. Root C at any non-terminal r, and contract internal nodes (other than r) of degree 2. For any internal node v of C, let $P(v)$ be the path from v to its rightmost child r_v, and then from r_v to some leaf terminal $\ell(v)$ using the leftmost possible path. Observe that paths $P(v)$ are edge disjoint. Pick a value $q \in \{0, 1, \ldots, h - 1\}$ uniformly at random, and mark the nodes at level $\ell = q \pmod h$. Consider the partition of C into edge-disjoint subtrees T which is induced by the marked levels[5]. Finally, for each such subtree T, we append to each leaf v of T the path $P(v)$: this defines a component C_T.

Trees T have at most h^h leaves: hence components C_T contain at most h^h terminals each. Observe that the component graph remains a tree. In order to bound the power of components C_T, note that each node u in the original tree has in each component a power not larger than the original power $p(u)$: hence it is sufficient to bound the expected number μ_u of components a node u belongs to. Suppose u is contracted or a leaf node. Then u is contained in precisely the same components as some edge e. This edge belongs deterministically to one subtree T (hence to C_T), and possibly to another component $C_{T'}$ if the node v with $e \in P(v)$ is marked: the

---

[5] Subtrees consisting of one (leaf) node can be neglected.

latter event happens with probability $1/h$. Hence in this case $\mu_u \leq 1 + 1/h$. For each other node u, observe that u belongs to one subtree T if it is not marked, and to at most two such subtrees otherwise. Furthermore, it might belong to one extra component $C_{T'}$ if the node v with $ul_u \in P(v)$ is marked, where l_u is the leftmost child of u. Hence, $\mu_u \leq 1 + 2/h$ in this case. Altogether, the decomposition of C has power at most $(1 + 2/h)p(C)$ in expectation.

From the above discussion, there exists (deterministically) an h^h decomposition of power at most $\frac{\lceil h/2 \rceil + 1}{\lceil h/2 \rceil - 1}(1 + \frac{2}{h})p(S) \leq (1 + \frac{14}{h})p(S)$. $\square$

We remark that for both min-cost Steiner tree and min-power spanning tree (which are special cases of min-power Steiner tree), improved $(1 + \frac{O(1)}{h})$ approximate c^h decompositions, $c = O(1)$, are known [1,2]. Finding a similar result for min-power Steiner tree, if possible, is an interesting open problem in our opinion (even if it would not directly imply any improvement of our approximation factor).

3   An Iterative Randomized Rounding Algorithm

In this section we present an improved approximation algorithm for min-power Steiner tree. Our approach is highly indebted to [3]. We consider the following LP relaxation for the problem:

$$
\begin{aligned}
\min \quad & \sum_{(Q,s):s \in Q \subseteq R} p_Q \cdot x_{Q,s} && (LP_{pow}) \\
s.t. \quad & \sum_{\substack{(Q,s):s \in Q \subseteq R, \\ s \notin W, Q \cap W \neq \emptyset}} x_{Q,s} \geq 1, && \forall \emptyset \neq W \subseteq R - \{r\}; \\
& x_{Q,s} \geq 0, && \forall s \in Q \subseteq R.
\end{aligned}
$$

Here r is an arbitrary *root* terminal. There is a variable $x_{Q,s}$ for each subset of terminals Q and for each $s \in Q$: the associated coefficient p_Q is the power of a min-power component C_Q on terminals Q. In particular, $S^* = C_R$ induces a feasible integral solution (where the only non-zero variable is $x_{R,r} = 1$). Let $C_{Q,s}$ be the *directed component* which is obtained by directing the edges of C_Q towards s. For a fractional solution x, let us define a directed capacity reservation by considering each (Q,s), and increasing by $x_{Q,s}$ the capacity of the edges in $C_{Q,s}$ Then the cut constraints ensure that each terminal is able to send one unit of (splittable) flow to the root without exceeding the mentioned capacity reservation. We remark that the authors of [3] consider essentially the same LP, the main difference being that p_Q is replaced by the cost c_Q of a min-cost component on terminals Q^6. In particular, the set of constraints in their LP is the same as in LP_{pow}. This allows us to reuse part of their results and techniques, which rely only on the properties of the set of constraints[7].

---

[6] Another technical difference w.r.t. [3] is that they consider only full components: this has no substantial impact on their analysis, and allows us to address the Steiner and spanning tree cases in a unified way.

[7] Incidentally, this observation might be used to address also other variants of the Steiner tree problem, with different objective functions.

Algorithm 1. An iterative randomized rounding approximation algorithm for min-power Steiner tree

---

(1) For $t = 1, 2, \ldots$
    (1a) Compute a $1+\varepsilon$ approximate solution x^t to LP_{pow} (w.r.t. the current instance).
    (1b) Sample one component C^t, where $C^t = C_Q$ with probability $\sum_{s \in Q} x^t_{Q,s} / \sum_{(Q',s')} x^t_{Q',s'}$. Set to zero the cost of the edges in C^t and update LP_{pow}.
    (1c) If there exists a Steiner tree of power zero, return it and halt.

Given Theorem 2, the proof of the following lemma follows along the same line as in [3].

Lemma 2. *For any constant $\varepsilon > 0$, a $1 + \varepsilon$ approximate solution to LP_{pow} can be computed in polynomial time.*

We exploit LP_{pow} within the iterative randomized rounding framework in [3]. Our algorithm (see also Algorithm 1) consists of a set of iterations. At each iteration t we compute a $(1+\varepsilon)$-approximate solution to LP_{pow}, and then sample one component $C^t = C_Q$ with probability proportional to $\sum_{s \in Q} x^t_{Q,s}$. We set to zero the cost of the edges of C^t in the graph, updating LP_{pow} consequently[8]. The algorithm halts when there exists a Steiner tree of cost (and power) zero: this halting condition can be checked in polynomial time.

Lemma 3. *Algorithm 1 halts in a polynomial number of rounds in expectation.*

4 The Approximation Factor

In this section we bound the approximation factor of Algorithm 1, both in the general and in the spanning tree case. Following [3], in order to simplify the analysis let us consider the following variant of the algorithm. We introduce a dummy variable $x_{r,r}$ with $p_r = 0$ corresponding to a dummy component containing the root only, and fix $x_{r,r}$ so that the sum of the x's is some fixed value M in all the iterations. For $M = n^{O(1)}$ large enough, this has no impact on the power of the solution nor on the behaviour of the algorithm (since sampling the dummy component has no effect). Furthermore, we let the algorithm run forever (at some point it will always sample components of power zero).

Let S^t be the min-power Steiner tree at the beginning of iteration t (in particular, $S^1 = S^*$). For a given sampled component C^t, we let $p(C^t)$ be its power in the considered iteration. We define similarly $p(S^t)$ and the corresponding cost $c(S^t)$. The expected approximation factor of the algorithm is bounded by:

---

[8] In the original algorithm in [3], the authors contract components rather than setting to zero the cost of their edges. Our variant has no substantial impact on their analysis, but it is crucial for us since contracting one edge (even if it has cost zero) can decrease the power of the solution.

$$\frac{1}{p(S^*)}\sum_t E[p(C^t)] = \frac{1}{p(S^*)}\sum_t \sum_{(Q,s)} E[\frac{x_{Q,s}^t}{M}p_Q] \le \frac{1+\varepsilon}{Mp(S^*)}\sum_t E[p(S^t)]. \quad (2)$$

Hence, it is sufficient to provide a good upper bound on $E[p(S^t)]$. We exploit the following high-level (ideal) procedure. We start from $\tilde{S} = S^*$, and at each iteration t we add the sampled component C^t to $\tilde{S}$ and delete some *bridge* edges B^t in $E(\tilde{S}) \cap E(S^*)$ in order to remove cycles (while maintaining terminal connectivity). By construction, $\tilde{S}$ is a feasible Steiner tree at any time. Furthermore, the power of $\tilde{S}$ at the beginning of iteration t is equal to the power $p(U^t)$ of the forest of non-deleted edges U^t of S^* at the beginning of the same iteration[9]. In particular, $p(S^t) \le p(U^t) = \sum_v p_{U^t}(v)$.

At this point our analysis deviates (and gets slightly more involved) w.r.t. [3]: in that paper the authors study the expected number of iterations before a given (single) edge is deleted. We rather need to study the behavior of collections of edges incident to a given node v. In more detail, let $e_v^1, \ldots, e_v^{d(v)}$ be the edges of S^* incident to v, in decreasing order of cost $c_v^1 \ge c_v^2 \ge, \ldots, \ge c_v^{d(v)}$ (breaking ties arbitrarily). Observe that $p_{U^t}(v) = c_v^i$ during the iterations when all edges $e_v^1, \ldots, e_v^{i-1}$ are deleted and e_v^i is still non-deleted. Define δ_v^i as the expected number of iterations before all edges $e_v^1, \ldots, e_v^i$ are deleted. For notational convenience, define also $\delta_v^0 = c_v^{d(v)+1} = 0$. Then

$$E[\sum_t p_{U^t}(v)] = \sum_{i=1}^{d(v)} c_v^i(\delta_v^i - \delta_v^{i-1}) = \sum_{i=1}^{d(v)} \delta_v^i(c_v^i - c_v^{i+1}). \quad (3)$$

We will provide a feasible upper bound δ^i on δ_v^i for all v (for a proper choice of the bridge edges B^t) with the following two properties for all i:

$$\text{(a) } \delta^i \le \delta^{i+1} \qquad \text{(b) } \delta^i - \delta^{i-1} \ge \delta^{i+1} - \delta^i.$$

In words, the δ^i's are increasing (which is intuitive since one considers larger sets of edges) but at decreasing speed. Consequently, from (3) one obtains

$$E[\sum_t p_{U^t}(v)] \le \delta^1 c_v^1 + \max_{i\ge 2}\{\delta^i - \delta^{i-1}\}\sum_{i=2}^{d(v)} c_v^i = \delta^1 c_v^1 + (\delta^2 - \delta^1)\sum_{i=2}^{d(v)} c_v^i. \quad (4)$$

Inspired by (4), we introduce the following classification of the edges of S^*. We say that the power of node v is *defined* by e_v^1. We partition the edges of S^* into the *heavy* edges H which define the power of both their endpoints, the *middle* edges M which define the power of exactly one endpoint, and the remaining *light* edges L which do not define the power of any node. Let $c(H) = \gamma_H c(S^*)$ and $c(M) = \gamma_M c(S^*)$. Observe that $p(S^*) = \alpha c(S^*)$ where $\alpha = 2\gamma_H + \gamma_M \in [1,2]$. Note also that in (4) heavy edges appear twice with coefficient δ^1, middle edges appear once with coefficient δ^1 and once with coefficient $\delta^2 - \delta^1$, and light edges appear twice with coefficient $\delta^2 - \delta^1$. Therefore one obtains

---

[9] Since edge weights of sampled components are set to zero, any bridge edge can be replaced by a path of zero cost edges which provides the same connectivity.

$$E[\sum_t p(U^t)] = \sum_v E[\sum_t p^t(v)] \le 2\delta^1 c(H) + (\delta^1 + \delta^2 - \delta^1)c(M) + 2(\delta^2 - \delta^1)c(L)$$

$$= (2\delta^1 \gamma_H + \delta^2 \gamma_M + 2(\delta^2 - \delta^1)(1 - \gamma_H - \gamma_M)) \cdot c(S^*)$$

$$= \left(2(\delta^2 - \delta^1) + (2\delta^1 - \delta^2)\alpha\right) \cdot \frac{p(S^*)}{\alpha} \overset{\alpha \ge 1}{\le} \delta^2 p(S^*). \tag{5}$$

Summarizing the above discussion, the approximation factor of the algorithm can be bounded by

$$\frac{1+\varepsilon}{Mp(S^*)} \sum_t E[p(S^t)] \le \frac{1+\varepsilon}{Mp(S^*)} \sum_t E[p(U^t)] \overset{(5)}{\le} \frac{(1+\varepsilon)\delta^2}{M}. \tag{6}$$

We next provide the mentioned bounds δ^i satisfying Properties (a) and (b): we start with the spanning tree case and then move to the more complex and technical general case.

4.1   The Spanning Tree Case

Observe that in this case the optimal solution $T^* := S^*$ is by definition a terminal spanning tree (i.e. a Steiner tree without Steiner nodes). Therefore we can directly exploit the following claim in [3].

Lemma 4. *[3] Let T^* be any terminal Steiner tree. Set $\tilde{T} := T^*$ and consider the following process. For $t = 1, 2, \ldots$: (a) Take any feasible solution x^t to LP_{pow}; (b) Sample one component $C^t = C_Q$ with probability proportional to variables $x^t_{Q,s}$; (c) Delete a subset of bridge edges B^t from $E(\tilde{T}) \cap E(T^*)$ so that all the terminals remain connected in $\tilde{T} - B^t \cup C^t$. There exists a randomized procedure to choose the B^t's so that any $W \subseteq E(T^*)$ is deleted after $M H_{|W|}$ iterations in expectation[10].*

By Lemma 4 with $W = \{e^1_v, \ldots, e^i_v\}$, we can choose $\delta^i = M \cdot H_i$. Observe that these δ^i's satisfy Properties (a) and (b) since $\frac{\delta^{i+1} - \delta^i}{M} = \frac{1}{i+1}$ is a positive decreasing function of i. Theorem 3 immediately follows by (6) since $\frac{(1+\varepsilon)\delta^2}{M} = \frac{(1+\varepsilon)M H_2}{M} = (1+\varepsilon) \cdot \frac{3}{2}$.

4.2   The Steiner Tree Case

Here we cannot directly apply Lemma 4 since S^* might not be a terminal spanning tree: w.l.o.g. assume that S^* consists of one full component. Following [3], we define a proper auxiliary terminal spanning tree T^*, the *witness tree*. We turn S^* into a rooted binary tree S^*_{bin} as follows: Split one edge, and root the tree at the newly created node r. Split internal nodes of degree larger than 3 by introducing dummy nodes and dummy edges of cost zero. We make the extra assumption[11], that we perform the latter step so that the i most expensive

---

[10] $H_q := \sum_{i=1}^q \frac{1}{i}$ is the q-th harmonic number.
[11] This is irrelevant for [3], but it is useful in the proof of Lemma 5.

edges incident to a given node appear in the highest possible (consecutive) levels of S^*_{bin}. Finally, shortcut internal nodes of degree 2. Tree T^* is constructed as follows. For each internal node v in S^*_{bin} with children u and z, mark uniformly at random exactly one of the two edges vu and vz. Given two terminals r' and r'', add $r'r''$ to T^* iff the path between r' and r'' in S^*_{bin} contains exactly one marked edge. We associate to each edge $f' \in E(S^*_{bin})$ a (non-empty) *witness set* $W(f')$ of edges of T^* as follows: $e = uv \in E(T^*)$ belongs to $W(f')$ iff the path between u and v in S^*_{bin} contains f'. There is a many-to-one correspondence from each $f \in E(S^*)$ to some $f' \in E(S^*_{bin})$: we let $W(f) := W(f')$.

We next apply the same deletion procedure as in Lemma 4 to T^*. When all the edges in $W(f)$ are deleted, we remove f from S^*: this process defines the bridge edges B^t that we remove from $\tilde{S}$ at any iteration. As shown in [3], the non-deleted edges U^t of S^* at the beginning of iteration t plus the components which are sampled in the previous iterations induce (deterministically) a feasible Steiner tree. Hence also in this case we can exploit the upper bound $p(S^t) \le p(U^t) = \sum_v p_{U^t}(v)$. Let us define $W^i(v) := \cup^i_{j=1} W(e^j_v)$. In particular, in order to delete all the edges $e^1_v, \ldots, e^i_v$ we need to delete $W^i(v)$ from T^*. The next technical lemma provides a bound on δ^i_v by combining Lemma 4 with an analysis of the distribution of $|W^i(v)|$. The crucial intuition here is that sets $W(e^j_v)$ are strongly correlated and hence $|W^i(v)|$ tends to be much smaller than $\sum^i_{j=1} |W(e^j_v)|$.

Lemma 5. $\delta^i_v \le \delta^i := \frac{1}{2^i} M H_i + (1 - \frac{1}{2^i}) \sum_{q \ge 1} \frac{1}{2^q} M H_{q+i}$.

Proof. Let us assume that v has degree $d(v) \ge 3$ and that $i < d(v)$, the other cases being analogous and simpler. Recall that we need to delete all the edges in $W^i(v)$ in order to delete $e^1_v, \ldots, e^i_v$, and this takes time $M H_{|W^i(v)|}$ in expectation by Lemma 4. Let us study the distribution of $|W^i(v)|$. Consider the subtree T' of S^*_{bin} given by (the edges corresponding to) $e^1_v, \ldots, e^i_v$ plus their sibling (possibly dummy) edges. Observe that, by our assumption on the structure of S^*_{bin}, this tree has $i + 1$ leaves and height i. We expand T' by appending to each leaf v of T' the only path of unmarked edges from v down to some leaf $\ell(v)$ of S^*_{bin}: let C' be the resulting tree (with $i + 1$ leaves). Each edge of T^* with both endpoints in (the leaves of) C' is a witness edge in $W^i(v)$. The number of these edges is at most i since the witness tree is acyclic: assume pessimistically that they are exactly i. Let r' be the root of T', and s' be the (only) leaf of T' such that the edges on the path from r' to s' are unmarked. Let also d' be the only leaf of T' which is not the endpoint of any e^j_v, $j \le i$ (d' is defined since $i < d(v)$). Observe that $Pr[s' = d'] = 1/2^i$ since this event happens only if the i edges along the path from r' to d' are unmarked. When $s' \ne d'$, there are at most $|E(P')| + 1$ extra edges in $W^i(v)$, where P' is a maximal path of unmarked edges starting from r' and going to the root of S^*_{bin}. If $h_{i,v}$ is the maximum value of $|E(P')|$, then $Pr[|E(P')| = q] = 1/2^{\min\{q+1, h_{i,v}\}}$ for $q \in [0, h_{i,v}]$. Altogether:

$$\delta^i_v \overset{Lem.4}{\le} \sum_{g \ge 1} Pr[|W^i(v)| = g] \cdot M H_g \le \frac{M H_i}{2^i} + (1 - \frac{1}{2^i}) \cdot \sum^{h_{i,v}}_{q=0} \frac{M H_{i+q+1}}{2^{\min\{q+1, h_{i,v}\}}} \le \delta^i. \qquad \square$$

The reader may check that the above δ^i's satisfy Properties (a) and (b) since

$$\frac{\delta^{i+1} - \delta^i}{M} = \frac{1}{(i+1)2^{i+1}} + (1 - \frac{1}{2^{i+1}}) \sum_{q \geq 1} \frac{1}{2^q(q+i+1)} + \frac{1}{2^{i+1}}(\sum_{q \geq 1} \frac{H_{q+i}}{2^q} - H_i)$$

is a positive decreasing function of i. Theorem 1 immediately follows from (6) and Lemma 5 since $\delta^2 = \frac{H_2}{4} + 3(\sum_{q \geq 1} \frac{H_q}{2^q} - \frac{H_1}{2} - \frac{H_2}{4}) = 3 \ln 4 - \frac{9}{4}$.

Acknowledgments. We thank Marek Cygan for reading a preliminary version of this paper and Zeev Nutov for very helpful and inspiring discussions.

References

1. Althaus, E., Calinescu, G., Mandoiu, I.I., Prasad, S.K., Tchervenski, N., Zelikovsky, A.: Power efficient range assignment for symmetric connectivity in static ad hoc wireless networks. Wireless Networks 12(3), 287–299 (2006)
2. Borchers, A., Du., D.-Z.: The k-Steiner ratio in graphs. SIAM Journal on Computing 26(3), 857–869 (1997)
3. Byrka, J., Grandoni, F., Rothvoß, T., Sanità, L.: An improved LP-based approximation for Steiner tree. In: STOC, pp. 583–592 (2010)
4. Calinescu, G.: Min-power Strong Connectivity. In: Serna, M., Shaltiel, R., Jansen, K., Rolim, J. (eds.) APPROX 2010, LNCS, vol. 6302, Springer, Heidelberg (2010)
5. Calinescu, G., Mandoiu, I.I., Zelikovsky, A.: Symmetric connectivity with minimum power consumption in radio networks. In: IFIP TCS, pp. 119–130 (2002)
6. Carmi, P., Katz, M.J.: Power assignment in radio networks with two power levels. Algorithmica 47(2), 183–201 (2007)
7. Clementi, A.E.F., Crescenzi, P., Penna, P., Rossi, G., Vocca, P.: On the Complexity of Computing Minimum Energy Consumption Broadcast Subgraphs. In: Ferreira, A., Reichel, H. (eds.) STACS 2001. LNCS, vol. 2010, pp. 121–131. Springer, Heidelberg (2001)
8. Clementi, A.E.F., Penna, P., Silvestri, R.: On the power assignment problem in radio networks. In: ECCC (2000)
9. Hajiaghayi, M.T., Immorlica, N., Mirrokni, V.S.: Power optimization in fault-tolerant topology control algorithms for wireless multi-hop networks. Transactions on Networking 15(6), 1345–1358 (2007)
10. Hajiaghayi, M.T., Kortsarz, G., Mirrokni, V.S., Nutov, Z.: Power optimization for connectivity problems. Mathematical Programming 110(1), 195–208 (2007)
11. Kirousis, L.M., Kranakis, E., Krizanc, D., Pelc, A.: Power consumption in packet radio networks. Theoretical Computer Science 243, 289–305 (2000)
12. Nutov, Z.: Survivable Network Activation Problems. In: Fernández-Baca, D. (ed.) LATIN 2012. LNCS, vol. 7256, pp. 594–605. Springer, Heidelberg (2012)
13. Nutov, Z., Yaroshevitch, A.: Wireless network design via 3-decompositions. Information Processing Letters 109(19), 1136–1140 (2009)
14. Prömel, H.J., Steger, A.: A new approximation algorithm for the Steiner tree problem with performance ratio 5/3. Journal of Algorithms 36(1), 89–101 (2000)
15. Wan, P.-J., Calinescu, G., Li, X.-Y., Frieder, O.: Minimum energy broadcast routing in static ad hoc wireless networks. In: INFOCOM, pp. 1162–1171 (2001)

Maximum Multicommodity Flows over Time without Intermediate Storage*

Martin Groß and Martin Skutella

Fakultät II – Mathematik und Naturwissenschaften,
Institut für Mathematik, Sekr. MA 5-2
Technische Universität Berlin, Straße des 17. Juni 136,
10623 Berlin, Germany
{gross,skutella}@math.tu-berlin.de

Abstract. Flows over time generalize classical "static" network flows by introducing a temporal dimension. They can thus be used to model non-instantaneous travel times for flow and variation of flow values over time, both of which are crucial characteristics in many real-world routing problems. There exist two different models of flows over time with respect to flow conservation: one where flow might be stored temporarily at intermediate nodes and a stricter model where flow entering an intermediate node must instantaneously progress to the next arc. While the first model is in general easier to handle, the second model is often more realistic since in applications like, e.g., road traffic, storage of flow at intermediate nodes is undesired or even prohibited. The main contribution of this paper is a fully polynomial time approximation scheme (FPTAS) for (min-cost) multi-commodity flows over time without intermediate storage. This improves upon the best previously known $(2+\varepsilon)$-approximation algorithm presented 10 years ago by Fleischer and Skutella (IPCO 2002).

1 Introduction

Two important characteristics of real-world network routing problems are the facts that flow along arcs varies over time and that flow does not arrive instantaneously at its destination but only after a certain delay. As none of these two characteristics are captured by classical network flows, the more powerful model of flows over time has been shifted into the focus of current research. Various interesting applications and examples can, for instance, be found in the surveys of Aronson [1] and Powell, Jailet and Odoni [15]. A more recent introduction to the area of flows over time is given in [16], and a recent application can, e.g., be found in [2].

Results from the literature. Network flows over time have first been studied by Ford and Fulkerson [7, 8], who developed a reduction of flow over time problems

---

* Supported by the DFG Research Center MATHEON "Mathematics for key technologies" in Berlin.

L. Epstein and P. Ferragina (Eds.): ESA 2012, LNCS 7501, pp. 539–550, 2012.

to static flow problems using *time-expanded networks*. This technique requires a discrete time model, but works for virtually all flow-over-time problems at the cost of a pseudo-polynomial blow-up in network size. Note that results for discrete time models often carry over to continuous time models (see, e. g., Fleischer and Tardos [6]). Furthermore, Ford and Fulkerson [7, 8] describe an efficient algorithm for the *maximum flow over time problem*, i. e., the problem of sending the maximum possible amount of flow from a source to a sink within a given time horizon. The algorithm performs a static minimum cost flow computation to find a static flow that is then sent repeatedly through the network.

Related to this problem is the *quickest flow problem*, which asks for the minimum amount of time necessary to send a given amount of flow from the source to the sink. Burkard, Dlaska and Klinz [3] describe a strongly polynomial algorithm for this problem by embedding Ford and Fulkersons' algorithm into Megiddo's parametric search framework [14]. A generalization of this problem, the *quickest transshipment problem*, asks for the minimum amount of time to fulfill given supplies and demands at the nodes. Hoppe and Tardos [11, 12] give a polynomial algorithm relying on submodular function minimization for this.

The *minimum cost flow over time problem* consists of computing a flow of minimum cost, fulfilling given supplies and demands within a specified time horizon. Klinz and Woeginger [13] show that this problem is weakly NP-hard; it can still be solved in pseudo-polynomial time using time-expanded networks.

The problems discussed above can all be generalized to the case of *multiple commodities*. In this setting, we either have a set of source-sink pairs between which flow is sent, or a set of supply-demand vectors specifying different types of supplies/demands that need to be routed in the network. Hall, Hippler, and Skutella [10] show that the multicommodity flow over time problem is NP-hard, even on series-parallel graphs. For multiple commodities, there are cases where storage of flow at intermediate nodes (i. e., nodes that are neither source nor sink nodes) can be necessary for obtaining an optimal solution. If storage at intermediate nodes and non simple flow paths are forbidden, the multicommodity flow over time problem is strongly NP-hard [10] and has no FPTAS, if P $\neq$ NP.

Fleischer and Skutella [4, 5] introduce the concept of *condensed time-expanded networks*, which rely on a rougher discretization of time, resulting in a blow-up in network size that is polynomially bounded. This, however, makes it necessary to increase the time-horizon slightly but still yields an FPTAS for the quickest multicommodity transshipment problem with intermediate storage. The result can be generalized to the quickest multicommodity flow over time problem with bounded cost, if costs on the arcs are non-negative. It is important to emphasize that the underlying technique critically relies on the fact that flow might be stored at intermediate nodes and that there are optimal solutions that send flow only along simple paths through the network. For the case where intermediate storage of flow is prohibited, there is a $(2 + \varepsilon)$-approximation algorithm based on entirely different ideas.

Contribution of this paper. There are numerous flow over time problems where non-simple flow paths are required for optimal solutions. Examples of such problems are multicommodity flows over time without intermediate storage, minimum cost flows over time with arbitrary (possibly negative) arc costs, and generalized flows over time. As in the first problem, the necessity of non-simple flow paths is often closely related to upper bounds on the amount of flow that can be stored at intermediate nodes. Such bounds do exist in most real-world applications and in many cases like, e. g., traffic networks intermediate storage of flow is forbidden completely.

Inspired by the work of Fleischer and Skutella, we present a new, more elaborate condensation technique for time-expanded networks whose analysis no longer requires that flow is being sent along simple paths only. Fleischer and Skutella use the number of arcs on a path to bound the rounding error, which does not yield good bounds with non-simple paths due to the potentially high number of arcs in them. To this end, we introduce a new type of LP formulation for flow over time problems that is somewhere in between an arc-based and a path-based formulation. This LP has a dual which can be approximately separated in polynomial time for most flow problems (more precisely, for flow problems that use standard flow conservation). We start by first studying the separation problem from a general point of view in Section 2. While this problem is NP-hard, we describe an FPTAS for it (which is sufficient for our purposes) that uses dynamic programming in conjunction with Meggido's parametric search framework [14]. In Section 3 we apply this result to the maximum multicommodity flow over time problem without intermediate storage and show that it yields an FPTAS for this NP-hard problem. Furthermore, in Section 4 we extend our techniques to the minimum cost multicommodity flow over time problem with non-negative arc costs.

2   The Restricted Minimum Cost s-t-Sequence Problem

In this section, we define the restricted minimum cost s-t-sequence problem. Informally speaking, this problem consists of finding an arc sequence of minimum cost between two nodes whose length is restricted to lie in a given interval. This length constraint implies that it can be necessary to incorporate cycles in an optimal solution. We present an FPTAS for this NP-hard problem. In Section 3 we make use of this result to solve a dual separation problem.

In order to define our problem formally, we need some notations for arc sequences first. Given a directed graph $G = (V, A)$, a v-w-sequence S is a sequence of arcs $(a_1, a_2, \ldots, a_{|S|})$, $a_i = (v_i, v_{i+1}) \in A$ with $v = v_1$, $w = v_{|S|+1}$. A v-w-path is a v-w-sequence with $v_i \neq v_j$ for all $i \neq j$. A v-cycle is a v-v-sequence with $v_i = v_j$ implying $i = j$ or $i, j \in \{1, |S| + 1\}$. Thus, our paths and cycles are simple arc sequences.

If an arc a is contained in a sequence S, we write $a \in S$. We refer to the number of occurrences of an arc a in a sequence S as $\#(a, S)$. By $S[\to a_i]$ we denote the subsequence $(a_1, a_2, \ldots, a_{i-1})$ of S. Furthermore, we extend the concept of incidence from arcs to sequences and write (by overloading notation) $S \in \delta_v^+$ if $a_1 \in \delta_v^+ := \{ a = (v, \cdot) \mid a \in A \}$ and $S \in \delta_v^- := \{ a = (\cdot, v) \mid a \in A \}$ if $a_{|S|} \in \delta_v^-$.

We also extend arc attributes $f_a \in \mathbb{R}, a \in A$ (e. g., costs, lengths, transit times, etc.) to sequences $S = (a_1, \ldots, a_{|S|})$ by defining $f_S := \sum_{i=1}^{|S|} f_{a_i}$.

Finally notice that we can decompose an s-t-sequence S into an s-t-path and cycles. With these notations, we can define our problem more formally.

Definition 1. *An instance of the* restricted minimum cost s-t-sequence problem *is given by a graph $G = (V, A)$ with costs $c_a \geq 0$ and lengths $\ell_a \geq 0$ on the arcs $a \in A$, two designated nodes $s, t \in V$ and a threshold value $\Delta > \ell^*$, with $\ell^* := \max_{a \in A} \ell_a$. The task is to find an s-t-sequence S of minimum cost $c_S := \sum_{a \in S} \#(a, S) \cdot c_a$ under the constraint that $\Delta - \ell^* \leq \ell_S \leq \Delta$.*

This problem is obviously NP-hard (e. g., by reduction of the length-bounded shortest path problem). The FPTAS that we will develop for this problem approximates the length constraint, i. e., it computes a solution at least as good as an optimal solution but with a relaxed length constraint $(1 - \varepsilon)\Delta - \ell^* \leq \ell_S \leq (1 + \varepsilon)\Delta$ for a given $\varepsilon > 0$. We begin by classifying the arcs into two groups.

Definition 2. *Given $\varepsilon > 0$, an arc $a \in A$ is* short *if $\ell_a \leq \varepsilon\Delta/n$, and* long *otherwise. A sequence is short, if all of its arcs are short, and long otherwise.*

It follows that a solution to the restricted minimum cost s-t-sequence problem can – due to the length constraint – contain at most n/ε long arcs. Since a long cycle needs to contain at least one long arc, it follows that at most n^2/ε arcs of the solution can be part of long cycles; the remaining arcs are part of an s-t-path or short cycles. Due to the fact that a path does not contain duplicate nodes, we make the following observation.

Observation 1. *Let S be a solution to the restricted minimum cost s-t-sequence problem. Then S contains at most $n + n^2/\varepsilon$ arcs that are not part of short cycles.*

For our FPTAS, we would like to round the arc lengths without introducing too much error. Sequences without short cycles are fine, because they have a nicely bounded number of arcs. However, we cannot simply restrict ourselves to sequences without short cycles. Therefore, we now define a *nice v-w-sequence S* to be a v-w-sequence that contains at most one short cycle, but is allowed to make several passes of this cycle. We will see that we can handle such sequences efficiently while losing not too much accuracy. Notice that we can ignore zero-length cycles, as these cycles do neither help us reach t nor can they improve the objective value. The proof of the following lemma can be found in the appendix.

Lemma 1. *For an s-t-sequence S with a path-cycle-decomposition, let C be a short cycle in the decomposition of S with maximal length-to-cost ratio $\frac{\sum_{a \in C} \ell_a}{\sum_{a \in C} c_a}$ and let ℓ_{short} be the total length of all short cycles in the decomposition of S. (Zero-cost cycles are assumed to have infinite ratio.) We define a corresponding nice sequence S' with cycle C by removing all short cycles and adding $\left\lfloor \frac{\ell_{short}}{\ell_C} \right\rfloor$ copies of C. Then $\ell_S - \varepsilon\Delta \leq \ell_S - \ell_C \leq \ell_{S'} \leq \ell_S$ and $c_{S'} \leq c_S$.*

Recall that we are looking for an s-t-sequence of lower or equal cost than the cost of an optimal solution, but with relaxed length constraints. Based on our observations, each optimal solution to the problem yields a corresponding nice sequence of lower or equal cost in the length interval between $\Delta - \ell^* - \varepsilon\Delta$ and Δ. Thus, it is sufficient to describe an algorithm that computes an optimal solution in the set of nice sequences within this length interval. In the following, we describe an algorithm that does so, but with a slightly larger set of allowed sequences. This can only improve the result, however.

Computing the best-ratio-cycles for each node. For each node $v \in V$, we compute the best-ratio short cycle containing v and refer to it as C_v. We can compute the cycle with the best ratio by using Megiddo's framework [14].

Defining rounded lengths. In order to compute the rest of the sequence, we will now introduce rounded lengths. Define $\delta := \varepsilon^2 \Delta/n^2$ and rounded arc lengths ℓ' by $\ell'_a := \lfloor \ell_a/\delta \rfloor \delta$ for all $a \in A$.

Computing the optimal sequences with regard to rounded lengths for each node. We use the rounded arc lengths in a dynamic program. Let $S(v, w, x, y)$ be a v-w-sequence of minimum cost with length x (with regard to ℓ') that contains y arcs. We denote the cost of $S(v, w, x, y)$ by $c(v, w, x, y)$. Notice that storing these values requires only $O(n \cdot n \cdot n^2 \varepsilon^{-2} \cdot (n + n^2 \varepsilon^{-1}))$ space, as there are only n nodes, $n^2 \varepsilon^{-2}$ different lengths and at most $n + n^2 \varepsilon^{-1}$ arcs in the sequences we are looking for (see Observation 1). We initialize the dynamic program with $c(v, w, x, y) := 0$ for $v = w$, $x = 0$, $y = 0$ and $c(v, w, x, y) := \infty$ for either $x = 0$ or $y = 0$. For $x \in \{0, \delta, \ldots, \lfloor n^2/\varepsilon^2 \rfloor \delta\}$ and $y \in \{0, 1, \ldots, \lfloor n + n^2 \varepsilon^{-1} \rfloor\}$ we compute

$$c(v, w, x, y) := \min \left\{ c(v, u, x - \ell'_a, y - 1) + c_a \mid a = (u, w) \in A, x - \ell'_a \geq 0 \right\} .$$

We now define $C(v, w, x) := \min \left\{ c(v, w, x, y) \mid y \in \{0, 1, \ldots, \lfloor n + n^2 \varepsilon^{-1} \rfloor\} \right\}$ as the minimum cost of a v-w-sequence with length x (with regard to ℓ') and at most y arcs. The sequences corresponding to the costs can be computed with appropriate bookkeeping. This dynamic program will either compute a sequence that does not use short cycles or at least a sequence that does not use more arcs than a nice sequence potentially would – which is also fine for our purposes.

Bounding the rounding error. The rounding error for an arc a is bounded by δ. Our sequences have at most $n + n^2/\varepsilon$ arcs. Therefore the rounding error for such a sequence S is at most $(n + n^2/\varepsilon)\delta = (\varepsilon + \varepsilon^2/n)\Delta$. Therefore we can bound the length of S with regard to the rounded lengths by $\ell_S - (\varepsilon + \varepsilon^2/n) \Delta \leq \ell'_S \leq \ell_S$.

Putting the parts together. For this computation, we test all possible start-nodes v for short cycles and all ways to distribute length between the s-v-, v-t-sequences and the short cycles. There are at most $O((n^2 \varepsilon^{-2})^2) = O(n^4 \varepsilon^{-4})$ ways to distribute the length. If the s-v- and v-t-sequence do not have sufficient length to match the length constraint, the remaining length is achieved by using the best-ratio cycle at v. We define $\mathbb{L} := \{0, \delta, \ldots, \lfloor n^2/\varepsilon^2 \rfloor \delta\}$ for brevity. Since we compute s-v- and v-t-sequences with regard to ℓ', we incur a rounding error of $2(\varepsilon + \varepsilon^2/n)$ there. Thus, we have to look for sequences with a length of at least

$L := (1 - 3\varepsilon - 2\varepsilon^2/n)\Delta - \ell^*$ with regard to ℓ' in order to consider all sequences with a length of at least $\Delta - \ell^* - \varepsilon\Delta$ with regard to ℓ. We can compute an s-t-sequence S with cost at most the cost of an optimal solution by

$$\min\left\{ C(s,v,\ell'_s) + c_{C_v}\left\lfloor\frac{L - \ell'_s - \ell'_t}{\ell_{C_v}}\right\rfloor + C(v,t,\ell'_t)\,\middle|\, v \in V, \ell'_s, \ell'_t \in \mathbb{L}, \ell'_s + \ell'_t \leq L\right\}$$

and using bookkeeping. The minimum can be computed in time $O(n^5\varepsilon^{-4})$. Thus, we can find a sequence S with cost at most the cost of an optimal solution and $\left(1 - 3\varepsilon - 2\varepsilon^2/n\right)\Delta - \ell^* \leq \ell_S \leq \left(1 + 2\varepsilon + 2\varepsilon^2/n\right)\Delta$.

Theorem 2. *For a given instance I of the restricted minimum cost s-t-sequence problem and $\varepsilon > 0$, we can compute in time polynomial in the input size and ε^{-1} an arc sequence whose cost is at most the cost of an optimal solution to I and whose length satisfies $(1 - \varepsilon)\Delta - \ell^* \leq \ell_S \leq (1 + \varepsilon)\Delta$.*

3   The Maximum Multicommodity Flow over Time Problem

We will now use this problem to approximate maximum multicommodity flows over time without intermediate storage.

3.1   Preliminaries

An instance of the *maximum multicommodity flow over time problem without intermediate storage* consists of a directed graph $G = (V, A)$ with capacities $u_a \geq 0$ and transit times $\tau_a \in \mathbb{N}_0$ on the arcs $a \in A$. Furthermore, we are given a set of commodities $K = \{1, \ldots, k\}$ with a source s_i and a sink t_i for all $i \in K$. Arc capacity u_a means that all commodities together can send at most u_a units of flow into arc a at any point in time. A transit time of τ_a means that flow entering arc a at time θ leaves the arc at time $\theta + \tau_a$. Finally, we are given a time horizon $T \in \mathbb{N}_0$ within which all flow needs to be sent. Let $n := |V|$, $m := |A|$, and $k := |K|$. Note that we will assume for every commodity $i \in K$, that there is a single source $s_i \in V$ and a single sink $t_i \in V$. Furthermore we assume that sources and sinks have no incoming and outgoing arcs, respectively (i.e., $\delta^-_{s_i} = \delta^+_{t_i} = \emptyset$ for all i). This can be achieved by adding k super-sources and super-sinks.

A *multicommodity flow over time without intermediate storage* $f : A \times K \times [0, T) \to \mathbb{R}_{\geq 0}$ assigns a flow rate $f(a, i, \theta)$ to each arc $a \in A$, each commodity $i \in K$ and all points in time $\theta \in [0, T)$. A flow rate $f(a, i, \theta)$ specifies how much flow of commodity i is sent into arc a at time θ. These flow rates need to fulfill flow conservation: $\sum_{a \in \delta^-_v} f(a, i, \theta - \tau_a) - \sum_{a \in \delta^+_v} f(a, i, \theta) = 0$ for all $i \in K, v \in V \setminus \{s_i, t_i\}, \theta \in [0, T)$. Such a flow is called feasible, if it obeys the capacity constraints as well: $\sum_{i \in K} f(a, i, \theta) \leq u_a$ for all $a \in A, \theta \in [0, T)$.

The objective is to compute a feasible multicommodity flow over time without intermediate storage that maximizes the amount of flow $|f|$ sent by all commodities within the time horizon, i.e., $|f| := \sum_{i \in K} \sum_{a \in \delta^-_{t_i}} \int_{\theta=0}^{T - \tau_a} f(a, i, \theta)$. For convenience, we define the flow rate of a flow over time to be zero on all arcs at all time points $\theta \notin [0, T)$.

For our setting of an integer time horizon and integral transit times, we can switch between the continuous time model of flows over time given above, where points in time are from a continuous interval $[0, T)$, and the discrete time model, where a discrete set of time points $\mathbb{T} := \{0, \dots, T-1\}$ is used. Such a flow over time $x : A \times K \times \mathbb{T} \to \mathbb{R}_{\geq 0}$ assigns flow rates to the arcs and commodities at the specified discrete time points only. It is assumed that flow is sent at this rate until the next time step. The viability of the switching is due to the following lemma (see, e. g., [5]).

Lemma 2. *Let f be a feasible multicommodity flow over time without intermediate storage in the continuous time model. Then there exists a feasible multicommodity flow over time x without intermediate storage in the discrete time model of the same value, and vice versa.*

Using the discrete time model, we can employ *time-expansion*, a standard technique from the literature first employed by Ford and Fulkerson [7, 8]. This technique is based on copying the network for each time step with arcs linking the copies based on their transit times. Furthermore, super-sources s_i' and -sinks t_i' are introduced for each commodity $i \in K$ and connected to the copies of the commodity's source and sink, respectively. A more thorough introduction can, for example, be found in [16]. For the sake of completeness, we give a formal definition of time-expanded networks in the appendix.

3.2   LP-Formulations

Using the discrete time model, we can formulate the problem as an LP in arc variables. A variable $x_{a,i,\theta}$ describes the flow rate of commodity i into arc a at time step θ which is equivalent to the flow on the copy of a starting in time layer θ in the time-expanded network. The only special cases for this time-expanded interpretation are the arcs of the form $(s_i', (s_i)_\theta)$ and $((t_i)_\theta, t_i')$, as these arcs are outside the copied original network. However, we can just consider the arcs $(s_i', (s_i)_\theta)$ being copies of each other belonging to time layer θ and the same for arcs $((t_i)_\theta, t_i')$.

$$
\begin{aligned}
\max \quad & \sum_{i \in K} \sum_{a \in \delta^-_{t_i'}} \sum_{\theta \in \mathbb{T}} x_{a,i,\theta}, \\
\text{s.t.} \quad & \sum_{i \in K} x_{a,i,\theta} && \leq u_a && \text{for all } a \in A, \theta \in \mathbb{T}, \\
& \sum_{a \in \delta^-_v} x_{a,i,\theta-\tau_a} - \sum_{a \in \delta^+_v} x_{a,i,\theta} && = 0 && \text{for all } i \in K, v \in V \setminus \{s_i', t_i'\}, \theta \in \mathbb{T}, \\
& x_{a,i,\theta} && \geq 0 && \text{for all } a \in A, i \in K, \theta \in \mathbb{T}.
\end{aligned}
$$

It is not difficult to see that, due to flow conservation, all flow contributing to the objective function is sent from a source to a sink along some sequence of arcs. We define $\mathcal{S}$ to be the set of all s_i'-t_i'-sequences for all $i \in K$. Then there is always an optimal solution to the maximum multicommodity flow over time problem that can be decomposed into sequences $\mathcal{S}^* \subseteq \mathcal{S}$. We want to split the sequences $S \in \mathcal{S}$ into smaller parts of nearly the same length $\Delta > 0$. Let τ^* be the maximum transit time of an arc. Then we can split S into subsequences of lengths between $\Delta - \tau^*$ and Δ. The length of the last subsequence can be anywhere between 0 and Δ, though, because we have to fit the remaining arcs there. This leads to the following definition.

$$\mathcal{S}^{\Delta}_{\Delta-\tau^*} := \{\, v\text{-}t'_i\text{-sequences } S \mid i \in K, v \in V, \tau_S \leq \Delta \,\}$$
$$\cup \{\, v\text{-}w\text{-sequences } S \mid i \in K, v, w \in V, \Delta - \tau^* \leq \tau_S \leq \Delta \,\}.$$

Since optimal solutions can be decomposed into sequences in $\mathcal{S}^{\Delta}_{\Delta-\tau^*}$, we can also formulate an LP based on variables for all sequences in $\mathcal{S}^{\Delta}_{\Delta-\tau^*}$. Variable $x_{S,i,\theta}$ describes the flow rate of commodity i into sequence S at time step θ. Therefore, the following LP, which we will refer to as the *split-sequence LP*, can be used to compute an optimal solution to our problem:

$$\max \sum_{i \in K} \sum_{S \in \delta^-_{t'_i}} \sum_{\theta \in \mathbb{T}} x_{S,i,\theta-\tau_S},$$

$$\text{s.t.} \quad \sum_{i \in K} \sum_{\substack{S \in \mathcal{S}^{\Delta}_{\Delta-\tau^*}}} \sum_{\substack{j=1,\dots,|S|:\\ a=a_j}} x_{S,i,\theta-\tau_{S[\to a_j]}} \;\leq\; u_a \quad \text{for all } a \in A, \theta \in \mathbb{T},$$

$$\sum_{S \in \delta^-_v} x_{S,i,\theta-\tau_S} - \sum_{S \in \delta^+_v} x_{S,i,\theta} \;=\; 0 \quad \text{for all } i \in K, v \in V \setminus \{s'_i, t'_i\}, \theta \in \mathbb{T},$$

$$x_{S,i,\theta} \;\geq\; 0 \quad \text{for all } S \in \mathcal{S}^{\Delta}_{\Delta-\tau^*}, i \in K, \theta \in \mathbb{T}.$$

This formulation of the problem has two advantages over the initial arc-based formulation. Firstly, all sequences in $\mathcal{S}^{\Delta}_{\Delta-\tau^*}$ – with the exception of those ending in a sink – have a guaranteed length of $\Delta - \tau^*$. Since the time horizon T is an upper bound for the length of any sequence in $\mathcal{S}$, we can conclude that a sequence $S \in \mathcal{S}$ can be split into at most $\lceil T/(\Delta - \tau^*) \rceil + 1$ segments. Secondly, when rounding up the transit times of the sequence-segments, we know that the error introduced by the rounding is bounded by τ^* for all segments save for the last one, which ends in a sink. These two advantages together allow us to obtain a strong bound on the rounding error.

3.3   Solving the Split-Sequence LP

Unfortunately, the split-sequence LP cannot be solved directly in polynomial time due to its enormous size. We will approach this problem in two steps.

1. We round transit times to generate a number of discrete time steps that is polynomially bounded in the input size and ε^{-1}. This introduces a rounding error, which we bound in Section 3.4.
2. The resulting LP has still exponentially many variables; therefore we will consider its dual, which has exponentially many constraints, but only polynomially many variables (in the input size and ε^{-1}). In Section 3.5 we argue that the dual separation problem can be approximated which yields an FPTAS for the original problem.

We begin by introducing a set $\mathcal{S}_L := \{S \text{ is a } v\text{-}w\text{-sequence with } \tau_S \geq L\}$ that be will necessary later due to the fact that we can only approximate the dual separation problem. We define $\overline{\mathcal{S}}_L^{\Delta} := \mathcal{S}_{\Delta-\tau^*}^{\Delta} \cup \mathcal{S}_L$ and replace $\mathcal{S}_{\Delta-\tau^*}^{\Delta}$ with $\overline{\mathcal{S}}_L^{\Delta}$ in the LP above. It is easy to see that this does not interfere with our ability to use this LP to solve our problem. We now round the transit times of the sequences in $\overline{\mathcal{S}}_L^{\Delta}$ up to the next multiple of Δ, i.e., $\tau_S' := \lceil \tau_S/\Delta \rceil \Delta$ for all $S \in \overline{\mathcal{S}}_L^{\Delta}$. Moreover, we define $T' := \lceil (1+\varepsilon)T/\Delta \rceil \Delta$ and $\mathbb{T}' := \{0, \Delta, \ldots, T' - \Delta\}$. Using these transit times yields the following LP, which we will refer to as the *sequence-rounded LP*:

$$\max \sum_{i \in K} \sum_{S \in \delta_{t_i'}^-} \sum_{\theta \in \mathbb{T}'} x_{S,i,\theta-\tau_S'},$$

$$\text{s.t.} \quad \sum_{i \in K} \sum_{S \in \overline{\mathcal{S}}_L^{\Delta}} \#(a,S) \cdot x_{S,i,\theta} \quad \leq \quad \Delta u_a \quad \text{for all } a \in A, \theta \in \mathbb{T}',$$

$$\sum_{S \in \delta_v^-} x_{S,i,\theta-\tau_S'} - \sum_{S \in \delta_v^+} x_{S,i,\theta} \quad = \quad 0 \qquad \text{for all } i \in K, v \in V \setminus \{s_i', t_i'\}, \theta \in \mathbb{T}',$$

$$x_{S,i,\theta} \quad \geq \quad 0 \qquad \text{for all } S \in \overline{\mathcal{S}}_L^{\Delta}, i \in K, \theta \in \mathbb{T}'.$$

3.4   Bounding the Error of the Sequence-Rounded LP

In this section, we will analyze the error introduced by using sequence-rounded transit times. The lemmas and proofs used in this subsection built on and extend techniques introduced by Fleischer and Skutella [5]. Due to space constraints, we have moved all proofs to the appendix. We begin by bounding the error introduced by switching from the given transit times to the sequence-rounded transit times.

Lemma 3. *Let $S \in \mathcal{S}$ be a source-sink sequence and let τ be the normal transit times and τ' the sequence-rounded transit times. Then $\tau_S \leq \tau_S' \leq \tau_S + \varepsilon^2 T$, for $\Delta := \frac{\varepsilon^2 T}{4n}$ and $L := (1 - 1/4\varepsilon^2/(1 + 1/4\varepsilon^2))\Delta$.*

This result enables us construct flows feasible with regard to the sequence-rounded transit times out of normal flows.

Lemma 4. *Let f be a multicommodity flow over time, sending $|f|$ flow units within a time horizon of T. Then, for $\varepsilon > 0$, there exists a multicommodity flow over time that is feasible with regard to the sequence-rounded transit times, sending $|f|/(1+\varepsilon)$ flow units within a time horizon of $(1+\varepsilon)T$.*

Furthermore, we can obtain a solution to the sequence-rounded LP out of any flow feasible with regard to the sequence-rounded transit times.

Lemma 5. *Let f be a feasible multicommodity flow over time with regard to the sequence-rounded transit times. Then there exists a solution x to the sequence rounded LP with equal value.*

Finally, a solution to the sequence-rounded LP can be turned into a normal flow.

Lemma 6. *Let x be a solution to the sequence-rounded LP. Then, for any $\varepsilon > 0$, one can compute a multicommodity flow over time without intermediate storage that sends $|x|/(1+\varepsilon)$ units of flow within time horizon $(1+\varepsilon)T$.*

3.5 Solving the Sequence-Rounded LP's Dual

It remains to show that we can solve the sequence-rounded LP efficiently.

Observation 3. *The dual of the sequence-rounded LP is*

$$min \sum_{a \in A} \sum_{\theta \in \mathbb{T}'} \Delta u_a \beta_{a,\theta},$$

$$s.t. \quad -\alpha_{v,i,\theta} + \alpha_{w,i,\theta+\tau'_S} + \sum_{a \in A} \#(a,S)\beta_{a,\theta} \geq 1 \quad for\ all\ S \in \overline{\mathcal{S}}^{\Delta}_L, i \in K, \theta \in \mathbb{T}',$$

$$\beta_{a,\theta} \qquad\qquad\qquad \geq 0 \quad for\ all\ a \in A, \theta \in \mathbb{T}'.$$

with $\alpha_{s'_i,i,\theta} = \alpha_{t'_i,i,\theta} = 0$ for all commodities $i \in K$ and $\theta \in \mathbb{T}'$, since these variables do not correspond to constraints in the primal LP and have just been added for notational brevity. The separation problem of this dual is to find a sequence $S \in \overline{\mathcal{S}}^{\Delta}_L$ with $\sum_{a \in A} \#(a,S) \cdot \beta_{a,\theta} < 1 + \alpha_{v,i,\theta} - \alpha_{w,i,\theta+\tau'_S}$ for some v,w,i,θ.

Notice that the number of different combinations of v,w,i,θ is polynomially bounded such that we can enumerate them efficiently. Thus, the separation problem is to find a v-w-sequence S of minimum cost with regard to β and $S \in \overline{\mathcal{S}}^{\Delta}_L$. This is the restricted minimum cost s-t-sequence problem introduced in Section 2 with β as the costs and τ as the lengths and Δ as the threshold. As we have seen there, we can only approximately solve this separation problem in the sense that we get a solution with a length between $(1 - \varepsilon)\Delta - \tau^*$ and $(1 + \varepsilon)\Delta$. Recall that we have defined $\overline{\mathcal{S}}^{\Delta}_L$ as $\mathcal{S}^{\Delta}_{\Delta - \tau^*} \cup \mathcal{S}_L$ for this purpose. Until this point, choosing $\mathcal{S}_L = \emptyset$ would have been completely fine, we have not needed it so far. However, using the equivalence of separation and optimization [9], we can find a solution to a dual problem where $\mathcal{S}_L$ contains the additional sequences of length at least $(1 - \varepsilon)\Delta - \tau^*$ found by the ellipsoid method. This leads to $L := (1 - \varepsilon)\Delta - \tau^*$ which is also in line with the value of L we have used in the analysis before. We conclude this section by stating our main result.

Theorem 4. *Let I be an instance of the maximum multicommodity flow over time problem without intermediate storage and let OPT be the value of an optimal solution to it. There exists an FPTAS that, for any given $\varepsilon > 0$, computes a solution with value at least $OPT/(1 + \varepsilon)$ and time horizon at most $(1 + \varepsilon)T$.*

We mention that for the multicommodity flow over time problems with storage at intermediate nodes considered in [5], Fleischer and Skutella obtained a stronger approximation result where only the time horizon is increased by a factor $1 + \varepsilon$ but the original, optimal flow values are sent. In our setting, however, we also need to approximate the flow values and can only send a $1/(1 + \varepsilon)$ fraction. This is due to the fact that for the case where intermediate storage is allowed, increasing the time horizon by a factor δ allows to increase flow values by a factor of δ (Lemma 4.8 in [5]). Such a result does not hold for our setting, though (see Example 1 below). In this context, it is also interesting to mention that hardly any results on approximating the flow value for a fixed time horizon are known and almost all approximation algorithms that guarantee an optimal flow value at the price of slightly increasing the time horizon rely on [5, Lemma 4.8].

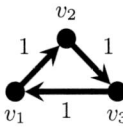 A gadget where increasing the time horizon does not increase the amount of flow that can be sent. The arc labels specify transit times, the capacities are one on all arcs. There are three commodities, the first having v_1 as a source and v_3 as a sink, the second with v_2 and v_1, the third with v_3 and v_2.

Fig. 1. Counterexample to Lemma 4.8 in [5] in our setting

Example 1. Consider the network in Figure 1. Within a time horizon of $T = 3$, three flow units can be sent by routing exactly one flow unit of each commodity. There is only a single source-sink-path for each commodity, and all paths have length 2. Therefore it is easy to see that this is indeed a maximal flow for this instance. Now consider time horizon $T = 4$. We can still not send more than three flow units, as we can only send flow into the three paths in the interval $[0, 2)$, but flow sent into one path during $[0, 1)$ blocks flow send into the next path during $[1, 2)$.

4   Extensions

Our approach above is based on the fact that we can efficiently approximate the separation problem of the dual LP. Therefore, we can extend this approach to related problems whose dual LP is also efficiently approximable.

The *multicommodity flow over time problem without intermediate storage* is similar to the maximum multicommodity flow over time problem without intermediate storage, but instead of a source and a sink for each commodity, we are given supplies and demands $b_{i,v}, i \in K, v \in V$ that specify how much flow of each commodity can be sent from or to a node. The task is to find a flow that satisfies all supplies and demands within a time horizon.

The *minimum cost multicommodity flow over time problem without intermediate storage* receives the same input as an instance of the multicommodity flow over time problem without intermediate storage together with arc costs $c_a \in \mathbb{R}$. The task is to find a feasible multicommodity flow over time of minimum cost fulfilling all supplies and demands within the given time horizon. Formal problem definitions and more details can be found in the appendix.

Theorem 5. *Let I be an instance of the minimum cost multicommodity flow over time problem with non-negative arc costs and let OPT be the value of an optimal solution to it. There exists an FPTAS that, for any given $\varepsilon > 0$, finds a solution of value at most OPT, with time horizon at most $(1 + \varepsilon)T$ that fulfills a $(1 + \varepsilon)^{-1}$ fraction of the given supplies and demands.*

Conclusion. We conclude by stating an open problem that might stimulate further research in this direction: Multicommodity flow over time problems without intermediate storage that are restricted to simple flow paths. It is known that these problems are strongly NP-hard [10] and thus, unless P=NP, do not have an FPTAS. On the positive side, there is a $(2 + \varepsilon)$-approximation algorithm known [4, 5].

Acknowledgments. The authors thank Melanie Schmidt and the anonymous referees for their helpful comments.

References

[1] Aronson, J.E.: A survey of dynamic network flows. Annals of Operations Research 20, 1–66 (1989)

[2] Braun, M., Winter, S.: Ad hoc solution of the multicommodity-flow-over-time problem. IEEE Transactions on Intelligent Transportation Systems 10(4), 658–667 (2009)

[3] Burkard, R.E., Dlaska, K., Klinz, B.: The quickest flow problem. ZOR – Methods and Models of Operations Research 37, 31–58 (1993)

[4] Fleischer, L., Skutella, M.: The Quickest Multicommodity Flow Problem. In: Cook, W.J., Schulz, A.S. (eds.) IPCO 2002. LNCS, vol. 2337, pp. 36–53. Springer, Heidelberg (2002)

[5] Fleischer, L., Skutella, M.: Quickest flows over time. SIAM Journal on Computing 36, 1600–1630 (2007)

[6] Fleischer, L.K., Tardos, É.: Efficient continuous-time dynamic network flow algorithms. Operations Research Letters 23, 71–80 (1998)

[7] Ford, L.R., Fulkerson, D.R.: Flows in Networks. Princeton University Press (1962)

[8] Ford, L.R., Fulkerson, D.R.: Constructing maximal dynamic flows from static flows. Operations Research 6, 419–433 (1987)

[9] Grötschel, M., Lovász, L., Schrijver, A.: Geometric Algorithms and Combinatorial Optimization. Springer (1988)

[10] Hall, A., Hippler, S., Skutella, M.: Multicommodity flows over time: Efficient algorithms and complexity. Theoretical Computer Science 379, 387–404 (2007)

[11] Hoppe, B.: Efficient dynamic network flow algorithms. PhD thesis, Cornell University (1995)

[12] Hoppe, B., Tardos, É.: The quickest transshipment problem. Mathematics of Operations Research 25, 36–62 (2000)

[13] Klinz, B., Woeginger, G.J.: Minimum cost dynamic flows: The series parallel case. Networks 43, 153–162 (2004)

[14] Megiddo, N.: Combinatorial optimization with rational objective functions. Mathematics of Operations Research 4, 414–424 (1979)

[15] Powell, W.B., Jaillet, P., Odoni, A.: Stochastic and dynamic networks and routing. In: Ball, M.O., Magnanti, T.L., Monma, C.L., Nemhauser, G.L. (eds.) Network Routing. Handbooks in Operations Research and Management Science, ch. 3, vol. 8, pp. 141–295. North–Holland, Amsterdam (1995)

[16] Skutella, M.: An introduction to network flows over time. In: Research Trends in Combinatorial Optimization, pp. 451–482. Springer (2009)

Approximating Earliest Arrival Flows
in Arbitrary Networks*

Martin Groß[1], Jan-Philipp W. Kappmeier[1],
Daniel R. Schmidt[2], and Melanie Schmidt[3]

[1] TU Berlin, Germany
{gross,kappmeier}@math.tu-berlin.de
[2] Universität zu Köln, Germany
schmidt@informatik.uni-koeln.de
[3] TU Dortmund, Germany
melanie.schmidt@udo.edu

Abstract. The earliest arrival flow problem is motivated by evacuation planning. It asks for a flow over time in a network with supplies and demands that maximizes the satisfied demands at every point in time. Gale [1959] has shown the existence of such flows for networks with a single source and sink. For multiple sources and a single sink the existence follows from work by Minieka [1973] and an exact algorithm has been presented by Baumann and Skutella [FOCS '06]. If multiple sinks are present, it is known that earliest arrival flows do not exist in general.

We address the open question of approximating earliest arrival flows in arbitrary networks with multiple sinks and present constructive approximations of time and value for them. We give tight bounds for the best possible approximation factor in most cases. In particular, we show that there is always a 2-value-approximation of earliest arrival flows and that no better approximation factor is possible in general. Furthermore, we describe an FPTAS for computing the best possible approximation factor (which might be better than 2) along with the corresponding flow for any given instance.

1 Introduction

Flows over time (also called *dynamic flows*) extend static network flows, introduced by Ford and Fulkerson [7]. For an overview on flows over time, see [17]. In contrast to static network flows, they incorporate a time aspect and are thus better suited to model real-world phenomena such as traffic, production flows or *evacuations*. For the latter, *quickest* flows over time are the model of choice. They capture the idea that each of a given number of individuals shall leave a dangerous area as quickly as possible [3, 6]. An evacuation scenario is modeled by a network, where nodes correspond to locations. Nodes containing evacuees are denoted as *sources* while the network's *sinks* model safe areas or exits. For networks with multiple

---

* This work was supported by DFG Research Center MATHEON "Mathematics for key technologies" in Berlin and by DFG grant SO 514/4-3.

L. Epstein and P. Ferragina (Eds.): ESA 2012, LNCS 7501, pp. 551–562, 2012.

sources and sinks, quickest flows are also denoted as *quickest transshipments* [10] and the problem to compute them is also refered to as the *evacuation problem* [18].

Earliest arrival flows are special quickest flows that maximize the number of evacuated people – i. e., the number of flow units that have reached a sink – at each point in time simultaneously. In particular, if multiple sinks are present, we are only interested in the total amount of flow that has arrived at the sinks but not at which specific sink flow has arrived. It is not clear that these flows exist in general, and indeed their existence depends on the number of sinks. Earliest arrival flows always exist if only one sink is present, as was first proven by Gale [8]. An alternative proof by Minieka [13] is based on the existence of (static) lexicographically maximal flows, and it can be extended to the case of multiple sources. Using discrete time expansion, it only holds in the so-called *discrete time model.* Philpott [14] shows that earliest arrival flows still exist in the *continuous time model* in networks with one sink. An alternative proof by Baumann and Skutella [1] works for multiple sources and in both time models.

Earliest arrival flows in networks with one sink have been intensely studied. For the discrete time model and networks with one source (and one sink), Minieka [13] and Wilkinson [19] develop pseudo-polynomial time algorithms, while Hoppe and Tardos [11] design a fully polynomial time approximation scheme. Fleischer and Tardos [6] develop an algorithm for the continuous time model and networks with one sink. For networks with multiple sources and a single sink, Baumann and Skutella [1] present an algorithm that is polynomial in the sum of input and output size. It is based on a strongly polynomial time algorithm for the quickest flow problem given by Hoppe [12] and works in both the discrete and the continuous time model. A fully polynomial-time approximation scheme is presented by Fleischer and Skutella [5].

In networks with multiple sinks, earliest arrival flows do not necessarily exist [1]. In the general case, we are not aware of any research on algorithms that test if an earliest arrival flow exists in a given network, or that classifies networks where this is always the case. In the special case of networks with zero transit times, earliest arrival flows exist in networks with either one source *or* one sink, but not in networks with multiple sources and sinks [4]. Schmidt and Skutella [16] give a characterization of the class of all networks with zero transit times that always allow for earliest arrival flows, regardless of the number of individuals, the capacity of passages and the capacity of the sinks.

There are no approximation results for earliest arrival flows in networks with multiple sources and sinks and only few related results exist. Baumann and Köhler [2] study approximation in networks with flow-dependend transit times. In this related scenario also, no earliest arrival flows exist in general. Their algorithm guarantees that for each point in time t, the amount of flow that has reached the sink is at least equal to the maximum amount of flow that can reach the sink up to time $t/4$. This type of approximation, which we call *time-approximation*, also occurs in a work by Fleischer and Skutella [5] to speed up the computation of quickest flows over time. Here, time is stretched by a factor of $(1 + \varepsilon)$, i. e., the resulting flow is an FPTAS. Their technique was also successfully applied to the computation of generalized flows over time [9].

Table 1. An overview of existence results for approximate earliest arrival flows

Approx. Type	$|S^+|$	$|S^-|$		$\tau \in \mathbb{R}^{|E|}$		$\tau \equiv 0$
Time (α) and	1	1	$\alpha = \beta = 1$	[8, 13]	$\alpha = \beta = 1$	[8, 13]
Flow Value (β)	2+	1	$\alpha = \beta = 1$	[1, 13]	$\alpha = \beta = 1$	[1, 13]
Flow Value (β)	1	2+	$\beta = 2$	Theorem 2,1	$\beta = 1$	[16]
	2+	2+	$\beta = 2$	Theorem 2,1	$\beta = 2$	Theorem 3, 1
Time (α)	1	2+	$\alpha \in [2,4]$	Theorem 5 and 4, [2]	$\alpha = 1$	[13, 16]
	2+	2+	$\alpha = T$	Section 4	$\alpha = T$	Lemma 5

Our Contribution. Networks with multiple sources and sinks arise naturally if evacuation scenarios include life boats or pick-up bus stations in an urban evacuation. The fact that these networks may not allow for earliest arrival flows is unsettling, and even an algorithm testing whether this is the case would not be completely satisfying. What flow over time do we use if the network does not allow for an earliest arrival flow?

We study the existence and computation of approximate earliest arrival flows in networks with multiple sources and sinks. To our knowledge, this has not been studied before. For simplicities sake, we assume that flow can be sent from any source to any sink, i. e., we are dealing with single commodity flows. However, the techniques described in this paper would apply to multicommodity flows as well. In this case, only the settings with multiple sources and sinks are interesting, because otherwise we have a single commodity flow, essentially. Since single commodity is a special case of multicommodity, all lower bounds carry over directly. The upper bound of Theorem 1 can be carried over by exchanging its maximum flow algorithm by a multicommodity maximum flow algorithm. Our main results are on *value-approximate* earliest arrival flows that send a β-fraction of the maximum flow value at each point in time. We show that for arbitrary networks, $\beta = 2$ is the best possible approximation factor, and give an algorithm that computes 2-value-approximate earliest arrival flows. While the algorithm is pseudopolynomial in the general case, we show that it is efficient for the special case of networks with zero transit times, and we also show that $\beta = 2$ is best possible even in this restricted scenario. We are not aware of any result on value-approximate earliest arrival flows over time in the literature.

We also study *time-approximate* earliest arrival flows and show that despite its usefulness in [2, 5, 9], time-approximation is (nearly) impossible for arbitrary networks with multiple sources and sinks, making value-approximation the method of choice. Additionally, we give some bounds for best possible time-approximation guarantees for networks with one source and multiple sinks.

Table 1 shows our existence results for approximate earliest arrival flows (see Section 2 for definitions of the notations used there). Notice that the upper bound $\beta \leq 2$ only holds in the discrete time model or if the transit times are zero. In the continuous model with nonzero transit times, we show that there is always flow that is $(1 + \varepsilon)$-time and 2-value approximate.

In reality, i. e., for evacuating a *specific* ship or building, a constructive algorithm for a 2-approximate earliest arrival flow is certainly more useful than a non-existence result. Still, given the stakes, the best possible approximation for the given *specific instance* is what is really needed. We give a fully polynomial time approximation scheme (FPTAS) for computing the best possible approximation factor for a given instance (which might be better than 2), which is based on techniques by Fleischer and Skutella [5]. For an arbitrary $\varepsilon > 0$, it incurs additional $(1 + \varepsilon)$-factors for both time and value.

Techniques and Outlook. Even though earliest arrival flow problems have been studied for over sixty years, it is not even known whether the problem is contained in $\mathcal{NP}$ or whether an algorithm with runtime polynomial in the input size exists. Known algorithms are either pseudo-polynomial, depend on factors like the number of breakpoints of the earliest arrival pattern ([1]), or rely on a special graph structure like series-parallel graphs ([15]).

We present tight bounds and constructive approximation algorithms for many settings, providing valuable aid in judging trade-offs in evacuation planning. In particular, we describe a technique for 2-value-approximations that is well suited for extension by additional side-constraints or for use with different models. Its main prerequisite is the existence of an iterative maximum flow algorithm. Notice that while we give a mostly complete overview about possible approximations, a few interesting cases remain open still.

Furthermore, we present a dual existence criterion for earliest arrival flow approximations with zero transit-times that is essentially a cut problem with side constraints. This opens new opportunities for proving lower approximation bounds in many settings and potentially provides insights in developing a combinatorial approximation algorithm for them as well.

Finally, we adapt the concept of geometrically condensed time-expanded networks from [5] to find the best possible approximation factor for any given instance. For applications with an unknown time horizon, as commonly found in evacuation settings, this allows us to obtain the best possible result for a given evacuation scenario. At the same time, our adaption is powerful enough to work in more general settings, e. g., when a time horizon is not completely unknown, but lies in a specified interval.

Due to space constrains, many proofs are omitted or have been concisely written. A full version will appear as a MATHEON preprint.

2   Preliminaries

We use $\mathcal{N} = (V, E, u, \tau, S^+, S^-, b)$ as a unified notation for both static networks and networks over time with slightly different interpretations. In both cases, V is a set of nodes, $E \subseteq V \times V$ is a set of edges, and $S^+ \subseteq V$ and $S^- \subseteq V$ denote disjoint sets of *sources* and *sinks*. For all $v \in V$, we denote the set of edges leaving a node v by $\delta^+(v)$ and the set of edges entering v by $\delta^-(v)$.

In the static case, $u : E \to \mathbb{Z}_{\geq 0}$ assigns a *capacity* u_e to each edge in the network, while in the dynamic case, u_e is interpreted as the maximal *inflow rate* of edge e. The function $\tau : E \to \mathbb{Z}_{\geq 0}$ is left undefined in the static case and denotes the transit time τ_e of an edge e in the dynamic case. Flow entering an edge e with a transit time of τ_e at time θ leaves e at time $\theta + \tau_e$. In both cases, $b : V \to \mathbb{Z}$ can be undefined; if it is defined, then $b_v > 0$ iff $v \in S^+$ and $b_v < 0$ iff $v \in S^-$. A positive b_v denotes the *supply* b_v of a source v, a negative b_v denotes a *demand* $-b_v$ of a sink v. We assume that $\sum_{v \in V} b_v = 0$.

We need three different types of flows during the paper: *static* flows, *flows over time in the discrete time model* and *flows over time in the continuous time model.* We also refer to the latter two cases as *dynamic flows.* All of them are mappings defined on different domains. Let $\mathcal{N} = (V, E, u, -, S^+, S^-, b)$ be a static network and let τ be a transit time function, i. e., $(V, E, u, \tau, S^+, S^-, b)$ is a network over time. A static flow $f : E \to \mathbb{R}_{\geq 0}$ assigns a flow value to each edge. Flows over time define flow rates instead of flow values because they additionally have a temporal dimension in their domain, which is $\{1, \ldots, T\}$ in the discrete and $[0, T)$ in the continuous case, i. e., discrete flows over time only change the flow rate after one time step, while continuous flows specify the flow rate for every moment in time. Now, a flow over time in the discrete time model is a mapping $f : E \times \{1, \ldots, T\} \to \mathbb{R}_{\geq 0}$, and a flow over time in the continuous time model is a Lebesque-integrable function $f : E \times [0, T) \to \mathbb{R}_{\geq 0}$. Both have to satisfy that no flow is left in the edges after the time horizon T, i. e., $f(e, \theta) = 0$ for all $\theta \geq T - \tau_e$.

All three types of flows have to satisfy *capacity constraints*, i. e., $f(e) \leq u_e$ for all $e \in E$ in the static case and $f(e, \theta) \leq u_e$ for all $e \in E$ and $\theta \in \{1, \ldots, T\}$ or $\theta \in [0, T)$, respectively, in the dynamic cases. Furthermore, they have to satisfy *flow conservation constraints*. For brevity, we define the *excess* in each node as the difference between the flow reaching the node and leaving it, i. e., $\mathrm{ex}_f(v) :=$ $\sum_{e \in \delta^-(v)} f(e) - \sum_{e \in \delta^+(v)} f(e)$ in the static case, $\mathrm{ex}_f(v, \theta) := \sum_{e \in \delta^-(v)} \sum_{\xi=0}^{\theta - \tau_e} f(e, \xi) - \sum_{e \in \delta^+(v)} \sum_{\xi=0}^{\theta} f(e, \xi)$ in the discrete dynamic case and $\mathrm{ex}_f(v, \theta) :=$ $\sum_{e \in \delta^-(v)} \int_0^{\theta - \tau_e} f(e, \xi)\, d\xi - \sum_{e \in \delta^+(v)} \int_0^{\theta} f(e, \xi)\, d\xi$ in the continuous dynamic case. Additionally, we define $\mathrm{ex}(v) := \mathrm{ex}(v, T)$ in both dynamic cases.

Now, both static and dynamic flows have to satisfy that $\mathrm{ex}(v) = 0$ for $v \in V \backslash (S^+ \cup S^-)$, $\mathrm{ex}(v) \leq 0$ for $v \in S^+$ and $\mathrm{ex}(v) \geq 0$ for $v \in S^-$. If a function b is defined, then they additionally have to satisfy that $\mathrm{ex}(v) \geq -b_v$ for all $v \in S^+$ and $\mathrm{ex}(v) \leq -b_v$ for all $v \in S^-$. Dynamic flows additionally have to satisfy $\mathrm{ex}(v, \theta) \geq 0$ for $v \in V \backslash S^-$ and $\mathrm{ex}(v, \theta) \leq 0$ for $v \in V \backslash S^+$ for all $\theta \in \{1, \ldots, T\}$ or $\theta \in [0, T)$, respectively.

In all three cases, a *maximum flow* is a flow that maximizes the *flow value* $|f| := \sum_{s^- \in S^-} \mathrm{ex}_f(s^-)$. In the dynamic cases, we denote the amount of flow sent to the sinks until time θ by a flow over time f as $|f|_\theta := \sum_{s^- \in S^-} \mathrm{ex}_f(s^-, \theta)$ with $\theta \in \{1, \ldots, T\}$ and $\theta \in [0, T)$, respectively. Let f_θ^* be a maximum flow over time horizon θ, then we define $p(\theta) := |f_\theta^*|_\theta$ for $\theta \in \{1, \ldots, T\}$ and $\theta \in [0, T)$, respectively. We refer to the values $p(\theta)$ as the *earliest arrival pattern* for $\mathcal{N}$ with supplies and demands b. An *earliest arrival flow* is a flow over time which simultaneously satisfies $|f|_\theta = p(\theta)$ for all points in time $\theta \in \{1, \ldots, T\}$ and $\theta \in [0, T)$, respectively. Notice that earliest arrival flows in networks with multiple sources / sinks associated with supplies / demands are sometimes also called earliest arrival *transshipments*, but we stick to the easier term flow. Now we define the two approximate versions of earliest arrival flows that we study.

Definition 1 (Approximate Earliest Arrival Flows). *An α-time-approx-imate earliest arrival flow is a flow over time f that achieves at every point in time $\theta \in \{1, \ldots, T\}$ and $\theta \in [0, T)$, respectively, at least as much flow value as possible at time θ/α, i. e., $|f|_\theta \geq p\left(\frac{\theta}{\alpha}\right)$. A β-value-approximate earliest arrival flow is a flow over time f that achieves at every point in time $\theta \in \{1, \ldots, T\}$ and $\theta \in [0, T)$, respectively, at least a β-fraction of the maximum flow value at time θ, i.e. $|f|_\theta \geq \frac{p(\theta)}{\beta}$.*

Any flow in the continuous time model induces a flow in the discrete time model that has an identical flow value at each integer point in time. Thus, we have:

Lemma 1. *A network that does not allow for a α-time-approximate earliest arrival flow in the discrete time model does also not allow for one in the continuous time model. The same holds for β-value-approximate earliest arrival flows.*

3   Value-Approximate Earliest Arrival Flows

Value-approximation has not been studied so far, although it is the more intuitive form of approximation: At each point in time, we wish to approximate the maximum amount of flow at this time. We show that this notion of approximation is surprisingly strong, as it allows for 2-approximations in the general case of multiple sinks *and* sources, in contrast to time-approximation.

Theorem 1. *Let $\mathcal{N} = (V, E, u, \tau, S^+, S^-, b)$ be a dynamic network (with possibly multiple sources and sinks). Then there exists a 2-value-approximative earliest arrival flow in $\mathcal{N}$ in the discrete time model.*

Proof. First, we use the standard concept of *time-expanded networks*, which was first described by Ford and Fulkerson [7] (see also [17]). A time expanded network $\mathcal{N}_\theta$ consists of θ time layers from 1 to θ. We denote the copy of node $v \in V$ in time layer i by v_i. If there is an edge $e = (v, w) \in E$, we insert the edge $(v_i, w_{i+\tau_e})$ for all $i = 1, \ldots, \theta - \tau_e$, where τ_e is the transit time of e. For each $s \in S^+$, there is a source s^*, and there are θ edges (s^*, s_i) for the θ copies of s. Similarly, there is a sink t^* for each $t \in S^-$ and θ edges (t_i, t^*). Finally, we add a supersource σ, $|S^+|$ edges (σ, s^*) with capacity $b(s^*)$, a super sink ω and $|S^-|$ edges (t^*, ω) with capacity $-b(t^*)$. A maximum flow with time horizon θ then corresponds to a maximum static flow from σ to ω in $\mathcal{N}_\theta$.

Consider the following algorithm. Start by computing a maximum flow f_1 in $\mathcal{N}_1$. Then, assume that so far the flow f_i was computed and that this flow is a feasible flow in $\mathcal{N}_i$ (which is true for f_1). Based on the static residual network $\mathcal{R}_{i+1}$ (which is induced by considering the flow f_i in the network $\mathcal{N}_{i+1}$), we define a new network $\mathcal{R}'_{i+1}$. It arises from $\mathcal{R}_{i+1}$ by deleting all edges of the form (t^*, t_j) for $j = 1, \ldots, i$ and all $t \in S^-$. These edges can only be present if they are backward edges as they are not part of $\mathcal{N}_{i+1}$. We augment f_i by σ-ω-paths in $\mathcal{R}'_{i+1}$ as long as such paths exist. Observe that if we did so in $\mathcal{R}_{i+1}$ instead, we would get a maximum flow from σ to ω in $\mathcal{N}_{i+1}$ and thus a maximum transshipment with time horizon $i+1$. Denote the resulting flow by f_{i+1}.

We show that f_i is a 2-value-approximate earliest arrival flow with time horizon i for all $i = 1, \ldots, \theta$. Let $f_{i,\max}$ be a maximum flow with time horizon i. Define the flow $g := |f_{i,\max} - f_i|$ by sending $(f_{i,\max} - f_i)(e)$ flow on e if this value is positive and sending $-(f_{i,\max} - f_i)$ on the reverse edge of e otherwise. Denote the value of a flow f by $|f|$; then $|f_{i,\max}| = |f_i| + |g|$. The flow g is valid in $\mathcal{R}_i$, but not necessarily in $\mathcal{R}'_i$. That is, the path decomposition of g can only contain σ-ω-paths that use a deleted edge (otherwise, the path would be present in $\mathcal{R}'_i$, but we augmented until there were no further σ-ω-paths). However, the value sent on these paths is bounded by the sum of the capacities of the deleted backward edges, and this sum can be at most $|f_i|$. Thus, $|g| \leq |f_i|$ and $|f_{i,\max}| \leq |f_i| + |f_i|$ and f_i is in fact a 2-value-approximate earliest arrival flow. □

The algorithm described in Theorem 1 has a pseudo-polynomial running time. However, we are only interested in an existence result here and will later use this result in conjunction with the FPTAS described in Section 5 to obtain an efficient approximation. Obtaining an efficient approximation directly is more complicated due to the fact that the earliest arrival pattern might have an exponential number of breakpoints (see [1] for a discussion).

Notice that the above algorithm only works in the discrete time model. For the continuous model, we conclude that there always exists a $(1 + \varepsilon)$-time and 2-value-approximate earliest arrival flow due to the following Lemma.

Lemma 2. *Let f be a α-time-β-value-approximate earliest arrival flow in the discrete time model, with ε as the length of a time step, in a network which has no flow arriving before time 1. Then there exists a $(1 + \varepsilon)\alpha$-time-β-value-approximation in the continuous time model.*

For the lower bound, we generalize a typical example for a network that does not allow for an earliest arrival flow to a family of networks which do not allow for β'-value-approximate earliest arrival flows, where β' converges to 2. Consider the network in Figure 1a. Let f be a flow in this network in the discrete time model and let $\mathcal{P}$ be a set of paths decomposing f. For $P \in \mathcal{P}_f$ we denote by $\sigma(P)$ the point in time where f starts to send flow along P and by $\ell(P)$ the length of P w.r.t. the transit times. If f sends B units of flow, then the *weighted sum of arrival times* Ξ_f of f is at least $\lfloor B \rfloor^2 + \lfloor B \rfloor$. Additionally, if f is β-value-approximate, then it can be modified to satisfy $\Xi_f \leq (n^2 + n)/(2\beta + 3n)$ without losing the approximation guarantee. This implies a bound which then leads to the following theorem.

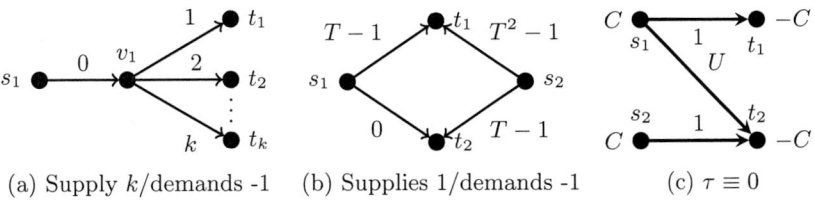

(a) Supply k/demands -1   (b) Supplies 1/demands -1   (c) $\tau \equiv 0$

Fig. 1. Counterexamples. In (a) and (b), edges have unit capacities.

Theorem 2. *For every $\beta < 2$, there exists an instantiation of the network depicted in Figure 1a that does not allow for a β-value-approximate earliest arrival flow in the discrete and continuous time model.*

We conclude with the interesting special case of zero-transit times, i.e., we assume $\tau \equiv 0$. This case has applications in transportation problems. Here, earliest arrival flows always exist iff our network has only one source or only one sink [4]. For a characterization of all networks with zero-transit times that allow for earliest arrival flows for all choices of supplies / demands and capacities, see [16]. For zero-transit times, our 2-value-approximation algorithm in Theorem 1 can be implemented by computing static max flows and repeating them until a terminal runs out of demand / supply. As this can not happen too often, this algorithm is polynomial. For zero transit times, the approximation also works in the continuous time model.

Lemma 3. *For networks with zero transit times, a 2-value-approximate earliest arrival flow can be computed with at most t static maximum flow computations in the continuous time model and at most $t \log b_{max}$ static maximum computations in the discrete time model, with $t := |S^+ \cup S^-|$ being the number of terminals and $b_{max} := \max_{v \in S^+ \cup S^-} |b_v|$ being the highest supply or demand.*

In the following, we show that $\beta = 2$ is best possible even in the case of zero transit times. For this purpose, we again use time-expanded networks. Recall from Theorem 1 that in a time-expanded network, we have copies of the network in each time step and super-terminals that send flow into the layers or recieve flow from the layers, respectively. Notice that due to zero-transit times, there are no edges between time layers. We need some additional notations. We define $\mathcal{P}^\theta$ to be the set of all s-t-paths in N_T containing nodes v_θ. $\mathcal{P}^\theta$ and $\mathcal{P}^{\theta'}$ are disjoint for $\theta \neq \theta'$ since no s-t-path can contain nodes v_θ and $w_{\theta'}$ for $\theta \neq \theta'$. Furthermore, we define $\mathcal{P} = \bigcup_{\theta=1}^{T} \mathcal{P}^\theta$. Recall that p denotes the earliest arrival pattern of N. We now define $p' : \{1, \dots, T\} \to \mathbb{R}$ to be the increase in the earliest arrival pattern between two time steps, i.e., $p'(1) := p(1)$ and $p'(\theta) := p(\theta) - p(\theta - 1)$ for all $\theta \in \{2, \dots, T\}$. Now, setting up the LP for calculating earliest arrival flows and dualizing it leads to the following existence criterion and then to the subsequent theorem.

Lemma 4. *A β-value-approximation in a network $\mathcal{N} = (V, E, u, \tau, S^+, S^-, b)$ with zero transit times exists if and only if the following (dual) LP is unbounded:*

$$
\begin{aligned}
min \quad & \sum_{e \in \mathcal{N}_T} y_e u_e - \sum_{\theta=1}^{T} \frac{p'(\theta)}{\beta} z_\theta, \\
s.t. \quad & \sum_{e \in P} y_e - z_\theta && \geq 0 && \text{for all } \theta \in \{1, \dots, T\}, P \in \mathcal{P}^\theta, \\
& y_e && \geq 0 && \text{for all } e \in E^T, \\
& z_\theta && \geq 0 && \text{for all } \theta \in \{1, \dots, T\}.
\end{aligned}
$$

Theorem 3. *For every $\beta < 2$, there exists a network with zero transit times which does not allow for a β-value-approximate earliest arrival flow.*

Proof. The dual program is a cut problem with side constraints. Imagine an integer solution for the dual program. Then, the variables y_e tell whether (and how often) an edge is 'cut'. For each cut of an edge, the dual program pays the capacity of the edge. It improves its objective function by increasing the variables z_θ. A variable z_θ tells how often each path in $\mathcal{P}^\theta$ has to be cut. Thus, increasing z_θ means that paths have to be cut more often. Finding values for the z_θ such that cutting every path appropriately often is cheaper than $\sum_{\theta=1}^{T} \frac{p'(\theta)}{\beta} z_\theta$ is sufficient to show that no β-flow-approximate earliest arrival flow exists, as multiplying all variables leads to arbitrary small objective function values.

Consider the network in Figure 2. The networks are constructed so that maximizing the flow for different time steps requires flow to be sent from different sources to different sinks, making it impossible to extend a maximum flow for one time step to be (nearly) maximal for all time steps. The networks belong to a family of networks $N_{\ell,k} = (V_{\ell,k}, E_{\ell,k}, u, \tau, S^+, S^-, b)$ with $V_{\ell,k} := \{s_i, t_i \mid i = 0, \ldots, \ell\}$, $E_{\ell,k} := \bigcup_{i=0}^{\ell} \{(s_i, t_j) \mid j = i, \ldots, \ell\}$, $S^+ := \bigcup_{i=0}^{\ell} s_i$, $S^- := \bigcup_{i=0}^{\ell} t_i$, capacities $u_{(s_i,t_j)} := k^{j-i}$ for all $(s_i, t_j) \in E$ and balances $b_{s_i} := k^\ell$ for $s_i \in S^+$, and $b_{t_i} := -k^\ell$ for $t_i \in S^-$. As time horizon, we set $T := k^\ell$.

We define values z_θ such that we can find a cut with cost less than $\sum_{\theta=1}^{k^\ell} z_\theta \cdot p'(\theta)$ that ensures that for all θ, every path through the θth copy of our network, i.e., every path in $\mathcal{P}^\theta$ is cut at least z_θ times. We define $z_\theta := \ell + 1 - \lceil \log_k \theta \rceil$ for $\theta \in \{1, \ldots, k^\ell\}$. This means that $z_1 = \ell + 1$, $z_2 = \cdots = z_k = \ell$, et cetera.

Notice that in every network $N_{\ell,k}$, we can always send k^ℓ units of flow in the first time step using the edge (s_0, t_ℓ). Furthermore, at a time k^{r-1} for $r \in \{1, \ldots, \ell+1\}$, we can use all edges (s_i, t_j) with $j - i = r$ to send $r \cdot k^\ell$ units of flow, thus $p(k^{r-1}) \geq r \cdot k^\ell$ for $r \in \{1, \ldots, \ell+1\}$ and it holds that $\sum_{\theta=1}^{k^\ell} z_\theta \cdot p'(\theta) = \sum_{r=1}^{\ell+1} \sum_{\theta=1}^{k^{r-1}} p'(\theta) = \sum_{r=1}^{\ell+1} p(k^{r-1}) \geq \sum_{r=1}^{\ell+1} r \cdot k^\ell = \frac{(\ell+1)(\ell+2)}{2} k^\ell$. It remains to define a cut cheaper than this term. Cutting edges from and to super terminals has the advantage of simultaneously cutting paths in different copies. We set $y_{(s,s_r)} := \frac{\ell+1}{2} - r$ for all $r \in \{0, \ldots, \frac{\ell-1}{2}\}$ and $y_{(t_r,t)} := r - \frac{\ell-1}{2}$ for all $r \in \{\frac{\ell-1}{2} + 1, \ldots, \ell\}$, incurring a cost of $\frac{(\ell+1)(\ell+3)}{2} k^\ell$. The remaining edges from and to super terminals are not cut. Notice that this choice means that a path through the edge $(s^i_{r_1}, t^i_{r_2})$ for $i \in \{1, \ldots, k^{\ell-m}\}$, $r_1 \leq r_2$, is already cut at least $r_2 - r_1 + 1$ times. Thus, only paths going through an $e = (s^i_{r_1}, t^i_{r_2})$ with $r_2 - r_1 \leq m - 1$ are not cut sufficiently often. For these, it holds $u_e = k^{r_2-r_1} \leq k^{m-1}$. Thus, we set $y_e = m+1$ for all these edges, inducing a cost less than $\ell^3 \cdot k^{m-1} \cdot k^{\ell-m} = \ell^3 \cdot k^{\ell-1}$

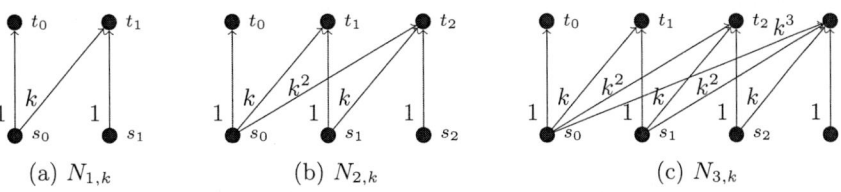

(a) $N_{1,k}$　　　　　　(b) $N_{2,k}$　　　　　　(c) $N_{3,k}$

Fig. 2. The networks $N_{\ell,k}$ for $\ell = 1, 2, 3$, supplies / demands are all k^ℓ

($m \leq \ell$ and less than ℓ^2 edges within each copy). The total cost of the cut is less than $\frac{(\ell+1)(\ell+3)}{2}k^\ell + \mathcal{O}(\ell^3 k^{\ell-1})$, and $\beta \sum_{i=1}^{k^\ell} z_i \cdot p'(i)$ is smaller than this if $\beta < \frac{1}{2} + \frac{1/2}{\ell+2} + \mathcal{O}\left(\frac{\ell}{k}\right)$. Thus, for $\beta > \frac{1}{2}$, there exist values for k and ℓ such that the network has no earliest arrival flow – for $\beta' = \frac{1}{2} + \epsilon$, first set ℓ such that $\frac{1/2}{\ell+2} < \frac{\epsilon}{2}$, and then set k such that $c'\frac{\ell}{k} < \frac{\epsilon}{2}$ for the constant c' hidden in $\mathcal{O}\left(\frac{\ell}{k}\right)$.   □

4   Time-Approximate Earliest Arrival Flows

It is a natural conjecture that time-approximation is a useful tool for our scenario and it is rather surprising that it is not. Notice that there is always a T-time-approximate earliest arrival flow in the discrete and continuous time model, where $T+1$ is the time horizon of the quickest transshipment, when one assumes that no flow arrives before time 1. However, this is basically all that can be achieved. In the network in Figure 1b, it is possible to have two units of flow arriving at time T or one at time 1 and one at time T^2, implying that an α-approximate earliest arrival flow has to satisfy $f(\lceil\alpha\rceil) = 1$ and $f(\lceil\alpha T\rceil) = 2$ (in the discrete time model). This is impossible unless $\alpha \geq T$. The same holds in the continuous time model by Lemma 1.

So, in the general case time-approximation is not as useful as expected or as useful as value-approximation. For the case of only one source, we give a lower and an upper bound on the best possible time-approximation guarantee. The upper bound follows from the work of Baumann and Köhler [2]: They describe a technique that splits the time horizon into exponentially growing intervals in which they compute independent flows over time. Demands and supply vectors create a dependence between the intervals (which is why this does not work for multiple sources *and* sinks), but for the case of only one source, this dependence does not worsen the approximation guarantee compared to the original result.

Theorem 4. *For networks with one source s, there is always a 4-time-approximate earliest arrival flow in the discrete time model. The same is true in the continuous time model if the shortest travel time from s to any sink is positive. For networks with one source and any $\epsilon > 0$, there is a $(4, 1 + \epsilon)$-time-value-approximate earliest arrival transshipment in the continuous time model.*

For the lower bound, we reconsider the network in Figure 1a that also provided the lower bound for the value-approximation. We already know that the weighted sum of arrival times Ξ_f satisfies $\Xi_f \geq \lfloor B\rfloor^2 + \lfloor B\rfloor$ if f is a flow in this network with flow value B. Additionally, every α-time-approximate earliest arrival flow f satisfies $\Xi_f \leq \sum_{i=1}^{k}\lceil\alpha(i+1)\rceil \leq 2k + \sum_{i=1}^{k}\alpha i$, implying the following theorem.

Theorem 5. *For every $\alpha < 2$, there exists an instantiation of the network in Figure 1a that does not allow for an α-time-approximate earliest arrival flow. This holds in the discrete and continuous time model.*

We conclude the section by stating that even in the case of zero-transit times, time-approximation is hopeless for multiple sources and sinks. Let $T(\mathcal{N})$ be the time horizon of a quickest flow in $\mathcal{N}$. In the discrete time model, there exists always a $\alpha = T(\mathcal{N})$-time-approximate earliest arrival flow, because the quickest flow has this approximation guarantee as there is no flow arriving before time one (this also holds in the continuous time model if no flow can arrive before time one). This cannot be improved much, as we see in the next lemma, and the arbitrary continuous case does not allow for any approximation.

Lemma 5. *Let $T(\mathcal{N})$ be the time horizon of a quickest flow for a given network $\mathcal{N}$. For every $\alpha \leq T(\mathcal{N})/2$, there exists a $C \in \mathbb{R}$ such that the zero transit time network $I_{C,C} =: T(\mathcal{N})$ (see Figure 1c) does not allow for an α-time-approximate earliest arrival flow in the discrete time model. For every finite α, there exist $C, U \in \mathbb{R}$ such that the zero transit time network $I_{C,U}$ does not allow for an α-time-approximate earliest arrival flow in the continuous time model.*

Notice that the examples can be modified so that all sources can reach all sinks by adding the edge (s_2, t_1) with capacity C^{-1}. The existence of this edge makes the analysis slightly more involved but does not change the result.

5   Optimal Value-Approximate Earliest Arrival Flows

In Section 3 we described algorithms for computing value-approximate earliest arrival flows and gave instances which allow no better approximation. Thus, our algorithms are best possible in the sense that no better approximation guarantee is possible in general. However, for a specific instance, a better value-approximation might be achievable. We now focus on approximating the optimal value-approximation factor in the continuous setting under the relaxation that flow may arrive later by a factor $(1+\varepsilon)$. This is equivalent to computing an earliest arrival flow approximation that is $(1+\varepsilon)$-time approximate and $(1+\varepsilon)\beta$-value approximate, where β is the best possible value approximation factor. We present a pseudopolynomial exact algorithm in the discrete time model and a fully polynomial time approximation scheme (FPTAS) for both time models.

Lemma 6. *In the discrete time model, an α-time-approximate earliest arrival flow or a β-value-approximate earliest arrival flow can be computed using an LP of pseudopolynomial size, where α/β are the best time/value-approximations possible for the given instance.*

Unfortunately, this does not work directly in the continuous time model, since the flow rate into an edge can change at infinitely many points in time. However, we can discretize time into intervals in which flow rates are considered to be constant, and use *geometrically condensed time-expanded networks*, a technique which was introduced by Fleischer and Skutella [5]. The combination of Lemma 6 and this technique yields the following theorem.

Theorem 6. *For arbitrary $\varepsilon > 0$ with $1/\varepsilon$ integral, a $(1 + O(\varepsilon))\alpha$-time- and $(1+\varepsilon)$-value-approximate earliest arrival flow or a $(1+O(\varepsilon))$-time- and $(1+\varepsilon)\beta$-value-approximate earliest arrival flow can be computed in running time polynomial in the input size and ε^{-1}, where α/β are the best time/value-approximations possible for the given instance. This holds in the continuous time model.*

Acknowledgments. We thank José Verschae and the anonymous reviewers for their helpful comments.

References

[1] Baumann, N., Skutella, M.: Solving evacuation problems efficiently: Earliest arrival flows with multiple sources. Mathematics of Operations Research 34(2), 499–512 (2009)

[2] Baumann, N., Köhler, E.: Approximating earliest arrival flows with flow-dependent transit times. Discrete Applied Mathematics 155(2), 161–171 (2007)

[3] Burkard, R.E., Dlaska, K., Klinz, B.: The quickest flow problem. Mathematical Methods of Operations Research 37, 31–58 (1993)

[4] Fleischer, L.K.: Faster algorithms for the quickest transshipment problem. SIAM Journal on Optimization 12(1), 18–35 (2001)

[5] Fleischer, L.K., Skutella, M.: Quickest flows over time. SIAM Journal on Computing 36(6), 1600–1630 (2007)

[6] Fleischer, L.K., Tardos, É.: Efficient continuous-time dynamic network flow algorithms. Operations Research Letters 23(3-5), 71–80 (1998)

[7] Ford, L.R., Fulkerson, D.R.: Flows in Networks. Princeton University Press (1962)

[8] Gale, D.: Transient flows in networks. Michigan Mathematical Journal 6, 59–63 (1959)

[9] Groß, M., Skutella, M.: Generalized maximum flows over time. In: Proceedings of the 9th WAOA, pp. 247–260 (to appear, 2012)

[10] Hoppe, B.: Efficient dynamic network flow algorithms. Ph.D. thesis, Cornell University (1995)

[11] Hoppe, B., Tardos, É.: Polynomial time algorithms for some evacuation problems. In: Proceedings of the 5th SODA, pp. 433–441 (1994)

[12] Hoppe, B., Tardos, É.: The quickest transshipment problem. Mathematics of Operations Research 25, 36–62 (2000)

[13] Minieka, E.: Maximal, lexicographic, and dynamic network flows. Operations Research 21, 517–527 (1973)

[14] Philpott, A.B.: Continuous-time flows in networks. Mathematics of Operations Research 15(4), 640–661 (1990)

[15] Ruzika, S., Sperber, H., Steiner, M.: Earliest arrival flows on series-parallel graphs. Networks 57(2), 169–173 (2011)

[16] Schmidt, M., Skutella, M.: Earliest arrival flows in networks with multiple sinks. Discrete Applied Mathematics (2011),
http://dx.doi.org/10.1016/j.dam.2011.09.023

[17] Skutella, M.: An introduction to network flows over time. In: Cook, W., Lovász, L., Vygen, J. (eds.) Research Trends in Combinatorial Optimization, pp. 451–482. Springer (2009)

[18] Tjandra, S.A.: Dynamic network optimization with application to the evacuation problem. Ph.D. thesis, Technical University of Kaiserslautern (2003)

[19] Wilkinson, W.L.: An algorithm for universal maximal dynamic flows in a network. Operations Research 19, 1602–1612 (1971)

Resource Buying Games

Tobias Harks[1] and Britta Peis[2]

[1] Department of Quantitative Economics, Maastricht University, The Netherlands
t.harks@maastrichtuniversity.nl
[2] Department of Mathematics, TU Berlin, Germany
peis@math.tu-berlin.de

Abstract. In *resource buying games* a set of players jointly buys a subset of a finite resource set E (e.g., machines, edges, or nodes in a digraph). The cost of a resource e depends on the number (or load) of players using e, and has to be paid completely by the players before it becomes available. Each player i needs at least one set of a predefined family $\mathcal{S}_i \subseteq 2^E$ to be available. Thus, resource buying games can be seen as a variant of congestion games in which the load-dependent costs of the resources can be shared arbitrarily among the players. A strategy of player i in resource buying games is a tuple consisting of one of i's desired configurations $S_i \in \mathcal{S}_i$ together with a payment vector $p_i \in \mathbb{R}_+^E$ indicating how much i is willing to contribute towards the purchase of the chosen resources. In this paper, we study the existence and computational complexity of pure Nash equilibria (PNE, for short) of resource buying games. In contrast to classical congestion games for which equilibria are guaranteed to exist, the existence of equilibria in resource buying games strongly depends on the underlying structure of the families $\mathcal{S}_i$ and the behavior of the cost functions. We show that for marginally non-increasing cost functions, matroids are exactly the right structure to consider, and that resource buying games with marginally non-decreasing cost functions always admit a PNE.

1 Introduction

We introduce and study *resource buying games* as a means to model selfish behavior of players jointly designing a resource infrastructure. In a resource buying game, we are given a finite set N of players and a finite set of resources E. We do not specify the type of the resources, they can be just anything (e.g., edges or nodes in a digraph, processors, trucks, etc.). In our model, the players jointly buy a subset of the resources. Each player $i \in N$ has a predefined family of subsets (called *configurations*) $\mathcal{S}_i \subseteq 2^E$ from which player i needs at least one set $S_i \in \mathcal{S}_i$ to be available. For example, the families $\mathcal{S}_i$ could be the collection of all paths linking two player-specific terminal-nodes s_i, t_i in a digraph $G = (V, E)$, or $\mathcal{S}_i$ could stand for the set of machines on which i can process her job on. The cost c_e of a resource $e \in E$ depends on the number of players using e, and needs to be paid completely by the players before it becomes available. As usual, we assume that the cost functions c_e are non-decreasing and normalized in the

L. Epstein and P. Ferragina (Eds.): ESA 2012, LNCS 7501, pp. 563–574, 2012.

sense that c_e never decreases with increasing load, and that c_e is zero if none of the players is using e. In a weighted variant of resource buying games, each player has a specific weight (demand) d_i, and the cost c_e depends on the sum of demands of players using e. In resource buying games, a strategy of player i can be regarded as a tuple (S_i, p_i) consisting of one of i's desired sets $S_i \in \mathcal{S}_i$, together with a payment vector $p_i \in \mathbb{R}_+^E$ indicating how much i is willing to contribute towards the purchase of the resources. The goal of each player is to pay as little as possible by ensuring that the bought resources contain at least one of her desired configurations. A *pure strategy Nash equilibrium* (PNE, for short) is a strategy profile $\{(S_i, p_i)\}_{i \in N}$ such that none of the players has an incentive to switch her strategy given that the remaining players stick to the chosen strategy. A formal definition of the model will be given in Section 2.

Previous Work. As the first seminal paper in the area of resource buying games, Anshelevich et al. [5] introduced *connection games* to model selfish behavior of players jointly designing a network infrastructure. In their model, one is given an undirected graph $G = (V, E)$ with non-negative (fixed) edge costs $c_e, e \in E$, and the players jointly design the network infrastructure by buying a subgraph $H \subseteq G$. An edge e of E is bought if the payments of the players for this edge cover the cost c_e, and, a subgraph H is bought if every $e \in H$ is bought. Each player $i \in N$ has a specified source node $s_i \in V$ and terminal node $t_i \in V$ that she wants to be connected in the bought subgraph. A strategy of a player is a payment vector indicating how much she contributes towards the purchase of each edge in E. Anshelevich et al. show that these games have a PNE if all players connect to a common source. They also show that general connection games might fail to have a PNE (see also Section 1 below). Several follow-up papers (cf. [3, 4, 6, 7, 9, 11, 12]) study the existence and efficiency of pure Nash and strong equilibria in connection games and extensions of them. In contrast to these works, our model is more general as we assume load-dependent congestion costs and weighted players. Load-dependent cost functions play an important role in many real-world applications as, in contrast to fixed cost functions, they take into account the intrinsic coupling between the quality or cost of the resources and the resulting demand for it. A prominent example of this coupling arises in the design of telecommunication networks, where the installation cost depends on the installed bandwidth which in turn should match the demand for it.

Hoefer [13] studied resource buying games for load-dependent non-increasing marginal cost functions generalizing fixed costs. He considers unweighted congestion games modeling cover and facility location problems. Among other results regarding approximate PNEs and the price of anarchy/stability, he gives a polynomial time algorithm computing a PNE for the special case, where every player wants to cover a single element.

First Insights. Before we describe our results and main ideas in detail, we give two examples motivating our research agenda.

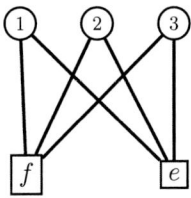
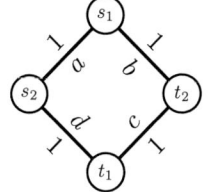

Fig. 1. (a)   **Fig. 1.** (b)

Consider the scheduling game illustrated in Fig.1(a) with two resources (machines) $\{e, f\}$ and three players $\{1, 2, 3\}$ each having unit-sized jobs. Any job fits on any machine, and the processing cost of machines e, f is given by $c_j(\ell_j(S))$, where $\ell_j(S)$ denotes the number of jobs on machine $j \in \{e, f\}$ under schedule S. In our model, each player chooses a strategy which is a tuple consisting of one of the two machines, together with a payment vector indicating how much she is willing to pay for each of the machines. Now, suppose the cost functions for the two machines are $c_e(0) = c_f(0) = 0$, $c_e(1) = c_f(1) = 1$, $c_e(2) = c_f(2) = 1$ and $c_e(3) = c_f(3) = M$ for some large $M > 0$. One can easily verify that there is no PNE: If two players share the cost of one machine, then a player with positive payments deviates to the other machine. By the choice of M, the case that all players share a single machine can never be a PNE. In light of this quite basic example, we have to restrict the set of feasible cost functions. Although the cost functions c_e and c_f of the machines in this scheduling game are monotonically non-decreasing, their marginal cost function is neither non-increasing, nor non-decreasing, where we call cost function $c_e : \mathbb{N} \to \mathbb{R}_+$ *marginally non-increasing [non-decreasing]* if

$$c_e(x + \delta) - c_e(x) \geq \ [\leq] \ c_e(y + \delta) - c_e(y) \quad \forall x \leq y; \ x, y, \delta \in \mathbb{N}. \qquad (1)$$

Note that cost functions with non-increasing marginal costs model economies of scale and include fixed costs as a special case. Now suppose that marginal cost functions are non-increasing and consider scheduling games on restricted machines with uniform jobs. It is not hard to establish a simple polynomial time algorithm to compute a PNE for this setting: Sort the machines with respect to the costs evaluated at load one. Iteratively, let the player whose minimal cost among her available resources is maximal exclusively pay for that resource, drop this player from the list and update the cost on the bought resource with respect to a unit increment of load.

While the above algorithm might give hope for obtaining a more general existence and computability result for PNEs for non-increasing marginal cost functions, we recall a counter-example given by [5]. Consider the connection game illustrated in Fig.1(b), where there are two players that want to establish an s_i-t_i path for $i = 1, 2$. Any strategy profile (*state*) of the game contains two paths, one for each player, that have exactly one edge e in common. In a PNE, no player would ever pay a positive amount for an edge that is not on her chosen path. Now, a player paying a positive amount for e (and at least one such player exists) would have an incentive to switch strategies as she could use the edge that is exclusively used (and paid) by the other player for free. Note that this example uses fixed costs which are marginally non-increasing.

Our Results and Outline. We study unweighted and weighted resource buying games and investigate the existence and computability of pure-strategy Nash equilibria (PNEs, for short). In light of the examples illustrated in Fig.1, we find that equilibrium existence is strongly related to two key properties of the game: the monotonicity of the marginal cost functions and the combinatorial structure of the allowed strategy spaces of the players.

We first consider non-increasing marginal cost functions and investigate the combinatorial structure of the strategy spaces of the players for which PNEs exist. As our main result we show that *matroids* are exactly the right structure to consider in this setting: In Section 3, we present a polynomial-time algorithm to compute a PNE for unweighted matroid resource buying games. This algorithm can be regarded as a far reaching, but highly non-trivial extension of the simple algorithm for scheduling games described before: starting with the collection of matroids, our algorithm iteratively makes use of deletion and contraction operations to minor the matroids, until a basis together with a suitable payment vector for each of the players is found. The algorithm works not only for fixed costs, but also for the more general marginally non-increasing cost functions. Matroids have a rich combinatorial structure and include, for instance, the setting where each player wants to build a spanning tree in a graph. In Section 4, we study weighted resource buying games. We prove that for non-increasing marginal costs and matroid structure, every (socially) optimal configuration profile can be obtained as a PNE. The proof relies on a complete characterization of configuration profiles that can appear as a PNE. We lose, however, polynomial running time as computing an optimal configuration profile is NP-hard even for simple matroid games with uniform players. In Section 5, we show that our existence result is "tight" by proving that the matroid property is also the maximal property of the configurations of the players that leads to the existence of a PNE: For every two-player weighted resource buying game having non-matroid set systems, we construct an isomorphic game that does not admit a PNE.

We finally turn in Section 6 to resource buying games having non-decreasing marginal costs. We show that every such game possesses a PNE regardless of the strategy space. We prove this result by showing that an optimal configuration profile can be obtained as a PNE. We further show that one can compute a PNE efficiently whenever one can compute a best response efficiently. Thus, PNE can be efficiently computed even in multi-commodity network games.

Connection to Classical Congestion Games. We briefly discuss connections and differences between resource buying games and classical congestion games. Recall the congestion game model: the strategy space of each player $i \in N$ consists of a family $\mathcal{S}_i \subseteq 2^E$ of a finite set of resources E. The cost c_e of each resource $e \in E$ depends on the number of players using e. In a classical congestion game, each player i chooses one set $S_i \in \mathcal{S}_i$ and needs to pay the *average* cost of every resource in S_i. Rosenthal [14] proved that congestion games always have a PNE. This stands in sharp contrast to resource buying games for which PNE need not exist even for unweighted singleton two-player games with non-decreasing costs, see Fig.1(a). For congestion games with weighted players, Ackermann et al. [2] showed that for non-decreasing marginal cost functions matroids are the maximal combinatorial structure of strategy spaces admitting PNE. In contrast, Theorem 5 shows that resource buying games with non-decreasing marginal cost functions always have a PNE *regardless* of the strategy space. Our characterization of matroids as the maximal combinatorial structure admitting PNE for resource buying games with non-increasing marginal costs is also different to the one of Ackermann et al. [2] for classical weighted matroid congestion games with non-decreasing marginal costs. Ackermann et al. prove the existence of PNE by using a potential function approach. Our existence result relies on a complete characterization of PNE implying that there exist payments so that the optimal profile becomes a PNE. For unweighted matroid congestion games, Ackermann et al. [1] prove polynomial convergence of best-response by using a (non-trivial) potential function argument. Our algorithm and its proof of correctness are completely different relying on matroid minors and cuts.

These structural differences between the two models become even more obvious in light of the computational complexity of computing a PNE. In classical network congestion games with non-decreasing marginal costs it is PLS-hard to compute a PNE [1,10] even for unweighted players. For network games with weighted players and non-decreasing marginal costs, Dunkel and Schulz [8] showed that it is NP-complete to decide whether a PNE exists. In resource buying (network) games with non-decreasing marginal costs one can compute a PNE in polynomial time even with weighted players (Theorem 6).

2   Preliminaries

The Model. A tuple $\mathcal{M} = (N, E, \mathcal{S}, (d_i)_{i \in N}, (c_r)_{r \in E})$ is called a *congestion model*, where $N = \{1, \ldots, n\}$ is the set of players, $E = \{1, \ldots, m\}$ is the set of *resources*, and $\mathcal{S} = \times_{i \in N} \mathcal{S}_i$ is a set of states (also called *configuration profiles*). For each player $i \in N$, the set $\mathcal{S}_i$ is a non-empty set of subsets $S_i \subseteq E$, called *the configurations of i*. If $d_i = 1$ for all $i \in N$ we obtain an *unweighted* game, otherwise, we have a *weighted* game. We call a configuration profile $S \in \mathcal{S}$ *(socially) optimal* if its total cost $c(S) = \sum_{e \in E} c_e(S)$ is minimal among all $S \in \mathcal{S}$.

Given a state $S \in \mathcal{S}$, we define $\ell_e(S) = \sum_{i \in N : e \in S_i} d_i$ as the total load of e in S. Every resource $e \in E$ has a *cost function* $c_e : \mathcal{S} \to \mathbb{N}$ defined as $c_e(S) = c_e(\ell_e(S))$. In this paper, all cost functions are non-negative, non-decreasing and normalized in the sense that $c_e(0) = 0$. We now obtain a *weighted resource buying game* as the (infinite) strategic game $G = (N, \mathcal{S} \times \mathcal{P}, \pi)$, where $\mathcal{P} = \times_{i \in N} \mathcal{P}_i$ with $\mathcal{P}_i = \mathbb{R}_+^{|E|}$ is the set of feasible payments for the players. Intuitively, each player chooses a configuration $S_i \in \mathcal{S}_i$ and a payment vector p_i for the resources. We say that a resource $e \in E$ is *bought* under strategy profile (S, p), if $\sum_{i \in N} p_i^e \geq c_e(\ell_e(S))$, where p_i^e denotes the payment of player i for resource e. Similarly, we say that a subset $T \subseteq E$ is bought if every $e \in T$ is bought. The private cost function of each player $i \in N$ is defined as $\pi_i(S) = \sum_{e \in E} p_i^e$ if S_i is bought, and $\pi_i(S) = \infty$, otherwise. We are interested in the existence of pure Nash equilibria, i.e., strategy profiles that are resilient against unilateral deviations. Formally, a strategy profile (S, p) is a *pure Nash equilibrium*, PNE for short, if $\pi_i(S, p) \leq \pi_i((S_i', S_{-i}), (p_i', p_{-i}))$ for all players $i \in N$ and all strategies $(S_i, p_i) \in S_i \times P_i$. Note that for PNE, we may assume w.l.o.g that a pure strategy (S_i, p_i) of player i satisfies $p_i^e \geq 0$ for all $e \in S_i$ and $p_i^e = 0$, else.

Matroid Games. We call a weighted resource buying game a *matroid (resource buying) game* if each configuration set $S_i \subseteq 2^{E_i}$ with $E_i \subseteq E$ forms the base set of some matroid $\mathcal{M}_i = (E_i, S_i)$. As it is usual in matroid theory, we will throughout write $\mathcal{B}_i$ instead of S_i, and $\mathcal{B}$ instead of S, when considering matroid games. Recall that a non-empty anti-chain[1] $\mathcal{B}_i \subseteq 2^{E_i}$ is the base set of a matroid $\mathcal{M}_i = (E_i, \mathcal{B}_i)$ on resource (*ground*) set E_i if and only if the following *basis exchange property* is satisfied: whenever $X, Y \in \mathcal{B}_i$ and $x \in X \setminus Y$, then there exists some $y \in Y \setminus X$ such that $X \setminus \{x\} \cup \{y\} \in \mathcal{B}_i$. For more about matroid theory, the reader is referred to [15].

3   An Algorithm for Unweighted Matroid Games

Let $M = (N, E, \mathcal{B}, (c_e)_{e \in E})$ be a model of an unweighted matroid resource buying game. Thus, $\mathcal{B} = \times_{i \in N} \mathcal{B}_i$ where each $\mathcal{B}_i$ is the base set of some matroid $\mathcal{M}_i = (E_i, \mathcal{B}_i)$, and $E = \bigcup_{i \in N} E_i$. In this section, we assume that the cost functions c_e, $e \in E$ are marginally non-increasing.

Given a matroid $\mathcal{M}_i = (E_i, \mathcal{B}_i)$, we will denote by $\mathcal{I}_i = \{I \subseteq E \mid I \subseteq B \text{ for some } B \in \mathcal{B}_i\}$ the collection of *independent sets* in $\mathcal{M}_i$. Furthermore, we call a set $C \subseteq E_i$ a *cut* of matroid $\mathcal{M}_i$ if $E_i \setminus C$ does not contain a basis of $\mathcal{M}_i$. Let $\mathcal{C}_i(\mathcal{M}_i)$ denote the collection of all inclusion-wise minimal cuts of $\mathcal{M}_i$. We will need the following basic insight at several places.

Lemma 1. *[15, Chapters 39 – 42] Let $\mathcal{M}$ be a weighted matroid with weight function $w : E \to \mathbb{R}_+$. A basis B is a minimum weight basis of $\mathcal{M}$ if and only if there exists no basis B^* with $|B \setminus B^*| = 1$ and $w(B^*) < w(B)$.*

In a strategy profile (B, p) of our game with $B = (B_1, \ldots, B_n) \in \mathcal{B}$ (and $n = |N|$) players will jointly buy a subset of resources $\bar{B} \subseteq E$ with $\bar{B} = B_1 \cup \ldots \cup B_n$. Such a strategy profile (B, p) is a PNE if none of the players $i \in N$ would need to pay less by switching to some other basis $B_i' \in \mathcal{B}_i$, given that all other players $j \neq i$ stick to their chosen strategy (B_j, p_j). By Lemma 1, it suffices to consider bases $\hat{B}_i \in \mathcal{B}_i$ with $\hat{B}_i = B_i - g + f$ for some $g \in B_i \setminus \hat{B}_i$ and $f \in \hat{B}_i \setminus B_i$. Note that by switching from B_i to $\hat{B}_i$, player i would need to pay the additional

---

[1] Recall that $\mathcal{B}_i \subseteq 2^{E_i}$ is an *anti-chain* (w.r.t. $(2^{E_i}, \subseteq)$) if $B, B' \in \mathcal{B}_i$, $B \subseteq B'$ implies $B = B'$.

marginal cost $c_f(l_f(B)+1) - c_f(l_f(B))$, but would not need to pay for element g. Thus, (B,p) is a PNE iff for all $i \in N$ and all $\hat{B}_i \in \mathcal{B}_i$ with $\hat{B}_i = B_i - g + f$ for some $g \in B_i \setminus \hat{B}_i$ and $f \in \hat{B}_i \setminus B_i$ holds $p_i^g \leq c_f(l_f(B)+1) - c_f(l_f(B))$.

We now give a polynomial time algorithm (see Algorithm 1 below) computing a PNE for unweighted matroid games with marginally non-increasing costs. The idea of the algorithm can roughly be described as follows: In each iteration, for each player $i \in N$, the algorithm maintains some independent set $B_i \in \mathcal{I}_i$, starting with $B_i = \emptyset$, as well as some payment vector $p_i \in \mathbb{R}_+^E$, starting with the all-zero vector. It also maintains a current matroid $\mathcal{M}_i' = (E_i', \mathcal{B}_i')$ that is obtained from the original matroid $\mathcal{M}_i = (E_i, \mathcal{B}_i)$ by deletion and contraction operations (see e.g., [15] for the definition of deletion and contraction in matroids.) The algorithm also keeps track of the current marginal cost $c_e' = c_e(\ell_e(B)+1) - c_e(\ell_e(B))$ for each element $e \in E$ and the current sequence $B = (B_1, \ldots, B_n)$. Note that c_e' denotes the amount that needs to be paid if some additional player i selects e into its set B_i. In each iteration, while there exists at least one player i such that B_i is not already a basis, the algorithm chooses among all cuts in $\mathcal{C} = \{C \in \mathcal{C}_i(\mathcal{M}_i') \mid \text{for some } i \in N\}$ an inclusion-wise minimal cut C^* whose bottleneck element (i.e., the element of minimal current weight in C^*) has maximal c'-weight (step 3). (We assume that some fixed total order $(E, \preceq)$ is given to break ties, so that the choices of C^* and e^* are unique.) It then selects the bottleneck element $e^* \in C^*$ (step 4), and some player i^* with $C^* \in \mathcal{C}_i(\mathcal{M}_i')$ (step 5). In an update step, the algorithm lets player i^* pay the marginal cost c_{e^*}' (step 7), adds e^* to B_{i^*} (step 8), and contracts element e^* in matroid $\mathcal{M}_{i^*}'$ (step 12). If B_{i^*} is a basis in the original matroid $\mathcal{M}_{i^*}$, the algorithm drops i^* from the player set N (step 10). Finally, the algorithm deletes the elements in $C^* \setminus \{e^*\}$ in all matroids $\mathcal{M}_i'$ for $i \in N$ (step 16), and iterates until $N = \emptyset$, i.e., until a basis has been found for all players.

Obviously, the algorithm terminates after at most $|N| \cdot |E|$ iterations, since in each iteration, at least one element e^* is dropped from the ground set of one of the players. Note that the inclusion-wise minimal cut C^* whose bottleneck element e^* has maximal weight (step 3), as well as the corresponding player i^* and the bottleneck element e^*, can be efficiently found, see the full version for a corresponding subroutine.

It is not hard to see that Algorithm 1 corresponds exactly to the procedure described in Section 1 to solve the scheduling game (i.e., the matroid game on uniform matroids) with non-increasing marginal cost functions. We show that the algorithm returns a pure Nash equilibrium also for general matroids. As a key Lemma, we show that the current weight of the chosen bottleneck element monotonically decreases.

Theorem 1. *The output (B,p) of the algorithm is a PNE.*

Proof. Obviously, at termination, each set B_i is a basis of matroid $\mathcal{M}_i$, as otherwise, player i would not have been dropped from N, in contradiction to the stopping criterium $N = \emptyset$. Thus, we first need to convince ourselves that the

---

Algorithm 1. COMPUTING PNE IN MATROIDS

Input: $(N, E, \mathcal{M}_i = (E_i, \mathcal{B}_i), c)$
Output: PNE (B, p)
1: Initialize $\mathcal{B}_i' = \mathcal{B}_i, E_i' = E_i, B_i = \emptyset, p_i^e = 0, t_e = 1$, and $c_e' = c_e(1)$ for each $i \in N$
   and each $e \in E$;
2: **while** $N \neq \emptyset$ **do**
3:     choose $C^* \leftarrow \operatorname{argmax}\{\min\{c_e' : e \in C\} \mid C \in \mathcal{C}$ inclusion-wise minimal$\}$
       where $\mathcal{C} = \{C \in \mathcal{C}_i(\mathcal{M}_i') \mid$ for some player $i \in N\}$;
4:     choose $e^* \leftarrow \operatorname{argmin}\{c_e' \mid e \in C^*\}$;
5:     choose i^* with $C^* \in \mathcal{C}_{i^*}(\mathcal{M}_{i^*}')$;
6:     $p_{i^*}^{e^*} \leftarrow c_{e^*}'$;
7:     $c_{e^*}' \leftarrow c_{e^*}(t_{e^*} + 1) - c_{e^*}(t_{e^*})$;
8:     $B_{i^*} \leftarrow B_{i^*} + e^*$;
9:     **if** $B_{i^*} \in \mathcal{B}_{i^*}$ **then**
10:        $N \leftarrow N - i^*$;
11:    **end if**
12:    $\mathcal{B}_{i^*}' \leftarrow \mathcal{B}_{i^*}'/e^* = \{B \subseteq E_{i^*}' \setminus \{e^*\} \mid B + e^* \in \mathcal{B}_{i^*}'\}$;
13:    $E_{i^*}' \leftarrow E_{i^*}' \setminus \{e^*\}$;
14:    $t_{e^*} \leftarrow t_{e^*} + 1$;
15:    **for all** players $i \in N$ **do**
16:        $\mathcal{B}_i' \leftarrow \mathcal{B}_i' \setminus (C^* \setminus \{e^*\}) = \{B \subseteq E_i' \setminus (C^* \setminus \{e^*\}) \mid B \in \mathcal{B}_i'\}$
17:        $E_i' \leftarrow E_i' \setminus (C^* \setminus \{e^*\})$;
18:    **end for**
19: **end while**
20: $B = (B_1, \ldots, B_n), p = (p_1, \ldots, p_n)$;
21: **Return** (B, p)

---

algorithm terminates, i.e., constructs a basis B_i for each matroid $\mathcal{M}_i$. However, this follows by the definition of contraction and deletion in matroids:

To see this, we denote by $N^{(k)}$ the current player set, and by $B_i^{(k)}$ and $\mathcal{M}_i^{(k)} = (E_i^{(k)}, \mathcal{B}_i^{(k)})$ the current independent set and matroid of player i at the beginning of iteration k. Suppose that the algorithm now chooses e^* in step 4 and player i^* in step 5. Thus, it updates $B_{i^*}^{(k+1)} \leftarrow B_{i^*}^{(k)} + e^*$ in step 8 and considers the base set $\mathcal{B}_{i^*}^{(k)}/e^*$ of the contracted matroid $\mathcal{M}_{i^*}^{(k)}/e^*$. Note that for each $B \in \mathcal{B}_{i^*}^{(k)}/e^*$, the set $B + e^*$ is a basis in $\mathcal{B}_{i^*}^{(k)}$, and, by induction, $B + B_{i^*}^{(k+1)}$ is a basis in the original matroid $\mathcal{M}_{i^*}$. Thus, $B_{i^*}^{(k+1)}$ is a basis in $\mathcal{M}_{i^*}$ (and i^* is dropped from $N^{(k)}$) if and only if $\mathcal{B}_{i^*}^{(k)}/e^* = \{\emptyset\}$.

Now consider any other player $i \neq i^*$ with $\mathcal{B}_i^{(k)} \neq \{\emptyset\}$ (and thus $i \in N^{(k)}$). Then, for the new base set $\mathcal{B}_i^{(k+1)} = \mathcal{B}_i^{(k)} \setminus (C^* \setminus \{e^*\})$ we still have $\mathcal{B}_i^{(k+1)} \neq \{\emptyset\}$, since otherwise $C^* \setminus \{e^*\}$ is a cut in matroid $\mathcal{M}_i^{(k)}$, in contradiction to the choice of C^*. Thus, since the algorithm only terminates when $N^{(k)} = \emptyset$ for the current iteration k, it terminates with a basis B_i for each player i.

Note that throughout the algorithm it is guaranteed that the current payment vectors $p = (p_1, \ldots, p_n)$ satisfy $\sum_{i \in N} p_i^e = c_e(\ell_e(B))$ for each $e \in E$ and the current independent sets $B = (B_1, \ldots, B_n)$. This follows, since the payments are only modified in step 7, where the marginal payment $p_{i^*}^{e^*} = c_{e^*}(\ell_{e^*}(B) + 1) - c_{e^*}(\ell_{e^*}(B))$ is assigned just before e^* was selected into the set B_{i^*}. Since we assumed the c_e's to be non-decreasing, this also guarantees that each component p_i^e is non-negative, and positive only if $e \in B_i$.

It remains to show that the final output (B,p) is a PNE. Suppose, for the sake of contradiction, that this were not true, i.e., that there exists some $i \in N$ and some basis $\hat{B}_i \in \mathcal{B}_i$ with $\hat{B}_i = B_i - g + f$ for some $g \in B_i \setminus \hat{B}_i$ and $f \in \hat{B}_i \setminus B_i$ such that $p_i^g > c_f(l_f(B+1)) - c_f(l_f(B))$. Let k be the iteration in which the algorithm selects the element g to be paid by player i, i.e., the algorithm updates $B_i^{(k+1)} \leftarrow B_i^{(k)} + g$. Let $C^* = C(k)$ be the cut for matroid $\mathcal{M}_i^{(k)} = (E_i^{(k)}, \mathcal{B}_i^{(k)})$ chosen in this iteration. Thus, the set $E_i^{(k)} \setminus C^*$ contains no basis in $\mathcal{B}_i^{(k)}$, i.e., no set $B \subseteq E_i^{(k)} \setminus C^*$ with $B + B_i^{(k)} \in \mathcal{B}_i$. Note that the final set B_i contains no element from C^* other than g, as all elements in $C^* \setminus \{g\}$ are deleted from matroid $\mathcal{M}_i^{(k)}/g$. We distinguish the two cases where $f \in C^*$, and where $f \notin C^*$.

In the first case, if $f \in C^*$, then, since the algorithm chooses g of minimal current marginal weight, we know that $p_i^g = c_g(l_g(B^{(k)}) + 1) - c_g(l_g(B^{(k)})) \leq c_f(l_f(B^{(k)}+1)) - c_f(l_f(B^{(k)}))$. Thus, the marginal cost of f must decrease at some later point in time, i.e., $c_f(l_f(B+1)) - c_f(l_f(B)) < c_f(l_f(B^{(k)}+1)) - c_f(l_f(B^{(k)}))$. But this cannot happen, since f is deleted from all matroids for which the algorithm has not found a basis up to iteration k.

However, also the latter case cannot be true: Suppose $f \notin C^*$. If $f \in E_i^{(k)}$, then $\hat{B}_i \setminus B_i^{(k)} \subseteq E_i^{(k)} \setminus C^*$, but $\hat{B}_i = \hat{B}_i \setminus B_i^{(k)} + B_i^{(k)} \in \mathcal{B}_i$, in contradiction to C^* being a cut in $\mathcal{M}_i^{(k)}$. Thus, f must have been dropped from E_i in some iteration l prior to k by either some deletion or contraction operation. We show that this is impossible (which finishes the proof): A contraction operation of type $\mathcal{M}_i^{(l)} \rightarrow \mathcal{M}_i^{(l)}/e_l$ drops only the contracted element e_l from player i's ground set $E_i^{(l)}$, after e_l has been added to the current set $B_i^{(l)} \subseteq B_i$. Thus, since $f \notin B_i$, f must have been dropped by the deletion operation in iteration l. Let $C(l)$ be the chosen cut in iteration l, and e_l the bottleneck element. Thus, $f \in C(l) - e_l$. Now, consider again the cut $C^* = C(k)$ of player i which was chosen in iteration k. Recall that the bottleneck element of $C(k)$ in iteration k was g. Note that there exists some cut $C' \supseteq C(k)$ such that C' is a cut of player i in iteration l and $C(k)$ was obtained from C' by the deletion and contraction operations in between iterations l and k. Why did the algorithm choose $C(l)$ instead of C'? The only possible answer is, that the bottleneck element a of C' has current weight $c_a^{(l)} \leq c_{e_l}^{(l)} \leq c_f^{(l)}$. On the other hand, if f was dropped in iteration l, then $c_f^{(l)} = c_f(l_f(B+1)) - c_f(l_f(B))$. Thus, by our assumption, $c_f^{(l)} < p_i^g = c_g^{(k)}$. However, since the cost function c_g is the marginally non-increasing, it follows that $c_g^{(k)} \leq c_g^{(l)}$. Summarizing, we yield $c_a^{(l)} \leq c_{e_l}^{(l)} \leq c_f^{(l)} < c_g^{(k)} \leq c_g^{(l)}$, and, in particular, $c_{e_l}^{(l)} < c_g^{(k)}$, in contradiction to Lemma 2 below (proven in the full version).

Lemma 2. *Let $\hat{c}_k$ denote the current weight of the bottleneck element chosen in step 4 of iteration k. Then this weight monotonically decreases, i.e., $l < k$ implies $\hat{c}_l \geq \hat{c}_k$ for all $l, k \in \mathbb{N}$.*

4   Weighted Matroid Games

For proving the existence of PNE in *weighted* matroid games with non-increasing marginal costs our algorithm presented before does not work anymore. We prove, however, that there exists a PNE in matroid games with non-increasing marginal costs even for weighted demands. To obtain our existence result, we now derive a complete characterization of configuration profiles $B \in \mathcal{B}$ in weighted matroid games $(N, E, \mathcal{B}, d, c)$ that can be obtained as a PNE. For our characterization, we need a few definitions: For $B \in \mathcal{B}$, $e \in E$ and $i \in N_e(B) := \{i \in N \mid e \in B_i\}$ let $\mathrm{ex}_i^e := \{f \in E - e \mid B_i - e + f \in \mathcal{B}_i\} \subseteq E$ denote the set of all resources f such that player i could exchange the resources e and f to obtain an alternative basis $B_i - e + f \in \mathcal{B}_i$. Note that $\mathrm{ex}_i(e)$ might be empty, and that, if $\mathrm{ex}_i(e)$ is empty, the element e lies in every basis of player i (by the matroid basis exchange property). Let $F := \{e \in E \mid e \text{ lies in each basis of } i \text{ for some } i \in N\}$ denote the set of elements that are "fixed" in the sense that they must lie in one of the players' chosen basis. Furthermore, we define for all $e \in E - F$ and all $i \in N_e(B)$ and all $f \in \mathrm{ex}_i(e)$ the value $\Delta_i(B; e \to f) := c_f(\ell_f(B_i + f - e, B_{-i})) - c_f(\ell_f(B))$ which is the marginal amount that needs to be paid in order to buy resource f if i switches from B_i to $B_i - e + f$. Finally, let $\Delta_i^e(B)$ be the minimal value among all $\Delta_i(B; e \to f)$ with $f \in \mathrm{ex}_i(e)$. The proof of the following characterization can be found in the full version.

Theorem 2. *Consider a weighted matroid resource buying game $(N, E, \mathcal{B}, d, c)$. There is a payment vector p such that the strategy profile (B, p) with $B \in \mathcal{B}$ is a PNE if and only if*

$$c_e(B) \leq \sum_{i \in N_e(B)} \Delta_i^e(B) \quad \text{for all } e \in E \setminus F. \tag{2}$$

Note that the above characterization holds for arbitrary non-negative and non-decreasing cost functions. In particular, if property (2) were true, it follows from the constructive proof that the payment vector p can be efficiently computed. The following Theorem3 states that matroid games with non-increasing marginal costs and weighted demands always possess a PNE. We prove Theorem 3 by showing that any socially optimal configuration $B \in \mathcal{B}$ satisfies (2).

Theorem 3. *Every weighted matroid resource buying game with marginally non-increasing cost functions possesses a PNE.*

Note that the above existence result does not imply an efficient algorithm for computing a PNE: It is straightforward to show that computing a socially optimal configuration profile is NP-hard even for unit demands and singleton strategies.

5 Non-matroid Strategy Spaces

In the previous section, we proved that for weighted matroid congestion games with non-negative, non-decreasing, marginally non-increasing cost functions, there always exists a PNE. In this section, we show that the matroid property of the configuration sets is also the maximal property needed to guarantee the existence of a PNE for *all* weighted resource buying games with marginally non-increasing costs (assuming that there is no a priori combinatorial structure how the strategy spaces are interweaved). This result and its proof (in the full version) is related to one of Ackermann et al. in [2] for the classical weighted matroid congestion games with average cost sharing and marginally non-decreasing cost functions.

Theorem 4. *For every non-matroid anti-chain S on a set of resources E, there exists a weighted two-player resource buying game $G = (\tilde{E}, (S_1 \times S_2) \times P, \pi)$ having marginally non-increasing cost functions, whose strategy spaces S_1 and S_2 are both isomorphic to S, so that G does not possess a PNE.*

6 Non-decreasing Marginal Cost Functions

In this section, we consider non-decreasing marginal cost functions on weighted resource buying games in general, i.e., $S = \times_{i \in N} S_i$ is not necessarily the cartesian product of matroid base sets anymore. We prove that for every socially optimal state S^* in a congestion model with non-decreasing marginal costs, we can define *marginal cost* payments p^* that result in a PNE. Formally, for a given socially optimal configuration profile $S^* \in S$ and a fixed order $\sigma = 1, \ldots, n$ of the players, we let $N_e(S^*) := \{i \in N \mid e \in S_i^*\}$ denote the players using e in S^*, $N_e^j(S^*) := \{i \in N_e(S^*) \mid i \leq_\sigma j\}$ denote the players in $N_e(S^*)$ prior or equal to j in σ, and $\ell_e^{\leq j}(S^*) = \sum_{i \in N_e^j(S^*)} d_i$ denote the load of these players on e in S^*. Given these definitions, we allocate the cost $c_e(\ell_e(S^*))$ for each resource $e \in E$ among the players in $N_e(S^*)$ by setting $p_i^e = 0$ if $e \notin S_i^*$ and $p_e^i = c_e(\ell_e^{\leq j}(S^*)) - c_e(\ell_e^{\leq j-1}(S^*))$ if player i is the j-th player in $N_e(S^*)$ w.r.t. σ. Let us call this payment vector *marginal cost pricing*.

Theorem 5. *Let S^* be a socially optimal solution. Then, marginal cost pricing induces a PNE.*

We now show that there is a simple polynomial time algorithm computing a PNE whenever we are able to efficiently compute a best-response. By simply inserting the players one after the other using their current best-response with respect to the previously inserted players, we obtain a PNE. It follows that for (multi-commodity) network games we can compute a PNE in polynomial time.

Theorem 6. *For multi-commodity network games with non-decreasing marginal costs, there is a polynomial time algorithm computing a PNE.*

The proof is straight-forward: Because payments of previously inserted players do not change in later iterations and marginal cost functions are non-decreasing, the costs of alternative strategies only increase as more players are inserted. Thus, the resulting strategy profile is a PNE.

7   Conclusions and Open Questions

We presented a detailed study on the existence and computational complexity of pure Nash equilibria in resource buying games. Our results imply that the price of stability is always one for both, games with non-decreasing marginal costs and games with non-increasing marginal costs and matroid structure. Regarding the price of anarchy, even on games with matroid structure it makes a difference whether cost functions are marginally non-increasing or marginally non-decreasing. For non-increasing marginal costs it is known that the price of anarchy is n (the lower bound even holds for singleton games). On the other hand, for non-decreasing marginal costs we can show that the price of anarchy for uniform and partition matroids is exactly n, while it is unbounded for graphical matroids even on instances with only two players. Convergence of best-response dynamics has not been addressed so far and deserves further research.

References

1. Ackermann, H., Röglin, H., Vöcking, B.: On the impact of combinatorial structure on congestion games. J. ACM 55, 1–25 (2008)
2. Ackermann, H., Röglin, H., Vöcking, B.: Pure Nash equilibria in player-specific and weighted congestion games. Theor. Comput. Sci. 410(17), 1552–1563 (2009)
3. Anshelevich, E., Caskurlu, B.: Exact and Approximate Equilibria for Optimal Group Network Formation. In: Fiat, A., Sanders, P. (eds.) ESA 2009. LNCS, vol. 5757, pp. 239–250. Springer, Heidelberg (2009)
4. Anshelevich, E., Caskurlu, B.: Price of Stability in Survivable Network Design. In: Mavronicolas, M., Papadopoulou, V.G. (eds.) SAGT 2009. LNCS, vol. 5814, pp. 208–219. Springer, Heidelberg (2009)
5. Anshelevich, E., Dasgupta, A., Tardos, É., Wexler, T.: Near-optimal network design with selfish agents. Theory of Computing 4(1), 77–109 (2008)
6. Anshelevich, E., Karagiozova, A.: Terminal backup, 3d matching, and covering cubic graphs. SIAM J. Comput. 40(3), 678–708 (2011)
7. Cardinal, J., Hoefer, M.: Non-cooperative facility location and covering games. Theor. Comput. Sci. 411, 1855–1876 (2010)
8. Dunkel, J., Schulz, A.: On the complexity of pure-strategy Nash equilibria in congestion and local-effect games. Math. Oper. Res. 33(4), 851–868 (2008)
9. Epstein, A., Feldman, M., Mansour, Y.: Strong equilibrium in cost sharing connection games. Games Econom. Behav. 67(1), 51–68 (2009)
10. Fabrikant, A., Papadimitriou, C., Talwar, K.: The complexity of pure Nash equilibria. In: Proc. of the 36th STOC, pp. 604–612 (2004)
11. Hoefer, M.: Non-cooperative tree creation. Algorithmica 53, 104–131 (2009)
12. Hoefer, M.: Strategic Cooperation in Cost Sharing Games. In: Saberi, A. (ed.) WINE 2010. LNCS, vol. 6484, pp. 258–269. Springer, Heidelberg (2010)
13. Hoefer, M.: Competitive cost sharing with economies of scale. Algorithmica 60, 743–765 (2011)
14. Rosenthal, R.: A class of games possessing pure-strategy Nash equilibria. Internat. J. Game Theory 2(1), 65–67 (1973)
15. Schrijver, A.: Combinatorial Optimization: Polyhedra and Efficiency. Springer (2003)

Succinct Data Structures for Path Queries[*]

Meng He[1], J. Ian Munro[2], and Gelin Zhou[2]

[1] Faculty of Computer Science, Dalhousie University, Canada
mhe@cs.dal.ca
[2] David R. Cheriton School of Computer Science, University of Waterloo, Canada
{imunro,g5zhou}@uwaterloo.ca

Abstract. Consider a tree T on n nodes, each having a weight drawn from $[1..\sigma]$. In this paper, we design succinct data structures to encode T using $nH(W_T) + o(n \lg \sigma)$ bits of space, such that we can support path counting queries in $O(\frac{\lg \sigma}{\lg \lg n} + 1)$ time, path reporting queries in $O((occ+1)(\frac{\lg \sigma}{\lg \lg n}+1))$ time, and path median and path selection queries in $O(\frac{\lg \sigma}{\lg \lg \sigma})$ time, where $H(W_T)$ is the entropy of the multiset of the weights of the nodes in T. Our results not only improve the best known linear space data structures [15], but also match the lower bounds for these path queries [18,19,16] when $\sigma = \Omega(n/\mathrm{polylog}(n))$.

1 Introduction

As fundamental structures in computer science, trees are widely used in modeling and representing different types of data in numerous applications. In many scenarios, objects are modeled as nodes of trees, and properties of objects are stored as weights or labels on the nodes. Thus researchers have studied the problem of preprocessing a weighted tree in which each node is assigned a weight, in order to support different path queries, where a path query computes a certain function over the weights of the nodes along a given path in the tree [1,7,13,17,15]. As an example, Chazelle [7] studied how to sum up weights along a query path efficiently.

In this paper, we design succinct data structures to maintain a weighted tree T such that we can support several path queries. We consider *path counting, path reporting, path median,* and *path selection* queries, where the first two kinds of queries are also called *path search* queries. The formal definitions of these queries are listed below. For the sake of simplicity, let $P_{u,v}$ be the set of nodes on the path from u to v, where u, v are nodes in T. Also, let $R_{p,q}$ be the set of nodes in T that have a weight in the range $[p, q]$.

- Path Counting Query: Given two nodes u and v, and a range $[p, q]$, return the cardinality of $P_{u,v} \cap R_{p,q}$;
- Path Reporting Query: Given two nodes u and v, and a range $[p, q]$, return the nodes in $P_{u,v} \cap R_{p,q}$;

[*] This work was supported by NSERC and the Canada Research Chairs Program.

L. Epstein and P. Ferragina (Eds.): ESA 2012, LNCS 7501, pp. 575–586, 2012.
© Springer-Verlag Berlin Heidelberg 2012

- Path Selection Query: Given two nodes u and v, and an integer $1 \leq k \leq |P_{u,v}|$, return the k-th smallest weight in the multiset of the weights of the nodes in $P_{u,v}$. If k is fixed to be $\lceil |P_{u,v}|/2 \rceil$, then path selection queries become path median queries.

As mentioned in [15], these queries generalize two-dimensional range counting, two-dimensional range reporting, range median, and range selection queries.

We represent our tree T as an *ordinal (ordered) tree*. This does not significantly impact the space cost, which is dominated by storing the weights of nodes. In addition, we assume that the weights are drawn from $[1..\sigma]$, or rank space. Thus a query range is an integral one, denoted by $[p..q]$ in the rest of this paper.

1.1   Previous Work

The path counting problem was studied by Chazelle [7], whose formulation is different from ours: Weights are assigned to edges instead of nodes, and a query asks for the number of edges on a given path that have a weight in a given range. Chazelle designed a linear space data structure to support queries in $O(\lg n)$ time. His technique, relying on tree partition, requires $O(\lg n)$ query time for an arbitrary set of weights. He et al. [15] slightly improved Chazelle's result. Under the standard word RAM model, their linear space data structure requires $O(\lg \sigma)$ query time when the weights are drawn from a set of σ distinct values.

The path reporting problem was proposed by He et al. [15], who obtained two solutions under the the word RAM model: One requires $O(n)$ words of space and $O(\lg \sigma + occ \lg \sigma)$ query time, the other requires $O(n \lg \lg \sigma)$ words of space but only $O(\lg \sigma + occ \lg \lg \sigma)$ query time, where occ is the size of output, and σ is the size of the set of weights.

The path median problem was first studied by Krizanc et al. [17], who gave two solutions for this problem. The first one, occupying $O(n \lg^2 n)$ words of space, supports queries in $O(\lg n)$ time. The second one requires $O(b \lg^3 n / \lg b)$ query time and $O(n \lg_b n)$ space, for any $2 \leq b \leq n$. Moreover, a linear space data structure that supports queries in $O(n^\epsilon)$ time can be obtained by setting $b = n^\epsilon / \lg^2 n$ for some small constant ϵ. He et al. [15] significantly improved these results. Under the standard word RAM model, their linear space data structure requires $O(\lg \sigma)$ query time only, where σ is the size of the set of weights.

Succinct Representation of Static Trees. The problem of encoding a static tree succinctly has been studied extensively. For unlabeled trees, a series of succinct representations have been designed [12,14,8,9,21]. For labeled trees, Geary et al. [12] presented a data structure to encode a tree on n nodes, each having a label drawn from an alphabet of size σ, supporting a set of navigational operations in constant time. The overall space cost is $n(\lg \sigma + 2) + O(\sigma n \lg \lg \lg n / \lg \lg n)$ bits, which is much more than the information-theoretic lower bound of $n \lg \sigma + 2n - O(\lg n)$ bits when σ is large. Ferragina et al. [10] and Barbay et al. [2,3] designed data structures for labeled trees using space close to the information-theoretic minimum, but supporting a more restricted set of operations.

Table 1. Best known linear space data structures and new results for path queries. We assume that occ is the size of output. All the results listed in the table are obtained under the standard word RAM model with word size $w = \Omega(\lg n)$. Note that $H(W_T)$ is at most $\lg \sigma$, which is $O(w)$.

Path Query Type	Best Known		new	
	Query Time	Space	Query Time	Space
Counting	$O(\lg \sigma)$	$O(n)$ words	$O(\frac{\lg \sigma}{\lg \lg n} + 1)$	$nH(W_T) + o(n \lg \sigma)$ bits
Reporting	$O((occ + 1) \lg \sigma)$	$O(n)$ words	$O((occ + 1)(\frac{\lg \sigma}{\lg \lg n} + 1))$	$nH(W_T) + o(n \lg \sigma)$ bits
Median / Selection	$O(\lg \sigma)$	$O(n)$ words	$O(\frac{\lg \sigma}{\lg \lg \sigma})$	$nH(W_T) + o(n \lg \sigma)$ bits

Succinct Data Structures for Range Queries. For two-dimensional orthogonal range search queries, the best known result is due to Bose et al. [5]. They presented a data structure that encodes a set of n points in an $n \times n$ grid using $n \lg n + o(n \lg n)$ bits to support range counting queries in $O(\frac{\lg n}{\lg \lg n})$ time, and range reporting queries in $O((occ + 1) \cdot \frac{\lg n}{\lg \lg n})$ time, where occ is the size of output. For range median and range selection queries, Brodal et al. [6] claimed that their data structure achieving $O(\frac{\lg n}{\lg \lg n})$ query time uses $O(n)$ words of space. A careful analysis reveals that their results are even better than claimed: in rank space, their data structure occupies $n \lg n + o(n \lg n)$ bits of space only, which is succinct.

1.2 Our Results

In this paper, we design succinct data structures for path queries over an ordinal tree T on n nodes, each having a weight drawn from $[1..\sigma]$. Let W_T be the multiset consisting of the weights of the nodes in T. Our data structures occupy $nH(W_T) + o(n \lg \sigma)$ bits of space, which is close to the information-theoretic lower bound ignoring lower-order terms, achieving faster query time compared to the best known $\Theta(n \lg n)$-bit data structures [15], which are not succinct. We summarize our main results in Table 1, along with the best known results [15].

It is noteworthy that our data structures for path counting, path median and path selection queries match the lower bounds for related range queries [18,19,16] when $\sigma = \Omega(n/\text{polylog}(n))$.

2 Preliminaries

2.1 Succinct Indexes of Bit Vectors and Integer Sequences

The bit vector is a key structure in many applications. Given a bit vector $B[1..n]$, we consider the following operations. For $\alpha \in \{0, 1\}$, $\text{RANK}_\alpha(B, i)$ returns the number of α-bits in $B[1..i]$, and $\text{SELECT}_\alpha(B, i)$ returns the position of the i-th α-bit in B. The following lemma addresses the problem of succinct representations of bit vectors.

Lemma 1 ([20]). *Let $B[1..n]$ be a bit vector with m 1-bits. B can be represented in $\lg \binom{n}{m} + O(n \lg \lg n / \lg n)$ bits to support* RANK, SELECT, *and the access to each bit in constant time.*

The RANK/SELECT operations can be generalized to sequences of integers. Given an integer sequence $S[1..n]$ in which the elements are drawn from $[1..s]$, we define $\text{RANK}_\alpha(S, i)$ and $\text{SELECT}_\alpha(S, i)$ by extending the definitions of these operations on bit vectors, letting $\alpha \in [1..s]$. Ferragina et al. [11] presented the following lemma:

Lemma 2 (Theorem 3.1 in [11]). *Let $S[1..n]$ be a sequence of integers in $[1..s]$, where $1 \leq s \leq \sqrt{n}$. S can be represented in $nH_0(S) + O(s(n \lg \lg n) / \log_s n)$ bits, where $H_0(S)$ is the zeroth empirical entropy of S, to support* RANK, SE-LECT, *and the access to any $O(\log_s n)$ consecutive integers in constant time.*

Note that the last operation is not explicitly mentioned in the original theorem. However, recall that S is divided into blocks of size $\lfloor \frac{1}{2} \log_s n \rfloor$, and the entry of table E that corresponds to some block G can be found in constant time. We can explicitly store in each entry the content of the block that corresponds to this entry. Thus the content of any block can be retrieved in constant time. It is easy to verify that the extra space cost is $o(n)$ bits.

When s is sufficiently small, say $s = O(\lg^\epsilon n)$ for some constant $0 < \epsilon < 1$, the second term in the space cost is upper bounded by $O(s(n \lg \lg n) / \log_s n) = O(n(\lg \lg n)^2 / \lg^{1-\epsilon} n) = o(n)$ bits.

The RANK operations can be further generalized to COUNT operations, where $\text{COUNT}_\beta(S, i)$ is defined to be $\sum_{\alpha=1}^{\beta} \text{RANK}_\alpha(S, i)$. Bose et al. [5] presented the following lemma:

Lemma 3 (Lemma 3 in [5]). *Let $S[1..n]$ be a sequence of integers in $[1..t]$, where $t = O(\lg^\epsilon n)$ for some constant $0 < \epsilon < 1$. Providing that any $\lceil \log^\lambda n \rceil$ consecutive integers in S can be retrieved in $O(1)$ time for some constant $\lambda > \epsilon$, S can be indexed using $o(n)$ additional bits to support* COUNT *in constant time.*

2.2   Properties of Ordinal Trees and Forests

In this subsection, we review and extend the discussion of ordinal trees by He et al. [15]. For any two nodes u and v in a given ordinal tree T, they denoted by $P_{u,v}$ the set of nodes on the path from u to v. For any node u and its ancestor w, they denoted by $A_{u,w}$ the set of nodes on the path from u to w, excluding the top node w. That is, $A_{u,w} = P_{u,w} - \{w\}$.

We further consider ordinal forests. An ordinal forest F is a left-to-right ordered list of ordinal trees. The preorder traversal sequence of F is defined to be the left-to-right ordered concatenation of the preorder traversal sequences of the ordinal trees in F. Nodes in the same ordinal tree or the same ordinal forest are identified by their preorder ranks, i.e., their positions in the preorder traversal sequence.

He et al. [15] made use of the deletion operation of tree edit distance [4]. They performed this operation on non-root nodes only. Let u be a non-root node, and let v be its parent. To delete u, its children are inserted in place of u into the list of children of v, preserving the original left-to-right order. We also perform this operation on roots. Let u be a root. An ordinal forest of trees rooted at the children of u occurs after deleting u, where the trees are ordered by their roots.

To support path queries, He et al. [15] defined notation in terms of weights. Let T be an ordinal tree, whose weights of nodes are drawn from $[1..\sigma]$. For any range $[a..b] \subseteq [1..\sigma]$, they defined $R_{a,b}$ to be the set of nodes in T that have a weight in $[a..b]$. They also defined $anc_{a,b}(T,x)$ to be the lowest ancestor of x that has a weight in $[a..b]$. Here we slightly modify their definition. We assume that $anc_{a,b}(T,x) =$ NULL if x has no ancestor that has a weight in $[a..b]$. To be consistent, we assume that the depth of the NULL node is 0. In addition, we define $F_{a,b}$, which is similar to $T_{a,b}$ in [15], to be the ordinal forest after deleting all the nodes that are not in $R_{a,b}$ from T, where the nodes are deleted from bottom to top. Note that there is a one-to-one mapping between the nodes in $R_{a,b}$ and the nodes in $F_{a,b}$. As proved in [15], the nodes in $R_{a,b}$ and the nodes in $F_{a,b}$ that correspond to them have the same relative positions in the preorder of T and the preorder of $F_{a,b}$.

2.3 Succinct Ordinal Tree Representation Based on Tree Covering

In this subsection, we briefly summarize the tree-covering based representation of ordinal trees [12,14,8,9], which we extend in Section 3 to support more powerful operations such as COUNT and SUMMARIZE.

The tree-covering based representation was proposed by Geary et al. [12] to represent an ordinal tree succinctly to support navigational operations. Let T be an ordinal tree on n nodes. A tree cover of T with a given parameter M is essentially a set of $O(n/M)$ *cover elements*, each being a connected subtree of size $O(M)$. These subtrees, being either disjoint or joined at the common root, cover all the nodes in T.

Geary et al. [12] proposed an algorithm to cover T with *mini-trees* or *tier-1 subtrees* for $M = \max\{\lceil(\lg n)^4\rceil, 2\}$. Again, they apply the algorithm to cover each mini-tree with *micro-trees* or *tier-2 subtrees* for $M' = \max\{\lceil(\lg n)/24\rceil, 2\}$. For $k = 1, 2$, the nodes that are roots of the tier-k subtrees are called *tier-k roots*. Note that a tier-k root can be the root node of multiple tier-k subtrees, and a tier-1 root must also be a tier-2 root.

To use tree-covering based representation to support more navigational operations, He et al. [14, Definition 4.22] proposed the notion of tier-k *preorder segments*, where the preorder segments are defined to be maximal substrings of nodes in the preorder sequence that are in the same mini-tree or micro-tree. Farzan and Munro [8] further modified the tree-covering algorithm. Their algorithm computes a tree cover such that nodes in a cover element are distributed into a constant number of preorder segments, as proved by Farzan et al. [9]. These results are summarized in the following lemma:

Lemma 4 (Theorem 1 in [8] and Lemma 2 in [9]). *Let T be an ordinal tree on n nodes. For a parameter M, the tree decomposition algorithm in [8] covers the nodes in T by $\Theta(n/M)$ cover elements of size at most $2M$, all of which are pairwise disjoint other than their root nodes. In addition, nodes in one cover element are distributed into $O(1)$ preorder segments.*

The techniques in [12,14,8,9] encode an unlabeled ordinal tree on n node in $2n + o(n)$ bits to support in constant time a set of operations related to nodes, tier-k subtrees and tier-k roots. For example, given an arbitrary node x, we can compute its depth, and find its i-th ancestor. Given two nodes x and y, we can compute their lowest common ancestor (LCA).

For $k = 1$ or 2, the tier-k subtrees are ordered and specified by their ranks. The following operations can be performed in constant time: For each tier-k subtree, we can find its root node, compute the preorder segments contained in it, and select the i-th node in preorder that belongs to this subtree. For each micro-tree, we can compute the encoding of its structure. For each node that is not a tier-k root, we can find the tier-k subtree to which it belongs, and its relative preorder rank in this tier-k subtree.

Similarly, for $k = 1$ or 2, the tier-k roots are ordered in preorder and specified by their ranks. Let r_i^1/r_i^2 denote the i-th tier-1/tier-2 root. Given a node x, we can compute its rank if x is a tier-k root, or determine that x is not a tier-k root. Conversely, given the rank of a tier-k root, we can compute its preorder rank.

3   Succinct Ordinal Trees over Alphabets of Size $O(\lg^\epsilon n)$

In this section, we present a succinct data structure to encode an ordinal tree T on n weighted nodes, in which the weights are drawn from $[1..t]$, where $t = O(\lg^\epsilon n)$ for some constant $0 < \epsilon < 1$. This succinct representation occupies $n(H_0(PLS_T) + 2) + o(n)$ bits of space to support a set of operations in constant time, where PLS_T is the preorder label sequence of T. As the entropy of W_T only depends on the frequencies of labels, we have $H_0(PLS_T) = H(W_T)$. Thus, the space cost can also be represented as $n(H(W_T) + 2) + o(n)$ bits. However, we use $H_0(PLS_T)$ in this section to facilitate space analysis. The starting points of our succinct representation are Geary et al.'s [12] succinct ordinal tree, and Bose et al.'s [5] data structure for orthogonal range search on a narrow grid.

In the rest of this section, we consider the operations listed below. For the sake of convenience, we call a node x an α-node or an α-ancestor if the weight of x is α. Also, we assume that nodes x, y are contained in T, α and β are in $[1..t]$, a node precedes itself in preorder, and a node is its own 0-th ancestor.

- PRE-RANK$_\alpha(T, x)$: Return the number of α-nodes that precede x in preorder;
- PRE-SELECT$_\alpha(T, i)$: Return the i-th α-node in preorder;
- PRE-COUNT$_\beta(T, x)$: Return the number of nodes preceding x in preorder that have a weight $\leq \beta$;
- DEPTH$_\alpha(T, x)$: Return the number of α-ancestors of x (excluding the root);
- LOWEST-ANC$_\alpha(T, x)$: Return the lowest α-ancestor of x if such an α-ancestor exists, otherwise return the root of T;
- COUNT$_\beta(T, x)$: Return the number of nodes on the path from x to the root of T (excluding the root) that have a weight $\leq \beta$;
- SUMMARIZE(T, x, y): Given that node y is an ancestor of x, this operation returns t bits, where the α-th bit is 1 if and only if there exists an α-node on the path from x to y (excluding y), for $1 \leq \alpha \leq t$.

We first compute the mini-micro tree cover of T using Lemma 4 for $M = \lceil \lg^2 n \rceil$ and $M' = \lceil \lg^\lambda n \rceil$ for some $\max\{\epsilon, \frac{1}{2}\} < \lambda < 1$. It is easy to verify that, with our choice of M and M', the operations in Subsection 2.3, which are related to tree nodes, tier-k subtrees and tier-k roots, can still be supported in constant time, using $2n + o(n)$ bits of space. Let n_1 and n_2 be the numbers of tier-1 roots and tier-2 roots. By Lemma 4, they are bounded by $O(n/M)$ and $O(n/M')$, respectively. To store the weights, we encode $PLS_T[1..n]$ using Lemma 2, occupying $nH_0(PLS_T) + o(n)$ bits of space. These are our main data structures that encode the structure of tree and the weights of nodes. We design auxiliary data structures of $o(n)$ bits to support operations.

Lemma 5. SUMMARIZE *can be supported in constant time, using $o(n)$ additional bits of space.*

Proof. To support SUMMARIZE, we need the following operations. For simplicity, we assume that x is a tier-1 root, and y is a tier-2 root.

- MINI-DEPTH(T, x): Return the number of tier-1 roots on the path from x to the root of T;
- MINI-ANC(T, x, i): Return the $(i + 1)$-th tier-1 root on the path from x to the root of T, providing that $0 \leq i < $ MINI-DEPTH(T, x);
- MICRO-DEPTH(T, y): Return the number of tier-2 roots on the path from y to the root of T;
- MICRO-ANC(T, y, i): Return the $(i + 1)$-th tier-2 root on the path from y to the root of T, providing that $0 \leq i < $ MICRO-DEPTH(T, y).

We only show how to support the first two operations. The other operations can be supported in the same way. We compute tree T' by deleting all the nodes other than the tier-1 roots from T. Clearly T' has n_1 nodes. To convert the nodes in T' and the tier-1 roots in T, we construct and store a bit vector $B_1[1..n]$, in which $B_1[i] = 1$ iff the i-th node in preorder of T is a tier-1 root. By Lemma 1, B_1 can be encoded in $o(n)$ bits to support RANK and SELECT in constant time. Note that the i-th node in T' corresponds to the tier-1 root of preorder rank j in T, providing that $B_1[j]$ is the i-th 1-bit in B_1. Thus the conversion can be done in constant time using RANK and SELECT. Applying the techniques in Subsection 2.3, we encode T' in $2n_1 + o(n_1) = o(n)$ bits to support MINI-DEPTH and MINI-ANC directly.

We then construct the following auxiliary data structures:

- A two-dimensional array $D[1..n_1, 0..\lceil \lg n \rceil]$, in which $D[i, j]$ stores a bit vector of length t whose α-th bit is 1 iff there exists an α-node on the path from r_i^1 to MINI-ANC$(T, r_i^1, 2^j)$.
- A two-dimensional array $E[1..n_2, 0..3\lceil \lg \lg n \rceil]$, in which $E[i, j]$ stores a bit vector of length t whose α-th bit is 1 iff there exists an α-node on the path from r_i^2 to MICRO-ANC$(T, r_i^2, 2^j)$.
- A table F that stores for every possible weighted tree p on $\leq 2M'$ nodes whose weights are drawn from $[1..t]$, every integer i, j in $[1..2M']$, a bit vector of length t whose α-th bit is 1 iff there exists an α-node on the path from the i-th node in preorder of p to the j-th node in preorder of p.
- All these paths do not include the top nodes.

Now we analyze the space cost. D occupies $n_1 \times (\lceil \lg n \rceil + 1) \times t = O(n/\lg^{1-\epsilon} n) = o(n)$ bits. E occupies $n_2 \times (3\lceil \lg \lg n \rceil + 1) \times t = O(n \lg \lg n / \lg^{\lambda - \epsilon} n) = o(n)$ bits. F has $O(n^{1-\delta})$ entries for some $\delta > 0$, each occupying t bits. Therefore, D, E, F occupy $o(n)$ bits in total.

Let r_a^1 and r_b^2 be the roots of the mini-tree and the micro-tree containing x, respectively. Also, let r_c^1 and r_d^2 be the roots of the mini-tree and the micro-tree containing y, respectively. Suppose that $a \neq c$ (the case that $a = c$ can be handled similarly). We define the following two nodes

$$r_e^1 = \text{Mini-Anc}(T, r_a^1, \text{Mini-Depth}(T, r_a^1) - \text{Mini-Depth}(T, r_c^1) - 1),$$
$$r_f^2 = \text{Micro-Anc}(T, r_e^1, \text{Micro-Depth}(T, r_e^1) - \text{Micro-Depth}(T, r_d^2) - 1).$$

To answer operation $\text{Summarize}(T, x, y)$, we split the path from x to y into $x - r_b^2 - r_a^1 - r_e^1 - r_f^2 - y$. We compute for each part a bit vector of length t whose α-th bit is 1 iff there exists an α-node on this part. The answer to the Summarize operation is the result of bitwise OR operation on these bit vectors.

The first part and the last part are paths in the same micro-tree. Thus their bit vectors can be computed by table lookups on F. We only show how to compute the bit vector of the third part; the bit vectors of the second and the fourth part can be computed in a similar way. Setting $w = \text{Mini-Depth}(T, r_a^1) - \text{Mini-Depth}(T, r_e^1)$, $j = \lfloor \lg w \rfloor$, tier-2 root $r_g^1 = \text{Mini-Anc}(T, r_a^1, w - 2^j)$, the path from r_a^1 to r_e^1 can be represented as the union of the path from r_a^1 to $\text{Mini-Anc}(T, r_a^1, 2^j)$ and path from r_g^1 to $\text{Mini-Anc}(T, r_g^1, 2^j)$. Therefore the vector of this part is equal to $D[a, j]$ OR $D[g, j]$. □

Due to space limitations, the details of supporting all the other operations listed in this subsection are omitted. To analyze the space cost of our data structure, we observe that the main cost is due to the sequence PLS_T and Subsection 2.3, which occupy $n(H_0(PLS_T) + 2) + o(n)$ bits in total. The other auxiliary data structures occupy $o(n)$ bits of space only. We thus have the following lemma.

Lemma 6. *Let T be an ordinal tree on n node, each having a weight drawn from $[1..t]$, where $t = O(\lg^\epsilon n)$ for some constant $0 < \epsilon < 1$. T can be represented in $n(H_0(PLS_T) + 2) + o(n)$ bits of space to support Pre-Rank, Pre-Select, Depth, Lowest-Anc, Count, Summarize, and the navigational operations described in Subsection 2.3 in constant time.*

4   Supporting Path Queries

We now describe the data structures for general path search queries. The basic idea is to build a conceptual range tree on $[1..\sigma]$ with the branching factor $t = \lceil \lg^\epsilon n \rceil$ for some constant $0 < \epsilon < 1$. In this range tree, a range $[a..b]$ corresponds to the nodes in T that have a weight in $[a..b]$, which are also said to be contained in the range $[a..b]$. We say that a range $[a..b]$ is empty if $a > b$. In addition, the length of a range $[a..b]$ is defined to be $b - a + 1$. Empty ranges have non-positive lengths.

We describe this range tree level by level. At the beginning, we have the top level only, which contains the root range $[1..\sigma]$. Starting from the top level, we keep splitting each range in current lowest level into t child ranges, some of which might be empty. Presuming that $a \leq b$, a range $[a..b]$ will be split into t child ranges $[a_1..b_1], [a_2..b_2], \cdots, [a_t..b_t]$ such that, for $a \leq j \leq b$, the nodes that have a weight j are contained in the child range of subscript $\lceil \frac{t(j-a+1)}{b-a+1} \rceil$. Doing a little math, we have that

$$a_i = \min \left\{ a \leq j \leq b \Big| \Big\lceil \frac{t(j-a+1)}{b-a+1} \Big\rceil = i \right\} = \Big\lfloor \frac{(i-1)(b-a+1)}{t} \Big\rfloor + a;$$

$$b_i = \max \left\{ a \leq j \leq b \Big| \Big\lceil \frac{t(j-a+1)}{b-a+1} \Big\rceil = i \right\} = \Big\lfloor \frac{i(b-a+1)}{t} \Big\rfloor + a - 1.$$

This procedure stops when all the ranges in current lowest level have length 1, which form the bottom level of the conceptual range tree.

We list all the levels from top to bottom. The top level is the first level, and the bottom level is the h-th level, where $h = \lceil \log_t \sigma \rceil + 1$ is the height of the conceptual range tree. It is clear that each value in $[1..\sigma]$ occurs in exactly one range at each level, and each leaf range corresponds to a single value in $[1..\sigma]$.

For each level in the range tree other than the bottom one, we create and explicitly store an auxiliary tree. Let T_l denote the auxiliary tree created for the l-th level. To create T_l, we list all the non-empty ranges at the l-th level in increasing order. Let them be $[a_1..b_1], [a_2..b_2], \cdots, [a_m..b_m]$. We have that $a_1 = 1, b_m = \sigma$, and $b_i = a_{i+1} - 1$ for $i = 1, 2, \cdots, m - 1$. Initially, T_l contains only a root node, say r. For $i = 1, 2, \cdots, m$, we compute forest F_{a_i,b_i} as described in Subsection 2.2, and then insert the trees in F_{a_i,b_i} into T_l in the original left-to-right order. That is, for $i = 1, 2, \cdots, m$, from left to right, we append the root node of each tree in F_{a_i,b_i} to the list of children of r. Thus, for $l = 1, 2, \cdots, h-1$, there exists a one-to-one mapping between the nodes in T and the non-root nodes in T_l, and the root of T_l corresponds to the NULL node.

We assign weights to the nodes in T_l. The root of T_l is assigned 1. For each node x in T, we denote by x_l the node at level l that corresponds to x. By the construction of the conceptual range tree, the range containing x at level $l + 1$ has to be a child range of the range containing x at level l. We assign a weight α to x_l if the range at level $l + 1$ is the α-th child range of the range at level l. As the result, we obtain an ordinal tree T_l on $n + 1$ nodes, each having a weight drawn from $[1..t]$. We explicitly store T_l using Lemma 6.

Equations 1 and 2 capture the relationship between x_l and x_{l+1}, for each node x in T and $l \in [1..h-1]$. Remind that nodes are identified by their preorder ranks. Presuming that node x is contained in a range $[a..b]$ at level l, and contained in a range $[a_\gamma..b_\gamma]$ at level $l + 1$, which is the γ-th child range of $[a..b]$, we have

$$rank = \text{PRE-RANK}_\gamma(T_l, |R_{1,a-1}| + 1)$$

$$x_{l+1} = |R_{1,a_\gamma-1}| + 1 + \text{PRE-RANK}_\gamma(T_l, x_l) - rank \tag{1}$$

$$x_l = \text{PRE-SELECT}_\gamma(T_l, x_{l+1} - |R_{1,a_\gamma-1}| - 1 + rank) \tag{2}$$

With these data structures, we can support path counting and reporting.

Theorem 1. *Let T be an ordinal tree on n nodes, each having a weight drawn from $[1..\sigma]$. Under the word RAM model with word size $w = \Omega(\lg n)$, T can be encoded in (a) $n(H(W_T) + 2) + o(n)$ bits when $\sigma = O(\lg^\epsilon n)$ for some constant $0 < \epsilon < 1$, or (b) $nH(W_T) + O(\frac{n \lg \sigma}{\lg \lg n})$ bits otherwise, supporting path counting queries in $O(\frac{\lg \sigma}{\lg \lg n} + 1)$ time, and path reporting queries in $O((occ+1)(\frac{\lg \sigma}{\lg \lg n} + 1))$ time, where $H(W_T)$ is the entropy of the multiset of the weights of the nodes in T, and occ is the output size of the path reporting query.*

Proof. Let $P_{u,v} \cap R_{p,q}$ be the query. $P_{u,v}$ can be partitioned into $A_{u,w}$, $A_{v,w}$ (see Subsection 2.2 for the definitions of $A_{u,w}$ and $A_{v,w}$) and $\{w\}$, where $w = LCA(u, v)$. It is trivial to compute $\{w\} \cap R_{p,q}$. We only consider how to compute $A_{u,w} \cap R_{p,q}$ and its cardinality. The computation of $A_{v,w} \cap R_{p,q}$ is similar.

Our recursive algorithm is shown in Algorithm 1. Providing that the range $[a..b]$ is at the l-th level of the conceptual range tree, $c = |R_{1,a-1}|+1, d = |R_{1,b}|+1$, and x, z are the nodes in T_l that corresponds to $anc_{a,b}(T, u)$ and $anc_{a,b}(T, w)$, the procedure SEARCH($[a..b], l, c, d, x, z, [p..q]$) returns the cardinality of $A_{u,w} \cap R_{a,b} \cap R_{p,q}$, and reports the nodes in the intersection if the given query is a path reporting query. To compute $A_{u,w} \cap R_{p,q}$ and its cardinality, we only need call SEARCH($[1..\sigma], 1, 1, n + 1, u + 1, w + 1, [p..q]$).

Now let us analyze the procedure SEARCH. In line 4, we maximize $[e'..f'] \subseteq [e..f]$ such that $[e'..f']$ can be partitioned into child ranges of subscripts α to β. To compute α (the computation of β is similar), let $\alpha' = \lceil \frac{t(e-a+1)}{b-a+1} \rceil$, the subscript of the child range containing the nodes that have a weight e. α is set to be α' if $e \le a_{\alpha'}$, otherwise α is set to be $\alpha'+1$. If $\alpha \le \beta$, we accumulate $|A_{u,w} \cap R_{a_\alpha,b_\beta}|$ in line 6, where COUNT$_{\alpha,\beta}(T, x)$ is defined to be COUNT$_\beta(T, x) -$ COUNT$_{\alpha-1}(T, x)$. We still need compute $A_{u,w} \cap (R_{e,f} - R_{a_\alpha,b_\beta}) = A_{u,w} \cap R_{e,f} \cap (R_{a_{\alpha-1},b_{\alpha-1}} \cup R_{a_{\beta+1},b_{\beta+1}})$. In the loop starting at line 14, for $\gamma \in \{\alpha - 1, \beta + 1\}$, we check whether $[a_\gamma..b_\gamma] \cap [e..f] \ne \phi$. If the intersection is not empty, then we compute $A_{u,w} \cap R_{e,f} \cap R_{a_\gamma,b_\gamma}$ and accumulate its cardinality by a recursive call in line 19. In lines 16 to 18, we adjust the parameters to satisfy the restriction of the procedure SEARCH. We increase c and d as shown in lines 16 and 17, because for $i = 1$ to t, the nodes in T that have a weight in $[a_i..b_i]$ correspond to the nodes in T_l that have a weight i and a preorder rank in $[c + 1..d]$. In line 18, for $\delta = \{x, z\}$, we set $\delta_\gamma = $ LOWEST-ANC$_\gamma(T_l, \delta)$. If δ_γ is the root of T_l, say $\delta_\gamma = 1$, then δ_γ is set to be 1. Otherwise, we compute δ_γ by Equation 1.

For path reporting queries, in lines 7 to 12, we report all nodes on the path $A_{u,w}$ that have a weight in $[a_\alpha..b_\beta]$. In line 8, we compute sum, a bit vector of length t, whose i-th bit is one iff $A_{u,w} \cap R_{a_i,b_i} \ne \phi$. Since $t = o(\lg n)$, using word operations, we can enumerate each 1-bit whose index is between α and β in constant time per 1-bit. For each 1-bit, assuming its index is i, we find all the nodes in $A_{u,w} \cap R_{a_i,b_i}$ using LOWEST-ANC operations repeatedly. At this moment, we only know the preorder ranks of the nodes to report in T_l. We have to convert it to the preorder rank in T_1 by using Equation 2 $l - 1$ times per node.

We now analyze the time cost of our algorithm. By the construction of the conceptual range tree, any query range $[p..q]$ will be partitioned into $O(h)$ ranges in the range tree. It takes constant time to access a range in the range tree, since, by

Subsection 2.3 and Lemma 6, all the operations on T_l take $O(1)$ time. Hence, it takes $O(h) = O(\frac{\lg \sigma}{\lg \lg n} + 1)$ time to compute $|A_{u,w} \cap R_{p,q}|$. For path reporting queries, it takes $O(h)$ time to report each node in $A_{u,w} \cap R_{p,q}$. Each 1-bit we enumerate in line 9 corresponds to one or more node to report, so the cost of enumeration is dominated by other parts of the algorithm. Therefore, it takes $O(h + |A_{u,w} \cap R_{p,q}|h) = O((|A_{u,w} \cap R_{p,q}| + 1)(\frac{\lg \sigma}{\lg \lg n} + 1))$ time to compute $A_{u,w} \cap R_{p,q}$.

Due to space limitations, the analysis of the space cost is omitted.   □

Algorithm 1. The algorithm for path search queries on $A_{u,w}$.

```
 1: procedure SEARCH([a..b], l, c, d, x, z, [p..q])
 2:     Let [a₁..b₁], [a₂..b₂], ⋯, [aₜ..bₜ] be the child ranges of [a..b];
 3:     count ← 0, [e..f] = [a..b] ∩ [p..q];
 4:     α = min{1 ≤ i ≤ t|e ≤ aᵢ ≤ bᵢ}, β = max{1 ≤ i ≤ t|aᵢ ≤ bᵢ ≤ f};
 5:     if α ≤ β then
 6:         count ← count + COUNT_{α,β}(T_l, x) − COUNT_{α,β}(T_l, z);
 7:         if the given query is a path reporting query then
 8:             sum ← SUMMARIZE(T_l, x, z);
 9:             for each i ∈ [α..β] that the i-th bit of sum is one do
10:                 report all i-nodes on the path from x to z (excluding z);
11:             end for
12:         end if
13:     end if
14:     for γ ∈ {α − 1, β + 1} do
15:         if 1 ≤ γ ≤ t and [a_γ..b_γ] ∩ [e..f] ≠ φ then
16:             c_γ ← c + PRE-COUNT_{γ−1}(T_l, d) − PRE-COUNT_{γ−1}(T_l, c);
17:             d_γ ← c + PRE-COUNT_γ(T_l, d) − PRE-COUNT_γ(T_l, c);
18:             δ_γ ← the preorder rank of LOWEST-ANC_γ(T_l, δ) in T_{l+1}, for δ ∈ {x, z};
19:             count ← count + SEARCH([a_γ..b_γ], l + 1, c_γ, d_γ, x_γ, z_γ, [p..q]);
20:         end if
21:     end for
22:     return count;
23: end procedure
```

To support path median and path selection queries, we apply the techniques we have developed in this section to generalize the linear space data structure of Brodal et al.'s [6] for range selection queries. We have the following theorem, and leave the proof in the full version of this paper.

Theorem 2. *Let T be an ordinal tree on n nodes, each having a weight drawn from $[1..\sigma]$. Under the word RAM model with word size $w = \Omega(\lg n)$, T can be represented in (a) $n(H(W_T) + 2) + o(n)$ bits when $\sigma = O(1)$, or (b) $nH(W_T) + O(\frac{n \lg \sigma}{\lg \lg \sigma})$ bits otherwise, supporting path median and path selection queries in $O(\frac{\lg \sigma}{\lg \lg \sigma})$ time.*

References

1. Alon, N., Schieber, B.: Optimal preprocessing for answering on-line product queries. Tech. rep., Tel Aviv University (1987)

2. Barbay, J., Golynski, A., Munro, J.I., Rao, S.S.: Adaptive Searching in Succinctly Encoded Binary Relations and Tree-Structured Documents. In: Lewenstein, M., Valiente, G. (eds.) CPM 2006. LNCS, vol. 4009, pp. 24–35. Springer, Heidelberg (2006)

3. Barbay, J., He, M., Munro, J.I., Rao, S.S.: Succinct indexes for strings, binary relations and multilabeled trees. ACM Transactions on Algorithms 7(4), 52 (2011)

4. Bille, P.: A survey on tree edit distance and related problems. Theor. Comput. Sci. 337(1-3), 217–239 (2005)

5. Bose, P., He, M., Maheshwari, A., Morin, P.: Succinct Orthogonal Range Search Structures on a Grid with Applications to Text Indexing. In: Dehne, F., Gavrilova, M., Sack, J.-R., Tóth, C.D. (eds.) WADS 2009. LNCS, vol. 5664, pp. 98–109. Springer, Heidelberg (2009)

6. Brodal, G.S., Gfeller, B., Jørgensen, A.G., Sanders, P.: Towards optimal range medians. Theor. Comput. Sci. 412(24), 2588–2601 (2011)

7. Chazelle, B.: Computing on a free tree via complexity-preserving mappings. Algorithmica 2, 337–361 (1987)

8. Farzan, A., Munro, J.I.: A Uniform Approach Towards Succinct Representation of Trees. In: Gudmundsson, J. (ed.) SWAT 2008. LNCS, vol. 5124, pp. 173–184. Springer, Heidelberg (2008)

9. Farzan, A., Raman, R., Rao, S.S.: Universal Succinct Representations of Trees? In: Albers, S., Marchetti-Spaccamela, A., Matias, Y., Nikoletseas, S., Thomas, W. (eds.) ICALP 2009, Part I. LNCS, vol. 5555, pp. 451–462. Springer, Heidelberg (2009)

10. Ferragina, P., Luccio, F., Manzini, G., Muthukrishnan, S.: Compressing and indexing labeled trees, with applications. J. ACM 57(1) (2009)

11. Ferragina, P., Manzini, G., Mäkinen, V., Navarro, G.: Compressed representations of sequences and full-text indexes. ACM Transactions on Algorithms 3(2) (2007)

12. Geary, R.F., Raman, R., Raman, V.: Succinct ordinal trees with level-ancestor queries. ACM Transactions on Algorithms 2(4), 510–534 (2006)

13. Hagerup, T.: Parallel preprocessing for path queries without concurrent reading. Inf. Comput. 158(1), 18–28 (2000)

14. He, M., Munro, J.I., Rao, S.S.: Succinct Ordinal Trees Based on Tree Covering. In: Arge, L., Cachin, C., Jurdziński, T., Tarlecki, A. (eds.) ICALP 2007. LNCS, vol. 4596, pp. 509–520. Springer, Heidelberg (2007)

15. He, M., Munro, J.I., Zhou, G.: Path Queries in Weighted Trees. In: Asano, T., Nakano, S.-I., Okamoto, Y., Watanabe, O. (eds.) ISAAC 2011. LNCS, vol. 7074, pp. 140–149. Springer, Heidelberg (2011)

16. Jørgensen, A.G., Larsen, K.G.: Range selection and median: Tight cell probe lower bounds and adaptive data structures. In: SODA, pp. 805–813 (2011)

17. Krizanc, D., Morin, P., Smid, M.H.M.: Range mode and range median queries on lists and trees. Nord. J. Comput. 12(1), 1–17 (2005)

18. Pǎtraşcu, M.: Lower bounds for 2-dimensional range counting. In: STOC, pp. 40–46 (2007)

19. Pǎtraşcu, M.: Unifying the landscape of cell-probe lower bounds. SIAM J. Comput. 40(3), 827–847 (2011)

20. Raman, R., Raman, V., Rao, S.S.: Succinct indexable dictionaries with applications to encoding k-ary trees, prefix sums and multisets. ACM Transactions on Algorithms 3(4) (2007)

21. Sadakane, K., Navarro, G.: Fully-functional succinct trees. In: SODA, pp. 134–149 (2010)

Approximation of Minimum Cost Homomorphisms

Pavol Hell[1], Monaldo Mastrolilli[2], Mayssam Mohammadi Nevisi[1],
and Arash Rafiey[2,3]

[1] School of Computing Science, SFU, Burnaby, Canada
{pavol,maysamm}@sfu.ca[*]
[2] IDSIA, Lugano, Switzerland
{monaldo,arash}@idsia.ch[**]
[3] Informatics Department, University of Bergen, Norway
arash.rafiey@ii.uib.no[* * *]

Abstract. Let H be a fixed graph without loops. We prove that if H is a co-circular arc bigraph then the minimum cost homomorphism problem to H admits a polynomial time constant ratio approximation algorithm; otherwise the minimum cost homomorphism problem to H is known to be not approximable. This solves a problem posed in an earlier paper. For the purposes of the approximation, we provide a new characterization of co-circular arc bigraphs by the existence of min ordering. Our algorithm is then obtained by derandomizing a two-phase randomized procedure. We show a similar result for graphs H in which all vertices have loops: if H is an interval graph, then the minimum cost homomorphism problem to H admits a polynomial time constant ratio approximation algorithm, and otherwise the minimum cost homomorphism problem to H is not approximable.

1 Introduction

We study the approximability of the minimum cost homomorphism problem, introduced below. A *c-approximation algorithm* produces a solution of cost at most c times the minimum cost. A *constant ratio* approximation algorithm is a c-approximation algorithm for some constant c. When we say a problem has a c-approximation algorithm, we mean a polynomial time algorithm. We say that a problem is *not approximable* if it there is no polynomial time approximation algorithm with a multiplicative guarantee unless $P = NP$.

The minimum cost homomorphism problem was introduced in [8]. It consists of minimizing a certain cost function over all homomorphisms of an input graph G to a fixed graph H. This offers a natural and practical way to model many optimization problems. For instance, in [8] it was used to model a problem of minimizing the cost of a repair and maintenance schedule for large machinery.

---

[*] Supported by NSERC Canada; IRMACS facilities gratefully acknowledged.
[**] Supported by the Swiss National Science Foundation project 200020-122110/1 "Approximation Algorithms for Machine Scheduling III".
[* * *] Supported by ERC advanced grant PREPROCESSING, 267959.

L. Epstein and P. Ferragina (Eds.): ESA 2012, LNCS 7501, pp. 587–598, 2012.
© Springer-Verlag Berlin Heidelberg 2012

It generalizes many other problems such as list homomorphism problems (see below), retraction problems [6], and various optimum cost chromatic partition problems [10,15,16,17]. (A different kind of the minimum cost homomorphism problem was introduced in [1].) Certain minimum cost homomorphism problems have polynomial time algorithms [7,8,9,14], but most are NP-hard. Therefore we investigate the approximability of these problems. Note that we approximate the cost over real homomophisms, rather than approximating the maximum weight of satisfied constraints, as in, say, MAXSAT.

We call a graph *reflexive* if every vertex has a loop, and *irreflexive* if no vertex has a loop. An *interval graph* is a graph that is the intersection graph of a family of real intervals, and a *circular arc graph* is a graph that is the intersection graph of a family of arcs on a circle. We interpret the concept of an intersection graph literally, thus any intersection graph is automatically reflexive, since a set always intersects itself. A bipartite graph whose complement is a circular arc graph, will be called a *co-circular arc bigraph*. When forming the complement, we take all edges that were not in the graph, including loops and edges between vertices in the same colour. In general, the word *bigraph* will be reserved for a bipartite graph with a fixed bipartition of vertices; we shall refer to *white* and *black* vertices to reflect this fixed bipartition. Bigraphs can be conveniently viewed as directed bipartite graphs with all edges oriented from the white to the black vertices. Thus, by definition, interval graphs are reflexive, and co-circular arc bigraphs are irreflexive. Despite the apparent differences in their definition, these two graph classes exhibit certain natural similarities [2,3]. There is also a concept of an *interval bigraph H*, which is defined for two families of real intervals, one family for the white vertices and one family for the black vertices: a white vertex is adjacent to a black vertex if and only if their corresponding intervals intersect.

A reflexive graph is a *proper interval graph* if it is an interval graph in which the defining family of real intervals can be chosen to be inclusion-free. A bigraph is a *proper interval bigraph* if it is an interval bigraph in which the defining two families of real intervals can be chosen to be inclusion-free. It turns out [11] that proper interval bigraphs are a subclass of co-circular arc bigraphs.

A *homomorphism* of a graph G to a graph H is a mapping $f : V(G) \to V(H)$ such that for any edge xy of G the pair $f(x)f(y)$ is an edge of H.

Let H be a fixed graph.

The *list homomorphism problem* to H, denoted ListHOM(H), seeks, for a given input graph G and lists $L(x) \subseteq V(H), x \in V(G)$, a homomorphism f of G to H such that $f(x) \in L(x)$ for all $x \in V(G)$. It was proved in [3] that for irreflexive graphs, the problem ListHOM(H) is polynomial time solvable if H is a co-circular arc bigraph, and is NP-complete otherwise. It was shown in [2] that for reflexive graphs H, the problem ListHOM(H) is polynomial time solvable if H is an interval graph, and is NP-complete otherwise.

The *minimum cost homomorphism problem* to H, denoted MinHOM(H), seeks, for a given input graph G and vertex-mapping costs $c(x, u), x \in V(G), u \in V(H)$, a homomorphism f of G to H that minimizes total cost $\sum_{x \in V(G)} c(x, f(x))$.

It was proved in [9] that for irreflexive graphs, the problem MinHOM(H) is polynomial time solvable if H is a proper interval bigraph, and it is NP-complete otherwise. It was also shown there that for reflexive graphs H, the problem MinHOM(H) is polynomial time solvable if H is a proper interval graph, and it is NP-complete otherwise.

In [20], the authors have shown that MinHOM(H) is not approximable if H is a graph that is not bipartite or not a co-circular arc graph, and gave randomized 2-approximation algorithms for MinHOM(H) for a certain subclass of co-circular arc bigraphs H. The authors have asked for the exact complexity classification for these problems. We answer the question by showing that the problem MinHOM(H) in fact has a $|V(H)|$-approximation algorithm for **all** co-circular arc bigraphs H. Thus for an irreflexive graph H the problem MinHOM(H) has a constant ratio approximation algorithm if H is a co-circular arc bigraph, and is not approximable otherwise. We also prove that for a reflexive graph H the problem MinHOM(H) has a constant ratio approximation algorithm if H is an interval graph, and is not approximable otherwise. We use the method of randomized rounding, a novel technique of randomized shifting, and then a simple derandomization.

A *min ordering* of a graph H is an ordering of its vertices $a_1, a_2, \ldots, a_n$, so that the existence of the edges $a_i a_j, a_{i'} a_{j'}$ with $i < i', j' < j$ implies the existence of the edge $a_i a_{j'}$. A *min-max ordering* of a graph H is an ordering of its vertices $a_1, a_2, \ldots, a_n$, so that the existence of the edges $a_i a_j, a_{i'} a_{j'}$ with $i < i', j' < j$ implies the existence of the edges $a_i a_{j'}, a_{i'} a_j$. For bigraphs, it is more convenient to speak of two orderings, and we define a *min ordering* of a bigraph H to be an ordering $a_1, a_2, \ldots, a_p$ of the white vertices and an ordering $b_1, b_2, \ldots, b_q$ of the black vertices, so that the existence of the edges $a_i b_j, a_{i'} b_{j'}$ with $i < i', j' < j$ implies the existence of the edge $a_i b_{j'}$; and a *min-max ordering* of a bigraph H to be an ordering of $a_1, a_2, \ldots, a_p$ of the white vertices and an ordering $b_1, b_2, \ldots, b_q$ of the black vertices, so that the existence of the edges $a_i b_j, a_{i'} b_{j'}$ with $i < i', j' < j$ implies the existence of the edges $a_i b_{j'}, a_{i'} b_j$. (Both are instances of a general definition of min ordering for directed graphs [13].)

In Section 2 we prove that co-circular arc bigraphs are precisely the bigraphs that admit a min ordering. In the realm of reflexive graphs, such a result is known about the class of interval graphs (they are precisely the reflexive graphs that admit a min ordering) [12]. In Section 3 we discuss a linear program that computes a solution to MinHOM(H) when H has a min-max ordering. In [9], the authors used a network flow problem equivalent to this linear program, to solve to MinHOM(H) when H admits a min-max ordering. In Section 4 we recall that MinHOM(H) is not approximable when H does not have min ordering, and describe a $|V(H)|$-approximation algorithm when H is a bigraph that admits a min ordering. Finally, in Section 5 we extend our results to reflexive graphs and suggest some future work.

2   Co-circular Bigraphs and Min Ordering

A reflexive graph has a min ordering if and only if it is an interval graph [12]. In this section we prove a similar result about bigraphs. Two auxiliary concepts from [3,5] are introduced first.

An *edge asteroid* of a bigraph H consists of $2k + 1$ disjoint edges $a_0b_0, a_1b_1,$ $\ldots, a_{2k}b_{2k}$ such that each pair a_i, a_{i+1} is joined by a path disjoint from all neighbours of $a_{i+k+1}b_{i+k+1}$ (subscripts modulo $2k + 1$).

An *invertible pair* in a bigraph H is a pair of white vertices a, a' and two pairs of walks $a = v_1, v_2, \ldots, v_k = a'$, $a' = v'_1, v'_2, \ldots, v'_k = a$, and $a' = w_1, w_2, \ldots, w_m = a$, $a = w'_1, w'_2, \ldots, w'_m = a'$ such that v_i is not adjacent to v'_{i+1} for all $i = 1, 2, \ldots, k$ and w_j is not adjacent to w'_{j+1} for all $j = 1, 2, \ldots, m$.

Theorem 1. *A bigraph H is a co-circular arc graph if and only if it admits a min ordering.*

Proof. Consider the following statements for a bigraph H:

1. H has no induced cycles of length greater than three and no edge asteroids
2. H is a co-circular-arc graph
3. H has a min ordering
4. H has no invertible pairs

$1 \Rightarrow 2$ is proved in [3].

$2 \Rightarrow 3$ is seen as follows: Suppose H is a co-circular arc bigraph; thus the complement $\overline{H}$ is a circular arc graph that can be covered by two cliques. It is known for such graphs that there exist two points, the *north pole* and the *south pole*, on the circle, so that the white vertices u of H correspond to arcs A_u containing the north pole but not the south pole, and the black vertices v of H correspond to arcs A_v containing the south pole but not the north pole. We now define a min ordering of H as follows. The white vertices are ordered according to the clockwise order of the corresponding clockwise extremes, i.e., u comes before u' if the clockwise end of A_u precedes the clockwise end of $A_{u'}$. The same definition, applied to the black vertices v and arcs A_v, gives an ordering of the black vertices of H. It is now easy to see from the definitions that if $uv, u'v'$ are edges of H with $u < u'$ and $v > v'$, then A_u and $A_{v'}$ must be disjoint, and so uv' is an edge of H.

$3 \Rightarrow 4$ is easy to see from the definitions (see, for instance [5]).

$4 \Rightarrow 1$ is checked as follows: If C is an induced cycle in H, then C must be even, and any two of its opposite vertices together with the walks around the cycle form an invertible pair of H. In an edge-asteroid $a_0b_0, \ldots, a_{2k}b_{2k}$ as defined above, it is easy to see that, say, a_0, a_k is an invertible pair. Indeed, there is, for any i, a walk from a_i to a_{i+1} that has no edges to the walk $a_{i+k}, b_{i+k}, a_{i+k}, b_{i+k}, \ldots, a_{i+k}$ of the same length. Similarly, a walk $a_{i+1}, b_{i+1}, a_{i+1}, b_{i+1}, \ldots, a_{i+1}$ has no edges to a walk from a_{i+k} to a_{i+k+1} implied by the definition of an edge-asteroid. By composing such walks we see that a_0, a_k is an invertible pair.   □

We note that it can be decided in time polynomial in the size of H, whether a graph H is a (co-)circular arc bigraph [19].

3   An Exact Algorithm

If H is a fixed bigraph with a min-max ordering, there is an exact algorithm for the problem MinHOM(H). Suppose H has the white vertices ordered $a_1, a_2, \cdots, a_p$, and the black vertices ordered $b_1, b_2, \cdots, b_q$. Define $\ell(i)$ to be the smallest subscript j such that b_j is a neighbour of a_i (and $\ell'(i)$ to be the smallest subscript j such that a_j is a neighbour of b_i) with respect to the ordering. Suppose G is a bigraph with white vertices u and black vertices v. We seek a minimum cost homomorphism of G to H that preserves colours, i.e., maps white vertices of G to white vertices of H and similarly for black vertices.

We define a set of variables $x_{u,i}, x_{v,j}$ for all vertices u and v of G and all $i = 1, 2, \ldots, p+1, j = 1, 2, \ldots, q+1$, and the following linear system $\mathcal{S}$.

For all vertices u (respectively v) in G and $i = 1, \ldots, p$ (respectively $j = 1, \ldots, q$)

- $x_{u,i} \geq 0$ (respectively $x_{v,j} \geq 0$)
- $x_{u,1} = 1$ (respectively $x_{v,1} = 1$)
- $x_{u,p+1} = 0$ (respectively $x_{v,q+1} = 0$)
- $x_{u,i+1} \leq x_{u,i}$ (respectively $x_{v,j+1} \leq x_{v,j}$).

For all edges uv of G and $i = 1, 2, \ldots, p, \ j = 1, 2, \ldots, q$

- $x_{u,i} \leq x_{v,\ell(i)}$
- $x_{v,j} \leq x_{u,\ell'(j)}$

Theorem 2. *There is a one-to-one correspondence between homomorphisms of G to H and integer solutions of $\mathcal{S}$. Furthermore, the cost of the homomorphism is equal to* $\sum\limits_{u,i} c(u,i)(x_{u,i} - x_{u,i+1}) + \sum\limits_{v,j} c(v,j)(x_{v,j} - x_{v,j+1})$.

Proof. If $f : G \to H$ is a homomorphism, we set the value $x_{u,i} = 1$ if $f(u) = a_t$ for some $t \geq i$, otherwise we set $x_{u,i} = 0$; and similarly for $x_{v,j}$. Now all the variables are non-negative, we have all $x_{u,1} = 1$, $x_{u,p+1} = 0$, and $x_{u,i+1} \leq x_{u,i}$; and similarly for $x_{v,j}$. It remains to show that $x_{u,i} \leq x_{v,\ell(i)}$ for any edge uv of G and any subscript i. (The proof of $x_{v,j} \leq x_{u,\ell'(j)}$ is analogous.) Suppose for a contradiction that $x_{u,i} = 1$ and $x_{v,\ell(i)} = 0$, and let $f(u) = a_r$, $f(v) = b_s$. This implies that $x_{u,r} = 1, x_{u,r+1} = 0$, whence $i \leq r$; and that $x_{v,s} = 1$, whence $s < \ell(i)$. Since both $a_i b_{\ell(i)}, a_r b_s$ are edges of H, the fact that we have a min ordering implies that $a_i b_s$ must also be an edge of H, contradicting the definition of $\ell(i)$.

Conversely, if there is an integer solution for $\mathcal{S}$, we define a homomorphism f as follows: we let $f(u) = a_i$ when i is the largest subscript with $x_{u,i} = 1$ (and similarly, $f(v) = b_j$ when j is the largest subscript with $x_{v,j} = 1$). Clearly, every vertex of G is mapped to some vertex of H, of the same colour. We prove that this is indeed a homomorphism by showing that every edge of G is mapped to an edge of H. Let $e = uv$ be an edge of G, and assume $f(u) = a_r, f(v) = b_s$. We will show that $a_r b_s$ is an edge of H. Observe that $1 = x_{u,r} \leq x_{v,\ell(r)} \leq 1$ and $1 = x_{v,s} \leq x_{u,\ell'(s)} \leq 1$, so we must have $x_{u,\ell'(s)} = x_{v,\ell(r)} = 1$. Also observe that $x_{u,i} = 0$ for all $i > r$, and $x_{v,j} = 0$ for all $j > s$. Thus, $\ell(r) \leq s$ and $\ell'(s) \leq r$.

Since $a_r b_{\ell(r)}$ and $a_{\ell'(s)} b_s$ are edges in H, we must have the edge $a_r b_s$, as we have a min-max ordering.

Furthermore, $f(u) = a_i$ if and only if $x_{u,i} = 1$ and $x_{u,i+1} = 0$, so, $c(u,i)$ contributes to the sum if and only if $f(u) = a_i$ (and similarly, if $f(v) = b_j$). $\square$

We have translated the minimum cost homomorphism problem to an integer linear program: minimize the objective function in Theorem 2 over the linear system $\mathcal{S}$. In fact, this linear program corresponds to a minimum cut problem in an auxiliary network, and can be solved by network flow algorithms [9,20]. We shall enhance the above system $\mathcal{S}$ to obtain an approximation algorithm for the case H is only assumed to have a min ordering.

4   An Approximation Algorithm

In this section we describe our approximation algorithm for MinHOM(H) in the case the fixed bigraph H has a min ordering, i.e., is a co-circular arc bigraph, cf. Theorem 1. We recall that if H is not a co-circular arc bigraph, then the list homomorphism problem ListHOM(H) is NP-complete [3], and this implies that MinHOM(H) is not approximable for such graphs H [20]. By Theorem 1 we conclude the following.

Theorem 3. *If a bigraph H has no min ordering, then MinHOM(H) is not approximable.*

Our main result is the following converse: if H has a min ordering (is a co-circular arc bigraph), then there exists a constant ratio approximation algorithm. (Since H is fixed, $|V(H)|$ is a constant.)

Theorem 4. *If H is a bigraph that admits a min ordering, then MinHOM(H) has a $|V(H)|$-approximation algorithm.*

Proof. Suppose H has a min ordering with the white vertices ordered $a_1, a_2, \cdots, a_p$, and the black vertices ordered $b_1, b_2, \cdots, b_q$. Let E' denote the set of all pairs $a_i b_j$ such that $a_i b_j$ is not an edge of H, but there is an edge $a_i b_{j'}$ of H with $j' < j$ and an edge $a_{i'} b_j$ of H with $i' < i$. Let $E = E(H)$ and define H' to be the graph with vertex set $V(H)$ and edge set $E \cup E'$. (Note that E and E' are disjoint sets.)

Observation 1. The ordering $a_1, a_2, \cdots, a_p$, and $b_1, b_2, \cdots, b_q$ is a min-max ordering of H'.

We show that for every pair of edges $e = a_i b_{j'}$ and $e' = a_{i'} b_j$ in $E \cup E'$, with $i' < i$ and $j' < j$, both $f = a_i b_j$ and $f' = a_{i'} b_{j'}$ are in $E \cup E'$.

If both e and e' are in E, $f \in E \cup E'$ and $f' \in E$.

If one of the edges, say e, is in E', there is a vertex $b_{j''}$ with $a_i b_{j''} \in E$ and $j'' < j'$, and a vertex $a_{i''}$ with $a_{i''} b_{j'} \in E$ and $i'' < i$. Now, $a_{i'} b_j$ and $a_i b_{j''}$ are both in E, so $f \in E \cup E'$. We may assume that $i'' \neq i'$, otherwise $f' = a_{i''} b_{j'} \in E$. If $i'' < i'$, then $f' \in E \cup E'$ because $a_{i'} b_{j''} \in E$; and if $i'' > i'$, then $f' \in E$ because $a_{i'} b_j \in E$.

If both edges e, e' are in E', then the earlier neighbours of a_i and b_j in E imply that $f \in E \cup E'$, and the earlier neighbours of $a_{i'}$ and $b_{j'}$ in E imply that $f' \in E \cup E'$.

Observation 2. Let $e = a_i b_j \in E'$. Then a_i is not adjacent in E to any vertex after b_j, or b_j is not adjacent in E to any vertex after a_i.

This easily follows from the fact that we have a min ordering.

Our algorithm first constructs the graph H' and then proceeds as follows. Consider an input bigraph G. Since H' has a min-max ordering, we can form the system $\mathcal{S}$ of linear inequalities for H'. By Theorem 2, homomorphisms of G to H' are in a one-to-one correspondence with integer solutions of $\mathcal{S}$. However, we are interested in homomorphisms of G to H, not H'. Therefore we shall add further inequalities to $\mathcal{S}$ to ensure that we only admit homomorphisms of G to H, i.e., avoid mapping edges of G to the edges in E'.

For every edge $e = a_i b_j \in E'$ and every edge $uv \in E(G)$, two of the following inequalities will be added to $\mathcal{S}$.

- if a_s is the first neighbour of b_j after a_i, we add the inequality

$$x_{v,j} \leq x_{u,s} + \sum_{a_t b_j \in E, \, t<i} (x_{u,t} - x_{u,t+1})$$

- else if b_j has no neighbours after a_i, we add the inequality

$$x_{v,j} \leq x_{v,j+1} + \sum_{a_t b_j \in E, \, t<i} (x_{u,t} - x_{u,t+1})$$

- if b_s is the first neighbour of a_i after b_j, we add the inequality

$$x_{u,i} \leq x_{v,s} + \sum_{a_i b_t \in E, \, t<j} (x_{v,t} - x_{v,t+1})$$

- else if a_i has no neighbour after b_j, we add the inequality

$$x_{u,i} \leq x_{u,i+1} + \sum_{a_i b_t \in E, \, t<j} (x_{v,t} - x_{v,t+1}).$$

Claim: There is a one-to-one correspondence between homomorphisms of G to H and integer solutions of the expanded system $\mathcal{S}$.

The correspondence between the integer solutions and the homomorphisms is defined as before. Thus we have a homomorphism of G to H' if and only if the old inequalities are satisfied. We shall show that the additional inequalities are also satisfied if and only if each edge of G is mapped to an edge in E, i.e., we have a homomorphism to H.

Suppose f is a homomorphism of G to H', obtained from an integer solution for $\mathcal{S}$, and, for some edge uv of G, let $f(u) = a_i$, $f(v) = b_j$. We have $x_{u,i} = 1$, $x_{u,i+1} = 0$, $x_{v,j} = 1$, $x_{v,j+1} = 0$, and for all $a_t b_j \in E$ with $t < i$ we have $x_{u,t} - x_{u,t+1} = 0$. If a_s is the first neighbour of b_j after a_i, then we will also have $x_{u,s} = 0$, and so the first inequality fails. Else if b_j is not adjacent to any vertex after a_i, and the second inequality fails. The remaining two other cases are similar.

Conversely, suppose f is a homomorphism of G to H (i.e., f maps the edges of G to the edges in E). For a contradiction, assume that the first inequalities fails (the other inequalities are similar). This means that for some edge

$uv \in E(G)$ and some edge $a_i b_j \in E'$, we have $x_{v,j} = 1$, $x_{u,s} = 0$, and the sum of $(x_{u,t} - x_{u,t+1}) = 0$, summed over all $t < i$ such that a_t is a neighbour of b_j. The latter two facts easily imply that $f(u) = a_i$. Since b_j has a neighbour after a_i, Observation 2 tells us that a_i has no neighbours after b_j, whence $f(v) = b_j$ and thus $a_i b_j \in E$, contradicting the fact that $a_i b_j \in E'$. This proves the Claim.

At this point, our algorithm will minimize the cost function over S in polynomial time using a linear programming algorithm. This will generally result in a fractional solution. (Even though the original system S is known to be totally unimodular [20] and hence have integral optima, we have added inequalities, and hence lost this advantage.) We will obtain an integer solution by a randomized procedure called *rounding*. We choose a random variable $X \in [0,1]$, and define the rounded values $x'_{u,i} = 1$ when $x_{u,i} \geq X$, and $x'_{u,i} = 0$ otherwise; and similarly for $x'_{v,j}$. It is easy to check that the rounded values satisfy the original inequalities, i.e., correspond to a homomorphism f of G to H'.

Now the algorithm will once more modify the solution f to become a homomorphism of G to H, i.e., to avoid mapping edges of G to the edges in E'. This will be accomplished by another randomized procedure, which we call *shifting*. We choose another random variable $Y \in [0,1]$, which will guide the shifting. Let F denote the set of all edges in E' to which some edge of G is mapped by f. If F is empty, we need no shifting. Otherwise, let $a_i b_j$ be an edge of F with maximum sum $i + j$ (among all edges of F). By the maximality of $i + j$, we know that $a_i b_j$ is the last edge of F from both a_i and b_j. Since $F \subseteq E'$, Observation 2 implies that $e = a_i b_j$ is also the last edge of E from a_i or from b_j. Suppose e is the last edge of E from a_i. (The shifting process is similar in the other case.) So a_i does not have any edges of F or of E after $a_i b_j$. (There could be edges of $E' - F$, but since no edge of G is mapped to such edges, they don't matter.) We now consider, one by one, vertices u in G such that $f(u) = a_i$ and u has a neighbour v in G with $f(v) = b_j$. (Such vertices u exist by the definition of F.) For such a vertex u, consider the set of all vertices a_t with $t < i$ such that $a_t b_j \in E$. This set is not empty, since e is in E' because of two edges of E. Suppose the set consists of a_t with subscripts t ordered as $t_1 < t_2 < \dots t_k$. The algorithm now selects one vertex from this set as follows. Let $P_{u,t} = \frac{x_{u,t} - x_{u,t+1}}{P_u}$, where

$$P_u = \sum_{a_t b_j \in E,\ t<i} (x_{u,t} - x_{u,t+1}).$$

Then a_{t_q} is selected if $\sum_{p=1}^{q} P_{u,t_p} < Y \leq \sum_{p=1}^{q+1} P_{u,t_p}$. Thus a concrete a_t is selected with probability $P_{u,t}$, which proportional to the difference of the fractional values $x_{u,t} - x_{u,t+1}$.

When the selected vertex is a_t, we shift the image of the vertex u from a_i to a_t. This modifies the homomorphism f, and hence the corresponding values of the variables. Namely, $x'_{u,t+1}, \dots, x'_{u,i}$ are reset to 0, keeping all other values the same. Note that these modified values still satisfy the original constraints, i.e., the modified mapping is still a homomorphism.

We repeat the same process for the next u with these properties, until $a_i b_j$ is no longer in F (because no edge of G maps to it). This ends the iteration on $a_i b_j$, and we proceed to the next edge $a_{i'} b_{j'}$ with the maximum $i' + j'$ for the next iteration. Each iteration involves at most $|V(G)|$ shifts. After at most $|E'|$ iterations, the set F is empty and we no longer need to shift.

We now claim that because of the randomization, the cost of this homomorphism is at most $|V(H)|$ times the minimum cost of a homomorphism. We denote by w the value of the objective function with the fractional optimum $x_{u,i}, x_{v,j}$, and by w' the value of the objective function with the final values $x'_{u,i}, x'_{v,j}$, after the rounding and all the shifting. We also denote by w^* the minimum cost of a homomorphism of G to H. Obviously we have $w \le w^* \le w'$.

We now show that the expected value of w' is at most a constant times w. We focus on the contribution of one summand, say $x'_{u,t} - x'_{u,t+1}$, to the calculation of the cost. (The other case, $x'_{v,s} - x'_{v,s+1}$, is similar.)

In any integer solution, $x'_{u,t} - x'_{u,t+1}$ is either 0 or 1. The probability that $x'_{u,t} - x'_{u,t+1}$ contributes to w' is the probability of the event that $x'_{u,t} = 1$ and $x'_{u,t+1} = 0$. This can happen in the following situations.

1. u is mapped to a_t by rounding, and is not shifted away. In other words, we have $x'_{u,t} = 1$ and $x'_{u,t+1} = 0$ after rounding, and these values don't change by shifting.

2. u is first mapped to some $a_i, i > t$, by rounding, and then re-mapped to a_t by shifting. This happens if there exist j and v such that uv is an edge of G mapped to $a_i b_j \in F$, and then the image of u is shifted to a_t, where $a_t b_j \in E$. In other words, we have $x'_{u,i} = x'_{v,j} = 1$ and $x'_{u,i+1} = x'_{v,j+1} = 0$ after rounding; and then u is shifted from a_i to a_t.

For the situation in 1, we compute the expectation as follows. The values $x'_{u,t} = 1, x'_{u,t+1} = 0$ are obtained by rounding if $x_{u,t+1} < X \le x_{u,t}$, i.e., with probability $x_{u,t} - x_{u,t+1}$. The probability that they are not changed by shifting is at most 1, whence this situation occurs with probability at most $x_{u,t} - x_{u,t+1}$, and the expected contribution is at most $c(u,t)(x_{u,t} - x_{u,t+1})$.

For the situation in 2, we first compute the contribution for a fixed i (for which there exist j and v as described above). The values $x'_{u,i} = x'_{v,j} = 1$ and $x'_{u,i+1} = x'_{v,j+1} = 0$ are obtained by rounding if X satisfies $\max\{x_{u,i+1}, x_{v,j+1}\} < X \le \min\{x_{u,i}, x_{v,j}\}$, i.e., with probability $\min\{x_{u,i}, x_{v,j}\} - \max\{x_{u,i+1}, x_{v,j+1}\} \le x_{v,j} - x_{u,i+1} \le x_{v,j} - x_{u,s} \le P_u$. In the last two inequalities above we have assumed that a_s is the first neighbour of b_j after a_i, and used the first inequality added above the Claim. If b_j has no neighbours after a_i, the proof is analogous, using the second added inequality. When uv maps to $a_i b_j$, we shift u to a_t with probability $P_{u,t} = \frac{(x_{u,t} - x_{u,t+1})}{P_u}$, so the overall probability is also at most $x_{u,t} - x_{u,t+1}$, and the expected contribution for a fixed i (with its j and v) is also at most $c(u,t)(x_{u,t} - x_{u,t+1})$.

Let r denote the number of vertices of H, of the same colour as a_t, that are incident with some edges of E'. Clearly the situation in 2 can occur at for at most r different values of i. Therefore a fixed u in G contributes at most

$(1+r)c(u,t)(x_{u,t} - x_{u,t+1})$ to the expected value of w'. Thus the expected value of w' is at most

$$(1+r) \left(\sum_{u,i} c(u,i)(x_{u,i} - x_{u,i+1}) + \sum_{v,j} c(v,j)(x_{v,j} - x_{v,j+1}) \right) \leq (1+r)w.$$

Since we have $w \leq w^*$, this means that the expected value of w' is at most $(1+r)w^*$. Note that $1 + r \leq 1 + |E'|$, and also $1 + r < |V(H)|$ because a_1 (and b_1) are not incident with any edges of E' by definition.

At this point we have proved that our two-phase randomized procedure produces a homomorphism whose expected cost is at most $(1+r)$ times the minimum cost. It can be transformed to a deterministic algorithm as follows. There are only polynomially many values $x_{u,t}$ (at most $|V(G)||V(H)|$). When X lies anywhere between two such consecutive values, all computations will remain the same. Thus we can derandomize the first phase by trying all these values of X and choosing the best solution. Similarly, there are only polynomially many values of the partial sums $\sum_{p=1}^{q} P_{u,t_p}$ (again at most $|V(G)||V(H)|$), and when Y lies between two such consecutive values, all computations remain the same. Thus we can also derandomize the second phase by trying all possible values and choosing the best. Since the expected value is at most $(1+r)$ times the minimum cost, this bound also applies to this best solution.   □

Corollary 1. *Let H be a co-circular arc bigraph in which at most r vertices of either colour are incident to edges of E', and let $c \geq 1 + r$ be any constant.*
   Then the problem MinHOM(H) has a c-approximation algorithm.

Note that c can be taken to be $|V(H)|$, or $1+|E'|$, as noted above. For $c = 1+|E'|$, we have an approximation with best bound when E' is small, in particular, an exact algorithm when E' is empty.

Finally, we conclude the following classification for the complexity of approximation of minimum cost homomorphism problems.

Corollary 2. *Let H be an irreflexive graph.*
   Then the problem MinHOM(H) has a constant ratio approximation algorithm if H is a co-circular arc bigraph, and is not approximable otherwise.

5   Extensions and Future Work

Interestingly, all our steps can be repeated verbatim for the case of reflexive graphs. If H is a reflexive graph with min-max ordering $a_1, a_2, \ldots, a_n$, we again define $\ell(i)$ as the smallest subscript j, with respect to this ordering, such that a_j is a neighbour of a_i. For an input graph G, we again define variables $x_{u,i}$, for $u \in V(G), i = 1, 2, \ldots, n$, and a the same system of linear inequalities $\mathcal{S}$ (restricted to the u's), and obtain an analogue of Theorem 2. Provided H has a min ordering, we can again add edges E' as above to produce a reflexive graph H' with a min-max ordering, with the analogous properties expressed in

Observations 1 and 2. We can add the corresponding inequalities to S as above, and there will again be a one-to-one correspondence between homomorphisms of G to H and the integer solutions to the system. Finally, we can define the approximation by the same sequence of rounding and shifting. Everything works exactly as before because it only depends on the definition of min (and min-max) ordering, which are the same. We leave the details to the reader. Finally, we use the fact that a reflexive graph has a min ordering if and only if it is an interval graph [2,12], and the fact that the list homomorphism problem ListHOM(H) is NP-complete if the reflexive graph H is not an interval graph [2]. The last facts implies, as in [20], that the problem MinHOM(H) is not approximable if H is a reflexive graph that is not an interval graph.

Theorem 5. *Let H be a reflexive graph.*
The problem MinHOM(H) has a $|V(H)|$-approximation algorithm if H is an interval graph, and is not approximable otherwise.

We leave open the problem of approximability of MinHOM problems for general graphs, i.e., graphs with loops allowed (some vertices have loops while other don't). It should be noted that the complexity of both ListHOM and MinHOM problems for general graphs has been classified in [4,9] respectively.

We also leave open the problem of approximability of MinHOM problems for directed graphs. The complexity of ListHOM and MinHOM problems for directed graphs has been classified in [13,14] respectively.

It would be particularly interesting to see (in both of the open cases) whether the complexity classification for constant ratio approximability again coincides with the complexity classification for list homomorphisms.

The most interesting open question is whether the approximation ratio can be bounded by a constant independent of H. Our algorithm is both a $|V(H)|$-approximation and a $1 + |E'|$-approximation algorithm. These are constants independent of the input G, but very much dependent on the fixed graph H. For many bipartite graphs H (including the bipartite tent, net, or claw), one can choose $|E'| = 1$, thus obtaining a 2-approximation algorithm. With a bit more effort it can be shown that a 2-approximation algorithm exists for the so-called doubly convex bigraphs. We have not excluded the possibility that there exist polynomial time 2-approximation algorithms (or k-approximation algorithms, for some absolute constant k) for all co-circular arc bigraphs H. Until such a possibility is excluded, there is not much interest in making slight improvements to the approximation ratio. However, we do have a more complicated d-approximation algorithm for co-circular arc bigraphs (and reflexive interval graphs) H with maximum degree d. We have not included it here but will be happy to communicate it to interested readers upon request.

We have recently learned that Benoit Larose and Adrian Lemaitre have also characterized bipartite graphs with a min ordering [18].

References

1. Aggarwal, G., Feder, T., Motwani, R., Zhu, A.: Channel assignment in wireless networks and classiffcation of minimum graph homomorphisms. Electronic Colloq. on Comput. Complexity (ECCC) TR06-040 (2006)
2. Feder, T., Hell, P.: List Homomorphism to reflexive graphs. J. Combin. Theory B 72, 236–250 (1998)
3. Feder, T., Hell, P., Huang, J.: List homomorphisms and circular arc graphs. Combinatorica 19, 487–505 (1999)
4. Feder, T., Hell, P., Huang, J.: Bi-arc graphs and the complexity of list homomorphisms. J. Graph Th. 42, 61–80 (2003)
5. Feder, T., Hell, P., Huang, J., Rafiey, A.: Interval graphs, adjusted interval digraphs, and reflexive list homomorphisms. Discrete Appl. Math. 160(6), 697–707 (2012)
6. Feder, T., Hell, P., Jonsson, P., Krokhin, A., Nordh, G.: Retractions to pseudo-forests. SIAM J. on Discrete Math. 24, 101–112 (2010)
7. Gupta, A., Hell, P., Karimi, M., Rafiey, A.: Minimum Cost Homomorphisms to Reflexive Digraphs. In: Laber, E.S., Bornstein, C., Nogueira, L.T., Faria, L. (eds.) LATIN 2008. LNCS, vol. 4957, pp. 182–193. Springer, Heidelberg (2008)
8. Gutin, G., Rafiey, A., Yeo, A., Tso, M.: Level of repair analysis and minimum cost homomorphisms of graphs. Discrete Appl. Math. 154, 881–889 (2006)
9. Gutin, G., Hell, P., Rafiey, A., Yeo, A.: A dichotomy for minimum cost graph homomorphisms. European J. Combin. 29, 900–911 (2008)
10. Halldórsson, M.M., Kortsarz, G., Shachnai, H.: Minimizing Average Completion of Dedicated Tasks and Interval Graphs. In: Goemans, M.X., Jansen, K., Rolim, J.D.P., Trevisan, L. (eds.) RANDOM 2001 and APPROX 2001. LNCS, vol. 2129, pp. 114–126. Springer, Heidelberg (2001)
11. Hell, P., Huang, J.: Interval bigraphs and circular arc graphs. J. Graph Theory 46, 313–327 (2004)
12. Hell, P., Nešetřil, J.: Graphs and homomorphisms. Oxford University Press (2004)
13. Hell, P., Rafiey, A.: The dichotomy of list homomorphisms for digraphs. In: SODA 2011 (2011)
14. Hell, P., Rafiey, A.: Duality for min-max orderings and dichotomy for minimum cost homomorphisms, arXiv:0907.3016v1 [cs.DM]
15. Jansen, K.: Approximation results for the optimum cost chromatic partition problem. J. Algorithms 34, 54–89 (2000)
16. Jiang, T., West, D.B.: Coloring of trees with minimum sum of colors. J. Graph Theory 32, 354–358 (1999)
17. Kroon, L.G., Sen, A., Deng, H., Roy, A.: The Optimal Cost Chromatic Partition Problem for Trees and Interval Graphs. In: D'Amore, F., Marchetti-Spaccamela, A., Franciosa, P.G. (eds.) WG 1996. LNCS, vol. 1197, pp. 279–292. Springer, Heidelberg (1997)
18. Larose, B., Lemaitre, A.: List-homomorphism problems on graphs and arc consistency (2012) (manuscript)
19. McConnell, R.M.: Linear-time recognition of circular-arc graphs. Algorithmica 37, 93–147 (2003)
20. Mastrolilli, M., Rafiey, A.: On the approximation of minimum cost homomorphism to bipartite graphs. Discrete Applied Mathematics (June 22, 2011) (in press), http://dx.doi.org/10.1016/j.dam.2011.05.002

Property Testing in Sparse Directed Graphs: Strong Connectivity and Subgraph-Freeness

Frank Hellweg and Christian Sohler[⋆]

Department of Computer Science, Technische Universität Dortmund
{frank.hellweg,christian.sohler}@tu-dortmund.de

Abstract. We study property testing in directed graphs in the bounded degree model, where we assume that an algorithm may only query the outgoing edges of a vertex, a model proposed by Bender and Ron [4]. As our first main result, we we present the first property testing algorithm for strong connectivity in this model, having a query complexity of $\mathcal{O}(n^{1-\epsilon/(3+\alpha)})$ for arbitrary $\alpha > 0$; it is based on a reduction to estimating the vertex indegree distribution. For subgraph-freeness we give a property testing algorithm with a query complexity of $\mathcal{O}(n^{1-1/k})$, where k is the number of connected componentes in the queried subgraph which have no incoming edge. We furthermore take a look at the problem of testing whether a weakly connected graph contains vertices with a degree of least 3, which can be viewed as testing for freeness of all orientations of 3-stars; as our second main result, we show that this property can be tested with a query complexity of $\mathcal{O}(\sqrt{n})$ instead of, what would be expected, $\Omega(n^{2/3})$.

1 Introduction

Property testing is a technique for solving decision problems that sacrifices some accuracy for the benefit of a sublinear time complexity. The sacrifice of accuracy is twofold: On the one hand, we allow property testing algorithms to accept a small margin of inputs that do not have the queried property Π but are similar to some inputs that have Π. More formally, for a proximity parameter $\epsilon < 1$, we say that an input is ϵ-*far* from having the property Π, if one must modify an ϵ-fraction of the input's description in order to construct an input that has Π. We only require a property testing algorithm for Π to give a reliable answer for inputs that either have the property Π or are ϵ-far from it.

The second relaxation in accuracy is due to the randomized nature of property testing algorithms: All those algorithms are *Monte Carlo* algorithms, which means that they are allowed to have a small constant error probability.

The most important measure for the performance of a property testing algorithm is its *query complexity*, which is the worst-case number of accesses to the input that it needs for inputs of a given size. We aim for algorithms that have a query complexity of $o(n)$ or even $\mathcal{O}(1)$.

[⋆] Research partly supported by DFG grant SO 514/3-2.

L. Epstein and P. Ferragina (Eds.): ESA 2012, LNCS 7501, pp. 599–610, 2012.

In this paper we are particularly interested in property testing for sparse directed graphs. Such graphs are assumed to be stored in adjacency list representation and have both an in- and an outdegree of at most some constant d; we require the adjacency lists to only contain the outgoing edges of a vertex, a model which has been introduced in [4]. This is a quite natural model for directed graphs: For example, the webgraph or, typically, graphs of social networks are sparse graphs which have directed links; in particular, the incoming edges of a vertex of these graphs might not be visible, for example in case of the incoming links of a website during a web crawl. To gain this knowledge, basically the whole graph has to be explored, and since these graphs are typically very large, this may be inappropriate. Property testing algorithms for this graph model can be useful to gain information about the structure of such graphs while exploring only a small portion of it.

Property testing has been introduced by Rubinfeld and Sudan [12], while Goldreich, Goldwasser, and Ron [8] have initiated the study of graph properties. In this paper the authors introduced property testing in the dense graph model, where graphs are assumed to be stored as an adjacency matrix. Furthermore, Goldreich and Ron have introduced property testing in the sparse graph model [9]. Since then a large variety of graph properties has been studied, including [5,7,10,11] in the sparse graph model and [2] in the dense graph model. These papers aim for identifying classes of testable properties: For the sparse graph model, the above series of papers shows that every hyperfinite graph property is testable, as well as every property in hyperfinite graphs; in the dense graph model, a graph property is testable if and only if it can be reduced to a problem of testing for satisfaction of one of a finite number of Szemerédi-Partitions.

Property testing in directed graphs can also be subdivided into property testing in the dense graph and the sparse graph models. In the dense graph model, Alon and Shapira have studied the property of subgraph-freeness [3]. Bender and Ron have studied the property of acyclicity in both the sparse graphs and the dense graph model and the property of strong connectivity for sparse graphs [4]. In the sparse graph model, they show that if a property testing algorithm is only allowed to query the outgoing edges of a vertex, there are no such algorithms with a query complexity of $o(n^{1/3})$ for acyclicity and $o(n^{1/2})$ for strong connectivity, where n is the number of vertices of the input graph. The assumption that only the outgoing edges of a vertex may be queried makes testing strong connectivity much harder: As Bender and Ron show, there is a one-sided error property testing algorithm with a query complexity of $\tilde{\mathcal{O}}(1/\epsilon)$ for strong connectivity without this constraint. Finally, Yoshida and Ito give a constant-time property testing algorithm for k-edge connectivity of directed graphs [13], which also relies on the visibility of incoming edges.

Our Results. In this paper we further study property testing in sparse directed graphs where only the outgoing edges of a vertex may be queried. In particular, we give the first property testing algorithm in this model for strong connectivity, which has a query complexity of $\mathcal{O}(n^{1-\epsilon/(3+\alpha)})$ for arbitrary $\alpha > 0$; this is the first main result of the paper. The algorithm is based on a reduction of the strong connectivity problem to a problem of estimating the vertex indegrees of a graph: We show that it is possible to define a locally computable partitioning of the input graph, such that small connected components that have no incoming edges become their own partitions; one can then construct a metagraph in which every partition of the original graph becomes a vertex. If the input graph is far from strongly connected, then the metagraph will contain many vertices with an indegree of 0, which can indeed be tested by statistics of the vertex indegrees.

The second property we study is subgraph-freeness, i.e., to test whether a graph H does not occur as a subgraph of a graph G. Let k be the number of connected components of H that have no incoming edge from another part of H: Then our algorithm has a query complexity of $\mathcal{O}(n^{1-1/k})$.

A problem connected to subgraph-freeness is testing whether a weakly connected graph is free of vertices with a degree of at least 3. This is equivalent to testing freeness of orientations of 3-stars. Birthday-paradox type arguments would imply a query complexity of $\Omega(n^{2/3})$ for this problem, but we can give an algorithm with one of $\mathcal{O}(n^{1/2})$, which is the second main result of the paper. This algorithm makes use of two facts: The first is that the above mentioned class of forbidden subgraphs induces some strong properties for graphs that are free of them; the second is that, when sampling edges, the probability of hitting a vertex twice as the target vertex of two different edges is disproportionally high if it has many incoming edges. This allows the algorithm to compute a ratio of two estimators, which will be considerably larger if the input graph has many vertices with a degree of at least 3.

2   Preliminaries

The graph model studied in this paper is the sparse graph model. If not explicitly stated else, all graphs in this paper are directed graphs whose vertices have an outdegree which is bounded by a constant d, as well as the indegree; this follows the notion in [4]. The graphs are assumed to be stored as adjacency lists. We at first define the notion of ϵ-farness:

Definition 1. *Let G, H be directed graphs as above, both having n vertices. We say that G is ϵ-far from H, if one has to change more than ϵdn entries of the adjacency lists of G to obtain a graph that is isomorphic to H.*

Let Π be a graph property. We say that G is ϵ-far from Π, if it is ϵ-far from any graph in Π.

Note that graphs as defined above have at most dn edges. This implies that changing ϵdn entries of adjacency lists means changing an ϵ-fraction of the graph's description. We can now define the way property testing algorithms get access to an input graph:

Definition 2. *Let $G = (V, E)$ be a directed graph with each vertex having an outdegree of at most $d \in \mathbb{N}$. We define $f_G : V \times \mathbb{N} \to V \cup \{+\}$ to be a function that for querying $f(v, i)$ returns the i-th neighbour of vertex $v \in V$ in the adjacency list representation of G, or $+$, if v has less than i neighbours.*

Property testing algorithms get access to f_G to gain knowledge about the input graph. A call to f_G takes $\mathcal{O}(1)$ time.

Definition 3. *Let $\mathcal{A}$ be an algorithm that has parameters f_G, ϵ and n. We define the* query complexity *of $\mathcal{A}$ as the worst case number of calls to f_G it performs for any graph G with n vertices. $\mathcal{A}$ is a property testing algorithm for a graph property Π, if:*

1. *The query complexity of $\mathcal{A}$ is sublinear in n.*
2. *$\mathcal{A}$ accepts every graph $G \in \Pi$ with a probability of at least $\frac{2}{3}$.*
3. *$\mathcal{A}$ rejects every graph G that is ϵ-far from Π with a probability of at least $\frac{2}{3}$.*

If $\mathcal{A}$ accepts every $G \in \Pi$ with probability 1, we say it has 1-sided error, else we say it has 2-sided error.

Finally, we define some graph properties that we will need throughout the rest of this paper. Let $G = (V, E)$ and $H = (V', E')$ be directed graphs.

We call H a *subgraph* of G, if there exists an injective mapping $g : V' \to V$ such that $(g(u), g(v)) \in E$ for all $(u, v) \in E'$; we say that G is H-*free*, if H is not a subgraph of G.

We call G *(strongly) connected*, if for all pairs of vertices $u, v \in V$ there is a directed path from u to v in G (we also say that v can be *reached* from u). We call G *weakly connected*, if for all $u, v \in V$ there is an undirected path between u and v. $U \subseteq V$ is a connected component of G, if the subgraph of G induced by U is strongly connected and there is no set of vertices $W \subseteq V - U$ such that the subgraph of G induced by $U \cup W$ is strongly connected; i.e., U is maximal.

We have to distinguish between several types of connected components of a graph that are witnesses to it being not strongly connected: Let A be a connected component of G. Then we call A a source component, if there is no edge (u, v) with $u \in V - A$ and $v \in A$; we call A a sink component, if there is no edge (u, v) with $u \in A$ and $v \in V - A$. In either case we call A a dead end.

To simplify the analysis of our algorithms, we use a sampling technique that deviates from the usual sampling of vertices (respectively, edges) with replacement. Instead, we sample each vertex (edge) of the input graph with a certain probability p. If a fixed number of vertex (edge) samples is exceeded, the algorithm aborts by returning an arbitrary answer; the probability of this event will be small. Note that in our analyses the case that the sample limit is exceeded seperately is considered seperately from the rest of the particular analysis; after that we use a union bound to bound the total error. Thus, in the rest of the analyses we can assume that each vertex (edge) is independently sampled with probability p.

3   Testing Strong Connectivity

In this section we study the graph property of strong connectivity. Bender and Ron have shown a lower bound of $\Omega(\sqrt{n})$ for algorithms with both one-sided and two-sided error for testing strong connectivity in the graph model we study

[4]. Our main result is a property testing algorithm with two-sided error and a query complexity of $\mathcal{O}(n^{1-\frac{\epsilon}{3+\alpha}})$ for an arbitrarily small constant $0 < \alpha < 1$. Algorithms with one-sided error cannot be faster than $\Omega(n)$: This can be shown by considering the graph family Bender and Ron used for their two sided error lower bound; the proof can be found in the appendix.

We will need the following lemma, which has been proved in [4]:

Lemma 1 ([4]). *Let $G = (V, E)$ be a directed graph with a vertex degree bounded by $d \in \mathbb{N}$. Assume that G is ϵ-far from being strongly connected. Then G contains more than $\epsilon dn/3$ dead ends.*

At first observe that if the input graph has many sink components, then a property testing algorithm can easily find a witness for not being connected; this follows directly from the analysis of a property testing algorithm in the model where directed edges can be accessed from both sides in [4]. However, in our model source components can not be handled in this way, as they cannot be identified by local graph traversals.

We will solve this problem by reducing it to a problem of measuring the distribution of the indegrees of a graph's vertices; this problem can then be solved in sublinear query complexity. The reduction has two parts: At first we need a locally computable reduction function $C : V \to \mathcal{P}(V)$ that suitably groups vertices of G to vertices of a metagraph. Furthermore, we need an algorithm that estimates the number of vertices of such a metagraph, since we need this value for the degree distribution estimation. We start by defining the reduction function C. For this purpose we need to define a criterion that we will use to group certain sets of vertices; $\alpha < 1$ is a small positive constant:

Definition 4. *Let $G = (V, E)$ as above. $U \subseteq V$ is called* compact component *if*

1. $|U| \le \frac{3+\alpha}{\epsilon d}$;
2. *the subgraph of G induced by U is strongly connected;*
3. *there are no vertices $v \in V - U$ and $u_1, u_2 \in U$ such that there are paths from u_1 to v and from v to u_1, each with a length of at most $\frac{3+\alpha}{\epsilon d}$.*

Note that it is possible that a vertex does not belong to any compact component; in this case, we define that its compact component only contains the vertex itself. The next two lemmas show that compact components are well-defined.

Lemma 2. *Let $G = (V, E)$ as above and $v \in V$. Then v belongs to at most one compact component.*

Proof. Assume that there are two distinct compact components U and W such that $v \in U$ and $v \in W$. Without loss of generality we assume that there exists a vertex $u \in U - W$. Since U is a compact component, the subgraph of G induced by U is strongly connected and has a size of at most $\frac{3+\alpha}{\epsilon d}$. Thus, because $v \in U$ and $u \in U$, there is a path from v to u and a path from u to v, both with a length of at most $\frac{3+\alpha}{\epsilon d}$; hence u violates the third condition for compact components for W, which is a contradiction to the assumption that W is a compact component. □

Since every vertex can only be in one compact component, we can denote the compact component of $v \in V$ as $C(v)$; if v has no compact component, then we define $C(v) = \{v\}$.

Lemma 3. *Let $G = (V, E)$ as above and $v \in V$. Then $C(u) = C(v)$ for every vertex $u \in C(v)$.*

Proof. For $|C(v)| = 1$, the lemma holds trivially. For $|C(v)| > 1$, assume that there exists a vertex $u \in C(v)$ such that $C(u) \neq C(v)$. Assume at first that there exists a vertex $w \in C(u) - C(v)$. Then, analogously to Lemma 3, there are paths from u to w and from w to u that violate the third condition for $C(v)$, which is a contradiction to $C(v)$ being a compact component.

Now assume $w \in C(v) - C(u)$; note that w might be v. By considering the paths from u to w and w to u we e get a contradiction to $C(u)$ being a compact component in the same way as above.   □

We denote by $C(G)$ be the metagraph of G that substitutes every compact component of G by a single vertex; the incoming and outgoing edges of the involved vertices are replaced by edges of their representative vertex, whereas edges between vertices in the same compact component are deleted. Both the in- and the outdegree of such a representative vertex can be at most $\frac{3+\alpha}{\epsilon d}(d-1) \leq \frac{3+\alpha}{\epsilon}$ (if $\frac{3+\alpha}{\epsilon d} \leq 1$, we have $\epsilon d \geq 3$ and no graph can be ϵ-far from strongly connected, since by changing $3n$ edges any possible input graph can be transformed into a strongly connected graph).

The interesting fact about $C(G)$ is that it replaces small source components of G with single vertices with an indegree of 0. Thus, if G has many small source components, then there will be many such vertices in $C(G)$, which should be easy to identify. Since our model does not allow to directly verify whether a vertex has an indegree of 0, this is a problem of measuring the distribution of vertex indegrees in $C(G)$.

Observation 1. *Let $G = (V, E)$ be as above and let $U \subseteq V$ be the set of vertices of a source component, $|U| \leq \frac{3+\alpha}{\epsilon d}$. Then U is a compact component.*

Proof. Since U is a source component, it is strongly connected by definition; additionally, the assumption guarantees us $|U| \leq \frac{3+\alpha}{\epsilon d}$. For the third condition for U we observe that U has no incoming edges, since it is a source component. Thus, there is no vertex $v \in V - U$ that has a path to some vertex in U.   □

It remains to show how sampling can be realized in $C(G)$. We do this by sampling in G instead and then computing the compact component of the sampled vertex v. We accept v only with probability $\frac{1}{|C(v)|}$, ensuring that each vertex of $C(G)$ has the same sample probability. Similarly, we sample an edge in $C(G)$ by sampling an edge in G and determining the compact components of both its vertices. If they belong to the same component, we throw the sample away. If they belong to distinct components which have m edges in the same direction as the sampled edge in between them, we accept the sample with probability $\frac{1}{m}$.

EstimateNumberOfVertices(n, G, d, ϵ)
$\quad x \leftarrow 0$; sample $t = \frac{3}{2\epsilon^2 d^2}$ vertices of G u.i.d.
$\quad$**foreach** sampled vertex v **do** $x \leftarrow x + \frac{1}{|C(v)|}$
$\quad$**return** $\hat{n} = \frac{n}{t} x$

For a sampled vertex v, its compact component can be computed as follows: Perform a breadth-first search in G that starts at v and aborts when it has explored all vertices at a depth of at most $\frac{3+\alpha}{\epsilon d}$. The largest strongly connected set of vertices that we find and that includes v is the candidate set W for the compact component; if it has a size of more than $\frac{3+\alpha}{\epsilon d}$, then v does not belong to a compact component. Now perform a second breadth-first search in G with all the vertices of W as starting vertices; abort it at after exploring all vertices at a depth of $\frac{6+2\alpha}{\epsilon d}$. If the subgraph of G explored by this traversal contains a structure that violates the third condition for W, v has no compact component; else, W is the compact component of v.

As mentioned above, we also need an algorithm that estimates the number of vertices in $C(G)$ in sublinear query complexity. This task can be done by the algorithm *EstimateNumberOfVertices*, which has an approach similar to the estimation algorithm for the number of connected components in [6]. It can be shown by applying a Hoeffding bound that $\tilde{n} - \epsilon d n \leq \hat{n} \leq \tilde{n} + \epsilon d n$ holds with probability at least $1 - 2e^{-3}$ for the value $\hat{n}$ which is returned by this algorithm; the number of queries to G is $\mathcal{O}(e^{-3} d^{(3+\alpha)/\epsilon d})$.

The last missing piece is the algorithm *EstimateReachableVertices*, which solves the reduced problem by estimating the distribution of vertex indegrees on $C(G)$. It uses an estimation of a given graph's vertex indegree distribution to estimate the number of vertices that are reached by at least one edge; together with *EstimateNumberOfVertices* this will allow us to estimate the number of vertices in $C(G)$ that have no incoming edge. We set $a_i := a^{d/i}$ for all $i = 1, \ldots, d$.

EstimateReachableVertices(n, G, d, ϵ)
$\quad$**for** $i \leftarrow d$ **downto** 1 **do**
$\qquad$Sample each edge of G with a probability of $p_i = a_i n^{-1/i}$,
$\qquad\qquad$aborting if more than $s_i = 8 d a_i n^{1-1/i}$ edges are sampled
$\qquad c_i \leftarrow$ number of vertices with exactly i incoming edges sampled
$\qquad \tilde{n}_i \leftarrow \frac{c_i}{p_i^i} - \sum_{i < j \leq d} \binom{j}{i} \tilde{n}_j (1 - p_i)^{j-i}$
$\quad$**return** $\tilde{m} = \frac{\epsilon d n}{16} + \sum_{1 \leq i \leq d} \tilde{n}_i$

The following lemma states that, with high probability, all the estimators $\tilde{n}_i$ in *EstimateReachableVertices* are tight to their expectation, which can be proved by induction over i and for each $\tilde{n}_i$ summing up all sources of error and bounding them by standard methods; the proof can be found in the appendix:

Lemma 4. *Let $G = (V, E)$ be a directed graph; let n_i be the number of vertices with exactly i incoming edges for $i = 1, \ldots, d$. Consider EstimateReachableVertices without the 3rd line, i.e., it does not abort when sampling too many edges. Then, with probability at least $7/8$, $|\tilde{n}_i - n_i| \leq \frac{\epsilon}{2^{i+3} d^{2i-2}} n)$ for all i.*

We use Lemma 4 to prove the correctness of *EstimateReachableVertices*:

Lemma 5. *Given a directed graph $G = (V, E)$ with a vertex degree bounded by $d \in \mathbb{N}$ and $m \leq n$ vertices that have incoming edges, EstimateReachableVertices returns a value $\tilde{m}$ such that $m \leq \tilde{m} \leq m + \frac{\epsilon d n}{8}$ with probability at least $3/4$ and with a query complexity of $\mathcal{O}(\frac{d^{8d+15}\log d}{\epsilon^3} n^{1-1/d})$.*

Proof. Let m be the number vertices in G that have at least one incoming edge; it holds $m = \sum_{1 \leq i \leq d} n_i$. Assume that $|\tilde{n}_i - n_i| \leq \frac{\epsilon}{2^{i+3}d^{2i-2}} n \leq \frac{\epsilon}{24} n$ for all $i = 1, \ldots, d$, then

$$\tilde{m} = \frac{\epsilon d n}{16} + \sum_{1 \leq i \leq d} \tilde{n}_i \leq \frac{\epsilon d n}{16} + \sum_{1 \leq i \leq d} n_i + |\tilde{n}_i - n_i| \leq m + \frac{\epsilon d n}{16} + \sum_{1 \leq i \leq d} \frac{\epsilon}{24} n = m + \frac{\epsilon d n}{8};$$

analogously, $\tilde{m} \geq m$.

The probability that the algorithm aborts due to sampling too many edges can be bounded to $\frac{1}{8d}$ for every single iteration by the Markov inequality; the union bound gives us a bound of $\frac{1}{8}$ for the probability that too many edges are sampled in at least one iteration. If this does not happen, the probability that at least one of the $\tilde{n}_i$ is not tight to n_i is at most $\frac{1}{8}$ due to Lemma 4. Thus, we have an overall failure probability of at most $\frac{1}{4}$.

The query complexity of every single iteration of the outer loop is bounded by $\mathcal{O}(da_i n^{1-1/i})$. Thus, the total query complexity of *EstimateReachableVertices* is at most $\mathcal{O}(\frac{d^{8d+15}\log d}{\epsilon^3} n^{1-1/d})$. $\qquad\square$

TestStrongConnectivity(n, G, ϵ)
   if $\epsilon d \geq 3 + \alpha$ **then return true**
   sample $s = \frac{72 + 24\alpha}{\alpha \epsilon d}$ vertices of G u.i.d.
   foreach sampled vertex v
      perform a BFS traversal in G starting at v, aborting after
         having explored $\frac{36 + 12\alpha}{\alpha \epsilon d}$ vertices
      if BFS identifies a sink component **then return false**
   $\hat{n} \leftarrow$ **EstimateNumberOfVertices**$(n, G, d, \frac{\alpha}{24(3+\alpha)}\epsilon)$
   $\tilde{m} \leftarrow$ **EstimateReachableVertices**$(n, C(G), \frac{3+\alpha}{\epsilon}, \frac{\alpha \epsilon^2 d}{6(3+\alpha)^2})$
   if $\tilde{m} < \hat{n} - \frac{\alpha}{12(3+\alpha)}\epsilon d n$ **then return false; else return true**

We can now state the main theorem of this chapter.

Theorem 1. *Let $G = (V, E)$ be a directed graph with both the maximum outdegree and the maximum indegree bounded to $d \in \mathbb{N}$; for a given vertex of G, only its outgoing edges are visible. Then for any $0 < \alpha < 1$ the property of G being strongly connected can be tested with two-sided error with a query complexity of $\mathcal{O}(\alpha^{-3} d^{4/\epsilon}(4/\epsilon)^{32/\epsilon + 22} \log \frac{1}{\epsilon} \cdot n^{1-\epsilon/(3+\alpha)})$.*

Proof. We can assume that $\epsilon d < 3 + \alpha$, elsewise *TestStrongConnectivity* returns the correct answer in the first line, since no graph can be more than $3/d$-far from being strongly connected.

At first, consider the case that G is ϵ-far from being connected. Then, due to Lemma 1, there are at more than $\frac{\epsilon dn}{3}$ dead ends in G. Assume that at least $\frac{\alpha\epsilon dn}{18+6\alpha}$ of them are sink components: Then at least $\frac{\alpha\epsilon dn}{36+12\alpha}$ of them have a size of at most $\frac{36+12\alpha}{\alpha\epsilon d}$ vertices, since G has n vertices. The probability that the sampling procedure in *TestStrongConnectivity* hits none of those is at most

$(1 - \frac{\alpha\epsilon dn}{(36+12\alpha)n})^s \le e^{-\alpha\epsilon ds/(36+12\alpha)} = e^{-2}$. If one of the small sink components is hit, it is completely explored by the corresponding BFS traversal and therefore identified; thus, the probability that G is accepted inadvertently is at most e^{-2} in this case.

Now assume that less than $\frac{\alpha\epsilon dn}{18+6\alpha}$ of the dead ends are sink components; thus, at least $\frac{\epsilon dn}{3} - \frac{\alpha\epsilon dn}{18+6\alpha} = \frac{6+\alpha}{6(3+\alpha)}\epsilon dn$ of them are source components. Since G has n vertices, at least $\frac{\alpha}{6(3+\alpha)}\epsilon dn$ of the source components have at most $\frac{3+\alpha}{\epsilon d}$ vertices and thus are compact components. Hence, $C(G)$ has at least $\frac{\alpha}{6(3+\alpha)}\epsilon dn$ vertices without an incoming edge. Let m be the number of vertices in $C(G)$ that have at least one incoming edge and let $\tilde{n}$ be the total number of vertices in $C(G)$; we know that $m \le \tilde{n} - \frac{\alpha}{6(3+\alpha)}\epsilon dn$. The call of *EstimateNumberOfVertices* guarantees us $\tilde{n} \le \hat{n} + \frac{\alpha\epsilon^2 d}{24(3+\alpha)^2} \cdot \frac{3+\alpha}{\epsilon} \cdot n = \hat{n} + \frac{\alpha}{24(3+\alpha)}\epsilon dn$ with probability at least $1 - 2e^{-3}$ by Lemma 5. Thus, the call of *EstimateReachableVertices* returns $\hat{m}$ such that $\hat{m} < m + \frac{\alpha}{24(3+\alpha)}\epsilon dn \le \hat{n} - \frac{\alpha}{12(3+\alpha)}\epsilon dn$ with probability at least $1 - 2e^{-3} - \frac{3}{4} \ge \frac{2}{3}$, causing *TestStrongConnectivity* to reject the input.

The second case is that G is strongly connected. G can only inadvertently be rejected if $\hat{m} < \hat{n} - \frac{\alpha}{12(3+\alpha)}\epsilon dn$; however, due to Lemma 5, $\hat{m}$ will be at least m with probability at least $\frac{3}{4}$, and *EstimateNumberOfVertices* guarantees us $\tilde{n} \ge \hat{n} - \frac{\alpha}{24(3+\alpha)}\epsilon dn$ with probability at least $1 - 2e^{-3}$. Since there are no source components in G, every vertex has an incoming edge, and thus $\hat{m} \ge m = \tilde{n} \ge \hat{n} - \frac{\alpha}{24(3+\alpha)}\epsilon dn$ with probability at least $\frac{2}{3}$, causing the algorithm to accept G. We conclude that the overall failure probability of *TestStrongConnectivity* is bounded by $\frac{1}{3}$.

The query complexity of the algorithm is dominated by the query complexity of the call of *EstimateReachableVertices*. The number of queries made to $C(G)$ in this call is $\mathcal{O}(\alpha^{-3}(\frac{3+\alpha}{\epsilon})^{(24+8\alpha)/\epsilon+21} \log \frac{1}{\epsilon} \cdot n^{1-\epsilon/(3+\alpha)})$. Since every query to $C(G)$ causes at most $\mathcal{O}(\epsilon^{-1}d^{(6+2\alpha)/\epsilon d})$ queries to G, this yields an overall query complexity of $\mathcal{O}(\epsilon^{-1}\alpha^{-3}d^{(6+2\alpha)/\epsilon d}(\frac{3+\alpha}{\epsilon})^{(24+8\alpha)/\epsilon+21} \log \frac{1}{\epsilon} \cdot n^{1-\epsilon/(3+\alpha)}) = \mathcal{O}(\alpha^{-3}d^{4/\epsilon}(4/\epsilon)^{32/\epsilon+22} \log \frac{1}{\epsilon} \cdot n^{1-\epsilon/(3+\alpha)})$. $\qquad\square$

4   Testing Subgraph-Freeness

In this chapter we study the problem of subgraph-freeness. We will only briefly present our results, see the appendix for a complete discussion. Let G and H be directed graphs with a vertex outdegree bounded by $d \in \mathbb{N}$; note that we do not require the vertex indegree to be bounded as well. Let n be the number of vertices of G and let m be the number of vertices of H. Let k be the number of source components of H. Let $\epsilon < 1$ be a proximity parameter.

Assume that G is ϵ-far from H-free. Then, G contains more than $\epsilon n/m$ edge-disjoint copies of H: If the maximum number of disjoint occurences was at most $\epsilon n/m$, then one could delete all edges that are incident to one of their vertices. In this way every occurence of H would loose at least one edge, and we would have constructed a H-free graph by removing at most ϵdn edges from G.

Our algorithm works as follows: Sample each vertex of G with a probability of $p = (3m/\epsilon n)^{1/k}$; starting from each of the sampled vertices, run a breadth-first search that aborts after having explored all vertices at a depth of at most m. If the subgraph induced by the explored vertices contains H as subgraph, return false, else true. By a birthday-paradox type argument we can show the following theorem for weakly connected H:

Theorem 2. *G can be tested for H-freeness with one-sided error and a running time of $\mathcal{O}\left((m/\epsilon)^{1/k} d^m n^{1-1/k}\right)$.*

By a more careful analysis of the number of occurences of H if G is ϵ-far from H-free, we get the following corollary for disconnected subgraphs H:

Corollary 1. *Assume that H consist of l weakly connected components and let k_{max} be the maximum and k_{min} be the minimum number of source components among them. Then G can be tested for H-freeness with one-sided error and in query complexity $\mathcal{O}\left((3/2)^l (m/\epsilon)^{1/k_{min}} d^m n^{1-1/k_{max}}\right)$.*

5   Testing 3-Star-Freeness

In this section we will study the problem of testing whether a weakly connected graph has vertices with a degree of at least 3 or not. This might seem trivial at first glance, but since algorithms are only able to see the outgoing edges of a vertex, there are graphs where it is hard to gain much information by typical exploration techniques like breadth-first searches; in fact, if all vertices with a degree of at least 3 in a graph have exclusively incoming edges, a breadth-first search will never reveal more than one edge of such a vertex. By a birthday-paradox type argument one might now assume that $\Omega(n^{2/3})$ queries are needed to solve this problem, which would ensure that, with high probability, 3 incoming edges of at least one of the vertices get sampled. However, we will show that this property can be tested with a query complexity of roughly $\mathcal{O}(n^{1/2})$ in weakly connected graphs.

Sketch of Test3StarFreeness(n, G, ϵ)

Handle cases where G has many 3-stars that have outgoing edges

$\hat{k} \leftarrow$ Estimate of the number of outgoing 2-stars

Sample each edge with a probability of $p = \frac{256d}{\epsilon^{3/2}\sqrt{n}}$

$\hat{c} \leftarrow \frac{1}{p^2}$ times the number of 2-way collisions of sampled edges

if $r := \frac{\hat{c}}{\hat{k}} \geq 1 + \frac{\epsilon}{24}$ **then return** false

else return true

This problem is related to testing subgraph-freeness: In fact, we are testing for a class of forbidden subgraphs which contains all orientations of 3-stars, where a 3-star is a graph consisting of a central vertex which is connected to 3 other vertices. This class of subgraphs induces some structural constraints on graphs that are free of them, which we can exploit. We will only briefly discuss our techniques at this point; see the appendix for a full presentation.

Let $G = (V, E)$ be a directed graph whose total vertex degree is bounded by $d \in \mathbb{N}$. We call the difference between the number vertices with an indegree of at least 2 and the number of vertices with an outdegree of at least 2 the *balance* $\mathcal{B}(G)$ of G.

If G is ϵ-far from 3-star-free, then it contains more than ϵn vertices with a degree of at least 3: For any such vertex, at most $d - 2$ edges have to be removed such that it gets a degree of less than 3 and thus is not a central vertex of a 3-star anymore; thus, if G had at most ϵn vertices with a degree of at least 3, less than ϵdn edges would have to be removed to eliminate all 3-stars, and hence G would not be ϵ-far from 3-star-free.

At first observe that the problem can be solved by standard property testing techniques if there are many (say, at least $\epsilon n/8d$) vertices with a degree of at least 3 that have at least one outgoing edge: Sampling at most 2 of the incoming edges of one of them would suffice to discover a 3-star and thus reject G; by a birthday-paradox type argument this will happen with high probability when sampling $\Theta(\sqrt{n})$ edges. The remaining edges can then be found by examining the adjacency list of the corresponding vertex. Thus, for the remainder of this section we only have to distinguish two cases: G is 3-star-free or G is ϵ-far from 3-star-free and has at least $\epsilon n - \epsilon n/8d = \Theta(\epsilon n)$ vertices with an indegree of at least 3 and no outgoing edges.

The basic idea for this is as follows: As we will see, the balance of G is relatively small, especially if G is 3-star-free. Thus, the number of vertices with an outdegree of at least 2 is a good estimate of the number vertices with an indegree of at least 2. We estimate both of these values: the first by a standard sampling approach, where we sample vertices and check whether they have 2 outgoing edges; the second by measuring the number of common target vertices of a set of randomly sampled edges, i.e., the number of collisions of the edges sampled in line 3 of the algorithm sketch.

For a graph that is 3-star-free, the ratio r between both of these values indeed is not much larger than 1 with high probability: The balance of G can be shown to be at most 1 in this case, and appropriately bounding the probability that $\hat{k}$ is much smaller than its expectation or that $\hat{c}$ is much larger gives the result.

However, if G has at least $\epsilon n - \epsilon n/8d = \Theta(\epsilon n)$ vertices with an indegree of at least 3, $\hat{c}$ will be larger with high probability: This is because the probability of sampling two of the edges of a vertex with indegree k is $\binom{k}{2}p^2(1 - p)^{k-2} \approx \frac{k(k-1)}{2}p^2$, which is quadratic in k; note that, for sufficiently large n, $(1 - p)^{k-2}$ is close to 1, since k is bounded by d. On the other hand, the balance of G can be proved to roughly increase by at most $k - 2$ for every such vertex, which is linear; thus, the factor by which $\hat{c}$ will increase for any vertex with indegree k

in G is considerably larger than the factor by which $\hat{k}$ will increase. This causes the expected number of collisions to be measurably higher than in the case that G is 3-star-free, since there are many vertices with an indegree of at least 3 in G: They form at least a $\Theta(\epsilon)$-fraction of the total number of vertices with an indegree of at least 2 (which is, trivially, at most n). Combining these facts, we can indeed show that r is at least $1 + \epsilon/24$ with probability 5/6 in this case.

The query complexity of *Test3StarFreeness* is dominated by the number of samples taken in line 3, which is $\mathcal{O}(pn) = \mathcal{O}(\frac{d\sqrt{n}}{\epsilon^{3/2}})$. For small constant n we solve the problem deterministically in time $\mathcal{O}(\frac{d^5}{\epsilon^3})$. This leads to the following theorem:

Theorem 3. *Let G be a weakly connected directed graph with a vertex degree bounded by $d \in \mathbb{N}$, and let $0 < \epsilon \leq 1$. Then* Test3StarFreeness *returns true if G is 3-star-free and false if G is ϵ-far from 3-star-free, each with probability at least $\frac{5}{6}$. Its query complexity is $\mathcal{O}(\max\{\frac{d\sqrt{n}}{\epsilon^{3/2}}, \frac{d^5}{\epsilon^3}\})$.*

References

1. Alon, N.: Testing subgraphs in large graphs. Random Struct. Algorithms 21(3-4), 359–370 (2002)
2. Alon, N., Fischer, E., Newman, I., Shapira, A.: A combinatorial characterization of the testable graph properties: it's all about regularity. SIAM Journal on Computing 39(1), 143–167 (2009)
3. Alon, N., Shapira, A.: Testing subgraphs in directed graphs. In: Proc. of the 35th ACM Symp. on the Theory of Computing (STOC), pp. 700–709 (2003)
4. Bender, M.A., Ron, D.: Testing properties of directed graphs: acyclicity and connectivity. Random Structures & Algorithms 20(2), 184–205 (2002)
5. Benjamini, I., Schramm, O., Shapira, A.: Every minor-closed property of sparse graphs is testable. In: Proc. of the 40th ACM Symp. on the Theory of Computing (STOC), pp. 393–402 (2008)
6. Chazelle, B., Rubinfeld, R., Trevisan, L.: Approximating the Minimum Spanning Tree Weight in Sublinear Time. SIAM Journal on Computing 34(6), 1370–1379 (2005)
7. Czumaj, A., Shapira, A., Sohler, C.: Testing hereditary properties of non-expanding bounded-degree graphs. SIAM Journal on Computing 38(6), 2499–2510 (2009)
8. Goldreich, O., Goldwasser, S., Ron, D.: Property Testing and its Connection to Learning and Approximation. J. of the ACM 45(4), 653–750 (1998)
9. Goldreich, O., Ron, D.: Property Testing in Bounded Degree Graphs. In: Proc. of the 29th ACM Symp. on the Theory of Computing (STOC), pp. 406–415 (1997)
10. Hassidim, A., Kelner, J.A., Nguyen, H.N., Onak, K.: Local graph partitions for approximation and testing. In: Proc. of the 50th IEEE Symp. on Foundations of Computer Science (FOCS), pp. 22–31 (2009)
11. Newman, I., Sohler, C.: Every property of hyper nite graphs is testable. In: Proc. of the 43rd ACM Symp. on the Theory of Computing (STOC), pp. 675–684 (2011)
12. Rubinfeld, R., Sudan, M.: Robust Characterizations of Polynomials with Applications to Program Testing. SIAM Journal on Computing 25(2), 252–271 (1996)
13. Yoshida, Y., Ito, H.: Testing k-edge-connectivity of digraphs. Journal of System Science and Complexity 23(1), 91–101 (2010)

Improved Implementation of Point Location in General Two-Dimensional Subdivisions*

Michael Hemmer, Michal Kleinbort, and Dan Halperin

Tel-Aviv University, Israel

Abstract. We present a major revamp of the point-location data structure for general two-dimensional subdivisions via randomized incremental construction, implemented in CGAL, the Computational Geometry Algorithms Library. We can now guarantee that the constructed directed acyclic graph $\mathcal{G}$ is of linear size and provides logarithmic query time. Via the construction of the Voronoi diagram for a given point set S of size n, this also enables nearest-neighbor queries in guaranteed $O(\log n)$ time. Another major innovation is the support of general unbounded subdivisions as well as subdivisions of two-dimensional parametric surfaces such as spheres, tori, cylinders. The implementation is exact, complete, and general, i.e., it can also handle non-linear subdivisions. Like the previous version, the data structure supports modifications of the subdivision, such as insertions and deletions of edges, after the initial preprocessing. A major challenge is to retain the expected $O(n \log n)$ preprocessing time while providing the above (deterministic) space and query-time guarantees. We describe efficient preprocessing algorithms, which explicitly verify the length $\mathcal{L}$ of the longest query path. However, instead of using $\mathcal{L}$, our implementation is based on the depth $\mathcal{D}$ of $\mathcal{G}$. Although we prove that the worst case ratio of $\mathcal{D}$ and $\mathcal{L}$ is $\Theta(n/\log n)$, we conjecture, based on our experimental results, that this solution achieves expected $O(n \log n)$ preprocessing time.

1   Introduction

Birn et al. [1] presented a structure for planar nearest-neighbor queries, based on Delaunay triangulations, named Full Delaunay Hierarchies (FDH). The FDH is a very simple, and thus light, data structure that is also very easy to construct. It outperforms many other methods in several scenarios, but it does not have a worst-case optimal behavior. However, it is claimed [1] that methods that do have this behavior are too cumbersome to implement and thus not available. We got challenged by this claim.

* This work has been supported in part by the 7th Framework Programme for Research of the European Commission, under FET-Open grant number 255827 (CGL— Computational Geometry Learning), by the Israel Science Foundation (grant no. 1102/11), and by the Hermann Minkowski–Minerva Center for Geometry at Tel Aviv University.

L. Epstein and P. Ferragina (Eds.): ESA 2012, LNCS 7501, pp. 611–623, 2012.

In this article we present an improved version of CGAL's planar point location that implements the famous incremental construction (RIC) algorithm as introduced by Mulmuley [2] and Seidel [3]. The algorithm constructs a linear size data structure that guarantees a logarithmic query time. It enables nearest-neighbor queries in guaranteed $O(\log n)$ time via planar point location in the Voronoi Diagram of the input points. In Section 4 we compare our revised implementation for point location, applied to nearest neighbor search, against the FDH. Naturally, this is only a byproduct of our efforts as planar point location is a very fundamental problem in Computational Geometry. It has numerous applications in a variety of domains including computer graphics, motion planning, computer aided design (CAD) and geographic information systems (GIS).

Previous Work. Most solutions can only provide an expected query time of $O(\log n)$ but cannot guarantee it, in particular, those that only require $O(n)$ space. Some may be restricted to static scenes that do not change, while others can only support linear geometry.

Triangulation-based point location methods, such as the approaches by Kirkpatrick [4] and Devillers [5] combine a logarithmic hierarchy with some walk strategy. Both require only linear space and Kirkpatrick can even guarantee logarithmic query time. However, both are restricted to linear geometry, since they build on a triangulation of the actual input.

Many methods can be summarized under the model of the trapezoidal search graph as pointed out by Seidel and Adamy [6]. Conceptually, the initial subdivision is further subdivided into trapezoids by emitting vertical rays (in both directions) at every endpoint of the input, which is the fundamental search structure. In principal, all these solutions can be generalized to support input curves that are decomposable into a finite number of x-monotone pieces.

The *slabs method* of Dobkin and Lipton [7] is one of the earliest examples. Every endpoint induces a vertical wall giving rise to $2n + 1$ vertical slabs. A point location is performed by a binary search to locate the correct slab and another search within the slab in $O(\log n)$ time. Preparata [8] introduced a method that avoids the decomposition into $n + 1$ slabs reducing the required space from $O(n^2)$ to $O(n \log n)$. Sarnak and Tarjan [9] went back to the slabs of Dobkin and Lipton and added the idea of persistent data structures, which reduced the space consumption to $O(n)$. Another example for this model is the separating chains method of Lee and Preparata [10]. Combining it with fractional cascading, Edelsbrunner et al. [11], achieved $O(\log n)$ query time as well. For other methods and variants the reader is referred to a comprehensive overview given in [12].

An asymptotically optimal solution is the randomized incremental construction (RIC), which was introduced by Mulmuley [2] and Seidel [3]. In the static setting, it achieves $O(n \log n)$ preprocessing time, $O(\log n)$ query time and $O(n)$ space, all in expectancy. As pointed out in [13], the latter two can even be worst-case guaranteed. It is also claimed there that one can achieve these worst-case bounds in an expected preprocessing time of $O(n \log^2 n)$, but no concrete proof is given. The approach is able to handle dynamic scenes; that is, it is possible to add or delete edges later on. This method is discussed in more detail in Section 2.

Contribution: We present here a major revision of the trapezoidal-map random incremental construction algorithm for planar point location in CGAL. As the previous implementation, it provides a linear size data structure for non-linear subdivisions that can handle static as well as dynamic scenes. The new version is now able to guarantee $O(\log n)$ query time and $O(n)$ space. Following recent changes in the "2D Arrangements" package [14], the implementation now also supports unbounded subdivisions as well as ones that are embedded on two-dimensional parametric surfaces. After a review of the RIC in Section 2, we discuss, in Section 3, the difference between the length $\mathcal{L}$ of the longest search path and the depth $\mathcal{D}$ of the DAG. We prove that the worst-case ratio of $\mathcal{D}$ and $\mathcal{L}$ is $\Theta(n/\log n)$. Moreover, we describe two algorithms for the preprocessing stage that achieve guaranteed $O(n)$ size and $O(\log n)$ query time. Both are based on a verification of $\mathcal{L}$ after the DAG has been constructed: An implemented one that runs in expected $O(n\log^2 n)$ time, and a more efficient one that runs in expected $O(n\log n)$ time. The latter is a very recent addition that was not included in the reviewed submission. However, the solution that is integrated into CGAL is based on a verification of $\mathcal{D}$. Based on our experimental results, we conjecture that it also achieves expected $O(n\log n)$ preprocessing time. Section 4 demonstrates the performance of the new implementation by comparing our point location in a Voronoi Diagram with the nearest neighbor implementation of the FDH and others. Section 5 presents more details on the new implementation. To the best of our knowledge, this is the only available implementation for guaranteed logarithmic query time point location in general two-dimensional subdivisions.

2   Review of the RIC for Point Location

We review here the random incremental construction (RIC) of an efficient point location structure, as introduced by [2,3] and described in [13,15]. For ease of reading we discuss the algorithm in case the input is in general position. Given an arrangement of n pairwise interior disjoint x-monotone curves, a random permutation of the curves is inserted incrementally, constructing the Trapezoidal Map, which is obtained by extending vertical walls from each endpoint upward and downward until an input curve is reached or the wall extends to infinity. During the incremental construction, an auxiliary search structure, a directed acyclic graph (DAG), is maintained. It has one root and many leaves, one for every trapezoid in the trapezoidal map. Every internal node is a binary decision node, representing either an endpoint p, deciding whether a query lies to the left or to the right of the vertical line through p, or a curve, deciding if a query is above or below it. When we reach a curve-node, we are guaranteed that the query point lies in the x-range of the curve. The trapezoids in the leaves are interconnected, such that each trapezoid knows its (at most) four neighboring

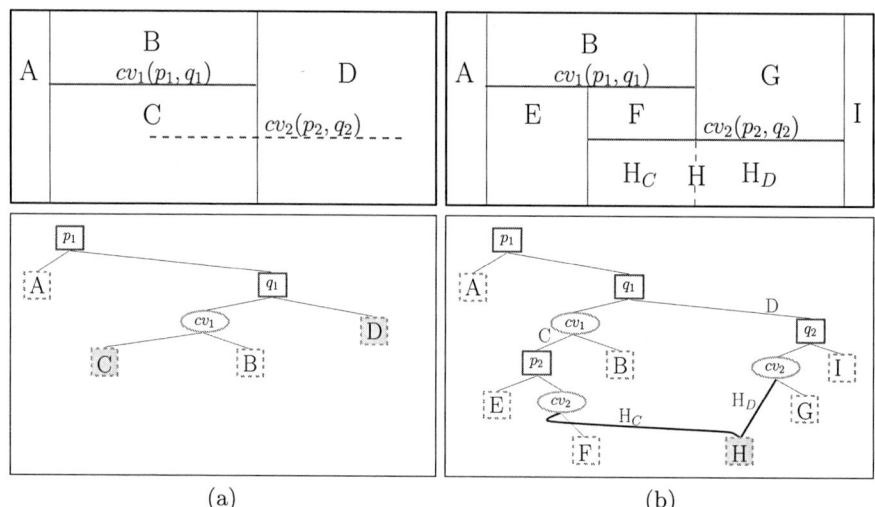

Fig. 1. Trapezoidal decomposition and the constructed DAG for two segments cv_1 and cv_2: (a) before and (b) after the insertion of cv_2. The insertion of cv_2 splits the trapezoids C, D into E, F, H_C and G, I, H_D, respectively. H_C and H_D are merged into H, as they share the same top (and bottom) curves.

trapezoids, two to the left and two to the right. In particular, there are no common x-coordinates for two distinct endpoints[1].

When a new x-monotone curve is inserted, the trapezoid containing its left endpoint is located by a search from root to leaf. Then, using the connectivity mechanism described above, the trapezoids intersected by the curve are gradually revealed and updated. Merging new trapezoids, if needed, takes time that is linear in the number of intersected trapezoids. The merge makes the data structure become a DAG (as illustrated in Figure 1) with expected $O(n)$ size, instead of an $\Omega(n \log n)$ size binary tree [6]. For an unlucky insertion order the size of the resulting data structure may be quadratic, and the longest search path may be linear. However, due to the randomization one can expect $O(n)$ space, $O(\log n)$ query time, and $O(n \log n)$ preprocessing time.

3   On the Difference between Paths and Search Paths

As shown in [13], one can build a data structure, which guarantees $O(\log n)$ query time and $O(n)$ size, by monitoring the size and the length of the longest search path $\mathcal{L}$ during the construction. The idea is that as soon as one of the values becomes too large, the structure is rebuilt using a different random insertion order. It is shown that only a small constant number of rebuilds is expected.

---

[1] In the general case all endpoints are lexicographically compared; first by the x-coordinate and then by the y-coordinate. This implies that two covertical points produce a virtual trapezoid, which has a zero width.

However, in order to retain the expected construction time of $O(n \log n)$, both values must be efficiently accessible. While this is trivial for the size, it is not clear how to achieve this for $\mathcal{L}$. Hence, we resort to the depth $\mathcal{D}$ of the DAG, which is an upper bound on $\mathcal{L}$ as the set of all possible search paths is a subset of all paths in the DAG. Thus, the resulting data structure still guarantees a logarithmic query time.

The depth $\mathcal{D}$ can be made accessible in constant time by storing the depth of each leaf in the leaf itself, and maintaining the maximum depth in a separate variable. The cost of maintaining the depth can be charged to new nodes, since existing nodes never change their depth value. This is not possible for $\mathcal{L}$ while retaining linear space, since each leaf would have to store a non-constant number of values, i.e., one for each valid search path that reaches it. In fact the memory consumption would be equivalent to the data structure that one would obtain without merging trapezoids, namely the trapezoidal search tree, which for certain scenarios requires $\Omega(n \log n)$ memory as shown in [6]. In particular, it is necessary to merge as (also in practice) the sizes of the resulting search tree and the resulting DAG considerably differ.

In Section 3.1 we show that for a given DAG its depth $\mathcal{D}$ can be linear while $\mathcal{L}$ is still logarithmic, that is, such a DAG would trigger an unnecessary rebuild. It is thus questionable whether one can still expect a constant number of rebuilds when relying on $\mathcal{D}$. Our experiments in Subsection 3.3 show that in practice the two values hardly differ, which indicates that it is sufficient to rely on $\mathcal{D}$. However, a theoretical proof to consolidate this is still missing. Subsection 3.2 provides efficient preprocessing solutions for the static scenario (where all segments are given in advance). As such, we see it as a concretization of, and an improvement over, the claim mentioned in [13].

3.1 Worst Case Ratio of Depth and Longest Search Path

The figure to the right shows the DAG of Figure 1 after inserting a third segment. There are two paths that reach the trapezoid N (black and gray arrows). However, the gray path is not a valid search path, since all points in N are to the right of q_1; that is, such a search would never visit the left child of q_1. It does, however, determine the depth of N, since it is the longer path of the two. In the sequel we use this observation 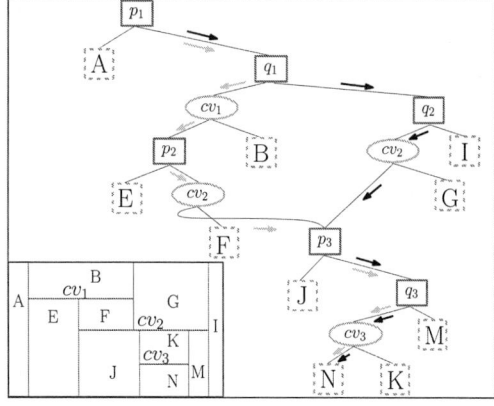 to construct an example that shows that the ratio between $\mathcal{D}$ and $\mathcal{L}$ can be as large as $\Omega(n/\log n)$. Moreover, we will show that this bound is tight.

We start by constructing a simple-to-demonstrate lower bound that achieves $\Omega(\sqrt{n})$ ratio between $\mathcal{D}$ and $\mathcal{L}$. Assuming that $n = k^2 \in \mathbb{N}$, the construction consists of k blocks, each containing k horizontal segments. The blocks are arranged as depicted in the figure to the right. Segments are inserted from top to bottom. A block starts with a large

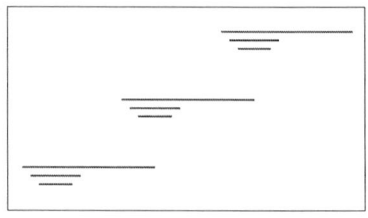

segment at the top, which we call the *cover segment*, while the other segments successively shrink in size. Now the next block is placed to the left and below the previous block. Only the cover segment of this block extends below the previous block, which causes a merge as illustrated in Figure 2. All $k = \sqrt{n}$ blocks are placed in this fashion. This construction ensures that each newly inserted segment intersects the trapezoid with the largest depth, which increases $\mathcal{D}$. The largest depth of $\Omega(n)$ is finally achieved in the trapezoid below the lowest segment. However, the actual search path into this trapezoid has only $O(\sqrt{n})$ length, since for each previous block it only passes through one node in order to skip it and $O(\sqrt{n})$ in the last block.

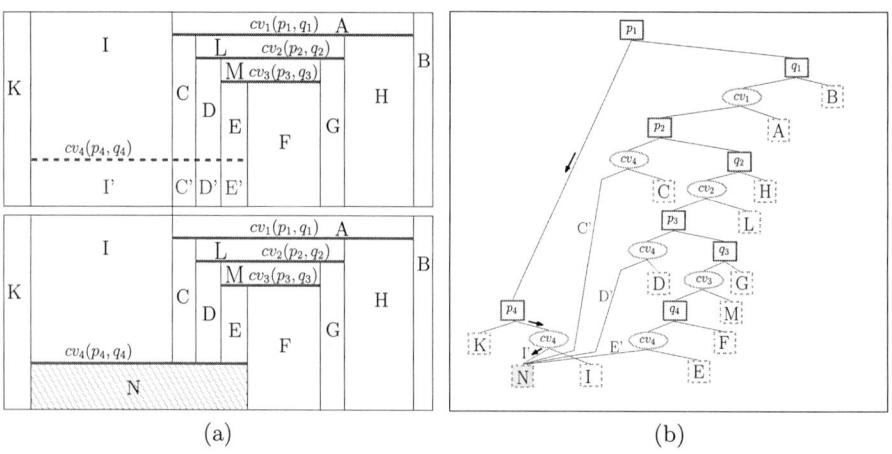

(a)   (b)

Fig. 2. (a) The trapezoidal-map after inserting cv_4. The map is displayed before and after the merge of I', C', D', and E' into N, in the top and bottom illustrations, respectively. A query path to the region of I' in N will take 3 steps, while the depth of N in this example is 11.

The following construction, which uses a recursive scheme, establishes the lower bound $\Omega(n/\log n)$ for $\mathcal{D}/\mathcal{L}$. Blocks are constructed and arranged in a similar fashion as in the previous construction. However, this time we have $\log n$ blocks, where block i contains $n/2^i$ segments. Within each block we then apply

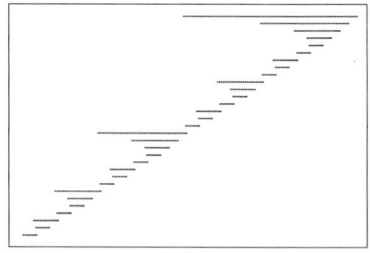

the same scheme recursively as depicted in the figure to the right. Again segments are inserted top to bottom such that the depth of $\Omega(n)$ is achieved in the trapezoid below the lowest segment. The fact that the lengths of all search paths are logarithmic can be proven by the following argument. By induction we assume that the longest path within a block of size $n/2^i$ is some constant times $(\log_2 n - i)$. Obviously this is true for a block containing only one segment. Now, in order to reach block i with $n/2^i$ segments, we require $i - 1$ comparisons to skip the $i - 1st$ preceding blocks. Thus in total the search path is of logarithmic length.

Theorem 1. *The $\Omega(n/\log n)$ worst-case lower bound on $\mathcal{D}/\mathcal{L}$ is tight.*

Proof. Obviously, $\mathcal{D}$ of $O(n)$ is the maximal achievable depth, since by construction each segment can only appear once along *any* path in the DAG. It remains to show that for any scenario with n segments there is no DAG for which $\mathcal{L}$ is smaller than $\Omega(\log n)$. Since there are n segments, there are at least n different trapezoids having these segments as their top boundary. Let T be a decision tree of the optimal search structure. Each path in the decision tree corresponds to a valid search path in the DAG and vice versa. The depth of T must be larger than $\log_2 n$, since it is only a binary tree. We conclude that the worst case ratio of $\mathcal{D}$ and $\mathcal{L}$ is $\Theta(n/\log n)$.   □

3.2   Efficient Solutions for Static Subdivisions

We first describe an algorithm for static scenes that runs in expected $O(n \log^2 n)$ time, constructing a DAG of linear size in which $\mathcal{L}$ is $O(\log n)$. The result is based on the following lemma. The proof is given in the full version of this paper [16].

Lemma 1. *Let S be a planar subdivision induced by n pairwise interior disjoint x-monotone curves. The expected size of the trapezoidal search tree $\mathcal{T}$, which is constructed as the RIC above but without merges, is $O(n \log n)$.*

The following algorithm *compute_max_search_path_length* computes $\mathcal{L}$ in expected $O(n \log^2 n)$ time. Starting at the root it descends towards the leaves in a recursive fashion. Taking the history of the current path into account, each recursion call maintains the interval of x values that are still possible. Thus, if an x-coordinate of a point node is not contained in the interval the recursion does not need to split. This means that the algorithm essentially mimics $\mathcal{T}$ (as it would have been constructed), since the recursion only follows possible search paths. By Lemma 1 the expected number of leaves of $\mathcal{T}$, and thus of search paths, is $O(n \log n)$. Since the expected length of a query is $O(\log n)$ this algorithm takes expected $O(n \log^2 n)$ time.

Definition 1. *$f(n)$ denotes the time it takes to verify that, in a linear size DAG constructed over a planar subdivision of n x-monotone curves, $\mathcal{L}$ is bounded by $c \log n$ for a constant c.*

Theorem 2. *Let S be a planar subdivision with n x-monotone curves. A point location data structure for S, which has $O(n)$ size and $O(\log n)$ query time in the worst case, can be built in $O(n \log n + f(n))$ expected time, where $f(n)$ is as defined above.*

Proof. The construction of a DAG with some random insertion order takes expected $O(n \log n)$ time. The linear size can be verified trivially on the fly. After the construction the algorithm *compute_max_search_path_length* is used to verify that $\mathcal{L}$ is logarithmic. The verification of the size and $\mathcal{L}$ may trigger rebuilds with a new random insertion order. However, according to [13], one can expect only a constant number of rebuilds. Thus, the overall expected runtime remains expected $O(n \log n + f(n))$.   □

The verification process described above takes expected $O(n \log^2 n)$ time. However, one can do better as we briefly sketch next. Let T be the collection of *all* the trapezoids created during the construction of the DAG, including intermediate trapezoids that are later killed by the insertion of later segments. Let $\mathcal{A}(T)$ denote the arrangement of the trapezoids. The *depth* of a point p in the arrangement is defined as the number of trapezoids in T that cover p. The key to the improved algorithm is the following observation by Har-Peled.

Observation 1. *The length of a path in the DAG for a query point p is at most three times the depth of p in $\mathcal{A}(T)$.*

It follows that we need to verify that the maximum depth of a point in $\mathcal{A}(T)$ is some constant $c_1 \log n$. We do this as follows. We first transform the collection T into a collection R of axis-parallel rectangles such that the maximum depth in $\mathcal{A}(R)$ is the same as the maximum depth in $\mathcal{A}(T)$. This can be done in $O(n log n)$ time since the input segments in S are interior pairwise disjoint, and we rely on the separation property of the segments in S stemming from [17]. We then apply an algorithm by Alt and Scharf [18], which detects the maximum depth in an arrangement of n rectangles in $O(n \log n)$ time. Notice that we only apply this verification algorithm on DAGs of linear size. Putting everything together we obtain:

Theorem 3. *Let S be a planar subdivision with n x-monotone curves. A point location data structure for S, which has $O(n)$ size and $O(\log n)$ query time in the worst case, can be built in $O(n \log n)$ expected time.* [2]

3.3   Experimental Results

Since $\mathcal{D}$ is an upper bound on $\mathcal{L}$ and since $\mathcal{D}$ is accessible in constant time our implementation explores an alternative that monitors $\mathcal{D}$ instead of $\mathcal{L}$. Though this may cause some additional rebuilds, the experiments in this section give strong evidence that one can still expect $O(n \log n)$ preprocessing time. We compared $\mathcal{D}$

---

[2] Theorem 3 has been added to the paper after the ESA review process. The main result of this section in the original submission was Theorem 2 with $f(n) = O(n \log^2 n)$.

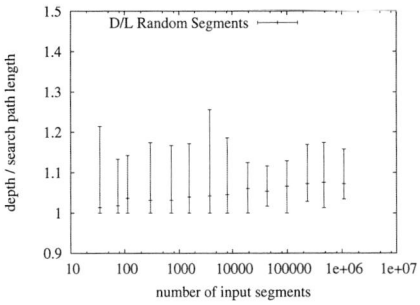 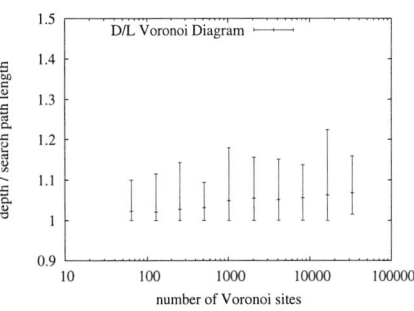

Fig. 3. Ratio of $\mathcal{D}$ and $\mathcal{L}$ in two scenarios: random segments (left), Voronoi diagram of random points (right). Plots show average value with error bars.

and $\mathcal{L}$ in two different scenarios: random non-intersecting line segments and Voronoi diagram for random sites.[3] Each scenario was tested with an increasing number of subdivision edges, with several runs for each input. Figure 3 displays the average $\mathcal{D}/\mathcal{L}$ ratio, and also the minimal and maximal ones. Obviously, the average ratio is close to 1 and never exceeded a value of 1.3.

These experimental results indicate that replacing the test for the length of the longest path $\mathcal{L}$ by the depth $\mathcal{D}$ of the DAG in the randomized incremental construction essentially does not harm the runtime. However, the following conjecture remains to be proven.

Conjecture 1. There exists a constant $c > 0$ such that the runtime of the randomized incremental algorithm, modified such that it rebuilds in case the depth $\mathcal{D}$ of the DAG becomes larger than $c \log n$, is expected $O(n \log n)$, i.e., the number of expected rebuilds is still constant.

4   Nearest Neighbor Search in Guaranteed $O(\log n)$ Time

As stated in the Introduction, we were challenged by the claim of Birn et al. [1] that guaranteed logarithmic nearest-neighbor search can be achieved via efficient point location on top of the Voronoi Diagram of the input points, but that this approach *"does not seem to be used in practice"*. With this section we would like to emphasize that such an approach is available and that it should be considered for use in practice. Using the RIC planar point location, the main advantage would be that query times are stable and independent of the actual scenario.

4.1   Nearest Neighbor Search via Voronoi Diagram

Given a set P of n points, which we wish to preprocess for efficient point location queries, we first create a Delaunay triangulation (DT) which takes $O(n \log n)$ expected time. The Voronoi diagram (VD) is then obtained by dualizing. Using a

---

[3] Additional experimental results that include also the scenarios constructed in Section 3.1 can be found at http://acg.cs.tau.ac.il/projects/lppl.

sweep, the arrangement representing the VD, which has at most $3n - 6$ edges, can be constructed in $O(n \log n)$ time. However, taking advantage of the spatial coherence of the edges, we use a more efficient method that directly inserts VD edges while crawling over the DT. The resulting arrangement is then further processed by our RIC implementation. If Conjecture 1 is true then this takes expected $O(n \log n)$ time. Alternatively, it would have been possible to implement the solution presented in Subsection 3.2 (for which we can prove expected $O(n \log n)$ preprocessing time).

4.2   Nearest Neighbor Search via Full Delaunay Hierarchy

The full Delaunay hierarchy (FDH) presented in [1] is based on the fact that one can find the nearest neighbor by performing a greedy walk on the edges of the Delaunay triangulation (DT). The difference is that the FDH keeps all edges that appear during the randomized construction [19] of the DT in a flattened n-level hierarchy structure, where level i contains the DT of the first i points. Thus, a walk that starts at the first point is accelerated due to long edges that appeared at an early stage of the construction process while the DT was still sparse. The FDH is a very light, easy to implement, and fast data structure with expected $O(n \log n)$ construction time that achieves an expected $O(\log n)$ query path length. However, a query may take $O(n)$ time since the degree of nodes can be linear. For the experiments we used two exact variants: a basic exact version (EFDH) and a (usually faster) version (FFDH) that first performs a walk using inexact floating point arithmetic and then continues with an exact walk.

4.3   Experimental Results

We compared our implementation for nearest-neighbor search using the RIC point location on the Voronoi-diagram (ENNRIC) to the following exact methods: EFDH, FFDH, CGAL's Delaunay hierarchy (CGAL_DH) [5], and CGAL's kd-tree (CGAL_KD).[4]

All experiments have been executed on a Intel(R) Core(TM) i5 CPU M 450 with 2.40GHz, 512 kB cache and 4GB RAM memory, running Ubuntu 10.10. Programs were compiled using g++ version 4.4.5 optimized with -O3 and -DNDEBUG. The left plot of Figure 4 displays the total query time in a random scenario, in which both input points and query points are randomly chosen within the unit square. Clearly, all methods have logarithmic query time, however due to larger constants ENNRIC is slower. The other plot presents a combined scenario of $(n - \lfloor \log n \rfloor)$ equally spaced input points on the unit circle and $\lfloor \log n \rfloor$ random outliers. The queries are random points in the same region. In this experiment the CGAL_KD and ENNRIC are significantly faster and maintain a stable query time. A similar scenario that was tested contains equally spaced input points on a circle and a point in the center with random query points inside the circle. The differences there are even more significant than in the previous scenario.

---

[4] Due to similar performance we elided the kd-tree implementation in ANN [20].

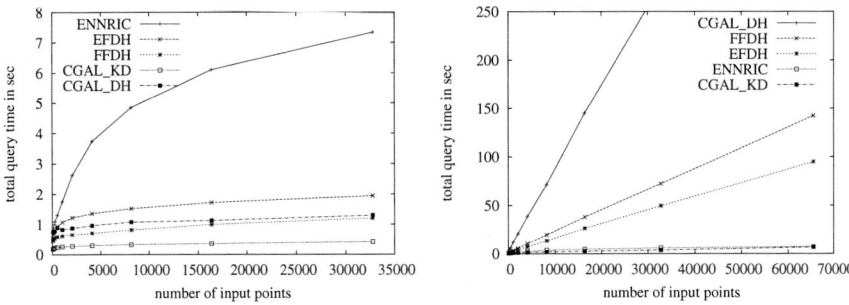

Fig. 4. Performance of 500k nearest-neighbor queries for different methods on two scenarios: (left) random points; (right) circle with outliers

As for the preprocessing time in all tested scenarios, obviously ENNRIC cannot compete with the fast construction time of the other methods.

5 CGAL's New RIC Point Location

With this article we announce our revamp of CGAL's implementation of planar point location via the randomized incremental construction of the trapezoidal map, which is going to be available in the upcoming CGAL release 4.1.

Like the previous implementation by Oren Nechushtan [21], it is part of the "2D Arrangements" package [22] of CGAL. It allows both insertions and deletions of edges. The implementation is exact and covers all degenerate cases. Following the *generic-programming paradigm* [23] it can be easily applied to linear geometry but also to non-linear geometry such as algebraic curves or Bézier curves. The main new feature, and this is what triggered this major revision, is the support for unbounded curves, as it was introduced for the "2D Arrangements" package in [14], enabling point location on two-dimensional parametric surfaces (e.g., spheres, tori, etc.) as well.

In addition we did a major overhaul of the code basis. In particular, we maintain the depth $\mathcal{D}$ of the DAG as described in Section 3 such that $\mathcal{D}$ is accessible in constant time. Thus we can now guarantee logarithmic query time after every operation. Moreover, the data structure now operates directly on the entities of the arrangement. In particular, it avoids copying of geometric data which can significantly reduce the amount of additional memory that is used by the search structure. This is important, since due to the generic nature of the code it is not clear whether the geometric types (user provided) are referenced.

To the best of our knowledge, this is the only available implementation of a point location method with a guaranteed logarithmic query time that can handle two-dimensional subdivisions to this generality. Furthermore, it is the fastest available point location method, in terms of query time, for CGAL arrangements.[5]

---

[5] See http://acg.cs.tau.ac.il/projects/lppl for a comparison to CGAL Landmarks point location [24].

6 Open Problem

Prove Conjecture 1, that is, prove that it is possible to rely on the depth $\mathcal{D}$ of the DAG and still expect only a constant number of rebuilds. This solution would not require any changes to the current implementation.

Acknowledgement. The authors thank Sariel Har-Peled for sharing Observation 1, which is essential to the expected $O(n \log n)$ time algorithm for producing a worst-case linear-size and logarithmic-time point-location data structure.

References

1. Birn, M., Holtgrewe, M., Sanders, P., Singler, J.: Simple and fast nearest neighbor search. In: Workshop on Algorithm Engineering and Experiments, pp. 43–54 (2010)
2. Mulmuley, K.: A fast planar partition algorithm. i. J. Symb. Comput. 10(3/4), 253–280 (1990)
3. Seidel, R.: A simple and fast incremental randomized algorithm for computing trapezoidal decompositions and for triangulating polygons. J. Comput. Geom. 1, 51–64 (1991)
4. Kirkpatrick, D.G.: Optimal search in planar subdivisions. SIAM J. Comput. 12(1), 28–35 (1983)
5. Devillers, O.: The Delaunay hierarchy. Int. J. Found. Comput. Sci. 13(2), 163–180 (2002)
6. Seidel, R., Adamy, U.: On the exact worst case query complexity of planar point location. J. Algorithms 37(1), 189–217 (2000)
7. Dobkin, D.P., Lipton, R.J.: Multidimensional searching problems. SIAM J. Comput. 5(2), 181–186 (1976)
8. Preparata, F.P.: A new approach to planar point location. SIAM J. Comput. 10(3), 473–482 (1981)
9. Sarnak, N., Tarjan, R.E.: Planar point location using persistent search trees. Commun. ACM 29(7), 669–679 (1986)
10. Lee, D.T., Preparata, F.P.: Location of a point in a planar subdivision and its applications. In: ACM Symposium on Theory of Computing. STOC 1976, pp. 231–235. ACM, New York (1976)
11. Edelsbrunner, H., Guibas, L.J., Stolfi, J.: Optimal point location in a monotone subdivision. SIAM J. Comput. 15(2), 317–340 (1986)
12. Snoeyink, J.: Point location. In: Goodman, J.E., O'Rourke, J. (eds.) Handbook of Discrete and Computational Geometry, pp. 767–785. CRC Press LLC, Boca Raton (2004)
13. de Berg, M., van Kreveld, M., Overmars, M., Schwarzkopf, O.: Computational Geometry: Algorithms and Applications, 3rd edn. Springer (2008)
14. Berberich, E., Fogel, E., Halperin, D., Melhorn, K., Wein, R.: Arrangements on parametric surfaces I: General framework and infrastructure. Mathematics in Computer Science 4, 67–91 (2010)
15. Mulmuley, K.: Computational geometry - an introduction through randomized algorithms. Prentice Hall (1994)
16. Hemmer, M., Kleinbort, M., Halperin, D.: Improved implementation of point location in general two-dimensional subdivisions. CoRR abs/1205.5434 (2012)

17. Guibas, L.J., Yao, F.F.: On translating a set of rectangles. In: STOC, pp. 154–160 (1980)
18. Alt, H., Scharf, L.: Computing the depth of an arrangement of axis-aligned rectangles in parallel. In: Proceedings of the 26th European Workshop on Computational Geometry (EuroCG), Dortmund, Germany, pp. 33–36 (March 2010)
19. Amenta, N., Choi, S., Rote, G.: Incremental constructions con brio. In: Symposium on Computational Geometry, pp. 211–219 (2003)
20. Mount, D.M., Arya, S.: Ann: A library for approximate nearest neighbor searching, http://www.cs.umd.edu/~mount/ANN/
21. Flato, E., Halperin, D., Hanniel, I., Nechushtan, O., Ezra, E.: The design and implementation of planar maps in CGAL. ACM Journal of Experimental Algorithmics 5, 13 (2000)
22. Wein, R., Berberich, E., Fogel, E., Halperin, D., Hemmer, M., Salzman, O., Zukerman, B.: 2D arrangements. In: CGAL User and Reference Manual, 4.0 edn., CGAL Editorial Board (2012)
23. Austern, M.H.: Generic Programming and the STL. Addison-Wesley (1999)
24. Haran, I., Halperin, D.: An experimental study of point location in planar arrangements in CGAL. ACM Journal of Experimental Algorithmics 13 (2008)

Parameterized Complexity of Induced
H-Matching on Claw-Free Graphs*

Danny Hermelin[1], Matthias Mnich[2], and Erik Jan van Leeuwen[3]

[1] Max Planck Institute for Informatics, Saarbrücken, Germany
hermelin@mpi-inf.mpg.de
[2] Cluster of Excellence "Multimodal Computing and Interaction", Saarland
University, Saarbrücken, Germany
mmnich@mmci.uni-saarland.de
[3] Dept. Computer, Control, Managm. Eng., Sapienza University of Rome, Italy
E.J.van.Leeuwen@dis.uniroma1.it

Abstract. The INDUCED H-MATCHING problem asks to find k disjoint, induced subgraphs isomorphic to H in a given graph G such that there are no edges between vertices of different subgraphs. This problem generalizes amongst others the classical INDEPENDENT SET and INDUCED MATCHING problems. We show that INDUCED H-MATCHING is fixed-parameter tractable in k on claw-free graphs when H is a fixed connected graph of constant size, and even admits a polynomial kernel when H is a clique. Both results rely on a new, strong algorithmic structure theorem for claw-free graphs. To show the fixed-parameter tractability of the problem, we additionally apply the color-coding technique in a nontrivial way. Complementing the above two positive results, we prove the W[1]-hardness of INDUCED H-MATCHING for graphs excluding $K_{1,4}$ as an induced subgraph. In particular, we show that INDEPENDENT SET is W[1]-hard on $K_{1,4}$-free graphs.

1 Introduction

A graph is *claw-free* if no vertex in the graph has three pairwise nonadjacent neighbors, i.e. if it does not contain a copy of $K_{1,3}$ as an induced subgraph. The class of claw-free graphs contains several well-studied graph classes such as line graphs, unit interval graphs, de Bruijn graphs, the complements of triangle-free graphs, and graphs of several polyhedra and polytopes. Consequently, claw-free graphs have attracted much interest, and are by now the subject of numerous research papers (see e.g. the survey [9]).

Our understanding of claw-free graphs and their structure was greatly extended with the recently developed theory of Chudnovsky and Seymour. This highly technical and detailed claw-free structure theory is contained in a sequence of seven papers called Claw-free graphs I–VII (see [4] for an accessible survey), and culminates in several variants of *claw-free decomposition theorems*. Each of these decomposition theorems shows that every connected claw-free graph can

---

* Supported by ERC StG project PAAl no. 259515. A full version of this paper can be found on the arXiv.

L. Epstein and P. Ferragina (Eds.): ESA 2012, LNCS 7501, pp. 624–635, 2012.

be constructed by starting with a collection of "basic claw-free graphs", and then gluing these basic graphs together in some controlled manner.

Recently, Chudnovsky and Seymour's claw-free structure theory has been deployed for algorithmic purposes. For example, it was used to develop approximation algorithms for CHROMATIC NUMBER on quasi-line graphs [3,17]. Faenza et al. [8] gave an algorithm to find a decomposition of claw-free graphs, and used this algorithm to improve on the fastest previously known algorithm for INDEPENDENT SET on claw-free graphs. Their decomposition, although similar in nature, does not rely on Chudnovsky and Seymour's structure theorem. Independently, the authors of this paper, along with Woeginger, gave an algorithm for the Chudnovsky-Seymour decomposition theorem, and used this to show that DOMINATING SET is fixed-parameter tractable on claw-free graphs, and admits a polynomial kernel as well [15]. (Fixed-parameter tractability was shown independently by Cygan et al. [5] using different methods.) Recently, Golovach et al. [14] used the algorithmic decomposition theorem in [15] to give a parameterized algorithm for the INDUCED DISJOINT PATHS problem on claw-free graphs.

In this paper we extend this line of research by considering the INDUCED H-MATCHING problem on claw-free graphs. This problem generalizes several important and well-studied problems, such as INDEPENDENT SET and INDUCED MATCHING. It can be defined as follows. We are given a claw-free graph G and an integer k, and the goal is to determine whether there is an induced occurrence of $k \cdot H$ in G; that is, whether there is a set of induced subgraphs $M = \{H_1, \ldots, H_k\}$ in G, each isomorphic to H, that are pairwise vertex-disjoint and satisfy that $(u,v) \notin E(G)$ for any pair of vertices $u \in V(H_i)$ and $v \in V(H_j)$ with $i \neq j$. As discussed further below, the study of this problem among other things requires a significantly stronger decomposition theorem for claw-free graphs compared to those developed previously.

The INDUCED H-MATCHING problem is closely related to the equally well-studied H-MATCHING problem. In H-MATCHING, the goal is to find induced subgraphs $H_1, \ldots, H_k$ isomorphic to H that are pairwise vertex-disjoint. Observe that when H has minimum degree at least 3, H-MATCHING reduces to INDUCED H-MATCHING by subdividing all the edges of G and H. On the other hand, on line graphs (a subclass of claw-free graphs), a reduction exists in the other direction. Recall that the *line graph* $L(G)$ of a graph G consists of a vertex v_e for each $e \in E(G)$ and there is an edge (v_e, v_f) in $L(G)$ if and only if e and f are incident to the same vertex in G. The graph G is known as the *pre-image* of $L(G)$. It can be easily seen that there is a bijection between the set of induced H-matchings on $L(G)$ and the set of H-matchings on G. Hence INDUCED H-MATCHING on line graphs inherits all known complexity results on H-MATCHING on general graphs. In particular, this implies that INDUCED H-MATCHING is NP-complete on line graphs if H is not edgeless [12,18,20].

In the context of parameterized complexity, the H-MATCHING problem has received significant attention. In particular, the problem is known to be fixed-parameter tractable by the size of the matching k [11,19] and even has a polynomial kernel (when H is fixed) [22]. Recently, tight lower bounds on the kernel size for specific graphs H were obtained in [6,16]. Note again that these results immediately carry over to INDUCED H-MATCHING on line graphs.

In general graphs, the INDUCED H-MATCHING problem is W[1]-hard for any clique H on $h \geq 1$ vertices when parameterized by the matching size k [7,23]. Marx [21] showed that the INDUCED H-MATCHING for a graph H on a single vertex (i.e. the INDEPENDENT SET problem) is W[1]-hard on $K_{1,5}$-free graphs. Another related result by Cameron and Hell [2] shows that on certain graph classes the problem can be reduced to an instance of INDEPENDENT SET on a graph in that same graph class, provided that the set of all occurrences of H are given. Their results, however, do not apply to claw-free graphs.

Our Results. The main result of this paper is that INDUCED H-MATCHING is fixed-parameter tractable on claw-free graphs when parameterized by k for fixed connected graphs H of constant size. It is important to note that requiring H to be of constant size is essential, since the problem is W[1]-hard with respect to $k + |V(H)|$ even for line graphs and co-bipartite graphs [13,14]. In the special case that H is a fixed clique, we also show that the problem admits a polynomial kernel. In contrast, we prove that the problem becomes W[1]-hard on $K_{1,4}$-free graphs. These results both complement and tighten the above-mentioned known hardness results on the problem.

Techniques. The main difference between this work and previous works on algorithms using the Chudnosky-Seymour structure theorem is that the so-called *twins* and *W-joins* seem hard to get rid of in INDUCED H-MATCHING. Hence we require a stronger version of the previously proposed decomposition theorems, one that does not require that such structures are not present in the given claw-free graph. We provide such a decomposition theorem in Section 2, and show that this decomposition can be found in polynomial time.

The new decomposition theorem that we develop has the advantage that it is simpler to state. In particular, it decomposes the input claw-free graph into only two graph classes. Yet, for one of these graph classes, namely the class of fuzzy circular-arc graphs (a superclass of proper circular-arc graphs), the INDUCED H-MATCHING problem was not previously known to be polynomial-time solvable for fixed H. We show that in fact it is. In the full version we exhibit stronger results on circular-arc graphs, and give evidence why these results are tight.

After obtaining our refined decomposition theorem, we solve INDUCED H-MATCHING on claw-free graphs by applying the color-coding technique [1] in a nontrivial way. To give some intuition behind this approach, we recall that to solve INDUCED H-MATCHING on line graphs G and H, we need to find pairwise vertex-disjoint induced subgraphs $H_1, \ldots, H_k$ in the pre-image of G such that H_i is isomorphic to the pre-image of H. In particular, we can find these H_i in our structural decomposition for G. However, if G and H are claw-free, there is no notion of pre-image and therefore no immediate relation between H and the H_i that we would want to find. Instead, we show that it is sufficient to find, in a coordinated manner and using color-coding, isomorphic copies of H itself.

Thus, in some sense, color-coding allows us to reduce INDUCED H-MATCHING to H-MATCHING. Complete details of this are given in Section 3.

For obtaining our polynomial kernel when H is a fixed clique of order h, we reduce the size of the decomposition-structure to $O(h^4 k^2)$. We then construct a kernel that mimics an easy algorithm for INDUCED H-MATCHING on such reduced decomposition-structures. This kernel actually reduces the problem to an equivalent instance of WEIGHTED INDEPENDENT SET of size polynomial in k, which in turn can be reduced to an equivalent instance of INDUCED H-MATCHING on claw-free graphs. This approach substantially simplifies the approach that was used for the polynomial kernel for DOMINATING SET on claw-free graphs [15].

2   Preliminaries

Let G be a graph. We call $I \subseteq V(G)$ an *independent set* if no two vertices of I are adjacent, and use $\alpha(G)$ to denote the size of a maximum independent set of G. The neighborhood of a vertex $v \in V(G)$ is denoted by $N(v) = \{u \in V(G) \mid (u,v) \in E(G)\}$, and the closed neighborhood of v is $N[v] = N(v) \cup \{v\}$. This notation extends to subsets of vertices. Given $X \subseteq V(G)$, the subgraph induced by X is $G[X] = (X, E(G) \cap (X \times X))$, and the graph $G - X$ is the subgraph induced by $V(G) \setminus X$. Then G is *claw-free* if $\alpha(G[N(v)]) \leq 2$ for any $v \in V(G)$.

We next define a variant of circular-arc graphs that will play an important role in the paper. Given a set $\mathcal{A}$ of objects, a graph G is the *intersection graph* of $\mathcal{A}$ if each vertex corresponds to an object of $\mathcal{A}$ and there is an edge between two vertices if and only if the corresponding objects intersect. In a *fuzzy intersection graph* G of $\mathcal{A}$, there is an edge between two vertices if the corresponding objects intersect in more than one point, and there is no edge if the corresponding objects do not intersect. The fuzziness stems from what happens if the two objects intersect in precisely one point: the graph may have an edge or not. We call $\mathcal{A}$ a *representation* of G. Now consider a set of arcs on a circle (throughout, we assume that arcs are not a single point). Then the set is called *proper* if no arc is a subset of another, *almost proper* if any arc that is a subset of another is in fact equal to the other, and *almost strict* if, for any maximal set of arcs with the same endpoints, at most one of these endpoints is also an endpoint of an arc outside the set. A graph is a *proper circular-arc graph* if it is the intersection graph of a set of proper arcs on a circle. A graph is a *fuzzy circular-arc graph* if it is the fuzzy intersection graph of an almost-proper, almost-strict set of arcs on a circle. Note that any intersection graph of an almost-proper set of arcs on a circle is a proper circular-arc graph; hence a fuzzy circular-arc graph without any fuzziness is just a proper circular-arc graph. It is known that any fuzzy circular-arc graph is claw-free [24].

A *hypergraph* $\mathcal{R}$ consists of a set of vertices $V(\mathcal{R})$ and a multi-set of edges $E(\mathcal{R})$ such that each $e \in E(\mathcal{R})$ is a subset of $V(\mathcal{R})$. We explicitly allow empty edges. Note that if $|e| \leq 2$ for every $e \in E(\mathcal{R})$, then $\mathcal{R}$ is a multi-graph, possibly with self-loops. A *strip-structure* of a graph G consists of a hypergraph $\mathcal{R}$ and for each edge $e \in E(\mathcal{R})$ a strip (J_e, Z_e) such that

- for each $e \in E(\mathcal{R})$, J_e is a claw-free graph and $Z_e \subseteq V(J_e)$ is an independent set;
- $V(J_e) \setminus Z_e \neq \emptyset$ for any $e \in E(\mathcal{R})$, and the sets $V(J_e) \setminus Z_e$ over all $e \in E(\mathcal{R})$ partition $V(G)$;
- for each $e \in E(\mathcal{R})$ and for each $r \in e$, there is a unique $z_e^r \in Z_e$ and $Z = \bigcup_{r \in e} z_e^r$;
- for each $r \in V(\mathcal{R})$, the union of $N(z_e^r)$ over all $e \in E(\mathcal{R})$ for which $r \in e$ induces a clique $C(r)$ in G;
- if $u, v \in V(G)$ are adjacent, then $u, v \in V(J_e)$ for some $e \in E(\mathcal{R})$ or $u, v \in C(r)$ for some $r \in V(\mathcal{R})$;
- $J_e[V(J_e) \setminus Z_e]$ equals $G[V(J_e) \setminus Z_e]$.

Given a strip (J, Z), the set $N(z)$ for $z \in Z$ is referred to as a *boundary* of (J, Z), while the (possibly empty) graph $J - N[Z]$ is called the *interior* of (J, Z). We call (J, Z) a *spot* if J is a three-vertex path and Z consists of its ends, and a *stripe* if no vertex of J is adjacent to more than one $z \in Z$.

We prove the following theorem in the full version of this paper.

Theorem 1. *Let G be a connected claw-free graph that is not a fuzzy circular-arc graph and has $\alpha(G) > 4$. Then G admits a strip-structure such that each strip is either a spot, or a stripe (J, Z) with $1 \leq |Z| \leq 2$ that is a fuzzy circular-arc graph or satisfies $\alpha(J) \leq 4$. Moreover, we can find such a strip-structure in polynomial time.*

3   Fixed-Parameter Algorithm

In this section, we prove the following theorem.

Theorem 2. *Let H be a fixed connected graph. Then INDUCED H-MATCHING is fixed-parameter tractable on claw-free graphs when parameterized by the size of the matching k.*

Throughout the section, we use H to denote a fixed connected graph of order h, and G to denote a claw-free graph on n vertices given as input to the INDUCED H-MATCHING problem. Observe that H is part of the problem description rather than part of the problem instance, and so its size is constant. We may assume that G is connected as the extension to the disconnected case is immediate.

The algorithm we use to prove Theorem 2 deploys the claw-free decomposition theorem stated in Theorem 1. Note that in order to apply this decomposition theorem, we need to first handle two cases: the case where $\alpha(G)$ is small, and the case where G is a fuzzy circular-arc graph.

Lemma 1. INDUCED H-MATCHING *on graphs G with $\alpha(G) \leq 4$ can be solved in polynomial time.*

This lemma is immediate from the observation that if $\alpha(G) \leq 4$, then any induced H-matching of G will have size at most 4. The proof of the following lemma is omitted and can be found in the full version.

Lemma 2. INDUCED H-MATCHING *on a fuzzy circular-arc graph can be solved in polynomial time.*

3.1 Guessing the Induced Strip-Graph

The first step of our algorithm in case G is not a fuzzy circular-arc graph and has $\alpha(G) > 4$ is to apply Theorem 1 to obtain a strip-structure of G.

Assume that G has an induced H-matching M of size k. Our goal from now on will be to discover M. We will call an edge of $\mathcal{R}$ *covered* if its corresponding strip contains a vertex of M. Similarly, we call a vertex r of $\mathcal{R}$ *covered* if the clique $C(r)$ corresponding to r contains a vertex of M. With these definitions, we can talk about the subgraph of $\mathcal{R}$ induced by covered edges and vertices, which we call the *covered subgraph* of $\mathcal{R}$.

As a first step, we guess what the covered subgraph of $\mathcal{R}$ looks like. Obviously, we cannot guess which vertices of $\mathcal{R}$ will be in the covered subgraph, as this would yield an $n^{O(hk)}$-time algorithm. However, we can guess the graph that the covered subgraph will be isomorphic to, which we will call the *base graph*. To bound the size of the base graph, note that each edge of $\mathcal{R}$ corresponds to at least one vertex of an occurrence of H. Hence the base graph has at most hk edges and at most $2hk$ vertices. As the base graph can be seen as a multigraph, there are at most $2^{O((hk)^3)}$ possibilities for this graph. (A tighter bound might be obtainable here, but the above bound is sufficient for our purposes.) Let B denote the base graph induced by M. Since we go through all possible choices of the base graph, we can assume that our guess is correct and isomorphic to B.

As a second step, we guess how the k occurrences of H are 'spread' over B. That is, for each edge e of B, we guess exactly what the intersection of M with the hypothetical strip corresponding to e would look like. In particular, we need to guess the intersection of M with the boundaries of the strip. We call this the *B-assignment*. To this end, we simply try all possible assignments of the hk vertices of M to the edges of B. For each edge of B, we have one bucket to indicate that the vertex should be in the interior of the strip and a bucket for each boundary of the strip. If we guess that the strip is a spot, then there is only a single bucket. Since there are at most three buckets per strip, this takes $O((3hk)^{hk})$ time. (Again, a better bound may be possible here, but this bound is sufficient for our purposes.)

For each assignment, we verify whether it is proper. That is, we need to verify whether the assignment could possibly correspond to an induced H-matching of size k. First, if two vertices of H are adjacent, then in each occurrence of H they must be assigned to the same edge of B or to two different edges of B that share an endpoint. In the latter case, they must also be assigned to the boundary buckets corresponding to that endpoint. Secondly, for each vertex b of B, the boundary buckets corresponding to b of the edges incident to b can contain vertices of at most one occurrence of H. Moreover, those vertices must induce a clique in H.

As we go through all possible B-assignments, we may assume that our guess corresponds to the actual way M is spread out over B. Thus, our goal now is to ensure the existence of a subgraph $\mathcal{R}_B$ of $\mathcal{R}$ which is isomorphic to B such that the k occurrences of H are spread in $\mathcal{R}_B$ according to the correct B-assignment. Obviously, we cannot actually find $\mathcal{R}_B$ in FPT-time, since SUBGRAPH ISOMORPHISM is known to be W[1]-hard, but we show that this is not necessary to reach the above goal.

3.2   Color-Coding

We apply the technique of color-coding [1] in a non-trivial way. We use a set of colors C_B containing colors for each vertex of B and for each bucket of each edge of B. Before we apply color-coding, we need to describe the elements that we wish to color. We will have an element for each vertex of $\mathcal{R}$ (a *vertex-element*), and for each edge of $\mathcal{R}$ which corresponds to a strip, we create an element for the interior of the strip and for each boundary of the strip (called *edge-elements*). If however the strip is a spot, then we create only a single element. Note that this corresponds directly to the creation of buckets as we did for the B-assignment. We denote the set of elements by $\mathcal{E}$.

We now color the elements of $\mathcal{E}$ with the colors of C_B. For this, we use a $|C_B|$-perfect hash family of colorings $\mathcal{F}$ where we are guaranteed that for any subset of C_B vertices and edges in $\mathcal{R}$, there is a subset of colorings in $\mathcal{F}$ that will assign distinct colors to this subset in all possible ways. According to [1] there exists such a family $\mathcal{F}$ of size at most $f(|C_B|) \cdot n^{O(1)} = f'(k) \cdot n^{O(1)}$ for computable functions f and f', and this family can be constructed in FPT-time with respect to $|C_B|$ (and hence with respect to k). We call a coloring *nice* if $\mathcal{R}_B$ is colored in a correct way, i.e. if the colors of C_B indicate the isomorphism between B and $\mathcal{R}_B$. Observe that the set of colorings $\mathcal{F}$ above assures us that if M is indeed an induced subgraph of G, one of the colorings in $\mathcal{F}$ will be nice. Thus, to complete the description of our algorithm, all that remains is to show the lemma below.

Lemma 3. *Given a nice coloring σ of $\mathcal{E}$, the induced H-matching M can be found in G in FPT-time with respect to k.*

Proof. First, we check whether the coloring of the vertices and edges is consistent. That is, each vertex-element should have received a vertex-color, each edge-element should have received an edge-color of the right type, each of the at most three edge-elements of the same edge should have received edge-colors that correspond to the coloring of an edge of B, and for each edge e its incident vertices have received colors that correspond to the coloring of the vertices of B incident to the edge colored in the way that e is colored. Any vertex or edge that is inconsistently colored is blanked, i.e. we remove its color. Note that if the fourth check above fails, we blank the elements of the edge and not of its incident vertices.

Second, we consider all strips (J, Z) in turn for which the corresponding edge was not blanked. Through the current B-assignment and using the coloring σ, we know what we need to find in this strip. In particular, consider the occurrences of H assigned to the boundaries of the strip. Recall that for each boundary, vertices of at most one occurrence can be assigned. Denote the vertices of the at most two occurrences assigned to the boundaries of the strip by H^{b_i}. We then use exhaustive enumeration to try all possible realizations X of the vertices of each H^{b_i} in J that are consistent with the current B-assignment. Then we remove $N[X]$ from $J \setminus Z$ and, if no vertices are assigned by the B-assignment to a boundary $N(z)$, we remove $N[z]$ from J as well. Call the resulting graph J'. From Theorem 1, we know that $\alpha(J') \leq 4$ or J is a fuzzy circular-arc graph. Hence, it follows from Lemma 1 and Lemma 2 that we can solve INDUCED H-MATCHING on J' in polynomial time. Denote the solution for X by $M^e(X)$. Let $X^e_{\max}$ denote a set X such that $|M^e(X)|$ is maximal. We then verify that $|M^e(X)|$ is at least the number of occurrences of H prescribed by the B-assignment; otherwise, we blank all elements of the edge.

After the second step, we have actually found all k' occurrences of H that are in the interior of a strip according to M and the B-assignment. It remains to find occurrences of H that span multiple strips or affect a strip boundary. For this, we color the vertices of G with a set of colors $C_{k''} = \{0, \ldots, k''\}$ where $k'' = k - k'$. For each edge e that is not blanked, consider $X^e_{\max}$. Recall that $X^e_{\max}$ consists of vertices of at most two occurrences of H, say occurrences $i, j \in \{1, \ldots, k''\}$. We first color the vertices of $X^e_{\max}$ (in G) by colors i and j from $C_{k''}$ in the obvious way (i.e. vertices from occurrence i receive color i), and color all remaining vertices of $V(J) \setminus Z$ with color 0. The crucial observation is that, by construction, any pair of vertices from different color classes in $C_{k''}$ will be independent. It is thus sufficient to find a copy of H in each color class of $C_{k''}$. This takes $n^{O(h)}$ time in total, where $h = |V(H)|$. By combining these copies of H with the copies in $M^e(X^e_{\max})$ for each $e \in E(\mathcal{R})$, we find an induced H-matching of size at least k.   □

4   Polynomial Kernel

The main result of this section is the following theorem.

Theorem 3. *Let H be a fixed clique. Then* INDUCED H-MATCHING *has a polynomial kernel with respect to the parameter k.*

We now prove Theorem 3. Throughout, we assume that $\alpha(G) > 4$ and that G is not a fuzzy-circular arc graph. Otherwise, we can solve INDUCED H-MATCHING in polynomial time by Lemmas 1 and 2, and reduce to a trivial "yes"- or "no"-instance. We can then also assume that we have used Theorem 1 and computed a strip-structure with a corresponding strip-graph $\mathcal{R}$.

First, we show that we can assume that $\mathcal{R}$ is a hypergraph with a bounded number of edges. The approach is based on one in [25] and [6] for H-MATCHING.

Lemma 4. *Either the strip-graph has $O(h^4 k^2)$ edges, or we can reduce (G, k), or G has an induced H-matching of size k.*

To obtain an intuition of the kernel, we give the following description of a simple algorithm for INDUCED H-MATCHING of H is a fixed clique. We first reduce the strip-graph $\mathcal{R}$ using Lemma 4. Then we observe that if we know the behavior of the induced H-matching on $C(r)$ for each $r \in \mathcal{R}$, then we can reduce to polynomial-time solvable instances of induced H-matching on individual strips. Since the number of vertices of $\mathcal{R}$ is bounded, this gives a fixed-parameter algorithm. The kernel will mimic this algorithm by reducing to an instance of WEIGHTED INDEPENDENT SET on general graphs of size polynomial in k. By using a Karp-reduction, we can obtain an instance of INDUCED H-MATCHING on claw-free graphs again.

For our purposes, we define WEIGHTED INDEPENDENT SET as the problem of given a graph G', a weight function w', and integers k', K', to decide whether G' has an independent set of size at least k' and weight at least K'.

Theorem 4. *Let H be a fixed clique. In polynomial time, we can reduce an instance of* INDUCED H-MATCHING *on claw-free graphs with parameter k to an equivalent instance of* WEIGHTED INDEPENDENT SET *with $O(2^h h^{4h+4} k^{2h+2})$ vertices, maximum vertex-weight at most k, k' bounded by $O(h^4 k^2)$, and K' bounded by k.*

Proof. First, we reduce the strip-graph using Lemma 4. Then, for each edge $e \in E(\mathcal{R})$, and each pair of integers $i, j \in \{-1, 0, \ldots, h\}$, we define the following two functions. Suppose that $|e| = 2$, and $e = \{x, y\}$. Then the function $w(e, i, j)$ is the size of a maximum induced H-matching on $J_e \setminus Z_e$ that,

- if $i = -1$, contains at least one vertex of $N(z_e^x)$;
- if $0 \le i \le h$, contains no vertices of $N[X \cup \{z_e^x\}]$ for at least one $X \subseteq N(z_e^x)$ of size i,
- if $j = -1$, contains at least one vertex of $N(z_e^y)$;
- if $j \ge 0$, contains no vertices of $N[Y \cup \{z_e^y\}]$ for at least one $Y \subseteq N(z_e^y)$ of size j.

The function w evaluates to $-\infty$ if an induced H-matching with the given constraints does not exist. It follows immediately from Lemmas 1 and 2 that the function w can be computed in polynomial time. We define a similar function $w(e, i)$ if $|e| = 1$, and compute its values in polynomial time. Below, we will largely ignore edges e with $|e| = 1$, as they are dealt with in essentially the same manner as edges with $|e| = 2$.

We now describe the instance of WEIGHTED INDEPENDENT SET that we will construct. For each edge $e \in E(\mathcal{R})$ and each $i, j \in \{-1, 0, \ldots, h\}$, we create a vertex $v_{i,j}^e$ of weight $w(e, i, j)$. For any $e \in E(\mathcal{R})$, we construct a clique on the vertices $v_{i,j}^e$. This expresses that for each strip, we can choose at most one solution.

It remains to coordinate the solutions of the strips. Let $\mathcal{U}$ denote the set of all sets U of integers such that $\sum_{i \in U} i = h$ and $\min_{i \in U} i \ge 1$ for each $U \in \mathcal{U}$. We also add the set $\{0\}$ to $\mathcal{U}$. Let E_r denote the set of edges in $\mathcal{R}$ containing a vertex $r \in V(\mathcal{R})$ and let $d_r = |E_r|$. Finally, we construct the set $\mathcal{P}_r$. Each

$P \in \mathcal{P}_r$ will be an assignment of the numbers of some $U \in \mathcal{U}$ to the edges of E_r. If $|U| < |E_r|$, then all remaining edges will be assigned the number zero. The set of all such assignments over all $U \in \mathcal{U}$ will be the set $\mathcal{P}_r$.

We create two types of vertices for each $r \in \mathcal{R}$. Whenever we refer to an edge e containing r, we assume that the second parameter of the function ω corresponds to the vertex z_r^e. Now, first, we create a vertex v_P^r for each $P \in \mathcal{P}_r$. The weight of v_P^r is set to 1, unless P assigns zero to every edge e, in which case the weight is set to zero. If P assigns i' to edge e, then we make v_P^r adjacent to $v_{i,j}^e$ for all $i \neq i'$ and for all j. This expresses that if an occurrence of H is a subset of $C(r)$, how the vertices of this occurrences are distributed over $C(r)$. Second, we create a vertex v_e^r for each e containing r and give it weight zero. We make this adjacent to $v_{i,j}^e$ for each $i \geq 0$ and for each j, and to $v_{i,j}^{e'}$ for each $e' \neq e$ containing r, each $i \neq 0$, and each j. This expresses that if an occurrence of H contains at least one vertex of an occurrence of H (but less than h), in which strip incident to r these vertices must be found. Finally, we construct a clique on all the vertices created for r.

This completes the construction of the instance of WEIGHTED INDEPENDENT SET. It follows from Lemmas 1 and 2 that this instance can be constructed in polynomial time. Denote the graph by G' and the weight function by w'. We claim that G' has $O(2^h h^{4h+2} k^{2h+2})$ vertices. To see this, we note that for each edge of $\mathcal{R}$, we create $(h+2)^2$ vertices. By Lemma 4, this gives a total of $O(h^6 k^2)$ vertices. We can count the number of vertices created for each vertex r of $\mathcal{R}$ as follows. Observe that $|\mathcal{U}| \leq 2^{h+1}$ (more precise bounds are known, but this will suffice). We can assume that the numbers of each $U \in \mathcal{U}$ are ordered from small to large. To bound $|\mathcal{P}_r|$, we note that the number of ordered subsets of $|E_r|$ of size at most h is at most $|E_r|^h$. For each ordered subset X, we create a bijection to each $U \in \mathcal{U}$ with $|U| = |X|$ by assigning the i-th element of U to the i-th element of X. We then extend this assignment to E_r by assigning zero to any elements of $E_r \setminus X$. This shows that $|\mathcal{P}_r|$ is $O(2^h h^{4h} k^{2h})$. The number of vertices created for each r thus is $O(2^h h^{4h} k^{2h})$. The claim follows.

To bound the weights, observe that if $\omega(e, i, j) > k$ for some e, i, j, then the instance of INDUCED H-MATCHING is a "yes"-instance and we can easily reduce to a trivial "yes"-instance of WEIGHTED INDEPENDENT SET. Hence all the weights are at most k. Finally, let k' be equal to the number of vertices and edges in $\mathcal{R}$, and let $K' = k$. Note that k' is bounded by $O(h^4 k^2)$ by Lemma 4.

We claim that INDUCED H-MATCHING on G, k is a "yes"-instance if and only if WEIGHTED INDEPENDENT SET on G', w', k', K' is a "yes"-instance. For the one direction, consider an induced H-matching M of size k. Now construct an independent set I of G' as follows. Initially, mark all vertices $v_{i,j}^e$ over all i, j as *potentials*. For each $r \in V(\mathcal{R})$, check if an occurrence of H in M is fully contained in $C(r)$. If so, find the assignment P of $\mathcal{P}_r$ induced by this occurrence. Then we add v_P^r to I. Moreover, we remove the potential mark from all neighbors of v_P^r. We proceed similarly if no vertex of M is contained in $C(r)$. Otherwise, a vertex of an occurrence of H in M is contained in $C(r)$, but there is also a vertex of this occurrence in $J_e \setminus N[z_r^e]$ for some e containing r. Then we add v_e^r to I and remove the potential mark from all its neighbors. All remaining potential vertices form an independent set by construction of G'. Moreover, for each edge e exactly one vertex $v_{i,j}^e$ is a potential vertex. This gives the required independent set. It has weight at least K' as M has size k.

For the converse, let I be an independent set of G' of size at least k' and weight at least K'. Observe that by construction of G' and k', $|I| = k'$. We now use the inverse of the above process to construct an induced H-matching M. Since $K' = k$, M will have size k. The theorem follows.   □

5   Discussion

In this section, we discuss the possibility of extending the work of this paper. In particular, we give some open problems.

Since we showed that INDUCED H-MATCHING on $K_{1,3}$-free graphs is fixed-parameter tractable for fixed connected graphs H, and even admits a polynomial kernel if H is a clique, it is natural to ask whether these results extend to $K_{1,4}$-free graphs. However, the following theorem shows that this is unlikely to be possible. We omit the proof due to space restrictions.

Theorem 5. INDUCED H-MATCHING *for any fixed clique H, and in particular* INDEPENDENT SET, *on $K_{1,4}$-free graphs is W[1]-hard.*

We also note that the running times of our results contain a term $n^{O(h)}$, where $h = |V(H)|$. Unfortunately, it has been shown that extending to the case where h is also a parameter is unlikely to succeed, as INDUCED H-MATCHING becomes W[1]-hard even on line graphs and co-bipartite graphs [13,14].

It is unclear however whether the results of this paper extend to graphs H with multiple components. This seems to be difficult even for fuzzy circular-arc graphs, as the natural reduction to an INDEPENDENT SET problem fails [10].

Another interesting open question is whether the polynomial kernel for INDUCED H-MATCHING on claw-free graphs extends beyond the case when H is a fixed clique. The proof of Theorem 3 actually extends to this more general case. However, the obstacle to extend our current techniques is that we are unable to find a lemma similar to Lemma 4 if H is not a clique.

References

1. Alon, N., Yuster, R., Zwick, U.: Color-coding. J. ACM 42(4), 844–856 (1995)
2. Cameron, K., Hell, P.: Independent packings in structured graphs. Math. Programming 105(2-3), 201–213 (2006)
3. Chudnovsky, M., Ovetsky, A.: Coloring quasi-line graphs. J. Graph Theory 54(1), 41–50 (2007)
4. Chudnovsky, M., Seymour, P.D.: The structure of claw-free graphs. In: Surveys in Combinatorics, vol. 327, pp. 153–171. Cambridge University Press (2005)
5. Cygan, M., Philip, G., Pilipczuk, M., Pilipczuk, M., Wojtaszczyk, J.O.: Dominating set is fixed parameter tractable in claw-free graphs. Theoretical Computer Science 412(50), 6982–7000 (2011)

6. Dell, H., Marx, D.: Kernelization of packing problems. In: Proc. SODA 2012, pp. 68–81 (2012)
7. Downey, R.G., Fellows, M.R.: Parameterized complexity. Springer (1999)
8. Faenza, Y., Oriolo, G., Stauffer, G.: An algorithmic decomposition of claw-free graphs leading to an $O(n^3)$-algorithm for the weighted stable set problem. In: Proc. SODA 2011, pp. 630–646 (2011)
9. Faudree, R.J., Flandrin, E., Ryjáček, Z.: Claw-free graphs - A survey. Discrete Math. 164(1-3), 87–147 (1997)
10. Fellows, M.R., Hermelin, D., Rosamond, F.A., Vialette, S.: On the parameterized complexity of multiple-interval graph problems. Theoretical Computer Science 410(1), 53–61 (2009)
11. Fellows, M.R., Knauer, C., Nishimura, N., Ragde, P., Rosamond, F.A., Stege, U., Thilikos, D.M., Whitesides, S.: Faster fixed-parameter tractable algorithms for matching and packing problems. Algorithmica 52(2), 167–176 (2008)
12. Garey, M., Johnson, D.: Computers and Intractability: A Guide to the Theory of NP-Completeness. Freeman (1979)
13. Golovach, P.A., Paulusma, D., van Leeuwen, E.J.: Induced Disjoint Paths in AT-Free Graphs. In: Fomin, F.V., Kaski, P. (eds.) SWAT 2012. LNCS, vol. 7357, pp. 153–164. Springer, Heidelberg (2012)
14. Golovach, P.A., Paulusma, D., van Leeuwen, E.J.: Induced Disjoint Paths in Claw-Free Graphs. In: Epstein, L., Ferragina, P. (eds.) ESA 2012. LNCS, vol. 7501, pp. 515–526. Springer, Heidelberg (2012)
15. Hermelin, D., Mnich, M., van Leeuwen, E.J., Woeginger, G.J.: Domination When the Stars Are Out. In: Aceto, L., Henzinger, M., Sgall, J. (eds.) ICALP 2011. LNCS, vol. 6755, pp. 462–473. Springer, Heidelberg (2011)
16. Hermelin, D., Wu, X.: Weak compositions and their applications to polynomial lower bounds for kernelization. In: Proc. SODA 2012, pp. 104–113 (2012)
17. King, A.D., Reed, B.A.: Bounding χ in terms of ω and Δ for quasi-line graphs. J. Graph Theory 59(3), 215–228 (2008)
18. Kirkpatrick, D., Hell, P.: On the complexity of general graph factor problems. SIAM J. Computing 12(3), 601–609 (1983)
19. Kneis, J., Mölle, D., Richter, S., Rossmanith, P.: Divide-and-Color. In: Fomin, F.V. (ed.) WG 2006. LNCS, vol. 4271, pp. 58–67. Springer, Heidelberg (2006)
20. Kobler, D., Rotics, U.: Finding maximum induced matchings in subclasses of claw-free and p_5-free graphs, and in graphs with matching and induced matching of equal maximum size. Algorithmica 37(4), 327–346 (2003)
21. Marx, D.: Efficient Approximation Schemes for Geometric Problems? In: Brodal, G.S., Leonardi, S. (eds.) ESA 2005. LNCS, vol. 3669, pp. 448–459. Springer, Heidelberg (2005)
22. Moser, H.: A Problem Kernelization for Graph Packing. In: Nielsen, M., Kučera, A., Miltersen, P.B., Palamidessi, C., Tůma, P., Valencia, F. (eds.) SOFSEM 2009. LNCS, vol. 5404, pp. 401–412. Springer, Heidelberg (2009)
23. Moser, H., Thilikos, D.M.: Parameterized complexity of finding regular induced subgraphs. In: Broersma, H., Dantchev, S.S., Johnson, M., Szeider, S. (eds.) Proc. ACiD 2006. Texts in Algorithmics, vol. 7, pp. 107–118. King's College, London (2006)
24. Oriolo, G., Pietropaoli, U., Stauffer, G.: On the recognition of fuzzy circular interval graphs. Discrete Math. 312(8), 1426–1435 (2012)
25. Prieto, E., Sloper, C.: Looking at the stars. Theoretical Computer Science 351(3), 437–445 (2006)

Solving Simple Stochastic Games with Few Coin Toss Positions*

Rasmus Ibsen-Jensen and Peter Bro Miltersen

Department of Computer Science
Aarhus University

Abstract. Gimbert and Horn gave an algorithm for solving simple stochastic games with running time $O(r!n)$ where n is the number of positions of the simple stochastic game and r is the number of its coin toss positions. Chatterjee *et al.* pointed out that a variant of strategy iteration can be implemented to solve this problem in time $4^r n^{O(1)}$. In this paper, we show that an algorithm combining value iteration with retrograde analysis achieves a time bound of $O(r2^r(r \log r + n))$, thus improving both time bounds. We also improve the analysis of Chatterjee *et al.* and show that their algorithm in fact has complexity $2^r n^{O(1)}$.

1 Introduction

Simple stochastic games is a class of two-player zero-sum games played on graphs that was introduced to the algorithms and complexity community by Condon [6]. A simple stochastic game is given by a directed finite (multi-)graph $G = (V, E)$, with the set of vertices V also called *positions* and the set of arcs E also called *actions*. There is a partition of the positions into V_1 (positions belonging to player Max), V_2 (positions belonging to player Min), V_R (coin toss positions), and a special terminal position GOAL. Positions of V_1, V_2, V_R have exactly two outgoing arcs, while the terminal position GOAL has none. We shall use r to denote $|V_R|$ (the number of coin toss positions) and n to denote $|V|-1$ (the number of non-terminal positions) throughout the paper. Between moves, a pebble is resting at one of the positions k. If k belongs to a player, this player should strategically pick an outgoing arc from k and move the pebble along this arc to another node. If k is a position in V_R, Nature picks an outgoing arc from k uniformly at random and moves the pebble along this arc. The objective of the game for player Max is to reach GOAL and should play so as to maximize his probability of doing so. The objective for player Min is to minimize player Max's probability of reaching GOAL.

* The authors acknowledge support from the Danish National Research Foundation and The National Science Foundation of China (under the grant 61061130540) for the Sino-Danish Center for the Theory of Interactive Computation, within which this work was performed. The authors also acknowledge support from the Center for Research in Foundations of Electronic Markets (CFEM), supported by the Danish Strategic Research Council.

L. Epstein and P. Ferragina (Eds.): ESA 2012, LNCS 7501, pp. 636–647, 2012.
© Springer-Verlag Berlin Heidelberg 2012

A *strategy* for a simple stochastic game is a (possibly randomized) procedure for selecting which arc or action to take, given the history of the play so far. A *positional strategy* is the very special case of this where the choice is deterministic and only depends on the current position, i.e., a positional strategy is simply a map from positions to actions. If player Max plays using strategy x and player Min plays using strategy y, and the play starts in position k, a random *play* $p(x, y, k)$ of the game is induced. We let $u^k(x, y)$ denote the probability that player Max will eventually reach GOAL in this random play. A strategy x^* for player Max is said to be *optimal* if for all positions k it holds that

$$\inf_{y \in S_2} u^k(x^*, y) \geq \sup_{x \in S_1} \inf_{y \in S_2} u^k(x, y), \tag{1}$$

where S_1 (S_2) is the set of strategies for player Max (Min). Similarly, a strategy y^* for player Min is said to be optimal if

$$\sup_{x \in S_1} u^k(x, y^*) \leq \inf_{y \in S_2} \sup_{x \in S_1} u^k(x, y). \tag{2}$$

A general theorem of Liggett and Lippman ([13], fixing a bug of a proof of Gillette [9]) restricted to simple stochastic games, implies that:

- Positional optimal strategies x^*, y^* for both players exist.
- For such optimal x^*, y^* and for all positions k,

$$\min_{y \in S_2} u^k(x^*, y) = \max_{x \in S_1} u^k(x, y^*).$$

This number is called the *value* of position k. We shall denote it $\mathrm{val}(G)_k$ and the vectors of values $\mathrm{val}(G)$.

In this paper, we consider *quantitatively solving* simple stochastic games, by which we mean computing the values of all positions of the game, given an explicit representation of G. Once a simple stochastic game has been quantitatively solved, optimal strategies for both players can be found in linear time [2]. However, it was pointed out by Anne Condon twenty years ago that no worst case polynomial time algorithm for quantitatively solving simple stochastic games is known. By now, finding such an algorithm is a celebrated open problem. Gimbert and Horn [10] pointed out that the problem of solving simple stochastic games parametrized by $r = |V_R|$ is *fixed parameter tractable*. That is, simple stochastic games with "few" coin toss positions can be solved efficiently. The algorithm of Gimbert and Horn runs in time $r! n^{O(1)}$. The next natural step in this direction is to try to find an algorithm with a better dependence on the parameter r. Thus, Dai and Ge [8] gave a *randomized* algorithm with *expected* running time $\sqrt{r!} n^{O(1)}$. Chatterjee *et al.* [4] pointed out that a variant of the standard algorithm of *strategy iteration* devised earlier by the same authors [5] can be applied to find a solution in time $4^r n^{O(1)}$ (they only state a time bound of $2^{O(r)} n^{O(1)}$, but a slightly more careful analysis yields the stated bound). The dependence on n in this bound is at least quadratic. The main result of this paper is an algorithm running in time $O(r 2^r (r \log r + n))$, thus improving all of the above bounds. More precisely, we show:

Theorem 1. *Assuming unit cost arithmetic on numbers of bit length up to $\Theta(r)$, simple stochastic games with n positions out of which r are coin toss positions, can be quantitatively solved in time $O(r2^r(r\log r + n))$.*

The algorithm is based on combining a variant of *value iteration* [16,7] with *retrograde analysis* [3,1]. We should emphasize that the time bound of Theorem 1 is valid *only* for simple stochastic games as originally defined by Condon. The algorithm of Gimbert and Horn (and also the algorithm of Dai and Ge, though this is not stated in their paper) actually applies to a generalized version of simple stochastic games where coin toss positions are replaced with *chance positions* that are allowed arbitrary out-degree and where a not-necessarily-uniform distribution is associated to the outgoing arcs. The complexity of their algorithm for this more general case is $O(r!(|E| + p))$, where p is the maximum bit-length of a transition probability (they only claim $O(r!(n|E| + p))$, but by using retrograde analysis in their Proposition 1, the time is reduced by a factor of n). The algorithm of Dai and Ge has analogous expected complexity, with the $r!$ factor replaced with $\sqrt{r!}$. While our algorithm and the strategy improvement algorithm of Chatterjee *et al.* can be generalized to also work for these generalized simple stochastic games, the dependence on the parameter p would be much worse - in fact exponential in p. It is an interesting open problem to get an algorithm with a complexity polynomial in 2^r as well as p, thereby combining the desirable features of the algorithms based on strategy iteration and value iteration with the features of the algorithm of Gimbert and Horn.

1.1   Organization of Paper

In Section 2 we present the algorithm and show how the key to its analysis is to give upper bounds on the difference between the value of a given simple stochastic game and the value of a *time bounded version* of the same game. In Section 3, we then prove such upper bounds. In fact, we offer two such upper bounds: One bound with a relatively direct proof, leading to a variant of our algorithm with time complexity $O(r^2 2^r(n + r\log r))$ and an *optimal* bound on the difference in value, shown using techniques from extremal combinatorics, leading to an algorithm with time complexity $O(r2^r(n + r\log r))$. In the Conclusion section, we briefly sketch how our technique also yields an improved upper bound on the time complexity of the strategy iteration algorithm of Chatterjee *et al.*

2   The Algorithm

2.1   Description of the Algorithm

Our algorithm for solving simple stochastic games with few coin toss positions is the algorithm of Figure 1.

In this algorithm, the vectors v and v' are real-valued vectors indexed by the positions of G. We assume the GOAL position has the index 0, so $v = (1, 0, ..., 0)$ is the vector that assigns 1 to the GOAL position and 0 to all other positions.

Function SolveSSG(G)

$v \leftarrow (1, 0, ..., 0)$;

for $i \in \{1, 2, \ldots, 10 \max\{r, 6\}2^{\max\{r,6\}}\}$ **do**

    $v \leftarrow \texttt{SolveDGG}(G, v)$;

    $v' \leftarrow v$;

    $v_k \leftarrow (v_j' + v_\ell')/2$, for all $k \in V_R$, v_j and v_ℓ being the two successors of v_k;

    Round each value v_k *down* to $7r$ binary digits;

$v \leftarrow \texttt{SolveDGG}(G, v)$;

$v \leftarrow \texttt{KwekMehlhorn}(v, 4^r)$;

return v

Fig. 1. Algorithm for solving simple stochastic games

SolveDGG is the retrograde analysis based algorithm from Proposition 1 in Andersson *et al.* [1] for solving *deterministic graphical games*. Deterministic graphical games are defined in a similar way as simple stochastic games, but they do not have coin toss positions, and arbitrary real payoffs are allowed at terminals. The notation $\texttt{SolveDGG}(G, v')$ means solving the deterministic graphical game obtained by replacing each coin toss position k of G with a terminal with payoff v_k', and returning the value vector of this deterministic graphical game. Finally, KwekMehlhorn is the algorithm of Kwek and Mehlhorn [12]. $\texttt{KwekMehlhorn}(v, q)$ returns a vector where each entry v_i in the vector v is replaced with the smallest fraction a/b with $a/b \geq v_i$ and $b \leq q$.

The complexity analysis of the algorithm is straightforward, given the analyses of the procedures SolveDGG and KwekMehlhorn from [1,12]. There are $O(r2^r)$ iterations, each requiring time $O(r \log r + n)$ for solving the deterministic graphical game. Finally, the Kwek-Mehlhorn algorithm requires time $O(r)$ for each replacement, and there are only r replacements to be made, as there are only r different entries different from $1, 0$ in the vector v, corresponding to the r coin toss positions, by standard properties of deterministic graphical games [1].

2.2 Proof of Correctness of the Algorithm

We analyse our main algorithm by first analysing properties of a simpler non-terminating algorithm, depicted in Figure 2. We shall refer to this algorithm as *modified value iteration*.

Let v^t be the content of the vector v immediately after executing SolveDGG in the $(t + 1)$'st iteration of the loop of ModifiedValueIteration on input G. To understand this variant of value iteration, we may observe that the v^t vectors can be given a "semantics" in terms of the value of a time bounded game.

Definition 1. *Consider the "timed modification" G^t of the game G defined as follows. The game is played as G, except that play stops and player Max loses when the play has encountered $t + 1$ (not necessarily distinct) coin toss positions. We let $\mathrm{val}(G^t)_k$ be the value of G^t when play starts in position k.*

Procedure ModifiedValueIteration(G)

$v \leftarrow (1, 0, ..., 0)$;
while *true* **do**
  $\quad$ $v \leftarrow$ SolveDGG(G, v);
  $\quad$ $v' \leftarrow v$;
  $\quad$ $v_k \leftarrow (v'_j + v'_\ell)/2$, for all $k \in V_R$, v_j and v_ℓ being the two successors of v_k;

Fig. 2. Modified value iteration

Lemma 1. $\forall k, t : v_k^t = \text{val}(G^t)_k$.

Proof. Straightforward induction on t ("backwards induction").

From Lemma 1 we immediately have $\forall k, t : \text{val}(G^t)_k \leq \text{val}(G^{t+1})_k$. Futhermore, it is true that $\lim_{t \to \infty} \text{val}(G^t) = \text{val}(G)$, where $\text{val}(G)$ is the value vector of G. This latter statement is very intuitive, given Lemma 1, but might not be completely obvious. It may be established rigorously as follows:

Definition 2. *For a given game G, let the game $\bar{G}^t$ be the following. The game is played as G, except that play stops and player Max loses when the play has encountered $t + 1$ (not necessarily distinct) positions. We let $\text{val}(\bar{G}^t)_k$ be the value of $\bar{G}^t$ when play starts in position k.*

(We note that $\text{val}(\bar{G}^t)$ is the valuation computed after t iterations of *unmodified* value iteration [7].) A very general theorem of Mertens and Neyman [14] linking the value of an infinite game to the values of its time limited versions implies that $\lim_{t \to \infty} \text{val}(\bar{G}^t) = \text{val}(G)$. Also, we immediately see that for any k, $\text{val}(\bar{G}^t)_k \leq \text{val}(G^t)_k \leq \text{val}(G)_k$, so we also have $\lim_{t \to \infty} \text{val}(G^t)_k = \text{val}(G)_k$.

To relate SolveDGG of Figure 1 to modified value iteration of Figure 2, it turns out that we want to upper bound the smallest t for which

$$\forall k : \text{val}(G)_k - \text{val}(G^t)_k \leq 2^{-5r}.$$

Let $T(G)$ be that t. We will bound $T(G)$ using two different approaches. The first, in Subsection 3.1, is rather direct and is included to show what may be obtained using completely elementary means. It shows that $T(G) \leq 5 \ln 2 \cdot r^2 \cdot 2^r$, for any game G with r coin toss positions (Lemma 3).

The second, in Subsection 3.2, identifies an *extremal* game (with respect to convergence rate) with a given number of positions and coin toss positions. More precisely:

Definition 3. *Let $S_{n,r}$ be the set of simple stochastic games with n positions out of which r are coin toss positions. Let $G \in S_{n,r}$ be given. We say that G is t-extremal if*

$$\max_k (\text{val}(G)_k - \text{val}(G^t)_k) = \max_{H \in S_{n,r}} \max_k (\text{val}(H)_k - \text{val}(H^t)_k).$$

We say that G is extremal if it is t-extremal for all t.

It is clear that t-extremal games exist for any choice of n, r and t. (That extremal games exist for any choice of n and r is shown later in the present paper.) To find an extremal game, we use techniques from extremal combinatorics. By inspection of this game, we then get a better upper bound on the convergence rate than that offered by the first approach. We show using this approach that $T(G) \leq 10 \cdot r \cdot 2^r$, for any game $G \in S_{n,r}$ with $r \geq 6$ (Lemma 7).

Assuming that an upper bound on $T(G)$ is available, we are now ready to finish the proof of correctness of the main algorithm. We will only do so explicitly for the bound on $T(G)$ obtained by the second approach from Subsection 3.2 (the weaker bound implies correctness of a version of the algorithm performing more iterations of its main loop). From Lemma 7, we have that for any game $G \in S_{n,r}$ with $r \geq 6$, $\mathrm{val}(G^t)_k$ approximates $\mathrm{val}(G)_k$ within an additive error of 2^{-5r} for $t \geq 10r \cdot 2^r$ and k being any position. SolveSSG differs from ModifiedValueIteration by rounding down the values in the vector v in each iteration. Let $\tilde{v}^t$ be the content of the vector v immediately after executing SolveDGG in the t'th iteration of the loop of SolveSSG. We want to compare $\mathrm{val}(G^t)_k$ with $\tilde{v}_k^t$ for any k. As each number is rounded down by less than 2^{-7r} in each iteration of the loop and recalling Lemma 1, we see by induction that

$$\mathrm{val}(G^t)_k - t2^{-7r} \leq \tilde{v}_k^t \leq \mathrm{val}(G^t)_k.$$

In particular, when $t = 10r2^r$, we have that $\tilde{v}_k^t$ approximates $\mathrm{val}(G)_k$ within $2^{-5r} + 10r2^r2^{-7r} < 2^{-4r}$, for any k, as we can assume $r \geq 6$ by the code of SolveSSG of Figure 1.

Lemma 2 of Condon [6] states that the value of a position in a simple stochastic game with n non-terminal positions can be written as a fraction with integral numerator and denominator at most 4^n. As pointed out by Chatterjee et al. [4], it is straightforward to see that her proof in fact gives an upper bound of 4^r, where r is the number of coin toss positions. It is well-known that two distinct fractions with denominator at most $m \geq 2$ differ by at least $\frac{1}{m(m-1)}$. Therefore, since $\tilde{v}_k^t$ approximates $\mathrm{val}(G)_k$ within $2^{-4r} < \frac{1}{4^r \cdot (4^r - 1)}$ from below, we in fact have that $\mathrm{val}(G)_k$ is the smallest rational number p/q so that $q \leq 4^r$ and $p/q \geq \tilde{v}_k^t$. Therefore, the Kwek-Mehlhorn algorithm applied to $\tilde{v}_k^t$ correctly computes $\mathrm{val}(G)_k$, and we are done.

We can not use the bound on $T(G)$ obtained by the first direct approach (in Subsection 3.1) to show the correctness of SolveSSG, but we can show the correctness of the version of it that runs the main loop an additional factor of $O(r)$ times, that is, i should range over $\{1, 2, \ldots, 5 \ln 2 \cdot r^2 2^r\}$ instead of over $\{1, 2, \ldots, 10r2^r\}$.

3   Bounds on the Convergence Rate

3.1   A Direct Approach

Lemma 2. *Let $G \in S_{n,r}$ be given. For all positions k and all integers $i \geq 1$, we have*

$$\mathrm{val}(G)_k - \mathrm{val}(G^{i \cdot r})_k \leq (1 - 2^{-r})^i.$$

Proof. If $\text{val}(G)_k = 0$, we also have $\text{val}(G^{i \cdot r})_k = 0$, so the inequality holds. Therefore, we can assume that $\text{val}(G)_k > 0$.

Fix some positional optimal strategy, x, for Max in G. Let y be any pure (i.e., deterministic, but not necessarily positional) strategy for Min with the property that y guarantees that the pebble will not reach GOAL after having been in a position of value 0 (in particular, any best reply to any strategy of Max, including x, clearly has this property).

The two strategies x and y together induce a probability space σ_k on the set of plays of the game, starting in position k. Let the probability measure on plays of G^t associated with this strategy be denoted Pr_{σ_k}. Let W_k be the event that this random play reaches GOAL. We shall also consider the event W_k to be a set of plays. Note that any position occurring in any play in W_k has non-zero value, by definition of y.

Claim. There is a play in W_k where each position occurs at most once.

Proof of Claim. Assume to the contrary that for all plays in W_k, some position occurs at least twice. Let y' be the modification of y where the second time a position, v, in V_2 is entered in a given play, y takes the same action as was used the first time v occurred. Let W' be the set of plays generated by x and y' for which the pebble reaches GOAL. We claim that W' is in fact the empty set. Indeed, if W' contains any play q, we can obtain a play in W' where each position occurs only once, by removing all transitions in q occurring between repetitions of the same position. Such a play is also an element of W_k, contradicting the assumption that all plays in W_k has a position occurring twice. The emptiness of W' shows that the strategy x does not guarantee that GOAL is reached with positive probability, when play starts in k. This contradicts either that x is optimal or that $\text{val}(G)_k > 0$. We therefore conclude that our assumption is incorrect, and that there is a play q in W_k where each position occurs only once, as desired. *This completes the proof of the claim.*

Consider a fixed play where each coin toss position occurs only once. The probability that this particular play occurs is at least 2^{-r}, according to the probability measure σ_k.

Let W_k^i be the set of plays in W_k that contains at most i occurrences of coin toss positions (and also let W_k^i denote the corresponding event with respect to the measure σ_k). Since the above claim holds for any position k of non-zero value and plays in W_k only visit positions of non-zero value, we see that $\text{Pr}_{\sigma_k}[\neg W_k^{i \cdot r} | W_k] \leq (1 - 2^{-r})^i$, for any i. Since x is optimal, we also have $\text{Pr}_{\sigma_k}[W_k] \geq \text{val}(G)_k$. Therefore,

$$\Pr_{\sigma_k}[W_k^{i \cdot r}] = \Pr_{\sigma_k}[W_k] - \Pr_{\sigma_k}[\neg W_k^{i \cdot r} | W_k] \Pr_{\sigma_k}[W_k]$$
$$\geq \text{val}(G)_k - (1 - 2^{-r})^i$$

The above derivation is true for any y guaranteeing that no play can enter a position of value 0 and then reach GOAL, and therefore it is also true for y being the optimal strategy in the time-limited game, $G^{i \cdot r}$. In that case, we have $\text{Pr}_{\sigma_k}[W_k^{i \cdot r}] \leq \text{val}(G^{i \cdot r})_k$. We can therefore conclude that $\text{val}(G^{i \cdot r})_k \geq \text{val}(G)_k - (1 - 2^{-r})^i$, as desired.

Lemma 3. *Let $G \in S_{n,r}$ be given.*

$$T(G) \leq 5 \ln 2 \cdot r^2 \cdot 2^r$$

Proof. We will show that for any $t \geq 5 \ln 2 \cdot r^2 \cdot 2^r$ and any k, we have that $\mathrm{val}(G)_k - \mathrm{val}(G^t)_k < 2^{-5r}$.

From Lemma 2 we have that $\forall i, k : \mathrm{val}(G)_k - \mathrm{val}(G^{i \cdot r})_k < (1 - 2^{-r})^i$. Thus,

$$\mathrm{val}(G)_k - \mathrm{val}(G^t)_k < (1 - 2^{-r})^{t/r} = ((1 - 2^{-r})^{2^r})^{\frac{t}{r \cdot 2^r}} < e^{-\frac{t}{r \cdot 2^r}} \leq e^{-\frac{5 \ln 2 r^2 2^r}{r \cdot 2^r}} = 2^{-5r}.$$

3.2 An Extremal Combinatorics Approach

The game of Figure 3 is a game in $S_{n,r}$. We will refer to this game as $E_{n,r}$. $E_{n,r}$ consists of no Max-positions. Each Min-position in $E_{n,r}$ has GOAL as a successor twice. The i'th coin toss position, for $i \geq 2$, has the $(i-1)$'st and the r'th coin toss position as successors. The first coin toss position has GOAL and the r'th coin toss position as successors. The game is very similar to a simple stochastic game used as an example by Condon [7] to show that unmodified value iteration converges slowly.

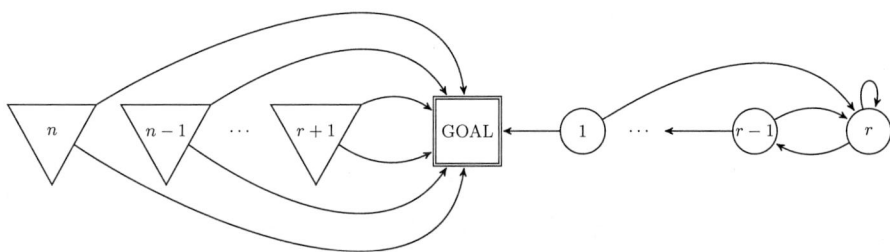

Fig. 3. The extremal game $E_{n,r}$. Circle nodes are coin toss positions, triangle nodes are Min positions and the node labeled GOAL is the GOAL position.

In this subsection we will show that $E_{n,r}$ is an extremal game in the sense of Definition 3 and upper bound $T(E_{n,r})$, thereby upper bounding $T(G)$ for all $G \in S_{n,r}$.

The first two lemmas in this subsection concerns assumptions about t-extremal games we can make without loss of generality.

Lemma 4. *For all n, r, t, there is a t-extremal game in $S_{n,r}$ with $V_1 = \emptyset$, i.e., without containing positions belonging to Max.*

Proof. Take any t-extremal game $G \in S_{n,r}$. Let x be a positional optimal strategy for Max in this game. Now we change the game, replacing each position belonging to Max with a position belonging to Min with both outgoing arcs making the choice specified by x. Call the resulting game H. We claim that H is also t-extremal. First, clearly, each position k of H has the same value as is has in G, i.e., $val(G)_k = val(H)_k$. Also, if we compare the values of the positions of the games H^t and G^t defined in the statement of Lemma 1, we see that $val(H^t)_k \leq val(G^t)_k$, since the only difference between H^t and G^t is that player Max has more options in the latter game. Therefore, $val(H)_k - val(H^t)_k \geq val(G)_k - val(G^t)_k$ so H must also be t-extremal.

Lemma 5. *For all n, r, t, there exists a t-extremal game in $S_{n,r}$, where all positions have value one and where no positions belong to player Max.*

Proof. By Lemma 4, we can pick a t-extremal game G in $S_{n,r}$ where no positions belong to player Max. Suppose that not all positions in G have value 1. Then, it is easy to see that the set of positions of value 0 is non-empty. Let this set be N. Let H be the game where all arcs into N are redirected to GOAL. Clearly, all positions in this game have value 1. We claim that H is also t-extremal.

Fix a position k. We shall show that $val(H)_k - val(H^t)_k \geq val(G)_k - val(G^t)_k$ and we shall be done. Let σ_k be a (not necessarily positional) optimal strategy for player Min in G^t for plays starting in k and let the probability measure on plays of G^t associated with this strategy be denoted $\Pr_{\sigma_k}$. We can extend σ_k to a strategy for G by making arbitrary choices after the coin toss budget runs out, so we have $\Pr_{\sigma_k}[\text{Play does not reach } N] \geq val(G)_k$. Also, by definition, $\Pr_{\sigma_k}[\text{Play reaches GOAL}] = val(G^t)_k$. That is,

$$\Pr_{\sigma_k}[\text{Play reaches neither GOAL nor } N] \geq val(G)_k - val(G^t)_k.$$

Let $\bar{\sigma}_k$ be an optimal strategy for plays starting in k for player Min in H^t. This strategy can also be used in G^t. Let the probability distribution on plays of G^t associated with this strategy be denoted $\Pr_{\bar{\sigma}_k}$. Note that plays reaching GOAL in H^t correspond to those plays reaching either GOAL or N in G^t. Thus, by definition, $\Pr_{\bar{\sigma}_k}[\text{Play reaches neither GOAL nor } N] = 1 - val(H^t)_k$. As σ_k can be used in H^t where $\bar{\sigma}_k$ is optimal, we have

$$1 - val(H^t)_k \geq val(G)_k - val(G^t)_k.$$

But since $val(H)_k = 1$, this is the desired inequality

$$val(H)_k - val(H^t)_k \geq val(G)_k - val(G^t)_k.$$

The following lemma will be used several times to change the structure of a game while keeping it extremal, eventually making the game into the specific game $E_{n,r}$ (in the context of extremal combinatorics, this is a standard technique pioneered by Moon and Moser [15]).

Lemma 6. *Given a game G. Let c be a coin toss position in G and let k be an immediate successor position k of c. Also, let a position k' with the following property be given: $\forall t : val(G^t)_{k'} \leq val(G^t)_k$. Let H be the game where the arc from c to k is redirected to k'. Then, $\forall t, j : val(H^t)_j \leq val(G^t)_j$.*

Proof. In this proof we will throughout use Lemma 1 and refer to the properties of ModifiedValueIteration. We show by induction on t that $\forall j, t : \text{val}(H^t)_j \leq \text{val}(G^t)_j$. For $t = 0$ we have $\text{val}(H^t)_j = \text{val}(G^t)_j$ by inspection of the algorithm. Now assume that the inequality holds for all values smaller than t and for all positions i and we will show that it holds for t and all positions j. Consider a fixed position j. There are three cases.

1. *The position j belongs to Max or Min.* In this case, we observe that the function computed by SolveDGG to determine the value of position j in ModifiedValueIteration is a monotonously increasing function. Also, the deterministic graphical game obtained when replacing coin toss positions with terminals is the same for G and for H. By the induction hypothesis, we have that for all i, $\text{val}(H^{t-1})_i \leq \text{val}(G^{t-1})_i$. So, $\text{val}(H^t)_j \leq \text{val}(G^t)_j$.

2. *The position j is a coin toss position, but not c.* In this case, we have

$$\text{val}(G^t)_j = \frac{1}{2}\text{val}(G^{t-1})_a + \frac{1}{2}\text{val}(G^{t-1})_b,$$

and

$$\text{val}(H^t)_j = \frac{1}{2}\text{val}(H^{t-1})_a + \frac{1}{2}\text{val}(H^{t-1})_b$$

where a and b are the successors of j. By the induction hypothesis, $\text{val}(H^{t-1})_a \leq \text{val}(G^{t-1})_a$ and $\text{val}(H^{t-1})_b \leq \text{val}(G^{t-1})_b$. Again, we have $\text{val}(H^t)_j \leq \text{val}(G^t)_j$.

3. *The position j is equal to c.* In this case, we have

$$\text{val}(G^t)_c = \frac{1}{2}\text{val}(G^{t-1})_a + \frac{1}{2}\text{val}(G^{t-1})_k$$

where a and k are the successors of c in G while

$$\text{val}(H^t)_c = \frac{1}{2}\text{val}(H^{t-1})_a + \frac{1}{2}\text{val}(H^{t-1})_{k'}.$$

By the induction hypothesis we have $\text{val}(H^{t-1})_a \leq \text{val}(G^{t-1})_a$. We also have that $\text{val}(H^{t-1})_{k'} \leq \text{val}(G^{t-1})_{k'}$ which is, by assumption, at most $\text{val}(G^{t-1})_k$. So, we have $\text{val}(H^t)_c \leq \text{val}(G^t)_c$.

Theorem 2. *$E_{n,r}$ is an extremal game in $S_{n,r}$.*

Due to space constraints, the proof does not appear here, but can be found in the full version of the paper [11]. The proof first uses Lemma 5 to restrict attention to games G without Max positions and with all positions having value 1. Then, Lemma 6 is used iteratively to change G into $E_{n,r}$. Having identified the extremal game $E_{n,r}$, we next estimate $T(E_{n,r})$.

Lemma 7. *For $\epsilon > 0$ and $t \geq 2(\ln 2\epsilon^{-1})2^r$, we have $\text{val}(E_{n,r})_k - \text{val}(E_{n,r}^t) \leq \epsilon$.*

Proof. We observe that for the purposes of estimating $T(E_{n,r})$, we can view $E_{n,r}$ as a game containing r coin toss positions only, since all Min-positions have transitions to GOAL only. Also, when modified value iteration is applied to a game G containing only coin toss positions, Lemma 1 implies that $\mathrm{val}(G^t)_k$ can be reinterpreted as the probability that the absorbing Markov process starting in state k is absorbed within t steps. By the structure of $E_{n,r}$, this is equal to the probability that a sequence of t fair coin tosses contains r consecutive tails. This is known to be exactly $1 - F_{t+2}^{(r)}/2^t$, where $F_{t+2}^{(r)}$ is the $(t+2)$'nd Fibonacci r-step number, i.e. the number given by the linear homogeneous recurrence $F_m^{(r)} = \sum_{i=1}^{r} F_{m-i}^{(r)}$ and the boundary conditions $F_m^{(k)} = 0$, for $m \leq 0$, $F_1^{(r)} = F_2^{(r)} = 1$. Asymptotically solving this linear recurrence, we have that $F_m^{(r)} \leq (\phi_r)^{m-1}$ where ϕ_r is the root near 2 to the equation $x + x^{-r} = 2$. Clearly, $\phi_r < 2 - 2^{-r}$, so

$$F_{t+2}^{(r)}/2^t < \frac{(2 - 2^{-r})^{t+1}}{2^t} = 2(1 - 2^{-r-1})^{t+1} < 2(1 - 2^{-r-1})^t.$$

Therefore, the probability that the chain is not absorbed within $t \geq 2(\ln 2\epsilon^{-1})2^r$ steps is at most $2(1 - 2^{-r-1})^{2(\ln 2\epsilon^{-1})2^r} \leq 2e^{-\ln 2\epsilon^{-1}} = \epsilon$.

4   Conclusions

We have shown an algorithm solving simple stochastic games obtaining an improved running time in the worst case compared to previous algorithms, as a function of its number of coin toss positions. It is relevant to observe that the *best* case complexity of the algorithm is strongly related to its *worst* case complexity, as the number of iterations of the main loop is fixed in advance.

As mentioned in the introduction, our paper is partly motivated by a result of Chatterjee *et al.* [4] analysing the *strategy iteration* algorithm of the same authors [5] for the case of simple stochastic games. We can in fact improve their analysis, using the techniques of this paper. We combine three facts:

1. ([5, Lemma 8]) For a game $G \in S_{n,r}$, after t iterations of the strategy iteration algorithm [5] applied to G, the (positional) strategy computed for Max guarantees a probability of winning of at least $\mathrm{val}(\bar{G}^t)_k$ against any strategy of the opponent when play starts in position k, where $\bar{G}^t$ is the game defined from G in Definition 2 of this paper.
2. For a game $G \in S_{n,r}$ and all k, t, $\mathrm{val}(\bar{G}^{t(n-r+1)})_k \geq \mathrm{val}(G^t)_k$. This is a direct consequence of the definitions of the two games, and the fact that in an optimal play, either a coin toss position is encountered at least after every $n - r + 1$ moves of the pebble, or never again.
3. Lemma 7 of the present paper.

These three facts together imply that the strategy iteration algorithm after $10(n - r + 1)r2^r$ iterations has computed a strategy that guarantees the values of the game within an additive error of 2^{-5r}, for $r \geq 6$. As observed by Chatterjee *et al.* [4], such a strategy is in fact optimal. Hence, we conclude that their strategy iteration algorithm terminates in time $2^r n^{O(1)}$. This improves their analysis of the algorithm significantly, but still yields a bound on its worst case running time inferior to the worst case running time of the algorithm presented here. On the other hand, unlike the algorithm presented in this paper, their algorithm has the desirable property that it may terminate faster than its worst case analysis suggests.

References

1. Andersson, D., Hansen, K.A., Miltersen, P.B., Sørensen, T.B.: Deterministic Graphical Games Revisited. In: Beckmann, A., Dimitracopoulos, C., Löwe, B. (eds.) CiE 2008. LNCS, vol. 5028, pp. 1–10. Springer, Heidelberg (2008)
2. Andersson, D., Miltersen, P.B.: The Complexity of Solving Stochastic Games on Graphs. In: Dong, Y., Du, D.-Z., Ibarra, O. (eds.) ISAAC 2009. LNCS, vol. 5878, pp. 112–121. Springer, Heidelberg (2009)
3. Bellman, R.E.: On the application of dynamic programming to the determination of optimal play in chess and checkers. Procedings of the National Academy of Sciences of the United States of America 53, 244–246 (1965)
4. Chatterjee, K., de Alfaro, L., Henzinger, T.A.: Termination criteria for solving concurrent safety and reachability games. In: Proceedings of the Twentieth Annual ACM-SIAM Symposium on Discrete Algorithms, SODA 2009, pp. 197–206 (2009)
5. Chatterjee, K., de Alfaro, L., Henzinger, T.A.: Strategy improvement for concurrent reachability games. In: Third International Conference on the Quantitative Evaluation of Systems. QEST 2006, pp. 291–300. IEEE Computer Society (2006)
6. Condon, A.: The complexity of stochastic games. Information and Computation 96, 203–224 (1992)
7. Condon, A.: On algorithms for simple stochastic games. In: Advances in Computational Complexity Theory. DIMACS Series in Discrete Mathematics and Theoretical Computer Science, vol. 13, pp. 51–73. American Mathematical Society (1993)
8. Dai, D., Ge, R.: New Results on Simple Stochastic Games. In: Dong, Y., Du, D.-Z., Ibarra, O. (eds.) ISAAC 2009. LNCS, vol. 5878, pp. 1014–1023. Springer, Heidelberg (2009)
9. Gillette, D.: Stochastic games with zero stop probabilities. In: Dresher, M., Tucker, A.W., Wolfe, P. (eds.) Contributions to the Theory of Games III. Annals of Mathematics Studies, vol. 39, pp. 179–187. Princeton University Press (1957)
10. Gimbert, H., Horn, F.: Simple Stochastic Games with Few Random Vertices Are Easy to Solve. In: Amadio, R.M. (ed.) FoSSaCS 2008. LNCS, vol. 4962, pp. 5–19. Springer, Heidelberg (2008)
11. Ibsen-Jensen, R., Miltersen, P.B.: Solving simple stochastic games with few coin toss positions, http://arxiv.org/abs/1112.5255
12. Kwek, S., Mehlhorn, K.: Optimal search for rationals. Inf. Process. Lett. 86(1), 23–26 (2003)
13. Liggett, T.M., Lippman, S.A.: Stochastic games with perfect information and time average payoff. SIAM Review 11(4), 604–607 (1969)
14. Mertens, J.F., Neyman, A.: Stochastic games. International Journal of Game Theory 10, 53–66 (1981)
15. Moon, J., Moser, L.: On cliques in graphs. Israel Journal of Mathematics 3, 23–28 (1965), doi:10.1007/BF02760024
16. Shapley, L.S.: Stochastic games. Proc. Nat. Acad. Science 39, 1095–1100 (1953)

Efficient Communication Protocols
for Deciding Edit Distance⋆

Hossein Jowhari

MADALGO, University of Aarhus
hjowhari@madalgo.au.dk

Abstract. In this paper we present two communication protocols on computing edit distance. In our first result, we give a one-way protocol for the following Document Exchange problem. Namely given $x \in \Sigma^n$ to Alice and $y \in \Sigma^n$ to Bob and integer k to both, Alice sends a message to Bob so that he learns x or truthfully reports that the edit distance between x and y is greater than k. For this problem, we give a randomized protocol in which Alice transmits at most $\tilde{O}(k \log^2 n)$ bits and each party's time complexity is $\tilde{O}(n \log n + k^2 \log^2 n)$.

Our second result is a simultaneous protocol for edit distance over permutations. Here Alice and Bob both send a message to a third party (the referee) who does not have access to the input strings. Given the messages, the referee decides if the edit distance between x and y is at most k or not. For this problem we give a protocol in which Alice and Bob run a $O(n \log n)$-time algorithm and they transmit at most $\tilde{O}(k \log^2 n)$ bits. The running time of the referee is bounded by $\tilde{O}(k^2 \log^2 n)$. To our knowledge, this result is the first upper bound for this problem.

Our results are obtained through mapping strings to the Hamming cube. For this, we use the Locally Consistent Parsing method of [5,6] in combination with the Karp-Rabin fingerprints. In addition to yielding non-trivial bounds for the edit distance problem, this paper suggest a new conceptual framework and raises new questions regarding the embeddability of edit distance into the Hamming cube which might be of independent interest.

1 Introduction

For integers m and n, let $ed(x, y)$ denote the standard *edit distance* between two strings x and y from the alphabet $[m] = \{1, \ldots, m\}$ where it is defined as the minimum number of substitutions, insertions and deletions of characters that is required to convert x to y. In this paper, we also consider two variants of this metric. The *Ulam* metric is a submetric of edit distance restricted to sequences with no character repetitions. The *edit distance with moves* [5,6], denoted by $ed_M(x, y)$, is defined similar to $ed(x, y)$ with addition of a block move operation. Namely moving the entire substring $x[i, j]$ to any location is considered a single operation. To simplify the presentation, throughout this paper we assume the

---

⋆ Part of this research was done while the author was a visiting student at MIT.

L. Epstein and P. Ferragina (Eds.): ESA 2012, LNCS 7501, pp. 648–658, 2012.

alphabet size m is $O(n^c)$ constant c, otherwise we can always use random hashing to make this happen.

Edit distance has been studied in various computational and mathematical models. In this paper we focus on computing edit distance in the context of two communication models. In the first problem, here denoted by DE_k and known as the *Document Exchange* problem [6], there are two communicating parties Alice and Bob respectively holding input strings x and y. We are interested in a one-way communication protocol where Alice sends a message to Bob and Bob either learns x or truthfully reports that $ed(x, y) > k$. Here in this paper, in addition to optimizing the total number of transmitted bits, we are interested in protocols where Alice and Bob both run poly(k, n)-time algorithms.

Protocols for the document exchange problem are of considerable practical importance in communicating data over noisy channels. Consider the following recurring scenario where A transmits a large file x to B over an unreliable link with possible errors of insertion, deletion and substitution of bits. Roughly speaking, using a protocol for the above problem, A can supplement his message with $A(x)$ and avoid retransmissions in the case of small amount of corruption. In a similar situation, these protocols are useful for minimizing communication in the consolidation of distributed data sites. For instance, instead of transmitting their repository, coordinated servers can instead exchange their differences through a document exchange protocol.

For the DE_k problem, we give a communication protocol that transmits $\tilde{O}(k \log^2 n)$ bits while Alice and Bob's running time is bounded by $\tilde{O}(n \log n + k^2 \log^2 n)$. To our knowledge this is the first time-efficient 1-way protocol for this problem [1].

In the second problem, denoted by $ED^{\uparrow}_{k,k+1}$, we are interested in *sketching* protocols for deciding edit distance. Namely, we would like to have an efficiently computable mapping $h_k : X \to \{0, 1\}^t$ with $t = \text{poly}(k, \log n)$, such that having access to only $h_k(x)$ and $h_k(y)$, there is an efficient algorithm that decides if $d(x, y) \leq k$ or not. Equivalently, defined in terms of the *simultaneous* protocols [11], Alice holds x, and Bob holds y but they are only allowed to send one message to a third party named the Referee whose duty is to decide $ed(x, y) \leq k$ without any access to the input strings. Likewise here we are interested in protocols that run time-efficient computations. In this problem (and in DE_k as well), we assume the communicating parties have access to a shared source of randomness.

The sketching protocols in addition to their application in data communication, have implications in classification and data structures for Nearest Neighbour search. In particular consider a scenario where we would like to find two strings with edit distance bounded by k in a collection of m strings. Using a sketching protocol for the above problem, one can first build sketches of the strings and then compare the short sketches instead of running an edit distance algorithm over long strings.

---

[1] In fact a time-efficient 1-way protocol that transmits $O(k \log k \log(n/k))$ bits can inferred from the work of Irmak et al. [9] which uses a different method. The author was not aware of this result in time of the submission of this paper.

For the problem of $ED^\uparrow_{k,k+1}$ restricted to non-repeating strings, we give a protocol that transmits at most $\tilde{O}(k \log^2 n)$ bits. The running time of all communicating parties is bounded by $\hat{O}(k^2 \log^2 n)$. Unfortunately our method does not yield an upper bound for standard edit distance. It remains an open problem whether there exists a simultaneous protocol for deciding edit distance over general strings.

Previous and Related Works. Also known as the *Remote File Synchronization* [9], the Document Exchange problem DE_k has been studied extensively by Cormode *et al.* [6]. Conditioning on $ed(x,y) \leq k$, it has been shown there is a deterministic one-way protocol for DE_k, resulting from the graph coloring method of [14], that transmits only $O(k \log n)$ bits. However in this protocol, Bob's running time is exponential in k. Similarly in a randomized protocol for this problem, Alice builds a random hash $h(x)$ and send it to Bob. Bob, using the same source of randomness, compares $h(x)$ with the hash value of all the strings within distance k from y. If he finds a match, x has been learned otherwise he can declare that $ed(x,y) > k$. It can be shown that random hashes of $O(k \log n)$ bit length are enough so that Bob succeeds with high probability. Once again Bob's running time will be exponential in k as he should try $n^{O(k)}$ number of strings.

However if Alice and Bob are allowed to use a multi-round protocol there is a time-efficient protocol. In [6] Cormode *et al* have given a time-efficient solution that transmits $O(k \log(n/k) \log k)$ bits in total and runs in $O(\log n)$ rounds. This protocols works through recursively dividing the string into $O(k)$ blocks and detecting the differences through comparing the fingerprints of each block. Building on similar ideas, Irmak *et al.* [9] have a shown a single round $O(k \log(n/k) \log n)$ protocol for the document exchange problem under edit distance with moves.

To our knowledge, prior to our work, there was no non-trivial upper bound on the sketching complexity of deciding edit distance. This was even true assuming computationally unbounded communicating parties. On the lower bound side, it is known that any protocol for DE_k needs to transmit $\Omega(k)$ bits. This follows from a lower bound for Hamming distance (see the concluding discussions in [3]) This fact remains true when Alice and Bob's inputs are non-repeating strings [8,17]. Needless to say, this lower bound also applies to the sketching variant since Bob can take up the role of the referee.

Considering the L_p metric and in particular the Hamming metric, both the document exchange and the sketching problem are fairly well-understood. In fact solutions for both of these problems can be derived from any exact sparse recovery scheme (see [7] for an introduction to sparse recovery). Here the L_p differences between x and y can be inferred from the measurements Ax and Ay. Considering that there are sparse recovery matrices with $O(k \log(n/k))$ number of rows, this leads to an $O(k \log(n/k) \log n)$ bits bound for our problems. There are however non-linear methods that give better bounds. In particular for Hamming distance we use the following result in our work.

Lemma 1. *[16] Let x and y be two points in $[m]^n$. There exists a randomized mapping $s_k : [m]^n \to \{0,1\}^{O(k \log n)}$ such that given $s_k(x)$ and $s_k(y)$, there is an algorithm that outputs all the triples $\{(x_i, y_i)\}$ satisfying $x_i \neq y_i$ in $O(k \log n)$ time. For l-sparse vectors x, the sketch $s_k(x)$ can be constructed in $O(l \log n)$ time and space in one pass.*

The best exact algorithm for edit distance runs in $O(n^2 / \log^2 n)$ time [13]. For the decision version of the problem, there is an algorithm by [12] that runs in $O(n + k^2)$ time. In contrast with little progress over the exact complexity, there is a long sequence of results on approximating edit distance. The best near-linear time approximation algorithm for edit distance [2] runs in $O(n^{1+\epsilon})$ time and outputs an $O(\log^{O(1/\epsilon)} n)$ approximation. The sketching complexity of the gap questions regarding edit distance has been studied directly or implicitly in [4,3,1]. In particular [3] gives a $O(1)$-size sketch for distinguishing between k versus $(kn)^{2/3}$. We refer the reader to [2] for a detailed history on these results.

Techniques and Ideas. In this paper we propose the following general framework that is applicable to all distance functions. Given $d : [m]^n \times [m]^n \to \mathbb{R}^+$, we design a mapping f from distance d to Hamming. Formally let $f : [m]^n \to \{0,1\}^{n'}$ having the following properties.

1. n' is bounded by $n^{O(1)}$ and for all x, $f(x)$ can be computed efficiently.

2. There exists $1 \leq e_f \ll n$ such that for distinct u and v, we have $0 < \mathcal{H}(f(u), f(v)) \leq e_f d(u, v)$ with probability at least $7/8$.

3. There is a polynomial time algorithm R_f where given $f(x)$ obtains x exactly with probability at least $7/8$.

Using the mapping f and the reconstruction procedure R_f, we propose the following protocol. In the beginning, Alice and Bob both compute $f(x)$ and $f(y)$, and then they run a document exchange protocol under Hamming distance over $f(x)$ and $f(y)$ with parameter $k' = O(e_f k)$ (see Lemma 1). It follows that if $d(x, y) \leq k$, Bob will learn $f(x)$ and subsequently using R_f he reconstructs x. Considering the bounds from Lemma 1, communication complexity of this solution is bounded by $O(e_f \cdot k \cdot \log n)$ while the running time depends on the construction time of $f(x)$ and the time complexity of R_f.

We remark that, we do not need to use a mapping with low contraction as apposed to the standard metric embeddings. As long as distinct strings are mapped to different vectors and they are efficiently retrievable, the contraction of distances is not important. Naturally deriving such mappings is considerably easier in comparison with the standard low distortion embeddings. In fact, as we see in Section 2, we almost directly obtain a mapping from the Cormode-Muthukrishnan's embedding of *edit distance with moves* into L_1 [5]. This works because edit distance with moves is upper bounded by edit distance and moreover the expansion factor of its mapping is $O(\log n \log^* n)$ which is considerably better than the existing bounds for edit distance [2]. The CM's embedding is

---

[2] The best low distortion embedding for edit distance, due to Ostrovsky and Rabani [15], has $2^{\Omega(\sqrt{\log n})}$ as both the expansion and contraction factor.

quite efficient. It constructs a parsing tree over the input in near-linear time and encodes the substrings that are marked in the process into $f(x)$. As we shall see, to complete the picture we just need to equip our mapping with an efficient reconstruction procedure.

Roughly speaking, our reconstruction algorithm, having collected the Rabin-Karp fingerprints of the substrings of x that were obtained from the encoding $f(x)$, it rebuilds the original parsing tree that was used to create $f(x)$. As we shall see, this can be done in a fairly straightforward manner. Once the parsing tree is reconstructed the original string x can be inferred from the labels of the leaves of this tree.

In our sketching result for deciding Ulam, similar to the solution of the document exchange, Alice and Bob use the mapping f to create $f(x)$ and $f(y)$. Granted $ed(x, y)$ is bounded, the referee will learn $f(x) - f(y)$ and from this difference he will be able to construct partial parsing trees corresponding to x and y. It will be shown that the partial trees are enough to decide if $ed(x, y) \leq k$ or not. This last step of deciding the edit distance from partial recovery of x and y relies on the non-repetitiveness of the input sequences.

2   A 1-Way Protocol for Document Exchange

Before presenting our solution for edit distance, as a warm-up and along the lines we described above, we present a simple solution for the *Ulam metric*. The mapping f_U for Ulam distance is defined as follows. Let π be a non-repeating string. Corresponding to each pair (i, j) where $i, j \in \Sigma$, $f_U(\pi)$ has a coordinate. We set $f_U(\pi)_{i,j} = 1$ if j appears right after i in π otherwise we set $f_U(\pi)_{i,j} = 0$. It can be verified easily that the expansion factor of f_U is bounded by a constant ($e_{f_U} \leq 8$). Moreover the reconstruction procedure R_{f_U} is straightforward. This gives us the following lemma.

Lemma 2. *There is a one-way protocol for DE_k restricted to permutations that transmits $O(k \log n)$ bits and its running time is bounded by $O(n \log n)$.*

To define our mapping for general strings, we use the following result by Cormode and Muthukrishnan [5].

Lemma 3. *There is a mapping $g : \Sigma^n \to [n]^l$, such that for every pair $u, v \in \Sigma^n$, we have*

$$\frac{1}{2} ed_M(u, v) \leq \| g(u) - g(v) \|_1 \leq O(\log n \log^* n) ed_M(u, v).$$

Moreover the number of non-zero entries in $g(u)$ is bounded by $2n$.

Unfortunately the mapping g generates exponentially long vectors, i.e $l = 2^{O(n \log m)}$, and thus it is unsuitable for a direct application. However since the generated vectors are very sparse, we deploy random hashing to reduce its dimensionality. In the following we proceed with a brief background on g and then we describe the random hash function (supplemented with some extra information) that gives us the desired mapping f. Finally we present the reconstruction algorithm R_f.

The ESP tree. The embedding of [5] uses a procedure called *Edit Sensitive Parsing* (ESP) to build a rooted tree $T_x = (V_x, E)$ over the input string x where the leaves of T_x are the single characters of x. Each node $v \in V_x$ represents a unique substring $x[i, j]$. This means the characters $x(i)x(i + 1)\ldots x(j)$ are at leaves of the subtree rooted at v. See Figure 3 for a pictorial representation. We let $s(v)$ denote the substring that v represents. Every non-leaf node of T_x has either 2 or 3 children. We introduce the following notations. We denote the left-hand child of v by v_l, the middle child (if exists) by v_m and the right-hand child by v_r.

We need the following fact regarding the properties of the ESP tree; its complete proof can be inferred from the details in [5].

Lemma 4. *The number of structurally distinct subtrees in T_x that represent the same (but arbitrary) $\alpha \in \Sigma^*$ is bounded by $2^{O(\log^* n)}$.*

Proof. Let $v \in V_x$ and let $x[i, j]$ be the substring represented by v. From the procedure ESP, we have that the tree rooted at v depends only on the content of a substring $x[i', j']$ where $i' = i - O(1)$ and $j' = j + O(\log^* n)$.   □

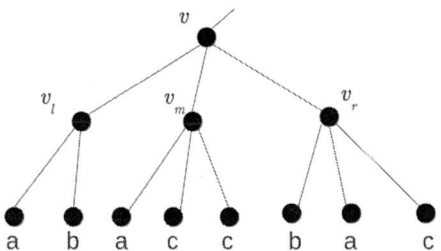

Fig. 1. A pictorial presentation of a part of an ESP tree

Given T_x, the vector $g(x)$ is defined as follows. Corresponding to each string α of length n, $g(x)$ has a coordinate. Let $g(x)[\alpha]$ represent the coordinate corresponding to α. We set $g(x)[\alpha]$ to be the number of nodes in T_x that represent α. This finishes the description of g.

Definition of f. We use a linear hash function to map the coordinates of g to polynomially bounded numbers. For the choice of our hash function we use the classical Rabin-Karp fingerprinting method [10]. Let $q > 4n^4$ be a prime and $r \in \mathbb{Z}_q^*$ be randomly selected from $[q]$. For $\alpha \in \Sigma^*$, the Rabin-Karp fingerprint of α is defined as $\Phi(\alpha) = \sum_{i=1}^{|\alpha|} \alpha[i] \cdot r^{i-1} \pmod{q}$. The following facts are well-known and the reader is referred to [10] for the proofs.

(F1) $\Phi(\alpha)$ can be computed in one pass over α using $O(\log n)$ bits of space.

(F2) Let $\beta \neq \alpha$ be a string of length at most n. $\Pr_r[\Phi(\alpha) = \Phi(\beta)] \leq \frac{1}{n^3}$.

(F3) We have $\Phi(\alpha \circ \beta) = \Phi(\alpha) + r^{|\alpha|+1}\Phi(\beta) \pmod{q}$ where the arithmetic operations are done in $\mathbb{Z}_q$.

Note 1. Since with high probability there will be no collision between the fingerprint of different strings, to simplify the presentation, in the rest of this paper we will assume that is the case.

Now we are ready to introduce $f : \Sigma^n \to [n] \times [q]^4$. Given $x \in \Sigma^n$, we set the coordinates of $f(x)$ as follows. For a non-leaf $v \in T_x$ with three children v_l, v_m and v_r, we set $f(x)[|s(v)|, \Phi(s(v)), \Phi(s(v_l)), \Phi(s(v_m)), \Phi(s(v_r))]$ to be the number of subtrees in T_x that represent $s(v)$ and their root substring is partitioned into three blocks of $s(v_l)$, $s(v_m)$ and $s(v_r)$. We do the same for the nodes with two children except that we regard the middle block $s(v_m)$ as the empty string with $\Phi(s(v_m)) = 0$. For the leaf $v \in T_x$, we set $f(x)[1, \Phi(s(v)), 0, 0, 0]$ to be the number of the occurrences of the character $s(v)$ in x. The rest of the coordinates in $f(x)$ are set to zero.

Proposition 1. *Let $u, v \in \Sigma^n$. We have $\mathcal{H}(f(u), f(v)) \leq 2^{O(\log^* n)} \| g(u) - g(v) \|_1$.*

Proof. Let S_u and S_v be the set of substrings that are represented by the nodes in the trees T_u and T_v respectively. Assuming the mapping Φ produces distinct values over $S_u \cup S_v$ and by Lemma 4, it follows there are at most $2^{O(\log^* n)}$ non-zero coordinates in f corresponding to a substring α. This completes the proof of the lemma. $\square$

The above proposition and Lemma 3, give us the following bound on the expansion factor of f.

Corollary 1. $e_f = O(\log n \cdot 2^{O(\log^* n)} \cdot \log^* n)$.

Reconstruction algorithm R_f. The description of this algorithm is given in Figure 2. Our procedure R_f works by constructing a tree T'_x that has the same sequence of (labeled) leaves as T_x does and as result we reconstruct x. The tree is constructed in a top-down fashion starting from the root of T_x and expanding the leaves by finding the encodings of its children in the vector $f(x)$. We point out that T'_x might not be structurally equivalent with T_x but it is guaranteed that the leaves are in the same labeling order.

Theorem 1. *There is a randomized one-way protocol for DE_k that transmits $\tilde{O}(k \log^2 n)$ bits and succeeds with high probability. The running time of each party is bounded by $O(n \log n + k^2 \log^2 n)$.*

Proof. (Sketch) The correctness of the result follows from Lemma 1, the general framework that we introduced in the introduction and the description of the the reconstruction procedure. The running time of creating $f(x)$ is bounded by

---

The Reconstruction Algorithm
Input: $f(x)$, **Output:** x

Initialization: Let S_x denote the set of non-zero coordinates in $f(x)$, *i.e*

$$S_x = \{ (l, j_0, j_1, j_2, j_3) \mid f(x)[l, j_0, j_1, j_2, j_3] \neq 0 \}.$$

Begin with an empty tree $T'_x = (V', E')$. Find $v \in S_x$ with $l = n$ and add it to V'. Let $S_x = S_x/\{v\}$. Note that such a node is unique since there is only one string with length n. This is going to be the root of T'_x. In the following, abusing the notation, we let $\Phi(v)$ denote the second coordinate of v.

Iteration: Repeat until S_x is empty.

1. Pick an arbitrary leaf $v \in T'_x$ which does not correspond to a single character.
2. Given the non-zero fingerprints j_1, j_2 and j_3 corresponding to the children of v, find the elements in v_l, v_m and v_r in S_x where $\Phi(v_l) = j_1$, $\Phi(v_m) = j_2$ and $\Phi(v_r) = j_3$. Break the ties arbitrarily.
3. Let $S_x = S_x/\{v_l, v_m, v_r\}$ and add the excluded elements to T'_x as the children of v. Note that if $j_2 = 0$, we would only add v_l and v_r.

Finalization: Counting from left to right, let x_i be the character that is represented by the i-th leaf of T'_x.

Fig. 2. Description of the reconstruction algorithm R_f

$O(n \log n)$ since the vector has at most $O(n)$ non-zero coordinates. The reconstruction procedure performed in a naive manner takes $O(n^2)$ time as finding each child of node might take linear time. To expedite this procedure, we do not build T'_x from the scratch. Since the trees T_x and T_y differ on at most $\tilde{O}(k \log n)$ nodes, we can start from T_y (after pruning its disagreements with T_x) and as result we need to process at most $\tilde{O}(k \log n)$ entries. We defer details of this improvement to the long version of this paper.

Having obtained x, Bob can decide $ed(x, y) \leq k$ in $O(n + k^2)$ time using the algorithm from [12]. This finishes the proof. □

3 A Sketching Result for Ulam

Let S_n be the set of permutations over $\{1, \ldots, n\}$. In this section we present our sketching result for edit distance over S_n. The following lemma is the heart of this result.

Lemma 5. *Let $x, y \in S_n$. Let f be the mapping described as before. Given the non-zero coordinates of $\Delta_{x,y} = f(x) - f(y)$, there is a $\tilde{O}(k^2 \log^2 n)$ algorithm that decides whether $ed(x, y) \leq k$ or not.*

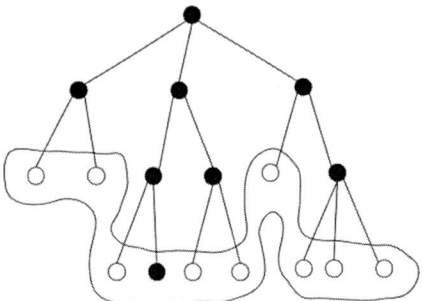

Fig. 3. A pictorial presentation of the tree $T(\mathcal{V}_X)$ shown in black nodes. The curved area shows the nodes in Γ_X.

Proof. First we observe that since x has no repetitions, $f(x)$ is a zero-one vector. Let

$$X = \{\, i \,|\, \Delta_{x,y}(i) = 1 \}.$$

The set X points to the substrings of T_x that are not represented by T_y. Similarly we define the set Y where it corresponds to the negative coordinates in $\Delta_{x,y}$.

Let $\mathcal{V}_X$ be the set of nodes in T_x that corresponds to X. Clearly $root(T_x) \in \mathcal{V}_X$. Now let $T(\mathcal{V}_X)$ be the rooted tree that is obtained by attempting to reconstruct T_x starting from $root(T_x)$. To build $T(\mathcal{V}_X)$, starting from the root, we try to expand every node until it reaches down to a single character (*i.e.* to a leaf of T_x) or its children are entirely missing in $\mathcal{V}_X$ because they have identical counterparts in T_y. Note that it is possible that one or two children are missing in $\mathcal{V}_X$. In that case, since we have supplemented every node with the fingerprint of its children, a node can still be expanded. It is also possible that some nodes in $\mathcal{V}_X$ are left unpicked after the end of this process. Those nodes represent substrings whose parents exist in both T_x and T_y but they are partitioned differently and hence have been mapped to different coordinates in $f(x)$ and $f(y)$. These nodes will be neglected.

Let Γ_X represent the set of vertices in $T_x/\mathcal{V}_X$ that are the immediate neighbors of the nodes in $T(\mathcal{V}_X)$. We also include the leaves of $T(\mathcal{V}_X)$ which represent single characters into Γ_X. Note that by the non-repetitiveness of x, $T(\mathcal{V}_X)$ is indeed a partial subtree of T_x rooted at $root(T_x)$ and hence Γ_X is well-defined. Also since for each node we have stored the information regarding its children, the set Γ_X can be computed. It should be clear that Γ_X gives a non-overlapping partitioning of x into $\tilde{O}(k \log n)$ blocks. By the definition, every block in Γ_X is identical to a represented substring in T_y. We perform the same process for string y and obtain Γ_Y.

We cannot claim that there exists a perfect matching between Γ_X and Γ_Y, however we can find a perfect matching between sets of consecutive blocks. To see this, let C be the longest block in $\Gamma_X \cup \Gamma_Y$ (breaking the ties arbitrarily) and w.l.o.g assume $C \in \Gamma_X$. Since C has an identical match in T_y it must match a set of consecutive blocks in Γ_Y. We pick this matching and continue with finding a matching for the longest block in the remained unmatched ones. It follows from the definition of Γ_X and Γ_Y and fact that there are no repeating substrings, every block will be matched at the end. Moreover such a mapping can be found in $O(k^2 \log^2 n)$ time.

Naturally, the mapping between the consecutive blocks in Γ_X and Γ_Y defines a perfect matching between the indices in x and y. Let $h(i)$ denote the index where x_i is mapped to. We create two permutations x' and y' so that x'_i and $y'_{h(i)}$ receive the same label. Clearly $ed(x', y') = ed(x, y)$ since relabeling does not affect the edit distance. It follows that we can compute the edit distance between x and y. This last step performed in a naive manner takes $O(n \log n)$ time but considering the fact that in an optimal alignment between x' and y' a block is either entirely aligned or it is deleted, we can find an optimal alignment using a dynamic programming for a weighted encoding of the problem. Therefore this can also be performed in $\tilde{O}(k^2 \log^2 n)$ time.   □

The above lemma combined with Lemma 1 give us the following result.

Theorem 2. *There is a randomized mapping* $u_k : S_n \to \{0,1\}^{O(k \log^2 n)}$ *such that given* $u_k(x)$ *and* $u_k(y)$ *for* $x, y \in S_n$, *there is an* $\tilde{O}(k^2 \log^2 n)$-*time algorithm that decides whether* $ed(x, y) \le k$ *or not.*

4 Concluding Remarks

1. In our protocol for the Document Exchange problem, we used a randomized mapping $f : (ed, [m]^n) \to (\text{Hamming}, \{0,1\}^{n'})$ with polynomially large n' and $e_f = \tilde{O}(\log n)$. Is there a similar mapping with $e_f = o(\log n)$? Such a mapping equipped with a polynomial time reconstruction procedure results in an improved protocol for the DE_k problem. On the other hand, given that such a mapping exists for the Ulam metric (the mapping f_U in Section 2), showing the impossibility of a similar fact for the edit distance will result in proving an interesting seperation theorem between the two metrics. From the angel of the low distortion embeddings, seperating Ulam from edit distance over repetitive strings has yet remained an open question.

2. In our sketching result for the Ulam metric, we have not used the simpler mapping f_U of Section 2. This is because it does not preserve the edit distance. In other words, there are pairs of strings (x_1, y_1) and (x_2, y_2) such that $f_U(x_1) - f_U(y_1)$ and $f_U(x_2) - f_U(y_2)$ are identical while $ed(x_1, y_1)$ and $ed(x_2, y_2)$ are arbitrarily far apart.

3. The sketching result for Ulam can be generalized to the case when only one of the strings is a permutation. This is true since we can still relabel the content of each mapped block with arbitrary characters. Also we may not have a perfect matching but the content of the blocks that aren't mapped will be revealed. In the case of general strings, we can also obtain a mapping between the blocks of the input strings. However, because of repetitions, it is not clear how we can use this information to learn the edit distance.

4. The sketching algorithm of Section 3 can be adapted to work as a streaming algorithm over interleaving input sequences. This follows from the fact that Lemma 1 is a streaming result. Moreover since ESP tree can be built in a streaming fashion (see Section 4.3 in [5]), we are able to derive a streaming algorithm.

Acknowledgement. The author would like to thank Ronitt Rubinfeld for kindly hosting his visit to MIT. Thanks also to Cenk Sahinalp, Mert Sağlam, Qin Zhang and Djamal Belazzougui for useful discussions. In particular thanks to Djamal Belazzougui for bringing [9] to the author's attention.

References

1. Andoni, A., Krauthgamer, R.: The computational hardness of estimating edit distance [extended abstract]. In: FOCS, pp. 724–734 (2007)
2. Andoni, A., Krauthgamer, R., Onak, K.: Polylogarithmic approximation for edit distance and the asymmetric query complexity. In: FOCS, pp. 377–386 (2010)
3. Bar-Yossef, Z., Jayram, T.S., Krauthgamer, R., Kumar, R.: Approximating edit distance efficiently. In: FOCS, pp. 550–559 (2004)
4. Batu, T., Ergün, F., Kilian, J., Magen, A., Raskhodnikova, S., Rubinfeld, R., Sami, R.: A sublinear algorithm for weakly approximating edit distance. In: STOC, pp. 316–324 (2003)
5. Cormode, G., Muthukrishnan, S.: The string edit distance matching problem with moves. ACM Transactions on Algorithms 3(1) (2007)
6. Cormode, G., Paterson, M., Sahinalp, S.C., Vishkin, U.: Communication complexity of document exchange. In: SODA, pp. 197–206 (2000)
7. Gilbert, A., Indyk, P.: Sparse recovery using sparse matrices. Proceedings of IEEE (2010)
8. Gopalan, P., Jayram, T.S., Krauthgamer, R., Kumar, R.: Estimating the sortedness of a data stream. In: SODA, pp. 318–327 (2007)
9. Irmak, U., Mihaylov, S., Suel, T.: Improved single-round protocols for remote file synchronization. In: INFOCOM, pp. 1665–1676 (2005)
10. Karp, R.M., Rabin, M.O.: Efficient randomized pattern-matching algorithms. IBM Journal of Research and Development 31(2), 249–260 (1987)
11. Kushilevitz, E., Nisan, N.: Communication complexity. Cambridge University Press (1997)
12. Landau, G.M., Myers, E.W., Schmidt, J.P.: Incremental string comparison. SIAM J. Comput. 27(2), 557–582 (1998)
13. Masek, W.J., Paterson, M.: A faster algorithm computing string edit distances. J. Comput. Syst. Sci. 20(1), 18–31 (1980)
14. Orlitsky, A.: Interactive communication: Balanced distributions, correlated files, and average-case complexity. In: FOCS, pp. 228–238 (1991)
15. Ostrovsky, R., Rabani, Y.: Low distortion embeddings for edit distance. J. ACM 54(5) (2007)
16. Porat, E., Lipsky, O.: Improved Sketching of Hamming Distance with Error Correcting. In: Ma, B., Zhang, K. (eds.) CPM 2007. LNCS, vol. 4580, pp. 173–182. Springer, Heidelberg (2007)
17. Sun, X., Woodruff, D.P.: The communication and streaming complexity of computing the longest common and increasing subsequences. In: SODA, pp. 336–345 (2007)

Approximation Algorithms for Wireless Link Scheduling with Flexible Data Rates

Thomas Kesselheim*

Department of Computer Science,
RWTH Aachen University, Germany

Abstract. We consider scheduling problems in wireless networks with respect to flexible data rates. That is, more or less data can be transmitted per time depending on the signal quality, which is determined by the signal-to-interference-plus-noise ratio (SINR). Each wireless link has a utility function mapping SINR values to the respective data rates. We have to decide which transmissions are performed simultaneously and (depending on the problem variant) also which transmission powers are used.

In the capacity-maximization problem, one strives to maximize the overall network throughput, i.e., the summed utility of all links. For arbitrary utility functions (not necessarily continuous ones), we present an $O(\log n)$-approximation when having n communication requests. This algorithm is built on a constant-factor approximation for the special case of the respective problem where utility functions only consist of a single step. In other words, each link has an individual threshold and we aim at maximizing the number of links whose threshold is satisfied. On the way, this improves the result in [Kesselheim, SODA 2011] by not only extending it to individual thresholds but also showing a constant approximation factor independent of assumptions on the underlying metric space or the network parameters.

In addition, we consider the latency-minimization problem. Here, each link has a demand, e.g., representing an amount of data. We have to compute a schedule of shortest possible length such that for each link the demand is fulfilled, that is, the overall summed utility (or data transferred) is at least as large as its demand. Based on the capacity-maximization algorithm, we show an $O(\log^2 n)$-approximation for this problem.

1 Introduction

The performance of wireless communication is mainly limited by the fact that simultaneous transmissions interfere. In the presence of interference, the connection quality can deteriorate or successful communication can even become impossible. In engineering, the common way to measure the quality of a wireless connection is the signal-to-interference-plus-noise ratio (SINR). This is the ratio of the strength of the intended signal and the sum of all other signal strengths

---

* This work has been supported by the UMIC Research Centre, RWTH Aachen University.

L. Epstein and P. Ferragina (Eds.): ESA 2012, LNCS 7501, pp. 659–670, 2012.

plus ambient noise. The higher it is, the better different symbols in a transmission can be distinguished and therefore the more information can be transmitted per time.

Existing algorithmic research on wireless networks mainly focused on transmissions using fixed data rates. This way successful reception becomes a binary choice: Either the quality is high enough and the transmission is successfully received or nothing is received due to too much interference. Interference constraints are, for example, given by thresholds on the SINR or by a graph whose edges model mutual conflicts. Typical optimization problems are similar to independent set or coloring problems. That is, the most important aspect is a discrete choice, making the problems usually non-convex and NP-hard.

This threshold assumption does not reflect the ability of wireless devices to adapt their data rates to different interference conditions. For example, a file transfer can be slowed down on a poor-quality connection but still be carried out. For this reason, a different perspective has been taken in some works on power control. Here, it is assumed that devices are perfectly able to adapt to the current conditions. This is reflected in a utility function, which is assumed to be a concave and differentiable function of the SINR. Under these assumptions, one can design distributed protocols for selecting transmission powers that converge to local optima of the summed utility [15]. Applying further restrictions on the utility functions, the summed utility becomes equivalent to a convex optimization problem and can thus be solved in polynomial time [2]. These algorithms solve a purely continuous optimization problem as they select transmission powers for each link from a continuous set and do not make a decision which transmissions are carried out. Requiring a certain minimal SINR for each transmission to be successful is thus only possible if the resulting set of feasible solutions is not empty. Furthermore, only under relatively strict assumptions, one is able to find a global optimum. It is even not clear if it is appropriate to assume continuity of utility functions as today's standards only support a fixed number of data rates.

In this paper, we take an approach that generalizes both perspectives. We assume each communication request has a utility function that depends arbitrarily on the SINR. That is, we are given a function for each transmission mapping each SINR to the throughput that can be achieved under these conditions. By selecting which transmissions are carried out and which powers are used, we maximize the overall network throughput. As we allow for non-continuous utility functions, this generalizes both the threshold and the convex objective functions. We only assume that the utility functions and their inverse can be evaluated efficiently.

More formally, we consider the following *capacity-maximization problem with flexible data rates*. One is given n communication requests, each being a link, that is, a pair of a sender and a receiver. For each link ℓ we are given a utility function u_ℓ, quantifying the value that each possible SINR γ_ℓ has. The task is to select a subset of these links and possibly transmission powers with the objective of maximizing $\sum_\ell u_\ell(\gamma_\ell)$, that is, the overall network's throughput.

1.1   Our Contribution

We present $O(\log n)$-approximations for the capacity-maximization problem with flexible data rates, for the case that our algorithm has to specify the transmission powers as well as for a given power assignment. In both cases, the algorithm is built on one for the respective capacity-maximization problem with individual thresholds. Here, utility functions are step functions with $u_\ell(\gamma_\ell) = 1$ for $\gamma_\ell \geq \beta_\ell$ and 0 otherwise. That is, we assume that each link has an individual threshold β_ℓ and we maximize the number of links whose threshold is satisfied. For this special case, we present constant-factor approximations for both variable and fixed transmission powers.

For the case of variable transmission powers, this extends the result in [16] in two ways. On the one hand, the algorithm in [16] relies on the fact that all thresholds are equal. In order to guarantee feasible solutions under individual thresholds, the links have to be processed in a different order and further checks have to be introduced. Furthermore, in contrast to [16], we are able to show a constant approximation factor without any further assumptions on the metric space respectively the model parameters.

For fixed transmission powers, we extend a greedy algorithm by Halldórsson and Mitra [13] that works with monotone, (sub-) linear power assignments. By modifying the processing order, we are able to prove a constant approximation factor independent of the thresholds. Furthermore, we present a simplified analysis.

In addition to capacity-maximization problems, we also consider latency minimization. That is, for each link there is a fixed amount of data that has to be transmitted in shortest possible time. We use the capacity-maximization algorithms repeatedly with appropriately modified utility functions. This way, we can achieve an $O(\log^2 n)$-approximation for both variants.

1.2   Related Work

In a seminal work on power control, Foschini and Miljanic [8] give a very simple distributed algorithm for finding a power assignment that satisfies the SINR targets of all links. They show that it converges from any starting point under the assumption that the set of feasible solutions is not empty. In subsequent works, more sophisticated techniques have been presented (for an overview see [20]). Very recently, this problem has also been considered from an algorithmic point of view [18,3] deriving bounds on how the network size or parameters determine the convergence time. While in these problems typically no objective function is considered, Huang et al. [15] present a game-theoretic approach to maximize the sum of link utilities, where the utility of each link is an increasing and strictly concave function of the SINR. They show that their algorithm converges to local optima of the sum-objective function. Chiang et al. [2] present an approach to compute the global optimum for certain objective functions by the means of geometric programming in a centralized way. All these algorithms have in common that they solve a continuous optimization problem. That is, transmission powers are chosen from a continuous set in order to maximize a continuous function. When requiring a minimum SINR, the mentioned algorithms only work under the assumption that all links can achieve this minimum SINR simultaneously. Since this is in general not true, one may have to solve an additional scheduling problem. Many heuristics have been presented for these scheduling problems but recently a number of approximation algorithms were studied as well. Most of them assume a common, constant threshold β for all links. Usually, for the approximation factors this β is considered constant, which is not appropriate in our case.

In a number of independent papers, the problem of finding a maximum feasible set under uniform transmission powers has been tackled. For example, Goussevskaia et al. [9] present an algorithm that computes a set that is at most a constant factor smaller than the optimal one under uniform powers. In contrast to this, Andrews and Dinitz [1] compare the set they compute to the optimal one using an arbitrary power assignment. Their approximation factor is $O(\log \Delta)$, where Δ is the ratio between the largest and the smallest distance between a sender and its corresponding receiver. This bound is tight as uniform power assignments, in general, cannot achieve better results. This is different for square-root power assignments [6,11], which choose powers proportional to $\sqrt{d^\alpha}$ for a sender-receiver pair of distance d. The best bound so far by Halldórsson and Mitra [13] shows that one can achieve an $O(\log \log \Delta + \log n)$-approximation this way. However, for large values of Δ the approximation factors can get as bad as $\Omega(n)$. In general, it is better to choose transmission powers depending on the selected link set. In [16,21] a constant-factor approximation for the combined problem of scheduling and power control was presented. While the analysis in the mentioned papers only proves this approximation factor for *fading metrics* (i.e. α is greater than the doubling dimension than the metric), we show in this paper that this result actually holds for all metric spaces.

Apart from these capacity-maximization problems and the online variants [5], one has focused on latency minimization, that is scheduling all transmission requests within shortest possible time. For this problem, distributed ALOHA-like algorithms have been analyzed [7,17,12].

The mentioned results only consider that signals propagate deterministically, neglecting short-term effects such as scattering, which are typically modeled stochastically. Dams et al. [4] present a black-box transformation to transfer algorithmic results to Rayleigh-fading conditions. Plugging in the results in this paper, we get an $O(\log n \cdot \log^* n)$-approximation for the flexible-rate problems and an $O(\log^* n)$-approximation for the respective threshold variants.

The only approximation algorithm for individual thresholds was given by Halldórsson and Mitra [14]. It is a constant-factor approximation for capacity maximization for uniform transmission powers. Santi et al. [19] consider latency minimization with flexible data rates. This approach, however, only considers quite restricted utility functions that have to be the same for all links. Furthermore, they only consider uniform transmission powers.

2   Formal Problem Statements

We identify the network devices by a set of nodes V in a metric space. If some sender transmits at a power level p then this signal is received at a strength of p/d^α by nodes whose distance to the sender is d. The constant α is called path-loss exponent. Given a set $\mathcal{L} \subseteq V \times V$ and a power assignment $p\colon \mathcal{L} \to \mathbb{R}_{\geq 0}$, the SINR of link $\ell = (s, r) \in \mathcal{L}$ is given by

$$\gamma_\ell(\mathcal{L}, p) = \frac{\frac{p(\ell)}{d(s,r)^\alpha}}{\sum_{\ell'=(s',r')\in\mathcal{L},\ell'\neq\ell} \frac{p(\ell')}{d(s',r)^\alpha} + N} \ .$$

Here, N denotes the constant ambient noise. To avoid ambiguities such as divisions by 0, we assume it to be strictly larger than 0. However, it may be arbitrarily small.

In the *capacity-maximization problem with flexible data rates*, we are given a set $\mathcal{R} \subseteq V \times V$ of pairs of nodes of a metric space. For each link $\ell \in \mathcal{R}$, we are given a utility function $u_\ell\colon [0, \infty) \to \mathbb{R}_{\geq 0}$. Furthermore, we are given a maximum transmission power $p_{\max} \in \mathbb{R}_{>0} \cup \{\infty\}$. The task is to select a subset $\mathcal{L}$ of $\mathcal{R}$ and a power assignment $p\colon \mathcal{L} \to [0, p_{\max}]$ such that $\sum_{\ell\in\mathcal{L}} u_\ell(\gamma_\ell(\mathcal{L}, p))$ is maximized.

We do not require the utility functions u_ℓ to be continuous. It neither has to be represented explicitly. We only assume that two possible queries on the utilities can be carried out. On the one hand, we need to access the maximum utility $u_\ell(p_{\max}/N)$ of a single link ℓ. Note that this value could be infinite if $p_{\max} = \infty$. We ignore this case as the optimal solution is not well-defined in this case. On the other hand, we assume that for each link ℓ given a value B no larger than its maximum utility, we can determine the smallest SINR γ_ℓ such that $u_\ell(\gamma_\ell) \geq B$. Both kinds of queries can be carried out in polynomial time for common cases of utility functions such as logarithmic functions or explicitly given step functions. As a technical limitation, we assume that $u_\ell(\gamma_\ell) = 0$ for $\gamma_\ell < 1$. This is not a weakness of our analysis but rather of the approaches studied so far. In the full version of this paper, we show that neither greedy nor ALOHA-like algorithms can dig the potential of SINR values smaller than 1.

In addition to this problem, we also consider the variant with fixed transmission powers. That is, we are given a set $\mathcal{R} \subseteq V \times V$ and a power assignment $p\colon \mathcal{R} \to \mathbb{R}_{\geq 0}$. We have to select a set $\mathcal{L} \subseteq \mathcal{R}$ maximizing $\sum_{\ell\in\mathcal{L}} u_\ell(\gamma_\ell(\mathcal{L}, p))$. We assume that the respective assumptions on the utility functions apply.

3   Capacity Maximization with Thresholds and Unlimited Transmission Powers

As a first step towards the final algorithm, we consider the following simplified problem. We are given a set $\mathcal{R}$ of links and a threshold $\beta(\ell) \geq 1$ for each $\ell \in \mathcal{R}$. We have to find a subset $\mathcal{L} \subseteq \mathcal{R}$ of maximum cardinality and a power assignment $p\colon \mathcal{L} \to \mathbb{R}_{\geq 0}$ such that $\gamma_\ell \geq \beta(\ell)$ for all $\ell \in \mathcal{L}$. In [16] an algorithm was presented that solves the special case where all $\beta(\ell)$ are identical. To guarantee feasibility, it inherently requires that all thresholds are equal. In consequence, to cope with individual thresholds some fundamental changes in the algorithm are necessary, that will be presented in this section.

Our algorithm iterates over the links ordered by their *sensitivity* $\beta(\ell)Nd(\ell)^\alpha$, which is the minimum power necessary to overcome ambient noise and to have $\gamma_\ell \geq \beta(\ell)$ in the absence of interference. Like the algorithm in [16] it selects a link if the currently considered link is far enough apart from all previously selected links. In this condition, we now have to take the individual thresholds into account and introduce further checks. Afterwards, like in [16], we determine the power assignment by iterating over links in reversed order.

More formally, let π be the ordering of the links by decreasing values of $\beta(\ell)d(\ell)^\alpha$ with ties broken arbitrarily. That is, if $\pi(\ell) < \pi(\ell')$ then $\beta(\ell)d(\ell)^\alpha \geq \beta(\ell')d(\ell')^\alpha$. Based on this ordering, define the following (directed) weight between two links ℓ and ℓ'. If $\pi(\ell) > \pi(\ell')$, we set

$$w(\ell, \ell') = \min\left\{1, \beta(\ell)\beta(\ell')\frac{d(s,r)^\alpha d(s',r')^\alpha}{d(s,r')^\alpha d(s',r)^\alpha} + \beta(\ell)\frac{d(s,r)^\alpha}{d(s,r')^\alpha} + \beta(\ell)\frac{d(s,r)^\alpha}{d(s',r)^\alpha}\right\} ,$$

otherwise we set $w(\ell, \ell') = 0$. For notational convenience we furthermore set $w(\ell, \ell) = 0$ for all $\ell \in \mathcal{R}$.

Our algorithm now works as follows: It iterates over all links in order of decreasing π values, i.e., going from small to large values of $\beta(\ell)d(\ell)^\alpha$. It adds ℓ' to the set $\mathcal{L}$ if $\sum_{\ell \in \mathcal{L}} w(\ell, \ell') \leq \tau$ for $\tau = 1/6\cdot3^\alpha+2$.

Afterwards, powers are assigned iterating over all links in order of increasing π values, i.e., going from large to small values of $\beta(\ell)d(\ell)^\alpha$. The power assigned to link ℓ' is set to

$$p(\ell') = 2\beta(\ell')Nd(s',r')^\alpha + 2\beta(\ell') \sum_{\substack{\ell=(s,r)\in\mathcal{L}\\ \pi(\ell)<\pi(\ell')}} \frac{p(\ell)}{d(s,r')^\alpha}d(s',r')^\alpha .$$

If there were only links of smaller π values, this power would yield an SINR of exactly $2\beta(\ell)$. When showing feasibility, we prove that due to the greedy selection condition, also taking the other links into account the SINR of link ℓ is at least $\beta(\ell)$. Due to space limitation, this proof can only be found in the full version of this paper.

3.1 Approximation Factor

In this section, we show that the algorithm achieves a constant approximation factor. In contrast to the analysis in [16] our analysis does not rely on any assumptions on the parameter α. We only require that the nodes are located in a metric space.

The central result our analysis builds on is a characterization of *admissible sets*. A set of links $\mathcal{L}$ is called admissible if there is some power assignment p such that $\gamma_\ell(\mathcal{L}, p) \geq \beta(\ell)$ for all $\ell \in \mathcal{L}$.

Lemma 1. *For each admissible set $\mathcal{L}$ there is a subset $\mathcal{L}' \subseteq \mathcal{L}$ with $|\mathcal{L}'| = \Omega(|\mathcal{L}|)$ and $\sum_{\ell' \in \mathcal{L}'} w(\ell, \ell') = O(1)$ for all $\ell \in \mathcal{R}$.*

That is, for each admissible set $\mathcal{L}$ there is a subset having the following property. If we take some further link ℓ, that does not necessarily belong to $\mathcal{L}$, the outgoing weight of this link to all links in the subset is bounded by a constant. Before coming to the proof of this lemma, let us first show how the approximation factor can be derived.

Theorem 1. *The algorithm is a constant-factor approximation.*

Proof. Let ALG be the set of links selected by our algorithm. Let furthermore be OPT the set of links in the optimal solution and $\mathcal{L}' \subseteq$ OPT be the subset described in Lemma 1. That is, we have $\sum_{\ell' \in \mathcal{L}'} w(\ell, \ell') \leq c$ for all $\ell \in \mathcal{R}$ for some suitable constant c. It now suffices to prove that $|\text{ALG}| \geq \frac{\tau}{c} \cdot |\mathcal{L}' \setminus \text{ALG}|$.

All $\ell' \in \mathcal{L}' \setminus \text{ALG}$ were not selected by the algorithm since the greedy condition was violated. That is, we have $\sum_{\ell \in \text{ALG}} w(\ell, \ell') > \tau$. Taking the sum over all $\ell' \in \mathcal{L}' \setminus \text{ALG}$, we get $\sum_{\ell' \in \mathcal{L}' \setminus \text{ALG}} \sum_{\ell \in \text{ALG}} w(\ell, \ell') > \tau \cdot |\mathcal{L}' \setminus \text{ALG}|$.

Furthermore, by our definition of $\mathcal{L}'$, we have $\sum_{\ell \in \text{ALG}} \sum_{\ell' \in \mathcal{L}'} w(\ell, \ell') \leq c \cdot |\text{ALG}|$. In combination this yields

$$|\text{ALG}| \geq \frac{1}{c} \sum_{\ell \in \text{ALG}} \sum_{\ell' \in \mathcal{L}'} w(\ell, \ell') \geq \frac{1}{c} \sum_{\ell' \in \mathcal{L}' \setminus \text{ALG}} \sum_{\ell \in \text{ALG}} w(\ell, \ell') > \frac{\tau}{c} \cdot |\mathcal{L}' \setminus \text{ALG}| \ .$$

3.2   Proof of Lemma 1 (Outline)

Due to space limitations the formal proof of Lemma 1 can only be found in the full version. Here, we only present the major steps, highlighting the general ideas.

In the first step, we use the fact that we can scale the thresholds by constant factors at the cost of decreasing the size of $\mathcal{L}$ by a constant factor. Considering such an appropriately scaled set $\mathcal{L}' \subseteq \mathcal{L}$, we use the SINR constraints and the triangle inequality to show

$$\frac{1}{|\mathcal{L}'|} \sum_{\ell = (s,r) \in \mathcal{L}'} \sum_{\substack{\ell' = (s',r') \in \mathcal{L}' \\ \pi(\ell') > \pi(\ell)}} \frac{\beta(\ell')d(s',r')^\alpha}{d(s',r)^\alpha} = O(1) \ .$$

In the next step, we use the following property of admissible set. After reversing the links (i.e. swapping senders and receivers) a subset of a constant fraction of the links is also admissible. Combining this insight with the above bound and Markov inequality, we can show that for each admissible set $\mathcal{L}$ there is a subset $\mathcal{L}' \subseteq \mathcal{L}$ with $|\mathcal{L}'| = \Omega(|\mathcal{L}|)$ and

$$\sum_{\ell' = (s',r') \in \mathcal{L}'} \frac{\min\{\beta(\ell)d(s,r)^\alpha, \beta(\ell')d(s',r')^\alpha\}}{\min\{d(s',r)^\alpha, d(s,r')^\alpha\}} = O(1) \qquad \text{for all } \ell = (s,r) \in \mathcal{L}'.$$

This subset has the property that not only for $\ell \in \mathcal{L}'$ but for all $\ell = (s, r) \in \mathcal{R}$

$$\sum_{\substack{\ell' = (s', r') \in \mathcal{L}' \\ \pi(\ell') < \pi(\ell)}} \min\left\{1, \frac{\beta(\ell) d(s, r)^\alpha}{d(s', r)^\alpha}\right\} + \min\left\{1, \frac{\beta(\ell) d(s, r)^\alpha}{d(s, r')^\alpha}\right\} = O(1) \ .$$

In the last step, we show that there is also a subset $\mathcal{L}'$ with $|\mathcal{L}'| = \Omega(|\mathcal{L}|)$ and

$$\sum_{\substack{\ell' = (s', r') \in \mathcal{L}' \\ \pi(\ell') < \pi(\ell)}} \min\left\{1, \beta(\ell)\beta(\ell') \frac{d(s, r)^\alpha d(s', r')^\alpha}{d(s', r)^\alpha d(s, r')^\alpha}\right\} = O(1) \quad \text{for all } \ell = (s, r) \in \mathcal{R} \ .$$

This is shown by decomposing the set $\mathcal{L}$. For one part we can use the result from above. For the remaining links we can show that among these links there has to be an exponential growth of the sensitivity and the distance to ℓ. In combination, the quotients added up in the sum decay exponentially, allowing us to bound the sum by a geometric series.

4   Capacity Maximization with Thresholds and Fixed Transmission Powers

In this section, we consider the case that a power assignment $p \colon \mathcal{R} \to \mathbb{R}_{\geq 0}$ is given. We assume that this power assignment is (sub-) linear and monotone in the link sensitivity. In our case of individual thresholds this means that if for two links $\ell, \ell' \in \mathcal{R}$ we have $\beta(\ell) d(s, r)^\alpha \leq \beta(\ell') d(s', r')^\alpha$, then $p(\ell) \leq p(\ell')$ and $\frac{1}{\beta(\ell)} \frac{p(\ell)}{d(s, r)^\alpha} \geq \frac{1}{\beta(\ell')} \frac{p(\ell')}{d(s', r')^\alpha}$. This condition is fulfilled in particular if all transmission powers are the same or if they are chosen by a (sub-) linear, monotone function depending on the sensitivity, e.g., linear power assignments $(p(\ell) \sim \beta(\ell) \cdot d(\ell)^\alpha)$ and square-root power assignments $(p(\ell) \sim \sqrt{\beta(\ell) \cdot d(\ell)^\alpha})$.

For the case of identical thresholds, Halldórsson and Mitra [13] present a constant-factor approximation that can be naturally extended as follows. Given two links $\ell = (s, r)$ and $\ell' = (s', r')$, and a power assignment p, we define the *affectance* of ℓ on ℓ' by $a_p(\ell, \ell') = \min\left\{1, \beta(\ell') \frac{p(\ell)}{d(s, r')^\alpha} \Big/ \left(\frac{p(\ell')}{d(s', r')^\alpha} - \beta(\ell')N\right)\right\}$. Algorithm 1 iterates over all $\ell' \in \mathcal{R}$ and adds a link to the tentative solution if the incoming and outgoing affectance to the previously selected links is at most $1/2$. At the end, only the feasible links are returned.

Theorem 2. *Algorithm 1 yields a constant-factor approximation.*

Algorithm 1: Capacity Maximization with Thresholds and Fixed Powers

---

initialize $\mathcal{L} = \emptyset$;
for $\ell' \in \mathcal{R}$ *in decreasing order of* π *values* **do**
  if $\sum_{\ell \in \mathcal{L}} a_p(\ell, \ell') + a_p(\ell', \ell) \leq \frac{1}{2}$ **then**
    add ℓ' to $\mathcal{L}$;

return $\{\ell' \in \mathcal{L} \mid \sum_{\ell \in \mathcal{L}} a_p(\ell, \ell') < 1\}$;

The analysis in [13] builds on a very involved argument, using a so-called *Red-Blue Lemma*. Essentially the idea is to match links in the computed solution and the optimal one. If the algorithm's solution was much smaller than the optimal one, one link would be left unmatched and thus taken by the algorithm. Our proof in contrast is much simpler and uses the same structure as the one for Theorem 1. Again, it is most important to characterize optimal solutions.

Lemma 2. *For each feasible set $\mathcal{L}$ there is a subset $\mathcal{L}' \subseteq \mathcal{L}$ with $|\mathcal{L}'| = \Omega(|\mathcal{L}|)$ and $\sum_{\ell' \in \mathcal{L}', \pi(\ell') < \pi(\ell)} a_p(\ell, \ell') + a_p(\ell', \ell) = O(1)$ for all $\ell \in \mathcal{R}$.*

Using this lemma, we can adopt the proof that the set $\mathcal{L}$ is at most a constant-factor smaller than the optimal solution literally from the one of Theorem 1. Using Markov inequality, one can see that the final solution has size at least $|\mathcal{L}|/2$. The formal proofs can be found in the full version.

5 Capacity Maximization with Thresholds and Limited Transmission Powers

Having found algorithms for the threshold problem that chooses powers from an unbounded set and for the one with fixed transmission powers, we are now ready to combine these two approaches to an algorithm that chooses transmission powers from a bounded set $[0, p_{\max}]$. The general idea has already been presented by Wan et al. [21].

We decompose the set $\mathcal{R}$ into two sets $\mathcal{R}_1 = \{\ell \in \mathcal{R} \mid \beta(\ell) N d(\ell)^\alpha \leq p_{\max}/4\}$; $\mathcal{R}_2 = \mathcal{R} \setminus \mathcal{R}_1$. On the set $\mathcal{R}_1$, we run a slightly modified algorithm for unlimited transmission powers. Due to the definition of the set $\mathcal{R}_1$ and the algorithm, it is guaranteed that all assigned transmission powers are at most $p_{\max}$. On the set $\mathcal{R}_2$, we run the fixed-power algorithm setting $p(\ell) = p_{\max}$ for all $\ell \in \mathcal{R}_2$. In the end, we return the better one of the two solutions.

This algorithm yields a constant-factor approximation as well. For the formal description and analysis, see the full version.

6 Capacity Maximization with General Utilities

In the previous sections we presented constant-factor approximations for the different variants of the capacity maximization problem with individual thresholds. In order to solve the flexible-rate problem with general utility function, we use the respective threshold algorithm as a building block. Inspired by an approach by Halldórsson to solve maximum weighted independent set in graphs [10], we perform the following steps.

For each link $\ell \in \mathcal{R}$, we determine the maximum utility that it can achieve, referred to as $u_\ell^{\max}$. This value is achieved if only this link is selected and (in case we can select powers) transmits at maximum power. Let the maximum of all maximum link utilities be called B. It is a lower bound on the value of the optimal solution. The optimal solution is in turn upper bounded by $n \cdot B$. For all $i \in \{0, 1, 2, \ldots, \lceil \log n \rceil\}$, we run the following procedure. For each link $\ell \in \mathcal{R}$, we query each utility function for the minimum SINR necessary to have utility at least $2^{-i} \cdot B$. This value is taken as the individual threshold of the respective link. On these thresholds, we run the respective algorithm from the previous sections. It returns a set of links and possibly (depending on the problem variant) also a power assignment. As the output, we take the best one among these $\lceil \log n \rceil + 1$ solutions.

Algorithm 2: Capacity Maximization with Flexible Rates

---

determine $B := \max_{\ell \in \mathcal{R}} u_\ell^{\max}$;
for $i \in \{0, 1, 2, \ldots, \lceil \log n \rceil\}$ **do**
  set $\beta(\ell) = \min\{\gamma_\ell \mid u_\ell(\gamma_\ell) \geq 2^{-i} \cdot B\}$ for all $\ell \in \mathcal{R}$;
  run respective algorithm with these thresholds, let S_i be the solution;
return the best one of the solutions S_i;

Theorem 3. *Algorithm 2 computes an $O(\log n)$-approximation.*

Proof. Consider the set $\mathcal{L}$ of links selected in the optimal solution and let γ_ℓ be the SINR of link $\ell \in \mathcal{L}$ in this solution. That is, the optimal solution has value $u(\text{OPT}) = \sum_{\ell \in \mathcal{L}} u_\ell(\gamma_\ell)$. By definition, we have $u(\text{OPT}) \geq B$.

Furthermore, for $i \in \{0, 1, 2, \ldots, \lceil \log n \rceil\}$, let $\mathcal{L}_i = \{\ell \in \mathcal{L} \mid u_\ell(\gamma_\ell) \geq 2^{-i} \cdot B\}$, i.e., the set of links having utility at least $2^{-i} \cdot B$ in the optimal solution. Some links might not be contained in any of these sets. Let therefore be $X = \mathcal{L} \setminus \mathcal{L}_{\lceil \log n \rceil + 1}$. We have that $\sum_{\ell \in X} u_\ell(\gamma_\ell) \leq n \cdot 2^{-\lceil \log n \rceil - 1} \cdot B \leq u(\text{OPT})/2$. So these links hardly contribute to the value of the solution and we leave them out of consideration.

Considering only the sets $\mathcal{L}_i$ defined above, we get $\sum_i 2^{-i} \cdot B \cdot |\mathcal{L}_i| \geq u(\text{OPT})/4$. Furthermore, solution S_i approximates the set $\mathcal{L}_i$ by a constant factor in size. Expressing this in terms of the utility we have $u(S_i) = \Omega(2^{-i} \cdot B \cdot |\mathcal{L}_i|)$.

Taking the sum over all i, we get $\sum_i u(S_i) = \Omega(\sum_i 2^{-i} \cdot B \cdot |\mathcal{L}_i|) = \Omega(u(\text{OPT}))$ and therefore $\max_i u(S_i) = \Omega(u(\text{OPT})/ \log n)$. ☐

7   Latency Minimization

Apart from capacity maximization, the other common problem considered in optimization with respect to SINR models is latency minimization. In the standard threshold case this means finding a schedule of shortest possible length in which transmission is carried out successfully at least once. For the case of fixed transmission powers, there are very simple distributed approximation algorithms [17,12]. With results shown above, not only these results but also the ones repeatedly applying a capacity-maximization algorithm can be easily transferred to the setting of individual thresholds.

Studying flexible data rates, this is different. Let us consider the following generalization. As before, we assume that each link has a utility function u_ℓ expressing the amount of data that can be transmitted in a time slot depending on the SINR. Furthermore there is a demand δ_ℓ representing the amount of data that has to be transmitted via this link. The task is to find a schedule fulfilling these demands, i.e. a sequence of subsets and possibly a power assignment for each step, such that $\sum_t u_\ell(\gamma_\ell^{(t)}) \geq \delta_\ell$. Here, $\gamma_\ell^{(t)}$ denotes the SINR of link ℓ in step t.

In order to solve this problem, we consider Algorithm 3. It repeatedly applies the flexible-rate algorithm on the remaining demands and uses each set for a time slot. The crucial part is which utility functions are considered. The idea is to modify the utility functions by scaling and rounding such that in each round reasonable progress is guaranteed. For our algorithm, we use two differently modified utility functions and we take the shorter one of the two schedules.

Algorithm 3: Latency Minimization

---

for $\ell \in \mathcal{R}$ set $u_{1,\ell}(\gamma_\ell) = \frac{1}{2n} \left\lfloor \frac{2n \cdot u_\ell(\gamma_\ell)}{\delta_\ell} \right\rfloor$, $\delta_{1,\ell} = 1$, $u_{2,\ell}(\gamma_\ell) = \frac{u_\ell(\gamma_\ell)}{u_\ell^{\max}}$, $\delta_{2,\ell} = \frac{\delta_\ell}{u_\ell^{\max}}$;

    for $i \in \{1,2\}$ do

        while $\sum_{\ell \in \mathcal{R}} \delta_{i,\ell} > 0$ do

            run Algorithm 2 w.r.t. $u'_{i,\ell}(\gamma_\ell) = \min\{\delta_{i,\ell}, u_{i,\ell}(\gamma_\ell)\}$;

            update $\delta_{i,\ell} := \delta_{i,\ell} - u'_{i,\ell}(\gamma_\ell)$ for all ℓ, where γ_ℓ is the achieved SINR;

    return the shorter schedule

Theorem 4. *Algorithm 3 computes an $O(\log^2 n)$-approximation.*

8   Conclusion

In this paper, we presented the flexible-rate capacity maximization problem and a first approximation algorithm. Maybe surprisingly, understanding the threshold variants brings about a reasonable approximation factor for this case as well.

The most striking limitation is that these approaches are not able to deal with SINR values smaller than 1 respectively a constant. Indeed, by the use of forward error-correcting codes, this case can actually occur in practice. However, we could find evidence that one will have to come up with new techniques to solve this problem.

The other major open issue is to improve the approximation factor. As a first step, one could ask whether there is a constant-factor approximation for *weighted* capacity maximization with thresholds. That is, each link has a threshold (or there is a global one) and a weight. The task is to select a feasible set of maximum weight. For linear power assignments, a constant-factor approximation is known [14] but all other approaches are only able to give $O(\log n)$ as well.

Another interesting topic for future research could include different objective functions. So far, we only maximized the sum of all link utilities. While it is trivial to maximize the minimal utility in a centralized way, it still remains an open problem to define further fairness aspects and solve the problem accordingly.

References

1. Andrews, M., Dinitz, M.: Maximizing capacity in arbitrary wireless networks in the SINR model: Complexity and game theory. In: Proc. 28th INFOCOM, pp. 1332–1340 (2009)

2. Chiang, M., Tan, C.W., Palomar, D., O'Neill, D., Julian, D.: Power control by geometric programming. IEEE Transactions on Wireless Communication 6(7), 2640–2651 (2007)

3. Dams, J., Hoefer, M., Kesselheim, T.: Convergence Time of Power-Control Dynamics. In: Aceto, L., Henzinger, M., Sgall, J. (eds.) ICALP 2011, Part II. LNCS, vol. 6756, pp. 637–649. Springer, Heidelberg (2011)

4. Dams, J., Hoefer, M., Kesselheim, T.: Scheduling in wireless networks with Rayleigh-fading interference. In: Proc. 24th SPAA, pp. 327–335 (2012)

5. Fanghänel, A., Geulen, S., Hoefer, M., Vöcking, B.: Online capacity maximization in wireless networks. In: Proc. 22nd SPAA, pp. 92–99 (2010)

6. Fanghänel, A., Kesselheim, T., Räcke, H., Vöcking, B.: Oblivious interference scheduling. In: Proc. 28th PODC, pp. 220–229 (2009)

7. Fanghänel, A., Kesselheim, T., Vöcking, B.: Improved algorithms for latency minimization in wireless networks. Theoretical Computer Science 412(24), 2657–2667 (2011)

8. Foschini, G.J., Miljanic, Z.: A simple distributed autonomous power control algorithm and its convergence. IEEE Transactions on Vehicular Technology 42(4), 641–646 (1992)

9. Goussevskaia, O., Wattenhofer, R., Halldórsson, M.M., Welzl, E.: Capacity of arbitrary wireless networks. In: Proc. 28th INFOCOM, pp. 1872–1880 (2009)

10. Halldórsson, M.M.: Approximations of weighted independent set and hereditary subset problems. Journal of Graph Algorithms and Applications 4, 1–16 (2000)

11. Halldórsson, M.M.: Wireless Scheduling with Power Control. In: Fiat, A., Sanders, P. (eds.) ESA 2009. LNCS, vol. 5757, pp. 361–372. Springer, Heidelberg (2009)

12. Halldórsson, M.M., Mitra, P.: Nearly Optimal Bounds for Distributed Wireless Scheduling in the SINR Model. In: Aceto, L., Henzinger, M., Sgall, J. (eds.) ICALP 2011, Part II. LNCS, vol. 6756, pp. 625–636. Springer, Heidelberg (2011)

13. Halldórsson, M.M., Mitra, P.: Wireless capacity with oblivious power in general metrics. In: Proc. 22nd SODA, pp. 1538–1548 (2011)

14. Halldórsson, M.M., Mitra, P.: Wireless capacity and admission control in cognitive radio. In: Proc. 31st INFOCOM, pp. 855–863 (2012)

15. Huang, J., Berry, R., Honig, M.: Distributed interference compensation for wireless networks. IEEE Journal on Sel. Areas in Comm. 24(5), 1074–1084 (2006)

16. Kesselheim, T.: A constant-factor approximation for wireless capacity maximization with power control in the SINR model. In: Proc. 22nd SODA, pp. 1549–1559 (2011)

17. Kesselheim, T., Vöcking, B.: Distributed Contention Resolution in Wireless Networks. In: Lynch, N.A., Shvartsman, A.A. (eds.) DISC 2010. LNCS, vol. 6343, pp. 163–178. Springer, Heidelberg (2010)

18. Lotker, Z., Parter, M., Peleg, D., Pignolet, Y.A.: Distributed power control in the SINR model. In: Proc. 30th INFOCOM, pp. 2525–2533 (2011)

19. Santi, P., Maheshwari, R., Resta, G., Das, S., Blough, D.M.: Wireless link scheduling under a graded SINR interference model. In: Proc. 2nd FOWANC, pp. 3–12 (2009)

20. Singh, V., Kumar, K.: Literature survey on power control algorithms for mobile ad-hoc network. Wireless Personal Communications, 1–7 (2010)

21. Wan, P.-J., Ma, C., Tang, S., Xu, B.: Maximizing Capacity with Power Control under Physical Interference Model in Simplex Mode. In: Cheng, Y., Eun, D.Y., Qin, Z., Song, M., Xing, K. (eds.) WASA 2011. LNCS, vol. 6843, pp. 84–95. Springer, Heidelberg (2011)

Extending Partial Representations of Function Graphs and Permutation Graphs[*]

Pavel Klavík[1], Jan Kratochvíl[1], Tomasz Krawczyk[2], and Bartosz Walczak[2]

[1] Department of Applied Mathematics,
Faculty of Mathematics and Physics, Charles University
{klavik,honza}@kam.mff.cuni.cz
[2] Theoretical Computer Science Department,
Faculty of Mathematics and Computer Science, Jagiellonian University
{krawczyk,walczak}@tcs.uj.edu.pl

Abstract. Function graphs are graphs representable by intersections of continuous real-valued functions on the interval $[0, 1]$ and are known to be exactly the complements of comparability graphs. As such they are recognizable in polynomial time. Function graphs generalize permutation graphs, which arise when all functions considered are linear.

We focus on the problem of extending partial representations, which generalizes the recognition problem. We observe that for permutation graphs an easy extension of Golumbic's comparability graph recognition algorithm can be exploited. This approach fails for function graphs. Nevertheless, we present a polynomial-time algorithm for extending a partial representation of a graph by functions defined on the entire interval $[0, 1]$ provided for some of the vertices. On the other hand, we show that if a partial representation consists of functions defined on subintervals of $[0, 1]$, then the problem of extending this representation to functions on the entire interval $[0, 1]$ becomes NP-complete.

1 Introduction

Geometric representations of graphs have been studied as part of graph theory from its very beginning. Euler initiated the study of graph theory by studying planar graphs in the setting of three-dimensional polytopes. The theorem of Kuratowski [16] provides the first combinatorial characterization of planar graphs and can be considered as the start of modern graph theory.

In this paper we are interested in intersection representations, which assign geometric objects to the vertices of graphs and the edges are encoded by intersections of objects. Formally, an intersection representation of G is a mapping $\phi : V(G) \to \mathcal{S}$ of the vertices of G to a class $\mathcal{S}$ of objects (sets) such that $\phi(u) \cap \phi(v) \neq \emptyset$ if and only if $uv \in E(G)$. This way, for different classes $\mathcal{S}$ we obtain various classes of representable graphs. Classic examples include interval

---

[*] Supported by ESF EuroGIGA project GraDR, the first two authors by Czech Science Foundation as grant No. GIG/11/E023, the last two authors by Ministry of Science and Higher Education of Poland as grant No. 884/N-ESF-EuroGIGA/10/2011/0.

L. Epstein and P. Ferragina (Eds.): ESA 2012, LNCS 7501, pp. 671–682, 2012.
© Springer-Verlag Berlin Heidelberg 2012

graphs, circle graphs, permutation graphs, string graphs, convex graphs, and function graphs [9,21]. As seen from these two monographs, geometric intersection graphs are intensively studied for their applied motivation, algorithmic properties, but also as a source of many interesting theoretical results that sometimes stimulate our curiosity (as one example we mention that string graphs requiring exponential number of crossing points in every representation are known, and yet the recognition problem is in NP).

Naturally, the recognition problem is the first one to consider. For most of the intersection defined classes the complexity of their recognition is known. For example, interval graphs can be recognized in linear time [3], while recognition of string graphs is NP-complete [14,20]. Our goal is to study the easily recognizable classes and explore if the recognition problem becomes harder when extra conditions are given with the input.

Partial Representations. A recent paper of Klavík et al. [13] introduced a question of extending partial representations for classes of intersection graphs. A *partial representation* of G is a representation $\phi : R \to S$ of the induced subgraph $G[R]$ for a set $R \subseteq V(G)$. The problem $\text{REPEXT}(\mathcal{G})$ of partial representation extension for a class $\mathcal{G}$ represented in a class S is defined as follows: given a graph $G \in \mathcal{G}$ and a partial representation $\phi : R \to S$ of G, decide whether there is a representation $\psi : V(G) \to S$ that *extends* ϕ, that is, such that $\psi|_R = \phi$.

The paper [13] investigates the complexity of the problem for intervals graphs (intersection graphs of intervals on a line) and presents an $O(n^2)$ algorithm for extending interval representations and an $O(nm)$ algorithm for extending proper interval representations of graphs with n vertices and m edges. A recent result of Bläsius and Rutter [4] solves the problem of extending interval representations in time $O(n + m)$, but the algorithm is involved.

A related problem of simultaneous graph representations was recently introduced by Jampani and Lubiw [11,12]: given two graphs G and H sharing common vertices $I = V(G) \cap V(H)$, decide whether there are representations ϕ of G and ψ of H such that $\phi|_I = \psi|_I$. In many cases we can use this problem to solve partial representation extension by introducing an additional graph and putting $I = R$. On the other hand, if $|I|$ is small, then we can test all essentially different possible representations of I and try to extend them to $V(G)$ and $V(H)$, which can give us a polynomial-time algorithm for fixed parameter $|I|$.

Several other problems have been considered in which a partial solution is given and the task is to extend it. For example, every k-regular bipartite graph is k-edge-colorable, but if some edges are pre-colored, the extension problem becomes NP-complete even for $k = 3$ [6], and even when the input is restricted to planar graphs [17]. For planar graphs, partial representation extension is solvable in linear time [1]. Every planar graph admits a straight-line drawing, but extending such representation is NP-complete [19].

Permutation and Function Graphs. In this paper, we consider two classes of intersection graphs: the class FUN of *function graphs*—intersection graphs of continuous functions $[0, 1] \to \mathbb{R}$, and the class PERM of *permutation graphs*, which is a subclass of FUN, represented the same way by linear functions.

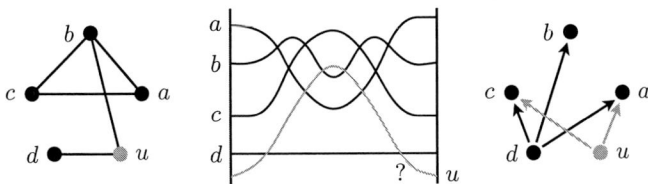

Fig. 1. A function graph G with a partial representation that is not extendable: $\phi(u)$ in order to intersect $\phi(b)$ and $\phi(d)$ must also intersect $\phi(a)$ or $\phi(c)$. The corresponding partial orientation of the comparability graph $\overline{G}$ is extendable.

A graph is a *comparability graph* if it is possible to orient its edges transitively. An orientation is *transitive* if $u \to v$ and $v \to w$ imply $u \to w$. Thus the relation $\to$ in a transitively oriented graph is a strict partial order. We denote the class of comparability graphs by CO. A *partial orientation* of a comparability graph is a transitive orientation of some of its edges. The problem ORIENTEXT is to decide whether we can orient the remaining edges to get a transitive orientation of the entire graph.

By coCO we denote the class of complements of comparability graphs. We have the following relations: FUN = coCO [10] and PERM = CO∩coCO [5]. We derive a transitive ordering from a function graphs as follows: if two functions do not intersect, then one is on top of the other; thus we can order the functions from bottom to top. For permutation graphs, we use the fact that PERM = coPERM.

Our Results. By a straightforward modification of the recognition algorithms of Golumbic [8,9] and by the property PERM = CO∩coCO, we get the following.

Proposition 1. *The problem* ORIENTEXT *can be solved in time* $O((n+m)\Delta)$ *for graphs with n vertices, m edges, and maximum degree Δ.*

Proposition 2. *The problem* REPEXT(PERM) *can be solved in time* $O(n^3)$ *for graphs with n vertices.*

Our first main result is a polynomial-time algorithm for REPEXT(FUN). Here the straightforward generalization of the recognition algorithm does not work. Even though FUN = coCO, the problems REPEXT(FUN) and ORIENTEXT are different, see Fig. 1. This is similar to what happens for the classes of proper and unit interval graphs: they are known to be equal, but their partial representation extension problems are different [13].

Theorem 1. *The problem* REPEXT(FUN) *can be solved in polynomial time.*

The second main result concerns partial representations by *partial functions* $f : [a, b] \to \mathbb{R}$ with $[a, b] \subseteq [0, 1]$, which generalize ordinary partial representations by functions. The problem REPEXT*(FUN) is to decide, for a given graph, whether a given partial representation by partial functions can be extended to a representation of the whole graph so that all partial functions are extended to functions defined on the entire $[0, 1]$.

Theorem 2. *The problem* REPEXT*(FUN) *is NP-complete.*

2 Extending Partial Orientation

We show how to modify the recognition algorithm of Golumbic [8,9] to obtain an algorithm for extending partial orientation of a comparability graph. The recognition algorithm repeats the following until the whole graph G is oriented. We pick an arbitrary unoriented edge and orient it in any direction. This may force several other edges to be oriented, according to the following rules:

- $u \to v$ and $v \to w$ force $u \to w$,
- $u \to v$, $vw \in E(G)$, and $uw \notin E(G)$ force $w \to v$.

If we find an edge that we are forced to reorient (change its direction), then we stop and answer that G is not a comparability graph. Otherwise, we finally orient all edges and obtain a transitive orientation of G. The running time of the algorithm is $O((n+m)\Delta)$.

Now, we adapt this algorithm to the problem ORIENTEXT. Since the algorithm processes edges in an arbitrary order, we can choose an ordering $e_1 < \ldots < e_m$ of the edges and always pick the first non-oriented edge in this ordering. Suppose that the first k edges $e_1, \ldots, e_k$ are preoriented by ϕ. If we pick an edge e_i, then we orient it according to ϕ if $i \leqslant k$ or arbitrarily otherwise. The algorithm additionally fails if it is forced to orient an edge e_i with $i \leqslant k$ in the opposite direction to the one forced by ϕ. In such a case, this orientation is forced by the orientation of $e_1, \ldots, e_{i-1}$, and thus the partial orientation is indeed not extendible. The running time of the algorithm is again $O((n+m)\Delta)$, which proves Proposition 1. The idea behind the proof of Proposition 2 is similar.

3 Extending Partial Representation of a Poset

Before we deal with function representations of graphs, we study representations of posets by continuous functions $[0,1] \to \mathbb{R}$. By a poset we mean a transitively oriented graph. We write $u <_P v$ to denote that there is an edge from u to v in a poset P. Since we are interested in algorithmic problems, we have to choose some discrete description of the functions, and the particular choice does not matter as long as we can convert from and to other descriptions in polynomial time. Here we restrict our attention to piecewise linear continuous functions. Specifically, each function $f : [0,1] \to \mathbb{R}$ that we consider is described by a tuple $(x_0, f(x_0)), \ldots, (x_k, f(x_k))$ of points in $[0,1] \times \mathbb{R}$ with $0 = x_0 < \ldots < x_k = 1$ so that f is linear on every interval $[x_i, x_{i+1}]$. We denote the family of such functions by $\mathcal{F}$. Note that every representation by continuous functions $[0,1] \to \mathbb{R}$ can be approximated by an equivalent representation by functions from $\mathcal{F}$. We define a natural order $<$ on $\mathcal{F}$ by setting $f < g$ whenever $f(x) < g(x)$ holds for every $x \in [0,1]$. A *representation* of a poset P is a mapping $\phi : V(P) \to \mathcal{F}$ such that

$$\forall u, v \in V(P) \colon (u <_P v \iff \phi(u) < \phi(v)).$$

It is worth to note that every poset has a representation of this kind, see [10]. A *partial representation* of a poset P is a mapping $\phi : R \to \mathcal{F}$ which is a representation of the subposet $P[R]$ induced on a set $R \subseteq V(P)$.

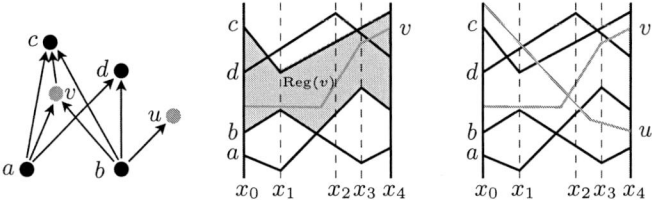

Fig. 2. A poset P and its partial representaion $\phi : \{a, b, c, d\} \to \mathcal{F}$. The diagram in the middle shows $\mathrm{Reg}(v)$ and a feasible $\psi(v)$ for a representaion ψ of P extending ϕ. The diagram to the right shows a representaion ψ of P extending ϕ.

In this section we provide a polynomial-time algorithm solving the following problem: given a poset P and its partial representation $\phi : R \to \mathcal{F}$, decide whether ϕ is extendable to a representation of P. Thus for the remainder of this section we assume that P is a poset, $R \subseteq V(P)$, and $\phi : R \to \mathcal{F}$ is a partial representation of P.

For a function $f \in \mathcal{F}$ we define sets

$$f{\uparrow} = \{(x, y) \in [0, 1] \times \mathbb{R} : y > f(x)\},$$
$$f{\downarrow} = \{(x, y) \in [0, 1] \times \mathbb{R} : y < f(x)\}.$$

For every vertex u of P we define a set $\mathrm{Reg}(u) \subseteq [0, 1] \times \mathbb{R}$, called the *region* of u, as follows. If $u \in R$, then $\mathrm{Reg}(u) = \phi(u)$. Otherwise,

$$\mathrm{Reg}(u) = \bigcap \{\phi(a){\downarrow} : a \in R \text{ and } a >_P u\} \cap \bigcap \{\phi(a){\uparrow} : a \in R \text{ and } a <_P u\}.$$

It follows that the function representing u in any representation of P extending ϕ must be contained entirely within $\mathrm{Reg}(u)$. See Fig. 2 for an illustration.

Lemma 1. *There is a representation of P extending ϕ if and only if any two incomparable vertices u and v of P satisfy $\mathrm{Reg}(u) \cap \mathrm{Reg}(v) \neq \emptyset$.*

Lemma 1 directly yields a polynomial-time algorithm for deciding whether P has a representation extending ϕ. Indeed, the lower or upper boundary of $\mathrm{Reg}(u)$ (if exists) is a function from $\mathcal{F}$ whose description can be easily computed from the descriptions of the functions $\phi(a)$ with $a \in R$ and $a <_P u$ or $a >_P u$, respectively. Having the descriptions of the lower and upper boundaries of all regions, we can easily check whether the intersection of any two of them is empty.

4   Modular Decomposition

The main tool that we use for constructing a polynomial-time algorithm for extending partial representations of function graphs is modular decomposition, also known as substitution decomposition. In this section we briefly discuss this concept and its connection to transitive orientations of graphs.

A graph is *empty* if it has no edges. A graph is *complete* if it has all possible edges. A non-empty set $M \subseteq V(G)$ is a *module* of G if every vertex in $V(G) - M$ is adjacent to either all or none of the vertices in M. The singleton sets and the whole $V(G)$ are the *trivial* modules of G. A non-empty graph is *prime* if it has no modules other than the trivial ones. A module M is *strong* if every module N satisfies $N \subseteq M$, $M \subseteq N$, or $M \cap N = \emptyset$. We denote the family of non-singleton strong modules of G by $\mathcal{M}(G)$. A strong module $M \subsetneq V(G)$ is *maximal* if there is no strong module N with $M \subsetneq N \subsetneq V(G)$. When G is a graph with at least two vertices, the maximal strong modules of G form a partition of $V(G)$, which we denote by $\mathcal{C}(G)$. It is easy to see that a set $M \subsetneq V(G)$ is a strong module of G if and only if M is a strong module of $G[N]$ for some $N \in \mathcal{C}(G)$. Applying this observation recursively, we see that the strong modules of G form a rooted tree, called the *modular decomposition* of G, in which

- $V(G)$ is the root;
- $\mathcal{C}(G[M])$ are the children of every $M \in \mathcal{M}(G)$;
- the singleton modules are the leaves.

In particular, G has at most $2|V(G)| - 1$ strong modules in total.

Any two distinct strong modules $M, N \subsetneq V(G)$ can be either *adjacent*, which means that any two vertices $u \in M$ and $v \in N$ are adjacent in G, or *non-adjacent*, which means that no two vertices $u \in M$ and $v \in N$ are adjacent in G. When M and N are two adjacent strong modules of G and P is a transitive orientation of G, we write $M <_P N$ to denote that $u <_P v$ for all $u \in M$ and $v \in N$.

Theorem 3 (Gallai [7]). *Let M and N be two adjacent strong modules of G. Every transitive orientation P of G satisfies either $M <_P N$ or $M >_P N$.*

For a module $M \in \mathcal{M}(G)$, we call the adjacency graph of $\mathcal{C}(G[M])$ the *quotient* of M and denote it by $G[M]/\mathcal{C}(G[M])$, and we call a transitive orientation of $G[M]/\mathcal{C}(G[M])$ simply a *transitive orientation* of M.

Theorem 4 (Gallai [7]). *The transitive orientations of G and the tuples of transitive orientations of non-singleton strong modules of G are in a one-to-one correspondence $P \leftrightarrow (P_M)_{M \in \mathcal{M}(G)}$ given by $M_1 <_{P_M} M_2 \iff M_1 <_P M_2$ for any $M \in \mathcal{M}(G)$ and $M_1, M_2 \in \mathcal{C}(M)$. In particular, G is a comparability graph if and only if $G[M]/\mathcal{C}(G[M])$ is a comparability graph for every $M \in \mathcal{M}(G)$.*

Theorem 5 (Gallai [7]). *Let M be a non-singleton strong module of G.*

1. *If $G[M]$ is not connected, then the maximal strong modules of $G[M]$ are the connected components of $G[M]$ and $G[M]/\mathcal{C}(G[M])$ is an empty graph.*
2. *If $\overline{G[M]}$ is not connected, then the maximal strong modules of $G[M]$ are the connected components of $\overline{G[M]}$ and $G[M]/\mathcal{C}(G[M])$ is a complete graph.*
3. *If $G[M]$ and $\overline{G[M]}$ are connected, then $G[M]/\mathcal{C}(G[M])$ is a prime graph.*

Theorem 5 allows us to classify non-singleton strong modules into three types. Namely, a non-singleton strong module M of G is

- *parallel* when $G[M]/\mathcal{C}(G[M])$ is empty;
- *serial* when $G[M]/\mathcal{C}(G[M])$ is complete;
- *prime* when $G[M]/\mathcal{C}(G[M])$ is prime.

Every parallel module has just one transitive orientation—there is nothing to orient in an empty quotient. Every serial module with k children has exactly $k!$ transitive orientations corresponding to the $k!$ permutations of the children. Finally, for prime modules we have the following.

Theorem 6 (Gallai [7]). *Every prime module of a comparability graph has exactly two transitive orientations, one being the reverse of the other.*

Golumbic [8,9] showed that the problems of computing the modular decomposition of a graph, computing the two transitive orientations of a prime comparability graph, and deciding whether a graph is a comparability graph are polynomial-time solvable. Actually, the first two of these problems can be solved in linear time [18].

5   Extending Partial Represenation of a Function Graph

In this section we provide a polynomial-time algorithm for extending partial representation of function graphs. However, for convenience, instead of function graphs we deal with their complements—comparability graphs. A *representation* of a comparability graph G is a representation of a transitive orientation of G, defined as in Section 3. A *partial representation* of G is a representation of an induced subgraph of G.

   Specifically, we prove that the following problem is polynomial-time solvable: given a comparability graph G and its partial representation $\phi : R \to \mathcal{F}$, decide whether ϕ is extendable to a representation of G. Thus for the remainder of this section we assume that G is a comparability graph, $R \subseteq V(G)$, and $\phi : R \to \mathcal{F}$ is a partial representation of G.

   A transitive orientation P of G *respects* ϕ if ϕ is a partial representation of P. The idea of the algorithm is to look for a transitive orientation P of G that respects ϕ and satisfies $\operatorname{Reg}_P(u) \cap \operatorname{Reg}_P(v) \neq \emptyset$ for any two adjacent vertices u and v of G, where by $\operatorname{Reg}_P(u)$ we denote the region of u with respect to P. By Lemma 1, such a transitive orientation exists if and only if ϕ is extendable. We make use of the modular decomposition of G and Theorem 4 to identify all transitive orientations of G. We apply to G a series of reductions, which ensure that every non-singleton strong module of G has exactly one or two transitive orientations respecting ϕ, while not changing the answer. Finally, after doing all these reductions, we express the existence of a requested transitive orientation of G by an instance of 2-SAT.

   A strong module M of G is *represented* if $M \cap R \neq \emptyset$. Any vertex from $M \cap R$ is a *representant* of M. Clearly, if M is represented, then all ancestors of M in the modular decomposition of G are represented as well.

   The first step of the algorithm is to compute the modular decomposition of G, which can be done in polynomial time as commented at the end of the previous section. Then, we apply three kinds of reductions, which modify G and its modular decomposition but do not affect R and ϕ:

1. If there is a non-singleton non-represented module M of G, then choose any vertex $u \in M$, remove $M - \{u\}$ (with all incident edges) from G and from the nodes of the modular decomposition, and replace the subtree rooted at M by the singleton module $\{u\}$ in the modular decomposition.
2. If there is a serial module M of G with two or more non-represented children, then we choose any non-represented child N of M and remove from G and from the modular decomposition all other non-represented children of M.
3. If there is a serial module M of G with two or more represented children and some non-represented children, then we remove from G and from the modular decomposition all non-represented children of M.

Lemma 2. *The graph G has a representation extending ϕ if and only if the graph G' obtained from G by reductions 1–3 has a representation extending ϕ.*

We apply reductions 1–3 in any order until none of them is applicable any more, that is, we are left with a graph G such that

– every non-singleton strong module of G is represented,
– every serial module of G has at most one non-represented child,
– every serial module of G with at least two represented children has no non-represented child.

For such G we have the following.

Lemma 3. *Let M be a non-singleton strong module of G. If M is*

– *a serial module with a non-represented child,*
– *a prime module with no two adjacent represented children,*

then M has exactly two transitive orientations, one being the reverse of the other, both respecting ϕ. Otherwise, M has just one transitive orientation respecting ϕ.

Lemma 4. *Let u be a non-represented vertex of G and M be the parent of $\{u\}$ in the modular decomposition of G. For transitive orientations P of G respecting ϕ, the set $\mathrm{Reg}_P(u)$ is determined only by the transitive orientation of M induced by P.*

Proof. Let a be a represented vertex of G adjacent to u. We show that the orientation of the edge au either is the same for all transitive orientations of G respecting ϕ or depends only on the transitive orientation of M. This suffices for the conclusion of the lemma, as the set $\mathrm{Reg}_P(u)$ is determined by the orientations of edges connecting u with represented vertices of G.

If $a \in M$, then clearly the orientation of the edge au depends only on the transitive orientation of M. Thus suppose $a \notin M$. Let b be a representant of M. Since M is a module, a is adjacent to b as well. By Theorem 3, the orientations of the edges au and ab are the same for every transitive orientation of G. The orientation of ab and thus of au in any transitive orientation of G respecting ϕ is fixed by ϕ. Therefore, all transitive orientations of G respecting ϕ yield the same orientation of the edge au. □

By Lemmas 3 and 4, for every $u \in V(G)$, all transitive orientations P of G respecting ϕ yield at most two different regions $\mathrm{Reg}_P(u)$. We can compute them following the argument in the proof of Lemma 4. Namely, if $u \in R$ then $\mathrm{Reg}_P(u) = \phi(u)$, otherwise we find the neighbors of u in R that bound $\mathrm{Reg}_P(u)$ from above and from below, depending on the orientation of M, and compute the geometric representation of $\mathrm{Reg}_P(u)$ as for the poset problem in Section 3.

Now, we describe a reduction of the problem to 2-SAT. For every $M \in \mathcal{M}(G)$ with two transitive orientations respecting ϕ, we introduce a boolean variable x_M. The two valuations of x_M represent the two transitive orientations of M. We write a formula of the form $\alpha = \alpha_1 \wedge \ldots \wedge \alpha_m$, where each clause α_j is a literal or an alternative of two literals of the form x_M or $\neg x_M$, as follows. By Lemma 4, the set $\mathrm{Reg}_P(u)$ for any vertex u is either the same for all valuations or determined by the valuation of just one variable. Therefore, for any two non-adjacent vertices u and v, whether $\mathrm{Reg}_P(u) \cap \mathrm{Reg}_P(v) \neq \emptyset$ depends on the valuation of at most two variables. For every valuation that yields $\mathrm{Reg}_P(u) \cap \mathrm{Reg}_P(v) = \emptyset$ we write a clause forbidding this valuation. Clearly, the resulting formula α is satisfiable if and only if G has a transitive orientation P respecting ϕ and such that $\mathrm{Reg}_P(u) \cap \mathrm{Reg}_P(v) \neq \emptyset$ for any non-adjacent $u, v \in V(G)$, which by Lemma 1 holds if and only if ϕ is extendable to a representation of G. We can test whether α is satisfiable in polynomial time by a classic result of Krom [15] (see also [2] for a linear-time algorithm).

6   Extending Partial Represenation of a Function Graph by Partial Functions

Let $\mathcal{F}^\circ$ denote the family of piecewise linear continuous functions $I \to \mathbb{R}$ with I being a closed subinterval of $[0, 1]$. We describe such a function f by a tuple $(x_0, f(x_0)), \ldots, (x_k, f(x_k))$ of points in $I \times \mathbb{R}$ with $x_0 < \ldots < x_k$ and $[x_0, x_k] = I$ so that f is linear on every interval $[x_i, x_{i+1}]$. We denote the interval I that is the domain of f by $\mathrm{dom}\, f$. For convenience, we also put the empty function (with empty domain) to $\mathcal{F}^\circ$. We say that a mapping $\psi : U \to \mathcal{F}$ extends a mapping $\phi : U \to \mathcal{F}^\circ$ if we have $\psi(u)|_{\mathrm{dom}\, \phi(u)} = \phi(u)$ for every $u \in U$.

We define the notions of a partial representation of a poset or graph by partial functions, which generalize partial representations by functions defined on the entire interval $[0, 1]$ and discussed earlier in the paper. A mapping $\phi : V(P) \to \mathcal{F}^\circ$ is a *partial representation* of a poset P if the following is satisfied for any $u, v \in V(P)$: if $u <_P v$, then $\phi(u)(x) < \phi(v)(x)$ for every $x \in \mathrm{dom}\, \phi(u) \cap \mathrm{dom}\, \phi(v)$. A mapping $\phi : V(G) \to \mathcal{F}^\circ$ is a *partial representation* of a comparability graph G if ϕ is a partial representation of some transitive orientation of G. The domain of ϕ is the whole set of vertices, as we may map non-represented vertices to the empty function.

In this section we prove that the following problem is NP-complete: given a comparability graph G and a partial representation $\phi : V(G) \to \mathcal{F}^\circ$ of G, decide whether ϕ is extendable to a representation $\psi : V(G) \to \mathcal{F}$ of G.

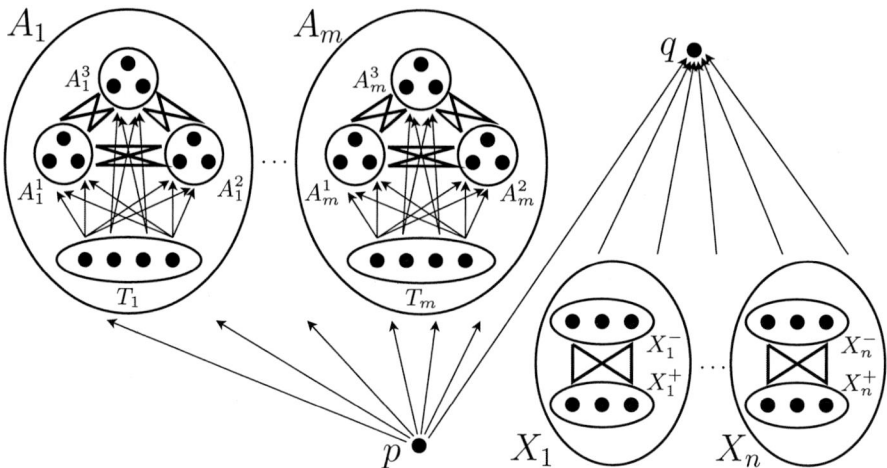

Fig. 3. The graph G and its modular decomposition. The edges whose orientation is fixed by ϕ are drawn directed.

For a function $f \in \mathcal{F}^\circ$ we define sets

$$f^\star = f \cup (([0,1] - \operatorname{dom} f) \times \mathbb{R}),$$
$$f\!\uparrow = \{(x,y) \in (\operatorname{dom} f \times \mathbb{R}) : y > f(x)\} \cup (([0,1] - \operatorname{dom} f) \times \mathbb{R}),$$
$$f\!\downarrow = \{(x,y) \in (\operatorname{dom} f \times \mathbb{R}) : y < f(x)\} \cup (([0,1] - \operatorname{dom} f) \times \mathbb{R}).$$

Let P be a poset and $\phi : V(P) \to \mathcal{F}^\circ$ be a partial representation of P. For every vertex u of P we define a set $\operatorname{Reg}(u) \subseteq [0,1] \times \mathbb{R}$, called the *region* of u, by

$$\operatorname{Reg}(u) = \phi(u)^\star \cap \bigcap \{\phi(a)\!\downarrow : a >_P u\} \cap \bigcap \{\phi(a)\!\uparrow : a <_P u\}.$$

It follows that the function representing u in any representation of P extending ϕ must be contained entirely within $\operatorname{Reg}(u)$. Lemma 1 generalizes verbatim to representations by partial functions.

Lemma 5. *There is a representation of P extending ϕ if and only if any two incomparable vertices u and v of P satisfy $\operatorname{Reg}(u) \cap \operatorname{Reg}(v) \neq \emptyset$.*

Lemma 5 shows that the problem of deciding whether a partial representation of a poset by partial functions is extendable is in P: a polynomial-time algorithm just tests whether $\operatorname{Reg}(u) \cap \operatorname{Reg}(v) \neq \emptyset$ for any two incomparable vertices u and v of the poset. It follows that the problem of deciding whether a partial representation of a comparability graph by partial functions is extendable is in NP: a non-deterministic polynomial-time algorithm can guess a transitive orientation and solve the resulting poset problem.

To prove that the latter problem is NP-hard, we show a polynomial-time reduction from 3-SAT. Let $\alpha = \alpha_1 \wedge \ldots \wedge \alpha_m$ be a boolean formula over variables $x_1, \ldots, x_n$, where each clause α_j is of the form $\alpha_j = \alpha_j^1 \vee \alpha_j^2 \vee \alpha_j^3$ with

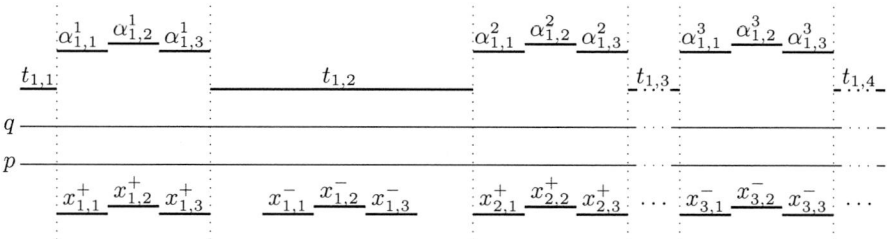

Fig. 4. The partial functions representing the vertices from X_1 and A_1, for a clause $\alpha_1 = \alpha_1^1 \vee \alpha_1^2 \vee \alpha_1^3$ with $\alpha_1^1 = x_1$, $\alpha_1^2 = x_2$, and $\alpha_1^3 = \neg x_3$

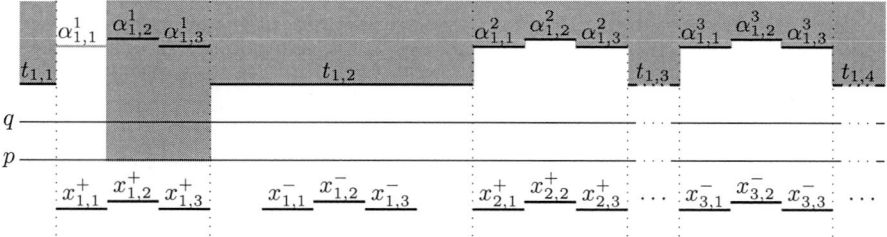

Fig. 5. $\mathrm{Reg}_P(\alpha_{1,1}^1)$ for a clause $\alpha_1 = \alpha_1^1 \vee \alpha_1^2 \vee \alpha_1^3$ with $\alpha_1^1 = x_1$, $\alpha_1^2 = x_2$, and $\alpha_1^3 = \neg x_3$ and an orientation of the module A_1 such that $A_1^1 >_P A_1^2, A_1^3$

$\alpha_j^k \in \{x_1, \neg x_1, \ldots, x_n, \neg x_n\}$. We construct a comparability graph G and its partial representation $\phi : V(G) \to \mathcal{F}^\circ$ that is extendable if and only if α is satisfiable. The vertex set of G consists of groups X_i of six vertices corresponding to variables, groups A_j of thirteen vertices corresponding to clauses, and two special vertices p and q. The edges and the modular decomposition of G are illustrated in Fig. 3. The partial representation ϕ is illustrated in Fig. 4, and the orientations of edges that are common for all transitive orientations of G respecting ϕ are shown again in Fig. 3. Every valuation satisfying α corresponds to a transitive orientation P of G that respects ϕ and satisfies $\mathrm{Reg}_P(u) \cap \mathrm{Reg}_P(v) \neq \emptyset$ for any two incomparable vertices u and v, as follows: x_i is true if and only if $X_i^+ <_P X_i^-$, and α_j^k is satisfied if A_j^k is maximal among A_j^1, A_j^2, A_j^3 with respect to $<_P$. This together with Lemma 5 implies that α is satisfiable if and only if ϕ is extendable. Figures 5 and 6 illustrate how the regions of the vertices depend on the chosen orientation.

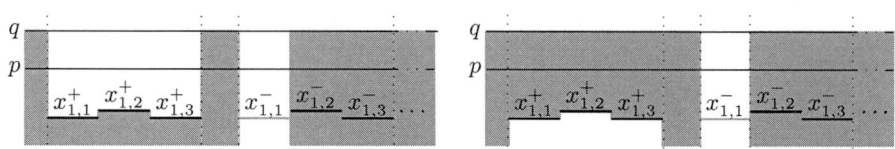

Fig. 6. $\mathrm{Reg}_P(x_{1,1}^-)$ for $X_1^- <_P X_1^+$ (to the left) and $X_1^+ <_P X_1^-$ (to the right)

References

1. Angelini, P., Di Battista, G., Frati, F., Jelínek, V., Kratochvíl, J., Patrignani, M., Rutter, I.: Testing planarity of partially embedded graphs. In: SODA 2010, pp. 202–221 (2010)
2. Aspvall, B., Plass, M.F., Tarjan, R.E.: A linear-time algorithm for testing the truth of certain quantified boolean formulas. Inform. Process. Lett. 8, 121–123 (1979)
3. Booth, K.S., Lueker, G.S.: Testing for the consecutive ones property, interval graphs, and planarity using PQ-tree algorithms. J. Comput. Sys. Sci. 13, 335–379 (1976)
4. Bläsius, T., Rutter, I.: Simultaneous PQ-ordering with applications to constrained embedding problems (2011), http://arxiv.org/abs/1112.0245
5. Even, S., Pnueli, A., Lempel, A.: Permutation graphs and transitive graphs. J. ACM 19, 400–410 (1972)
6. Fiala, J.: NP-completeness of the edge precoloring extension problem on bipartite graphs. J. Graph Theory 43, 156–160 (2003)
7. Gallai, T.: Transitiv orientierbare Graphen. Acta Mathematica Academiae Scientiarum Hungaricae 18, 25–66 (1967)
8. Golumbic, M.C.: The complexity of comparability graph recognition and coloring. Computing 18, 199–208 (1977)
9. Golumbic, M.C.: Algorithmic Graph Theory and Perfect Graphs. Academic Press (1980)
10. Golumbic, M.C., Rotem, D., Urrutia, J.: Comparability graphs and intersection graphs. Discrete Math. 43, 37–46 (1983)
11. Jampani, K.R., Lubiw, A.: Simultaneous Interval Graphs. In: Cheong, O., Chwa, K.-Y., Park, K. (eds.) ISAAC 2010. LNCS, vol. 6506, pp. 206–217. Springer, Heidelberg (2010)
12. Jampani, K.R., Lubiw, A.: The simultaneous representation problem for chordal, comparability and permutation graphs. Graph Algorithms Appl. 16, 283–315 (2012)
13. Klavík, P., Kratochvíl, J., Vyskočil, T.: Extending Partial Representations of Interval Graphs. In: Ogihara, M., Tarui, J. (eds.) TAMC 2011. LNCS, vol. 6648, pp. 276–285. Springer, Heidelberg (2011)
14. Kratochvíl, J.: String graphs. II. recognizing string graphs is NP-hard. J. Combin. Theory Ser. B 52, 67–78 (1991)
15. Krom, M.R.: The decision problem for a class of first-order formulas in which all disjunctions are binary. Zeitschrift für Mathematische Logik und Grundlagen der Mathematik 13, 15–20 (1967)
16. Kuratowski, K.: Sur le problème des courbes gauches en topologie. Fund. Math. 15, 217–283 (1930)
17. Marx, D.: NP-completeness of list coloring and precoloring extension on the edges of planar graphs. J. Graph Theory 49, 313–324 (2005)
18. McConnell, R.M., Spinrad, J.P.: Modular decomposition and transitive orientation. Discrete Math. 201, 189–241 (1999)
19. Patrignani, M.: On Extending a Partial Straight-Line Drawing. In: Healy, P., Nikolov, N.S. (eds.) GD 2005. LNCS, vol. 3843, pp. 380–385. Springer, Heidelberg (2006)
20. Schaefer, M., Sedgwick, E., Štefankovič, D.: Recognizing string graphs in NP. In: STOC 2002, pp. 1–6 (2002)
21. Spinrad, J.P.: Efficient Graph Representations. Field Institute Monographs (2003)

A Fast and Simple Subexponential Fixed Parameter Algorithm for One-Sided Crossing Minimization

Yasuaki Kobayashi and Hisao Tamaki

Meiji University, Kawasaki, Japan 214-8571
{yasu0207,tamaki}@cs.meiji.ac.jp

Abstract. We give a subexponential fixed parameter algorithm for one-sided crossing minimization. It runs in $O(3^{\sqrt{2k}} + n)$ time, where n is the number of vertices of the given graph and parameter k is the number of crossings. The exponent of $O(\sqrt{k})$ in this bound is asymptotically optimal assuming the Exponential Time Hypothesis and the previously best known algorithm runs in $2^{O(\sqrt{k}\log k)} + n^{O(1)}$ time. We achieve this significant improvement by the use of a certain interval graph naturally associated with the problem instance and a simple dynamic program on this interval graph. The linear dependency on n is also achieved through the use of this interval graph.

1 Introduction

A *two-layer drawing* of a bipartite graph G with bipartition (X, Y) of $V(G)$ places vertices in X on one line and those in Y on another line parallel to the first and draws edges as straight line segments between these two lines. We call these parallel lines *layers* of the drawing. A *crossing* in a two-layer drawing is a pair of edges that intersect each other at a point not representing a vertex. Note that the set of crossings in a two-layer drawing of G is completely determined by the order of the vertices in X on one layer and the order of the vertices in Y on the other layer. We consider the following problem.

OSCM(One-Sided Crossing Minimization)

Instance: $(G, X, Y, <, k)$, where G is a bipartite graph on $X \cup Y$ with $E(G) \subseteq X \times Y$, $<$ is a total order on X, and k is a positive integer.

Question: Is there a total order $<'$ on Y such that the two-layer drawing of G in which the vertices in X are ordered by $<$ in one layer and those in Y are ordered by $<'$ in the other layer has k or fewer crossings?

OSCM is a key subproblem in a popular approach to multi-layer graph drawing, called the "Sugiyama approach" [19], which repeatedly solves OSCM for two adjacent layers as it sweeps the layers from top to bottom and vice versa, in hope of reducing the total number of crossings in the entire drawing.

L. Epstein and P. Ferragina (Eds.): ESA 2012, LNCS 7501, pp. 683–694, 2012.

OSCM is known to be NP-complete [8], even for sparse graphs [18]. On the positive side, Dujmović and Whitesides [7] showed that OSCM is fixed parameter tractable [5], that is, it can be solved in $f(k)n^{O(1)}$ time for some function f. More specifically, the running time of their algorithm is $O(\psi^k \cdot n^2)$, where $n = |V(G)|$ and $\psi \sim 1.6182$ is the golden ratio. This result was later improved by Dujmović, Fernau, and Kaufmann [6] who gave an algorithm with running time $O(1.4656^k + kn^2)$. Very recently, Fernau et al. [9] reduced this problem to weighted FAST (feedback arc sets in tournaments) and, using the algorithm of Alon, Lokshtanov, and Saurabh [2] for weighted FAST, gave a subexponential time algorithm that runs in $2^{O(\sqrt{k}\log k)} + n^{O(1)}$ time. Karpinski and Schudy [14] considered a different version of weighted FAST proposed in [1], which imposes certain restrictions called probability constraints on the instances, and gave a faster algorithm that runs in $2^{O(\sqrt{OPT})} + n^{O(1)}$ time where OPT is the cost of an optimal solution. However, reducing OSCM to this version of FAST seems difficult: a straightforward reduction produces an instance that does not satisfy the required probability constraints. Nagamochi gave a polynomial time 1.4664-approximate algorithm [17] and $(1.2964 + 12/(\delta - 4))$-approximate algorithm when the minimum degree δ of a vertex in Y is at least 5 [16].

Our main result in this paper is the following.

Theorem 1. *OSCM can be solved in $O(3^{\sqrt{2k}} + n)$ time, assuming that G is given in the adjacency list representation and X is given in a list sorted in the total order $<$.*

Our algorithm is faster than any of the previously known parameterized algorithms. Both the dependency $O(3^{\sqrt{2k}})$ on k and the dependency $O(n)$ on n are strictly better than the algorithms cited above. In particular, the exponent $\sqrt{2k}$ does not contain the $\log k$ factor or any hidden constant as in the exponent $O(\sqrt{k}\log k)$ of [9], the only previously known subexponential algorithm for OSCM. Note that the running time of our algorithm is linear in n as long as $k \leq \frac{\log_3^2 n}{2} + O(1)$. The improvement is not only of theoretical but also of practical importance: the range of k for which the problem can be practically solvable is significantly extended.

Moreover, the exponent of $O(\sqrt{k})$ in our bound is asymptotically optimal under the Exponential Time Hypothesis (ETH) [11], a well-known complexity assumption which states that, for each $k \geq 3$, there is a positive constant c_k such that k-SAT cannot be solved in $O(2^{c_k n})$ time where n is the number of variables. ETH has been used to derive lower bounds on parameterized and exact computation (see [15] for a survey). The proof of the following theorem is in Section 6.

Theorem 2. *There is no $2^{o(\sqrt{k})}n^{O(1)}$ time algorithm for OSCM unless ETH fails.*

Another advantage of our algorithm over the previous algorithms is simplicity. The algorithm in [7] involves several reduction rules for kernelization and the improvement in [6] is obtained by introduction of additional reduction rules which entail more involved analysis. The algorithm in [9] relies on the algorithm in [2] for the more general problem of FAST. Our result suggests that OSCM is significantly easier than FAST in that it does not require any advanced algorithmic techniques or sophisticated combinatorial structures used in the algorithm of [2] for FAST, in deriving a subexponential algorithm.

Our algorithm is along the lines of earlier work [7,6]. We emphasize that our improvement does not involve any complications but rather comes with simplifications. Our algorithm does not require any kernelization. It is a straightforward dynamic programming algorithm on an interval graph associated with each OSCM instance. This interval graph is implicit in the earlier work [7,6], but is neither made explicit nor fully exploited in the previous work. Once we recognize the key role this interval graph plays in the problem, the design and analysis of an efficient algorithm becomes rather straightforward. Below we sketch how this works.

Fix an OSCM instance $(G, X, Y, <, k)$. For each vertex $y \in Y$, let l_y (r_y, resp.) denote the smallest (largest, resp.) $x \in X$ adjacent to y, with respect to the given total order $<$. We denote the half-open interval $[l_y, r_y) = \{x \in X \mid l_y \leq x < r_y\}$ in the ordered set $(X, <)$ by I_y and denote the system of intervals $\{I_y \mid y \in Y\}$ by $\mathcal{I}$. For simplicity, we assume here that the degree of each vertex y in Y is at least 2 so that the interval I_y is non-empty. Our formal treatment in Section 3 does not need this assumption. A key observation in [7] (see Lemma 2 in the present paper), is that if $r_u \leq l_v$ for distinct vertices $u, v \in Y$ then u precedes v in any optimal ordering of Y. Therefore, to determine the optimal ordering on Y, we only need to determine the pairwise order for each pair $\{u, v\}$ such that $l_v < r_u$ and $l_u < r_v$, that is, such that the intervals I_u and I_v intersect each other. Thus, the problem can be viewed as that of orienting edges of the interval graph defined by the interval system $\mathcal{I}$. The fact exploited in earlier work [7,6] to obtain fixed parameter algorithms for OSCM is that, in our terminology, this interval graph has at most k edges in feasible instances of OSCM, as each pair of u and v such that I_u and I_v intersect each other contributes at least one crossing to the drawing no matter which ordering of this pair in Y is chosen. Our interval graph view tells us more: the clique size of this interval graph for a feasible instance is at most $\sqrt{2k} + 1$, as otherwise it has more than k edges, and hence it has a path-decomposition of width at most $\sqrt{2k}$ (see [3], for example, for interval graphs and their path-decompositions). Our algorithm is a natural dynamic programming algorithm based on this path-decomposition and runs in time exponential in the width of the decomposition.

We remark that the interval system $\mathcal{I}$ also plays an important role in reducing the dependency of the running time on n to $O(n)$. See Section 4 for details.

The rest of this paper is organized as follows. In Section 2, we give preliminaries of the problem and outline our entire algorithms. In Section 3, we describe the construction of the interval systems used in our algorithm. In Section 4, we describe a preprocessing phase of our algorithm. In Section 5, we describe our dynamic programming algorithm. Finally, in Section 6, we give a proof of Theorem 2.

2   Preliminaries and Outline of the Algorithm

In this section, we give some preliminaries and outline our algorithm claimed in Theorem 1. Throughout the remainder of this paper, $(G, X, Y, <, k)$ will always be the given instance of OSCM. We assume that G does not have any parallel edges or isolated vertices. We denote the number of vertices $|V(G)|$ by n and the number of edges $|E(V)|$ by m. For each $v \in X \cup Y$, we denote the set of neighbors of v in G by $N(v)$ and its degree $|N(v)|$ by $d(v)$. We assume that $N(v)$ is given as a list, together with its length $d(v)$. We also assume that X is given as a list in which the vertices are ordered by $<$.

For each pair of distinct vertices $u, v \in Y$, we denote by $c(u, v)$ the number of pairs (x, x') with $x \in N(u)$, $x' \in N(v)$, and $x' < x$. Note that $c(u, v)$ is the number of crossings between the edges incident with u and those incident with v when the position of u precedes that of v in the layer for Y. We extend this notation for sets: for each disjoint subsets U and V of Y, we define $c(U, V) = \sum_{u \in U, v \in V} c(u, v)$.

We represent total orderings by permutations in our algorithm. Let U be a finite set. A *permutation* on U, in this paper, is a sequence of length $|U|$ in which each member of U appears exactly once. We denote the set of all permutations on U by $\Pi(U)$. Let $\pi \in \Pi(U)$. We define the total order $<_\pi$ on U naturally induced by π: for $u, v \in U$, $u <_\pi v$ if and only if u appears before v in π. When U and V are disjoint finite sets, $\pi \in \Pi(U)$, and $\sigma \in \Pi(V)$, we denote by $\pi + \sigma$ the permutation on $U \cup V$ that is a concatenation of π and σ, the sequence consisting of π followed by σ.

For each subset U of Y and a permutation π on U, we denote by $c(\pi)$ the number of crossings among the edges incident with U when the vertices in U is ordered by π, that is,

$$c(\pi) = \sum_{u,v \in U, u <_\pi v} c(u, v).$$

For each subset U of Y, we define $\text{opt}(U) = \min\{c(\pi) \mid \pi \in \Pi(U)\}$. The goal of our algorithm is to decide if $\text{opt}(Y) \leq k$.

We need the following simple observation to bound the number of edges in feasible instances of OSCM. The proof of the following lemma is omitted due to space limitations.

Lemma 1. *If G has a two-layer drawing with at most k crossings then $|E(G)| \leq |V(G)| + k - 1$.*

We also need the following lemma due to Dujmović and Whitesides [7].

Lemma 2. (Lemma 1 in [7]) *Suppose u and v are distinct vertices in Y such that $c(u, v) = 0$. Then we have $u <_\pi v$ in every optimal permutation on Y, unless we also have $c(v, u) = 0$.*

Motivated by this lemma, let us call an unordered pair $\{u, v\}$ of distinct vertices in Y *forced to* (u, v) if $c(u, v) = 0$ and $c(v, u) > 0$. We say that it is *forced* if it is forced either to (u, v) or to (v, u). We say such an unordered pair is *orientable* if $c(u, v) > 0$ and $c(v, u) > 0$; *free* if $c(u, v) = 0$ and $c(v, u) = 0$. We use the above lemma in the following form.

Corollary 1. *Let π be an optimal permutation on Y and let u, v be distinct vertices in Y. If $\{u, v\}$ is forced to (u, v) then we have $u <_\pi v$. If $\{u, v\}$ is free, then the permutation π' obtained from π by swapping the positions of u and v is also optimal.*

Proof. The first part is a restatement of Lemma 2. Suppose pair $\{u, v\}$ is free. This means that $N(u) = N(v) = \{x\}$ for some $x \in X$. Clearly, u and v are indistinguishable in our problem. Therefore the second part holds.   □

Since each orientable pair contributes at least one crossing in any ordering of Y, the following is obvious.

Proposition 1. *Assuming that the given OSCM instance is feasible, the number of orientable pairs is at most k.*

The following is an outline of our algorithm.

1. If $m \geq n + k$ then stop with "No".
2. Construct the interval system $\mathcal{I}$ described in the introduction and another interval system $\mathcal{J}$, which inherits the property of $\mathcal{I}$ that each intersecting pair of intervals contributes at least one crossing in the drawing and is designed to allow degree-1 vertices and to facilitate dynamic programming. The construction of these interval systems can be done in $O(m)$ time. See Section 3.
3. If $\mathcal{J}$ contains more than k intersecting pairs, stop with "No".
4. Precompute $c(u, v)$ and $c(v, u)$ for all orientable pairs of vertices $u, v \in Y$. This can be done in $O(n + k)$ total time. If infeasibility is detected during this precomputation, stop immediately with "No". See Section 4 for details of this step.
5. Compute $\mathrm{opt}(Y)$ by a dynamic programming algorithm based on the interval system $\mathcal{J}$. In this computation, the values of $c(u, v)$ are needed only for orientable pairs. If infeasibility is detected during this computation, stop immediately with "No". If the computation is successful and $\mathrm{opt}(Y) \leq k$ then answer "Yes"; otherwise answer "No". This step can be performed in $O(3^{\sqrt{2k}} + n)$ time. See Section 5.

The total running time of the algorithm is dominated by the dynamic programming part and is $O(3^{\sqrt{2k}} + n)$.

It is straightforward to augment the dynamic programming tables so that, when the last step is complete, an optimal permutation on Y can be constructed. We note that this optimal solution is correct even if $\mathrm{opt}(Y) > k$, as long as the dynamic programming computation is completed.

3   Interval Systems

We refer to the interval system $\mathcal{I} = \{I_y \mid y \in Y\}$ defined in the introduction as the *naive interval system*. Recall $I_y = [l_y, r_y)$, where l_y is the smallest neighbor of y and r_y is the largest neighbor of y, with respect to the total order $<$ on X. The construction of $\mathcal{I}$ can be done in $O(m)$ time: we scan X in the given total order $<$ and, as we scan $x \in X$, we do necessary book-keeping to record l_y and r_y for each $y \in N(x)$.

We need another system $\mathcal{J} = \{J_y \mid y \in Y\}$ of intervals which is slightly more complicated than the naive system. This complication comes from the need to deal with vertices in Y of degree 1 and to facilitate dynamic programming. The system $\mathcal{J}$ will satisfy the following conditions. Let $J_y = [a_y, b_y]$ for each $y \in Y$.

J1. For each y, a_y and b_y are integers satisfying $1 \leq a_y < b_y \leq 2|Y|$.
J2. For each t, $1 \leq t \leq 2|Y|$, there is a unique vertex $y \in Y$ such that $a_y = t$ or $b_y = t$.
J3. If $b_u < a_v$ for $u, v \in Y$, then $c(u, v) = 0$.

Conditions J1 and J2 are for the sake of the ease of dynamic programming described in the next section, while condition J3 is the essential property that $\mathcal{J}$ shares with the naive interval system. The construction of $\mathcal{J}$ based on $\mathcal{I}$ can be done in $O(m)$ time. The details are omitted in this version.

We restate Corollary 1 using our interval system $\mathcal{J}$. We say that a permutation π on $U \subseteq Y$ is *consistent with* $\mathcal{J}$ if $b_u < a_v$ implies $u <_\pi v$ for every pair $u, v \in U$.

Lemma 3. *Let U be an arbitrary subset of Y. There is an optimal permutation π on U that is consistent with $\mathcal{J}$.*

Proof. Let π be an optimal permutation on U. For each $x \in X$, let U_x denote the set of vertices in U that are adjacent to x but no other vertices in X. For each x, each pair of distinct vertices in U_x is free and therefore, applying Corollary 1 to the instance where Y is replaced by U, we may assume that π restricted on U_x is consistent with $\mathcal{J}$. Now, let u, v be arbitrary vertices in U and suppose $b_u < a_v$. By property J3 of $\mathcal{J}$, we have $c(u, v) = 0$. If $c(v, u) > 0$ then $\{u, v\}$ is forced to (u, v) and we must have $u <_\pi v$. Otherwise, $\{u, v\}$ is free and $u, v \in U_x$ for some $x \in X$. By the assumption on π above, we have $u <_\pi v$ in this case as well.   □

4   Computing the Crossing Numbers

Dujmović and Whitesides [7] give an algorithm for computing the crossing numbers $c(u, v)$ for all pairs $\{u, v\}$ in $O(kn^2)$ time. We spend $O(n + k)$ time for precomputing $c(u, v)$ for all orientable pairs, ignoring forced and free pairs.

We use the naive interval system $\mathcal{I} = \{I_y \mid y \in Y\}$, where $I_y = [l_y, r_y)$, in this computation.

For each $y \in Y$ and $x \in X$, let $d^{<x}(y) = |\{z \in N(y) \mid z < x\}|$ and $d^{\leq x}(y) = |\{z \in N(y) \mid z \leq x\}|$. Then, we have $c(u,v) = \sum_{x \in N(u)} d^{<x}(v)$.

It turns out helpful to decompose the above sum as follows.

$$c(u,v) = \left[\sum_{x \in N(u), l_v < x \leq r_v} d^{<x}(v) \right] + d(v) \cdot (d(u) - d^{\leq r_v}(u)). \tag{1}$$

For each $x \in X$, let $Y_x = \{y \in Y \mid l_y < x < r_y\}$ be the set of vertices in Y whose corresponding intervals strictly contain x.

In the following, we call an ordered pair (u,v) *orientable* if the corresponding unordered pair is orientable. We evaluate these sums simultaneously for all orientable pairs (u,v), using a counter $c[u,v]$ for each pair. We represent these counters by a $|Y| \times |Y|$ two-dimensional array. Since we cannot afford to initialize all of its elements, we initialize $c[u,v]$ to 0 only for orientable pairs (u,v). Our algorithm proceeds as follows.

1. Scan X in the total order $<$, maintaining Y_x as we scan x. When we scan $x \in X$, we initialize $c[u,v]$ to 0 for each $u \in N(x)$ and each $v \in Y_x$.
2. Scan X again in the total order $<$, maintaining Y_x and $d^{<x}(y)$ for each $y \in Y$, as we scan x. Suppose we are scanning $x \in X$. For each $u \in N(x)$ and each $v \in Y_x$, we add $d^{<x}(v)$, the summand in (1), to $c[u,v]$. Moreover, for each $u \in Y_x$ and $v \in N(x)$ such that $r_v = x$, we add $d(v) \cdot (d(u) - d^{\leq x}(u))$, the second term in (1), to $c[u,v]$.

The proof of the following lemma is omitted in this version.

Lemma 4. *Assuming that the given OSCM instance is feasible, the running time of the above algorithm is $O(n+k)$.*

To control the running time for infeasible instances, we count the number of times the initialization of a counter occurs in the first scan. As soon as the number exceeds $2k$, we stop the computation and report infeasibility.

5 Dynamic Programming

In this section, we describe our dynamic programming algorithm for computing $\mathrm{opt}(Y)$. Owing to the previous section, we assume in this section that $c(u,v)$ and $c(v,u)$ are available for all orientable pairs $\{u,v\}$.

We use the interval system $\mathcal{J} = \{J_y \mid y \in Y\}$ we have defined in Section 3, where $J_y = [a_y, b_y]$.

A standard dynamic programming approach (see [4,10], for example) gives us the following exponential upper bound on the complexity of computing $\mathrm{opt}(U)$, which we need for small subproblems.

Lemma 5. *Let $V \subseteq Y$ and assume that $c(u,v)$ is available in $O(1)$ time for each pair of distinct vertices $u,v \in V$. Then, $\mathrm{opt}(U)$ for all $U \subseteq V$ can be computed in $O(h2^h)$ total time where $h = |V|$.*

For each t, $1 \leq t \leq 2|Y|$, let $L_t = \{y \in Y \mid b_y \leq t\}$, $M_t = \{y \in Y \mid a_y \leq t < b_y\}$, and $R_t = \{y \in Y \mid t < a_y\}$. Note that

1. if $t = a_y$ for some $y \in Y$ then $L_t = L_{t-1}$, $M_t = M_{t-1} \cup \{y\}$, and $R_t = R_{t-1} \setminus \{y\}$;
2. if $t = b_y$ for some $y \in Y$ then $L_t = L_{t-1} \cup \{y\}$, $M_t = M_{t-1} \setminus \{y\}$, and $R_t = R_{t-1}$.

In other words, when interval J_y opens at t, y is moved from the "right set" to the "middle set"; when it closes at t, y is moved from the "middle set" to the "left set".

For each integer t, $1 \leq t \leq 2|Y|$, we compute the following and store the results in a table: (1) $c(L_t, \{y\})$, for each $y \in M_t$; (2) $\mathrm{opt}(L_t \cup S)$, for each $S \subseteq M_t$.

The recurrences for (1) are straightforward. The base case is $c(L_1, \{y\}) = 0$, where $L_1 = \emptyset$ and y is the unique element of M_1. Let $2 \leq t \leq 2|Y|$ and suppose first that $t = a_y$ for some $y \in Y$. Note that $L_t = L_{t-1}$ and $M_t \setminus M_{t-1} = \{y\}$. Therefore, for $v \in M_t \setminus \{y\}$, we have $c(L_t, \{v\}) = c(L_{t-1}, \{v\})$. Since $b_u < a_y$ for each $u \in L_t = L_{t-1}$, we have $c(L_t, \{y\}) = 0$. Suppose next that $t = b_y$ for some $y \in Y$. Note that $L_t \setminus L_{t-1} = \{y\}$ and $M_{t-1} \setminus M_t = \{y\}$ in this case. For each $v \in M_t$, we have $c(L_t, \{v\}) = c(L_{t-1} \cup \{y\}, \{v\}) = c(L_{t-1}, \{v\}) + c(y, v)$. Note that pair (y, v) is orientable, as $y, v \in M_{t-1}$, and hence $c(y, v)$ is available. Thus, in either case, the table entries of type (1) for t can be computed from the entries for $t - 1$ in $O(h)$ time, where $h = |M_t|$.

We now turn to the recurrences for type (2) entries. Since $L_0 = \emptyset$, the base case $\mathrm{opt}(L_0 \cup \emptyset) = 0$ is trivial. To facilitate the induction steps, we define, for each $y \in Y$ and disjoint subsets Y_1, Y_2 of $Y \setminus \{y\}$,

$$\mathrm{opt}(Y_1, y, Y_2) = \min\{c(\pi) \mid \pi \in \Pi(Y_1 \cup \{y\} \cup Y_2), Y_1 <_\pi \{y\} <_\pi Y_2\},$$

where, by $U <_\pi V$, we mean $u <_\pi v$ for every $u \in U$ and every $v \in V$. In other words, $\mathrm{opt}(Y_1, y, Y_2)$ is the cost of the optimal permutation on $Y_1 \cup \{y\} \cup Y_2$ subject to the condition that it is of the form $\pi_1 + y + \pi_2$, where $\pi_1 \in \Pi(Y_1)$ and $\pi_2 \in \Pi(Y_2)$. Note that $\mathrm{opt}(Y_1, y, Y_2)$ can be computed by

$$\mathrm{opt}(Y_1, y, Y_2) = \mathrm{opt}(Y_1) + c(Y_1, Y_2) + c(Y_1, \{y\}) + c(\{y\}, Y_2) + \mathrm{opt}(Y_2).$$

Lemma 6. *Let $1 \leq t \leq 2|Y|$ and suppose $a_y = t$ for some $y \in Y$. For each $S \subseteq M_t$ with $y \in S$, we have*

$$\mathrm{opt}(L_t \cup S) = \min\{\mathrm{opt}(L_{t-1} \cup T, y, U) \mid T \cup U = S \setminus \{y\}, T \cap U = \emptyset\}.$$

Proof. We first show that the left-hand side of the equality is upper-bounded by the right-hand side. Let (T^*, U^*) be a bipartition (T, U) of $S \setminus \{y\}$ that minimizes $\mathrm{opt}(L_{t-1} \cup T, y, U)$. Let σ_1 be an optimal permutation on $L_t \cup T^*$, and σ_2 an optimal permutation on U^*. Define $\sigma \in \Pi(L_t \cup S)$ by $\sigma = \sigma_1 + y + \sigma_2$.

Then, we have

$$\text{opt}(L_t \cup S) \leq c(\sigma)$$
$$= c(\sigma_1) + c(L_t \cup T^*, U^*) + c(L_t \cup T^*, \{y\}) + c(\{y\}, U^*) + c(\sigma_2)$$
$$= \text{opt}(L_t \cup T^*, y, U^*)$$
$$= \text{opt}(L_{t-1} \cup T^*, y, U^*)$$
$$= \min\{\text{opt}(L_{t-1} \cup T, y, U) \mid T \cup U = S \setminus \{y\}, T \cap U = \emptyset\}.$$

To show the inequality in the other direction, let π be an optimal permutation on $L_t \cup S$. By Corollary 1, we may assume that π is consistent with $\mathcal{J}$. Let $U_0 = \{u \in L_t \cup S \mid y <_\pi u\}$. Since $L_t = L_{t-1}$, $y \in R_{t-1}$, and π is consistent with $\mathcal{J}$, we have $U_0 \subseteq S$. Let $T_0 = S \setminus (U_0 \cup \{y\})$ and write $\pi = \pi_1 + y + \pi_2$, where $\pi_1 \in \Pi(L_t \cup T_0)$ and $\pi_2 = \Pi(U_0)$. Since π is optimal, π_1 must be optimal in $L_t \cup T_0$ and π_2 must be optimal in U_0. Thus, we have

$$\text{opt}(L_t \cup S) = c(\pi)$$
$$= \text{opt}(L_{t-1} \cup T_0) + c(L_t \cup T_0, U_0)$$
$$+ c(L_t \cup T_0, \{y\}) + c(\{y\}, U_0) + \text{opt}(U_0)$$
$$= \text{opt}(L_{t-1} \cup T_0, y, U_0)$$
$$\geq \min\{\text{opt}(L_{t-1} \cup T, y, U) \mid T \cup U = S \setminus \{y\}, T \cap U = \emptyset\}. \qquad \square$$

The dynamic programming gives us the optimal solution $\text{opt}(Y)$ since $L_{2|Y|} = Y$.

Lemma 7. *Let $1 \leq t \leq 2|Y|$ and let $h = |M_t|$. Given a table that lists the values of $c(L_{t-1}, \{v\})$ for every $v \in M_{t-1}$ and $\text{opt}(L_{t-1} \cup S)$ for every $S \subseteq M_{t-1}$, we can compute in $O(3^h)$ time the values of $c(L_t, \{v\})$ for all $v \in M_t$ and the values of $\text{opt}(L_t \cup S)$ for all $S \subseteq M_t$.*

Proof. We have observed that all the values of $c(L_t, \{v\})$ for $v \in M_t$ can be computed from the table entries in $O(h)$ time. We estimate the time for computing $\text{opt}(L_t \cup S)$ for $S \subseteq M_t$. Before computing this for individual S, we first compute $\text{opt}(U)$ and $c(L_{t-1}, U)$ for all $U \subseteq M_t$ in $O(h2^h)$ time, using Lemma 5 and the table entries for $c(L_{t-1}, \{v\})$ where $v \in M_{t-1}$ respectively. Moreover, we compute $c(T, U)$ for all disjoint subset $T, U \subset S$ as follows. For the base of this computation, $c(\{v\}, U)$, for all pair $U \subseteq M_t$ and $v \in (M_t \setminus U)$, can be computed in $O(h2^h)$ time. $c(T, U)$ can be calculated by the recurrence $c(T, U) = c(T \setminus \{v\}, U) + c(\{v\}, U)$ for some $v \in T$. We evaluate this recurrence for all disjoint subset $T, U \subseteq M_t$ using dynamic programming in $O(3^h)$ time. We store all those values, opt and c, in a table.

Suppose $t = b_y$ for some $y \in Y$. Then $L_t = L_{t-1} \cup \{y\}$. Since $(S \cup \{y\}) \subseteq M_{t-1}$, for each $S \subseteq M_t$, $\text{opt}(L_t \cup S) = \text{opt}(L_{t-1} \cup (S \cup \{y\}))$ is available in the table. Suppose $t = a_y$ for some $y \in Y$. Then, we have $L_t = L_{t-1}$ and $M_t = M_{t-1} \cup \{y\}$. Let $S \subseteq M_t$. If $y \notin S$, then $\text{opt}(L_t \cup S) = \text{opt}(L_{t-1} \cup S)$ is available in the table, since $S \subseteq M_{t-1}$. Suppose $y \in S$. Then, by Lemma 6,

$$\text{opt}(L_t \cup S) = \min\{\text{opt}(L_{t-1} \cup T, y, U) \mid T \cup U = S \setminus \{y\}, T \cap U = \emptyset\}.$$

For each bipartition (U, T) of $S \setminus \{y\}$, we have

$$\text{opt}(L_{t-1} \cup T, y, U) = \text{opt}(L_{t-1} \cup T) + c(L_{t-1} \cup T, U)$$
$$+ c(L_{t-1} \cup T, \{y\}) + c(\{y\}, U) + \text{opt}(U)$$
$$= \text{opt}(L_{t-1} \cup T) + c(L_{t-1}, U) + c(T, U)$$
$$+ c(L_{t-1}, \{y\}) + c(T, \{y\}) + c(\{y\}, U) + \text{opt}(U).$$

All the terms are in the tables. It takes $O(1)$ time for each bipartition (T, U) of $S \setminus \{y\}$ and hence $O(2^{|S|-1})$ time for all such bipartitions. The computation for all $S \subseteq M_t$ with $y \in S$ takes $\sum_{S \subseteq M_t} O(2^{|S|-1}) = O(3^h)$ time. □

Each pair of vertices in M_t contributes at least one crossing in any ordering of Y. Therefore, for the given instance to be feasible, we have $h(h-1)/2 \leq k$ and hence $h \leq \sqrt{2k}+1$, where $h = |M_t|$. Using this bound and an observation that $|M_t| \geq 2$ for at most k values of t, it is straightforward to derive a bound of $O(k3^{\sqrt{2k}} + n)$ on the running time of the entire dynamic programming computation. For a tighter analysis, we need the following lemma.

Lemma 8. *Assume that $\mathcal{J}$ has at most k intersecting pairs of intervals. Let $H = \lceil \sqrt{2k} \rceil + 1$ and, for $2 \leq h \leq H$, let c_h denote the number of values of t with $|M_t| = h$. Then, we have $c_h \leq 2^{H-h+2}$ for $2 \leq h \leq H$.*

Proof. Fix h, $2 \leq h \leq H$. Let $t_1 < t_2 < \ldots < t_{c_h}$ be the members of $\{t \mid |M_t| = h\}$. The first set M_{t_1} contains $h(h-1)/2$ paris of vertices each corresponding to an intersecting pair of intervals. We claim that M_{t_i} for each $2 \leq i \leq c_h$ contributes at least $h-1$ new intersecting pairs of intervals. This is obvious if $M_{t_i} \neq M_{t_{i-1}}$. If $M_{t_i} = M_{t_{i-1}}$, then $M_{t_{i-1}+1}$ must contain a vertex not in $M_{t_{i-1}}$ that contributes h new intersecting pairs. Therefore we have $h(h-1)/2 + (c_h - 1)(h-1) \leq k$. Solving this inequality for c_h, we have $c_h \leq k/(h-1) - h/2 + 1$.

If $H - h + 2 \geq \log_2 k$, the claimed bound $c_h \leq 2^{H-h+2}$ is obvious since then $2^{H-h+2} \geq k$. So suppose $j = H - h + 2 < \log_2 k$. Then, we have $(H-j+1)(H+2j) = H^2 + H + j(H - 2j + 2) > H^2 > 2k$ since $H \geq \sqrt{2k} + 1 \geq 2(\log_2 k - 1)$. Therefore, we have $k/(h-1) = k/(H-j+1) < (H+2j)/2 = H/2 + j$ and hence

$$c_h < H/2 + j - (H - j + 2)/2 + 1 = \frac{3j}{2} \leq 2^j.$$

This is the claimed bound. □

Lemma 9. *Assume that the given OSCM instance is feasible, the total running time of the dynamic programming algorithm based on Lemma 7 is $O(3^{\sqrt{2k}} + n)$.*

Proof. Let H and c_h, $2 \leq h \leq H$, be as in Lemma 8. Then, from Lemma 7, it follows that the total running time of the dynamic programming algorithm is $O(\sum_{2 \leq h \leq H} c_h 3^h + n)$. By Lemma 8, the first term is bounded by

$$\sum_{2 \leq h \leq H} c_h 3^h \leq \sum_{2 \leq h \leq H} 2^{H-h+2} 3^h = \sum_{2 \leq h \leq H} \left(\frac{2}{3}\right)^{H-h+2} 3^{H+2} = O(3^H),$$

and therefore we have the claimed bound. □

To control the running time for infeasible instances, we compute c_h for each $2 \leq h \leq H$ and, if c_h exceeds the proved bound, we immediately stop the computation as we have detected infeasibility.

6   Proof of Theorem 2

Impagliazo, Paturi, and Zane [12] have shown that, under ETH, there is no $2^{o(m)}$ time algorithm for 3-SAT where m is the number of clauses. We confirm in the Lemmas below that the standard chain of reductions from 3-SAT to OSCM showing the NP-completeness of sparse OSCM is such that the number r of edges in the resulting OSCM instance is linear in the number m of clauses of the input 3-SAT instance. Theorem 2 follows, since a $2^{o(\sqrt{k})}n^{O(1)}$ time algorithm for OSCM would imply a $2^{o(m)}$ time algorithm for 3-SAT as $k \leq r^2 = O(m^2)$ in our reduction and hence would violate ETH.

Lemma 10. [13] *There is a polynomial time algorithm that, given a 3-CNF formula ϕ with n variables and m clauses, computes a graph G with $O(n + m)$ vertices and $O(n + m)$ edges such that ϕ is satisfiable if and only if G has a vertex cover of size at most $n + 2m$.*

Lemma 11. [13] *There is a polynomial time algorithm that, given a graph G with n vertices and m edges, and integer k, computes a directed graph D with $O(n + m)$ vertices and $O(n + m)$ arcs such that G has a vertex cover of size at most k if and only if D has a feedback arc set of size at most k.*

Lemma 12. (Section 4 in [18]) *There is a polynomial time algorithm that, given a directed graph D with n vertices and m edges, and integer k, computes an OSCM instance $I = (G, X, Y, <, k')$ with $O(n + m)$ vertices, $O(n + m)$ edges, and $k' = \sum_{u,v \in Y} \min(c(u,v), c(v,u)) + 2k$ such that D has a feedback arc set of size at most k if and only if I is feasible.*

Acknowledgement. We deeply appreciate valuable comments from anonymous referees on our earlier submission, especially the one suggesting the optimality of our result under ETH.

References

1. Ailon, N., Charikar, M., Newman, A.: Aggregating inconsistent information:Ranking and clustering. Journal of the ACM 55(5): Article No. 23 (2008)
2. Alon, N., Lokshtanov, D., Saurabh, S.: Fast FAST. In: Albers, S., Marchetti-Spaccamela, A., Matias, Y., Nikoletseas, S., Thomas, W. (eds.) ICALP 2009, Part I, LNCS, vol. 5555, pp. 49–58. Springer, Heidelberg (2009)
3. Bodlaender, H.: A Tourist Guide through Treewidth. Acta Cybernetica 11, 1–23 (1993)
4. Bodlaender, H., Fomin, F., Kratsch, D., Thilikos, D.: A Note on Exact Algorithms for Vertex Ordering Problems on Graphs. Theory of Computing Systems 50(3), 420–432 (2012)

5. Downey, R.G., Fellows, M.R.: Parameterized Complexity. Springer, Heidelberg (1998)
6. Dujmović, V., Fernau, H., Kaufmann, M.: Fixed parameter algorithms for one-sided crossing minimization revisited. Journal of Discrete Algorithms 6(2), 313–323 (2008)
7. Dujmović, V., Whitesides, S.: An Efficient Fixed Parameter Tractable Algorithm for 1-Sided Crossing Minimization. Algorithmica 40(1), 15–31 (2004)
8. Eades, P., Wormald, N.C.: Edge crossings in drawings of bipartite graphs. Algorithmica 11(4), 379–403 (1994)
9. Fernau, H., Fomin, F.V., Lokshtanov, D., Mnich, M., Philip, G., Saurabh, S.: Ranking and Drawing in Subexponential Time. In: Iliopoulos, C.S., Smyth, W.F. (eds.) IWOCA 2010. LNCS, vol. 6460, pp. 337–348. Springer, Heidelberg (2011)
10. Held, M., Karp, R.M.: A dynamic programming approach to sequencing problems. Journal of the Society for Industrial and Applied Mathematics 10, 196–210 (1962)
11. Impagliazzo, R., Paturi, R.: On the complexity of k-SAT. Journal of Computer and System Sciences 62, 367–375 (2001)
12. Impagliazzo, R., Paturi, R., Zane, F.: Which problems have strongly exponential complexity? Journal of Computer and System Sciences 63, 512–530 (2001)
13. Karp, R.M.: Reducibility among combinatorial problems. In: Complexity of Computer Computations, pp. 85–103. Plenum Press (1972)
14. Karpinski, M., Schudy, W.: Faster Algorithms for Feedback Arc Set Tournament, Kemeny Rank Aggregation and Betweenness Tournament. In: Cheong, O., Chwa, K.-Y., Park, K. (eds.) ISAAC 2010, Part I, LNCS, vol. 6506, pp. 3–14. Springer, Heidelberg (2010)
15. Lokshtanov, D., Marx, D., Saurabh, S.: Lower bounds based on the Exponential Time Hypothesis, The Complexity Column by Arvind V. Bulletin of the EATCS, pp. 41–72 (2011)
16. Nagamochi, H.: On the one-sided crossing minimization in a bipartite graph with large degree. Theoretical Computer Science 332, 417–446 (2005)
17. Nagamochi, H.: An improved bound on the one-sided minimum crossing number in two-layered drawings. Discrete and Computational Geometry 33(4), 569–591 (2005)
18. Muñoz, X., Unger, W., Vrt'o, I.: One Sided Crossing Minimization Is NP-Hard for Sparse Graphs. In: Mutzel, P., Jünger, M., Leipert, S. (eds.) GD 2001. LNCS, vol. 2265, pp. 115–123. Springer, Heidelberg (2002)
19. Sugiyama, K., Tagawa, S., Toda, M.: Methods for visual understanding of hierarchical system structures. IEEE Transactions on Systems, Man, and Cybernetics 11(2), 109–125 (1981)

Minimum Average Distance Triangulations

László Kozma

Universität des Saarlandes, Saarbrücken, Germany
kozma@cs.uni-saarland.de

Abstract. We study the problem of finding a triangulation T of a planar point set S such as to minimize the expected distance between two points x and y chosen uniformly at random from S. By distance we mean the length of the shortest path between x and y along edges of T, with edge weights given as part of the problem. In a different variant of the problem, the points are vertices of a simple polygon and we look for a triangulation of the interior of the polygon that is optimal in the same sense. We prove that a general formulation of the problem in which the weights are arbitrary positive numbers is strongly NP-complete. For the case when all weights are equal we give polynomial-time algorithms. In the end we mention several open problems.

1 Introduction

The problem addressed in this paper is a variant of the classical *network design* problem. In many applications, the average routing cost between pairs of nodes is a sensible network characteristic, one that we seek to minimize. If costs are *additive* (e.g., time delay) and *symmetric*, an edge-weighted, undirected graph G is a suitable model of the connections between endpoints. The task is then to find a spanning subgraph T of G that minimizes the average distance. Johnson *et al.* [1] study the problem when the total edge weight of T is required to be less than a given budget constraint. They prove this problem to be NP-complete, even in the special case when all weights are equal and the budget constraint forces the solution to be a *spanning tree*.

Here we study the problem in a planar embedding: vertices of G are points in the plane, edges of G are straight segments between the points and weights are given as part of the problem. Instead of limiting the total edge weight of the solution, we require the edges of T to be non-intersecting. From a theoretical point of view this turns out to be an essential difference: the problem now has a geometric structure that we can make use of. As an application we could imagine that we wanted to connect n cities with an optimal railroad network using straight line connections and no intersections. We now give a more precise definition of the problem.

Given a set of points $S = \{p_1, \ldots, p_n\} \subset \mathbb{R}^2$, and weights $w : S^2 \to \mathbb{R}$, having $w(x, x) = 0$ and $w(x, y) = w(y, x)$, for all $x, y \in S$, we want to find a *geometric*, *crossing-free* graph T with vertex set S and edge weights given by w, such that the expected distance between two points chosen uniformly at random from S is as small as possible. By *distance* we mean the length of the shortest path

L. Epstein and P. Ferragina (Eds.): ESA 2012, LNCS 7501, pp. 695–706, 2012.
© Springer-Verlag Berlin Heidelberg 2012

in T and we denote it by d_T. Since adding an edge cannot increase a distance, it suffices to consider *maximal* crossing-free graphs, i.e., *triangulations*. We call this problem MINIMUM AVERAGE DISTANCE TRIANGULATION (MADT).

The previous formulation, if we omit the normalizing factor, amounts to finding a triangulation T that minimizes the following quantity:

$$\mathcal{W}(T) = \sum_{1 \le i < j \le n} d_T(p_i, p_j).$$

Similarly, we ask for the triangulation T of the interior of a polygon with n vertices that has the minimum value $\mathcal{W}(T)$. In this case the triangulation consists of all boundary edges and a subset of the diagonals of the polygon.

We note that in mathematical chemistry, the quantity $\mathcal{W}(T)$ is a widely used characteristic of molecular structures, known as Wiener index [2,3]. Efficient computation of the Wiener index for special graphs, as well as its combinatorial properties have been the subject of significant research [4,5,6].

Optimal triangulations. Finding optimal triangulations with respect to various criteria has been intensively researched in the past decades [7,8]. One particularly well-studied problem is *minimum weight triangulation* (MWT). For polygons, the solution of MWT is found by the $\mathcal{O}(n^3)$ algorithm due to Gilbert [9] and Klincsek [10], a classical example of dynamic programming. For point sets and Euclidean weights MWT was proven to be NP-hard by Mulzer and Rote [11]. For unit weights MWT is trivial since all triangulations have the same cost.

In contrast to both MWT and budgeted network design [1], MADT is interesting even for unit weights. In case of simple polygons, the problem is neither trivial, nor NP-hard. The algorithm we give in § 2.2 uses dynamic programming but it is much more involved than the $\mathcal{O}(n^3)$ algorithm for MWT. Surprisingly, the ideas of the MWT algorithm do not seem to directly carry over to the MADT problem, which, in fact, remains open for Euclidean weights, even for polygons. What makes our criterion of optimality difficult is that it is highly nonlocal. It is nontrivial to decompose the problem into smaller parts and known techniques do not seem to help.

Our results. We study triangulations of point sets and of polygons. In the case of equal weights on all allowed edges, we assume w.l.o.g. that the weights are equal to one and we refer to the distance as *link distance*. Using link distance, the solution is easily obtained when one point or one vertex can be connected to all the others. This is shown in § 2.1. For the more general case of simple polygons (when no vertex can be connected to all other vertices) in § 2.2 we give an algorithm with a running time of $\mathcal{O}(n^{11})$ that uses dynamic programming. Our approach exploits the geometric structure of the problem, making a decomposition possible in this case.

For general point sets and arbitrary positive, symmetric weights (not necessarily obeying the triangle inequality), in § 2.3 we prove the problem to be strongly NP-complete, ruling out the existence of an efficient exact algorithm or of a fully polynomial time approximation scheme (FPTAS), unless P=NP. The hardness proof is a gadget-based reduction from PLANAR3SAT. Again, the nonlocality of the cost function makes the reduction somewhat difficult, requiring a careful balancing between the magnitudes of edge weights and the size of the construction. For lack of space we omit some of the details of the reduction, for the full proofs we refer the reader to the preprint version of this paper [12].

We leave the problem open in the case of Euclidean weights but we present the results of computer experiments for certain special cases in § 3.

2 Results

2.1 Link Distance with One-Point-Visibility

We call a point set S *one-point-visible* if one of the points $p \in S$ can be connected to all the points $q \in S$, where $p \neq q$, using straight segments that do not contain a point of S, except at endpoints. This condition is less restrictive than the usual *generality* (no three points collinear). Similarly, we call a polygon *one-vertex-visible* if one of the vertices can be connected to all others with diagonals or boundary edges of the polygon. The set of *one-vertex-visible* polygons includes all convex polygons. A *fan* is a triangulation in which one point or vertex (called the *fan handle*) is connected to all other points or vertices.

Theorem 1. *For a one-vertex-visible polygon every fan triangulation has the same average distance and this is the smallest possible. For a one-point-visible point set every fan triangulation has the same average distance and this is the smallest possible.*

Proof. The smallest possible distance of 1 is achieved for exactly $2n - 3$ pairs of vertices in polygons and $3n - h - 3$ pairs of points in point sets (with h points on the convex hull), for every triangulation (these are the pairs that are connected with an edge and all triangulations of the same polygon or point set have the same number of edges). In a fan triangulation, all remaining pairs are at distance 2 from each other: the path between two vertices not connected with an edge can go via the fan handle.   □

2.2 Link Distance in Simple Polygons

We now look at polygons that do not admit a fan triangulation. It would be desirable to decompose the problem and deal with the parts separately. The difficulty lies in the fact that when we are triangulating a smaller piece of the polygon, the decisions affect not just the distances within that piece but also the distances between external vertices. We need to do some bookkeeping of these global distances, but first we make some geometric observations.

Assume that an optimum triangulation T has been found. We use a clockwise ordering of the vertices p_1 to p_n and we denote by p_d be the third vertex of the triangle that includes $p_1 p_n$. Let us visit the vertices from p_1 to p_d in clockwise order (if $p_1 p_d$ is a boundary edge, then we have no other vertices in between). Let p_a be the last vertex in this order such that $d_T(p_a, p_1) < d_T(p_a, p_d)$. There has to be such a vertex, since p_1 itself has this property and p_d does not. Let p_c be the first vertex for which $d_T(p_c, p_d) < d_T(p_c, p_1)$. Again, such a vertex clearly exists (in a degenerate case we can have $p_1 = p_a$ or $p_c = p_d$ or both). Let p_b denote the vertex (other than p_n) that is connected to both p_1 and p_d (unless $p_1 p_d$ is on the boundary).

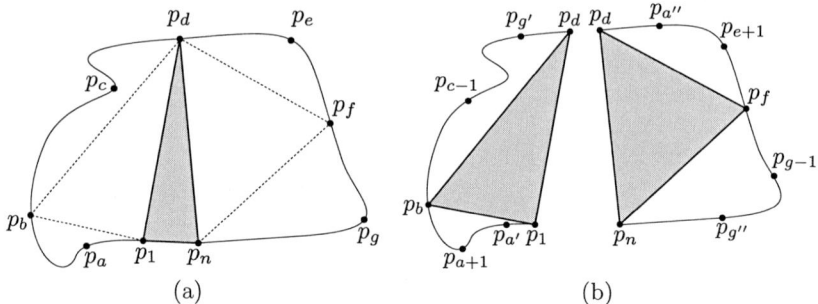

Fig. 1. (a) Special vertices of the polygon. (b) Splitting up the polygon.

On the other side of the triangle $p_1 p_d p_n$ we similarly visit the vertices from p_d to p_n and assign the label p_e to the last vertex such that $d_T(p_e, p_d) < d_T(p_e, p_n)$ and p_g to the first vertex such that $d_T(p_g, p_n) < d_T(p_g, p_d)$ and we let p_f be the vertex connected to both p_d and p_n (Fig. 1(a)). Now we can observe some properties of these vertices.

Lemma 1. *Let $1 \le k \le d$. Then the following hold (analogous statements hold for $d \le k \le n$):*
(a) $d_T(p_k, p_1) < d_T(p_k, p_d)$ iff $1 \le k \le a$.
(b) $d_T(p_k, p_1) > d_T(p_k, p_d)$ iff $c \le k \le d$.
(c) $d_T(p_k, p_1) = d_T(p_k, p_d)$ iff $a < k < c$. In particular, if p_b exists, then $a < b < c$. Otherwise $a = 1$, $c = d = 2$, and $p_1 p_2$ is on the boundary.

Proof. (a) The largest index k for which $d_T(p_k, p_1) < d_T(p_k, p_d)$ is $k = a$ by the definition of p_a. For the converse, observe that for all intermediary vertices p_k on a shortest path between p_1 and p_a we have $d_T(p_k, p_1) < d_T(p_k, p_d)$. Now suppose there is a vertex p_l, with $1 \le l \le a$, such that $d_T(p_l, p_d) \le d_T(p_l, p_1)$. Such an inequality also holds for all intermediary vertices on the shortest path between p_l and p_d. Since the shortest path between p_l and p_d intersects the shortest path between p_a and p_1, the common vertex has to be at the same time strictly closer to p_1 and closer or equal to p_d, a contradiction.

(b) Similar argument as for *(a)*.

(c) First, observe that $a < c$, otherwise some vertex would have to be strictly closer to both p_1 and p_d, a contradiction. Then, since for $1 \le k < c$ we have $d_T(p_k, p_1) \le d_T(p_k, p_d)$ and for $a < k \le d$ we have $d_T(p_k, p_d) \le d_T(p_k, p_1)$, it follows that for indices in the intersection of the two intervals ($a < k < c$) we have $d_T(p_k, p_d) = d_T(p_k, p_1)$. The converse follows from (a) and (b). Also, we have $d_T(p_b, p_1) = d_T(p_b, p_d) = 1$. □

Equipped with these facts, we can split the distance between two vertices on different sides of the $p_1 p_d p_n$ triangle into locally computable components.

Let $1 \leq x \leq d$. Consider the shortest path between p_x and p_d. Clearly, for all vertices p_k on this path $1 \leq k \leq d$ holds, otherwise the path would go via the edge $p_1 p_n$ and it could be shortened via $p_1 p_d$. Similarly, given $d \leq y \leq n$, for all vertices p_k on the shortest path between p_y and p_n, we have $d \leq k \leq n$. We conclude that $d_T(p_x, p_d)$ and $d_T(p_y, p_n)$ only depend on the triangulations of $(p_1, \ldots, p_d)$ and $(p_d, \ldots, p_n)$ respectively. We now express the global distance $d_T(p_x, p_y)$ in terms of these two local distances.

Lemma 2. Let $p_1, \ldots, p_n$ defined as before, $1 \leq x \leq d$, and $d \leq y \leq n$, and let $\phi = d_T(p_x, p_d) + d_T(p_y, p_n)$. Then the following holds, covering all possible values of x and y:

$$d_T(p_x, p_y) = \begin{cases} \phi - 1 & \text{if } d \leq y \leq e; \\ \phi + 1 & \text{if } g \leq y \leq n \text{ and } a < x \leq d; \\ \phi & \text{otherwise} . \end{cases}$$

Proof. In each of the cases we use Lemma 1 to argue about the possible ways in which the shortest path can cross the triangle $p_1 p_d p_n$. For example, if $d \leq y \leq e$, the shortest path goes through p_d, therefore we have $d_T(p_x, p_y) = d_T(p_x, p_d) + d_T(p_y, p_d)$. Since $d_T(p_y, p_d) = d_T(p_y, p_n) - 1$, we obtain $d_T(p_x, p_y) = \phi - 1$. The other cases use similar reasoning and we omit them for brevity.   □

Lemma 2 allows us to decompose the problem into parts that can be solved separately. We proceed as follows: we *guess* a triangle $p_1 p_d p_n$ that is part of the optimal triangulation and we use it to split the polygon in two. We also *guess* the special vertices p_a, p_c, p_e, p_g. We recursively find the optimal triangulation of the two smaller polygons with vertices $(p_1, \ldots, p_d)$ and $(p_d, \ldots, p_n)$. Besides the distances *within* the subpolygons we also need to consider the distances *between* the two parts. Using Lemma 2 we can split these distances into a distance to p_d in the left part, a distance to p_n in the right part and a constant term.

We now formulate an extended cost function $\mathcal{W}_{\mathrm{EXT}}$, that has a second term for accumulating the distances to endpoints that result from splitting up global distances. The coefficient $\alpha \in \mathbb{N}$ will be uniquely determined by the sizes of the polygons, which in turn are determined by the choice of the index d. We express this new cost function for a general subpolygon $(p_i, \ldots, p_j)$:

$$\mathcal{W}_{\mathrm{EXT}}(T, \alpha)\Big|_i^j = \sum_{i \leq x < y \leq j} d_T(p_x, p_y) + \alpha \sum_{i \leq x \leq j} d_T(p_x, p_j).$$

Observe that minimizing $\mathcal{W}_{\mathrm{EXT}}(T, 0)\Big|_1^n$ solves the initial problem. We can split the sums and rewrite the distances using Lemma 2, until we can express $\mathcal{W}_{\mathrm{EXT}}$ recursively in terms of smaller polygons and the indices of the special vertices a, c, e, g. Note that p_a, p_c, p_e, p_g play the same role as in the earlier discussion, but now the endpoints are p_i and p_j instead of p_1 and p_n.

$$\mathcal{W}_{\mathrm{EXT}}(T,\alpha)\Big|_i^j = \sum_{i\le x<y\le d} d_T(p_x,p_y) + \sum_{d\le x<y\le j} d_T(p_x,p_y) + \sum_{\substack{i\le x\le d\\ d\le y\le j}} d_T(p_x,p_y)$$

$$- \sum_{i\le x\le j} d_T(p_x,p_d) + \alpha \sum_{i\le x\le d} d_T(p_x,p_j) + \alpha \sum_{d\le x\le j} d_T(p_x,p_j) - \alpha\cdot d_T(p_d,p_j)$$

$$= \mathcal{W}_{\mathrm{EXT}}(T,\alpha+j-d)\Big|_i^d + \mathcal{W}_{\mathrm{EXT}}(T,\alpha+d-i)\Big|_d^j$$

$$+ (\alpha+j-g+1)(d-a-1) + (e-d+1)(i-d).$$

How can we make sure that the constraints imposed by the choice of the special vertices p_a, p_c, p_e, p_g are respected by the recursive subcalls? If the left side of the triangle is on the boundary $(d=i+1)$, it follows that $a=i$ and $c=d$ and it is trivially true that $d_T(p_a,p_i) < d_T(p_a,p_d)$. Similarly, if $d=j-1$, it follows that $e=d$ and $g=j$, therefore $d_T(p_e,p_d) < d_T(p_e,p_j)$. The general case remains, when one of the sides of the triangle is not on the boundary. The following lemma establishes a necessary and sufficient condition for the constraints to hold. We write it only for the side p_ip_d and special indices a and c, a symmetric argument works for the side p_dp_j and special indices e and g.

Lemma 3. *Let $p_i,\dots,p_j$ be a triangulated polygon. Assume that the triangulation contains the triangles $p_ip_dp_j$ and $p_ip_bp_d$. Then the following hold:*

(a) *We have a as the largest index $(i \le a \le d)$ for which $d_T(p_a,p_i) < d_T(p_a,p_d)$ iff $a+1$ is the smallest index $(i \le a+1 \le b)$ such that $d_T(p_{a+1},p_b) < d_T(p_{a+1},p_i)$.*

(b) *We have c as the smallest index $(i \le c \le d)$ for which $d_T(p_c,p_d) < d_T(p_c,p_i)$ iff $c-1$ is the largest index $(b \le c-1 \le d)$ such that $d_T(p_{c-1},p_b) < d_T(p_{c-1},p_d)$.*

Proof. (a) If a is the largest index such that $d_T(p_a,p_i) < d_T(p_a,p_d)$ then $a+1$ is the smallest index such that $d_T(p_{a+1},p_d) \le d_T(p_{a+1},p_i)$. Since $a+1 \le b$, the shortest path between p_{a+1} and p_d contains p_b, therefore $d_T(p_{a+1},p_b) < d_T(p_{a+1},p_i)$. To see that $a+1$ is the smallest index with this property, we need to prove that $d_T(p_k,p_b) \ge d_T(p_k,p_i)$ for all $i \le k \le a$. This inequality follows from $d_T(p_k,p_d) > d_T(p_k,p_i)$ and $d_T(p_k,p_d) = d_T(p_k,p_b)+1$.

For the converse, assume $a+1$ to be the smallest index such that $d_T(p_{a+1},p_b) < d_T(p_{a+1},p_i)$. Then $d_T(p_a,p_b) \ge d_T(p_a,p_i)$. Since $d_T(p_a,p_d) = d_T(p_a,p_b)+1$, it follows that $d_T(p_a,p_i) < d_T(p_a,p_d)$. To see that a is the largest index with this property, we need $d_T(p_k,p_i) \ge d_T(p_k,p_d)$ for all $a < k \le b$ (for $k > b$ the inequality clearly holds). Again, this follows from $d_T(p_k,p_i) > d_T(p_k,p_b)$ and $d_T(p_k,p_d) = d_T(p_k,p_b)+1$.

(b) Similarly. □

Procedure EXT (Fig. 2) returns the cost $\mathcal{W}_{\mathrm{EXT}}$ of a triangulation that minimizes this cost. We can modify our procedure without changing the asymptotic running time, such as to return the actual triangulation achieving minimum cost. The results are then merged to form the full solution.

Lemma 3 tells us that in all recursive calls two of the four special vertices are fixed and we only need to guess the remaining two. We label these new vertices as p'_a, p'_g and p''_a, p''_g. They play the same role in their respective small polygons as a and g in the large polygon (see Fig. 1(b) for illustration). The notation $p \leftrightarrow q$ indicates the condition that vertices p and q see each other within the polygon.

---

procedure EXT $\big((p_i, \ldots, p_j),\ p_a, p_c, p_e, p_g,\ \alpha\big)$:
    if $(a = i)$ and $(c = e = i + 1)$ and $(g = j = i + 2)$:
        return $3 + 2\alpha$;   /* *the polygon has only three vertices* */
    else:

$$\textbf{return} \quad \min_{\substack{p_d, p'_a, p'_g, p''_a, p''_g: \\ i \le a' \le a+1 \le c-1 \le g' \le d \\ d \le a'' \le e+1 \le g-1 \le g'' \le j \\ p_i \leftrightarrow p_d \leftrightarrow p_j}} \Big\{ \text{EXT}\big((p_i, \ldots, p_d),\ p'_a, p_{a+1}, p_{c-1}, p'_g,\ \alpha + j - d\big)$$
$$+ \text{EXT}\big((p_d, \ldots, p_j),\ p''_a, p_{e+1}, p_{g-1}, p''_g,\ \alpha + d - i\big)$$
$$+ (\alpha + j - g + 1)(d - a - 1) + (e - d + 1)(i - d)\Big\};$$

Fig. 2. Procedure for finding the triangulation that minimizes $\mathcal{W}_{\text{EXT}}(T, \alpha)$

The process terminates, since every subcall is on polygons of smaller sizes and we exit the recursion on triangles. Clearly every triangulation can be found by the algorithm and the correctness of the decomposition is assured by Lemma 2. The fact that indices a, c, e, g indeed fulfill their necessary role (and thus the expressions in the cost-decomposition are correct) is guaranteed by Lemma 3.

For the cases when there is no suitable vertex that can be connected to the required endpoints and whose index fulfills the required inequalities, we adopt the convention that the minimum of the empty set is $+\infty$, thus abandoning those branches in the search tree.

Theorem 2. *The running time of* EXT *on a polygon with n vertices is $\mathcal{O}(n^{11})$.*

Proof. Observe that all polygons in the function calls have contiguous indices, therefore we can encode them with two integers between 1 and n. Furthermore, if the initial call has $\alpha = 0$, then it can be shown that on each recursive call the parameter α becomes "n minus the number of vertices in the polygon". For this reason it is superfluous to pass α as an argument. There are four remaining parameters which can all take n different values. We can build up a table containing the return values of all $\mathcal{O}(n^6)$ possible function calls, starting from the smallest and ending with the full polygon. In computing one entry we take the minimum of $\mathcal{O}(n^5)$ values, giving a total running time of $\mathcal{O}(n^{11})$.   □

2.3   Arbitrary Positive Weights

We now prove that the decision version of MADT for point sets is NP-complete if the edge weigths w are arbitrary positive numbers, such that $w(p_i, p_j) = 0$ iff $i = j$ and $w(p_i, p_j) = w(p_j, p_i)$ for all i, j. Such a weight function is called a *semimetric*. We leave open the status of the problem for *metric* weights (that obey the triangle inequality). For lack of space, we omit many of the details of the proofs in this section and we refer the reader to the preprint version of the paper [12]. Where possible, we give a short, intuitive explanation.

MADT (decision version): *For given $\mathcal{W}^\star \in \mathbb{R}$, is there a triangulation T of a given point set S and weights w, such that $\mathcal{W}(T) \leq \mathcal{W}^\star$?*

The problem is clearly in NP, since for a given triangulation T, we can use an all-pairs shortest path algorithm to compute $\mathcal{W}(T)$ and compare it with $\mathcal{W}^\star$ in polynomial time.

We prove NP-hardness via a reduction from PLANAR3SAT [13]. In 3SAT, given a 3-CNF formula, we ask whether there exists an assignment of truth values such that each clause has at least one *true* literal. PLANAR3SAT restricts the question to planar formulae: those that can be represented as a planar graph in which vertices represent both variables and clauses of the formula and there is an edge between clause C and variable x iff C contains either x or $\neg x$.

Knuth and Ragunathan [14] observed that PLANAR3SAT remains NP-complete if it is restricted to formulae embedded in the following fashion: variables are arranged on a horizontal line with three-legged clauses on the two sides of the line. Clauses and their three legs are properly nested, i.e., none of the legs cross each other. We can clearly have an actual embedding in which all three legs of a clause are straight lines and the "middle" leg is perpendicular to the line of the variables. For simplicity, let us call such an embedding of a formula a *planar circuit*.

We put two extra conditions on the admissible planar circuits, such that PLANAR3SAT remains NP-complete when restricted to planar circuits obeying these conditions: (**R1**) no variable appears more than once in the same clause, and (**R2**) every variable appears in at least two clauses.

Lemma 4. *Given a planar circuit ϕ_1, we can transform it into a planar circuit ϕ_2 that obeys R1 and R2, such that ϕ_2 has a satisfying assignment iff ϕ_1 does.*

Proof. We examine every possible way in which R1 or R2 can be violated and give transformations that remove the violations while preserving the planar embedding, as well as the satisfiability of the circuit. □

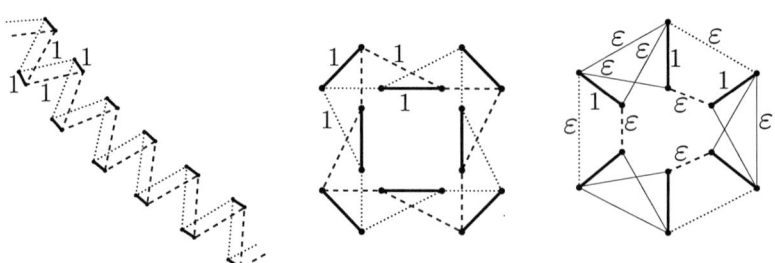

Fig. 3. (a) Wire gadget.   (b) Simplified variable gadget.   (c) Clause gadget.

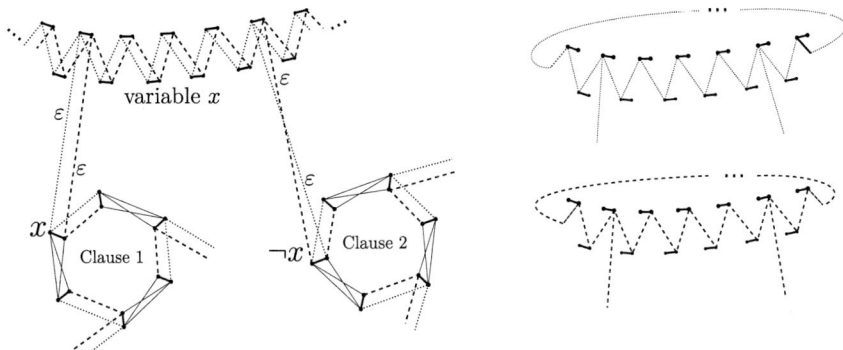

Fig. 4. (a) Bridge between variable and clause.   (b) Pure triangulations of a variable.

The gadgets used in the reduction are shown in Fig. 3. They consist of points in the plane and weights for each pair of points. A weight can take one of three values: the value 1, a small value ε and a large value σ (higher than any distance in the triangulation). We call edges with weight σ *irrelevant* and we do not show them in the figures. Including irrelevant edges in a triangulation never decreases the cost $\mathcal{W}$, therefore we can safely ignore them. The values ε and σ depend on the problem size (number of clauses and variables).

The basic building block is the *wire*, shown in Fig. 3(a). The thick solid edges are part of all triangulations. From each pair of intersecting dotted and dashed edges exactly one is included in any triangulation. The weights of all non-irrelevant edges in a wire are 1. The wire-piece can be bent and stretched freely as long as we do not introduce new crossings or remove a crossing between two edges (irrelevant edges do not matter).

The *variable* is a wire bent into a loop, with the corresponding ends glued together. We illustrate this in Fig. 3(b) using a wire-piece with 16 vertices. The construction works with any $4k$ vertices for $k \geq 3$ and we will use *much* more than 16 vertices for each variable. The *clause* gadget and the weights of its edges are shown in Fig. 3(c).

A *bridge* is a pair of edges that links a variable to a clause. A clause gadget has three fixed "places" where bridges are connected to it. We use *parallel* or *crossing* bridges, as seen in Fig. 4(a). Given a PLANAR3SAT instance (a planar circuit), we transform it into an instance of MADT as follows: we replace the vertices of the planar circuit by variable- or clause gadgets and we replace the edge between clause C and variable x by a parallel bridge if C contains x and by a crossing bridge if C contains $\neg x$.

Lemma 5. *Using the gadgets and transformations described above we can represent any planar circuit as a MADT instance.*   □

Since the gadgets allow a large amount of flexibility, the proof is straightforward. Now we can formulate our main theorem:

Theorem 3. *We can transform any planar circuit ϕ into a* MADT *instance consisting of a point set S in the plane, a* semimetric *weight function $w : S^2 \to \mathbb{R}$ and a threshold $\mathcal{W}^\star$, such that S admits a triangulation T with $\mathcal{W}(T) \leq \mathcal{W}^\star$ iff ϕ has a satisfying assignment. All computations can be done in polynomial time and the construction is of polynomial size, as are all parameters.*

Corollary 1. MADT *with semimetric weights is strongly NP-complete.*

The proof of Theorem 3 relies on a sequence of lemmas. The high level idea is the following: we call a triangulation of the construction *pure*, if every variable gadget, together with its associated bridges contains either only dashed edges or only dotted edges (besides the thick solid edges), see Fig. 4(b). First we show that we only need to consider *pure* triangulations and thus we can use the pure states of the gadgets to encode an assignment of truth values, with the convention *dotted* $\to$ (*true*), and *dashed* $\to$ (*false*). Then, we prove that satisfying assignments lead to triangulations with the smallest cost. Finally, we bound the difference in cost between different satisfying assignments and we show how to generate a *baseline* triangulation, with cost not far from the cost of a satisfying assignment (if one exists). The cost of the baseline triangulation can be computed in polynomial time.

We denote the number of variables in our planar circuit by n_v and the number of clauses by n_c. Due to condition R2, we have $n_v \leq 1.5n_c$. We denote the number of vertices in a variable gadget between two bridges (not including the bridge endpoints) by N (the same value for all variables). In Fig. 4(a), for instance, we have $N = 14$. The proof requires a careful balancing of the parameter N describing the size of the construction, and the weight ε.

Lemma 6. *If $N > 5 \cdot 10^5 n_c{}^3$, for any impure triangulation T_{impure} of the construction we can find a pure triangulation T_{pure}, with $\mathcal{W}(T_{pure}) < \mathcal{W}(T_{impure})$.*

Proof. The main idea is that if a variable is impure then the loop of the gadget is necessarily broken in some place and this leads to a penalty in cost that cannot be offset by any other change in the triangulation. □

Lemma 7. *Let T_{SAT} be a triangulation corresponding to a satisfying assignment of a planar circuit (assuming such an assignment exists) and let T_{nonSAT} be the triangulation with smallest cost $\mathcal{W}(T_{nonSAT})$ among all triangulations corresponding to nonsatisfying assignments. Then, for $N > 5 \cdot 10^5 n_c{}^3$ and $\varepsilon < \frac{1}{N^2}$, we have $\mathcal{W}(T_{nonSAT}) - \mathcal{W}(T_{SAT}) \geq \frac{N^2}{32}$.* □

Proof. We count the distances that can be smaller in T_{nonSAT} than in T_{SAT} and we bound their contribution to $\mathcal{W}$. Then we count those distances that are provably smaller in T_{SAT} than in T_{nonSAT} and we add up the differences. The crucial fact that makes the proof possible is the following: T_{nonSAT} has at least one clause with all three literals *false*. For this clause, crossing the gadget from one bridge to another costs at least $1 + 2\varepsilon$. In T_{SAT} clause crossings cost at most 4ε (Fig. 5). As each clause crossing participates in $\Omega(N^2)$ shortest paths, given bounds on N and ε we obtain the result for the difference between costs. □

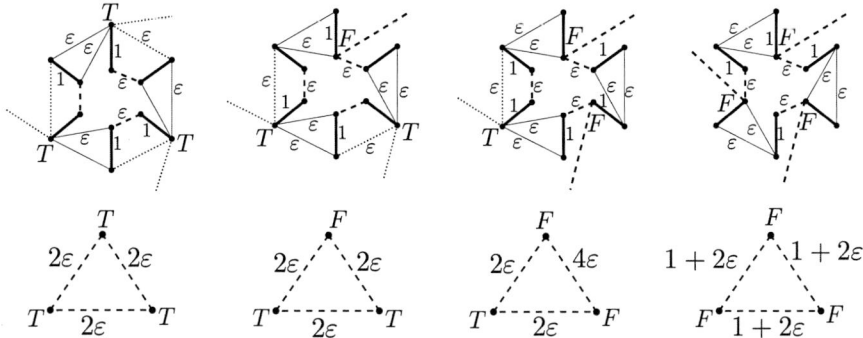

Fig. 5. (top) Optimal triangulation of a clause.   (bottom) Cost of crossing a clause.

Lemma 8. *If T_{SAT1} and T_{SAT2} are two triangulations corresponding to different satisfying assignments, then, given previous bounds on N and ε, we have that $|\mathcal{W}(T_{SAT1}) - \mathcal{W}(T_{SAT2})| \leq 150n_c{}^2 N$.*   □

Lemma 9. *For any $T_{baseline}$ and any T_{SAT}, given previous bounds, we have $|\mathcal{W}(T_{baseline}) - \mathcal{W}(T_{SAT})| \leq 150n_c{}^2 N$.*   □

We can now generate the threshold as $\mathcal{W}^{\star} = \mathcal{W}(T_{\text{baseline}}) + 300n_c{}^2 N + 1$. In accordance with our previous constraints we set $N = 10^6 n_c{}^3$, $\varepsilon = \frac{1}{2N^2}$, and this ensures that $\mathcal{W}^{\star}$ falls in the gap between satisfying and nonsatisfying triangulations. We note that all constructions and computations can be performed in polynomial time and all parameters have polynomial magnitude. This concludes the proof of NP-completeness.   □

3 Open Questions

The main unanswered question is the status of the problem for metric, in particular, for Euclidean weights. For polygons we have not succeeded in establishing results similar to Lemma 1, that would enable a dynamic programming approach. For point sets we suspect that the problem remains NP-hard, but we have not found a reduction. The Euclidean-distance problem remains open even for the special case of regular polygons with n vertices (in this case the boundary of the polygon gives a $\frac{\pi}{2}$-approximation). Computer simulation shows that the exact solution is the fan for n up to 18, *except* for the cases $n = 7$ and $n = 9$, where the solutions are shown in Fig. 6(a). We conjecture these to be the only counterexamples. As another special case, Fig. 6(b) shows the solutions obtained for a 2-by-n grid, for n up to 16.

With unit weights, the problem is unsolved for point sets not admitting a fan, even in special cases such as a 3-by-n grid. For the NP-hard variant, the question remains whether reasonable approximation schemes exist (an FPTAS is impossible, unless P=NP). One could also place other constraints on the allowed geometric graphs *besides* non-crossing, such as a total budget on the sum of edge weights or bounded degree. Further variants include allowing Steiner points or the problem of maximizing the average distance, instead of minimizing it.

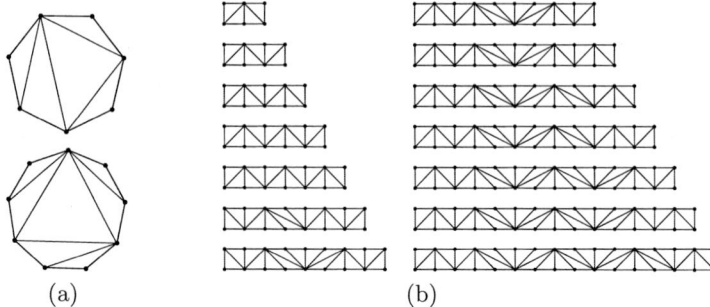

(a)   (b)

Fig. 6. (a) Solution for regular polygons.   (b) Solution for 2-by-n grids.

Acknowledgement. I thank my advisor, Raimund Seidel, for mentioning the problem at the INRIA-McGill-Victoria Workshop on Computational Geometry (2011) at the Bellairs Research Institute, and for valuable comments. I also thank several anonymous reviewers for pointing out errors and suggesting improvements.

References

1. Johnson, D.S., Lenstra, J.K., Kan, A.H.G.R.: The complexity of the network design problem. Networks 8(4), 279–285 (1978)
2. Wiener, H.: Structural Determination of Paraffin Boiling Points. J. of the Am. Chem. Soc. 69(1), 17–20 (1947)
3. Rouvray, D.H.: Predicting chemistry from topology. Sci. Am. 255 (1986)
4. Mohar, B., Pisanski, T.: How to compute the Wiener index of a graph. J. of Mathematical Chemistry 2, 267–277 (1988)
5. Dobrynin, A.A., Entringer, R., Gutman, I.: Wiener index of trees: Theory and applications. Acta Applicandae Mathematicae 66, 211–249 (2001)
6. Nilsen, C.W.: Wiener index and diameter of a planar graph in subquadratic time (2009)
7. Aurenhammer, F., Xu, Y.: Optimal triangulations. In: Encyclopedia of Optimization, pp. 2757–2764 (2009)
8. Bern, M.W., Eppstein, D.: Mesh generation and optimal triangulation. In: Computing in Euclidean Geometry, pp. 23–90 (1992)
9. Gilbert, P.D.: New results in planar triangulations. Report R-850, Coordinated Sci. Lab., Univ. Illinois, Urbana, IL (1979)
10. Klincsek, G.T.: Minimal triangulations of polygonal domains. Discr. Math. 9, 121–123 (1980)
11. Mulzer, W., Rote, G.: Minimum-weight triangulation is NP-hard. J. ACM 55, 11: 1–11 (2008)
12. Kozma, L.: Minimum average distance triangulations (2012), http://arxiv.org/abs/1112.1828
13. Lichtenstein, D.: Planar formulae and their uses. SIAM J. Comput. 11(2), 329–343 (1982)
14. Knuth, D.E., Raghunathan, A.: The problem of compatible representatives. SIAM J. Discr. Math. 5, 422–427 (1992)

Colouring AT-Free Graphs

Dieter Kratsch[1] and Haiko Müller[2]

[1] Université de Lorraine, LITA
57045 Metz, Cedex 01, France
kratsch@univ-metz.fr
[2] University of Leeds, School of Computing
Leeds, LS2 9JT, United Kingdom
h.muller@leeds.ac.uk

Abstract. A vertex colouring assigns to each vertex of a graph a colour
such that adjacent vertices have different colours. The algorithmic complexity of the COLOURING problem, asking for the smallest number of
colours needed to vertex-colour a given graph, is known for a large number of graph classes. Notably it is NP-complete in general, but polynomial
time solvable for perfect graphs. A triple of vertices of a graph is called an
asteroidal triple if between any two of the vertices there is a path avoiding all neighbours of the third one. Asteroidal triple-free graphs form
a graph class with a lot of interesting structural and algorithmic properties. Broersma *et al.* (ICALP 1997) asked to find out the algorithmic
complexity of COLOURING on AT-free graphs. Even the algorithmic complexity of the k-COLOURING problem, which asks whether a graph can
be coloured with at most a fixed number k of colours, remained unknown
for AT-free graphs. First progress was made recently by Stacho who presented an $O(n^4)$ time algorithm for 3-colouring AT-free graphs (ISAAC
2010). In this paper we show that k-COLOURING on AT-free graphs is in
XP, *i.e.* polynomial time solvable for any fixed k. Even more, we present
an algorithm using dynamic programming on an asteroidal decomposition which, for any fixed integers k and a, solves k-COLOURING on any
input graph G in time $\mathcal{O}(f(a,k) \cdot n^{g(a,k)})$, where a denotes the asteroidal
number of G, and $f(a,k)$ and $g(a,k)$ are functions that do not depend
on n. Hence for any fixed integer k, there is a polynomial time algorithm solving k-COLOURING on graphs of bounded asteroidal number.
The algorithm runs in time $\mathcal{O}(n^{8k+2})$ on AT-free graphs.

1   Introduction

Graph Colouring. One of the classical subjects in graph theory is vertex
colouring. It involves the labelling of the vertices of some given graph by colours
such that no two adjacent vertices receive the same colour. The COLOURING
problem asks to compute the smallest number of colours sufficient to colour a
graph. The related k-COLOURING problem asks to decide whether a graph can
be coloured with at most k colours. COLOURING as well as k-COLOURING for any
fixed $k \geq 3$ were among the first problems shown to be NP-complete [7]. Due

L. Epstein and P. Ferragina (Eds.): ESA 2012, LNCS 7501, pp. 707–718, 2012.
© Springer-Verlag Berlin Heidelberg 2012

to the importance of these problems in theory and applications there has been considerable interest in studying their algorithmic complexity when restricted to certain graph classes, see, *e.g.* [10]. An outstanding result in this respect is due to Grötschel, Lovász, and Schrijver [8], who show that COLOURING is polynomial time solvable for perfect graphs.

Asteroidal Triple-Free Graphs. Three vertices of a graph form an asteroidal triple (short AT) if between any two of them there is a path avoiding all neighbours of the third. This concept was introduced by Lekkerkerker and Boland in 1962 to characterise interval graphs as those chordal graphs having no AT [16]. A graph without AT is called asteroidal triple-free (short AT-free). AT-free graphs contain various classes of perfect graphs as *e.g.* permutation, interval, and cobipartite graphs. However not all AT-free graphs are perfect. For example, the C_5 (the cycle on 5 vertices) is AT-free and not perfect. For definitions and further information on graph classes, see, *e.g.* [1].

The study of the structure of AT-free graphs goes back to the nineties. It was initiated by Corneil, Olariu and Stewart in two fundamental papers [4,5]. While AT-free graphs seem to lack algorithmically useful representations like geometric intersection models or vertex orderings, their structure provides other algorithmically useful tools; dominating pairs and 2LexBFS, *i.e.* a modified lexicographic breadth-first search, are worth mentioning [4,5]. Walter introduced asteroidal sets as those independent sets of a graph in which every triple of vertices forms an AT [18]. The asteroidal number of a graph G, denoted by an(G), is the maximum cardinality of an asteroidal set in G. Note that an$(G) \leq 1$ iff G is complete, and that an$(G) \leq 2$ iff G is AT-free. In [13] Kloks *et al.* study the structure of graphs of bounded asteroidal number.

Related Work. The structural and algorithmic results of Corneil, Olariu and Stewart inspired the study of the algorithmic complexity of NP-complete problems on AT-free graphs. Polynomial time algorithms for AT-free graphs were presented for various well-known NP-complete problems among them INDEPENDENT SET, DOMINATING SET and FEEDBACK VERTEX SET [2,14,15]. Other problems remain NP-complete on AT-free graphs as, *e.g.* CLIQUE, PARTITION INTO CLIQUES and BANDWIDTH [2,12]. Various algorithmic results for AT-free graphs were generalised to graphs of bounded asteroidal number [2,3,15].

Fifteen years ago at ICALP 1997, Broersma *et al.* (see [2]) asked about the algorithmic complexity of COLOURING, HAMILTONIAN PATH and HAMILTONIAN CIRCUIT on AT-free graphs. Up to our knowledge, none of these questions has been answered to this day. The only progress made so far was achieved by Stacho who recently provided a polynomial time algorithm for 3-colouring AT-free graphs [17]. His algorithm runs in time $O(n^4)$ and is based on a sophisticated analysis of the structure of 3-colourable AT-free graphs.

Our Work. We present an algorithm using dynamic programming on an asteroidal decomposition of the connected input graph. For any fixed integer k, it decides in polynomial time whether a given AT-free graph can be coloured with at most k colours. Hence k-COLOURING on AT-free graphs is in XP. We present our algorithm in the following more general setting: for any fixed integers

k and a, our algorithm solves k-COLOURING on an input graph G with n vertices in time $\mathcal{O}(f(a,k) \cdot n^{g(a,k)})$ where $\text{an}(G) \leq a$ and both functions $f(a,k)$ and $g(a,k)$ do not depend on n. Hence for any fixed integer k, the algorithm solves k-COLOURING on graphs of bounded asteroidal number in polynomial time. On AT-free graphs it runs in time $\mathcal{O}(n^{8k+2})$.

The paper is organised as follows. Preliminaries on graphs and graphs of bounded asteroidal number are given in Section 2. In Section 3 we introduce the asteroidal decomposition of graphs. This new decomposition of graphs is of interest on its own. In Section 4 vertex colourings and their representations are studied. In Section 5 we present our algorithm, analyse its running time, and prove its correctness.

2 Preliminaries

All graphs we consider are finite, undirected and simple. Let $G = (V, E)$ be such a graph and let $n = |V|$ and $m = |E|$. We denote by $G[U]$ the subgraph of G induced by $U \subseteq V$. We write $G - U$ instead of $G[V \setminus U]$ for $U \subseteq V$, and $G - u$ instead of $G[V \setminus \{u\}]$ for $u \in V$. The subset of V formed by vertices adjacent to $v \in V$ is the *open neighbourhood* $\mathrm{N}_G(v) = \{u \mid \{u, v\} \in E\}$ of v in G. The *closed neighbourhood* of v is $\mathrm{N}_G[v] = \{v\} \cup \mathrm{N}_G(v)$. Both concepts generalise to subsets $U \subseteq V$. We define $\mathrm{N}_G[U] = \bigcup_{u \in U} \mathrm{N}[u]$ and $\mathrm{N}_G(U) = \mathrm{N}[U] \setminus U$. A *clique* in $G = (V, E)$ is a subset of V consisting of pairwise adjacent vertices. A set of pairwise non-adjacent vertices is *independent*. Indices in notations are omitted if there is no ambiguity. A set of vertices $U \subseteq V$ of a graph $G = (V, E)$ is called *connected* if $G[U]$ is connected. A maximal connected induced subgraph is called *(connected) component*. That is, $G[U]$ is a component of G if U is a maximal connected set in G. For two non-adjacent vertices u and w of G, we denote by $\mathrm{cc}(u, w)$ the maximal connected set of $G - \mathrm{N}[u]$ containing w.

The following definition is equivalent to the aforementioned one by Walter. A set $A \subseteq V$ is *asteroidal* in $G = (V, E)$ if for every $v \in A$ there is a component of $G - \mathrm{N}[v]$ containing all vertices of $A \setminus \{v\}$. Hence asteroidal sets are independent. An independent set A is asteroidal if $A \setminus \{v\} \subseteq \mathrm{cc}(v, u)$ holds for every pair of different vertices $u, v \in A$. An asteroidal set of size three is called *asteroidal triple*. The maximum size of an asteroidal set of G is the *asteroidal number* of G, denoted by $\text{an}(G)$. Graphs with asteroidal number at most two are called *AT-free*. Note that the following sets are asteroidal in every graph $G = (V, E)$: $\varnothing$, $\{v\}$ for all $v \in V$, and $\{u, v\}$ for all pairs of non-adjacent vertices u and v. For further information on graphs of bounded asteroidal number we refer to [13].

Definition 1. *For each vertex v of a graph $G = (V, E)$ we define a binary relation $\preccurlyeq_v$ on $\mathrm{N}(v)$ as follows:* $u \preccurlyeq_v w \iff \mathrm{N}(u) \setminus \mathrm{N}(v) \subseteq \mathrm{N}(w) \setminus \mathrm{N}(v)$.

The relation $\preccurlyeq_v$ is reflexive and transitive. The *width* of $(\mathrm{N}(v), \preccurlyeq_v)$ is the maximum size of an antichain, that is a subset of $\mathrm{N}(v)$ of pairwise $\preccurlyeq_v$-incomparable vertices. The above definition and the following lemma are important for our representations of colourings.

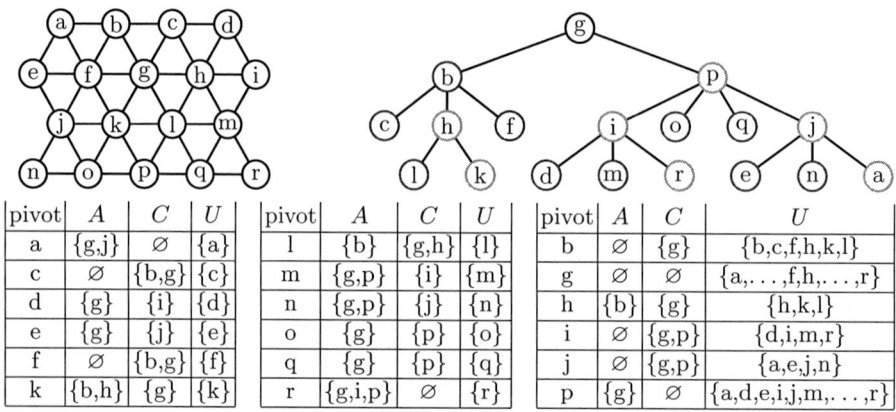

pivot	A	C	U
a	{g,j}	∅	{a}
c	∅	{b,g}	{c}
d	{g}	{i}	{d}
e	{g}	{j}	{e}
f	∅	{b,g}	{f}
k	{b,h}	{g}	{k}

pivot	A	C	U
l	{b}	{g,h}	{l}
m	{g,p}	{i}	{m}
n	{g,p}	{j}	{n}
o	{g}	{p}	{o}
q	{g}	{p}	{q}
r	{g,i,p}	∅	{r}

pivot	A	C	U
b	∅	{g}	{b,c,f,h,k,l}
g	∅	∅	{a,...,f,h,...,r}
h	{b}	{g}	{h,k,l}
i	∅	{g,p}	{d,i,m,r}
j	∅	{g,p}	{a,e,j,n}
p	{g}	∅	{a,d,e,i,j,m,...,r}

Fig. 1. An asteroidal decomposition of a triagonal grid. Red nodes are internal children, green nodes are external children. The nodes of the tree are labelled by pivots only.

Lemma 1 ([15]). *Let v be any vertex of a graph $G = (V, E)$ and let $S \subseteq N(v)$ be an independent set of G. If the asteroidal number of G is at most a then the width of the restriction $(S, \preceq_v)$ is at most $2a^2$.*

Remark 1. Lemma 1 can be strengthened as follows. If G is AT-free then the width of $(S, \preceq_v)$ is at most four [15].

3   Asteroidal Decomposition

In this section we introduce a new decomposition of graphs called asteroidal decomposition. It is inspired by techniques used to construct algorithms solving NP-complete problems for AT-free graphs, as *e.g.* INDEPENDENT SET [2] and FEEDBACK VERTEX SET [15].

Let us define an asteroidal decomposition of a connected graph $G = (V, E)$. Such a decomposition recursively decomposes G into connected induced subgraphs $G[U]$, which we call *blocks* of G, by applying the following operation while $|U| \geq 2$. To decompose a block $G[U]$, choose an arbitrary vertex $u \in U$ as *pivot*. Then $G[U]$ decomposes into the components of $G[U \setminus N[u]]$ called *external u-children* of $G[U]$ and into the components of $G[U \cap N(u)]$ called *internal u-children* of $G[U]$. Like in various graph decompositions, a fixed asteroidal decomposition of a connected graph $G = (V, E)$ can be described by means of a decomposition tree. We assign the graph G and its pivot $v \in V$ to the root of the tree T. If $G[U]$ and its pivot $u \in U$ is assigned to a node of T then its internal and external u-children are assigned to the children of this node. Finally to each leaf of T we assign $G[\{v\}]$ and v, which acts as pivot. Figure 1 gives an example.

Theorem 1. *Each decomposition tree T of an asteroidal decomposition of a connected graph $G = (V, E)$ has $n = |V|$ nodes. It can be computed in time $O(nm)$.*

Proof. Let $G[U]$ be a block of T with children $G[U_i]$ for $i = 1, \ldots, l$. The pivot $u \in U$ partitions $U \setminus \{u\}$ into $U_1, \ldots, U_l$. That is, each vertex in V is the pivot of exactly one block of T. Therefore T has n nodes. For each block $G[U]$ its external u-children are the components of $G[U \setminus N[u]]$ and its internal u-children are the components $G[U \cap N(u)]$. Using a linear time algorithm to compute the components of a graph gives the overall running time $O(nm)$.   □

It is crucial for our algorithmic application of the asteroidal decomposition that its blocks can actually be described as (A, C, U)-blocks defined as follows.

Definition 2. *Let $G = (V, E)$ be a graph, $A \subseteq V$ an asteroidal set of G and $C \subseteq V$ a clique of G. $G[U]$ is said to be an (A, C, U)-block of G if it is a component of $G[\bigcap_{c \in C} N(c)] - N[A]$.*

That is, U is a maximal subset of V such that all vertices in U belong to the same component of $G - N[A]$, all vertices in U are adjacent to all vertices in C, and $G[U]$ is connected. We use the following extended labelling of the decomposition tree. Each node of T corresponds to an (A, C, U)-block of G, and carries the label (A, C, U, u). Here $u \in U$ is the pivot of the block, *i.e.* the vertex chosen to decompose $G[U]$. Each leaf of T corresponds to a block of the form $(A, C, \{u\})$ for a vertex $u \in V$, which also acts as pivot. The entire graph G is the $(\varnothing, \varnothing, V)$-block of G and thus the label of the root of T is $(\varnothing, \varnothing, V, v)$, where v is the pivot of G. Let $u \in U$ be the vertex chosen to decompose $G[U]$ in the decomposition tree. Then the children of the (A, C, U)-block in the tree are blocks called either internal or external. An internal child of the (A, C, U)-block is an $(A, C \cup \{u\}, U')$-block with $U' \subseteq U \cap N(u)$, and an external child is an (A'', C, U'')-block with $U'' \subseteq U \setminus N[u]$ and $A'' \subseteq A \cup \{u\}$. Finally let us mention some properties of an (A, C, U)-block with pivot $u \in U$ which are easy consequences of the definition. (1) U uniquely determines C but not A. (2) A, C and u uniquely determine U. (3) A, C and u uniquely determine the internal and external children of $G[U]$. (4) $N(U) \subseteq C \cup N(A)$.

The following lemma shows that all blocks of an asteroidal decomposition are indeed (A, C, U)-blocks, and that corresponding labels (A, C, U, u) can be assigned in time $O(nm)$ per node of the decomposition tree.

Lemma 2. *There is an $O(n^2 m)$ algorithm computing for each block $G[U]$ of a given asteroidal decomposition of G an asteroidal set A and a clique C of G such that $G[U]$ is an (A, C, U)-block.*

Proof. Let T be an asteroidal decomposition tree of $G = (V, E)$. We consider the pivots on the path P in T from the root to the node with label (U, u). All pivots with an internal child on P form a clique C such that every vertex of U is adjacent to all vertices of C since all vertices of an internal w-child are adjacent to w. The pivots with an external child on P form an independent set S since no vertex of an external w-child is adjacent to w. Furthermore $G[U]$ is a component of $G[\bigcap_{c \in C} N(c)] - N[S]$.

For each $X \subseteq S$ we define a vertex $v \in X$ to be X-*essential* if it has a neighbour in $N(U) \setminus N(X \setminus \{v\})$. We compute $A \subseteq S$ as follows. First let $X = S$.

While there is a vertex $v \in X$ that is not X-essential we replace X by $X \setminus \{v\}$. The procedure stops when all vertices of X are X-essential. This set is called A.

For two different vertices $v, w \in A$ we have $w \in cc(v, u)$ since w is A-essential. That is, $A \setminus \{v\} \subseteq cc(v, u)$. Consequently, A is an asteroidal set of G. Whenever we update X, $G[U]$ remains a component of $G[\bigcap_{c \in C} N(c)] - N[X]$. Hence this is the case for $X = A$ as well.

Given G and T, the sets C and S can be constructed in linear time. In time $\mathcal{O}(nm)$ we find $N(U) \cap N(v)$ for all $v \in S$, and select $A \subseteq S$ in time $\mathcal{O}(n^2)$.   □

Our algorithm to decide whether a graph of bounded asteroidal number is k-colourable uses dynamic programming on an asteroidal decomposition tree of the connected input graph.

4   Partial Colourings and Their Representations

In this section we study partial k-colourings of graphs of bounded asteroidal number and representations of such colourings. To achieve polynomial running time for fixed k, we use representations which are based on combinatorial properties given in [15].

A *vertex colouring* of a graph $G = (V, E)$ is a partition of V into independent sets of G. A *partial colouring* of G is a packing of disjoint independent sets of G. In both cases we refer to these independent sets as *colour classes*. A *k-colouring* is a vertex colouring with at most k colour classes. We adopt the convention that a k-colouring is a set of exactly k colour classes where empty classes are allowed. Slightly abusing notation, we allow multiple empty sets. For a partial colouring $\mathcal{P}$ of G let $\bigcup \mathcal{P} = \{v \in V \mid \exists F \in \mathcal{P} \text{ s.t. } v \in F\}$ be the set of *coloured* vertices. All vertices in $V \setminus \bigcup \mathcal{P}$ are *uncoloured*. Let $\mathcal{P}$ and $\mathcal{Q}$ be partial colourings of $G = (V, E)$. The *restriction* of $\mathcal{P}$ to U is $\mathcal{P}|U = \{F \cap U \mid F \in \mathcal{P}\}$. If $\mathcal{P} = \mathcal{Q}|\bigcup \mathcal{P}$ then we call $\mathcal{P}$ a *restriction* of $\mathcal{Q}$ and we also call $\mathcal{Q}$ an *extension* of $\mathcal{P}$. A partial colouring $\mathcal{Q}$ is a *common extension* of the partial colourings $\mathcal{P}_1, \mathcal{P}_2, \mathcal{P}_3, \ldots, \mathcal{P}_l$ if $\bigcup \mathcal{Q} = \bigcup_{i=1}^{l} \bigcup \mathcal{P}_i$ and $\mathcal{P}_i = \mathcal{Q}|\bigcup \mathcal{P}_i$ for all $i = 1, 2, \ldots, l$.

Definition 3. *Let v be a vertex of G and let $\mathcal{P}$ and $\mathcal{R}$ be partial colourings of $G[N(v)]$. We say that $\mathcal{R}$ v-represents $\mathcal{P}$ if $\mathcal{R} = \mathcal{P}|\bigcup \mathcal{R}$ and for each vertex $u \in \bigcup \mathcal{P}$ there is a vertex $w \in \bigcup \mathcal{R}$ such that $u \preccurlyeq_v w$ and there is a colour class $F \in \mathcal{P}$ with $\{u, w\} \subseteq F$.*

The v-represents-relation on partial colourings of $G[N(v)]$ is transitive: If $\mathcal{R}$ v-represents $\mathcal{P}$ and $\mathcal{Q}$ v-represents $\mathcal{R}$, then $\mathcal{Q}$ v-represents $\mathcal{P}$ too. The following lemma is an immediate consequence of Lemma 1.

Lemma 3. *Let v be a vertex of G and $an(G) \leq a$. Every vertex colouring of $G[N(v)]$ can be v-represented by a partial colouring in which each colour class has size at most $2a^2$.*

Definition 4. *Let $\mathcal{P}$ and $\mathcal{R}$ be partial colourings of a graph $G = (V, E)$. For $u \in V$ we say $\mathcal{P}$ is u-consistent with $\mathcal{R}$ if there is a common extension $\mathcal{Q}$ of $\mathcal{P}$ and $\mathcal{R}$ such that the restriction $\mathcal{R}|N(u)$ u-represents $\mathcal{Q}|N(u)$.*

The correctness of our algorithm is based on the following lemma which shows that maintaining certain partial colourings is sufficient to verify k-colourability.

Lemma 4. *Let $G = (V, E)$ be a connected graph, $u \in V$. Let $U_1, \ldots, U_p$ induce the components of $G[N(u)]$, and $U_{p+1}, \ldots, U_q$ induce the components of $G - N[u]$. Let $k \geq 1$ and let $\mathcal{R}$ be a partial $(k-1)$-colouring of $G[N(u)]$. If*

- *for $i = 1, \ldots, p$ a vertex $(k-1)$-colouring $\mathcal{P}_i$ of $G[U_i]$ exists that is u-consistent with $\mathcal{R}$, and*
- *for $i = p+1, \ldots, q$ a vertex k-colouring $\mathcal{P}_i$ of $G[U_i]$ exists that is u-consistent with $\mathcal{R}$,*

then there is a vertex k-colouring of G.

Proof. Let us assume that $|V| \geq 2$ and $k > 1$; all other cases are trivial. We construct a common extension $\mathcal{Q}$ of $\mathcal{P}_1, \ldots, \mathcal{P}_q$ that is a vertex k-colouring of G.

Let us first give the idea of the construction. We permute the colour classes in $\mathcal{P}_1, \ldots, \mathcal{P}_p$ to match the colouring $\mathcal{R}$. This can be done independently on each component $G[U_i]$ of $G[N(u)]$ since there are no edges between these components. Similarly we permute the colour classes in $\mathcal{P}_{p+1}, \ldots, \mathcal{P}_q$ which can be done independently on each component $G[U_i]$ of $G - N[u]$ since there are no edges between these components. Finally we assign to u a colour not used on $N(u)$, which must exist since $\mathcal{P}_i$ is a $(k-1)$-colouring of $G[U_i]$ for each $i = 1, \ldots, p$. The resulting colouring is $\mathcal{Q}$.

Formally, for $i = 1, 2, \ldots, q$, let $\mathcal{Q}_i$ be the common extension of $\mathcal{P}_i$ and $\mathcal{R}$ that exists by definition. For each $F \in \mathcal{R}$, let F_i be the colour class in $\mathcal{Q}_i$ containing F. For $F = \{u\}$ let F_i be a (possibly empty) colour class of $\mathcal{Q}_i$ that is disjoint with $N(u)$. If $\mathcal{R}$ has $k-1$ non-empty colour classes then $\mathcal{Q} = \{\bigcup_{i=1}^{q} F_i \mid F \in \mathcal{R} \cup \{\{u\}\}\}$. Otherwise, for $i = p+1, \ldots, q$, the remaining colour classes $F_i \in \mathcal{Q}_i$ that are disjoint with $N(u)$, are joined arbitrarily to empty colour classes of $\mathcal{R}$ to form colour classes of $\mathcal{Q}$. Consequently $\mathcal{Q}$ is an extension of $\mathcal{P}_i$ for all $i = 1, \ldots, q$. Now we show that $\mathcal{Q}$ is a vertex k-colouring of G.

Covering: We have $V = \{u\} \cup \bigcup_{i=1}^{q} U_i$, and $N(u) \neq \varnothing$ because G is connected, consequently $q \geq p \geq 1$. The vertex u is coloured in $\mathcal{Q}$ by its construction. For each $i = 1, \ldots, q$, $\mathcal{Q}$ covers U_i because $\mathcal{Q}$ is an extension of $\mathcal{P}_i$.

Packing: The vertex u belongs to exactly one colour class $F \in \mathcal{Q}$, which is disjoint with $N(u)$. Now assume there is an index $i = 1, \ldots, q$ and a vertex $x_i \in U_i$ that belongs to two classes of $\mathcal{Q}$. This leads to a contradiction since $\mathcal{Q}|U_i = \mathcal{P}_i$ and $\mathcal{P}_i$ is a vertex colouring of $G[U_i]$.

Independence: Assume for the sake of a contradiction, there is an edge $\{x, y\}$ between two vertices in the same class $F \in \mathcal{Q}$. Since there are no edges between vertices in different components of $G[N(u)]$ or $G - N[u]$ we may assume $x \in U_i$ for some $i = 1, \ldots, p$, and $y \in U_j$ for some $j = p+1, \ldots, q$. Since $\mathcal{R}$ u-represents $\mathcal{P}_i$ there is a vertex $z \in N(u) \cap F$ with $x \preccurlyeq_u z$. This implies the existence of an edge $\{z, y\}$ with $y, z \in F$, which contradicts the fact that $\mathcal{Q}_j$ is a common extension of $\mathcal{P}_j$ and $\mathcal{R}$. Therefore each class of $\mathcal{Q}$ is an independent set of G.

Size: We have $|\mathcal{Q}| \leq k$ by construction.   □

5   The Algorithm

The inputs of our algorithm are a connected graph $G = (V, E)$ and fixed integers k and a. (Recall that a disconnected graph is k-colourable iff all its components are k-colourable.) The number of available colours is k, and it is fixed in advance. The asteroidal number of G is at most a, which can be checked in time $\mathcal{O}(n^{a+2})$; see, e.g. [11]; there is no need to know $an(G)$ explicitly. Contrary to the previous section, we assume in this one that for every partial colouring $\mathcal{P}$ of a graph $G = (V, E)$ the colour $\mathcal{P}(v) = i$ of each coloured vertex v is explicitly known. For all $i = 1, 2, \ldots, k$, let $\mathcal{P}_i = \{v \in V \mid \mathcal{P}(v) = i\}$ be the i-th colour class of $\mathcal{P}$.

Now we define candidate colourings of (A, C, U)-blocks and their validity, which play a major role in our algorithm. For any vertex s of G an s-colouring of G is a partial k-colouring of $G[N[s]]$ such that (1) s is coloured, (2) for each $u \in N(s)$ there is a coloured $w \in N(s)$ such that $u \preccurlyeq_s w$, and (3) each colour class other than $\{s\}$ is an antichain of $(N(s), \preccurlyeq_s)$. Two partial colourings $\mathcal{P}$ and $\mathcal{P}'$ are *consistent* if their union $\mathcal{P} \cup \mathcal{P}'$ is a partial colouring as well. That is, each vertex has at most one colour, and adjacent vertices do not have the same colour. For an independent set A and a clique C of G, both possibly empty, an *(A,C)-colouring* consists of $|A| + 1$ consistent partial colourings, namely a vertex colouring of $G[C]$ and an s-colouring for each vertex $s \in A$. A *candidate colouring* of an (A, C, U)-block is an (A, C)-colouring; note that no vertex of U is coloured. For an independent set B and a clique D with $B \cup D \subseteq A \cup C$, the *restriction* of an (A, C)-colouring $\mathcal{P}$ to (B, D) is the (B, D)-colouring that contains all s-colourings of $\mathcal{P}$ for $s \in B$. A candidate colouring $\mathcal{P}$ of an (A, C, U)-block of G is *valid* if there is a vertex k-colouring $\mathcal{Q}$ of $G[\bigcup \mathcal{P} \cup U]$ such that $\mathcal{P}$ is a restriction of $\mathcal{Q}$.

By Lemma 1 each of its colour classes contains at most $2a^2$ vertices. The algorithm encodes a candidate colouring $\mathcal{P}$ of an (A, C, U)-block by a sequence $(\mathrm{col}(C), \mathrm{col}(s_1), \mathrm{col}(s_2), \ldots, \mathrm{col}(s_{|A|}))$, called encoding of $\mathcal{P}$. (For simplicity we may assume that every coloured vertex is stored with its colour as a pair.) For each $s \in A$, $\mathrm{col}(s)$ denotes an s-colouring of $G[N[s]]$. Furthermore, every colour class of $\mathcal{P}$ may also contain one vertex of C. The total number of coloured vertices in such an encoding of a candidate colouring of an (A, C, U)-block is at most $2a^3k + k$. As pointed out above, not each such sequence $(\mathrm{col}(C), \mathrm{col}(s_1), \mathrm{col}(s_2), \ldots, \mathrm{col}(s_{|A|}))$ encodes indeed a candidate colouring; a vertex may have different colours and vertices of the same colour may be adjacent in G. Fortunately it is easy to find out whether all s_j-colourings, $j = 1, 2, \ldots, |A|$, and the colouring of C are pairwise consistent. This can be done by simply checking all vertices and edges of the graph induced by all coloured vertices of $\mathcal{P}$ in time $\mathcal{O}((2a^3k + k)^2) = \mathcal{O}(a^6 k^2)$ per sequence. All sequences of valid candidate colourings of a certain block will be stored in a dictionary allowing access in time logarithmic in the size of the dictionary.

Lemma 5. *An (A, C, U)-block of an asteroidal decomposition of a graph $G = (V, E)$ has at most $k^k n^{2a^3k}$ different candidate colourings. All candidate colourings of an (A, C, U)-block can be generated in time $\mathcal{O}(a^6 k^{k+2} n^{2a^3k+3})$.*

Proof. Consider an (A, C, U)-block. Let $\mathcal{R}$ be a candidate colouring of this block. Then for each $s \in A$, at most $2a^2k$ vertices are coloured by the s-colouring of $\mathcal{R}$. Hence $|A| \leq a$ and $|C| \leq k$ imply $|\bigcup \mathcal{R}| \leq 2a^3k + k$. To generate the encodings of all candidate colourings (and many other encodings) we assign the vertices to its colour starting with colour 1, then 2, and so on. For each colour i, $i = 1, 2, \ldots k$, we consecutively choose among the not yet assigned vertices one or no vertex of C to colour it i, and either only s_j is coloured i or we assign one by one at most $2a^2$ vertices of $\mathrm{N}(s_j)$ to the i-th colour class of the s_j-colouring of $\mathcal{R}$ for $j = 1, 2, \ldots, |A|$. In this way we generate all sequences that may encode a candidate colouring $\mathcal{R}$ of the (A, C, U)-block of G. Taking into account that there are at most $k^{|C|} \leq k^k$ colourings of C, we obtain the stated bound.

To generate all candidate colourings $\mathcal{R}$ of the (A, C, U)-block we guess all encodings of candidate colourings of this block as described above, e.g. in a quasilexicographic order based on some fixed linear order of the vertices. Clearly we only choose among vertices of $C \cup \mathrm{N}[A]$. For each encoding generated in this way, we need to verify whether it is indeed an encoding of an (A, C)-colouring. This is easy for vertices of C. For every vertex $s_j \in A$ it requires checking whether the encoding of the s_j-colouring, $j = 1, 2, \ldots, |A|$ fulfils conditions (1)–(3) of the definition on the previous page. Checking condition (1) is easy. It is not hard to check conditions (2) and (3) as soon as $(\mathrm{N}(s_j), \preccurlyeq_{s_j})$ is available. This can be computed in time $\mathcal{O}(n^3)$ for every j. Hence this part of the test can be done in time $\mathcal{O}(akn^3)$ per generated encoding. Finally we need to verify whether an encoding corresponds indeed to a partial k-colouring of the subgraph induced by the coloured vertices. As described above, this can be done in time $\mathcal{O}(a^6 k^2)$. □

Consequently both inserting and searching an encoding in the dictionary can be done in time $O(a^3 k \log n)$.

Now we describe our algorithm that uses dynamic programming on an asteroidal decomposition tree of the input graph G. As shown in Section 3, such a tree T with labels (A, C, U, u) assigned to its nodes can be computed in time $O(n^2 m)$, see Lemma 2. Note that in a graph G of asteroidal number at most a, $|A| \leq a$ for all asteroidal sets A of G. Furthermore, in every k-colourable graph, all cliques have size at most k. Thus if there is an (A, C, U)-block with $|C| > k$ in the asteroidal decomposition then G is not k-colourable and the algorithm stops. The fundamental idea of the algorithm is to compute in a bottom-up manner for each (A, C, U)-block of the decomposition tree the collection of its valid candidate colourings. If the algorithm detects an (A, C, U)-block without valid candidate colouring then it declares G not k-colourable and stops. Otherwise the algorithm outputs that G is k-colourable after finding that the unique candidate colouring of G, which does not colour any vertex of the $(\varnothing, \varnothing, V)$-block, is valid.

One may view our algorithm as producing a huge number of valid candidate colourings in all the leaves of the decomposition tree, and then verifying in a bottom-up manner which of them can be composed to valid candidate colourings of interior nodes of the tree T. It should be mentioned that the algorithm can be modified to output a k-colouring of G if there is one. To do this, store for every valid candidate colouring $\mathcal{P}$ of an (A, C, U)-block one (arbitrary) extension $\mathcal{Q}$ which is a vertex k-colouring of $G[\bigcup \mathcal{P} \cup U]$; finally output the extension stored for the unique valid candidate colouring of G which is a k-colouring of G.

Let us describe the dynamic programming on the asteroidal decomposition tree T in more detail. First consider a leaf of T. Let its label be $(A, C, \{v\}, v)$ with $v \in V$. Since $C \cup \{v\}$ is a clique the graph G has no k-colouring if $|C| \geq k$, and the algorithm rejects the input. Otherwise candidate colourings of this block are of the following form: for every $s \in A$ an s-colouring of $G[N[s]]$ is part of the candidate colouring; also every colour class contains at most one vertex of C. By Lemma 5, there are at most $k^k n^{2a^3 k}$ encodings of candidate colourings of the $(A, C, \{v\})$-block. Verifying whether such a candidate colouring is valid can be done by checking whether the graph induced by v and all coloured vertices is correctly k-coloured, since $U = \{v\}$ consists of one vertex only.

Now consider an interior node of T. Let its label be (A, C, U, u). By construction each child of the (A, C, U)-block is either internal or external u-child. Note that the correctness of this step is based on Lemma 4. Due to the bottom-up manner of the dynamic programming, the valid candidate colourings of each u-child have already been computed and each u-child has at least one. The valid candidate colourings of the (A, C, U)-block are computed as follows. We generate all candidate colourings of this block as already described and verify for each one whether it is valid. To check whether a fixed candidate colouring $\mathcal{P}$ is valid, for every u-colouring $\mathcal{P}'$ of $G[N[u]]$ which is consistent with $\mathcal{P}$, we add the encoding of $\mathcal{P}'$ to the one of $\mathcal{P}$. For each partial colouring of $G[U \cup C \cup N[A \cup \{u\}]]$ obtained in this way, we compute its restriction to the vertex set of all internal u-children and external u-children of the (A, C, U)-block. We check whether the restriction to the vertex set of the corresponding u-child is an already computed valid candidate colouring for it, which can be checked by dictionary look-up. Moreover, the candidate colourings of the internal u-children have to be u-represented by $\mathcal{P}'$. If this is the case for all u-children then the candidate colouring of the (A, C, U)-block under verification is indeed valid and will be stored as a valid candidate colouring of the (A, C, U)-block in the dictionary.

Theorem 2. *The algorithm has running time $O(a^9 k^{k+3} n^{2a^2(a+1)k+2})$.*

The proof of the theorem is omitted due to space restrictions.

Corollary 1. *The algorithm can be modified such that its running time on AT-free graphs is $\mathcal{O}(n^{8k+2})$ for any fixed k.*

Proof. Firstly by the remark to Lemma 1, colour classes of s-colourings on AT-free graphs can be chosen to have at most 4 vertices (instead of $2a^2 = 8$). Secondly, it can be shown that we can choose an asteroidal decomposition of an AT-free graph such that $|A| \leq 1$ holds for every (A, C, U)-block of the decomposition. $\square$

In addition to Lemma 4 which deals with u-consistent partial colourings we provide in Theorem 3 a proof of correctness showing that our algorithm computes exactly the valid candidate colourings of an (A, C, U)-block of G.

Theorem 3. *For every (A, C, U)-block of G, the algorithm correctly computes all its valid candidate colourings. Thus our algorithm correctly decides whether the input graph G is k-colourable.*

Proof. Let T be the decomposition tree of G computed by the algorithm. For the purpose of this proof we address a node of the decomposition tree T with pivot u as u-node and the label of the node as (A_u, C_u, U_u, u). Consider any edge $\{x, y\} \in E$. We show that x and y have different colours in any candidate colouring of an (A, C, U)-block satisfying $x, y \in U$ that the algorithm accepted as a valid one.

(1) Clearly for every block the algorithm checks the validity of all of its candidate colourings and it correctly computes all valid candidate colourings of those blocks assigned to a leaf of the decomposition tree.

(2) Assume that w.l.o.g. the y-node of T belongs to the path from the root to the x-node (which might be a leaf). Since x and y are adjacent in G, x belongs to an internal y-child. Consequently x and y must have different colours.

(3) Let the w-node of T be the least common ancestor of the x-node and the y-node in T. Since there are no edges between vertices of two internal (resp. two external) w-children w.l.o.g. we may assume that x belongs to an internal w-child and that y belongs to an external w-child. Note that $x, y \in U_z$ implies that the z-node is an ancestor of the w-node (including the w-node itself) in T. Thus it suffices to show that x and y have different colours in every valid candidate colouring of the (A_w, C_w, U_w)-block. To do this consider any candidate colouring of this block and any w-colouring chosen to decompose this block. In the w-colouring there is a coloured vertex x' having same colour as x such that $x \preccurlyeq_w x'$ ($x = x'$ is possible), and thus $\{x', y\} \in E$. Let the (A_v, C_v, U_v)-block be the external w-child containing y. This implies that the restriction of the candidate colouring and the w-colouring under inspection is a valid candidate colouring of the (A_v, C_v, U_v)-block; hence x' and y have different colours and also x and y have different colours in this k-colouring. Thus any candidate colouring of the (A_w, C_w, U_w)-block accepted as valid by the algorithm is indeed valid.

Finally recall that a valid candidate colouring of the $(\varnothing, \varnothing, V)$-block is a vertex k-colouring of G. $\qquad\square$

6 Conclusions

We presented an algorithm that for any fixed integers k and a, solves the k-COLOURING problem in time $\mathcal{O}(f(a, k) \cdot n^{g(a,k)})$ assuming the asteroidal number of the input graph is at most a. Hence for any fixed k, the algorithm solves k-COLOURING in polynomial time on graphs of bounded asteroidal number, and thus k-COLOURING for graphs of bounded asteroidal number is in XP. The running time of the algorithm on AT-free graphs is $\mathcal{O}(n^{8k+2})$. We mention that our algorithm can be modified to solve within the same running time the LIST

k-COLORING problem [6] which requires in addition that every vertex u must receive a colour from some given set $L(u) \subseteq \{1, \ldots, k\}$.

The following two related questions are challenging. What is the algorithmic complexity of COLOURING on AT-free graphs; is the problem NP-complete? Taking as parameter the number of available colours, is the COLOURING problem on AT-free graphs fixed-parameter tractable; is it W[1]-hard?

Very recently it was brought to our attention that COLOURING for graphs of asteroidal number at most three remains NP-complete [9].

References

1. Brandstädt, A., Bang Le, V., Spinrad, J.: Graph classes: a survey. SIAM (1999)
2. Broersma, H.-J., Kloks, T., Kratsch, D., Müller, H.: Independent sets in asteroidal triple-free graphs. SIAM Journal on Discrete Mathematics 12, 276–287 (1999)
3. Broersma, H.-J., Kloks, T., Kratsch, D., Müller, H.: A generalization of AT-free graphs and a generic algorithm for solving triangulation problems. Algorithmica 32, 594–610 (2002)
4. Corneil, D.G., Olariu, S., Stewart, L.: Asteroidal triple-free graphs. SIAM Journal on Discrete Mathematics 10, 399–430 (1997)
5. Corneil, D.G., Olariu, S., Stewart, L.: Linear time algorithms for dominating pairs in asteroidal triple-free graphs. SIAM Journal on Computing 28, 1284–1297 (1999)
6. Couturier, J.-F., Golovach, P.A., Kratsch, D., Paulusma, D.: List Coloring in the Absence of a Linear Forest. In: Kolman, P., Kratochvíl, J. (eds.) WG 2011. LNCS, vol. 6986, pp. 119–130. Springer, Heidelberg (2011)
7. Garey, M.R., Johnson, D.S.: Computers and Intractability: A guide to the Theory of NP-completeness. Freeman, New York (1979)
8. Grötschel, M., Lovász, L., Schrijver, A.: Polynomial algorithms for perfect graphs. Annals on Discrete Mathematics 21, 325–356 (1984)
9. Golovach, P.: Private communication
10. Johnson, D.S.: The NP-complete column: an ongoing guide. Journal of Algorithms 6, 434–451 (1985)
11. Kloks, T., Kratsch, D., Müller, H.: Asteroidal Sets in Graphs. In: Möhring, R.H. (ed.) WG 1997. LNCS, vol. 1335, pp. 229–241. Springer, Heidelberg (1997)
12. Kloks, T., Kratsch, D., Müller, H.: Approximating the bandwidth for asteroidal triple-free graphs. Journal of Algorithms 32, 41–57 (1999)
13. Kloks, T., Kratsch, D., Müller, H.: On the structure of graphs with bounded asteroidal number. Graphs and Combinatorics 17, 295–306 (2001)
14. Kratsch, D.: Domination and total domination on asteroidal triple-free graphs. Discrete Applied Mathematics 99, 111–123 (2000)
15. Kratsch, D., Müller, H., Todinca, I.: Feedback vertex set on AT-free graphs. Discrete Applied Mathematics 156, 1936–1947 (2008)
16. Lekkerkerker, C.G., Boland, J.C.: Representation of a finite graph by a set of intervals on the real line. Fundamenta Mathematicae 51, 45–64 (1962)
17. Stacho, J.: 3-Colouring AT-Free Graphs in Polynomial Time. In: Cheong, O., Chwa, K.-Y., Park, K. (eds.) ISAAC 2010, Part II. LNCS, vol. 6507, pp. 144–155. Springer, Heidelberg (2010)
18. Walter, J.R.: Representations of chordal graphs as subtrees of a tree. Journal of Graph Theory 2, 265–267 (1978)

Routing Regardless of Network Stability

Bundit Laekhanukit[1], Adrian Vetta[2], and Gordon Wilfong[3]

[1] McGill University, Montreal QC H3A 0G4, Canada
blaekh@cs.mcgill.ca
[2] McGill University, Montreal QC H3A 0G4, Canada
vetta@math.mcgill.ca
[3] Bell Laboratories, Alcatel-Lucent, Murray Hill NJ 07974, USA
gtw@research.bell-labs.com

Abstract. How effective are interdomain routing protocols, such as the *Border Gateway Protocol*, at routing packets? Theoretical analyses have attempted to answer this question by ignoring the packets and instead focusing upon protocol stability. To study stability, it suffices to model only the control plane (which determines the routing graph) – an approach taken in the *Stable Paths Problem*. To analyse packet routing, it requires modelling the interactions between the control plane and the forwarding plane (which determines when packets are forwarded), and our first contribution is to introduce such a model. We then examine the effectiveness of packet routing in this model for the broad class *next-hop preferences with filtering*. Here each node v has a *filtering list* $\mathcal{D}(v)$ consisting of nodes it does not want its packets to route through. Acceptable paths (those that avoid nodes in the filtering list) are ranked according to the *next-hop*, that is, the neighbour of v that the path begins with. On the negative side, we present a strong inapproximability result. For filtering lists of cardinality at most one, given a network in which an equilibrium is guaranteed to exist, it is NP-hard to approximate the maximum number of packets that can be routed to within a factor of $O(n^{1-\epsilon})$, for any constant $\epsilon > 0$. On the positive side, we give algorithms to show that in two fundamental cases *every* packet will eventually route with probability one. The first case is when each node's filtering list contains only itself, that is, $\mathcal{D}(v) = \{v\}$. Moreover, with positive probability every packet will be routed before the control plane reaches an equilibrium. The second case is when all the filtering lists are empty, that is, $\mathcal{D}(v) = \emptyset$. Thus, with probability one packets will route even when the nodes don't care if their packets cycle! Furthermore, with probability one every packet will route even when the control plane has *no* equilibrium at all. To our knowledge, these are the first results to guarantee the possibility that all packets get routed without stability. These positive results are tight – for the general case of filtering lists of cardinality one, it is not possible to ensure that every packet will eventually route.

1 Introduction

In the *Stable Paths Problem* (SPP) [1], we are given a directed graph $G = (V, A)$ and a sink (or destination) node r. Furthermore, each node v has a ranked list

L. Epstein and P. Ferragina (Eds.): ESA 2012, LNCS 7501, pp. 719–730, 2012.

of some of its paths to r and the lowest ranked entry in the list is the "empty path"[1]. This preference list is called v's list of *acceptable paths*. A set of paths, one path $\mathcal{P}(v)$ from each node v's list of acceptable paths, is termed *stable* if (i) they are *consistent*: if $u \in \mathcal{P}(v)$, then $\mathcal{P}(u)$ must be the subpath of $\mathcal{P}(v)$ beginning at u, and (ii) they form an *equilibrium*: for each node v, $\mathcal{P}(v)$ is the path ranked highest by v of the form $vP(w)$ where w is a neighbour of v. The stable paths problem asks whether a stable set of paths exists in the network. The SPP has risen to prominence as it is viewed as a *static* description of the problem that the Border Gateway Protocol (BGP) is trying *dynamically* to solve. BGP can be thought of as trying to find a set of stable routes to r so that routers can use these routes to send packets to r.

Due to the importance of BGP, both practical and theoretical aspects of the SPP have been studied in great depth. To avoid overloading the reader with practical technicalities and for reasons of space, we defer discussions on a sample of this vast literature and on the technical aspects of BGP to the full paper. Two observations concerning the SPP, though, are pertinent here and motivate our work:

(1) Even if a stable solution exists, the routing tree induced by a consistent set of paths might not be spanning. Hence, a stable solution may **not** actually correspond to a functioning network – there may be isolated nodes that cannot route packets to the sink! Disconnectivities arise because nodes may prefer the empty-path to any of the paths proffered by its neighbours; for example, a node might not trust certain nodes to handle its packets securely or in a timely fashion, so it may reject routes traversing such unreliable domains. This problem of non-spanning routing trees has quite recently been studied in the context of a version of BGP called iBGP [4]. In Section 3, we show that non-connectivity is a very serious problem (at least, from the theoretical side) by presenting an $O(n^{1-\epsilon})$ hardness result for the combinatorial problem of finding a *maximum cardinality stable subtree*.

(2) The SPP says nothing about the dynamic behaviour of BGP. Stable routings are significant for many practical reasons (e.g., network operators want to know the routes their packets are taking), but while BGP is operating at the control plane level, packets are being sent at the forwarding plane level without waiting for stability (if, indeed, stability is ever achieved). Thus, it is important to study network performance in the dynamic case. For example, what happens to the packets whilst a network is unstable? This is the main focus of our paper: to investigate packet routing under network dynamics.

Towards this goal, we define a distributed protocol, inspired by BGP, that stops making changes to the routing graph (i.e., becomes stable) if it achieves a stable solution to the underlying instance of SPP. The current routing graph itself is determined by the control plane but the movement of packets is determined by the *forwarding plane*. Thus, our distributed protocol provides a framework

---

[1] Clearly, the empty path is not a real path to the sink; we call it a path for clarity of exposition.

under which the control and forwarding planes interact; essentially, this primarily means that we need to understand the relative speeds at which links change and packets move.

Given this model we analyse the resulting trajectory of packets. In a stable solution, a node in the stable tree containing the sink would have its packets route whereas an isolated node would not. For unstable networks, or for stable networks that have not converged, things are much more complicated. Here the routes selected by nodes are changing over time and, as we shall see, this may cause the packets to cycle. If packets can cycle, then keeping track of them is highly non-trivial. Our main results, however, are that for two fundamental classes of preference functions (i.e., two ways of defining acceptable paths and their rankings) all packets will route with probability one in our model. That is, there is an execution of the our distributed protocol such that *every* packet in the network will reach the destination (albeit, possibly, slowly) even in instances where the network has no stable solution. (Note that we are ignoring the fact that in BGP packets typically have a *time-to-live* attribute meaning that after traversing a fixed number of nodes the packet will be dropped.) Furthermore, when the network does have a stable solution, we are able to guarantee packet routing even before the time when the network converges.

These positive results on the routing rate are to our knowledge, the first results to guarantee the possibility of packet routing without stability. The results are also tight in the sense that, for any more expressive class of preference function, our hardness results show that guaranteeing that all packet eventually route is not possible – thus, packets must be lost.

2   The Model and Results

We represent a network by a directed graph $G = (V, A)$ on n nodes. The destination node in the network is denoted by a distinguished node r called a *sink* node. We assume that, for every node $v \in V$, there is a directed path in G from v to the sink r, and the sink r has no outgoing arc. At any point in time t, each node v chooses at most one of its out-neighbours w as its *chosen next-hop*; thus, v selects one arc (v, w) or selects none. These arcs form a *routing graph* $\mathcal{R}_t$, each component of which is a *1-arborescence*, an *in-arborescence*[2] T plus possibly one arc (v, w) emanating from the root v of T, i.e, T and $T \cup \{(v, w)\}$ are both 1-arborescences. (If the root of a component does select a neighbour, then that component contains a unique cycle.) When the context is clear, for clarity of exposition, we abuse the term *tree* to mean a 1-arborescence, and we use the term *forest* to mean a set of trees. A component (tree) in a routing graph is called a *sink-component* if it has the sink r as a root; other components are called *non-sink components*.

Each node selects its outgoing arc according to its preference list of acceptable paths. We examine the case where these lists can be generated using two of the

---

[2] An *in-arborescence* is a graph T such that the underlying undirected graph is a tree and every node has a unique path to a root node.

most common preference criteria in practice: *next-hop preferences* and *filtering*. For next-hop preferences, each node $v \in V$ has a ranking on its *out-neighbours*, nodes w such that $(v, w) \in A$. We say that w is the *k-th choice* of v if w is an out-neighbour of v with the k-th rank. For $k = 1, 2, \ldots, n$, we define a set of arcs A_k to be such that $(v, w) \in A_k$ if w is the k-th choice of v, i.e., A_k is the set of *the k-th choice arcs*. Thus, $A_1, A_2, \ldots, A_n$ partition a set of arcs A, i.e., $A = A_1 \cup A_2 \cup \ldots A_n$. We call the entire graph $G = (V, A)$ an *all-choice* graph. A *filtering list*, $\mathcal{D}(v)$, is a set of nodes that v never wants its packets to route through. We allow nodes to use filters and otherwise rank routes via next-hop preferences, namely *next-hop preferences with filtering*.

To be able to apply these preferences, each node $v \in V$ is also associated with a path $\mathcal{P}(v)$, called v's *routing path*. The routing path $\mathcal{P}(v)$ may **not** be the same as an actual v, r-path in the routing graph. A routing path $\mathcal{P}(v)$ (resp., a node v) is *consistent* if $\mathcal{P}(v)$ is a v, r-path in the routing graph; otherwise, we say that $\mathcal{P}(v)$ (resp., v) is *inconsistent*. A node v is *clear* if the routing path $\mathcal{P}(v) \neq \emptyset$, i.e., v has a path to the sink; otherwise, v is *opaque*. (The node v will never keep a path that is not a v, r-path.) We say that a node w is *valid* for v or is a *valid choice* for v if w is clear and $\mathcal{P}(w)$ contains no nodes in the filtering list $\mathcal{D}(w)$. If w is a valid choice for v, and v prefers w to all other valid choices, then we say that w is the *best valid choice* of v. A basic step in the dynamic behaviour of BGP is that, at any time t, some subset V_t of nodes is *activated* meaning that every node $v \in V_t$ chooses the most highest ranked acceptable path $\mathcal{P}(v)$ that is consistent with one of its neighbours' chosen paths at time $t - 1$. The routing graph $\mathcal{R}_t$ consists of the first arc in each routing path at time t.

Protocol variations result from such things as restricting V_t so that $|V_t| = 1$, specifying the relative rates that nodes are chosen to be activated and allowing other computations to occur between these basic steps. In our protocol, we assume that activation orderings are *fair* in that each node activates exactly once in each time period – a *round* – the actual ordering however may differ in each round. While our protocol is not intended to model exactly the behaviour of BGP, we tried to let BGP inspires our choices and captures the essential coordination problem that makes successful dynamic routing hard. Again, a detailed discussion on these issues an on the importance of a fairness-type criteria is deferred to the full paper.

Procedure 1. Activate(v)

Input: A node $v \in V - \{r\}$.

1: **if** v has a valid choice **then**
2:     Choose a best valid choice w of v.
3:     Change the outgoing arc of v to (v, w).
4:     Update $\mathcal{P}(v) := v\mathcal{P}(w)$ (the concatenation of v and $\mathcal{P}(w)$).
5: **else**
6:     Update $\mathcal{P}(v) := \emptyset$.
7: **end if**

Procedure 2. Protocol($G,r,\mathcal{R}_0$)

Input: A network $G = (V, A)$, a sink node r and a routing graph $\mathcal{R}_0$

1: Initially, every node generates a packet.
2: **for** round $t = 1$ to ... **do**
3:     Generate a permutation π_t of nodes in $V - \{r\}$ using an external algorithm $\mathbb{A}$.
4:     **Control Plane:** Apply Activate(v) to activate each node in the order in π_t. This forms a routing graph $\mathcal{R}_t$.
5:     **Forwarding Plane:** Ask every node to forward packets it has, and wait until every packet is moved by at most n hops (forwarded n times) or gets to the sink.
6:     **Route-Verification:** Every node learns which paths it has in the routing graph, i.e., update $\mathcal{P}(v) := v, r$-path in $\mathcal{R}_t$.
7: **end for**

This entire mechanism can thus be described using two algorithms as follows. Once activated, a node v updates its routing path $\mathcal{P}(v)$ using the algorithm in Procedure 1. The generic protocol is described in Procedure 2. This requires an external algorithm $\mathbb{A}$ which acts as a *scheduler* that generates a permutation – an order in which nodes will be activated in each round. We will assume that these permutations are independent and randomly generated. Our subsequent routing guarantees will be derived by showing the existence of specific permutations that ensure all packets route. These permutations are different in each of our models, which differ only in filtering lists. We remark that our model is incorporated with a *route-verification* step, but this is not a feature of BGP (again, see the full version for a discussion).

With the model defined, we examine the efficiency of packet routing for the three cases of next-hop preferences with filtering:

- General Filtering. The general case where the filtering list $\mathcal{D}(v)$ of any node v can be an arbitrary subset of nodes.

- Not me! The subcase where the filtering list of node v consists only of itself, $\mathcal{D}(v) = \{v\}$. Thus, a node does not want a path through itself, but otherwise has no nodes it wishes to avoid.

- Anything Goes! The case where every filtering list is empty, $\mathcal{D}(v) = \emptyset$. Thus a node does not even mind if its packets cycle back through it!

We partition our analyses based upon the types of filtering lists. Our first result is a strong hardness result presented in Section 3. Not only can it be hard to determine if every packet can be routed but the maximum number of packets that can be routed cannot be approximated well even if the network can reach equilibrium. Specifically,

Theorem 1. *For filtering lists of cardinality at most one, it is NP-hard to approximate the maximum stable subtree to within a factor of $O(n^{1-\epsilon})$, for any constant $\epsilon > 0$.*

Corollary 1. *For filtering lists of cardinality at most one, given a network in which an equilibrium is guaranteed to exist, it is NP-hard to approximate the*

maximum number of packets that can be routed to within a factor of $O(n^{1-\epsilon})$, for any constant $\epsilon > 0$.

However, for its natural subcase where the filtering list of a node consists only of itself (that is, a node doesn't want to route via a cycle!), we obtain a positive result in Section 5.

Theorem 2. *If the filtering list of a node consists only of itself, then an equilibrium can be obtained in n rounds. However, every packet will be routed in $\frac{n}{3}$ rounds, that is, before stability is obtained!*

Interestingly, we can route every packet in the case $\mathcal{D}(v) = \emptyset$ for all $v \in V$; see Section 4. Thus, even if nodes don't care whether their packets cycle, the packets still get through!

Theorem 3. *If the filtering list is empty then every packet can be routed in 4 rounds, even when the network has no equilibrium.*

Theorems 2 and 3 are the first theoretical results showing that packet routing can be done in the absence of stability. For example, every packet will be routed even in the presence of dispute wheels. Indeed, packets will be routed even if some nodes *never* actually have paths to the sink. Note that when we say that every packet will route with probability one we mean that, assuming permutations are drawn at random, we will eventually get a fair activation sequence that routes every packet. It is a nice open problem to obtain high probability guarantees for fast packet routing under such an assumption.

3   General Filtering

Here we consider hardness results for packet routing with general filtering lists. As discussed, traditionally the theory community has focused upon the stability of $\mathcal{R}$ – the routing graph is stable if every node is selecting their best valid neighbour (and is consistent). For example, there are numerous intractability results regarding whether a network has an equilibrium. However, that the routing graph may be stable even if it is not spanning! There may be singleton nodes that prefer to stay disconnected rather than take any of the offered routes. Thus, regardless of issues such as existence and convergence, an equilibrium may not even route the packets. This can be particularly problematic when the nodes use filters. Consider our problem of maximising the number of nodes that can route packets successfully. We show that this cannot be approximated to within a factor of $n^{1-\epsilon}$, for any $\epsilon > 0$ unless P = NP. The proof is based solely upon a control plane hardness result: it is NP-hard to approximate the maximum-size stable tree to within a factor of $n^{1-\epsilon}$. Thus, even if equilibria exist, it is hard to determine if there is one in which the *sink-component* (the component of $\mathcal{R}$ containing the sink) is large.

  Formally, in the maximum-size stable tree problem, we are given a directed graph $G = (V, E)$ and a sink node r; each node $v \in V$ has a ranking of its neighbours and has a filtering list $\mathcal{D}(v)$. Given a tree (arborescence) $T \subseteq G$, we say that a node u is *valid* for a node v if $(u, v) \in E$ and a v, r-path in T does not contain any node of $\mathcal{D}(v)$. We say that T is *stable* if, for every arc (u, v) of T, v is valid for u, and u prefers v to any of its neighbours in G that are valid for u (w.r.t. T). Our goal is to find the stable tree (sink-component) with the maximum number of nodes. We will show that even when $|\mathcal{D}(v)| = 1$ for all nodes $v \in V$, the maximum-size stable tree problem cannot be approximated to within a factor of $n^{1-\epsilon}$, for any constant $\epsilon > 0$, unless P = NP.

The proof is based on the hardness of $3SAT$ [2]: given a CNF-formula on N variables and M clauses, it is NP-hard to determine whether there is an assignment satisfying all the clauses. Take an instance of 3SAT with N variables, $x_1, x_2, \ldots, x_N$ and M clauses $C_1, C_2, \ldots, C_M$. We now create a network $G = (V, A)$ using the following gadgets:

- VAR-GADGET: For each variable x_i, we have a gadget $H(x_i)$ with four nodes a_i, u_i^T, u_i^F, b_i. The nodes u_i^T and u_i^F have first-choice arcs (u_i^T, a_i), (u_i^F, a_i) and second-choice arcs (u_i^T, b_i), (u_i^F, b_i). The node a_i has two arcs (a_i, u_i^T) and (a_i, u_i^F); the ranking of these arcs can be arbitrary. Each node in this gadget has itself in the filtering list, i.e., $\mathcal{D}(v) = \{v\}$ for all nodes v in $H(x_i)$.

- CLAUSE-GADGET: For each clause C_j with three variables $x_{i(1)}, x_{i(2)}, x_{i(3)}$, we have a gadget $Q(C_j)$. The gadget $Q(C_j)$ has four nodes $s_j, q_{1,j}, q_{2,j}, q_{3,j}, t_j$. The nodes $q_{1,j}, q_{2,j}, q_{3,j}$ have first-choice arcs $(q_{1,j}, t_j)$, $(q_{2,j}, t_j)$, $(q_{3,j}, t_j)$. The node s_j has three arcs $(s_j, q_{1,j})$, $(s_j, q_{2,j})$, $(s_j, q_{3,j})$; the ranking of these arcs can be arbitrary, so we may assume that $(s_j, q_{z,j})$ is a zth-choice arc. Define the filtering list of s_j and t_j as $\mathcal{D}(s_j) = \{s_j\}$ and $\mathcal{D}(t_j) = \{d_0\}$. (The node d_0 will be defined later.) For $z = 1, 2, 3$, let $u_{i(z)}^T$ be a node in a Var-Gadget $H(x_{i(z)})$; the node $q_{z,j}$ has a filtering list $\mathcal{D}(q_{z,j}) = \{u_{i(z)}^T\}$, if assigning $x_{i(z)} = $ True satisfies the clause C_j; otherwise, $\mathcal{D}(q_{z,j}) = \{u_{i(z)}^F\}$.

To build G, we first add a sink node r and a *dummy sink* d_0; we connect d_0 to r by a first-choice arc (d_0, r). We arrange Var-Gadgets and Clause-Gadgets in any order. Then we add a first-choice arc from the node a_1 of the first Var-Gadget $H(x_1)$ to the sink r. For $i = 2, 3, \ldots, N$, we add a first-choice arc (b_i, a_{i-1}) joining gadgets $H(x_{i-1})$ and $H(x_i)$. We join the last Var-Gadget $H(x_N)$ and the first Clause-Gadget $Q(C_1)$ by a first-choice arc (t_1, a_N). For $j = 2, 3, \ldots, M$, we add a first-choice arc (t_j, s_{j-1}) joining gadgets $Q(C_{j-1})$ and $Q(C_j)$. This forms a line of gadgets. Then, for each node $q_{z,j}$ of each Clause-Gadget $Q(C_j)$, we add a second-choice arc $(q_{z,j}, d_0)$ joining $q_{z,j}$ to the dummy sink d_0. Finally, we add L *padding* nodes $d_1, d_2, \ldots, d_L$ and join each node d_i, for $i = 1, 2, \ldots, L$, to the last Clause-Gadget $Q(c_M)$ by a first-choice arc (d_i, s_M); the filtering list of each node d_i is $\mathcal{D}(d_i) = \{d_0\}$, for all $i = 0, 1, \ldots, L$. The parameter L can be any positive integer depending on a given parameter. Observe that the number of nodes in the graph G is $4N + 5M + L + 2$, and $|\mathcal{D}(v)| = 1$ for all nodes v of G. The reduction is illustrated in Figure 3(a).

The correctness of the reduction is proven in the next theorem. The proof is provided in the full version.

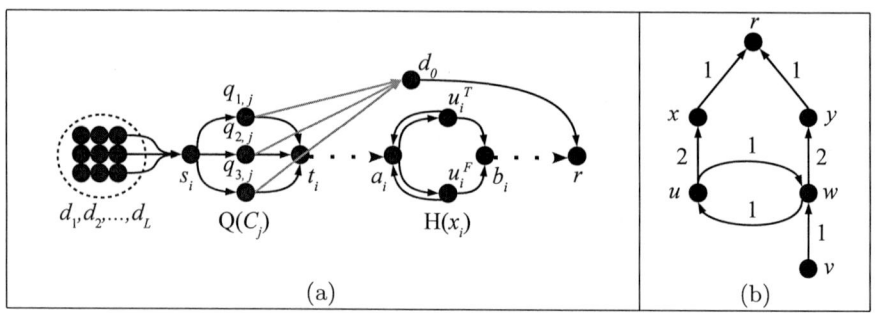

Fig. 1. (a) The reduction from 3SAT. (b) A network with no stable solution.

Theorem 4. *For any constant $\epsilon > 0$, given an instance of the maximum-size stable tree problems with a directed graph G on n nodes and filtering lists of cardinality $|\mathcal{D}(v)| = 1$ for all nodes v, it is NP-hard to distinguish between the following two cases of the maximum-size stable tree problem.*

- YES-INSTANCE: *The graph G has a stable tree spanning all the nodes.*
- NO-INSTANCE: *The graph G has no stable tree spanning n^ϵ nodes.*   □

From the perspective of the nodes, it is NP-hard to determine whether adding an extra node to its filtering list can lead to solutions where none of its packets ever route. In other words, it **cannot** avoid using intermediate node it dislikes!

4   Filtering: Anything-Goes!

Here we consider the case where every node has an empty filtering list. This case is conceptually simple but still contains many technical difficulties involved in tracking packets when nodes become mistaken in their connectivity assessments. In this case, networks with no stable solutions can exist (for example, see Figure 3(b)), and there can be fair activation sequences under which a node will never be in the sink-component. We show, however, that even in such circumstances, every packet still reaches the sink, and this is the case for all networks. Specifically, we present a fair activation sequence of four rounds that routes every packet, even when there is no equilibrium.

When filtering lists are empty, a node v only needs to known whether its neighbour u has a path to the sink since v does not have any node it dislikes. Thus, we can view each node as having two states: clear or opaque. A node is *clear* if it is in the routing-tree (the nomenclature derives from the fact that a packet at such a node will then reach the sink – that is, "clear"); otherwise, a node is *opaque*. Of course, as nodes update their chosen next-hop over time, they may be mistaken in their beliefs (inconsistent) as the routing graph changes. In other words, some clear nodes may not have "real" paths to the sink. After the learning step at the end of the round, these clear-opaque states are correct again.

Our algorithm and analysis are based on properties of the network formed by the first-choice arcs, called the *first class network*. We say that an arc (u, v) of G is a *first-choice* arc if v is the most preferred neighbour of u. We denote the first class network by $F = (V, A_1)$, where A_1 are the first-choice arcs. As in a routing graph $\mathcal{R}$, every node in F has one outgoing arc. Thus, every component of F is a 1-*arborescences*, a tree-like structure with either a cycle or a single node as a root. We denote the components of F by $F_0, F_1, \ldots, F_\ell$, where F_0 is the component containing the sink r. Observe that, when activated, every node in F_0 will always choose its neighbour in F_0. So, we may assume wlog that F_0 is a singleton. Each F_j has a unique cycle C_j, called a *first class cycle* (We may assume the directed cycle in F_0 is a self-loop at the sink r.) The routing graph at the beginning of Round t is denoted by $\mathcal{R}_t$. We denote by $\mathcal{K}_t$ and $\mathcal{O}_t$ the sets of clear and the set of opaque nodes at the start of Round t. Now, we will show that there is an activation sequence which routes every packet in four rounds.

The proof has two parts: a *coordination phase* and a *routing phase*. In the first phase, we give a coordination algorithm that generates a permutation that gives a red-blue colouring of the nodes with the following three properties: (i) For each F_j, every node in F_j has the same colour, i.e., the colouring is coordinated. (ii) If the first class cycle C_j of F_j contains a clear node then all nodes in F_j must be coloured blue. (iii) Subject to the first two properties the number of nodes coloured red is maximised. The usefulness of this colouring mechanism lies in the fact that the corresponding permutation is a fair activation sequence that will force the red nodes to lie in the sink-component and the blue nodes in non-sink-components. Moreover, bizarrely, running this coordination algorithm four times in a row ensures that every packet routes! So in the second phase, we run the coordination algorithm three more times.

Procedure 3. Coordinate($\mathcal{K}_t$)

Input: A set of clear nodes $\mathcal{K}_t$.
Output: A partition (R, B) of V.
 1: Let $B_0 := \bigcup_{i:V(C_j)\cap\mathcal{K}_t \neq \emptyset} V(F_i)$ be a set of nodes containing in an F-component whose first class cycle C_i has a clear node.
 2: Initialise $q := 0$.
 3: **repeat**
 4:    Update $q := q + 1$.
 5:    Initialise $B_q := B_{q-1}$, $R_q := \{r\}$ and $U := V - (R_q \cup B_q)$.
 6:    **while** $\exists$ a node $v \in U$ that prefers a node in R_q to nodes in $B_q \cup (U \cap K_t)$ **do**
 7:       Move v from U to R_q.
 8:    **end while**
 9:    **while** $\exists$ a node $v \in U$ that prefers a node in B_q to nodes in $R_q \cup (U \cap K_t)$ **do**
10:       Move v from U to B_q.
11:    **end while**
12:    Move $U \cap \mathcal{K}_t$ from U to B_q.
13: **until** $B_q = B_{q-1}$.
14: **return** (R_q, B_q).

4.1   Coordination Phase

The algorithm Coordinate($\mathcal{K}_t$) in Procedure 3 constructs a red-blue colour-ing of the nodes, i.e. the final partition (R, B) of V. At the termination of Coordinate($\mathcal{K}_t$), by the construction, any node $v \in R$ prefers some node in R to any node $w \in B$, and any node $v \in B$ prefers some node in B to any node $w \in R$.

Given a partition (R, B), we generate an activation sequence as follows. First, we greedily activate nodes in $R - \{r\}$ whenever their most-preferred *clear* neigh-bours are in R. (We activate nodes of R in the same order as we constructed R.) This forms a sink-component on R. Next, we activate nodes in B. We start by activating nodes in $B_0 = \bigcup_{i:C_i \cap K_i \neq \emptyset} V(F_i)$ – the components of F whose first class cycles contain at least one clear node. For each F_i with $V(F_i) \subseteq B_0$, take a clear node $v \in C_i \cap K_t$. Then activate the nodes of F_i (except v) in an increasing order of distance from v in F_i, and after that we activate v. This forms a non-sink-component F_i in the routing graph as every node can choose its first-choice. Finally, we activate nodes in $B - B_0$ whenever their most-preferred *clear* neighbours are in B (we use the same order as in the construction of B). This creates non-sink-components on B and implies the next lemma.

Lemma 1. *Let π_t be an activation sequence generated from (R, B) as above. At the end of the round, the following hold:*

- *The sink-component includes R and excludes B.*
- **Coordination:** *For each F_i, either all the nodes of F_i are in the sink-component or none of them are.*
- *Let $B_0 = \bigcup_{i:V(C_i) \cap K_t \neq \emptyset} V(F_i)$, and suppose $\mathcal{K}_t = B_0$. If a packet travels for n hops but does not reach the sink, then it must be at a node in $\mathcal{K}_t$.*

Proof. The first statement follows from the construction. For the second state-ment, it suffices to show that, for each F_i, either $V(F_i) \subseteq R$ or $V(F_i) \subseteq B$. Suppose not. Then there is a $V(F_i)$ crossing R or B. But, then some node in R (respectively, B) would have a first-choice in B (respectively, R), and this is not possible by the construction of (R, B).

For the third statement, note that a packet that travels for n hops but does not reach the sink must be stuck in some cycle. Consider the construction of (R, B). Since $\mathcal{K}_t = B_0$, we only add node to B whenever it prefers some node in B to any node in R. Because $U \cap \mathcal{K}_t = \emptyset$, nodes in $B - B_0$ cannot create a cycle on their own. Thus, the packet is stuck in a cycle that contains a clear node; the only such cycles are the first class cycles of B_0 since $\mathcal{K}_t = B_0$. □

The following lemma follows by the construction of a partition (R, B).

Lemma 2. *Let (R', B') be any partition generated from Coordinate(.), and let (R, B) be a partition obtained by Coordinate($\mathcal{K}_t$), where $\mathcal{K}_t \subseteq B'$. Then $R' \subseteq R$.*

Proof. Consider a partition (R_q, B_q) constructed during a call to Coordinate($\mathcal{K}_t$). Observe that $B_0 \subseteq B'$ because $B_0 = \bigcup_{i:V(C_j) \cap K_t \neq \emptyset} V(F_i)$ and Lemma 1 implies that each F_1 is contained entirely in R' or B'. By the construction of (R', B'), since $B_0 \subseteq B'$, every node of R' must have been added to R_1, i.e., $R' \subseteq R_1$. Inductively, if $R' \subseteq R_q$ for some $q \geq 1$, then $B_q \subseteq B'$ and thus $R' \subseteq R_{q+1}$ by the same argument. □

4.2   Routing Phase: A Complete Routing in Four Rounds

Running the coordination algorithm four times ensures every packet will have been in the sink-component at least once, and thus, every packet routes.

Theorem 5. *In four rounds every packet routes.*

Proof. The first round $t = 1$ is simply the coordination phase. We will use subscripts on R and B (e.g., R_t and B_t) to denote the final colourings output in each round and not the intermediate sets R_q/B_q used in Coordinate(.). Now, consider a packet generated by any node of V. First, we run Coordinate($\mathcal{K}_1$) and obtain a partition (R_1, B_1). By Lemma 1, if the packet is in R_1, then it is routed successfully, and we are done. Hence, we may assume that the packet does not reach the sink and the packet is in B_1. Note that, now, each F_i is either contained in R_1 or B_1 by Lemma 1.

We now run Coordinate($\mathcal{K}_2$) and obtain a partition (R_2, B_2). By Lemma 1, $\mathcal{K}_2 = R_1$. So, if the packet does not reach the sink, it must be in B_2. Since no F-component crosses R_1, we have $R_1 = \mathcal{K}_2 = \bigcup_{i:V(C_i) \cap \mathcal{K}_2 \neq \emptyset} V(F_i)$. So, $R_1 \subseteq B_2$ (since $\mathcal{K}_2 \subseteq B_2$) and $R_2 \subseteq B_1$, and Lemma 1 implies that the packet is in R_1.

Third, we run Coordinate($\mathcal{K}_3$) and obtain a partition (R_3, B_3). Applying the same argument as before, we have that the packet is in R_2 (or it is routed), $R_2 \subseteq B_3$ and $R_3 \subseteq B_2$. Now, we run Coordinate($\mathcal{K}_4$) and obtain a partition (R_4, B_4). Since $R_3 = \mathcal{K}_4 \subseteq B_2$, Lemma 2 implies that $R_2 \subseteq R_4$. Thus, the packet is routed successfully since R_4 is contained in the sink-component. □

5   Filtering: Not-Me!

In practice, it is important to try to prevent cycles forming in the routing graph of a network. To achieve this, *loop-detection* is implemented in the BGP-4 protocol [3]. The "Not-Me" filtering encodes loop-detection as in the BGP-4 protocol simply by having a filtering list $\mathcal{D}(v) = \{v\}$, for every node v. In contrast to Theorem 4, which says that it is NP-hard to determine whether we can route every packet, here we show that every packet will route. Moreover, we exhibit a constructive way to obtain a stable spanning tree via fair activation sequences. Interestingly, all of the packets will have routed before stability is obtained. In particular, we give an algorithm that constructs an activation sequence such that every packet routes successfully in $\frac{1}{3}n$ rounds, and the network itself becomes stable in n rounds. This is the most complex result in the paper; just the algorithm itself is long. So here we give a very high level overview and defer the algorithm and its proof of performance to the full paper.

When filtering lists are non-empty, a complication arises since even if w is the most preferred choice of v and w has non-empty routing path, v still may not be able to choose w. This makes the routing graph hard to manipulate. The key idea is to manipulate the routing graph a little-by-little in each round. To do this, we find a spanning tree with a *strong stability property* – a spanning tree S has the strong stability property on $\mathbb{O}$ if, for every node $v \in \mathbb{O}$, the most preferred choice of v is its parent w, even if it may choose any node except those that are descendants and in $\mathbb{O}$. Thus, if we activate nodes of S in increasing order of distance from the sink r, then every node $v \in \mathbb{O}$ will always choose w.

It is easy to find a *stable spanning tree*, a tree where no node wants to change its choice, and given a stable spanning tree S, it is easy to force opaque nodes in $\mathcal{O}_t$ to make the same choices as in S. But, this only applies to the set of opaque nodes, so it may not hold in the later rounds. The strong stability property allows us to make a stronger manipulation. Intuitively, the strong stability property says that once we force every node $v \in \mathbb{O}$ to make the same choice as in S, we can maintain these choices in all the later rounds. Moreover, in each round, if we cannot route all the packets, then we can make the strong stability property span three additional nodes; if so, the property spans one more node. Thus, in $\frac{1}{3}n$ rounds, every packet will route, but we need n rounds to obtain stability.

Theorem 6. *There is an activation sequence that routes every packet in $\lfloor n/3 \rfloor$ rounds and gives a stable spanning tree in n rounds.*

Acknowledgements. We thank Michael Schapira and Sharon Goldberg for interesting discussions on this topic.

References

1. Griffin, T., Bruce Shepherd, F., Wilfong, G.T.: The stable paths problem and interdomain routing. IEEE/ACM Trans. Netw. 10(2), 232–243 (2002)
2. Karp, R.M.: Reducibility among combinatorial problems. In: Complexity of Computer Computations, pp. 85–103 (1972)
3. Stewart III, J.W.: BGP4: Inter-Domain Routing in the Internet. Addison-Wesley Longman Publishing Co., Inc., Boston (1998)
4. Vissicchio, S., Cittadini, L., Vanbever, L., Bonaventure, O.: ibgp deceptions: More sessions, fewer routes. In: INFOCOM (2012)

The Simplex Tree: An Efficient Data Structure for General Simplicial Complexes

Jean-Daniel Boissonnat and Clément Maria

INRIA Sophia Antipolis-Méditerranée and École Normale Supérieure de Cachan
{jean-daniel.boissonnat,clement.maria}@inria.fr

Abstract. This paper introduces a new data structure, called simplex tree, to represent abstract simplicial complexes of any dimension. All faces of the simplicial complex are explicitly stored in a trie whose nodes are in bijection with the faces of the complex. This data structure allows to efficiently implement a large range of basic operations on simplicial complexes. We provide theoretical complexity analysis as well as detailed experimental results. We more specifically study Rips and witness complexes.

Keywords: simplicial complex, data structure, flag complex, Rips complex, witness complex, relaxed witness complex.

1 Introduction

Simplicial complexes are widely used in combinatorial and computational topology, and have found many applications in topological data analysis and geometric inference. A variety of simplicial complexes have been defined, for example the Čech complex, the Rips complex and the witness complex [7,6]. However, the size of these structures grows very rapidly with the dimension of the data set, and their use in real applications has been quite limited so far.

We are aware of only a few works on the design of data structures for general simplicial complexes. Brisson [4] and Lienhardt [9] have introduced data structures to represent d-dimensional cell complexes, most notably subdivided manifolds. While those data structures have nice algebraic properties, they are very redundant and do not scale to large data sets or high dimensions. Zomorodian [11] has proposed the tidy set, a compact data structure to simplify a simplicial complex and compute its homology. Since the construction of the tidy set requires to compute the maximal faces of the simplicial complex, the method is especially designed for flag complexes. Flag complexes are a special type of simplicial complexes (to be defined later) whose combinatorial structure can be deduced from its graph. In particular, maximal faces of a flag complex can be computed without constructing explicitly the whole complex. In the same spirit, Attali et al. [1] have proposed the skeleton-blockers data structure. Again, the representation is general but it requires to compute blockers, the simplices which are not contained in the simplicial complex but whose proper subfaces are. Computing the blockers is difficult in general and details on the construction are given

L. Epstein and P. Ferragina (Eds.): ESA 2012, LNCS 7501, pp. 731–742, 2012.
© Springer-Verlag Berlin Heidelberg 2012

only for flag complexes, for which blockers can be easily obtained. As of now, there is no data structure for general simplicial complexes that scales to dimension and size. Best implementations have been restricted to flag complexes.

Our approach aims at combining both generality and scalability. We propose a tree representation for simplicial complexes. The nodes of the tree are in bijection with the simplices (of all dimensions) of the simplicial complex. In this way, our data structure explicitly stores all the simplices of the complex but does not represent explicitly all the adjacency relations between the simplices, two simplices being adjacent if they share a common subface. Storing all the simplices provides generality, and the tree structure of our representation enables us to implement basic operations on simplicial complexes efficiently, in particular to retrieve incidence relations, *ie* to retrieve the faces that contain, or are contained, within a given simplex. Moreover, storing exactly one node per simplex ensures that the size of the structure adapts to the intrinsic complexity of the simplicial complex to be represented.

Background: A *simplicial complex* is a pair $\mathcal{K} = (V, S)$ where V is a finite set whose elements are called the *vertices* of $\mathcal{K}$ and S is a set of non-empty subsets of V that is required to satisfy the following two conditions : 1. $p \in V \Rightarrow \{p\} \in S$ and 2. $\sigma \in S, \tau \subseteq \sigma \Rightarrow \tau \in S$. Each element $\sigma \in S$ is called a *simplex* or a *face* of $\mathcal{K}$ and, if $\sigma \in S$ has precisely $s + 1$ elements ($s \geq -1$), σ is called an *s-simplex* and the dimension of σ is s. The dimension of the simplicial complex $\mathcal{K}$ is the largest k such that S contains a k-simplex. We define the j-skeleton, $j \geq 0$, of a simplicial complex $\mathcal{K}$ to be the simplicial complex made of the faces of $\mathcal{K}$ of dimension at most j. In particular, the 1-skeleton of $\mathcal{K}$ contains the vertices and the edges of $\mathcal{K}$. It has the structure of a graph, and we will equivalently talk about the graph of the simplicial complex, and use graph notations.

A *face* of a simplex $\sigma = \{p_0, \cdots, p_s\}$ is a simplex whose vertices form a subset of $\{p_0, \cdots, p_s\}$. A *proper* face is a face different from σ and the *facets* of σ are its proper faces of maximal dimension. A simplex $\tau \in \mathcal{K}$ admitting σ as a face is called a *coface* of σ. The subset of simplices consisting of all the cofaces of a simplex $\sigma \in \mathcal{K}$ is called the *star* of σ.

2   Simplex Tree

In this section, we introduce a new data structure which can represent any simplicial complex. This data structure is a trie [2] which will explicitly represent all the simplices and will allow efficient implementation of basic operations on simplicial complexes.

2.1   Simplicial Complexes and Trie

Let $\mathcal{K}$ be a simplicial complex of dimension k, V its vertex set. The vertices are labeled from 1 to $|V|$ and ordered accordingly. We can thus associate to each simplex of $\mathcal{K}$ a word on the alphabet $1 \cdots |V|$. Specifically, a j-simplex of $\mathcal{K}$ is

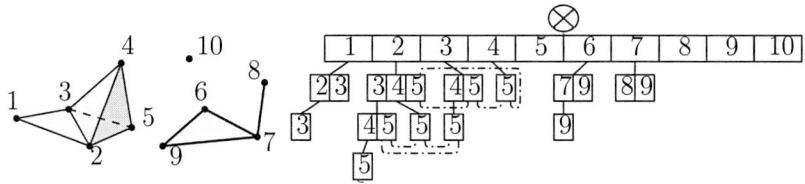

Fig. 1. A simplicial complex on 10 vertices and its simplex tree. The deepest node represents the tetrahedron of the complex. All the positions of a given label at a given depth are linked in a circular list, as illustrated in the case of label 5.

uniquely represented as the word of length $j+1$ consisting of the ordered set of the labels of its $j+1$ vertices. Formally, let simplex $\sigma = \{v_{\ell_0}, \cdots, v_{\ell_j}\}$, where $v_{\ell_i} \in V$, $\ell_i \in \{1, \cdots, |V|\}$ and $\ell_0 < \cdots < \ell_j$. σ is represented by the word $[\sigma] = [\ell_0, \cdots, \ell_j]$. The last label of the word representation of a simplex σ will be called the last label of σ and denoted by $last(\sigma)$.

The simplicial complex $\mathcal{K}$ can be defined as a collection of words on an alphabet of size $|V|$. To compactly represent the set of simplices of $\mathcal{K}$, we store the corresponding words in a tree satisfying the following properties:

1. The nodes of the simplex tree are in bijection with the simplices (of all dimensions) of the complex. The root is associated to the empty face.
2. Each node of the tree, except the root, stores the label of a vertex. Specifically, a node N associated to simplex $\sigma \neq \emptyset$ stores the label of vertex $last(\sigma)$.
3. The vertices whose labels are encountered along a path from the root to a node N, associated to a simplex σ, are the vertices of σ. Labels are sorted by increasing order along such a path, and each label appears exactly once.

We call this data structure the *Simplex Tree* of $\mathcal{K}$. It may be seen as a trie [2] on the words representing the simplices of the complex (Figure 1). The depth of the root is 0 and the depth of a node is equal to the dimension of the simplex it represents plus one.

In addition, we augment the data structure so as to quickly locate all the instances of a given label in the tree. Specifically, all the nodes at a same depth j which contain a same label ℓ are linked in a circular list $L_j(\ell)$, as illustrated in Figure 1 for label $\ell = 5$. We also attach to each set of sibling nodes a pointer to their parent so that we can access a parent in constant time.

The children of the root of the simplex tree are called the *top nodes*. The top nodes are in bijection with the elements of V, the vertices of $\mathcal{K}$. Nodes which share the same parent (e.g. the top nodes) will be called *sibling nodes*.

We give a constructive definition of the simplex tree. Starting from an empty tree, we insert the words representing the simplices of the complex in the following manner. When inserting the word $[\sigma] = [\ell_0, \cdots, \ell_j]$ we start from the root, and follow the path containing successively all labels $\ell_0, \cdots, \ell_i$, where $[\ell_0, \cdots, \ell_i]$ denotes the longest prefix of $[\sigma]$ already stored in the simplex tree. We then append to the node representing $[\ell_0, \cdots, \ell_i]$ a path consisting of the nodes storing labels $\ell_{i+1}, \cdots, \ell_j$. It is easy to see that the three properties above are satisfied.

Hence, if $\mathcal{K}$ consists of $|\mathcal{K}|$ simplices (including the empty face), the associated simplex tree contains exactly $|\mathcal{K}|$ nodes.

We use dictionaries for searching, inserting and removing elements among a set of sibling nodes. As the size of a dictionary is linear in the number of elements it stores, these additional structures do not change the asymptotic complexity of the simplex tree. For the top nodes, we simply use an array since the set of vertices V is known and fixed. Let $d^o_{\max}$ denote the maximal outdegree of a node in the tree distinct from the root. Remark that $d^o_{\max}$ is at most the maximal degree of a vertex in the graph of the simplicial complex. In the following, we will denote by D_m the maximal number of operations needed to perform a search, an insertion or a removal in a dictionary of maximal size $d^o_{\max}$ (for example, with red-black trees $D_m = O(\log(d^o_{\max}))$).

2.2   Operations on a Simplex Tree

- SEARCH/INSERT/REMOVE-SIMPLEX use the previous top-down traversal to search for the node representing a simplex of dimension j in $O(jD_m)$ operations. Then return, insert or remove a node from a dictionary.
- LOCATE-FACETS uses j searches for locating the facets of a simplex of dimension j, in $O(j^2 D_m)$ operations.
- LOCATE-COFACES locates the cofaces of a simplex τ using the lists which link the nodes storing a same label. This allows to locate the nodes which represent supersets of τ and are potential cofaces.

These elementary operations allow to implement more operations, like inserting a simplex and its subfaces, removing a simplex and its cofaces, perform elementary collapses and edge contractions. We refer to [3] for details on these operations, their implementation and their complexity.

3   Construction of Simplicial Complexes

In this section, we detail how to implement two important types of simplicial complexes, the flag and the witness complexes, using simplex trees.

3.1   Flag Complexes

A flag complex is a simplicial complex whose combinatorial structure is entirely determined by its 1-skeleton. Specifically, a simplex is in the flag complex if and only if its vertices form a clique in the graph of the simplicial complex.

Given the 1-skeleton of a flag complex, we call *expansion of order k* the operation which reconstructs the k-skeleton of the flag complex. If the 1-skeleton is stored in a simplex tree, the expansion of order k consists in successively inserting all the simplices of the k-skeleton into the simplex tree.

Let $G = (V, E)$ be the graph of the simplicial complex, where V is the set of vertices and $E \subseteq V \times V$ is the set of edges. For a vertex $v_\ell \in V$, we denote by

$$\mathcal{N}^+(v_\ell) = \{\ell' \in \{1, \cdots, |V|\} \mid (v_\ell, v_{\ell'}) \in E \wedge \ell' > \ell\}$$

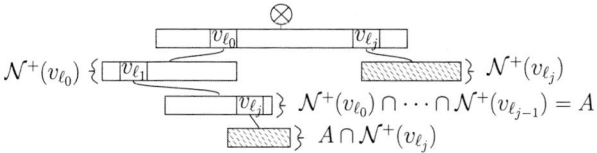

Fig. 2. Representation of a set of sibling nodes as intersection of neighborhoods

the set of labels of the neighbors of v_ℓ in G that are bigger than ℓ. Let N_{ℓ_j} be a node in the tree that stores label ℓ_j and represents the word $[\ell_0, \cdots, \ell_j]$. The children of N_{ℓ_j} store the labels in $\mathcal{N}^+(v_{\ell_0}) \cap \cdots \cap \mathcal{N}^+(v_{\ell_j})$. Indeed, the children of N_{ℓ_j} are neighbors in G of the vertices v_{ℓ_i}, $0 \le i \le j$, (by definition of a clique) and must have a bigger label than $\ell_0, \cdots, \ell_j$ (by construction of the simplex tree). Consequently, the sibling nodes of N_{ℓ_j} are exactly the nodes that store the labels in $A = \mathcal{N}^+(v_{\ell_0}) \cap \cdots \cap \mathcal{N}^+(v_{\ell_{j-1}})$, and the children of N_{ℓ_j} are exactly the nodes that store the labels in $A \cap \mathcal{N}^+(v_{\ell_j})$. See Figure 2.

For every vertex v_ℓ, we have an easy access to $\mathcal{N}^+(v_\ell)$ since $\mathcal{N}^+(v_\ell)$ is exactly the set of labels stored in the children of the top node storing label ℓ. We easily deduce an in-depth expansion algorithm. The time complexity for the expansion algorithm depends on our ability to compute fast intersections of the type $A \cap \mathcal{N}^+(v_{\ell_j})$. In practice, we have observed that the time taken by the expansion algorithm depends linearly on the size of the output simplicial complex.

The Rips Complex: Rips complexes are geometric flag complexes which are popular in computational topology due to their simple construction and their good approximation properties. Given a set of vertices V in a metric space and a parameter $r > 0$, the Rips graph is defined as the graph whose set of vertices is V and two vertices are joined by an edge if their distance is at most r. The Rips complex is the flag complex defined on top of this graph. We will use this complex for our experiments on the construction of flag complexes.

3.2 Witness Complexes

The Witness Complex: has been first introduced in [6]. Its definition involves two given sets of points in a metric space, the landmarks L and the witnesses W. We say that a witness $w \in W$ *witnesses* a simplex $\sigma \subseteq L$ iff $\forall x \in \sigma$ and $\forall y \in L \setminus \sigma$ we have $d(w,x) \le d(w,y)$; or, equivalently, the vertices of σ are the $|\sigma|$ nearest neighbors of w in L.

The *witness complex* $\mathrm{Wit}(W,L)$ is the maximal simplicial complex, with vertex set L, whose faces admit a witness in W. Equivalently, a simplex belongs to the witness complex if and only if it is witnessed and all its facets belong to the witness complex. Such simplex will be called *fully witnessed*.

Construction Algorithm: We suppose that L and W are finite sets and give them labels $\{1, \cdots, |L|\}$ and $\{1, \cdots, |W|\}$ respectively. We describe how to construct the k-skeleton of the witness complex, where k is any integer in $\{1, \cdots, |L| - 1\}$.

Our construction algorithm is incremental, from lower to higher dimensions. At step j it inserts in the simplex tree the j-dimensional fully witnessed simplices.

During the construction of the k-skeleton of the witness complex, we need to access the nearest neighbors of the witnesses, in L. To do so, we compute the $k + 1$ nearest neighbors of all the witnesses in a preprocessing phase, and store them in a $|W| \times (k + 1)$ matrix. Given an index $j \in \{0, \cdots, k\}$ and a witness $w \in W$, we can then access in constant time the $(j + 1)^{\text{th}}$ nearest neighbor of w, noted s_j^w. We maintain a list of *active witnesses*, initialized with W.

We insert the vertices of $\text{Wit}(W, L)$ in the simplex tree. For each witness $w \in W$ we insert a top node storing the label of the nearest neighbor of w in L, if no such node already exists. w is initially an active witness and we make it point to the node mentionned above, representing the 0-simplex it witnesses.

We maintain the following loop invariants: 1. at the beginning of iteration j, the simplex tree contains the $(j - 1)$-skeleton of the witness complex $\text{Wit}(W, L)$; 2. the active witnesses are the elements of W that witness a $(j - 1)$-simplex of the complex; each active witness w points to the node representing the $(j - 1)$-simplex in the tree it witnesses.

At iteration $j \geq 1$, we traverse the list of active witnesses. Let w be an active witness. We first compute the $(j + 1)^{\text{th}}$ nearest neighbor s_j^w of w using the nearest neighbors matrix (Step 1). Let σ_j be the j-simplex witnessed by w and let us decompose the word representing σ_j into $[\sigma_j] = [\sigma'].[s_j^w].[\sigma'']$ ("." denotes the concatenation of words). We then look for the location in the tree where σ_j might be inserted (Step 2). To do so, we start at the node N_w which represents the $(j-1)$-simplex witnessed by w. Observe that the word associated to the path from the root to N_w is exactly $[\sigma'].[\sigma'']$. We walk $||[\sigma'']||$ steps up from N_w, reach the node representing $[\sigma']$ and then search downwards for the word $[s_w^j] . [\sigma'']$ The cost of this operation is $O(jD_m)$.

If the node representing σ_j exists, σ_j has already been inserted; we update the pointer of w and return. If the simplex tree contains neither this node nor its father, σ_j is not fully witnessed because the facet represented by its longest prefix is missing. We consequently remove w from the set of active witnesses. Lastly, if the node is not in the tree but its father is, we check whether σ_j is fully witnessed. To do so, we search for the $j + 1$ facets of σ_j in the simplex tree (Step 3). The cost of this operation is $O(j^2 D_m)$ using SEARCH-FACETS. If σ_j is fully witnessed, we insert σ_j in the simplex tree and update the pointer of the active witness w. Else, we remove w from the list of active witnesses.

It is easily seen that the loop invariants are satisfied at the end of iteration j.

Complexity: The cost of accessing a neighbor of a witness using the nearest neighbors matrix is $O(1)$. We access a neighbor (Step 1) and locate a node in the simplex tree (Step 2) at most $k|W|$ times. In total, the cost of Steps 1 and 2 together is $O((kD_m + 1)k|W|)$. In Step 3, either we insert a new node in the simplex tree, which happens exactly $|\mathcal{K}|$ times (the number of faces in the complex), or we remove an active witness, which happens at most $|W|$ times. The total cost of Step 3 is thus $O((|\mathcal{K}| + |W|)k^2 D_m)$. In conclusion, constructing the k-skeleton of the witness complex takes time:

$$O((|\mathcal{K}| + |W|)k^2 D_m + k|W|) = O((|\mathcal{K}| + |W|)k^2 D_m).$$

Landmark Insertion: We can also insert a new landmark x proceeding to a local variant of the witness complex construction where, by "local", we mean that we reconstruct only the star of vertex x. Constructing $\text{Wit}(W, L \cup \{x\})$ from the simplex tree representation of $\text{Wit}(W, L)$ takes time $O((|W^x| + C_{\{x\}})k^2 D_m)$, where $C_{\{x\}}$ is the size of the star of x and W^x the set of the *reverse neighbors* of x. We refer to [3] for more details on the algorithm and its analysis.

Relaxed Witness Complex: Given a relaxation parameter $\rho \geq 0$ we define the *relaxed witness complex* [6]. We say that a witness $w \in W$ ρ-*witnesses* a simplex $\sigma \subseteq L$ iff $\forall x \in \sigma$ and $\forall y \in L \setminus \sigma$ we have $d(w, x) \leq d(w, y) + \rho$.

The *relaxed witness complex* $\text{Wit}^\rho(W, L)$ with parameter ρ is the maximal simplicial complex, with vertex set L, whose faces admit a ρ-witness in W. For $\rho = 0$, the relaxed witness complex is the standard witness complex.

We resort to the same incremental algorithm as above. At each step j, we insert, for each witness w, the j-simplices which are ρ-witnessed by w. Differently from the standard witness complex, there may be more than one j-simplices that are witnessed by a given witness $w \in W$. Consequently, we do not maintain a pointer from each active witness to the last inserted simplex it witnesses. We use simple top-down insertions from the root of the simplex tree.

Given a witness w and a dimension j, we generate all the j-simplices which are ρ-witnessed by w. We suppose we are given the sorted list of nearest neighbors of w in L, noted $\{z_0 \cdots z_{|L|-1}\}$, and their distance to w, noted $m_i = d(w, z_i)$, with $m_0 \leq \cdots \leq m_{|L|-1}$, breaking ties arbitrarily. Note that if one wants to construct only the k-skeleton of the complex, it is sufficient to know the list of neighbors of w that are at distance at most $m_k + \rho$ from w. We preprocess this list of neighbors for all witnesses. For $i \in \{0, \cdots, |L| - 1\}$, we define the set A_i of landmarks z such that $m_i \leq d(w, z) \leq m_i + \rho$. For $i \leq j + 1$, w ρ-witnesses all the j-simplices that contain $\{z_0, \cdots, z_{i-1}\}$ and a $(j + 1 - i)$-subset of A_i, provided $|A_i| \geq j + 1 - i$. It is easy to see that all j-simplices that are ρ-witnessed by w are obtained this way, and exactly once, when i ranges from 0 to $j + 1$.

For all $i \in \{0, \cdots, j + 1\}$, we compute A_i and generate all the simplices which contain $\{z_0, \cdots, z_{i-1}\}$ and a subset of A_i of size $(j + 1 - i)$. In order to easily update A_i when i is incremented, we maintain two pointers to the list of neighbors, one to z_i and the other to the end of A_i. We check in constant time if A_i contains more than $j + 1 - i$ vertices, and compute all the subsets of A_i of cardinality $j + 1 - i$ accordingly.

Complexity. If R_j is the number of j-simplices ρ-witnessed by w, generating all those simplices takes $O(j + R_j)$ operations.

4   Experiments

In this section, we report on the performance of our algorithms on both real and synthetic data, and compare them to existing software. More specifically, we

Data	$|\mathcal{P}|$	D	d	r	k	T_g	$|E|$	T_{Rips}	$|\mathcal{K}|$	T_{tot}	$T_{\mathrm{tot}}/|\mathcal{K}|$
Bud	49,990	3	2	0.11	3	1.5	1,275,930	104.5	354,695,000	104.6	$3.0 \cdot 10^{-7}$
Bro	15,000	25	?	0.019	25	0.6	3083	36.5	116,743,000	37.1	$3.2 \cdot 10^{-7}$
Cy8	6,040	24	2	0.4	24	0.11	76,657	4.5	13,379,500	4.61	$3.4 \cdot 10^{-7}$
Kl	90,000	5	2	0.075	5	0.46	1,120,000	68.1	233,557,000	68.5	$2.9 \cdot 10^{-7}$
S4	50,000	5	4	0.28	5	2.2	1,422,490	95.1	275,126,000	97.3	$3.6 \cdot 10^{-7}$

Data	$|L|$	$|W|$	D	d	ρ	k	T_{nn}	T_{Wit^ρ}	$|\mathcal{K}|$	T_{tot}	$T_{\mathrm{tot}}/|\mathcal{K}|$
Bud	10,000	49,990	3	2	0.12	3	1.	729.6	125,669,000	730.6	$12 \cdot 10^{-3}$
Bro	3,000	15,000	25	?	0.01	25	9.9	107.6	2,589,860	117.5	$6.5 \cdot 10^{-3}$
Cy8	800	6,040	24	2	0.23	24	0.38	161	997,344	161.2	$23 \cdot 10^{-3}$
Kl	10,000	90,000	5	2	0.11	5	2.2	572	109,094,000	574.2	$5.7 \cdot 10^{-3}$
S4	50,000	200,000	5	4	0.06	5	25.1	296.7	163,455,000	321.8	$1.2 \cdot 10^{-3}$

Fig. 3. Data, timings (in s.) and statistics for the construction of Rips complexes (TOP) and relaxed witness complexes (BOTTOM)

benchmark the construction of Rips complexes, witness complexes and relaxed witness complexes. Our implementations are in C++. We use the ANN library (http://www.cs.umd.edu/~mount/ANN/) to compute the 1-skeleton graph of the Rips complex, and to compute the lists of nearest neighbors of the witnesses for the witness complexes. All timings are measured on a Linux machine with 3.00 GHz processor and 32 GB RAM. Timings are provided by the clock function from the Standard C Library, and zero means that the measured time is below the resolution of the clock function. For its efficiency and flexibility, we use the map structure from the Standard Template Library for storing sets of sibling nodes, except for the top nodes which are stored in an array.

We use a variety of both real and synthetic datasets. **Bud** is a set of points sampled from the surface of the *Stanford Buddha* in $\mathbb{R}^3$. **Bro** is a set of 5×5 *high-contrast patches* derived from natural images, interpreted as vectors in $\mathbb{R}^{25}$, from the Brown database (with parameter $k = 300$ and cut 30%) [5]. **Cy8** is a set of points in $\mathbb{R}^{24}$, sampled from the space of conformations of the cyclo-octane molecule [10], which is the union of two intersecting surfaces. **Kl** is a set of points sampled from the surface of the figure eight Klein Bottle embedded in $\mathbb{R}^5$. Finally **S4** is a set of points uniformly distributed on the unit 4-sphere in $\mathbb{R}^5$. Datasets are listed in Figure 3 with details on the sets of points $\mathcal{P}$ or landmarks L and witnesses W, their size $|\mathcal{P}|$, $|L|$, $|W|$, the ambient dimension D, the intrinsic dimension d of the object the sample points belong to (if known), the parameter r or ρ, the dimension k up to which we construct the complexes, the time T_g to construct the Rips graph or the time T_{nn} to compute the lists of nearest neighbors of the witnesses, the number of edges $|E|$, the time for the construction of the Rips complex T_{Rips} or for the construction of the witness complex T_{Wit^ρ}, the size of the complex $|\mathcal{K}|$, and the total construction time T_{tot} and average construction time per face $T_{\mathrm{tot}}/|\mathcal{K}|$.

We test the performance of our algorithms on these datasets, and compare them to the JPlex (http://comptop.stanford.edu/u/programs/jplex/) library which is a Java software package which can be used with Matlab. JPlex is

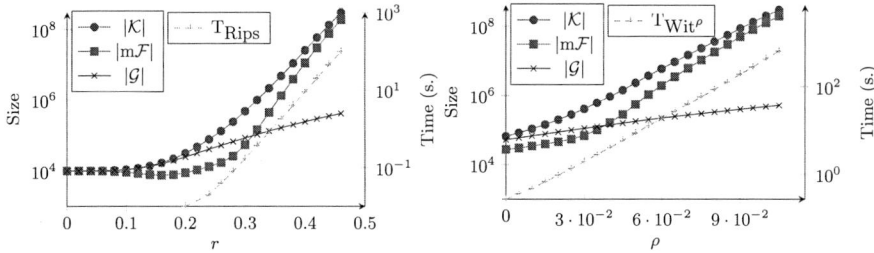

Fig. 4. Statistics and timings for the Rips complex (Left) and the relaxed witness complex (Right) on 10000 points from **S4**

widely used to construct simplicial complexes and to compute their homology. We also provide an experimental analysis of the memory performance of our data structure compared to other representations. Unless mentioned otherwise, all simplicial complexes are computed up to the embedding dimension. All timings are averaged over 10 independent runs. Due to the lack of space, we cannot report on the performance of each algorithm on each dataset but the results presented are a faithful sample of what we have observed on other datasets.

As illustrated in Figure 3, we are able to construct and represent both Rips and relaxed witness complexes of up to several hundred million faces in high dimensions, on all datasets.

4.1   Memory Performance of the Simplex Tree

In order to represent the combinatorial structure of an arbitrary simplicial complex, one needs to mark all maximal faces. Indeed, from the definition of a simplicial complex, we cannot infer the higher dimensional faces from the lower dimensional ones. Moreover, the number of maximal simplices of a k-dimensional simplicial complex is at least $|V|/k$. In the case, considered in this paper, where the vertices are identified by their labels, a minimal representation of the maximal simplices would then require at least $\Omega(\log |V|)$ bits per maximal face. The simplex tree uses $O(\log |V|)$ memory bits per face *of any dimension*.

The following experiment compares the memory performance of the simplex tree with the minimal representation described above, and with the representation of the 1-skeleton. Figure 4 shows results for both Rips and relaxed witness complexes associated to 10000 points from **S4** and various values of, respectively, the distance threshold r and the relaxation parameter ρ. The figure plots the total number of faces $|\mathcal{K}|$, the number of maximal faces $|\text{m}\mathcal{F}|$, the size of the 1-skeleton $|\mathcal{G}|$ and the construction times T_{Rips} and T_{Wit^ρ}. The quantities $|\mathcal{K}|$, $|\text{m}\mathcal{F}|$ and $|\mathcal{G}|$ stand, respectively, for the asymptotic size of the simplex tree, the size of the optimal representation and of the size of the representation of the 1-skeleton.

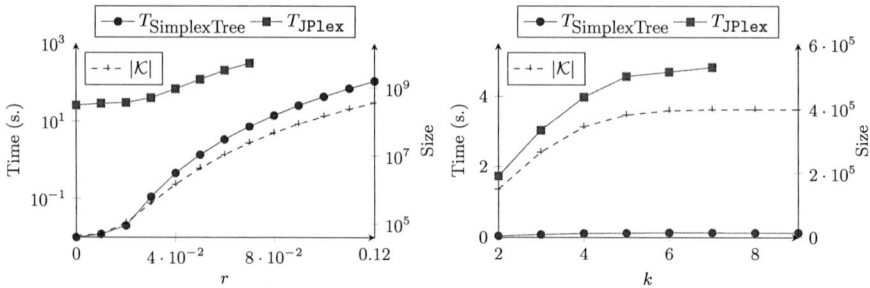

Fig. 5. Statistics and timings for the construction of the Rips complex on (Left) **Bud** and (Right) **Cy8**

As expected, the 1-skeleton is significantly smaller than the two other representations. However, as explained earlier, a representation of the graph of the simplicial complex is only well suited for flag complexes.

As shown on the figure, the total number of faces and the number of maximal faces remain close along the experiment. Interestingly, we catch the topology of **S4** when $r \approx 0.4$ for the Rips complex and $\rho \approx 0.08$ for the relaxed witness complex. For these "good" values of the parameters, the total number of faces is not much bigger than the number of maximal faces. Specifically, the total number of faces of the Rips complex is less than 2.3 times bigger than the number of maximal faces, and the ratio is less than 2 for the relaxed witness complex.

4.2 Construction of Rips Complexes

We test our algorithm for the construction of Rips complexes. In Figure 5 we compare the performance of our algorithm with JPlex along two directions.

In the first experiment, we build the Rips complex on 45000 points from the dataset **Bud**. Our construction is at least 43 times faster than JPlex along the experiment, and several hundred times faster for small parameter r. Moreover, JPlex is not able to handle the full dataset **Bud** nor big simplicial complexes due to memory allocation issues, whereas our method has no such problems. In our experiments, JPlex is not able to compute complexes of more than 23 million faces ($r = 0.07$) while the simplex tree construction runs successfully until $r = 0.12$, resulting in a complex of 368 million faces.

In the second experiment, we construct the Rips complex on the 6040 points from **Cy8**, with threshold $r = 0.31$, for different dimensions k. Again, our method outperforms JPlex, by a factor 14 to 20 (from small to big k). Again, JPlex cannot compute complexes of dimension higher than 7.

The simplex tree and the expansion algorithm we have described are, by design, output sensitive. For example, when we construct the Rips complex on **Cy8** for dimensions k higher than 5, the size of the output complex is constant, and so is the time for the Rips complex construction using a simplex tree. This is not the case for JPlex. Even further, as shown by our experiments, the construction

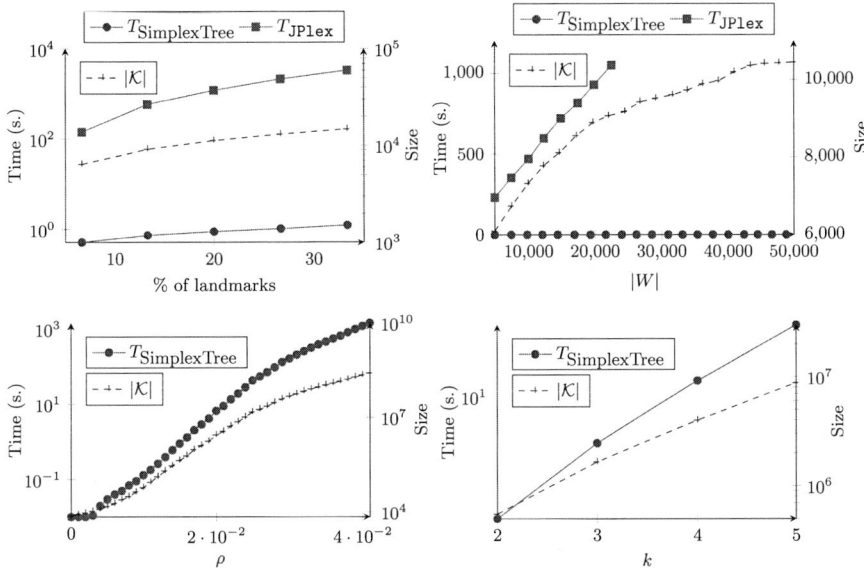

Fig. 6. Statistics and timings for the construction of: (TOP) the witness complex and (BOTTOM) the relaxed witness complex, on datasets (Left) **Bro** and (Right) **Kl**

time using a simplex tree depends linearly on the size of the output complex. Indeed, when the Rips graphs are dense enough so that the time for the expansion dominates the full construction, we observe that the average construction time per face is constant and equal to 2.9×10^{-7} seconds for the first experiment, and 5.4×10^{-7} seconds for the second experiment (with standard errors 0.2% and 1.3% respectively).

4.3 Construction of Witness Complexes

JPlex does not provide an implementation of the relaxed witness complex as defined in this paper. Consequently, we were only able to compare the algorithms on the construction of the witness complex. Figure 6 (top) shows the results of two experiments on the full construction of the witness complex.

The first one compares the performance of the simplex tree algorithm and of JPlex on the dataset **Bro** consisting of 15000 points in dimension $\mathbb{R}^{25}$. Subsets of different size of landmarks are selected at random among the sample points. Our algorithm is from several hundred to several thousand times faster than JPlex (from small to big subsets of landmarks). We stopped the experiment when JPlex became too slow. Moreover, the simplex tree algorithm spends more than 99% of the time to compute the nearest neighbors of the witnesses.

In the second experiment, we construct the witness complex on 2500 landmarks from **Kl**, and sets of witnesses of different size. The simplex tree algorithm outperforms JPlex, being tens of thousands times faster. We stopped the experiment for `JPlex` when it became too slow; differently, the simplex tree algorithm stayed under 0.1 second all along the experiment. Moreover, the simplex tree algorithm spends about 90% of the time to compute the nearest neighbors of the witnesses.

Finally we test the full construction of the relaxed witness complex along two directions, as illustrated in Figure 6 (bottom). In the first experiment, we compute the 5-skeleton of the relaxed witness complex on **Bro**, with 15000 witnesses and 1000 landmarks selected randomly, for different values of the parameter ρ. In the second experiment, we construct the k-skeleton of the relaxed witness complex on **Kl** with 10000 landmarks, 100000 witnesses and fixed parameter $\rho = 0.07$, for various k. We are able to construct and store complexes of up to 260 million faces. In both cases the construction time is linear in the size of the output complex, with a contruction time per face equal to 4.9×10^{-6} seconds in the first experiment, and 4.0×10^{-6} seconds in the second experiment (with standard errors 1.6% and 6.3% respectively).

Acknowledgments. The authors thanks A.Ghosh, S. Hornus, D. Morozov and P. Skraba for discussions that led to the idea of representing simplicial complexes by tries [8]. They also thank S. Martin and V. Coutsias for providing the cyclo-octane data set. This research has been partially supported by the 7th Framework Programme for Research of the European Commission, under FET-Open grant number 255827 (CGL Computational Geometry Learning).

References

1. Attali, D., Lieutier, A., Salinas, D.: Efficient data structure for representing and simplifying simplicial complexes in high dimensions. In: SoCG 2011, pp. 501–509 (2011)
2. Bentley, J.L., Sedgewick, R.: Fast algorithms for sorting and searching strings. In: SODA 1997, pp. 360–369 (1997)
3. Boissonnat, J.-D., Maria, C.: The Simplex Tree: An Efficient Data Structure for General Simplicial Complexes. Rapport de recherche RR-7993, INRIA (June 2012)
4. Brisson, E.: Representing geometric structures in d dimensions: topology and order. In: SCG 1989, pp. 218–227 (1989)
5. Carlsson, G., Ishkhanov, T., Silva, V., Zomorodian, A.: On the local behavior of spaces of natural images. Int. J. Comput. Vision 76, 1–12 (2008)
6. de Silva, V., Carlsson, G.: Topological estimation using witness complexes. In: Eurographics Symposium on Point-Based Graphics (2004)
7. Edelsbrunner, H., Harer, J.: Computational Topology - an Introduction. American Mathematical Society (2010)
8. Hornus, S.: Private communication
9. Lienhardt, P.: N-dimensional generalized combinatorial maps and cellular quasi-manifolds. Int. J. Comput. Geometry Appl. 4(3), 275–324 (1994)
10. Martin, S., Thompson, A., Coutsias, E.A., Watson, J.: Topology of cyclo-octane energy landscape. J Chem. Phys. 132(23), 234115 (2010)
11. Zomorodian, A.: The tidy set: a minimal simplicial set for computing homology of clique complexes. In: SCG 2010, pp. 257–266 (2010)

Succinct Posets

J. Ian Munro and Patrick K. Nicholson[*]

David R. Cheriton School of Computer Science, University of Waterloo
{imunro,p3nichol}@cs.uwaterloo.ca

Abstract. We describe an algorithm for compressing a partially ordered set, or *poset*, so that it occupies space matching the information theory lower bound (to within lower order terms), in the worst case. Using this algorithm, we design a succinct data structure for representing a poset that, given two elements, can report whether one precedes the other in constant time. This is equivalent to succinctly representing the transitive closure graph of the poset, and we note that the same method can also be used to succinctly represent the transitive reduction graph. For an n element poset, the data structure occupies $n^2/4 + o(n^2)$ bits, in the worst case, which is roughly half the space occupied by an upper triangular matrix. Furthermore, a slight extension to this data structure yields a succinct oracle for reachability in arbitrary directed graphs. Thus, using roughly a quarter of the space required to represent an arbitrary directed graph, reachability queries can be supported in constant time.

1 Introduction

Partially ordered sets, or *posets*, are useful for modelling relationships between objects, and appear in many different areas, such as natural language processing, machine learning, and database systems. As problem instances in these areas are ever-increasing in size, developing more space efficient data structures for representing posets is becoming an increasingly important problem.

When designing a data structure to represent a particular type of combinatorial object, it is useful to first determine how many objects there are of that type. By a constructive enumeration argument, Kleitman and Rothschild [11] showed that the number of n element posets is $2^{n^2/4+O(n)}$. Thus, the information theory lower bound indicates that representing an arbitrary poset requires $\lg(2^{n^2/4+O(n)}) = n^2/4 + O(n)$ bits[1]. This naturally raises the question of how a poset can be represented using only $n^2/4 + o(n^2)$ bits, *and* support efficient query operations. Such a representation, that occupies space matching the information theoretic lower bound to within lower order terms while supporting efficient query operations, is called a *succinct data structure* [9].

The purpose of this paper is to answer this question by describing the first succinct representation of arbitrary posets. We give a detailed description of our

---

[*] This research was funded in part by NSERC of Canada, the Canada Research Chairs program, and an NSERC PGS-D Scholarship.
[1] We use $\lg n$ to denote $\lceil \log_2 n \rceil$.

L. Epstein and P. Ferragina (Eds.): ESA 2012, LNCS 7501, pp. 743–754, 2012.

results in Section 4, but first provide some definitions in Section 2 and then highlight some of the previous work related to this problem in Section 3.

2   Definitions

A (finite) poset P, is a reflexive, antisymmetric, transitive binary relation $\preceq$ on a set S of n elements, denoted $P = (S, \preceq)$. Let a and b be two elements in S. If $a \preceq b$, we say a *precedes* b. We refer to queries of the form, "Does a precede b?" as *precedence queries*. If neither $a \preceq b$ or $b \preceq a$, then a and b are *incomparable*, and we denote this as $a \not\prec b$. For convenience we write $a \prec b$ if $a \preceq b$ and $a \neq b$.

Each poset $P = (S, \preceq)$ is uniquely described by a directed acyclic graph, or *DAG*, $G_c = (S, E_c)$, where $E_c = \{(a, b) : a \prec b\}$ is the set of edges. The DAG G_c is the *transitive closure graph* of P. Note that a precedence query for elements a and b is equivalent to the query, "Is the edge (a, b) in E_c?" Alternatively, let $G_r = (S, E_r)$ be the DAG such that $E_r = \{(a, b) : a \prec b, \nexists_{c \in S}, a \prec c \prec b\}$, i.e., the minimal set of edges that imply all the edges in E_c by transitivity. The DAG G_r also uniquely describes P, and is called the *transitive reduction graph* of P.

Posets are also sometimes illustrated using a *Hasse diagram*, which displays all the edges in the transitive reduction, and indicates the direction of an edge (a, b) by drawing element a above b. We refer to elements that have no outward edges in the transitive reduction as *sinks*, and elements that have no inward edges in the transitive reduction as *sources*. See Figure 1 for an example. Since all these concepts are equivalent, we may freely move between them when discussing a poset, depending on which representation is the most convenient.

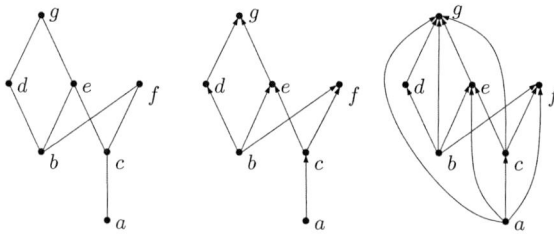

Fig. 1. A Hasse diagram of a poset (left), the transitive reduction (centre), and the transitive closure (right). Elements a and b are sources, and elements f and g are sinks.

A *linear extension* $L = \{a_1, ..., a_n\}$ is a total ordering of the elements in S such if $a_i \prec a_j$ for some $i \neq j$, then $i < j$. However, note that the converse is not necessarily true: we cannot determine whether $a_i \prec a_j$ unless we know that a_i and a_j are comparable elements. A *chain* of a poset, $P = (S, \preceq)$, is a total ordering $C = \{a_1, ..., a_k\}$ on a subset of k elements from S such that $a_i \prec a_j$ iff $i < j$, for $1 \leq i < j \leq k$. An *antichain* is a subset $A = \{a_1, ..., a_k\}$ of k elements from S, such that each a_i and a_j are incomparable, for $1 \leq i < j \leq k$. The *height* of a poset is the number of elements in its maximum chain, and the *width* of a poset is the number of elements in its maximum antichain.

For a graph $G = (V, E)$, we use $E(H)$ to denote the set of edges $\{(a, b) : (a, b) \in E, a \in H, b \in H\}$, where $H \subseteq V$. Similarly, we use $G(H)$ to denote the subgraph of G induced by H, i.e., the subgraph with vertex set H and edge set $E(H)$. Finally, if $(a, b) \in E$, or $(b, a) \in E$, we say that b is a *neighbour* of a in G.

3 Previous Work

Previous work in the area of succinct data structures includes representations of arbitrary undirected graphs [6], planar graphs [1], and trees [14]. There has also been interest in developing reachability oracles for planar directed graphs [18], as well as approximate distance oracles for undirected graphs [19]. For restricted classes of posets, such as lattices [17] and distributive lattices [7], space efficient representations have been developed, though they are not succinct.

One way of storing a poset is by representing either its transitive closure graph, or transitive reduction graph, using an adjacency matrix. If we topologically order the vertices of this graph, then we can use an upper triangular matrix to represent the edges, since the graph is a DAG. Such a matrix occupies $\binom{n}{2}$ bits, and can, in a single bit probe, be used to report whether an edge exists in the graph between two elements. Thus, this simple approach achieves a space bound that is roughly two times the information theory lower bound. An alternative representation, called the *ChainMerge* structure, was proposed by Daskalakis et al. [4], and occupies $O(nw)$ *words* of space, where w is the width of the poset. The ChainMerge structure, like the adjacency matrix of the transitive closure graph, supports precedence queries in $O(1)$ time.

Recently, Farzan and Fischer [5] presented a data structure that represents a poset using $2nw(1 + o(1)) + (1 + \varepsilon)n \lg n$ bits, where w is the width of the poset, and $\varepsilon > 0$ is an arbitrary positive constant. This data structure supports precedence queries in $O(1)$ time, and many other operations in time proportional to the width of the poset. These operations are best expressed in terms of the transitive closure and reduction graphs, and include: reporting all neighbours of an element in the transitive closure in $O(w + k)$ time, where k is the number of reported elements; reporting all neighbours of an element in the transitive reduction in $O(w^2)$ time; reporting an arbitrary neighbour of an element in the transitive reduction in $O(w)$ time; reporting whether an edge exists between two elements in the transitive reduction in $O(w)$ time; reporting all elements that, for two elements a and b, are both preceded by a and precede b in $O(w+k)$ time; among others. The basic idea of their data structure is to encode the ChainMerge structure of Daskalakis et al. [4] using bit sequences, and answer queries using rank and select operations on these bit sequences.

Since the data structure of Farzan and Fischer [5] is adaptive on width, it is appropriate for posets where the width is small relative to n. However, if we select a poset of n elements uniformly at random from the set of all possible n element posets, then it will have width $n/2 + o(n)$ with high probability [11]. Thus, this representation may occupy as many as $n^2 + o(n^2)$ bits, which is roughly *four times* the information theory lower bound. Furthermore, with the exception of precedence queries, all other operations take linear time for such a poset.

4 Our Results

Our results hold in the word-RAM model of computation with word size $\Theta(\lg n)$ bits. Our main result is summarized in the following theorem:

Theorem 1. *Let* $P = (S, \preceq)$ *be a poset, where* $|S| = n$. *There is a succinct data structure for representing* P *that occupies* $n^2/4 + O((n^2 \lg \lg n)/ \lg n)$ *bits, and can support precedence queries in* $O(1)$ *time: i.e., given elements* $a, b \in S$, *report whether* $a \preceq b$.

The previous theorem implies that we can, in $O(1)$ time, answer queries of the form, "Is the edge (a, b) in the transitive closure graph of P?" In fact, we can also apply the same representation to support, in $O(1)$ time, queries of the form, "Is the edge (a, b) in the transitive reduction graph of P?" However, at present it seems as though we can only support efficient queries in one or the other, *not both* simultaneously. For this reason we focus on the closure, since it is likely more useful, but state the following theorem:

Theorem 2. *Let* $G_r = (S, E_r)$ *be the transitive reduction graph of a poset, where* $|S| = n$. *There is a succinct data structure for representing* G_r *that occupies* $n^2/4 + O((n^2 \lg \lg n)/ \lg n)$ *bits, and, given elements* $a, b \in S$, *can report whether* $(a, b) \in E_r$ *in* $O(1)$ *time.*

Reachability in Directed Graphs: For an arbitrary DAG, the *reachability relation* between vertices is a poset: i.e., given two vertices, a and b, the relation of whether there a directed path from a to b in the DAG. As a consequence, Theorem 1 implies that there is a data structure that occupies $n^2/4 + o(n^2)$ bits, and can support reachability queries in a DAG, in $O(1)$ time. As far as we are aware, this is the first reachability oracle for arbitrary DAGs that uses strictly less space than an upper triangular matrix, and supports reachability queries in $O(1)$ time. We can even strengthen this observation by noting that for an *arbitrary directed graph* G, the *condensation* of G— the graph that results by contracting each strongly connected component into a single vertex [3, Section 22.5]— is a DAG. Given vertices a and b, if a and b are in the same strongly connected component, then b is reachable from a. Otherwise, we can apply Theorem 1 to the condensation of G. Thus, we get the following corollary:

Corollary 1. *Let* G *be a directed graph. There is a data structure that occupies* $n^2/4 + o(n^2)$ *bits and, given two vertices of* G, a *and* b, *can report whether* b *is reachable from* a *in* $O(1)$ *time.*

Note that the space bound of the previous corollary is roughly a quarter of the space required to represent an arbitrary directed graph! Switching back to the terminology of order theory, the previous corollary generalizes Theorem 1 to the larger class of binary relations known as *quasi-orders*: i.e., binary relations that are reflexive and transitive, but not necessarily antisymmetric. In fact, reflexivity does not restrict the binary relation very much, so we can further generalize Theorem 1 to arbitrary *transitive binary relations*; we defer the details to the full version.

Overview of the data structure: The main idea behind our succinct data structure is to develop an algorithm for compressing a poset so that it occupies space matching the information theory lower bound (to within lower order terms), in the worst case. The difficulty is ensuring that we are able to query the compressed structure efficiently. Our first attempt at designing a compression algorithm was essentially a reverse engineered version of an enumeration proof by Kleitman and Rothschild [10]. However, though the algorithm achieved the desired space bound, there was no obvious way to answer queries on the compressed data due to one crucial compression step. Though there are several other enumeration proofs (cf., [11,2]), they all appeal to a similar strategy, making them difficult to use as the basis for a data structure. This led us to develop an alternate compression algorithm, that borrows techniques from extremal graph theory.

We believe it is conceptually simpler to present our algorithm as having two steps. In the first step, we preprocess the poset, removing edges in its transitive closure graph, to create a new poset where the height is not too large. We refer to what remains as a *flat* poset. We then make use of the fact that, in a flat poset, either balanced biclique subgraphs of the transitive closure graph— containing $\Omega(\lg n / \lg \lg n)$ elements— must exist, or the poset is relatively sparsely connected. In the former case, the connectivity between these balanced biclique subgraphs and the remaining elements is shown to be space efficient to encode using the fact that all edges implied by transitivity are in the transitive closure graph. In the latter case, we can directly apply techniques from the area of succinct data structures to compress the poset.

5   Succinct Data Structure

In this section we describe a succinct data structure for representing posets. In order to refer to the elements in the poset, we assume each element has a label. Since our goal is to design a data structure that occupies $n^2/4 + o(n^2)$ bits, we are free to assign arbitrary $O(\lg n)$-bit labels to the elements, as such a labeling will require only $O(n \lg n)$ bits. Thus, we can assume each element in our poset has a distinct integer label, drawn from the range $[1, n]$. Our data structure always refers to elements by their labels, so often when we refer to "element a", it means "the element in S with label a", depending on context.

5.1   Preliminary Data Structures

Given a bit sequence $B[1..n]$, we use $\texttt{access}(B, i)$ to denote the i-th bit in B, and $\texttt{rank}(S, i)$ to denote the number of 1 bits in the prefix $B[1..i]$. We make use of the following lemma, which can be used to support access and rank operations on bit sequences, while compressing the sequence to its 0th-order empirical entropy.

Lemma 1 (Raman, Raman, Rao [16]). *Given a bit sequence B of length n, of which β bits are 1, there is a data structure that can represent B using $\lg \binom{n}{\beta} + O(n \lg \lg n / \lg n)$ bits, and support* $\texttt{access}$*, and* $\texttt{rank}$ *on B in $O(1)$ time.*

5.2   Flattening a Poset

Let $\gamma > 0$ be a parameter, to be fixed later; the reader would not be misled by thinking that we will eventually set $\gamma = \lg n$. We call a poset γ-*flat* if it has height no greater than γ. In this section, we describe a preprocessing algorithm for posets that outputs a data structure of size $O(n^2/\gamma)$ bits, that transforms a poset into a γ-flat poset, without losing any information about its original structure. After describing this preprocessing algorithm, we develop a compression algorithm for flat posets. Using the preprocessing algorithm together with the compression algorithm yields a succinct data structure for posets.

Let $P = (S, \preceq)$ be an arbitrary poset with transitive closure graph $G_c = (S, E_c)$. We decompose the elements of S into antichains based on their height within P. Let $\mathcal{H}(P)$ denote the height of P. All the sources in S are of height 1, and therefore are assigned to the same set. Each non-source element $a \in S$ is assigned a height equal to the number of elements in the maximum length path from a source to a. We use U_h to denote the set of all the elements of height h, $1 \leq h \leq \mathcal{H}(P)$, and $\mathcal{U}$ to denote the set $\{U_1, ..., U_{\mathcal{H}(P)}\}$. Furthermore, each set U_h is an antichain, since if $a \prec b$ then b has height strictly greater than a.

Next, we compute a linear extension $\mathcal{L}$ of the poset P in the following way, using $\mathcal{U}$. The linear extension $\mathcal{L}$ is ordered such that all elements in U_i come before U_{i+1} for all $1 \leq i < \mathcal{H}(P)$, and the elements within the same U_i are ordered arbitrarily within $\mathcal{L}$. Given any subset $S' \subseteq S$, we use the notation $S'(x)$ to denote the element ranked x-th according to $\mathcal{L}$, among the elements in the subset S'. We illustrate these concepts in Figure 2. Later, this particular linear extension will be used extensively, when we output the structure of the poset as a bit sequence.

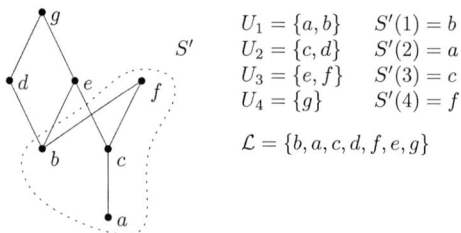

$$U_1 = \{a, b\} \quad S'(1) = b$$
$$U_2 = \{c, d\} \quad S'(2) = a$$
$$U_3 = \{e, f\} \quad S'(3) = c$$
$$U_4 = \{g\} \quad S'(4) = f$$

$$\mathcal{L} = \{b, a, c, d, f, e, g\}$$

Fig. 2. The antichain decomposition of the poset from Figure 1. The set S' is the set of elements surrounded by the dotted line. Note that $\mathcal{L}$ is only one of many possible linear extensions.

We now describe a preprocessing algorithm to transform an arbitrary poset P into a γ-flat poset $\tilde{P}$. We assume P is not γ-flat, otherwise we are done. Given two consecutive antichains U_i and U_{i+1}, we define a *merge step* to be the operation of replacing U_i and U_{i+1} by a new antichain $U_i' = U_i \cup U_{i+1}$, and outputting and removing all the edges between elements in U_i and U_{i+1} in the transitive closure of P, i.e., $E_c(U_i \cup U_{i+1})$. We say that U_{i+1} is the *upper antichain*, U_i is the *lower antichain*, and refer to the new antichain U_i' as the *merged antichain*. Each antichain U_j where $j > i+1$ becomes antichain U_{j-1}' in the *residual decomposition*, after the merge step. To represent the edges, let B be a bit sequence, storing $|U_i||U_{i+1}|$ bits. The bit sequence B is further subdivided into sections, denoted B^x, for each $x \in [1, |U_i|]$, where the bit $B^x[y]$ represents whether there is an edge from $U_i(x)$ to $U_{i+1}(y)$; or equivalently, whether $U_i(x) \prec U_{i+1}(y)$. We say that antichain U_{i+1} is *associated* with B, and vice versa. The bit sequence B is represented using the data structure of Lemma 1, which compresses it to its 0th-order empirical entropy[2]. Note that, after the merge step, the elements in merged antichain U_i' are ordered, in the linear extension $\mathcal{L}$, such that $U_i'(x) = U_i(x)$ for $1 \le x \le |U_i|$ and $U_i'(y + |U_i|) = U_{i+1}(y)$ for $1 \le y \le |U_{i+1}|$.

Algorithm FLATTEN$(\mathcal{U}, i)$: where i is the index of an antichain in $\mathcal{U}$.

---

if $i \ge |\mathcal{U}|$ **then**
    EXIT
end if
if $|U_i| + |U_{i+1}| \le 2n/\gamma$ **then**
    Perform a merge step on U_i and U_{i+1}
else
    $i \leftarrow i + 1$
end if
FLATTEN$(\mathcal{U}, i)$

There are many possible ways that we could apply merge steps to the poset in order to make it γ-flat. The method we choose, presented in algorithm FLATTEN, has the added benefit that accessing the output bit sequences is straightforward. Let $\tilde{\mathcal{U}}$ be the residual antichain decomposition that remains after executing FLATTEN$(\mathcal{U}, 1)$, and $\tilde{P}$ be the resulting poset. The number of antichains in $\tilde{\mathcal{U}}$ is at most γ, and therefore the remaining poset $\tilde{P}$ is γ-flat. We make the following further observation:

Lemma 2. FLATTEN$(\mathcal{U}, 1)$ *outputs* $O(n^2/\gamma)$ *bits.*

Proof. Consider the decomposition $\mathcal{U}$ and let $m = \mathcal{H}(P) = |\mathcal{U}|$. Let $n_1, ..., n_m$ denote the number of elements in $U_1, ..., U_m$, and $n_{s,t}$ to denote $\sum_{i=s}^{t} n_i$. We use the fact that the expression $\sum_{i=s}^{t-1} ((\sum_{j=s}^{i} n_j) n_{i+1}) \le n_{s,t}(n_{s,t} - 1)/2$, where $1 \le s < t \le m$. For each of the at most γ antichains in $\tilde{\mathcal{U}}$, there are two cases. Either the antichain was not created as the result of merge steps, or the antichain has size at most $2n/\gamma$, and is the result of some sequence of merge steps. Thus, the previous inequality implies that FLATTEN outputs no more than $O(n_{s,t}^2)$ bits for such an antichain, where $n_{s,t} = O(n/\gamma)$. Therefore, the total number of bits output during the merging steps is $O((n/\gamma)^2 \gamma) = O(n^2/\gamma)$. $\square$

---

[2] We note that for our purposes in this section, compression of the bit sequence is not required to achieve the desired asymptotic space bounds. However, the fact that Lemma 1 compresses the bit sequence will indeed matter in Section 5.3.

We now show how to use the output of FLATTEN to answer precedence queries on elements that end up in the same antichain in the residual decomposition:

Lemma 3. *There is a data structure of size $O(n^2/\gamma)$ bits that, given elements a and b can determine in $O(1)$ time whether a precedes b, if both a and b belong to the same antichain in the residual antichain decomposition $\tilde{\mathcal{U}}$.*

Proof. We add additional data structures to the output of FLATTEN in order to support queries. Since the labels of elements in S are in the range $[1, n]$, we can treat elements as array indices. Thus, it is trivial to construct an $O(n \lg n)$ bit array that, given elements $a, b \in S$, returns values i, i', j, j', x, x', y and y' in $O(1)$ time such that $U_i(x) = a$, $U_j(y) = b$, $U_{i'}(x') = a$, $U_{j'}(y') = b$, where $U_i, U_j \in \mathcal{U}$ and $U_{i'}, U_{j'} \in \tilde{\mathcal{U}}$. We additionally store the length of each antichain in $\mathcal{U}$, as well as an array A containing $|\mathcal{U}|$ *records*. For each antichain $U_k \in \mathcal{U}$, if U_k is the upper antichain during a merge step[3], then: $A[k]$.pnt points to the start of the sequence, B, associated with U_k, and; $A[k]$.len stores the length of the lower antichain. Recall that after the merge step, the element $U_k(z)$ has rank $z + A[k]$.len in the merged antichain. Thus, $A[k]$.len is the *offset* of the ranks of the elements of U_k within the merged antichain. These extra data structures occupy $O(n \lg n)$ bits and are dominated by the size of the output of FLATTEN, so the claimed space bound holds by Lemma 2.

We now discuss answering queries. Given $a, b \in S$, if $i' \neq j'$, then we return "different antichains". Otherwise, if $i = j$, then we return "$a \not\preceq b$", unless $a = b$, in which case we return "$a \preceq b$". Otherwise, assume without loss of generality that $i < j$. Thus, U_j is the upper antichain, and $A[j]$.pnt is a pointer to a sequence B, whereas U_i is a subset of the lower antichain $\hat{U}_k$ with which U_j is merged at some step, and $A[i]$.len is the offset of U_i's elements within $\hat{U}_k$. Let $\Delta = x + A[i]$.len, and return "$a \prec b$" if $B^\Delta[y] = 1$ and "$a \not\prec b$" otherwise. Section B^Δ begins at the $((\Delta - 1)|U_j|)$-th bit of B so we can access $B^\Delta[y]$ in $O(1)$ time. □

5.3   Compressing Flat Posets

In this section we describe a compression algorithm for flat posets that, in the worst case, matches the information theory lower bound to within lower order terms. We begin by stating the following lemma, which is a constructive deterministic version of a well known theorem by Kővári, Sós, and Turán [12]:

Lemma 4 (Mubayi and Turán [13]). *There is a constant c_{min} such that, given a graph with $|V| \geq c_{min}$ vertices and $|E| \geq 8|V|^{3/2}$ edges, we can find a balanced biclique $K_{q,q}$, where $q = \Theta(\lg |V| / \lg(|V|^2/|E|))$, in time $O(|E|)$.*

Let $\tilde{P}$ be a $(\lg n)$-flat poset, $G_c = (S, E_c)$ be its transitive closure, and $\tilde{\mathcal{U}} = \{U_1, ..., U_m\}$ be its antichain decomposition (discussed in the last section), which

---

[3] Note that, with the exception of the first merge step, $U_k \in \mathcal{U}$ is *not* the k-th antichain in the decomposition when the merge step occurs, but we will store records for the index k rather than some intermediate index.

contains $m \leq \lg n$ antichains. We now prove our key lemma, which is crucial for the compression algorithm.

Lemma 5 (Key Lemma). *Consider the subgraph $G_\Upsilon = G_c(U_i \cup U_{i+1})$ for some $1 \leq i < m$, and ignore the edge directions so that G_Υ is undirected. Suppose G_Υ contains a balanced biclique subgraph with vertex set D, and $|D| = \tau$. Then there are at most $2^{\tau/2+1} - 1$ ways that the vertices in D can be connected to each vertex in $S \setminus (U_i \cup U_{i+1})$.*

Proof. Each vertex $v \in S \setminus (U_i \cup U_{i+1})$ is in U_j, where, either $j > i + 1$ or $j < i$. Without loss of generality, consider the case where $j > i+1$. If any vertex $u \in D \cap U_{i+1}$ has an edge to v, then *all* vertices in $D \cap U_i$ have an edge to v. Thus, the vertices in $D \cap U_{i+1}$ can have edges to v in $2^{\tau/2} - 1$ ways, or the vertices in $D \cap U_i$ can have edges to v in $2^{\tau/2} - 1$ ways, or the vertices in D can have no edges to v. In total, there are $2^{\tau/2+1} - 1$ ways to connect v to D.   □

Algorithm COMPRESS-FLAT$(\hat{P}, \hat{n}, \hat{\mathcal{U}}, \hat{m})$: where $\hat{P} = (\hat{S}, \preceq)$ is a $(\lg n)$-flat poset of $\hat{n} \leq n$ elements, and $\hat{\mathcal{U}} = \{\hat{U}_1, ..., \hat{U}_{\hat{m}}\}$ is a decomposition of the elements in $\hat{P}$ into $\hat{m}$ antichains.

1: **if** $\hat{m} = 1$ **then**
2:    EXIT
3: **else if** $|\hat{U}_1 \cup \hat{U}_2| \geq c_{\min}$ and $|E_c(\hat{U}_1 \cup \hat{U}_2)| \geq (\hat{n}/\lg \hat{n})^2$ **then**
4:    Apply Lemma 4 to the subgraph $G_c(\hat{U}_1 \cup \hat{U}_2)$. This computes a balanced biclique with vertex set $D \subset \hat{U}_1 \cup \hat{U}_2$ such that $\tau = |D| = \Omega(\lg \hat{n}/\lg \lg \hat{n}) = O(\lg \hat{n})$.
5:    Let $H \leftarrow \hat{S} \setminus (\hat{U}_1 \cup \hat{U}_2)$. Output an array of integers Y, where $Y[k] \in [0, 2^{\tau/2+1} - 1]$ and indicates how $H(k)$ is connected to D (see Lemma 5).
6:    Let $V_1 \leftarrow \hat{U}_1 \cap D$, $V_2 \leftarrow \hat{U}_2 \cap D$, $\hat{U}_1 \leftarrow \hat{U}_1 \setminus V_1$, and $\hat{U}_2 \leftarrow \hat{U}_2 \setminus V_2$.
7:    For each $a \in V_1$ output a sequence W_a of $|\hat{U}_2|$ bits, where $W_a[k] = 1$ iff $a \prec \hat{U}_2(k)$.
8:    For each $b \in V_2$ output a sequence W_b of $|\hat{U}_1|$ bits, where $W_b[k] = 1$ iff $\hat{U}_1(k) \prec b$.
9:    COMPRESS-FLAT$(\hat{P} \setminus D, \hat{n} - \tau, \hat{\mathcal{U}}, \hat{m})$
10: **else**
11:    Perform a merge step on $\hat{U}_1$ and $\hat{U}_2$
12:    Set $\hat{m} \leftarrow \hat{m} - 1$
13:    COMPRESS-FLAT$(\hat{P}, \hat{n}, \hat{\mathcal{U}}, \hat{m})$
14: **end if**

Consider the algorithm COMPRESS-FLAT. The main idea is to apply Lemma 4 to the bottom two consecutive antichains the antichain decomposition, if they have many edges— defined on line 3— between them in the transitive closure graph. If they have few edges between them, then we apply a merge step. The algorithm terminates when only one antichain remains. We refer to the case on lines 4-9 as the *dense case*, and the case on lines 11-13 as the *sparse case*. We now prove that the size of the output of the compression algorithm matches the information theory lower bound to within lower order terms.

Lemma 6. *The output of* COMPRESS-FLAT$(\tilde{P}, n, \tilde{U}, m)$ *is no more than* $n^2/4 + O((n^2 \lg \lg n)/\lg n)$ *bits.*

Proof (Sketch). In the base case (line 2), the lemma trivially holds since nothing is output. Next we show that the total output from all the sparse cases cannot exceed $O((n^2 \lg \lg n)/\lg n)$ bits. Recall that the representation of Lemma 1 compresses to $\lg\lceil\binom{t}{\beta}\rceil + O(t \lg \lg t / \lg t)$ bits, where t is the length of the bit sequence, and β is the number of 1 bits. We use the fact that $\lg\lceil\binom{t}{\beta}\rceil \le \beta \lg(et/\beta) + O(1)$ [8, Section 4.6.4]. For a single pass through the sparse case, the total number of bits represented by B is $t = O(n^2)$, and $\beta = O((n/\lg n)^2)$ bits are 1's. Thus, the first term in the space bound to represent B using Lemma 1 (applying the inequality) is $O((n^2 \lg \lg n)/\lg^2 n)$ bits. Since we can enter the sparse case *at most* $\lg n$ times before exiting on line 2, the total number of bits occupied by the first term is bounded by $O((n^2 \lg \lg n)/\lg n)$. With a more careful analysis, it can be shown that the second term $(O(t \lg \lg t / \lg t))$ does not dominate the cost by using inequality from Lemma 2.

We now prove the lemma by induction for the dense case. Let $\mathcal{S}(n)$ denote the number of bits output by COMPRESS-FLAT$(\tilde{P}, n, \tilde{U}, m)$. We can assume $\mathcal{S}(n_0) \le n_0^2/4 + c_0(n_0^2 \lg \lg n_0)/\lg n_0$ for all $3 \le n_0 < n$, where $c_0 > 0$ is a sufficiently large constant. All the additional self-delimiting information— for example, storing the length of the sequences output on lines 5-8— occupies no more than $c_1 \lg n$ bits for some constant $c_1 > 0$. Finally, recall that $c_2 \lg n / \lg \lg n \le \tau \le c_3 \lg n$ for some constants $c_2, c_3 > 0$. We have:

$$\mathcal{S}(n) = \frac{\tau}{2}(|U_1| + |U_2| - \tau) + (n - (|U_1| + |U_2|)) \lg(2^{\tau/2+1}) + c_1 \lg n + \mathcal{S}(n - \tau)$$

$$< \left(\frac{\tau}{2} + 1\right)(n - \tau) + c_1 \lg n + \left(\frac{1}{4} + \frac{c_0 \lg \lg n}{\lg(n-\tau)}\right)(n^2 - 2n\tau + \tau^2)$$

$$< \frac{n^2}{4} + \frac{c_0 n^2 \lg \lg n}{\lg(n-\tau)} - (c_5 - c_4)n \quad (c_5 < c_0 c_2, \; c_4 > 1)$$

Note that through our choice of c_0 and c_4, we can ensure that $c_5 - c_4$ is a positive constant. If $\lg(n - \tau) = \lg n$, then the inductive step clearly holds. The alternative case can only happen when $n \ge 2^k$, for some positive integer k, and $n - \tau < 2^k$, due to the ceiling function on $\lg$. Thus, the alternative case only occurs once every $\Omega(n/\lg n)$ times, since $\tau = O(\lg n)$. By charging this extra cost to the rightmost negative term, the induction holds.   □

We now sketch how to support precedence queries on a $(\lg n)$-flat poset. As in the previous section, if element a is removed in the dense case, we say a is *associated* with the output on lines 6-9. Similarly, for each antichain $U_i \in \tilde{U}$ involved in a merge step as the upper antichain in the sparse case, we say that U is *associated* with the output of that merge step, and vice versa.

Lemma 7. *Let $\tilde{P}$ be a $(\lg n)$-flat poset on n elements, with antichain decomposition $\tilde{U} = \{U_1, ..., U_m\}$. There is a data structure of size $n^2/4 + O((n^2 \lg \lg n)/\lg n)$ bits that, given elements a and b, can report whether a precedes b in $O(1)$ time.*

Proof (Sketch). We augment the output of COMPRESS-FLAT with additional data structures in order to answer queries efficiently. We denote the first set of elements removed in a dense case as D_1, the second set as D_2 and so on. Let D_r denote the last set of elements removed in a dense case, for some $r = O(n \lg \lg n / \lg n)$. Let $S_\ell = S/(\cup_{i=1}^{\ell} D_i)$, for $1 \leq \ell \leq r$. We define $M_\ell(x)$ to be the number of elements $a \in S_\ell$ such that $S(y) = a$, and $y \leq x$. We now discuss how to compute $M_\ell(x)$ in $O(1)$ time using a data structure of size $O(n^2 \lg \lg n / \lg n)$ bits. Define M'_ℓ to be a bit sequence, where $M'_\ell[x] = 1$ iff $S(x) \in S_\ell$, for $x \in [1, n]$. We represent M'_ℓ using the data structure of Lemma 1, for $1 \leq \ell \leq r$. Overall, these data structures occupy $O(n^2 \lg \lg n / \lg n)$ bits, since $r = O((n \lg \lg n)/\lg n)$, and each bit sequence occupies $O(n)$ bits by Lemma 1. To compute $M_\ell(x)$ we return $\mathtt{rank}_1(M'_\ell, x)$, which requires $O(1)$ time by Lemma 1. By combining the index just described with techniques similar in spirit to those used in Lemma 3, we can support precedence queries in $O(1)$ time. The idea is to find the output associated with first query element that was removed during the execution of the compression algorithm, and find the correct bit in the output to examine using the index just described. □

Theorem 1 follows by combining Lemmas 3 (with γ set to $\lg n$) and 7.

6   Concluding Remarks

We have presented the first succinct data structure for arbitrary posets. For a poset of n elements, our data structure occupies $n^2/4 + o(n^2)$ bits and supports precedence queries in $O(1)$ time. This is equivalent to supporting $O(1)$ time queries of the form, "Is the edge (a, b) in the transitive closure graph of P?"

We close with a few brief remarks, and defer additional details to the full version. Our first remark is that if we want to support edge queries on the transitive reduction instead of the closure, a slightly simpler data structure can be used. The reason for this simplification is that for the transitive reduction, our key lemma does not require the antichains containing the biclique to be consecutive, and, furthermore, we can "flatten" the transitive reduction in a simpler way than by using Lemma 3. Our second remark is that many of the bit sequences we use in our data structure do not require the rank operation, and there are alternative data structures with smaller lower order terms for these situations; see Raman, Raman, Rao [16] for alternatives, as well as Okanohara and Sadakane [15] for implementations. Finally, we remark that we can report the neighbours of an arbitrary element in the transitive closure graph efficiently, without asymptotically increasing the space bound of Theorem 1. This is done by encoding the neighbours using a bit sequence, if there are few of them, and checking all $n - 1$ possibilities via queries to the data structure of Theorem 1, if there are many.

References

1. Barbay, J., Aleardi, L.C., He, M., Munro, J.I.: Succinct representation of labeled graphs. Algorithmica 62(1-2), 224–257 (2012)
2. Brightwell, G., Jurgen Promel, H., Steger, A.: The average number of linear extensions of a partial order. J. Comb. Theo., Series A 73(2), 193–206 (1996)
3. Cormen, T.H., Leiserson, C.E., Rivest, R.L., Stein, C.: Introduction to Algorithms, 2nd edn. The MIT Press (2001)
4. Daskalakis, C., Karp, R.M., Mossel, E., Riesenfeld, S., Verbin, E.: Sorting and selection in posets. In: Proc. SODA, pp. 392–401. SIAM (2009)
5. Farzan, A., Fischer, J.: Compact Representation of Posets. In: Asano, T., Nakano, S.-i., Okamoto, Y., Watanabe, O. (eds.) ISAAC 2011. LNCS, vol. 7074, pp. 302–311. Springer, Heidelberg (2011)
6. Farzan, A., Munro, J.I.: Succinct Representations of Arbitrary Graphs. In: Halperin, D., Mehlhorn, K. (eds.) ESA 2008. LNCS, vol. 5193, pp. 393–404. Springer, Heidelberg (2008)
7. Habib, M., Nourine, L.: Tree structure for distributive lattices and its applications. Theoretical Computer Science 165(2), 391 (1996)
8. He, M.: Succinct Indexes. Ph.D. thesis, University of Waterloo (2007)
9. Jacobson, G.: Space-efficient static trees and graphs. In: Proc. FOCS, pp. 549–554 (1989)
10. Kleitman, D.J., Rothschild, B.L.: The number of finite topologies. Proceedings of the American Mathematical Society 25, 276 (1970)
11. Kleitman, D.J., Rothschild, B.L.: Asymptotic enumeration of partial orders on a finite set. Transactions of the American Mathematical Society 205, 205–220 (1975)
12. Kővári, T., Sós, V.T., Turán, P.: On a problem of Zarankiewicz. Coll. Math. 3, 50–57 (1954)
13. Mubayi, D., Turán, G.: Finding bipartite subgraphs efficiently. Information Processing Letters 110(5), 174–177 (2010)
14. Munro, J.I., Raman, V.: Succinct representation of balanced parentheses and static trees. SIAM J. Comput. 31(3), 762–776 (2001)
15. Okanohara, D., Sadakane, K.: Practical entropy-compressed rank/select dictionary. In: ALENEX (2007)
16. Raman, R., Raman, V., Rao, S.S.: Succinct indexable dictionaries with applications to encoding k-ary trees and multisets. In: Proc. SODA, pp. 233–242. SIAM (2002)
17. Talamo, M., Vocca, P.: An efficient data structure for lattice operations. SIAM J. on Comp. 28(5), 1783–1805 (1999)
18. Thorup, M.: Compact oracles for reachability and approximate distances in planar digraphs. J. ACM 51(6), 993–1024 (2004)
19. Thorup, M., Zwick, U.: Approximate distance oracles. J. ACM 52(1), 1–24 (2005)

Polynomial-Time Approximation Schemes for Shortest Path with Alternatives

Tim Nonner

IBM Research - Zurich
tno@zurich.ibm.com

Abstract. Consider the generic situation that we have to select k alternatives from a given ground set, where each element in the ground set has a random arrival time and cost. Once we have done our selection, we will greedily select the first arriving alternative, and the total cost is the time we had to wait for this alternative plus its random cost. Our motivation to study this problem comes from public transportation, where each element in the ground set might correspond to a bus or train, and the usual user behavior is to greedily select the first option from a given set of alternatives at each stop. We consider the arguably most natural arrival time distributions for such a scenario: exponential distributions, uniform distributions, and distributions with mon. decreasing linear density functions. For exponential distributions, we show how to compute an optimal policy for a complete network, called a shortest path with alternatives, in $\mathcal{O}(n(\log n + \delta^3))$ time, where n is the number of nodes and δ is the maximal outdegree of any node, making this approach practicable for large networks if δ is relatively small. Moreover, for the latter two distributions, we give PTASs for the case that the distribution supports differ by at most a constant factor and only a constant number of hops are allowed in the network, both reasonable assumptions in practice. These results are obtained by combining methods from low-rank quasi-concave optimization with fractional programming. We finally complement them by showing that general distributions are NP-hard.

1 Introduction

Due to its wide application in route planning, finding a shortest path in a given network is a cornerstone problem in combinatorial optimization, which spawned algorithms like Dijkstra and Bellman-Ford [3,8] present in any textbook. The basic idea of route planning is to provide a fixed path that indicates which arc to follow at each node in order to reach the destination in a minimum amount of time. In public transportation, for instance, each arc might correspond to a bus line connecting two stops. In such a scenario, it is necessary to wait at each stop until a bus of the selected line arrives. If the exact arrival times of all buses are known, then we can find an optimal path by using a time-expanded network [8] that simulates arrival times at the expense of adding additional nodes

L. Epstein and P. Ferragina (Eds.): ESA 2012, LNCS 7501, pp. 755–765, 2012.
© Springer-Verlag Berlin Heidelberg 2012

and arcs. However, an ongoing study of the Dublin bus system[1] revealed that arrival times are sometimes more a rule of thumb, which provide some guidance how frequently a bus arrives (e.g. every 5 or 10 minutes). In this case of stochastic arrival times, selecting a single line is suboptimal, since there might be another line arriving earlier that is also a good choice, possibly with a different next stop, yielding a smaller total transit time. Instead, it is then more convenient to provide some alternatives from which we should greedily pick the one with smallest realized arrival time. In other words, the one that comes first. Therefore, given distributions of the arrival times and (possibly random) transit times of the arcs, we consider the problem of optimally selecting alternatives, say at most k at each node, such that the expected time to reach the destination is minimized, which is the sum of all waiting and transit times. Note that, using linearity of expectation, this problem contains the classical problem of finding a shortest path as a special case for $k = 1$ or if all arrival times are zero.

For example, consider the network depicted in Subfigure 1(a) with source $\mathfrak{s}$ and destination $\mathfrak{t}$. The solid arcs represent a solution for $k = 2$ with two alternatives leaving source $\mathfrak{s}$. Hence, depending on the realized arrival times, we either pick the arc leading to node v, or the arc leading to node u. At node v we have no choice, but there are again two alternatives at node u. In any case, we end up at destination $\mathfrak{t}$.

To gain a better understanding, consider now a simple network consisting of only two nodes, source $\mathfrak{s}$ and destination $\mathfrak{t}$, and δ parallel arcs connecting $\mathfrak{s}$ and $\mathfrak{t}$, as depicted in Subfigure 1(b). In this case, we can think of the arcs labeled $1, 2, \ldots, \delta$ as an abstract ground set, where each element i in this ground set has a random arrival time A_i and random cost/transit time T_i. Hence, once we have done our selection of at most k alternatives, we have to pay the random cost of the first arriving alternative, a quite generic scenario. However, because our motivation comes from shortest paths and public transportation, we refer to an element in this ground set as an arc throughout this paper. Note that even if $k = \infty$, it might not be optimal to include all arcs in order to minimize the waiting time, since there might be an arc with a very small expected arrival time, but an enormous transit time. In this case, it is very likely that we pick this arc, spending a long time in transit. Therefore, finding an optimal set of alternatives is non-trivial even for such simple networks. Indeed, we show that this problem is NP-hard even if all transit times are zero.

Motivated by the NP-hardness of general arrival times, even for simple networks, we consider the arguably most natural arrival times for this scenario: *exponential arrival times*, that is, exponentially distributed arrival times, *uniform arrival times*, that is, arrival times which are uniformly distributed between 0 and some given positive parameter l, and *linear arrival times*, that is, arrival times whose density function f has the form $f(t) = b - a \cdot t$ for $t \leq l$ and $f(t) = 0$ for $t > l$ for some parameters $a, b, l \geq 0$ such that $\int_{t=0}^{\infty} f(t)\, dt = 1$. Note that uniform arrival times are the most natural ones in public transportation

---

[1] For instance, for many bus lines listed on http://www.dublinbus.ie/en/ , there are only the times listed when the buses leave the terminal.

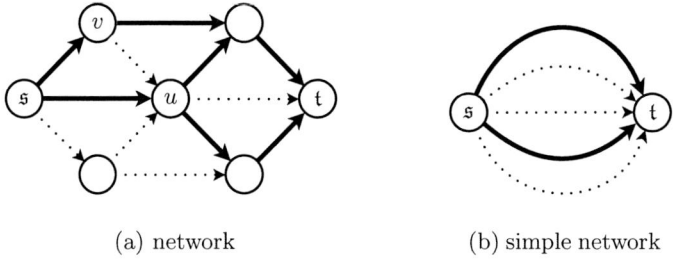

(a) network   (b) simple network

Fig. 1. Example alternatives

system with random but periodic service times. For instance, if a bus arrives at about every 5 minutes at some stop, then this is naturally modeled by a uniform distribution with $l = 5$ minutes.

Contributions and Outline. First, we formally present the objective for simple networks in Section 2. Using this, we obtain an $\mathcal{O}(\delta^2)$ time algorithm for simple networks with exponential arrival times in Section 3. By applying a Dijkstra-type procedure in Section 5, we extend this result to general networks, which gives an algorithm with running time $\mathcal{O}(n(\log n + \delta^3))$, where n is the number of nodes and δ is the maximal outdegree of any node. Next, we give a PTAS for simple networks with uniform arrival times in Section 4. Specifically, for any sufficiently small $\epsilon > 0$, this algorithm yields a $1 + \epsilon$-approximation in time $\delta^{1/\epsilon^{\mathcal{O}(1)}}$. The only requirement is that the l-parameters of the uniform arrival times differ by at most a constant factor. It is worth mentioning here that straight-forward approaches like geometric rounding of the l-parameters do not work, even if this restriction is satisfied, since the objective derived in Section 2 is quite sensitive with respect to even small changes of the input. Finally, we extend this PTAS to the linear arrival times described above. Both PTASs are extended to general networks in Section 5 for the case that we restrict the number of hops to constantly many. This makes sense especially in public transportation, since it is not convenient to switch buses very often, but only a small number of times. Complementing these results, we give an NP-hardness proof for simple networks with arbitrary arrival times in Section 6.

Techniques for Exponential Arrival Times. To construct the algorithm for simple networks with exponential arrival times, we show that, in this case, finding an optimal set of alternatives is equivalent to solving a fractional knapsack problem with uniform weights, that is, minimizing a fraction of linear functions subject to a cardinality constraint. It is a classic result of Megiddo [13] that such fractional programming problems can be solved in polynomial time if the underlying linear problem can be solved in polynomial time. The main idea, which has been extended many times, for instance to transfer it to approximate solutions [10,2,4], is to transform the fractional objective into a parametric linear objective, and then perform a local search on the parameter. Using this, our fractional knapsack problem can be clearly solved in $\mathcal{O}(\delta^2 \log^2 \delta)$ time. Another approach for such problems arises from the context of quasi-concave/convex optimization. Specifically, we can also interpret our knapsack objective as a quasi-concave function of rank two [12,14]. Thus, motivated by a similar approach of Nikolova et al. [14], we find that an optimal solution corresponds to a vertex of the shadow of the considered polytope, and since we can prove that there are only polynomially many such vertices, this yields an enumeration approach running in $\mathcal{O}(\delta^2 \log \delta)$ time, which we sketch in Subsection 3.1. Finally, by combining this enumeration approach with the fractional programming approach of Megiddo, we are able to improve by a logarithmic factor to $\mathcal{O}(\delta^2)$ running time in Subsection 3.2. Our main argument here is that, having an unsorted list of all vertices of the shadow polytope, we are able to replace a sorting step by an iterative median computation. To the best of our knowledge, this simple but powerful combination has not been used before to improve the running time of algorithms dealing with fractional objectives. Replacing a sorting step by a median computation is also an argument in [2], but we do not see any connection to our case.

Techniques for Uniform Arrival Times. We obtain the PTAS for simple networks with uniform arrival times by applying a well-known technique [15]: partition the search space into *small sets* (Subsection 4.1) and *large sets* (Subsection 4.2), where the maximum size of the small sets depends on the required precision $\epsilon > 0$. Then enumerate all small sets and provide a different approach for the large sets. The main insight here is that uniform arrival times start to behave like exponential arrival times for large sets. This results in a reduction to a quasi-concave optimization problem, which can be solved using the enumeration approach from Subsection 3.1.

Related Work. The problem most related to ours is the stochastic shortest path problem described by Bertsekas and Tsitsiklis [1]. Here, at each node, we need to select a probability distribution over the outgoing arcs. The network is then randomly traversed according to these distributions, and, similar to our problem, the goal is to minimize the expected time until the destination is reached. However, since this problem only requires the selection of a single distribution at each node, it is still much closer related to the traditional shortest path problem. In contrast, the difficulty of our problem mainly stems from the fact that we need to select a set of alternatives instead of a single arc.

Starting with a paper of Dreyfus [6], there is a line of research considering the case of time-dependent transit times, see for instance [9] and the references therein. Waiting plays a different role in such a scenario, since it might be advantageous to wait at a node for the decrease of some transit times.

Waiting is a common theme in on-line algorithms. For instance, in the famous secretary problem [7], a sequence of secretaries of different quality arrives, and we have to wait for the right secretary, that is, stop this process when we think that the current candidate is good enough. Other well-known examples are the the TCP-acknowledgment problem [5] or the ski rental problem [11]. However, our problem is an off-line problem, since we can only indirectly influence the waiting times by selecting the right alternatives.

Formal Problem Definition. We are given a network modeled by a directed graph $G = (V, E)$ with source and destination nodes $\mathfrak{s} \in V$ and $\mathfrak{t} \in V$, respectively. Let $n := |V|$, and let δ be the maximum outdegree of any node in G. Each arc $e = (v, u) \in E$ has a random arrival time A_e, which is the time we have to wait until we may use this arc after arriving at node v. Once we select this arc, it takes time T_e to transit it to node u, where T_e might be a random variable. We allow parallel arcs, and we assume that all arrival and transit times are positive. The goal is to specify a subgraph G' with a single source $\mathfrak{s}$, a single sink $\mathfrak{t}$, and maximum outdegree k, where k is an additionally given positive integer. The *alternatives* at each node v in G' are then the set of (at most k) outgoing arcs of v, and the cost of G' is the expected time to travel from $\mathfrak{s}$ to $\mathfrak{t}$ under the assumption that, at each node v, the alternative with minimum realized arrival time is selected. Finally, a *simple network* is a network with $V = \{\mathfrak{s}, \mathfrak{t}\}$. In this case, we may assume that there are no arcs from $\mathfrak{t}$ to $\mathfrak{s}$, and hence δ is the number of arcs from $\mathfrak{s}$ to $\mathfrak{t}$.

2   Objective in Simple Networks

In simple networks with only two nodes, assume that the parallel arcs connecting the source $\mathfrak{s}$ with the destination $\mathfrak{t}$ are labeled $1, 2, \ldots, \delta$ with arrival times $A_1, A_2, \ldots, A_\delta$ and transit times $T_1, T_2, \ldots, T_\delta$. For each arc i, let $f_i(t)$ denote the density function and $F_i(t) := \Pr[A_i \leq t] = \int_{s=0}^{t} f_i(s)\, ds$ the cumulative distribution of its arrival time A_i. For a given subset of arcs $\sigma \subseteq \{1, \ldots, \delta\}$, let $\text{cost}(\sigma)$ denote the combined expected waiting and transit time. Therefore, our goal is to find a set σ that minimizes

$$\text{cost}(\sigma) := \int_{t=0}^{\infty} \sum_{i \in \sigma} f_i(t)(t + \mathbb{E}[T_i]) \prod_{j \in \sigma: j \neq i} (1 - F_j(t))\, dt \tag{1}$$

$$= \int_{t=0}^{\infty} \prod_{i \in \sigma} (1 - F_i(t))\, dt + \int_{t=0}^{\infty} \sum_{i \in \sigma} f_i(t)\mathbb{E}[T_i] \prod_{j \in \sigma: j \neq i} (1 - F_j(t))\, dt.$$

3   Exponential Arrival Times

In this section, we consider simple networks with exponential arrival times. Specifically, the arrival time A_i of each arc i is exponentially distributed with some parameter $\beta_i > 0$, i.e., $f_i(t) = \beta_i e^{-t\beta_i}$. Using standard arithmetic and Equation (1), we obtain the fractional programming objective

$$\text{cost}(\sigma) = \frac{1 + \sum_{i \in \sigma} \mathbb{E}[T_i]\, \beta_i}{\sum_{i \in \sigma} \beta_i}, \tag{2}$$

which we will use throughout this section. Note that if $k = \infty$, then it is easy to minimize this objective by using the observation that if there is an arc $i \in \{1, 2, \ldots, \delta\} \backslash \sigma$ with $\mathbb{E}[T_i] < \mathrm{cost}(\sigma)$, then $\mathrm{cost}(\sigma \cup \{i\}) < \mathrm{cost}(\sigma)$. This implies that if we first sort the arcs such that $\mathbb{E}[T_1] \leq \mathbb{E}[T_2] \leq \ldots \leq \mathbb{E}[T_\delta]$, then an optimal set of alternatives has the form $\sigma = \{1, 2, \ldots, i\}$ for some $1 \leq i \leq \delta$, and can hence be found in linear time after the sorting. However, we need to deal with the case of a general k.

3.1   Enumeration

Let $a \in \mathbb{R}^{\delta+1}$ be the vector with $a_i := \mathbb{E}[T_i]\beta_i$ for $1 \leq i \leq \delta$ and additionally $a_{\delta+1} := 1$, and let $b \in \mathbb{R}^{\delta+1}$ be the vector with $b_i := \beta_i$ for $1 \leq i \leq \delta$ and additionally $b_{\delta+1} := 0$. Having a $(\delta+1)$th component is for the sake of exposition. Moreover, for any $0 \leq \lambda \leq 1$, let $c^\lambda := (\lambda - 1)a + \lambda b$ be a linear combination of the vectors a and b. Then, for each component $1 \leq i \leq \delta$, there is exactly one parameter $0 \leq \lambda_i \leq 1$ such that $c_i^\lambda < 0$ for all $0 \leq \lambda < \lambda_i$ and $c_i^\lambda \geq 0$ for all $\lambda_i \leq \lambda \leq 1$. Moreover, for any pair of components $1 \leq i < j \leq \delta$, we either have that one of these components always dominates the other one, i.e., $c_i^\lambda \leq c_j^\lambda$ for all $0 \leq \lambda \leq 1$ or $c_j^\lambda \leq c_i^\lambda$ for all $0 \leq \lambda \leq 1$, or there is exactly one parameter $0 \leq \lambda_{ij} \leq 1$ such that $c_i^{\lambda_{ij}} = c_j^{\lambda_{ij}}$. Hence, in the latter case, the domination of components i and j swaps at λ_{ij}. Define then the set

$$\Lambda_{ab} := \{\lambda_i \mid 1 \leq i \leq \delta\} \cup \{\lambda_{ij} \mid 1 \leq i < j \leq \delta : i \text{ and } j \text{ swap dom.}\} \cup \{0\}.$$

which has quadratic size $\mathcal{O}(\delta^2)$. Note that each vector $v \in \{0,1\}^{\delta+1}$ that satisfies the cardinality constraint $1 \leq \sum_{i=1}^{\delta} v_i \leq k$ describes a set of alternatives σ as characteristic vector, i.e., for each $1 \leq i \leq \delta$, $i \in \sigma$ if $v_i = 1$ and $i \notin \sigma$ if $v_i = 0$. Finally, for any $0 \leq \lambda \leq 1$, let $v^\lambda \in \{0,1\}^{\delta+1}$ be the vector with $1 \leq \sum_{i=1}^{\delta} v_i^\lambda \leq k$ and $v_{\delta+1}^\lambda = 1$ that maximizes $(c^\lambda)^t v^\lambda$. We use these definitions in the following algorithm.

Algorithm $\mathcal{A}^{\mathrm{enum}}$ Input: a, b, k

---

1. Compute Λ_{ab}.
2. For each $\lambda \in \Lambda_{ab}$, compute v^λ.
3. Return the set σ which corresponds to some vector v^λ computed in the last step that minimizes $\mathrm{cost}(\sigma)$.

The following theorem analyzes this algorithm[2].

Theorem 1. *For simple networks with exponential arrival times, algorithm $\mathcal{A}^{\mathrm{enum}}$ returns an optimal set of alternatives. It can be implemented such that it runs in time $\mathcal{O}(\delta^2 \log \delta)$.*

---

[2] All proofs are deferred to the full version of this paper.

3.2   Combining Fractional Programming and Enumeration

The arguably best implementation of algorithm $\mathcal{A}^{\text{enum}}$ initially sorts the $\mathcal{O}(\delta^2)$ many elements in Λ_{ab}, which takes superquadratic time. To avoid this step, consider the following median-based algorithm.

Algorithm $\mathcal{A}^{\text{median}}$ Input: a, b, k

---

1. Compute Λ_{ab} and set $\Lambda' := \Lambda_{ab}$.
2. While $|\Lambda'| > 2$:
   (a) Compute a median $\lambda' \in \Lambda'$ such that the sizes of the sets $\Lambda'_{\geq} := \{\lambda \in \Lambda' \mid \lambda \geq \lambda'\}$ and $\Lambda'_{<} := \{\lambda \in \Lambda' \mid \lambda < \lambda'\}$ differ by at most one.
   (b) Compute $v^{\lambda'}$.
   (c) If $(c^{\lambda'})^t v^{\lambda'} \leq 0$, then set $\Lambda' := \Lambda'_{\geq}$, and otherwise set $\Lambda' := \Lambda'_{<}$.
3. Let $\{\lambda_1, \lambda_2\} := \Lambda'$ with $\lambda_1 < \lambda_2$.
4. Return v^{λ} for some $\lambda_1 < \lambda < \lambda_2$.

The following theorem gives the correctness.

Theorem 2. *For simple networks with exponential arrival times, algorithm $\mathcal{A}^{median}$ returns an optimal set of alternatives. It can be implemented in $\mathcal{O}(\delta^2)$ time.*

4   Uniform Arrival Times

In this section, we consider simple networks with uniform arrival times. Hence, each arc i has a parameter l_i such that its arrival time is uniformly distributed in interval $[0, l_i]$, i.e., $f_i(t) = 1/l_i$ for $1 \leq t \leq l_i$ and $f_i(t) = 0$ for $t > l_i$. It follows that $F_i(t) = t/l_i$ is a linear function for $t \leq l_i$, and $F_i(t) = 1$ for $t > l_i$. For a set of arcs $\sigma \subseteq \{1, \ldots, \delta\}$, define $l_\sigma := \min_{i \in \sigma} l_i$ and $l_\sigma^{\max} := \max_{i \in \sigma} l_i$. We assume throughout this section that the l-parameters of the uniform arrival times differ by at most a constant factor, which implies that $l_\sigma^{\max}/l_\sigma = \mathcal{O}(1)$ for any set σ. Note that the realized arrival time of a set σ is bounded by l_σ. Using this and Equation (1), we get the objective $\text{cost}(\sigma) = \int_{t=0}^{l_\sigma} g_\sigma(t) \, dt$, where, for each $0 \leq t \leq l_\sigma$, the function g_σ is defined as

$$g_\sigma(t) := \prod_{i \in \sigma} \left(1 - \frac{t}{l_i}\right) + \sum_{i \in \sigma} \frac{\mathbb{E}[T_i]}{l_i} \prod_{j \in \sigma: j \neq i} \left(1 - \frac{t}{l_j}\right).$$

We use this objective throughout this section. To find a $1 + \epsilon$-approximation for this objective with respect to a given precision $\epsilon > 0$, we distinguish two cases: *small sets*, that is, sets of size $\leq \tau$, and *large sets*, that is, sets of size $> \tau$, where τ is a constant threshold depending on ϵ that will be defined later on. The case of small sets is considered in Subsection 4.1, and the case of large sets is considered in Subsection 4.2. Combining both cases finally yields a PTAS in Subsection 4.2.

4.1   Small Sets

If there is an optimal set σ which is small, then we can find it in polynomial time by enumerating $\mathcal{O}(\delta^\tau)$ many subsets. However, for each such subset, we need to approximately compute $\mathrm{cost}(\sigma)$ in order to return the one which minimizes this objective. We can do this by numerically computing this integral via interpolating the function g_σ at the positions $t_s := (1 - \epsilon)^s l_\sigma$ for $0 \leq s \leq r$ for some fixed r depending on ϵ. Specifically, this gives the value $\mathrm{cost}_r(\sigma) := t_r g_\sigma(t_r) + \sum_{s=1}^r (t_{s-1} - t_s) g_\sigma(t_{s-1}) = t_r g_\sigma(t_r) + \sum_{s=1}^r \epsilon t_{s-1} g_\sigma(t_{s-1})$. The following lemma shows that a linear r suffices to obtain a sufficient approximation, which implies quadratic time $\mathcal{O}(\delta^2)$ to evaluate $\mathrm{cost}_r(\sigma)$.

Lemma 1. *For any $\epsilon > 0$, we obtain that $(1 - \epsilon)\mathrm{cost}(\sigma) \leq \mathrm{cost}_r(\sigma) \leq \mathrm{cost}(\sigma)$ for some $r = \mathcal{O}(\delta)$.*

Lemma 1 immediately implies the following lemma.

Lemma 2. *For any $\epsilon > 0$ and any threshold τ, if there is a small optimal set with respect to τ, then the enumeration approach described above yields a $1 + \epsilon$-approximation in time $\mathcal{O}(\delta^{2+\tau})$.*

4.2   Large Sets

For large sets, assume that we know an arc $i^* \in \sigma^*$ with $l_{i^*} := l_{\sigma^*}$ for an optimal set σ^*. This is no restriction, since we can test all possible arcs i^* in a linear number of iterations. Note that fixing i^* restricts the arcs i to consider to the ones with $l_i \geq l_{i^*}$. Therefore, for simplicity, assume w.l.o.g. that $l_{i^*} = \min_{1 \leq i \leq \delta} l_i$. We then use algorithm $\mathcal{A}^{\mathrm{enum}}$ described in Section 3 with modified input vectors $a \in \mathbb{R}^{\delta+1}$ with $a_i := \mathbb{E}[T_i]/l_i$ for $1 \leq i \leq \delta$ and additionally $a_{\delta+1} := 1$, and $b \in \mathbb{R}^{\delta+1}$ with $b_i := l_{i^*}/l_i$ for $1 \leq i \leq \delta$ and additionally $b_{\delta+1} := 0$. Moreover, in Step 2, let v^λ be the vector that maximizes $(c^\lambda)^t v^\lambda$ subject to $v_{i^*}^\lambda = 1$, which ensures that i^* is contained in the returned set σ. Finally, we use $\mathrm{cost}(\sigma)$ adapted to uniform arrival times in Step 3. To approximately compute $\mathrm{cost}(\sigma)$ in this case, we can for instance use the method described in Subsection 4.1 which runs in quadratic time $\mathcal{O}(\delta^2)$. The following lemma states that the approximation ratio of this modification of algorithm $\mathcal{A}^{\mathrm{enum}}$ becomes arbitrarily good for a sufficiently large threshold τ.

Lemma 3. *For any $\epsilon > 0$, if there is an optimal set which is large with respect to some threshold $\tau = 1/\epsilon^{\mathcal{O}(1)}$, then the modified algorithm $\mathcal{A}^{\mathrm{enum}}$ finds a $1 + \epsilon$-approximation in $\delta^{\mathcal{O}(1)}$ time.*

Combining Lemma 3 with Lemma 2 gives the final theorem. Note here that the running time of this combined approach is dominated by the enumeration of the small sets.

Theorem 3. *There is a PTAS for simple networks with uniform arrival times if the l-parameters differ by at most a constant factor. The running time is $\delta^{1/\epsilon^{\mathcal{O}(1)}}$.*

Using the fact that a linear arrival time can be modeled as the minimum of two uniform arrival times, we obtain the following theorem.

Theorem 4. *There is a PTAS for simple networks with linear arrival times if the l-parameters differ by at most a constant factor. The running time is $\delta^{1/\epsilon^{\mathcal{O}(1)}}$.*

5   Extension to General Networks

Using an adaption of the well-known method of Dijkstra, we obtain the following theorem.

Theorem 5. *For any family of arrival times, any algorithm for simple networks with running time $T(\delta)$ yields an algorithm for general networks with running time $\mathcal{O}(n(\log n + \delta T(\delta)))$. If the algorithm for simple networks has approximation ratio α, then the algorithm for general networks has approximation ratio α^Δ, where Δ is the number of allowed hops.*

Theorem 5 gives the following corollaries in combination with Theorem 2 and Theorem 4.

Corollary 1. *In general networks with exponential arrival times, we can compute an optimal set of alternatives in $\mathcal{O}(n(\log n + \delta^3))$ time.*

Corollary 2. *There is a PTAS for general networks with uniform or linear arrival times if the l-parameters differ by at most a constant factor and only a constant number of hops are allowed. The running time is $\mathcal{O}(n(\log n + \delta^{1/\epsilon^{\mathcal{O}(1)}}))$.*

6   Hardness of Simple Networks

In this section, we prove that the simple network case is NP-hard, even if all transit times are zero. Specifically, we prove the following quite general theorem.

Theorem 6. *Given random variables $A_1, A_2, \ldots, A_\delta$ and a positive integer k, it is NP-hard to find a subset $\sigma \subseteq \{1, \ldots, \delta\}$ of at most size k such that $\mathbb{E}\left[\min_{i \in \sigma} A_i\right]$ is minimized.*

7   Conclusion

This paper considers the problem of finding a shortest path with alternatives. We think that, especially for simple networks, the problem setting is quite generic and worth considering in its own right. Three different arrival time distributions are considered: exponential arrival times, uniform arrival times, and linear arrival times. Uniform arrival are arguably the first choice for stochastic public transportation networks. However, it is interesting that although these arrival times differ significantly, they essentially lead to the same algorithm: enumerate a quadratic number of candidate solutions and take the best one.

We consider an off-line scenario where all alternatives need to be selected before the journey starts, whereas in many practical scenarios, it might be possible to update the travel plan according to newly obtained measurements and to submit the next bus to pick in real-time. However, if the movements of buses are partly stochastic, and hence there is no accurate forecast for the complete planning horizon of the journey, it still might be helpful to use our methods as a tactical tool to select directions which are robust with respect to future stochastic disturbances. Combining our methods with the standard tool of using a time-expanded network in order to include more precise information for the near future also seems to be a natural enhancement. The author is currently working on such an extension which includes a test with real data.

To obtain the PTASs for uniform and linear arrival times, we require several times the restriction that the l-parameters differ by at most a constant factor, a reasonable assumption in practice. An interesting open problem is to get rid of this restriction. Moreover, our NP-hardness proof only works for general arrival time distributions, but we conjecture that already uniform arrival times are NP-hard.

Although there are different approaches from literature to solve the fractional programming problem considered in the context of exponential arrival times, namely the approach of Megiddo [13] and an enumeration approach [14], we show that combining both improves the running time by a logarithmic factor. We think that the used ideas might be helpful to get rid of logarithmic factors in other fractional programming problems.

References

1. Bertsekas, D.P., Tsitsiklis, J.N.: An analysis of stochastic shortest path problems. Math. Oper. Res. 16, 580–595 (1991)
2. Billionnet, A.: Approximation algorithms for fractional knapsack problems. Operations Research Letters 30(5), 336–342 (2002)
3. Cormen, T.H., Leiserson, C.E., Rivest, R.L.: Introduction to Algorithms. MIT Press (1990)
4. Correa, J., Fernandes, C., Wakabayashi, Y.: Approximating a class of combinatorial problems with rational objective function. Mathematical Programming 124, 255–269 (2010)
5. Dooly, D.R., Goldman, S.A., Scott, S.D.: TCP dynamic acknowledgment delay: Theory and practice (extended abstract). In: Proceedings of the 30th Annual ACM Symposium on the Theory of Computing (STOC 1998), pp. 389–398 (1998)
6. Dreyfus, S.E.: An appraisal of some shortest-path algorithms. Operations Research 17(3), 395–412 (1969)
7. Dynkin, E.B.: The optimum choice of the instant for stopping a Markov process. Soviet Math. Dokl. 4 (1963)
8. Ford, L., Fulkerson, D.: Flows in Networks. Princeton University Press (1962)
9. Foschini, L., Hershberger, J., Subhash, S.: On the complexity of time-dependent shortest paths. In: Proceedings of the 22nd Annual ACM-SIAM Symposium on Discrete Algorithms (SODA 2011), pp. 327–341 (2011)

10. Hashizume, S., Fukushima, M., Katoh, N., Ibaraki, T.: Approximation algorithms for combinatorial fractional programming problems. Mathematical Programming 37, 255–267 (1987)
11. Karlin, A.R., Manasse, M.S., Rudolph, L., Sleator, D.D.: Competitive snoopy caching. Algorithmica 3, 77–119 (1988)
12. Kelner, J.A., Nikolova, E.: On the hardness and smoothed complexity of quasi-concave minimization. In: Proceedings of the 48th Annual IEEE Symposium on Foundations of Computer Science (FOCS 2007), pp. 472–482 (2007)
13. Megiddo, N.: Combinatorial optimization with rational objective functions. In: Proceedings of the 10th annual ACM symposium on Theory of computing (STOC 1978), New York, NY, USA, pp. 1–12 (1978)
14. Nikolova, E., Kelner, J.A., Brand, M., Mitzenmacher, M.: Stochastic Shortest Paths Via Quasi-convex Maximization. In: Azar, Y., Erlebach, T. (eds.) ESA 2006. LNCS, vol. 4168, pp. 552–563. Springer, Heidelberg (2006)
15. Shachnai, H., Tamir, T.: Handbook of Approximation Algorithms and Metaheuristics. In: Gonzalez, T.F. (ed.). Chapman and Hall/CRC Computer and Information Science Series (2007)

On Computing Straight Skeletons
by Means of Kinetic Triangulations[*]

Peter Palfrader[1], Martin Held[1], and Stefan Huber[2]

[1] FB Computerwissenschaften, Universität Salzburg, A-5020 Salzburg, Austria
[2] FB Mathematik, Universität Salzburg, A-5020 Salzburg, Austria
{ppalfrad,held,shuber}@cosy.sbg.ac.at

Abstract. We study the computation of the straight skeleton of a planar straight-line graph (PSLG) by means of the triangulation-based wavefront propagation proposed by Aichholzer and Aurenhammer in 1998, and provide both theoretical and practical insights. As our main theoretical contribution we explain the algorithmic extensions and modifications of their algorithm necessary for computing the straight skeleton of a general PSLG within the entire plane, without relying on an implicit assumption of general position of the input, and when using a finite-precision arithmetic. We implemented this extended algorithm in C and report on extensive experiments. Our main practical contribution is (1) strong experimental evidence that the number of flip events that occur in the kinetic triangulation of real-world data is linear in the number n of input vertices, (2) that our implementation, Surfer, runs in $\mathcal{O}(n \log n)$ time on average, and (3) that it clearly is the fastest straight-skeleton code currently available.

1 Introduction

1.1 Motivation

The straight skeleton of a simple polygon is a skeletal structure similar to the generalized Voronoi diagram, but comprises straight-line segments only. It was introduced to computational geometry by Aichholzer et al. [1], and later generalized to planar straight-line graphs (PSLGs) by Aichholzer and Aurenhammer [2]. Currently, the most efficient straight-skeleton algorithm for PSLGs, by Eppstein and Erickson [7], has a worst-case time and space complexity of $\mathcal{O}(n^{17/11+\epsilon})$ for any $\epsilon > 0$. For a certain class of simple polygons with holes Cheng and Vigneron [6] presented a randomized algorithm that runs in $\mathcal{O}(n\sqrt{n}\log^2 n)$ time. However, both algorithms employ elaborate data structures in order to achieve these complexities and are not suitable for implementation.

The first comprehensive straight-skeleton code was implemented by Cacciola [4] and is shipped with the CGAL library [5]. It handles polygons with holes as input, and requires $\mathcal{O}(n^2 \log n)$ time and $\mathcal{O}(n^2)$ space for real-world datasets,

---

[*] The authors would like to thank Willi Mann for valuable discussions and comments. Work supported by Austrian FWF Grant L367-N15.

L. Epstein and P. Ferragina (Eds.): ESA 2012, LNCS 7501, pp. 766–777, 2012.

see [10] for an explanation and experimental analysis. The code Bone by Huber and Held [10] has been, until now, the fastest implementation. It handles PSLGs as input and runs in $\mathcal{O}(n \log n)$ time and $\mathcal{O}(n)$ space in practice.

The algorithm by Aichholzer and Aurenhammer [2] propagates a wavefront by means of kinetic triangulations and it can handle PSLGs as input. No better complexity bound than $\mathcal{O}(n^3 \log n)$ has been established so far, due to only a trivial $\mathcal{O}(n^3)$ upper bound on the number of flip events that might occur in the triangulation in the worst case, even though no input is known that requires more than $\mathcal{O}(n^2 \log n)$ time. The basic algorithm is suitable for implementation, but no full implementation has been published. In fact, if general position cannot be assumed then one needs to fill in algorithmic gaps prior to an actual implementation. For instance, the basic algorithm of [2] may loop if multiple concurrent events are not processed properly, even if exact arithmetic is used.

Our contribution. We thoroughly investigate Aichholzer and Aurenhammer's triangulation-based approach [2] both from a theoretical and a practical point of view. We start with a discussion of our extensions to their basic algorithm, which are necessary in order to handle arbitrary PSLGs. In particular, we present a procedure for resolving a loop of events in the kinetic triangulation, we report on technical details concerning unbounded triangles in order to triangulate the entire plane, and we discuss how to handle parallel edges in G that lead to infinitely fast vertices in the kinetic triangulation. Besides these extensions, we address a major open question raised by Aichholzer and Aurenhammer [2]: How many flip events shall we expect to occur in the kinetic triangulation of practical data?

We implemented the basic algorithm and our extensions in C, and present extensive statistics based on test runs for about 20 000 industrial and synthetic datasets of different characteristics. As first important practical contribution we provide strong experimental evidence for Aichholzer and Aurenhammer's conjecture that only $\mathcal{O}(n)$ flip events suffice for all practical datasets.

Our code, Surfer, can be run with two arithmetic back-ends: (i) with double-precision floating-point arithmetic, and (ii) with the extended-precision library MPFR [8]. Tests clearly show that Surfer runs in $\mathcal{O}(n \log n)$ time for all our datasets. Furthermore, it is reliable enough to handle datasets of a few million vertices with floating-point arithmetic. In comparison with the code provided by CGAL, Surfer has several advantages: (i) it is by a linear factor faster in practice, (ii) it can handle PSLGs as input, and (iii) its space complexity is in $\mathcal{O}(n)$. Furthermore, even though the worst-case complexity of Surfer is worse than that of Bone [10], our code turns out to be faster by a factor of 10 in runtime tests.

1.2 Preliminaries and Basic Definitions

Definition of the straight skeleton. Aichholzer et al. [1] defined a straight skeleton $\mathcal{S}(G)$ of a PSLG G based on a so-called wavefront propagation process. The idea is that every edge of G sends out two wavefront copies that move in a parallel fashion and with constant speed on either side, see Fig. 1. Consider a non-terminal vertex v of G: As the wavefront progresses, wavefront vertices, i.e., copies of v, move along angular bisectors defined by pairs of consecutive edges (in the cyclic incidence order) that are incident at v. At a terminal vertex v of G an additional wavefront edge orthogonal to the incident input edge is emanated such that the wavefront forms a rectangular cap at v. (It is assumed that G contains no isolated vertices.)

We denote by $\mathcal{W}_G(t)$ the wavefront at time t and interpret $\mathcal{W}_G(t)$ as a 2-regular graph which has the shape of a mitered offset curve of G. During the propagation of $\mathcal{W}_G$ topological changes occur:

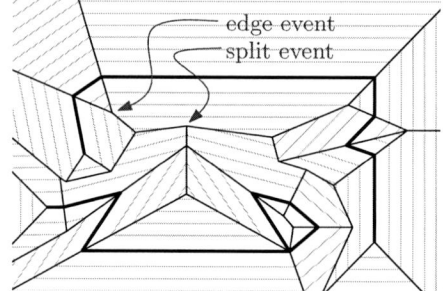

- An *edge event* occurs when a wavefront edge shrinks to zero length and vanishes.
- A *split event* occurs when a reflex wavefront vertex meets a wavefront edge and splits the wavefront into parts. We call a wavefront vertex *reflex* if the angle of the incident wavefront edges on the propagation side is reflex.

Fig. 1. A PSLG G, wavefronts $\mathcal{W}_G(t)$ in gray for different times t, and the straight skeleton $\mathcal{S}(G)$

The straight skeleton $\mathcal{S}(G)$ is then defined as the union of loci that are traced out by the wavefront vertices, see Fig. 1. Every vertex of $\mathcal{S}(G)$ is due to an event.

The triangulation-based algorithm. The idea that drives the triangulation-based algorithm by Aichholzer and Aurenhammer [2] is to keep for any $t \geq 0$ the area $\mathbb{R}^2 \setminus \bigcup_{t' < t} \mathcal{W}_G(t')$ triangulated. In other words, they maintain a kinetic triangulation of those parts of the plane that have not yet been swept by the wavefront $\mathcal{W}_G$. Every edge event and every split event is indicated by the collapse of a triangle as either (i) a wavefront edge collapsed to zero length or (ii) a wavefront vertex met a wavefront edge. Hence, the topological changes of the wavefront are indicated by the topological changes of the kinetic triangulation. However, some topological changes of the kinetic triangulation do not correspond to a wavefront event, namely when a reflex wavefront vertex meets an inner triangulation edge. One needs to flip this edge in order to maintain a valid triangulation. Hence, these events are called *flip events*. We follow the notation of Aichholzer and Aurenhammer and call an inner triangulation diagonal a *spoke* in the remainder of this paper.

Their algorithm starts with a constrained triangulation of G and adapts it to a triangulation for the initial wavefront $\mathcal{W}_G(0)$ by duplicating the edges of G and inserting zero-length edges that are emanated at terminal vertices of G. For every triangle a collapse time is computed. If a triangle collapses in finite time it is put into a priority queue Q prioritized by its collapse time. Then one event after the other is fetched from Q in chronological order, and the necessary topological changes are applied to the kinetic triangulation. Eventually, Q is empty as no further triangle is collapsing in finite time. At that point all straight-skeleton nodes were computed.

Note that there are $\mathcal{O}(n)$ edge and split events as $\mathcal{S}(G)$ is of linear size. An edge and split event locally changes the topology of the wavefront and adapts the velocities of some vertices. Consequently, the collapse times of all incident triangles need to be recomputed. Hence, a single edge or split event may require $\mathcal{O}(n \log n)$ time, which leads to $\mathcal{O}(n^2 \log n)$ as total time complexity for all edge and split events.

The signed area of a triangle can be expressed as quadratic polynomial in t and, hence, a single triangle can collapse at most at two single points in time. As there are at most $\binom{n}{3}$ combinatorial possibilities for triangles among n vertices over the entire simulation time, there are at most $\mathcal{O}(n^3)$ flip events. This is the best known bound. A single flip event requires $\mathcal{O}(1)$ modifications in Q and, thus, can be handled in $\mathcal{O}(\log n)$ time. In total, the algorithm has a worst-case complexity of $\mathcal{O}((n^2 + k) \log n) \subseteq \mathcal{O}(n^3 \log n)$, where $k \in \mathcal{O}(n^3)$ denotes the number of flip events. However, no input is known that causes more than $\mathcal{O}(n^2)$ flip events.

2   Handling Unbounded Triangles

If the entire straight skeleton of a general PSLG G is to be computed, one has to ensure that the initial triangulation covers a portion of the plane that is large enough to contain all nodes of $\mathcal{S}(G)$. The natural idea of computing $\mathcal{S}(G)$ inside of a "large" box or triangle is difficult to cast into a reliable implementation for two reasons: First, no efficient method is known for computing a good upper bound on the maximum distance of a node of $\mathcal{S}(G)$ from G. Second, picking a truly large box might be a heuristic attempt to ensure coverage for all practically relevant needs but it will result in lots of very skinny triangles. These triangles are difficult to process correctly on a finite-precision arithmetic, and they place a burden on floating-point filters used to speed up exact geometric computing.

Therefore we construct a triangulation of the entire plane as follows: First, we compute a constrained Delaunay triangulation of G inside of its convex hull, $CH(G)$. Then we attach an *unbounded triangle* to every edge of $CH(G)$. Such an unbounded triangle has one finite edge on $CH(G)$ and two unbounded edges. These unbounded edges are thought to meet at infinity.

While computing the collapse time of a (finite) triangle with three vertices moving at constant speed amounts to solving a quadratic equation in one variable, it is less obvious how to deal with unbounded triangles. We note that it is not sufficient to regard an unbounded triangle as collapsed only if its finite edge has shrunk to zero length. This would allow unbounded triangles to move to the interior of the wavefront, causing us to miss events that change the topology of the wavefront later on: The left part of Fig. 2 depicts a portion of a PSLG G (in bold) such that the input edges, or wavefronts, w_1, w_2, w_3 and the spokes s_1, s_2 lie on $CH(G)$. If the angles at the reflex wavefront vertices are chosen

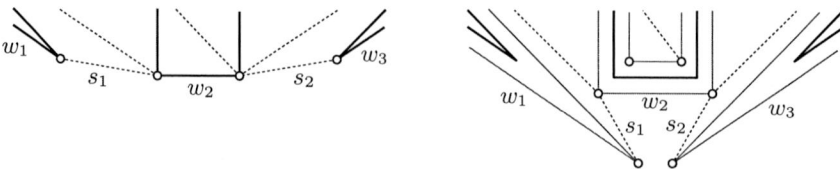

Fig. 2. A crash of wavefronts might be missed if unbounded triangles are handled naively

appropriately then the wavefronts will collide while no triangle collapse serves as a witness of the event, see the right part.

In order to prevent such problems we proceed as follows: We consider the stereographic projection of the plane $\mathbb{R}^2$ to the sphere S^2, which maps the origin to the south pole of S^2 and the north pole represents all points at infinity. Every triangle of our triangulation, including the unbounded triangles, is mapped to a spherical triangle on S^2. The infinite edges of unbounded triangles are arcs of great circles supporting the north and south pole of S^2.

We now regard an unbounded triangle of $\mathbb{R}^2$ as collapsed when its spherical counterpart collapsed, i.e., when its three vertices lie on a great circle. A collapse of an unbounded triangle indicates either an edge event or a flip event, and both events can be handled in a similar fashion as for bounded triangles. Of course, it would be quite cumbersome for an actual implementation to compute collapse times of spherical triangles. Fortunately this can be avoided: The spherical counterpart of an unbounded triangle Δ collapses if the two finite vertices v_1, v_2 and the north pole lie on a great circle. This is the case if and only if v_1, v_2 and the south pole lie on a great circle. Hence Δ collapses if the two finite vertices and the origin are collinear in the plane, i.e., if the triangle formed by v_1, v_2 and the origin collapses.

3   Handling Input without General Position Assumed

In this section we describe the nuts and bolts required for turning the basic algorithm into an implementation that can handle real-world data while using standard floating-point arithmetic. Problems arise because (1) the (usually implicit) assumption of general position (GPA) is not warranted for real-world applications, and (2) finite-precision arithmetic does not guarantee to process all events in the correct order. Note that working with finite precision arithmetic precludes approaches like symbolic perturbation.

3.1   Vertices Moving at Infinite Speed

Consider the c-shaped PSLG shown in Fig. 3(a) together with a wavefront and a part of the triangulation. As the wavefront progresses, the shaded triangle Δ_1 will collapse since the edge e between vertices v_1 and v_2 will shrink to zero

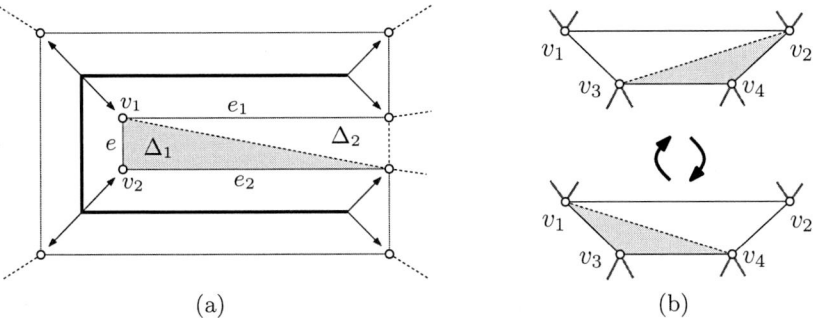

Fig. 3. Without general position assumed. (a) The edge event for the shaded triangle Δ_1 will create a vertex moving at infinite speed. (b) A loop of concurrent flip events.

length. (Triangle Δ_2 and some other triangles will collapse at the same time, but this is irrelevant at the moment.)

The standard procedure for an edge event is to replace the two vertices v_1, v_2 with a new vertex that moves along the angular bisector of the two incident wavefront edges, at the speed required to follow the wavefront propagation. In the situation shown in Fig. 3(a), the two wavefront edges e_1, e_2 are parallel and overlap at the time of the event. This means that the new vertex just created moves along the supporting line of e_1, e_2, but it has to move at infinite speed to "keep up" with the wavefront propagation since it runs perpendicular to the direction of the wavefront propagation.

This problem is resolved by introducing triangulation vertices that are marked as *infinitely fast*. Like any other vertex, such a vertex v is the central vertex of a triangle fan of one or more triangles, $\Delta_1, \Delta_2, \ldots, \Delta_n$. This fan is enclosed at v by two overlapping wavefronts. (Otherwise v would not be infinitely fast.) This implies that all triangles of the fan are collapsing at the time the infinitely fast moving vertex comes into existence. Among all these triangles we choose either Δ_1 or Δ_n, depending on which has the shorter wavefront edge to v. Let v' be the vertex next to v on the shorter wavefront edge.

In the chosen triangle we process an edge event as if v had become coincident with v': We add the path from v to v' as a straight skeleton arc, and v and v' merge into a new kinetic vertex, leaving behind a straight skeleton node.

3.2   Infinite Loops of Flip Events

At first sight, it appears evident that the basic triangulation-based algorithm terminates as there are only finitely many events to be processed. For edge and split events an even simpler argument can be applied: Every edge and split event reduces the number of triangles and, thus, only $O(n)$ many edge and split events can occur. While flip events do not result in a reduction of the triangle count, we still make progress in the wavefront propagation if no two flip events occur at the same time.

However, if general position is not assumed and, thus, two or more (flip) events may occur at the same time then this standard argument for the termination of the basic triangulation-based algorithm fails. Fig. 3(b) shows two wavefront vertices v_1, v_2 that move downwards and two wavefront vertices v_3, v_4 that move upwards. Now assume that all four vertices will become collinear at some future point in time. Then the two triangles shown will collapse at the same time. Hence, we have two choices on how to proceed: We can either flip the spoke (v_1, v_2) or the spoke (v_2, v_3). If we chose to flip (v_1, v_2) and subsequently (v_2, v_3), then we would achieve progress as all four vertices could proceed with their movement. If, however, we chose to flip (v_2, v_3) then no progress would be achieved, and a subsequent flip of (v_1, v_4) would get us into an infinite loop. Modifying the set-up of Fig. 3(b) by regarding (v_1, v_2) as an edge of the wavefront yields an example for a possible event loop that involves a split event and a flip event.

This simple example can be made more complex in order to incorporate multiple concurrent split and flip events. We emphasize that such a loop of events can occur even if exact arithmetic is used for computing all event times without numerical error. That is, this is a genuine algorithmic problem once we allow inputs that trigger multiple events at the same time.

Indications for a flip event loop. One might suspect that processing an event twice is a clear indication that we encountered the same triangulation a second time and that the event processing ended up in a loop. However, this is not correct, as demonstrated in Fig. 4. Assume that the vertices v_1 and v_6 move downwards while the vertices $v_2, \ldots, v_5$ move upwards such that all six vertices become collinear at some point t in time, lined up in the order $v_1, \ldots, v_6$, with no two vertices coinciding. Hence, all four triangles depicted have collapse events scheduled for time t. In Fig. 4, the collapse event chosen is indicated by shading, and the next triangulation is obtained by flipping the dashed spoke. The triangle Δ_1 is processed twice and still all eight triangulations are different and, thus, we did not end up in a loop. However, if in (viii) we continue with Δ_3 instead of, say, Δ_2, then we obtain the same triangulation as in (v) and we are indeed caught in a loop.

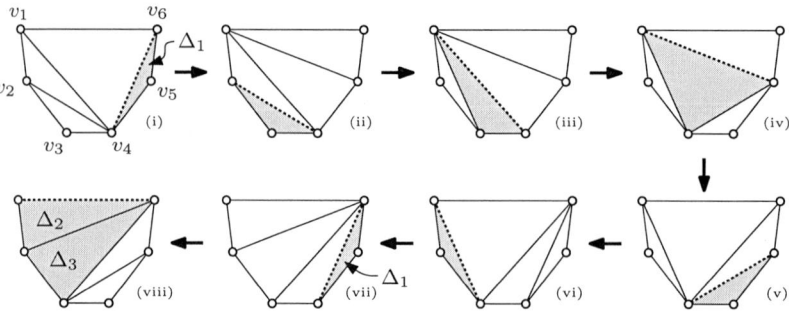

Fig. 4. Encountering an event twice need not imply a loop nor the same triangulation

In general, it is a save to handle non-flip events prior to flip events, as non-flip events reduce the number of triangles. However, preferring concurrent non-flip events over flip events requires that one notice the event times are identical, which is prone to errors when using finite-precision arithmetic. And, of course, we were still left with the problem of detecting and handling event loops that consist only of concurrent flip events. Summarizing, two questions need to be addressed: (i) how to detect event loops and (ii) how to cope with them. We pay particular attention to handling event loops on finite-precision arithmetic, since our implementation operates with double-precision or MPFR-based extended precision arithmetic.

Handling flip event loops. In order to detect and handle event loops we maintain a history H that saves for every processed event a triple (t^*, t, Δ), where t^* denotes the number of triangles remaining in the triangulation, Δ denotes the sorted vertex-triple of the collapsed triangle, and t denotes its collapse time. These triples are stored in the same sequential order as the corresponding events are processed. In addition, we maintain a search data structure S that stores the triples (t^*, t, Δ) in lexicographical order. Note that every non-flip event decrements t^*. Thus, t^* measures the progress of the wavefront propagation in a discrete manner, and we can empty both data structures whenever t^* is decreased.

Whenever an event is to be processed, we check whether the corresponding triple is already stored in S. If we ended up in a loop then we are guaranteed to find that triple already stored in S. As explained in Fig. 4, the opposite conclusion need not be true. Nonetheless, once we observe that the triple (t^*, t, Δ) was already handled, we apply the following method in order to resolve a potential loop. (No harm is caused in case of a false alarm.)

First of all, as t^* has not changed, the entire potential loop comprises flip events only. With exact arithmetic, all triples between the first occurrence T_1 and second occurrence T_2 of (t^*, t, Δ) in H have the identical value for t. With finite-precision arithmetic, we declare all triples between T_1 and T_2 to have happened at the same time, even though there may be slight deviations in time for the triples between T_1 and T_2.

Next, we trace back the triples in H from T_2 to T_1 and mark all triangles that are known to have collapsed due to their role in a flip event. An inductive argument shows that the union of these triangles forms one or more polygons that have collapsed to straight-line segments. Let P denote the one polygon that contains Δ. We now roll back all triples between T_2 and T_1 that flip edges in P, including T_2. That is, we start at T_2 and visit all triples in H until we reach T_1, and if the corresponding triangle is in P we undo the flip and remove the triple from H and S. (Recall that only flips occurred between T_1 and T_2.)

We now adapt the triangulation for P and for the triangles Δ_e opposite to edges e of P as follows, see Fig. 5. Let $v_1, \ldots, v_k$ denote the vertices of P in sorted order with respect to the line to which P collapsed. First, we ensure that $v_i v_{i+1}$ is a diagonal for all $1 \leq i < k$. That is, the path $v_1, \ldots, v_k$ is part of the triangulation. This path tessellates P into cells $C(e)$ that are bounded by two or more edges of the path and a single edge e of P. We denote by v_e the opposite

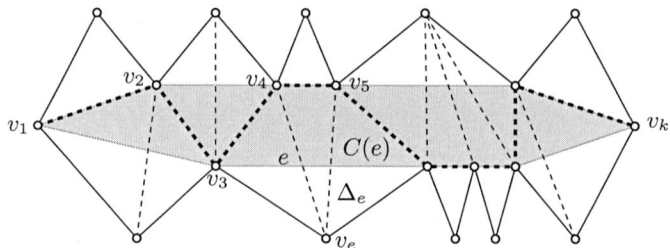

Fig. 5. An entire polygon P (shaded) collapsed to a line. We adapt the triangulation within P and the triangles Δ_e attached to edges e of P by first inserting a monotone chain $v_1, \ldots, v_k$ (bold dashed) and subsequently triangulating the resulting cells of P and the triangles Δ_e.

vertex of e in Δ_e, see Fig. 5. Then we triangulate for each cell $C(e)$ the area $C(e) \cup \Delta_e$ such that every diagonal is incident to v_e.

Let T^* denote the last triple in H after the rollback. We use an algorithm by Hanke et al. [9] to append after T^* a sequence of triples that transfers the original triangulation (at the state just after T^*) into the new triangulation explained above. Similarly, those triples are inserted into S. Each of the new triples of H is also furnished with a pointer that points to T_1. The idea behind this pointer is that the time values of all triples between T_1 and triples that point to T_1 are considered to be equal. (With exact arithmetic they all are indeed equal.) After this reconfiguration of the triangulation and of H and S we proceed as usual.

Assume now that one of the triangles Δ_e has zero area as v_e was collinear with the vertices of P, too. If our loop-detection method later reports another potential loop because a triple T_3, which is in H, would be processed twice and T_3 contains a pointer back to T_1 then we repeat our resolution method presented above. However, the only difference is now that we consider all events between T_1 and T_3 to happen concurrently. That is, the rollback phase does not stop at T_3 but is extended back to T_1. As a consequence, the resulting new polygon includes the old polygon P, and we detected even more vertices that are collinear at that particular time. As there are only finitely many triangles, at some point in time, we obtain a largest polygon P. Hence, it is guaranteed that the above method also terminates on finite-precision arithmetic.

Of course, we make sure to compute all numerical values used by our algorithm in a canonical way. In particular, different computations of the collapse time of the same triangle are guaranteed to yield precisely the same numerical value. Also, note that we carefully avoid the use of precision thresholds for judging whether the collapse times of two triangles are identical: On a finite-precision arithmetic this would be prone to errors and one could not guarantee that the algorithm will terminate.

Sorting vertices along the collapse line. Finally, we discuss a subtle detail when sorting the vertices $v_1, \ldots, v_k$ along the line to which P collapsed. First, we determine the minimum $t_{\min}$ and the maximum $t_{\max}$ among the collapse times of the triangles in P. Next, we determine a fitting straight line $L_{\min}$ ($L_{\max}$, resp.) of the vertices $v_1, \ldots, v_k$ at time $t_{\min}$ ($t_{\max}$, resp.) by means of a least-square fitting. Then we sort $v_1, \ldots, v_k$ at time $t_{\min}$ with respect to $L_{\min}$ and obtain the sequence $v_{i_1}, \ldots, v_{i_k}$; likewise for $t_{\max}$ and $L_{\max}$. If we obtain the same order then we proceed with it as described above. If, however, a vertex v_{i_j} has a different position j' in the sorted sequence with respect to $t_{\max}$ then we declare the vertices $v_{i_{j'}}$ and v_{i_j} to coincide. Furthermore, we enforce a spoke e between those two vertices in the triangulation and handle one of both non-flip events that correspond to the collapse of the two triangles sharing e. Consequently, t^* is decremented and we again have a guaranteed progress of our algorithm.

4 Experimental Results

We implemented the full wavefront-propagation algorithm in C. The resulting code, Surfer, can be run with two arithmetic back-ends: (i) with double-precision floating-point arithmetic, and (ii) with the arbitrary-precision library MPFR [8].

We tested Surfer on about twenty thousand polygons and PSLGs, with up to 2.5 million vertices per data. Both real-world and contrived data of different characteristics was tested, including CAD/CAM designs, printed-circuit board layouts, geographic maps, space filling curves, star-shaped polygons, and random polygons generated by RPG [3], as well as sampled spline curves, families of offset curves, font outlines, and fractal curves. Some datasets contain also circular arcs, which we approximated by polygonal chains in a preprocessing step.

4.1 Number of Flip Events

The best upper bound on the number of flip events is $\mathcal{O}(n^3)$, but no input is known to cause more than $\mathcal{O}(n^2)$ flip events. While we provide no new theoretical insight on the maximum number of flip events, our tests provide strong experimental evidence that we can indeed expect a linear number of flip events for all practical data. Fig. 6 shows the number of flip events per input vertex (y-axis), for different input

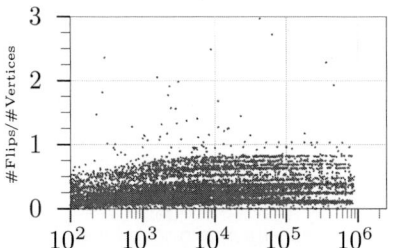

Fig. 6. The number of flip events is linear in practice. (input size on the x-axis).

sizes n arranged on the x-axis. On average, our algorithm had to deal with a total of $n/4$ flip events. Over the entire set of twenty-thousand inputs only a dozen cases, mostly sampled arcs, required more than $2n$ flip events. This clearly demonstrates the linear nature of this number in practice. It is interesting to note the clusters in this plot. Some clusters, but not all, correspond to different types of input. For instance, a closer inspection of the test results revealed that synthetic "random" polygons generated by RPG [3] require significantly more flips than random axis-aligned polygons.

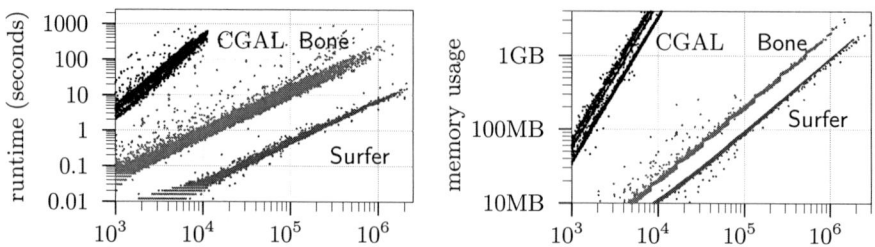

Fig. 7. Runtime and memory usage behavior of CGAL, Bone, and Surfer for inputs of different sizes (x-axis). Bone and Surfer use their IEEE 754 double precision backend.

4.2   Runtime Performance Statistics

The following tests were conducted on an Intel Core i7-980X CPU clocked at 3.33 GHz, with Ubuntu 10.04. Surfer was compiled by GCC 4.4.3.

By default, Surfer uses standard IEEE 754 double-precision floating-point arithmetic, but it can be built to use the MPFR library [8], enabling extended-precision floating-point operations. When using floating-point arithmetic it computes the straight skeleton of inputs with a million vertices in about ten seconds. In particular, our tests confirm an $\mathcal{O}(n \log n)$ runtime for practical data, including any time spent handling degenerate cases.

Comparison to other skeletonizers. We compared the runtime of Surfer against both Bone, the fastest other known implementation of a straight skeleton algorithm by Huber and Held [10], and against Cacciola's implementation [4] that is shipped with the CGAL library, version 4.0 [5]. Input to the latter was confined to polygonal data as the implementation cannot handle generalized PSLGs.

As can be seen in the left plot of Fig. 7, Surfer consistently outperforms Bone by a factor of about ten. Furthermore, it is by a linear factor faster than the CGAL code. In particular, for inputs with 10^4 vertices CGAL already takes well over one hundred seconds whereas Surfer runs in a fraction of one second. Note, though, that the CGAL code uses an exact-predicates-inexact-constructors kernel and, thus, could be expected to be somewhat slower. However, its timings do not improve substantially when run with an inexact kernel. Further analysis revealed an average runtime (in seconds) of $5.8 \cdot 10^{-7} n \log n$ for Surfer, $1.5 \cdot 10^{-5} n \log n$ for Bone, and $4.5 \cdot 10^{-7} n^2 \log n$ for the CGAL code.

Measurements of memory use, shown in the right plot of Fig. 7, confirm the expected linear memory footprint of Surfer. Its memory consumption is similar to that of Bone, while the CGAL code exhibits quadratic memory requirements.

As stated, Surfer can use the MPFR library for extended-precision floating-point operations. Obviously, extended-precision arithmetic incurs a penalty both in runtime and space requirements. Our tests with MPFR version 3.0.0 showed a decrease in speed by a factor of roughly $9.64 \cdot 10^{-4} p \sqrt{p} + 9$, where p denotes the MPFR precision. In particular, running Surfer with an MPFR precision of 100 takes about ten times as long as running it in IEEE 754 mode; at a precision of

1000 the slow-down factor is already 40. (Likely, the slow-down follows a $p\sqrt{p}$ law due to the increased complexity of doing multiplications with a larger number of digits.) The memory requirement increases linearly as the MPFR precision is increased. Roughly, the blow-up factor is modeled by $3.66 + 9.72 \cdot 10^{-3}p$.

5   Conclusion

We explain how the triangulation-based straight-skeleton algorithm by Aichholzer and Aurenhammer can be extended to make it handle real-world data on a finite-precision arithmetic. While the basic algorithm is simple to implement, all the subtle details discussed in this paper increase its algorithmic and implementational complexity. However, extensive tests clearly demonstrate that the resulting new code Surfer is the fastest straight-skeleton code currently available. In particular, our tests provide experimental evidence that only a linear number of flips occurs in the kinetic triangulation of practical data, allowing Surfer to run in $\mathcal{O}(n \log n)$ time in practice, despite of an $\mathcal{O}(n^3 \log n)$ theoretical worst-case complexity.

References

1. Aichholzer, O., Alberts, D., Aurenhammer, F., Gärtner, B.: Straight Skeletons of Simple Polygons. In: Proc. 4th Internat. Symp. of LIESMARS, Wuhan, P.R. China, pp. 114–124 (1995)
2. Aichholzer, O., Aurenhammer, F.: Straight Skeletons for General Polygonal Figures in the Plane. In: Samoilenko, A.M. (ed.) Voronoi's Impact on Modern Science, Book 2. Institute of Mathematics of the National Academy of Sciences of Ukraine, Kiev, Ukraine, pp. 7–21 (1998)
3. Auer, T., Held, M.: Heuristics for the Generation of Random Polygons. In: Proc. Canad. Conf. Comput. Geom. (CCCG 1996), pp. 38–44. Carleton University Press, Ottawa (1996)
4. Cacciola, F.: 2D Straight Skeleton and Polygon Offsetting. In: CGAL User and Reference Manual. CGAL Editorial Board, 4.0 edition (2012)
5. CGAL. Computational Geometry Algorithms Library, http://www.cgal.org/
6. Cheng, S.-W., Vigneron, A.: Motorcycle Graphs and Straight Skeletons. Algorithmica 47, 159–182 (2007)
7. Eppstein, D., Erickson, J.: Raising Roofs, Crashing Cycles, and Playing Pool: Applications of a Data Structure for Finding Pairwise Interactions. Discrete Comput. Geom. 22(4), 569–592 (1999)
8. GNU. The GNU MPFR Library, http://www.mpfr.org/
9. Hanke, S., Ottmann, T., Schuierer, S.: The Edge-Flipping Distance of Triangulations. J. Universal Comput. Sci. 2, 570–579 (1996)
10. Huber, S., Held, M.: Theoretical and Practical Results on Straight Skeletons of Planar Straight-Line Graphs. In: Proc. 27th Annu. ACM Sympos. Comput. Geom., Paris, France, pp. 171–178 (June 2011)

A Self-adjusting Data Structure
for Multidimensional Point Sets*

Eunhui Park and David M. Mount

University of Maryland, College Park MD 20742, USA
{ehpark,mount}@cs.umd.edu

Abstract. A data structure is said to be *self-adjusting* if it dynamically reorganizes itself to adapt to the pattern of accesses. Efficiency is typically measured in terms of *amortized complexity*, that is, the average running time of an access over an arbitrary sequence of accesses. The best known example of such a data structure is Sleator and Tarjan's splay tree. In this paper, we introduce a self-adjusting data structure for storing multidimensional point data. The data structure is based on a quadtree-like subdivision of space. Like a quadtree, the data structure implicitly encodes a subdivision of space into cells of constant combinatorial complexity. Each cell is either a quadtree box or the set-theoretic difference of two such boxes. Similar to the traditional splay tree, accesses are based on an splaying operation that restructures the tree in order to bring an arbitrary internal node to the root of the tree. We show that many of the properties enjoyed by traditional splay trees can be generalized to this multidimensional version.

Keywords: Geometric data structures, quadtrees, self-adjusting data structures, amortized analysis.

1 Introduction

The development of efficient data structures for geometric search and retrieval is of fundamental interest in computational geometry. The continued interest in geometric data structures is motivated by the tension between the various desirable properties such a data structure should possess, including efficiency (in space and query time), generality, simplicity, efficient handling of updates, and accuracy (when approximation is involved). Partition trees, including the quadtree and its many variants, are among the most popular geometric data structures. The quadtree defines a hierarchical subdivision of space into hypercube-shaped cells [24]. The simplicity and general utility of quadtree-based data structures is evident in the wide variety of problems to which they have been applied, both in theory and practice.

Our focus here stems from the study of dynamic data structures for storing multidimensional point sets. Through path compression (see e.g., [5,9,16]), compressed quadtrees can store an arbitrary set of n points in $\mathbb{R}^d$ in linear space.

---

* This work has been supported by the National Science Foundation under grant CCF-1117259 and the Office of Naval Research under grant N00014-08-1-1015.

L. Epstein and P. Ferragina (Eds.): ESA 2012, LNCS 7501, pp. 778–789, 2012.

(Throughout, we assume that d is a constant.) Since a compressed quadtree may have height $\Theta(n)$, it is desirable to consider ways of maintaining balance within the tree. Examples of such approaches include the randomized *skip quadtree* of Eppstein *et al.* [14] and our own *quadtreap* data structure, which, respectively, generalize the skip list data structure [23] and the treap data structure [25] to a multidimensional setting. It also includes Chan's data structures [7,8], which are based on ordering the points according to the Morton ordering. These data structures all support efficient insertion and deletion and can answer approximate proximity queries, such as approximate nearest neighbor searching and approximate range searching.

The above data structures guarantee $O(\log n)$ time (possibly randomized) to access each node of the tree, and so are efficient in the worst case. When access patterns may be highly skewed, however, it is of interest to know whether the data structure can achieve even higher degrees of efficiency by exploiting the fact that some nodes are much more likely to be accessed than others. A natural standard for efficiency in such cases is based on the entropy of the access distribution. Entropy-optimal data structures for point location have been proposed by Arya *et al.* [1,2] and Iacono [19]. These data structures are static, however, and cannot readily be generalized to other types of search problems.

A commonly studied approach for adaptive efficiency for one-dimensional data sets involves the notion of a *self-adjusting data structure*, that is, a data structure that dynamically reorganizes itself to fit the pattern of data accesses. Although many self-adjusting data structures have been proposed, the best-known example is the splay tree of Sleator and Tarjan [26,29]. This is a binary search tree that stores no internal balance information, but nonetheless provides efficient dynamic dictionary operations (search, insert, delete) with respect to *amortized time*, that is, the average time per operation over a sequence of operations. Sleator and Tarjan demonstrated that (or in some cases conjectured that) the splay tree adapts well to skewed access patterns. For example, rather than relating access time to the logarithm of the number of points in the set, it is related to a parameter that characterizes the degree of locality in a query. Examples of measures of locality include the working set bound [26], static and dynamic finger bounds [10,11,26], the unified bound [6,17,26], key-independent optimality [20], and the BST model [12]. Numerous papers have been devoted to the study of dynamic efficiency of data structures (see, e.g., [13,15,22,28,30]).

Given the fundamental importance of the splay-tree to the storage of 1-dimensional point sets and the numerous works it has inspired, it is remarkable that, after decades of study, there are no comparable self-adjusting data structures for storing multidimensional point sets. In this paper we propose such a data structure, called the *splay quadtree*. Like the splay tree, the splay quadtree achieves adaptability through a restructuring operation that moves an arbitrary internal node to the root of the tree. (In our tree, data is stored exclusively at the leaf nodes, and so it is not meaningful to bring a leaf node to the root.) Each node of our tree is associated with a region of space, called a cell, which is either a quadtree box or the set-theoretic difference of two quadtree boxes. As seen in other variants of the quadtree, such as the BBD tree [3, 4] and quadtreap [21], this generalization makes it possible to store an arbitrary set of n points in a tree of linear size and logarithmic height.

Although there are complexities present in the multidimensional context that do not arise in the 1-dimensional context, we show how to define a generalization of the splaying operation of [26] to the splay quadtree. We show that the fundamental properties of the standard splay tree apply to our data structure as well. In particular, we show that the amortized time of insertions and deletions into a tree of size n is $O(\log n)$. We also present multidimensional generalizations of many of the results of [26], including the Balance Theorem, the Static Optimality Theorem, the Working Set theorem, and variants of the Static Finger Theorem for a number of different proximity queries, including box queries (which generalize point-location queries), approximate nearest neighbor queries, and approximate range counting queries.

The rest of the paper is organized as follows. In Section 2, we present some background material on the BD tree, the variant of the quadtree on which the splay quadtree is based. In Section 3, we present the splaying operation and analyze its amortized time. In Section 4, we present and analyze algorithms for point insertion and deletion. Finally, in Section 5, we present the various search and optimality results. Due to space limitations, a number of proofs and figures have been omitted. They will appear in the full version of the paper.

2   Preliminaries

We begin by recalling some of the basic elements of quadtrees and BD-trees. A *quadtree* is a hierarchical decomposition of space into d-dimensional hypercubes, called *cells*. The root of the quadtree is associated with a unit hypercube $[0, 1]^d$, and we assume that (through an appropriate scaling) all the points fit within this hypercube. Each internal node has 2^d children corresponding to a subdivision of its cell into 2^d disjoint hypercubes, each having half the side length.

It will be convenient to consider a binary variant of the quadtree. Each decomposition step splits a cell by an axis-orthogonal hyperplane that bisects the cell's longest side. If there are ties, the side with the smallest coordinate index is selected. Although cells are not hypercubes, each cell has constant aspect ratio. Henceforth, we use the term *quadtree box* in this binary context.

If the point distribution is highly skewed, a compressed quadtree may have height $\Theta(n)$. One way to deal with this is to introduce a partitioning mechanism that allows the algorithm to "zoom" into regions of dense concentration. In [3, 4] a subdivision operation, called *shrinking*, was proposed to achieve this. The resulting data structure is called a *box-decomposition tree* (*BD-tree*). This is a binary tree in which the cell associated with each node is either a quadtree box or the set theoretic difference of two such boxes, one enclosed within the other. Thus, each cell is defined by an *outer box* and an optional *inner box*. Although cells are not convex, they have bounded aspect ratio and are of constant combinatorial complexity.

The spay quadtree is based on a decomposition method that combines shrinking and splitting at each step. We say that a cell of a BD-tree is *crowded* if it contains two or more points or if it contains an inner box and at least one point. Let B denote this cell's outer box, and let C be any quadtree box contained

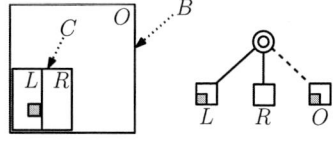

within B, such that, if the cell has an inner box, C contains it. We generate a cell $O = B \setminus C$. We then split C by bisecting its longest side forming two cells L and R. By basic properties of quadtree boxes, the inner box (if it exists) lies entirely inside of one of the two cells. We make the convention, called the *inner-left convention*, that the inner box (if it exists) lies within L. We generate a node with three children called the *left*, *right*, and *outer* child, corresponding to the cells L, R, and O, respectively. We allow for the possibility that $C = B$, which implies that the outer child's cell is empty, which we call a *vacuous leaf*. Our algorithms will maintain the invariant that the left and right children separate at least two entities (either two points or a point and inner box). It follows from the analysis given in [21] that the size of the tree is linear in the number of points.

As shown in [21], the tree can be restructured through a local operation, called *promotion*, which is analogous to rotation in binary trees. The operation is given any non-root internal node x and is denoted by *promote(x)*. Let y denote x's parent. There are three cases depending on which child x is (see Fig. 1). Note that promotion does not alter the underlying subdivision of space.

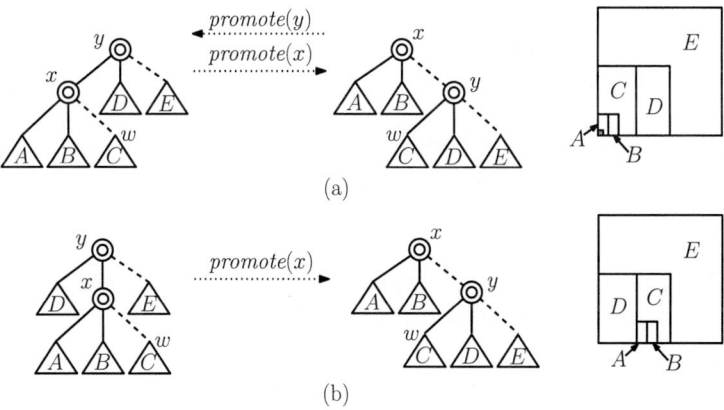

(a)

(b)

Fig. 1. (a) Left and outer promotions (b) right promotion

Left Child: This makes y the outer child of x, and x's old outer child becomes the new left child of y (see Fig. 1(a)).

Outer Child: This is the inverse of the left-child promotion, where the roles of x and y are reversed (see Fig. 1(a)).

Right Child: To maintain the convention that a cell has only one inner box, this operation is defined only if x's left sibling node does not have an inner box (node D in Fig. 1(b)). Letting y denote x's parent, $promote(x)$ swaps x with its left sibling and then performs $promote(x)$ (which is now a left-child promotion).

It is easily verified that promotion can be performed in $O(1)$ time and preserves the inner-left convention. We say that a promotion is *admissible*, if it is either a left or outer promotion or if it is a right promotion under the condition that the left child has no inner box.

3   Self-adjusting Quadtrees

Like the standard splay tree [26], our self-adjusting quadtree, or *splay quadtree*, maintains no balance information and is based on an operation, called *splaying*, that moves an arbitrary internal node to the root through a series of promotions. This operation is described in terms of primitives, called *basic splaying steps*.

3.1   Basic Splaying Steps

Let us begin with some definitions. Let x be internal, non-root node, and let $p(x)$ denote its parent. If $p(x)$ is not the root, let $g(x)$ denote x's grandparent. Define $rel(p(x), x)$ to be one of the symbols L, R, or O depending on whether x is the left, right, or outer child of $p(x)$, respectively. If $p(x)$ is not the root, we also define $str(x)$ to be a two-character string over the alphabet $\{L, R, O\}$ whose first character is $rel(g(x), p(x))$ and whose second character is $rel(p(x), x)$. The basic splaying step is defined as follows.

Basic Splay(x):
  zig: If $p(x)$ is the root, perform $promote(x)$ (see Fig. 2(a)).
  zig-zag: If $p(x)$ is not the root, and $str(x) \in \{LO, RO, OL\}$, do $promote(x)$, and then $promote(x)$ (see Fig. 2(b) for the case $str(x) = LO$).
  zig-zig: Otherwise, perform $promote(p(x))$ and then perform $promote(x)$ (see Fig. 2(c) for the case $str(x) = LL$).

Note that basic splaying may not be applicable to a node because it would result in an inadmissible promotion. We say that a basic splaying step is *admissible* if all of its promotions are admissible. We will show later that our splaying algorithm performs only admissible promotions.

3.2   Splaying Operation

As mentioned earlier, the splay quadtree is based on an operation that modifies the tree so that a given internal node becomes the root. We start with a definition. Let T be a splay quadtree, let u be a node of T, and let t be T's root.

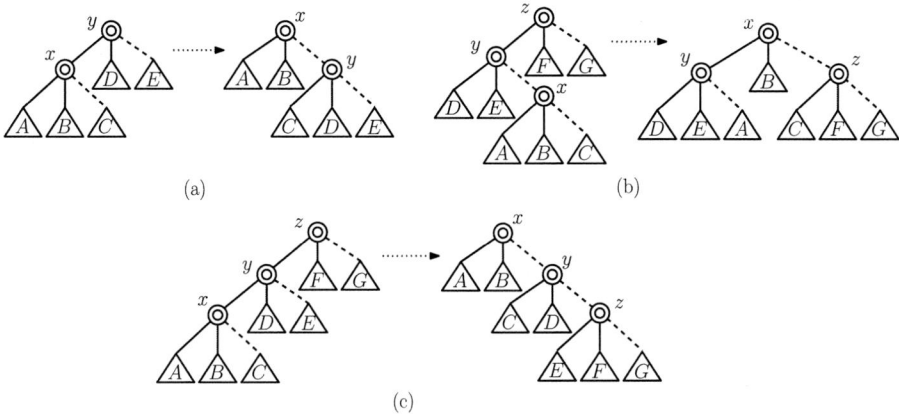

(a)   (b)

(c)

Fig. 2. The basic splaying operations: (a) *zig*, (b) *zig-zag*, and (c) *zig-zig*, where (before promotion) $y = p(x)$ and $z = g(x)$

Letting $\langle u = u_1, u_2, \ldots, u_n = t \rangle$ denote the path from u to t, the *path sequence* $seq_T(u)$ is defined

$$seq_T(u) = rel(u_n, u_{n-1}) \ldots rel(u_3, u_2) rel(u_2, u_1).$$

Let ε denote the empty string. Using the terminology of regular expressions, this sequence can be thought of as a string in the language $\{L, R, O\}^*$. Thus, after splaying u, we have $seq_T(u) = \varepsilon$.

In the standard binary splay tree [26], rotations are performed bottom-up along the access path. In our case, we will need to take care to avoid inadmissible promotions. We perform the splaying operation in three phases, each of which transforms the path sequence into successively more restricted form. Due to space limitations, the details of the algorithm and its analysis are presented in the full version. Here is a high-level overview of the process.

Phase 1: Initially, the path sequence is arbitrary, that is, $seq_T(u) \in \{L, R, O\}^*$, which can be represented by $\{\{O, L\}^*R\}^*\{O, L\}^*$. This phase compresses each substring of the form $\{O, L\}^*$ to a substring consisting of at most one such character by an appropriate sequence of *zig-zig* or *zig-zag* steps. The output has the form $\{\{O, L, \varepsilon\}R\}^*\{O, L, \varepsilon\}$.

Phase 2: In this phase, we transform $seq_T(u)$ such that the sibling node of any right child on the path does not have an inner box. After Phase 1, we have $seq_T(u) \in \{\{O, L, \varepsilon\}R\}^*\{O, L, \varepsilon\}$, which can be equivalently represented as $\{OR, LR, R\}^*\{O, L, \varepsilon\}$. Here, we consider the case where the sequence starts with OR, that is, $str(x) = OR$. Thus, since x's parent, $p(x)$ is an outer child, and the left child of $p(x)$, which is a sibling node of x, has an inner box, we cannot perform a right-child promotion at x. To solve this, we transform $seq_T(u)$ to $\{LR, R\}^*\{OL^*, O, L, \varepsilon\}$ by removing the OR substrings through *zig-zig* steps.

Fig. 3 shows an example of a *zig-zig* step for such an OR substring. Let y denote x's parent. First, we perform *promote(y)*. Before *promote(y)*, x's left sibling (labeled w_1 in Fig. 3) may have an inner box. But, after *promote(y)*, x's new sibling node, z, does not have an inner box. Thus, now *promote(x)* is admissible. We perform *promote(x)*. By the right-child property, prior to this *zig-zig* step, z's children, A and B, and x's children, E and F do not have inner boxes. After the operation, the right-child property still holds since x's children, E and F, and z's children, A and B, do not have inner boxes. The inner-left convention is still preserved, since prior to the operation, the inner box consisting of the union of A and B's cells in Fig 3 lies in y's left child (labeled w_1 in Fig. 3), and after the operation, the inner box consisting of the union of E and F's cells lies in y's left child (labeled w_2 in Fig. 3).

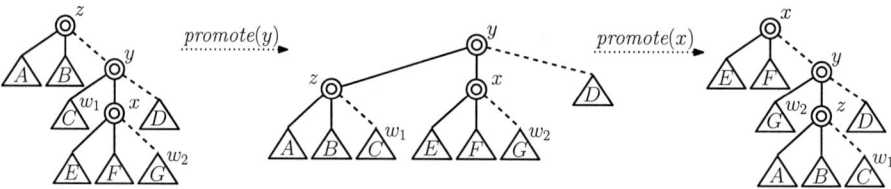

Fig. 3. A *zig-zig* splaying for an OR substring

Note that x's old outer child (labeled w_2 in Fig. 3) becomes to the left child of x's new outer child, y after the *zig-zig* step. Thus, if $str(x)$ is OR, and u is x's outer child, the value of $str(u)$ after *zig-zig* step is OL. This occurs recursively. Therefore, $seq(u) = \{OR, LR, R\}^*\{O, L, \varepsilon\}$ is transformed to a sequence of the form $\{LR, R\}^*\{OL^*, O, L, \varepsilon\}$ by removing OR.

Phase 3: After Phase 2, the path sequence has been modified so that all right-child transitions precede outer-child transitions. Since inner boxes are generated only by outer-child transitions, it follows that promotions performed on any right-child nodes of the current path are admissible. This phase completes the process (thus brining u to the root) by a series of basic splaying steps (*zig-zag* if $str(x) \in \{RO, OL\}$ and *zig-zag* otherwise).

Let us now analyze the amortized time of splaying. As in [26], we will employ a potential-based argument. We assign each node x a non-negative weight $w(x)$. We define x's *size*, $s(x)$, to be the sum of the weights of its descendants, including x itself, and we define its *rank*, $r(x)$, to be $\log s(x)$. (Unless otherwise specified, all logs are taken base 2.) We also define the *potential* Φ of the tree to be the sum of the ranks of all its nodes. The *amortized time*, a, of an operation is defined to be $t + \Phi' - \Phi$, where t is the actual time of the operation, Φ is the potential before the operation, and Φ' is the potential after the operation. It is easily verified that our splay operations satisfy the general splay properties described by Subramanian [27], and therefore, by the analysis presented there we have:

Lemma 1. *The amortized time to splay a tree with root t at a node u is at most $O(r(t) - r(u)) = O(\log s(t)/s(u))$.*

4   Updates to Splay Quadtrees

The insertion of a point q into a splay quadtree can be performed by a straight-forward modification of the incremental insertion process presented in [21]. This algorithm first determines the leaf node x containing q by a simple descent of the tree. It then generates a new node u to separate q from x's current contents and replaces x with u. The leaf containing q becomes the right child of u. After insertion, we splay u.

The deletion process is complimentary but more complex. We first find the leaf node containing q. Let u be this node's parent. As in [21], we apply a restructuring procedure to the subtree rooted at u so that the tree structure would be the same *as if* the point q had been the last point to be inserted. After restructuring, we delete q by simply "undoing" the insertion process. We then splay q's parent node. The analysis is similar to that of [26]. Details will appear in the full version.

Lemma 2. *Given a splay quadtree T containing an n-element points set, insertion or deletion of a point can be performed in $O(\log n)$ amortized time.*

5   Search and Optimality Results

In this section, we present theorems related to the complexity of performing various types of geometric searches on splay quadtrees. We provide generalizations to a number of optimality results for the standard splay tree, including the Balance Theorem, the Static Optimality Theorem, the Working Set theorem, and variants of the Static Finger Theorem. As in [26], our approach is based on applying Lemma 1 under an appropriate assignment of weights to the nodes.

We first discuss theorems related to the search time for accessing points. These theorems are essentially the same as those proved in [26] for 1-dimensional splay trees. Given a query point q, an *access query* returns the leaf node of T whose cell contains q. (Thus, it can be thought of as a point location query for the subdivision of space induced by T.) An access query is processed by traversing the path from the root to the leaf node containing q, and splaying the parent of this leaf.

The working set property states that once an item is accessed, accesses to the same item in the near future are particularly efficient [26]. Through the application of an appropriate weight assignment, similar to one used in [26] we obtain the following.

Theorem 1. (Working Set Theorem) *Consider a set P of n points in $\mathbb{R}^d$. Let $q_1, \ldots, q_m$ be a sequence of access queries. For each access q_j $(1 \leq j \leq m)$, let t_j be the number of different cells accessed before q_j since its previous access, or since the beginning of the sequence if this is q_j's first access. The total time to answer m queries is $O(\sum_{j=1}^{m} \log(t_j + 1) + m + n \log n)$.*

Two additional properties follow as consequences of the above theorem (see, e.g., Iacono [18]):

Theorem 2. (Balance Theorem) *Consider a set P of n points in $\mathbb{R}^d$ and a splay quadtree T for P. Let $q_1, \ldots, q_m$ be a sequence of access queries. The total time to answer these queries is $O((m + n) \log n + m)$.*

Theorem 3. (Static Optimality Theorem) *Given a subdivision Z, and the empirical probability p_z for each cell $z \in Z$, the total time to answer m access queries is $O(m \cdot entropy(Z))$.*

In traditional splay trees, the *static finger theorem* states that the running times of a series of accesses can be bounded by the sum of logarithms of the distances of each accessed item to a given fixed key, called a *finger*. In the 1-dimensional context, the notion of distance is based on the number of keys that lie between two items [26]. Intuitively, this provides a measure of the degree of locality among a set of queries with respect to a static key. Generalizing this to a geometric setting is complicated by the issue of how to define an appropriate notion of distance based on *betweenness*. In this section, we present a number of static finger theorems for different queries in the splay quadtree.

In general, our notion of distance to a static finger is based on the number of relevant objects (points or quadtree cells) that are closer to the finger than the accessed object(s). Define the distance from a point p to a geometric object Q, denoted $dist(p, Q)$, to be the minimum distance from p to any point of Q. Let $b(f, r)$ denote a ball of radius r centered at a point f. Our first static-finger result involves access queries.

Theorem 4. *Consider a set P of n points in $\mathbb{R}^d$, for some constant d. Let $q_1, \ldots, q_m$ be a sequence of access queries, where each q_j $(1 \leq j \leq m)$ is a point of P. If f is any fixed point in $\mathbb{R}^d$, the total time to access points $q_1, \ldots, q_m$ is $O\left(\sum_{j=1}^{m} \log N_f(q_j) + m + n \log n\right)$, where $N_f(q_j)$ is the number of points of P that are closer to f than q_j is.*

This theorem extends the static finger theorem for the traditional splay trees of [26] to a multidimensional setting. However, it is limited to the access of points that are in the given set. In a geometric setting, it is useful to extend this to points lying outside the set. The simplest generalization is to point location queries in the quadtree subdivision, that is, determining the leaf cell that contains a given query point. More generally, given a quadtree box Q, a *box query* returns the smallest outer box of any node of a splay quadtree that contains Q. (Point location queries arise as a special case corresponding to an infinitely small box.)

Our next result establishes a static finger theorem to box queries (and so applies to point location queries as well). Given a query box Q, starting from the root of the tree, we descend to the child node that contains Q until no such child exists. We then return the outer box of the current node, and if the current node is an internal node, we splay it, otherwise, we splay its parent node.

The following theorem establishes the static finger theorem for box queries.

Theorem 5. *Consider a set P of n points in $\mathbb{R}^d$ and a splay quadtree T for P. Let $Q_1, \ldots, Q_m$ be a sequence of box queries. If f is any fixed point in $\mathbb{R}^d$, the total time to answer these queries is $O\left(\sum_{j=1}^{m} \log N_f(Q_j) + m + n \log n\right)$, where $N_f(Q_j)$ is the number of leaf cells overlapping the ball $b(f, dist(f, Q_j))$ in the subdivision induced by T.*

Theorem 5 expresses the time for a series of box queries based on the number of leaf cells. In some applications, the induced subdivision is a merely a byproduct of a construction involving a set of P of points. In such case, it is more natural to express the time in terms of points, not quadtree cells. Given a set of box queries that are local to some given finger point f, we wish to express total access time as the function of the number of points of P in an appropriate neighborhood of f. As before, our analysis is based on quadtree leaf cells that are closer to f than the query box. The problem is that such leaf cells may generally contain points are very far from f, relative to the query box. To handle this, we allow a constant factor expansion in the definition of local neighborhood. Consider a ball containing the closest points of all given boxes from f (see Fig. 4(a)). Given a point set P, a finger point f, any constant c, and the box queries $Q_1, \ldots, Q_m$, the *working set* W_{box} with respect to P, f, c, and the box queries is defined:

$$W_{\text{box}}(P, f, c, Q_1, \ldots, Q_m) = \{p : p \in b(f, r_{\text{box}}(1+c)) \cap P, \ r_{\text{box}} = \max_j dist(f, Q_j)\}.$$

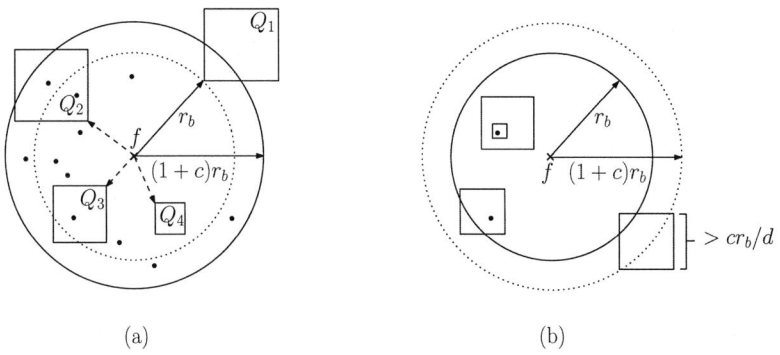

(a)   (b)

Fig. 4. (a) Working set, (b) the leaf cells overlapping a ball

Theorem 6. *Let $Q_1, \ldots, Q_m$ be the sequence of box queries. For any point $f \in \mathbb{R}^d$ and any positive constant c, the total time to answer these queries is $O(m \log(|W_{\text{box}}| + (\frac{1}{c})^{d-1}) + n \log n)$.*

We will also analyze the time for answering approximate nearest neighbor. First, we consider approximate nearest neighbor queries using splay quadtrees. Let q be the query point, and let $\varepsilon > 0$ be the approximating factor. We apply a variant of the algorithm of [4]. We use a priority queue U and maintain the closest point

p to q. Initially, we insert the root of the splay quadtree into U, and set p to a point infinitely far away. Then we repeatedly carry out the following process. First, we extract the node v with the highest priority from the queue, that is, the node closest to the query point. If the distance from v's cell to q exceeds $dist(q,p)/(1+\varepsilon)$, we stop and return p. Since no subsequent point in any cell to be encountered can be closer to q than $dist(q,p)/(1+\varepsilon)$, p is an ε-approximate nearest neighbor. Otherwise, we descend v's subtree to visit the leaf node closest to the query point. As we descend the path to this leaf, for each node u that is visited, we compute the distance to the cell associated with u's siblings and then insert these siblings into U. If the visited leaf node contains a point, and the distance from q to this point is closer than p, we update p to this point. Finally, we splay the parent of this leaf node.

Consider any sequence of m approximate nearest neighbor queries, $q_1,\ldots,q_m$. The working set for these queries can be defined to be

$$W_{\text{ann}}(P,f,c,q_1,\ldots,q_m)$$
$$= \{p : p \in b(f,r_{\text{ann}}(1+c)) \cap P,\ r_{\text{ann}} = \max_j (dist(f,q_j) + dist(q_j,\text{NN}(q_j)))\}.$$

The following theorem shows that the time to answer these sequence of queries can be related to the size of this working set.

Theorem 7. *Let $q_1,\ldots,q_m$ be a sequence of ε-approximate nearest neighbor queries. For any point $f \in \mathbb{R}^d$ and any positive constant c, the total time to answer all these queries is $O\big(m\big(\frac{1}{\varepsilon}\big)^d \log\big(|W_{\text{ann}}| + \big(\frac{1}{c}\big)^{d-1}\big) + n\log n\big)$.*

References

1. Arya, S., Malamatos, T., Mount, D.M.: A simple entropy-based algorithm for planar point location. ACM Trans. Algorithms, 3, article: 17 (2007)
2. Arya, S., Malamatos, T., Mount, D.M., Wong, K.-C.: Optimal expected-case planar point location. SIAM J. Comput. 37, 584–610 (2007)
3. Arya, S., Mount, D.M.: Approximate range searching. Comput. Geom. Theory Appl. 17, 135–163 (2001)
4. Arya, S., Mount, D.M., Netanyahu, N.S., Silverman, R., Wu, A.: An optimal algorithm for approximate nearest neighbor searching. J. Assoc. Comput. Mach. 45, 891–923 (1998)
5. Bern, M.: Approximate closest-point queries in high dimensions. Inform. Process. Lett. 45, 95–99 (1993)
6. Bădoiu, M., Cole, R., Demaine, E.D., Iacono, J.: A unified access bound on comparison-based dynamic dictionaries. Theo. Comp. Sci. 382, 86–96 (2007)
7. Chan, T.: A minimalist's implementation of an approximate nearest neighbor algorithm in fixed dimensions (2006),
http://www.cs.uwaterloo.ca/~tmchan/pub.html (unpublished manuscript)
8. Chan, T.M.: Closest-point problems simplified on the ram. In: Proc. 13th Annu. ACM-SIAM Sympos. Discrete Algorithms, pp. 472–473 (2002)
9. Clarkson, K.L.: Fast algorithms for the all nearest neighbors problem. In: Proc. 24th Annu. IEEE Sympos. Found. Comput. Sci., pp. 226–232 (1983)

10. Cole, R.: On the dynamic finger conjecture for splay trees. Part 2 The proof. SIAM J. Comput. 30, 44–85 (2000)
11. Cole, R., Mishra, B., Schmidt, J., Siegel, A.: On the dynamic finger conjecture for splay trees. Part 1: Splay sorting log n-block sequences. SIAM J. Comput. 30, 1–43 (2000)
12. Demaine, E.D., Harmon, D., Iacono, J., Pǎtraşcu, M.: The geometry of binary search tree. In: Proc. 20th Annu. ACM-SIAM Sympos. Discrete Algorithms, pp. 496–505 (2009)
13. Elmasry, A.: On the sequential access theorem and deque conjecture for splay trees. Theo. Comp. Sci. 314, 459–466 (2004)
14. Eppstein, D., Goodrich, M.T., Sun, J.Z.: The skip quadtree: A simple dynamic data structure for multidimensional data. Internat. J. Comput. Geom. Appl. 18, 131–160 (2008)
15. Georgakopoulos, G.F.: Chain-splay trees, or, how to achieve and prove $\log\log N$-competitiveness by splaying. Inform. Process. Lett. 106, 37–43 (2008)
16. Har-Peled, S.: Geometric Approximation Algorithms. American Mathematical Society, Providence (2011)
17. Iacono, J.: Alternatives to splay trees with $O(\log n)$ worst-case access times. In: Proc. 12th Annu. ACM-SIAM Sympos. Discrete Algorithms, pp. 516–522 (2001)
18. Iacono, J.: Distribution-sensitive data structures. PhD thesis, Rutgers, The state University of New Jersey, New Brunswick, New Jersey (2001)
19. Iacono, J.: Expected asymptotically optimal planar point location. Comput. Geom. Theory Appl. 29, 19–22 (2004)
20. Iacono, J.: Key-independent optimality. Algorithmica 42, 3–10 (2005)
21. Mount, D.M., Park, E.: A dynamic data structure for approximate range searching. In: Proc. 26th Annu. Sympos. Comput. Geom., pp. 247–256 (2010)
22. Pettie, S.: Splay trees, Davenport-Schinzel sequences, and the deque conjecture. In: Proc. 19th Annu. ACM-SIAM Sympos. Discrete Algorithms, pp. 1115–1124 (2008)
23. Pugh, W.: Skip lists: A probabilistic alternative to balanced trees. Commun. ACM 33, 668–676 (1990)
24. Samet, H.: Foundations of Multidimensional and Metric Data Structures. Morgan Kaufmann, San Francisco (2006)
25. Seidel, R., Aragon, C.: Randomized search trees. Algorithmica 16, 464–497 (1996)
26. Sleator, D.D., Tarjan, R.E.: Self-adjusting binary search trees. J. Assoc. Comput. Mach. 32, 652–686 (1985)
27. Subramanian, A.: An explanation of splaying. J. Algorithms 20(3), 512–525 (1996)
28. Sundar, R.: On the deque conjecture for the splay algorithm. Combinatorica 12, 95–124 (1992)
29. Tarjan, R.E.: Sequential access in splay trees takes linear time. Combinatorica 5, 367–378 (1985)
30. Wang, C.C., Derryberry, J., Sleator, D.D.: $o(\log\log n)$-competitive dynamic binary search trees. In: Proc. 17th Annu. ACM-SIAM Sympos. Discrete Algorithms, pp. 374–383 (2006)

TSP Tours in Cubic Graphs: Beyond 4/3

José R. Correa, Omar Larré, and José A. Soto

Universidad de Chile, Santiago, Chile
correa@uchile.cl, olarre@dim.uchile.cl, jsoto@dim.uchile.cl

Abstract. After a sequence of improvements Boyd, Sitters, van der Ster, and Stougie proved that any 2-connected graph whose n vertices have degree 3, i.e., a cubic graph, has a Hamiltonian tour of length at most $(4/3)n$, establishing in particular that the integrality gap of the subtour LP is at most 4/3 for cubic graphs and matching the conjectured value of the famous 4/3 conjecture. In this paper we improve upon this result by designing an algorithm that finds a tour of length $(4/3 - 1/61236)n$, implying that cubic graphs are among the few interesting classes for which the integrality gap of the subtour LP is strictly less than 4/3.

1 Introduction

The traveling salesman problem (TSP) in metric graphs is a landmark problem in combinatorial optimization and theoretical computer science. Given a graph in which edge-distances form a metric the goal is to find a tour of minimum length visiting each vertex at least once. In particular, understanding the approximability of the TSP has attracted much attention since Christofides [7] designed a 3/2-approximation algorithm for the problem. Despite the great efforts, Christofides' algorithm continues to be the current champion, while the best known lower bound, proved by Papadimitriou and Vempala [15], states that the problem is NP-hard to approximate with a factor better than 220/219. A key lower bound to study the approximability of the problem is the so-called subtour linear program which has long been known to have an integrality gap of at most 3/2 [20]. Although no progress has been made in decreasing the integrality gap of the subtour LP, its value is conjectured to be 4/3 (see e.g., Goemans [9]).

Very recently there have been several improvements for important special cases of the metric TSP. Oveis Gharan, Saberi, and Singh [14] design a $3/2 - \epsilon$-approximation algorithm for the case of graph metrics, while Mömke and Svensson [11] improve that to 1.461, using a different approach. Mucha [12] then showed that the approximation guarantee of the Mömke and Svensson algorithm is 13/9. These results in particular show that the integrality gap of the subtour LP is below 3/2 in case the metric comes from a graph. Another notable recent result, due to An, Shmoys, and Kleinberg [2], states that there is a $(1 + \sqrt{5})/2$-approximation algorithm for the s-t path version of the TSP on arbitrary metrics, improving on the natural extension of Christofides' heuristic.

This renewed interest in designing algorithms for the TSP in graph metrics has also reached the case when we further restrict to graph metrics induced by

L. Epstein and P. Ferragina (Eds.): ESA 2012, LNCS 7501, pp. 790–801, 2012.

special classes of graphs. Gamarnik, Lewenstein, and Sviridenko [8] show that in a 3-connected cubic graph on n vertices there is always a TSP tour –visiting each vertex at least once– of length at most $(3/2 - 5/389)n$, improving upon Christofides' algorithm for this graph class. A few years later, Aggarwal, Garg, and Gupta [1] improved the result obtaining a bound of $(4/3)n$ while Boyd et al. [5] showed that the $(4/3)n$ bound holds even if only 2-connectivity assumed. Finally, Mömke and Svensson [11] approach prove that the $(4/3)n$ bound holds for subcubic 2-connected graphs. Interestingly, the latter bound happens to be tight and thus it may be tempting to conjecture that there are cubic graphs on n vertices for which no TSP tour shorter than $(4/3)n$ exists. In this paper we show that this is not the case. Namely, we prove that any 2-connected cubic graph on n vertices has a TSP tour of length $(4/3 - \epsilon)n$, for $\epsilon = 1/61236 > 0.000016$.

On the other hand, Qian, Schalekamp, Williamson, and van Zuylen [16] show that the integrality gap of the subtour LP is strictly less than 4/3 for metrics where all the distances are either 1 or 2. Their result, based on the work of Schalekamp, Williamson, and van Zuylen [17], constitutes the first relevant special case of the TSP for which the integrality gap of the subtour LP is strictly less than 4/3. Our result implies in particular that the integrality gap of the subtour LP is also strictly less than 4/3 in 2-connected cubic graphs.

From a graph theoretic viewpoint, our result can also be viewed as a step towards resolving Barnette's [4] conjecture, stating that every bipartite, planar, 3-connected, cubic graph is Hamiltonian (a similar conjecture was formulated by Tait [18], refuted by Tutte [19], reformulated by Tutte and refuted by Horton (see [6]), and finally reformulated by Barnette more than 40 years ago). For Barnette's graphs (i.e., those with the previous properties) on n vertices it is easy to construct TSP tours of length at most $(4/3)n$, however no better bounds were known. Very recently, Kawarabayashi and Ozeki [10] proved that the $(4/3)n$ bound holds even for 3-connected planar graphs. In the full version of this paper we improve this bound for Barnette's graphs to $(4/3 - 1/18)n < 1.28n$.

Our Approach. An *Eulerian subgraph cover* (or simply a *cover*) is a collection $\Gamma = \{\gamma_1, \ldots, \gamma_j\}$ of connected multi-subgraphs of G, called *components*, satisfying that (i) every vertex of G is covered by exactly one component, (ii) each component is Eulerian and (iii) no edge appears more than twice in the same component. Every cover Γ can be transformed into a TSP tour $T(\Gamma)$ of the entire graph by contracting each component, adding a doubled spanning tree in the contracted graph (which is connected) and then uncontracting the components. Boyd et al [5] defined the *contribution* of a vertex v in a cover Γ, as $z_\Gamma(v) = \frac{(\ell+2)}{h}$, where ℓ and h are the number of edges and vertices of the component of Γ in which v lies. The extra 2 in the numerator is added for the cost of the double edge used to connect the component to the other in the spanning tree mentioned above, so that $\sum_{v \in V} z_\Gamma(v)$ equals the number of edges in the final TSP tour $T(\Gamma)$ plus 2. Let $\mathcal{D} = \{(\Gamma_i, \lambda_i)\}_{i=1}^k$, be a distribution over covers of a graph. This is, each Γ_i is a cover of G and each λ_i is a positive number so that $\sum_{i=1}^k \lambda_i = 1$. The *average contribution* of a vertex v with respect to distribution $\mathcal{D}$ is defined as $z_\mathcal{D}(v) = \sum_{i=1}^k z_{\Gamma_i}(v)\lambda_i$.

Given a 2-connected cubic graph G, Boyd et al. [5] find a TSP tour T of G with at most $\frac{4}{3}|V(G)|-2$ edges. Their approach has two phases. In the first phase, they transform G into a simpler cubic 2-connected graph H not containing certain ill-behaved structures (called p-rainbows, for $p \geq 1$). In the second phase, they use a linear programming approach to find a polynomial collection of perfect matchings for H such that a convex combination of them gives every edge a weight of $1/3$. Their complements induce a distribution over cycle covers of H. By performing local operations on each cover, they get a distribution of Eulerian subgraph covers having average vertex contribution bounded above by $4/3$. They use this to find a TSP tour for H with at most $\frac{4}{3}|V(H)|-2$ edges, which can be easily transformed into a TSP tour of G having the desired guarantee.

In this paper, we improve Boyd et al.'s technique and show that every 2-connected cubic graph G admits a TSP tour with at most $(4/3 - \epsilon)|V(G)| - 2$ edges. The first difference between approaches, described in Section 2, is that our simplification phase is more aggressive. Specifically, we set up a framework to eliminate large families of structures that we use to get rid of all chorded 6-cycles. This clean-up step can very likely be extended to larger families and may ultimately lead to improved results when combined with an appropriate second phase. The second difference, described in Section 3, is that we extend the set of local operations of the second phase by allowing the alternation of 6-cycles of an Eulerian subgraph cover. Again, it is likely that one can further exploit this idea to get improved guarantees. Mixing these new ideas appropriately requires significant work but ultimately leads us to find a distribution $\mathcal{D}$ of covers of the simplified graph H for which $\sum_{v \in V(H)} z_{\mathcal{D}}(v) \leq (\frac{4}{3} - \epsilon)n - 2$. From there, we obtain a TSP tour of G with the improved guarantee.

2   Graph Simplification Phase

Our algorithm starts by reducing the input 2-connected cubic graph G into a simpler 2-connected cubic graph H which does not contain a cycle of length six with one or more chords as subgraph. In addition our reduction satisfies that if H has a TSP tour of length at most $(4/3 - \epsilon)|V(H)| - 2$ then G has a TSP tour of length at most $(4/3 - \epsilon)|V(G)| - 2$, where $V(H)$ and $V(G)$ denote the vertex sets of H and G respectively. In what follows, we describe the main steps of the graph reduction process without discussing the tedious, but straightforward, case analysis, which is left to the full version.

The first step to achieve this is to remove the 6-cycles having two chords. Let γ be such a cycle and consider the two vertices, v_1 and v_2 connecting γ to the rest of G. Our procedure replaces the whole 6-cycle by a 4-cycle with a chord, identifying v_1 and v_2 with the vertices of degree 2 in the chorded 4-cycle. It is easy to observe that a TSP tour in the reduced graph can be turned into a TSP tour in the original graph with exactly 2 extra edges (and also 2 more vertices). To see this note that in any 6-cycle with two chords there is a Hamiltonian path between v_1 and v_2 and also is obviously Hamiltonian. This procedure in particular removes the p-rainbow structure in Boyd et al. [5]. Note that, at least theoretically, we could apply this procedure to any 2-cut in G

satisfying that there is a Hamiltonian path between the two cut edges and that it has a Hamiltonian cycle, but testing this cannot be done in polynomial time.

The second step is to consider 6-cycles having only one chord. Let γ be such a cycle and consider the four edges, $e_1, e_2, e_3,$ and e_4 connecting γ to the rest of G. Letting w_i be the endpoint of e_i outside γ, we distinguish three cases. If only two of the w_i's are distinct, we proceed as in the previous case, replacing the structure by a chorded 4-cycle. If three of the w_i's are distinct we replace the 7 vertex structure formed by γ and the w_i adjacent to two vertices in γ by a triangle (identifying the degree two vertices in the structure with those in the triangle). One can check that a TSP tour in the reduced graph can be turned into one in the original one with at most 5 extra edges (and 4 more vertices), and therefore if the tour in the reduced graph had length $\alpha n - 2$ with $5/4 \leq \alpha \leq 4/3$, the new one will as well. The final case is when all four w_i's are distinct. Assume, without loss of generality that the $w_i's$ are indexed in the cyclic order induced by γ. In this case we replace the 6-cycle by an edge e and we either connect w_1, w_2 to one endpoint of e and w_3, w_4 to the other, or we connect w_1, w_4 to one endpoint and w_2, w_3 to the other. The previous choice can always be made so that e is not a bridge. Again we check that a TSP tour in the reduced graph can be turned into a TSP tour in the original one with at most 5 extra edges (and 4 more vertices).

Finally note that all the above reduction steps strictly decrease the number of vertices in the graph and therefore only a linear number of these needs to be done. Also each step requires only polynomial time, not only to find the desired structure, but also to recover the TSP tour in the original graph. Thus this graph simplification phase runs in polynomial time.

3   Eulerian Subgraph Cover Phase

For every edge set $F \subseteq E$, the vector $\chi^F \in \{0,1\}^E$ given by $\chi_e^F = 1$ if $e \in F$ and $\chi_e^F = 0$ otherwise, is known as the *incident vector* of F. We say that a matching M is *3-cut perfect* if M is a perfect matching intersecting every 3-cut in exactly one edge. Based on ideas by Naddef and Pulleyblank [13], Boyd et al. [5] prove:

Lemma 1. *Let $G = (V, E)$ be a 2-connected cubic graph. Then, the vector $\frac{1}{3}\chi^E$ can be expressed as a convex combination of incidence vectors of 3-cut perfect matchings of G. This is, there are 3-cut perfect matchings $\{M_i\}_{i=1}^k$ and positive real numbers $\lambda_1, \lambda_2, \ldots, \lambda_k$ such that*

$$\sum_{i=1}^k \lambda_i = 1 \quad (1) \qquad and \qquad \frac{1}{3}\chi^E = \sum_{i=1}^k \lambda_i \chi^{M_i}. \quad (2)$$

Furthermore, Barahona [3] provides an algorithm to find a convex combination of $\frac{1}{3}\chi^E$ having $k \leq 7n/2 - 1$ in $O(n^6)$ time.

Consider a graph G that is cubic, 2-connected and *reduced*. That is, no 6-cycle in G has chords. We also assume that $n \geq 10$ as every cubic 2-connected graph on less than 10 vertices is Hamiltonian.

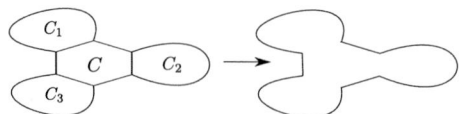

Fig. 1. Operation (U1)

Let $\{M_i\}_{i=1}^k$ and $\{\lambda_i\}_{i=1}^k$ be the 3-cut matchings and coefficients guaranteed by Lemma 1. Let $\{\mathcal{C}_i\}_{i=1}^k$ be the family of *cycle covers* associated to the matchings $\{M_i\}_{i=1}^k$. This is, $\mathcal{C}_i$ is the collection of cycles induced by $E \setminus M_i$. Since each matching M_i is 3-cut perfect, the corresponding cycle cover $\mathcal{C}_i$ does not contain 3-cycles. Furthermore every 5-cycle in $\mathcal{C}_i$ is induced (i.e., it has no chord in G).

In what follows we define three local operations, (U1), (U2) and (U3) that will be applied iteratively to the current family of covers. Each operation is aimed to reduce the contribution of each component of the family. We stress here that operations (U2) and (U3) are exactly those used by Boyd et al., but for reader's convenience we explain each one completely. We start with operation (U1).

(U1) Consider a cycle cover $\mathcal{C}$ of the current family. If C_1, C_2 and C_3 are three disjoint cycles of $\mathcal{C}$, that intersect a fixed 6-cycle C of G, then we merge them into the simple cycle obtained by taking their symmetric difference with C. This is, the new cycle in $V(C_1) \cup V(C_2) \cup V(C_3)$ having edge set $(E(C_1) \cup E(C_2) \cup E(C_3)) \Delta E(C)$.

An example of (U1) is depicted in Figure 1. We apply (U1) as many times as possible to get a new cycle cover $\{\mathcal{C}_i^{\text{U1}}\}_{i=1}^k$. Then we apply the next operation.

(U2) Consider a cycle cover $\mathcal{C}$ of the current family. If C_1 and C_2 are two disjoint cycles of $\mathcal{C}$, that intersect a fixed 4-cycle C of G, then we merge them into a simple cycle obtained by taking their symmetric difference with C. This is, the new cycle in $V(C_1) \cup V(C_2)$ having edge set $(E(C_1) \cup E(C_2)) \Delta E(C)$.

We apply (U2) as many times as possible to obtain a new cycle cover $\{\mathcal{C}_i^{\text{U2}}\}_{i=1}^k$ of G. The next operation transforms a cycle cover $\mathcal{C}$ of the current family into an Eulerian subgraph cover Γ, having components that are not necessarily cycles.

(U3) Let Γ be an Eulerian subgraph cover of the current family. If γ_1 and γ_2 are two components of Γ, each one having at least 5 vertices, whose vertex set intersect a fixed 5-cycle C of G, then combine them into a single component, by adding at most 1 extra edge.

To explain how we combine the components in operation (U3) we need two lemmas that are almost identical to results by Boyd et al. [5].

Lemma 2. *Let H_1 and H_2 be two connected Eulerian multi-subgraphs of a cubic graph G having at least two vertices in common and let H_3 be the sum of H_1 and H_2, i.e., the union of their vertices and the sum of their edges (allowing multiple parallel edges). Then we can remove (at least) two edges from H_3 such that it stays connected and Eulerian.*

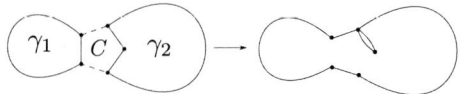

Fig. 2. Sketch of operation (U3)

Lemma 3. *If v belongs to a component γ of any of the covers Γ considered by the algorithm, then at least two of its 3 neighbors are in the same component.*

Observe that if γ is a component of a cover in the current family, and C is an arbitrary cycle of G containing a vertex of γ then, by the cubicity of G and Lemma 3, C and γ must share at least two vertices. In particular, if γ_1 and γ_2 are the two components intersecting a 5-cycle C considered by operation (U3), then one of them, say γ_1, must contain exactly 2 vertices of C and the other one must contain the other 3 vertices (note that they cannot each share 2 vertices, since then a vertex of C would not be included in the cover). To perform (U3) we first merge γ_1 and C using Lemma 2 removing 2 edges, and then we merge the resulting component with γ_2, again removing 2 edges. Altogether, we added the 5 edges of C and removed 4 edges. Finally, we remove 2 edges from each group of triple or quadruple edges that may remain, so that each edge appears at most twice in each component. Figure 2 shows an example of (U3).

Remark 1. Operation (U3) generates components having at least 10 vertices. Therefore, any component having 9 or fewer vertices must be a cycle. Furthermore, all the cycles generated by (U1) or (U2) contain at least 10 vertices (this follows from the fact that G is reduced, and so operation (U2) always involve combine 2 cycles of length at least 5). From here we observe that any component having 9 or fewer vertices must be in the original cycle cover $\{C_i\}_{i=1}^k$.

We say that a 4-cycle C with a chord is isolated if the two edges incident to it are not incident to another chorded 4-cycle. The following is the main result of this section. Before proving it we show it implies the main result of the paper.

Proposition 1 (Main Proposition). *Let $\{\Gamma_i\}_{i=1}^k$ be the family of Eulerian subgraph covers at the end of the algorithm (that is, after applying all operations), and let $z(v) = z_{\mathcal{D}}(v)$ be the average contribution of vertex v for the distribution $\mathcal{D} = \{(\Gamma_i, \lambda_i)\}_{i=1}^k$. Furthermore, let γ_i be the component containing v in Γ_i and $\Gamma(v) = \{\gamma_i\}_{i=1}^k$. We have the following.*

(P1) *If v is in an isolated chorded 4-cycle then $z(v) \leq 4/3$.*
(P2) *If v is in a non-isolated chorded 4-cycle of G then $z(v) \leq 13/10$.*
(P3) *Else, if there is an induced 4-cycle $\gamma \in \Gamma(v)$, then $z(v) \leq 4/3 - 1/60$.*
(P4) *Else, if there is an induced 5-cycle $\gamma \in \Gamma(v)$, then $z(v) \leq 4/3 - 1/60$.*
(P5) *Else, if there is an induced 6-cycle $\gamma \in \Gamma(v)$, then we have both $z(v) \leq 4/3$ and $\sum_{w \in V(\gamma)} z(w) \leq 6 \cdot (4/3 - 1/729)$.*
(P6) *In any other case $z(v) \leq 13/10$.*

Theorem 1. *Every 2-connected cubic graph $G = (V, E)$ admits a TSP tour of length at most $(4/3 - \epsilon)|V| - 2$, where $\epsilon = 1/61236$. This tour can be computed in polynomial time.*

Proof. From Section 2, we can assume that G is also reduced and so the Main Proposition holds. Let B be the union of the vertex sets of all isolated chorded 4-cycles of G. We say a vertex is *bad* if it is in B, and *good* otherwise. We claim that the proportion of bad vertices in G is bounded above by $6/7$. To see this, construct the auxiliary graph G' from G by replacing every isolated chorded 4-cycle with an edge between its two neighboring vertices. Since G' is cubic, it contains exactly $2|E(G')|/3$ vertices, which are good in G. Hence, for every bad vertex there are at least $(1/4) \cdot (2/3) = 1/6$ good ones, proving the claim.

The Main Proposition guarantees that every bad vertex v contributes a quantity $z(v) \leq 4/3$. Now we show that the average contribution of all the good vertices is at most $(4/3 - \delta)$ for some δ to be determined. To do this, define $\mathcal{H} = \{\gamma \in \bigcup_i \Gamma_i : |V(\gamma)| = 6\}$ as the collection of all 6-cycles appearing in some cover of the final family, and let $H = \bigcup_{\gamma \in \mathcal{H}} V(\gamma)$ be the vertices included in some 6-cycle of $\mathcal{H}$. It is easy to check that B and H are disjoint. Furthermore, the Main Proposition guarantees that if $v \in V \backslash (B \cup H)$ then $z(v) \leq (4/3 - 1/60)$. So we focus on bounding the contribution of the vertices in H.

For every $v \in H$, let $f(v)$ be the number of distinct 6-cycles in $\mathcal{H}$ containing v. Since G is cubic, there is an absolute constant K, such that $f(v) \leq K$. By the main proposition, $z(v) \leq 4/3$ for $v \in H$ and for every $\gamma \in \mathcal{H}$, $\sum_{v \in V(\gamma)} z(v) \leq 6 \cdot (4/3 - \epsilon')$, where $\epsilon' = 1/729$. Putting all together we have:

$$K \cdot \sum_{v \in H} \left[z(v) - \left(\frac{4}{3} - \frac{\epsilon'}{K} \right) \right] = |H|\epsilon' + K \sum_{v \in H} \left(z(v) - \frac{4}{3} \right)$$

$$\leq 6|\mathcal{H}|\epsilon' + \sum_{v \in H} f(v) \left(z(v) - \frac{4}{3} \right) = 6|\mathcal{H}|\epsilon' + \sum_{\gamma \in \mathcal{H}} \sum_{v \in V(\gamma)} \left(z(v) - \frac{4}{3} \right)$$

$$\leq 6|\mathcal{H}|\epsilon' - \sum_{\gamma \in \mathcal{H}} 6\epsilon' = 0.$$

It follows that $\frac{1}{|H|} \sum_{v \in H} z(v) \leq (4/3 - \epsilon'/K)$. Since $\epsilon'/K \leq 1/60$, we get

$$\sum_{v \in V} z(v) \leq \sum_{v \in B} z(v) + \sum_{v \in H} z(v) + \sum_{v \in V \backslash (B \cup H)} z(v)$$

$$\leq \frac{4}{3}|B| + \left(\frac{4}{3} - \frac{\epsilon'}{K} \right) (|V| - |B|) = |V| \left(\frac{4}{3} - \frac{\epsilon'}{7K} \right).$$

We conclude that there is an index i such that $\sum_{v \in V} z_i(v) \leq |V| (4/3 - \epsilon'/(7K))$. By adding a double spanning tree of $G/E(\Gamma_i)$ we transform Γ_i into a TSP tour T of length $|V| (4/3 - \epsilon'/(7K)) - 2$. Noting that $K \leq 12$ and $\epsilon' = 1/729$ we obtain the desired bound[1]. Clearly, all operations can be done in polynomial time.   □

---

[1] As the graph is cubic, v has at most 6 vertices at distance 2. These can be joined through another vertex in $\binom{6}{2} = 15$ ways to form 6-cycles, but 3 of these combinations clearly do not create a 6-cycle, namely those corresponding to two vertices at distance 2 of v coming from the same neighbor of v, for a total of 15-3=12 possible 6-cycles.

3.1 Proof of the Main Proposition

We start by a lemma, whose proof is the same as that of [5, Observation 1].

Lemma 4 ([5]). *For each vertex $v \in V$, and each $i \in \{1, ..., k\}$, the contribution $z_i(v) := z_{\Gamma_i}(v)$ is*

(a) at most $\frac{h+2}{h}$, where $h = \min\{t, 10\}$ and v is on a t-cycle belonging to one of the cycle covers $\mathcal{C}_i$, $\mathcal{C}_i^{U1}$ and $\mathcal{C}_i^{U2}$.
(b) at most $\frac{13}{10}$ if operation (U3) modified the component containing v.

We will also use the following notation in our proof. For any subset J of indices in $[k] := \{1, ..., k\}$, define $\lambda(J) = \sum_{i \in J} \lambda_i$.

The proofs of parts (P1) through (P4) are similar to the arguments used by Boyd et al. [5] to show that $z(v) \leq 4/3$ when v is a 4-cycle or 5-cycle. By using the fact that G is reduced (i.e. it contains no chorded 6-cycles) we obtain a better guarantee in (P2), (P3) and (P4). The details of these parts are deferred to the full version of the paper. To prove part (P5) we heavily use the fact that operation (U1) is applied to the initial cycle cover (recall that this operation was not used in [5]).

Proof of Part (P5). Let $\gamma \in \Gamma(v)$ be an induced 6-cycle containing v. By Remark 1, γ is in some initial cycle cover $\mathcal{C}_i$. We can assume that no 4-cycle shares exactly one edge with γ, as otherwise operations (U1) or (U2) would have modified γ and so, by the end of the algorithm γ would not be a 6-cycle.

We can also assume that γ does not intersect the 5-cycles contained in an initial cycle cover. Indeed, if this was not the case, define $S_5 = \{w \in V(\gamma) : w$ is in some 5-cycle C of an initial cycle cover$\}$. If $w \notin S_5$ then in every initial cover, the cycle containing w is of length at least 6; using Lemma 4, part (P4) of the Main Proposition, and the fact that $S_5 \neq \emptyset$ implies $|S_5| \geq 2$, we conclude that $\sum_{w \in V(\gamma)} z(w) \leq |S_5| \left(\frac{4}{3} - \frac{1}{60}\right) + |V(C) \setminus S_5|\frac{4}{3} \leq 6\left(\frac{4}{3} - \frac{1}{180}\right)$, and also that $z(w) \leq 4/3$ for all $w \in V$.

Under the assumptions above, all the components containing v in the final family of covers have length at least 6. Using Lemma 4 we conclude not only that $z(v) \leq \max\{13/10, 8/6\} = 4/3$ (which proves the first statement of P5) but also that $z(w) \leq 4/3$ for the 6 vertices $w \in V(\gamma)$.

Let us continue with the proof. Denote the edges of γ as $a_1, ..., a_6$ and the 6 edges incident to γ as $e_1, ..., e_6$, as in Figure 3.

We now define some sets of indices according on how γ intersects the matchings $M_1, ..., M_k$. For every symbol $Z \in \{X_0\} \cup \{X_1^q\}_{q=1}^6 \cup \{X_2^q\}_{q=1}^3 \cup \{Y_2^q\}_{q=1}^6 \cup \{X_3^q\}_{q=1}^2$, we define Z as the set of indices i for which the matching M_i contains the bold edges indicated in Figure 4. For example, $X_0 = \{i: \{e_1, ..., e_6\} \in M_i\}$, $X_1^1 = \{i: \{a_1, a_3, a_5\} \in M_i\}$, and so on. Let also $x_0 = \lambda(X_0)$, $x_i^q = \lambda(X_i^q)$ and $y_2^q = \lambda(Y_i^q)$ for every i and q and define

$$x_1 = \sum_{q=1}^{6} x_1^q, \quad x_2 = \sum_{q=1}^{3} x_2^q, \quad y_2 = \sum_{q=1}^{6} y_2^q, \quad x_3 = \sum_{q=1}^{2} x_3^q, \quad \overline{x}_2 = x_2 + y_2.$$

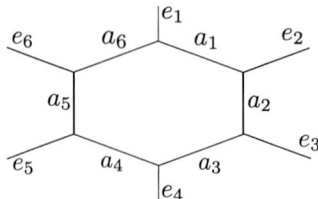

Fig. 3. Induced 6-cycle γ

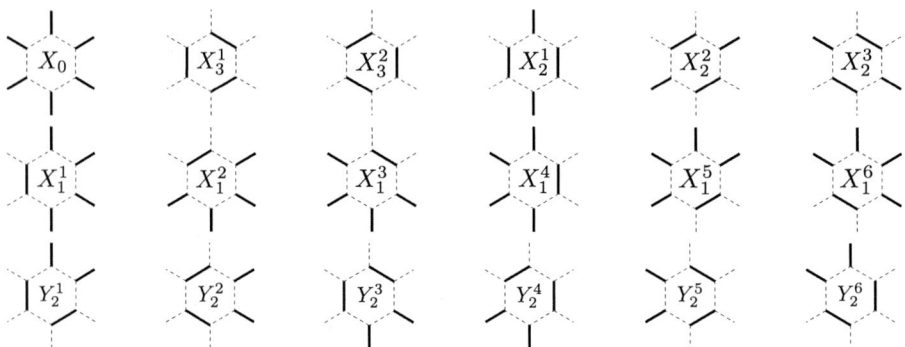

Fig. 4. Ways in which a matching can intersect γ. The orientation is as in Figure 3.

Equation (1) implies that $x_0 + x_1 + \overline{x}_2 + x_3 = 1$. Equation (2) applied to the set $\{e_1, \ldots, e_6\}$ of edges incident to γ implies that $6x_0 + 4x_1 + 2\overline{x}_2 = 6/3$. Hence, $3x_0 + 2x_1 + \overline{x}_2 = 1$. It follows that

$$2x_0 + x_1 = x_3. \tag{3}$$

Recall that there are no 4-cycles in G and no 5-cycles in an initial cycle cover intersecting γ in exactly one edge. Consider $w \in V(\gamma)$ and $i \in [k]$.

If $i \in X_0$ (i.e., M_i shares no edge with γ) then $w \in V(\gamma)$ and $\gamma \in C_i$. By Lemma 4 we have, $z_i(w) \leq 8/6$. If $i \in X_1 := \cup_{q=1}^6 X_1^q$ (i.e., M_i contains exactly one edge of γ) then, as no 4-cycle shares exactly one edge with γ, w must be in a cycle $C \in C_i$ of length at least 9; therefore, $z_i(w) \leq 11/9$. If $i \in X_3 := \cup_{q=1}^2 X_3^q$ (i.e., M_i contains three edges of γ) then we have two cases. The first case is that γ is intersected by 1 or 3 cycles of C_i. Then, by the end of operation (U1), w must be in a cycle of C_i^{U1} of length at least 9 and so $z_i(w) \leq 11/9$. The second case is that γ is intersected by 2 cycles of C_i. One of them shares exactly 2 edges with γ, thence it must be of length at least 8. The other cycle shares exactly one edge with γ and so it must be of length at least 6. Therefore, in this case, 4 of the vertices w of γ satisfy $z_i(w) \leq 10/8$ and the remaining 2 satisfy $z_i(w) \leq 8/6$.

We still need to analyze the indices $i \in X_2 := \cup_{q=1}^3 X_2^q$ and $i \in Y_2 := \cup_{q=1}^6 Y_2^q$ (i.e., those for which M_i shares two edges with γ). Let $0 < \delta \leq 1$ be a constant to be determined. We divide the rest of the proof in two scenarios.

Scenario 1: If x_3 (which equals $\max\{x_0, x_1, x_3\}$ by (3)) is at least δ.

If $i \in X_2 \cup Y_2$, then every vertex $w \in \gamma$ is in a cycle $C \in \mathcal{C}_i$ of length at least 6; therefore $z_i(w) \le 8/6$ and

$$\sum_{w \in V(\gamma)} z(w) \le 6 \cdot (x_0 8/6 + x_1 11/9 + \overline{x}_2 8/6) + x_3 \left(2 \cdot \frac{8}{6} + 4 \cdot \frac{10}{8} \right)$$

$$\le 6 \cdot \left((1 - x_3)4/3 + x_3 \left(\frac{4}{3} - \frac{1}{18} \right) \right) \le 6 \cdot (4/3 - \delta/18). \qquad (4)$$

Scenario 2: If x_3 (which equals $\max\{x_0, x_1, x_3\}$ by (3)) is at most δ.

We start by stating a lemma whose proof we defer to the full version.

Lemma 5. *Define* $\beta := 1/9 - \delta$*. Then at least one of the following cases hold:*

- **Case 1:** $x_2^1, x_2^2, x_2^3 \ge \beta$.
- **Case 2:** $x_2^1, y_2^2, y_2^5 \ge \beta$.
- **Case 3:** $x_2^2, y_2^3, y_2^6 \ge \beta$.
- **Case 4:** $x_2^3, y_2^1, y_2^4 \ge \beta$.
- **Case 5:** $y_2^1, y_2^4, y_2^2, y_2^5 \ge \beta$.
- **Case 6:** $y_2^2, y_2^5, y_2^3, y_2^6 \ge \beta$.
- **Case 7:** $y_2^1, y_2^4, y_2^3, y_2^6 \ge \beta$.

Denote an index $i \in X_2 \cup Y_2$ as *long* if there are at least 2 vertices of $V(\gamma)$ contained in a single cycle of $\mathcal{C}_i^{\mathrm{U}1}$ of length at least 7, otherwise denote it as *short*. A set $Z \subseteq [k]$ is called long if Z contains only long indices.

Consider a short index $i \in X_2 \cup Y_2$. Since the matching M_i contains two edges of γ, we must be in the case where γ intersects exactly two cycles of $\mathcal{C}_i^{\mathrm{U}1}$ and both of them are 6-cycles (we assumed at the beginning of the proof of this part that no cycle in $\mathcal{C}_i$ of length at most 5 intersects γ). The next lemma complements what happens in each of the cases introduced in Lemma 5.

Lemma 6.

(1) If X_2^1, X_2^2 and X_2^3 are non-empty then at least one of them is long.
(2) If X_2^1, Y_2^2 and Y_2^5 are non-empty then at least one of them is long.
(3) If X_2^2, Y_2^1 and Y_2^4 are non-empty then at least one of them is long.
(4) If X_2^3, Y_2^3 and Y_2^6 are non-empty then at least one of them is long.
(5) If Y_2^1, Y_2^4, Y_2^2 and Y_2^5 are non-empty then at least one of them is long.
(6) If Y_2^2, Y_2^5, Y_2^3 and Y_2^6 are non-empty then at least one of them is long.
(7) If Y_2^1, Y_2^4, Y_2^3 and Y_2^6 are non-empty then at least one of them is long.

Proof. We only prove items 1, 2 and 5, since the proofs for the rest are analogous.

(1) Assume for contradiction that there are short indices $i_1 \in X_2^1$, $i_2 \in X_2^2$ and $i_3 \in X_3^3$. In particular, every vertex of γ is in a 6-cycle of $\mathcal{C}_{i_p}^{\mathrm{U}1}$ (and thus, of $\mathcal{C}_{i_p}$) for $p = 1, 2, 3$. From this, we deduce that the neighborhood of γ in G is exactly as depicted in Figure 5. Now focus on the short index $i_1 \in X_2^1$. Since G is as in Figure 5, there are three cycles of $\mathcal{C}_{i_1}$ sharing each one edge with a 6-cycle of G. But then, as Figure 6 shows, operation (U1) would have merge them into a unique cycle C in $\mathcal{C}_{i_1}^{\mathrm{U}1}$ of length at least 16, contradicting the fact that i_1 is short.

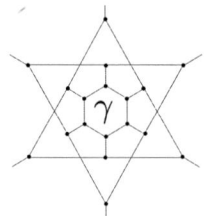

Fig. 5. 6-cycle γ for the case in which X_2^1, X_2^2 and X_2^3 are nonempty and not long

Fig. 6. Operation (U1) applied to cycles in $\mathcal{C}_{i_1}$, where i_1 is a short index of X_2^1

 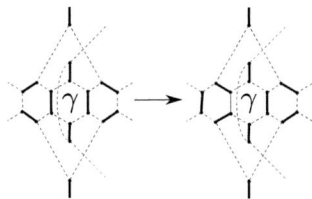 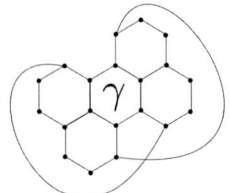

Fig. 7. 6-cycle γ for the case X_2^1, Y_2^2, Y_2^5 are nonempty and not long

Fig. 8. Operation (U1) applied to cycles in $\mathcal{C}_{i_1}$, where i_1 a short index of X_2^1

Fig. 9. 6-cycle γ for the case Y_2^1, Y_2^4, Y_2^2, Y_2^5 are non-empty and not long

(2) Assume for contradiction that there are short cycles $i_1 \in X_2^i$ $i_2 \in Y_2^2$ and $i_3 \in Y_2^5$. In particular, every vertex of γ is in a 6-cycle of $\mathcal{C}_{i_p}^{\mathrm{U1}}$ (and thus, of $\mathcal{C}_{i_p}$) for $p = 1, 2, 3$. From this, we deduce that the neighborhood of γ in G is exactly as depicted in Figure 7,
   Focus on the short index $i_1 \in X_2^1$. Since G is as in Figure 7, there are three cycles of $\mathcal{C}_{i_1}$ that share one edge each with a 6-cycle of G. But in this case, as Figure 8 shows, operation (U1) would have merge them into a unique cycle C in $\mathcal{C}_{i_1}^{\mathrm{U1}}$ of length at least 16, contradicting the fact that i_1 is short.

(5) Assume for contradiction that there are short indices $i_1 \in Y_2^1$, $i_2 \in Y_2^4$, $i_3 \in Y_2^2$ and $i_4 \in Y_2^5$. In particular every vertex of γ is in a 6-cycle of $\mathcal{C}_{i_p}^{\mathrm{U1}}$ (and thus, of $\mathcal{C}_{i_p}$), for $p = 1, 2, 3, 4$. Then, the neighborhood of γ in G is exactly as depicted in Figure 7. But this structure shows a contradiction, as matching M_{i_1} cannot be completed to the entire graph.   □

Using Lemmas 5 and 6 we conclude that there is a long set of indices $Z \subseteq X_2 \cup Y_2$ for which $\lambda(Z) \geq \beta$. In particular, using Lemma 4, we conclude that for every $i \in Z$, there are 2 vertices w in γ with $z_i(w) \leq 9/7$, and for the remaining four vertices of γ, $z_i(w) \leq 4/3$. Altogether, $\sum_{w \in V(\gamma)} z(w)$ is at most

$$6 \cdot \left(x_0 \frac{8}{6} + x_1 \frac{11}{9} + (\bar{x}_2 - \beta)\frac{8}{6} \right) + \beta \left(2 \cdot \frac{9}{7} + 4 \cdot \frac{8}{6} \right) + x_3 \left(2 \cdot \frac{8}{6} + 4 \cdot \frac{10}{8} \right)$$

$$\leq 6(1 - \beta)\frac{4}{3} + \beta \left(2 \cdot \frac{9}{7} + 4 \cdot \frac{8}{6} \right) = 6 \cdot \left(\frac{4}{3} - \frac{1/9 - \delta}{63} \right). \tag{5}$$

Setting $\delta = 2/81$, so that $(1/9 - \delta)/63 = \delta/18 = 1/729$, and using (4) and (5) we conclude that in any scenario, $\sum_{w \in V(\gamma)} z(w) \leq 6 \cdot (4/3 - 1/729)$.

Proof of Part (P6). If none of the cases indicated by the previous parts hold then there are no 4, 5 and 6-cycles in $\Gamma(v)$. In other words, all the components containing v in the final family of covers have length at least 7. Using Lemma 4 we conclude that $z(v) \leq \max\{13/10, 9/7\} = 13/10$.

Acknowledgements. This work was supported by Núcleo Milenio Información y Coordinación en Redes ICM/FIC P10-024F. The second author was also supported by a Conicyt fellowship, and the third author by CMM-Basal grant.

References

1. Aggarwal, N., Garg, N., Gupta, S.: A 4/3-approximation for TSP on cubic 3-edge-connected graphs. arXiv:1101.5586v1 (2011)
2. An, H.-C., Kleinberg, R., Shmoys, D.B.: Improving Christofides' Algorithm for the s-t Path TSP. In: STOC 2012 (2012)
3. Barahona, F.: Fractional packing of T-joins. SIAM J. Disc. Math. 17, 661–669 (2004)
4. Barnette, D.: Conjecture 5. In: Tutte, W.T. (ed.) Recent Progress in Combinatorics: Proceedings of the Third Waterloo Conference on Combinatorics (1968)
5. Boyd, S., Sitters, R., van der Ster, S., Stougie, L.: TSP on Cubic and Subcubic Graphs. In: Günlük, O., Woeginger, G.J. (eds.) IPCO 2011. LNCS, vol. 6655, pp. 65–77. Springer, Heidelberg (2011)
6. Bondy, J.A., Murty, U.S.R.: Graph Theory With Applications. Macmillan, London (1976)
7. Christofides, N.: Worst-case analysis of a new heuristic for the travelling salesman problem, Report 388, Graduate School of Industrial Administration, CMU (1976)
8. Gamarnik, D., Lewenstein, M., Sviridenko, M.: An improved upper bound for the TSP in cubic 3-edge-connected graphs. Oper. Res. Lett. 33(5), 467–474 (2005)
9. Goemans, M.X.: Worst-case comparison of valid inequalities for the TSP. Math. Program. 69, 335–349 (1995)
10. Kawarabayashi, K., Ozeki, K.: Spanning closed walks and TSP in 3-connected planar graphs. In: SODA 2012 (2012)
11. Mömke, T., Svensson, O.: Approximating graphic TSP by matchings. In: FOCS 2011 (2011)
12. Mucha, M.: 13/9-approximation for graphic TSP. In: STACS 2012 (2012)
13. Naddef, D., Pulleyblank, W.: Matchings in regular graphs. Discrete Math. 34, 283–291 (1981)
14. Oveis Gharan, S., Saberi, A., Singh, M.: Worst case analysis of a new heuristic for the traveling salesman problem. In: FOCS 2011 (2011)
15. Papadimitriou, C.H., Vempala, S.: On the approximability of the Traveling Salesman Problem. Combinatorica 26(1), 101–120 (2006)
16. Qian, J., Schalekamp, F., Williamson, D.P., van Zuylen, A.: On the Integrality Gap of the Subtour LP for the 1,2-TSP. In: Fernández-Baca, D. (ed.) LATIN 2012. LNCS, vol. 7256, pp. 606–617. Springer, Heidelberg (2012)
17. Schalekamp, F., Williamson, D.P., van Zuylen, A.: A proof of the Boyd-Carr conjecture. In: SODA 2012 (2012)
18. Tait, P.G.: Listing's Topologie. Philosophical Magazine 17, 30–46 (1884)
19. Tutte, W.T.: On hamiltonian circuits. J. London Math. Soc. 7, 169–176 (1946)
20. Wolsey, L.A.: Heuristic analysis, linear programming and branch and bound. Mathematical Programming Studies 13, 121–134 (1980)

FPT Algorithms for Domination in Biclique-Free Graphs

Jan Arne Telle and Yngve Villanger

Department of Informatics, University of Bergen, Norway
{telle,yngvev}@ii.uib.no

Abstract. A class of graphs is said to be biclique-free if there is an integer t such that no graph in the class contains $K_{t,t}$ as a subgraph. Large families of graph classes, such as any nowhere dense class of graphs or d-degenerate graphs, are biclique-free. We show that various domination problems are fixed-parameter tractable on biclique-free classes of graphs, when parameterizing by both solution size and t. In particular, the problems k-DOMINATING SET, CONNECTED k-DOMINATING SET, INDEPENDENT k-DOMINATING SET and MINIMUM WEIGHT k-DOMINATING SET are shown to be FPT, when parameterized by $t + k$, on graphs not containing $K_{t,t}$ as a subgraph. With the exception of CONNECTED k-DOMINATING SET all described algorithms are trivially linear in the size of the input graph.

1 Introduction

The k-dominating set problem is one of the most well-studied NP-complete problems in algorithmic graph theory. Given a graph G and an integer k, we ask if G contains a set S of at most k vertices such that every vertex of G is either in S or adjacent to a vertex of S. To cope with the intractability of this problem it has been studied both in terms of approximability [15] (relaxing the optimality) and fixed-parameter tractability (relaxing the runtime). In this paper we consider also weighted k-domination and the variants asking for a connected or independent k-dominating set.

The k-dominating set problem is notorious in the theory of fixed-parameter tractability (see [10,22,12] for an introduction to parameterized complexity). It was the first problem to be shown $W[2]$-complete [10], and it is hence unlikely to be FPT, i.e. unlikely to have an algorithm with runtime $f(k)n^c$ for f a computable function, c a constant and n the number of vertices of the input graph. However, by restricting the class of input graphs, say to planar graphs, we can obtain FPT algorithms [2], even if the problem remains NP-complete on planar graphs [13]. In the race to find the boundary between FPT and W-hardness one typically wants the weakest possible restriction when proving FPT and the strongest possible restriction when proving W-hardness. In this paper, we push the tractability frontier forward by considering the above variants of k-dominating set on t-biclique free graphs and showing that they are FPT when parameterized by $k + t$. The t-biclique free graphs are those that do not contain

L. Epstein and P. Ferragina (Eds.): ESA 2012, LNCS 7501, pp. 802–812, 2012.
© Springer-Verlag Berlin Heidelberg 2012

$K_{t,t}$ as a subgraph, and to the best of our knowledge, they form the largest class of graphs for which FPT algorithms are known for k-dominating set. Our algorithms are simple and rely on results from extremal graph theory that bound the number of edges in a t-biclique free graph, see the Bollobás book [4].

The parameterized complexity of the dominating set problem has been heavily studied with the tractability frontier steadily pushed forward by enlarging the class of graphs under consideration. One such line of improvements for k-dominating set consists of the series of FPT algorithms starting with planar graphs by Alber et al. [2], followed by bounded genus graphs by Ellis et al. [11], H-minor free graphs by Demaine et al. [8], bounded expansion graphs by Nesetril and Ossona de Mendez [19], and culminating in the FPT algorithm for nowhere dense classes of graphs by Dawar and Kreutzer [7]. See Figure 1. Alon and Gutner [3] have shown that k-dominating set on d-degenerate graphs parameterized by $k + d$ is FPT. Nowhere dense classes and d-degenerate graph classes are incomparable and Dawar and Kreutzer [7] mention that: 'it would be interesting to compare nowhere dense classes of graphs to graph classes of bounded degeneracy'. In this paper we base such a comparison on the fact that any nowhere dense class of graphs or d-degenerate class of graphs is a t-biclique-free class of graphs, for some t. See Section 2. Let us remark that also for a nowhere dense class $\mathcal{C}$ of graphs there is a parameter analogous to the d in d-degenerate graphs and the t in t-biclique-free graphs, with Dawar and Kreutzer [7] mentioning that: '...the exact parameter dependence of the algorithms depends on the function h, which is determined by the class of structures $\mathcal{C}$'. A relation between the different graph class properties can be seen in Figure 1.

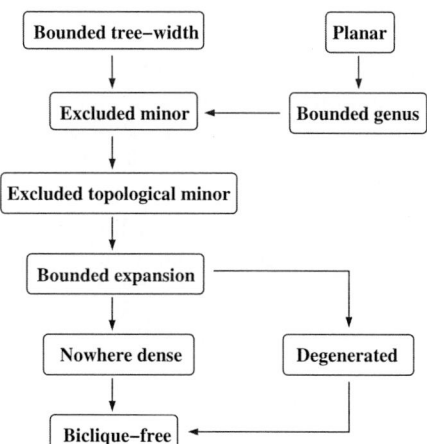

Fig. 1. Inclusion relations between some of the mentioned graph class properties. We refer to Nesetril and Ossona de Mendez [20] for a more refined view.

Raman and Saurabh [24] have shown that k-dominating set is $W[2]$-hard on K_3-free graphs and FPT on graphs not containing $K_{2,2}$ as a subgraph (i.e. 2-biclique-free graphs). Philip et al. [23] have shown that k-dominating set on graphs not containing $K_{i,j}$ as a subgraph (which they call $K_{i,j}$-free) has a polynomial kernel, which is stronger than simply saying that it is FPT. However, their algorithm is parameterized by k only and considers $i + j$ to be a constant. They mention explicitly that: 'Another challenge is to...get a running time of the form $O(n^c)$ for $K_{i,j}$-free graphs where c is independent of i and j.' In this paper we do not directly meet this challenge but instead do something related. By showing that k-dominating set on t-biclique free graphs is FPT when parameterized by $k + t$, we generalize all FPT results for k-domination on restricted graph classes that we have found in the literature. Note that we could not expect to meet the challenge of a polynomial kernel when parameterizing by $k + t$, as Dom et al. [9] have shown that the k-Dominating Set problem on d-degenerate graphs (a subclass of $(d + 1)$-biclique-free graphs) does not have a kernel of size polynomial in both d and k unless the polynomial hierarchy collapses to the third level.

Our result extends to showing that connected k-domination and independent k-domination on t-biclique-free graphs are both FPT when parameterized by $k + t$. Note that Cygan et al. [6] have shown that connected k-domination has no polynomial kernel on graphs of degeneracy 2 (a subclass of 3-biclique-free graphs) unless the polynomial hierarchy collapses to the third level. For connected k-domination we use a subroutine developed by Misra et al. [18] for the Group Steiner Tree problem. The FPT borderline for connected k-domination and independent k-domination prior to our work resembled the one for k-domination, with Dawar and Kreutzer [7] showing that both problems are FPT on nowhere dense classes of graphs and Golovach and Villanger [14] showing that both problems are FPT on d-degenerate graphs when parameterized by $k + d$. Our algorithm generalizes these results.

Our result extends also to weighted k-domination. Alon and Gutner [3] show that weighted k-domination on d-degenerate graphs parameterized by $d + k$ is FPT but fixed-parameter tractability was not known for nowhere dense classes of graphs prior to our result for t-biclique free graphs.

A famous open problem in parameterized complexity is the question if there is an FPT algorithm deciding if a graph G is k-biclique-free and in the Conclusion section we briefly mention this open problem in light of our algorithms.

2   Graph Classes and Problems

We use standard graph theory terminology. For a graph $G = (V, E)$ and $S \subseteq V$ we denote by $N[S]$ the vertices that are either in S or adjacent to a vertex of S, and we denote by $G[S]$ the subgraph of G induced by S. We denote always $|V| = n$ and $|E| = m$. The distance between two vertices is the number of edges in a shortest path linking them. Let us consider some classes of graphs.

Definition 1 (Degenerate classes). *A class of graphs $\mathcal{C}$ is said to be degenerate if there is an integer d such that every induced subgraph of any $G \in \mathcal{C}$ has a vertex of degree at most d.*

Many interesting families of graphs are degenerate. For example, graphs embeddable on some fixed surface, degree-bounded graphs and non-trivial minor-closed families of graphs. Another broad property of graph classes, recently introduced by Nesetril and Ossona de Mendez [21], is the property of being nowhere dense. There are several equivalent definitions, we use the following based on the concept of a shallow minor. The radius of a connected graph G is the minimum over all vertices v of G of the maximum distance between v and another vertex. For non-negative integer r a graph H is a shallow minor at depth r of a graph G if there exists a subgraph X of G whose connected components have radius at most r, such that H is a simple graph obtained from G by contracting each component of X into a single vertex and then taking a subgraph.

Definition 2 (Nowhere dense classes). *A class of graphs C is said to be nowhere dense if there is a function f such that for every $r \geq 0$ the graph $K_{f(r)}$ is not a shallow minor at depth r of any $G \in C$.*

Many interesting families of graphs are nowhere dense, like graphs of bounded expansion and graphs locally excluding a minor. We now consider a class of graphs which was shown by Philip et al. [23] to strictly contain the degenerate classes of graphs and show that it also contains the nowhere dense classes. We denote by $K_{t,t}$ the complete bipartite graph with t vertices on each side of the bipartition.

Definition 3 (Biclique-free classes). *A class of graphs C is said to be t-biclique-free, for some $t > 0$, if $K_{t,t}$ is not a subgraph of any $G \in C$, and it is said to be biclique-free if it is t-biclique-free for some t.*

Fact 1. *Any degenerate or nowhere dense class of graphs is biclique-free, but not vice-versa.*

Proof. For completeness we give a full proof. We first show that if a class of graphs C is degenerate then it is biclique-free. Assume that every induced subgraph of any $G \in C$ has a vertex of degree at most d. Then $K_{d+1,d+1}$ cannot be a subgraph of any $G \in C$ since its vertices would induce a subgraph where every vertex has degree larger than d.

We next show that if a class of graphs C is nowhere dense then it is biclique-free. Assume a function f such that for every $r \geq 0$ the graph $K_{f(r)}$ is not a shallow minor at depth r of any $G \in C$. Then $K_{f(1)-1,f(1)-1}$ cannot be a subgraph of any $G \in C$ since we in such a subgraph could contract a matching of $f(1) - 2$ edges crossing the bipartition and get $K_{f(1)}$ as a shallow minor at depth 1.

Finally, we show a biclique-free class of graphs C that is neither degenerate nor nowhere dense. For any value of $k \geq 2$ there exists a k-regular graph of girth 5, call it R_k, see e.g. [1]. Since $K_{2,2}$ is a 4-cycle the class $\{R_k : k \geq 2\}$ is biclique-free but not degenerate. Let S_k be the graph obtained by subdividing once each edge of K_k. Since S_k contains K_k as a shallow minor at depth 1 the class $\{S_k : k \geq 2\}$ is biclique-free but not nowhere dense. Let the class C contain, for each value of $k \geq 2$, the graph we get by taking one copy of S_k and one copy of R_k and adding a single edge between some vertex of S_k and some vertex of R_k to make it a connected graph. The class $\mathcal{C}$ is biclique-free but it is neither degenerate nor nowhere dense.

A k-dominating set of a graph $G = (V, E)$ is a set $S \subseteq V$ with $|S| = k$ and $N[S] = V$. We will be considering parameterized versions of several domination-type problems in biclique-free classes of graphs. In each case we ask for a dominating set of size exactly k but note that an algorithm for this problem can also be used to find the smallest k.

k-DOMINATING SET
Input: Integers k, t and a t-biclique-free graph G.
Parameter: $k + t$
Question: Is there a k-dominating set in G?

CONNECTED k-DOMINATING SET
Input: Integers k, t and a t-biclique -free graph G.
Parameter: $k + t$
Question: Is there a k-dominating set S in G with $G[S]$ connected?

INDEPENDENT k-DOMINATING SET
Input: Integers k, t and a t-biclique -free graph G.
Parameter: $k + t$
Question: Is there a k-dominating set S in G with $G[S]$ having no edges?

WEIGHTED k-DOMINATING SET
Input: Integers k, t and a t-biclique -free graph G with positive vertex weights.
Parameter: $k + t$
Output: A k-dominating set S in G with minimum sum of weights, if it exists.

3   Simple Algorithms for Domination in Biclique-Free Graph Classes

When studying the boundary between W-hardness and fixed parameter tractability one tries to find the strongest restriction such that the problem remains hard and the weakest restriction such that the problem becomes fixed parameter tractable. In most cases the arguments on both sides become more and more involved as one approaches the boundary. In this section we give fairly simple algorithms for variants of the k-dominating set problem on the class of biclique-free graphs.

3.1   Extremal Combinatorics and High Degree Vertices

Before starting to describe the algorithm we need some tools from extremal combinatorics.

Bollobás in his book "Extremal Graph Theory"[4] discusses the so called Zarankiewicz Problem of giving an upper bound for the number of edges in graphs where $K_{s,t}$ is forbidden as a subgraph for integers s, t. It is worthwhile to point out that there is a significant difference between forbidding a K_t and a $K_{s,t}$, as a graph without K_t may contain $\Omega(n^2)$ edges while a graph without $K_{s,t}$ contains $O(n^{2-1/t})$ edges [4]. Another difference that we have already mentioned is that the k-dominating set problem is $W[2]$-hard on K_3 free graphs [24] while it is fixed parameter tractable on t-biclique-free graphs. The proposition below turns out to be very useful when studying graphs without $K_{s,t}$ as a subgraph.

Proposition 1 (Bollobás [4] VI.2). *For integers s, t let $G = (V_1, V_2, E)$ be a bipartite graph not containing $K_{s,t}$ as a subgraph where $|V_1| = n_1$ and $|V_2| = n_2$. Then for $2 \leq s \leq n_2$ and $2 \leq t \leq n_1$ we have that $|E| < (s-1)^{\frac{1}{t}}(n_2 - t + 1)n_1^{1-\frac{1}{t}} + (t-1)n_1$.*

A convenient consequence of Proposition 1 is that we can use it to say something about the number of high degree vertices in graphs that are t-biclique-free. For ease of notation let $f(k,t) = 2k(t + 1 + (4k)^t)$.

Lemma 1. *Let k and t be positive integers and let G be a t-biclique-free graph on n vertices where $f(k,t) \leq n$. Then there are less than $(4k)^t$ vertices of G with degree at least $\frac{n-k}{k}$.*

Proof. On the contrary let us assume that there exists a vertex set $X \subset V$ where $(4k)^t = |X|$ and each vertex $v \in X$ has degree at least $\frac{n-k}{k}$ in G. Clearly such a vertex set also exists if there are more then $(4k)^t$ vertices of degree at least $\frac{n-k}{k}$. Let $Y = V \setminus X$ and define $x = |X|$ and thus $|Y| = n - x$. There are now at least $x(\frac{n-k}{k} - x)$ edges in G between vertex sets X and Y.

As G is a t-biclique-free graph we know by Proposition 1 that the number of edges between X and Y in G is less than $(t-1)^{\frac{1}{t}}(n-x-t+1)x^{1-\frac{1}{t}} + (t-1)x$ which is trivally at most $2(n-x)x^{1-\frac{1}{t}} + tx$. As $x = (4k)^t$ we aim for a contradiction by starting from the observation that,

$$x(\tfrac{n-k}{k} - x) < 2(n-x)x^{1-\frac{1}{t}} + tx$$
$$(\tfrac{n-k}{k} - x) < 2(n-x)x^{-\frac{1}{t}} + t$$
$$n/k < 2nx^{-\frac{1}{t}} - 2x^{1-\frac{1}{t}} + t + 1 + x$$
$$1 < \tfrac{2k}{x^{1/t}} - (2kx^{1-\frac{1}{t}})/n + k(t+1+x)/n$$
$$1 < \tfrac{2k}{(4k)^{t/t}} - (2k(4k)^{t(1-\frac{1}{t})})/n + k(t+1+(4k)^t)/n$$
$$1 < \tfrac{1}{2} + k(t+1+(4k)^t)/n$$
$$\tfrac{1}{2} < k(t+1+(4k)^t)/n$$

The assumption was that $f(k,t) = 2k(t + 1 + (4k)^t) \leq n$ which means that $k(t+1+(4k)^t)/n \leq \frac{1}{2}$ and we get the contradiction.

3.2   Enumeration of Partial Dominating Sets

A simple way to decide if a graph has a dominating set of size k is to enumerate all inclusion minimal dominating sets and check if one of them is of size at most

k. If the goal is an FPT algorithm this approach fails already for planar graphs as they may contain $O(n^k)$ minimal dominating sets of size $k + 1$.[1] Our way around this obstruction is to build the dominating sets in stages by enumerating only some subsets of each dominating set of size at most k in such a way that all remaining vertices can be classified into a "small" number of equivalence classes. This provides a framework where several variants of domination can be discussed and compared.

Like before let $f(k,t) = 2k(t + 1 + (4k)^t)$.

Lemma 2. *For positive integers k and t let G be a t-biclique-free graph on n vertices where $f(k,t) \leq n$. Then there exists an algorithm that in time $O((n + m)k \cdot (4k)^{tk})$ outputs a family of vertex subsets $\mathcal{F}$ such that $|\mathcal{F}| \leq (4k)^{tk}$ and for any vertex set S where $|S| \leq k$ and $V = N[S]$ there is $X \in \mathcal{F}$ with $X \subseteq S$ such that X dominates at least $n - f(k,t)$ vertices of G, i.e. $|N[X]| \geq n - f(k,t)$.*

Proof. For a graph G let us say that a family $\mathcal{F}$ of vertex subsets satisfies invariant $\mathcal{D}$ (for Domination) if for every $S \subseteq V$ such that $|S| \leq k$ and $N[S] = V$ there exists $X \in \mathcal{F}$ such that $X \subseteq S$. A family $\mathcal{F}$ of vertex subsets is defined to be of branch depth i if it holds for every $X \in \mathcal{F}$ that if $|V \setminus N[X]| > f(k,t)$ then $|X| \geq i$. For short we will denote a family $\mathcal{F}$ of vertex subsets satisfying invariant $\mathcal{D}$ of branch depth i as $\mathcal{F}_{\mathcal{D},i}$. Note that to prove the lemma it suffices to find in $O((n+m)k \cdot (4k)^{tk})$ time a family $\mathcal{F}_{\mathcal{D},k}$ since by the Domination invariant there is for every S with $|S| \leq k$ and $N[S] = V$ some $X \in \mathcal{F}_{\mathcal{D},k}$ with $X \subseteq S$, and since this is at branch depth k we know that if $|X| < k$ then $|N[X]| \geq n - f(k,t)$ and if $|X| \geq k$ then necessarily $X = S$ so that $|N[X]| = n$.

We will prove the lemma by induction on i, where the induction hypothesis is that a family $\mathcal{F}_{\mathcal{D},i}$ of cardinality at most $(4k)^{ti}$ can be computed in $O((n+m)i \cdot (4k)^{ti})$ time. The base case is obtained by simply observing that $\{\emptyset\}$ satisfies invariant $\mathcal{D}$ and is of branch depth 0, i.e. we can take $\mathcal{F}_{\mathcal{D},0} = \{\emptyset\}$.

Now for the induction step. Let us assume that a family $\mathcal{F}_{\mathcal{D},i-1}$ of cardinality $(4k)^{t(i-1)}$ is provided and let us argue how $\mathcal{F}_{\mathcal{D},i}$ can be obtained in $\mathcal{O}((n + m) \cdot (4k)^t)$ time. Let X be one of the at most $(4k)^{t(i-1)}$ elements in $\mathcal{F}_{\mathcal{D},i-1}$ of cardinality $i - 1$ where $|V \setminus N[X]| > f(k,t)$. Every dominating set $S \supset X$ with $|S| \leq k$ has to dominate all vertices of $V \setminus N[X]$. Thus, at least one of the vertices of $S \setminus X$ has to dominate at least $\frac{|V \setminus N[X]|}{|S \setminus X|}$ of these vertices. Let Z be the set of vertices in $V \setminus X$ that dominates at least $\frac{|V \setminus N[X]|}{|S \setminus X|}$ of the vertices in $V \setminus N[X]$. By Lemma 1 $|Z| \leq (4k)^t$ and it is a trivial task to obtain Z in $O(n + m)$ time provided G and X.

Now update $\mathcal{F}_{\mathcal{D},i-1}$ by removing X and adding set $X \cup \{w\}$ for every $w \in Z$. As every dominating set $S \supset X$ with $|S| \leq k$ contains at least one vertex of Z

---

[1] Consider an independent set of size n that is partitioned into k colours of size n/k. For each colour add one vertex that is adjacent to all vertices of this colour and add one additional vertex u adjacent to all n vertices in the independent set. We obtain $(n/k)^k$ minimal dominating sets by selecting u and exactly one vertex from each colour class.

the thus updated set $\mathcal{F}_{\mathcal{D},i-1}$ satisfies the invariant $\mathcal{D}$. Repeat this procedure for every set $X \in \mathcal{F}_{\mathcal{D},i-1}$ such that $|X| = i - 1$ and $|V \setminus N[X]| > f(k,t)$. As there are at most $(4k)^{t(i-1)}$ such sets of size $i-1$ and each of these are replaced by at most $(4k)^t$ sets the resulting family of sets is of size at most $(4k)^t \cdot |\mathcal{F}_{\mathcal{D},i-1}|$ and is of branching depth i. This completes the proof.

Each element $X \in \mathcal{F}_{\mathcal{D},k}$ will now be used to define an equivalence relation on $V \setminus X$ based on the possible neighbors among the undominated vertices $V \setminus N[X]$.

Definition 4. *For a graph $G = (V,E)$ and a vertex set $X \subset V$ let $W = V \setminus N[X]$. Let $\equiv_X$ be the binary relation on $V \setminus X$ with $u \equiv_X v$ if $N[u] \cap W = N[v] \cap W$. This is an equivalence relation and we say that $v \in V \setminus X$ belongs to the equivalence class $\mathcal{E}$ corresponding to $N[v] \cap W$.*

Definition 5. *For a graph $G = (V,E)$ and a vertex set $X \subset V$ a set $A = \{\mathcal{E}_1, \mathcal{E}_2, \ldots, \mathcal{E}_r\}$ of equivalence classes of $\equiv_X$ is called dominating if mapping each element $\mathcal{E}_i \in A$ to an arbitrary vertex $v_i \in \mathcal{E}_i$ (it does not matter which since all vertices in $\mathcal{E}_i$ have the same closed neighborhoods in W) we have $V = N[X] \cup \bigcup_{i=1}^{r} N[v_i]$. Let $\mathcal{A}_{X,r}$ be defined as the set of all equivalence classes of cardinality r that are dominating.*

Lemma 3. *For positive integers k and t let G be a t-biclique-free graph on n vertices where $f(k,t) \leq n$. Let X be an element of the family $\mathcal{F}_{\mathcal{D},k}$ and let $r \leq k - |X|$. Then $|\mathcal{A}_{X,r}| \leq 2^{r \cdot f(k,t)}$ and $\mathcal{A}_{X,r}$ can be computed in $O((n+m) \cdot 2^{r \cdot f(k,t)})$ time.*

Proof. Let $W = V \setminus N[X]$. Note $|W| \leq f(k,t)$ and hence there are at most $2^{f(k,t)}$ subsets of W. For each vertex $v \in V \setminus X$ compute $N[v] \cap W$ and add v to its equivalence class. Note there are at most $2^{f(k,t)}$ equivalence classes. For each of the $(2^{f(k,t)})^r$ possible subsets $A = \{\mathcal{E}_1, \mathcal{E}_2, \ldots, \mathcal{E}_r\}$ of equivalence classes add A to $\mathcal{A}_{X,r}$ if A is dominating. The running time for this is $O((n+m) \cdot 2^{r \cdot f(k,t)})$.

3.3 Various Domination Problems on Biclique-Free Graph Classes

This subsection combines the previous combinatorial results into algorithms for different domination problems.

Theorem 1. *Given a t-biclique-free graph $G = (V,E)$ the following problems, as defined in Section 2, are fixed parameter tractable when parameterizing by $k + t$:*

1. *k-DOMINATING SET, $O((n+m) \cdot 2^{O(tk^2(4k)^t)})$ time,*
2. *CONNECTED k-DOMINATING SET,*
3. *INDEPENDENT k-DOMINATING SET,*
4. *WEIGHTED k-DOMINATING SET.*

Proof. In all cases, if $|V| < f(k,t) = 2k(t+1+(4k)^t)$ we can simply enumerate all $\binom{f(k,t)}{k}$ vertex subsets of size k and test in $O((n+m) \cdot k^2)$ time if the specific properties for the problem is satisfied. Otherwise we first enumerate the family $\mathcal{F}_{\mathcal{D},k}$ containing at most $(4k)^{tk}$ elements in $O((n+m)k \cdot (4k)^{tk})$ time using Lemma 2. For each element $X \in \mathcal{F}_{\mathcal{D},k}$ we apply Lemma 3 and compute $\mathcal{A}_{X,r}$ which is of size at most $2^{r \cdot f(k,t)}$ in time $O((n+m) \cdot 2^{r \cdot f(k,t)})$. The value $r \leq k - |X|$ for which this is computed will depend on the problem to be solved in the following way:

1. By definition, there is a k-DOMINATING SET in G if and only if there is some $X \in \mathcal{F}_{\mathcal{D},k}$ for which $\mathcal{A}_{X,k-|X|} \neq \emptyset$. Applying Lemma 3 $(4k)^{tk}$ times give a total running time of $O((n+m) \cdot 2^{O(tk^2(4k)^t)})$.

2. In Misra et al. [18] an FPT algorithm is given for the following Group Steiner Tree problem:

   - Given a graph H, subsets of vertices $T_1, T_2, ..., T_l$, and an integer p, with parameter $l \leq p$, does there exist a subgraph of H on p vertices that is a tree T and includes at least one vertex from each $T_i, 1 \leq i \leq l$?

   We claim that our input graph G has a connected k-dominating set if and only if there exists some $X \in \mathcal{F}_{\mathcal{D},k}$ and some $r \in \{0, 1, ..., k - |X|\}$ and some $A \in \mathcal{A}_{X,r}$ such that the following Group Steiner Tree problem has a Yes-answer:

   - Let $X = \{v_1, v_2, ..., v_{|X|}\}$ and $A = \{\mathcal{E}_1, \mathcal{E}_2, ..., \mathcal{E}_r\}$. Set $H = G$ and set the vertices of the equivalence class $\mathcal{E}_i$ to form a subset T_i, for each $1 \leq i \leq r$ and additionally $T_i = \{v_i\}$ for $r+1 \leq i \leq r+|X|$. We thus have $l = r + |X|$ and set $p = k$.

   Let us argue for the claim. For one direction assume that S is a connected k-dominating set of G. We then know there exists $X \subseteq S$ with $X \in \mathcal{F}_{\mathcal{D},k}$. Let $A = \{\mathcal{E}_1, \mathcal{E}_2, ..., \mathcal{E}_r\}$ be the equivalence classes of $\equiv_X$ containing at least one vertex of S. Note that $0 \leq r \leq k - |X|$ and $A \in \mathcal{A}_{X,r}$. The Group Steiner Tree problem we formed from this X and A has a Yes-answer, for example by taking any subgraph of $G[S]$ inducing a tree. For the other direction, if the Group Steiner Tree problem formed by some X and A has a Yes-answer by some tree T then the set of vertices of this tree T will necessarily be a connected k-dominating set of G.

   The total running time will be FPT in $k+t$ as the possible choices of X and A are a function of $k + t$ only and the Group Steiner Tree problem is FPT.

3. For every $X \in \mathcal{F}_{\mathcal{D},k}$ such that $G[X]$ is an independent set, check by brute force in $O((n+m) \cdot 2^{f(k,t)})$ if there exists an independent dominating set of size $k - |X|$ in the graph $G[V \setminus N[X]]$. The runtime will be linear FPT.

4. For each vertex $v \in V$ let $w(v)$ be the weight of the vertex. The weight $w(A)$ of a set $A = \{\mathcal{E}_1, \mathcal{E}_2, ..., \mathcal{E}_r\}$ of equivalence classes is defined as $\sum_{i=1}^{r} w(v_i)$ where class $\mathcal{E}_i$ is mapped to the minimum weight vertex $v_i \in \mathcal{E}_i$. The k-dominating set of minimum weight is then obtained by minimizing over all $X \in \mathcal{F}_{\mathcal{D},k}$ such that $\mathcal{A}_{X,k-|X|} \neq \emptyset$ the value of $\min_{A \in \mathcal{A}_{X,k-|X|}} w(A) + \sum_{v \in X} w(v)$. The runtime will be linear FPT.

4   Conclusion

In this paper we have pushed forward the FPT boundary for the k-dominating set problem, and some of its variants, to the case of t-biclique-free graphs parameterized by $k + t$. This generalizes all FPT algorithms for k-dominating set on special graph classes we have found in the literature, in particular the algorithms for the uncomparable classes d-degenerate and nowhere dense. By the result of Dom et al. [9] this problem does not have a polynomial kernel unless the polynomial hierarchy collapses to the third level.

The basic idea for our algorithm is to branch on vertices of sufficiently high degree until the remaining vertices can be partitioned into a small number of equivalence classes. Unsurprisingly, the applied techniques have some similarities with the algorithms for the d-degenerate classes and nowhere dense classes. Usually the algorithms become more complicated as they apply to more general classes of graphs. In this paper our generalized algorithm is still fairly simple with a running time that can trivially be summed up to be $O((n + m) \cdot 2^{O(tk^2(4k)^t)})$.

The described algorithm resolves the k-dominating set problem on all t-biclique-free graphs. As the algorithm only uses extremal combinatorial properties of the graph class there is no need to verify that the input graph indeed is t-biclique-free. Consider the following somewhat strange parameterized problem:

k-DOMINATING SET OR k-BICLIQUE
Input: Graph G and integer k.
Parameter: k
Output: Either 'G is not k-biclique-free' or 'G has a k-dominating set' or 'G does not have a k-dominating set'

By using our algorithm for k-dominating set on an arbitrary graph G, with $k = t$, we actually get an FPT algorithm for k-DOMINATING SET OR k-BICLIQUE, since we either conclude by Lemma 1 that G is not k-biclique-free or we have few high degree vertices and are able to decide if the graph has a dominating set of size k. This gives us a somewhat strange result, resembling the situation for the Ramsey problem asking if a graph has either a clique of size k or an independent set of size k [17,16]. The k-dominating set problem is $W[2]$-hard on general graphs and Bulatov and Marx [5] gives reason to believe that deciding if a graph G is k-biclique-free is not FPT on general graphs. Nevertheless, one of the two problems can always be resolved in FPT time.

References

1. Abreu, M., Funk, M., Labbate, D., Napolitano, V.: A family of regular graphs of girth 5. Discrete Mathematics 308, 1810–1815 (2008)
2. Alber, J., Bodlaender, H.L., Fernau, H., Kloks, T., Niedermeier, R.: Fixed parameter algorithms for dominating set and related problems on planar graphs. Algorithmica 33, 461–493 (2002)
3. Alon, N., Gutner, S.: Linear time algorithms for finding a dominating set of fixed size in degenerated graphs. Algorithmica 54, 544–556 (2009)

4. Bollobás, B.: Extremal graph theory, Dover Books on Mathematics. Dover Publications (2004)
5. Bulatov, A.A., Marx, D.: Constraint Satisfaction Parameterized by Solution Size. In: Aceto, L., Henzinger, M., Sgall, J. (eds.) ICALP 2011, Part I, LNCS, vol. 6755, pp. 424–436. Springer, Heidelberg (2011)
6. Cygan, M., Pilipczuk, M., Pilipczuk, M., Wojtaszczyk, J.O.: Kernelization Hardness of Connectivity Problems in d-Degenerate Graphs. In: Thilikos, D.M. (ed.) WG 2010. LNCS, vol. 6410, pp. 147–158. Springer, Heidelberg (2010)
7. Dawar, A., Kreutzer, S.: Domination problems in nowhere-dense classes. In: Kannan, R., Kumar, K.N. (eds.). FSTTCS LIPIcs, Schloss Dagstuhl - Leibniz-Zentrum fuer Informatik, vol. 4, pp. 157–168 (2009)
8. Demaine, E.D., Fomin, F.V., Hajiaghayi, M.T., Thilikos, D.M.: Subexponential parameterized algorithms on bounded-genus graphs and h-minor-free graphs. J. ACM 52, 866–893 (2005)
9. Dom, M., Lokshtanov, D., Saurabh, S.: Incompressibility through Colors and IDs. In: Albers, S., Marchetti-Spaccamela, A., Matias, Y., Nikoletseas, S., Thomas, W. (eds.) ICALP 2009, Part I, LNCS, vol. 5555, pp. 378–389. Springer, Heidelberg (2009)
10. Downey, R.G., Fellows, M.R.: Parameterized Complexity. Springer (1999)
11. Ellis, J.A., Fan, H., Fellows, M.R.: The dominating set problem is fixed parameter tractable for graphs of bounded genus. J. Algorithms 52, 152–168 (2004)
12. Flum, J., Grohe, M.: Parameterized Complexity Theory. Springer (2006)
13. Garey, M.R., Johnson, D.S.: Computers and Intractability: A Guide to the Theory of NP-Completeness. W. H. Freeman (1979)
14. Golovach, P.A., Villanger, Y.: Parameterized Complexity for Domination Problems on Degenerate Graphs. In: Broersma, H., Erlebach, T., Friedetzky, T., Paulusma, D. (eds.) WG 2008. LNCS, vol. 5344, pp. 195–205. Springer, Heidelberg (2008)
15. Johnson, D.S.: Approximation algorithms for combinatorial problems. J. Comput. Syst. Sci. 9, 256–278 (1974)
16. Khot, S., Raman, V.: Parameterized complexity of finding subgraphs with hereditary properties. Theor. Comput. Sci. 289, 997–1008 (2002)
17. Kratsch, S.: Co-nondeterminism in compositions: a kernelization lower bound for a ramsey-type problem. In: Rabani, Y. (ed.) SODA, pp. 114–122. SIAM (2012)
18. Misra, N., Philip, G., Raman, V., Saurabh, S., Sikdar, S.: FPT Algorithms for Connected Feedback Vertex Set. In: Rahman, M. S., Fujita, S. (eds.) WALCOM 2010. LNCS, vol. 5942, pp. 269–280. Springer, Heidelberg (2010)
19. Nesetril, J., de Mendez, P.O.: Structural properties of sparse graphs. Building bridges between Mathematics and Computer Science, vol. 19. Springer (2008)
20. Nesetril, J., de Mendez, P.O.: First order properties on nowhere dense structures. J. Symb. Log. 75, 868–887 (2010)
21. Nesetril, J., de Mendez, P.O.: On nowhere dense graphs. Eur. J. Comb. 32, 600–617 (2011)
22. Niedermeier, R.: Invitation to fixed-parameter algorithms. Oxford Lecture Series in Mathematics and Its Applications. Oxford University Press (2006)
23. Philip, G., Raman, V., Sikdar, S.: Solving Dominating Set in Larger Classes of Graphs: FPT Algorithms and Polynomial Kernels. In: Fiat, A., Sanders, P. (eds.) ESA 2009. LNCS, vol. 5757, pp. 694–705. Springer, Heidelberg (2009)
24. Raman, V., Saurabh, S.: Short cycles make w-hard problems hard: Fpt algorithms for w-hard problems in graphs with no short cycles. Algorithmica 52, 203–225 (2008)

Maximum Flow Networks for Stability Analysis of LEGO® Structures

Martin Waßmann and Karsten Weicker

HTWK Leipzig, IMN, Postfach 301166, 04251 Leipzig, Germany
http://portal.imn.htwk-leipzig.de/fakultaet/weicker/

Abstract. To determine the stability of LEGO® structures is an interesting problem because the special binding mechanism prohibits the usage of methods of structural frame design or dynamic physics engines. We propose a new two-phase approach where instances of maximum-flow networks are constructed. In a first phase, the distribution of compressive and tensile forces is computed which is used in a second phase to model the moments within the structure. By solving the maximum-flow networks we can use the resulting flow as a sufficient criterion for the stability of the structure. The approach is demonstrated for two exemplary structures which outperform previous results using a multi-commodity flow network.

Keywords: Buildable objects, physics simulation, maximum flow, push-relabel algorithm.

1 Introduction

The toy construction set with LEGO® bricks has been introduced in 1949 and proved to be a world-wide success not only for children. In the field of automated design using evolutionary algorithms, LEGO® is used as an interesting application domain. In the late 1990s, the buildable objects experiments of Funes and Pollack [3,4,5] gained attention, where impressively long crane arms are constructed.

In an attempt to both reproduce the results of Funes and Pollack and apply more advanced techniques for the representation and optimisation of LEGO® structures, we examined the existing heuristic for the analysis of the stability. However, the error rate, with which stable structures are dismissed, appears to be to high.

As a consequence, the stability analysis is the major challenge: The physics of structural frame design cannot be applied since LEGO® does not follow the rules of mechanics with punctiform conjunctions. Also existing dynamic physics engine cannot be applied since the successive and isolated consideration of the various forces within a structure leads to insufficient exactness.

Therefore, we model the forces within the LEGO® structure as two maximum-flow networks that are solved using the push-relabel algorithm [6] with the gap relabelling heuristic and global relabelling [1]. The resulting maximum flow indicates whether the structure is stable.

L. Epstein and P. Ferragina (Eds.): ESA 2012, LNCS 7501, pp. 813–824, 2012.

The paper is organised as follows. After a short review of the related literature in Section 2 we present a physical model of the forces within a LEGO® structure in Section 3. Section 4 is devoted to the new approach using maximum-flow networks. After a short discussion of a few results gained with the technique in Section 5 we conclude in Section 6.

2   Related Work

The analysis of the stability of LEGO® structures was realised by Funes and Pollack [3,4,5] by isolating the single forces and moments. For each force (caused by gravitation or an external load) and each brick within the structure, the resulting moment on the brick is computed. Now these moments and their counteracting moments are used to formulate a flow network for each force in which the moments are propagated towards the bearing. As a consequence, each force may generate a transported moment at the joint of any two bricks. However, the sum of all those transported moments (over all forces) must be below the maximum moment capacity for each joint. Formally, this model corresponds to a multicommodity network flow problem. This approach is problematic in two ways: first, the problem is NP-hard which is the reason for approximation algorithms only, and, second, this approach overestimates the moments within the structures since the isolation of forces does not consider compensation of opposing moments.

In our approach, we use maximum-flow networks $N = (G, c, s, t)$ where $G = (V, E)$ is a directed graph, $c : E \rightarrow \mathbb{R}$ the capacity, $s \in V$ the source vertex, and $t \in V$ the target vertex. The simple Edmonds-Karp algorithm [2] solves the problem with time complexity $\mathcal{O}(|E|^2 \cdot |V|)$. More competitive is the push-relabel algorithm by Goldberg and Tarjan [6] with time complexity $\mathcal{O}(|V|^2 \cdot \sqrt{|E|})$. We use in addition the version of the push-relabel algorithm by Cherkassky and Goldberg [1] where two heuristics are used: The gap relabelling heuristic simplifies the used label values in the case where certain vertices cannot reach the target vertex which may be detected if a label value is found not to be in use. For the gap relabelling heuristic, a special data structure, the residual graph, is introduced and the label values are initialised using a reverse breadth-first search.

To a certain degree, our approach is the reverse counterpart to the technique of network flow programming where certain instances of flow network problems may be transformed into instances of integer programming. In this paper, we derive the physical model of the forces within a LEGO® structure and transform the resulting system of equations into two maximum-flow network problem instances. By solving these instances we get an approximation on the stability of the LEGO® structure.

3   Physics of LEGO® Structures

Figure 1a shows an exemplary LEGO® brick. Within our problem domain, all forces act vertically, i.e. transmitted forces affect bricks at the horizontal positions defined by knobs of the brick whereas external loads are applied at the

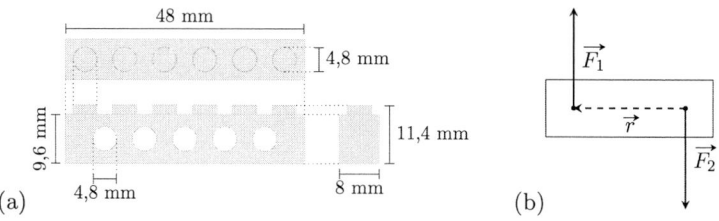

Fig. 1. (a) shows a LEGO® brick and (b) two parallel forces resulting in a moment

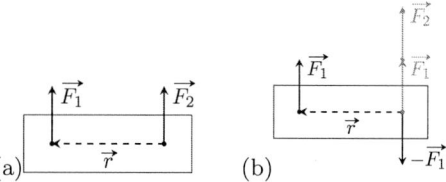

Fig. 2. (a) shows two forces that can be viewed in (b) as a compound of a single force and a moment

holes. The horizontal width of "one knob" is given by the constant $c_x = 8$mm. Figure 1b shows two parallel forces in opposite directions that can be combined as a moment iff $\overrightarrow{F_1} = \overrightarrow{F_2}$.

The combination of two forces into the same direction as shown in Figure 2a is more complicated. By adding the forces $\overrightarrow{F_1}$ and $-\overrightarrow{F_1}$ which equalise each other, we can view the two forces as a single tensile force $\overrightarrow{F_1} + \overrightarrow{F_2}$ and a pair of forces resulting in a moment (shown in Figure 2b).

We want to show with the small example in Figure 3 how we can express the forces and counteracting forces within a LEGO® structure as system of equations. The structure is given as upright projection and as graph where the vertices are triples giving the x- and y-coordinate on a grid and the length of the respective brick.

In Figure 4, we show exemplary how a single brick (with number 2), for which all forces and moments must be balanced, may be isolated.

First, we get an equation for the tensile and compressive forces:

$$\overrightarrow{F_{g2}} + \overrightarrow{F_{3,2}} + \overrightarrow{F_{4,2}} + \overrightarrow{F_R} = \overrightarrow{0}$$

where the force $\overrightarrow{F_{gx}}$ denotes the gravitational force of brick x and the force $\overrightarrow{F_{x,y}}$ is the force between bricks x and y. Given the moments and the forces we can identify the condition for the balance of the moments:

$$\overrightarrow{M_R} + \overrightarrow{M_{3,2}} + \overrightarrow{M_{4,2}} + \overrightarrow{r_{g2}} \times \overrightarrow{F_{g2}} + \overrightarrow{r_{3,2}} \times \overrightarrow{F_{3,2}} + \overrightarrow{r_{4,2}} \times \overrightarrow{F_{4,2}} = \overrightarrow{0}$$

where the moments $\overrightarrow{M_{x,y}}$ are those moments that are passed from brick x to brick y. The r-vectors give the distance between the respective force and its

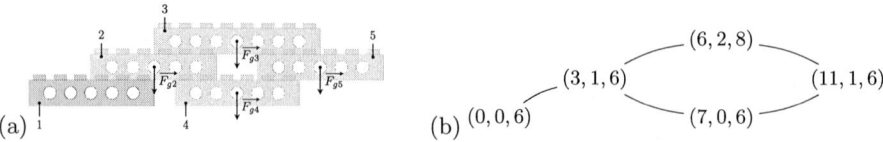

Fig. 3. Modelling a LEGO® structure: (a) shows the structure, the number of the bricks, and the forces given by the bricks' weight; (b) shows the corresponding graph representation where in each vertex the first two values are the position and the third value is the size of the brick.

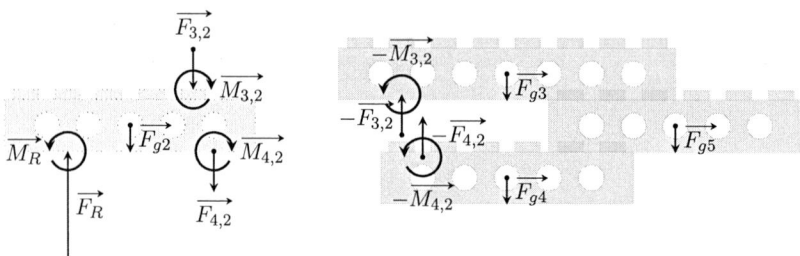

Fig. 4. By isolating brick 2, the forces affecting this brick become obvious and can be put into balance

equal yet opposite counterpart – moreover, we assume that the forces between bricks are applied in the middle position of the common knobs. In our example: $\overrightarrow{r_{g2}} = (1.5 \cdot c_x, 0, 0)^T$, $\overrightarrow{r_{3,2}} = (3 \cdot c_x, 0, 0)^T$, and $\overrightarrow{r_{4,2}} = (3.5 \cdot c_x, 0, 0)^T$.

These two equations, describing the equilibrium of forces and moments, set up for every brick of a structure and complemented by the following auxiliary constraints create a system of equation. The structure is stable iff this system can be solved. The following auxiliary constraints are derived from the physics of the LEGO® bricks and their connections.

1. The compressive forces could exceed the material strength of a LEGO® brick. This effect can be neglected because the weight of approximately 1.5 million bricks would be needed. Let $F_C = 5.44$kN be the maximal possible compressive force per knob.
2. The possible tensile forces depend on the number $k(u, v)$ of involved knobs in a joint of two bricks u and v and the possible maximal force per knob F_B. Then the maximal tensile force is $k(u, v) \cdot F_B$. The value for F_B depends on the age and condition of the bricks – for new bricks more than 3N are reasonable. We use $F_B = 2.5$N.
3. The main reason for unstable structures are moments that exceed the maximal moment capacity of a connection because of leverages. The maximum moment capacity of a connection with $k(u, v)$ knobs is given as $\frac{k(u,v)^2}{2} \cdot F_B$.

Generally, such a system can not be solved trivially because it may be underdetermined. Experiments using constraint satisfaction solvers showed an unacceptable runtime which is not a big surprise in the face of the exponential asymptotic runtime (with n being the number of forces).

The task is to analyse a given LEGO$^{\circledR}$ structure and to decide whether the structure will be stable. Basically, we are satisfied with an approximation algorithm. However there should only be false negatives and no false positives and the error rate should be smaller than in the approach of Funes and Pollack [3]. In addition, a reasonable wall-clock runtime is an important factor because usually the algorithm is executed within an evolutionary algorithm more than 20 million times.

4   Maximum-Flow Approach

We decided to go the inverse direction of flow network programming. We turn the system of equation into two maximum-flow networks that are solved successively:

1. Only the compressive and tensile forces are considered as if each brick is fixed in its orientation and cannot be tilted. If all gravitational and external forces can flow to the bearings, we get a possible distribution of the forces within the structure.
2. Now, the effective forces from the first maximum-flow network are used within a model for the moments. The maximum-flow corresponds to the moments that are passed between the bricks. The capacities within the flow network correspond to the maximum moment capacities.

This approach was developed within the unpublished Master's Thesis [7].

In the remainder of the section, we will discuss the construction of the two networks in detail.

4.1   Network of Forces in Phase 1

Let $G = (B \cup L, E)$ be the undirected graph corresponding to the LEGO$^{\circledR}$ structure where the vertices consist of the placeable bricks B and the bearings L. Both are annotated by the position and the size of the brick. The edges in E reflect the connections between bricks.

Then, we construct the maximum-flow network

$$N_{\mathrm{f}} = (G_{\mathrm{f}}, c_{\mathrm{f}}, s, t)$$

where the graph $G_{\mathrm{f}} = (V_{\mathrm{f}}, E_{\mathrm{f}})$ contains the vertices $V_{\mathrm{f}} = B \cup L \cup \{s, t\}$. The edge set E_{f} contains all edges in E (and transforms each edge into two directed edges). Moreover, for each force (from gravitation or external loads) an edge is inserted from s to the vertices in B. Each vertex in L is connected to t because the bearings are able to counteract the forces.

$$E_{\mathrm{f}} = \big\{(b_1, b_2) \big| \{b_1, b_2\} \in E\big\} \cup \big\{(s, b) \big| b \in B\big\} \cup \big\{(\ell, t) \big| \ell \in L\big\}$$

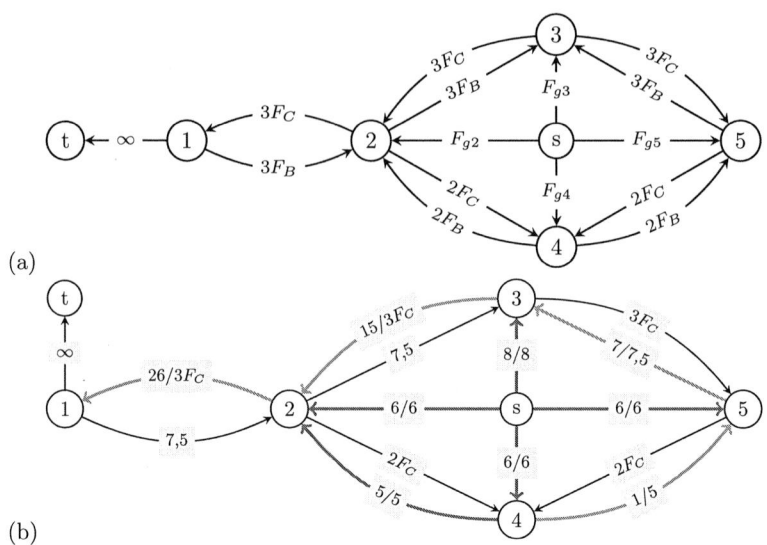

(a)

(b)

Fig. 5. The maximum-flow network (for tensile and compressive forces) constructed for the example of Figure 3. Subfigure (a) shows the constructed network and (b) shows the resulting maximum-flow. The example uses a modified gravitational constant.

The capacity function c_f is determined as follows:

$$c_f((u,v)) = \begin{cases} k(u,v) \cdot F_C & \text{iff } \{u,v\} \in E \text{ and } u \text{ is atop of } V \\ k(u,v) \cdot F_B & \text{iff } \{u,v\} \in E \text{ and } u \text{ is below } V \\ g \cdot weight(v) & \text{iff } u = s \text{ and } v \in B \\ \infty & \text{iff } v = t \text{ and } u \in L \end{cases}$$

where g is the gravitational constant and $weight(v)$ is the weight in kg. The network is constructed in such a way that the stability conditions concerning tensile and compressive forces are fulfilled if the maximum-flow uses all edges starting in s to their full capacity.

Figure 5 shows the resulting network for the simple example of Figure 3. The computed maximum-flow in subfigure (b) demonstrates that the conditions for the tensile and compressive forces are met and the full weight of the stones is transfered to the bearing. For the sake of simplicity, we adapted in our example the gravitational constant in such a way that a brick with 6 knobs corresponds to a gravitational force of 6N.

4.2   Network of Moments in Phase 2

From the forces between the bricks determined in phase 1 we can compute the effective moment for a brick a as follows:

$$M_a = \sum_{\{a,b\} \in E} d_m(b,a) \cdot (-f(b,a))$$

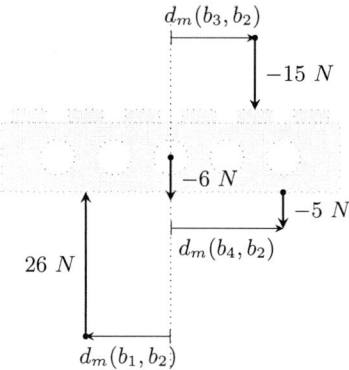

Fig. 6. For brick 2 the forces from phase 2 are shown that affect the moment

where $f(b, a)$ is the flow from vertex b to vertex a computed in phase 1 and $d_m(b, a)$ is the distance on the x-axis between the mass centre of the LEGO® brick a and the position of the force from the other brick. This is shown for brick 2 in Figure 6: Using the distances $d_m(b_1, b_2) = -1.5 \cdot c_x$, $d_m(b_3, b_2) = 1.5 \cdot c_x$, and $d_m(b_4, b_2) = 2 \cdot c_x$ we get

$$M_{b_2} = \underbrace{((-1.5) \cdot (-26\text{N}) + 1.5 \cdot 15\text{N} + 2 \cdot 5\text{N})}_{=71.5\text{N}} \cdot \underbrace{c_x}_{=8\text{mm}} = 0.572\text{Nm}.$$

If the moments are known for all bricks, we must again balance the moments by passing reactive moments along the joints between bricks. However, the situation is different from phase 1 because instead of passing all the forces towards the bearings positive and negative moments can equalise each other. As a consequence, the flow network model is quite different: We introduce a new source vertex s that induces all the positive moments into the network; in the same manner the target vertex t induces the negative moments. The flow within the network balances all positive and negative moments.

However, the bearings need special handling. The overall counteracting moment

$$M_R = -\sum_{b \in B} M_b$$

must be balanced by the bearings. To achieve this, we introduce a new vertex ℓ that combines all bearings, i.e. ℓ is connected with capacity ∞ to all $\ell_i \in L$. Depending on the sign of M_R, ℓ is connected with the source or the target vertex. This is shown in Figure 7a. Now, the vertex ℓ may be used to balance two counteracting equal valued moments as shown in Figure 7b.

Because of the infinite capacity between ℓ and all $\ell_i \in L$, we can omit the ℓ_i and substitute them completely by ℓ.

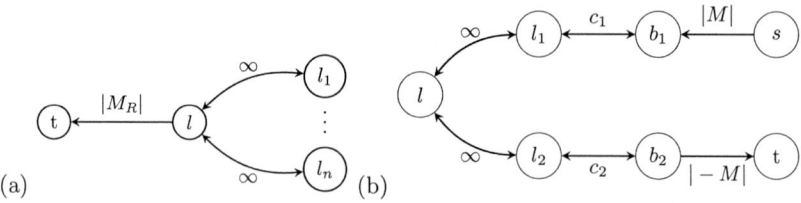

Fig. 7. A new vertex ℓ is introduced for all bearings in (a). Subfigure (b) shows how a positive and a negative moment may be balanced using the substitute vertex ℓ.

The maximum-flow network

$$N_\mathrm{m} = ((V_\mathrm{m}, E_\mathrm{m}), c_\mathrm{m}, s, t)$$

results with $V_\mathrm{m} = B \cup \{s, t, \ell\}$. And the edge set contains again all edges between bricks from the underlying graph $G = (V, E)$ as well as the edges described above.

$$
\begin{aligned}
E_\mathrm{m} = &\{(b_1, b_2) \mid \{b_1, b_2\} \in E \wedge b_1, b_2 \in B\} \\
&\cup \{(s, b) \mid b \in B \wedge M_b > 0\} \\
&\cup \{(b, t) \mid b \in B \wedge M_b < 0\} \\
&\cup \{(\ell, t) \mid \sum_{b \in B} M_b < 0\} \\
&\cup \{(s, \ell) \mid \sum_{b \in B} M_b > 0\} \\
&\cup \{(b, \ell) \mid \{b, \ell'\} \in E \wedge b \in B \wedge \ell' \in L\}
\end{aligned}
$$

The capacity function c_m is determined as follows (using the maximum moment capacity described in Section 3):

$$
c_\mathrm{m}((u, v)) = \begin{cases}
\frac{k(u,v)^2}{2} \cdot F_\mathrm{B}, & \text{iff } \{u, v\} \in E \\
|M_v|, & \text{iff } u = s \text{ and } v \in B \\
|M_u|, & \text{iff } u \in B \text{ and } v = t \\
|\sum_{b \in B} M_b|, & \text{iff } (u, v) = (\ell, t) \text{ or } (u, v) = (s, \ell) \\
0, & \text{otherwise}
\end{cases}
$$

To continue the example introduced in Figure 3, we get the maximum-flow network in Figure 8a. Because the moments for all bricks are positive all $b \in B$ are connected with the source vertex s. If we solve the given flow problem as shown in Figure 8b, we see that not all moments can be balanced within the structure. The capacity between ℓ and vertex 2 is too small. As a consequence, the structure would not be stable (under the circumstances of the artificially increased gravitational constant as described in the previous section).

The approach with two phases enables us to use single-commodity flow networks where the previous technique of Funes and Pollack [3] needed multi-commodity flow networks (where they are not able to balance moments within the structure).

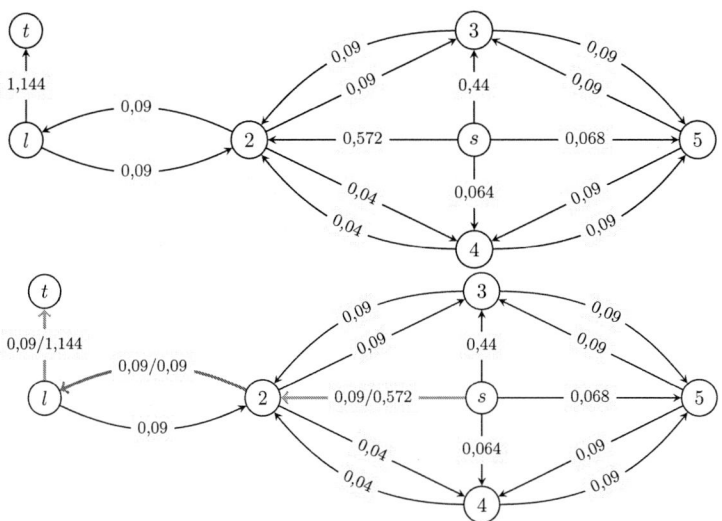

Fig. 8. Model for the moments in the example (a) and the resulting solution of the maximum-flow network in (b)

However, the presented technique is still a heuristic since only one possible distribution of forces is computed in the first phase. As a consequence an unfortunately chosen distribution may lead to a maximum-flow network in phase 2 that cannot be solved. The structure is declared instable, even though there might exist a force distribution for which the second phase succeeds.

In addition, this approach does not consider additional factors that affect the stability of LEGO® structures, e.g., if LEGO® bricks are placed side by side, tilting of a brick is inhibited to a certain degree. In such a case moments may not only be transmitted across connections, as in our model, but also to adjacent bricks, resulting in a more complex flow network with a higher overall capacity.

We restricted the description of the approach to 2D LEGO® structures but it can be extended easily to 3D by separating the moments in x- and y-direction and solving two (independent) maximum-flow networks instead of one. First 3D results using our implementation are available but omitted in this paper.

5   Results

The proposed technique has been used within an evolutionary algorithm to construct LEGO® structures that fulfil certain requirements. So far we considered cranes that are able to carry a load at the end of the crane and bridges where the driving surface must support an evenly distributed load.[1]

---

[1] The source code is available for download at
http://portal.imn.htwk-leipzig.de/fakultaet/weicker/forschung/download

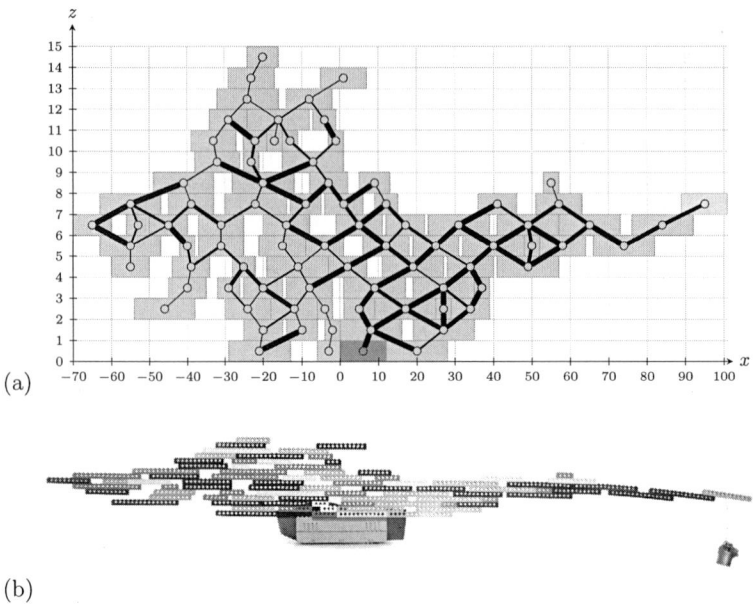

(a)

(b)

Fig. 9. Exemplary LEGO® structure that serves as a crane. (a) shows the solution of the maximum-flow network problem – the thickness of the edges indicates the strength of the transmitted moment. (b) shows the result of a practical evaluation whether the structure holds.

The concept and the details of the evolutionary algorithm are described in detail in [8]. In this section, we present two examples of the generated structures with the aim to demonstrate the high quality of the presented approach to analyse the structures' stability.

The crane problem was previously already considered by Funes and Pollack [3,4]. The crane is tied to one brick that serves as a bearing and the aim is to maximise both the length of the crane and the load that can be carried by the crane. Figure 9 shows a structure where a load of 0.235 kg is supported in a distance of 0.664 m.

Table 1 compares the best cranes reported in the literature. The results that are produced with our presented stability analysis outperform the results of Funes and Pollack both in length and in the load capacity. It is difficult to

Table 1. Comparison of the LEGO® cranes

Funes/Pollack [4]		Waßmann/Weicker [8]	
length	load	length	load
0.5 m	0.148 kg	0.496 m	0.482 kg
		0.664 m	0.236 kg
0.16 m	0.5 kg	0.256 m	1.0 kg

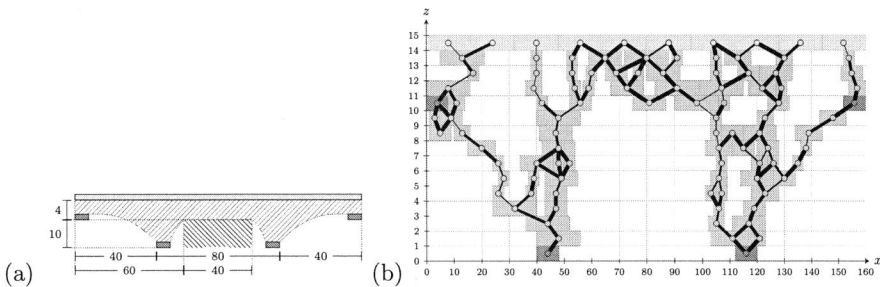

Fig. 10. LEGO® bridge. (a) shows the specified requirements for the evolution of the bridge. The plan in (b) corresponds to one of the constructed bridges.

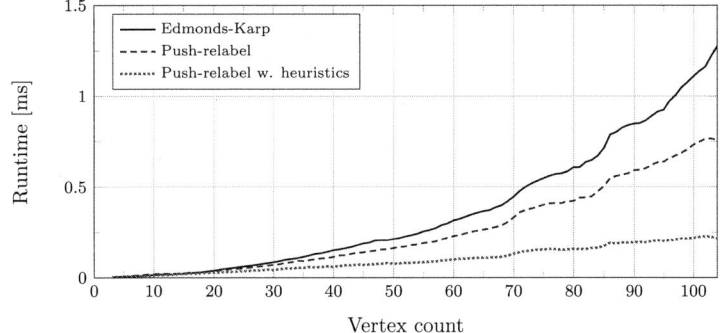

Fig. 11. Resulting runtime for the used max-flow algorithms

determine the exact share of the success for the physics simulation because we used a different algorithm as Funes and Pollack additionally.

The ability to handle more than one bearing is shown with the construction of a bridge as shown in Figure 10. The four bearings in the example are displayed as bricks in dark gray.

During our extensive experiments, we measured the runtime for the analysis of the stability. We tested the Edmonds-Karp algorithm [2], the Push-Relabel algorithm [6], and the Push-Relabel algorithm with heuristics [1]. Figure 11 show the runtime for LEGO® structures with up to 104 bricks. For more than 60 bricks, the Push-Relabel algorithm with heuristics is approximately 4–5 times faster than the Edmonds-Karp algorithm. With less than 20 bricks there is almost no observable difference between the three algorithms.

6   Conclusion

We have presented a new approach for the evaluation of LEGO® structures' stability. The core idea of transforming the involved physical equations into flow networks is not a new one. The known technique by Funes and Pollack [3] considered the moments only resulting in a multi-commodity network flow problem. In our approach, the physical equations are considered in two stages: first, a maximum-flow network for the compressive and tensile forces and, second, a maximum-flow network for the moments. Three primary advantages come along with our technique: We substituted the NP-hard multi-commodity network flow problem by two maximum-flow network problems that are solvable in polynomial time. Furthermore, the new flow network model for the moments is constructed in such a way that positive and negative moments may already balance each other within the structure. This leads to an improved predictive quality of the network model. And last, our approach is able to handle multiple bearings.

The results that have been constructed by an evolutionary algorithm outperform the results presented in [3,4]. This result is stressed even more if the number of structures is considered that are examined by the respective algorithms. Where [3] mentions 220,000 or 390,000 evaluated structures within one evolution of a LEGO® crane we used 50,400 evaluations to create the presented results.

For clarity and brevity, we restricted the presentation of the approach to 2D structures. However, the approach works for 3D structures analogously and similar results are available, e.g., for 3D cranes. The 3D experiments could be extended to additional classes of problems – 3D bridges could be an interesting application because our stability evaluation is able to handle more than one bearing.

Another direction of future research could be an extension of the physical model by external horizontal forces. This would enable the analysis of tower-like structures.

References

1. Cherkassky, B.V., Goldberg, A.V.: On implementing the push-relabel method for the maximum flow problem. Algorithmica 19, 390–410 (1997)
2. Edmonds, J., Karp, R.M.: Theoretical improvements in algorithmic efficiency for network flow problems. Journal of the ACM 19(2), 248–264 (1972)
3. Funes, P., Pollack, J.B.: Computer evolution of buildable objects. In: Husbands, P., Harvey, I. (eds.) Fourth European Conf. on Artificial Life, pp. 358–367. MIT Press, Cambridge (1997)
4. Funes, P., Pollack, J.B.: Computer evolution of buildable objects. In: Bentley, P.J. (ed.) Evolutionary Design by Computers, pp. 387–403. Morgan Kaufmann, San Francisco (1999)
5. Funes, P.J.: Buildable evolution. SIG Evolution 2(3), 6–19 (2007)
6. Goldberg, A.V., Tarjan, R.E.: A new approach to the maximum flow problem. Journal of the ACM 35, 921–940 (1988)
7. Waßmann, M.: Physiksimulation und Evolution von LEGO®-Strukturen. Master's thesis, HTWK Leipzig, Leipzig, Germany (2011)
8. Waßmann, M., Weicker, K.: Buildable objects revisited. To appear in PPSN 2012 – 12th International Conference on Parallel Problem Solving from Nature. Springer, Heidelberg (2012)

Average Case Analysis of Java 7's Dual Pivot Quicksort*

Sebastian Wild and Markus E. Nebel

Fachbereich Informatik, Technische Universität Kaiserslautern
{s_wild,nebel}@cs.uni-kl.de

Abstract. Recently, a new Quicksort variant due to Yaroslavskiy was chosen as standard sorting method for Oracle's Java 7 runtime library. The decision for the change was based on empirical studies showing that on average, the new algorithm is faster than the formerly used classic Quicksort. Surprisingly, the improvement was achieved by using a dual pivot approach, an idea that was considered not promising by several theoretical studies in the past. In this paper, we identify the reason for this unexpected success. Moreover, we present the first precise average case analysis of the new algorithm showing e. g. that a random permutation of length n is sorted using $1.9n \ln n - 2.46n + \mathcal{O}(\ln n)$ key comparisons and $0.6n \ln n + 0.08n + \mathcal{O}(\ln n)$ swaps.

1 Introduction

Due to its efficiency in the average, Quicksort has been used for decades as general purpose sorting method in many domains, e. g. in the C and Java standard libraries or as UNIX's system sort. Since its publication in the early 1960s by Hoare [1], classic Quicksort (Algorithm 1) has been intensively studied and many modifications were suggested to improve it even further, one of them being the following: Instead of partitioning the input file into two subfiles separated by a single pivot, we can create s partitions out of $s - 1$ pivots.

Sedgewick considered the case $s = 3$ in his PhD thesis [2]. He proposed and analyzed the implementation given in Algorithm 2. However, this dual pivot Quicksort variant turns out to be clearly inferior to the much simpler classic algorithm. Later, Hennequin studied the comparison costs for any constant s in his PhD thesis [3], but even for arbitrary $s \geq 3$, he found no improvements that would compensate for the much more complicated partitioning step.[1] These negative results may have discouraged further research along these lines.

Recently, however, Yaroslavskiy proposed the new dual pivot Quicksort implementation as given in Algorithm 3 at the Java core library mailing list.[2] He

---

* This research was supported by DFG grant NE 1379/3-1.

[1] When s depends on n, we basically get the Samplesort algorithm from [4]. [5], [6] or [7] show that Samplesort can beat Quicksort if hardware features are exploited. [2] even shows that Samplesort is asymptotically optimal with respect to comparisons. Yet, due to its inherent intricacies, it has not been used much in practice.

[2] The discussion is archived at http://permalink.gmane.org/gmane.comp.java. openjdk.core-libs.devel/2628.

L. Epstein and P. Ferragina (Eds.): ESA 2012, LNCS 7501, pp. 825–836, 2012.

Algorithm 1. Implementation of classic Quicksort as given in [8] (see [2], [9] and [10] for detailed analyses).
Two pointers i and j scan the array from left and right until they hit an element that does not belong in their current subfiles. Then the elements $A[i]$ and $A[j]$ are exchanged. This "crossing pointers" technique is due to Hoare [11], [1].

QUICKSORT(A, *left, right*)

      // Sort the array A in index range *left*, ..., *right*. We assume a sentinel $A[0] = -\infty$.
1 **if** *right* − *left* ≥ 1
2        $p := A[right]$ // Choose rightmost element as pivot
3        $i := left - 1;$ $j := right$
4   **do**
5   **do** $i := i + 1$ **while** $A[i] < p$ **end while**
6   **do** $j := j - 1$ **while** $A[j] > p$ **end while**
7   **if** $j > i$ **then** Swap $A[i]$ and $A[j]$ **end if**
8   **while** $j > i$
9   Swap $A[i]$ and $A[right]$ // Move pivot to final position
10   QUICKSORT(A, *left* , $i - 1$)
11   QUICKSORT(A, $i + 1$, *right*)
12 **end if**

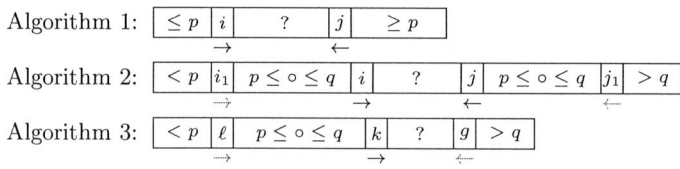

Algorithm 1:
Algorithm 2:
Algorithm 3:

Fig. 1. Comparison of the partitioning schemes of the three Quicksort variants discussed in this paper. The pictures show the invariant maintained in partitioning.

initiated a discussion claiming his new algorithm to be superior to the runtime library's sorting method at that time: the widely used and carefully tuned variant of classic Quicksort from [12]. Indeed, Yaroslavskiy's Quicksort has been chosen as the new default sorting algorithm in Oracle's Java 7 runtime library after extensive empirical performance tests.

In light of the results on multi-pivot Quicksort mentioned above, this is quite surprising and asks for explanation. Accordingly, since the new dual pivot Quicksort variant has not been analyzed in detail, yet[3], corresponding average case results will be proven in this paper. Our analysis reveals the reason why dual pivot Quicksort can indeed outperform the classic algorithm and why the partitioning method of Algorithm 2 is suboptimal. It turns out that Yaroslavskiy's partitioning method is able to take advantage of certain asymmetries in the

---

[3] Note that the results presented in http://iaroslavski.narod.ru/quicksort/ DualPivotQuicksort.pdf provide wrong constants and thus are insufficient for our needs.

Algorithm 2. Dual Pivot Quicksort with Sedgewick's partitioning as proposed in [2] (Program 5.1). This is an equivalent Java-like adaption of the original ALGOL-style program.

DUALPIVOTQUICKSORTSEDGEWICK(A, *left*, *right*)

      // Sort the array A in index range *left*, ..., *right*. We assume a sentinel $A[0] = -\infty$.

  1 **if** *right* $-$ *left* ≥ 1

  2      $i := left$; $i_1 := left$; $j := right$; $j_1 := right$; $p := A[left]$; $q := A[right]$

  3   **if** p > q **then** Swap p and q **end if**

  4   **while** *true*

  5          $i := i + 1$

  6   **while** $A[i] \leq q$

  7   **if** $i \geq j$ **then break** outer while **end if** // pointers have crossed

  8   **if** $A[i] < p$ **then** $A[i_1] := A[i]$; $i_1 := i_1 + 1$; $A[i] := A[i_1]$ **end if**

  9              $i := i + 1$

10   **end while**

11          $j := j - 1$

12   **while** $A[j] \geq p$

13   **if** $A[i] > q$ **then** $A[j_1] := A[j]$; $j_1 := j_1 - 1$; $A[j] := A[j_1]$ **end if**

14   **if** $i \geq j$ **then break** outer while **end if** // pointers have crossed

15              $j := j - 1$

16   **end while**

17          $A[i_1] := A[j]$; $A[j_1] := A[i]$

18          $i_1 := i_1 + 1$; $j_1 := j_1 - 1$

19          $A[i] := A[i_1]$; $A[j] := A[j_1]$

20   **end while**

21      $A[i_1] := p$; $A[j_1] := q$

22   DUALPIVOTQUICKSORTSEDGEWICK(A, *left* , $i_1 - 1$)

23   DUALPIVOTQUICKSORTSEDGEWICK(A, $i_1 + 1$, $j_1 - 1$)

24   DUALPIVOTQUICKSORTSEDGEWICK(A, $j_1 + 1$, *right*)

25 **end if**

---

outcomes of key comparisons. Algorithm 2 fails to utilize them, even though being based on the same abstract algorithmic idea.

2   Results

In this paper, we give the first precise average case analysis of Yaroslavskiy's dual pivot Quicksort (Algorithm 3), the new default sorting method in Oracle's Java 7 runtime library. Using these original results, we compare the algorithm to existing Quicksort variants: The classic Quicksort (Algorithm 1) and a dual pivot Quicksort as proposed by Sedgewick in [2] (Algorithm 2).

    Table 1 shows formulæ for the expected number of key comparisons and swaps for all three algorithms. In terms of comparisons, the new dual pivot Quicksort by Yaroslavskiy is best. However, it needs more swaps, so whether it can outperform the classic Quicksort, depends on the relative runtime contribution of swaps and

---

Algorithm 3. Dual Pivot Quicksort with Yaroslavskiy's partitioning method

---

DUALPIVOTQUICKSORTYAROSLAVSKIY(A, $left$, $right$)
    // Sort the array A in index range $left, \ldots, right$. We assume a sentinel $A[0] = -\infty$.
1 **if** $right - left \geq 1$
2      $p := A[left]$; $q := A[right]$
3   **if** $p > q$ **then** Swap p and q **end if**
4      $\ell := left + 1$; $g := right - 1$; $k := \ell$
5   **while** $k \leq g$
6   **if** $A[k] < p$
7   Swap $A[k]$ and $A[\ell]$
8              $\ell := \ell + 1$
9   **else**
10   **if** $A[k] > q$
11   **while** $A[g] > q$ **and** $k < g$ **do** $g := g - 1$ **end while**
12   Swap $A[k]$ and $A[g]$
13                 $g := g - 1$
14   **if** $A[k] < p$
15   Swap $A[k]$ and $A[\ell]$
16                     $\ell := \ell + 1$
17   **end if**
18   **end if**
19   **end if**
20         $k := k + 1$
21   **end while**
22     $\ell := \ell - 1$; $g := g + 1$
23   Swap $A[left]$ and $A[\ell]$ // Bring pivots to final position
24   Swap $A[right]$ and $A[g]$
25   DUALPIVOTQUICKSORTYAROSLAVSKIY(A, $left$, $\ell - 1$)
26   DUALPIVOTQUICKSORTYAROSLAVSKIY(A, $\ell + 1$, $g - 1$)
27   DUALPIVOTQUICKSORTYAROSLAVSKIY(A, $g + 1$, $right$)
28 **end if**

Table 1. Exact expected number of comparisons and swaps of the three Quicksort variants in the random permutation model. The results for Algorithm 1 are taken from [10, p. 334] (for $M = 1$). $\mathcal{H}_n = \sum_{i=1}^{n} \frac{1}{i}$ is the nth harmonic number, which is asymptotically $\mathcal{H}_n = \ln n + 0.577216\ldots + \mathcal{O}(n^{-1})$ as $n \to \infty$.

	Comparisons	Swaps
Classic Quicksort (Algorithm 1)	$2(n+1)\mathcal{H}_{n+1} - \frac{8}{3}(n+1)$ $\approx 2n \ln n - 1.51n + \mathcal{O}(\ln n)$	$\frac{1}{3}(n+1)\mathcal{H}_{n+1} - \frac{7}{9}(n+1) + \frac{1}{2}$ $\approx 0.33n \ln n - 0.58n + \mathcal{O}(\ln n)$
Sedgewick (Algorithm 2)	$\frac{32}{15}(n+1)\mathcal{H}_{n+1} - \frac{856}{225}(n+1) + \frac{3}{2}$ $\approx 2.13n \ln n - 2.57n + \mathcal{O}(\ln n)$	$\frac{4}{5}(n+1)\mathcal{H}_{n+1} - \frac{19}{25}(n+1) - \frac{1}{4}$ $\approx 0.8n \ln n - 0.30n + \mathcal{O}(\ln n)$
Yaroslavskiy (Algorithm 3)	$\frac{19}{10}(n+1)\mathcal{H}_{n+1} - \frac{711}{200}(n+1) + \frac{3}{2}$ $\approx 1.9n \ln n - 2.46n + \mathcal{O}(\ln n)$	$\frac{3}{5}(n+1)\mathcal{H}_{n+1} - \frac{27}{100}(n+1) - \frac{7}{12}$ $\approx 0.6n \ln n + 0.08n + \mathcal{O}(\ln n)$

comparisons, which in turn differ from machine to machine. Section 4 shows some running times, where indeed Algorithm 3 was fastest.

Remarkably, the new algorithm is significantly better than Sedgewick's dual pivot Quicksort in both measures. Given that Algorithms 2 and 3 are based on the same algorithmic idea, the considerable difference in costs is surprising. The explanation of the superiority of Yaroslavskiy's variant is a major discovery of this paper. Hence, we first give a qualitative teaser of it. Afterwards, Section 3 gives a thorough analysis, making the arguments precise.

2.1   The Superiority of Yaroslavskiy's Partitioning Method

Let $p < q$ be the two pivots. For partitioning, we need to determine for every $x \notin \{p, q\}$ whether $x < p$, $p < x < q$ or $q < x$ holds by comparing x to p *and/or* q. Assume, we first compare x to p, then averaging over all possible values for p, q and x, there is a $1/3$ chance that $x < p$ – in which case we are done. Otherwise, we still need to compare x and q. The expected number of comparisons for one element is therefore $1/3 \cdot 1 + 2/3 \cdot 2 = 5/3$. For a partitioning step with n elements including pivots p and q, this amounts to $5/3 \cdot (n-2)$ comparisons in expectation.

In the random permutation model, knowledge about an element $y \neq x$ does not tell us whether $x < p$, $p < x < q$ or $q < x$ holds. Hence, one could think that any partitioning method should need at least $5/3 \cdot (n-2)$ comparisons in expectation. But this is not the case.

The reason is the independence assumption above, which only holds true for algorithms that do comparisons at exactly *one location in the code*. But Algorithms 2 and 3 have several compare-instructions at different locations, and how often those are reached *depends* on the pivots p and q. Now of course, the number of elements smaller, between and larger p and q, directly depends on p and q, as well! So if a comparison is executed often if p is large, it is clever to first check $x < p$ there: The comparison is done more often than on average if and only if the probability for $x < p$ is larger than on average. Therefore, the expected number of comparisons can drop below the "lower bound" $5/3$ for this element!

And this is exactly, where Algorithms 2 and 3 differ: Yaroslavskiy's partitioning always evaluates the "better" comparison first, whereas in Sedgewick's dual pivot Quicksort this is not the case. In Section 3.3, we will give this a more quantitative meaning based on our analysis.

3   Average Case Analysis of Dual Pivot Quicksort

We assume input sequences to be random permutations, i.e. each permutation π of elements $\{1, \ldots, n\}$ occurs with probability $1/n!$. The first and last elements are chosen as pivots; let the smaller one be p, the larger one q.

Note that all Quicksort variants in this paper fulfill the following property:

Property 1. Every key comparison involves a pivot element of the current partitioning step.

3.1   Solution to the Dual Pivot Quicksort Recurrence

In [13], Hennequin shows that Property 1 is a sufficient criterion for *preserving randomness* in subfiles, i.e. if the whole array is a (uniformly chosen) random permutation of its elements, so are the subproblems Quicksort is recursively invoked on. This allows us to set up a recurrence relation for the expected costs, as it ensures that all partitioning steps of a subarray of size k have the same expected costs as the initial partitioning step for a random permutation of size k.

The expected costs C_n for sorting a random permutation of length n by any dual pivot Quicksort with Property 1 satisfy the following recurrence relation:

$$C_n = \sum_{1 \le p < q \le n} \Pr[\text{pivots } (p,q)] \cdot (\text{partitioning costs} + \text{recursive costs})$$

$$= \sum_{1 \le p < q \le n} \frac{2}{n(n-1)} (\text{partitioning costs} + C_{p-1} + C_{q-p-1} + C_{n-q}) ,$$

for $n \ge 3$ with base cases $C_0 = C_1 = 0$ and $C_2 = d$.[4]

We confine ourselves to linear expected partitioning costs $a(n+1) + b$, where a and b are constants depending on the kind of costs we analyze. The recurrence relation can then be solved by standard techniques – the detailed calculations can be found in Appendix A. The closed form for C_n is

$$C_n = \tfrac{6}{5}a \cdot (n+1) \left(\mathcal{H}_{n+1} - \tfrac{1}{5}\right) + \left(-\tfrac{3}{2}a + \tfrac{3}{10}b + \tfrac{1}{10}d\right) \cdot (n+1) - \tfrac{1}{2}b,$$

which is valid for $n \ge 4$ with $\mathcal{H}_n = \sum_{i=1}^{n} \frac{1}{i}$ the nth harmonic number.

3.2   Costs of One Partitioning Step

In this section, we analyze the expected number of swaps and comparisons used in the first partitioning step on a random permutation of $\{1, \ldots, n\}$. The results are summarized in Table 2. To state the proofs, we need to introduce some notation.

Table 2. Expected costs of the first partitioning step for the two dual pivot Quicksort variants on a random permutation of length n (for $n \ge 3$)

	Comparisons	Swaps
Sedgewick (Algorithm 2)	$\frac{16}{9}(n+1) - 3 - \frac{2}{3}\frac{1}{n(n-1)}$	$\frac{2}{3}(n+1) + \frac{1}{2}$
Yaroslavskiy (Algorithm 3)	$\frac{19}{12}(n+1) - 3$	$\frac{1}{2}(n+1) + \frac{7}{6}$

---

[4] d can easily be determined manually: For Algorithm 3, it is 1 for comparisons and $\frac{5}{2}$ for swaps and for Algorithm 2 we have $d = 2$ for comparisons and $d = \frac{5}{2}$ for swaps.

Notation. Let S be the set of all elements smaller than both pivots, M those in the middle and L the large ones, i. e.

$$S := \{1,\ldots,p-1\}, \quad M := \{p+1,\ldots,q-1\}, \quad L := \{q+1,\ldots,n\}.$$

Then, by Property 1 the algorithm cannot distinguish $x \in C$ from $y \in C$ for any $C \in \{S,M,L\}$. Hence, for analyzing partitioning costs, we replace all non-pivot elements by s, m or l when they are elements of S, M or L, respectively. Obviously, all possible results of a partitioning step correspond to the same word $s \cdots s\, p\, m \cdots m\, q\, l \cdots l$. The following example will demonstrate these definitions.

Example 1.   Example permutation before ...   ... and after partitioning.

$$
\begin{array}{l}
pq \\
\boxed{2}\ 4\ 7\ 8\ 1\ 6\ 9\ 3\ \boxed{5} \qquad\qquad 1\ \boxed{2}\ 4\ 3\ \boxed{5}\ 6\ 9\ 8\ 7\\
p\ m\ l\ l\ s\ l\ l\ m\ q s\ p\ m\ m\ q\ l\ l\ l\ l
\end{array}
$$

Next, we define position sets $\mathcal{S}$, $\mathcal{M}$ and $\mathcal{L}$ as follows:

$$
\begin{aligned}
\mathcal{S} &:= \{2,\ldots,p\}, \\
\mathcal{M} &:= \{p+1,\ldots,q-1\}, \\
\mathcal{L} &:= \{q,\ldots,n-1\}.
\end{aligned}
$$

in the example:
$$
\begin{array}{ccccccc}
\mathcal{S} & \mathcal{M} & \mathcal{M} & \mathcal{L} & \mathcal{L} & \mathcal{L} & \mathcal{L} \\
\boxed{2}\ 4 & 7 & 8 & 1 & 6 & 9 & 3\ \boxed{5}\\
\scriptstyle 1\ \scriptstyle 2 & \scriptstyle 3 & \scriptstyle 4 & \scriptstyle 5 & \scriptstyle 6 & \scriptstyle 7 & \scriptstyle 8\ \scriptstyle 9
\end{array}
$$

Now, we can formulate the main quantities occurring in the analysis below: For a given permutation, $c \in \{s,m,l\}$ and a set of positions $\mathcal{P} \subset \{1,\ldots,n\}$, we write $c @ \mathcal{P}$ for the number of c-type elements occurring *at positions* in $\mathcal{P}$ of the permutation. In our last example, $\mathcal{M} = \{3,4\}$ holds. At these positions, we find elements 7 and 8 (before partitioning), both belonging to L. Thus, $l @ \mathcal{M} = 2$, whereas $s @ \mathcal{M} = m @ \mathcal{M} = 0$.

Now consider a *random* permutation. Then $c @ \mathcal{P}$ becomes a random variable. In the analysis, we will encounter the conditional expectation of $c @ \mathcal{P}$ *given* that the random permutation induces the pivots p and q, i. e. the first and last element of the permutation are p and q *or* q and p, respectively. We abbreviate this quantity as $\mathbb{E}\,[c @ \mathcal{P} \,|\, p,q]$. As the number $\#c$ of c-type elements only depends on the pivots, not on the permutation itself, $\#c$ is a fully determined constant in $\mathbb{E}\,[c @ \mathcal{P} \,|\, p,q]$. Hence, given pivots p and q, $c @ \mathcal{P}$ is a hypergeometrically distributed random variable: For the c-type elements, we draw their $\#c$ positions out of $n-2$ possible positions via sampling without replacement. Drawing a position in $\mathcal{P}$ is a 'success', a position not in $\mathcal{P}$ is a 'failure'.

Accordingly, $\mathbb{E}\,[c @ \mathcal{P} \,|\, p,q]$ can be expressed as the mean of this hypergeometric distribution: $\mathbb{E}\,[c @ \mathcal{P} \,|\, p,q] = \#c \cdot \frac{|\mathcal{P}|}{n-2}$. By the law of total expectation, we finally have

$$
\begin{aligned}
\mathbb{E}\,[c @ \mathcal{P}] &= \sum_{1 \le p < q \le n} \mathbb{E}\,[c @ \mathcal{P} \,|\, p,q] \cdot \Pr[\text{pivots }(p,q)] \\
&= \frac{2}{n(n-1)} \sum_{1 \le p < q \le n} \#c \cdot \frac{|\mathcal{P}|}{n-2}.
\end{aligned}
$$

Comparisons in Algorithm 3. Algorithm 3 contains five places where key comparisons are used, namely in lines 3, 6, 10, 11 and 14. Line 3 compares the two pivots and is executed exactly once. Line 6 is executed once per value for k except for the last increment, where we leave the loop before the comparison is done. Similarly, line 11 is run once for every value of g except for the last one.

The comparison in line 10 can only be reached, when line 6 made the 'else'-branch apply. Hence, line 10 causes as many comparisons as k attains values with $A[k] \geq p$. Similarly, line 14 is executed once for all values of g where $A[g] \leq q$.[5]

At the end, q gets swapped to position g (line 24). Hence we must have $g = q$ there. Accordingly, g attains values $\mathcal{G} = \{n-1, n-2, \ldots, q\} = \mathcal{L}$ at line 11. We always leave the outer while loop with $k = g+1$ or $k = g+2$. In both cases, k (at least) attains values $\mathcal{K} = \{2, \ldots, q-1\} = \mathcal{S} \cup \mathcal{M}$ in line 11. The case "$k = g+2$" introduces an additional term of $3 \cdot \frac{n-q}{n-2}$; see Appendix B for the detailed discussion.

Summing up all contributions yields the conditional expectation $c_n^{p,q}$ of the number of comparisons needed in the first partitioning step for a random permutation, given it implies pivots p and q:

$$c_n^{p,q} = 1 + |\mathcal{K}| + |\mathcal{G}| + \big(\mathbb{E}\left[m\,@\,\mathcal{K}\,|\,p,q\right] + \mathbb{E}\left[l\,@\,\mathcal{K}\,|\,p,q\right]\big)$$
$$+ \big(\mathbb{E}\left[s\,@\,\mathcal{G}\,|\,p,q\right] + \mathbb{E}\left[m\,@\,\mathcal{G}\,|\,p,q\right]\big)$$
$$+ 3 \cdot \tfrac{n-q}{n-2}$$
$$= n - 1 + \big((q-p-1) + (n-q)\big)\frac{q-2}{n-2}$$
$$+ \big((p-1) + (q-p-1)\big)\frac{n-q}{n-2}$$
$$+ 3 \cdot \tfrac{n-q}{n-2}$$
$$= n - 1 + (n-p-1)\frac{q-2}{n-2} + (q+1)\frac{n-q}{n-2}\,.$$

Now, by the law of total expectation, the expected number of comparisons in the first partitioning step for a random permutation of $\{1, \ldots, n\}$ is

$$c_n := \mathbb{E}\,c_n^{p,q} = \tfrac{2}{n(n-1)} \sum_{p=1}^{n-1} \sum_{q=p+1}^{n} c_n^{p,q}$$
$$= n - 1 + \tfrac{2}{n(n-1)(n-2)} \sum_{p=1}^{n-1}(n-p-1) \sum_{q=p+1}^{n}(q-2)$$
$$+ \tfrac{2}{n(n-1)(n-2)} \sum_{q=2}^{n}(n-q)(q+1) \sum_{p=1}^{q-1} 1$$
$$= n - 1 + \big(\tfrac{5}{12}(n+1) - \tfrac{4}{3}\big) + \tfrac{1}{6}(n+3) = \tfrac{19}{12}(n+1) - 3\,.$$

---

[5] Line 12 just swapped $A[k]$ and $A[g]$. So even though line 14 literally says "$A[k] < p$", this comparison actually refers to an element first reached as $A[g]$.

Swaps in Algorithm 3. Swaps happen in Algorithm 3 in lines 3, 7, 12, 15, 23 and 24. Lines 23 and 24 are both executed exactly once. Line 3 once swaps the pivots if needed, which happens with probability $1/2$. For each value of k with $A[k] < p$, one swap occurs in line 7. Line 12 is executed for every value of k having $A[k] > q$. Finally, line 15 is reached for all values of g where $A[g] < p$ (see footnote 5).

Using the ranges $\mathcal{K}$ and $\mathcal{G}$ from above, we obtain $s_n^{p,q}$, the conditional expected number of swaps for partitioning a random permutation, given pivots p and q. There is an additional contribution of $\frac{n-q}{n-2}$ when k stopps with $k = g+2$ instead of $k = g+1$. As for comparisons, its detailed discussion is deferred to Appendix B.

$$
\begin{aligned}
s_n^{p,q} &= \tfrac{1}{2} + 1 + 1 + \mathbb{E}\left[s\,@\,\mathcal{K}\,|\,p,q\right] + \mathbb{E}\left[l\,@\,\mathcal{K}\,|\,p,q\right] + \mathbb{E}\left[s\,@\,\mathcal{G}\,|\,p,q\right] + \tfrac{n-q}{n-2} \\
&= \tfrac{5}{2} + (p-1)\frac{q-2}{n-2} + (n-q)\frac{q-2}{n-2} + (p-1)\frac{n-q}{n-2} + \tfrac{n-q}{n-2} \\
&= \tfrac{5}{2} + (n+p-q-1)\frac{q-2}{n-2} + p\cdot\frac{n-q}{n-2} \ .
\end{aligned}
$$

Averaging over all possible p and q again, we find

$$
\begin{aligned}
s_n := \mathbb{E}\, s_n^{p,q} &= \tfrac{5}{2} + \tfrac{2}{n(n-1)(n-2)} \sum_{q=2}^{n}(q-2)\sum_{p=1}^{q-1}(n+p-q-1) \\
&\quad + \tfrac{2}{n(n-1)(n-2)} \sum_{q=2}^{n}(n-q)\sum_{p=1}^{q-1}p \\
&= \tfrac{5}{2} + \left(\tfrac{5}{12}(n+1) - \tfrac{4}{3}\right) + \tfrac{1}{12}(n+1) \;=\; \tfrac{1}{2}(n+1) + \tfrac{7}{6} \ .
\end{aligned}
$$

Comparisons in Algorithm 2. Key comparisons happen in Algorithm 2 in lines 3, 6, 8, 12 and 13. Lines 6 and 12 are executed once for every value of i respectively j (without the initialization values *left* and *right* respectively). Line 8 is reached for all values of i with $A[i] \le q$ except for the last value. Finally, the comparison in line 13 gets executed for every value of j having $A[j] \ge p$.

The value-ranges of i and j are $\mathcal{I} = \{2, \ldots, \hat{\imath}\}$ and $\mathcal{J} = \{n-1, n-2, \ldots, \hat{\imath}\}$ respectively, where $\hat{\imath}$ depends on the positions of m-type elements. So, lines 6 and 12 together contribute $|\mathcal{I}| + |\mathcal{J}| = n-1$ comparisons. For lines 8 and 13, we get additionally

$$
\left(\mathbb{E}\left[s\,@\,\mathcal{I}'\,|\,p,q\right] + \mathbb{E}\left[m\,@\,\mathcal{I}'\,|\,p,q\right]\right) + \left(\mathbb{E}\left[m\,@\,\mathcal{J}\,|\,p,q\right] + \mathbb{E}\left[l\,@\,\mathcal{J}\,|\,p,q\right]\right)
$$

many comparisons (in expectation), where $\mathcal{I}' := \mathcal{I} \setminus \hat{\imath}$. As i and j cannot meet on an m-type element (both would not stop), $m\,@\,\{\hat{\imath}\} = 0$, so

$$
\mathbb{E}\left[m\,@\,\mathcal{I}'\,|\,p,q\right] + \mathbb{E}\left[m\,@\,\mathcal{J}\,|\,p,q\right] = q - p - 1 \ .
$$

Positions of m-type elements do not contribute to $s\,@\,\mathcal{I}'$ (and $l\,@\,\mathcal{J}$) by definition. Hence, it suffices to determine the number of non-m-elements located

Table 3. $\mathbb{E}\left[c @ \mathcal{P}\right]$ for $c = s, m, l$ and $\mathcal{P} = \mathcal{S}, \mathcal{M}, \mathcal{L}$

	$\mathcal{S}$	$\mathcal{M}$	$\mathcal{L}$
s	$\frac{1}{6}(n-1)$	$\frac{1}{12}(n-3)$	$\frac{1}{12}(n-3)$
m	$\frac{1}{12}(n-3)$	$\frac{1}{6}(n-1)$	$\frac{1}{12}(n-3)$
l	$\frac{1}{12}(n-3)$	$\frac{1}{12}(n-3)$	$\frac{1}{6}(n-1)$

at positions in $\mathcal{I}'$. A glance at Figure 1 suggests to count non-m-type elements left of (and including) the last value of i_1, which is p. So, the first $p-1$ of all $(p-1) + (n-q)$ non-m-positions are contained in $\mathcal{I}'$, thus $\mathbb{E}\left[s @ \mathcal{I}' \mid p, q\right] = (p-1)\frac{p-1}{(p-1)+(n-q)}$. Similarly, we can show that $l @ \mathcal{J}$ is the number of l-type elements right of i_1's largest value: $\mathbb{E}\left[l @ \mathcal{J} \mid p, q\right] = (n-q)\frac{n-q}{(p-1)+(n-q)}$. Summing up all contributions, we get

$$c_n'^{p,q} = n - 1 + q - p - 1 + (p-1)\frac{p-1}{(p-1)+(n-q)} + (n-q)\frac{n-q}{(p-1)+(n-q)}.$$

Taking the expectation over all possible pivot values yields

$$c_n' = \frac{2}{n(n-1)} \sum_{p=1}^{n-1} \sum_{q=p+1}^{n} c_n'^{p,q} = \frac{16}{9}(n+1) - 3 - \frac{2}{3}\frac{1}{n(n-1)}.$$

This is not a linear function and hence does not directly fit our solution of the recurrence from Section 3.1. The exact result given in Table 1 is easily proven by induction. Dropping summand $-\frac{2}{3}\frac{1}{n(n-1)}$ and inserting the linear part into the recurrence relation, still gives the correct leading term; in fact, the error is only $\frac{1}{90}(n+1)$.

Swaps in Algorithm 2. The expected number of swaps has already been analyzed in [2]. There, it is shown that Sedgewick's partitioning step needs $\frac{2}{3}(n+1)$ swaps, on average – excluding the pivot swap in line 3. As we count this swap for Algorithm 3, we add $\frac{1}{2}$ to the expected value for Algorithm 2, for consistency.

3.3 Superiority of Yaroslavskiy's Partitioning Method – Continued

In this section, we abbreviate $\mathbb{E}\left[c @ \mathcal{P}\right]$ by $E_c^{\mathcal{P}}$ for conciseness. It is quite enlightening to compute $E_c^{\mathcal{P}}$ for $c = s, m, l$ and $\mathcal{P} = \mathcal{S}, \mathcal{M}, \mathcal{L}$, see Table 3: There is a remarkable *asymmetry*, e. g. averaging over all permutations, *more than half* of all l-type elements are located at positions in $\mathcal{L}$. Thus, if we *know* we are looking at a position in $\mathcal{L}$, it is much more advantageous to first compare with q, as with probability $> \frac{1}{2}$, the element is $> q$. This results in an expected number of comparisons $< \frac{1}{2} \cdot 2 + \frac{1}{2} \cdot 1 = \frac{3}{2} < \frac{5}{3}$. Line 11 of Algorithm 3 is exactly of this type. Hence, Yaroslavskiy's partitioning method exploits the knowledge about the different position sets comparisons are reached for. Conversely, lines 6 and 12 in Algorithm 2 are of the opposite type: They check the unlikely outcome first.

We can roughly approximate the expected number of comparisons in Algorithms 2 and 3 by expressing them in terms of the quantities from Table 3 (using $\mathcal{K} = \mathcal{S} \cup \mathcal{M}$, $\mathcal{G} \approx \mathcal{L}$ and $E_s^{\mathcal{I}'} + E_l^{\mathcal{J}} \approx E_s^{\mathcal{S}} + E_l^{\mathcal{L}} + E_s^{\mathcal{M}}$):

$$
\begin{aligned}
c_n' &= n - 1 + \quad\quad \mathbb{E}\,\#m \quad\quad + \quad\quad E_s^{\mathcal{I}'} + E_l^{\mathcal{J}} \\
&\approx \quad n \;\; + \left(E_m^{\mathcal{S}} + E_m^{\mathcal{M}} + E_m^{\mathcal{L}}\right) + \left(E_s^{\mathcal{S}} + E_l^{\mathcal{L}} + E_s^{\mathcal{M}}\right) \\
&\approx \left(1 + 3 \cdot \tfrac{1}{12} + 3 \cdot \tfrac{1}{6}\right)n \;\approx\; 1.75n \quad\quad (\text{exact: } 1.78n - 1.22 + o(1))
\end{aligned}
$$

$$
\begin{aligned}
c_n &= n + \quad\quad E_m^{\mathcal{K}} \quad\quad + \quad\quad E_l^{\mathcal{K}} \quad\quad + E_s^{\mathcal{G}} + E_m^{\mathcal{G}} \\
&\approx n + \left(E_m^{\mathcal{S}} + E_m^{\mathcal{M}}\right) + \left(E_l^{\mathcal{S}} + E_l^{\mathcal{M}}\right) + E_s^{\mathcal{L}} + E_m^{\mathcal{L}} \\
&\approx \left(1 + 5 \cdot \tfrac{1}{12} + 1 \cdot \tfrac{1}{6}\right)n \;\approx\; 1.58n \quad\quad (\text{exact: } 1.58n - 0.75)
\end{aligned}
$$

Note that both terms involve six '$E_c^{\mathcal{P}}$-terms', but Algorithm 2 has *three* 'expensive' terms, whereas Algorithm 3 only has *one* such term.

4   Some Running Times

Extensive performance tests have already been done for Yaroslavskiy's dual pivot Quicksort. However, those were based an optimized implementation intended for production use. In Figure 2, we provide some running times of the basic variants as given in Algorithms 1, 2 and 3 to directly evaluate the algorithmic ideas, complementing our analysis.

Note: This is not intended to replace a thorough performance study, but merely to demonstrate that Yaroslavskiy's partitioning method performs well – at least on our machine.

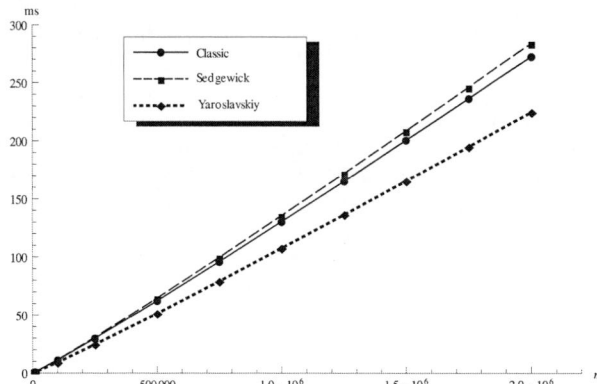

Fig. 2. Running times of Java implementations of Algorithms 1, 2 and 3 on an Intel Core 2 Duo P8700 laptop. The plot shows the average running time of 1000 random permutations of each size.

5 Conclusion and Future Work

Having understood how the new Quicksort saves key comparions, there are plenty of future research directions. The question if and how the new Quicksort can compensate for the many extra swaps it needs, calls for further examination. One might conjecture that comparisons have a higher runtime impact than swaps. It would be interesting to see a closer investigation – empirically or theoretically.

In this paper, we only considered the most basic implementation of dual pivot Quicksort. Many suggestions to improve the classic algorithm are also applicable to it. We are currently working on the effect of selecting the pivot from a larger sample and are keen to see the performance impacts.

Being intended as a standard sorting method, it is not sufficient for the new Quicksort to perform well on random permutations. One also has to take into account other input distributions, most notably the occurrence of equal keys or biases in the data. This might be done using Maximum Likelihood Analysis as introduced in [14], which also helped us much in discovering the results of this paper. Moreover, Yaroslavskiy's partitioning method can be used to improve Quickselect. Our corresponding results are omitted due to space constraints.

References

1. Hoare, C.A.R.: Quicksort. The Computer Journal 5(1), 10–16 (1962)
2. Sedgewick, R.: Quicksort. Phd thesis, Stanford University (1975)
3. Hennequin, P.: Analyse en moyenne d'algorithme, tri rapide et arbres de recherche. Ph.d. thesis, Ecole Politechnique, Palaiseau (1991)
4. Frazer, W.D., McKellar, A.C.: Samplesort: A Sampling Approach to Minimal Storage Tree Sorting. Journal of the ACM 17(3), 496–507 (1970)
5. Sanders, P., Winkel, S.: Super Scalar Sample Sort. In: Albers, S., Radzik, T. (eds.) ESA 2004. LNCS, vol. 3221, pp. 784–796. Springer, Heidelberg (2004)
6. Leischner, N., Osipov, V., Sanders, P.: GPU sample sort. In: 2010 IEEE International Symposium on Parallel Distributed Processing IPDPS, pp. 1–10. IEEE (2009)
7. Blelloch, G.E., Leiserson, C.E., Maggs, B.M., Plaxton, C.G., Smith, S.J., Zagha, M.: A comparison of sorting algorithms for the connection machine CM-2. In: Proceedings of the Third Annual ACM Symposium on Parallel Algorithms and Architectures - SPAA 1991, pp. 3–16. ACM Press, New York (1991)
8. Sedgewick, R.: Implementing Quicksort programs. Comm. ACM 21(10), 847–857 (1978)
9. Sedgewick, R.: Quicksort with Equal Keys. SIAM Journal on Computing 6(2), 240–267 (1977)
10. Sedgewick, R.: The analysis of Quicksort programs. Acta Inf. 7(4), 327–355 (1977)
11. Hoare, C.A.R.: Algorithm 63: Partition. Comm. 4(7), 321 (1961)
12. Bentley, J.L.J., McIlroy, M.D.: Engineering a sort function. Software: Practice and Experience 23(11), 1249–1265 (1993)
13. Hennequin, P.: Combinatorial analysis of Quicksort algorithm. Informatique Théorique et Applications 23(3), 317–333 (1989)
14. Laube, U., Nebel, M.E.: Maximum likelihood analysis of algorithms and data structures. Theoretical Computer Science 411(1), 188–212 (2010)

Author Index

Abed, Fidaa   12
Abraham, Ittai   24
Abu-Affash, A. Karim   36
Afshani, Peyman   48
Ahmed, Mahmuda   60
Ajwani, Deepak   72
Albers, Susanne   84
Arge, Lars   96
Aumüller, Martin   108

Babka, Martin   121
Bansal, Manisha   133
Bansal, Nikhil   145
Barman, Siddharth   157
Batz, Gernot Veit   169
Belazzougui, Djamal   181
Belovs, Aleksandrs   193
Boissonnat, Jean-Daniel   731
Bonichon, Nicolas   205
Brodal, Gerth Stølting   217
Buchin, Kevin   229
Buchin, Maike   229
Bulánek, Jan   121

Cabello, Sergio   241
Caragiannis, Ioannis   253
Cardinal, Jean   241
Carmi, Paz   36
Chan, T.-H. Hubert   265, 277
Chang, Jessica   289
Chatterjee, Krishnendu   301
Chawla, Shuchi   157
Chazelle, Bernard   313
Chechik, Shiri   325
Chen, Fei   265
Childs, Andrew M.   337
Correa, José R.   790
Čunát, Vladimír   121
Cygan, Marek   349, 361

Davis, James M.   373
Davoodi, Pooya   217
de Berg, Mark   383
de Keijzer, Bart   395
Deleuran, Lasse   96

Delling, Daniel   24, 407
Díaz, Josep   419
Dietzfelbinger, Martin   108
Dutta, Debojyoti   431

Fisikopoulos, Vissarion   443
Fogel, Efi   455
Fomin, Fedor V.   467
Fusco, Emanuele G.   479

Gabow, Harold N.   289
Garg, Naveen   133
Gavoille, Cyril   205
Georgiadis, Loukas   491
Giesen, Joachim   503
Goldberg, Andrew V.   24
Golovach, Petr A.   515
Grandoni, Fabrizio   349, 527
Groß, Martin   539, 551
Gupta, Neelima   133

Halperin, Dan   455, 611
Hanusse, Nicolas   205
Harks, Tobias   563
He, Meng   575
Held, Martin   766
Hell, Pavol   587
Hellweg, Frank   599
Hellwig, Matthias   84
Hemmer, Michael   455, 611
Henzinger, Monika   301
Hermelin, Danny   624
Huang, Chien-Chung   12
Huber, Stefan   766

Ibsen-Jensen, Rasmus   636
Italiano, Giuseppe F.   491

Jaggi, Martin   503
Jowhari, Hossein   648

Kaklamanis, Christos   253
Kanellopoulos, Panagiotis   253
Kappmeier, Jan-Philipp W.   551
Kapralov, Michael   431

Katz, Matthew J.   36
Kesselheim, Thomas   659
Khuller, Samir   289
Kimmel, Shelby   337
Klavík, Pavel   671
Kleinbort, Michal   611
Kobayashi, Yasuaki   683
Kortsarz, Guy   361
Kothari, Robin   337
Koucký, Michal   121
Kozma, László   695
Kratochvíl, Jan   671
Kratsch, Dieter   707
Krawczyk, Tomasz   671
Krinninger, Sebastian   301
Kyropoulou, Maria   253

Laekhanukit, Bundit   719
Langerman, Stefan   241
Larré, Omar   790
Laue, Sören   503
Laura, Luigi   491
Leonardi, Stefano   349
Lewenstein, Moshe   217

Maria, Clément   731
Mastrolilli, Monaldo   587
Matias, Yossi   1
Meulemans, Wouter   229
Meyer, Ulrich   72
Miltersen, Peter Bro   636
Mnich, Matthias   624
Mølhave, Thomas   96
Mount, David M.   778
Müller, Haiko   707
Mulzer, Wolfgang   313
Munro, J. Ian   575, 743

Nanongkai, Danupon   301
Navarro, Gonzalo   181
Nebel, Markus E.   825
Nevisi, Mayssam Mohammadi   587
Nicholson, Patrick K.   743
Ning, Li   265
Nonner, Tim   755
Nutov, Zeev   361

Palfrader, Peter   766
Park, Eunhui   778
Paulusma, Daniël   515

Peis, Britta   563
Pelc, Andrzej   479
Peñaranda, Luis   443
Perković, Ljubomir   205
Pilipczuk, Marcin   349
Porat, Asaf   455
Post, Ian   431
Pottonen, Olli   419
Pruhs, Kirk   145

Rafiey, Arash   587
Raman, Rajeev   217
Reichardt, Ben W.   193
Revsbæk, Morten   96
Roeloffzen, Marcel   383

Saks, Michael   121
Sanders, Peter   169
Sankowski, Piotr   349
Santaroni, Federico   491
Saurabh, Saket   467
Schäfer, Guido   395
Schmidt, Daniel R.   551
Schmidt, Melanie   551
Serna, Maria   419
Sgall, Jiří   2
Shi, Elaine   277
Shinde, Rajendra   431
Skutella, Martin   539
Sohler, Christian   599
Song, Dawn   277
Soto, José A.   790
Speckmann, Bettina   229, 383
Srinivasa Rao, Satti   217

Tamaki, Hisao   683
Telle, Jan Arne   802
Trabelsi, Yohai   36
Truelsen, Jakob   96

Umboh, Seeun   157

van Leeuwen, Erik Jan   419, 515, 624
Veith, David   72
Vetta, Adrian   719
Villanger, Yngve   467, 802

Walczak, Bartosz   671
Waßmann, Martin   813
Weicker, Karsten   813

Wenk, Carola   60

Werneck, Renato F.   24, 407

Wild, Sebastian   825

Wilfong, Gordon   719

Williamson, David P.   373

Woelfel, Philipp   108

Zeh, Norbert   48

Zhou, Gelin   575